AF869703

CAMBRIDGE MONOGRAPHS ON APPLIED AND COMPUTATIONAL MATHEMATICS

Series Editors
M. ABLOWITZ, S. DAVIS, J. HINCH,
A. ISERLES, J. OCKENDON, P. OLVER

41

Spectral and Spectral Element Methods for Fractional Ordinary and Partial Differential Equations

The *Cambridge Monographs on Applied and Computational Mathematics* series reflects the crucial role of mathematical and computational techniques in contemporary science. The series publishes expositions on all aspects of applicable and numerical mathematics, with an emphasis on new developments in this fast-moving area of research.

State-of-the-art methods and algorithms as well as modern mathematical descriptions of physical and mechanical ideas are presented in a manner suited to graduate research students and professionals alike. Sound pedagogical presentation is a prerequisite. It is intended that books in the series will serve to inform a new generation of researchers.

A complete list of books in the series can be found at www.cambridge.org/mathematics. Recent titles include the following:

25. Algebraic geometry and statistical learning theory, *Sumio Watanabe*
26. A practical guide to the invariant calculus, *Elizabeth Louise Mansfield*
27. Difference equations by differential equation methods, *Peter E. Hydon*
28. Multiscale methods for Fredholm integral equations, *Zhongying Chen, Charles A. Micchelli & Yuesheng Xu*
29. Partial differential equation methods for image inpainting, *Carola-Bibiane Schönlieb*
30. Volterra integral equations, *Hermann Brunner*
31. Symmetry, phase modulation and nonlinear waves, *Thomas J. Bridges*
32. Multivariate approximation, *V. Temlyakov*
33. Mathematical modelling of the human cardiovascular system, *Alfio Quarteroni, Luca Dede', Andrea Manzoni & Christian Vergara*
34. Numerical bifurcation analysis of maps, *Yuri A. Kuznetsov & Hil G.E. Meijer*
35. Operator-adapted wavelets, fast solvers, and numerical homogenization, *Houman Owhadi & Clint Scovel*
36. Spaces of measures and their applications to structured population models, *Christian Düll, Piotr Gwiazda, Anna Marciniak-Czochra & Jakub Skrzeczkowski*
37. Quasi-interpolation, *Martin Buhmann & Janin Jäger*
38. The Christoffel–Darboux kernel for data analysis, *Jean Bernard Lasserre, Edouard Pauwels & Mihai Putinar*
39. Geometry of the phase retrieval problem, *Alexander H. Barnett, Charles L. Epstein, Leslie Greengard & Jeremy Magland*
40. Discrete variational problems with interfaces, *Roberto Alicandro, Andrea Braides, Marco Cicalese & Margherita Solci*

The first (ever in the history of the USA) International Symposium on Fractional PDEs: Theory, Numerics and Applications, organized by George Em Karniadakis and Jan S. Hesthaven in Newport, RI, June 3–5, 2013, supported by the US Air Force Office of Scientific Research, US Army Research Office, and US Department of Energy/Pacific Northwest National Lab. This book recollects the loving memory of Mark M. Meerschaert and Rudolf Gorenflo in the group photo. Source: the website of the department of applied mathematics at Brown University.

Spectral and Spectral Element Methods for Fractional Ordinary and Partial Differential Equations

MOHSEN ZAYERNOURI
Michigan State University

LI-LIAN WANG
Nanyang Technological University, Singapore

JIE SHEN
Eastern Institute of Technology, Ningbo, China

GEORGE EM KARNIADAKIS
Brown University, Rhode Island

CAMBRIDGE
UNIVERSITY PRESS

CAMBRIDGE
UNIVERSITY PRESS

Shaftesbury Road, Cambridge CB2 8EA, United Kingdom

One Liberty Plaza, 20th Floor, New York, NY 10006, USA

477 Williamstown Road, Port Melbourne, VIC 3207, Australia

314–321, 3rd Floor, Plot 3, Splendor Forum, Jasola District Centre, New Delhi – 110025, India

103 Penang Road, #05–06/07, Visioncrest Commercial, Singapore 238467

Cambridge University Press is part of Cambridge University Press & Assessment, a department of the University of Cambridge.

We share the University's mission to contribute to society through the pursuit of education, learning and research at the highest international levels of excellence.

www.cambridge.org
Information on this title:www.cambridge.org/9781108490993

DOI: 10.1017/9781108867160

© Mohsen Zayernouri, Li-Lian Wang, Jie Shen, and George Em Karniadakis 2024

This publication is in copyright. Subject to statutory exception and to the provisions of relevant collective licensing agreements, no reproduction of any part may take place without the written permission of Cambridge University Press & Assessment.

When citing this work, please include a reference to the DOI 10.1017/9781108867160

First published 2024

A catalogue record for this publication is available from the British Library

A Cataloging-in-Publication data record for this book is available from the Library of Congress

ISBN 978-1-108-49099-3 Hardback

Cambridge University Press & Assessment has no responsibility for the persistence or accuracy of URLs for external or third-party internet websites referred to in this publication and does not guarantee that any content on such websites is, or will remain, accurate or appropriate.

To our families

Contents

Preface

The mathematical theory of fractional calculus is almost as old as the classical calculus. However, it took the scientific and engineering community much longer to discover and truly appreciate the beauty (i.e., power, flexibility, and natural expressibility) of fractional-order modeling in a wide range of crucial applications. The main reason for that can be sought in the very nature of such generalizing-order models, emerging as nonlocal/memory-dependent. Over the past decades, an extensive amount of work has been devoted to developing efficient numerical methods that can make such a new and interesting way of thinking in mathematical modeling tractable across multiple disciplines.

It is fair to say that the first attempts to solve such problems were based on Fourier and Laplace transforms, rendering the given fractional *differential* problem *algebraic*, in which the final solution eventually can be yielded as an inverse Fourier/Laplace of (often complicated) functions in the frequency/wave-number domains. Since our knowledge to obtain such inverses had been quite limited to some particular class of functions, their general/practical use remained focused on rather elementary/linear problems and usually in one spatial or temporal dimension. This motivated the numerical analysis community to overcome this challenge by introducing *discretized fractional calculus*, hence, developing (local) finite-difference and later (local) finite-volume methods to essentially *discretize* the underlying fractional-order operators as initial-/boundary-value problems, which began in the early 1980s. There are now 40+ years of new developments on that front, where particularly over the past two decades, more efficient finite-difference/volume schemes have been developed for solving fractional problems with more desirable degrees of applicability, i.e., accuracy and cost; this is still an ongoing field of study!

This book, however, is focused on the use of *global* numerical methods, namely *spectral methods* in single-domains (in any dimension) and *spectral element methods* in multi-domains, to solve fractional ordinary and partial differential equations (FODEs/FPDEs). The main driving force and philosophy for this contribution is the underlying *global nature* of such operators, which would allow one to rather spectrally/globally approximate the history/nonlocality (instead of marching/surfing through time/space). Developing spectral numerical methods for fractional models began in the early 1990s, employing smooth basis functions (i.e., Fourier, Legendre, Chebychev, etc. polynomials). However, the first USA-based International Symposium on Fractional PDEs: Theory, Numerics & Applications on June 3–5 of 2013 (Newport, RI) can be marked as the time and place of birth for a new class of spectral approaches for solving FODEs/FPDEs. The new theory of regular/singular fractional Sturm–Liouville eigenproblems with explicit and orthogonal eigenfunctions, known as *Jacobi poly-fractonomials*, was born, which was later extended to the family of *generalized Jacobi functions*. This book aims to methodically present more than a decade of sustained research and development on the theory and practice of FODEs/FPDEs, mainly accomplished by the authors in addition to citing and acknowledging important contributions made by others in this field to the best of our ability. We understand that it is nearly impossible to cite all the existing works, and we do apologize to our respected colleagues all over the world if their works have not been cited here. This book consists of eight chapters, each serving as a standing pillar to establish step by step our story of dedication to this field.

In Chapter 1, we learn how fractional calculus establishes an interconnecting bridge at the interface between differential equations, mathematical physics, and probability theory. We begin by introducing and classifying the most commonly used types of fractional-order differential operators in the context of a variety of *diffusion models* and based on *random walks*. This way, not only does the reader learn about the stochastic interpretation of fractional diffusion models but she/he also gradually acquainted with the preliminary definitions in fractional calculus as well as their physical and mathematical meanings. In Chapter 2, we focus on the spectral expansions and related approximation properties. We introduce a series of spectral theories including fractional Sturm–Liouville eigenproblems and four new families of orthogonal/log orthogonal functions, i.e., Jacobi poly-fractonomials, tempered Jacobi poly-fractonomials, generalized Jacobi functions, and generalized log-normal functions, which will serve us as new spectral (nonsmooth) basis functions. In Chapters 3 and 4, we present the corresponding spectral and spectral element methods for FODEs and FPDEs, respectively, based on the new bases introduced in Chapter 2 and other existing ones that we introduce in due time. A number of variational/non-variational approaches are employed here, including: Galerkin methods (where the trial and test spaces coincide), Petrov–Galerkin methods (where the trial and test spaces do not necessarily coincide), and spectral collocation methods. Chapter 6 serves as a unique place to gain a deeper understanding of the *meaning of fractional Laplacian* in *bounded domain* with a variety of boundary conditions. We employ different definitions of fractional Laplacian and treat them via spectral and/or spectral element methods, and we also present all the required and careful comparisons between their predictions and performances. Chapter 5 examines the *integral* definition of fractional Laplacian in *unbounded* domains with a number of interesting applications. We devote Chapter 7 to introducing and reviewing around 40 years of developments and major contributions to developing time-integration methods for fractional models and systems. We subsequently introduce a new fractional Adams family of schemes, which seamlessly generalize the standard integer-order Adams schemes, introducing the additional notion of *history/memory* and its efficient numerical treatment. We do this by introducing fast methods and enriching the solution space by adding correction terms. We end this chapter with recent methods that can capture/control several random singularities at the initial condition level, making them even more practical. Finally in Chapter 8, we introduce a number of interesting applications in which FPDEs naturally emerge as the right (if not parsimonious!) models to describe the complex behavior of nature.

Acknowledgments

The present book is the product of dedicated, enthusiastic, and inclusive research, carried out by the authors and their diverse research groups on the subject over the past decade. To make this happen, we owe a great deal of acknowledgments to our families and the US federal funding agencies that supported us, in addition to our esteemed collaborators, postdoc associates, and Ph.D. students.

First and foremost, we would like to give hearty thanks to our families for their unconditionally stabilizing love and support all along. Moreover, we wish to give our best acknowledgments to the US Army Research Office, the US Army Research Lab, the US Air Force Office of Scientific Research, the US Department of Energy, Pacific Northwest National Lab, the US National Science Foundation, and Sandia National Lab for their strong support. Especially, we thank Joseph Myers, Bruce West, Radhakrishnan Balu, Luis Bravo Robles, and Anindya Ghoshal at the US ARO/ARL. Moreover, our warm appreciation goes to Fariba Fahroo and Jean-Luc Cambier at the US Air Force Office of Scientific Research and Matthew Munson at the United States Air Force Academy, in addition to Lee Jameson at the US National Science Foundation, William Spotz and Ali Ghassemian at the US Department of Energy, and Michael Parks at the Sandia National Lab.

We also wish to acknowledge our collaborators as well as our colleagues, postdoc research associates, and Ph.D. students who were with us on this journey and took part in this collective cutting-edge research effort, in alphabetical order: Mark Ainsworth, Sheng Chen, Marta D'Elia, Christian Glusa, Mamikon Gulian, Ben-Yu Guo, Jan S. Hesthaven, Ehsan Kharazmi, Zhiping Mao, Mark M. Meerschaert, Eduardo de Moraes, Guofei Pang, Ali Akhavan Safaei, Mehdi Samiee, Seyed Hadi Seyedi, Changtao Sheng, Fangying Song, Jorge L. Suzuki, Hong Wang, Fanhai Zeng, Handy (Zhongqiang) Zhang, and Xuan Zhao.

Finally, we acknowledge Cambridge University Press and warmly thank Kaitlin Leach and Arman Chowdhury for their editorial/publishing support.

1
Fractional Calculus and Anomalous Transport

We first present the essential concepts of fractional calculus that will form a proper mathematical language for understanding and modeling *anomalous transport* phenomena. There are different ways of motivating and introducing the subject. For instance as the following historical review suggests, fractional calculus was developed shortly after the advent of the standard integer-order calculus and for the purpose of generalizing the notion of classical differentiation and integration to any non-integer order. While this school of thought is mathematically interesting, we find the multi-scale link between the fractional calculus and probability theory particularly fascinating and insightful. Hence in this chapter, we will have a step-by-step generalization of the standard/normal transport (diffusion) model to a variety of anomalous transport models, governed by fractional differential operators at the macroscopic level in addition to their direct connection to their underlying anomalous stochastic processes, occurring at the microscopic scale.

We initially introduce the standard (normal) diffusion model that solves the probability density function of the stochastic Brownian motion/process, satisfying the normal scaling property. Our first generalization of the standard diffusion happens then through a new definition of the process increments, where they no longer are drawn from a normal (Gaussian) distribution, leading to α-stable Lévy flights at the microscopic level and correspondingly an anomalous diffusion model with a *fractional Laplacian* at the macroscopic scale that forms the first anomalous transport model we present in our book. Next, we show how the Riemann–Liouville fractional derivatives emerge in another anomalous diffusion model corresponding to the asymmetric α-stable Lévy flights at small scales. Subsequently, we introduce the notion of subdiffusion stochastic processes, in which the Caputo time-fractional derivative appears in the anomalous subdiffusion fractional model. Interestingly we combine the previous two cases, and construct *continuous-time random walks*, where a space-time fractional diffusion model will solve the evolution of the probability density function of the stochastic process. In the rest of the chapter, we motivate and introduce many other types of fractional derivatives that will code more complexity and variability at micro-to-macroscopic scales, including fractional material derivatives, time-variable diffusivity for the fractional Brownian motion, tempered/variable-order/distributed-order/vector fractional calculus, etc. We will end this chapter with a clarifying discussion on the notion of global versus local numerical methods for solving fractional differential equations, and why we are mainly regarding global (i.e., spectral and spectral element methods) for solving an array of fractional models for anomalous transport phenomena throughout the present book.

1.1 A Short Historical Review

The theory of classical *integer-order* calculus was independently developed by Isaac Newton and Gottfried Wilhelm Leibniz in seventeenth-century Europe, while its initial notion and basic ideas had in fact appeared in ancient Egypt, Greece, then in China and the Middle East, and later on, in medieval Europe and in India. Yet the notion of *fractional calculus*, as a generalizing theory dealing with derivatives/integrals of any (real or complex) arbitrary orders, first appeared in a letter written to Guillaume de l'Hôpital by Gottfried Wilhelm Leibniz in 1695, which can be marked as the birth certificate of fractional calculus. Shortly after the advent of standard calculus, Leibniz himself wrote to one of the Bernoulli brothers describing the similarity between the binomial theorem and the Leibniz rule for the fractional derivative of a product of two functions, breaking new mathematical ground for the development of a *non-integer order* theory. Subsequently, the term "fractional calculus" was officially introduced in Niels Henrik Abel's early work in Abel (1823), where all the fundamental elements of such generalizing calculus were constructed. Interestingly and independently, the foundations of the subject were laid by Liouville (1832a,b), where along his line of thought, Bernhard Riemann established a rather more solid approach to construct the fractional operators by introducing new crucial mathematical building blocks. It should be noted that in his work (which was published after his death in Riemann (1876)), Riemann faced some serious technical problems with the notion of "an arbitrary function of integration" (issues like the constant of integration in evaluation of standard indefinite integrals, or anti-derivatives, or primitive functions). Consequently, he could not continue this approach. It was in fact A. V. Letnikov, a Russian mathematician, who later obtained the first rigorous and comprehensive construction of the theory of the fractional integro-differentiation, which was later called the "*Riemann–Liouville integral and derivative*". An extended description of his results can be found in Letnikov and Chernykh (2011).

1.2 Why Anomalous Transport and Fractional Modeling?

The classical calculus leads to the formulation of differential/integral/integro-differential operators, all integer orders being associated with their underlying conventional statistical physics. However such models cannot fully describe/predict the realistic nonlocal and complex nature of the *anomalous transport* phenomena due to their inherently local character in space-time. Nature is abundant with such processes, in which for instance a cloud of particles spreads in a different manner from traditional diffusion. This class of physical phenomena is characterized by fascinating processes that exhibit non-Markovian (long-range memory) effects, non-Fickian (nonlocal in space) interactions, non-ergodic statistics, and non-equilibrium dynamics. The phenomena of anomalous transport have been observed in a wide variety of complex, multi-scale, and multi-physics systems such as: *sub-/superdiffusion in subsurface transport, kinetic plasma turbulence, aging polymers, glassy materials, in addition to amorphous semiconductors, biological cells, heterogeneous tissues, and fractal disordered media*, see e.g., Klages et al. (2008) and references therein. From the scientific point of view, better understanding and effective exploiting of anomalous transport phenomena via fractional operators will continue to improve and perhaps transform our perspectives towards future discoveries.

1.3 Anomalous Transport and Stochastic Processes

Fractional calculus is a natural mathematical language for representing anomalous transport from *micro-to-macro* phenomena. In fact, fractional calculus establishes an interconnecting bridge at the interface between statistical physics, differential equations, and probability theory. Therefore, we find it effective to begin by introducing and classifying the most commonly used types of fractional-order differential operators in the context of *diffusion models* based on *random walks*. In this regard, and for simplicity, we restrict our discussion to one spatial dimension except for a few remarks, in which the extension to higher dimensions is touched upon. For other pedagogical reasons, we aim to describe the underlying stochastic processes in terms of their discretizations, thinking of them as sequences of random variables $X_{n\Delta t}$, defined as the cumulative sum of jumps at time $n\Delta t$, where $\Delta t > 0$ denotes the time-step of the random walker and the integer n represents the number index of the jump. We will show how fractional partial differential equations (FPDEs) govern the statistical properties of *continuous-time* random walks, which are appropriate scaling (or long-time) limits of the discrete random walks.

Before we begin our introduction to the stochastic processes, we provide a summary of symbols used in Chapter 1, in addition to their descriptions, and the referral to their definitions or first appearance in Table 1.1.

1.3.1 Standard (Normal) Diffusion

The connection between Brownian motion B_t and the standard diffusion model was studied in seminal works by Louis Bachelier (1900) and Albert Einstein (1905). The diffusion equation is posed in

$$\begin{cases} \dfrac{\partial u}{\partial t}(x,t) = k^2 \Delta u(x,t), & x \in \mathbb{R}, \quad t > 0, \\ u(x,t=0) = u_0(x), & x \in \mathbb{R}, \end{cases} \tag{1.3.1}$$

as a mixed initial/boundary value problem in which $k^2 > 0$ is the diffusion coefficient and $\Delta u = \partial^2 u/\partial x^2$ denotes the standard (integer-order) Laplacian. Brownian motion B_t is a continuous-time stochastic process defined for $t \geq 0$, whose discretized version using the time-step Δt has the property that $B_{t=0} = 0$ and

$$B_{t+\Delta t} = B_t + \Delta B; \quad \Delta B \sim \mathcal{N}(\mu = 0, \sigma = k\sqrt{\Delta t}), \tag{1.3.2}$$

where ΔB's denote independent and identically distributed (i.i.d.) increments. The notation $\sim$ indicates that the increment ΔB at each time-step of Δt is drawn from a normal distribution $\mathcal{N}(0, k\sqrt{\Delta t})$ with mean $\mu = 0$ and standard deviation $\sigma = k\sqrt{\Delta t}$, where the corresponding normal probability density function (PDF) is explicitly given by

$$p_{\mathcal{N}}(x; \mu, \sigma) = \frac{1}{\sigma\sqrt{2\pi}} e^{-\frac{1}{2}\left(\frac{x-\mu}{\sigma}\right)^2}. \tag{1.3.3}$$

The rule (1.3.2) for sampling a path of B_t at times $m\Delta t$, $m = 0, 1, 2, ...,$ is an example of a discrete stochastic differential equation (SDE), and is referred to as the *Euler–Maruyama* discretization of Brownian motion (Kloeden and Platen, 1992)[1]. The discrete process $B_{m\Delta t}$ should be thought of as tracing

[1] Another frequently used discrete random walk that leads to Brownian motion simply involves steps of fixed length to the left or right with probability 1/2 each; see Lawler (2010). In the

Table 1.1. *Symbols used in Chapter 1, their description, and the referral to their equation/section.*

Symbol	Description	Equation/Section
$\frac{\partial}{\partial t}$	Partial derivative in time	(1.3.1)
Δ	Laplacian	(1.3.1)
$\mathcal{N}(\mu, \sigma)$	Normal random variable	(1.3.2)
B_t	Brownian motion	(1.3.2)
$\langle \cdot \rangle$	Mean of a random variable	(1.3.4)
α	Fractional order of spatial derivative	Sec. 1.3.2
S_α	Stable random variable	Sec. 1.3.2
f_α	Stable density function	(1.3.13)
X_t^α	Stable Lévy walk	(1.3.15)
$(-\Delta)^{\alpha/2}$	Fractional negative Laplacian	(1.3.17)
Γ	Gamma function	(1.3.18)
$\mathcal{F}$	Fourier transform	(1.3.20)
${}_{a}^{\mathrm{RL}}D_x^\alpha$	Left-sided Riemann–Liouville derivative	(1.3.24)
${}_{x}^{\mathrm{RL}}D_b^\alpha$	Right-sided Riemann–Liouville derivative	(1.3.25)
$X_t^{\alpha,p}$	Asymmetric Lévy walk	(1.3.29)
$(-\Delta)_M^{\alpha/2}$	Directional fractional Laplacian	(1.3.32)
β	Fractional order of temporal derivative	Sec. 1.3.4
ψ_β	Standard stable subordinator density function	(1.3.34)
$\mathcal{L}$	Laplace transform	(1.3.36)
$B_{\tau(t)}$	Brownian motion with waiting times	Sec. 1.3.4
${}_{a}^{C}D_t^\beta$	Caputo derivative	(1.3.40)
E_θ	One-parameter Mittag-Leffler function $E_{\theta,1}$	(1.3.42)
$B_H(t)$	Fractional Brownian motion	(1.3.61)
H	Hurst index	(1.3.62)
$\mathbb{E}[B_H(t)B_H(s)]$	Covariance structure of a fractional Brownian motion	(1.3.64)
$\frac{\partial^{\alpha,\lambda}}{\partial x^\alpha}$	Tempered fractional derivative	(1.3.65)
$\mathfrak{L}^{\alpha(\cdot,\cdot)}$	Variable-order fractional Laplacian	(1.3.67)
${}_{-\infty}^{RL}D_t^{\alpha(\cdot,\cdot)}$	Variable-order left-sided Riemann–Liouville derivative	(1.3.68)
${}_{-\infty}^{C}D_t^{\alpha(\cdot,\cdot)}$	Variable-order left-sided Caputo fractional derivative	(1.3.69)
${}^{D}\mathcal{D}_\phi$	Distributed-order fractional derivative	(1.3.70)
${}^{\phi}\mathcal{H}(\mathbb{R})$	Distributed fractional Sobolev space on $\mathbb{R}$	(1.3.71)
$\mathcal{J}_M^{1-\beta}$	Meerschaert et al.'s fractional integral	(1.3.74)
∇_M^β	Meerschaert et al.'s fractional gradient	(1.3.75)
$\mathrm{div}_M^\beta V$	Meerschaert et al.'s fractional divergence	(1.3.76)
$\mathrm{curl}_M^\beta V$	Meerschaert et al.'s fractional curl	(1.3.77)
Grad_W^α	Tarasov's fractional gradient	(1.3.78)
Div_W^α	Tarasov's fractional divergence	(1.3.79)
Curl_W^α	Tarasov's fractional curl	(1.3.80)

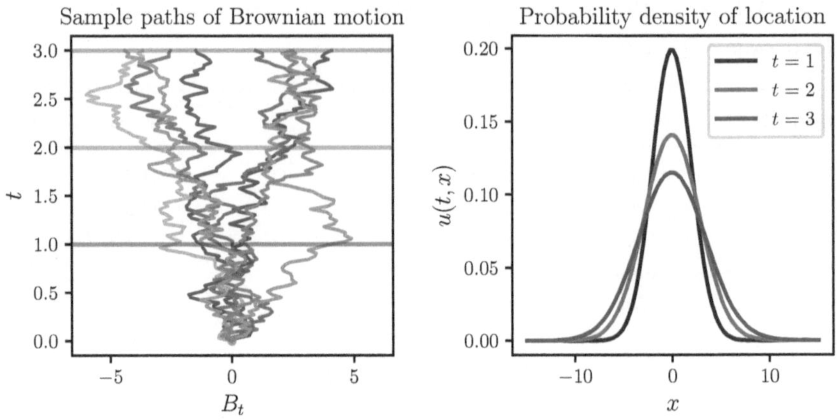

Figure 1.1 *(Left)* Eight independent sample paths of Brownian motion representing the path of a particle starting at the origin and stepping according to the rule (1.3.2). *(Right)* For $t = 1, 2, 3$, the probability density of the location of the particle, i.e., the fundamental solution to the classical heat equation (1.3.1). Source: Suzuki et al. (2023).

a path in $\mathbb{R}$ of a particle undergoing "jumps" in a random direction at time intervals of size Δt. At each time t, the position B_t of the particle is a *random variable*. It can be shown that the paths of the continuous-time process B_t are almost surely continuous in time (Revuz and Yor, 2013). From (1.3.2) and the central limit theorem (CLT), it follows that Brownian motion satisfies the *linear scaling* property (in time), that is,

$$\langle B_t^2 \rangle = 2k^2 t, \tag{1.3.4}$$

where the left-hand side denotes the *variance* of the random variable B_t. Given an initial PDF of particles $u_0(x)$ in $\mathbb{R}$ at $t = 0$, which subsequently undergo Brownian motion, *the corresponding PDF of the cloud of particles in* $\mathbb{R}$ *at each instance, i.e.,* $u(x,t)$*, is governed by* (1.3.1). In other words, diffusing particles (random walkers) described at a *microscopic scale* by Brownian motion, i.e., by (1.3.2) in discrete time, have their PDF in space – a *macroscopic property* – governed by the diffusion (heat) equation (Meerschaert and Sikorskii, 2011, Section 1.1). This is illustrated in Figure 1.1.

The consistency between this macroscopic description and the microscopic model is illustrated by scaling properties. A necessary property of such a Brownian motion model is the second-moment condition (1.3.4), which states that on average, particles travel a distance $k\sqrt{t}$ from their initial position after

long-time limit, all such discrete walks that draw increments from a finite-variance distribution lead to Brownian motion, due to the central limit theorem (Zaburdaev et al., 2015) time t. This is reflected by the fact that the solution of (1.3.1) with initial condition $u_0(x) = \delta(x)$, being the Dirac delta function, is

$$u(x,t) = \frac{1}{\sqrt{4\pi t}} e^{-x^2/4k^2 t}, \tag{1.3.5}$$

which is the normal density (1.3.3) with standard deviation $k\sqrt{t}$. Note that this solution has the property

$$u(x,t_2) = \left(\frac{t_2}{t_1}\right)^{-1/2} u\left(\frac{x}{(t_2/t_1)^{-1/2}}, t_1\right), \quad t_2 > t_1 > 0. \tag{1.3.6}$$

Thus, the distribution of a plume of particles in this diffusion model spreads out as $(t_2/t_1)^{1/2}$ as time elapses from t_1 to t_2, consistent with (1.3.4). The model for normal diffusion reviewed here is also referred to as *Fickian* diffusion. The heat equation (1.3.1) can be derived from the mass conservation with flux term J,

$$\frac{\partial u}{\partial t} + \frac{\partial J}{\partial x} = 0 \tag{1.3.7}$$

under Fick's law $J = \nabla u$. As discussed by Schumer et al. (2001), the fractional diffusion equations we introduce below follow from mass conservation with non-Fickian fluxes.

1.3.2 α-Stable Lévy Flights and Fractional Laplacian

There are important physical systems that exhibit diffusive behavior but do not satisfy the scaling property (1.3.4) (Klafter and Sokolov, 2005). This type of diffusion is referred to as *anomalous diffusion* as it cannot be adequately described by (1.3.2) with normally distributed increments. We need to develop a microscopic model that generalizes Brownian motion B_t, and a corresponding macroscopic model that generalizes the diffusion equation (1.3.1).

The first model we propose remains in the framework of a discrete SDE with (i.i.d.) increments,

$$X_{t+\Delta t} = X_t + \Delta X, \quad X_0 = 0, \tag{1.3.8}$$

but the increments ΔX are no longer drawn from a normal distribution. It follows from the central limit theorem (CLT) that the only way to obtain a microscopic model in this framework that is statistically distinct from B_t, i.e., not equivalent in distribution, is to draw step sizes from a PDF with *infinite* variance (Zaburdaev et al., 2015; Meerschaert and Sikorskii, 2011).

We introduce the isotropic α-stable random variable $S_\alpha(\gamma, \sigma, \mu)$. This family of random variables is defined by their *characteristic function*, defined below[2]. For a general random variable X, the characteristic function φ_X is related to the PDF p_X by

$$\varphi_X(\xi) = \int e^{i\xi x} p_X(x)dx = \mathbb{E}[e^{i\xi x}], \tag{1.3.9}$$

being the mathematical expectation of the exponential function $e^{i\xi x}$. For instance, the characteristic function of the normal random variable is $e^{i\xi\mu-\sigma^2\xi^2/2}$. Moreover a straightforward calculation reveals that the α-stable random variable has the following characteristic function

$$\varphi_\alpha(\xi;\gamma,\sigma,\mu) = e^{i\xi\mu \,-\, |\sigma\xi|^\alpha(1-i\,\gamma\,\mathrm{sgn}(\xi)\,\Phi)}, \tag{1.3.10}$$

where

$$\Phi = \begin{cases} \tan\left(\frac{\pi\alpha}{2}\right), & \text{if } \alpha \neq 1, \\ -\frac{2}{\pi}\log(|\sigma\xi|), & \text{otherwise.} \end{cases} \tag{1.3.11}$$

The parameter $\alpha \in (0,2]$ is referred to as the *stability parameter* of the distribution, $\mu \in \mathbb{R}$ as the center of distribution, in addition, $\gamma \in [-1,1]$ denotes the skewness, and $\sigma \in (0,\infty)$ represents the scale. The *isotropic* or *symmetric* α-stable distribution $S_\alpha(\gamma = 0,\sigma,\mu = 0)$ therefore has the characteristic function

$$\varphi_\alpha(\xi;\gamma = 0,\sigma,\mu = 0) = e^{-\sigma|\xi|^\alpha}, \tag{1.3.12}$$

generalizing the characteristic function of the normal distribution with mean $\mu = 0$ and standard deviation $\sigma/\sqrt{2}$ and reducing to it when $\alpha = 2$. By the Fourier inversion theorem, the PDF of $S_\alpha(\gamma,\sigma,\mu)$ can be written

$$f_\alpha(x;\gamma,\sigma,\mu) = \frac{1}{2\pi}\int e^{-i\xi x}\varphi_\alpha(\xi;\gamma,\sigma,\mu)d\xi. \tag{1.3.13}$$

In general, the α-stable density does not admit a closed-form expression[3], but in the symmetric case where $\gamma = \mu = 0$, it has the property that

$$f_\alpha(x;\gamma = 0,\sigma,\mu = 0) \sim \frac{1}{|\sigma x|^{1+\alpha}} \quad \text{for large } x, \quad 0 < \alpha < 2, \tag{1.3.14}$$

as discussed in, e.g., Cont and Tankov (2003) or Nolan (2020). In other words, the density exhibits *power-law* (also called *Paretian*) tails. This is in contrast to the rapidly decaying square-exponential tails of the normal distribution. In

[2] Several parametrizations of the α-stable characteristic function exist. The parametrization (1.3.10) is due to Samorodnitsky and Taqqu (1994). See Nolan (1998, 2020) for discussions of alternate forms.

[3] Special cases are $\alpha = 2$ corresponding to the normal distribution, $\alpha = 1$ and $\gamma = 0$ corresponding to the Cauchy distribution, and $\alpha = 1/2$ and $\gamma = 1$ corresponding to the Lévy distribution.

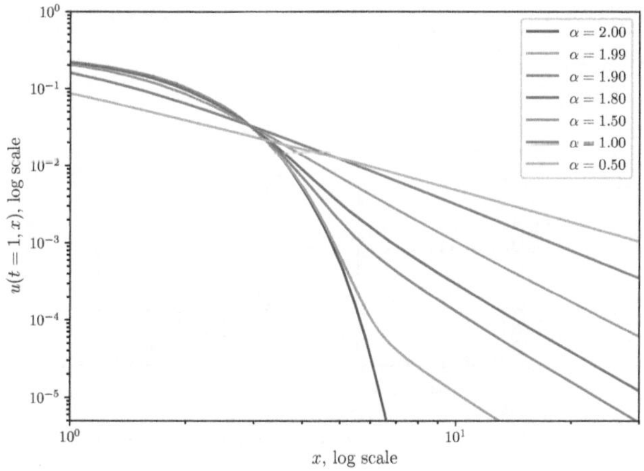

Figure 1.2 A plot of the α-stable densities in a log-log scale that illustrates the tail behavior asserted in (1.3.14). While α-stable densities do not have a closed form expression for all x, their simple, asymptotic inverse power-law behavior is an important heuristic. Source: Suzuki et al. (2023).

many settings, such tails are informally referred to as being examples of heavy tails or fat tails (Adler et al., 1998; Haas and Pigorsch, 2009), see Figure 1.2.

Using the isotropic distribution given above, we introduce the isotropic α-stable Lévy *flight* X_t^α by providing the corresponding discrete stochastic process. This is given for $t = k\Delta t$ with integer k by $X_0^\alpha = 0$ and the rule (Meerschaert and Sikorskii, 2011)

$$X_{t+\Delta t}^\alpha = X_t^\alpha + \Delta X^\alpha; \quad \Delta X^\alpha \sim S_\alpha(\gamma = 0, \sigma = k(\Delta t)^{1/\alpha}, \mu = 0). \quad (1.3.15)$$

The continuous-time stochastic process X_t^α for $t \geq 0$ can be thought of as a scaling limit as $\Delta t \to 0$ of the above random walk, and enjoys several theoretical properties such as stability and an extended central limit theorem (Meerschaert and Scheffler, 2001). However, it also has the property that for $\alpha < 2$, the paths of X_t^α are almost surely *discontinuous*, in contrast to Brownian motion; hence, the name Lévy "*flight*".

Given an initial distribution $u_0(x)$ of particles in $\mathbb{R}$ which undergo α-stable Lévy flight, the evolution of the distribution $u(x,t)$ for $t > 0$ is governed by the *space-fractional* diffusion equation (Meerschaert and Sikorskii, 2011, Section 1.2)

$$\begin{aligned} \frac{\partial u}{\partial t}(x,t) &= -k^\alpha(-\Delta)^{\alpha/2} u(x,t), \\ u(x,0) &= u_0(x), \end{aligned} \quad (1.3.16)$$

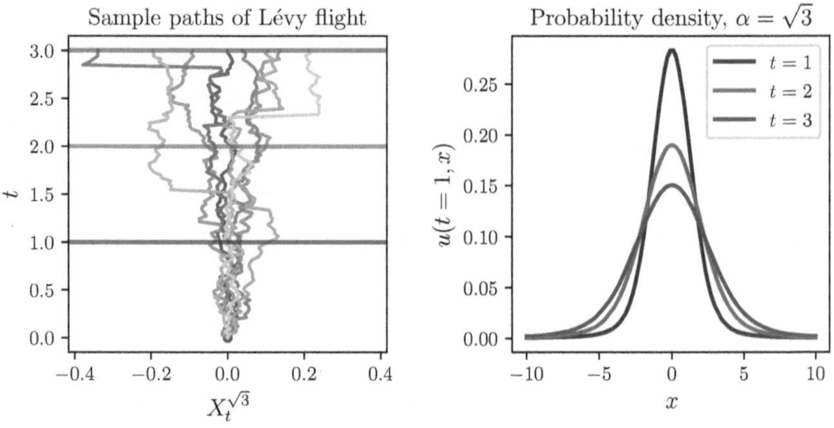

Figure 1.3 *(Left)* Eight independent sample paths of symmetric α-stable Lévy flight with $\alpha = \sqrt{3}$ representing the path of a particle starting at the origin and stepping according to the rule (1.3.15). *(Right)* For $t = 1, 2, 3$, the probability density of the location of the particle given by (1.3.13), i.e., the fundamental solution to the fractional diffusion equation (1.3.16). Compared to Figure 1.1, note that despite some qualitative similarity between the shapes of the density functions, the presence of long jumps signifies a striking difference between the paths of a particle undergoing Lévy flight versus Brownian motion. Source: Suzuki et al. (2023).

as illustrated in Figure 1.3. The fractional power of the negative Laplacian $(-\Delta)^{\alpha/2}$ is defined for $0 < \alpha < 2$ and for any dimension d as

$$(-\Delta)^{\alpha/2}u(x) = C_{d,\alpha} \text{ p.v.} \int_{\mathbb{R}^d} \frac{u(x) - u(y)}{|x - y|^{d+\alpha}} dy, \quad x \in \mathbb{R}^d, \tag{1.3.17}$$

with

$$C_{d,\alpha} = \frac{4^{\alpha/2}\Gamma\left(\alpha/2 + \frac{d}{2}\right)}{\pi^{d/2}|\Gamma(-\alpha/2)|}; \tag{1.3.18}$$

see Lischke et al. (2020), where "p.v." denotes the principal value of the singular integral. We have defined this operator in any dimension for future reference, although our present discussion is only restricted to the case $d = 1$. Perhaps the simplest characterization of the fractional Laplacian is the Fourier representation,

$$\mathcal{F}[(-\Delta)^{\alpha/2}u](\xi) = |\xi|^{\alpha}\mathcal{F}[u](\xi), \tag{1.3.19}$$

where the Fourier transform is

$$\mathcal{F}[u](\xi) = \int e^{-i\xi x}u(x)dx. \tag{1.3.20}$$

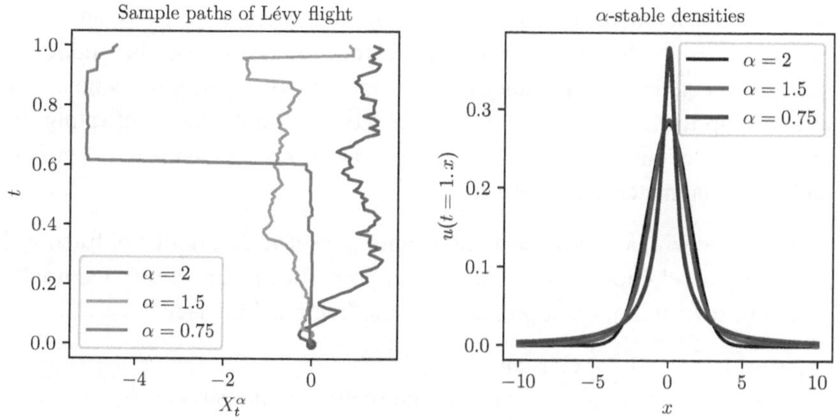

Figure 1.4 The seemingly innocuous heavy tails of the α-stable density, signifying non-vanishing probability of long jumps, are responsible for the striking properties of α-stable Lévy flights. As α decreases from 2, more mass in the middle region of the density is lost and is transferred towards the tails and the center, so that the relative probability of very small movements and very long movements increases (*right*). This is evident in the sample paths of the process (*left*). Source: Suzuki et al. (2023).

The simplest case of (1.3.16) is the initial condition $u_0(x) = \delta(x)$, in which case the solution is

$$u(x,t) = f_\alpha(x; \gamma = 0, \sigma = kt^{1/\alpha}, \mu = 0). \qquad (1.3.21)$$

This is known as the fundamental solution. Although this solution cannot be written in closed form, it satisfies

$$u(x,t_2) = \left(\frac{t_2}{t_1}\right)^{-1/\alpha} u\left(\frac{x}{(t_2/t_1)^{-1/\alpha}}, t_1\right), \quad t_2 > t_1 > 0, \qquad (1.3.22)$$

as shown in Meerschaert and Sikorskii (2011, Section 1.2). This illustrates that a plume of particles undergoing isotropic α-stable Lévy flight spreads by a factor of $(t_2/t_1)^{1/\alpha}$ as time elapses from t_1 to t_2, a faster rate when $\alpha < 2$ than the normal rate $t^{1/2}$. Thus, the α-stable Lévy flight is an example of *superdiffusion*. The dependence of the above solution as well as sample paths on α is shown in Figure 1.4.

Remark 1.1 Since $\alpha > 0$, the tail behavior of the isotropic α-stable density implies that the second moment of X_t^α diverges for $\alpha < 2$,

$$\langle X_t^\alpha \rangle^2 = \infty, \quad 0 < \alpha < 2. \qquad (1.3.23)$$

with the first moment (the mean) diverging also when $\alpha \leq 1$ (Nolan, 2020; Zaburdaev et al., 2015). This implies that the variance of α-stable motion is not a useful statistic for parametrizing α-stable Lévy flight; it bears no useful relationship to α. This aspect can be tackled in several ways, motivating the introduction of further fractional-order operators, such as tempered operators and fractional material derivatives.

While we will later provide a rather comprehensive description of fractional Laplacian in addition to its meaning and properties in Chapter 6, we briefly point out several important properties of the fractional Laplacian:

(1) $(-\Delta)^{\alpha/2}c = 0$, for constant c.
(2) The fractional Laplacian satisfies the *semigroup* property $(-\Delta)^{\alpha/2}(-\Delta)^{\beta/2} = (-\Delta)^{(\alpha+\beta)/2}$ (Samko et al., 1993).
(3) Unlike integer-order derivatives, the fractional Laplacian is indeed a *nonlocal* operator, i.e., the value of $(-\Delta)^{\alpha/2}u(x)$ depends on the values of u in all of $\mathbb{R}$ (or $\mathbb{R}^d$, for $d > 1$). In contrast, the value of any integer-order derivative of u at x depends only on the values of u in an infinitesimal neighborhood of x.

1.3.3 The Riemann–Liouville Fractional Derivatives and Asymmetric α-Stable Lévy Flight

The fractional Laplacian (1.3.17) was introduced in the previous section as a symmetric or rotation invariant operator for describing the symmetric or isotropic α-stable Lévy flight. This model involved a stability parameter $0 < \alpha \leq 2$ allowing it to generalize normal diffusion, with the scale σ and center μ playing similar roles as the standard deviation and mean of the normal distribution. However, the stable distribution also allows for a skewness parameter $\gamma \in [-1, 1]$, with $\beta = 0$ in the symmetric case, which has no analogue in the normal distribution or for Brownian motion. This is due to the CLT, which states that the use of any finite-variance distribution for the i.i.d. increments ΔX in (1.3.8), no matter how asymmetric, leads to X_t being normally distributed, so that the density is necessarily symmetric about the mean.

In this section, we introduce the one-sided Riemann–Liouville fractional derivatives as appropriate operators for modeling asymmetric α-stable Lévy flights, which are defined by (1.3.15) with $\Delta X^\alpha \sim S_\alpha(\gamma, \sigma = k(\Delta t)^{1/\alpha}, \mu = 0)$ for nonzero β. The left-sided and right-sided Riemann–Liouville derivatives in $\mathbb{R}$ are defined, for $n = \lceil \alpha \rceil$, as

$$ {}^{\mathrm{RL}}_{\ a}D^{\alpha}_{x}u(x) = \frac{1}{\Gamma(n-\alpha)}\left[\frac{d^n}{dz^n}\int_a^z \frac{u(y)}{|z-y|^{\alpha-n+1}}dy\right]_{z=x}, \tag{1.3.24} $$

$$ {}^{\mathrm{RL}}_{\ x}D^{\alpha}_{b}u(x) = \frac{(-1)^n}{\Gamma(n-\alpha)}\left[\frac{d^n}{dz^n}\int_z^b \frac{u(y)}{|z-y|^{\alpha-n+1}}dy\right]_{z=x}. \tag{1.3.25} $$

The books of Podlubny (1999), Oldham and Spanier (1974), and Meerschaert and Sikorskii (2011) discussed these operators in detail. These derivatives are frequently used in models with $a = -\infty$ and $b = \infty$. In connection with initial-value problems (IVPs), the left-sided Riemann–Liouville derivative in time, ${}^{\mathrm{RL}}_{\ 0}D^{\alpha}_{t}u(t)$, is sometimes used with $a = 0$. We have written the definitions (1.3.24) and (1.3.25) to avoid ambiguities in notation, and clearly show that substitution of the variable x occurs after integration and differentiation. An alternative approach is to define the Riemann–Liouville fractional integrals separately, as in the right-hand sides of (1.3.24) and (1.3.25); see Samko et al. (1993).

One quirk of the notation for Riemann–Liouville derivatives in (1.3.24) and (1.3.25) is the subtle writing of the upper and lower limits of integration $[a, x]$ and $[b, x]$, respectively, as subscripts. While this is suggestive, the result is that the variable of evaluation x occurs twice in the notation for each operator. If these derivatives are evaluated at any fixed value of x, this value should be substituted in both locations; thus, ${}^{\mathrm{RL}}_{\ a}D^{\alpha}_{5}u(5)$ represents a valid evaluation of the derivatives, but ${}^{\mathrm{RL}}_{\ a}D^{\alpha}_{x}u(5)$ and ${}^{\mathrm{RL}}_{\ a}D^{\alpha}_{5}u(x)$ do not.

With $a = -\infty$ and $b = \infty$, the Riemann–Liouville derivatives can be represented by the following Fourier representations

$$ \mathcal{F}[{}^{\mathrm{RL}}_{-\infty}D^{\alpha}_{x}u](\xi) = (-i\xi)^{\alpha}\mathcal{F}[u](\xi), \tag{1.3.26} $$

$$ \mathcal{F}[{}^{\mathrm{RL}}_{\ x}D^{\alpha}_{\infty}u](\xi) = (i\xi)^{\alpha}\mathcal{F}[u](\xi). \tag{1.3.27} $$

In one dimension, these can be used in the asymmetric diffusion model

$$ \begin{aligned} &\frac{\partial u}{\partial t}(x,t) = \frac{-k^{\alpha}}{\cos(\pi\alpha/2)}\left[p\ {}^{\mathrm{RL}}_{-\infty}D^{\alpha}_{x}u(x,t) + (1-p)\,{}^{\mathrm{RL}}_{\ x}D^{\alpha}_{\infty}u(x,t)\right],\\ &u(x,t=0) = u_0(x), \end{aligned} \tag{1.3.28} $$

which describes anomalous diffusion of independent particles. Here, the positions of each particle at time-steps of $k\Delta t$ for integer k are governed by (1.3.8) with increments ΔX being drawn from the *asymmetric* α-stable distribution

$$ \Delta X \sim S_{\alpha}(\gamma = 2p-1, \sigma = k(\Delta t)^{1/\alpha}, \mu = 0). \tag{1.3.29} $$

The resulting random variable given by the sum of k increments is denoted by $X_t^{\alpha,p}$, for $t = k\Delta t$. Thus, the skewness ranges from $\gamma = -1$ when $p = 0$ to $\gamma = 1$ when $p = 1$. The fundamental solution of (1.3.28) cf. (1.3.21) is

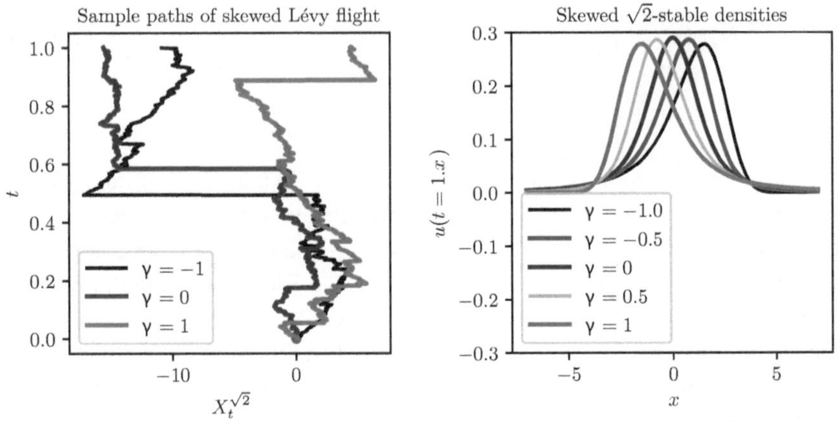

Figure 1.5 α-stable Lévy flights allow for asymmetric diffusion, which has no analogue within the classical diffusion framework. The α-stable density (1.3.29) admits a skewness parameter β, ranging from −1 to 1, which can adjust the relative probability of long jumps in a given direction (*right*), a statistical property that is evident in the sample paths (*left*). Such models are governed by the fractional-order diffusion equation involving Riemann–Liouville derivatives, as in (1.3.28). Source: Suzuki et al. (2023).

$$u(x,t) = f_\alpha(x; \gamma = 2p - 1, \sigma = kt^{1/\alpha}, \mu). \tag{1.3.30}$$

Sample paths of the process $X_t^{\alpha,p}$ are illustrated in Figure 1.5. Note that when $p = 1/2$, the distribution reverts to the symmetric α-stable distribution, and it can be shown in this case that (1.3.28) reduces to (1.3.16); more specifically,

$$\frac{1}{\cos(\pi\alpha/2)}\left[\frac{1}{2}\left({}^{\mathrm{RL}}_{-\infty}D^{\alpha}_{x}u(x)\right) + \frac{1}{2}\left({}^{\mathrm{RL}}_{x}D^{\alpha}_{\infty}u(x)\right)\right] = (-\Delta)^{\alpha/2}u(x). \tag{1.3.31}$$

The Fourier representation (1.3.26) suggests that the left-sided Riemann–Liouville derivative ${}^{\mathrm{RL}}_{-\infty}D^{\alpha}_{x}u$ should be thought of as a fractional power of the operator $\partial/\partial x$. However, the correspondence between (1.3.28) and (1.3.29) makes it clear that to obtain a complete description of α-stable Lévy flights in one dimension necessitates two operators, a left-sided and a right-sided operator, which agree with one another when $\alpha = 2$. Our interest in these models lies in the fact that an extended CLT holds for processes with i.i.d. increments drawn from distributions with infinite variance, but for which the tails of the density function satisfy Pareto-type conditions as in (1.3.14). For such processes, α-stable distributions play an analogous role to the normal distribution in the classical central limit theorem; unlike the classical theorem, for full generality, skewed α-stable distributions must be included in such a result. See Meerschaert and Scheffler (2001) or Meerschaert and Sikorskii (2011) for a treatment of these results.

We note that the Riemann–Liouville derivative can be utilized in dimensions $d > 1$. An anisotropic diffusion operator was introduced by Meerschaert et al. (1999) and Benson et al. (2000) as

$$-(-\Delta)_M^{\alpha/2} u(x) = C_{\alpha,d} \int_{|\theta|=1} D_\theta^\alpha u(x) M(d\theta), \tag{1.3.32}$$

$$C_{\alpha,d} = \frac{\Gamma\left(\frac{1-\alpha}{2}\right)\Gamma\left(\frac{d+\alpha}{2}\right)}{2\pi^{\frac{1+d}{2}}},$$

where $M(d\theta)$ denotes a non-negative measure on the angle θ in the unit sphere $\{|\theta| = 1\}$ in $\mathbb{R}^d$, and the Riemann–Liouville directional derivative is given by

$$D_\theta^\alpha u(x) = {}^{\mathrm{RL}}_{-\infty}D_t^\alpha v(t)\big|_{t=0}, \quad \text{where} \quad v(t) = u(x + t\theta).$$

Benson et al. (2000) showed that when the measure M is uniform, the operator (1.3.32) reduces to the fractional Laplacian (1.3.17). In higher dimensions and for general measures M, the operator (1.3.32) plays an analogous role to the operator in the right-hand side of (1.3.28), which is in fact a special case of it for $d = 1$. As such, it is used in models of anistropic multivariate α-stable Lévy diffusion.

1.3.4 Subdiffusion and Caputo Fractional Derivatives

The superdiffusive model introduced above, in which a plume of particles spreads out in space with rate $t^{1/\alpha}$ for $0 < \alpha < 2$, raises the question of whether a process can be constructed which results in diffusion *slower* than the Brownian rate $t^{1/2}$. In this section, we introduce such a model, constructed as Brownian motion with random waiting times drawn from a skewed stable distribution, supported over positive real numbers with a power-law tail. Here, we step away from the framework of the SDE given by (1.3.8). Rather than being defined by a simple time-stepping scheme with i.i.d. increments, the paths of the process are defined by a transformation or *post-processing* of Brownian paths B_t.

We introduce Brownian motion with waiting times, denoted by $B_{\tau(t)}$. The intuition is that the particle paths traced out in space by a discretization of $B_{\tau(t)}$ are paths of discretized Brownian motion B_t, but the particles wait at each point of the path for a random time drawn from the totally skewed stable distribution. The operational time $\tau(t)$, which introduces waiting and replaces linear time t, is an *inverse stable subordinator*. This is a stochastic process in the variable t, although we write $\tau(t)$ rather than using a subscript for typographical reasons. This process is constructed by first defining the *stable subordinator* $D(t)$, and defining $\tau(t)$ to be the inverse process of $D(\tau)$[4]. Both $D(t)$ and $\tau(t)$ are nondecreasing processes with time. In terms of paths, $\tau(t)$ arises from $D(t)$ as

$$\tau(t) = \inf\{\tau \text{ such that } D(\tau) > t\}. \tag{1.3.33}$$

Intuitively, $D(t)$ represents a cumulative waiting time process, keeping track of the total time waited by a particle throughout a path, while the inverse $\tau(t)$ represents an operational time, i.e., the time spent traveling. The increments of $D(t)$ represent the time waited at each location of a particle before the jump to the next location. More specifically, $D(t)$ is a totally skewed β-stable Lévy process (1.3.29) with stability index $\beta \in (0, 1)$, $\gamma = 1$, scale $\sigma = \cos(\pi\beta/2)$, and center $\mu = 0$; see Meerschaert and Sikorskii (2011), Example 5.14. The construction of sample paths of $B_{\tau(t)}$ is demonstrated in Figure 1.6. The resulting probability density function of $D(t)$,

$$\psi_\beta(t) = f_\beta(t; \gamma = 1, \sigma = \cos(\pi\beta/2), \mu = 0) \tag{1.3.34}$$

for waiting times is supported in non-negative real numbers. Due to the non-negative support of the waiting time density, the characteristic function (1.3.10) yields the Laplace transform of the waiting time density as

$$\mathcal{L}\left[\psi_\beta\right](s) = e^{-s^\beta}, \tag{1.3.35}$$

where the Laplace transform is defined as

$$\mathcal{L}[u](s) = \int_0^\infty e^{-st} u(t) dt. \tag{1.3.36}$$

See Meerschaert and Sikorskii (2011), p. 108 and p. 156 for a discussion. The variance of the process $B_{\tau(t)}$ is given by

$$\langle B_{\tau(t)} \rangle^2 = \frac{2}{\Gamma(\beta+1)} t^\beta, \quad 0 < \beta < 1, \tag{1.3.37}$$

which is the desired subdiffusive property. Note that the finiteness of the variance does not imply that the normal CLT applies to $B_{\tau(t)}$, which is not equal in distribution to Brownian motion nor to any Lévy process. In fact, $B_{\tau(t)}$ is not a Markov process.

[4] The definition of τ in terms of D is an example of a right-continuous inverse of an increasing function. Paths of D, thought of as functions of t, are nondecreasing, so that each path of τ constructed in this way is a continuous-from-the-right inverse of the parent path of D used to construct it.

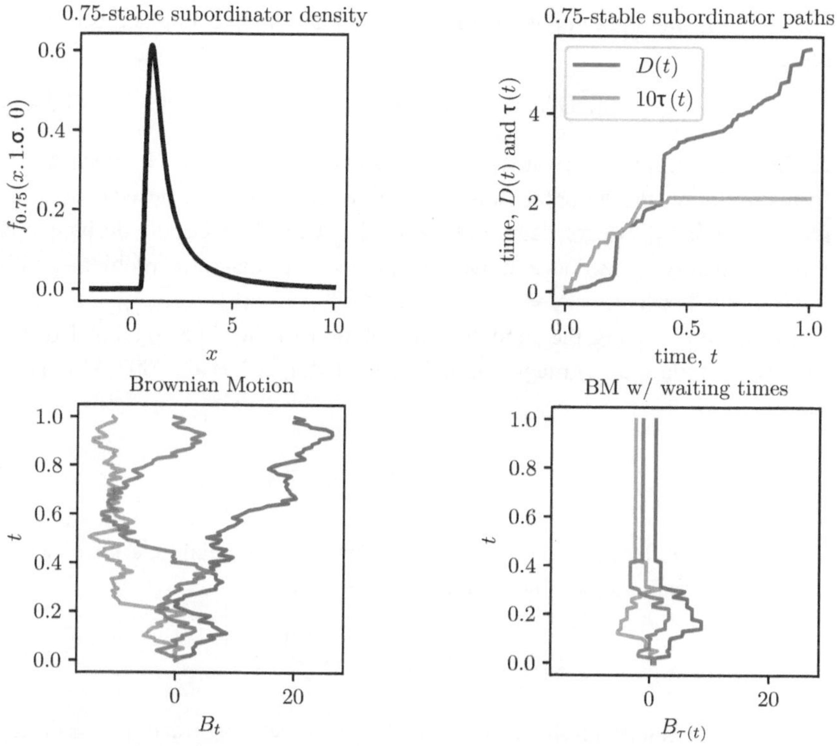

Figure 1.6 (*Top left*) Example of an α-stable subordinator density function, representing the density for random waiting times for the processes corresponding to the time-fractional diffusion equation (1.3.38). (*Top right*) Sample path of the subordinator (cumulative waiting time) $D(t)$, the parent path, and the inverse subordinator (operational time) $\tau(t)$ given by (1.3.33). Note that as t increases, $\tau(t)$ need not advance. (*Bottom left*) Three sample paths of Brownian motion. (*Bottom right*) Three sample paths of Brownian motion with waiting times, constructed from the Brownian paths in the bottom left panel. The particles trace out the same Brownian paths in space, but now wait for potentially several time-steps at each location, as specified by the operation time $\tau(t)$. Source: Suzuki et al. (2023).

The probability density of Brownian motion with waiting times $B_{\tau(t)}$ is governed by the time-fractional diffusion equation,

$$ {}_0^{\mathrm{C}}D_t^{\beta} u(x,t) = k^2 \Delta u(x,t), \tag{1.3.38} $$

$$ u(x, t = 0) = u_0(x). \tag{1.3.39} $$

Here, the Caputo derivative is defined for $0 < \beta < 1$ by

$$ {}_a^{\mathrm{C}}D_t^{\beta} u(t) = \frac{1}{\Gamma(1-\beta)} \int_a^t \frac{u'(s)}{|s-t|^{\beta}} ds. \tag{1.3.40} $$

For $a = 0$, this operator is characterized by the simple Laplace transform representation

$$\mathcal{L}[{}_0^{\mathrm{C}}D_t^{\beta}u](s) = s^{\beta}\mathcal{L}[u](s) - s^{\beta-1}u(0). \tag{1.3.41}$$

Higher-order Caputo derivatives can be defined, although the Laplace transforms of the resulting operators involve initial conditions for derivatives of u; see Section 2.3 of Meerschaert and Sikorskii (2011). The Caputo derivative is most frequently utilized as a derivative in time for initial-value problems, with the fractional order $0 < \alpha < 1$.

Before introducing the fundamental solution to the time-fractional diffusion, we introduce the Mittag-Leffler function (Mainardi et al., 2007; Mainardi, 2020)

$$E_{\theta}(z) = \sum_{\ell=0}^{\infty} \frac{z^{\ell}}{\Gamma(\theta\ell+1)}, \quad \theta > 0. \tag{1.3.42}$$

This Mittag-Leffler function $E_{\theta}(z)$ reduces to the exponential function e^z when $\theta = 1$, and has the following Laplace transform property

$$\mathcal{L}[E_{\theta}(-k^2t^{\theta})](s) = \frac{s^{\theta-1}}{s^{\theta}+k^2}, \tag{1.3.43}$$

which immediately implies that $E_{\beta}(-k^2t^{\beta})$ solves the fractional ordinary differential equation

$${}_0^{\mathrm{C}}D_t^{\beta}u = k^2u. \tag{1.3.44}$$

Returning to the diffusion equation (1.3.38) with initial condition $u(x,t=0) = \delta(x)$, applying the Fourier transform in space implies that

$$\mathcal{F}[u(\cdot,t)]\,(\xi) = E_{\beta}(-k^2\xi^2t^{\beta}), \tag{1.3.45}$$

which, as shown by Mainardi et al. (2007), yields a solution that can be written as

$$u(x,t) = t^{-\beta/2}U(|x|/t^{\beta/2}) \tag{1.3.46}$$

with

$$U(x) = \frac{1}{2}\sum_{k=0}^{\infty} \frac{(-x)^k}{k!\,\Gamma[-(\beta/2)k+1-(\beta/2)]} \tag{1.3.47}$$

being a special case of the Fox–Wright function. Note that $U(x) = u(x,t=1)$. While the fundamental solution above is transcendental, it has the following properties: for $\alpha = 1$, it reduces to the solution (1.3.5) of the classical diffusion equation; for $0 < \alpha < 1$, the solution decays faster than exponential and slower than Gaussian; and the second moment of the solution is

$$\sigma^2(t) = \frac{2}{\Gamma(\alpha+1)} t^{\beta}. \tag{1.3.48}$$

Note that the t^{β} scaling of this second moment is consistent with the scaling of the fundamental solution above.

1.3.5 Continuous-Time Random Walks and Space-Time Fractional Diffusion

Both the α-stable Lévy flight $X_t^{\alpha,p}$, which led to the space-fractional diffusion equation discused in Section 1.3.3, and the Brownian motion with β-stable subordinator operational time $B_{T^{\beta}(t)}$, which led to the time-fractional diffusion equation discussed in Section 1.3.4, are examples of *continuous-time random walks*. A continuous-time random walk (CTRW) allows for a general family of processes in space to be time-changed by a general family of waiting-time processes. To illustrate this concept, we consider the process $X_{T^{\beta}(t)}^{\alpha,p}$, which is α-stable Lévy flight $X_t^{\alpha,p}$ defined at the discrete level by (1.3.29) time-changed by the β-stable subordinator process $t \mapsto T^{\beta}(t)$ introduced in Section 1.3.4. This models a particle that performs independent jumps drawn from the α-stable process, waiting at each point for a random time drawn independently from the β-stable subordinator process. As shown by, e.g., Meerschaert and Sikorskii (2011) (Section 4.5), the probability density of this particle position is then governed by a differential equation that is fractional in both time and space,

$$\begin{aligned} &{}_0^C D_t^{\beta} u(x,t) = \frac{-k^{\alpha}}{\cos(\pi\alpha/2)} \left[p \left({}_{-\infty}^{RL} D_x^{\alpha} u(x,t) \right) + (1-p) \left({}_x^{RL} D_{\infty}^{\alpha} u(x,t) \right) \right], \\ &u(x, t = 0) = u_0(x). \end{aligned} \tag{1.3.49}$$

While intuitive, this result deserves a more detailed discussion within the general theory of CTRWs. In the standard CTRW model, particles wait at a location for time drawn from a density function ψ, and jump to a new location by an increment drawn from a density function ϕ. The waiting time and jump samples are assumed to be i.i.d., and uncoupled from each other (Zaburdaev et al., 2015; Metzler and Klafter, 2000; Scalas et al., 2004; Torrejon and Emelianenko, 2018). Thus, the densities ψ and ϕ completely determine the CTRW. From the waiting time density ψ, the probability that a particle will remain at any given position for time t is

$$\Psi(t) = 1 - \int_0^t \psi(\tau)d\tau; \tag{1.3.50}$$

this is referred to as the *survival probability* of a CTRW particle. Then, given an initial probability density of a particle $u_0(x) = u(x, t = 0)$, which can also be thought of as an initial distribution of an ensemble of independent particles, the following equation was derived by Montroll and Weiss (1965) for the density at later time

$$u(x,t) = \Psi(t)u_0(x) - \int_0^t \psi(t-\tau) \int_{-\infty}^{\infty} \phi(y)u(x-y,\tau)dyd\tau. \tag{1.3.51}$$

This equation is central to the CTRW theory[5]. Taking the Laplace transform in time, the Fourier transform in space, and solving for $\mathcal{F}[\mathcal{L}[u]]\,(\xi, s)$ yields the Montroll–Weiss equation (Montroll and Weiss, 1965),

$$\mathcal{F}[\mathcal{L}[u]]\,(\xi, s) = \frac{1 - \mathcal{L}[\psi](s)}{s} \frac{\mathcal{F}[u_0](\xi)}{1 - \mathcal{L}[\psi](s)\mathcal{F}[\phi](\xi)}. \tag{1.3.52}$$

If ϕ is the α-stable density (1.3.29) and ψ is the β-stable subordinator density (1.3.34), then $\mathcal{F}[\phi]$ is given by the analytical formula (1.3.10) and $\mathcal{L}[\psi]$ by (1.3.35), so that the Montroll–Weiss equation represents a closed-form solution of u in (ξ, s)-space. Unsurprisingly, it is impossible to perform inverse transforms and obtain u itself analytically, but u can be shown to satisfy (1.3.49) using the representations (1.3.41) and (1.3.26).

1.3.6 Lévy Walks and Fractional Material Derivatives

The superdiffusive α-stable Lévy flight exhibits infinite mean-squared displacement (MSD), which is a drawback for certain applications. Related to this is the infinite speed of propagation intrinsic to Lévy flights, i.e., the fact that particles have a nonzero probability of traveling an arbitrary large distance in a unit of time. Brownian motion also suffers from this feature, although this probability of large excursions is so low that MSD remains finite. A prototypical model of the superdiffusion that cannot be described by a Lévy flight is ballistic motion, in which particles simply move from an initial configuration in fixed random directions with speed v, for all time t. A ballistic particle travels a distance vt in time t from an initial position x_0. If reorientations are allowed, then the positions of these so-called *sub-ballistic particles* in space-time are confined to a ballistic cone

[5] This equation was also derived by Scher and Lax (1973a,b) and is referred to as the CTRW equation of Scher and Lax by Klafter and Silbey (1980). Other authors, such as Torrejon and Emelianenko (2018) refer to this as the *master equation* of a CTRW.

$$\{(x,t) : x \in [x_0 - vt, x_0 + vt], t \geq 0\}. \qquad (1.3.53)$$

Because the density function of the particle positions is compactly supported, all moments of the position are finite. Such a process cannot be described by Lévy flights.

To capture such a behavior, we introduce the *Lévy walk model*, following Zaburdaev et al. (2015). Such models are based on continuous-in-time motion of particles, rather than instantaneous jumps. A speed v of particles in a medium is specified; each particle moves with speed v in a chosen direction, before a reorientation event occurs, in which the direction changes instantaneously and the particle continues to move with speed v in the next direction. Assuming the direction at reorientation is sampled uniformly on the unit sphere, such a walk is determined by a PDF for the duration of movement $\psi(\tau)$. This leads to a survival probability $\Psi(t)$ given by (1.3.50), with ψ now representing the duration density. Thus, $\Psi(t)$ returns the probability that a particle has persisted in a given direction for time τ, i.e., has *not* experienced reorientation for time τ.

Similar to the CTRW case, a master equation can be derived for the probability density $u(x,t)$ of the location of the particle in the Laplace–Fourier space:

$$\mathcal{F}[\mathcal{L}[u]]\,(\xi, s) = \frac{\mathcal{L}[\Psi](s + iv\xi) + \mathcal{L}[\Psi](s - iv\xi)}{2 - \mathcal{L}[\psi](s + iv\xi) + \mathcal{L}[\psi](s - iv\xi)} \mathcal{F}[u_0](\xi). \quad (1.3.54)$$

Unlike the master equation for CTRWs, this equation exhibits coupling in Fourier and Laplace variables, representing coupling in space-time[6]. This results in governing equations that are considerably more complex than those of a standard CTRW. For a Lévy walk, ψ is taken to be a Pareto-type distribution,

$$\psi(\tau) = \frac{1}{\tau_0} \frac{\gamma}{(1 + \tau/\tau_0)^{\gamma}}, \qquad \tau_0 > 0, \gamma > 0. \qquad (1.3.55)$$

An asymptotic expansion of $\mathcal{L}[\psi]$ and $\mathcal{L}[\Psi]$ substituted in (1.3.54) yields the following approximation for the evolution of the density function of a Lévy walk in Fourier–Laplace space:

$$\mathcal{F}[\mathcal{L}[u]]\,(\xi, s) \approx \frac{(s + iv\xi)^{\gamma - 1} + (s - iv\xi)^{\gamma - 1}}{(s + iv\xi)^{\gamma} + (s - iv\xi)^{\gamma}} \mathcal{F}[u_0](\xi). \qquad (1.3.56)$$

Given v and u_0, this equation can be inverted to compute $u(x,t)$, but obtaining a governing equation in (x,t) is less straightforward from this point on, due

[6] A Lévy walk may be compared to a non-standard CTRW in which waiting times prior to jumps are correlated to the jump length, e.g., proportional to the jump length, so that long excursions are penalized by long waiting times, see Zaburdaev et al. (2015).

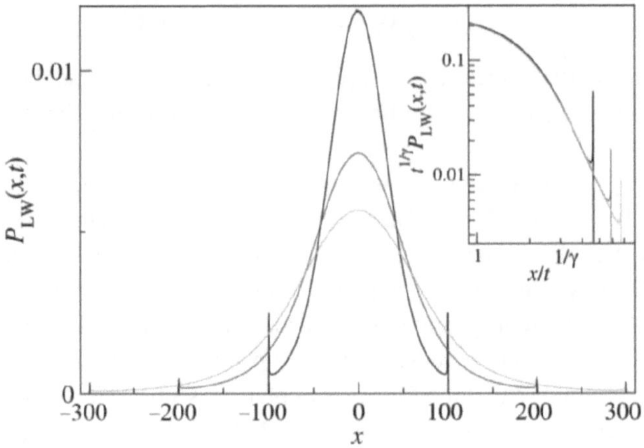

Figure 1.7 The evolution of PDF (denoted P_{LW} in the figure) of a Lévy walk, reproduced from (Zaburdaev et al., 2015). Here, $\gamma = 3/2$ and the density is plotted for $t = 100$ (black), $t = 200$ (darker gray), and $t = 300$ (lighter gray). The density mimics the density of a γ-stable Lévy flight in an interior region of the ballistic cone, scaling outwards as $t^{1/\gamma}$, supported inside the ballistic front (consisting of two points in one dimension) that scales outwards as t. Source: Suzuki et al. (2023).

to space-time coupling. Sokolov and Metzler (2003) suggested defining a fractional *material* or *substantial* derivative

$$(v^{-1}\partial_t \pm \partial_x)^{1/\gamma} u := \mathcal{F}^{-1}\mathcal{L}^{-1}\left[(s + iv\xi)^{\gamma} + (s - iv\xi)^{\gamma}\mathcal{F}[u_0](\xi)\right], \quad (1.3.57)$$

in order to obtain a governing equation for $u(x,t)$. Some works, such as those of Chen and Deng (2015), have explored numerical discretizations for these operators.

Despite the greater mathematical difficulties related to governing equations, as compared to other fractional models, Lévy walks have been widely used due to the physical nature of finite speed of propagation and finite MSD; see Zaburdaev et al. (2015) for a survey. When $1 < \gamma < 2$, by numerical approximations, it can be seen that $u(x,t)$ evolves from a δ-distribution with "a central part of the profile approximated by the Lévy distribution sandwiched between two ballistic peaks" that propogates at speed v, with an MSD and self-similarity property for large t that features a superdiffusive scale factor of $t^{1/\gamma}$ (Zaburdaev et al., 2015), see Figure 1.7.

Finally, the comparison of normal diffusion, superdiffusion, and subdiffusion via MSD of the particle models and the scaling-in-time of the fundamental

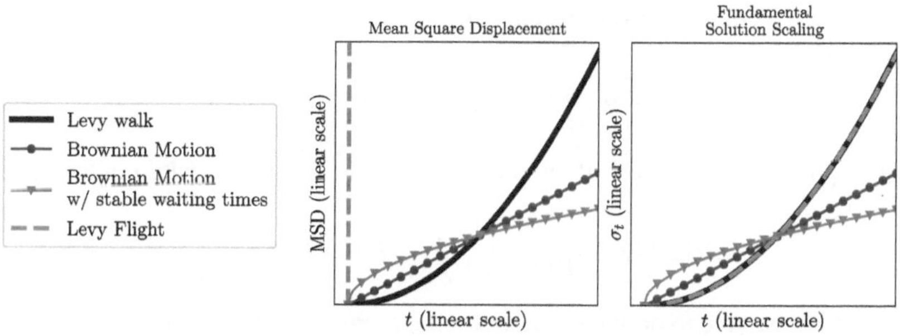

Figure 1.8 Brownian motion exhibits both MSD and scaling factor that are *linear* in time. Superdiffusive Lévy flight exhibits an infinite MSD and a fundamental solution scaling factor t^{α} for $\alpha > 1$, while the superdiffusive Lévy walk exhibits the same scaling of the fundamental solution as well as a finite MSD that scales as t^{α}. The subdiffusive Brownian motion with waiting times exhibits sublinear MSD and fundamental solution scaling, proportional to $t^{1/\alpha}$ for $\alpha < 1$. Source: Suzuki et al. (2023).

solution to the diffusion equations governing the density functions are illustrated in Figure 1.8.

1.3.7 Fractional Brownian Motion

In Section 1.3.1, we introduced the standard diffusion model as an integer-order PDE that governs the PDF of a cloud of particles spreading around according to Brownian motion. In earlier sections, we learned that by generalizing the construction of such underlying microscopic motions to incorporate new anomalous effects (e.g., heavy-tailed jumps, stochastic waiting times, etc.), we could arrive at a generalizing time-/space-fractional PDE in each case, governing the PDF evolution of the underlying process, being either *subdiffusive* or *superdiffusive*.

Interestingly, it is possible to construct yet another class of important stochastic processes directly via Brownian motion without obtaining any fractional PDE in the limit, which alternatively can exhibit sub-super-diffusive behavior in the process. We will show that the corresponding PDF equation of such a process called *fractional Brownian motion* emerges as an *integer-order* PDE yet with a time-dependent diffusion coefficient.

Derivation of Fractional Brownian Motion

Fractional Brownian motion is the fractional derivative of a Brownian motion (Biagini et al., 2008; Meerschaert and Sikorskii, 2011). Suppose that B_t is a standard Brownian motion, defined in Section 1.3.1, with the characteristic function

$$\mathbb{E}[e^{ikB_t}] = e^{-tk^2/2}, \forall t \geq 0.$$

The Caputo fractional derivative of order $0 < \alpha < 1$ can be written in the form

$$_{-\infty}^{\ C}D_x^{\alpha} f(x) = \frac{1}{\Gamma(1-\alpha)} \int_{-\infty}^{x} f'(u)(x-u)^{-\alpha} du.$$

We hope to be able to define the fractional Brownian motion by replacing $f(x)$ by the Brownian motion B_t, but the derivative $B'(s)$ does not even exist since the paths of a Brownian motion are nowhere differentiable (almost surely). However, for a continuous function $g(s)$, we can define

$$\int_a^b g(s)B(ds) \approx \sum_{i=1}^{n} g(s_i)B(\Delta s_i), \tag{1.3.58}$$

where

$$\Delta s = (b-a)/n, \quad s_j = a + j\Delta s, \quad B(\Delta s_j) = B(s_j) - B(s_{j-1}),$$

for $j = 0, 1, \ldots, n$, and the approximating sum on the right converges (*in probability*) to the stochastic integral on the left as $n \to \infty$. Note that $B(\Delta s_j)$ is normal with mean zero and variance $s_j - s_{j-1}$, and that $B(\Delta s_1), \ldots, B(\Delta s_n)$ are independent, since B_t has independent increments. Then $\sum_j g(s_j)B(\Delta s_j)$ is normal with mean zero and variance $\sum_j g^2(s_j)\Delta s$, and it follows by taking the limit that

$$\int_a^b g(s)B(ds) \sim \mathcal{N}\left(0, \int_a^b |g(s)|^2 ds\right), \tag{1.3.59}$$

assuming that $|g(s)|^2$ is integrable over $a < s \leq b$. As a common practice in mathematics, the improper integral is defined as a limit of proper integrals

$$\int_{-\infty}^b g(s)B(ds) = \lim_{a\to-\infty} \int_a^b g(s)B(ds) \quad \text{in probability,}$$

and then

$$\int_{-\infty}^b g(s)B(ds) \sim \mathcal{N}\left(0, \int_{-\infty}^b |g(s)|^2 ds\right), \tag{1.3.60}$$

assuming that $|g(s)|^2$ is integrable over $-\infty < s \leq b$. Now we intend to define a *fractional derivative of the Brownian motion* by the formula

$$I(t) = \frac{1}{\Gamma(1-\alpha)} \int_{-\infty}^{t} (t-s)^{-\alpha} B(ds),$$

but the challenge is that since $g(s) = (t - s)^{-\alpha}$, we have

$$\int_{-\infty}^{t} |g(s)|^2 ds = \infty.$$

To work around this, we first define

$$I_a(t) = \frac{1}{\Gamma(1-\alpha)} \int_a^t (t-s)^{-\alpha} B(ds)$$

and consider the difference

$$\begin{aligned} B_H(t) &= \lim_{a \to -\infty} I_a(t) - I_a(0) \\ &= \frac{1}{\Gamma(1-\alpha)} \int_{-\infty}^{\infty} \left[(t-s)_+^{-\alpha} - (0-s)_+^{-\alpha} \right] B(ds), \end{aligned} \tag{1.3.61}$$

where $(x)_+ = x, \forall x > 0$ and vanishes otherwise. Here, we adopt the convention $0^0 = 0$.

Remark 1.2 Since $\int |g(s)|^2 ds < \infty$, $g(s)$ being the integrand of the above stochastic integral, for any $-1/2 < \alpha < 1/2$, this above stochastic integral is defined only for this admissible range of $\alpha \in (-1/2, 1/2)$.

Hurst Index H

The *Hurst index* $H = 1/2 - \alpha$ for $0 < H < 1$ governs the *self-similarity* of the fractional Brownian motion, defined in (1.3.61), that is,

$$B_H(ct) = c^H B_H(t). \tag{1.3.62}$$

To see why this is indeed the case, first note that the random (Brownian) measure $B(ds)$ in (1.3.61) has a scaling property $B(c\,ds) \sim c^{1/2} B(ds)$, since for an interval $V = [a, b]$, we have

$$B(V) \sim \mathcal{N}(0, |V|)$$

and

$$B(cV) \sim \mathcal{N}(0, |cV|) \sim c^{1/2} B(V).$$

Then a change of variables $s = cs'$ yields (1.3.62).

Remark 1.3 (Special Case: $H = 1/2$) The fractional Brownian motion $B_H(t)$ is a Brownian motion B_t on $t \in \mathbb{R}$ when $H = 1/2$. Indeed since in the special case $H = 1/2$, we have $\alpha = 0$, and then for $t \geq 0$, we get

$$B_H(t) = \int_{-\infty}^{\infty} I(0 \leq s < t)\, B(ds) = B_t - B_0 = B_t,$$

while for $t < 0$ we get

$$B_H(t) = \int_{-\infty}^{\infty} I(t \le s < 0)\, B(ds) = B_0 - B_t = -B_1(t) \sim B_t.$$

It follows from self-similarity $B_H(ct) \sim c^H B_H(t)$ that a fractional Brownian motion satisfies $B_H(t) \sim t^H B_H(1)$, $\forall t \in \mathbb{R}$, where the stochastic integral $B_H(1)$ is normal with mean zero.

The PDF of Fractional Brownian Motion

The PDF $p(x,t)$ of a fractional Brownian motion $B_H(t)$ satisfies a diffusion equation with variable coefficients

$$\frac{\partial}{\partial t} p(x,t) = \left(2HDt^{2H-1}\right) \frac{\partial^2}{\partial x^2} p(x,t), \quad \forall t > 0, \tag{1.3.63}$$

where the case $1/2 < H < 1$ is associated with a *superdiffusive* process, and $0 < H < 1/2$ represents a *subdiffusive* fractional Brownian motion. In fact, (1.3.63) is an *integer-order* diffusion model with time-variable diffusivity $2HDt^{2H-1}$, which grows nonlinearly with time (except when $H = 1$).

To see why (1.3.63) holds, we note that $B_H(t)$ has a PDF $p(x,t)$ with Fourier transform

$$\hat{p}(k,t) = \mathbb{E}[e^{-ikB_H(t)}] = e^{-Dt^{2H}k^2}, \quad \forall t > 0,$$

for some constant $D > 0$. Then, by taking the first time-derivative on both sides we have

$$\frac{d}{dt}\hat{p}(k,t) = 2HDt^{2H-1}(ik)^2\hat{p}(k,t),$$

where its inverse Fourier transform yields (1.3.63).

The increments of the fractional Brownian motion are *not* independent; however, its increments are *stationary*, i.e.,

$$B_H(t_2) - B_H(t_1) \sim B_H(t_2 - t_1).$$

Moreover, the fractional Brownian motion $B_H(t)$ with $H \neq 1/2$ is *not* a Lévy process, since it does not have independent increments. There are many Gaussian stochastic processes whose PDFs solve (1.3.63); however, fractional Brownian motion is *the only self-similar Gaussian process with stationary increments.*

Fractal Sample Paths and Covariance Structure of $B_H(t)$

The sample path of a fractional Brownian motion $B_H(t)$ is a random fractal with dimension $d = 2 - H$ (Falconer, 2003, Theorem 16.7). As the Hurst index

H increases from $1/2$ to 1, we are applying a fractional integral of increasing order, so the path of $B_H(t)$ becomes smoother. Moreover, the covariance structure of a fractional Brownian motion $B_H(t)$ is given by

$$\mathbb{E}[B_H(t)B_H(s)] = \frac{C}{2}\left[|t|^{2H} + |s|^{2H} - |t-s|^{2H}\right]. \tag{1.3.64}$$

This can be easily computed employing two important properties of $B_H(t)$, the self-similarity and stationary increments. When $0 < s < t$, since $B_H(t) \sim t^H B_H(1)$, we have $\mathbb{E}[B_H(t)^2] = t^{2H}C$, where $C = \mathbb{E}[B_H(1)^2]$. Now one can write

$$(B_H(t) - B_H(s))^2 = B_H(t)^2 + B_H(s)^2 - 2B_H(t)B_H(s),$$

and take expectations to get

$$C(t-s)^{2H} = Ct^{2H} + Cs^{2H} - 2\mathbb{E}[B_H(t)B_H(s)].$$

That yields

$$\mathbb{E}[B_H(t)B_H(s)] = \frac{C}{2}\left[t^{2H} + s^{2H} - (t-s)^{2H}\right].$$

The case $0 < t < s$ is similar, and we can combine these two cases to obtain (1.3.64). Note that the case for $t < 0$ or $s < 0$ is again similar, and leads to the same result.

1.3.8 Tempered Fractional Calculus

Consider a CTRW with independent power-law jumps each having probability density function

$$f(x) = C\,\alpha x^{-\alpha-1}\mathbf{1}_{[C^{1/\alpha},\infty]}(x).$$

The second moment of such distribution is divergent that accounts for the anomalous jumps, called *Lévy flights*. We learned that the limit of this sum is an α-stable Lévy motion, whose probability density $p(t,x)$ satisfies the fractional diffusion

$$\frac{\partial p(t,x)}{\partial t} = k^2 \frac{\partial^\alpha p(t,x)}{\partial x^\alpha}$$

with diffusion coefficient k^2. In practice, however, many physical processes take place in *bounded domains* in *finite times* and have *finite moments*. Therefore, the divergent second moments may not be applicable to such processes. There exist different techniques to address this barrier (Sokolov et al., 2004; Mantegna and Stanley, 1994), among which the most popular one is a new variation of the fractional calculus, where power-laws are *tempered*

Table 1.2. *Tempered fractional derivatives of order $\alpha \in (0,1]$, $\lambda \geq 0$ and $x \in [a,b]$.*

	Left-Sided, ${}^{*}_{a}D^{\alpha,\lambda}_{x} f(x),\ x \geq a$	Right-Sided, ${}^{*}_{x}D^{\alpha,\lambda}_{b} f(x),\ x \leq b$
C	$\frac{e^{-\lambda x}}{\Gamma(1-\alpha)} \int_a^x (x-s)^{-\alpha} \frac{d\, e^{\lambda s} f(s)}{ds} ds$	$\frac{e^{\lambda x}}{\Gamma(1-\alpha)} \int_x^b (x-s)^{-\alpha} \frac{d\, e^{-\lambda s} f(s)}{ds} ds$
RL	$\frac{e^{-\lambda x}}{\Gamma(1-\alpha)} \frac{d}{dx} \int_a^x (x-s)^{-\alpha} e^{\lambda s} f(s) ds$	$\frac{e^{\lambda x}}{\Gamma(1-\alpha)} \frac{d}{dx} \int_x^b (x-s)^{-\alpha} e^{-\lambda s} f(s) ds$

by an exponential factor (Sabzikar et al., 2015; Baeumer and Meerschaert, 2010; Cartea and del Castillo-Negrete, 2007; Beghin, 2015; Wu et al., 2016; Liemert et al., 2017; Boniece et al., 2018). This exponential tempering has both mathematical and practical advantages (Sabzikar et al., 2015; Beghin, 2015, and references therein). Tempered fractional diffusion applies an exponential tempering factor to the particle jump density, i.e.,

$$f_\epsilon(x) = C_\epsilon^{-1} x^{-\alpha-1} e^{\lambda x} \mathbf{1}_{[\epsilon,\infty]}(x),$$

where

$$C_\epsilon = \int_\epsilon^\infty x^{-\alpha-1} e^{\lambda x} dx.$$

Theorems 2.1 and 2.2 in Sabzikar et al. (2015) prove that a CTRW with tempered density and an independent Poisson process N_t^ϵ converges to a tempered stable Lévy motion, whose probability density $p(t,x)$ is the point source solution to the tempered space-fractional diffusion equation

$$\frac{\partial p(t,x)}{\partial t} = \mp k^2 \frac{\partial^{\alpha,\lambda} p(t,x)}{\partial x^\alpha} \tag{1.3.65}$$

('$-$' is for $0 < \alpha < 1$ and '$+$' is for $1 < \alpha < 2$). This equation models transient superdiffusion, where a particle plume initially spreads faster than what the traditional diffusion equation predicts, but later relaxes to a typical diffusion profile, see Sabzikar et al. (2015) for more details. The corresponding tempered fractional derivatives of order α with tempering parameter $\lambda \geq 0$ in Caputo and Riemann–Liouville sense are given in Table 1.2.

1.3.9 Variable-Order Fractional Derivatives

Given the physical meaning within stochastic models of the fractional order α in derivatives such as (1.3.17), (1.3.24), and (1.3.40), it is reasonable to expect that these parameters may vary in space and/or time. Variable-order fractional models are convenient to describe anomalous diffusion in the case of heterogeneous materials or media, or, more generally, when the nature of the diffusion process (subdiffusive, superdiffusive, and classical) changes with space and/or time. While models with constant fractional order are the simplest and most widely used, some of the model descriptions we have discussed are enriched by the use of variable fractional orders as *field variables*.

More flexible models, characterized by variable fractional orders, have been introduced for both space- and time-fractional differential operators (Razminia et al., 2012; Antil and Rautenberg, 2019; Zheng and Wang, 2020b; Darve et al., 2021; D'Elia and Glusa, 2021) and several discretization methods have been designed (Schneider et al., 2010; Chen et al., 2015; Zheng and Wang, 2020a). The improved descriptive power of variable-order fractional operators has been demonstrated in some recent works on parameter estimation (Pang et al., 2019, 2020a; Zheng et al., 2020).

Given a function

$$\alpha : \mathbb{R}^d \times \mathbb{R}^+ \to \mathbb{R}, \tag{1.3.66}$$

i.e., a function $\alpha(\boldsymbol{x}, t)$ of space and time, we define variable-order operators as follows. For a function $u(\boldsymbol{x}, t)$ with $\boldsymbol{x} \in \mathbb{R}^d$ and $t \in \mathbb{R}$, we define the variable-order fractional Laplacian[7] as

$$\mathfrak{L}^{\alpha(\cdot,\cdot)} u(\boldsymbol{x}, t) = C_{d,\alpha(\boldsymbol{x},t)} \text{ p.v.} \int_{\mathbb{R}^d} \frac{u(\boldsymbol{x}, t) - u(\boldsymbol{y}, t)}{|\boldsymbol{x} - \boldsymbol{y}|^{d+\alpha(\boldsymbol{x},t)}} dy. \tag{1.3.67}$$

Here, $\alpha(\boldsymbol{x}, t)$ is restricted to take values in $(0, 2)$. Note that for constant α, $\mathfrak{L}^\alpha = (-\Delta)^{\alpha/2}$. For $d = 1$ and $\alpha(x, t)$ restricted to $(0, 1)$, we define the variable-order left-sided Riemann–Liouville fractional derivative as

$${}_{-\infty}^{\text{RL}}D_t^{\alpha(\cdot,\cdot)} u(x, t) = \frac{1}{\Gamma(1 - \alpha(x, t))} \frac{\partial}{\partial x} \int_{-\infty}^{x} \frac{u(y, t)}{|x - y|^{\alpha(x,t)}} dy. \tag{1.3.68}$$

The right-sided Riemann–Liouville may be defined for variable order in an analogous way. We define the variable-order Caputo fractional derivative, again for $\alpha(x, t)$ taking values in $(0, 1)$, as

$${}_{-\infty}^{\text{C}}D_t^{\alpha(\cdot,\cdot)} u(x, t) = \frac{1}{\Gamma(1 - \alpha(x, t))} \int_{-\infty}^{t} \frac{\partial u}{\partial t}(x, s) \frac{1}{|s - t|^{\alpha(x,t)}} ds. \tag{1.3.69}$$

1.3.10 Distributed-Order Fractional Calculus

There are physical systems that demonstrate a non-scaling behavior corresponding to crossover between different power-laws or even non-power-law

[7] For more recent works and novel definitions of variable-order fractional Laplacians, we refer the reader to Antil and Rautenberg (2019) and D'Elia and Glusa (2021).

behavior. Distributed-order models offer a tool for mathematical modeling of such *multi-scale* phenomena. In this case, the differential order is distributed over a range of values rather than being just a fixed fraction as it is in fractional operators. Caputo (1995, 2001) proposed the use of distributed-order fractional derivatives to generalize the stress–strain relation of inelastic media and Fick's law. The connection of such operators in subdiffusion with the corresponding CTRW was first established in Chechkin et al. (2002), where the authors proposed diffusion-like equations with time and space fractional derivatives of distributed-order for the kinetic description of anomalous diffusion and relaxation phenomena. They studied the "natural form" of the corresponding distributed-order diffusion equation, which describes an underlying random process that may lead to retarded subdiffusion (super slow diffusion) or accelerating superdiffusion. Later, in Chechkin et al. (2008), they discussed the "modified form" of distributed-order equations with both temporal and spatial fractional derivatives and showed the process describing anomalous relaxation and diffusion phenomena getting less anomalous in the course of time evolution (accelerating subdiffusion and decelerating superdiffusion). For more detailed discussions see also Chechkin et al. (2003); Umarov and Steinberg (2006); Kochubei (2008); Chen and Kumagai (2008); Kochubei (2011); Jiao et al. (2012); Sandev et al. (2015, 2018) and the references therein. The application of distributed-order fractional derivatives is also discussed in rheological models (Lorenzo and Hartley, 2002; Atanackovic et al., 2009), economic processes with distributed memory fading (Tarasova and Tarasov, 2018), time domain analysis of control, filtering and signal processing (Li et al., 2011; Li and Chen, 2014), vibration and optimal control (Zaky and Machado, 2017; Duan and Baleanu, 2018), and frequency domain analysis (Bagley and Torvik, 2000). There is also a rapidly growing interest in the use of distributed-order fractional derivatives in the construction of mathematical models in the field of uncertainty quantification as the inherent uncertainty of experimental data can be directly incorporated into the differential operators (Kharazmi et al., 2017). The distributed-order fractional derivative over $x \in [a, b]$ with order $\alpha \in [\alpha_{min}, \alpha_{max}]$ can be expressed in the general form as

$$ {}^{D}\mathcal{D}_{\phi} u(x) = \int_{\alpha_{min}}^{\alpha_{max}} \phi(\alpha) \; {}^{*}_{a}D^{\alpha}_{x} u(x) \, d\alpha, \quad \forall x > a, \tag{1.3.70} $$

in which the prescript $*$ stands for any type of fractional derivative, and $\phi(\alpha)$ can be thought of as a distribution function, where $\alpha \mapsto \phi(\alpha)$ is a continuous mapping in $[\alpha_{min}, \alpha_{max}]$. We note that α_{min} and α_{max} are only the theoretical lower and upper terminals in the definition of distributed-order derivatives. However, in general, the distribution function $\phi(\alpha)$ can arbitrarily confine the domain of integration in each realization of practical problems, and thus requires different regularity of underlying function space. In Kharazmi et al. (2017), the authors defined the *distributed fractional Sobolev space* on $\mathbb{R}$, denoted by ${}^{\phi}\mathcal{H}(\mathbb{R})$, as a compliant underlying function space for the distributed-order derivatives. Let $\phi \in L^1([\alpha_{min}, \alpha_{max}])$, $0 \leq \alpha_{min} < \alpha_{max}$, be non-negative. Then,

$$
{}^{\phi}\mathcal{H}(\mathbb{R}) = \left\{ v \in L^2(\mathbb{R}) : \int_{\alpha_{min}}^{\alpha_{max}} [\phi(\alpha)(1 + |\omega|^2)^{\alpha}]^{\frac{1}{2}} \, \mathcal{F}(v)(\omega) \, d\alpha \in L^2(\mathbb{R}) \right\}, \tag{1.3.71}
$$

endowed with norm

$$
\| \cdot \|_{\phi, I} = \inf_{\tilde{v} \in {}^{\phi}\mathcal{H}(\mathbb{R}), \tilde{v}|_I = (\cdot)} \|\tilde{v}\|_{\phi, \mathbb{R}}.
$$

Subsequently, the *distributed* fractional Sobolev space on the finite closed interval I, denoted by ${}^{\phi}\mathcal{H}(I)$, is defined as

$$
{}^{\phi}\mathcal{H}(I) = \{ v \in L^2(I) : \ \exists \tilde{v} \in {}^{\phi}\mathcal{H}(\mathbb{R}) \ \text{s.t.} \ \tilde{v}|_I = v \},
$$

equipped with the equivalent left- and right-sided norms

$$
\| \cdot \|^2_{l, \phi, I} = \| \cdot \|^2_{L^2(I)} + \int_{\alpha_{min}}^{\alpha_{max}} \phi(\alpha) \| \, {}^{RL}_{x_L} D^{\alpha}_{x} (\cdot) \|^2_{L^2(I)} \, d\alpha, \tag{1.3.72}
$$

$$
\| \cdot \|^2_{r, \phi, I} = \| \cdot \|^2_{L^2(I)} + \int_{\alpha_{min}}^{\alpha_{max}} \phi(\alpha) \| \, {}^{RL}_{x} D^{\alpha}_{x_R} (\cdot) \|^2_{L^2(I)} \, d\alpha. \tag{1.3.73}
$$

We note that when $\phi > 0$ is continuous in I, ${}^{\phi}\mathcal{H}(\mathbb{R})$ is equivalent to $H^{\alpha_{max}}(\mathbb{R})$. However, in general, the choice of ϕ can arbitrarily confine the domain of integration in practice. In other words, α_{min} and α_{max} are only the theoretical lower and upper terminals in the definition of distributed-order fractional derivatives. For instance, in a distributed subdiffusion problem, the temporal derivative is associated with $\alpha_{min} = 0$ and $\alpha_{max} = 1$, and in a superdiffusion problem, the theoretical upper terminal $\alpha_{max} = 2$. We particularly can define ϕ in any possible subset of the interval $[\alpha_{min}, \alpha_{max}]$. Hence, in each realization of a physical process (e.g., sub- or superdiffusion) ϕ can be obtained from data, where the theoretical setting of the problem remains invariant yet requiring the solution to have less regularity (since ${}^{\phi}\mathcal{H}(\mathbb{R}) \supset H^{\alpha_{max}}(\mathbb{R})$ in general, see Figure 1.9).

1.3.11 Fractional Vector Calculus

The history of fractional vector calculus (FVC) is not as long as fractional calculus (Adda, 1998; Engheta, 1998; Naqvi and Abbas, 2004; Meerschaert

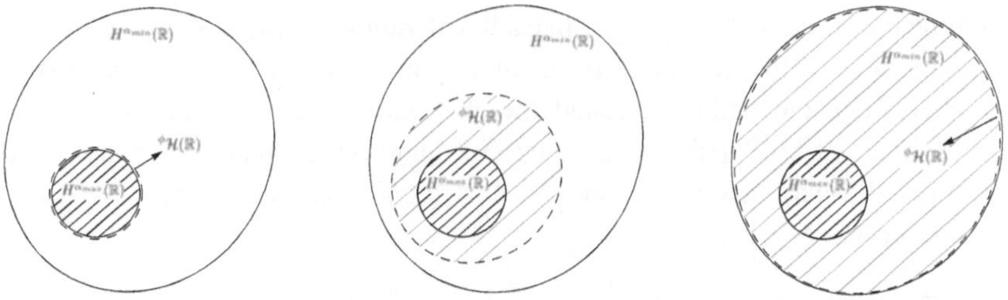

Figure 1.9 Schematic of distributed fractional Sobolev space $^{\phi}\mathcal{H}(\mathbb{R})$: (left) $\phi = \delta(\alpha - \alpha_{max})$ hence $^{\phi}\mathcal{H}(\mathbb{R}) = H^{\alpha_{max}}(\mathbb{R})$; (middle) ϕ defined on a compact support in $[\alpha_{min}, \alpha_{max}]$, hence, $^{\phi}\mathcal{H}(\mathbb{R}) \supset H^{\alpha_{max}}(\mathbb{R})$; (right) $\phi = \delta(\alpha - \alpha_{min})$, where $^{\phi}\mathcal{H}(\mathbb{R}) = H^{\alpha_{min}}(\mathbb{R})$. Here δ represents the Dirac delta function/distribution. Source: Kharazmi et al. (2017). Copyright ©2017 Society for Industrial and Applied Mathematics. Reprinted with permission. All rights reserved.

et al., 2006). Some of the main approaches in formulating FVC have been briefly described in Tarasov (2008). In particular, we review the two major contributions by Meerschaert et al. (2006) and Tarasov (2008).

• **Meerschaert et al.'s Approach:** In this approach (Meerschaert et al., 2006), the fractional integral operator is defined in a general form as

$$\mathcal{J}_M^{1-\beta}[\cdot] = \int_{\|\theta\|=1} \theta\, D_\theta^{\beta-1} \theta^T [\cdot]\, M(d\theta), \quad \theta \in \mathbb{R}^d, \tag{1.3.74}$$

where $D_\theta^{\beta-1}(\cdot)$ denotes the inverse Fourier transform of $(i\mathbf{k} \cdot \theta)^{\beta-1}\mathcal{F}[\cdot]$, $0 < \beta \le 1$, the symbol θ^T denotes transpose of column vector θ, and $M(d\theta)$ represents a positive finite-measure on the set of unit vectors. Given the integral operator (1.3.74), the fractional gradient, fractional divergence, and fractional curl operators are defined as follows

$$\nabla_M^\beta u(\boldsymbol{x}) = \mathcal{J}_M^{1-\beta} \nabla u(\boldsymbol{x}) = \int_{\|\theta\|=1} \theta\, D_\theta^\beta u(\boldsymbol{x})\, M(d\theta), \tag{1.3.75}$$

$$\mathrm{div}_M^\beta \boldsymbol{V}(\boldsymbol{x}) = \nabla \cdot \mathcal{J}_M^{1-\beta} \boldsymbol{V}(\boldsymbol{x}) = \int_{\|\theta\|=1} D_\theta^\beta \boldsymbol{V}(\boldsymbol{x}) \cdot \theta\, M(d\theta), \tag{1.3.76}$$

$$\mathrm{curl}_M^\beta \boldsymbol{V}(\boldsymbol{x}) = \nabla \times \mathcal{J}_M^{1-\beta} \boldsymbol{V}(\boldsymbol{x}) = \int_{\|\theta\|=1} \nabla \times \theta\, D_\theta^{\beta-1} \theta \cdot \boldsymbol{V}(\boldsymbol{x})\, M(d\theta), \tag{1.3.77}$$

where the operation symbols $\cdot$ and $\times$ denote inner and outer products of two vectors, respectively, and $D_\theta^\beta \boldsymbol{V}(\mathbf{x})$ represents the component-wise directional fractional derivative for vector field $\boldsymbol{V}$.

Note that in the case of vector field $V(\mathbf{x})$, the fractional integration operator $\mathcal{J}_M^{1-\beta}$ transforms the original field to another field $V^\tau = \mathcal{J}_M^{1-\beta} V$ in a nonlocal fashion. Therefore, the classical divergence and curl theorems for the resulting vector field V^τ will yield the fractional counterparts for the vector field V after substituting $V^\tau = \mathcal{J}_M^{1-\beta} V$. This is, however, a point of controversy as discussed in the following.

• **Tarasov's Approach:** In Tarasov (2008), Tarasov argued that although Meerschaert et al.'s approach (Meerschaert et al., 2006) allows the use of usual (integer) integral Stokes' and Green's theorems, it cannot be considered as a fractional generalization of vector calculus as the proposed curl operator can be presented as the usual (integer) curl operator. In return, he proposed to replace the Riemann–Liouville derivative with the Caputo one to overcome the failure of the Newton–Leibniz formula for the fractional generalization of Green's formula. Thus, he proved the *fundamental theorem of fractional calculus (FTFC)* by considering the fractional Riemann–Liouville integral and Caputo derivative as its antiderivative operator. Recall ${}_a\mathcal{I}_x^\alpha$ and ${}_a^{\mathrm{C}}D_x^\alpha$ being the fractional integral and Caputo derivative operators, respectively, which act on a real-valued function $f(x) \in L_1[a,b]$ for $\alpha > 0$, and $f(x) \in AC^n[a,b]$ for $n-1 < \alpha < n$. Here $AC^n[a,b]$ stands for the space of functions with absolutely continuous n-th derivatives in $[a,b]$. Then, FTFC can be represented as

$$
{}_a^{\mathrm{C}}D_x^\alpha \, {}_a\mathcal{I}_{x'}^\alpha f(x'') = f(x), \quad (\alpha > 0),
$$

$$
{}_a\mathcal{I}_x^\alpha \, {}_a^{\mathrm{C}}D_{x'}^\alpha f(x'') = f(x) - f(a), \quad (0 < \alpha < 1),
$$

which implies that the Caputo derivative of the fractional integral of a function returns the function itself when the orders of derivative and integral are the same. Thus, the Caputo derivative is the corresponding antiderivative operator of the fractional integral.

Let W be a domain of $\mathbb{R}^3$. Let $f(\mathbf{x})$ and $F(\mathbf{x})$ be real-valued scalar and vector fields, respectively, that have continuous derivatives up to order $n-1$ on W, such that the $n-1$ derivatives are absolutely continuous, i.e., $f, F \in AC^n[W]$. Then, a fractional generalization of the nabla operator is defined by

$$
\nabla_W^\alpha = {}^C\mathbf{D}_W^\alpha = \mathbf{e}_1 \, {}^C D_W^\alpha[x] + \mathbf{e}_2 \, {}^C D_W^\alpha[y] + \mathbf{e}_3 \, {}^C D_W^\alpha[z], \quad n-1 < \alpha < n,
$$

and hence, the fractional gradient, divergence, and curl operators are defined as

$$\mathrm{Grad}_W^\alpha f = {}^C\mathbf{D}_W^\alpha f = \mathbf{e}_l \, {}^C D_W^\alpha[x_l] f(x,y,z), \tag{1.3.78}$$

$$\mathrm{Div}_W^\alpha F = ({}^C\mathbf{D}_W^\alpha, F) = {}^C D_W^\alpha[x_l] F_l(x,y,z), \tag{1.3.79}$$

$$\mathrm{Curl}_W^\alpha F = [{}^C\mathbf{D}_W^\alpha, F] = \mathbf{e}_l \, \varepsilon_{lmk} \, {}^C D_W^\alpha[x_m] F_k(x,y,z), \tag{1.3.80}$$

respectively, using Einstein's notation and ε_{lmk} as Levi-Civita symbol. Other relations for fractional differential and integral vector operations, such as

$$\mathrm{Div}_W^\alpha \, \mathrm{Grad}_W^\alpha = ({}^C\mathbf{D}_W^\alpha)^2 = {}^C D_W^\alpha[x_l] \, {}^C D_W^\alpha[x_l], \tag{1.3.81}$$

and fractional circulation, flux, and volume integral are given in Tarasov (2008). The extension to fractional Green's theorem for a rectangle, fractional Stokes' theorem, and fractional Gauss' theorem for a parallelepiped are also proved in Tarasov (2008).

1.4 Global vs. Local Numerical Methods

Computational approaches appear as suitable tools for solving FODEs/FPDEs since exact solutions are only available for a small class of linear problems in unbounded domains. Here, we do not aim to provide a complete description of all existing numerical methods. We rather define the two main classes of computational approaches extensively employed for solving fractional-order problems in the present book: (i) *hp*-methods, including spectral methods (also known as *p-methods*) and spectral element methods (also known as *hp-methods*), and (ii) finite-difference methods.

In *hp*-methods, instead of solving the strong (original) form of the fractional model, we discretize the corresponding weak (variational) form of the problem. Such a weak form is obtained by multiplying the governing equation by another proper function, called *test function*, integrating over the computational domain, and then using integration-by-parts to transfer the derivative load from the solution to test function. After the weak/variational form is derived, we represent/approximate the solution in terms of a finite linear combination of some particular functions, called *basis/trial functions*. By plugging the solution expansion into the weak form and testing the problem on as many as the number of expansion terms, a linear system for the corresponding unknown solution coefficients is constructed. In spectral methods (p-methods), the basis/test functions are nontrivially defined over the whole computational domain and the whole domain is treated as a single element. Moreover in spectral methods, the numerical solution converges to the true solution by increasing the number of terms in the basis expansion (i.e., increasing the maximum basis order in the solution expansion). In spectral element methods (hp-method), we partition the domain into a finite number of non-overlapping patches, called *elements*, over which the basis/test functions are locally defined. In the formulation of spectral element methods, the degree of bases in each element can arbitrarily change such that the method can be understood as a spectral method-like in each element. Therefore, the solution in a spectral element method converges to the exact solution as the number of elements and/or the number of expansion terms (i.e., the basis order) in each element increases. The proper mathematical type of the convergence will be later rigorously determined.

In the finite-difference methods, however, we approximate the solution of a fractional model at certain temporal/spatial grid points, and then, we satisfy the governing equation at each of those grid points, replacing all the underlying fractional derivatives in the governing equation (i.e., the strong form) with their finite-difference approximations. This gives a finite system of algebraic equations to be solved in place of the corresponding differential equation. Therefore, the numerical solution converges point-wise to the true solution by increasing the number of grid points in space and/or time.

Remark 1.4 Given the aforementioned descriptions, finite-difference methods are *"local"* discretization schemes intrinsically; however, the nature of spectral and spectral element methods is indeed *"global (nonlocal)"* as the true solution is always approximated via basis *functions* (rather than grid *points*) over the entire global domain/sub-domains.

Why Global Numerical Methods?

The local finite-difference schemes are generally easy to implement; however, such numerical methods impose history/nonlocal calculations in the context of FODEs/FPDEs, which contain multiple summations over the solution history or spatial neighborhood. Hence, they can be computationally expensive as the inherent memory-dependent/nonlocal nature of fractional operators increases the complexity of arithmetic operations. Hence, such global hp-methods naturally fit the nonlocal processes, described by FODEs/FPDEs.

To address the aforementioned computational challenges, this book introduces the spectral theories on fractional and tempered fractional Sturm–Liouville problems in both regular and singular forms. The corresponding eigenfunctions and their generalizations will establish new classes of orthogonal basis/test functions with attractive spectral properties, which make them suitable for constructing efficient spectral and spectral element methods for fractional-order problems. Moreover, recalling the Askey scheme of orthog-

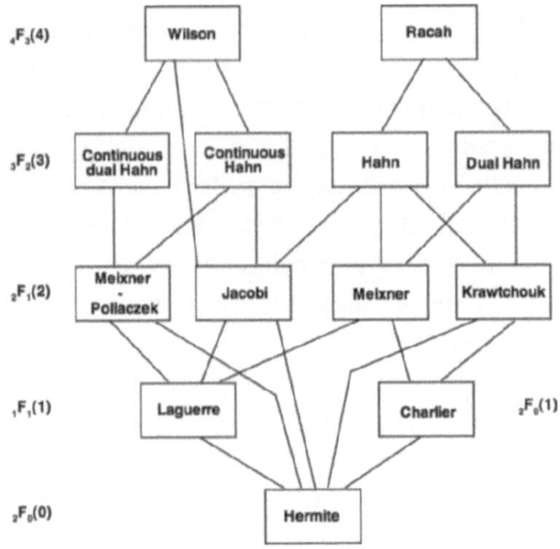

Figure 1.10 The Askey scheme of classical orthogonal polynomials. This big family of orthogonal polynomials is only a subset of the new extended family of poly-fractonomials and can be recovered for the special case of fractional order $\nu = 2$. Source: Xiu and Karniadakis (2002). Copyright ©2002 Society for Industrial and Applied Mathematics. Reprinted with permission. All rights reserved.

onal polynomials, shown in Figure 1.10, such an extension from Jacobi polynomials of integer order to their fractional order, introduced in Chapter 2, will create new families of orthogonal functions that can be used as basis functions to represent solutions of FODEs and FPDEs, encoding the nonlocal processes with heavy-tail measures.

Nonetheless, the time-integration of nonlinear and/or stiff FPDEs, where the characteristic time scales of problems strongly vary, will be inevitably treated by finite-difference methods. In Chapter 7, we will introduce the generalization of the well-known multi-step implicit-explicit (IMEX) methods in addition to the fast and accurate computation of the corresponding history terms for carrying out stable long-time integrations.

2

Spectral Expansions and Related Approximations

We first review the classic Jacobi polynomials as the eigenfunctions of the Sturm–Liouville problems, being extensively employed in spectral and spectral element methods for standard (integer-order) problems. Next, we present the need for developing fractional-order spectral theories, which can explicitly yield rather *non-polynomial* yet orthogonal eigensolutions, in order to effectively represent the singularities in solutions to FODEs/FPDEs. To this end, we present the regular and singular theories of fractional Sturm–Liouville eigenproblems. We call the corresponding explicit eignefunctions of these problems *Jacobi poly-fractonomials* (JPFs), being a linear combination of *fractonomials* (rather than monomials of integer powers). We demonstrate their attractive properties including their analytic fractional derivatives/integrals, three-term recursions, special values, function approximability, etc. Subsequently by introducing the generalized Jacobi polynomials and their degeneracy, we introduce the notion of *generalized Jacobi poly-fractonomials* (GJPFs), expanding the range of admissible parameters also allowing function singularities of negative indices at both ends. Next, we present a rigorous approximation theory for GJPFs with numerical examples. We further generalize our fractional Sturm–Liouville theories to regular/singular tempered fractional Sturm–Liouville eigenproblems, where a new exponentially tempered family of fractional orthogonal basis functions emerges. In the last section, we introduce a variant of orthogonal basis functions suitable for anomalous transport that occurs over significantly longer time-periods. The logarithmic orthogonal functions and their generalization will provide efficient basis/test functions to solve such slowly evolving and long-range nonlocal FODEs.

In Table 2.1, we provide a summary of the most relevant symbols in addition to their description and the place of the first appearance in Chapter 2.

Table 2.1. *Symbols used in Chapter 2, their description, and the referral to their equation, definition, or section.*

Symbol	Description	Equation/ Section
$P_n^{(\alpha,\beta)}(x)$	Jacobi polynomials	(2.1.1)
${}_2F_1(a,b;c;z)$	Hypergeomeric function	Sec. 2.1
$\mathcal{L}_i^{\mu}$	Regular fractional Sturm–Liouville operator of kind-I	(2.2.19)
$\Phi_n^{(1)}(x)$	Regular Jacobi poly-fractonomial of kind-I	(2.2.33)
$\lambda_n^{(1)}$	Eigenvalues of RFSLP-I	(2.2.34)
$\Phi_n^{(2)}(x)$	Regular Jacobi poly-fractonomial of kind-II	(2.2.35)
$\lambda_n^{(2)}$	Eigenvalues of RFSLP-II	(2.2.36)
$t^{k+\mu}$	Fractonomial	(2.6)
$\mathcal{F}_e^{n+\mu}$	Fractonomial expansion set	(2.2.43)
$\mathcal{L}_i^{\alpha,\beta;\mu}$	Singular fractional Sturm–Liouville operator of kind-I	(2.2.55)
${}^{(1)}\mathcal{P}_n^{\alpha,\beta,\mu}(x)$	Singular Jacobi poly-fractonomial of kind-I	(2.2.60)
$\Lambda_n^{(1)}$	Eigenvalues of SFSLP-I	(2.2.61)
${}^{(2)}\mathcal{P}_n^{\alpha,\beta,\mu}(x)$	Singular Jacobi poly-fractonomial of kind-II	(2.2.62)
$\Lambda_n^{(2)}$	Eigenvalues of SFSLP-II	(2.2.63)
$\mathcal{Q}_n^{(\alpha,\beta)}(x)$	Jacobi poly-fractonomials associated with Riesz potential	(2.3.20)
$\mathcal{P}_n^{(\alpha,\beta)}(x)$	Generalized Jacobi poly-fractonomial (GJPF)	(2.4.1)
${}^{l}\mathcal{P}_n^{(\alpha,-\beta)}(x)$	Left-sided GJPF	(2.4.8)
${}^{r}\mathcal{P}_n^{(-\alpha,\beta)}(x)$	Right-sided GJPF	(2.4.9)
$\mathcal{J}_n^{-\mu,-\nu}(x)$	Two-sided GJPF	(2.4.28)
$\pi_N^{\alpha,\beta}$	Jacobi-weighted $L^2_{\omega^{(\alpha,\beta)}}$-orthogonal projection	Sec. 2.5
$B^m_{\alpha,\beta}(\Omega)$	Jacobi-weighted Sobolev space	(2.5.4)
${}^{l}\mathbb{F}_N^{(\alpha,-\beta)}(\Lambda)$	Left-sided poly-fractonomial space	(2.5.15)
${}^{r}\mathbb{F}_N^{(-\alpha,\beta)}(\Lambda)$	Right-sided poly-fractonomial space	(2.5.28)
${}^{l}\pi_N^{(\alpha,-\beta)}$	Left-sided $L^2_{\omega^{(\alpha,-\beta)}}$-orthogonal projection	(2.5.22)
${}^{r}\pi_N^{(-\alpha,\beta)}$	Right-sided $L^2_{\omega^{(-\alpha,\beta)}}$-orthogonal projection	(2.5.31)
$\mathcal{B}_s^m(\Lambda)$	Space of two-sided Riesz GJPFs	(2.5.3)
$\Pi_N^{(-\alpha,-\alpha)}$	Two-sided $L^2_{\omega^{(-\alpha,-\alpha)}}$-orthogonal projection	(2.5.40)
${}^{\mathrm{RL}}_{x_L}\mathbb{I}_x^{\mu,\tau}$	Left-sided tempered Riemann–Liouville fractional integral	(2.6.1)
${}^{\mathrm{RL}}_{x}\mathbb{I}_{x_R}^{\mu,\tau}$	Right-sided tempered Riemann–Liouville integral	(2.6.2)
${}^{\mathrm{RL}}_{x_L}\mathbb{D}_x^{\mu,\tau}$	Left-sided tempered Riemann–Liouville derivative	(2.6.3)
${}^{\mathrm{RL}}_{x}\mathbb{D}_{x_R}^{\mu,\tau}$	Right-sided tempered Riemann–Liouville derivative	(2.6.4)
${}^{\mathrm{C}}_{x_L}\mathbb{D}_x^{\mu,\tau}$	Left-sided tempered Caputo derivative	(2.6.5)
${}^{\mathrm{C}}_{x}\mathbb{D}_{x_R}^{\mu,\tau}$	Right-sided tempered Caputo derivative	(2.6.6)
$\mathcal{L}_i^{\mu,\tau}$	Regular tempered fractional Sturm–Liouville operator	(2.6.17)
$F_n^{(1)}(x)$	Regular tempered Jacobi poly-fractonomial of kind-I	(2.6.21)
$F_n^{(2)}(x)$	Regular tempered Jacobi poly-fractonomial of kind-II	(2.6.23)

Table 2.1. *(cont.)*

Symbol	Description	Equation/ Section
$\mathbb{P}_n^{(1)}(x)$	Singular tempered Jacobi poly-fractonomial of kind-I	(2.6.37)
$\mathbb{P}_n^{(2)}(x)$	Singular tempered Jacobi poly-fractonomial of kind-II	(2.6.39)
$\mathcal{S}_n^{(\alpha,\beta)}(t)$	Logarithmic orthogonal function	(2.7.3)
$\mathcal{P}_K^{\log t}$	Finite-dimensional space of log orthogonal functions	(2.7.11)
$\hat{\pi}_N^{\alpha,\beta}$	Orthogonal projection onto $\mathcal{P}_N^{\log t}$	(2.7.12)
$\mathcal{S}_n^{(\alpha,\beta,\lambda)}(t)$	Generalized log orthogonal functions (GLOFs)	Definition 2.59
$\mathcal{P}_N^{\gamma,\log t}$	Finite-dimensional space of GLOFs	(2.7.36)
$\pi_N^{\alpha,\beta,\lambda}$	Projection operator onto $\mathcal{P}_N^{\frac{\beta-\lambda}{2},\log t}$	(2.7.38)

2.1 Jacobi Polynomials

Jacobi polynomials, occasionally called *hypergeometric* polynomials, were introduced by Carl Gustav Jacob Jacobi (1804–1851). They have played a crucial role in the development of theory and practice of polynomial spectral methods, see e.g., Gottlieb and Orszag (1977); Karniadakis et al. (2005); Hesthaven et al. (2007); Canuto et al. (2007); Shen et al. (2011a). Jacobi polynomials $P_n^{(\alpha,\beta)}(x)$, $n \geq 0$, are a class of classical orthogonal polynomials with respect to the weight $(1-x)^\alpha(1+x)^\beta$ on the interval $(-1,1)$. There are several ways to define the Jacobi polynomials (Szegő, 1975). Here we adopt the definition through hypergeometric functions as shown below. For $\alpha, \beta > -1$, the (classical) Jacobi polynomial of degree $n \in \mathbb{N}_0$ is defined by

$$\begin{aligned} P_n^{(\alpha,\beta)}(x) &= \frac{(\alpha+1)_n}{n!}\,{}_2F_1\Big(-n, n+\alpha+\beta+1; \alpha+1; \frac{1-x}{2}\Big) \\ &= (-1)^n \frac{(\beta+1)_n}{n!}\,{}_2F_1\Big(-n, n+\alpha+\beta+1; \beta+1; \frac{1+x}{2}\Big) \end{aligned} \tag{2.1.1}$$

for $x \in (-1,1)$, where

$${}_2F_1(a,b;c;z) = \sum_{k=0}^{\infty} \frac{(a)_k (b)_k}{(c)_k \Gamma(k+1)} z^k \tag{2.1.2}$$

for $|z| < 1$ and given $(a)_k = \frac{\Gamma(a+k)}{\Gamma(a)}$, $b > 0$. They are mutually orthogonal with respect to the Jacobi weight function: $\omega^{(\alpha,\beta)}(x) = (1-x)^\alpha(1+x)^\beta$, namely,

$$\int_{-1}^{1} P_n^{(\alpha,\beta)}(x) P_{n'}^{(\alpha,\beta)}(x) \omega^{(\alpha,\beta)}(x)\, dx = \gamma_n^{(\alpha,\beta)} \delta_{nn'}, \tag{2.1.3}$$

where $\delta_{nn'}$ is the Kronecker delta symbol, and

$$\gamma_n^{(\alpha,\beta)} = \frac{2^{\alpha+\beta+1}\Gamma(n+\alpha+1)\Gamma(n+\beta+1)}{(2n+\alpha+\beta+1)n!\ \Gamma(n+\alpha+\beta+1)}. \tag{2.1.4}$$

They satisfy the three-term recurrence relation

$$\begin{aligned} &P_{n+1}^{(\alpha,\beta)}(x) = \big(a_n^{(\alpha,\beta)} x - b_n^{(\alpha,\beta)}\big) P_n^{(\alpha,\beta)}(x) - c_n^{(\alpha,\beta)} P_{n-1}^{(\alpha,\beta)}(x), \quad n \ge 1, \\ &P_0^{(\alpha,\beta)}(x) = 1, \quad P_1^{(\alpha,\beta)}(x) = \frac{1}{2}(\alpha+\beta+2)x + \frac{1}{2}(\alpha-\beta), \end{aligned} \tag{2.1.5}$$

where

$$a_n^{(\alpha,\beta)} = \frac{(2n+\alpha+\beta+1)(2n+\alpha+\beta+2)}{2(n+1)(n+\alpha+\beta+1)}, \tag{2.1.6a}$$

$$b_n^{(\alpha,\beta)} = \frac{(\beta^2-\alpha^2)(2n+\alpha+\beta+1)}{2(n+1)(n+\alpha+\beta+1)(2n+\alpha+\beta)}, \tag{2.1.6b}$$

$$c_n^{(\alpha,\beta)} = \frac{(n+\alpha)(n+\beta)(2n+\alpha+\beta+2)}{(n+1)(n+\alpha+\beta+1)(2n+\alpha+\beta)}. \tag{2.1.6c}$$

Moreover, we have

$$P_n^{(\alpha,\beta)}(-x) = (-1)^n P_n^{(\beta,\alpha)}(x), \quad P_n^{(\alpha,\beta)}(1) = \frac{(\alpha+1)_n}{n!}. \tag{2.1.7}$$

The Jacobi polynomials are the eigenfunctions of the singular Sturm–Liouville problem (SLP)

$$\mathscr{L}_{\alpha,\beta}[P_n^{(\alpha,\beta)}](x) = \lambda_n^{(\alpha,\beta)} P_n^{(\alpha,\beta)}(x), \tag{2.1.8}$$

with the corresponding eigenvalues

$$\lambda_n^{(\alpha,\beta)} = n(n+\alpha+\beta+1), \tag{2.1.9}$$

where the singular Sturm–Liouville operator is

$$\begin{aligned} \mathscr{L}_{\alpha,\beta}[u](x) :&= -(1-x)^{-\alpha}(1+x)^{-\beta}\big[(1-x)^{\alpha+1}(1+x)^{\beta+1}u'(x)\big]' \\ &= (x^2-1)u''(x) + \big(\alpha-\beta+(\alpha+\beta+2)x\big)u'(x), \end{aligned} \tag{2.1.10}$$

in which the superscripts $'$ and $''$ denote the first and second (ordinary) derivatives of $u(x)$ with respect to x, respectively.

Why Fractional Eigenproblems?

The Sturm–Liouville theory has been the keystone for the development of spectral methods and the theory of self-adjoint operators. For many applications, the Sturm–Liouville problems (SLPs) are studied as boundary-value problems. However prior to 2013, mostly integer-order differential operators in SLPs have been employed, and such operators did not include any fractional differential operators with any known *explicit* eigensolutions. In most of the fractional Sturm–Liouville formulations presented recently, the ordinary derivatives in a traditional Sturm–Liouville problem were replaced with fractional derivatives, and the resulting problems were solved using a variety of numerical approaches. However, in such numerical approaches, round-off errors and the pseudo-spectra, introduced in approximating the infinite-dimensional boundary-value problem as a finite-dimensional eigen-value problem, prohibited computing more than a handful of eigenvalues and eigenfunctions with the desired precision. Furthermore, these papers did not really examine the common properties of fractional Sturm–Liouville problems (FSLPs) such as orthogonality of the eigenfunctions of fractional operators in addition to the matter of eigensolutions being real-valued as the eigensolutions of the RFSLP often appeared to be complex-valued, see e.g., Jin and William (2012).

Making the aforementioned fundamental properties for FSLPs available is certainly crucial in constructing proper numerical methods for efficient treatment of fractional operators, which is the central theme of the present book.

2.2 Jacobi Poly-fractonomials (JPFs)

A regular fractional Sturm–Liouville problem (RFSLP) of two types has been defined in Klimek and Agrawal (2012), where it has been shown that the eigenvalues of the problem are real-valued, and the eigenfunctions corresponding to distinct eigenvalues are orthogonal. In addition, the spectral properties of a regular FSLP for diffusion operators were studied in Bas and Metin (2012) demonstrating that the fractional diffusion operator is indeed self-adjoint. However, the eigensolutions had not been obtained explicitly and no approximation theories had been published prior to 2013.

In the rest of this chapter, we present a series of spectral theories for the regular and singular fractional Sturm–Liouville problems (RFSLP and SFSLP) and demonstrate its utility by constructing explicitly proper bases for spectral approximations of fractional functions.

In Sections 2.2.1 and 2.2.2, we first consider a regular problem of two kinds, i.e., RFSLP-I and -II. Then, we obtain the analytical eigensolutions to these problems explicitly for the first time. We show that the explicit eigenvalues to RFSLP-I and -II are real, discrete and simple. In addition, we demonstrate that the corresponding eigenfunctions are of non-polynomial form, called Jacobi *poly-fractonomials*. We also show that these eigenfunctions are orthogonal and dense in $L^2_\omega(-1,1)$, forming a complete basis in the Hilbert space. We subsequently extend the regular problem to a singular fractional Sturm–Liouville problem again of two kinds SFSLP-I and -II, and prove that the eigenvalues of these singular problems are real and the eigenfunctions corresponding to distinct eigenvalues are orthogonal; these too are computed analytically. We show that the eigensolutions to such singular problems share many fundamental properties with their regular counterparts, with the explicit eigenfunctions of SFSLP-I and -II completing the family of the Jacobi poly-fractonomials. Finally, we complete the spectral theory for the regular and singular FSLPs by analyzing the approximation properties of the eigenfunctions of RFSLPs and SFSLPs, which are employed as basis functions in approximation theory. Our numerical tests verify the theoretical exponential convergence in approximating non-polynomial functions in $L^2_w(-1,1)$. We compare with the standard polynomial basis functions (such as Legendre polynomials) demonstrating the fast exponential convergence of the poly-fractonomial bases.

Review of the Preliminary Definitions for FSLPs

For $x \in \Omega = (0,1)$ and $\rho \in \mathbb{R}^+$, the left and right Riemann–Liouville fractional integrals are respectively defined as

$$ {}^{\mathrm{RL}}_{0}\mathcal{I}^{\rho}_{x} f(x) = \frac{1}{\Gamma(\rho)} \int_0^x \frac{f(\tau)}{(x-\tau)^{1-\rho}} \mathrm{d}\tau, \quad {}^{\mathrm{RL}}_{x}\mathcal{I}^{\rho}_{1} f(x) = \frac{1}{\Gamma(\rho)} \int_x^1 \frac{f(\tau)}{(\tau-x)^{1-\rho}} \mathrm{d}\tau, \tag{2.2.1} $$

where $\Gamma(\cdot)$ represents the Euler gamma function. Moreover, for real $\rho \in [k-1,k)$ with $k \in \mathbb{N}$, the Riemann–Liouville fractional derivatives are defined by

$$ {}^{\mathrm{RL}}_{0}D^{\rho}_{x} f(x) = \frac{\mathrm{d}^k}{\mathrm{d}x^k} \left\{ {}^{\mathrm{RL}}_{0}\mathcal{I}^{k-\rho}_{x} f(x) \right\}, \quad {}^{\mathrm{RL}}_{x}D^{\rho}_{1} f(x) = (-1)^k \frac{\mathrm{d}^k}{\mathrm{d}x^k} \left\{ {}^{\mathrm{RL}}_{x}\mathcal{I}^{k-\rho}_{1} f(x) \right\}. \tag{2.2.2} $$

The Caputo fractional derivatives of order s are defined by

$$ {}^{\mathrm{C}}_{0}D^{\rho}_{x} f(x) = {}^{\mathrm{RL}}_{0}\mathcal{I}^{k-\rho}_{x} \left\{ \frac{\mathrm{d}^k}{\mathrm{d}x^k} f(x) \right\}, \quad {}^{\mathrm{C}}_{x}D^{\rho}_{1} f(x) = (-1)^k \, {}^{\mathrm{RL}}_{x}\mathcal{I}^{k-\rho}_{1} \left\{ \frac{\mathrm{d}^k}{\mathrm{d}x^k} f(x) \right\}. \tag{2.2.3} $$

It is easy to see that

- ${}_0^{\mathrm{C}}D_x^{\rho}\, c = 0,\ \rho > 0$, where c is a constant.
- If $f(0) = 0$, then

$$ {}_0^{\mathrm{C}}D_x^{\rho} f(x) = {}_0^{\mathrm{RL}}D_x^{\rho} f(x), \quad 0 < \rho < 1. $$

- For any function $f \in L^1(\Omega)$ (i.e., space of integrable functions in Ω),

$$ {}_0^{\mathrm{RL}}\mathcal{I}_x^{\rho}\, {}_0^{\mathrm{RL}}\mathcal{I}_x^{r} f(x) = {}_0^{\mathrm{RL}}\mathcal{I}_x^{\rho+r} f(x), \quad \rho, r \geq 0. $$

- Let $\rho > 0$ and $r > -1$. Then we have

$$ {}_0^{\mathrm{RL}}\mathcal{I}_x^{\rho}\, x^r = \frac{\Gamma(r+1)}{\Gamma(r+1+\rho)} x^{r+\rho}, \quad {}_0^{\mathrm{RL}}D_x^{\rho}\, x^r = \frac{\Gamma(r+1)}{\Gamma(r+1-\rho)} x^{r-\rho}, \quad r - \rho > -1. $$

We next extend the above definitions to a more general interval. Let $\Omega = (x_L, x_R)$. The corresponding left-sided and right-sided Riemann–Liouville integrals of order $\mu \in (0,1)$, are defined, respectively, as

$$ {}_{x_L}^{\mathrm{RL}}\mathcal{I}_x^{\mu} f(x) = \frac{1}{\Gamma(\mu)} \int_{x_L}^{x} \frac{f(s)}{(x-s)^{1-\mu}} \mathrm{d}s, \quad x > x_L, \tag{2.2.4} $$

and

$$ {}_{x}^{\mathrm{RL}}\mathcal{I}_{x_R}^{\mu} f(x) = \frac{1}{\Gamma(\mu)} \int_{x}^{x_R} \frac{f(s)}{(s-x)^{1-\mu}} \mathrm{d}s, \quad x < x_R. \tag{2.2.5} $$

The corresponding inverse operators, i.e., the left-sided and right-sided fractional derivatives of order $\mu \in (0,1)$, are then defined based on (2.2.4) and (2.2.5), as

$$ {}_{x_L}^{\mathrm{RL}}D_x^{\mu} f(x) = \frac{\mathrm{d}}{\mathrm{d}x} \left\{ {}_{x_L}^{\mathrm{RL}}\mathcal{I}_x^{1-\mu} f(x) \right\} = \frac{1}{\Gamma(1-\mu)} \frac{\mathrm{d}}{\mathrm{d}x} \int_{x_L}^{x} \frac{f(s)}{(x-s)^{\mu}} \mathrm{d}s, \quad x > x_L, \tag{2.2.6} $$

and

$$ {}_{x}^{\mathrm{RL}}D_{x_R}^{\mu} f(x) = -\frac{\mathrm{d}}{\mathrm{d}x} \left\{ {}_{x}^{\mathrm{RL}}\mathcal{I}_{x_R}^{1-\mu} f(x) \right\} = \frac{1}{\Gamma(1-\mu)} \left(-\frac{\mathrm{d}}{\mathrm{d}x}\right) \int_{x}^{x_R} \frac{f(s)}{(s-x)^{\mu}} \mathrm{d}s, \quad x < x_R. \tag{2.2.7} $$

Furthermore, the corresponding left-sided and right-sided Caputo derivatives of order $\mu \in (0,1)$ are obtained as

$$ {}_{x_L}^{\mathrm{C}}D_x^{\mu} f(x) = {}_{x_L}^{\mathrm{RL}}\mathcal{I}_x^{1-\mu} \left\{ \frac{\mathrm{d}f}{\mathrm{d}x}(x) \right\} = \frac{1}{\Gamma(1-\mu)} \int_{x_L}^{x} \frac{f'(s)}{(x-s)^{\mu}} \mathrm{d}s, \quad x > x_L, \tag{2.2.8} $$

and

$$ {}_{x}^{\mathrm{C}}D_{x_R}^{\mu} f(x) = {}_{x}^{\mathrm{RL}}\mathcal{I}_{x_R}^{1-\mu} \left\{ -\frac{\mathrm{d}f}{\mathrm{d}x}(x) \right\} = \frac{1}{\Gamma(1-\mu)} \int_{x_L}^{x} \frac{-f'(s)}{(x-s)^{\mu}} \mathrm{d}s, \quad x < x_R. \tag{2.2.9} $$

The two definitions of the left-sided and right-sided fractional derivatives of both Riemann–Liouville and Caputo types are linked by the following relationship, which can be derived by direct calculation, see e.g., Podlubny (1999) as

$$ {}^{\mathrm{RL}}_{x_L}\mathcal{D}^{\mu}_{x} f(x) = \frac{f(x_L)}{\Gamma(1-\mu)(x+x_L)^{\mu}} + {}^{\mathrm{C}}_{x_L}\mathcal{D}^{\mu}_{x} f(x), \tag{2.2.10} $$

and

$$ {}^{\mathrm{RL}}_{x}\mathcal{D}^{\mu}_{x_R} f(x) = \frac{f(x_R)}{\Gamma(1-\mu)(x_R-x)^{\mu}} + {}^{\mathrm{C}}_{x}\mathcal{D}^{\mu}_{x_R} f(x). \tag{2.2.11} $$

Note that the Riemann–Liouville fractional derivative acting on fractional integral of the same side and order μ is an identical operator:

$$ {}^{\mathrm{RL}}_{x_L}\mathcal{D}^{\mu}_{x}\, {}^{\mathrm{RL}}_{x_L}\mathcal{I}^{\mu}_{x} f(x) = f(x), \quad {}^{\mathrm{RL}}_{x}\mathcal{D}^{\mu}_{x_R}\, {}^{\mathrm{RL}}_{x}\mathcal{I}^{\mu}_{x_R} f(x) = f(x). \tag{2.2.12} $$

Moreover, the fractional integration-by-parts for the aforementioned fractional derivatives is obtained, see e.g., Klimek and Agrawal (2012) as

$$ \int_{x_L}^{x_R} f(x)\, {}^{\mathrm{RL}}_{x}\mathcal{D}^{\mu}_{x_R} g(x)\mathrm{d}x = \int_{x_L}^{x_R} g(x)\, {}^{\mathrm{C}}_{x_L}\mathcal{D}^{\mu}_{x} f(x)\mathrm{d}x - f(x)\, {}^{\mathrm{RL}}_{x}\mathcal{I}^{\mu}_{x_R} g(x)\Big|_{x=x_L}^{x_R}, \tag{2.2.13} $$

and

$$ \int_{x_L}^{x_R} f(x)\, {}^{\mathrm{RL}}_{x_L}\mathcal{D}^{\mu}_{x} g(x)\mathrm{d}x = \int_{x_L}^{x_R} g(x)\, {}^{\mathrm{C}}_{x}\mathcal{D}^{\mu}_{x_R} f(x)\mathrm{d}x + f(x)\, {}^{\mathrm{RL}}_{x_L}\mathcal{I}^{\mu}_{x} g(x)\Big|_{x=x_L}^{x_R}. \tag{2.2.14} $$

Finally, we recall a useful property of the Riemann–Liouville fractional derivatives. Assuming $0 < p, q < 1$ and $f(x_L) = 0$, we have that for $x > x_L$,

$$ {}^{\mathrm{RL}}_{x_L}\mathcal{D}^{p+q}_{x} g(x) = {}^{\mathrm{RL}}_{x_L}\mathcal{D}^{p}_{x} \left\{ {}^{\mathrm{RL}}_{x_L}\mathcal{D}^{q}_{x} g(x) \right\} = {}^{\mathrm{RL}}_{x_L}\mathcal{D}^{q}_{x} \left\{ {}^{\mathrm{RL}}_{x_L}\mathcal{D}^{p}_{x} g(x) \right\}. \tag{2.2.15} $$

While the notation introduced above is mainly associated with the Riemann–Liouville and Caputo fractional derivatives, we will use the notation $\frac{\partial^{\alpha}(\cdot)}{\partial x^{\alpha}}$ and $\frac{\partial^{\alpha}(\cdot)}{\partial(-x)^{\alpha}}$ to represent the left-sided (positive) and right-sided (negative) fractional derivative of Grünwald–Letnikov sense.

2.2.1 Regular Fractional Sturm–Liouville Problems of Kind-I and -II

We consider a regular fractional Sturm–Liouville problem (RFSLP) of order $\nu \in (0, 2)$ (Klimek and Agrawal, 2012), where the differential part contains both left-sided and right-sided fractional derivatives, each of order $\mu = \nu/2 \in (0, 1)$ as

$$
{}^{\mathrm{RL}}D^{\mu}\left[p_i(x)\,{}^{\mathrm{C}}D^{\mu}\,\Phi_{\lambda}^{(i)}(x)\right]+q_i(x)\Phi_{\lambda}^{(i)}(x)+\lambda w_i(x)\Phi_{\lambda}^{(i)}(x)=0,\quad x\in(x_L,x_R),
\tag{2.2.16}
$$

where $i\in\{1,2\}$, with $i=1$ denoting RFSLP of the first kind, where ${}^{\mathrm{RL}}D^{\mu}\equiv {}^{\mathrm{RL}}_{\ x}D^{\mu}_{x_R}$ (i.e., right-sided Riemann–Liouville fractional derivative of order μ) and ${}^{\mathrm{C}}D^{\mu}\equiv {}^{\ \mathrm{C}}_{x_L}D^{\mu}_{x}$ (i.e., left-sided Caputo fractional derivative of order μ), and $i=2$ corresponding to RFSLP of the second kind in which ${}^{\mathrm{RL}}D^{\mu}\equiv {}^{\mathrm{RL}}_{x_L}D^{\mu}_{x}$ and ${}^{\mathrm{C}}D^{\mu}\equiv {}^{\mathrm{C}}_{x}D^{\mu}_{x_R}$ (i.e., respectively, left-sided Riemann–Liouville and right-sided Caputo fractional derivative of order μ). In such a setting, $\mu\in(0,1)$, $p_i(x)\neq 0$, $w_i(x)$ is a non-negative weight function, and $q_i(x)$ is a potential function. Also, p_i, q_i and w_i are real-valued continuous functions in the interval (x_L,x_R).

The eigenproblem (2.2.16) is subject to the boundary conditions

$$
a_1\Phi_{\lambda}^{(i)}(x_L)+a_2\,{}^{\mathrm{RL}}\mathcal{I}^{1-\mu}\left[p_i(x)\,{}^{\mathrm{C}}D^{\mu}\,\Phi_{\lambda}^{(i)}(x)\right]\Big|_{x=x_L}=0, \tag{2.2.17}
$$

$$
b_1\Phi_{\lambda}^{(i)}(x_R)+b_2\,{}^{\mathrm{RL}}\mathcal{I}^{1-\mu}\left[p_i(x)\,{}^{\mathrm{C}}D^{\mu}\,\Phi_{\lambda}^{(i)}(x)\right]\Big|_{x=x_R}=0, \tag{2.2.18}
$$

where

$$
a_1^2+a_2^2\neq 0,\quad b_1^2+b_2^2\neq 0.
$$

In this notation, ${}^{\mathrm{RL}}\mathcal{I}^{1-\mu}\equiv {}^{\mathrm{RL}}_{\ x}\mathcal{I}^{1-\mu}_{x_R}$ (i.e., right-sided Riemann–Liouville fractional integration of order $1-\mu$) when $i=1$ for RFSLP of the first kind, while ${}^{\mathrm{RL}}\mathcal{I}^{1-\mu}\equiv {}^{\mathrm{RL}}_{x_L}\mathcal{I}^{1-\mu}_{x}$ (i.e., left-sided Riemann–Liouville fractional integration of order $1-\mu$) when $i=2$ for RFSLP of the second kind.

The problem of finding the eigenvalues λ such that the boundary-value problem (2.2.16)–(2.2.18) has nontrivial solutions yields the eigenfunctions of the regular fractional Sturm–Liouville eigenvalue problem. The following theorem characterizes the eigensolutions.

Theorem 2.1 (see Klimek and Agrawal, 2012) *The eigenvalues of* (2.2.16) *are real, and the eigenfunctions corresponding to distinct eigenvalues are orthogonal with respect to the weight functions $w_i(x)$ where $i\in\{1,2\}$.*

Regular Boundary-Value Problem Definition

In this section, we shall solve two particular forms of RFSLP (2.2.16)–(2.2.18) denoted by RFSLP-I when $i=1$ and RFSLP-II when $i=2$ of order $\nu=2\mu\in(0,2)$, where the potential functions $q_i(x)=0$ in both cases. To this end, the following nonlocal differential operator is defined

$$
\mathcal{L}_i^{\mu}:={}^{\mathrm{RL}}D^{\mu}\left[\mathcal{K}\ {}^{\mathrm{C}}D^{\mu}(\cdot)\right], \tag{2.2.19}
$$

where $\mathcal{K}$ is constant, and by the notation we introduced

$$\mathcal{L}_1^{\mu} := {}^{\text{RL}}_{\ x}D^{\mu}_{x_R}\left[\mathcal{K}\ {}^{\text{C}}_{x_L}D^{\mu}_{x}(\cdot)\right]$$

in RFSLP-I (i.e., first we take the left-sided μ-th order Caputo derivative of the function multiplied by a constant, and then we take the right-sided Riemann–Liouville derivative of order μ), and for the case of RFSLP-II, we reverse the order of the right-sided and left-sided derivatives for the inner and outer fractional derivatives in the operator, i.e.,

$$\mathcal{L}_2^{\mu} := {}^{\text{RL}}_{x_L}D^{\mu}_{x}\left[\mathcal{K}\ {}^{\text{C}}_{\ x}D^{\mu}_{x_R}(\cdot)\right],$$

where $\mu \in (0,1)$. In fact, we have set $p_i(x) = \mathcal{K}$, a nonzero constant, referred to as *stiffness* constant, which yields the regularity character to the boundary-value problem. That being defined, we consider the RFSLP (-I and -II) as

$$\mathcal{L}_i^{\mu}\Phi_{\lambda}^{(i)}(x) + \lambda(1-x)^{-\mu}(1+x)^{-\mu}\Phi_{\lambda}^{(i)}(x) = 0, \quad i \in \{1,2\},\ x \in (-1,1). \tag{2.2.20}$$

We shall solve (2.2.20) subject to a homogeneous Dirichlet and a homogeneous *integro-differential* boundary condition to the problems RFSLP-I and RFSLP-II, respectively, as

$$\begin{aligned} &\Phi_{\lambda}^{(1)}(-1) = 0, \\ &{}^{\text{RL}}_{\ x}\mathcal{I}^{1-\mu}_{1}\left[\mathcal{K}\ {}^{\text{C}}_{-1}D^{\mu}_{x}\,\Phi_{\lambda}^{(1)}(x)\right]\Big|_{x=1} = 0, \end{aligned} \tag{2.2.21}$$

and

$$\begin{aligned} &\Phi_{\lambda}^{(2)}(1) = 0, \\ &{}^{\text{RL}}_{-1}\mathcal{I}^{1-\mu}_{x}\left[\mathcal{K}\ {}^{\text{C}}_{\ x}D^{\mu}_{1}\,\Phi_{\lambda}^{(2)}(x)\right]\Big|_{x=-1} = 0. \end{aligned} \tag{2.2.22}$$

The boundary conditions (2.2.21) and (2.2.22) are natural in nonlocal calculus and fractional differential equations, and they are motivated by the fractional integration-by-parts (2.2.13) and (2.2.14). In fact, the fundamental properties of eigensolutions in the theory of classical Sturm–Liouville problems are connected with the integration-by-parts formula and the choice of the boundary conditions. For instance, the continuity or discreteness of the eigen-spectrum in boundary-value problems is highly dependent on the type of boundary conditions enforced. In the setting chosen here, we shall show that the eigen-spectra of RFSLP-I and RFSLP-II are simple and fully discrete.

Remark 2.2 Having defined the zero boundary conditions in (2.2.21) and (2.2.22), and due to the relationship between Caputo and Riemann–Liouville fractional derivatives, the above fractional differential operators $\mathcal{L}_1^{\mu}$ and $\mathcal{L}_2^{\mu}$ respectively are *identical* to

$$\mathcal{L}_1^{\mu} := {}^{\mathrm{RL}}_{\ \ x}D^{\mu}_{x_R}\left[\mathcal{K}\ {}^{\mathrm{RL}}_{x_L}D^{\mu}_{x}(\cdot)\right]$$

and

$$\mathcal{L}_2^{\mu} := {}^{\mathrm{RL}}_{x_L}D^{\mu}_{x}\left[\mathcal{K}\ {}^{\mathrm{RL}}_{\ \ x}D^{\mu}_{x_R}(\cdot)\right]$$

alternatively. In general, this setting and dealing only with Riemann–Liouville fractional derivatives would define the eigenvalue problem in a bigger function space compared to the case where the middle fractional derivatives are defined of Caputo sense. However, we keep the notation as before just to make use of the fractional integration-by-parts easier to understand in this setting.

Analytical Eigensolutions to RFSLP-I and -II

Here, we intend to derive the analytical solution $\Phi_{\lambda}^{(i)}(x)$ to RFSLP-I and -II (2.2.20), subject to the homogeneous Dirichlet and integro-differential boundary conditions (2.2.21)–(2.2.22).

We first recall the following integral representations of the standard Jacobi polynomials $P_n^{(\alpha,\beta)}(x)$.

Lemma 2.3 (Askey and Fitch, 1969) *For* $\mu > 0,\ \alpha, \beta > -1,\ n \in \mathbb{N}_0$ *and* $x \in (-1, 1)$, *we have*

$$(1-x)^{\alpha+\mu}\frac{P_n^{(\alpha+\mu,\beta-\mu)}(x)}{P_n^{(\alpha+\mu,\beta-\mu)}(1)} = \frac{\Gamma(\alpha+\mu+1)}{\Gamma(\alpha+1)\Gamma(\mu)}\int_x^1 \frac{(1-y)^{\alpha}}{(y-x)^{1-\mu}}\frac{P_n^{(\alpha,\beta)}(y)}{P_n^{(\alpha,\beta)}(1)}\,\mathrm{d}y, \tag{2.2.23}$$

and

$$(1+x)^{\beta+\mu}\frac{P_n^{(\alpha-\mu,\beta+\mu)}(x)}{P_n^{(\alpha-\mu,\beta+\mu)}(-1)} = \frac{\Gamma(\beta+\mu+1)}{\Gamma(\beta+1)\Gamma(\mu)}\int_{-1}^x \frac{(1+y)^{\beta}}{(x-y)^{1-\mu}}\frac{P_n^{(\alpha,\beta)}(y)}{P_n^{(\alpha,\beta)}(-1)}\,\mathrm{d}y. \tag{2.2.24}$$

By the definitions of Riemann–Liouville fractional integrals in (2.2.4)–(2.2.5) and evaluating the special values of Jacobi polynomials via (2.1.7), we can rewrite the two formulas in the above lemma respectively as

$${}^{\mathrm{RL}}_{\ \ x}I_1^{\mu}\left\{(1-x)^{\alpha}P_n^{(\alpha,\beta)}(x)\right\} = \frac{\Gamma(n+\alpha+1)}{\Gamma(n+\alpha+\mu+1)}(1-x)^{\alpha+\mu}P_n^{(\alpha+\mu,\beta-\mu)}(x), \tag{2.2.25}$$

and

$${}^{\mathrm{RL}}_{-1}I_x^{\mu}\left\{(1+x)^{\beta}P_n^{(\alpha,\beta)}(x)\right\} = \frac{\Gamma(n+\beta+1)}{\Gamma(n+\beta+\mu+1)}(1+x)^{\beta+\mu}P_n^{(\alpha-\mu,\beta+\mu)}(x). \tag{2.2.26}$$

In view of (2.2.12), we obtain from (2.2.25)–(2.2.26) immediately that

$$ {}^{\mathrm{RL}}_{\ x}D^{\mu}_{1}\{(1-x)^{\alpha+\mu}P_n^{(\alpha+\mu,\beta-\mu)}(x)\} = \frac{\Gamma(n+\alpha+\mu+1)}{\Gamma(n+\alpha+1)}(1-x)^{\alpha}P_n^{(\alpha,\beta)}(x), \tag{2.2.27} $$

and

$$ {}^{\mathrm{RL}}_{-1}D^{\mu}_{x}\{(1+x)^{\beta+\mu}P_n^{(\alpha-\mu,\beta+\mu)}(x)\} = \frac{\Gamma(n+\beta+\mu+1)}{\Gamma(n+\beta+1)}(1+x)^{\beta}P_n^{(\alpha,\beta)}(x). \tag{2.2.28} $$

For $\mu \in (0,1)$, taking $\alpha = \mp\mu$ or $\beta = \pm\mu$ in (2.2.25) or (2.2.26), we obtain the following useful formulas

$$ {}^{\mathrm{RL}}_{\ x}I^{\mu}_{1}\{(1-x)^{-\mu}P_n^{(-\mu,\mu)}(x)\} = \frac{\Gamma(n-\mu+1)}{n!}P_n(x), \tag{2.2.29} $$

and

$$ {}^{\mathrm{RL}}_{-1}I^{\mu}_{x}\{(1+x)^{-\mu}P_n^{(\mu,-\mu)}(x)\} = \frac{\Gamma(n-\mu+1)}{n!}P_n(x), \tag{2.2.30} $$

where $P_n(x) = P_n^{(0,0)}(x)$ represents the Legendre polynomial of degree n.

On the other hand, we can set $\alpha = \beta = 0$ in (2.2.27)–(2.2.28) to obtain

$$ {}^{\mathrm{RL}}_{\ x}D^{\mu}_{1}\{(1-x)^{\mu}P_n^{(\mu,-\mu)}(x)\} = \frac{\Gamma(n+\mu+1)}{n!}P_n(x), \tag{2.2.31} $$

and

$$ {}^{\mathrm{RL}}_{-1}D^{\mu}_{x}\{(1+x)^{\mu}P_n^{(-\mu,\mu)}(x)\} = \frac{\Gamma(n+\mu+1)}{n!}P_n(x) \tag{2.2.32} $$

for $\mu \in (0,1)$.

The formulas (2.2.29)–(2.2.32) are the key to proving the following theorem.

Theorem 2.4 *Let $\mu \in (0,1)$ and $n \in \mathbb{N}$. The exact eigenfunctions to* (2.2.20) *with $i = 1$, i.e., RFSLP-I, subject to* (2.2.21) *are given by*

$$ \Phi_n^{(1)}(x) = (1+x)^{\mu}\, P_{n-1}^{(-\mu,\mu)}(x), \tag{2.2.33} $$

and the corresponding distinct eigenvalues are given by

$$ \lambda_n^{(1)} = -\mathcal{K}\frac{\Gamma(n+\mu)}{\Gamma(n-\mu)}. \tag{2.2.34} $$

Moreover, the exact eigenfunctions to (2.2.20) *with $i = 2$, i.e., RFSLP-II, subject to* (2.2.22) *are*

$$ \Phi_n^{(2)}(x) = (1-x)^{\mu}\, P_{n-1}^{(\mu,-\mu)}(x), \tag{2.2.35} $$

and the corresponding distinct eigenvalues are

$$\lambda_n^{(2)} = \lambda_n^{(1)} = -\mathcal{K}\,\frac{\Gamma(n+\mu)}{\Gamma(n-\mu)}. \tag{2.2.36}$$

Proof We carry out the proof in three steps.

Step 1: First, we prove (2.2.33) and (2.2.34). From (2.2.33), it is clear that $\Phi_n^{(1)}(-1) = 0$. To check the other boundary condition, since $\Phi_n^{(1)}(-1) = 0$, by the property (2.2.10), we can replace ${}_{-1}^{\ C}\mathcal{D}_x^{\mu}$ by ${}_{-1}^{\mathrm{RL}}\mathcal{D}_x^{\mu}$. Thus by (2.2.32) and

$$\begin{aligned}\Big\{{}_{x}^{\mathrm{RL}}\mathcal{I}_1^{1-\mu}\big[\mathcal{K}\ {}_{-1}^{\ C}\mathcal{D}_x^{\mu}\,\Phi_n^{(1)}(x)\big]\Big\}\Big|_{x=1} &= \Big\{{}_{x}^{\mathrm{RL}}\mathcal{I}_1^{1-\mu}\big[\mathcal{K}\ {}_{-1}^{\mathrm{RL}}\mathcal{D}_x^{\mu}\,\Phi_n^{(1)}(x)\big]\Big\}\Big|_{x=1}\\ &= \Big\{{}_{x}^{\mathrm{RL}}\mathcal{I}_1^{1-\mu}\big[\mathcal{K}\ {}_{-1}^{\mathrm{RL}}\mathcal{D}_x^{\mu}\big((1+x)^{\mu}\ P_{n-1}^{(-\mu,\mu)}(x)\big)\big]\Big\}\Big|_{x=1}\\ &\overset{(2.2.32)}{=} \mathcal{K}\,\frac{\Gamma(n+\mu)}{(n-1)!}\Big\{{}_{x}^{\mathrm{RL}}\mathcal{I}_1^{1-\mu}\,P_{n-1}(x)\Big\}\Big|_{x=1},\end{aligned}$$

which vanishes, as by (2.2.25) with $\alpha = \beta = 0$ and $\mu \to 1-\mu$, we have

$${}_{x}^{\mathrm{RL}}\mathcal{I}_1^{1-\mu}\,P_{n-1}(x) = \frac{(n-1)!}{\Gamma(n+1-\mu)}(1-x)^{1-\mu}\,P_{n-1}^{(1-\mu,\mu-1)}(x). \tag{2.2.37}$$

Thus, $\Phi_n^{(1)}(x)$ satisfies the second boundary condition in (2.2.21).

Now, we need to show that $\Phi_n^{(1)}(x)$ indeed satisfies (2.2.20) with $i = 1$, and the corresponding eigenvalues are given by (2.2.34). To verify this, we take the fractional integral of order μ on both sides of (2.2.20) with $i = 1$, leading to

$$\mathcal{K}\ {}_{-1}^{\ C}\mathcal{D}_x^{\mu}\,\Phi_n^{(1)}(x) = -\lambda\ {}_{x}^{\mathrm{RL}}\mathcal{I}_1^{\mu}\big\{(1-x^2)^{-\mu}\,\Phi_n^{(1)}(x)\big\}. \tag{2.2.38}$$

Substituting (2.2.33) and replacing the Caputo derivative by the Riemann–Liouville one, thanks to the homogeneous Dirichlet boundary condition, we obtain

$$\mathcal{K}\ {}_{-1}^{\mathrm{RL}}\mathcal{D}_x^{\mu}\big\{(1+x)^{\mu}\ P_{n-1}^{(-\mu,\mu)}(x)\big\} = -\lambda\,{}_{x}^{\mathrm{RL}}\mathcal{I}_1^{\mu}\big\{(1-x)^{-\mu}\,P_{n-1}^{(-\mu,\mu)}(x)\big\}.$$

Using (2.2.32) and (2.2.29), respectively, we can work out the fractional derivative on the left-hand side and the fractional integration on the right-hand side that yields

$$\mathcal{K}\,\frac{\Gamma(n+\mu)}{\Gamma(n)}\ P_{n-1}(x) = -\lambda\frac{\Gamma(n-\mu)}{\Gamma(n)}\ P_{n-1}(x).$$

Thus we have

$$\lambda \equiv \lambda_n^{(1)} = -\mathcal{K}\,\frac{\Gamma(n+\mu)}{\Gamma(n-\mu)}, \quad n \geq 1,$$

which shows that the eigenvalues of RFSLP-I are real-valued and distinct. The eigenpairs give the explicit solutions claimed in Theorem 2.1. Moreover, the orthogonality of the eigenfunctions (2.2.33) with respect to the weight function $w(x) = (1-x^2)^{-\mu}$ can be shown, in view of

$$\begin{aligned}\int_{-1}^{1} \Phi_k^{(1)}(x)\,\Phi_j^{(1)}(x)\,w(x)\,\mathrm{d}x &= \int_{-1}^{1} (1+x)^{2\mu}\,P_{k-1}^{(-\mu,\mu)}(x)\,P_{j-1}^{(-\mu,\mu)}(x)w(x)\,\mathrm{d}x \\ &= \int_{-1}^{1} P_{k-1}^{(-\mu,\mu)}(x)\,P_{j-1}^{(-\mu,\mu)}(x)\,(1-x)^{-\mu}(1+x)^{\mu}\,\mathrm{d}x \\ &= \gamma_{k-1}^{(-\mu,\mu)}\delta_{kj},\end{aligned}$$

where we used the orthogonality of Jacobi polynomials in (2.1.3)–(2.1.4).

Step 2: The proof of the eigensolution to RFSLP-II in (2.2.35)–(2.2.36) follows the same lines as in **Step 1**, so we just sketch the derivation below. It is clear that $\Phi_n^{(2)}(1)=0$. In view of the property (2.2.11), we can replace ${}^{\mathrm{C}}_{x}D_1^{\mu}$ by ${}^{\mathrm{RL}}_{x}D_1^{\mu}$ in the second boundary condition given in (2.2.22). Hence, substituting it into (2.2.35) and working out the inner fractional derivative using (2.2.31), we can obtain

$$\begin{aligned}\left\{{}^{\mathrm{RL}}_{x}I_1^{1-\mu}\left[\mathcal{K}\ {}^{\mathrm{C}}_{x}D_1^{\mu}\Phi_n^{(2)}(x)\right]\right\}\Big|_{x=-1} &= \mathcal{K}\,\frac{\Gamma(n)}{\Gamma(n+\mu)}\left\{{}^{\mathrm{RL}}_{-1}\mathcal{I}_x^{1-\mu}P_{n-1}(x)\right\}\Big|_{x=-1} \\ &= \mathcal{K}\left\{(1+x)^{\mu}P_{n-1}^{(-\mu,\mu)}(x)\right\}\Big|_{x=-1}=0,\end{aligned}$$

where we used (2.2.25) with $\alpha=\beta=0$ to evaluate the above fractional integral.

To check if $\Phi_n^{(2)}(x)$ given in (2.2.35) satisfies (2.2.20) with $i=2$, we substitute (2.2.35) into (2.2.20) and carry out the fractional integration of order μ on both sides using (2.2.26). Then, by working out the fractional derivative on the left-hand side using (2.2.29), we can verify that (2.2.35) satisfies the boundary-value problem, provided that (2.2.36) are the real-values distinct eigenvalues of RFSLP-II.

Finally, the orthogonality of the eigenfunctions (2.2.35) with respect to the weight function $w(x)=(1-x^2)^{-\mu}$ can be shown as

$$\begin{aligned}\int_{-1}^{1} \Phi_k^{(2)}(x)\,\Phi_j^{(2)}(x)\,w(x)\,\mathrm{d}x &= \int_{-1}^{1} P_{k-1}^{(\mu,-\mu)}(x)\,P_{j-1}^{(\mu,-\mu)}(x)\,(1-x)^{\mu}(1+x)^{-\mu}\,\mathrm{d}x \\ &= \gamma_{k-1}^{(\mu,-\mu)}\delta_{kj},\end{aligned}$$

where we used the orthogonality of Jacobi polynomials in (2.1.3)–(2.1.4).

Step 3: It remains to prove that the set $\{\Phi_n^{(i)}(x) : n=1,2,\ldots\}$ forms a basis for the infinite-dimensional Hilbert space $L_w^2(-1,1)$, and $\lambda_n^{(i)}$, the corresponding eigenvalue for each n, is simple. For any $f(x)\in L_w^2(-1,1)$, it is clear that

$$g^{(i)}(x)=(1\pm x)^{-\mu}f(x)\in L^2_{w^{(i)}}(-1,1),\quad i\in\{1,2\},$$

with $\mu\in(0,1)$ and

$$w^{(1)}(x)=(1-x)^{-\mu}(1+x)^{\mu},\quad w^{(2)}(x)=(1-x)^{\mu}(1+x)^{-\mu}.$$

Hereafter, the upper signs are corresponding to the case $i = 1$, i.e., RFSLP-I, and the lower signs are corresponding to the case $i = 2$, i.e., RFSLP-II. Hence, given any real sequence $\{a_n^{(i)}\}_{n=1}^N$, we have

$$\begin{aligned}
\Big\| \sum_{n=1}^{N} a_n^{(i)} \Phi_n^{(i)} - f \Big\|_{L^2_w} &= \Big\| \sum_{n=1}^{N} a_n^{(i)} (1 \pm x)^{\mu} P_{n-1}^{(\mp\mu,\pm\mu)} - f \Big\|_{L^2_w} \\
&= \Big\| (1 \pm x)^{\mu} \Big(\sum_{n=1}^{N} a_n^{(i)} P_{n-1}^{(\mp\mu,\pm\mu)} - (1 \pm x)^{-\mu} f \Big) \Big\|_{L^2_w} \\
&= \Big\| (1 \pm x)^{\mu} \Big(\sum_{n=1}^{N} a_n^{(i)} P_{n-1}^{(\mp\mu,\pm\mu)} - g^{(i)} \Big) \Big\|_{L^2_w} \\
&= \Big\| \sum_{n=1}^{N} a_n^{(i)} P_{n-1}^{(\mp\mu,\pm\mu)} - g^{(i)} \Big\|_{L^2_{w^{(i)}}}.
\end{aligned}$$

We choose $\{a_n^{(i)}\}$ to be the expansion coefficients of $g^{(i)}(x)$ in terms of the Jacobi polynomials $\{P_{n-1}^{(\mp\mu,\pm\mu)}(x)\}_{n=1}^N$, i.e.,

$$a_n^{(i)} = \frac{1}{\gamma_{n-1}^{(\mp\mu,\pm\mu)}} \int_{-1}^{1} g^{(i)}(x)\, P_{n-1}^{(\mp\mu,\pm\mu)}(x)\, w^{(i)}(x)\, dx.$$

We conclude from Toepler's theorem on approximation by orthogonal systems that

$$\lim_{N\to\infty} \Big\| \sum_{n=1}^{N} a_n^{(i)} P_{n-1}^{(\mp\mu,\pm\mu)} - g^{(i)} \Big\|_{L^2_{w^{(i)}}} \to 0, \quad i \in \{1, 2\}.$$

Thus

$$\sum_{n=1}^{N} a_n^{(i)} \Phi_n^{(i)}(x) \xrightarrow{L^2_w} f(x),$$

which implies $\{\Phi_n^{(i)} : n = 1, 2, \ldots\}$ is dense in $L^2_w(-1, 1)$ and it forms a basis for this Hilbert space.

To show the simplicity of the eigenvalues, assume that corresponding to the eigenvalue $\lambda_j^{(i)}$, there exists another eigenfunction $\Phi_j^{(i)*}(x) \in L^2_w(-1, 1)$ in addition to $\Phi_j^{(i)}(x)$, which is by Theorem 2.1 orthogonal to the rest of the eigenfunctions $\Phi_n^{(i)}(x)$, $n \neq j$. By the density of the eigenfunctions set, we can represent $\Phi_j^{(i)*}(x)$ as

$$\Phi_j^{(i)*}(x) = \sum_{n=1}^{\infty} b_n^{(i)} \Phi_n^{(i)}(x), \tag{2.2.39}$$

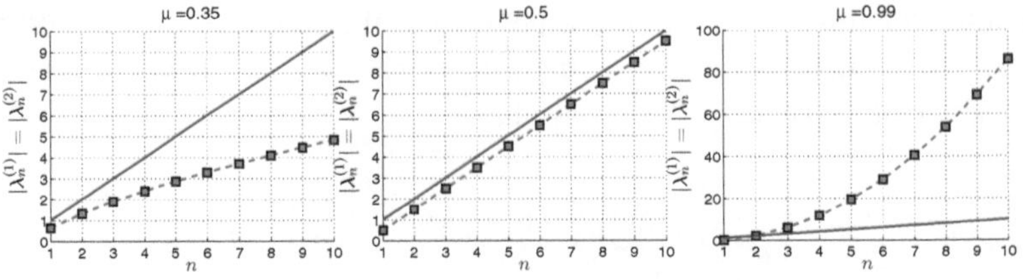

Figure 2.1 Magnitude of the eigenvalues of RFSLP-I and RFSLP-II, $|\lambda_n^{(1)}| = |\lambda_n^{(2)}|$, versus n, corresponding to $\mu = 0.35$, left: sublinear growth, $\mu = 0.5$, middle: linear growth, and $\mu = 0.99$, right: superlinear-subquadratic growth. The solid line denotes the linear growth. Reprinted from Zayernouri and Karniadakis (2013) with permission from Elsevier.

where there exists at least one coefficient $b_k^{(i)} \neq 0$ with $k \neq j$. Now, by multiplying both sides by $\Phi_k^{(i)}(x)$ and integrating with respect to the weight function $w(x)$, we obtain

$$\int_{-1}^{1} \Phi_j^{(i)*}(x)\, \Phi_k^{(i)}(x)\, w(x)\, \mathrm{d}x = \sum_{n=1}^{\infty} b_n^{(i)} \int_{-1}^{1} \Phi_n^{(i)}(x)\, \Phi_k^{(i)}(x)\, w(x)\, \mathrm{d}x$$
$$= b_k^{(i)} \gamma_{k-1}^{(\mp\mu,\pm\mu)} \neq 0,$$

which contradicts Theorem 2.1. Therefore, the eigenvalues $\{\lambda_n^{(i)}\}$ are simple. This completes the proof. □

The growth of the magnitude of the eigenvalues of RFSLP-I and -II, $|\lambda_n^i|$, $i \in \{1, 2\}$, is plotted in Figure 2.1, corresponding to three values of $\mu = 0.35$, $\mu = 0.5$, and $\mu = 0.99$. We observe that there are two growth modes, depending on either $\mu \in (0, 1/2)$, where a sublinear growth in $|\lambda_n^1| = |\lambda_n^2|$ is observed, or $\mu \in (1/2, 1)$, where a superlinear-subquadratic growth mode is noticed. The case $\mu = 1/2$ leads to an exactly linear growth mode.

In order to visually get more sense of how the eigensolutions look, we plot in Figure 2.2 the eigenfunctions of RFSLP-I, $\Phi_n^{(1)}(x)$ of different orders and corresponding to different values of μ used in Figure 2.1. In each plot, we compare the eigensolutions with the corresponding standard Jacobi polynomials $P_n^{(-\mu,\mu)}(x)$. In a similar fashion, we plot the eigenfunctions of RFSLP-II, $\Phi_n^{(2)}(x)$, of different orders compared to $P_n^{(\mu,-\mu)}(x)$ in Figure 2.3.

So far, the eigenfunctions have been defined in the interval $(-1, 1)$. The following lemma provides a useful shifted definition of $\Phi_n^{(i)}(x)$, which is not only more convenient to work with but also helps exploit some interesting properties.

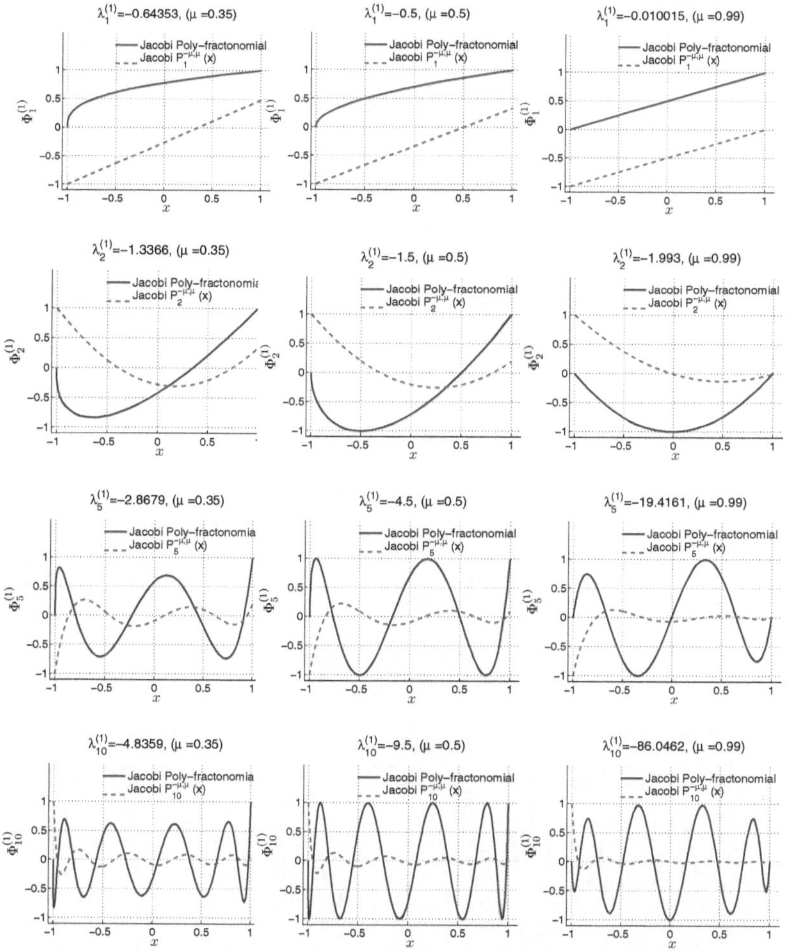

Figure 2.2 Eigenfunctions of RFSLP-I, $\Phi_n^{(1)}$, versus x, for $n = 1$ (first row), $n = 2$ (second row), $n = 5$ (third row), and $n = 10$ (last row), corresponding to the fractional order $\mu = \nu/2 = 0.35$ (left column), $\mu = \nu/2 = 0.5$ (middle column), and $\mu = \nu/2 = 0.99$ (right column). Reprinted from Zayernouri and Karniadakis (2013) with permission from Elsevier.

Lemma 2.5 *The shifted eigensolutions to RFSLP-I and -II, denoted by* $\widetilde{\Phi}_n^{(i)}(t)$, $i \in \{1, 2\}$, *are given by*

$$\begin{aligned}\widetilde{\Phi}_n^{(1)}(t) &= \Phi_n^{(1)}(2t-1)\\ &= 2^{\mu}\sum_{j=0}^{n-1}(-1)^{n+j-1}\binom{n-1+j}{j}\binom{n-1+\mu}{n-1-j}t^{j+\mu},\end{aligned} \tag{2.2.40}$$

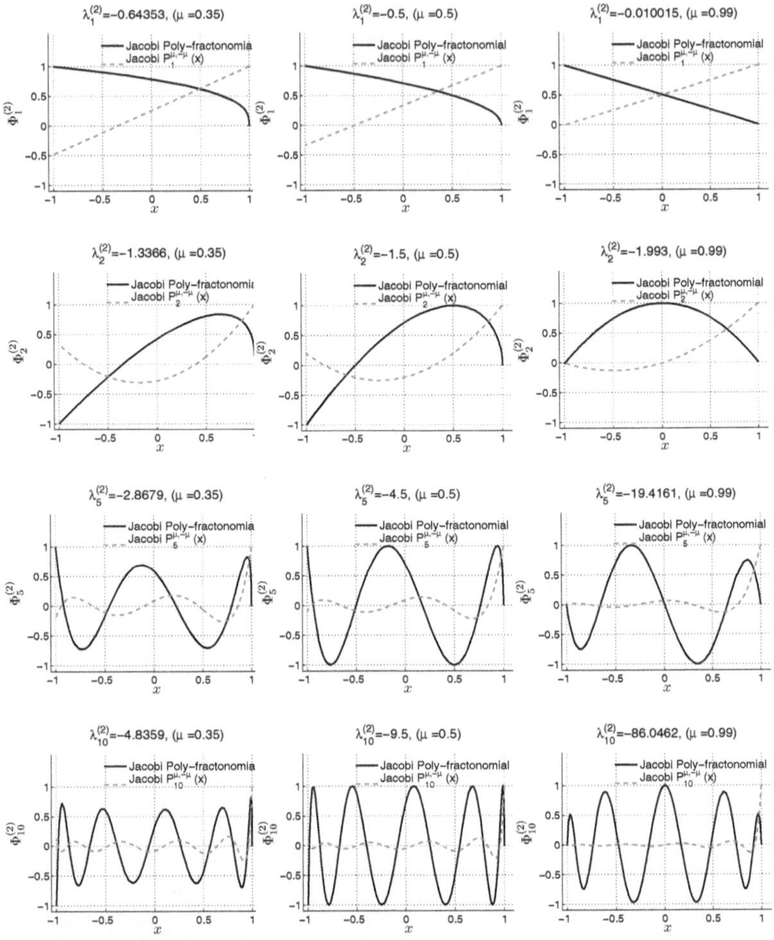

Figure 2.3 Eigenfunctions of RFSLP-II, $\Phi_n^{(2)}$, versus x, for $n = 1$ (first row), $n = 2$ (second row), $n = 5$ (third row), and $n = 10$ (last row), corresponding to the fractional order $\mu = \nu/2 = 0.35$ (left column), $\mu = \nu/2 = 0.5$ (middle column), and $\mu = \nu/2 = 0.99$ (right column). Reprinted from Zayernouri and Karniadakis (2013) with permission from Elsevier.

and

$$\begin{aligned}\widetilde{\Phi}_n^{(2)}(t) &= \Phi_n^{(2)}(1-2t) = (-1)^{n-1}\widetilde{\Phi}_n^{(1)}(t)\\ &= 2^{\mu}\sum_{j=0}^{n-1}(-1)^j\binom{n-1+j}{j}\binom{n-1+\mu}{n-1-j}t^{j+\mu},\end{aligned} \tag{2.2.41}$$

where $t \in (0,1)$.

Proof We first obtain the shifted RFSLP-I by performing an affine mapping from interval $[-1, 1]$ to $[0, 1]$. To do so, we recall the power series expansion of the standard Jacobi polynomial $P_n^{(\alpha,\beta)}(x)$ for $x \in [-1, 1]$:

$$\begin{aligned} P_n^{(\alpha,\beta)}(x) &= \sum_{j=0}^{n} \binom{n+\alpha+\beta+j}{j} \binom{n+\alpha}{n-j} \left(\frac{x-1}{2}\right)^j \\ &= \sum_{j=0}^{n} (-1)^{n+j} \binom{n+\alpha+\beta+j}{j} \binom{n+\beta}{n-j} \left(\frac{x+1}{2}\right)^j. \end{aligned} \tag{2.2.42}$$

With the change of variable $x = 2t - 1$, we obtain from (2.2.33) and (2.2.42) that

$$\begin{aligned} \widetilde{\Phi}_n^{(1)}(t) &= \Phi_n^{(1)}(x) = (1+x)^{\mu} \; P_{n-1}^{(-\mu,\mu)}(x) \\ &= 2^{\mu} \sum_{j=0}^{n-1} (-1)^{n+j-1} \binom{n-1+j}{j} \binom{n-1+\mu}{n-1-j} t^{j+\mu}. \end{aligned}$$

Similarly, we derive from (2.2.35) and (2.2.42) with the change of variable $x = 1 - 2t$ that

$$\begin{aligned} \widetilde{\Phi}_n^{(2)}(t) &= \Phi_n^{(2)}(x) = (1-x)^{\mu} \; P_{n-1}^{(\mu,-\mu)}(x) \\ &= 2^{\mu} \sum_{j=0}^{n-1} (-1)^{j} \binom{n-1+j}{j} \binom{n-1+\mu}{n-1-j} t^{j+\mu}. \end{aligned}$$

This verifies (2.2.41) and completes the proof. □

Definition 2.6 A *fractonomial* is defined as a function $f : \mathbb{C} \to \mathbb{C}$ of non-integer power, denoted as $t^{k+\mu}$, where $k \in \mathbb{Z}^+$ and $\mu \in (0, 1)$, in which the power can be represented as a sum of an integer and non-integer number. Moreover, denoted by $\mathcal{F}_e^{n+\mu}$ is the *fractal expansion set*, which is defined as the set of all fractonomials of order less than or equal $n + \mu$ as

$$\mathcal{F}_e^{n+\mu} = \text{span}\{t^{k+\mu} \; : \; \mu \in (0, 1) \text{ and } k = 0, 1, \ldots, n\}. \tag{2.2.43}$$

Remark 2.7 All fractonomials are zero-valued at $t = 0$. Moreover, asymptotically, when $\mu \to 0$, a fractonomial of order $n+\mu$ approaches the monomial t^n.

Definition 2.8 A *poly-fractonomial* of order $n + \mu < \infty$ with $n \in \mathbb{N}_0$ and $\mu \in (0, 1)$, is defined as a linear combination of the elements in the fractal expansion set $\mathcal{F}_e^{n+\mu}$, as

$$F_{n+\mu}(t) = a_0 \, t^{\mu} + a_1 \, t^{1+\mu} + \cdots + a_n \, t^{n+\mu},$$

where $a_j \in \mathbb{C}, \; j \in \{0, 1, \ldots, n\}$ are constants. Moreover, denoted by $\mathbb{F}^{n+\mu}$ is the space of all poly-fractonomials up to order $n + \mu$. By Remark 2.7, all elements in $\mathbb{F}^{n+\mu}$ asymptotically approach the corresponding standard polynomial of order n with coefficients $\{a_j\}$.

Remark 2.9 It is observed that $\mathbb{F}^{n+\mu} \subset L^2_w$ since $\mu \in (0,1)$, and hence, all poly-fractonomials in $\mathbb{F}^{n+\mu}$ can be represented as an infinite sum in terms of the shifted eigenfunctions of RFSLP-I and -II. It is true by the density of the eigenfunctions in L^2_w, shown in **Step 3** of the proof in Theorem 2.4.

Lemma 2.10 *The fractional Caputo derivative of order* $\mu \in (0,1)$ *of any polynomial in* $\mathbb{P}_n$ *with* $n \geq 1$ *lies in the space of poly-fractonomials* $\mathbb{F}^{n-1+\tilde{\mu}}$, *where* $\tilde{\mu} = 1 - \mu \in (0,1)$.

Proof For any $f \in \mathbb{P}_n$, we write $f(t) = \sum_{j=0}^{n} a_j t^j$. From Podlubny (1999) and for $\mu \in (0,1)$, we have

$$ {}^{C}_{0}D^{\mu}_{t} t^k = \begin{cases} 0, & k = 0, \\ \frac{\Gamma(k+1)}{\Gamma(k+1-\mu)} t^{k-\mu}, & 0 < \mu \leq k. \end{cases} \tag{2.2.44} $$

Thus,

$$ \begin{aligned} {}^{C}_{0}D^{\mu}_{t} f(t) &= \sum_{j=0}^{n} a_j \, {}^{C}_{0}D^{\mu}_{t} t^j = \sum_{j=1}^{n} a_j \frac{\Gamma(j+1)}{\Gamma(j+1-\mu)} t^{j-\mu} \\ &= \sum_{j=0}^{n-1} a_{j+1} \frac{\Gamma(j+2)}{\Gamma(j+2-\mu)} t^{j+(1-\mu)} \in \mathbb{F}^{n-1+\tilde{\mu}}, \end{aligned} $$

where $\tilde{\mu} = 1 - \mu \in (0,1)$. □

Theorem 2.11 *The shifted eigensolutions to* (2.2.20), $\widetilde{\Phi}^{(i)}_n(t)$, $n \in \mathbb{N}$, *form a complete hierarchical basis for the finite-dimensional space of poly-fractonomials* $\mathbb{F}^{n-1+\mu}$, *where* $\mu \in (0,1)$.

Proof From Lemma 2.5 and Definition 2.8, it is clear that

$$ \mathbb{F}^{n-1+\mu} = \text{span}\{\widetilde{\Phi}^{(i)}_k : k = 1, 2, \ldots, n\}. \tag{2.2.45} $$

Moreover, we can rewrite the two expansions in Lemma 2.5 in the matrix-vector multiplication form as

$$ \vec{\Phi}^{(i)} = \mathbb{T}^{(i)} \, \vec{t}, \tag{2.2.46} $$

where

$$\vec{\Phi}^{(i)} = \begin{pmatrix} \widetilde{\Phi}_1^{(i)}(t) \\ \widetilde{\Phi}_2^{(i)}(t) \\ \vdots \\ \widetilde{\Phi}_n^{(i)}(t) \end{pmatrix}, \quad \vec{t} = \begin{pmatrix} t^{\mu} \\ t^{1+\mu} \\ \vdots \\ t^{n-1+\mu} \end{pmatrix},$$

and $\mathbb{T}^{(i)} = \{T_{kj}^{(i)}\}_{k,j=1,\dots,n}$ is an $n \times n$ matrix with entries given by

$$T_{kj}^{(1)} = 2^{\mu}(-1)^{k+j}\binom{k+j-2}{j-1}\binom{k-1+\mu}{k-j}, \quad \text{if } 1 \le j \le k,$$

and $T_{kj}^{(1)} = 0$, if $k < j \le n$; and $T_{kj}^{(2)} = (-1)^{k-1} T_{kj}^{(1)}$. Thus for each $i \in \{1,2\}$, $\mathbb{T}^{(i)}$ is a *lower-triangular* matrix. Thanks to the orthogonality of $\Phi_n^{(i)}$, the eigenfunctions are linearly independent, so the matrix $\mathbb{T}^{(i)}$ is invertible. Let $\mathcal{T}^{(i)}$ be the inverse matrix of $\mathbb{T}^{(i)}$, which is also lower-triangular. Hence,

$$\vec{t} = \mathcal{T}^{(i)}\, \vec{\Phi}^{(i)}. \tag{2.2.47}$$

In other words, each element in the poly-fractonomial space $\mathbb{F}^{n-1+\mu}$, say $t^{m+\mu}$, $0 \le m \le n-1$, can be uniquely represented through the following expansion

$$t^{m+\mu} = \sum_{j=1}^{n} c_j\, \widetilde{\Phi}_j^{(i)}(t) = \sum_{j=1}^{n} \{\mathcal{T}_{mj}^{(i)}\}\, \widetilde{\Phi}_j^{(i)}(t) = \sum_{j=1}^{m} \{\mathcal{T}_{mj}^{(i)}\}\, \widetilde{\Phi}_j^{(i)}(t), \tag{2.2.48}$$

where the last equality holds since $\mathcal{T}^{(i)}$ is a lower-triangular matrix. As seen from (2.2.43), the fractal expansion set $\mathcal{F}_e^{n+\mu} \subset \mathcal{F}_e^{n+1+\mu}$, which indicates that the shifted eigensolutions $\{\widetilde{\Phi}_n^{(i)}\}$ form a *hierarchical* expansion basis set. □

2.2.2 Singular Fractional Sturm–Liouville Problems of Kind-I and -II

In the sequel, we extend the study of the regular fractional Sturm–Liouville problem to the singular FSLP. We begin with the definitions of the singular fractional Sturm–Liouville of first kind (SFSLP-I) and second kind (SFSLP-II) of order $\nu = 2\mu \in (0,2)$ with the parameters

$$-1 < \alpha < 2-\mu, \quad -1 < \beta < \mu - 1 \tag{2.2.49}$$

in SFSLP-I ($i = 1$), and

$$-1 < \alpha < \mu - 1, \quad -1 < \beta < 2-\mu \tag{2.2.50}$$

in SFSLP-II ($i = 2$). More precisely, for $x \in (-1, 1)$, we consider

$$ {}^{\rm RL}D^{\mu}\{\omega^{(\alpha+1,\beta+1)}(x)\ {}^{\rm C}D^{\mu}\,\mathcal{P}^{(i)}(x)\} + \Lambda^{(i)}\omega^{(\alpha+1-\mu,\beta+1-\mu)}(x)\,\mathcal{P}^{(i)}(x) = 0, \tag{2.2.51} $$

subject to the boundary conditions

$$ \mathcal{P}^{(i)}((-1)^i) = 0, \tag{2.2.52} $$

$$ \big\{{}^{\rm RL}\mathcal{I}^{1-\mu}\big[\omega^{(\alpha,\beta)}(x)\,{}^{\rm C}D^{\mu}\,\mathcal{P}^{(i)}(x)\big]\big\}\big|_{x=(-1)^{i+1}} = 0, \tag{2.2.53} $$

where $\omega^{(\alpha,\beta)}(x) = (1-x)^{\alpha}(1+x)^{\beta}$ is the Jacobi weight function and the fractional operators in the above for two kinds are taken to be

$$ \big\{{}^{\rm RL}D^{\mu},\ {}^{\rm C}D^{\mu},\ {}^{\rm RL}\mathcal{I}^{1-\mu}\big\} \equiv \begin{cases} \big\{{}^{\rm RL}_{\ x}D^{\mu}_{1},\ {}^{\rm C}_{-1}D^{\mu}_{x},\ {}^{\rm RL}_{\ x}I^{1-\mu}_{1}\big\}, & i = 1, \\ \big\{{}^{\rm RL}_{-1}D^{\mu}_{x},\ {}^{\rm C}_{1}D^{\mu}_{x},\ {}^{\rm RL}_{-1}I^{1-\mu}_{x}\big\}, & i = 2. \end{cases} \tag{2.2.54} $$

Theorem 2.12 *The eigenvalues of SFSLP-I and -II* (2.2.49)–(2.2.54) *are real-valued, and the eigenfunctions corresponding to distinct eigenvalues are orthogonal with respect to the weight function* $w(x) = (1-x)^{\alpha+1-\mu}(1+x)^{\beta+1-\mu}$.

Proof We first show that all eigenvalues are real. For simplicity, we introduce the fractional differential operator of order 2μ associated with (2.2.51):

$$ \mathcal{L}_i^{\alpha,\beta;\mu} := {}^{\rm RL}D^{\mu}\{(1-x)^{\alpha+1}(1+x)^{\beta+1}\ {}^{\rm C}D^{\mu}(\cdot)\}. \tag{2.2.55} $$

Assume that $\Lambda^{(i)}$ is an eigenvalue of (2.2.51)–(2.2.54) with the corresponding eigenfunction $\eta^{(i)}(x)$, where $i \in \{1, 2\}$. Taking the complex conjugate on (2.2.51) leads to

$$ \mathcal{L}_i^{\alpha,\beta;\mu}\bar{\eta}^{(i)}(x) + \bar{\Lambda}^{(i)}w(x)\bar{\eta}^{(i)}(x) = 0, \tag{2.2.56} $$

with the corresponding boundary conditions from (2.2.52):

$$ \bar{\eta}^{(i)}((-1)^i) = 0, \quad \big\{{}^{\rm RL}\mathcal{I}^{1-\mu}\big[\omega^{(\alpha+1,\beta+1)}(x)\ {}^{\rm C}D^{\mu}\bar{\eta}^{(i)}(x)\big]\big\}\big|_{x=(-1)^{i+1}} = 0. $$

Multiplying (2.2.51) with $\mathcal{P}^{(i)} = \eta^{(i)}$ by $\bar{\eta}^{(i)}$ and (2.2.56) by $\eta^{(i)}$ and then subtracting the resulting equations, leads to

$$ \begin{aligned} &(\Lambda^{(i)} - \bar{\Lambda}^{(i)})w(x)\eta^{(i)}(x)\bar{\eta}^{(i)}(x) \\ &\quad = \eta^{(i)}(x)\mathcal{L}_i^{\alpha,\beta;\mu}\bar{\eta}^{(i)}(x) - \bar{\eta}^{(i)}(x)\mathcal{L}_i^{\alpha,\beta;\mu}\eta^{(i)}(x). \end{aligned} \tag{2.2.57} $$

Integrating both its sides over the interval $(-1, 1)$ and utilizing the fractional integration-by-parts (2.2.13)–(2.2.14), we obtain from the fact that the fractional operators in (2.2.54) are properly paired up that

$$
\begin{aligned}
&(\Lambda^{(i)} - \bar{\Lambda}^{(i)}) \int_{-1}^{1} |\eta^{(i)}(x)|^2 \, w(x) \, dx \\
&= -\eta^{(i)}((-1)^{i+1}) \{ {}^{\mathrm{RL}}\mathcal{I}^{1-\mu} [\omega^{(\alpha+1,\beta+1)}(x) \, {}^{\mathrm{C}}D^{\mu} \eta^{(i)}(x)] \} |_{x=(-1)^{i+1}} \\
&+ \bar{\eta}^{(i)}((-1)^{i+1}) \{ {}^{\mathrm{RL}}\mathcal{I}^{1-\mu} [\omega^{(\alpha+1,\beta+1)}(x) \, {}^{\mathrm{C}}D^{\mu} \bar{\eta}^{(i)}(x)] \} |_{x=(-1)^{i+1}} \\
&+ \eta^{(i)}((-1)^{i}) - \bar{\eta}((-1)^{i}).
\end{aligned} \tag{2.2.58}
$$

Now, by applying the boundary conditions for $\eta^{(i)}$ and $\bar{\eta}^{(i)}$, we obtain

$$
(\Lambda^{(i)} - \bar{\Lambda}^{(i)}) \int_{-1}^{1} |\eta^{(i)}(x)|^2 \, w(x) \, dx = 0, \tag{2.2.59}
$$

which implies $\Lambda^{(i)} = \bar{\Lambda}^{(i)}$, as $\eta^{(i)}(x)$ is a nontrivial solution of the problem, and $w(x)$ is non-negative in the interval $(-1, 1)$. Thus the eigenvalues are all real.

We now prove that the eigenfunctions corresponding to distinct eigenvalues are orthogonal with respect to the weight function $w(x)$. Assume that $\eta_1^{(i)}(x)$ and $\eta_2^{(i)}(x)$ are two eigenfunctions corresponding to two distinct eigenvalues $\Lambda_1^{(i)}$ and $\Lambda_2^{(i)}$, respectively. Following the same lines as the derivation of (2.2.59) but with $\{\Lambda_1^{(i)}, \eta_1^{(i)}\}, \{\Lambda_2^{(i)}, \eta_2^{(i)}\}$ in place of $\{\Lambda^{(i)}, \eta^{(i)}\}, \{\bar{\Lambda}^{(i)}, \bar{\eta}^{(i)}\}$, respectively, we can derive the following identity:

$$
(\Lambda_1^{(i)} - \Lambda_2^{(i)}) \int_{-1}^{1} \eta_1^{(i)}(x) \eta_2^{(i)}(x) \, w(x) \, dx = 0.
$$

As $\Lambda_1^{(i)} \neq \Lambda_2^{(i)}$, the corresponding eigenfunctions are mutually orthogonal. □

Theorem 2.13 *The exact eigenfunctions of SFSLP-I, i.e.,* (2.2.49)–(2.2.54) *with* $i = 1$*, are given by*

$$
\mathcal{P}_n^{(1)}(x) = {}^{(1)}\mathcal{P}_n^{(\alpha,\beta,\mu)}(x) = (1+x)^{-\beta+\mu-1} \; P_{n-1}^{(\alpha-\mu+1, -\beta+\mu-1)}(x), \tag{2.2.60}
$$

and the corresponding distinct eigenvalues are

$$
\Lambda_n^{(1)} = {}^{(1)}\Lambda_n^{(\alpha,\beta,\mu)} = -\frac{\Gamma(n-\beta+\mu-1)\Gamma(n+\alpha+1)}{\Gamma(n-\beta-1)\Gamma(n+\alpha-\mu+1)}. \tag{2.2.61}
$$

Moreover, the exact eigenfunctions to SFSLP-II, i.e., (2.2.49)–(2.2.54) *with* $i = 2$*, are given by*

$$
\mathcal{P}_n^{(2)}(x) = {}^{(2)}\mathcal{P}_n^{(\alpha,\beta,\mu)}(x) = (1-x)^{-\alpha+\mu-1} \; P_{n-1}^{(-\alpha+\mu-1, \beta-\mu+1)}(x), \tag{2.2.62}
$$

and the corresponding distinct eigenvalues are

$$
\Lambda_n^{(2)} = {}^{(2)}\Lambda_n^{(\alpha,\beta,\mu)} = -\frac{\Gamma(n-\alpha+2\mu-1)\Gamma(n+\beta+1)}{\Gamma(n-\alpha+\mu-1)\Gamma(n+\beta-\mu+1)}. \tag{2.2.63}
$$

Proof The proof is essentially similar to that of Theorem 2.4. Here we only consider SFSLP-I and prove (2.2.60)–(2.2.61).

As $-\beta+\mu-1>0$, it is clear that ${}^{(1)}\mathcal{P}_n^{(\alpha,\beta,\mu)}(-1)=0$. Now verify that ${}^{(1)}\mathcal{P}_n^{(\alpha,\beta,\mu)}(x)$ also satisfies the boundary condition (2.2.53) with $i=1$. Since ${}^{(1)}\mathcal{P}_n^{(\alpha,\beta,\mu)}(-1)=0$, we can replace ${}_{-1}^{\ C}D_x^{\mu}$ by ${}_{-1}^{\rm RL}D_x^{\mu}$ and obtain

$$
\begin{aligned}
&\Big\{{}_{x}^{\rm RL}I_1^{1-\mu}\Big[\omega^{(\alpha+1,\beta+1)}(x)\,{}_{-1}^{\ C}D_x^{\mu}\ {}^{(1)}\mathcal{P}_{n-1}^{(\alpha,\beta,\mu)}(x)\Big]\Big\}\Big|_{x=1}\\
&\quad=\Big\{{}_{x}^{\rm RL}I_1^{1-\mu}\Big[\omega^{(\alpha+1,\beta+1)}(x)\,{}_{-1}^{\rm RL}D_x^{\mu}\ {}^{(1)}\mathcal{P}_{n-1}^{(\alpha,\beta,\mu)}(x)\Big]\Big\}\Big|_{x=1}\\
&\quad=\Big\{{}_{x}^{\rm RL}I_1^{1-\mu}\Big[\omega^{(\alpha+1,\beta+1)}(x)\,{}_{-1}^{\rm RL}D_x^{\mu}\big\{(1+x)^{-\beta+\mu-1}\ P_{n-1}^{(\alpha-\mu+1,\,-\beta+\mu-1)}(x)\big\}\Big]\Big\}\Big|_{x=1}.
\end{aligned}
$$

Using (2.2.28) with $n\to n-1$, $\alpha\to 1+\alpha$, $\beta\to-\beta-1$, we find

$$
\begin{aligned}
&{}_{-1}^{\rm RL}D_x^{\mu}\big\{(1+x)^{-\beta+\mu-1}\ P_{n-1}^{(\alpha-\mu+1,\,-\beta+\mu-1)}(x)\big\}\\
&\quad=\frac{\Gamma(n-\beta+\mu-1)}{\Gamma(n-\beta-1)}(1+x)^{-1-\beta}\ P_{n-1}^{(1+\alpha,-1-\beta)}(x),
\end{aligned}
\tag{2.2.64}
$$

so we can continue the derivation and obtain

$$
\begin{aligned}
&\Big\{{}_{x}^{\rm RL}I_1^{1-\mu}\Big[\omega^{(\alpha+1,\beta+1)}(x)\,{}_{-1}^{\ C}D_x^{\mu}\ {}^{(1)}\mathcal{P}_{n-1}^{(\alpha,\beta,\mu)}(x)\Big]\Big\}\Big|_{x=1}\\
&\quad=\frac{\Gamma(n-1-\beta+\mu)}{\Gamma(n-1-\beta)}\Big\{{}_{x}^{\rm RL}I_1^{1-\mu}\big[\omega^{(\alpha+1,\beta+1)}(x)(1+x)^{-1-\beta}P_{n-1}^{(1+\alpha,-1-\beta)}(x)\big]\Big\}\Big|_{x=1}\\
&\quad=\frac{\Gamma(n-1-\beta+\mu)}{\Gamma(n-1-\beta)}\Big\{{}_{x}^{\rm RL}I_1^{1-\mu}\big[(1-x)^{1+\alpha}\ P_{n-1}^{(1+\alpha,-1-\beta)}(x)\big]\Big\}\Big|_{x=1}=0,
\end{aligned}
$$

where we used the formula (2.2.25) with $n\to n-1$, $\mu\to 1-\mu$, $\alpha\to 1+\alpha$, $\beta\to-\beta-1$, that is,

$$
\begin{aligned}
&{}_{x}^{\rm RL}I_1^{1-\mu}\big[(1-x)^{1+\alpha}\ P_{n-1}^{(1+\alpha,-1-\beta)}(x)\big]\\
&\quad=\frac{\Gamma(n+\alpha+1)}{\Gamma(n+\alpha+2-\mu)}(1-x)^{\alpha+2-\mu}\ P_{n-1}^{(\alpha+2-\mu,-\beta-2+\mu)}(x).
\end{aligned}
\tag{2.2.65}
$$

The next step is to show that the eigenpairs in (2.2.60)–(2.2.61) satisfy SFSLP (2.2.51). Firstly, we take a fractional integration of order μ on both sides of (2.2.51) and substitute (2.2.60) into the resulting equation. Then, again by replacing the Caputo derivative by the Riemann–Liouville derivation, due to the homogeneous Dirichlet boundary condition, we obtain

$$
\begin{aligned}
&(1-x)^{\alpha+1}(1+x)^{\beta+1}\,{}_{-1}^{\rm RL}D_x^{\mu}\big[(1+x)^{-\beta+\mu-1}\ P_{n-1}^{(\alpha-\mu+1,\,-\beta+\mu-1)}(x)\big]\\
&\qquad=-\Lambda^{(1)}\ {}_{x}^{\rm RL}I_1^{\mu}\big[(1-x)^{\alpha+1-\mu}\ P_{n-1}^{(\alpha-\mu+1,\,-\beta+\mu-1)}(x)\big].
\end{aligned}
$$

Working out the fractional derivative and integral by (2.2.64) and (2.2.25), we find

$$\frac{\Gamma(n-1-\beta+\mu)}{\Gamma(n-1-\beta)}(1-x)^{\alpha+1}\ P_{n-1}^{(\alpha+1,-\beta-1)}(x)$$
$$= -\Lambda^{(1)}\frac{\Gamma(n+\alpha+\mu+1)}{\Gamma(n+\alpha+1)}(1-x)^{\alpha+1}\ P_{n-1}^{(\alpha+1,-\beta-1)}(x),$$

which implies

$$\Lambda^{(1)} \equiv {}^{(1)}\Lambda_n^{\alpha,\beta,\mu} = -\frac{\Gamma(n-\beta+\mu-1)\Gamma(n+\alpha+1)}{\Gamma(n-\beta-1)\Gamma(n+\alpha-\mu+1)}.$$

Now, we need to check Theorem 2.12, to see if the eigenvalues of SFSLP-I are indeed real-valued and distinct, and the orthogonality of the eigenfunctions with respect to the Jacobi weight function $w(x) = (1-x)^{1+\alpha-\mu}(1+x)^{1+\beta-\mu}$:

$$\int_{-1}^{1} {}^{(1)}\mathcal{P}_k^{(\alpha,\beta,\mu)}(x)\ {}^{(1)}\mathcal{P}_j^{(\alpha,\beta,\mu)}(x)\, w(x)\, \mathrm{d}x$$
$$= \int_{-1}^{1} \left[(1+x)^{-\beta+\mu-1}\right]^2 P_{k-1}^{(\alpha-\mu+1,\,-\beta+\mu-1)}(x)\, P_{j-1}^{(\alpha-\mu+1,\,-\beta+\mu-1)}(x)\, w(x)\, \mathrm{d}x$$
$$= \int_{-1}^{1} P_{k-1}^{(\alpha^*,\beta^*)}(x)\, P_{j-1}^{(\alpha^*,\beta^*)}(x)\,(1-x)^{\alpha^*}(1+x)^{\beta^*}\mathrm{d}x = \gamma_{k-1}^{(\alpha^*,\beta^*)}\delta_{kj},$$

where $\alpha^* = \alpha-\mu+1$, $\beta^* = -\beta+\mu-1$, and we used the orthogonality of Jacobi polynomials.

Following the proof in **Step 3** of Theorem 2.4, we can show that the set $\{{}^{(1)}\mathcal{P}_n^{(\alpha,\beta,\mu)}(x) \ :\ n = 1,2,\dots\}$ forms a basis for the infinite-dimensional Hilbert space $L^2_w(-1,1)$, and $\Lambda_n^{(1)}$, the corresponding eigenvalue for each n, is simple.

The results (2.2.62)–(2.2.63) on the eigenpairs of SFLSP-II can be shown in a very similar fashion. □

Lemma 2.14 *The shifted eigenfunctions to SFSLP-I and -II in Theorem 2.13 are given by*

$${}^{(1)}\widetilde{\mathcal{P}}_n^{(\alpha,\beta,\mu)}(t) = {}^{(1)}\mathcal{P}_n^{(\alpha,\beta,\mu)}(2t-1)$$
$$= 2^{\tilde{\mu}^{(1)}}\sum_{j=0}^{n-1}(-1)^{n+j-1}\binom{n-1+j}{j}\binom{n+\tilde{\mu}^{(1)}-1}{n-1-j}t^{j+\tilde{\mu}^{(1)}}, \tag{2.2.66}$$

and

$$\begin{aligned} {}^{(2)}\widetilde{\mathcal{P}}_n^{(\alpha,\beta,\mu)}(t) &= {}^{(2)}\mathcal{P}_n^{(\alpha,\beta,\mu)}(1-2t) \\ &= 2^{\tilde{\mu}^{(2)}} \sum_{j=0}^{n-1} (-1)^j \binom{n-1+j}{j} \binom{n+\tilde{\mu}^{(2)}-1}{n-1-j} t^{j+\tilde{\mu}^{(2)}}, \end{aligned} \tag{2.2.67}$$

where $t \in (0,1)$ *and*

$$\tilde{\mu}^{(1)} = -\beta + \mu - 1, \quad \tilde{\mu}^{(2)} = -\alpha + \mu - 1.$$

Proof The expansion (2.2.66) is a direct consequence of (2.2.42) and (2.2.60). Noting the relation

$${}^{(2)}\mathcal{P}_n^{(\alpha,\beta,\mu)}(-x) = (-1)^{n-1}\, {}^{(1)}\mathcal{P}_n^{(\beta,\alpha,\mu)}(x), \quad x \in (-1,1),$$

we obtain (2.2.67) from (2.2.66) directly. □

Theorem 2.15 *The shifted eigensolutions to (2.2.51):* ${}^{(i)}\widetilde{\mathcal{P}}_n^{(\alpha,\beta,\mu)}(t)$*, form a complete hierarchical basis for the finite-dimensional space of poly-fractonomials* $\mathbb{F}^{n-1+\tilde{\mu}^{(i)}}$*, where* $\tilde{\mu}^{(1)} = -\beta + \mu - 1$ *and* $\tilde{\mu}^{(2)} = -\alpha + \mu - 1$*, where* $0 < \tilde{\mu}^{(1)} < \mu$ *and* $0 < \tilde{\mu}^{(2)} < \mu$*.*

Proof The proof follows that of Theorem 2.11. □

The growth of the magnitude in the eigenvalues of SFSLP-I, $|\Lambda_n^{(1)}|$, exhibits a similar behavior as one observed in RFSLP-I and -II. However, there are another two degrees of freedom in the choice of parameters α and β, which affect the magnitude of the eigenvalues. It turns out that in case of SFSLP-I ($i = 1$), the optimal highest magnitude is achieved when $\alpha \to 2 - \mu$ and $\beta \to -1$, $\forall \mu \in (0,1)$. The growth of the $|\Lambda_n^{(1)}|$ corresponding to three values of $\mu = 0.35$, $\mu = 0.5$, and $\mu = 0.99$ is shown in Figure 2.4. Again, we observe

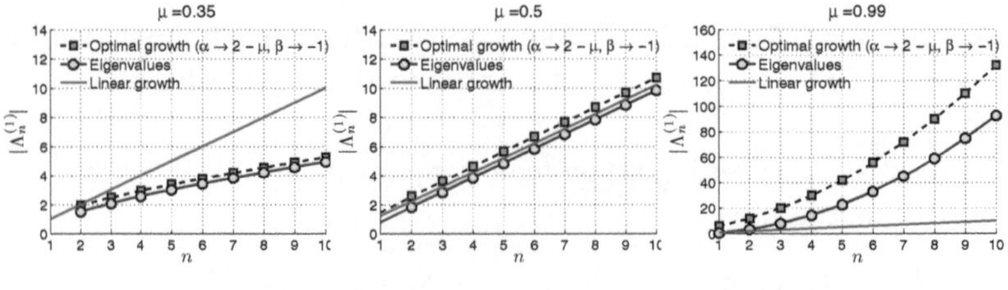

Figure 2.4 Magnitude of the eigenvalues of SFSLP-I, $|\Lambda_n^{(1)}|$, versus n, corresponding to $\alpha = 0$ and $\beta = -0.7$, corresponding to different fractional order $\mu = 0.35$, left: sublinear growth, $\mu = 0.5$, middle: linear growth, and $\mu = 0.99$, right: superlinear-subquadratic growth. Here we compare the growth of the eigenvalues to the optimal case when $\alpha \to 2 - \mu$ and $\beta \to -1$. Reprinted from Zayernouri and Karniadakis (2013) with permission from Elsevier.

about the two growth modes of $|\Lambda_n^{(1)}|$, depending on either $\mu \in (0, 1/2)$, where a sublinear growth in $|\Lambda_n^1|$ is observed, or, $\mu \in (1/2, 1)$, where a superlinear-subquadratic growth mode is valid; the case $\mu = 1/2$ leads to an exactly linear growth mode. Corresponding to the aforementioned fractional orders μ, in Figure 2.5, we plot the eigenfunctions of SFSLP-I, $\mathcal{P}_n^{(1)}(x)$, of different orders and corresponding to different values of μ used in Figure 2.4. In a similar fashion, we compare the eigensolutions with the corresponding standard Jacobi polynomials $P_n^{(\alpha-\mu+1,-\beta+\mu-1)}(x)$ in each plot.

In Figure 2.7, the growth of the magnitude in $\Lambda_n^{(2)}$, corresponding to three values of $\mu = 0.35$, $\mu = 0.5$, and $\mu = 0.99$ is plotted. In SFSLP-II ($i = 2$), the optimal highest magnitude in the eigenvalues is achieved when $\alpha \to -1$ and $\beta \to 2 - \mu$. Moreover, in Figure 2.6, we plot the eigenfunctions of SFSLP-II, $\mathcal{P}_n^{(2)}(x)$, of different fractional orders and corresponding to different μ used in Figure 2.7. This time, we compare the eigensolutions with the corresponding standard Jacobi polynomials $P_n^{(-\alpha+\mu-1,\beta-\mu+1)}(x)$ in each plot.

2.3 Degeneracy of Generalized Jacobi Polynomials

We now aim to introduce a broader family of non-polynomial weakly singular spectral bases to be employed in the subsequent chapters. To this end, we focus our attention on further generalizing the aforementioned regular Jacobi poly-fractonomials, $\Phi_n^{(i)}(x)$ in (2.2.33) and (2.2.35) and the singular Jacobi poly-fractonomials, ${}^{(i)}\mathcal{P}_n^{(\alpha,\beta,\mu)}(x)$ in (2.2.60) and (2.2.62), both in terms of the underlying *singular structures* and their admissible range of input parameters. However the backbone of such a development demands understanding of the generalized Jacobi "polynomials" and their "degeneracy."

2.3.1 Generalized Jacobi Polynomials and Their Degeneracy

According to Szegő (1975, (4.21.2)), the classical Jacobi polynomials with the parameters $\alpha, \beta > -1$ in (2.1.1) can be generalized to the Jacobi polynomials with parameters $\alpha, \beta \in \mathbb{R}$ through the same hypergeometric representation:

$$
\begin{aligned}
P_n^{(\alpha,\beta)}(x) &= \frac{(\alpha+1)_n}{n!}\,{}_2F_1\Big(-n, n+\alpha+\beta+1; \alpha+1; \frac{1-x}{2}\Big) \\
&= \frac{(\alpha+1)_n}{n!} + \sum_{j=1}^{n-1} \frac{(n+\alpha+\beta+1)_j(\alpha+j+1)_{n-j}}{j!\,(n-j)!}\Big(\frac{x-1}{2}\Big)^j \\
&\quad + \frac{(n+\alpha+\beta+1)_n}{n!}\Big(\frac{x-1}{2}\Big)^n, \quad n \ge 1,
\end{aligned}
\tag{2.3.1}
$$

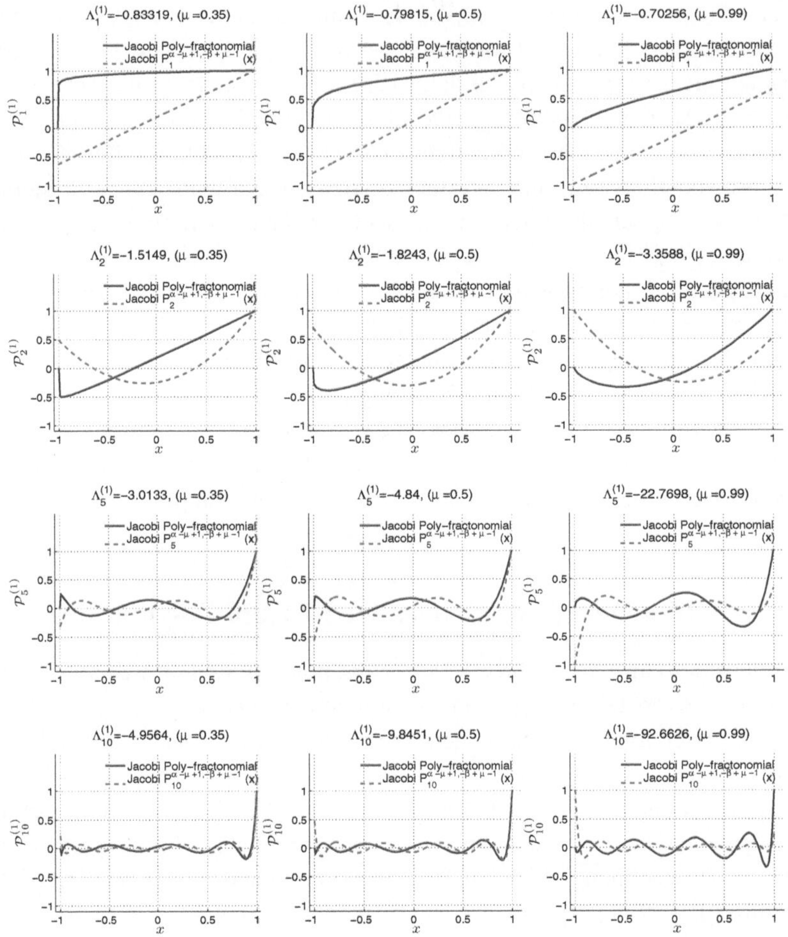

Figure 2.5 Eigenfunctions of SFSLP-I, $\mathcal{P}_n^{(1)}$, versus x, for $n = 1$ (first row), $n = 2$ (second row), $n = 5$ (third row), and $n = 10$ (last row), corresponding to the fractional order $\mu = \nu/2 = 0.35$ (left column), $\mu = \nu/2 = 0.5$ (middle column), and $\mu = \nu/2 = 0.99$ (right column). Here, we take the same values $\alpha = 0$ and $\beta = -0.7$, as shown in Figure 2.4. Reprinted from Zayernouri and Karniadakis (2013) with permission from Elsevier.

and $P_0^{(\alpha,\beta)}(x) \equiv 1$. Here, we adopt the normalization and formulation of the coefficients to avoid artificial singularities. Alternatively, we can define the generalized Jacobi polynomial such that

$$\begin{aligned} P_n^{(\alpha,\beta)}(x) &= (-1)^n \frac{(\beta+1)_n}{n!} {}_2F_1\Big(-n, n+\alpha+\beta+1; \beta+1; \frac{1+x}{2}\Big) \\ &= (-1)^n P_n^{(\beta,\alpha)}(-x), \quad x \in (-1,1). \end{aligned} \tag{2.3.2}$$

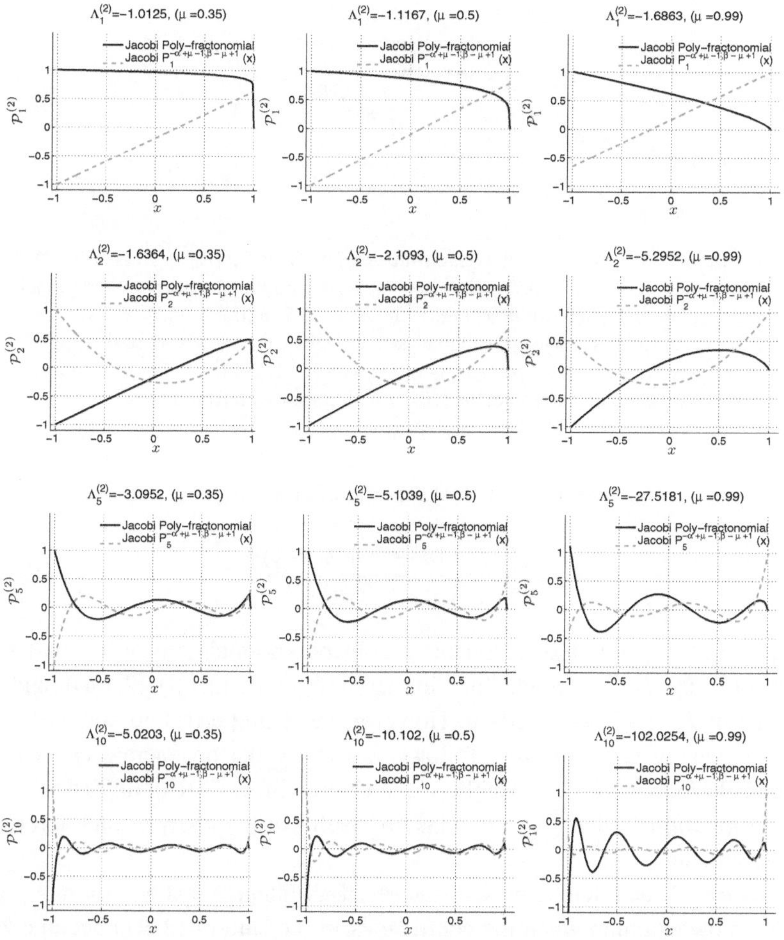

Figure 2.6 Eigenfunctions of SFSLP-II, $\mathcal{P}_n^{(2)}$, versus x, for $n = 1$ (first row), $n = 2$ (second row), $n = 5$ (third row), and $n = 10$ (last row), corresponding to the fractional order $\mu = \nu/2 = 0.35$ (left column), $\mu = \nu/2 = 0.5$ (middle column), and $\mu = \nu/2 = 0.99$ (right column). Here, we take the same values $\alpha = -0.7$ and $\beta = 0$, as shown in Figure 2.7. Reprinted from Zayernouri and Karniadakis (2013) with permission from Elsevier.

It is evident that $P_n^{(\alpha,\beta)}(x)$ is always a polynomial in x for all $\alpha, \beta \in \mathbb{R}$, and satisfies

$$P_n^{(\alpha,\beta)}(1) = \frac{(\alpha+1)_n}{n!}, \quad P_n^{(\alpha,\beta)}(-1) = (-1)^n \frac{(\beta+1)_n}{n!}. \tag{2.3.3}$$

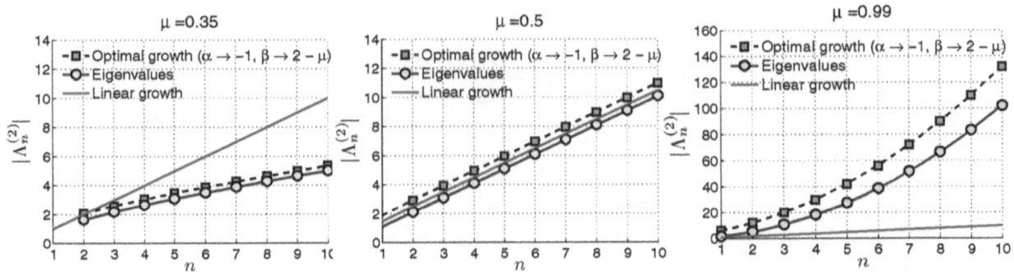

Figure 2.7 Magnitude of the eigenvalues of SFSLP-II, $|\Lambda_n^{(2)}|$, versus n, corresponding to $\alpha = -0.7$ and $\beta = 0$, corresponding to different fractional order $\mu = 0.35$, left: sublinear growth, $\mu = 0.5$, middle: linear growth, and $\mu = 0.99$, right: superlinear-subquadratic growth. Here we compare the growth of the eigenvalues to the optimal case when $\alpha \to -1$ and $\beta \to 2 - \mu$. Reprinted from Zayernouri and Karniadakis (2013) with permission from Elsevier.

Moreover, we observe from (2.3.1) that the coefficient of x^n is

$$l_n^{(\alpha,\beta)} := \frac{(n+\alpha+\beta+1)_n}{2^n n!}. \tag{2.3.4}$$

For $\alpha, \beta > -1$, the (classical) Jacobi polynomials are orthogonal with respect to the Jacobi weight function $\omega^{(\alpha,\beta)}(x)$ (see (2.1.3)–(2.1.4)), and the degree of $P_n^{(\alpha,\beta)}(x)$ is exactly n. However, these properties do not carry over to the general case. In what follows, we study the degenerate cases as in Szegő (1975, p.63–65) and Cagliero and Koornwinder (2015). In particular, the transformation (reflection) formulas are useful in the analysis and algorithm development.

Let $n \ge 1$, and denote by $\mathbb{Z}_{\le k}$ the set of all integers that are not more than k. We now examine when the coefficients in the sum of (2.3.1) become zero and which degeneracies can occur. For this purpose, it is necessary to consider the following two rising factorials.

- **Case I**: $(n+\alpha+\beta+1)_j = 0$ for some $j \in \{1, 2, \ldots, n\}$. Equivalently, $(n+\alpha+\beta+1)_n = 0$ or alternatively, this happens when

$$\alpha+\beta \in \mathbb{Z}_{\le -2}, \quad n+\alpha+\beta+1 \le 0 \quad \text{and} \quad 2n+\alpha+\beta \ge 0. \tag{2.3.5}$$

In other words, if $\alpha+\beta$ satisfies (2.3.5), then

$$(n+\alpha+\beta+1)_j = 0 \quad \text{for all} \quad j = -n-\alpha-\beta, \ldots, n, \tag{2.3.6}$$

so the corresponding coefficients of $P_n^{(\alpha,\beta)}(x)$ in (2.3.1) vanish.

- **Case II**: $(\alpha + j + 1)_{n-j} = 0$ for some $j \in \{0, 1, \ldots, n-1\}$. Equivalently, $(\alpha + 1)_n = 0$ or alternatively, this happens when

$$\alpha \in \mathbb{Z}_{\leq -1} \quad \text{and} \quad n + \alpha \geq 0. \tag{2.3.7}$$

  In other words, if α satisfies (2.3.7), then

$$(\alpha + j + 1)_{n-j} = 0 \quad \text{for all} \quad j = 0, \ldots, -\alpha - 1. \tag{2.3.8}$$

  so the corresponding coefficients of $P_n^{(\alpha,\beta)}(x)$ in (2.3.1) vanish.

  In view of (2.3.2), the same property holds with β in place of α in (2.3.7)–(2.3.8).

One verifies readily that if α, β meet the conditions for both cases, then the generalized Jacobi polynomial $P_n^{(\alpha,\beta)}(x)$ vanishes identically.

Proposition 2.16 *For $n \geq 1$, $P_n^{(\alpha,\beta)}(x) \equiv 0$, if and only if*

$$\alpha, \beta \in \mathbb{Z}_{\leq -1} \quad \textit{and} \quad \max\{-\alpha, -\beta\} \leq n \leq -\alpha - \beta - 1. \tag{2.3.9}$$

Case I causes the degeneracy that $P_n^{(\alpha,\beta)}(x)$ has degree lower than n in x. Indeed, it is seen from (2.3.4) that its leading coefficient vanishes.

Proposition 2.17 *Let $n \geq 1$, and assume $P_n^{(\alpha,\beta)}(x)$ does not vanish identically. Then $P_n^{(\alpha,\beta)}(x)$ has degree $< n$, if and only if (2.3.5) holds. Its degree is $-n - \alpha - \beta - 1$ and*

$$P_n^{(\alpha,\beta)}(x) = \frac{(-n-\beta)_{2n+\alpha+\beta+1}}{(-n-\alpha-\beta)_{2n+\alpha+\beta+1}} P_{-n-\alpha-\beta-1}^{(\alpha,\beta)}(x). \tag{2.3.10}$$

Case II causes the degeneracy that $P_n^{(\alpha,\beta)}(x)$ vanishes for $x = \pm 1$ with a certain multiplicity.

Proposition 2.18 *Let $n \geq 1$, and assume $P_n^{(\alpha,\beta)}(x)$ does not vanish identically.*

(a) *$P_n^{(\alpha,\beta)}(1) = 0$ if and only if $\alpha \in \mathbb{Z}_{\leq -1}$ and $n + \alpha \geq 0$. The zero $x = 1$ has a multiplicity $-\alpha$. Moreover, we have the transformation formula*

$$P_n^{(\alpha,\beta)}(x) = \frac{(n+\alpha+\beta+1)_{-\alpha}\,(n+\alpha)!}{n!}\Big(\frac{x-1}{2}\Big)^{-\alpha} P_{n+\alpha}^{(-\alpha,\beta)}(x). \tag{2.3.11}$$

(b) $P_n^{(\alpha,\beta)}(-1) = 0$ *if and only if* $\beta \in \mathbb{Z}_{\le -1}$ *and* $n + \beta \ge 0$. *The zero* $x = -1$ *has a multiplicity* $-\beta$. *Moreover, we have the transformation formula*

$$P_n^{(\alpha,\beta)}(x) = \frac{(n+\alpha+\beta+1)_{-\beta}(n+\beta)!}{n!}\left(\frac{x+1}{2}\right)^{-\beta} P_{n+\beta}^{(\alpha,-\beta)}(x). \quad (2.3.12)$$

(c) *For* $\alpha, \beta \in \mathbb{Z}_{\le -1}$ *and* $n + \alpha + \beta \ge 0$, *we have the transformation formula*

$$P_n^{(\alpha,\beta)}(x) = \left(\frac{x-1}{2}\right)^{-\alpha}\left(\frac{x+1}{2}\right)^{-\beta} P_{n+\alpha+\beta}^{(-\alpha,-\beta)}(x). \quad (2.3.13)$$

The transformation formula in (c) is a direct consequence of (a)–(b). One can verify (a)–(b) by using the hypergeometric representations in (2.3.1)–(2.3.2). Combining Proposition 2.17 with the case (a) or (b), one can derive some other transformation formulas as in Cagliero and Koornwinder (2015).

The generalized Jacobi polynomials $P_n^{(\alpha,\beta)}(x)$ with $\alpha, \beta \in \mathbb{Z}_{\le -1}$ play an important role in the construction and analysis of optimal spectral algorithms for boundary-value problems (see Guo et al., 2006b). We can apply the previous results to study their degeneracy.

Example 2.19 Let $\alpha = -k$, $\beta = -l$ with integers $k \ge l \ge 1$.

(i) We find from Proposition 2.16 that

$$P_n^{(-k,-l)}(x) \equiv 0, \quad k \le n \le k + l - 1. \quad (2.3.14)$$

(ii) In view of (2.3.13), we can transform $P_n^{(-k,-l)}(x)$ with $n \ge k+1$ back to the classical Jacobi polynomial via

$$P_n^{(-k,-l)}(x) = 2^{-k-l}(x-1)^k(x+1)^l P_{n-k-l}^{(k,l)}(x), \quad n \ge k + l. \quad (2.3.15)$$

(iii) The degeneracy does not occur when $n = 0, 1, \ldots, k-1$. Indeed, we have $P_0^{(-k,-l)}(x) \equiv 1$ and by (2.3.1),

$$P_1^{(-k,-l)}(x) = \frac{1}{2}(2-k-l)x + \frac{1}{2}(l-k).$$

Then $P_n^{(-k,-l)}(x)$ with $2 \le n \le k-1$ can be computed by the recurrence relation (**??**)–(**??**).

Remark 2.20 Observe from (2.3.15) that $\{P_n^{(-k,-l)}(x)\}_{n\geq k+l}$ are mutually orthogonal with the non-integrable Jacobi weight function $\omega^{(-k,-l)}(x) = (1-x)^{-k}(1+x)^{-l}$.

2.3.2 Integral Representations of Generalized Jacobi Polynomials

It is seen that the fractional integral formulas in Lemma 2.3 have played an important role in the study of fractional Sturm–Liouville problems. In fact, they are also valid for generalized Jacobi polynomials with α or $\beta \leq -1$, which can be shown by the Bateman's fractional integral formula (Bateman, 1909) of hypergeometric functions:

$$ {}_2F_1(a,b;c+\mu;x) = \frac{\Gamma(c+\mu)}{\Gamma(c)\Gamma(\mu)} x^{1-(c+\mu)} \int_0^x t^{c-1}(x-t)^{\mu-1} {}_2F_1(a,b;c;t)\, dt, \tag{2.3.16} $$

where real $\Re\{c\}, \Re\{\mu\} > 0$ and $|x| < 1$ if the series is infinite.

The following more general integral representations with real parameters α or β can be founded in Szegő (1975, p.96) and Andrews et al. (1999, p.314).

Theorem 2.21 *Let $\mu \in \mathbb{R}^+$, $n \in \mathbb{N}_0$ and $|x| < 1$.*

(i) *For $\alpha > -1$ and $\beta \in \mathbb{R}$, we have*

$$ (1-x)^{\alpha+\mu} \frac{P_n^{(\alpha+\mu,\beta-\mu)}(x)}{P_n^{(\alpha+\mu,\beta-\mu)}(1)} = \frac{\Gamma(\alpha+\mu+1)}{\Gamma(\alpha+1)\Gamma(\mu)} \int_x^1 \frac{(1-y)^\alpha}{(y-x)^{1-\mu}} \frac{P_n^{(\alpha,\beta)}(y)}{P_n^{(\alpha,\beta)}(1)}\, dy. $$

(ii) *For $\alpha \in \mathbb{R}$ and $\beta > -1$, we have*

$$ (1+x)^{\beta+\mu} \frac{P_n^{(\alpha-\mu,\beta+\mu)}(x)}{P_n^{(\beta+\mu,\alpha-\mu)}(1)} = \frac{\Gamma(\beta+\mu+1)}{\Gamma(\beta+1)\Gamma(\mu)} \int_{-1}^x \frac{(1+y)^\beta}{(x-y)^{1-\mu}} \frac{P_n^{(\alpha,\beta)}(y)}{P_n^{(\beta,\alpha)}(1)}\, dy. $$

Proof Taking $a = -n, b = n+\alpha+\beta+1, c = \alpha+1$ and setting $t = (1-y)/2$ and $x \to (1-x)/2$ in (2.3.16), we obtain the first formula from the definition of the Jacobi polynomial in (2.3.1). The second formula follows from the first one and the property: $P_n^{(\alpha,\beta)}(x) = (-1)^n P_n^{(\beta,\alpha)}(-x)$ in (2.3.2) straightforwardly. □

Moreover, we can derive from Bateman's formula and some relevant properties of hypergeometric functions the following integral identities.

Theorem 2.22 *Let $\mu \in \mathbb{R}^+$, $n \in \mathbb{N}_0$ and $x \in (-1,1)$.*

(i) *For $\alpha > -1$ and $\beta \in \mathbb{R}$,*

$$\frac{(1-x)^{\alpha+\mu}}{(1+x)^{n+\alpha+1}} \frac{P_n^{(\alpha+\mu,\beta)}(x)}{P_n^{(\alpha+\mu,\beta)}(1)} = \frac{2^\mu \Gamma(\alpha+\mu+1)}{\Gamma(\alpha+1)\Gamma(\mu)} \int_x^1 \frac{(1-y)^\alpha}{(1+y)^{n+\alpha+\mu+1}} \frac{P_n^{(\alpha,\beta)}(y)}{P_n^{(\alpha,\beta)}(1)} \times (y-x)^{\mu-1} dy.$$

(ii) *For $\alpha \in \mathbb{R}$ and $\beta > -1$,*

$$\frac{(1+x)^{\beta+\mu}}{(1-x)^{n+\beta+1}} \frac{P_n^{(\alpha,\beta+\mu)}(x)}{P_n^{(\beta+\mu,\alpha)}(1)} = \frac{2^\mu \Gamma(\beta+\mu+1)}{\Gamma(\beta+1)\Gamma(\mu)} \int_{-1}^x \frac{(1+y)^\beta}{(1-y)^{n+\beta+\mu+1}} \frac{P_n^{(\alpha,\beta)}(y)}{P_n^{(\beta,\alpha)}(1)} \times (x-y)^{\mu-1} dy.$$

Proof We first derive the integral identity

$$\begin{aligned} &x^{c+\mu-1}(1-x)^{a-c}{}_2F_1(a,b+\mu;c+\mu;x) \\ &= \frac{\Gamma(c+\mu)}{\Gamma(c)\Gamma(\mu)} \int_0^x (x-t)^{\mu-1}(1-t)^{a-c-\mu}t^{c-1}{}_2F_1(a,b;c;t)\,dt \end{aligned} \tag{2.3.17}$$

from Bateman's formula (2.3.16) and Pfaff's transformation (cf. Andrews et al., 1999, p. 68)

$${}_2F_1(a,b;c;x) = (1-x)^{-a}\,{}_2F_1\Big(a,c-b;c;\frac{x}{x-1}\Big). \tag{2.3.18}$$

Indeed, applying (2.3.18) to the two hypergeometric functions in (2.3.16) yields

$$\begin{aligned} &x^{c+\mu-1}(1-x)^{-a}{}_2F_1\Big(a,c+\mu-b;c+\mu;\frac{x}{x-1}\Big) \\ &= \frac{\Gamma(c+\mu)}{\Gamma(c)\Gamma(\mu)} \int_0^x t^{c-1}(x-t)^{\mu-1}(1-t)^{-a}\,{}_2F_1\Big(a,c-b;c;\frac{t}{t-1}\Big)\,dt. \end{aligned} \tag{2.3.19}$$

Changing the variables: $\hat{t} = t/(t-1)$ and $\hat{x} = x/(x-1)$ in (2.3.19), and noting that

$$t = \frac{\hat{t}}{\hat{t}-1},\quad 1-t = \frac{1}{1-\hat{t}},\quad x = \frac{\hat{x}}{\hat{x}-1},\quad 1-x = \frac{1}{1-\hat{x}},\quad x-t = \frac{\hat{t}-\hat{x}}{(1-\hat{x})(1-\hat{t})},$$

we obtain from (2.3.19) and direct calculation that

$$\begin{aligned} &\hat{x}^{c+\mu-1}(1-\hat{x})^{a-c}\,{}_2F_1(a,c+\mu-b;c+\mu;\hat{x}) \\ &= \frac{\Gamma(c+\mu)}{\Gamma(c)\Gamma(\mu)} \int_0^{\hat{x}} \hat{t}^{c-1}(1-\hat{t})^{a-c+\mu}(\hat{x}-\hat{t})^{\mu-1}\,{}_2F_1(a,c-b;c;\hat{t})\,d\hat{t}. \end{aligned}$$

With $x, t, b+c$ in place of $\hat{x}, \hat{t}, b$ in the above identity, we obtain (2.3.17).

Thus, taking $a = -n, b = n+\alpha+\beta+1, c = \alpha+1, t = (1-y)/2$ and $x \to (1-x)/2$ in (2.3.17), we obtain the formula in (i) from (2.3.1) and (2.3.3). The second formula follows from the first one and the property: $P_n^{(\alpha,\beta+\mu)}(x) = (-1)^n P_n^{(\beta+\mu,\alpha)}(-x)$ in (2.3.2) immediately. □

Moreover, we have the following spectral relationship for the classical Riesz potential, see Podlubny (1999, Theorems 6.5 & 6.7). For notational convenience, we denote

$$Q_n^{(\alpha,\beta)}(x) = (1-x)^\alpha(1+x)^\beta P_n^{(\alpha,\beta)}(x). \tag{2.3.20}$$

Theorem 2.23 *Let $\nu \in (0,1)$ and r, l be two integers.*

(i) *If $r, l > -\frac{\nu+1}{2}$, then for $n+r+l \ge 1$,*

$$\int_{-1}^{1} \frac{Q_n^{(\frac{\nu-1}{2}+r,\frac{\nu-1}{2}+l)}(t)}{|x-t|^\nu}dt = \frac{\pi(-1)^r 2^{r+l}\Gamma(n+\nu)}{n!\,\Gamma(\nu)\cos\frac{\nu\pi}{2}} P_{n+r+l}^{(\frac{\nu-1}{2}-r,\frac{\nu-1}{2}-l)}(x). \tag{2.3.21}$$

(ii) *If $r, l > -\frac{\nu}{2}-1$, then for $n+r+l+1 \ge 0$,*

$$\int_{-1}^{1} \frac{\operatorname{sign}(x-t)}{|x-t|^\nu} Q_n^{(\frac{\nu}{2}+r,\frac{\nu}{2}+k)}(t)dt = \frac{\pi(-1)^r 2^{r+l+1}\Gamma(n+\nu)}{n!\,\Gamma(\nu)\sin\frac{\pi\nu}{2}} P_{n+r+l+1}^{(\frac{\nu}{2}-r-1,\frac{\nu}{2}-l-1)}(x). \tag{2.3.22}$$

Theorem 2.24 *If $\alpha > -1, \beta > -1, 0 < \nu < 1$ and γ is an arbitrary real number, then we have*

$$\begin{aligned}\int_{-1}^{1} &\left(\operatorname{sign}(x-t) + \frac{\tan(\pi\gamma)}{\tan\frac{\nu\pi}{2}}\right)\frac{Q_m^{(\alpha,\beta)}(t)}{|x-t|^\nu}dt \\ &= \frac{\sin\pi\left(\gamma-\frac{\nu}{2}\right)\Phi_1(x) + \sin\pi\left(\gamma+\frac{\nu}{2}-\beta\right)\Phi_2(x)}{\Gamma(\nu)\sin\frac{\nu\pi}{2}\cos\gamma\pi},\end{aligned} \tag{2.3.23}$$

where

$$\begin{aligned}\Phi_1(x) &= \frac{\Gamma(m+\alpha+1)\Gamma(m+\nu)\Gamma(\beta-\nu+1)(-1)^m}{2^{-\alpha-\beta+\nu-1}\Gamma(m+\alpha+\beta-\nu+2)m!} \\ &\quad \times {}_2F_1\left(m+\nu, \nu-m-\alpha-\beta-1; -\beta+\nu; \frac{1+x}{2}\right),\end{aligned} \tag{2.3.24}$$

$$\begin{aligned}\Phi_2(x) &= \frac{\Gamma(m+\beta+1)\Gamma(\nu-\beta-1)(-1)^{m+1}}{2^{-\alpha}(1+x)^{\nu-\beta-1}} \\ &\quad \times {}_2F_1\left(m+\beta+1, -m-\alpha; \beta-\nu+2; \frac{1+x}{2}\right).\end{aligned} \tag{2.3.25}$$

As some important special cases, we have the following integral representations.

Theorem 2.25 *If* $0 < \nu < 1, \gamma$ *is an arbitrary real number and* r *and* k *are integer numbers such that* $r > -1 + \gamma - \frac{\nu}{2}, k > -1 - \gamma - \frac{\nu}{2}$*, then for* $-1 < x < 1$*, the following holds for* $m + r + k + 1 \geq 0$:

$$\int_{-1}^{1} \left(\text{sign}(x-t) + \frac{\tan(\pi\gamma)}{\tan\frac{\nu\pi}{2}} \right) \frac{Q_m^{(-\gamma+\frac{\nu}{2}+r,\gamma+\frac{\nu}{2}+k)}(t)}{|x-t|^\nu} dt$$
$$= \frac{\pi(-1)^{r+k+1} \sin\pi\left(\gamma - \frac{\nu}{2}\right) 2^{r+k+1} \Gamma(m+\nu)}{m!\,\Gamma(\nu) \sin\frac{\nu\pi}{2} \cos(\gamma\pi) \sin\pi\left(-\gamma + \frac{\nu}{2} - k\right)} P_{m+r+k+1}^{(\gamma+\frac{\nu}{2}-r-1,-\gamma+\frac{\nu}{2}-k-1)}(x). \tag{2.3.26}$$

Theorem 2.25 is a generalization of Popov's formula (Popov, 1982).

Theorem 2.26 (Spectral relationship for the classical Riesz potential) *If* $0 < \nu < 1$ *and* r *and* k *are integer numbers such that* $r > -\frac{\nu+1}{2}, k > -\frac{\nu+3}{2}$*, then for* $-1 < x < 1$*, the following holds:*

$$\int_{-1}^{1} \frac{Q_m^{(\frac{\nu-1}{2}+r,\frac{\nu+1}{2}+k)}}{|x-t|^\nu} dt = \frac{\pi(-1)^r 2^{r+k+1} \Gamma(m+\nu)}{m!\,\Gamma(\nu)\cos\frac{\nu\pi}{2}} P_{m+r+k+1}^{(\frac{\nu-1}{2}-r,\frac{\nu-3}{2}-k)}(x) \tag{2.3.27}$$

for $m + r + k + 1 > 0$.

In the general case, the Jacobi polynomials in the right-hand side of (2.3.26)–(2.3.27) are the Jacobi polynomials orthogonal with non-integrable weight function.

In order to consider classical Jacobi polynomials, we must have

$$\frac{\nu-1}{2} - r > -1, \quad \frac{\nu-3}{2} - k > -1, \tag{2.3.28}$$

in addition to the conditions of Theorem 2.26. There is only one pair of values for r and k which simultaneously satisfies the conditions of Theorem 2.26 and (2.3.28): $r = 0, k = -1$. In this case, the formula (2.3.27) takes the following simple form.

Theorem 2.27 *If* $0 < \nu < 1$*, then for* $-1 < x < 1$*, the following holds:*

$$\int_{-1}^{1} \frac{Q_m^{(\frac{\nu-1}{2},\frac{\nu-1}{2})}(t)}{|x-t|^\nu} dt = \frac{\pi\Gamma(m+\nu)}{m!\,\Gamma(\nu)\cos\frac{\nu\pi}{2}} P_m^{(\frac{\nu-1}{2},\frac{\nu-1}{2})}(x) \tag{2.3.29}$$

for $m = 0, 1, 2 \ldots.$

Theorem 2.28 *If* $0 < \nu < 1$ *and* r *and* k *are integer numbers such that* $r > -1 - \frac{\nu}{2}, k > -1 - \frac{\nu}{2}$, *then for* $-1 < x < 1$, *the following holds for* $m + r + k + 1 \geq 0$:

$$\int_{-1}^{1} \frac{\operatorname{sign}(x-t)}{|x-t|^{\nu}} Q_m^{(\frac{\nu}{2}+r, \frac{\nu}{2}+k)}(t)\mathrm{d}t = \frac{\pi(-1)^r \Gamma(m+\nu)}{2^{-r-k-1} m!\, \Gamma(\nu) \sin \frac{\pi\nu}{2}} P_{m+r+k+1}^{(\frac{\nu}{2}-r-1, \frac{\nu}{2}-k-1)}(x). \tag{2.3.30}$$

2.4 Generalized Jacobi Poly-fractonomials (GJPFs)

Generalized Jacobi poly-fractonomials (GJPFs), initially termed as the *generalized Jacobi functions* in Guo et al. (2009), generalize the structure of Jacobi poly-fractonomials and take the form

$$\mathcal{P}_n^{(\alpha,\beta)}(x) = (1-x)^{\hat{\alpha}}(1+x)^{\hat{\beta}}\, P_{n_1}^{(\bar{\alpha},\bar{\beta})}(x), \quad x \in (-1, 1), \tag{2.4.1}$$

for $\alpha, \beta \in \mathbb{R}$ and $n_1 = n - [\hat{\alpha}] - [\hat{\beta}] \geq 0$, where

$$\hat{\alpha} := \begin{cases} -\alpha, & \text{if } \alpha \leqslant -1, \\ 0, & \text{if } \alpha > -1, \end{cases} \quad \bar{\alpha} := \begin{cases} -\alpha, & \text{if } \alpha \leqslant -1, \\ \alpha, & \text{if } \alpha > -1. \end{cases} \tag{2.4.2}$$

Alternatively, we can formulate it in a more explicit form as

$$\mathcal{P}_n^{(\alpha,\beta)}(x) = \begin{cases} P_n^{(\alpha,\beta)}(x), & \text{if } \alpha, \beta > -1, \\ (1+x)^{-\beta} P_{n-[-\beta]}^{(\alpha,-\beta)}(x), & \text{if } \alpha > -1,\ \beta \leq -1, \\ (1-x)^{-\alpha} P_{n-[-\alpha]}^{(-\alpha,\beta)}(x), & \text{if } \alpha \leq -1,\ \beta > -1, \\ (1-x)^{-\alpha}(1+x)^{-\beta} P_{n-[-\alpha]-[-\beta]}^{(-\alpha,-\beta)}(x), & \text{if } \alpha, \beta \leq -1. \end{cases} \tag{2.4.3}$$

It is seen that GJPFs are polynomials under one of the four conditions: (i) $\alpha, \beta > -1$; (ii) $\alpha > -1$ and $-\beta \in \mathbb{N}$; (iii) $-\alpha \in \mathbb{N}$ and $\beta > -1$; or (iv) $-\alpha, -\beta \in \mathbb{N}$. In particular, in Cases (ii)–(iv), $\mathcal{P}_n^{(\alpha,\beta)}(x)$ satisfies the transformation formulas in Proposition 2.18, and they are mutually orthogonal with respect to the non-integral generalized Jacobi weight function $\omega^{(\alpha,\beta)}(x) = (1-x)^{\alpha}(1+x)^{\beta}$. We refer to Guo et al. (2006b) for their applications in the construction and analysis of optimal spectral methods for boundary-value problems. On the other hand, for non-integer $\alpha \leq -1$ or $\beta \leq -1$, $\mathcal{P}_n^{(\alpha,\beta)}(x)$ is a non-polynomial function.

2.4.1 Relationship between GJPFs and Jacobi Poly-fractonomials

To highlight the interconnection between the GJPFs and Jacobi poly-fractonomials, we first note that the power indices of the singular terms in GJPFs are greater than one, see (2.4.3), which at first might seem to disagree with the definition of Jacobi poly-fractonomials. However, to establish a suitable form of GJPFs comparable with the Jacobi poly-fractonomials, we rewrite the two middle cases in (2.4.3) by factorizing out the terms $(1+x)^{-\beta-1}$ and $(1-x)^{-\alpha-1}$, respectively, as

$$\mathcal{P}_n^{(\alpha,\beta)}(x) = \begin{cases} (1+x)^{-\beta-1}\big[(1+x)P_{n-[-\beta]}^{(\alpha,-\beta)}(x)\big], & \text{if } \alpha > -1,\ \beta \le -1, \\ (1-x)^{-\alpha-1}\big[(1-x)P_{n-[-\alpha]}^{(-\alpha,\beta)}(x)\big], & \text{if } \alpha \le -1,\ \beta > -1. \end{cases} \tag{2.4.4}$$

Next, we recall the following useful identities:

$$\begin{aligned} (1+x)P_n^{(a,b+1)}(x) &= \frac{2}{2n+a+b+2}\big((n+b+1)P_n^{(a,b)}(x) + (n+1)P_{n+1}^{(a,b)}(x)\big), \\ (1-x)P_n^{(a+1,b)}(x) &= \frac{2}{2n+a+b+2}\big((n+a+1)P_n^{(a,b)}(x) - (n+1)P_{n+1}^{(a,b)}(x)\big). \end{aligned} \tag{2.4.5}$$

Next, using (2.4.4) and (2.4.5), we can obtain the GJPFs as

$$\mathcal{P}_n^{(\alpha,\beta)}(x) = \begin{cases} c_1\,(1+x)^{-\beta-1}P_n^{(\alpha,-\beta-1)}(x) \ + \ c_2\,(1+x)^{-\beta-1}P_{n+1}^{(\alpha,-\beta-1)}(x), \\ (\text{if } \alpha > -1,\ \beta \le -1), \\ c_3\,(1-x)^{-\alpha-1}P_n^{(-\alpha-1,\beta)}(x) \ + \ c_4\,(1-x)^{-\alpha-1}P_{n+1}^{(-\alpha-1,\beta)}(x), \\ (\text{if } \alpha \le -1,\ \beta > -1), \end{cases} \tag{2.4.6}$$

where

$$c_1 = \frac{2(n-\beta)}{2n+\alpha-\beta+1}, \quad c_2 = \frac{2(n+1)}{2n+\alpha-\beta+1},$$
$$c_3 = \frac{2(n-\alpha)}{2n-\alpha+\beta+1}, \quad c_4 = -\frac{2(n+1)}{2n-\alpha+\beta+1}.$$

This new form of one-sided GJPFs consists of singular terms being consistent with the singularity structure of the regular Jacobi poly-fractonomials, $\Phi_n^{(i)}(x) = (1 \pm x)^{\mu} P_{n-1}^{(\mp\mu,\pm\mu)}(x)$, $\mu \in (0,1)$, $i = 1, 2$, with the singular power less than one given in (2.2.33) and (2.2.35). Hence,

$$
\mathcal{P}_n^{(\alpha,\beta)}(x) = \begin{cases} c_1\Phi_n^{(1)}(x) + c_2\Phi_{n+1}^{(1)}(x), \\ (\text{if } \alpha = -\mu > -1,\ \beta = -1-\mu \le -1), \\ \\ c_3\Phi_n^{(2)}(x) + c_4\Phi_{n+1}^{(2)}(x), \\ (\text{if } \alpha = -1-\mu \le -1,\ \beta = -\mu > -1). \end{cases} \tag{2.4.7}
$$

Note that a similar explicit relationship can be formulated between the GJPFs, $\mathcal{P}_n^{(\alpha,\beta)}(x)$, and the singular Jacobi poly-fractonomials, ${}^{(i)}\mathcal{P}_n^{(a,b,\mu)}(x), i = 1, 2$, given in (2.2.60) and (2.2.62).

Remark 2.29 Given the hierarchical structure of the Jacobi poly-fractonomials and their generalized counterparts, one can see that any span of one-sided GJPFs and Jacobi poly-fractonomials would yield identical function spaces with the same singularity features, using the same range of parameters in (2.4.6). We refer to (2.4.3) as "generalized Jacobi poly-fractonomials" since: (i) they mimic the left- and right-sided singular structure of the Jacobi poly-fractonomials, being linear combinations of *fractonomials* according to Definition 2.6 and 2.8, (ii) they broaden the range of parameters regardless of what eigen-boundary-value problems they can originate from whilst preserving useful properties, and (iii) they allow two-sided singularities, which further extends the notion of fractonomials to their two-sided counterparts.

In what follows, we aim to take advantage of the aforementioned generalization of the eigenfunctions and define a set of GJPFs that are suitable basis functions for solving fractional differential equations, allowing the singular terms to appear with *negative* indices when needed. We will demonstrate the effectiveness of such basis constructions in the next chapters.

2.4.2 GJPFs for One-Sided Fractional Derivatives

We modify the definition of GJPFs in (2.4.3) for the two cases as follows.

Definition 2.30 (One-sided generalized Jacobi poly-fractonomials) Let $n \in \mathbb{N}$ and $x \in (-1, 1)$.

(i) For $\beta > -1$ and $\alpha \in \mathbb{R}$, the left-sided GJF is defined by

$$
{}^{l}\mathcal{P}_n^{(\alpha,-\beta)}(x) := (1+x)^{\beta} P_n^{(\alpha,\beta)}(x). \tag{2.4.8}
$$

(ii) For $\alpha > -1$ and $\beta \in \mathbb{R}$, the right-sided GJF is defined by

$$
{}^{r}\mathcal{P}_n^{(-\alpha,\beta)}(x) := (1-x)^{\alpha} P_n^{(\alpha,\beta)}(x). \tag{2.4.9}
$$

In view of (2.3.2), one verifies readily that the so-defined one-sided GJPFs satisfy

$$ {}^{r}\mathcal{P}_n^{(-\alpha,\beta)}(-x) = (-1)^n \, {}^{l}\mathcal{P}_n^{(\beta,-\alpha)}(x). \tag{2.4.10} $$

Moreover, for $\alpha, \beta > -1$, one finds from (2.1.3)–(2.1.4) the orthogonality:

$$ \begin{aligned} &\int_{-1}^{1} {}^{r}\mathcal{P}_n^{(-\alpha,\beta)}(x)\, {}^{r}\mathcal{P}_m^{(-\alpha,\beta)}(x)\, \omega^{(-\alpha,\beta)}(x)\, dx \\ &\quad = \int_{-1}^{1} {}^{l}\mathcal{P}_n^{(\alpha,-\beta)}(x)\, {}^{l}\mathcal{P}_m^{(\alpha,-\beta)}(x)\, \omega^{(\alpha,-\beta)}(x)\, dx \\ &\quad = \int_{-1}^{1} P_n^{(\alpha,\beta)}(x) P_m^{(\alpha,\beta)}(x)\, \omega^{(\alpha,\beta)}(x)\, dx = \gamma_n^{(\alpha,\beta)} \delta_{mn}. \end{aligned} \tag{2.4.11} $$

In view of the definition of the Riemann–Liouville fractional integrals, we can work out the constants by (2.3.3) and then rewrite the formulas in Theorem 2.21 as follows.

Theorem 2.31 *Let $\mu \in \mathbb{R}^+$, $n \in \mathbb{N}_0$ and $x \in (-1, 1)$.*

(i) *For $\alpha > -1$ and $\beta \in \mathbb{R}$,*

$$ {}^{\mathrm{RL}}_{\;x}I_1^{\mu}\{{}^{r}\mathcal{P}_n^{(-\alpha,\beta)}\}(x) = \frac{\Gamma(n+\alpha+1)}{\Gamma(n+\alpha+\mu+1)}\, {}^{r}\mathcal{P}_n^{(-\alpha-\mu,\beta-\mu)}(x). \tag{2.4.12} $$

(ii) *For $\alpha \in \mathbb{R}$ and $\beta > -1$,*

$$ {}^{\mathrm{RL}}_{-1}I_x^{\mu}\{{}^{l}\mathcal{P}_n^{(\alpha,-\beta)}\}(x) = \frac{\Gamma(n+\beta+1)}{\Gamma(n+\beta+\mu+1)}\, {}^{l}\mathcal{P}_n^{(\alpha-\mu,-\beta-\mu)}(x). \tag{2.4.13} $$

Setting $\alpha = \beta = 0$ in (2.4.12)–(2.4.13) leads to the following important fractional integral formulas for the Legendre polynomials: for $\mu \ge 0$,

$$ \begin{aligned} {}^{\mathrm{RL}}_{\;x}I_1^{\mu} P_n(x) &= \frac{n!}{\Gamma(n+\mu+1)}\, {}^{r}\mathcal{P}_n^{(-\mu,-\mu)}(x) \\ &= \frac{n!}{\Gamma(n+\mu+1)} (1-x)^{\mu} P_n^{(\mu,-\mu)}(x) \end{aligned} \tag{2.4.14} $$

and

$$ {}^{\mathrm{RL}}_{-1}I_x^{\mu} P_n(x) = \frac{n!}{\Gamma(n+\mu+1)} (1+x)^{\mu} P_n^{(-\mu,\mu)}(x). \tag{2.4.15} $$

These two formulas have important implications:

- For any $\phi \in \mathbb{P}_n$, we have

$$ {}^{\mathrm{RL}}_{\;x}I_1^{\mu} \phi = (1-x)^{\mu} \psi_-, \quad {}^{\mathrm{RL}}_{-1}I_x^{\mu} \phi = (1+x)^{\mu} \psi_+, \quad \exists!\ \psi_{\pm} \in \mathbb{P}_n. \tag{2.4.16} $$

Indeed, we note from Proposition 2.17 that $\deg(P_n^{(\mu,-\mu)}) = \deg(P_n^{(-\mu,\mu)}) = n$ for any $\mu \ge 0$.

- For non-integer $\mu > 0$, ${}^{\rm RL}_{\;\;x}I_1^{\mu}$ (resp. ${}^{\rm RL}_{-1}I_x^{\mu}$) takes a smooth function to a function with limited regularity at the end-point $x = 1$ (resp. $x = -1$). It is a distinctive difference from the integration of integer order μ.
- For integer $\mu \in \mathbb{N}$, we find from the transformation formulas in Proposition 2.18 that both (2.4.14) and (2.4.15) reduce to

$$\begin{aligned} I^{\mu} P_n(x) &= \frac{(n+\mu)!\,(n-\mu)!}{2^{\mu}(n!)^2}(1-x^2)^{\mu} P_{n-\mu}^{(\mu,\mu)}(x) \\ &= (-1)^{\mu}\frac{2^{\mu}(n+\mu)!\,(n-\mu)!}{(n!)^2} P_{n-\mu}^{(-\mu,-\mu)}(x), \quad n \ge \mu. \end{aligned} \tag{2.4.17}$$

These relations motivated the introduction of GJPs that could lead to optimal spectral algorithms and simplified analysis as shown in Guo et al. (2006b) and Shen et al. (2011a).

We now obtain from Theorem 2.31 the following useful "inverse" formulas.

Theorem 2.32 *Let* $s \in \mathbb{R}^+$, $n \in \mathbb{N}_0$ *and* $x \in (-1, 1)$.

- *For* $\alpha > s - 1$ *and* $\beta \in \mathbb{R}$,

$${}^{\rm RL}_{\;\;x}D_1^s\{{}^r\mathcal{P}_n^{(-\alpha,\beta)}\}(x) = \frac{\Gamma(n+\alpha+1)}{\Gamma(n+\alpha-s+1)}\,{}^r\mathcal{P}_n^{(-\alpha+s,\beta+s)}(x). \tag{2.4.18}$$

- *For* $\alpha \in \mathbb{R}$ *and* $\beta > s - 1$,

$${}^{\rm RL}_{-1}D_x^s\{{}^l\mathcal{P}_n^{(\alpha,-\beta)}\}(x) = \frac{\Gamma(n+\beta+1)}{\Gamma(n+\beta-s+1)}\,{}^l\mathcal{P}_n^{(\alpha+s,-\beta+s)}(x). \tag{2.4.19}$$

Observe that if $\alpha = 0$ in (2.4.18), the fractional derivative operator takes $(1-x)^s P_n^{(s,\beta-s)}(x)$ to the polynomial $P_n^{(0,\beta)}(x)$. Conversely, if $\alpha + s = k \in \mathbb{N}_0$, it takes the polynomial $(1-x)^k P_n^{(k,\beta-s)}(x)$ to $(1-x)^{k-s} P_n^{(k-s,\beta)}(x)$. Such remarkable properties are essential for efficient spectral algorithms to be developed later.

The analysis of GJPFs essentially relies on the orthogonality of fractional derivatives of GJPFs. Recall the derivative formula of the classical Jacobi polynomials (see, e.g., Shen et al. (2011a, P. 72)): for $\alpha, \beta > -1$ and $n \ge l$,

$$D^l P_n^{(\alpha,\beta)}(x) = \kappa_{n,l}^{(\alpha,\beta)} P_{n-l}^{(\alpha+l,\beta+l)}(x), \tag{2.4.20}$$

where $\kappa_{n,l}^{(\alpha,\beta)}$ is some constant. Noting that

$$ {}^{\mathrm{RL}}_{\;x}D_1^{s+l} = (-1)^l D^l \; {}^{\mathrm{RL}}_{\;x}D_1^{s}, \quad {}^{\mathrm{RL}}_{-1}D_x^{s+l} = D^l \; {}^{\mathrm{RL}}_{-1}D_x^{s} $$

for $s \in \mathbb{R}^+$ and $l \in \mathbb{N}$, we derive from (2.1.3) and (2.4.18)–(2.4.20) the following orthogonality relations:

- For $\alpha > 0$, $\alpha + \beta > -1$, and $n, n' \geq l \geq 0$,

$$ \int_{-1}^{1} {}^{\mathrm{RL}}_{\;x}D_1^{\alpha+l} \; {}^{r}\mathcal{P}_n^{(-\alpha,\beta)}(x) \; {}^{\mathrm{RL}}_{\;x}D_1^{\alpha+l} \; {}^{r}\mathcal{P}_{n'}^{(-\alpha,\beta)}(x) \, \omega^{(l,\alpha+\beta+l)}(x) \, dx = h_{n,l}^{(\alpha,\beta)} \delta_{nn'}, \tag{2.4.21} $$

where

$$ h_{n,l}^{(\alpha,\beta)} := \frac{2^{\alpha+\beta+1}\Gamma^2(n+\alpha+1)\Gamma(n+\alpha+\beta+l+1)}{(2n+\alpha+\beta+1)n!\,(n-l)!\,\Gamma(n+\alpha+\beta+1)}. \tag{2.4.22} $$

- For $\alpha + \beta > -1$, $\beta > 0$, and $n, n' \geq l \geq 0$,

$$ \int_{-1}^{1} {}^{\mathrm{RL}}_{-1}D_x^{\beta+l} \; {}^{l}\mathcal{P}_n^{(\alpha,-\beta)}(x) \; {}^{\mathrm{RL}}_{-1}D_x^{\beta+l} \; {}^{l}\mathcal{P}_{n'}^{(\alpha,-\beta)}(x) \, \omega^{(\alpha+\beta+l,l)}(x) \, dx = h_{n,l}^{(\beta,\alpha)} \delta_{nn'}. \tag{2.4.23} $$

2.4.3 GJPFs for Riesz Fractional Integrals and Derivatives

The Riesz fractional integrals and derivatives include both the left- and right-sided fractional integral and derivatives, which are defined as follows (see Samko et al. (1993, (12.44–45))).

Definition 2.33 (Riesz fractional integrals and derivatives) For real $\mu > 0$ and $\mu \neq 1, 3, 5, \ldots$, the Riesz potential of $u(x)$ is given by

$$ I_1^{\mu} u(x) = \frac{1}{2\Gamma(\mu)\cos\frac{\pi\mu}{2}} \int_{-1}^{1} \frac{u(t)}{|x-t|^{1-\mu}} dt = \frac{1}{2\cos\frac{\pi\mu}{2}} \left({}^{\mathrm{RL}}_{-1}I_x^{\mu} + {}^{\mathrm{RL}}_{\;x}I_1^{\mu} \right) u(x), \tag{2.4.24} $$

while for real $\mu > 0$ and $\mu \neq 2, 4, 6, \ldots$, the modified Riesz potential of $u(x)$ is defined by

$$ \begin{aligned} \hat{I}_1^{\mu} u(x) &= \frac{1}{2\Gamma(\mu)\sin\frac{\pi\mu}{2}} \int_{-1}^{1} \frac{\mathrm{sign}(x-t)}{|x-t|^{1-\mu}} u(t)\, dt \\ &= \frac{1}{2\sin\frac{\pi\mu}{2}} \left({}^{\mathrm{RL}}_{-1}I_x^{\mu} - {}^{\mathrm{RL}}_{\;x}I_1^{\mu} \right) u(x). \end{aligned} \tag{2.4.25} $$

Correspondingly, the Riesz fractional derivative is defined as

$$
\begin{aligned}
D^{\mu}_{|x|}u(x) &:= \frac{(-1)^{\left\lfloor \frac{\mu+1}{2} \right\rfloor}}{2\cos\frac{\pi\mu}{2}}\big({}^{\rm RL}_{-1}D^{\mu}_{x} + {}^{\rm RL}_{x}D^{\mu}_{1}\big)u(x) \\
&= \begin{cases} D^{k}(I^{k-\mu}_{1}u)(x), & k \text{ odd}, \\ D^{k}(\hat{I}^{k-\mu}_{1}u)(x), & k \text{ even}, \end{cases}
\end{aligned}
\tag{2.4.26}
$$

where $\lfloor a \rfloor$ is the largest integer not exceeding the real number a.

Remark 2.34 Note that for non-integer $\mu > 0$, the one-sided Riemann–Liouville fractional integrals can be represented as

$$
\begin{aligned}
{}^{\rm RL}_{-1}I^{\mu}_{x} &= \cos\big(\frac{\pi\mu}{2}\big)I^{\mu}_{1} + \sin\big(\frac{\pi\mu}{2}\big)\hat{I}^{\mu}_{1}, \\
{}^{\rm RL}_{x}I^{\mu}_{1} &= \cos\big(\frac{\pi\mu}{2}\big)I^{\mu}_{1} - \sin\big(\frac{\pi\mu}{2}\big)\hat{I}^{\mu}_{1}.
\end{aligned}
\tag{2.4.27}
$$

In view of the two-sided nature of Riesz fractional operators, the one-sided GJPFs introduced in the previous section are not suitable for approximation of FDEs with Riesz derivatives. Instead, we define a new class of two-sided GJPFs: for real $\mu, \nu > -1$, and $x \in (-1, 1)$,

$$
\mathcal{J}^{-\mu,-\nu}_{n}(x) = (1-x)^{\mu}(1+x)^{\nu}P^{(\mu,\nu)}_{n}(x), \quad n \in \mathbb{N}_0. \tag{2.4.28}
$$

We then derive from (2.1.3) that the general Jacobi functions $\mathcal{J}^{-\mu,-\nu}_{n}(x)$ are mutually orthogonal:

$$
\int_{-1}^{1} \mathcal{J}^{-\mu,-\nu}_{n}(x)\mathcal{J}^{-\mu,-\nu}_{m}(x)\omega^{(-\mu,-\nu)}(x) = \gamma^{(\mu,\nu)}_{n}\delta_{mn}, \tag{2.4.29}
$$

and they span the approximation space:

$$
\mathbb{F}^{-\mu,-\nu}_{N}(\Omega) := \big\{\mathcal{J}^{-\mu,-\nu}_{n}(x) : n = 0, 1, \ldots, N\big\}, \tag{2.4.30}
$$

where we denoted $\Omega := (-1, 1)$.

The following Riesz fractional calculus properties are useful in solving a certain class of fractional differential equations.

Theorem 2.35 *If $s \in (2k-1, 2k)$ with $k \in \mathbb{N}$, then*

$$
I^{2k-s}_{1}\mathcal{J}^{-\frac{s}{2},-\frac{s}{2}}_{m}(x) = (-1)^{k}\frac{\Gamma(m+s+1-2k)}{2^{-2k}m!}P^{(\frac{s}{2}-2k,\frac{s}{2}-2k)}_{m+2k}(x), \tag{2.4.31}
$$

and for $j = 0, 1, \ldots, 2k-1$,

$$
D^{s-j}_{|x|}\mathcal{J}^{-\frac{s}{2},-\frac{s}{2}}_{m}(x) = 2^{j}(-1)^{k}\frac{\Gamma(m-j+1+s)}{m!}P^{(\frac{s}{2}-j,\frac{s}{2}-j)}_{m+j}(x). \tag{2.4.32}
$$

These formulas can be derived from Theorem 2.23, which we leave as an exercise to the interested reader.

We point out in particular that with $j = 0$ in (2.4.32), we have

Corollary 2.36 *If $s \in (2k-1, 2k)$ with $k \in \mathbb{N}$, then*

$$D^s_{|x|} \mathcal{J}_m^{-\frac{s}{2},-\frac{s}{2}}(x) = (-1)^k \frac{\Gamma(m+1+s)}{m!} P_m^{(\frac{s}{2},\frac{s}{2})}(x). \tag{2.4.33}$$

2.4.4 GJPFs for Two-Sided Fractional Derivatives

For $1 < \beta < 2, 0 < \mu, \nu < \beta, \mu + \nu = \beta, 0 \le p \le 1$ and $p \neq 1/2$, we define the two-sided fractional integral operator

$$\mathcal{I}_p^{\mu,\nu,\rho} := C_\beta (p \, {}^{\mathrm{RL}}_{-1}I^\rho_x + (1-p) \, {}^{\mathrm{RL}}_{x}I^\rho_1), \tag{2.4.34}$$

where

$$C_\beta := C(\beta, \mu, \nu) = \frac{\sin(\pi\mu) + \sin(\pi\nu)}{\sin(\pi\beta)}.$$

Then for $s \in (k-1, k)$, we define the two-sided fractional derivative operator

$$\mathcal{D}_p^{\mu,\nu,s} := \frac{\mathrm{d}^k}{\mathrm{d}x^k} \mathcal{I}_p^{\mu,\nu,k-s}. \tag{2.4.35}$$

It turns out that suitable basis functions for dealing with two-sided FDEs with different coefficients are of the form $\mathcal{J}_n^{-\mu,-\nu}(x)$, where (μ, ν) are determined as follows (see Ervin et al. (2018); Mao and Karniadakis (2018)):

Theorem 2.37 *Given (p, β) such that $0 \le p \le 1$ and $1 < \beta < 2$, let (μ, ν) be determined from*

$$\mu + \nu = \beta, \qquad p \sin(\pi\mu) = (1-p)\sin(\pi\nu), \tag{2.4.36}$$

then for $n = 0, 1, 2, \ldots$, it holds that

$$\mathcal{I}_p^{\mu,\nu,2-\beta} \mathcal{J}_n^{-\mu,-\nu}(x) = 4 \frac{\Gamma(n+\beta-1)}{n!} P_{n+2}^{(\nu-2,\mu-2)}(x), \tag{2.4.37}$$

and for $k = 1, 2, \ldots, n+2$,

$$\mathcal{D}_p^{\mu,\nu,k+\beta-2} \mathcal{J}_n^{-\mu,-\nu}(x) = \frac{\Gamma(n+k+\beta-1)}{2^{k-2} n!} P_{n+2-k}^{(\nu-2+k,\mu-2+k)}(x). \tag{2.4.38}$$

In particular, for $k = 2$,

$$\mathcal{D}_p^{\mu,\nu,\beta} \mathcal{J}_n^{-\mu,-\nu}(x) = \frac{\Gamma(n+\beta+1)}{n!} P_n^{(\nu,\mu)}(x). \tag{2.4.39}$$

For the sake of simplicity, we shall denote $\mathcal{I}_p^{\mu,\nu,\rho}$ and $\mathcal{D}_p^{\mu,\nu,\rho}$ by $\mathcal{I}_p^{\rho}$ and $\mathcal{D}_p^{\rho}$, respectively.

By virtue of (2.1.3), a consequent result of equation (2.4.38) is the orthogonality of $\mathcal{D}_p^{\beta+l}$ for $l = -1, 0, 1, \ldots, \min\{m,n\}$. If (μ, ν) and (p, β) satisfy the conditions of Theorem 2.37, then we have

$$\int_{-1}^{1} \mathcal{D}_p^{\beta+l}\, \mathcal{J}_m^{-\mu,-\nu}(x) \mathcal{D}_p^{\beta+l}\, \mathcal{J}_n^{-\mu,-\nu}(x) \omega^{(\nu+l,\mu+l)}(x)\, \mathrm{d}x = 0, \quad \forall n \neq m, \tag{2.4.40}$$

and

$$\int_{-1}^{1} \mathcal{D}_{1-p}^{\beta+l}\, \mathcal{J}_m^{-\mu,-\nu}(x) \mathcal{D}_{1-p}^{\beta+l}\, \mathcal{J}_n^{-\mu,-\nu}(x) \omega^{(\nu+l,\mu+l)}(x)\, \mathrm{d}x = 0, \quad \forall n \neq m. \tag{2.4.41}$$

2.5 Approximation by Jacobi Polynomials and Generalized Jacobi Poly-fractonomials (GJPFs)

We present in this section basic approximation results employing Jacobi polynomials and generalized Jacobi poly-fractonomials.

2.5.1 Approximation by Jacobi Polynomials

Let $\Omega = (-1, 1)$ and let $\omega^{(\alpha,\beta)}(x) = (1-x)^{\alpha}(1+x)^{\beta}$ with $\alpha, \beta > -1$ be the Jacobi weight function as before. For any $u \in L^2_{\omega^{(\alpha,\beta)}}(\Omega)$, we write

$$u(x) = \sum_{n=0}^{\infty} \hat{u}_n^{(\alpha,\beta)} P_n^{(\alpha,\beta)}(x) \quad \text{with} \quad \hat{u}_n^{(\alpha,\beta)} = \frac{\left(u, P_n^{(\alpha,\beta)}\right)_{\omega^{(\alpha,\beta)}}}{\gamma_n^{(\alpha,\beta)}}. \tag{2.5.1}$$

Define the $L^2_{\omega^{(\alpha,\beta)}}$-orthogonal projection $\pi_N^{\alpha,\beta} : L^2_{\omega^{(\alpha,\beta)}}(\Omega) \to \mathbb{P}_N$ such that

$$\left(\pi_N^{\alpha,\beta} u - u,\ v\right)_{\omega^{(\alpha,\beta)}} = 0, \quad \forall v \in \mathbb{P}_N, \tag{2.5.2}$$

or equivalently,

$$(\pi_N^{\alpha,\beta} u)(x) = \sum_{n=0}^{N} \hat{u}_n^{(\alpha,\beta)} P_n^{(\alpha,\beta)}(x). \tag{2.5.3}$$

It is known that $\pi_N^{\alpha,\beta} u$ is the best polynomial approximation of u in the $L^2_{\omega^{(\alpha,\beta)}}$-norm.

For better characterization of the truncation error, we introduce the non-uniformly (or anisotropic) Jacobi-weighted Sobolev space:

$$B^m_{\alpha,\beta}(\Omega) := \{u : u^{(k)} \in L^2_{\omega^{(\alpha+k,\beta+k)}}(\Omega),\ \ 0 \le k \le m\}, \quad m \in \mathbb{N}, \tag{2.5.4}$$

equipped with the inner product, norm and semi-norm

$$\begin{aligned}(u, v)_{B^m_{\alpha,\beta}} &= \sum_{k=0}^{m} (u^{(k)}, v^{(k)})_{\omega^{(\alpha+k,\beta+k)}},\\ \|u\|_{B^m_{\alpha,\beta}} &= (u, u)^{1/2}_{B^m_{\alpha,\beta}}, \quad |u|_{B^m_{\alpha,\beta}} = \|u^{(m)}\|_{\omega^{(\alpha+m,\beta+m)}}.\end{aligned} \tag{2.5.5}$$

It is noteworthy that the derivatives of different order are weighted with different weight functions which should be distinguished from the usual weighted Sobolev space.

We have the following fundamental approximation results on the Jacobi polynomial expansion, which plays a fundamental role in Jacobi spectral approximation theory and Jacobi spectral methods (Shen et al., 2011a).

Theorem 2.38 *Let $\alpha, \beta > -1$. For any $u \in B^m_{\alpha,\beta}(\Omega)$,*

(i) *if $0 \le l \le m \le N + 1$, we have*

$$\begin{aligned}&\|(\pi^{\alpha,\beta}_N u - u)^{(l)}\|_{\omega^{(\alpha+l,\beta+l)}}\\ &\quad \le c\sqrt{\frac{(N-m+1)!}{(N-l+1)!}}(N+m)^{(l-m)/2}\|u^{(m)}\|_{\omega^{(\alpha+m,\beta+m)}},\end{aligned} \tag{2.5.6}$$

(ii) *if $0 \le l < N + 1$ and $m > N + 1$, we have*

$$\begin{aligned}&\|(\pi^{\alpha,\beta}_N u - u)^{(l)}\|_{\omega^{(\alpha+l,\beta+l)}}\\ &\quad \le c(2\pi N)^{-1/4}\Big(\frac{\sqrt{e/2}}{N}\Big)^{N-l+1}\|u^{(N+1)}\|_{\omega^{(\alpha+N+1,\beta+N+1)}},\end{aligned} \tag{2.5.7}$$

where $c \approx 1$ for $N \gg 1$.

The non-uniformly Jacobi-weighted Sobolev framework is essentially motivated by the derivative relation, the orthogonality and Parseval's identity. More precisely, we have the explicit derivative formula

$$\partial^k_x P^{(\alpha,\beta)}_n(x) = d^{\alpha,\beta}_{n,k} P^{(\alpha+k,\beta+k)}_{n-k}(x), \quad n \ge k, \tag{2.5.8}$$

where we denote $\partial^k_x = \frac{d^k}{dx^k}$ and

$$d^{\alpha,\beta}_{n,k} = \frac{\Gamma(n+k+\alpha+\beta+1)}{2^k\Gamma(n+\alpha+\beta+1)}. \tag{2.5.9}$$

As a result, we have from the orthogonality of Jacobi polynomials that

$$\int_{-1}^{1} \partial_x^k P_n^{(\alpha,\beta)}(x)\partial_x^k P_l^{(\alpha,\beta)}(x)\omega^{(\alpha+k,\beta+k)}(x)\mathrm{d}x = h_{n,k}^{\alpha,\beta}\delta_{nl}, \tag{2.5.10}$$

where for $n \geq k$, the constant

$$\begin{aligned} h_{n,k}^{\alpha,\beta} &= (d_{n,k}^{\alpha,\beta})^2\gamma_{n-k}^{(\alpha+k,\beta+k)} \\ &= \frac{2^{\alpha+\beta+1}\Gamma(n+\alpha+1)\Gamma(n+\beta+1)\Gamma(n+k+\alpha+\beta+1)}{(2n+\alpha+\beta+1)(n-k)!\,\Gamma^2(n+\alpha+\beta+1)}. \end{aligned} \tag{2.5.11}$$

Summing (2.5.10) for all $0 \leq k \leq m$, we find that the Jacobi polynomials are orthogonal in the Jacobi-weighted Sobolev space $B^m_{\alpha,\beta}(\Omega)$, namely,

$$\left(P_n^{(\alpha,\beta)}, P_l^{(\alpha,\beta)}\right)_{B^m_{\alpha,\beta}} = 0, \quad \text{if} \quad n \neq l. \tag{2.5.12}$$

Indeed, from Parseval's identity, we have

$$\|u^{(k)}\|^2_{\omega^{(\alpha+k,\beta+k)}} = \sum_{n=k}^{\infty} h_{n,k}^{\alpha,\beta}|\hat{u}_n^{(\alpha,\beta)}|^2, \quad k \geq 0. \tag{2.5.13}$$

Accordingly, the norm of the space $B^m_{\alpha,\beta}(\Omega)$ defined in (2.5.4) can be expressed in terms of the expansion coefficients.

This framework and the results can be extended to the generalized Jacobi polynomials with α and β being negative integers.

Theorem 2.39 *For any $k, l \in \mathbb{Z}$, and $u \in B^m_{k,l}(I)$, we have that for $0 \leq \mu \leq m \leq N+1$ and integers m, μ,*

$$\begin{aligned} &\|(\pi_N^{k,l}u - u)^{(\mu)}\|_{\omega^{(k+\mu,l+\mu)}} \\ &\leq c\sqrt{\frac{(N-m+1)!}{(N-\mu+1)!}}(N+m)^{(\mu-m)/2}\|u^{(m)}\|_{\omega^{(k+m,l+m)}}, \end{aligned} \tag{2.5.14}$$

where c is a positive constant independent of m, N and u.

Proof The proof is essentially the same as that of Theorem 2.38 (Shen et al., 2011a, Chapter 3) as sketched below. Using the derivative relation of Jacobi polynomials repeatedly yields

$$\partial_x^i P_n^{(k,l)}(x) = d_{n,i}^{k,l} P_{n-i}^{(k+i,l+i)}(x),$$

which implies

$$\int_{-1}^{1} \partial_x^i P_n^{(k,l)}(x)\partial_x^i P_m^{(k,l)}(x)\omega^{(k+i,l+i)}(x)\mathrm{d}x = h_{n,i}^{k,l}\delta_{mn}.$$

One can follow the same procedure as for the proof of Theorem 2.38 by working out the constants $d_{n,i}^{k,l}$ and $h_{n,i}^{k,l}$ to obtain the desired result. □

We refer to Shen et al. (2011a) for more approximation results on higher-order orthogonal projections, quadratures and interpolations.

2.5.2 Approximation by One-Sided GJPFs

We define the finite-dimensional left-sided poly-fractonomial space:

$$\begin{aligned} {}^{l}\mathbb{F}_N^{(\alpha,-\beta)}(\Omega) &= \{(1+x)^\beta \psi : \psi \in \mathbb{P}_N\} \\ &= \operatorname{span}\{{}^{l}\mathcal{P}_n^{(\alpha,-\beta)} : 0 \le n \le N\} \end{aligned} \tag{2.5.15}$$

for $\Omega = (-1,1)$. By the orthogonality (2.4.11), we can expand any $u \in L^2_{\omega^{(\alpha,-\beta)}}(\Omega)$ as

$$u(x) = \sum_{n=0}^{\infty} \hat{u}_n^{(\alpha,-\beta)} {}^{l}\mathcal{P}_n^{(\alpha,-\beta)}(x), \tag{2.5.16}$$

where

$$\hat{u}_n^{(\alpha,-\beta)} = \frac{1}{\gamma_n^{(\alpha,\beta)}} \int_{-1}^{1} u(x)\, {}^{l}\mathcal{P}_n^{(\alpha,-\beta)}(x) \omega^{(\alpha,-\beta)}(x)\, dx.$$

There holds the Parseval identity:

$$\|u\|^2_{\omega^{(\alpha,-\beta)}} = \sum_{n=0}^{\infty} \gamma_n^{(\alpha,\beta)} |\hat{u}_n^{(\alpha,-\beta)}|^2. \tag{2.5.17}$$

Consider the $L^2_{\omega^{(\alpha,-\beta)}}$-orthogonal projection onto ${}^{l}\mathbb{F}_N^{(\alpha,-\beta)}(\Omega)$:

$$\left({}^{l}\pi_N^{(\alpha,-\beta)} u - u, v_N\right)_{\omega^{(\alpha,-\beta)}} = 0, \quad \forall v_N \in {}^{l}\mathbb{F}_N^{(\alpha,-\beta)}(\Omega). \tag{2.5.18}$$

Then, it is easy to derive from (2.4.19) that for any $l \in \mathbb{N}_0$, we have

$$\left({}_{-1}^{\mathrm{RL}}D_x^{\beta+l}({}^{l}\pi_N^{(\alpha,-\beta)} u - u), D^l w_N\right)_{\omega^{(\alpha+\beta+l,l)}} = 0, \quad \forall w_N \in \mathbb{P}_N. \tag{2.5.19}$$

To describe the projection error, we define

$${}^{l}\mathcal{B}^m_{\alpha,\beta}(\Omega) := \left\{u \in L^2_{\omega^{(\alpha,-\beta)}}(\Omega) : {}_{-1}^{\mathrm{RL}}D_x^{\beta+l} u \in L^2_{\omega^{(\alpha+\beta+l,l)}}(\Omega) \text{ for } 0 \le l \le m\right\}. \tag{2.5.20}$$

For notational simplicity, we introduce the set

$$ {}^l\Sigma^{\alpha,\beta} := \big\{(\alpha,\beta) : \beta > 0, \ \ \alpha > -1\big\}. $$

From (2.4.11) and (2.5.16), we derive that for $(\alpha,\beta) \in {}^l\Sigma^{\alpha,\beta}$ and $l \in \mathbb{N}_0$,

$$ \| {}^{\rm RL}_{-1}D_x^{\beta+l} u\|_{\omega^{(\alpha+\beta+l,l)}} = \sum_{n=l}^{\infty} h_{n,l}^{(\beta,\alpha)} |\hat{u}_n^{(\alpha,-\beta)}|^2. \tag{2.5.21} $$

Theorem 2.40 *Let $(\alpha,\beta) \in {}^l\Sigma^{\alpha,\beta}$ and $u \in {}^l\mathcal{B}^m_{\alpha,\beta}(\Omega)$.*

(i) *For $0 \le l \le m$, we have*

$$ \big\| {}^{\rm RL}_{-1}D_x^{\beta+l}({}^l\pi_N^{(\alpha,-\beta)}u - u)\big\|_{\omega^{(\alpha+\beta+l,l)}} \le cN^{l-m} \big\| {}^{\rm RL}_{-1}D_x^{\beta+l}u\big\|_{\omega^{(\alpha+\beta+m,m)}}. \tag{2.5.22} $$

(ii) *For $0 \le m \le N$, we also have the $L^2_{\omega^{(\alpha,-\beta)}}$-estimate:*

$$ \big\|{}^l\pi_N^{(\alpha,-\beta)}u - u\big\|_{\omega^{(\alpha,-\beta)}} \le cN^{-(\beta+m)} \big\| {}^{\rm RL}_{-1}D_x^{\beta+m}u\big\|_{\omega^{(\alpha+\beta+m,m)}}. \tag{2.5.23} $$

Proof By (2.5.16), (2.5.18) and (2.5.21), we have

$$ \begin{aligned} \big\| {}^{\rm RL}_{-1}D_x^{\beta+l}({}^l\pi_N^{(\alpha,-\beta)}u - u)\big\|^2_{\omega^{(\alpha+\beta+l,l)}} &= \sum_{n=N+1}^{\infty} h_{n,l}^{(\beta,\alpha)} |\hat{u}_n^{(\alpha,-\beta)}|^2 \\ = \sum_{n=N+1}^{\infty} \frac{h_{n,l}^{(\beta,\alpha)}}{h_{n,m}^{(\beta,\alpha)}} h_{n,m}^{(\beta,\alpha)} |\hat{u}_n^{(\alpha,-\beta)}|^2 &\le \frac{h_{N+1,l}^{(\beta,\alpha)}}{h_{N+1,m}^{(\beta,\alpha)}} \big\| {}^{\rm RL}_{-1}D_x^{\beta+m}u\big\|^2_{\omega^{(\alpha+\beta+m,m)}}. \end{aligned} \tag{2.5.24} $$

We now estimate the constant factor. By (2.4.22) and a direct calculation, we find that for $0 \le l \le m \le N$,

$$ \begin{aligned} \frac{h_{N+1,l}^{(\beta,\alpha)}}{h_{N+1,m}^{(\beta,\alpha)}} &= \frac{\Gamma(N+\alpha+\beta+l+2)(N-m+1)!}{\Gamma(N+\alpha+\beta+m+2)(N-l+1)!} \\ &= \frac{1}{(N+\alpha+\beta+2+l)\cdots(N+\alpha+\beta+1+m)} \frac{(N-m+1)!}{(N-l+1)!} \\ &\le N^{l-m}\frac{(N-m+1)!}{(N-l+1)!}, \end{aligned} \tag{2.5.25} $$

where we used the fact: $\alpha+\beta > -1$. Thus, we obtain (2.5.22) from (2.5.24) and (2.5.25).

The $L^2_{\omega^{(\alpha,-\beta)}}$-estimates can be obtained by using the same argument. We sketch the derivation below. By (2.5.17) and (2.5.21),

$$\begin{aligned}\|{}^{l}\pi_N^{(\alpha,-\beta)}u-u\|^2_{\omega^{(\alpha,-\beta)}}&=\sum_{n=N+1}^{\infty}\gamma_n^{(\alpha,\beta)}|\hat u_n^{(\alpha,-\beta)}|^2\\&=\sum_{n=N+1}^{\infty}\frac{\gamma_n^{(\alpha,\beta)}}{h_{n,m}^{(\beta,\alpha)}}h_{n,m}^{(\beta,\alpha)}|\hat u_n^{(\alpha,-\beta)}|^2\le\frac{\gamma_{N+1}^{(\alpha,\beta)}}{h_{N+1,m}^{(\beta,\alpha)}}\|{}^{\rm RL}_{-1}D_x^{\beta+m}u\|^2_{\omega^{(\alpha+\beta+m,m)}}.\end{aligned}\tag{2.5.26}$$

Working out the constants by (2.1.4) and (2.4.22), we use the property of the Gamma function to derive that

$$\begin{aligned}\frac{\gamma_{N+1}^{(\alpha,\beta)}}{h_{N+1,m}^{(\beta,\alpha)}}&=\frac{\Gamma(N+\alpha+2)\Gamma(N+m+2)(N-m+1)!}{\Gamma(N+\beta+2)\Gamma(N+\alpha+\beta+m+2)(N+m+1)!}\\&\le cN^{-2(\beta+m)}.\end{aligned}\tag{2.5.27}$$

This ends the proof. □

We now analyze spectral approximation by the right-sided GJPFs. Similarly, we introduce the parameter set

$${}^{r}\Sigma^{\alpha,\beta}:=\{(\alpha,\beta)\,:\,\alpha>0,\ \ \beta>-1\},$$

and define the approximation right-sided poly-fractonomial space

$$\begin{aligned}{}^{r}\mathbb{F}_N^{(-\alpha,\beta)}(\Omega)&=\{(1-x)^{\alpha}\psi\,:\,\psi\in\mathbb{P}_N\}\\&=\operatorname{span}\{{}^{r}\mathcal{P}_n^{(-\alpha,\beta)}\,:\,0\le n\le N\}.\end{aligned}\tag{2.5.28}$$

By the orthogonality of GJPFs, we can expand any $u\in L^2_{\omega^{(-\alpha,\beta)}}(\Omega)$ as

$$u(x)=\sum_{n=0}^{\infty}\hat u_n^{(-\alpha,\beta)}\,{}^{r}\mathcal{P}_n^{(-\alpha,\beta)}(x),\tag{2.5.29}$$

where

$$\hat u_n^{(-\alpha,\beta)}=\frac{1}{\gamma_n^{(\alpha,\beta)}}\int_{-1}^{1}u(x)\,{}^{r}\mathcal{P}_n^{(-\alpha,\beta)}\omega^{(-\alpha,\beta)}(x)\,dx$$

and there holds the Parseval identity:

$$\|u\|^2_{\omega^{(-\alpha,\beta)}}=\sum_{n=0}^{\infty}\gamma_n^{(\alpha,\beta)}|\hat u_n^{(-\alpha,\beta)}|^2.\tag{2.5.30}$$

Consider the $L^2_{\omega^{(-\alpha,\beta)}}$-orthogonal projection upon ${}^{r}\mathbb{F}_N^{(-\alpha,\beta)}(\Omega)$ defined by

$$\big({}^{r}\pi_N^{(-\alpha,\beta)}u-u,v_N\big)_{\omega^{(-\alpha,\beta)}}=0,\quad\forall\,v_N\in{}^{r}\mathbb{F}_N^{(-\alpha,\beta)}(\Omega).\tag{2.5.31}$$

By definition, we have

$$({}^r\pi_N^{(-\alpha,\beta)}u)(x) = \sum_{n=0}^{N} \hat{u}_n^{(-\alpha,\beta)}\, {}^r\mathcal{P}_n^{(-\alpha,\beta)}(x). \tag{2.5.32}$$

Theorem 2.41 *Let* $(\alpha,\beta) \in {}^r\Sigma^{\alpha,\beta}$, *and let* $u \in {}^r\mathcal{B}^m_{\alpha,\beta}(\Omega)$ *with* $m \in \mathbb{N}_0$.

(i) *For* $0 \le l \le m \le N$,

$$\begin{aligned} &\big\| {}^{\mathrm{RL}}_{\ x}D_1^{\alpha+l}({}^r\pi_N^{(-\alpha,\beta)}u - u)\big\|_{\omega^{(l,\alpha+\beta+l)}} \\ &\quad\le N^{(l-m)/2}\sqrt{\frac{(N-m+1)!}{(N-l+1)!}}\, \big\| {}^{\mathrm{RL}}_{\ x}D_1^{\alpha+m}u\big\|_{\omega^{(m,\alpha+\beta+m)}}. \end{aligned} \tag{2.5.33}$$

In particular, if m is fixed, then

$$\big\| {}^{\mathrm{RL}}_{\ x}D_1^{\alpha+l}({}^r\pi_N^{(-\alpha,\beta)}u - u)\big\|_{\omega^{(l,\alpha+\beta+l)}} \le cN^{l-m}\big\| {}^{\mathrm{RL}}_{\ x}D_1^{\alpha+m}u\big\|_{\omega^{(m,\alpha+\beta+m)}}, \tag{2.5.34}$$

where $c \approx 1$ *for* $N \gg 1$.

(ii) *For* $0 \le m \le N$, *we also have the* $L^2_{\omega^{(-\alpha,\beta)}}$*-estimates:*

$$\big\|{}^r\pi_N^{(-\alpha,\beta)}u - u\big\|_{\omega^{(-\alpha,\beta)}} \le cN^{-\alpha}\sqrt{\frac{(N-m+1)!}{(N+m+1)!}}\, \big\| {}^{\mathrm{RL}}_{\ x}D_1^{\alpha+m}u\big\|_{\omega^{(m,\alpha+\beta+m)}}. \tag{2.5.35}$$

In particular, if m is fixed, then

$$\big\|{}^r\pi_N^{(-\alpha,\beta)}u - u\big\|_{\omega^{(-\alpha,\beta)}} \le cN^{-(\alpha+m)}\big\| {}^{\mathrm{RL}}_{\ x}D_1^{\alpha+m}u\big\|_{\omega^{(m,\alpha+\beta+m)}}, \tag{2.5.36}$$

where $c \approx 1$ *for* $N \gg 1$.

We can follow the same lines as for the proof of Theorem 2.40 to derive the above estimates, so we leave it as an exercise.

2.5.3 Approximation by Two-Sided GJPFs

We first consider the two-sided Riesz GJPFs defined in (2.4.28). Let $s \in (2k-1, 2k)$ with $k \in \mathbb{N}$, and $l \in \mathbb{N}_0$; we denote for any $m \in \mathbb{N}_0$,

$$\mathcal{B}^m_s(\Omega) := \Big\{u \in L^2_{\omega^{(-\frac{s}{2},-\frac{s}{2})}}(\Omega) : \ D^{s-k+l}_{|x|}u \in L^2_{\omega^{(\frac{s}{2}-k+l,\frac{s}{2}-k+l)}}(\Omega),\ 0 \le l \le m\Big\}.$$

In view of the orthogonality (2.4.29), we can expand any $u \in L^2_{\omega^{(-\alpha,-\alpha)}}(\Omega)$ as

$$u(x) = \sum_{n=0}^{\infty} \hat{u}_n^{(-\alpha,-\alpha)}\mathcal{J}_n^{\alpha,-\alpha}(x), \tag{2.5.37}$$

and there holds the Parseval identity:

$$\|u\|^2_{\omega^{(-\alpha,-\alpha)}} = \sum_{n=0}^{\infty} \gamma_n^{(\alpha,\alpha)} |\hat{u}_n^{(-\alpha,-\alpha)}|^2. \tag{2.5.38}$$

Moreover, for any $2\alpha \in (2k-1, 2k)$ with $k \in \mathbb{N}$, we have

$$\| D_{|x|}^{2\alpha-k+l} u \|^2_{\omega^{(\alpha-k+l,\alpha-k+l)}} = \sum_{n=l}^{\infty} h_{n,l}^{(\alpha,\alpha)} |\hat{u}_n^{(-\alpha,-\alpha)}|^2. \tag{2.5.39}$$

Consider the $L^2_{\omega^{(-\alpha,-\alpha)}}$-orthogonal projection upon $\mathbb{F}_N^{(-\alpha,-\alpha)}(\Omega)$ defined by

$$\left(\Pi_N^{(-\alpha,-\alpha)} u - u, v_N\right)_{\omega^{(-\alpha,-\alpha)}} = 0, \quad \forall\, v_N \in \mathbb{F}_N^{(-\alpha,-\alpha)}(\Omega). \tag{2.5.40}$$

By definition, we have

$$(\Pi_N^{(-\alpha,-\alpha)} u)(x) = \sum_{n=0}^{N} \hat{u}_n^{(-\alpha,-\alpha)} \mathcal{J}_n^{(-\alpha,-\alpha)}(x). \tag{2.5.41}$$

Theorem 2.42 *Assume $2\alpha \in (2k-1, 2k)$ and $k \in \mathbb{N}$, and let $u \in \mathcal{B}^m_{2\alpha}(\Omega)$. Then we have*

$$\|D_{|x|}^{2\alpha-k+l}(\Pi_N^{(-\alpha,-\alpha)} u - u)\|_{\omega^{(\alpha-k+l,\alpha-k+l)}} \le cN^{l-m} \|D_{|x|}^{2\alpha-k+m} u\|_{\omega^{(\alpha-k+m,\alpha-k+m)}}, \tag{2.5.42}$$

and

$$\|\Pi_N^{(-\alpha,-\alpha)} u - u\|_{\omega^{(-\alpha,-\alpha)}} \le cN^{k-(2\alpha+m)} \|D_{|x|}^{2\alpha-k+m} u\|_{\omega^{(\alpha-k+m,\alpha-k+m)}}. \tag{2.5.43}$$

Proof By (2.5.46) (with $\mu = \nu = \alpha$) and (2.5.39), we have

$$\begin{aligned} \left\| D_{|x|}^{2\alpha-k+l}(\Pi_N^{(-\alpha,-\alpha)} u - u)\right\|^2_{\omega^{(\alpha-k+l,\alpha-k+l)}} &= \sum_{n=N+1}^{\infty} h_{n,l}^{(\alpha,\alpha)} |\hat{u}_n^{(-\alpha,-\alpha)}|^2 \\ = \sum_{n=N+1}^{\infty} \frac{h_{n,l}^{(\alpha,\alpha)}}{h_{n,m}^{(\alpha,\alpha)}} h_{n,m}^{(\alpha,\alpha)} |\hat{u}_n^{(-\alpha,-\alpha)}|^2 &\le \frac{h_{N+1,l}^{(\alpha,\alpha)}}{h_{N+1,m}^{(\alpha,\alpha)}} \left\| D_{|x|}^{2\alpha-k+m} u\right\|^2_{\omega^{(\alpha-k+m,\alpha-k+m)}}. \end{aligned} \tag{2.5.44}$$

We now estimate the constant factor. Similar to the proof in Theorem 2.40, we find that for $0 \le l \le m \le N$,

$$\frac{h_{N+1,l}^{(\alpha,\alpha)}}{h_{N+1,m}^{(\alpha,\alpha)}} \le cN^{l-m} \frac{(N+k-m+1)!}{(N+k-l+1)!}. \tag{2.5.45}$$

Thus, we obtain (2.5.42) from (2.5.44) and (2.5.45).

The $L^2_{\omega^{(-\alpha,-\alpha)}}$-estimate can be obtained by using the same argument. We sketch the derivation below. By (2.5.38) and (2.5.39),

$$\|\Pi_N^{(-\alpha,-\alpha)}u - u\|^2_{\omega^{(-\alpha,-\alpha)}} = \sum_{n=N+1}^{\infty} \gamma_n^{(\alpha,\alpha)} |\hat{u}_n^{(-\alpha,-\alpha)}|^2$$

$$= \sum_{n=N+1}^{\infty} \frac{\gamma_n^{(\alpha,\alpha)}}{h_{n,m}^{(\alpha,\alpha)}} h_{n,m}^{(\alpha,\alpha)} |\hat{u}_n^{(\alpha,\alpha)}|^2 \le \frac{\gamma_{N+1}^{(\alpha,\alpha)}}{h_{N+1,m}^{(\alpha,\alpha)}} \|D_{|x|}^{2\alpha-k+m} u\|^2_{\omega^{(\alpha-k+m,\alpha-k+m)}}.$$

Similarly, by (2.1.4) and (2.4.22), we obtain that

$$\frac{\gamma_{N+1}^{(\alpha,\alpha)}}{h_{N+1,m}^{(\alpha,\alpha)}} = \frac{(N+1)!\,(N+k+m+1)!\,(N+k-m+1)!}{\Gamma(N+2\alpha+2)\Gamma(N+2\alpha-k+m+2)(N+k+m+1)!}$$
$$\le cN^{2k-(4\alpha+2m)}.$$

This completes the proof. □

Next, we consider the two-sided GJPFs with different coefficients. Define

$$\left(\Pi_N^{(-\mu,-\nu)}u - u, v_N\right)_{\omega^{(-\mu,-\nu)}} = 0, \quad \forall v_N \in \mathbb{F}_N^{-\mu,-\nu}(\Omega), \tag{2.5.46}$$

and for μ, ν satisfying (2.4.36), we denote

$$\tilde{\mathcal{B}}^m_{\beta,p}(\Omega) := \left\{u \in L^2_{\omega^{(-\mu,-\nu)}}(\Omega) : \mathcal{D}_p^{\beta+l}u \in L^2_{\omega^{(\nu+l,\mu+l)}}(\Omega),\ -1 \le l \le m\right\}. \tag{2.5.47}$$

Then we have the approximation results for the projection errors.

Theorem 2.43 *Assume* $1 < \beta < 2$ *and let* $u \in \tilde{\mathcal{B}}^m_{\beta,p}(\Omega)$ *with* $m \in \mathbb{N}$. *Then for a given* p, $0 \le p \le 1$, *if* $0 < \mu, \nu < \beta$, *and* μ, ν *satisfying* (2.4.36), *we have*

$$\|\mathcal{D}_p^{\beta+l}(\pi_N^{(-\mu,-\nu)}u - u)\|_{\omega^{(\nu+l,\mu+l)}} \le cN^{l-m}\|\mathcal{D}_p^{\beta+m}u\|_{\omega^{(\nu+m,\mu+m)}}, \tag{2.5.48}$$

and

$$\|\pi_N^{(-\mu,-\nu)}u - u\|_{\omega^{(-\mu,-\nu)}} \le cN^{-(\beta+m)}\|\mathcal{D}_p^{\beta+m}u\|_{\omega^{(\nu+m,\mu+m)}}. \tag{2.5.49}$$

The proof is similar to that of Theorem 2.42, so we omit it.

2.6 Tempered Jacobi Poly-fractonomials

As reviewed in Chapter 1, the continuum-time random walk can be regarded as a general model for particle kinetics, which allows for incorporating waiting times and/or non-Gaussian jump distributions with divergent second moments to account for Lévy flights. *Exponentially tempering* the probability distribution of the waiting times and the anomalously large displacements results in tempered-stable Lévy processes with *finite* moments, where the fluid (continuous) limit leads to the tempered fractional diffusion equation. The development of fast and accurate numerical schemes for such nonlocal problems requires a new spectral theory and suitable choice of basis functions.

In what follows, we introduce two classes of regular and singular tempered fractional Sturm–Liouville problems of two kinds (TFSLP-I and TFSLP-II) of order $\nu \in (0,2)$. The aforementioned tempered fractional differential operators are both of tempered Riemann–Liouville and tempered Caputo type of fractional order $\mu = \nu/2 \in (0,1)$. We show the well-posedness of the boundary-value problems and that the eigenvalues of the regular tempered problems are real-valued and the corresponding eigenfunctions are orthogonal. Then we obtain the explicit eigensolutions to the TFSLP-I and -II as non-polynomial functions, which we define as *tempered Jacobi poly-fractonomials*. These eigenfunctions are orthogonal with respect to the weight function associated with the TFSLP-I and -II, see Zayernouri et al. (2015a).

Notation and Preliminary Definitions for TFSLPs

We start with some preliminary definitions of fractional and tempered fractional calculus following Podlubny (1999); Cartea and del Castillo-Negrete (2007); Meerschaert and Sikorskii (2011).

Definition 2.44 (Tempered Fractional Integrals and Derivatives) The left-sided *tempered* Riemann–Liouville fractional integral of order $\mu \in (0,1)$ is defined as

$$\begin{aligned} {}^{\mathrm{RL}}_{x_L}\mathbb{I}^{\mu,\tau}_{x} f(x) &= e^{-\tau x}\ {}^{\mathrm{RL}}_{x_L}I^{\mu}_{x}\{e^{\tau x} f\}(x) \\ &= \frac{1}{\Gamma(\mu)}\int_{x_L}^{x} \frac{e^{-\tau(x-s)} f(s)}{(x-s)^{1-\mu}}\mathrm{d}s, \quad x > x_L, \end{aligned} \tag{2.6.1}$$

and the right-sided one is

$$\begin{aligned} {}^{\mathrm{RL}}_{x}\mathbb{I}^{\mu,\tau}_{x_R} f(x) &= e^{\tau x}\ {}^{\mathrm{RL}}_{x}I^{\mu}_{x_R}\{e^{-\tau x} f\}(x) \\ &= \frac{1}{\Gamma(\mu)}\int_{x}^{x_R} \frac{e^{-\tau(s-x)} f(s)}{(s-x)^{1-\mu}}\mathrm{d}s, \quad x < x_R. \end{aligned} \tag{2.6.2}$$

The corresponding *tempered* Riemann–Liouville fractional derivatives of order $\mu \in (0,1)$ with tempering parameter $\tau \geq 0$ are given by

$$\begin{aligned} {}^{\mathrm{RL}}_{x_L}\mathbb{D}^{\mu,\tau}_{x} f(x) &= e^{-\tau x}\ {}^{\mathrm{RL}}_{x_L}D^{\mu}_{x}\{e^{\tau x} f\}(x) \\ &= \frac{e^{-\tau x}}{\Gamma(1-\mu)}\frac{\mathrm{d}}{\mathrm{d}x}\int_{x_L}^{x} \frac{e^{\tau s} f(s)}{(x-s)^{\mu}}\mathrm{d}s, \quad x > x_L, \end{aligned} \tag{2.6.3}$$

and

$$
\begin{aligned}
{}^{\mathrm{RL}}_{\ x}\mathbb{D}^{\mu,\tau}_{x_R} f(x) &= e^{\tau x}\ {}^{\mathrm{RL}}_{\ x}D^{\mu}_{x_R}\{e^{-\tau x} f\}(x) \\
&= \frac{e^{\tau x}}{\Gamma(1-\mu)} \frac{\mathrm{d}}{\mathrm{d}x} \int_x^{x_R} \frac{e^{-\tau s} f(s)}{(s-x)^{\mu}} \mathrm{d}s, \quad x < x_R.
\end{aligned} \tag{2.6.4}
$$

An alternative approach in defining the tempered fractional derivatives is based on the left- and right-sided *tempered* Caputo derivatives of order $\mu \in (0,1)$, defined respectively, as

$$
\begin{aligned}
{}^{\ \mathrm{C}}_{x_L}\mathbb{D}^{\mu,\tau}_{x} f(x) &= e^{-\tau x}\ {}^{\ \mathrm{C}}_{x_L}D^{\mu}_{x}\{e^{\tau x} f\}(x) \\
&= \frac{e^{-\tau x}}{\Gamma(1-\mu)} \int_{x_L}^{x} \frac{[e^{\tau s} f(s)]'}{(x-s)^{\mu}} \mathrm{d}s, \quad x > x_L,
\end{aligned} \tag{2.6.5}
$$

and

$$
\begin{aligned}
{}^{\mathrm{C}}_{x}\mathbb{D}^{\mu,\tau}_{x_R} f(x) &= e^{\tau x}\ {}^{\mathrm{C}}_{x}D^{\mu}_{x_R}\{e^{-\tau x} f\}(x) \\
&= \frac{e^{\tau x}}{\Gamma(1-\mu)} \int_{x}^{x_R} \frac{[e^{-\tau s} f(s)]'}{(s-x)^{\mu}} \mathrm{d}s, \quad x < x_R.
\end{aligned} \tag{2.6.6}
$$

When $\tau = 0$, the tempered fractional integrals and derivatives reduce to the usual non-tempered ones discussed previously.

Similar to the non-tempered case, the relationship between the tempered Riemann–Liouville and tempered Caputo fractional derivatives when $\mu \in (0,1)$ is given by

$$
{}^{\mathrm{RL}}_{x_L}\mathbb{D}^{\mu,\tau}_{x} f(x) = \frac{f(x_L)}{\Gamma(1-\mu)(x-x_L)^{\mu}} + {}^{\ \mathrm{C}}_{x_L}\mathbb{D}^{\mu,\tau}_{x} f(x), \tag{2.6.7}
$$

and

$$
{}^{\mathrm{RL}}_{\ x}\mathbb{D}^{\mu}_{x_R} f(x) = \frac{f(x_R)}{\Gamma(1-\mu)(x_R-x)^{\mu}} + {}^{\mathrm{C}}_{x}\mathbb{D}^{\mu,\tau}_{x_R} f(x). \tag{2.6.8}
$$

These two tempered fractional derivatives coincide with each other when the functions vanish at the boundary.

Let $\Omega = (x_L, x_R)$. Similarly, we have the fractional integration-by-parts

$$
\left(f, {}^{\mathrm{RL}}_{\ x}\mathbb{D}^{\mu,\tau}_{x_R} g\right)_{\Omega} = \left(g, {}^{\ \mathrm{C}}_{x_L}\mathbb{D}^{\mu,\tau}_{x} f\right)_{\Omega} - f(x)\, {}^{\mathrm{RL}}_{\ x}\mathbb{I}^{\mu,\tau}_{x_R} g(x)\Big|_{x=x_L}^{x=x_R}, \tag{2.6.9}
$$

and

$$
\left(f, {}^{\mathrm{RL}}_{x_L}\mathbb{D}^{\mu,\tau}_{x} g\right)_{\Omega} = \left(g, {}^{\mathrm{C}}_{x}\mathbb{D}^{\mu,\tau}_{x_R} f\right)_{\Omega} + f(x)\, {}^{\mathrm{RL}}_{x_L}\mathbb{I}^{\mu,\tau}_{x} g(x)\Big|_{x=x_L}^{x=x_R}, \tag{2.6.10}
$$

where $(\cdot,\cdot)_{\Omega}$ represents the standard L^2-inner product.

2.6.1 Regular Tempered Fractional Problems of Kind-I and -II

After the above preparation and setting the underlying spaces, we now introduce the following regular TFSLP of order $\nu = 2\mu \in (0,2)$,

$$ {}^{\mathrm{RL}}\mathbb{D}^{\mu,\tau}\big[\, p_i(x)\; {}^{\mathrm{C}}\mathbb{D}^{\mu,\tau}\; F_\lambda^{(i)}(x) \,\big] + \lambda\, w_i(x)\, F_\lambda^{(i)}(x) = 0, \quad x \in (x_L, x_R), \tag{2.6.11} $$

where $i = 1,2$, with $i = 1$ denoting the regular TFSLP of the first kind, in which ${}^{\mathrm{RL}}\mathbb{D}^{\mu,\tau} \equiv {}^{\mathrm{RL}}_{\;x}\mathbb{D}^{\mu,\tau}_{x_R}$ and ${}^{\mathrm{C}}\mathbb{D}^{\mu,\tau} \equiv {}^{\;\mathrm{C}}_{x_L}\mathbb{D}^{\mu,\tau}_{x}$. Moreover, $i = 2$ corresponds to the regular TFSLP of the second kind where ${}^{\mathrm{RL}}\mathbb{D}^{\mu,\tau} \equiv {}^{\mathrm{RL}}_{x_L}\mathbb{D}^{\mu,\tau}_{x}$ and ${}^{\mathrm{C}}\mathbb{D}^{\mu,\tau} \equiv {}^{\mathrm{C}}_{x}\mathbb{D}^{\mu,\tau}_{x_R}$. In such setting, $\mu \in (0,1)$, $\tau \ge 0$, $p_i(x) \neq 0$, and $w_i(x)$ is a non-negative weight function. In addition, we assume that p_i and w_i are real-valued continuous functions in the interval (x_L, x_R). The tempered fractional boundary-value problem (2.6.11) is subject to the following boundary conditions

$$ a_1 F_\lambda^{(i)}(x_L) + a_2\, {}^{\mathrm{RL}}\mathbb{I}^{1-\mu,\,\tau}\big[\, p_i(x)\; {}^{\mathrm{C}}\mathbb{D}^{\mu,\tau} F_\lambda^{(i)}(x) \,\big]\big|_{x=x_L} = 0, \tag{2.6.12} $$

$$ b_1 F_\lambda^{(i)}(x_R) + b_2\, {}^{\mathrm{RL}}\mathbb{I}^{1-\mu,\,\tau}\big[\, p_i(x)\; {}^{\mathrm{C}}\mathbb{D}^{\mu,\tau} F_\lambda^{(i)}(x)\big]\big|_{x=x_R} = 0, \tag{2.6.13} $$

where $a_1^2 + a_2^2 \neq 0$, $b_1^2 + b_2^2 \neq 0$. Here, ${}^{\mathrm{RL}}\mathbb{I}^{1-\mu,\,\tau} \equiv {}^{\mathrm{RL}}_{\;x}\mathbb{I}^{1-\mu,\,\tau}_{x_R}$ when $i = 1$ for the regular TFSLP of the first kind, while, ${}^{\mathrm{RL}}\mathbb{I}^{1-\mu,\,\tau} \equiv {}^{\mathrm{RL}}_{x_L}\mathbb{I}^{1-\mu,\,\tau}_{x}$ when $i = 2$ for the regular TFSLP of the second kind.

Theorem 2.45 *The eigenvalues of the regular TFSLP of kind-I and -II (2.6.11) subject to the nonlocal boundary conditions* (2.6.12) *and* (2.6.13) *are real, and the eigenfunctions corresponding to distinct eigenvalues are orthogonal with respect to the weight functions $w_i(x)$.*

Proof We first consider TFSLP of the first kind (i.e., $i = 1$):

$$ {}^{\mathrm{RL}}_{\;x}\mathbb{D}^{\mu,\tau}_{x_R}\big[\, p_1(x)\; {}^{\;\mathrm{C}}_{x_L}\mathbb{D}^{\mu,\tau}_{x}\; F_\lambda^{(1)}(x)\big] + \lambda\, w_1(x)\, F_\lambda^{(1)}(x) = 0. $$

Now, by the definition of the tempered derivatives ${}^{\mathrm{RL}}_{\;x}\mathbb{D}^{\mu,\tau}_{x_R}$ and ${}^{\;\mathrm{C}}_{x_L}\mathbb{D}^{\mu,\tau}_{x}$, we can re-write the first problem as

$$ e^{\tau x}\, {}^{\mathrm{RL}}_{\;x}D^{\mu}_{x_R}\, e^{-\tau x}\big[\, p_i(x)\, e^{-\tau x}\; {}^{\;\mathrm{C}}_{x_L}D^{\mu}_{x}\, e^{\tau x}\; F_\lambda^{(1)}(x) \,\big] + \lambda\, w_1(x)\, F_\lambda^{(1)}(x) = 0, $$

where we multiply both sides by $e^{-\tau x}$ to obtain

$$ {}^{\mathrm{RL}}_{\;x}D^{\mu}_{x_R}\big[\, e^{-2\tau x}\, p_1(x)\; {}^{\;\mathrm{C}}_{x_L}D^{\mu}_{x}\big(e^{\tau x}\, F_\lambda^{(1)}(x)\big)\big] + \lambda\, e^{-2\tau x}\, w_1(x)\big(\, e^{\tau x}\, F_\lambda^{(1)}(x)\big) = 0. $$

By taking

$$ P_1(x) \equiv e^{-2\tau x}\, p_1(x), \quad W_1(x) \equiv e^{-2\tau x}\, w_1(x), \quad \Phi_\lambda^{(1)}(x) \equiv e^{\tau x}\; F_\lambda^{(1)}(x), $$

we can transform the tempered boundary-value problem of the first kind to

$$ {}^{\mathrm{RL}}_{\ x}D^{\mu}_{x_R}\big[\, P_1(x)\ {}^{\ \mathrm{C}}_{x_L}D^{\mu}_{x}\, \Phi^{(1)}_{\lambda}(x) \big] + \lambda\, W_1(x)\Phi^{(1)}_{\lambda}(x) = 0. \qquad (2.6.14) $$

Clearly, $P_1(x) \neq 0$ and $W_1(x)$ are continuous functions in (x_L, x_R). Moreover, we obtain the boundary conditions corresponding to this change of variables (functions) from (2.6.12) and (2.6.13) by employing the definitions of the tempered fractional integration as

$$ a_1\Phi^{(1)}_{\lambda}(x_L) + a_2\, {}^{\mathrm{RL}}_{\ x}I^{1-\mu}_{x_R}\big[\, P_1(x)\ {}^{\ \mathrm{C}}_{x_L}D^{\mu}_{x}\, \Phi^{(1)}_{\lambda}(x) \big]\big|_{x=x_L} = 0, \quad (2.6.15) $$

$$ b_1\Phi^{(1)}_{\lambda}(x_R) + b_2\, {}^{\mathrm{RL}}_{\ x}I^{1-\mu}_{x_R}\big[\, P_1(x)\ {}^{\ \mathrm{C}}_{x_L}D^{\mu}_{x}\, \Phi^{(1)}_{\lambda}(x) \big]\big|_{x=x_R} = 0. \quad (2.6.16) $$

Following Klimek and Agrawal (2012), the transformed eigen-problem has real-valued eigenvalues and eigenfunctions. Moreover, the corresponding eigenfunctions to distinct eigenvalues are orthogonal with respect to $W_1(x)$. Clearly, the transformed problem (2.6.14) subject to (2.6.15) and (2.6.16) shares the same eigenvalues with the original problem (2.6.11) subject to (2.6.12) and (2.6.13), when $i = 1$. Therefore, the eigenvalues of the regular TFSLP of the first kind are real-valued. Moreover, let $\xi^{(1)}_1$ and $\xi^{(1)}_2$ be eigenfunctions corresponding to two distinct eigenvalues λ_1 and λ_2. Then, by the orthogonality of the transformed problem, we have

$$ \int_{x_L}^{x_R} \xi^{(1)}_1(x)\, \xi^{(1)}_2(x)\, W_1(x)\, \mathrm{d}x = 0, $$

which can be re-written by the inverse transformation of the eigenfunctions as

$$ \int_{x_L}^{x_R} \big(e^{\tau x}\,\Xi^{(1)}_1(x)\big)\big(e^{\tau x}\,\Xi^{(1)}_2(x)\big)\, e^{-2\tau x}\, w_1(x)\, \mathrm{d}x = 0, $$

or

$$ \int_{x_L}^{x_R} \Xi^{(1)}_1(x)\ \Xi^{(1)}_2(x)\ w_1(x)\, \mathrm{d}x = 0, $$

where $\Xi^{(1)}_1(x)$ and $\Xi^{(1)}_2(x)$ are the corresponding real-valued eigenfunctions associated with the distinct eigenvalues λ_1 and λ_2.

When $i = 2$, we follow similar steps by taking $P_2(x) \equiv e^{2\tau x}p_2(x)$ and $W_2(x) \equiv e^{2\tau x}w_2(x)$, and $\Phi^{(2)}_{\lambda}(x) \equiv e^{-\tau x}\, F^{(2)}_{\lambda}(x)$ and through the transformation of the tempered boundary-value problem of the second kind, we complete the proof. □

Regular Tempered Eigenproblems

We specifically solve two regular TFSLPs, denoted by regular TFSLP-I and -II of order $\nu = 2\mu \in (0, 2)$, by choosing particular forms $p_i(x)$ and $w_i(x)$. To this end, we consider the following tempered nonlocal differential operator

$$\mathcal{L}_i^{\mu,\tau} := {}^{\mathrm{RL}}\mathbb{D}^{\mu,\tau}\big[e^{(-1)^{i+1}2\tau x}\ {}^{C}\mathbb{D}^{\mu,\tau}(\cdot)\big], \quad i = 1,2, \tag{2.6.17}$$

where

$$\mathcal{L}_1^{\mu} := {}^{\mathrm{RL}}_{\ x}\mathbb{D}_{x_R}^{\mu,\tau}\big[e^{2\tau x}\ {}^{\mathrm{C}}_{x_L}\mathbb{D}_{x}^{\mu,\tau}(\cdot)\big]$$

in the regular TFSLP-I, and for the case of the regular TFSLP-II, we reverse the order of the right-sided and left-sided tempered derivative for the inner and outer fractional derivatives in the operator, i.e.,

$$\mathcal{L}_2^{\mu} := {}^{\mathrm{RL}}_{x_L}\mathbb{D}_{x}^{\mu,\tau}\big[e^{-2\tau x}\ {}^{\mathrm{C}}_{\ x}\mathbb{D}_{x_R}^{\mu,\tau}(\cdot)\big],$$

where $\mu \in (0,1)$. We note that the term $e^{(-1)^{i+1}2\tau x} \neq 0$, $\forall x \in [x_L, x_R]$, yields the regularity character to the tempered boundary-value problem. That being defined, we consider the regular TFSLP (-I and -II) as

$$\mathcal{L}_i^{\mu,\tau} F_\lambda^{(i)}(x) + \lambda\, e^{(-1)^{i+1}2\tau x}\,(1-x^2)^{-\mu}\, F_\lambda^{(i)}(x) = 0. \tag{2.6.18}$$

We shall solve (2.6.18) subject to a homogeneous Dirichlet and a homogeneous fractional *integro-differential* boundary condition

$$F_\lambda^{(1)}(-1) = 0, \tag{2.6.19}$$
$${}^{\mathrm{RL}}_{\ x}\mathbb{I}_{1}^{1-\mu,\tau}\big[e^{2\tau x}\ {}^{\mathrm{C}}_{-1}\mathbb{D}_{x}^{\mu,\tau} F_\lambda^{(1)}(x)\big]\big|_{x=1} = 0,$$

and

$$F_\lambda^{(2)}(1) = 0, \tag{2.6.20}$$
$${}^{\mathrm{RL}}_{-1}\mathbb{I}_{x}^{1-\mu,\tau}\big[e^{-2\tau x}\ {}^{\mathrm{C}}_{\ x}\mathbb{D}_{1}^{\mu,\tau} F_\lambda^{(2)}(x)\big]\big|_{x=-1} = 0,$$

which are enforced on the regular TFSLP-I and TFSLP-II, respectively.

Explicit Eigensolutions to the Regular TFSLP-I and -II

In what follows, we derive the analytical solution $F_\lambda^{(i)}(x)$ to the regular TFSLP-I and -II (2.6.18), subject to the homogeneous Dirichlet and integro-differential boundary conditions (2.6.19) and (2.6.20).

Theorem 2.46 *The exact eigenfunctions to* (2.6.18)*, when* $i = 1$*, i.e., the regular TFSLP-I, subject to* (2.6.19) *are given by*

$$F_n^{(1)}(x) = e^{-\tau x}(1+x)^{\mu}\ P_{n-1}^{(-\mu,\mu)}(x), \quad \forall n \geq 1, \tag{2.6.21}$$

and the corresponding distinct eigenvalues are

$$\lambda_n^{(1)} = -\frac{\Gamma(n+\mu)}{\Gamma(n-\mu)}, \quad \forall n \geq 1. \tag{2.6.22}$$

Moreover, the exact eigenfunctions to (2.6.18), *when* $i = 2$, *i.e., the regular TFSLP-II subject to* (2.6.20), *are given as*

$$F_n^{(2)}(x) = e^{\tau x}(1-x)^{\mu} \; P_{n-1}^{(\mu,-\mu)}(x), \quad \forall n \geq 1 \tag{2.6.23}$$

with the corresponding distinct eigenvalues given by

$$\lambda_n^{(2)} = \lambda_n^{(1)} = -\frac{\Gamma(n+\mu)}{\Gamma(n-\mu)}, \quad \forall n \geq 1. \tag{2.6.24}$$

The eigenfunctions of the regular TFSLP-I and -II are *non-polynomial* functions due to their particular structure i.e., the multiplier $e^{\mp\tau x}\,(1 \pm x)^{\mu}$. These eigenfunctions in fact generalize the existing *Jacobi poly-fractonomial* functions, which were obtained to be $(1 \pm x)^{\mu} P_{n-1}^{(\mp\mu,\pm\mu)}(x)$, introduced in Section 2.2.1 as the exact eigenfunctions of the regular fractional (non-tempered) Sturm–Liouville problems (RFSLP-I and -II). We note that when the tempering parameter $\tau = 0$, they reduce to the Jacobi poly-fractonomials. Here, to distinguish the tempered eigenfunctions from them, we refer to $F_n^{(i)}(x)$ as *tempered Jacobi poly-fractonomial.* Clearly, when $\tau = 0$ and $\mu \to 1$ or 0, the tempered Jacobi poly-fractonomials (2.6.21) and (2.6.23) coincide with the well-known Jacobi polynomials, which are the eigenfunctions of the standard (integer-order) Sturm–Liouville problem.

Remark 2.47 The regular TFSLP-I and -II share the same eigenvalues with RFSLP-I and -II in Section 2.2.1. Note that the growth in the magnitude of eigenvalues with respect to n is dependent on the fractional derivative order μ. The asymptotic values are obtained as

$$|\lambda_n^{(i)}| = \begin{cases} n^2, & \mu \to 1, \\ n, & \mu \to 1/2, \\ 1, & \mu \to 0. \end{cases} \tag{2.6.25}$$

Hence, there are two modes of growth in the magnitude of $\lambda_n^{(i)}$, the *sublinear* mode corresponding to $0 < \mu < 1/2$, and *superlinear subquadratic* mode corresponding to $1/2 < \mu < 1$. In particular, when $\mu = 1/2$, the eigenvalues grow linearly with n.

Properties of Eigenfunctions of the Regular TFSLP-I and -II

Next, we list a number of important properties of the solutions to the regular TFSLP-I and -II in (2.6.18). Again the upper signs correspond to $i = 1$, solution to the regular TFSLP-I, and the lower signs correspond to the regular TFSLP-II when $i = 2$.

- **Recurrence relations**:

$$\begin{aligned} F_1^{(i)}(x) &= e^{\mp\tau x}(1 \pm x)^{\mu}, \\ F_2^{(i)}(x) &= e^{\mp\tau x}(1 \pm x)^{\mu}(x \mp \mu), \\ &\vdots \\ a_n F_{n+1}^{(i)}(x) &= b_n x F_n^{(i)}(x) - c_n F_{n-1}^{(i)}(x), \end{aligned} \tag{2.6.26}$$

where

$$a_n = 4n^2(n-1), \quad b_n = 2n(2n-1)(2n-2),$$
$$c_n = 4n(n-1 \mp \mu)(n-1 \pm \mu).$$

- **Orthogonality**:

$$\int_{-1}^{1} e^{(-1)^{i+1}2\tau x}(1-x)^{-\mu}(1+x)^{-\mu}\, F_k^{(i)}(x) F_m^{(i)}(x) \mathrm{d}x = \hat{\gamma}_k^{\alpha_i,\beta_i}\delta_{kj},$$

where

$$\hat{\gamma}_k^{\alpha_i,\beta_i} = \frac{2}{2k-1}\frac{\Gamma(k+\alpha_i)\Gamma(k+\beta_i)}{(k-1)!\ \Gamma(k)}, \tag{2.6.27}$$

and $(\alpha_1, \beta_1) = (-\mu, \mu), (\alpha_2, \beta_2) = (\mu, -\mu)$.

- **Fractional derivatives**:

$$ {}^{\mathrm{RL}}_{-1}\mathbb{D}_x^{\mu,\tau}\, F_n^{(1)}(x) = {}^{\mathrm{C}}_{-1}\mathbb{D}_x^{\mu,\tau}\, F_n^{(1)}(x) = \frac{\Gamma(n+\mu)}{\Gamma(n)} e^{-\tau x} P_{n-1}(x), \tag{2.6.28}$$

$$ {}^{\mathrm{RL}}_{x}\mathbb{D}_1^{\mu,\tau}\, F_n^{(2)}(x) = {}^{\mathrm{C}}_{x}\mathbb{D}_1^{\mu,\tau}\, F_n^{(2)}(x) = \frac{\Gamma(n+\mu)}{\Gamma(n)} e^{\tau x} P_{n-1}(x), \tag{2.6.29}$$

where $P_{n-1}(x)$ denotes the standard Legendre polynomial of degree $n-1$.

- **Special values**:

$$F_n^{(1)}(-1) = 0, \quad F_n^{(1)}(1) = e^{-\tau}\, 2^{\mu}\binom{n-1-\mu}{n-1}, \tag{2.6.30}$$

and

$$F_n^{(2)}(1) = 0, \quad F_n^{(2)}(-1) = e^{\tau}(-1)^{n-1} F_n^{(1)}(1). \tag{2.6.31}$$

2.6.2 Singular Tempered Fractional Problems of Kind-I and -II

In the regular setting $p_i(x)$ did not vanish in $(-1, 1)$, and here, we aim to generalize it to singular eigenproblems, in which

$$p_i(x) = e^{(-1)^{i+1}2\tau x}\ (1-x)^{\alpha+1}(1+x)^{\beta+1} \tag{2.6.32}$$

that vanishes at the end-points. We present the definition of the singular TFSLP-I and singular TFSLP-II of order $\nu = 2\mu \in (0, 2)$, associated this time with the additional parameters $-1 < \alpha < 2 - \mu$, and $-1 < \beta < \mu - 1$ in the singular TFSLP-I ($i = 1$), and $-1 < \alpha < \mu - 1$, and $-1 < \beta < 2 - \mu$ in the singular TFSLP-II ($i = 2$), for $x \in (-1, 1)$ as

$$\begin{aligned} &{}^{\mathrm{RL}}\mathbb{D}^{\mu,\tau}\{e^{(-1)^{i+1}2\tau x}\ (1-x)^{\alpha+1}(1+x)^{\beta+1}\ {}^{\mathrm{C}}\mathbb{D}^{\mu,\tau}\ \mathbb{P}^{(i)}(x)\} \\ &+\Lambda^{(i)}e^{(-1)^{i+1}2\tau x}\ (1-x)^{\alpha+1-\mu}(1+x)^{\beta+1-\mu}\ \mathbb{P}^{(i)}(x) = 0, \end{aligned} \tag{2.6.33}$$

where $\tau \geq 0$, $\mu \in (0, 1)$ and $i = 1, 2$. Similar to the regular setting, $i = 1$ denotes the singular TFSLP-I in which ${}^{\mathrm{RL}}\mathbb{D}^{\mu,\tau} \equiv {}^{\mathrm{RL}}_{\ \ x}\mathbb{D}_1^{\mu,\tau}$ and ${}^{\mathrm{C}}\mathbb{D}^{\mu,\tau} \equiv {}^{\mathrm{C}}_{-1}\mathbb{D}_x^{\mu,\tau}$, also $i = 2$ corresponds to the singular TFSLP-II where ${}^{\mathrm{RL}}\mathbb{D}^{\mu,\tau} \equiv {}^{\mathrm{RL}}_{-1}\mathbb{D}_x^{\mu,\tau}$ and ${}^{\mathrm{C}}\mathbb{D}^{\mu,\tau} \equiv {}^{\mathrm{C}}_{x}\mathbb{D}_1^{\mu,\tau}$. The singular problem (2.6.33) is subject to

$$\mathbb{P}^{(i)}(\ (-1)^i\ ) = 0, \tag{2.6.34}$$

$$\{\ {}^{\mathrm{RL}}\mathbb{I}^{1-\mu,\ \tau}[p_i(x)\ {}^{\mathrm{C}}\mathbb{D}^{\mu,\tau}\mathbb{P}^{(i)}(x)]\}\big|_{x=(-1)^{i+1}} = 0, \tag{2.6.35}$$

in which we follow the same definitions as in the regular problems i.e., ${}^{\mathrm{RL}}\mathbb{I}^{1-\mu,\ \tau} \equiv {}^{\mathrm{RL}}_{\ \ x}\mathbb{I}_1^{1-\mu,\ \tau}$ when $i = 1$ in the singular TFSLP-I, and ${}^{\mathrm{RL}}\mathbb{I}^{1-\mu,\ \tau} \equiv {}^{\mathrm{RL}}_{-1}\mathbb{I}_x^{1-\mu,\ \tau}$ in case of $i = 2$ in the singular TFSLP-II with $p_i(x)$ given in (2.6.32) and

$$w_i(x) = e^{(-1)^{i+1}2\tau x}\ (1-x)^{\alpha+1-\mu}(1+x)^{\beta+1-\mu} \tag{2.6.36}$$

in (2.6.33), which is a non-negative function.

Theorem 2.48 *The eigenvalues of the singular TFSLP-I and -II* (2.6.33)–(2.6.35) *are real-valued, and the eigenfunctions corresponding to distinct eigenvalues of* (2.6.33)–(2.6.35) *are orthogonal with respect to the weight function $w_i(x)$ given in* (2.6.36). *Moreover, the exact eigenfunctions of the singular TFSLP-I* (2.6.33)–(2.6.35), *when $i = 1$, are given by*

$$\begin{aligned} \mathbb{P}_n^{(1)}(x) &= {}^{(1)}\mathbb{P}_n^{(\alpha,\beta,\mu,\tau)}(x) \\ &= e^{-\tau x}(1+x)^{-\beta+\mu-1}\ P_{n-1}^{(\alpha-\mu+1,-\beta+\mu-1)}(x), \end{aligned} \tag{2.6.37}$$

and the corresponding distinct eigenvalues are

$$\Lambda_n^{(1)} = {}^{(1)}\Lambda_n^{\alpha,\beta,\mu} = -\frac{\Gamma(n-\beta+\mu-1)\Gamma(n+\alpha+1)}{\Gamma(n-\beta-1)\Gamma(n+\alpha-\mu+1)}. \tag{2.6.38}$$

In addition, the exact eigenfunctions to the singular TFSLP-II (2.6.33)–(2.6.35), *in case of $i = 2$, are given by*

$$\begin{aligned} \mathbb{P}_n^{(2)}(x) &= {}^{(2)}\mathbb{P}_n^{(\alpha,\beta,\mu,\tau)}(x) \\ &= e^{\tau x}(1-x)^{-\alpha+\mu-1}\ P_{n-1}^{(-\alpha+\mu-1,\beta-\mu+1)}(x), \end{aligned} \tag{2.6.39}$$

and the corresponding distinct eigenvalues are

$$\Lambda_n^{(2)} = {}^{(2)}\Lambda_n^{\alpha,\beta,\mu} = -\frac{\Gamma(n-\alpha+2\mu-1)\Gamma(n+\beta+1)}{\Gamma(n-\alpha+\mu-1)\Gamma(n+\beta-\mu+1)}. \quad (2.6.40)$$

Remark 2.49 In Section 2.2.1, the standard singular Sturm–Liouville problem was generalized to the fractional (non-tempered) Sturm–Liouville problems (SFSLP-I and -II) and the singular Jacobi poly-fractonomials of the first and second kind given in (2.2.61) and (2.2.63), which were explicitly obtained as the eigenfunctions. Here, the tempered Jacobi poly-fractonomials involve another parameter i.e., $\tau \geq 0$, which further generalizes the existing singular fractional eigenfunctions, and they in fact complete the whole family of tempered Jacobi poly-fractonomials. Moreover, we note that the singular TFSLP-I and -II also share the same eigenvalues, hence the same asymptotic values, with SFSLP-I and -II.

Properties of the Eigensolutions to the Singular TFSLP-I and -II

We list a number of properties of the eigensolutions to the singular TFSLP-I and -II as follows.

- **Recurrence relations**:
  A recurrence relation is obtained for the tempered Jacobi poly-fractonomials ${}^{(i)}\mathbb{P}_n^{(\alpha,\beta,\mu,\tau)}(x)$ of the first kind, when $i = 1$:

$$\begin{aligned}
{}^{(1)}\mathbb{P}_1^{(\alpha,\beta,\mu,\tau)}(x) &= e^{-\tau x}(1+x)^{-\beta+\mu-1}, \\
{}^{(1)}\mathbb{P}_2^{(\alpha,\beta,\mu,\tau)}(x) &= \frac{1}{2}e^{-\tau x}(1+x)^{-\beta+\mu-1}\left[\alpha+\beta-2\mu+2+(\alpha-\beta+2)x\right], \\
&\vdots
\end{aligned}$$

$$a_n\, {}^{(1)}\mathbb{P}_{n+1}^{(\alpha,\beta,\mu,\tau)}(x) = (b_n + c_n x)\, {}^{(1)}\mathbb{P}_n^{(\alpha,\beta,\mu,\tau)}(x) - d_n\, {}^{(1)}\mathbb{P}_{n-1}^{(\alpha,\beta,\mu,\tau)}(x),$$

  where

$$\begin{aligned}
a_n &= 2n(n+\alpha-\beta)(2n+\alpha-\beta-2), \\
b_n &= (2n-\alpha+\beta-1)(\alpha-\beta)(\alpha+\beta-2\mu+2), \\
c_n &= (2n-\alpha+\beta)(2n-\alpha+\beta-1)(2n-\alpha+\beta-2), \\
d_n &= 2(n-\alpha+\mu-2)(n+\beta-\mu)(2n-\alpha+\beta).
\end{aligned}$$

  The second kind when $i = 2$:

$${}^{(2)}\mathbb{P}_1^{(\alpha,\beta,\mu,\tau)}(x) = e^{\tau x}(1-x)^{-\alpha+\mu-1},$$

$$ {}^{(2)}\mathbb{P}_2^{(\alpha,\beta,\mu,\tau)}(x) = \frac{1}{2} e^{\tau x}(1-x)^{-\alpha+\mu-1}\left[-\alpha-\beta+2\mu-2+(-\alpha+\beta+2)x\right], $$

$$ \vdots $$

$$ a_n^* \, {}^{(2)}\mathbb{P}_{n+1}^{(\alpha,\beta,\mu,\tau)}(x) = (b_n^* + c_n^* x)\, {}^{(2)}\mathbb{P}_n^{(\alpha,\beta,\mu,\tau)}(x) - d_n^* \, {}^{(2)}\mathbb{P}_{n-1}^{(\alpha,\beta,\mu,\tau)}(x), $$

where

$$ \begin{aligned} a_n^* &= 2n(n-\alpha+\beta)(2n-\alpha+\beta-2), \\ b_n^* &= (2n-\alpha+\beta-1)(\alpha-\beta)(\alpha+\beta-2\mu+2), \\ c_n^* &= (2n+\alpha+\beta)(2n+\alpha+\beta-1)(2n+\alpha+\beta-2), \\ d_n^* &= 2(n+\alpha-\mu)(n-\beta+\mu-2)(2n+\alpha-\beta). \end{aligned} $$

- **Orthogonality**:

$$ \int_{-1}^{1} {}^{(i)}\mathbb{P}_k^{(\alpha,\beta,\mu,\tau)}(x)\, {}^{(i)}\mathbb{P}_j^{(\alpha,\beta,\mu,\tau)}(x)\, w_i(x)\, \mathrm{d}x = {}^{(i)}C_k^{\alpha,\beta,\mu}\, \delta_{kj}, \quad (2.6.41) $$

where

$$ {}^{(1)}C_k^{\alpha,\beta,\mu} = \frac{2^{\alpha-\beta+1}}{2k+\alpha-\beta-1} \frac{\Gamma(k+\alpha-\mu+1)\Gamma(k-\beta+\mu-1)}{(k-1)!\,\Gamma(k+\alpha-\beta)}, $$

$$ {}^{(2)}C_k^{\alpha,\beta,\mu} = \frac{2^{-\alpha+\beta+1}}{2k-\alpha+\beta-1} \frac{\Gamma(k-\alpha+\mu-1)\Gamma(k+\beta-\mu+1)}{(k-1)!\,\Gamma(k-\alpha+\beta)}. $$

- **Fractional derivatives**:

$$ \begin{aligned} {}^{RL}_{-1}\mathbb{D}_x^{-\beta+\mu+1,\tau}\left\{ {}^{(1)}\mathbb{P}_n^{(\alpha,\beta,\mu,\tau)}(x) \right\} &= {}^{C}_{-1}\mathbb{D}_x^{-\beta+\mu+1,\tau}\left\{ {}^{(1)}\mathbb{P}_n^{(\alpha,\beta,\mu,\tau)}(x) \right\} \\ &= \frac{\Gamma(n+\mu)}{\Gamma(n)} e^{\tau x} P_{n-1}^{(\alpha-\beta,0)}(x), \end{aligned} $$

and

$$ \begin{aligned} {}^{RL}_{x}\mathbb{D}_1^{-\alpha+\mu-1,\tau}\left\{ {}^{(2)}\mathbb{P}_n^{(\alpha,\beta,\mu,\tau)}(x) \right\} &= {}^{C}_{x}\mathbb{D}_1^{-\alpha+\mu-1,\tau}\left\{ {}^{(2)}\mathbb{P}_n^{(\alpha,\beta,\mu,\tau)}(x) \right\} \\ &= \frac{\Gamma(n+\mu)}{\Gamma(n)} e^{-\tau x} P_{n-1}^{(0,\beta-\alpha)}(x), \end{aligned} $$

where $P_{n-1}^{(\alpha-\beta,0)}(x)$ and $P_{n-1}^{(0,\beta-\alpha)}(x)$ denote the standard Jacobi polynomials.

- **Special values**:

$$ {}^{(1)}\mathbb{P}_n^{(\alpha,\beta,\mu,\tau)}(-1) = 0, \quad {}^{(1)}\mathbb{P}_n^{(\alpha,\beta,\mu,\tau)}(1) = 2^{-\beta+\mu-1}\binom{n+\alpha-\mu}{n-1} e^{-\tau}, $$

$$ {}^{(2)}\mathbb{P}_n^{(\alpha,\beta,\mu,\tau)}(1) = 0, \quad {}^{(2)}\mathbb{P}_n^{(\alpha,\beta,\mu,\tau)}(-1) = 2^{-\alpha+\mu-1}\binom{n+\beta-\mu}{n-1} e^{-\tau}. $$

2.6.3 Approximability of the Tempered Eigenfunctions

Following Zayernouri et al. (2015a), we introduce the tempered eigenfunctions (regular and singular) as new basis functions in $L^2_{w_i}(-1,1)$. Then, we discuss how fast the expansion coefficients of the approximation would decay with N.

Theorem 2.50 *The set of regular eigenfunctions* $\{F_n^{(i)}(x) : n = 1,2,\ldots\}$ *and the singular eigenfunctions* $\{{}^{(i)}\mathbb{P}_n^{(\alpha,\beta,\mu,\tau)}(x) : n = 1,2,\ldots\}$ *each form a basis for the infinite-dimensional Hilbert space* $L^2_{w_i}(-1,1)$, *and the corresponding eigenvalues* $\lambda_n^{(i)}$ *and* $\Lambda_n^{(i)}$ *are simple.*

Proof It suffices to prove the regular part as the singular part follows similar steps. Let $f(x) \in L^2_{w_i}(-1,1)$ and then clearly $g(x) = e^{\pm\tau x}(1 \pm x)^{-\mu} f(x) \in L^2_{w_i}(-1,1)$ as well when $\mu \in (0,1)$. Considering the upper signs to correspond to the regular TFSLP-I, $i = 1$, and the lower signs to correspond to the case $i = 2$, i.e., the regular TFSLP-II, we have

$$
\begin{aligned}
&\Big\| \sum_{n=1}^{N} a_n^{(i)} F_n^{(i)}(x) - f(x) \Big\|_{L^2_{w_i}} \\
&\quad = \Big\| e^{\mp\tau x}(1 \pm x)^{\mu} \Big(\sum_{n=1}^{N} a_n^{(i)} P_{n-1}^{(\mp\mu,\pm\mu)}(x) - e^{\pm\tau x}(1 \pm x)^{-\mu} f(x) \Big) \Big\|_{L^2_{w_i}} \\
&\quad = \Big\| e^{\mp\tau x} \Big(\sum_{n=1}^{N} a_n^{(i)} P_{n-1}^{(\mp\mu,\pm\mu)}(x) - g(x) \Big) \Big\|_{L^2_{\tilde{w}_i}} \\
&\quad \le \| e^{\mp\tau x} \|_{L^2_{\tilde{w}_i}} \Big\| \sum_{n=1}^{N} a_n^{(i)} P_{n-1}^{(\mp\mu,\pm\mu)}(x) - g(x) \Big\|_{L^2_{\tilde{w}_i}} \\
&\quad \le c \Big\| \sum_{n=1}^{N} a_n^{(i)} P_{n-1}^{(\mp\mu,\pm\mu)}(x) - g(x) \Big\|_{L^2_{\tilde{w}_i}},
\end{aligned}
\tag{2.6.42}
$$

where we used the Cauchy–Schwarz inequality and the weight functions are

$$
\tilde{w}_1(x) = (1-x)^{-\mu}(1+x)^{\mu}, \quad \tilde{w}_2(x) = (1-x)^{\mu}(1+x)^{-\mu}.
$$

Following **Step 3** in the proof of Theorem 2.4, we take $\{a_n^{(i)}\}$ to be the Jacobi expansion coefficients of $g(x)$ in terms of $\{P_{n-1}^{(\mp\mu,\pm\mu)}(x)\}$. Then we can show

$$
\sum_{n=1}^{N} a_n^{(i)} F_n^{(i)}(x) \xrightarrow{L^2_{w_i}} f(x).
$$

Therefore, the set of tempered eigenfunctions $\{F_n^{(i)}(x) : n = 1, 2, \ldots\}$ is dense in the Hilbert space $L^2_{w_i}(-1,1)$ and forms a basis for this space.

To show the simplicity of the eigenvalues, assume that corresponding to the eigenvalue $\lambda_j^{(i)}$, there exists another eigenfunction $F_j^{(i)*}(x) \in L^2_{w_i}(-1,1)$ in addition to $F_j^{(i)}(x)$, which is by Theorem 2.45 orthogonal to the rest of the eigenfunctions $F_n^{(i)}(x)$, $n \neq j$. By the density of the set of eigenfunctions, we can represent $F_j^{(i)*}(x)$ as

$$F_j^{(i)*}(x) = \sum_{n=1}^{\infty} b_n^{(i)} F_n^{(i)}(x), \tag{2.6.43}$$

where there exists at least one $b_k^{(i)} \neq 0$ with $k \neq j$. Now, by multiplying both sides by $F_k^{(i)}(x)$, and integrating with respect to the weight function $w_i(x)$, we obtain

$$\begin{aligned}\int_{-1}^{1} F_j^{(i)*}(x) F_k^{(i)}(x)\, w_i(x)\, \mathrm{d}x &= \sum_{n=1}^{\infty} b_n^{(i)} \int_{-1}^{1} F_n^{(i)}(x) F_k^{(i)}(x)\, w_i(x)\, \mathrm{d}x \\ &= b_k^{(i)} C_k \neq 0,\end{aligned} \tag{2.6.44}$$

which contradicts Theorem 2.45, and this completes the proof. □

Spectral Approximation Using Singular Tempered Basis

Next, we study the approximation properties of the family of Jacobi poly-fractonomials ${}^{(i)}\mathbb{P}_n^{(\alpha,\beta,\mu,\tau)}(x)$ by representing $f(x) \in L^2_{w_i}(-1,1)$ with w_i given in (2.6.36) as

$$f(x) \approx \mathcal{S}_N^{\alpha,\beta,\tau}[f](x) := \sum_{n=1}^{N} \hat{f}_n \; {}^{(i)}\mathbb{P}_n^{(\alpha,\beta,\mu,\tau)}(x), \tag{2.6.45}$$

where by the orthogonality (2.6.41), the expansion coefficients are

$$\hat{f}_n = \frac{1}{C_n^{\alpha,\beta,\mu}} \int_{-1}^{1} f(x) \; {}^{(i)}\mathbb{P}_n^{(\alpha,\beta,\mu,\tau)}(x)\, w_i(x)\, \mathrm{d}x. \tag{2.6.46}$$

Here, we need to understand how fast the expansion coefficients $\hat{f}_n$ decay in n. Denoting the fractional part of the Sturm–Liouville problem (2.6.33) by $\mathcal{L}_i^{\alpha,\beta;\mu,\tau}$, we have

$$\mathcal{L}_i^{\alpha,\beta;\mu,\tau}\{ {}^{(i)}\mathbb{P}_n^{(\alpha,\beta,\mu,\tau)}(x)\} = -\Lambda_n^{(i)} w_i(x) \; {}^{(i)}\mathbb{P}_n^{(\alpha,\beta,\mu,\tau)}(x). \tag{2.6.47}$$

Thus we can rewrite (2.6.46) as

$$\hat{f}_n = -\frac{1}{C_n^{\alpha,\beta,\mu}\,\Lambda_n^{(i)}}\int_{-1}^{1} f(x)\,\mathcal{L}_i^{\alpha,\beta;\mu,\tau}\{\,{}^{(i)}\mathbb{P}_n^{(\alpha,\beta,\mu,\tau)}(x)\}\,\mathrm{d}x$$
$$= -\frac{1}{C_n^{\alpha,\beta,\mu}\,\Lambda_n^{(i)}}\int_{-1}^{1} f(x)\,{}^{\mathrm{RL}}\mathbb{D}^{\mu,\tau}\{\omega^{(\alpha+1,\beta+1)}(x)\;{}^{\mathrm{C}}\mathbb{D}^{\mu,\tau}\,\{{}^{(i)}\mathbb{P}_n^{(\alpha,\beta,\mu,\tau)}(x)\}\}\,\mathrm{d}x.$$

Using fractional integration-by-parts formulas (2.6.9) and (2.6.10), we obtain

$$\hat{f}_n = -\frac{1}{C_n^{\alpha,\beta,\mu}\,\Lambda_n^{(i)}}\int_{-1}^{1} \omega^{(\alpha+1,\beta+1)}(x)\;{}^{\mathrm{C}}\mathbb{D}^{\mu,\tau}\,\{{}^{(i)}\mathbb{P}_n^{(\alpha,\beta,\mu,\tau)}(x)\}\,{}^{\mathrm{C}}\mathbb{D}^{\mu,\tau} f(x)\,\mathrm{d}x$$
$$= -\frac{1}{C_n^{\alpha,\beta,\mu}\,\Lambda_n^{(i)}}\int_{-1}^{1} {}^{(i)}\mathbb{P}_n^{(\alpha,\beta,\mu,\tau)}(x)\,{}^{\mathrm{RL}}\mathbb{D}^{\mu,\tau}\{\omega^{(\alpha+1,\beta+1)}(x)\;{}^{\mathrm{C}}\mathbb{D}^{\mu,\tau}\,\{f(x)\}\}\,\mathrm{d}x$$
$$= -\frac{1}{C_n^{\alpha,\beta,\mu}\,\Lambda_n^{(i)}}\int_{-1}^{1} {}^{(i)}\mathbb{P}_n^{(\alpha,\beta,\mu,\tau)}(x)\,\mathcal{L}_i^{\alpha,\beta;\mu,\tau}\{f(x)\}\,\mathrm{d}x$$
$$= -\frac{1}{C_n^{\alpha,\beta,\mu}\,\Lambda_n^{(i)}}\int_{-1}^{1} {}^{(i)}\mathbb{P}_n^{(\alpha,\beta,\mu,\tau)}(x)\;f_{(1)}(x)w_i(x)\,\mathrm{d}x,$$

where we denoted $f_{(1)}(x) := w_i^{-1}(x)\mathcal{L}_i^{\alpha,\beta;\mu,\tau}\{f(x)\}$. This implies

$$\hat{f}_n = -\frac{1}{\Lambda_n^{(i)}}(\widehat{f_{(1)}(x)})_n. \tag{2.6.48}$$

We further define

$$f_{(m)}(x) := w_i^{-1}(x)\mathcal{L}_i^{\alpha,\beta;\mu,\tau}\{f_{m-1}(x)\}, \quad m \geq 1,$$

with $f_0(x) \equiv f(x)$. By carrying out the fractional integration-by-parts another $(m-1)$ times, we can obtain

$$\hat{f}_n = \frac{(-1)^m}{(\Lambda_n^{(i)})^m}(\widehat{f_{(m)}(x)})_n. \tag{2.6.49}$$

Thus, by Parseval's identity, we find that if $f_{(m)}(x) \in L^2_{w_i}(-1,1)$ for $m \geq 1$, then

$$
\begin{aligned}
\|\mathcal{S}_N^{\alpha,\beta,\tau}[f]-f\|_{L^2_{w_i}}^2 &= \sum_{n=N+1}^{\infty} C_n^{\alpha,\beta,\mu}|\hat{f}_n|^2 \\
&= \sum_{n=N+1}^{\infty} \frac{C_n^{\alpha,\beta,\mu}}{|\Lambda_n^{(i)}|^{2m}}|(\widehat{f_{(m)}(x)})_n|^2 \\
&\le |\Lambda_{N+1}^{(i)}|^{-2m} \sum_{n=N+1}^{\infty} C_n^{\alpha,\beta,\mu}|(\widehat{f_{(m)}(x)})_n|^2 \\
&\le |\Lambda_N^{(i)}|^{-2m} \|f_{(m)}\|_{L^2_{w_i}}^2 .
\end{aligned}
$$

In summary, if $f_{(m)}(x) \in L^2_{w_i}(-1,1)$ for $m = 0, 1, \ldots$, then we have

$$
\|\mathcal{S}_N^{\alpha,\beta,\tau}[f]-f\|_{L^2_{w_i}} \le |\Lambda_N^{(i)}|^{-m} \|f_{(m)}\|_{L^2_{w_i}} . \qquad (2.6.50)
$$

Consequently, when $m \to \infty$, and $f_{(m)}(x) \in L^2_{w_i}(-1,1)$, we have the spectral decay of the expansion error.

2.7 Log Orthogonal Functions and Their Generalizations

It is well known that classical spectral methods can provide high accuracy for problems with smooth solutions, but may not have any advantage for problems with nonsmooth solutions. On the other hand, solutions of many practical applications involve weakly singular solutions, such as in nonsmooth domains, with non-matching boundary conditions, in integral equations with singular/weakly singular kernels, and in fractional differential equations.

For problems which exhibit weakly singular behaviors at the initial time for initial-value problems or at one end-point for boundary-value problems, we construct special classes of orthogonal functions, through a suitable log mapping to Laguerre functions, which are capable of resolving weak singularities. We shall develop basic approximation results for two new classes of orthogonal functions, log orthogonal functions (LOFs) and generalized log orthogonal functions (GLOFs). In particular, these results indicate that approximation by the LOFs and GLOFs to functions behaving like $t^r(-\log t)^k$ near $t = 0$ will converge exponentially for any real $r \ge 0$, $k \in \mathbb{N}_0$. This set of basis functions can approximate regular polynomials t^k ($k \in \mathbb{N}_0$) and weakly singular functions like $t^r(-\log t)^k$ ($0 < r < 1$, $k \in \mathbb{N}_0$) with exponential accuracy. Thus, LOFs and GLOFs are particularly suitable for problems whose solutions exhibit weak singularities behaving like $\sum_i c_i t^{r_i}(-\log t)^{k_i}$ near $t = 0$ for small $r_i > 0$. In particular, solutions of time fractional differential equations and boundary-value problems with one-sided fractional derivatives fall into this category. Hence, the spectral methods using GLOFs that we propose in this paper can be used to deal with a large class of fractional differential equations having weak singularities at the initial time or at one end-point.

2.7.1 Log Orthogonal Functions

To fix the idea, we consider the canonical time interval $I = (0, 1)$ and introduce the mapping

$$y(t) := -(\beta + 1) \log t, \quad \beta > -1, \quad t \in I, \tag{2.7.1}$$

which maps I to $\mathbb{R}^+$. We shall make use of the Laguerre polynomial $\mathscr{L}_n^{(\alpha)}(y)$, $\alpha > -1$ and $y \in \mathbb{R}^+$, satisfying

$$\int_0^\infty \mathscr{L}_n^{(\alpha)}(y)\, \mathscr{L}_m^{(\alpha)}(y)\, y^\alpha e^{-y}\, dy = \gamma_n^{(\alpha)} \delta_{mn}, \tag{2.7.2}$$

where

$$\gamma_n^{(\alpha)} = \frac{\Gamma(n + \alpha + 1)}{\Gamma(n + 1)}.$$

Definition 2.51 (LOFs) For $\alpha, \beta > -1$, we define the logarithmic orthogonal functions as

$$\mathcal{S}_n^{(\alpha,\beta)}(t) = \mathscr{L}_n^{(\alpha)}(y(t)) = \mathscr{L}_n^{(\alpha)}(-(\beta + 1)\log t), \quad n = 0, 1, \ldots. \tag{2.7.3}$$

From the properties of Laguerre polynomials (see Shen et al. (2011a)) and the following relations

$$y = -(\beta + 1) \log t, \quad dy = -(\beta + 1)t^{-1}dt, \quad \partial_t = -(\beta + 1)t^{-1}\partial_y, \tag{2.7.4}$$

we can easily derive the following properties of the LOFs.

Lemma 2.52 *The LOFs satisfy the following properties.*

(i) *Three-term recurrence relation*

$$\mathcal{S}_{n+1}^{(\alpha,\beta)}(t) = \frac{2n + \alpha + 1 + (\beta + 1)\log t}{n + 1}\mathcal{S}_n^{(\alpha,\beta)}(t) - \frac{n + \alpha}{n + 1}\mathcal{S}_{n-1}^{(\alpha,\beta)}(t),$$
$$\mathcal{S}_0^{(\alpha,\beta)}(t) = 1, \quad \mathcal{S}_1^{(\alpha,\beta)}(t) = (\beta + 1)\log t + \alpha + 1. \tag{2.7.5}$$

(ii) *Derivative relations*

$$(\beta + 1)^{-1} t\partial_t \mathcal{S}_n^{(\alpha,\beta)}(t) = \mathcal{S}_{n-1}^{(\alpha+1,\beta)}(t) = \sum_{l=0}^{n-1} \mathcal{S}_l^{(\alpha,\beta)}(t), \quad n \geq 1. \tag{2.7.6}$$

(iii) *Orthogonality*

$$\int_0^1 \mathcal{S}_n^{(\alpha,\beta)}(t)\, \mathcal{S}_m^{(\alpha,\beta)}(t)\, (-\log t)^\alpha\, t^\beta\, dt = \gamma_n^{(\alpha,\beta)} \delta_{mn}, \tag{2.7.7}$$

where

$$\gamma_n^{(\alpha,\beta)} := \frac{\Gamma(n+\alpha+1)}{(\beta+1)^{\alpha+1}\Gamma(n+1)}.$$

(iv) *Sturm–Liouville problem*

$$(-\log t)^{-\alpha} t^{-\beta} \partial_t\big((-\log t)^{\alpha+1} t^{\beta+2} \partial_t \mathcal{S}_n^{(\alpha,\beta)}(t)\big) + n(\beta+1)\mathcal{S}_n^{(\alpha,\beta)}(t) = 0. \tag{2.7.8}$$

(v) *Gauss-LOFs quadrature*
Let $\{y_j^{(\alpha)}, \omega_j^{(\alpha)}\}_{j=0}^N$ *be the Gauss-node and -weight of* $\mathcal{L}_{n+1}^{(\alpha)}(y)$. *Denote*

$$\big\{t_j^{(\alpha,\beta)} := e^{-(\beta+1)^{-1} y_j^{(\alpha)}}, \quad \chi_j^{(\alpha,\beta)} = (\beta+1)^{-\alpha-1}\omega_j^{(\alpha)}\big\}_{j=0}^N. \tag{2.7.9}$$

Then we have the exactness of the quadrature

$$\int_0^1 p(t)(-\log t)^\alpha t^\beta dt = \sum_{j=0}^N p(t_j^{(\alpha,\beta)})\chi_j^{(\alpha,\beta)}, \quad \forall p \in \mathcal{P}_{2N+1}^{\log t}, \tag{2.7.10}$$

where

$$\mathcal{P}_K^{\log t} := \text{span}\{1, \log t, (\log t)^2, \ldots, (\log t)^K\}. \tag{2.7.11}$$

Proof The three-term recurrence relation (2.7.5) is a straightforward result from the three-term recurrence relation of the Laguerre polynomials with the variable transform (2.7.4). The formula (2.7.6) can be obtained from (2.7.4) and the derivative relation of the Laguerre polynomials:

$$(\beta+1)^{-1} t\partial_t\{\mathcal{S}_n^{(\alpha,\beta)}(t)\} = -\partial_y \mathcal{L}_n^{(\alpha)}(y) = \mathcal{L}_{n-1}^{(\alpha+1)}(y) = \mathcal{S}_{n-1}^{(\alpha+1,\beta)}(t).$$

The orthogonality (2.7.7) is a direct consequence of (2.7.2), since with the change of variable $y = -(\beta+1)\log t$, we have

$$\int_0^1 \mathcal{S}_n^{(\alpha,\beta)}(t)\, \mathcal{S}_m^{(\alpha,\beta)}(t)\, (-\log t)^\alpha t^\beta\, dt = \frac{1}{(\beta+1)^{\alpha+1}} \int_0^\infty \mathcal{L}_n^{(\alpha)}(y)\, \mathcal{L}_m^{(\alpha)}(y)\, y^\alpha e^{-y}\, dy.$$

The property (2.7.8) is valid as using

$$y = -(\beta+1)\log t, \quad \partial_y = -(\beta+1)^{-1} t\, \partial_t, \quad \mathcal{L}_n^{(\alpha)}(y) = \mathcal{S}_n^{(\alpha,\beta)}(t)$$

leads to

$$y^{-\alpha}e^{y}\partial_y\big(y^{\alpha+1}e^{-y}\partial_y\mathscr{L}_n^{(\alpha)}(y)\big)$$
$$=\frac{(-\log t)^{-\alpha}t^{-\beta}}{\beta+1}\partial_t\big((-\log t)^{\alpha+1}t^{\beta+2}\partial_t\mathcal{S}_n^{(\alpha,\beta)}(t)\big).$$

Finally, setting $t=e^{-(\beta+1)^{-1}y}$, we can obtain (2.7.9) from the Laguerre–Gauss quadrature:

$$\int_0^1 p(t)(-\log t)^{\alpha}t^{\beta}\mathrm{d}t=\int_{\mathbb{R}^+}p(x(y))\frac{y^{\alpha}e^{-y}}{(\beta+1)^{\alpha+1}}\,\mathrm{d}y$$
$$=\sum_{j=0}^{N}p(x(y_j^{(\alpha)}))\frac{\omega_j^{(\alpha)}}{(\beta+1)^{\alpha+1}}=\sum_{j=0}^{N}p(t_j^{(\alpha,\beta)})\chi_j^{(\alpha,\beta)}.$$

The proof is completed. □

Remark 2.53 We used two parameters α and β to provide better flexibility, e.g., they allow us to effectively deal with problems with weight $(-\log t)^{\alpha}t^{\beta}$. In the applications considered in this paper, the log term does not appear so we can take $\alpha=0$. On the other hand, taking $\beta=0$ offers good approximation properties for problems with weight $(-\log t)^{\alpha}$, which cannot be well approximated by classical orthogonal polynomials.

2.7.2 Approximation by Log Orthogonal Functions

For $\alpha,\beta>-1$, we define the weight function $\chi^{\alpha,\beta}(t):=(-\log t)^{\alpha}t^{\beta}$. Consider the orthogonal projection $\hat{\pi}_N^{\alpha,\beta}u$ of $u\in L^2_{\chi^{\alpha,\beta}}(I)$ onto $\mathcal{P}_N^{\log t}$, defined by

$$(u-\hat{\pi}_N^{\alpha,\beta}u,v)_{\chi^{\alpha,\beta}}=\int_0^1\{u-\hat{\pi}_N^{\alpha,\beta}u\}(t)\,v(t)\,\chi^{\alpha,\beta}(t)\mathrm{d}t=0, \tag{2.7.12}$$

for all $v\in\mathcal{P}_N^{\log t}$. Thanks to the orthogonality of the LOFs $\{\mathcal{S}_n^{(\alpha,\beta)}\}_{n=0}^{\infty}$, we have

$$(\hat{\pi}_N^{\alpha,\beta}u)(t)=\sum_{n=0}^{N}\hat{u}_n^{\alpha,\beta}\,\mathcal{S}_n^{(\alpha,\beta)}(t), \tag{2.7.13}$$

where

$$\hat{u}_n^{\alpha,\beta}=\frac{1}{\gamma_n^{(\alpha,\beta)}}\int_0^1 u(t)\,\mathcal{S}_n^{(\alpha,\beta)}(t)\,\chi^{\alpha,\beta}(t)\mathrm{d}t.$$

To better describe the projection error $\hat{\pi}_N^{\alpha,\beta}u$, we define a pseudo-derivative

$$\widehat{\partial}_t u:=t\partial_t u \tag{2.7.14}$$

and introduce the non-uniformly weighted Sobolev space

$$A^k_{\alpha,\beta}(I) := \{v \in L^2_{\chi^{\alpha,\beta}}(I) : \widehat{\partial}_t^{\,j} v \in L^2_{\chi^{\alpha+j,\beta}}(I),\ j = 1,2,\ldots,k\}, \quad (2.7.15)$$

equipped with the semi-norm and norm

$$|v|_{A^m_{\alpha,\beta}} := \|\widehat{\partial}_t^{\,m} v\|_{\chi^{\alpha+m,\beta}}, \quad \|v\|_{A^m_{\alpha,\beta}} := \Big(\sum_{k=0}^{m} |v|^2_{A^k_{\alpha,\beta}}\Big)^{1/2}.$$

Theorem 2.54 *Let m, N, $k \in \mathbb{N}$ and $\alpha, \beta > -1$. For any $u \in A^m_{\alpha,\beta}(I)$ and $0 \le k \le \widetilde{m} := \min\{m, N+1\}$, we have*

$$\|\widehat{\partial}_t^{\,k}(u - \hat{\pi}_N^{\alpha,\beta} u)\|_{\chi^{\alpha+k,\beta}} \le \sqrt{(\beta+1)^{k-\widetilde{m}} \frac{(N-\widetilde{m}+1)!}{(N-k+1)!}}\, \|\widehat{\partial}_t^{\,\widetilde{m}} u\|_{\chi^{\alpha+\widetilde{m},\beta}}. \quad (2.7.16)$$

In particular, fixing $\alpha = \beta = k = 0$ and $m < N+1$, we have

$$\|u - \hat{\pi}_N^{0,0} u\| \le c N^{-\frac{m}{2}} \|\widehat{\partial}_t^{\,m} u\|_{\chi^{m,0}}. \quad (2.7.17)$$

Proof For any $u \in A^m_{\alpha,\beta}(I)$, we obtain from (2.7.6) and (2.7.14) that

$$\widehat{\partial}_t^{\,l} \mathcal{S}_n^{(\alpha,\beta)}(t) = (\beta+1)^l \mathcal{S}_{n-l}^{(\alpha+l,\beta)}(t), \quad l \le n. \quad (2.7.18)$$

Thus the orthogonality (2.7.7) implies that

$$\|\widehat{\partial}_t^{\,l} u\|^2_{\chi^{\alpha+l,\beta}} = \sum_{n=l}^{\infty} (\beta+1)^{2l} \gamma_{n-l}^{(\alpha+l,\beta)} |\hat{u}_n^{\alpha,\beta}|^2, \quad l \ge 1.$$

Therefore,

$$\begin{aligned}
\|\widehat{\partial}_t^{\,k}(u - \hat{\pi}_N^{\alpha,\beta} u)\|^2_{\chi^{\alpha+k,\beta}} &= \sum_{n=N+1}^{\infty} (\beta+1)^{2k} \gamma_{n-k}^{(\alpha+k,\beta)} |\hat{u}_n^{\alpha,\beta}|^2 \\
&\le \max\left\{\frac{\gamma_{n-k}^{(\alpha+k,\beta)}}{\gamma_{n-\widetilde{m}}^{(\alpha+\widetilde{m},\beta)}}\right\} \sum_{n=N+1}^{\infty} (\beta+1)^{2k} \gamma_{n-\widetilde{m}}^{(\alpha+\widetilde{m},\beta)} |\hat{u}_n^{\alpha,\beta}|^2 \\
&\le (\beta+1)^{2(k-\widetilde{m})} \frac{\gamma_{N+1-k}^{(\alpha+k,\beta)}}{\gamma_{N+1-\widetilde{m}}^{(\alpha+\widetilde{m},\beta)}} \|\widehat{\partial}_t^{\,\widetilde{m}} u\|^2_{\chi^{\alpha+\widetilde{m},\beta}} \\
&\le (\beta+1)^{k-\widetilde{m}} \frac{(N-\widetilde{m}+1)!}{(N-k+1)!} \|\widehat{\partial}_t^{\,\widetilde{m}} u\|^2_{\chi^{\alpha+\widetilde{m},\beta}}.
\end{aligned}$$

Finally the special case can be proved by the following useful result: for any constant $a, b \in \mathbb{R}$, $n \in \mathbb{N}$, $n + a > 1$ and $n + b > 1$ (see Zhao et al. (2013, Lemma 2.1)),

$$\frac{\Gamma(n+a)}{\Gamma(n+b)} \le \nu_n^{a,b} n^{a-b}, \quad (2.7.19)$$

where

$$\nu_n^{a,b} = \exp\Big(\frac{a-b}{2(n+b-1)} + \frac{1}{12(n+a-1)} + \frac{(a-b)^2}{n}\Big). \tag{2.7.20}$$

□

Remark 2.55 The essential difference between approximations by LOFs and traditional polynomials can be explained by the estimate (2.7.17). In fact, since $\widehat{\partial}_t t^r = rt^r$, it is easy to check that $\|\widehat{\partial}_t^m t^r\|_{\chi^m} < \infty$ for all $r \geq 0$ and any positive integer m. So the LOFs can approximate a function whose singularity behaves as $\sum_i c_i t^{r_i}$ with an exponential convergence.

On the contrary, the polynomial approximation error depends on the regularity defined by the usual derivative. Specifically, for the classical polynomial projection $\Pi_N\colon L^2 \to P_N^t : = span\{1, t, \ldots, t^N\}$, it holds that

$$\|u - \Pi_N u\| \leq cN^{-m}\|\partial_t^m u\|.$$

Hence, functions behaving as $\sum_i c_i t^{r_i}$ with several small $r_i > 0$ cannot be well approximated by polynomials. □

We next analyze the LOF interpolation errors. Let $\{t_j^{(\alpha,\beta)}\}_{j=0}^N$ be the mapped Gauss points defined in (2.7.9). We define the mapped Lagrange functions

$$l_j(y(t)) = \frac{\prod\limits_{i\neq j}\big(y(t) - y(t_i^{(\alpha,\beta)})\big)}{\prod\limits_{i\neq j}\big(y(t_j^{(\alpha,\beta)}) - y(t_i^{(\alpha,\beta)})\big)} = \frac{\prod\limits_{i\neq j}\log(t_i^{(\alpha,\beta)}/t)}{\prod\limits_{i\neq j}\log(t_i^{(\alpha,\beta)}/t_j^{(\alpha,\beta)})}, \tag{2.7.21}$$

and the interpolation operator $\mathcal{I}_N^{\alpha,\beta} : C(I) \to P_N^{\log t}$ by

$$\mathcal{I}_N^{\alpha,\beta} v(t) = \sum_{j=0}^{N} v(t_j^{(\alpha,\beta)}) l_j(y(t)). \tag{2.7.22}$$

Obviously, we have $\mathcal{I}_N^{\alpha,\beta} v(t_j^{(\alpha,\beta)}) = v(t_j^{(\alpha,\beta)})$, $j = 0, 1, \ldots, N$.

We first establish a stability result.

Theorem 2.56 *For any $v \in C(I) \cap A^1_{\alpha,\beta}(I)$ and $\widehat{\partial}_t v \in L^2_{\chi^{\alpha,\beta}}(I)$, we have*

$$\|\mathcal{I}_N^{\alpha,\beta} v\|_{\chi^{\alpha,\beta}} \leq c\sqrt{(\beta+1)^{\alpha}}\big(c_1^{\beta} N^{-\frac{1}{2}}\|\widehat{\partial}_t v\|_{\chi^{\alpha,\beta}} + c_2^{\beta}\sqrt{\log N}\|v\|_{A^1_{\alpha,\beta}}\big), \tag{2.7.23}$$

where

$$c_1^{\beta} = (\beta+1)^{-\frac{1}{2}}, \quad c_2^{\beta} = 2\sqrt{\max\{1, \beta+1\}}.$$

Proof Let $t(y) = e^{-(\beta+1)^{-1}y}$ and $\tilde{v}(y) = v(t(y))$. Via relations (2.7.22) and (2.7.9), we have

$$\mathcal{I}_N^{\alpha,\beta} v(t) = \mathcal{I}_N^{\alpha} \tilde{v}(y) := \sum_{j=0}^{N} \tilde{v}(y_j^{(\alpha)}) l_j(y), \quad y \in \mathbb{R}^+.$$

Thanks to Guo et al. (2006a, (3.12) with $\beta = 1$), we have

$$\|\mathcal{I}_N^{\alpha} \tilde{v}\|_{y^\alpha e^{-x}} \le c\big(N^{-\frac{1}{2}} \sqrt{M_1^{\tilde{v}}} + 2\sqrt{\log N}\sqrt{M_2^{\tilde{v}}}\big),$$

where

$$M_1^{\tilde{v}} = \int_0^\infty (\partial_y \tilde{v}(y))^2\, y^\alpha e^{-y}\, \mathrm{d}y, \quad M_2^{\tilde{v}} = \int_0^\infty \big(\tilde{v}^2 + y(\partial_y \tilde{v})^2\big) y^\alpha e^{-y}\, \mathrm{d}y.$$

On the other hand, we have

$$\int_0^\infty (\partial_y \tilde{v}(y))^2\, y^\alpha e^{-y}\, \mathrm{d}y = \int_0^1 \Big(\frac{t}{\beta+1}\partial_t v(t)\Big)^2 (-(\beta+1)\log t)^\alpha t^{\beta+1}\, \frac{\beta+1}{t}\mathrm{d}t$$
$$= (\beta+1)^{\alpha-1} \int_0^1 (\widehat{\partial}_t v(t))^2\, (-\log t)^\alpha t^\beta \mathrm{d}t = (\beta+1)^{\alpha-1} \|\widehat{\partial}_t v\|^2_{\chi^{\alpha,\beta}},$$

and

$$\int_0^\infty \big(\tilde{v}^2 + y(\partial_y \tilde{v})^2\big) y^\alpha e^{-y}\, \mathrm{d}y$$
$$= (\beta+1)^{\alpha+1} \int_0^1 \Big(v^2 + \frac{(-\log t)}{\beta+1}(\widehat{\partial}_t v)^2\Big)\, (-\log t)^\alpha t^\beta \mathrm{d}t$$
$$\le (\beta+1)^\alpha \max\{1, (\beta+1)\} \|v\|^2_{A^1_{\alpha,\beta}}.$$

We can then derive (2.7.23) by combining the above relations. □

With the above stability result at our disposal, we are now ready to estimate the interpolation error.

Theorem 2.57 *For $\alpha, \beta > -1$, if $v \in C(I) \cap A^m_{\alpha,\beta}(I)$ and $\widehat{\partial}_t v \in A^{m-1}_{\alpha,\beta}(I)$ with integer $m \ge 1$, then we have*

$$\|\mathcal{I}_N^{\alpha,\beta} v - v\|_{\chi^{\alpha,\beta}} \le c\sqrt{\frac{(N+1-\widetilde{m})!}{(\beta+1)^{\widetilde{m}} N!}} \Big\{c_1^\beta \|\widehat{\partial}_t^{\widetilde{m}} v\|_{\chi^{\alpha+m-1,\beta}} + c_2^\beta \sqrt{\log N} \|\widehat{\partial}_t^{\widetilde{m}} v\|_{\chi^{\alpha+m,\beta}}\Big\}, \tag{2.7.24}$$

where

$$c_1^\beta = (\beta+1)^{-\frac{1}{2}}, \quad c_2^\beta = 2\sqrt{\max\{1, \beta+1\}}, \quad \widetilde{m} = \min\{m, N+1\}.$$

Proof By the triangle inequality, we have

$$\|\mathcal{I}_N^{\alpha,\beta} v - v\|_{\chi^{\alpha,\beta}} \le \|\mathcal{I}_N^{\alpha,\beta} v - \hat{\pi}_N^{\alpha,\beta} v\|_{\chi^{\alpha,\beta}} + \|\hat{\pi}_N^{\alpha,\beta} v - v\|_{\chi^{\alpha,\beta}}. \tag{2.7.25}$$

We only need to estimate the first term since the estimate for the second term is already available in Theorem 2.54. Thanks to (2.7.23),

$$\begin{aligned}\|\mathcal{I}_N^{\alpha,\beta} v - \hat{\pi}_N^{\alpha,\beta} v\|_{\chi^{\alpha,\beta}} &= \|\mathcal{I}_N^{\alpha,\beta}(v - \hat{\pi}_N^{\alpha,\beta} v)\|_{\chi^{\alpha,\beta}} \\ &\le c\sqrt{(\beta+1)^\alpha}\big(c_1^\beta N^{-\frac{1}{2}}\|\widehat{\partial}_t(v - \hat{\pi}_N^{\alpha,\beta} v)\|_{\chi^{\alpha,\beta}} \\ &\quad + c_2^\beta\sqrt{\log N}\|v - \hat{\pi}_N^{\alpha,\beta} v\|_{A^1_{\alpha,\beta}}\big).\end{aligned} \tag{2.7.26}$$

The error term $\|v - \hat{\pi}_N^{\alpha,\beta} v\|_{A^\mu_{\alpha,\beta}}$ $(\mu = 0, 1)$ can be estimated by Theorem 2.54. Moreover, we have

$$\|\widehat{\partial}_t(v - \hat{\pi}_N^{\alpha,\beta} v)\|_{\chi^{\alpha,\beta}} \le \|\widehat{\partial}_t v - \hat{\pi}_N^{\alpha,\beta}\{\widehat{\partial}_t v\}\|_{\chi^{\alpha,\beta}} + \|\hat{\pi}_N^{\alpha,\beta}\{\widehat{\partial}_t v\} - \widehat{\partial}_t\{\hat{\pi}_N^{\alpha,\beta} v\})\|_{\chi^{\alpha,\beta}}.$$

We now follow the procedure as in Bernardi and Maday (1997); Guo (1998). Note that by (2.7.6),

$$\begin{aligned}\widehat{\partial}_t v &= \sum_{n=0}^{\infty} \hat{v}_n^{\alpha,\beta}\widehat{\partial}_t \mathcal{S}_n^{(\alpha,\beta)} = \sum_{n=1}^{\infty} \hat{v}_n^{\alpha,\beta}\Big((\beta+1)\sum_{l=0}^{n-1} \mathcal{S}_l^{(\alpha,\beta)}\Big) \\ &= \sum_{l=0}^{\infty}\Big((\beta+1)\sum_{n=l+1}^{\infty} \hat{v}_n^{\alpha,\beta}\Big)\mathcal{S}_l^{(\alpha,\beta)},\end{aligned}$$

which implies

$$\hat{\pi}_N^{\alpha,\beta}\{\widehat{\partial}_t v\}(t) = \sum_{n=0}^{N} \hat{v}_{1,n}^{\alpha,\beta}\mathcal{S}_n^{(\alpha,\beta)}(t), \quad \hat{v}_{1,n}^{\alpha,\beta} := (\beta+1)\sum_{k=n+1}^{\infty} \hat{v}_k^{\alpha,\beta}.$$

Similarly, we have

$$\widehat{\partial}_t\{\hat{\pi}_N^{\alpha,\beta} v\}(t) = \sum_{n=0}^{N-1}\Big((\beta+1)\sum_{k=n+1}^{N} \hat{v}_k^{\alpha,\beta}\Big)\mathcal{S}_n^{(\alpha,\beta)}(t) = \sum_{n=0}^{N-1}(\hat{v}_{1,n}^{\alpha,\beta} - \hat{v}_{1,N}^{\alpha,\beta})\mathcal{S}_n^{(\alpha,\beta)}(t).$$

Hence using (2.7.7) leads to

$$\begin{aligned}\|\hat{\pi}_N^{\alpha,\beta}(\{\widehat{\partial}_t v\} - \widehat{\partial}_t\{\hat{\pi}_N^{\alpha,\beta} v\})\|^2_{\chi^{\alpha,\beta}} &= \sum_{n=0}^{N} \gamma_n^{(\alpha,\beta)}(\hat{v}_{1,N}^{\alpha,\beta})^2 \\ &= \gamma_N^{(\alpha,\beta)}(\hat{v}_{1,N}^{\alpha,\beta})^2\sum_{n=0}^{N}\gamma_n^{(\alpha,\beta)}(\gamma_N^{(\alpha,\beta)})^{-1}.\end{aligned} \tag{2.7.27}$$

We now estimate $s_N := \sum_{n=0}^{N}\gamma_n^{(\alpha,\beta)}(\gamma_N^{(\alpha,\beta)})^{-1}$. For any $\alpha \ge 0$, in view of the expression of $\gamma_n^{(\alpha,\beta)}$, we have $s_N \le N+1$. For $-1 < \alpha < 0$, we use the Stirling formula to deduce that for a large integer $M \le k \le N$,

$$\frac{\gamma_k^{(\alpha,\beta)}}{\gamma_N^{(\alpha,\beta)}} = \frac{\Gamma(N+1)\Gamma(k+\alpha+1)}{\Gamma(N+\alpha+1)\Gamma(k+1)} \sim N^{-\alpha}k^{\alpha}.$$

Therefore, there exists a constant c_M such that

$$s_N = \sum_{n=0}^{N} \gamma_n^{(\alpha,\beta)} (\gamma_N^{(\alpha,\beta)})^{-1} \le N^{-\alpha}(c_M + c\sum_{k=M}^{N} k^{\alpha}) \le cN. \tag{2.7.28}$$

Finally, combining (2.7.25)–(2.7.28) and Theorem 2.54 leads to (2.7.24). □

Remark 2.58 Let $\{t_j^{(\alpha,\beta)}\}_{j=0}^N$ and $\{\chi_j^{(\alpha,\beta)}\}_{j=0}^N$ be the same as in (2.7.9). Then, we have the following estimate for the quadrature error:

$$\Big| \int_0^1 v(t)\chi^{\alpha,\beta}(t)\mathrm{d}t - \sum_{j=0}^{N} v(t_j^{(\alpha,\beta)})\chi_j^{(\alpha,\beta)} \Big| \le \sqrt{\frac{\Gamma(\alpha+1)}{(\beta+1)^{\alpha+1}}}\, \|\mathcal{I}_N^{\alpha,\beta} v - v\|_{\chi^{\alpha,\beta}}.$$

Indeed, the above estimate can be derived from

$$\sum_{j=0}^{N} v(t_j^{(\alpha,\beta)})\chi_j^{(\alpha,\beta)} = \int_0^1 \mathcal{I}_N^{\alpha,\beta} v(t)\chi^{\alpha,\beta}(t)\mathrm{d}t,$$

and

$$\int_0^1 (-\log t)^{\alpha} t^{\beta}\mathrm{d}t = \frac{1}{(\beta+1)^{\alpha+1}}\int_0^{\infty} y^{\alpha}e^{-y}\mathrm{d}y = \frac{\Gamma(\alpha+1)}{(\beta+1)^{\alpha+1}}.$$

□

To provide some insights why the LOFs are suitable for approximating singular functions such as t^s, $s > 0$, we plot the distributions of the Gauss-LOFs quadrature nodes $\{t_j^{(\alpha,\beta)}\}_{j=0}^N$ with various N and α, β in Figures 2.8–2.9. We observe from Figure 2.8 that the nodes are clustered near zero with a much higher density near $t = 0$ than the usual Gauss–Radau points. Figure 2.9 illustrates the influence of the parameters (α, β) on the distributions of the nodes. We see that when α increases and β is fixed, the quadrature nodes move towards the left end-point; but when β increases and α is fixed, the quadrature nodes move away from zero.

We next demonstrate the accuracy of Gauss-LOFs quadrature for computing

$$\int_0^1 f(t)(-\log t)^{\alpha} t^{\beta}\mathrm{d}t, \quad \alpha, \beta > -1, \tag{2.7.29}$$

with the following functions $f(t) = \sin t,\ e^t,\ t^{-1/3}$ and $t^{1/10}$, respectively. The quadrature errors are shown in Figure 2.10 (left). We observe that

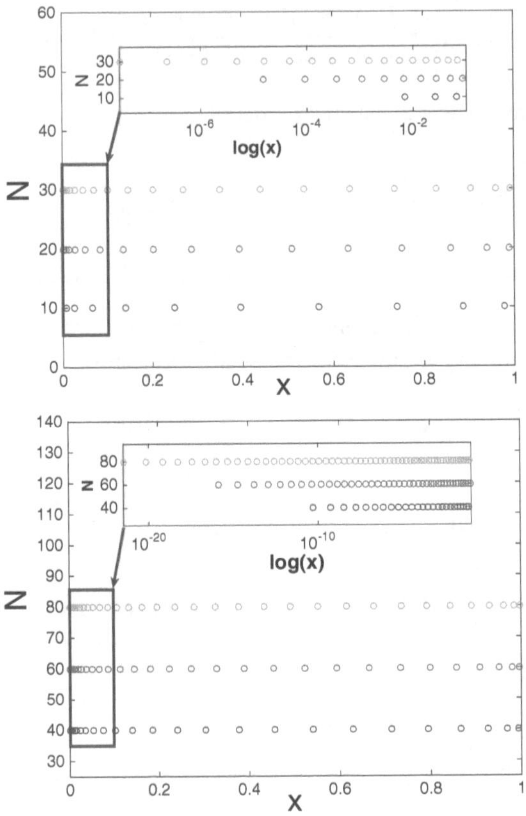

Figure 2.8 Zero distribution of $\mathcal{S}_n^{(\alpha,\beta)}(t)$: $\alpha = 0,\ \beta = 5$ with different N. Source: Chen and Shen (2022) by permission of Oxford University Press.

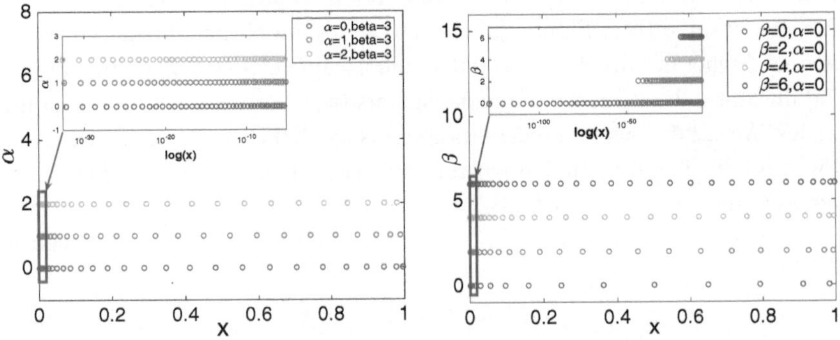

Figure 2.9 Zero distribution of $\mathcal{S}_n^{(\alpha,\beta)}(t)$: $N = 80$ with different $\alpha,\ \beta$. Source: Chen and Shen (2022) by permission of Oxford University Press.

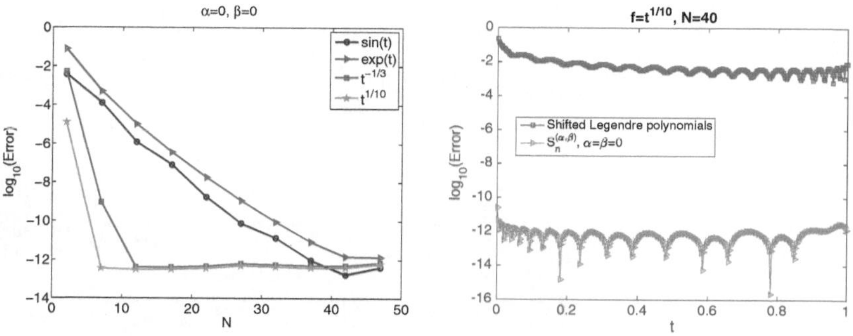

Figure 2.10 Left: Gauss-LOFs quadrature errors. Right: projection errors. Source: Chen and Shen (2022) by permission of Oxford University Press.

the errors decay exponentially in all cases. We note that $f(t) = t^{-1/3}$ is singular and cannot be computed efficiently by the classical Gauss quadrature. However, $t^{-1/3}$ is smooth with the norm defined through the pseudo-derivative (2.7.14), so we achieve exponential convergence for this case as well.

Next, we compute the projection error for $f(t) = t^{1/10}$ which is nonsmooth in the usual Sobolev space, but it is smooth with the norm defined through the pseudo-derivative. In Figure 2.10 (right), we plot the projection errors by using the shifted Legendre polynomial $L_n(2t - 1)$, $t \in I$ and LOFs for function $f(t) = t^{1/10}$ with the fixed degree of basis $N = 40$. We observe that the projection error by using LOFs is uniformly small for all $t \in [0, 1]$, while the error by using the shifted Legendre polynomial is very large.

2.7.3 Generalized Log Orthogonal Functions

The LOFs introduced in the last section are capable of resolving a certain type of singularities at $t = 0$, but LOFs $\mathcal{S}_n^{(\alpha,\beta)}(t)$ consist of $\{(-\log t)^k\}_{k=0}^n$, so grow very fast near $t = 0$ (see Figure 2.11 (left)). This behavior may affect the accuracy in many situations. In addition, derivatives of LOFs involve the singular term t^{-1}, which may not be favorable for numerical solutions of PDEs or fractional PDEs.

Therefore, we shall introduce below the generalized log orthogonal functions (GLOFs) which are more suitable for numerical approximations of functions with weak singularities at one end-point.

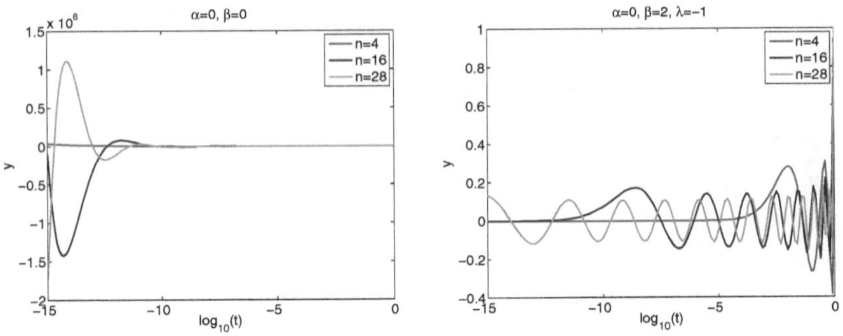

Figure 2.11 Left: Graphs of $y = \mathcal{S}_n^{(\alpha,\beta)}(t)$. Right: Graphs of $y = \mathcal{S}_n^{(\alpha,\beta,\Omega)}(t)$. Source: Chen and Shen (2022) by permission of Oxford University Press.

Definition 2.59 (GLOFs) Let $\alpha, \beta > -1$ and $\Omega \in \mathbb{R}$. We define the GLOFs by

$$\mathcal{S}_n^{(\alpha,\beta,\Omega)}(t) := t^{\frac{\beta-\Omega}{2}} \mathcal{S}_n^{(\alpha,\beta)}(t), \quad n \geq 0. \tag{2.7.30}$$

In particular, $\mathcal{S}_n^{(\alpha,\beta,\beta)}(t) = \mathcal{S}_n^{(\alpha,\beta)}(t)$.

The GLOFs enjoy similar properties as those listed in Lemma 2.52 for LOFs. It is evident from (2.7.7) that we have the orthogonality

$$\int_0^1 \mathcal{S}_n^{(\alpha,\beta,\Omega)}(t)\, \mathcal{S}_m^{(\alpha,\beta,\Omega)}(t)\, (-\log t)^\alpha t^\Omega \, dt = \gamma_n^{(\alpha,\beta)} \delta_{mn}, \tag{2.7.31}$$

where $\gamma_n^{(\alpha,\beta)}$ is the same as in (2.7.7).

Moreover, we derive from (2.7.6) and Definition 2.59 that

$$\begin{aligned}
\partial_t \mathcal{S}_n^{(\alpha,\beta,\Omega)}(t) &= \frac{\beta-\Omega}{2} t^{\frac{\beta-\Omega}{2}-1} \mathcal{S}_n^{(\alpha,\beta)}(t) + t^{\frac{\beta-\Omega}{2}} \partial_t \mathcal{S}_n^{(\alpha,\beta)}(t) \\
&= \frac{\beta-\Omega}{2} t^{\frac{\beta-\Omega-2}{2}} \mathcal{S}_n^{(\alpha,\beta)}(t) + (\beta+1) t^{\frac{\beta-\Omega-2}{2}} \mathcal{S}_{n-1}^{(\alpha+1,\beta)}(t) \\
&= \frac{\beta-\Omega}{2} \mathcal{S}_n^{(\alpha,\beta,\Omega+2)}(t) + (\beta+1) \mathcal{S}_{n-1}^{(\alpha+1,\beta,\Omega+2)}(t).
\end{aligned} \tag{2.7.32}$$

The pseudo-derivative corresponding to GLOFs should be defined as

$$\widehat{\partial}_{\gamma,t} u = t^{1+\gamma} \partial_t \{t^{-\gamma} u\}. \tag{2.7.33}$$

Then, thanks to the definition of $\mathcal{S}_n^{(\alpha,\beta,\Omega)}$ and (2.7.6), we have the following important derivative relation:

$$(\beta+1)^{-1} \widehat{\partial}_{\frac{\beta-\Omega}{2},t}\, \mathcal{S}_n^{(\alpha,\beta,\Omega)}(t) = \mathcal{S}_{n-1}^{(\alpha+1,\beta,\Omega)}(t) = \sum_{l=0}^{n-1} \mathcal{S}_l^{(\alpha,\beta,\Omega)}(t). \tag{2.7.34}$$

Let

$$t_j^{(\alpha,\beta,\Omega)} := t_j^{(\alpha,\beta)},\ \chi_j^{(\alpha,\beta,\Omega)} := (t_j^{(\alpha,\beta)})^{\Omega-\beta}\chi_j^{(\alpha,\beta)},\ j=0,1,\ldots,N, \tag{2.7.35}$$

and define

$$\mathcal{P}_N^{\gamma,\log t} := \{t^\gamma p(t) : \ p \in \mathcal{P}_N^{\log t}\}. \tag{2.7.36}$$

Then, we have the GLOF-Gauss quadrature:

$$\int_0^1 f(t)(-\log t)^\alpha t^\Omega \mathrm{d}t = \sum_{j=0}^N f(t_j^{(\alpha,\beta,\Omega)})\chi_j^{(\alpha,\beta,\Omega)}, \quad \forall f \in \mathcal{P}_{2N+1}^{\beta-\Omega,\log t}. \tag{2.7.37}$$

In addition, we derive from Definition 2.51 and the closed form of the Laguerre polynomial that

$$\mathcal{S}_n^{(\alpha,\beta,\Omega)}(t) = \sum_{k=0}^n \frac{(-1)^k}{k!}\binom{n+\alpha}{n-k} t^{\beta-\Omega/2}[-(\beta+1)\log t]^k, \quad t \in I := (0,1).$$

2.7.4 Approximation by Generalized Log Orthogonal Functions

Let $\alpha, \beta > -1$ and $\Omega \in \mathbb{R}$. We define the projection operator $\pi_N^{\alpha,\beta,\Omega}$: $L^2_{\chi^{\alpha,\Omega}}(I) \to \mathcal{P}_N^{\frac{\beta-\Omega}{2},\log t}$ by

$$(u - \pi_N^{\alpha,\beta,\Omega}u, v)_{\chi^{\alpha,\Omega}} = 0, \quad \forall v \in \mathcal{P}_N^{\frac{\beta-\Omega}{2},\log t}, \tag{2.7.38}$$

where the weight function $\chi^{\alpha,\Omega}(t) := (-\log t)^\alpha t^\Omega$. In view of the orthogonality of the GLOFs, we have

$$\pi_N^{\alpha,\beta,\Omega}u = \sum_{n=0}^N \hat{u}_n^{\alpha,\beta,\Omega}\, \mathcal{S}_n^{(\alpha,\beta,\Omega)}, \tag{2.7.39}$$

with

$$\hat{u}_n^{\alpha,\beta,\Omega} = \frac{1}{\gamma_n^{(\alpha,\beta)}}\int_0^1 u(t)\, \mathcal{S}_n^{(\alpha,\beta,\Omega)}(t)\, \chi^{\alpha,\Omega}(t)\mathrm{d}t.$$

We define non-uniformly weighted Sobolev spaces

$$A^k_{\alpha,\beta,\Omega}(I) := \big\{v \in L^2_{\chi^{\alpha,\Omega}}(I) : \ \widehat{\partial}^j_{\frac{\beta-\Omega}{2},t} v \in L^2_{\chi^{\alpha+j,\Omega}}(I),\ j=1,2,\ldots,k\big\}, \tag{2.7.40}$$

for integer $k \geq 1$, which is equipped with the semi-norm and norm

$$|v|_{A^m_{\alpha,\beta,\Omega}} := \|\widehat{\partial}^m_{\frac{\beta-\Omega}{2},t} v\|_{\chi^{\alpha+m,\Omega}}, \quad \|v\|_{A^m_{\alpha,\beta,\Omega}} := \Big(\sum_{k=0}^m |v|^2_{A^k_{\alpha,\beta,\Omega}}\Big)^{\frac{1}{2}}.$$

Theorem 2.60 *Let $m, N, k \in \mathbb{N}$, $\Omega \in \mathbb{R}$ and $\alpha, \beta > -1$. For any $u \in A^m_{\alpha,\beta,\Omega}(I)$ and $0 \le k \le \widetilde{m} = \min\{m, N+1\}$, we have*

$$\|\widehat{\partial}^k_{\frac{\beta-\Omega}{2},t}(u - \pi_N^{\alpha,\beta,\Omega}u)\|_{\chi^{\alpha+k,\Omega}} \le \sqrt{(\beta+1)^{k-\widetilde{m}}\frac{(N-\widetilde{m}+1)!}{(N-k+1)!}}\,\|\widehat{\partial}^{\widetilde{m}}_{\frac{\beta-\Omega}{2},t}u\|_{\chi^{\alpha+\widetilde{m},\Omega}},$$

where $\widehat{\partial}_{\frac{\beta-\Omega}{2},t}$ is the pseudo-derivative defined in (2.7.33).

Proof For any $u \in A^m_{\alpha,\beta,\Omega}(I)$, we can expand it as $u = \sum_{n=0}^{\infty}\hat{u}_n^{\alpha,\beta,\Omega}\mathcal{S}_n^{(\alpha,\beta,\Omega)}$. By (2.7.34),

$$\widehat{\partial}^l_{\frac{\beta-\Omega}{2},t}\mathcal{S}_n^{(\alpha,\beta,\Omega)}(t) = (\beta+1)^l\mathcal{S}_{n-l}^{(\alpha+l,\beta,\Omega)}(t), \quad l \le n,$$

which, together with (2.7.31), leads to

$$\|\widehat{\partial}^l_{\frac{\beta-\Omega}{2},t}u\|^2_{\chi^{\alpha+l,\Omega}} = \sum_{n=l}^{\infty}(\beta+1)^{2l}\gamma_{n-l}^{(\alpha+l,\beta)}|\hat{u}_n^{\alpha,\beta,\Omega}|^2, \quad l \ge 1.$$

Following the same procedure as in the proof of Theorem 2.54, we can obtain the desired result (2.60). □

We now turn to the interpolation error estimate. Let $\{t_j^{\alpha,\beta}\}_{j=0}^N$ be the same set of collocation points as for the LOFs. We define the interpolation operator $\mathcal{I}_N^{\alpha,\beta,\Omega} : C(I) \to \mathcal{P}_N^{\frac{\beta-\Omega}{2},\log t}$ by

$$(\mathcal{I}_N^{\alpha,\beta,\Omega}v)(t_j^{\alpha,\beta}) = v(t_j^{\alpha,\beta}), \quad j = 0, 1, \ldots, N.$$

Note that

$$(\mathcal{I}_N^{\alpha,\beta,\Omega}v)(t) = \sum_{j=0}^{N} v(t_j^{(\alpha,\beta)})l_j^{\beta,\Omega}(y(t)), \quad y(t) = -(\beta+1)\log t, \tag{2.7.41}$$

where $\{l_j^{\beta,\Omega}\}$ are the Lagrange interpolating basis defined by

$$l_j^{\beta,\Omega}(y(t)) = \frac{t^{\frac{\beta-\Omega}{2}}\prod_{i\neq j}\log(t_i^{(\alpha,\beta)}/t)}{(t_j^{(\alpha,\beta)})^{\frac{\beta-\Omega}{2}}\prod_{i\neq j}\log(t_i^{(\alpha,\beta)}/t_j^{(\alpha,\beta)})}. \tag{2.7.42}$$

In view of (2.7.22) and (2.7.21), we have

$$\mathcal{I}_N^{\alpha,\beta,\Omega}v(t) = t^{\frac{\beta-\Omega}{2}}\mathcal{I}_N^{\alpha,\beta}\{t^{\frac{\Omega-\beta}{2}}v(t)\} \in \mathcal{P}_N^{\frac{\beta-\Omega}{2},\log t}.$$

Hence, we can derive the following result from Theorem 2.57.

Theorem 2.61 *Let m and N be positive integers, $\alpha, \beta > -1$ and $\Omega \in \mathbb{R}$. For any $v \in C(I) \cap A^m_{\alpha,\beta,\Omega}(I)$ and $\widehat{\partial}_{\frac{\beta-\Omega}{2},t} v \in A^{m-1}_{\alpha,\beta,\Omega}(I)$, we have*

$$\|\mathcal{I}_N^{\alpha,\beta,\Omega} v - v\|_{\chi^{\alpha,\Omega}} \le c\sqrt{\frac{(N+1-\widetilde{m})!}{(\beta+1)^{\widetilde{m}-\alpha} N!}}\Big\{c_1^\beta \|\widehat{\partial}^{\widetilde{m}}_{\frac{\beta-\Omega}{2},t} v\|_{\chi^{\alpha+m-1,\Omega}} + c_2^\beta \sqrt{\log N}\|\widehat{\partial}^{\widetilde{m}}_{\frac{\beta-\Omega}{2},t} v\|_{\chi^{\alpha+m,\Omega}}\Big\},$$

where $c_1^\beta = (\beta+1)^{-\frac{1}{2}}$, $c_2^\beta = 2\sqrt{\max\{1, \beta+1\}}$ and $\widetilde{m} = \min\{m, N+1\}$.

Proof Since

$$\|\mathcal{I}_N^{\alpha,\beta,\Omega} v - v\|_{\chi^{\alpha,\Omega}} = \|\mathcal{I}_N^{\alpha,\beta}\{t^{\frac{\Omega-\beta}{2}} v\} - t^{\frac{\Omega-\beta}{2}} v\|_{\chi^{\alpha,\beta}}$$

and

$$\widehat{\partial}_t\{t^{\frac{\Omega-\beta}{2}} v\} = t^{\frac{\Omega-\beta}{2}}\widehat{\partial}_{\frac{\beta-\Omega}{2},t} v \implies \widehat{\partial}_t^{\widetilde{m}}\{t^{\frac{\Omega-\beta}{2}} v\} = t^{\frac{\Omega-\beta}{2}}\widehat{\partial}^{\widetilde{m}}_{\frac{\beta-\Omega}{2},t} v,$$

we can derive the desired result from the above relation and Theorem 2.57. □

Explicit Error Estimate for a Class of Weakly Singular Functions

The result in Theorem 2.61 is not easy to interpret for general functions, so we consider the following typical weakly singular functions

$$f(t) = t^r(-\log t)^k, \quad r \ge 0, \ k \in \mathbb{N}_0.$$

We first present a very useful integral identity of the Laguerre polynomials.

Lemma 2.62 *Let $s > 0$, $\alpha > -1$. For $k, n \in \mathbb{N}_0$ and $n > k$, there exists*

$$\int_0^\infty y^{\alpha+k} e^{-sy} \mathscr{L}_n^{(\alpha)}(y) dy = \Big(\frac{s-1}{s}\Big)^n \frac{k!}{s^{\alpha+k+1}} \sum_{j=0}^{k} \frac{\Gamma(n-j+k+\alpha+1)}{(j!)^2\,\Gamma(n-j+1)}\Big(\frac{s}{1-s}\Big)^j. \tag{2.7.43}$$

Proof The case $k = 0$ is a direct result of Gradshteyn and Ryzhik (2015, 7.414: 8), i.e.,

$$\int_0^\infty e^{-sy} y^\alpha \mathscr{L}_n^{(\alpha)}(y) dy = \frac{\Gamma(\alpha+n+1)}{\Gamma(n+1)}(1-1/s)^n s^{-\alpha-1}.$$

For $k > 0$, owing to Rodrigues' formula (cf. (5.1.5) in Szegő (1975)), we have that

$$\begin{aligned}\int_0^\infty y^k e^{-sy} y^\alpha \mathscr{L}_n^{(\alpha)}(y)\mathrm{d}y &= \frac{1}{n!}\int_0^\infty y^k e^{(1-s)y}\, \partial_y^n (y^{n+\alpha}e^{-y})\mathrm{d}y \\ &= \frac{(-1)^n}{n!}\int_0^\infty \partial_y^n (y^k e^{(1-s)y})\, y^{n+\alpha}e^{-y}\mathrm{d}y \\ &= \frac{(-1)^n}{n!}\sum_{j=0}^k \binom{n}{j}(1-s)^{n-j}\int_0^\infty \frac{k!}{j!}y^{k-j}\, y^{n+\alpha}e^{-sy}\mathrm{d}y \\ &= (-1)^n \sum_{j=0}^k \frac{k!\,\Gamma(n-j+k+\alpha+1)}{(j!)^2\Gamma(n-j+1)}\frac{(1-s)^{n-j}}{s^{n-j+\alpha+k+1}}.\end{aligned}$$

One can easily check the equivalence of the relations (2.7.43) and (2.7.4), which completes the proof. □

With the above lemma in hand, we have the following error estimate.

Theorem 2.63 *Consider* $f(t) = t^r(-\log t)^k$, $r \ge 0$, $k \in \mathbb{N}_0$, *and let* $\Omega > -1 - 2r$, $\alpha, \beta > -1$ *and* $\beta > \Omega$, *so that the constant*

$$R_{\alpha,\beta,\Omega} := \left|\frac{2r+\Omega-\beta}{2r+2+\Omega+\beta}\right| < 1.$$

Then $f \in L^2_{\chi^{\alpha,\Omega}}(I)$ *and*

$$\|f - \pi_N^{\alpha,\beta,\Omega} f\|_{\chi^{\alpha,\Omega}} \le c\,(k+1)!\; N^{\frac{\alpha+1}{2}+k}\,(R_{\alpha,\beta,\Omega})^N, \tag{2.7.44}$$

when

$$N > -\frac{2k+\alpha+2}{2\log(R_{r,\beta,\Omega})}.$$

Here the constant

$$c \approx \sqrt{\frac{2^{\alpha+1+k}(\beta+1)^{2\alpha+2-k}}{(\beta+\Omega+2r+2)^{\alpha+1+k}}}.$$

Proof Since $\Omega > -1 - 2r$, it is easy to check that $f \in L^2_{\chi^{\alpha,\Omega}}$ and $R_{\alpha,\beta,\Omega} < 1$. Thanks to the orthogonality of the basis $\{S_n^{(\alpha,\beta,\Omega)}\}_{n=0}^\infty$, we can write

$$f = \sum_{n=0}^{\infty} \hat{f}_n^{\alpha,\beta,\Omega} \mathcal{S}_n^{(\alpha,\beta,\Omega)}, \quad \pi_N^{\alpha,\beta,\Omega} f = \sum_{n=0}^{N} \hat{f}_n^{\alpha,\beta,\Omega} \mathcal{S}_n^{(\alpha,\beta,\Omega)}$$

with coefficients

$$\hat{f}_n^{\alpha,\beta,\Omega} = (\gamma_n^{(\alpha,\beta)})^{-1} \int_0^1 f(t)\, \mathcal{S}_n^{(\alpha,\beta,\Omega)}(t)\, \chi^{\alpha,\Omega}(t) \mathrm{d}t.$$

Using the change of variable $y(t) = -(\beta+1)\log t$ and $\mathrm{d}y = -(\beta+1)t^{-1}\mathrm{d}t$, we derive

$$\begin{aligned}\hat{f}_n^{\alpha,\beta,\Omega} &= \frac{(\beta+1)^{\alpha+1}\Gamma(n+1)}{\Gamma(n+\alpha+1)} \int_0^1 t^r(-\log t)^k\, t^{(\beta-\Omega)/2} \mathscr{L}_n^{(\alpha)}(y(t))\,(-\log t)^{\alpha} t^{\Omega} \mathrm{d}t \\ &= \frac{(\beta+1)^{-k}\Gamma(n+1)}{\Gamma(n+\alpha+1)} \int_0^{\infty} \exp\Big(-\frac{\beta+\Omega+2r+2}{2(\beta+1)} y\Big) y^{\alpha+k} \mathscr{L}^{(\alpha)}(y) \mathrm{d}y.\end{aligned}$$

Taking $s = (\beta+\Omega+2r+2)/(2\beta+2)$ in (2.7.43) leads to

$$\hat{f}_n^{\alpha,\beta,\Omega} = \Big(\frac{s-1}{s}\Big)^n \frac{\Gamma(n+1)\, k!\, (\beta+1)^{-k}}{\Gamma(n+\alpha+1)\, s^{\alpha+k+1}} \sum_{j=0}^{k} \frac{\Gamma(n-j+k+\alpha+1)}{(j!)^2\, \Gamma(n-j+1)} \Big(\frac{s-1}{s}\Big)^{-j}.$$

Owing to Zhao et al. (2013, Lemma 2.1), we have

$$\frac{\Gamma(n+a)}{\Gamma(n+b)} \le \nu_n^{a,b} n^{a-b}, \quad n+a>1 \text{ and } n+b>1, \tag{2.7.45}$$

where

$$\nu_n^{a,b} = \exp\Big(\frac{a-b}{2(n+b-1)} + \frac{1}{12(n+a-1)} + \frac{(a-b)^2}{n}\Big). \tag{2.7.46}$$

A combination of the above results and noting $R_{r,\beta,\Omega} < 1$, we obtain

$$\begin{aligned}\|f - \pi_N^{\alpha,\beta,\Omega} f\|_{\chi^{\alpha,\Omega}}^2 &= \sum_{N+1}^{\infty} |\hat{f}_n^{\alpha,\beta,\Omega}|^2 \gamma_n^{\alpha,\beta} \le C_{\alpha,\beta}^{\Omega,r} \sum_{N+1}^{\infty} \nu_n^{\alpha,0} n^{2k+\alpha} (R_{r,\beta,\Omega})^{2n-2k} \\ &\le C_{\alpha,\beta}^{\Omega,r} \nu_N^{\alpha,0} \int_N^{\infty} (R_{r,\beta,\Omega})^{2x-2k} x^{2k+\alpha} \mathrm{d}x,\end{aligned}$$

where

$$C_{\alpha,\beta}^{\Omega,r} = \frac{2^{\alpha+1+k}(\beta+1)^{2\alpha+2-k}}{(\beta+\Omega+2r+2)^{\alpha+1+k}} \big((k+1)!\big)^2.$$

Finally, as $(R_{r,\beta,\Omega})^{2x-2k} x^{2k+\alpha+2}$ is a decreasing function of x when $N > -\frac{2k+\alpha+2}{2\log(R_{r,\beta,\Omega})}$, we conclude that

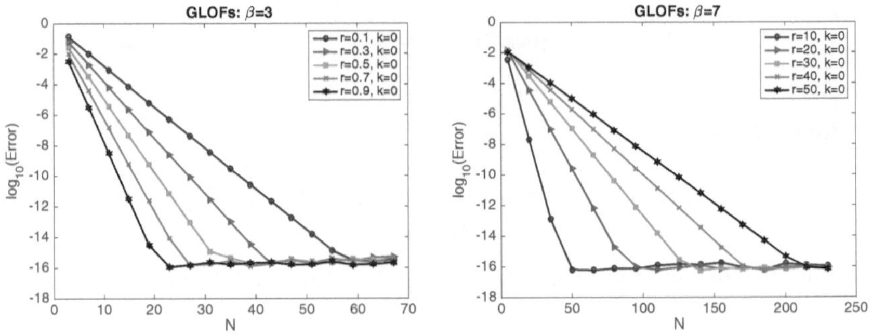

Figure 2.12 GLOFs Approximation: $\mathcal{S}_n^{(\alpha,\beta,\Omega)}$, $\alpha = \Omega = 0$. Source: Chen and Shen (2022) by permission of Oxford University Press.

$$\|f - \pi_N^{\alpha,\beta,\Omega} f\|_{\chi^{\alpha,\Omega}} \le \sqrt{C_{\alpha,\beta}^{\Omega,r}}\; N^{\frac{\alpha+1}{2}+k}\; (R_{r,\beta,\Omega})^{N-k}.$$

The proof is complete. □

The above theorem provides an accurate estimate for the GLOFs to a large class of singular functions. In particular, by setting $\alpha = \Omega = 0$, we have an estimate in L^2-norm.

Corollary 2.64 *For $f(t) = t^r(-\log t)^k$, $r \ge 0$, $k \in \mathbb{N}$, it holds that*

$$\|f - \pi_N^{0,\beta,0} f\| \le \sqrt{2}^k (\beta+1)^{-k} k!\, N^k \sqrt{2(\beta+1)N} \left|\frac{2r-\beta}{2r+\beta+2}\right|^{N-k}. \tag{2.7.47}$$

In particular, for $f = t^r$, $r \ge 0$, we have

$$\|f - \pi_N^{0,\beta,0} f\| \le \sqrt{2(\beta+1)N} \left|\frac{2r-\beta}{2r+\beta+2}\right|^{N}. \tag{2.7.48}$$

In order to verify the above theoretical results, we plot the error curves for the GLOFs approximation to $f(t) = t^r$ with various r in Figure 2.12 (left) with $r \in (0,1)$ and in Figure 2.12 (right) with r being integers. We observe exponential convergence for all $r \ge 0$. We also plot the error curves for the GLOFs approximation to $f(t) = t(-\log t)^k$ and $f(t) = t^2(-\log t)^k$ in Figure 2.13. We observe the exponential convergence in all cases. All of these numerical results are consistent with the theoretical results in Corollary 2.64.

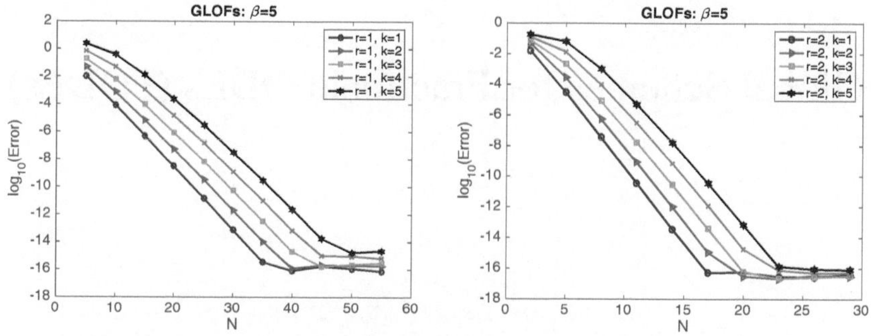

Figure 2.13 GLOFs Approximation: $\mathcal{S}_n^{(\alpha,\beta,\Omega)}$, $\alpha = \Omega = 0$. Source: Chen and Shen (2022) by permission of Oxford University Press.

3
Global Schemes for Fractional ODEs (FODEs)

This chapter provides a comprehensive presentation of global numerical methods for solving fractional ODEs (FODEs) employing the polynomial and non-polynomial bases, introduced in Chapter 2. We have already provided the meaning of the term "global" and explained why global numerical methods are suitable for solving the nonlocal fractional models in Section 1.4. The FODEs of interest will be initial-/boundary-value problems, posed using a variety of fractional derivatives (e.g., Caputo, Riemann–Liouville, Riesz, one-sided, two-sided, variable-order, distributed-order, etc.), introduced in Chapters 1 and 2. We devote the first two sections in this chapter to introducing a series of variational and non-variational spectral methods in *single domains*, where the solution singularities can occur at the initial or boundary points. In a variational formulation of an FODE, one first obtains the weak (variational) form of the given equation, where the highest derivative order is reduced using integration-by-parts, and then one solves the variational formulation by constructing the corresponding (finite-dimensional) *solution* and *test* subspaces. In non-variational problems, one rather directly solves the strong (original) FODE, hence assuming a higher regularity in the solution. All the spectral methods introduced in Sections 3.1 and 3.2 can only treat initial-/boundary-point singularities. In Section 3.3, we therefore introduce a spectral element method (SEM) for FODEs in *multiple domains* for the main purpose of capturing possible *interior/boundary* singularities.

More specifically in Section 3.1, we first present the variational Galerkin and Petrov–Galerkin spectral methods in single domains. In these problems, after introducing the weak formulation of FODE, we construct the underlying solution and test subspaces using the span of bases/test functions introduced in Chapter 2. When these two subspaces coincide, the variational formulation is called the *Galerkin* method, otherwise the method would be of *Petrov–Galerkin* sense. Petrov–Galerkin methods appear naturally in the numerical solution of FODEs due to the appearance of the left- and right-sided fractional derivatives and/or integrals in the variational form. Therefore, the "suitable" basis and test functions can belong to different subspaces with different boundary/initial conditions and/or different end-singularities. We provide a rigorous treatment of FODEs with constant/variable coefficients of fixed-to-distributed order problems.

In Section 3.2, we begin with the classical Jacobi–Gauss–Lobatto interpolation, and subsequently we formulate fractional pseudo-spectral differentiation matrices for the Riemann–Liouville, Caputo and modified Riemann–Liouville, in addition to their Birkhoff interpolation differentiation counterparts to be examined in the context of FODEs with the corresponding error analyses. Next, we further highlight the idea of *fractional Lagrange interpolants* and introduce fractional Lagrange bases based on Gauss–Radau interpolation, obtaining the corresponding fractional differentiation matrices, followed with rigorous error analyses. For effective treatment of variable-order FODEs, we subsequently present a tunable collocation method, where Caputo and Riemann–Liouville fractional differentiation matrices are constructed in a stable fashion, followed with numerous numerical examples with only one-sided singularities. For treating different singularities at all end-points, we finally formulate a generalized tunable collocation method, based on the generalized Jacobi poly-fractonomials, introduced in Chapter 2.

In Section 3.3, we present a Petrov–Galerkin spectral element method for FODEs in *multiple sub-domains*. We employ the standard (modal) spectral element basis and the Jacobi poly-fractonomials as the test functions. We show an effective procedure for assembling the global linear system from elemental (local) mass and stiffness matrices. Moreover, we efficiently obtain the nonlocal (history) stiffness matrices, in which the nonlocality is presented analytically for uniform grids. One interesting feature of this scheme is that we can construct non-uniform grids over the whole computational domain in order to capture singular solutions both at end-points and the interior domain.

3.1 Galerkin and Petrov–Galerkin Spectral Methods

3.1.1 Generalized Jacobi Poly-fractonomials as Spectral Bases

In this section, we present spectral methods using the generalized Jacobi poly-fractonomials (GJPFs), defined in the last chapter to solve several prototypical fractional differential equations. We first consider a fractional initial-value problem, followed by a fractional boundary-value problem involving one-sided fractional derivatives.

A Fractional Initial-Value Problem (FIVP)

As the first example, we consider the fractional initial-value problem of order $s \in (0,1)$:

$$ {}_0^{\mathrm{C}}D_t^s\, u(t) = f(t), \quad t \in I := (0,T); \quad u(0) = u_0. \tag{3.1.1}$$

For the non-homogeneous initial conditions $u(0) = u_0$, we first decompose the solution $u(t)$ into

$$u(t) = v(t) + u_0, \tag{3.1.2}$$

with $v(0) = 0$. As ${}_0^{\mathrm{C}}D_t^s\, u_0 = 0$, the FIVP (3.1.1) is equivalent to the following FIVP with Riemann–Liouville fractional derivative:

$$ {}_0^{\mathrm{RL}}D_t^s\, v(t) = f(t), \quad t \in I; \quad v(0) = 0. \tag{3.1.3}$$

A Petrov–Galerkin scheme for (3.1.3) is to find

$$v_N \in {}^{l}V_N^{(s)}(I) := \big\{ t^s \varphi(t) : \forall\, \varphi(t) \in \mathbb{P}_N,\ t \in I \big\}, \tag{3.1.4}$$

such that

$$({}_0^{\mathrm{RL}}D_t^s\, v_N, w_N) = (f, w_N), \quad \forall\, w_N \in \mathbb{P}_N. \tag{3.1.5}$$

We expand $f(t)$ as

$$f(t) = \sum_{n=0}^{\infty} \tilde{f}_n\, P_n(x(t)), \quad x(t) = (2t-T)/T \in (-1,1), \tag{3.1.6}$$

and write

$$v_N(t) = \sum_{n=0}^{N} \tilde{u}_n^{(s)}\, {}^{l}\mathcal{P}_n^{(-s,-s)}(x(t)) \in {}^{l}V_N^{(s)}(I). \tag{3.1.7}$$

Taking $w_N = P_l(x(t))$ in (3.1.5), we obtain from (2.4.19) and the orthogonality of Legendre polynomials that

$$\tilde{u}_n^{(s)} = \frac{n!}{\Gamma(n+s+1)} \tilde{f}_n, \quad 0 \le n \le N. \tag{3.1.8}$$

Therefore, we can obtain the numerical solution v_N without solving any algebraic equation, so the method is very efficient.

As for the error estimate, we have the following result (see Chen et al. (2016)).

Theorem 3.1 *Let v and v_N be the solution of* (3.1.3) *and* (3.1.5), *respectively. Then*

$$\| {}_0^{\mathrm{RL}}D_t^s\, (v - v_N) \| \le c N^{-m} \| f^{(m)} \|_{\omega^{(m-1,m-1)}}, \tag{3.1.9}$$

where $\omega^{(m-1,m-1)} = t^{m-1}(T-t)^{m-1}$.

Proof Let ${}^l\pi_N^{(-s,-s)}$ be as defined in (2.5.18) for $0 < s < 1$. By (2.4.19), we have

$$\big({}^{\rm RL}_{0}D_t^s\, {}^l\pi_N^{(-s,-s)} v - v, \psi\big) = 0, \quad \forall \psi \in \mathbb{P}_N.$$

Then by (3.1.3),

$$\big(f - {}^{\rm RL}_{0}D_t^s\, {}^l\pi_N^{(-s,-s)} v,\, \psi\big) = \big({}^{\rm RL}_{0}D_t^s\, v - {}^{\rm RL}_{0}D_t^s\, {}^l\pi_N^{(-s,-s)} v, \psi\big) = 0, \quad \forall \psi \in \mathbb{P}_N.$$

Let $\pi_N f$ be the L^2-orthogonal projection of f upon $\mathbb{P}_N$. We infer from the above that

$${}^{\rm RL}_{0}D_t^s\, {}^l\pi_N^{(-s,-s)} v = \pi_N f = {}^{\rm RL}_{0}D_t^s\, v_N.$$

Therefore,

$$\begin{aligned} \|\, {}^{\rm RL}_{0}D_t^s\, (v - v_N)\| &= \big\|\, {}^{\rm RL}_{0}D_t^s\, (v - {}^l\pi_N^{(-s,-s)} v_N)\big\| \\ &\le \big\|\, {}^{\rm RL}_{0}D_t^s (v - {}^l\pi_N^{(-s,-s)} v)\big\| + \|\pi_N f - f\|. \end{aligned} \tag{3.1.10}$$

It follows from Theorem 2.40 (with $\alpha = -\beta = s$ and $0 < s < 1$), and the Legendre approximation results that

$$\|\, {}^{\rm RL}_{0}D_t^s (v - v_N)\| \le cN^{-m}\big(\big\|\, {}^{\rm RL}_{0}D_t^{s+m} v\big\|_{\omega^{(m,m)}} + \|f^{(m)}\|_{\omega^{(m-1,m-1)}}\big). \tag{3.1.11}$$

We deduce from (3.1.1) that

$$\big\|\, {}^{\rm RL}_{0}D_t^{s+m} v\big\|_{\omega^{(m,m)}} \le c\big\| f^{(m)}\big\|_{\omega^{(m-1,m-1)}}.$$

This completes the proof. □

The analysis of this simple model shows that the accuracy of this Petrov–Galerkin method only depends on the regularity of the source term in the usual Jacobi-weighted Sobolev space. Consequently, if the source term $f(t)$ is smooth, it enjoys the spectral accuracy as typical spectral methods for integer-order differential equations.

One-Sided Fractional Boundary-Value Problems

Now we consider a model one-sided fractional boundary-value problem

$${}^{\rm RL}_{-1}D_x^{\nu}\, u(x) = f(x), \quad x \in \Omega := (-1, 1); \quad u(\pm 1) = 0, \tag{3.1.12}$$

where $\nu \in (1, 2)$. Let $s = \nu - 1$ and introduce the solution and test function spaces

$$\begin{aligned} U_s &:= \big\{u \in L^2_{\omega^{(-s,-1)}}(\Omega) \,:\, {}^{\rm RL}_{-1}D_x^s\, u \in L^2_{\omega^{(0,s-1)}}(\Omega)\big\}; \\ V_s &:= \big\{v \in L^2_{\omega^{(-1,-s)}}(\Omega) \,:\, Dv \in L^2_{\omega^{(0,1-s)}}(\Omega)\big\}, \end{aligned} \tag{3.1.13}$$

equipped with the norms

$$\begin{aligned} \|u\|_{U_s} &= \big(\|u\|^2_{\omega^{(-s,-1)}} + \|{}^{\rm RL}_{-1}D^s_x\, u\|^2_{\omega^{(0,s-1)}}\big)^{1/2}; \\ \|v\|_{V_s} &= \big(\|v\|^2_{\omega^{(-1,-s)}} + \|Dv\|^2_{\omega^{(0,1-s)}}\big)^{1/2}. \end{aligned} \tag{3.1.14}$$

For $u \in U_s$ and $v \in V_s$, we expand

$$\begin{aligned} u(x) &= \sum_{n=1}^{\infty} \hat{u}_n {}^l\mathcal{P}_n^{(-s,-1)}(x) = (1-x)^s(1+x)\sum_{n=1}^{\infty} \hat{u}_n P_{n-1}^{(s,1)}(x), \\ v(x) &= \sum_{n=1}^{\infty} \hat{v}_n {}^r\mathcal{P}_n^{(-1,-s)}(x) = (1-x)(1+x)^s\sum_{n=1}^{\infty} \hat{v}_n P_{n-1}^{(1,s)}(x). \end{aligned} \tag{3.1.15}$$

With the above setup, we can build in the homogeneous boundary conditions and also perform fractional integration-by-parts. Hence, a weak form of (3.1.12) is to find $u \in U_s$ such that

$$a(u,v) := ({}^{\rm RL}_{-1}D^s_x u,\, Dv) = (f,v), \quad \forall\, v \in V_s. \tag{3.1.16}$$

Let ${}^l\mathbb{F}_N^{(-s,-1)}(\Omega)$ and ${}^r\mathbb{F}_N^{(-1,-s)}(\Omega)$ be the finite-dimensional spaces as defined in the previous chapter. Then the GJF-Petrov–Galerkin scheme for (3.1.16) is to find $u_N \in {}^l\mathbb{F}_N^{(-s,-1)}(\Omega)$ such that

$$a(u_N, v_N) = \big({}^{\rm RL}_{-1}D^s_x u_N,\, Dv_N\big) = (f, v_N), \quad \forall\, v_N \in {}^r\mathbb{F}_N^{(-1,-s)}(\Omega). \tag{3.1.17}$$

We derive from (2.4.18) and ${}^{\rm RL}_{-1}D^1_x = \frac{\rm d}{{\rm d}x}$ that

$$a({}^l\mathcal{P}_n^{(-s,-1)}, {}^r\mathcal{P}_m^{(-1,-s)}) = C^s_{n,m}({}^l\mathcal{P}_n^{(0,s-1)}, {}^r\mathcal{P}_m^{(0,1-s)}) = 0, \quad \forall n \neq m. \tag{3.1.18}$$

Hence, we can obtain u_N directly without solving any algebraic equation. Moreover, we have the following estimates.

Theorem 3.2 *Let $s \in (0,1)$, and let u and u_N be the solutions of* (3.1.16) *and* (3.1.17)*, respectively. If $u \in U_s \cap {}^l\mathcal{B}^m_{s,-1}(\Omega)$ with $0 \le m \le N$, then we have*

$$\|u - u_N\|_{U_s} \le cN^{-m}\big\|{}^{\rm RL}_{-1}D^{s+m}_x u\big\|_{\omega^{(m,s-1+m)}}. \tag{3.1.19}$$

In particular, if $f^{(m-1)} \in L^2_{\omega^{(m,s-1+m)}}(\Omega)$ for $m \ge 1$, we have

$$\|u - u_N\|_{U_s} \le cN^{-m}\big\|f^{(m-1)}\big\|_{\omega^{(m,s-1+m)}}. \tag{3.1.20}$$

Here, c is a positive constant independent of u, N and m.

Proof We derive from (3.1.16) and (3.1.17) that

$$a(u - u_N, v_N) = 0, \quad \forall\, v_N \in {}^r\mathbb{F}_N^{(-1,-s)}(\Omega),$$

which, along with (3.1.18), implies immediately that $u_N = {}^{l}\pi_N^{(-s,-1)}u$. Since

$$ {}_{-1}^{\mathrm{RL}}D_x^{\nu}u = {}_{-1}^{\mathrm{RL}}D_x^{s+1}u = f, $$

the desired results follow from Theorem 2.40. □

Fractional Boundary-Value Problems with Riesz Derivatives

We consider first the so-called Riesz fractional differential equations, followed by a more general case with a zeroth-order term. We consider the Riesz fractional equation of order $2\alpha \in (2k-1, 2k)$ with $k \in \mathbb{N}$:

$$ \begin{cases} (-1)^k D_{|x|}^{2\alpha}u(x) = f(x), & x \in \Omega, \\ u^{(l)}(\pm 1) = 0, & l = 0, 1, \ldots, k-1. \end{cases} \tag{3.1.21} $$

A Petrov–Galerkin spectral method for (3.1.21) is to find $u_N \in \mathbb{F}_N^{(-\alpha,-\alpha)}(\Omega)$, defined in (2.4.30), such that

$$ (-1)^k (D_{|x|}^{2\alpha}u_N, v_N)_{\omega^{(\alpha,\alpha)}} = (f, v_N)_{\omega^{(\alpha,\alpha)}}, \quad \forall v_N \in \mathbb{P}_N. \tag{3.1.22} $$

Note that, in contrast to the Galerkin spectral method in which the trial (basis) space and the test space coincide, here in the Petrov–Galerkin spectral method, the trial and test spaces are different. The solution to this discrete problem can be found directly as follows. Given

$$ f(x) = \sum_{m=0}^{\infty} \hat{f}_m P_m^{(\alpha,\alpha)}(x), \tag{3.1.23} $$

write

$$ u_N(x) = \sum_{n=0}^{N} \hat{u}_n \mathcal{J}_n^{-\alpha,-\alpha}(x). \tag{3.1.24} $$

Plugging the above into (3.1.22), and using (2.4.33) and the orthogonality of $\{P_m^{(\alpha,\alpha)}\}$ in $L^2_{\omega^{(\alpha,\alpha)}}(\Omega)$, we find

$$ \hat{u}_n = \hat{f}_n \Big/ \Big(2\cos(\pi\alpha)\frac{\Gamma(n+1+2\alpha)}{n!}\Big), \quad \forall 0 \le n \le N. \tag{3.1.25} $$

Moreover, we have the following error estimates.

Theorem 3.3 *Assume that $f^{(j)} \in L^2_{\omega^{\alpha+j,\alpha+j}}(\Omega)$ for $0 \le j \le m$. Then we have*

$$ \|u - u_N\|_{\omega^{(-\alpha,-\alpha)}} \le cN^{-2\alpha-m}\|f^{(m)}\|_{\omega^{(\alpha+m,\alpha+m)}}, \tag{3.1.26} $$

and

$$ \|D_{|x|}^{2\alpha}(u - u_N)\|_{\omega^{(\alpha,\alpha)}} \le cN^{-m}\|f^{(m)}\|_{\omega^{(\alpha+m,\alpha+m)}}. \tag{3.1.27} $$

Proof Let $\pi_N^{(-\alpha,-\alpha)}$ be as defined in (2.5.2). It is clear from (3.1.25) that $u_N = \pi_N^{(-\alpha,-\alpha)} u$. Hence, by (2.5.43), we have

$$\begin{aligned}\|u - u_N\|_{\omega^{(-\alpha,-\alpha)}} &= \|u - \pi_N^{(-\alpha,-\alpha)} u\|_{\omega^{(-\alpha,-\alpha)}}\\ &\le cN^{-2\alpha-m}\|D_{|x|}^{2\alpha+m} u\|_{\omega^{(\alpha+m,\alpha+m)}}.\end{aligned}$$

On the other hand, we derive from (2.5.42) with $k = i = 0$ that

$$\begin{aligned}\|D_{|x|}^{2\alpha} u - u_N\|_{\omega^{(\alpha,\alpha)}} &= \|D_{|x|}^{2\alpha}(u - \pi_N^{(-\alpha,-\alpha)} u)\|_{\omega^{(\alpha,\alpha)}}\\ &\le cN^{-m}\|D_{|x|}^{2\alpha+m} u\|_{\omega^{(\alpha+m,\alpha+m)}}.\end{aligned}$$

Since $(-1)^k D_{|x|}^{2\alpha+m} u = f^{(m)}$, we obtain the desired results from the above two inequalities. □

A More General Case

For the previous examples, we have developed optimal spectral methods using GJPFs with the merits: (i) the numerical solution can be determined directly without solving any algebraic equation; and (ii) the error converges faster than any algebraic rate as long as the source term f is smooth, despite the fact that the solution is weakly singular at the end-point(s). However, it is not possible to construct such optimal spectral methods for more general FODEs. Nevertheless, using proper GJPFs still allows us to (i) deal with fractional derivatives efficiently, and (ii) resolve the leading singular term in the solution.

Consider for example

$$\begin{cases} -D_{|x|}^{2\alpha} u(x) + \rho u(x) = f(x), \quad x \in \Omega = (-1,1), \\ u(\pm 1) = 0, \end{cases} \tag{3.1.28}$$

where $\rho > 0$ and $2\alpha \in (1,2)$.

The GJPF spectral-Galerkin approximation to (3.1.28) is to find $u_N \in \mathbb{F}_N^{(-\alpha,-\alpha)}(\Omega)$ such that

$$a(u_N, v_N) := \rho(u_N, v_N) - (D_{|x|}^{2\alpha} u_N, v_N) = (f, v_N) \tag{3.1.29}$$

for all $v_N \in \mathbb{F}_N^{(-\alpha,-\alpha)}(\Omega)$. We write

$$u_N(x) = \sum_{n=0}^{N} \tilde{u}_n \mathcal{J}_n^{-\alpha,-\alpha}(x). \tag{3.1.30}$$

Set

$$\begin{aligned} &U = (\tilde{u}_0, \tilde{u}_1, \ldots, \tilde{u}_N)^T, \quad M = (M_{jk}), \quad S = (S_{jk}), \\ &M_{jk} = (\mathcal{J}_k^{\alpha,-\alpha}, \mathcal{J}_j^{\alpha,-\alpha}), \quad S_{jk} = -(D_{|x|}^{2\alpha} \mathcal{J}_k^{\alpha,-\alpha}, \mathcal{J}_j^{\alpha,-\alpha}), \\ &F = (\tilde{f}_0, \tilde{f}_1, \ldots, \tilde{f}_N)^T, \quad \tilde{f}_j = (f, \mathcal{J}_j^{\alpha,-\alpha}). \end{aligned} \tag{3.1.31}$$

Then (3.1.29) reduces to the following linear system:

$$(\rho M + S)\, U = F.$$

We know from (2.4.32) and (2.1.3) that S is a diagonal matrix with

$$S_{jk} = \frac{2^{2\alpha+1}\Gamma^2(k+\alpha+1)}{(k!)^2(2k+2\alpha+1)} \delta_{jk}. \tag{3.1.32}$$

The mass matrix M is full but its entries can be evaluated exactly. More precisely, $M_{jk} = 0$ for odd $j+k$, while for even $j+k$, we have

$$\begin{aligned} M_{jk} &= \int_{-1}^{1} (1-x^2)^\alpha P_j^{(\alpha,\alpha)}(x)(1-x^2)^\alpha P_k^{(\alpha,\alpha)}(x)\, dx \\ &= \frac{\sqrt{\pi}\Gamma(2\alpha+1)}{\Gamma(2\alpha+\frac{3}{2})} \frac{(-1)^{\frac{j-k}{2}}(j+k)!}{2^{j+k} j!\, k!} \frac{(-j-\alpha)_\nu(-k-\alpha)_\nu}{(2\alpha+\frac{3}{2})_\nu \nu!}, \end{aligned} \tag{3.1.33}$$

where

$$\nu := \frac{j+k}{2}, \quad (a)_\nu = \frac{\Gamma(a+\nu)}{\Gamma(a)}.$$

Lemma 3.4 *For any* $u \in \{u \in L^2_{\omega^{(-\alpha,-\alpha)}}(\Omega) : D_{|x|}^{2\alpha} u \in L^2(\Omega)\}$, *we have*

$$\|u\|^2_{\omega^{(-\alpha,-\alpha)}} \le -(D_{|x|}^{2\alpha} u, u). \tag{3.1.34}$$

Proof We expand $u(x) = \sum_{i=0}^\infty \tilde{u}_i \mathcal{J}_i^{\alpha,-\alpha}(x)$ and have

$$\|u\|^2_{\omega^{(-\alpha,-\alpha)}} = \sum_{i=0}^\infty \tilde{u}_i^2 \gamma_i^{(\alpha,\alpha)}.$$

By Theorem 2.35 and (2.1.3), we have

$$\begin{aligned} -(D_{|x|}^{2\alpha} u, u) &= \Big(-\sum_{i=0}^\infty \tilde{u}_i D_{|x|}^{2\alpha} \mathcal{J}_i^{\alpha,-\alpha}, \sum_{j=0}^\infty \tilde{u}_j \mathcal{J}_j^{\alpha,-\alpha}\Big) \\ &= \Big(\sum_{i=0}^\infty \tilde{u}_i \frac{\Gamma(i+2\alpha+1)}{i!} P_i^{(\alpha,\alpha)}, \sum_{j=0}^\infty \tilde{u}_j P_j^{(\alpha,\alpha)}\Big)_{\omega^{(\alpha,\alpha)}} \\ &= \sum_{i=0}^\infty \frac{\Gamma(i+2\alpha+1)}{i!} \tilde{u}_i^2 \gamma_i^{(\alpha,\alpha)}. \end{aligned}$$

We can easily obtain the desired result (3.1.34) by comparing the above results. □

In view of (3.1.34), we define the energy norm associated with (3.1.28) by

$$\|u\|_{B^\alpha}^2 = \rho(u,u) - (D_{|x|}^{2\alpha}u, u). \tag{3.1.35}$$

Theorem 3.5 *Assume* $2\alpha \in (1,2)$. *Let* $u \in \mathcal{B}_{2\alpha}^m(\Omega)$, *and let* u *and* u_N *be the solution of* (3.1.28) *and* (3.1.29), *respectively. Then we have*

$$\|u - u_N\|_{B^\alpha} \le cN^{1-m}\|D_{|x|}^{2\alpha-1+m}u\|_{\omega^{(\alpha-1+m,\alpha-1+m)}}. \tag{3.1.36}$$

Proof Let us denote $\tilde{u}_N := \pi_N^{(-\alpha,-\alpha)}u$ and $e_N := \tilde{u}_N - u_N$. We derive from (3.1.28) and (3.1.29) that for all $v_N \in \mathbb{F}_N^{(-\alpha,-\alpha)}(\Omega)$,

$$\begin{aligned} a(e_N, v_N) &= \rho(e_N, v_N) - (D_{|x|}^{2\alpha}e_N, v_N) \\ &= \rho(\tilde{u}_N - u, v_N) - (D_{|x|}^{2\alpha}(\tilde{u}_N - u), v_N). \end{aligned}$$

Taking $v_N = e_N$ in the above, we derive from (3.1.35) and the Cauchy–Schwarz inequality that

$$\begin{aligned} \|e_N\|_{B^\alpha}^2 &:= \rho(e_N, e_N) - (D_{|x|}^{2\alpha}e_N, e_N) \\ &\le \rho\|\pi_N^{(-\alpha,-\alpha)}u - u\|_{\omega^{(\alpha,\alpha)}}\|e_N\|_{\omega^{(-\alpha,-\alpha)}} \\ &\quad + \|D_{|x|}^{2\alpha}(\pi_N^{(-\alpha,-\alpha)}u - u)\|_{\omega^{(\alpha,\alpha)}}\|e_N\|_{\omega^{(-\alpha,-\alpha)}}. \end{aligned}$$

From the above and Lemma 3.4, we derive

$$\|e_N\|_{B^\alpha} \le \rho\|\pi_N^{(-\alpha,-\alpha)}u - u\|_{\omega^{(-\alpha,-\alpha)}} + \|D_{|x|}^{2\alpha}(\pi_N^{(-\alpha,-\alpha)}u - u)\|_{\omega^{(\alpha,\alpha)}}.$$

Then by Theorem 2.42, we obtain

$$\|\pi_N^{(-\alpha,-\alpha)}u - u\|_{\omega^{(-\alpha,-\alpha)}} \le cN^{1-(2\alpha+m)}\|D_{|x|}^{2\alpha-1+m}u\|_{\omega^{(\alpha-1+m,\alpha-1+m)}} \tag{3.1.37}$$

and

$$\|D_{|x|}^{2\alpha}(\pi_N^{(-\alpha,-\alpha)}u - u)\|_{\omega^{(\alpha,\alpha)}} \le cN^{1-m}\|D_{|x|}^{2\alpha-1+m}u\|_{\omega^{(\alpha-1+m,\alpha-1+m)}}, \tag{3.1.38}$$

which implies

$$\|e_N\|_{B^\alpha} \le cN^{1-m}\|D_{|x|}^{2\alpha-1+m}u\|_{\omega^{(\alpha-1+m,\alpha-1+m)}}. \tag{3.1.39}$$

On the other hand,

$$\begin{aligned} &-(D_{|x|}^{2\alpha}(u - \pi_N^{(-\alpha,-\alpha)}u), u - \pi_N^{(-\alpha,-\alpha)}u) \\ &\quad \le \|D_{|x|}^{2\alpha}(\pi_N^{(-\alpha,-\alpha)}u - u)\|_{\omega^{(\alpha,\alpha)}}\|u - \pi_N^{(-\alpha,-\alpha)}u\|_{\omega^{(-\alpha,-\alpha)}}, \end{aligned}$$

as

$$\|u - \pi_N^{(-\alpha,-\alpha)} u\| \le \|u - \pi_N^{(-\alpha,-\alpha)} u\|_{\omega^{(-\alpha,-\alpha)}}.$$

We find from the above that

$$\|u - \pi^{(-\alpha,-\alpha)} u\|_{B^\alpha} \le \rho \|u - \pi_N^{(-\alpha,-\alpha)} u\|_{\omega^{(-\alpha,-\alpha)}} + \|D_{|x|}^{2\alpha}(\pi_N^{(-\alpha,-\alpha)} u - u)\|_{\omega^{(\alpha,\alpha)}}.$$

Finally, since $u - u_N = u - \tilde{u}_N + e_N$, combining the above with (3.1.39), (3.1.37) and (3.1.38), we obtain the desired result. □

Remark 3.6 We emphasize that unlike previous examples, here we cannot easily bound the errors in terms of f. In particular, the smoothness of f does not imply that the right-hand side of (3.1.36) is bounded for any m. Hence, only an algebraic convergence rate can be achieved in this case.

Two-Sided FODEs with Different Coefficients

We consider the two-sided fractional differential equation

$$\begin{cases} \mathcal{D}_p^\beta u(x) := C_\beta (p \, {}_{-1}^{\mathrm{RL}} D_x^\beta u(x) + (1-p) \, {}_x^{\mathrm{RL}} D_1^\beta u(x)) = f(x), & \in \Omega, \\ u(\pm 1) = 0, \end{cases} \tag{3.1.40}$$

where $1 < \beta < 2$, $0 \le p \le 1$, C_β defined in (2.4.34) , and $f(x)$ is a given function.

Let μ, ν satisfy (2.4.36) and $0 < \mu, \nu < \beta$. A Petrov–Galerkin spectral method for (3.1.40) is to find $u_N \in \mathbb{F}_N^{(-\mu,-\nu)}$ such that

$$(\mathcal{D}_p^\beta u_N, v_N)_{\omega^{(\nu,\mu)}} = (f, v_N)_{\omega^{(\nu,\mu)}}, \quad \forall v_N \in \mathbb{P}_N. \tag{3.1.41}$$

The solution to this discrete problem can be found directly as follows. We expand $f(x)$ as

$$f(x) = \sum_{m=0}^{\infty} \hat{f}_m P_m^{(\nu,\mu)}(x), \tag{3.1.42}$$

and write

$$u_N(x) = \sum_{n=0}^{N} \hat{u}_n \mathcal{J}_n^{-\mu,-\nu}(x). \tag{3.1.43}$$

Plugging (3.1.43) and (3.1.42) into (3.1.41), and using (2.4.39) and the orthogonality of $\{P_m^{(\nu,\mu)}\}$ in $L^2_{\omega^{(\nu,\mu)}}(\Omega)$, we find

$$\hat{u}_n = \hat{f}_n \Big/ \Big(\frac{\Gamma(n+1+\beta)}{n!} \Big), \quad \forall\, 0 \le n \le N. \tag{3.1.44}$$

Moreover, we have the following error estimates.

Theorem 3.7 *Assuming $f^{(j)} \in L^2_{\omega^{(\nu+j,\mu+j)}}(\Omega)$ for $0 \le j \le m$, let u and u_N be the solution of* (3.1.40) *and* (3.1.41), *then we have*

$$\|u - u_N\|_{\omega^{(-\mu,-\nu)}} \le cN^{-\beta-m}\|f^{(m)}\|_{\omega^{(\nu+m,\mu+m)}}, \tag{3.1.45}$$

and

$$\|\mathcal{D}^{\beta}_{p}(u - u_N)\|_{\omega^{(\nu,\mu)}} \le cN^{-m}\|f^{(m)}\|_{\omega^{(\nu+m,\mu+m)}}. \tag{3.1.46}$$

Proof It is clear from (3.1.44) that $u_N = \pi_N^{(-\mu,-\nu)}u$. Hence, by (2.5.49), we have

$$\begin{aligned}\|u - u_N\|_{\omega^{(-\mu,-\nu)}} &= \|u - \pi_N^{(-\mu,-\nu)}u\|_{\omega^{(-\mu,-\nu)}}\\ &\le cN^{-\beta-m}\|\mathcal{D}^{\beta+m}_{p}u\|_{\omega^{(\nu+m,\mu+m)}}.\end{aligned}$$

On the other hand, we derive from (2.5.48) with $l = 0$ that

$$\begin{aligned}\|\mathcal{D}^{\beta}_{p}(u - u_N)\|_{\omega^{(\nu,\mu)}} &= \|\mathcal{D}^{\beta}_{p}(u - \pi_N^{(-\mu,-\nu)}u)\|_{\omega^{(\nu,\mu)}}\\ &\le cN^{-m}\|\mathcal{D}^{\beta+m}_{p}u\|_{\omega^{(\nu+m,\mu+m)}}.\end{aligned}$$

Since $\mathcal{D}^{\beta+m}_{p}u = f^{(m)}$, we obtain the desired results from the above. □

3.1.2 Petrov–Galerkin Spectral Method for Tempered FODEs

We now test the efficiency of the tempered Jacobi poly-fractonomials, developed in Section 2.6, in solving some tempered fractional differential equations (FDEs). To this end, we develop a Petrov–Galerkin spectral method for the following tempered FODE:

$$\begin{cases} {}^{\text{RL}}_{-1}\mathbb{D}^{2\mu,\tau}_{x}u(x) = f(x), & x \in \Omega,\\ u(-1) = 0, \end{cases} \tag{3.1.47}$$

where $\tau \ge 0$ and $2\mu \in (0, 1)$. We choose U_N to be the space of basis (trial) functions, defined in terms of the following eigenfunction of the first kind as

$$U_N = \text{span}\{F_n^{(1)}(x), \quad 1 \le n \le N\}, \tag{3.1.48}$$

whose elements i.e., any linear combinations of

$$F_n^{(1)}(x) = e^{-\tau x}(1 + x)^{\mu}\; P_{n-1}^{(-\mu,\mu)}(x),$$

satisfy the left-boundary condition. We define the space of test functions as

$$V_N = \text{span}\{F_n^{(2)}(x), \quad \forall 1 \le n \le N\}, \tag{3.1.49}$$

where we recall

$$F_n^{(2)}(x) = e^{\tau x}(1-x)^{\mu}\; P_{n-1}^{(\mu,-\mu)}(x).$$

It is straightforward to check that $\forall u \in U_N$ and $w \in V_N$, we have the following bilinear form

$$a(u,w) = \left({}^{\mathrm{RL}}_{-1}\mathbb{D}_x^{2\mu,\tau} u, w\right) = \left({}^{\mathrm{RL}}_{-1}\mathbb{D}_x^{\mu,\tau} u\,,\, {}^{\mathrm{RL}}_{x}\mathbb{D}_1^{\mu,\tau} w\right). \tag{3.1.50}$$

Hence, we consider the corresponding weak form of (3.1.47) that is to find $u_N \in U_N$ such that

$$a(u_N,w) := \left({}^{\mathrm{RL}}_{-1}\mathbb{D}_x^{\mu,\tau} u_N\,,\, {}^{\mathrm{RL}}_{x}\mathbb{D}_1^{\mu,\tau} w\right) = (f,w), \quad \forall w \in V_N, \tag{3.1.51}$$

which becomes equivalent to the strong form (3.1.47) when $u_N \approx u$ possesses sufficient smoothness. Next, we seek the approximate solution u_N of the form

$$u_N(x) = \sum_{n=1}^{N} b_n F_n^{(1)}(x), \tag{3.1.52}$$

where $\{b_n\}$ are the unknown expansion coefficients. By plugging (3.1.52) into (3.1.51), which enforces the residual

$$R_N(x) = {}^{\mathrm{RL}}_{-1}\mathbb{D}_x^{2\mu,\tau} u_N(x) - f(x)$$

to be L^2-orthogonal to all elements in V_N, we obtain

$$\sum_{n=1}^{N} b_n \left({}^{\mathrm{RL}}_{-1}\mathbb{D}_x^{\mu,\tau} F_n^{(1)}\,,\, {}^{\mathrm{RL}}_{x}\mathbb{D}_1^{\mu,\tau} F_k^{(2)}\right) = \left(f, F_k^{(2)}\right), \quad k = 1,2,\ldots,N,$$

which yields a *diagonal* stiffness matrix on the left-hand side, due to (2.6.28) and (2.6.29), whose diagonal entries are given by

$$\gamma_k = \left(\frac{\Gamma(k+\mu)}{\Gamma(k)}\right)^2 \frac{2}{2k-1}.$$

Consequently, we obtain the expansion coefficients as

$$b_k = \frac{1}{\gamma_k}\left(f\,,\, F_k^{(2)}\right). \tag{3.1.53}$$

In Figure 3.1, we demonstrate the convergence of the proposed Petrov–Galerkin scheme. We plot the log-log L^2-error versus N, the number of expansion terms in (3.1.52), corresponding to the limit fractional orders $2\mu = 1/10$ and $2\mu = 9/10$, considering two different exact solutions: (i) $u(x) = e^{-x}(1+x)^5$ (left), and (ii) $u(x) = e^{1+x+x^2}(1+x)^2$ (right). These plots show the spectral mode of convergence in the Petrov–Galerkin spectral method. In what follows, we further provide the stability and convergence analysis of the method.

Stability and Convergence Analysis

We carry out the discrete stability analysis given the pair of U_N and V_N. We represent u_N as

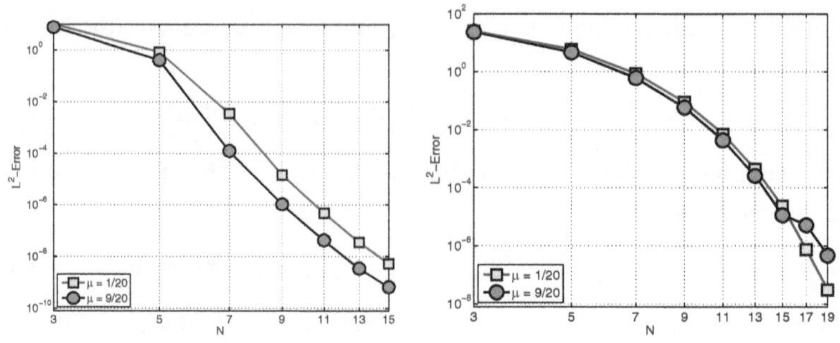

Figure 3.1 Petrov–Galerkin scheme for TFODE (3.1.47): L^2-error versus N, the number of expansion terms in (3.1.52), corresponding to the limit fractional orders $2\mu = 1/10$ and $2\mu = 9/10$. Here, the exact solutions are $u(x) = e^{-x}(1+x)^5$ (left) and $u(x) = e^{1+x+x^2}(1+x)^2$ (right). Source: Zayernouri et al. (2015a). Copyright ©2015 Society for Industrial and Applied Mathematics. Reprinted with permission. All rights reserved.

$$u_N(x) = \sum_{n=1}^{N} \hat{u}_n F_n^{(1)}(x), \tag{3.1.54}$$

and choose v_N to be the following linear combination of elements in V_N as

$$v_N(x) = \sum_{k=1}^{N} \hat{u}_k F_k^{(2)}(x). \tag{3.1.55}$$

Hence, we obtain

$$\begin{aligned} a(u_N, v_N) &= \left({}^{\text{RL}}_{-1}\mathbb{D}_x^{\mu,\tau} u_N \, , {}^{\text{RL}}_{x}\mathbb{D}_1^{\mu,\tau} v_N \right) \\ &= \sum_{n=1}^{N} \hat{u}_n \sum_{k=1}^{N} \hat{u}_k \int_{-1}^{1} {}^{\text{RL}}_{-1}\mathbb{D}_x^{\mu,\tau} F_n^{(1)}(x) \, {}^{\text{RL}}_{x}\mathbb{D}_1^{\mu} F_k^{(2)}(x) \, dx \\ &= \sum_{n=1}^{N} \hat{u}_n \frac{\Gamma(n+\mu)}{\Gamma(n)} \sum_{k=1}^{N} \hat{u}_k \frac{\Gamma(k+\mu)}{\Gamma(k)} \int_{-1}^{1} P_{n-1}(x) P_{k-1}(x) \, dx \\ &= \sum_{n=1}^{N} \left(\frac{\Gamma(n+\mu)}{\Gamma(n)} \right)^2 \frac{2}{2n-1} \hat{u}_n^2. \end{aligned} \tag{3.1.56}$$

Moreover, we have

$$\begin{aligned} \|v_N\|_{V_N}^2 = \| {}^{\text{RL}}_{x}\mathbb{D}_1^{\mu,\tau} v_N \|^2 &= \int_{-1}^{1} \Big(\sum_{k=1}^{N} \hat{u}_k \, {}^{\text{RL}}_{x}\mathbb{D}_1^{\mu,\tau} F_k^{(2)}(x) \Big)^2 dx \\ &= \int_{-1}^{1} \Big(\sum_{k=1}^{N} \hat{u}_k e^{\tau x} \frac{\Gamma(k+\mu)}{\Gamma(k)} P_{k-1}(x) \Big)^2 dx. \end{aligned} \tag{3.1.57}$$

From (3.1.56)–(3.1.57), we observe

$$\|v_N\|_{V_N}^2 \geq C_1^2 \int_{-1}^{1} \Big(\sum_{k=1}^{N} \hat{u}_k \frac{\Gamma(k+\mu)}{\Gamma(k)} P_{k-1}(x)\Big)^2 dx$$
$$= C_1^2\, a(u_N, v_N),$$

and

$$\|v_N\|_{V_N}^2 \leq C_2^2 \int_{-1}^{1} \Big(\sum_{k=1}^{N} \hat{u}_k \frac{\Gamma(k+\mu)}{\Gamma(k)} P_{k-1}(x)\Big)^2 dx$$
$$= C_2^2\, a(u_N, v_N),$$

where

$$C_1 = \min_{-1\leq x\leq 1} |e^{\tau x}| = e^{-\tau}, \quad C_2 = \max_{-1\leq x\leq 1} e^{\tau x} = e^{\tau}.$$

We can obtain similar results for $\|u_N\|_{U_N}^2$ as

$$C_1^2 a(u_N, v_N) \leq \|u_N\|_{U_N}^2 \leq C_2^2 a(u_N, v_N). \tag{3.1.58}$$

Hence, there exists a positive constant C such that

$$\|v_N\|_{V_N} \leq C \|u_N\|_{U_N},$$

by which and using the right inequality of (3.1.58), we obtain

$$\sup_{v_N \in V_N} \frac{a(u_N, v_N)}{\|v_N\|_{V_N}} \geq \frac{\frac{1}{C_2^2}\|u_N\|_{U_N}^2}{C\|u_N\|_{U_N}} = \frac{1}{C C_2^2}\|u_N\|_{U_N}, \quad \forall u_N \in U_N, \tag{3.1.59}$$

that is stability is guaranteed for $\beta = \frac{1}{C C_2^2}$. Therefore, Céa's lemma holds:

$$\|u - u_N\|_{U_N} \leq \Big(1 + \frac{M}{\beta}\Big)\|u - \tilde{u}_N\|_{U_N}, \quad \forall \tilde{u}_N \in U_N, \tag{3.1.60}$$

in which the continuity constant $M = 1$. Next, in order to obtain the corresponding (projection) error estimates, we expand the exact solution u, when $2\mu \in (0, 1)$, in terms of the following infinite series of tempered Jacobi poly-fractonomials

$$u(x) = \sum_{n=1}^{\infty} \hat{u}_n e^{-\tau x}(1+x)^{\mu} P_{n-1}^{(-\mu,\mu)}(x). \tag{3.1.61}$$

Here, we would like to bound $\|u - u_N\|_{U_N}$ in terms of higher-order derivatives. We first note that

$$
\begin{aligned}
{}^{\mathrm{RL}}_{-1}\mathbb{D}_x^{r+\mu,\tau} u(x) &= e^{-\tau x}[{}^{\mathrm{RL}}_{-1}D_x^{r+\mu}\, e^{-\tau x} u(x)] = e^{-\tau x}\Big[\frac{\mathrm{d}^r}{\mathrm{d}x^r}\, {}^{\mathrm{RL}}_{-1}D_x^{\mu}\, e^{-\tau x} u(x)\Big] \\
&= \sum_{n=1}^{\infty} \hat{u}_n \frac{\Gamma(n+\mu)}{\Gamma(n)} e^{-\tau x} \frac{\mathrm{d}^r}{\mathrm{d}x^r}[P_{n-1}(x)],
\end{aligned}
$$

where

$$
\frac{\mathrm{d}^r}{\mathrm{d}x^r}[P_{n-1}(x)] = \begin{cases} \frac{(n-1+r)!}{2r(n-1)!} P_{n-1-r}^{(r,r)}(x), & r \le n, \\ 0, & r > n. \end{cases}
$$

Therefore,

$$
{}^{\mathrm{RL}}_{-1}D_x^{r+\mu,\tau} u(x) = \sum_{n=r}^{\infty} \hat{u}_n \frac{\Gamma(n+\mu)}{\Gamma(n)} \frac{(n-1+r)!}{2r(n-1)!} e^{-\tau x} P_{n-1-r}^{(r,r)}(x).
$$

Hence,

$$
\begin{aligned}
&\|(1-x^2)^{r/2}\, {}^{\mathrm{RL}}_{-1}\mathbb{D}_x^{r+\mu,\tau} u\|^2 \\
&= \int_{-1}^{1} (1-x^2)^r \Big(\sum_{n=r}^{\infty} \hat{u}_n \frac{\Gamma(n+\mu)}{\Gamma(n)} \frac{(n-1+r)!}{2r(n-1)!} e^{-\tau x} P_{n-1-r}^{(r,r)}(x)\Big)^2 \mathrm{d}x \\
&\ge \sum_{n=r}^{\infty} \Big(\hat{u}_n \frac{\Gamma(n+\mu)}{\Gamma(n)} \frac{(n-1+r)!}{2r(n-1)!}\Big)^2 \int_{-1}^{1} e^{-2\tau x}(1-x^2)^r (P_{n-1-r}^{(r,r)}(x))^2 \mathrm{d}x \\
&\ge C \sum_{n=r}^{\infty} \Big(\hat{u}_n \frac{\Gamma(n+\mu)}{\Gamma(n)} \frac{(n-1+r)!}{2r(n-1)!}\Big)^2 \int_{-1}^{1} (1-x^2)^r (P_{n-1-r}^{(r,r)}(x))^2 \mathrm{d}x \\
&= C \sum_{n=r}^{\infty} \Big[\hat{u}_n \frac{\Gamma(n+\mu)}{\Gamma(n)}\Big]^2 \frac{2}{2n+1} \frac{(n-1+r)!}{(n-1-r)!},
\end{aligned}
$$

where $C = \min_{x\in(-1,1)} e^{-2\tau x} = e^{-2\tau}$. Noting that $\frac{(n-1+r)!}{(n-1-r)!}$ is minimized when $n = N+1$, we have

$$
\begin{aligned}
\|u - u_N\|_{U_N}^2 &\le \sum_{n=N+1}^{\infty} \Big[\hat{u}_n \frac{\Gamma(n+\mu)}{\Gamma(n)}\Big]^2 \\
&\le \sum_{n=N+1}^{\infty} \Big[\hat{u}_n \frac{\Gamma(n+\mu)}{\Gamma(n)}\Big]^2 \frac{(n-1+r)!}{(n-1-r)!} \frac{(N-r)!}{(N+r)!} \\
&\le \frac{1}{C} \frac{(N-r)!}{(N+r)!} \|\, {}^{\mathrm{RL}}_{-1}\mathbb{D}_x^{r+\mu,\tau} u\|_{L^2}^2 \\
&\le c N^{-2r} \|\, {}^{\mathrm{RL}}_{-1}\mathbb{D}_x^{r+\mu,\tau} u\|_{L^2}^2,
\end{aligned}
$$

where $r \ge 1$ and $2\mu \in (0,1)$.

3.1.3 Generalized Log Orthogonal Functions as Spectral Bases for Solving FODEs

In this section, we shall use generalized log orthogonal functions (GLOFs) as the basis functions to solve some types of fractional differential equations.

Application to a Fractional Initial-Value Problem (FIVP)

Given $g, q \in L^2(I)$ with $I := (0,1)$, we consider the following Caputo fractional differential equation of order $\nu \in (0,1)$:

$$ {}_0^{\mathrm{C}}D_t^{\nu} u(t) + q(t)u(t) = g(t), \quad t \in I; \quad u(0) = u_0. \tag{3.1.62}$$

We shall first homogenize the initial condition. Setting $u = v + u_0$ into the above equation, we find that the problem (3.1.62) is equivalent to

$$ {}_0^{\mathrm{C}}D_t^{\nu} v(t) + q(t)v(t) = g(t) - u_0 q(t), \quad v(0) = 0. \tag{3.1.63}$$

A main difficulty in obtaining an accurate approximate solution of (3.1.63) is that the solution of this problem is weakly singular at $t = 0$ even if q and g are smooth. To design an effective approach to deal with this difficulty, we need to understand the nature of this singularity.

Applying ${}_0^{\mathrm{RL}}I_t^{\nu}$ into both sides of (3.1.63) and using the fact that ${}_0^{\mathrm{RL}}I_t^{s}\,{}_0^{\mathrm{RL}}I_t^{r} = {}_0^{\mathrm{RL}}I_t^{s+r}$, we find

$$ v(t) + \frac{1}{\Gamma(\nu)} \int_0^t (t-\tau)^{\nu-1} q(\tau) v(\tau) d\tau = {}_0^{\mathrm{RL}}I_t^{\nu}\{g - u_0 q\}(t).$$

We then find from Cao et al. (2003, Theorem 2.1) that the solution near $t = 0$ behaves like

$$ v(t) = \sum_{i=0}^{\infty} \sum_{j=1}^{\infty} \tilde{v}_{ij}\, t^{i+j\nu}. \tag{3.1.64}$$

This is why the usual approximations based on global or piecewise polynomials cannot approximate $v(t)$ well. On the other hand, based on the analysis from the last section, the GLOFs are particularly suitable for this problem.

Let us define $X_N^0 = \{t^{\frac{\beta-\lambda}{2}} p : p \in \mathcal{P}_N^{\log t}, \beta > \lambda\}$. Then, the GLOF-Galerkin method for (3.1.63) is to find $v_N \in X_N^0$ such that

$$ ({}_0^{\mathrm{C}}D_t^{\nu} v_N, w) + (q v_N, w) = \big(\mathcal{I}_N^{\alpha,\beta,\lambda}\{g - u_0 q\}, w\big), \quad \forall w \in X_N^0. \tag{3.1.65}$$

Writing

$$ v_N = \sum_{n=0}^{N} \tilde{v}_n^{\alpha,\beta,\lambda} \mathcal{S}_n^{(\alpha,\beta,\lambda)}, \quad \bar{v} = (\tilde{v}_0^{\alpha,\beta,\lambda}, \tilde{v}_1^{\alpha,\beta,\lambda}, \ldots, \tilde{v}_N^{\alpha,\beta,\lambda})^T,$$

and setting

$$
\begin{aligned}
S_{kj} &= ({}_0^C D_t^{\nu} \mathcal{S}_j^{(\alpha,\beta,\lambda)}, \mathcal{S}_k^{(\alpha,\beta,\lambda)}), \quad S = (S_{kj}), \\
M_{kj} &= (q \mathcal{S}_j^{(\alpha,\beta,\lambda)}, \mathcal{S}_k^{(\alpha,\beta,\lambda)}), \quad M = (M_{kj}), \\
f_j &= (\mathcal{I}_N^{\alpha,\beta,\lambda}\{g - u_0 q\}, \mathcal{S}_j^{(\alpha,\beta,\lambda)}), \quad \bar{f} = (f_0, f_1, \dots, f_N)^T,
\end{aligned}
\tag{3.1.66}
$$

we can derive from (3.1.65) the following linear system:

$$
(S + M)\,\bar{v} = \bar{f}. \tag{3.1.67}
$$

The entries of M and $\bar{f}$ can be computed accurately by using the quadrature formula, but the computation of the stiffness matrix S needs special care.

Indeed, for any $v, w \in X_N^0$,

$$
\begin{aligned}
({}_0^C D_t^{\nu} v, w) &= \int_0^1 \frac{1}{\Gamma(1-\nu)} \int_0^t \frac{v'(s)}{(t-s)^{\nu}}\, \mathrm{d}s\, w(t)\mathrm{d}t \\
&\overset{s=t\tau}{=} \frac{1}{\Gamma(1-\nu)} \int_0^1 \int_0^1 \frac{v'(t\tau)}{(1-\tau)^{\nu}}\, \mathrm{d}\tau\, w(t) t^{1-\nu} \mathrm{d}t.
\end{aligned}
$$

Note that the integrand in the above is weakly singular as $t \to 0$ and $\tau \to 1$. In order to compute accurately the inner integral, we split it into two terms:

$$
\begin{aligned}
\int_0^1 \frac{v'(t\tau)}{(1-\tau)^{\nu}}\, \mathrm{d}\tau &= \int_0^{\frac{1}{2}} v'(t\tau)\,(1-\tau)^{-\nu}\, \mathrm{d}\tau + \int_{\frac{1}{2}}^1 v'(t\tau)\,(1-\tau)^{-\nu}\, \mathrm{d}\tau \\
&= \frac{1}{2} \int_0^1 v'(\frac{t\tau}{2})\,(1-\frac{\tau}{2})^{-\nu}\, \mathrm{d}\tau + \frac{1}{4^{1-\nu}} \int_{-1}^1 v'(\frac{t(\xi+3)}{4})\,(1-\xi)^{-\nu}\, \mathrm{d}\xi.
\end{aligned}
$$

Hence,

$$
\begin{aligned}
({}_0^C D_t^{\nu} v, w) &= \frac{1}{2\Gamma(1-\nu)} \int_0^1 \int_0^1 v'(\frac{t\tau}{2})\,(1-\frac{\tau}{2})^{-\nu}\, \mathrm{d}\tau\, w(t) t^{1-\nu} \mathrm{d}t \\
&\quad + \frac{1}{4^{1-\nu}\Gamma(1-\nu)} \int_0^1 \int_{-1}^1 v'(\frac{t(\xi+3)}{4})(1-\xi)^{-\nu}\, \mathrm{d}\xi\, w(t) t^{1-\nu} \mathrm{d}t.
\end{aligned}
\tag{3.1.68}
$$

The first term has weak singularity as $t, \tau \to 0$ while the second term has weak singularities as $t \to 0$ and $\xi \to 1$. Therefore, the first term can be computed by using the tensor product of Gauss-GLOF quadratures (in t and τ) which is effective with weak singularities as $t, \tau \to 0$, and the second term can be computed by using the tensor product of Gauss-GLOF quadrature in t and of Gauss–Jacobi quadrature in ξ with weight function $(1-\xi)^{-\nu}$. More precisely,

$$\left({}_0^C D_t^\nu v_N, w \right) \approx \frac{1}{2\Gamma(1-\nu)} \sum_{i=0}^{N_I} \sum_{j=0}^{N_I} v_N'\left(\frac{t_i\, t_j}{2}\right)\left(1-\frac{t_j}{2}\right)^{-\nu} w(t_i) t_i^{1-\nu} \chi_i \chi_j$$
$$+ \frac{1}{4^{1-\nu}\Gamma(1-\nu)} \sum_{i=0}^{N_I} \sum_{j=0}^{N_I} v_N'\left(\frac{t_i(\xi_j+3)}{4}\right) w(t_i) t_i^{1-\nu} \chi_i \eta_j,$$

where $N_I \geq N$ is a suitable number, $\{t_i, \chi_i\}_{i=0}^{N_I}$ are the Gauss-LOFs nodes with weight function $\chi^{0,0} \equiv 1$, and $\{\xi_i, \eta_i\}_{i=0}^{N_I}$ are the Gauss–Jacobi nodes with weight function $(1-\tau)^{-\nu}$.

We present below some numerical results. We consider

$$ {}_0^C D_t^\nu u(t) + \mathcal{K} u(t) = 0, \quad u(0) = 1, \tag{3.1.69}$$

whose solution is Diethelm (2010, Theorem 4.3) $u(t) = E_\nu(-\mathcal{K} t^\nu)$ where $E_\gamma(z)$ is the Mittag-Leffler function

$$E_\gamma(z) = \sum_{j=0}^{\infty} \frac{z^j}{\Gamma(\gamma j + 1)}. \tag{3.1.70}$$

We fix the parameters $\alpha = 0, \beta = 5, \lambda = 0$, and plot in the left of Figure 3.2 the convergence rates for various values of ν with $\mathcal{K} = 1$. It is clear that the solution is not smooth in the classical Sobolev space, but it is smooth in the space defined through the pseudo-derivative, so we still obtain an exponential convergence rate.

Next, we consider

$$ {}_0^C D_t^\nu u(t) + (1+\sin t) u(t) = \cos t, \quad u(0) = 1,$$

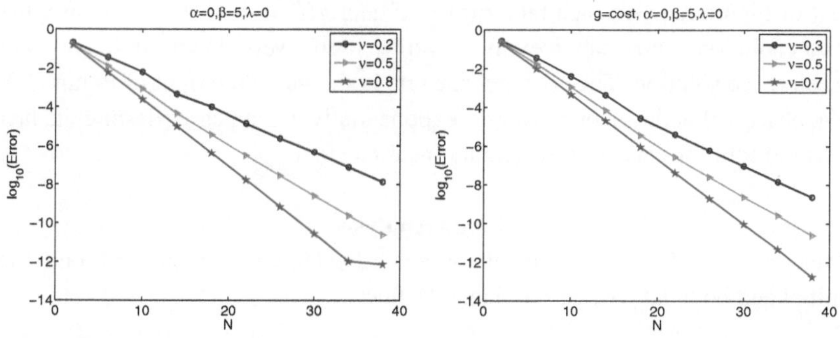

Figure 3.2 Left: (3.1.69) with $\mathcal{K} = 1$. Right: $g = \cos t$. Source: Chen and Shen (2022) by permission of Oxford University Press.

for which the exact solution is unknown. Although the coefficients are smooth, the exact solution is expected to be weakly singular near zero but smooth in the space defined through the pseudo-derivative. We fix the parameters $\alpha = 0$, $\beta = 5, \lambda = 0$, and plot in the right of Figure 3.2 the convergence rates for various values of ν. We obtain again exponential convergence rates.

A Boundary-Value Problem (BVP)

We consider

$$\begin{cases} -{}_0^C D_t^\mu u(t) + q(t)u(t) = g(t), & t \in I = (0,1), \\ u(0) = 0, \quad u(1) = 0, \end{cases} \tag{3.1.71}$$

where $\mu \in (1,2)$ and g, q are given functions.

Similar to the initial-value problem (3.1.62), the solution of the above problem is usually weakly singular even with smooth g and q. However, it can be approximated accurately by GLOFs since the solution is smooth in the space defined through the pseudo-derivative (2.7.33).

Let us denote

$$X_N^{0,0} := \text{span}\Big\{\phi_n = \frac{n}{n+\alpha}\mathcal{S}_n^{(\alpha,\beta,\lambda)} - \mathcal{S}_{n-1}^{(\alpha,\beta,\lambda)} \,:\, n = 1,2,\ldots,N,\ \beta > \lambda\Big\}.$$

Note that we have $\phi_n(0) = \phi_n(1) = 0$ for $n \ge 1$. The GLOF Galerkin method is: find $u_N \in X_N^{0,0}$ such that

$$-({}_0^C D_t^\mu u_N, w) + (qu_N, w) = (\mathcal{I}_N^{\alpha,\beta,\lambda} g, w), \quad \forall w \in X_N^{0,0}. \tag{3.1.72}$$

The stiffness and mass matrices of the above problem can be formulated as in the case of the initial-value problem considered above.

We now present some numerical results. We first take $q(t) = e^t$ and the exact solution to be $u(t) = t^{3/2}(1-t)$. The convergence rate is shown on the left of Figure 3.3. We then take $q(t) = e^t$ and $g(t) = t\sin t$. In this case, the exact solution is not known explicitly so we used a very fine mesh to compute a reference solution. The convergence rate is shown on the right of Figure 3.3. We observe that the error converges exponentially in both cases despite the fact that the solutions are weakly singular near $t = 0$.

Error Analysis

We carry out below error analysis for the GLOF Galerkin schemes for both the initial and boundary-value problems. Define

- for $0 \le s < \frac{1}{2}$,

$$H_0^s(I) := \{f \in L^2(I) :\ {}_0^C D_t^s f,\ {}_t^C D_1^s f \in L^2(I)\};$$

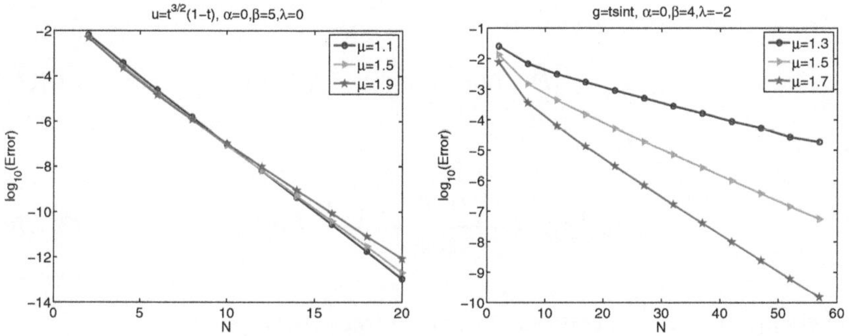

Figure 3.3 Left: $q(t) = e^t$ and $u = t^{3/2}(1-t)$. Right: $q(t) = e^t$ and $g = t\sin t$. Source: Chen and Shen (2022) by permission of Oxford University Press.

- for $\frac{1}{2} < s \le 1$,

$$H_0^s(I) := \{f \in L^2(I) : \, {}_0^{\mathrm{C}}D_t^s f, \, {}_t^{\mathrm{C}}D_1^s f \in L^2(I), \, f(0) = f(1) = 0\},$$

equipped with norm and semi-norm

$$\|f\|_s = \sqrt{\|f\|^2 + |f|_s^2} \quad \text{with} \quad |f|_s = \sqrt{\|{}_0^{\mathrm{C}}D_t^s f\|^2 + \|{}_t^{\mathrm{C}}D_1^s f\|^2}.$$

It can be shown that the above definitions coincide with the usual definitions by space interpolation.

To avoid repetition, we use the following weak formulation for both problems (3.1.63) and (3.1.71): find $P \in H_0^s(I)$ such that

$$a(P,w) := (-1)^{\sigma(s)}({}_0^{\mathrm{C}}D_t^s P, {}_t^{\mathrm{C}}D_1^s w) + (qP,w) = (Q,w), \quad \forall w \in H_0^s(I), \tag{3.1.73}$$

where

- for (3.1.63): $s = \nu/2, \quad \sigma(s) = 0, \quad P(t) = v(t), \quad Q(t) = \{g - u_0 q\}(t)$,
- for (3.1.71): $s = \mu/2, \quad \sigma(s) = 1, \quad P(t) = u(t), \quad Q(t) = g(t)$.

The error analysis follows similar procedures used in Ervin and Roop (2006) and Li and Xu (2009, 2010). We first recall some useful results.

Lemma 3.8 *Let $s, r \in [0,1]/\{\frac{1}{2}\}$ and $s \le r$. For any $f, h \in H_0^r(I)$, there exists*

- *Ervin and Roop (2006, Corollary 2.15)*

$$\|f\| \le c_1 |f|_s \le c_2 |f|_r, \tag{3.1.74}$$

- *Li and Xu (2010, Lemma 2.8)*

$$({}_0^C D_t^{2s} f, h) = ({}_0^C D_t^s f, {}_t^C D_1^s h), \tag{3.1.75}$$

- *Li and Xu (2010, Lemma 2.6)*

$$c_1 |f|_s^2 \le \frac{({}_0^C D_t^s f, {}_t^C D_1^s f)}{\cos(s\pi)} \le c_2 |f|_s^2, \tag{3.1.76}$$

where c_1, c_2 are two positive constants independent of function f.

Thanks to relation (3.1.75) and the identity

$$({}_0^C D_t^{\mu} u, w) = ({}_0^C D_t^s u, {}_t^C D_1^s w), \quad \mu = 2s,\ s \in (1/2, 1),$$

we can rewrite (3.1.65) and (3.1.72) as: find $P_N \in X_N$ such that

$$a(P_N, w) = (\mathcal{I}_N^{\alpha,\beta,\lambda} Q, w), \quad \forall w \in X_N, \tag{3.1.77}$$

where $X_N = X_N^0$ for (3.1.65), and $X_N = X_N^{0,0}$ for (3.1.72).

Lemma 3.9 *If $q(t)$ is bounded and $\min_{t\in[0,1]} q(t) \ge 0$, then there exist $c(s), c(q) > 0$ such that for any P, $\tilde{P} \in H_0^s(I)$, we have*

$$c(s)\|P\|_s^2 \le a(P, P), \quad a(P, \tilde{P}) \le c(q)\|P\|_s \|\tilde{P}\|_s. \tag{3.1.78}$$

Proof Due to the fact $(-1)^{\sigma(s)} \cos(s\pi) > 0$ and the relation (3.1.76), it is easy to derive that

$$c_1(-1)^{\sigma(s)} \cos(s\pi)|P|_s^2 \le (-1)^{\sigma(s)}({}_0^C D_t^s P, {}_t^C D_1^s P) \le a(P, P),$$

where c_1 is a constant independent of the function P. Then, we can then derive the first inequality in (3.1.78) from the generalized Poincaré inequality (3.1.74). The second inequality in (3.1.78) is a direct consequence of the Cauchy–Schwarz inequality. □

Thanks to (3.1.78), the existence-uniqueness of the weak formulation (3.1.73) and the schemes (3.1.77) follows immediately from the Lax–Milgram lemma.

As for the error estimate, we have the following result.

Theorem 3.10 *Let $-1 < \alpha \le 0$, $\lambda \le 0$ and $\beta > 1$. Let P and P_N be respectively the solution of (3.1.73) and (3.1.77) with $\min_{t\in[0,1]} q(t) \ge 0$. Then, we have*

$$\|P - P_N\|_s \le c\sqrt{(\beta-1)^{-\widetilde{m}}\frac{(N-\widetilde{m}+1)!}{(N+1)!}}\,\|\widehat{\partial}^{\widetilde{m}+1}_{\frac{\beta-\lambda-2}{2},t}P\|_{\chi^{\alpha+\widetilde{m},\lambda}}$$
$$+c\sqrt{\frac{(N+1-\widetilde{m})!}{(\beta+1)^{\widetilde{m}-\alpha}N!}}\Big\{c_1^{\beta}\|\widehat{\partial}^{\widetilde{m}}_{\frac{\beta-\lambda}{2},t}Q\|_{\chi^{\alpha+m-1,\lambda}} \qquad (3.1.79)$$
$$+c_2^{\beta}\sqrt{\log N}\,\|\widehat{\partial}^{\widetilde{m}}_{\frac{\beta-\lambda}{2},t}Q\|_{\chi^{\alpha+m,\lambda}}\Big\},$$

where

$$c_1^{\beta} = (\beta+1)^{-\frac{1}{2}}, \quad c_2^{\beta} = 2\sqrt{\max\{1,\beta+1\}}, \quad \widetilde{m} = \min\{m, N+1\}.$$

Proof For any $w_N \in X_N$, we derive from (3.1.73) and (3.1.77) that

$$a(P - P_N, w_N) = (Q - \mathcal{I}_N^{\alpha,\beta,\lambda}Q, w_N). \qquad (3.1.80)$$

Let $e_N = P_N - w_N$; we have

$$c(s)\|e_N\|_s^2 \le a(e_N, e_N) = a(P - w_N, e_N) + a(P_N - P, e_N). \qquad (3.1.81)$$

Taking $w_N = e_N$ in (3.1.80), we find

$$a(P_N - P, e_N) = (\mathcal{I}_N^{\alpha,\beta,\lambda}Q - Q, e_N) \le \|\mathcal{I}_N^{\alpha,\beta,\lambda}Q - Q\|\,\|e_N\|. \qquad (3.1.82)$$

We then derive from (3.1.81) and (3.1.78) that

$$c(s)\|w_N - P_N\|_s \le c(\|P - w_N\|_s + \|\mathcal{I}_N^{\alpha,\beta,\lambda}Q - Q\|),$$

which, along with (3.1.82), implies that

$$\|P - P_N\|_s \le \|P - w_N\|_s + \|w_N - P_N\|_s \le c\inf_{w_N\in X_N}\|P - w_N\|_s$$
$$+ c\|\mathcal{I}_N^{\alpha,\beta,\lambda}Q - Q\|. \qquad (3.1.83)$$

Next, we set

$$w_N(t) = -\int_t^1 \pi_N^{\alpha,\beta-2,\lambda}\{\partial_\tau P\}(\tau)d\tau.$$

Obviously, $w_N(1) = 0$. Writing

$$\pi_N^{\alpha,\beta-2,\lambda}\{\partial_\tau P\}(\tau) = \tau^{\frac{\beta-\lambda}{2}-1}\sum_{k=0}^{N}\tilde{c}_k(\log\tau)^k,$$

and integrating-by-parts, we find

$$\int_t^1 \pi_N^{\alpha,\beta-2,\lambda}\{\partial_\tau P\}(\tau)\mathrm{d}\tau = \int_t^1 \tau^{\frac{\beta-\lambda}{2}-1}\sum_{k=0}^{N}\tilde{c}_k(\log\tau)^k\mathrm{d}\tau$$
$$= t^{\frac{\beta-\lambda}{2}}\sum_{k=0}^{N}\tilde{d}_k(\log t)^k,$$

which implies $w_N(0) = 0$. Hence, $w_N \in X_N$. On the other hand, we have

$$\partial_t P - \partial_t w_N = (I - \pi_N^{\alpha,\beta-2,\lambda})\partial_t P.$$

We can then derive from the above relations and Poincaré inequality (3.1.74) that

$$\|P - w_N\|_s \le c\|\partial_t P - \partial_t w_N\| \le c\|\partial_t P - \partial_t w_N\|_{\chi^{\alpha,\lambda}}$$
$$= c\|\partial_t P - \pi_N^{\alpha,\beta-2,\lambda}\partial_t P\|_{\chi^{\alpha,\lambda}}.$$

Finally, combining the above and (3.1.83) and applying Theorems 2.60 and 2.61, we obtain the desired result. □

Remark 3.11 We can also show that for $P = t^r$ and $Q = t^q$, the estimate (3.1.79) leads to exponential convergence rate if r and q are within certain range.

3.1.4 Petrov–Galerkin Methods for Distributed-Order FODEs

Distributed-order fractional operators offer a rigorous tool for mathematical modeling of multi-physics phenomena. In this case, the differential order is distributed over a range of values rather than being just a fixed fraction as it is in standard/fractional ODEs/PDEs. There is a rapidly growing interest in the use of fractional derivatives in the construction of mathematical models, which contain distributed order, e.g., in the field of uncertainty quantification as the inherent uncertainty of experimental data can be directly incorporated into the differential operators. In Section 1.3.10, we introduced the *distributed Sobolev spaces* and their associated norms. We also show their equivalence to the defined left-side and right-side norms as well. By employing Riemann–Liouville derivatives, we define the corresponding distributed-order differential equation. Here, we aim to obtain its variational form to develop a Petrov–Galerkin spectral method following the theory of fractional Sturm–Liouville eigenproblems (FSLP) in Chapter 2 and employ the corresponding eigenfunctions (Jacobi poly-fractonomials) of the first kind as the bases and those of the second kind as test functions. We will develop a spectrally accurate

Gauss–Legendre quadrature rule in the construction of a linear system, where we investigate the stability and error analysis of the scheme.

In the following theorem, we prove the equivalence (shown by the notation $\sim$) of the aforementioned norms. We remark that this notation has exceptionally a different meaning in Chapter 1, where $\sim$ meant "*distributed as*" in the context of probability theory.

Theorem 3.12 *Let* $\phi \in L^1([\alpha_{min}, \alpha_{max}])$ *be non-negative. Given the definitions in* (1.3.72) *and* (1.3.73), *the norms* $\|\cdot\|_{\phi,I}$, $\|\cdot\|_{l,\phi,I}$, *and* $\|\cdot\|_{r,\phi,I}$ *are equivalent.*

Proof Let ϕ be bounded in $(\alpha_{min}, \alpha_{max})$. Then,

$$C_1\phi_{min}\, A \;\leq\; \|\cdot\|^2_{l,\phi,I} \;\leq\; C_2\phi_{max}\, A,$$
$$C_3\phi_{min}\, B \;\leq\; \|\cdot\|^2_{r,\phi,I} \;\leq C_4\phi_{max}\, B,$$

where

$$A = \|\cdot\|^2_{L^2(I)} + \int_{\alpha_{min}}^{\alpha_{max}} \|\, {}^{\mathrm{RL}}_{x_L}D^{\alpha}_{x}(\cdot)\,\|^2_{L^2(I)}\; d\alpha,$$
$$B = \|\cdot\|^2_{L^2(I)} + \int_{\alpha_{min}}^{\alpha_{max}} \|\, {}^{\mathrm{RL}}_{x}D^{\alpha}_{x_R}(\cdot)\,\|^2_{L^2(I)}\; d\alpha,$$

and C_1, C_2, C_3, and C_4 are positive constants. From Li and Xu (2009), we know that for any fixed $\alpha = s$, $\|\cdot\|_{l,s,I} \sim \|\cdot\|_{r,s,I}$, that is,

$$\|\, {}^{\mathrm{RL}}_{x_L}D^{s}_{x}(\cdot)\,\|^2_{L^2(I)} \sim \|\, {}^{\mathrm{RL}}_{x}D^{s}_{x_R}(\cdot)\,\|^2_{L^2(I)},$$

hence let $\vec{s} = \{s_1, s_2, \ldots, s_Q\}$, and similarly $\forall s = s_q$ the aforementioned equivalence holds. Therefore, any linear combination of

$$\sum_{q=1}^{Q} w_q \left\|\, {}^{\mathrm{RL}}_{x_L}D^{s_q}_{x}(\cdot)\right\|^2_{L^2(I)} \sim \sum_{q=1}^{Q} w_q \left\|\, {}^{\mathrm{RL}}_{x}D^{s_q}_{x_R}(\cdot)\right\|^2_{L^2(I)}.$$

Taking $Q \to \infty$ and assuming $\{w_q, s_q\}$ to be Riemann integral weights and points in $[\alpha_{min}, \alpha_{max}]$,

$$\int_{\alpha_{min}}^{\alpha_{max}} \|\, {}^{\mathrm{RL}}_{x_L}D^{s}_{x}(\cdot)\,\|^2_{L^2(I)}\, d\alpha \sim \int_{\alpha_{min}}^{\alpha_{max}} \|\, {}^{\mathrm{RL}}_{x}D^{s}_{x_R}(\cdot)\,\|^2_{L^2(I)}\, d\alpha.$$

By adding $\|\cdot\|^2_{L^2(I)}$ to both sides of the above equivalence, we obtain $A \sim B$; and by (3.1.84), $\|\cdot\|_{l,\phi,I} \sim \|\cdot\|_{r,\phi,I}$.

In addition, from Li and Xu (2009), we know that $\forall\alpha = s$ fixed, $\|\cdot\|^2_{s,I} \sim \|\cdot\|^2_{l,s,I}$. Let $\vec{s} = \{s_1, s_2, \ldots, s_Q\}$ thus $\forall s = s_q \in \vec{s}$, $\|\cdot\|^2_{s_q,I} \sim \|\cdot\|^2_{l,s_q,I}$. Therefore, for any linear combination of $\sum_{q=1}^{Q} w_q \parallel \cdot \parallel_{s_q,I}^2 \sim \sum_{q=1}^{Q} w_q \parallel \cdot \parallel_{l,s_q,I}^2$. Taking $Q \to \infty$, we obtain

$$\int_{\alpha_{min}}^{\alpha_{max}} \parallel \cdot \parallel_{\alpha,I}^2 \, d\alpha \sim \int_{\alpha_{min}}^{\alpha_{max}} \parallel \cdot \parallel_{l,\alpha,I}^2 \, d\alpha,$$

where the right-hand side of the equivalence is

$$\begin{aligned}\int_{\alpha_{min}}^{\alpha_{max}} \left(\parallel \cdot \parallel_{L^2(I)}^2 + \parallel {}^{\text{RL}}_{x_L}D_x^{\alpha}(\cdot) \parallel_{L^2(I)}^2 \right) d\alpha &= (\alpha_{max} - \alpha_{min}) \parallel \cdot \parallel_{L^2(I)}^2 \\ &+ \int_{\alpha_{min}}^{\alpha_{max}} \parallel {}^{\text{RL}}_{x_L}D_x^{\alpha}(\cdot) \parallel_{L^2(I)}^2 \, d\alpha \sim \parallel \cdot \parallel_{l,\phi,I}^2 \sim \parallel \cdot \parallel_{r,\phi,I}^2.\end{aligned}$$

Therefore,

$$\int_{\alpha_{min}}^{\alpha_{max}} \parallel \cdot \parallel_{\alpha,I}^2 \, d\alpha \sim \parallel \cdot \parallel_{l,\phi,I}^2 \sim \parallel \cdot \parallel_{r,\phi,I}^2. \tag{3.1.84}$$

We can also show that

$$\begin{aligned}\phi_{min} \int_{\alpha_{min}}^{\alpha_{max}} \parallel (1 + |\omega|^2)^{\frac{\alpha}{2}} \mathcal{F}(\cdot)(\omega) \parallel_{L^2(\mathbb{R})}^2 \, d\alpha &\leq \parallel \cdot \parallel_{\phi,\mathbb{R}}^2 \\ &\leq \phi_{max} \int_{\alpha_{min}}^{\alpha_{max}} \parallel (1 + |\omega|^2)^{\frac{\alpha}{2}} \mathcal{F}(\cdot)(\omega) \parallel_{L^2(\mathbb{R})}^2 \, d\alpha.\end{aligned}$$

Because of the non-negativity of the norms, we have

$$\begin{aligned}&\phi_{min} \inf_{\tilde{v} \in {}^{\phi}\mathcal{H}(\mathbb{R}), \tilde{v}|_I = (\cdot)} \int_{\alpha_{min}}^{\alpha_{max}} \parallel (1 + |\omega|^2)^{\frac{\alpha}{2}} \mathcal{F}(\tilde{v})(\omega) \parallel_{L^2(\mathbb{R})}^2 \, d\alpha \\ &\leq \inf_{\tilde{v} \in {}^{\phi}\mathcal{H}(\mathbb{R}), \tilde{v}|_I = (\tilde{v})} \parallel \cdot \parallel_{\phi,\mathbb{R}}^2 \\ &\leq \phi_{max} \inf_{\tilde{v} \in {}^{\phi}\mathcal{H}(\mathbb{R}), \tilde{v}|_I = (\cdot)} \int_{\alpha_{min}}^{\alpha_{max}} \parallel (1 + |\omega|^2)^{\frac{\alpha}{2}} \mathcal{F}(\tilde{v})(\omega) \parallel_{L^2(\mathbb{R})}^2 \, d\alpha.\end{aligned} \tag{3.1.85}$$

In general, ${}^{\phi}\mathcal{H}(\mathbb{R}) \subset H^{\alpha_{max}}(\mathbb{R})$, $\forall \alpha \in [\alpha_{min}, \alpha_{max}]$. Therefore, we have

$$\begin{aligned}&\inf_{\tilde{v} \in {}^{\phi}\mathcal{H}(\mathbb{R}), \tilde{v}|_I = (\cdot)} \int_{\alpha_{min}}^{\alpha_{max}} \parallel (1 + |\omega|^2)^{\frac{\alpha}{2}} \mathcal{F}(\tilde{v})(\omega) \parallel_{L^2(\mathbb{R})}^2 \, d\alpha, \\ &= \int_{\alpha_{min}}^{\alpha_{max}} \inf_{\tilde{v} \in {}^{\phi}\mathcal{H}(\mathbb{R}), \tilde{v}|_I = (\cdot)} \parallel (1 + |\omega|^2)^{\frac{\alpha}{2}} \mathcal{F}(\tilde{v})(\omega) \parallel_{L^2(\mathbb{R})}^2 \, d\alpha, \\ &\leq C \int_{\alpha_{min}}^{\alpha_{max}} \inf_{\tilde{v} \in H^{\alpha}(\mathbb{R}), \tilde{v}|_I = (\cdot)} \parallel (1 + |\omega|^2)^{\frac{\alpha}{2}} \mathcal{F}(\tilde{v})(\omega) \parallel_{L^2(\mathbb{R})}^2 \, d\alpha, \\ &= C \int_{\alpha_{min}}^{\alpha_{max}} \parallel \cdot \parallel_{\alpha,I}^2 \, d\alpha.\end{aligned}$$

However, for some choices of $\phi = \delta(\alpha - \alpha_{min})$ and thus ${}^{\phi}\mathcal{H}(\mathbb{R}) \supset H^{\alpha_{max}}(\mathbb{R})$, $\forall \alpha \in [\alpha_{min}, \alpha_{max}]$. Therefore,

$$\int_{\alpha_{min}}^{\alpha_{max}} \inf_{\tilde{v}\in {}^{\phi}\mathcal{H}(\mathbb{R}),\tilde{v}|_I=(\cdot)} \| (1+|\omega|^2)^{\frac{\alpha}{2}} \mathcal{F}(\tilde{v})(\omega) \|_{L^2(\mathbb{R})}^2 \, d\alpha,$$

$$\geq \widetilde{\mathcal{C}} \int_{\alpha_{min}}^{\alpha_{max}} \inf_{\tilde{v}\in H^{\alpha}(\mathbb{R}),\tilde{v}|_I=(\cdot)} \| (1+|\omega|^2)^{\frac{\alpha}{2}} \mathcal{F}(\tilde{v})(\omega) \|_{L^2(\mathbb{R})}^2 \, d\alpha,$$

$$= \widetilde{\mathcal{C}} \int_{\alpha_{min}}^{\alpha_{max}} \| \cdot \|_{\alpha,I}^2 \, d\alpha,$$

which by (3.1.85) and $\| \cdot \|_{s,I} = \inf_{\tilde{v}\in H^s(\mathbb{R}),\tilde{v}|_I=(\cdot)} \|\tilde{v}\|_{s,\mathbb{R}}$, we get

$$\int_{\alpha_{min}}^{\alpha_{max}} \| \cdot \|_{\alpha,I}^2 \, d\alpha \sim \| \cdot \|_{\phi,I}^2 \, . \tag{3.1.86}$$

Comparing (3.1.84) and (3.1.86), we have

$$\| \cdot \|_{\phi,I}^2 \sim \| \cdot \|_{l,\phi,I}^2 \sim \| \cdot \|_{r,\phi,I}^2 \, .$$

This ends the proof. □

Variational Formulation of the Distributed-Order Problem

In order to obtain the variational form we multiply the problem of strong form by a proper test function v (defined later) and integrate over the computational domain:

$$\int_{\Omega} {}^{D}\mathcal{D}_{\phi} u(t)\, v(t)\, d\Omega = \int_{\Omega} f(t;u)\, v(t)\, d\Omega. \tag{3.1.87}$$

Using the definition of distributed-order fractional derivatives defined in (1.3.70), we get

$$\int_{\alpha_{min}}^{\alpha_{max}} \phi(\alpha) \int_{\Omega} {}^{\mathrm{RL}}_{0}D_t^{\alpha} u(t)\, v(t)\, d\Omega\, d\alpha = \int_{\Omega} f(t;u)\, v(t)\, d\Omega, \tag{3.1.88}$$

where

$$\forall \alpha \in (\alpha_{min}, \alpha_{max}), \quad \int_{\Omega} {}^{\mathrm{RL}}_{0}D_t^{\alpha} u\, v\, d\Omega = \left({}^{\mathrm{RL}}_{0}D_t^{\alpha} u, v\right)_{\Omega}$$

denotes the well-known L^2-inner product. We define the bilinear form associated with $\alpha \in (\alpha_{min}, \alpha_{max})$ as

$$a^{(\alpha)}(u,v) = \left({}^{\mathrm{RL}}_{0}D_t^{\alpha} u, v\right)_{\Omega} = ({}^{\mathrm{RL}}_{0}D_t^{\alpha/2} u, {}^{\mathrm{RL}}_{t}D_T^{\alpha/2} v)_{\Omega}. \tag{3.1.89}$$

We choose v such that $v(T) = 0$ and ${}^{\mathrm{RL}}_{t}D_T^{\alpha/2} v$ is integrable in Ω for all $\alpha \in (\alpha_{min}, \alpha_{max})$. Moreover, let U be the solution space defined as

$$U = \left\{ u \in L^2(\Omega) : \sqrt{\int_{\alpha_{min}}^{\alpha_{max}} \phi(\alpha) \| {}^{\mathrm{RL}}_{0}D_t^{\alpha/2} u \|_{L^2(\Omega)}^2 d\alpha} < \infty \right.$$
$$\left. \text{s.t.} \quad \begin{array}{ll} u(0) = 0 & \text{if } \alpha_{max} \in (0,1] \\ u(0) = u'(t)|_{t=0} = 0 & \text{if } \alpha_{max} \in (1,2] \end{array} \right\},$$

and let V be the test function space given by

$$V = \left\{ v \in L^2(\Omega) : \sqrt{\int_{\alpha_{min}}^{\alpha_{max}} \phi(\alpha) \| {}^{\mathrm{RL}}_{t}D_{T}^{\alpha/2} v \|^2_{L^2(\Omega)} d\alpha} < \infty \;\; \text{s.t.} \;\; v(T) = 0 \right\}.$$

The problem thus reads as: find $u \in U$ such that

$$a(u, v) = l(v), \quad \forall v \in V,$$

where

$$a(u, v) := \int_{\alpha_{min}}^{\alpha_{max}} \phi(\alpha)\, a^{(\alpha)}(u, v)\, d\alpha \tag{3.1.90}$$

denotes the *distributed* bilinear form and $l(v) := (f, v)_{\Omega}$.

Petrov–Galerkin Method: Modal Expansion

In the Petrov–Galerkin (PG) spectral method, we follow the theory of fractional Sturm–Liouville eigenproblems (FSLP) in Chapter 2 and employ the corresponding eigenfunctions, Jacobi poly-fractonomials (of the first kind) given in the reference domain $(-1, 1)$ by

$$^{(1)}\mathcal{P}_n^{\mu}(\xi) := (1 + \xi)^{\mu} P_{n-1}^{(-\mu,\mu)}(\xi), \quad \xi \in (-1, 1), \tag{3.1.91}$$

as *non-polynomial* basis functions consisting of a fractional term multiplied by the Jacobi polynomial $P_{n-1}^{(-\mu,\mu)}(\xi)$, hence we employ (3.1.91) in construction of a basis to formulate a projection type scheme, namely *modal* expansion. We represent the solution in terms of the elements of the basis space U_N given as follows

$$U_N = \text{span}\left\{ {}^{(1)}\mathcal{P}_n^{\mu}(\xi), \;\; \xi \in (-1, 1), \;\; n = 1, 2, \ldots, N \right\} \tag{3.1.92}$$

via the poly-fractonomial *modal* expansion as

$$u_N(\xi) = \sum_{n=1}^{N} c_n \, {}^{(1)}\mathcal{P}_n^{\mu}(\xi), \tag{3.1.93}$$

in which μ is to be fixed as a fractional parameter *a priori* depending on the range of distribution order interval, i.e., $\mu \in (0, 1)$ if $\alpha_{max} \in (0, 1]$ and $\mu \in (1, 2)$ if $\alpha_{max} \in (1, 2]$. It can also be tuned to capture possible singularities in the exact solution if some knowledge about that is available.

Moreover, in the PG scheme, we employ another space of test functions V_N, however of the same dimension, given by

$$V_N = \text{span}\left\{ {}^{(2)}\mathcal{P}_k^{\mu}(\xi), \;\; \xi \in (-1, 1), \;\; k = 1, 2, \ldots, N \right\}, \tag{3.1.94}$$

in which

$$ {}^{(2)}\mathcal{P}_k^{\mu}(\xi):=(1-\xi)^{\mu} P_{k-1}^{(\mu,-\mu)}(\xi), \quad \xi \in (-1,1) $$

denotes the Jacobi poly-fractonomial of the second kind, which is the explicit eigenfunction of fractional Sturm–Liouville problem of the second kind.

Since $\phi(\alpha) \geq 0$ and $\int_{\alpha_{min}}^{\alpha_{max}} \phi(\alpha)\,d\alpha = 1$, it is not difficult to see that $U_N \subset U$ and $V_N \subset V$ when μ is chosen properly. Therefore, the bilinear form (3.1.89) reduces to the *discrete* bilinear form

$$ a_h^{\alpha}(u_N, v_N) = \left({}^{\mathrm{RL}}_{\;0}D_t^{\alpha/2} u_N, {}^{\mathrm{RL}}_{\;t}D_T^{\alpha/2} v_N\right)_{\Omega}, \tag{3.1.95} $$

and thus the problem reads as: find $u_N \in U_N$ such that

$$ a_h(u_N, v_N) = l_h(v_N), \quad \forall v_N \in V_N, \tag{3.1.96} $$

where

$$ a_h(u_N, v_N) := \int_{\alpha_{min}}^{\alpha_{max}} \phi(\alpha)\; a_h^{\alpha}(u_N, v_N)\,d\alpha $$

denotes the *discrete* distributed bilinear form and $l_h(v_N) := (f, v_N)_{\Omega}$ represents the load vector.

By substituting the expansion (3.1.93), choosing $v_N = {}^{(2)}\mathcal{P}_k^{\mu}(\xi) \in V_N$, $k = 1, 2, \ldots, N$, the discrete distributed bilinear form in (3.1.96) can be written as

$$ \int_{\alpha_{min}}^{\alpha_{max}} \phi(\alpha) \sum_{n=1}^{N} c_n \left(\frac{2}{T}\right)^{\alpha} \left({}^{\mathrm{RL}}_{-1}D_{\xi}^{\alpha/2}[{}^{(1)}\mathcal{P}_n^{\mu}(\xi)], {}^{\mathrm{RL}}_{\;\xi}D_1^{\alpha/2}[{}^{(2)}\mathcal{P}_k^{\mu}(\xi)]\right)_{\Omega} d\alpha. $$

We have

$$ {}^{\mathrm{RL}}_{-1}D_{\xi}^{\alpha/2}[{}^{(1)}\mathcal{P}_n^{\mu}(\xi)] = \frac{\Gamma(n+\mu)}{\Gamma(n+\eta)}\, {}^{(1)}\mathcal{P}_n^{\eta}(\xi), \tag{3.1.97} $$

$$ {}^{\mathrm{RL}}_{\;\xi}D_1^{\alpha/2}[{}^{(2)}\mathcal{P}_k^{\mu}(\xi)], = \frac{\Gamma(k+\mu)}{\Gamma(k+\eta)}\, {}^{(2)}\mathcal{P}_k^{\eta}(\xi), \tag{3.1.98} $$

where $\eta = \mu - \alpha/2$. Thus, by changing the order of summation, the above integral takes the form

$$ \begin{aligned} &\sum_{n=1}^{N} c_n \int_{\alpha_{min}}^{\alpha_{max}} \phi(\alpha) \left(\frac{2}{T}\right)^{\alpha} R_{n,k}^{\mu,\alpha} \left({}^{(1)}\mathcal{P}_n^{\eta}, {}^{(2)}\mathcal{P}_k^{\eta}\right)_{\Omega} d\alpha \\ &= \sum_{n=1}^{N} c_n \int_{\alpha_{min}}^{\alpha_{max}} \phi(\alpha) \left(\frac{2}{T}\right)^{\alpha} R_{n,k}^{\mu,\alpha} \int_{-1}^{1} {}^{(1)}\mathcal{P}_n^{\eta}(\xi)\; {}^{(2)}\mathcal{P}_k^{\eta}(\xi)\; d\xi\, d\alpha, \end{aligned} \tag{3.1.99} $$

where

$$ R_{n,k}^{\mu,\alpha} := \frac{\Gamma(n+\mu)}{\Gamma(n+\mu-\alpha/2)} \frac{\Gamma(k+\mu)}{\Gamma(k+\mu-\alpha/2)}. \tag{3.1.100} $$

Thus changing the order of integration yields

$$a_h(u_N, v_N) = \sum_{n=1}^{N} c_n \int_{-1}^{1} \left[\int_{\alpha_{min}}^{\alpha_{max}} \phi(\alpha) \left(\frac{2}{T}\right)^{\alpha} R_{n,k}^{\mu,\alpha} \; {}^{(1)}\mathcal{P}_n^{\eta}(\xi) \; {}^{(2)}\mathcal{P}_k^{\eta}(\xi) \; d\alpha \right] d\xi. \tag{3.1.101}$$

Theorem 3.13 (Spectrally Accurate Quadrature Rule in α-Direction) *We have the following properties and highly accurate quadrature rule in α-direction.*

(i) *For any fixed $\xi_0 \in (-1, 1)$ and $n \geq 1$, the Jacobi polynomial $P_n^{(\pm\alpha,\mp\alpha)}(\xi_0)$ is a polynomial of order n in α.*

(ii) *Let $\phi \in H^r(\alpha_{min}, \alpha_{max})$ with $r > 0$. Then for any $\mu \geq \alpha_{max}/2$,*

$$\left| \int_{\alpha_{min}}^{\alpha_{max}} \phi(\alpha) \left(\frac{2}{T}\right)^{\alpha} R_{n,k}^{\mu,\alpha} \; {}^{(1)}\mathcal{P}_n^{\eta}(\xi) \; {}^{(2)}\mathcal{P}_k^{\eta}(\xi) \; d\alpha - \sum_{q=1}^{Q} \widetilde{w}_q \, \phi_N(\alpha_q) \left(\frac{2}{T}\right)^{\alpha_q} R_{n,k}^{\mu,\alpha_q} \; {}^{(1)}\mathcal{P}_n^{\eta_q}(\xi) \; {}^{(2)}\mathcal{P}_k^{\eta_q}(\xi) \right| \leq \mathscr{C} \, Q^{-r} \, \|\phi\|_{H^r(\alpha_{min},\alpha_{max})}, \tag{3.1.102}$$

where the constant $\mathscr{C} > 0$, $\{\alpha_q, \widetilde{w}_q\}|_{q=1}^{Q}$ is the set of Gauss–Legendre quadrature points and weights, and

$$\eta_q = \mu - \alpha_q/2, \quad \phi_N(\alpha) = \sum_{n=0}^{N} \hat{\phi}_n \, P_n(\alpha),$$

with $\{\hat{\phi}_n\}$ being the Legendre polynomial expansion coefficients of $\phi(\alpha)$.

(iii) *If $\phi(\alpha)$ is smooth, the quadrature rule in α-direction is exponentially accurate.*

Proof (i) The Jacobi polynomials, $P_n^{(\alpha,\beta)}(\xi)$, can be constructed via the three-term recurrence relation. By letting $\beta = -\alpha$, the corresponding relation reduces to

$$P_{n+1}^{(\alpha,-\alpha)}(\xi) = \frac{2n+1}{n+1} \, \xi \, P_n^{(\alpha,-\alpha)}(\xi) - \frac{n - \alpha^2}{n(n+1)} \, P_{n-1}^{(\alpha,-\alpha)}(\xi). \tag{3.1.103}$$

Thus the Jacobi polynomials evaluated at $\xi = \xi_0 \in [-1, 1]$ are obtained in the following standard form

$$P_0^{(\alpha,-\alpha)}(\xi_0) = \mathfrak{P}_0(\alpha) = 1 \quad \text{(zeroth order in } \alpha\text{)},$$
$$P_1^{(\alpha,-\alpha)}(\xi_0) = \mathfrak{P}_1(\alpha) = \alpha + \xi_0 \quad \text{(linear in } \alpha\text{)},$$
$$P_2^{(\alpha,-\alpha)}(\xi_0) = \mathfrak{P}_2(\alpha) = \frac{1}{2}\,\alpha^2 + \frac{3}{2}\,\xi_0\,\alpha + \frac{3\,\xi_0^2 - 1}{2} \quad \text{(quadratic in } \alpha\text{)},$$
$$P_3^{(\alpha,-\alpha)}(\xi_0) = \mathfrak{P}_3(\alpha) = \frac{1}{6}\,\alpha^3 + \xi_0\,\alpha^2 + \frac{15\,\xi_0^2 - 4}{6}\,\alpha + \frac{5\,\xi_0^3 - 3\,\xi_0}{2} \quad \text{(cubic in } \alpha\text{)}.$$

Assume that $P_k^{(\alpha,-\alpha)}(\xi_0)$ and $P_{k-1}^{(\alpha,-\alpha)}(\xi_0)$ are respectively polynomials of order k and $k-1$ in α. Using (3.1.103) for $n = k+1$, we get

$$P_{k+1}^{(\alpha,-\alpha)}(\xi_0) = \frac{2k+1}{k+1}\,\xi_0\,P_k^{(\alpha,-\alpha)}(\xi_0) - \frac{k-\alpha^2}{k(k+1)}\,P_{k-1}^{(\alpha,-\alpha)}(\xi_0),$$

which is a polynomial of order $k+1$ in α due to the second term. Hence, by mathematical induction, $P_n^{(\alpha,-\alpha)}(\xi_0) = \mathfrak{P}_n(\alpha)$ is a polynomial of order n in α, for any fixed $\xi_0 \in [-1, 1]$.

Using the same argument, we can show that $P_n^{(-\alpha,\alpha)}(\xi_0) = \mathfrak{P}_n(-\alpha)$ is also a polynomial of order n in α, for any given $\xi_0 \in [-1, 1]$.

(ii) The inner integral of the discrete distributed bilinear form (3.1.101) can be written as

$$\int_{\alpha_{min}}^{\alpha_{max}} \phi(\alpha)\left(\frac{2}{T}\right)^{\alpha} R_{n,k}^{\mu,\alpha}\ {}^{(1)}\mathcal{P}_n^{\eta}(\xi)\ {}^{(2)}\mathcal{P}_k^{\eta}(\xi)\ d\alpha = \Gamma(n+\mu)\Gamma(k+\mu)$$
$$\times \int_{\alpha_{min}}^{\alpha_{max}} \phi(\alpha)\left(\frac{2}{T}\right)^{\alpha} \frac{(1-\xi^2)^{\eta}}{\Gamma(n+\eta)\Gamma(k+\eta)}\,P_{n-1}^{(-\eta,\eta)}(\xi)\,P_{k-1}^{(\eta,-\eta)}(\xi)\,d\alpha, \tag{3.1.104}$$

where $\eta = \mu - \alpha/2$.

From Property-(i), we know that $P_{n-1}^{(\frac{\alpha}{2}-\mu,\mu-\frac{\alpha}{2})}(\xi)$ and $P_{k-1}^{(\mu-\frac{\alpha}{2},\frac{\alpha}{2}-\mu)}(\xi)$ are polynomials in α of order $n-1$ and $k-1$, respectively, for any $\xi \in [-1, 1]$ and μ fixed. Thus, we can write

$$P_{n-1}^{(\frac{\alpha}{2}-\mu,\mu-\frac{\alpha}{2})}(\xi) = \sum_{r=0}^{n-1} \sigma_r\,P_r(\alpha), \tag{3.1.105}$$

$$P_{k-1}^{(\mu-\frac{\alpha}{2},\frac{\alpha}{2}-\mu)}(\xi) = \sum_{l=0}^{k-1} \widetilde{\sigma}_l\,P_l(\alpha). \tag{3.1.106}$$

By plugging (3.1.105) and (3.1.106) into (3.1.104), we obtain

$$\int_{\alpha_{min}}^{\alpha_{max}} \phi(\alpha) \left(\frac{2}{T}\right)^{\alpha} \frac{(1-\xi^2)^{\eta}}{\Gamma(n+\eta)\Gamma(k+\eta)} P_{n-1}^{(-\eta,\eta)}(\xi)\, P_{k-1}^{(\eta,-\eta)}(\xi)\, d\alpha$$
$$= \int_{\alpha_{min}}^{\alpha_{max}} \phi(\alpha)\, \mathcal{W}_{kn}^{\xi,\mu}(\alpha) \sum_{r=0}^{n-1} \sigma_r\, P_r(\alpha) \sum_{l=0}^{k-1} \widetilde{\sigma}_l\, P_l(\alpha)\, d\alpha, \tag{3.1.107}$$

where

$$\mathcal{W}_{kn}^{\xi,\mu}(\alpha) = \left(\frac{2}{T}\right)^{\alpha} \frac{(1-\xi)^{\mu-\alpha}}{\Gamma(n+\mu-\alpha)} \frac{(1+\xi)^{\mu-\alpha}}{\Gamma(k+\mu-\alpha)}$$

is *smooth* in any compact support in $[\alpha_{min}, \alpha_{max}]$ and its polynomial expansion

$$\mathcal{W}_{kn}^{\xi,\mu}\big|_N(\alpha) = \mathcal{W}_N(\alpha) = \sum_{q=0}^{N} \rho_q\, P_q(\alpha)$$

converges exponentially i.e.,

$$\left\|\mathcal{W}_{kn}^{\xi,\mu}(\alpha) - \mathcal{W}_N(\alpha)\right\| \le c_1 \exp(-c_2 N^{c_3}), \tag{3.1.108}$$

in which $\|.\|$ denotes the L^2-norm in $(\alpha_{min}, \alpha_{max})$. If the distribution function $\phi \in H^r(\alpha_{min}, \alpha_{max})$, $r > 0$, we have the projection error

$$\left\|\phi(\alpha) - \phi_N(\alpha)\right\| \le c_4\, N^{-r}\, \|\phi\|_{H^r(\alpha_{min},\alpha_{max})}, \tag{3.1.109}$$

where $\phi_N(\alpha) = \sum_{n=0}^{N} \hat{\phi}_n\, P_n(\alpha)$ as before. Consequently, the integrand in (3.1.107) can be well approximated via

$$\phi(\alpha)\, \mathcal{W}_{kn}^{\xi,\mu}(\alpha) \sum_{r=0}^{n-1} \sigma_r\, P_r(\alpha) \sum_{l=0}^{k-1} \widetilde{\sigma}_l\, P_l(\alpha)$$
$$\approx \phi_N(\alpha)\, \mathcal{W}_N(\alpha) \sum_{r=0}^{n-1} \sigma_r\, P_r(\alpha) \sum_{l=0}^{k-1} \widetilde{\sigma}_l\, P_l(\alpha). \tag{3.1.110}$$

Next, let

$$\mathcal{I} = \int_{\alpha_{min}}^{\alpha_{max}} \phi(\alpha)\, \mathcal{W}_{kn}^{\xi,\mu}(\alpha) \sum_{r=0}^{n-1} \sigma_r\, P_r(\alpha) \sum_{l=0}^{k-1} \widetilde{\sigma}_l\, P_l(\alpha)\, d\alpha, \tag{3.1.111}$$
$$\mathcal{I}_N = \int_{\alpha_{min}}^{\alpha_{max}} \phi_N(\alpha)\, \mathcal{W}_N(\alpha) \sum_{r=0}^{n-1} \sigma_r\, P_r(\alpha) \sum_{l=0}^{k-1} \widetilde{\sigma}_l\, P_l(\alpha)\, d\alpha,$$

where $\mathcal{I}_N$ can be *accurately* calculated via

$$\mathcal{I}_N = \sum_{q=1}^{Q} \widetilde{w}_q\, \phi_N(\alpha_q)\, \mathcal{W}_N(\alpha_q) \sum_{r=0}^{n-1} \sigma_r\, P_r(\alpha_q) \sum_{l=0}^{k-1} \widetilde{\sigma}_l\, P_l(\alpha_q), \tag{3.1.112}$$

employing a Gauss–Legendre quadrature rule, provided $Q = 2N$. Thus by the Cauchy–Schwarz inequality and Hölder inequality, we have

$$\begin{aligned}
|\mathcal{I} - \mathcal{I}_N| &\le \sqrt{\alpha_{min} - \alpha_{max}} \left\| (\phi\, \mathcal{W}_{kn}^{\xi,\mu} - \phi_N\, \mathcal{W}_N) \sum_{r=0}^{n-1} \sigma_r P_r \sum_{l=0}^{k-1} \widetilde{\sigma}_l P_l \right\| \\
&\le \sqrt{\alpha_{min} - \alpha_{max}} \left\| \sum_{r=0}^{n-1} \sigma_r P_r \sum_{l=0}^{k-1} \widetilde{\sigma}_l P_l \right\| \left\| \phi\, \mathcal{W}_{kn}^{\xi,\mu} - \phi_N\, \mathcal{W}_N \right\| \\
&\le c_5 \left\| \phi\, \mathcal{W}_{kn}^{\xi,\mu} - \phi_N\, \mathcal{W}_N \right\|.
\end{aligned} \tag{3.1.113}$$

Using the triangle inequality and (3.1.108)–(3.1.109), we can further obtain

$$\begin{aligned}
|\mathcal{I} - \mathcal{I}_N| &\le c_5 \left\| [(\phi - \phi_N) + \phi_N][(\mathcal{W}_{kn}^{\xi,\mu} - \mathcal{W}_N) + \mathcal{W}_N] - \phi_N\, \mathcal{W}_N \right\| \\
&= c_5 \big\| (\phi - \phi_N)(\mathcal{W}_{kn}^{\xi,\mu} - \mathcal{W}_N) + \phi_N(\mathcal{W}_{kn}^{\xi,\mu} - \mathcal{W}_N) \\
&\quad + (\phi - \phi_N)\mathcal{W}_N + \phi_N\, \mathcal{W}_N - \phi_N\, \mathcal{W}_N \big\| \\
&\le c_5 \big(\|(\phi - \phi_N)(\mathcal{W}_{kn}^{\xi,\mu} - \mathcal{W}_N)\| + \|\phi_N(\mathcal{W}_{kn}^{\xi,\mu} - \mathcal{W}_N)\| \\
&\quad + \|(\phi - \phi_N)\mathcal{W}_N\| \big) \\
&\le c_5 \big(\|\phi - \phi_N\| \|\mathcal{W}_{kn}^{\xi,\mu} - \mathcal{W}_N\| + \|\phi_N\| \|\mathcal{W}_{kn}^{\xi,\mu} - \mathcal{W}_N\| \\
&\quad + \|\phi - \phi_N\| \|\mathcal{W}_N\| \big) \\
&\le c_5 \big(c_4\, N^{-r} \|\phi\|_{H^r(\alpha_{min}, \alpha_{max})} \cdot c_1 \exp(-c_2 N^{c_3}) \\
&\quad + \|\phi_N\| c_1 \exp(-c_2 N^{c_3}) + \|\mathcal{W}_N\| c_4\, N^{-r} \|\phi\|_{H^r(\alpha_{min}, \alpha_{max})} \big) \\
&\le c_6\, N^{-r} \|\phi\|_{H^r(\alpha_{min}, \alpha_{max})}.
\end{aligned} \tag{3.1.114}$$

In view of this, we can obtain (3.1.102) from (3.1.105), (3.1.106), (3.1.111) and (3.1.112).

(iii) If $\phi(\alpha)$ is smooth, then the approximation $\phi_N(\alpha)$, in (3.1.109), converges with an exponential accuracy and so does the norm in (3.1.114). Thus,

$$|\mathcal{I} - \mathcal{I}_N| \le \mathscr{C}_1 \exp(-\mathscr{C}_2\, N^{\mathscr{C}_3}), \tag{3.1.115}$$

and therefore, the quadrature rule becomes exponentially accurate in Q. □

By Theorem 3.13 and performing an affine mapping from $\alpha \in [\alpha_{min}, \alpha_{max}]$ to the standard domain $\alpha_{st} \in (-1, 1)$, the inner integral in (3.1.101) can be evaluated with spectral accuracy by employing a Gauss–Legendre quadrature rule. Then by changing the order of summation and integral $a_h(u_N, v_N) = l_h(v_N)$ can be written as:

$$\sum_{n=1}^{N}\sum_{j=1}^{Q} c_n\, J_\alpha\, w_j\, \phi_j \left(\frac{2}{T}\right)^{\alpha_j} R_{n,k}^{\mu,\alpha_j} \int_{-1}^{1} {}^{(1)}\mathcal{P}_n^{\eta_j}(\xi)\; {}^{(2)}\mathcal{P}_k^{\eta_j}(\xi)\, d\xi \tag{3.1.116}$$
$$= \left(f, {}^{(2)}\mathcal{P}_k^{\mu}\right)_\Omega, \quad k = 1, 2, \ldots, N,$$

where

$$J_\alpha = \frac{\alpha_{max} - \alpha_{min}}{2}, \quad \alpha_j = \alpha(\alpha_{st_j}), \quad \phi_j = \phi(\alpha(\alpha_{st_j})), \quad \eta_j = \mu - \frac{\alpha_j}{2}$$

and α_{st_j}, w_j are the quadrature points and weights respectively.

The corresponding linear system is

$$\boldsymbol{S}\,\boldsymbol{c} = \boldsymbol{F}, \tag{3.1.117}$$

where the entries of the stiffness matrix S and force vector $\boldsymbol{F}$ are given by

$$\boldsymbol{S}_{kn} = J_\alpha \sum_{j=1}^{Q} w_j\, \phi_j \left(\frac{2}{T}\right)^{\alpha_j} R_{n,k}^{\mu,\alpha_j} \int_{-1}^{1} {}^{(1)}\mathcal{P}_n^{\eta_j}(\xi)\; {}^{(2)}\mathcal{P}_k^{\eta_j}(\xi)\, d\xi, \tag{3.1.118}$$

and

$$\boldsymbol{F}_k = \int_{-1}^{1} f(\xi)\, {}^{(2)}\mathcal{P}_k^{\mu}(\xi)\, d\xi. \tag{3.1.119}$$

Remark 3.14 For each fixed j and given the structure of ${}^{(1)}\mathcal{P}_n^{\eta_j}(\xi)$ and ${}^{(2)}\mathcal{P}_k^{\eta_j}(\xi)$, the above integrations take the form

$$\int_{-1}^{1} {}^{(1)}\mathcal{P}_n^{\eta_j}(\xi)\, {}^{(2)}\mathcal{P}_k^{\eta_j}(\xi) d\xi = \int_{-1}^{1} (1-\xi^2)^{\eta_j}\, P_{k-1}^{(\eta_j,-\eta_j)}(\xi)\, P_{n-1}^{(-\eta_j,\eta_j)}(\xi)\, d\xi,$$
$$\int_{-1}^{1} f(\xi)\, {}^{(2)}\mathcal{P}_k^{\mu}(\xi)\, d\xi = \int_{-1}^{1} (1-\xi)^{\mu}\, f(\xi)\, P_{k-1}^{(\mu,-\mu)}(\xi)\, d\xi.$$

Therefore, the full stiffness matrix $\boldsymbol{S}$ and $\boldsymbol{F}$ can be evaluated accurately using a proper Gauss–Lobatto–Jacobi rule corresponding to the weight functions $(1-\xi^2)^{\eta_j}$ and $(1-\xi)^{\eta_j}$, respectively.

Discrete Stability Analysis

In this section, we study the stability of the previous numerical scheme based on the aforementioned choice of solution and test function space considering the bilinear form in (3.1.96).

Theorem 3.15 *The scheme* (3.1.96) *is stable and the following inequality holds*

$$\inf_{u_N \in U_N} \sup_{v_N \in V_N} \frac{a_h(u_N, v_N)}{\|u_N\|_{U_N}\, \|v_N\|_{V_N}} \geq \beta.$$

Proof Recall from (3.1.90) that

$$a_h(u_N, v_N) = \int_{\alpha_{min}}^{\alpha_{max}} \phi(\alpha)\, a_h^{\alpha}(u_N, v_N)\, d\alpha,$$

where by the fractional derivative of poly-fractonomials,

$$a_h^{\alpha}(u_N, v_N) = a_h^{\alpha}\Big(\sum_{n=1}^{N} a_n\, {}^{(1)}\mathcal{P}_n^{\mu}, \sum_{k=1}^{N} a_k\, {}^{(2)}\mathcal{P}_k^{\mu}\Big) =$$
$$\sum_{n=1}^{N}\sum_{k=1}^{N} a_n a_k R_{n,k}^{\mu,\alpha} \int_{-1}^{1} (1-\xi)^{\eta}(1+\xi)^{-\eta}(1+\xi)^{2\eta} P_{k-1}^{(\eta,-\eta)}(\xi) P_{n-1}^{(-\eta,\eta)}(\xi) d\xi$$

for $\alpha \in (\alpha_{min}, \alpha_{max})$. As before, $\eta = \mu - \alpha/2$ and $R_{n,k}^{\mu,\alpha}$ is defined in (3.1.100). Note that here $(1+\xi)^{\eta}$ is replaced by $(1+\xi)^{-\eta}(1+\xi)^{2\eta}$. If $\eta > -1/2$, then the function $(1+\xi)^{2\eta}$ is non-negative, nondecreasing, continuous and integrable. Therefore,

$$a_h^{\alpha}(u_N, v_N) \geq C_0 \sum_{n=1}^{N}\sum_{k=1}^{N} a_n a_k R_{n,k}^{\mu,\alpha}$$
$$\times \int_{-1}^{1} (1-\xi)^{\eta}(1+\xi)^{-\eta} P_{k-1}^{(\eta,-\eta)}(\xi) P_{n-1}^{(-\eta,\eta)}(\xi)\, d\xi.$$

Moreover, the Jacobi polynomial $P_{n-1}^{(-\eta,\eta)}(\xi)$ can be expanded as

$$P_{n-1}^{(-\eta,\eta)}(\xi) = \sum_{j=0}^{n-1}(-1)^{j-n+1}\binom{n-1+j}{j}\binom{n-1+\eta}{n-1-j}\Big(\frac{1+\xi}{2}\Big)^{j}$$
$$= \sum_{j=0}^{n-1}(-1)^{j-n+1}\mathcal{A}_{n,j}^{\eta}\binom{n-1+j}{j}\binom{n-1-\eta}{n-1-j}\Big(\frac{1+\xi}{2}\Big)^{j},$$

where $\mathcal{A}_{n,j}^{\eta} = \binom{n-1+\eta}{n-1-j}/\binom{n-1-\eta}{n-1-j}$ is nondecreasing, positive and bounded for all n, j, η. Therefore, there exists $C_1 = C_1(n) > 0$ such that

$$a_h^{\alpha}(u_N, v_N) \geq C_0 \sum_{n=1}^{N}\sum_{k=1}^{N} a_n a_k\, R_{n,k}^{\mu,\alpha}\, C_1(n) \int_{-1}^{1} (1-\xi)^{\eta}(1+\xi)^{-\eta} P_{k-1}^{(\eta,-\eta)}(\xi) P_{n-1}^{(\eta,-\eta)}(\xi)\, d\xi$$
$$\geq C_0 \sum_{k=1}^{N} a_k^2\, R_{k,k}^{\mu,\alpha}\, C_1(k)\, \gamma_{k-1}^{(\eta,-\eta)},$$

where

$$\gamma_{k-1}^{(\eta,-\eta)} = \frac{2}{2k-1}\frac{\Gamma(k+\mu)\,\Gamma(k-\mu)}{(k-1)!\;\Gamma(k)}.$$

Hence,

$$a_h(u_N, v_N) \geq C_0 \int_{\alpha_{min}}^{\alpha_{max}} \phi(\alpha) \sum_{k=1}^{N} a_k^2 \, R_{k,k}^{\mu,\alpha} \, C_1(k) \, \gamma_{k-1}^{(\eta,-\eta)} \, d\alpha. \qquad (3.1.120)$$

Moreover, we have

$$\|v_N\|_{V_N}^2 = \int_{\alpha_{min}}^{\alpha_{max}} \phi(\alpha) \, \| {}^{\text{RL}}_{-1}D_x^{\alpha/2} v_N \|_{L^2(\Omega)}^2 d\alpha,$$

where taking

$$v_N(\xi) = \sum_{k=1}^{N} a_k \, {}^{(2)}\mathcal{P}_k^{\mu}(\xi),$$

we can write

$$\begin{aligned} \| {}^{\text{RL}}_{-1}D_x^{\alpha/2} v_N \|_{L^2(\Omega)}^2 &= \int_{-1}^{1} \Big(\sum_{k=1}^{N} a_k \, C_k^{\mu,\alpha} \, (1-\xi)^{\eta} P_{k-1}^{(\eta,-\eta)}(\xi) \Big)^2 d\xi \\ &\leq \int_{-1}^{1} \sum_{k=1}^{N} a_k^2 \, (C_k^{\mu,\alpha})^2 \, (1-\xi)^{2\eta} (P_{k-1}^{(\eta,-\eta)}(\xi))^2 d\xi, \end{aligned}$$

where we used Jensen's inequality and denoted

$$C_k^{\mu,\alpha} := \frac{\Gamma(k+\mu)}{\Gamma(k+\eta)} = \frac{\Gamma(k+\mu)}{\Gamma(k+\mu-\alpha/2)}.$$

By multiplying the integrand by $(1+\xi)^{-\eta}(1+\xi)^{\eta}$ and changing the order of summation and integration, we obtain

$$\begin{aligned} \| {}^{\text{RL}}_{-1}D_x^{\alpha/2} v_N \|_{L^2(\Omega)}^2 &\leq \sum_{k=1}^{N} a_k^2 \, (C_k^{\mu,\alpha})^2 \\ &\times \int_{-1}^{1} (1-\xi)^{\eta}(1+\xi)^{-\eta}(1-\xi^2)^{\eta} P_{k-1}^{(\eta,-\eta)}(\xi) \, P_{k-1}^{(\eta,-\eta)}(\xi) \, d\xi \\ &\leq \sum_{k=1}^{N} a_k^2 \, (C_k^{\mu,\alpha})^2 \, C_2 \\ &\times \int_{-1}^{1} (1-\xi)^{\eta}(1+\xi)^{-\eta} P_{k-1}^{(\eta,-\eta)}(\xi) \, P_{k-1}^{(\eta,-\eta)}(\xi) \, d\xi \end{aligned}$$

since $\eta > -1/2$ and consequently $(1-\xi^2)^{\eta}$ is non-negative and integrable in the domain of integration. By the orthogonality of Jacobi polynomials, we get

$$\| {}^{\text{RL}}_{-1}D_x^{\alpha/2} v_N \|_{L^2(\Omega)}^2 \leq C_2 \sum_{k=1}^{N} a_k^2 \, (C_k^{\mu,\alpha})^2 \, \gamma_{k-1}^{(\eta,-\eta)},$$

and thus

$$\|v_N\|_{V_N} \leq \sqrt{C_2 \int_{\alpha_{min}}^{\alpha_{max}} \phi(\alpha) \sum_{k=1}^{N} a_k^2 \; (C_k^{\mu,\alpha})^2 \gamma_{k-1}^{(\eta,-\eta)} \, d\alpha}. \qquad (3.1.121)$$

Similarly for $\|u_N\|_{U_N}^2$,

$$\|u_N\|_{U_N}^2 = \int_{\alpha_{min}}^{\alpha_{max}} \phi(\alpha) \; \| {}_{-1}^{\text{RL}}D_x^{\alpha/2} u_N \|_{L^2(\Omega)}^2 d\alpha,$$

where by Jensen's inequality,

$$\begin{aligned}
\| {}_{-1}^{\text{RL}}D_x^{\alpha/2} u_N \|_{L^2(\Omega)}^2 &= \int_{-1}^{1} \Big(\sum_{n=1}^{N} a_n \, C_n^{\mu,\alpha} \, (1+\xi)^{\eta} P_{n-1}^{(-\eta,\eta)}(\xi) \Big)^2 d\xi \\
&\leq \int_{-1}^{1} \sum_{n=1}^{N} a_n^2 \; (C_n^{\mu,\alpha})^2 \, (1+\xi)^{2\eta} (P_{n-1}^{(-\eta,\eta)}(\xi))^2 d\xi.
\end{aligned}$$

Following similar steps, and by multiplying the integrand by $(1-\xi)^{-\eta}(1-\xi)^{\eta}$ and changing the order of summation and integration, we obtain

$$\begin{aligned}
&\| {}_{-1}^{\text{RL}}D_x^{\alpha/2} u_N \|_{L^2(\Omega)}^2 \\
&\leq \sum_{n=1}^{N} a_n^2 \; (C_n^{\mu,\alpha})^2 \int_{-1}^{1} (1-\xi)^{-\eta}(1+\xi)^{\eta}(1-\xi^2)^{\eta} P_{n-1}^{(-\eta,\eta)}(\xi) \; P_{n-1}^{(-\eta,\eta)}(\xi) \; d\xi \\
&\leq \sum_{n=1}^{N} a_n^2 \; (C_n^{\mu,\alpha})^2 \, C_3 \int_{-1}^{1} (1-\xi)^{-\eta}(1+\xi)^{\eta} P_{n-1}^{(-\eta,\eta)}(\xi) \; P_{n-1}^{(-\eta,\eta)}(\xi) \; d\xi
\end{aligned}$$

since $(1-\xi^2)^{\eta}$ is non-negative and integrable in the domain of integral. Next, by the orthogonality of Jacobi polynomials,

$$\| {}_{-1}^{\text{RL}}D_x^{\alpha/2} u_N \|_{L^2(\Omega)}^2 \leq C_3 \sum_{n=1}^{N} a_n^2 \; (C_n^{\mu,\alpha})^2 \gamma_{n-1}^{(-\eta,\eta)}.$$

Therefore,

$$\|u_N\|_{U_N} \leq \sqrt{C_3 \int_{\alpha_{min}}^{\alpha_{max}} \phi(\alpha) \sum_{n=1}^{N} a_n^2 \; (C_n^{\mu,\alpha})^2 \gamma_{n-1}^{(\eta,-\eta)} \, d\alpha}, \qquad (3.1.122)$$

where $\gamma_{n-1}^{(-\eta,\eta)} = \gamma_{n-1}^{(\eta,-\eta)}$.

Therefore, using (3.1.120), (3.1.121) and (3.1.122), we obtain

$$\inf_{u_N \in U_N} \sup_{v_N \in V_N} \frac{a_h(u_N, v_N)}{\|u_N\|_{U_N} \; \|v_N\|_{V_N}} \geq \frac{C_0 \; C_{1_{min}}}{\sqrt{C_2 \; C_3}} = \beta.$$

This ends the proof. □

Projection Error Analysis

In this section, we analyze the projection errors on approximation by expansions of the defined set of basis functions.

Theorem 3.16 *Let*

$$\frac{d^r u}{dt^r} \in U \quad i.e., \quad \int_{\alpha_{min}}^{\alpha max} \phi(\alpha) \parallel {}^{RL}_{0}D_t^{r+\alpha/2} u \parallel^2_{L^2(\Omega)} d\alpha < \infty,$$

and let u_N be the projection of the exact solution u. Then we have

$$\|u - u_N\|_U^2 \le C N^{-2r} \int_{\alpha_{min}}^{\alpha_{max}} \phi(\alpha) \| {}^{DL}_{0}D_t^{r+\alpha/2} u \|^2_{L^2(\Omega)} d\alpha. \qquad (3.1.123)$$

Proof By performing an affine mapping from $t \in [0, T]$ to the standard domain $\xi \in (-1, 1)$, we expand the exact solution u in terms of the following infinite series of Jacobi poly-fractonomials

$$u(\xi) = \sum_{n=1}^{\infty} c_n \, {}^{(1)}\mathcal{P}_n^{\mu}(\xi). \qquad (3.1.124)$$

Then

$$\begin{aligned} {}^{RL}_{0}D_t^{r+\alpha/2} u(\xi(t)) &= \left(\frac{2}{T}\right)^{r+\alpha/2} {}^{RL}_{-1}\mathcal{I}_\xi^{\mu-\alpha/2} \frac{d^r}{d\xi^r} {}^{RL}_{-1}D_\xi^{\mu} u(\xi) \\ &= \left(\frac{2}{T}\right)^{r+\alpha/2} \sum_{n=1}^{\infty} c_n \frac{\Gamma(n+\mu)}{\Gamma(n)} {}^{RL}_{-1}I_\xi^{\mu-\alpha/2} \frac{d^r}{d\xi^r} P_{n-1}(\xi), \end{aligned}$$

where

$$\frac{d^r}{d\xi^r} P_{n-1}(\xi) = \begin{cases} \dfrac{(n-1+r)!}{2r(n-1)!} P_{n-1-r}^{(r,r)}, (\xi) & r < n, \\ 0, & r \ge n. \end{cases}$$

Thus by multiplying with a proper weight function,

$$w(\xi) = (1+\xi)^{r/2-\mu+\alpha/2} (1-\xi)^{r/2},$$

the right-hand side of (3.1.123) becomes

$$\begin{aligned} \int_{\alpha_{min}}^{\alpha_{max}} \phi(\alpha) \left(\frac{2}{T}\right)^{2r+\alpha/2} \Bigg\| w(\xi) \sum_{n=r+1}^{\infty} c_n \frac{\Gamma(n+\mu)}{\Gamma(n)} \frac{(n-1+r)!}{2r(n-1)!} \\ \times {}^{RL}_{-1}I_\xi^{\mu-\alpha/2} P_{n-1-r}^{(r,r)}(\xi) \Bigg\|^2_{L^2(\Omega)} d\alpha. \end{aligned} \qquad (3.1.125)$$

By expanding the Jacobi polynomial as

$$P_{n-1-r}^{(r,r)}(\xi) = \\ (-1)^{n-1-r} \sum_{j=0}^{n-1-r} (-1)^j \binom{n-1+r+j}{j} \binom{n-1}{n-1-r-j} \left(\frac{1+\xi}{2}\right)^j.$$

Changing the order of summation and the integration, we obtain the fractional integral as

$$\begin{aligned} {}^{\text{RL}}_{-1}\mathcal{I}_{\xi}^{\mu-\alpha/2} \, P_{n-1-r}^{(r,r)}(\xi) &= (-1)^{n-1-r} \, (1+\xi)^{\mu-\alpha/2} \\ &\times \sum_{j=0}^{n-1-r} \binom{n-1+r+j}{j} \binom{n-1}{n-1-r-j} \frac{\Gamma(1+j)}{\Gamma(1+j+\mu-\alpha/2)} (-2)^{-j} (1+\xi)^j \\ &= (-1)^{n-1-r} \, (1+\xi)^{\mu-\alpha/2} \sum_{q=0}^{n-1-r} \tilde{c}_q(\alpha) \; P_q^{(r,r)}(\xi), \end{aligned} \tag{3.1.126}$$

where the coefficient, $\tilde{c}_q(\alpha)$ can be obtained using the orthogonality of Jacobi polynomials. Hence by taking $C_n = c_n \, (-1)^{n-1} \frac{\Gamma(n+\mu)}{\Gamma(n)}$, (3.1.125) takes the form

$$\begin{aligned} I_\phi &:= \int_{\alpha_{min}}^{\alpha_{max}} \phi(\alpha) \left(\frac{2}{T}\right)^{2r+\alpha} \\ &\times \left\| (1-\xi^2)^{r/2} \sum_{n=r+1}^{\infty} C_n \frac{(n-1+r)!}{2r(n-1)!} (-1)^{-r} \sum_{q=0}^{n-1-r} \tilde{c}_q(\alpha) P_q^{(r,r)}(\xi) \right\|_{L^2(\Omega)}^2 d\alpha \\ &= \int_{\alpha_{min}}^{\alpha_{max}} \phi(\alpha) \left(\frac{2}{T}\right)^{2r+\alpha} \\ &\times \left[\int_{-1}^{1} (1-\xi^2)^r \Big(\sum_{n=r+1}^{\infty} C_n \frac{(n-1+r)!}{2r(n-1)!} \sum_{q=0}^{n-1-r} \tilde{c}_q(\alpha) \, P_q^{(r,r)}(\xi) \Big)^2 d\xi \right] d\alpha. \end{aligned} \tag{3.1.127}$$

Then we change the order of two summations and obtain

$$\begin{aligned} I_\phi &= \int_{\alpha_{min}}^{\alpha_{max}} \phi(\alpha) \left(\frac{2}{T}\right)^{2r+\alpha} \\ &\times \left[\int_{-1}^{1} (1-\xi^2)^r \sum_{q=0}^{\infty} \Big(\sum_{n=r+1+q}^{\infty} C_n \frac{(n-1+r)!}{2r(n-1)!} \Big)^2 \tilde{c}_q^2(\alpha) \, (P_q^{(r,r)}(\xi))^2 d\xi \right] d\alpha \\ &= \sum_{q=0}^{\infty} \Big(\sum_{n=r+1+q}^{\infty} C_n \frac{(n-1+r)!}{2r(n-1)!} \Big)^2 \\ &\times \int_{-1}^{1} (1-\xi^2)^r \, (P_q^{(r,r)}(\xi))^2 d\xi \int_{\alpha_{min}}^{\alpha_{max}} \phi(\alpha) \left(\frac{2}{T}\right)^{2r+\alpha} \tilde{c}_q^2(\alpha) \, d\alpha. \end{aligned}$$

Using the orthogonality of Jacobi polynomials, we further derive

$$
\begin{aligned}
I_{\phi} &= \sum_{q=0}^{\infty} \left(\sum_{n=r+1+q}^{\infty} C_n \frac{(n-1+r)!}{2r(n-1)!} \right)^2 \\
&\quad \times \frac{2^{2r+1}}{2q+2r+1} \frac{((q+1)!)^2}{q!\,(q+2r)!} \int_{\alpha_{min}}^{\alpha_{max}} \phi(\alpha) \left(\frac{2}{T}\right)^{2r+\alpha} \tilde{c}_q^2(\alpha)\, d\alpha \\
&= \frac{2^{2r+1}(\frac{2}{T})^{2r}}{(2r)^2} \sum_{q=0}^{\infty} \left(\sum_{n=r+1+q}^{\infty} C_n \frac{(n-1+r)!}{(n-1)!} \right)^2 \\
&\quad \times \frac{(q+1)^2\, q!}{(2q+2r+1)(q+2r)!} \int_{\alpha_{min}}^{\alpha_{max}} \phi(\alpha) \left(\frac{2}{T}\right)^{\alpha} \tilde{c}_q^2(\alpha)\, d\alpha.
\end{aligned} \tag{3.1.128}
$$

Moreover, using the approximation of the solution given in (3.1.93) and by multiplying with the proper weight functions, the left-hand side of (3.1.123) takes the form

$$
\begin{aligned}
\left\| (1+\xi)^{-\mu+\alpha/2} (u - u_N) \right\|_U^2 &= \int_{\alpha_{min}}^{\alpha_{max}} \phi(\alpha) \left(\frac{2}{T}\right)^{\alpha} \\
&\quad \times \left\| (1+\xi)^{-\mu+\alpha/2} \sum_{n=N+1}^{\infty} c_n \left(\frac{\Gamma(n+\mu)}{\Gamma(n)} \right) {}^{\mathrm{RL}}_{-1}\mathcal{I}_{\xi}^{\mu-\alpha/2} P_{n-1}(\xi) \right\|_{L^2(\Omega)}^2 d\alpha,
\end{aligned} \tag{3.1.129}
$$

where

$$
{}^{\mathrm{RL}}_{-1}\mathcal{D}_{\xi}^{\alpha/2} = {}^{\mathrm{RL}}_{-1}\mathcal{D}_{\xi}^{\alpha/2-\mu+\mu} = {}^{\mathrm{RL}}_{-1}\mathcal{I}_{\xi}^{\mu-\alpha/2}\, {}^{\mathrm{RL}}_{-1}\mathcal{D}_{\xi}^{\mu},
$$

and the fractional derivative is taken using the derivative of poly-fractonomials. By expanding the Legendre polynomial as

$$
P_{n-1}(\xi) = (-1)^{n-1} \sum_{j=0}^{n-1} \binom{n-1+j}{j} \binom{n-1}{n-1-j} (-12)^{-j} (1+\xi)^j,
$$

and following similar steps as in (3.1.126), we obtain the fractional integral as

$$
{}^{\mathrm{RL}}_{-1}\mathcal{I}_{\xi}^{\mu-\alpha/2} P_{n-1}(\xi) = (-1)^{n-1} (1+\xi)^{\mu-\alpha/2} \sum_{q=0}^{n-1} \tilde{a}_q(\alpha)\, P_q(\xi), \tag{3.1.130}
$$

where the coefficients $\{\tilde{a}_q(\alpha)\}$ can be obtained using the orthogonality of Legendre polynomials. Hence, (3.1.129) takes the form

$$
\begin{aligned}
\left\| (1+\xi)^{-\mu+\alpha/2} (u - u_N) \right\|_U^2 &= \int_{\alpha_{min}}^{\alpha_{max}} \phi(\alpha) \left(\frac{2}{T}\right)^{\alpha} \\
&\quad \times \int_{-1}^{1} \left(\sum_{n=N+1}^{\infty} C_n \sum_{q=0}^{n-1} \tilde{a}_q(\alpha)\, P_q(\xi) \right)^2 d\xi\, d\alpha,
\end{aligned} \tag{3.1.131}
$$

where $C_n = c_n\,(-1)^{n-1}\,\frac{\Gamma(n+\mu)}{\Gamma(n)}$. We change the order of two summations to use the orthogonality of Legendre polynomials and obtain

$$
\begin{aligned}
&\big\|(1+\xi)^{-\mu+\alpha/2}\,(u-u_N)\big\|_U^2 = \\
&\int_{\alpha_{min}}^{\alpha_{max}} \phi(\alpha)\left(\frac{2}{T}\right)^{\alpha} \int_{-1}^{1}\Big(\sum_{q=0}^{N}\sum_{n=q+1}^{\infty} C_n\,\tilde{a}_q(\alpha)\,P_q(\xi) \\
&\quad + \sum_{q=N+1}^{\infty}\sum_{n=q+1}^{\infty} C_n\,\tilde{a}_q(\alpha)\,P_q(\xi)\Big)^2 \mathrm{d}\xi\,\mathrm{d}\alpha \\
&= \int_{\alpha_{min}}^{\alpha_{max}} \phi(\alpha)\left(\frac{2}{T}\right)^{\alpha}\Big(\sum_{q=0}^{N}\Big(\sum_{n=q+1}^{\infty} C_n\Big)^2 \tilde{a}_q^2(\alpha)\int_{-1}^{1}(P_q(\xi))^2\,\mathrm{d}\xi \\
&\quad + \sum_{q=N+1}^{\infty}\Big(\sum_{n=q+1}^{\infty} C_n\Big)^2 \tilde{a}_q^2(\alpha)\int_{-1}^{1}(P_q(\xi))^2\,\mathrm{d}\xi\Big)\,\mathrm{d}\alpha \\
&\le \sum_{q=N+1}^{\infty}\Big(\sum_{n=q+1}^{\infty} C_n\Big)^2 \frac{2}{2q+1}\int_{\alpha_{min}}^{\alpha_{max}} \phi(\alpha)\left(\frac{2}{T}\right)^{\alpha}\tilde{a}_q^2(\alpha)\,\mathrm{d}\alpha \\
&\le C\left(\frac{N!}{(N-r)!}\right)^2 \sum_{q=N+1}^{\infty}\Big(\sum_{n=q+1}^{\infty} C_n\,\frac{(n-1+r)!}{(n-1)!}\Big)^2 \frac{2}{2q+1} \\
&\quad \times \frac{(q+1)^2\,q!}{(2q+2r+1)(q+2r)!}\int_{\alpha_{min}}^{\alpha_{max}} \phi(\alpha)\left(\frac{2}{T}\right)^{\alpha}\tilde{a}_q^2(\alpha)\,\mathrm{d}\alpha \\
&\le C\left(\frac{N!}{(N-r)!}\right)^2 \int_{\alpha_{min}}^{\alpha_{max}} \phi(\alpha)\left(\frac{2}{T}\right)^{2r+\alpha} \big\|w(\xi)\,{}^{\mathrm{RL}}_{-1}D_{\xi}^{r+\alpha/2}u\big\|_{L^2(\Omega)}^2\,\mathrm{d}\alpha \\
&\le C\left(\frac{N!}{(N-r)!}\right)^2 \int_{\alpha_{min}}^{\alpha_{max}} \phi(\alpha)\,\big\|{}^{\mathrm{RL}}_{0}D_{t}^{r+\alpha/2}u\big\|_{L^2(\Omega)}^2\,\mathrm{d}\alpha,
\end{aligned}
$$

where we denoted $w(\xi) = (1+\xi)^{r/2-\mu+\alpha/2}\,(1-\xi)^{r/2}$. Therefore,

$$
\begin{aligned}
\|u-u_N\|_U^2 &\le \big\|(1+\xi)^{-\mu+\alpha/2}\,(u-u_N)\big\|_U^2 \\
&\le C\,N^{-2r}\int_{\alpha_{min}}^{\alpha_{max}} \phi(\alpha)\big\|{}^{\mathrm{RL}}_{0}D_{t}^{r+\alpha/2}u\big\|_{L^2(\Omega)}^2\,\mathrm{d}\alpha.
\end{aligned}
$$

This completes the proof. □

Remark 3.17 Since the inf-sup condition holds (see Theorem 3.15), by the Banach–Nečas–Babuška theorem (Ern and Guermond, 2013), the error in the numerical scheme is less than or equal to a constant times the projection error. Choosing the projection u_N in Theorem 3.16, we infer the spectral accuracy of the scheme.

Numerical Tests

In order to examine the convergence of the scheme with *modal* expansions, we consider problems with smooth and nonsmooth solutions.

Smooth Solutions

Let $\alpha \in [0, 2]$ and consider the following two cases:

- Case I: $u^{ext} = t^5, \;\; \phi(\alpha) = \Gamma(6-\alpha)/5!\,, \;\; f(t) = \frac{t^5 - t^3}{\log(t)}$
- Case II: $u^{ext} = t^3, \;\; \phi(\alpha) = \Gamma(4-\alpha)\sinh(\alpha)$, and

$$f(t) = \frac{6t(t^2 - \cosh(2) - \sinh(2)\log(t))}{\log(t)^2 - 1}.$$

By taking the simulation time $T = 2$ and for different choices of μ, we provide the convergence study in L^∞-norm, L^2-norm, H^{μ_1}-norm and ${}^{\phi}\mathcal{H}$-norm using the PG scheme and in L^∞-norm using the collocation scheme. It is observed that the choice of μ has an important effect on the convergence behavior of the scheme. For instance, since the exact solution is a polynomial, as $\mu \to 1$, we recover the exponential convergence in capturing the exact solution.

Tables 3.1 and 3.2 show the convergence behavior of the simulation results based on the PG scheme for the two case-I and II respectively.

Nonsmooth Solutions

Since the exact solution is not always known and in contrast to the standard fractional ODEs where the forcing term gives some regularity information about the exact solution, in distributed-order problems such a prediction is rather difficult to make. Hence, the fractional parameter μ can play the role of a *fine-tuning knob* giving the possibility of searching for the best/optimal case, where the highest rate can be achieved with minimal degrees of freedom. Here, we let $\alpha \in [0, 1]$ and consider the following two cases of singular solution, where by the proper choice of μ we can easily capture the singularity of the solution.

- Case III: $u^{ext} = t^{\mu_0}, \;\; \phi(\alpha) = \Gamma(1 + \mu_0 - \alpha)/(\mu_0)!, \;\; \mu_0 = 1/10, 9/10,$
- Case IV: $u^{ext} = t^{\mu_0}\sin(t), \;\; \phi(\alpha), \;\; \mu_0 = 75/100, 25/100.$

In Case III, we are able to obtain the exact solution only with one term by choosing $\mu = \mu_0$. In case-IV, we take $\mu = \mu_0$ and expand $\sin(t)$ using Taylor series. Table 3.3 shows the L^2-norm convergence of the PG scheme for two different distribution functions.

Table 3.1. *Case-I; PG scheme convergence study in L^∞-norm, L^2-norm, H^{μ_1}-norm and ${}^\phi\mathcal{H}$-norm, where $T = 2$. Source: Kharazmi et al. (2017).* *Copyright ©2017 Society for Industrial and Applied Mathematics. Reprinted with permission. All rights reserved.*

		$\mu_1 = 1 + 10^{-4}$		
N	L^∞-Error	L^2-Error	H^{μ_1}-Error	${}^\phi\mathcal{H}$-Error
2	9.49784	3.38063	20.604	6.52507
4	0.163486	0.0823368	0.802757	0.187176
6	9.71043×10^{-8}	6.7433×10^{-8}	8.37613×10^{-7}	1.70551×10^{-7}
8	2.9053×10^{-9}	2.32457×10^{-9}	3.53574×10^{-8}	6.59486×10^{-9}
10	2.27748×10^{-10}	2.01002×10^{-10}	3.67074×10^{-9}	6.38469×10^{-10}
		$\mu_1 = 1.1$		
N	L^∞-Error	L^2-Error	H^{μ_1}-Error	${}^\phi\mathcal{H}$-Error
2	9.6776	3.2898	23.3004	6.38693
4	0.160434	0.0661304	0.872809	0.157957
6	0.0000947942	0.0000589784	0.00107458	0.00015822
8	3.10668×10^{-6}	2.19939×10^{-6}	0.0000507737	6.59429×10^{-6}
10	2.48519×10^{-7}	1.9822×10^{-7}	5.5753×10^{-6}	6.61409×10^{-7}
		$\mu_1 = 1.5$		
N	L^∞-Error	L^2-Error	H^{μ_1}-Error	${}^\phi\mathcal{H}$-Error
2	9.8476	3.10681	35.8457	5.96161
4	0.102534	0.0264974	0.949183	0.0718496
6	0.000584995	0.00015106	0.0117235	0.000524729
8	0.0000272655	7.37649×10^{-6}	0.000989158	0.0000306404
10	2.75271×10^{-6}	7.75346×10^{-7}	0.000158823	3.72512×10^{-6}

Condition Number

The condition number of the constructed linear system is obtained for different distribution functions, $\phi(\alpha)$. Table 3.4 shows the condition number of the constructed linear system for case-I and II based on PG for the aforementioned distribution functions.

Moreover, three cases for the distribution function are considered: 1) the distribution is more biased toward the left of domain, 2) the distribution is symmetric and 3) the distribution is more biased toward the right of domain, namely *left biased*, *symmetric* and *right biased* respectively. The distribution functions are well-known normal, exponential, log-normal, Cauchy, Laplace, Beta and Maxwell distributions; however, they are truncated and normalized,

Table 3.2. *Case-II; PG scheme convergence study in L^∞-norm, L^2-norm, H^{μ_1}-norm and ${}^\phi\mathcal{H}$-norm, where $T = 2$. Source: Kharazmi et al. (2017). Copyright ©2017 Society for Industrial and Applied Mathematics. Reprinted with permission. All rights reserved.*

		$\mu_1 = 1 + 10^{-4}$		
N	L^∞-Error	L^2-Error	H^{μ_1}-Error	${}^\phi\mathcal{H}$-Error
2	0.379134	0.325253	1.44392	1.86897
4	6.80222×10^{-7}	6.33141×10^{-7}	4.61395×10^{-6}	5.29606×10^{-6}
6	5.22608×10^{-8}	4.52071×10^{-8}	4.80236×10^{-7}	5.08899×10^{-7}
8	1.27547×10^{-8}	9.98313×10^{-9}	1.0532×10^{-7}	1.049×10^{-7}
10	7.31142×10^{-9}	7.21402×10^{-9}	3.44882×10^{-8}	3.39574×10^{-8}
		$\mu_1 = 1.1$		
N	L^∞-Error	L^2-Error	H^{μ_1}-Error	${}^\phi\mathcal{H}$-Error
2	0.369682	0.263829	1.45384	1.62458
4	0.000646557	0.000569995	0.00548608	0.00499413
6	0.0000458334	0.0000438926	0.000636023	0.000511403
8	7.74333×10^{-6}	7.36329×10^{-6}	0.000147177	0.000107932
10	2.02013×10^{-6}	1.84714×10^{-6}	0.000048212	0.0000327428
		$\mu_1 = 1.5$		
N	L^∞-Error	L^2-Error	H^{μ_1}-Error	${}^\phi\mathcal{H}$-Error
2	0.288508	0.114871	1.25471	0.848595
4	0.00403916	0.00163979	0.0511667	0.0190804
6	0.000406095	0.000169817	0.0106909	0.00268063
8	0.0000789352	0.0000336939	0.00358698	0.000671243
10	0.0000219275	9.49574×10^{-6}	0.00153771	0.000228446

Table 3.3. *Case-IV; PG scheme convergence study in L^2-norm, where $T = 2$. Source: Kharazmi et al. (2017). Copyright ©2017 Society for Industrial and Applied Mathematics. Reprinted with permission. All rights reserved.*

	$\mu = 75/100$		$\mu = 25/100$	
N	$\phi(\alpha) = 1$	$\phi(\alpha) =$ Normal	$\phi(\alpha) = 1$	$\phi(\alpha) =$ Normal
2	1.56682×10^{-1}	1.62765×10^{-1}	1.5773×10^{-1}	1.548×10^{-1}
4	3.13043×10^{-3}	3.3898×10^{-3}	3.4228×10^{-3}	3.28626×10^{-3}
6	2.55359×10^{-5}	2.81522×10^{-5}	2.8956×10^{-5}	2.76729×10^{-5}
8	1.13562×10^{-7}	1.2512×10^{-7}	4.24126×10^{-7}	1.40114×10^{-7}
10	2.60471×10^{-9}	7.84647×10^{-10}	3.9524×10^{-7}	5.49882×10^{-8}

Table 3.4. *Case-I and II; PG scheme condition number of the constructed linear system, where $T = 2$. Source: Kharazmi et al. (2017).* *Copyright ©2017 Society for Industrial and Applied Mathematics. Reprinted with permission. All rights reserved.*

	$\mu = 2 - 10^{-8}$		$\mu = 2 - 10^{-1}$		$\mu = 1 + 1/2$		$\mu = 1 + 1/10$	
N	Case-I	Case-II	Case-I	Case-II	Case-I	Case-II	Case-I	Case-II
6	29706.682	4863.50	14319.465	2168.87	661.70145	70.1081	51.928935	7.19267
10	240000.55	33494.8	90197.388	11817.9	1339.2855	130.925	42.754111	9.35597
14	882010.62	118283	279501.78	35395.6	1941.9838	190.309	47.335770	13.0337
18	2.2811229×10^6	301479	633307.06	79324.0	2505.5107	247.627	50.713428	16.7944

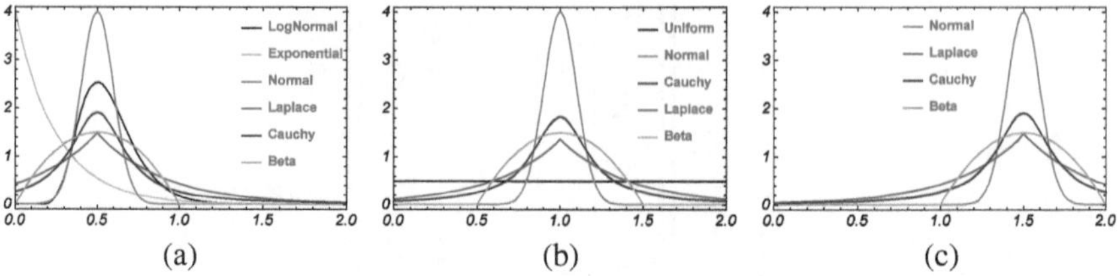

Figure 3.4 Distribution functions: (a) Left biased (b) Symmetric (c) Right biased. Source: Kharazmi et al. (2017). Copyright ©2017 Society for Industrial and Applied Mathematics. Reprinted with permission. All rights reserved.

see Figure 3.4. For these distributions, the condition number of the constructed linear system based on the two methods is computed and provided in Tables 3.5, 3.6 and 3.7.

3.2 Spectral Collocation Methods for FODEs

3.2.1 Jacobi–Gauss–Lobatto Interpolation

For $\alpha, \beta > -1$, let $\{x_j := x_{N,j}^{(\alpha,\beta)}, \omega_j := \omega_{N,j}^{(\alpha,\beta)}\}_{j=0}^{N}$ be the set of Jacobi–Gauss–Lobatto (JGL) quadrature nodes and weights, where the nodes are zeros of $(1-x^2)\partial_x P_N^{(\alpha,\beta)}(x)$ on $\Omega := (-1,1)$. Hereafter, we assume that $\{x_j\}$ are arranged in ascending order so that $x_0 = -1$ and $x_N = 1$. Moreover, to alleviate the burden of heavy notation, we sometimes drop the parameters α, β in the notation, whenever it is clear from the context. The JGL quadrature enjoys the exactness (see, e.g., Shen et al. (2011a, Ch. 3)):

$$\int_{-1}^{1} \phi(x)\omega^{(\alpha,\beta)}(x)\,dx = \sum_{j=0}^{N} \phi(x_j)\omega_j, \quad \forall\, \phi \in \mathbb{P}_{2N-1}, \tag{3.2.1}$$

where $\mathbb{P}_N$ is the set of all polynomials of degree at most N.

Let $\mathcal{I}_N u$ be the Lagrange polynomial interpolant of $u \in C(\bar{\Omega})$ at JGL points defined by

$$(\mathcal{I}_N u)(x) = \sum_{j=0}^{N} u(x_j) l_j(x) \in \mathbb{P}_N, \tag{3.2.2}$$

where the interpolating basis polynomials $\{l_j\}_{j=0}^{N}$ can be expressed by

$$l_j(x) = \sum_{n=0}^{N} t_{nj} P_n^{(\alpha,\beta)}(x), \quad 0 \le j \le N, \tag{3.2.3}$$

Table 3.5. *Left biased distribution function; PG scheme condition number of the constructed linear system, where $T = 2$. Source: Kharazmi et al. (2017). Copyright ©2017 Society for Industrial and Applied Mathematics. Reprinted with permission. All rights reserved.*

N	LogNormal	Exponential	Normal	Laplace	Cauchy	Beta
			$\mu = 2 - 1/10$			
6	62101.5	130227	85410.4	21714.7	31361	70458.4
10	1.28119×10^6	2.62266×10^6	2.15167×10^6	186527	294630	1.51681×10^6
14	9.84911×10^6	1.61563×10^7	1.98724×10^7	668178	1.04066×10^6	1.22216×10^6
18	4.48721×10^7	5.34428×10^7	1.0748×10^8	1.62018×10^6	2.41399×10^6	5.86944×10^6
			$\mu = 1 + 1/10$			
6	200.626	505.679	300.643	71.5348	100.504	233.849
10	654.259	1397.74	1309.12	91.8644	140.467	816.254
14	1322.63	1969.15	3437.72	64.4093	98.4193	1780.96
18	2145.61	2178	7037.99	70.7541	91.8767	3134.64

Table 3.6. *Symmetric distribution function; PG scheme condition number of the constructed linear system, where $T = 2$. Source: Kharazmi et al. (2017).* *Copyright ©2017 Society for Industrial and Applied Mathematics. Reprinted with permission. All rights reserved.*

N	Uniform	Normal	Cauchy	Laplace	Beta
			$\mu = 2 - 1/10$		
6	3104.6331	16261.6	9328.53	6969.22	13404.7
10	17244.219	220205	80110.2	51304.3	157767
14	52095.460	1.28827×10^6	308549	179737	817268
18	117338.89	4.86517×10^6	803362	441628	2.77821×10^6
			$\mu = 1 + 1/10$		
6	9.5451677	44.0857	27.6858	17.9872	37.4224
10	11.211269	85.3822	30.0092	21.361	64.2331
14	12.360897	127.844	30.9343	20.1997	87.2794
18	15.900925	172.888	34.2126	22.1982	107.403

Table 3.7. *Right biased distribution function; PG scheme condition number of the constructed linear system, where $T = 2$. Source: Kharazmi et al. (2017).* *Copyright ©2017 Society for Industrial and Applied Mathematics. Reprinted with permission. All rights reserved.*

N	Normal	Laplace	Cauchy	Beta
		$\mu = 2 - 1/10$		
6	2985.08	2274.39	2368.82	2105.03
10	23439.3	13948.9	15471.9	13474.7
14	92925.7	45201.4	52626.9	45325.7
18	259993	107082	129454	110813
		$\mu = 1 + 1/10$		
6	6.99801	6.60933	6.54308	6.18762
10	10.8049	9.81523	10.0861	10.1067
14	15.1563	14.2958	14.0265	14.6423
18	19.4415	18.4003	18.8445	18.8791

where

$$t_{nj} := \frac{\omega_j}{\tilde{\gamma}_n^{(\alpha,\beta)}} P_n^{(\alpha,\beta)}(x_j),$$

with

$$\tilde{\gamma}_n^{(\alpha,\beta)} = \gamma_n^{(\alpha,\beta)}, \quad 0 \le n \le N-1; \quad \tilde{\gamma}_N^{(\alpha,\beta)} = \Big(2 + \frac{\alpha+\beta+1}{N}\Big)\gamma_N^{(\alpha,\beta)}. \tag{3.2.4}$$

Connection Problems

In the spectral algorithms, the transform between Jacobi expansions with different parameters will be used frequently. This defines a *connection problem* that connects the representations of polynomials between two spaces:

$$\mathbb{P}_N = \mathrm{span}\big\{P_n^{(\alpha,\beta)} \,:\, 0 \le n \le N\big\} = \mathrm{span}\big\{P_l^{(a,b)} \,:\, 0 \le l \le N\big\}$$

for $\alpha, \beta, a, b > -1$. More precisely, given the Jacobi expansion coefficients $\{\hat{u}_n^{(\alpha,\beta)}\}$ of $u \in \mathbb{P}_N$, find the coefficients $\{\hat{u}_l^{(a,b)}\}$ such that

$$u(x) = \sum_{n=0}^{N} \hat{u}_n^{(\alpha,\beta)} P_n^{(\alpha,\beta)}(x) = \sum_{l=0}^{N} \hat{u}_l^{(a,b)} P_l^{(a,b)}(x). \tag{3.2.5}$$

This can be resolved by the transform:

$$\hat{\boldsymbol{u}}^{(a,b)} = {}^{(\alpha,\beta)}\boldsymbol{C}^{(a,b)}\, \hat{\boldsymbol{u}}^{(\alpha,\beta)}, \tag{3.2.6}$$

where $\hat{\boldsymbol{u}}^{(\alpha,\beta)}$ and $\hat{\boldsymbol{u}}^{(a,b)}$ are column-$(N+1)$ vectors of the coefficients, and ${}^{(\alpha,\beta)}\boldsymbol{C}^{(a,b)}$ is the connection matrix of the transform from $\big\{P_n^{(\alpha,\beta)}\big\}$ to $\big\{P_l^{(a,b)}\big\}$. One finds from the orthogonality (2.1.3) and (3.2.5) that the entries of ${}^{(\alpha,\beta)}\boldsymbol{C}^{(a,b)}$, i.e., the connection coefficients, are given by

$${}^{(\alpha,\beta)}\boldsymbol{C}_{ln}^{(a,b)} := \frac{1}{\gamma_l^{(a,b)}} \int_{-1}^{1} P_l^{(a,b)}(x) P_n^{(\alpha,\beta)}(x)\, \omega^{(a,b)}(x) dx. \tag{3.2.7}$$

Note that ${}^{(\alpha,\beta)}\boldsymbol{C}_{ln}^{(a,b)} = 0$, if $n < l$, and the nonzero entries can be computed efficiently by using the recurrence formulas below.

Proposition 3.18 *For $\alpha, \beta, a, b > -1$, the connection coefficients in* (3.2.7) *can be computed recursively as follows.*

(i) *For $n = 0$,*

$${}^{(\alpha,\beta)}\boldsymbol{C}_{00}^{(a,b)} = 1, \quad {}^{(\alpha,\beta)}\boldsymbol{C}_{l0}^{(a,b)} = 0, \quad l = 1, \ldots, N. \tag{3.2.8}$$

(ii) *For $n = 1$,*

$$\begin{aligned} {}^{(\alpha,\beta)}\boldsymbol{C}_{01}^{(a,b)} &= \frac{\alpha(b+1) - \beta(a+1) + b - a}{a+b+2}, \\ {}^{(\alpha,\beta)}\boldsymbol{C}_{11}^{(a,b)} &= \frac{\alpha+\beta+2}{a+b+2}, \end{aligned} \tag{3.2.9}$$

and ${}^{(\alpha,\beta)}\boldsymbol{C}_{l1}^{(a,b)} = 0$ for $l = 2, \ldots, N$.

(iii) *With the initials in* (3.2.8)–(3.2.9), *we compute*

$$\begin{aligned}{}^{(\alpha,\beta)}\mathbf{C}^{(a,b)}_{l,n+1} &= \breve{a}_l\hat{a}_n\,{}^{(\alpha,\beta)}\mathbf{C}^{(a,b)}_{l-1,n} + (\breve{b}_l\hat{a}_n - \hat{b}_n)\,{}^{(\alpha,\beta)}\mathbf{C}^{(a,b)}_{ln}\\ &+ \breve{c}_l\hat{a}_n\,{}^{(\alpha,\beta)}\mathbf{C}^{(a,b)}_{l+1,n} - \hat{c}_n\,{}^{(\alpha,\beta)}\mathbf{C}^{(a,b)}_{l,n-1}\end{aligned} \tag{3.2.10}$$

for $n = 1,\ldots,N-1$, $l = 0,\ldots,n+1$, *and set* ${}^{(\alpha,\beta)}\mathbf{C}^{(a,b)}_{l,n+1} = 0$ *for* $l = n+2,\ldots,N$. *Here, we define* ${}^{(\alpha,\beta)}\mathbf{C}^{(a,b)}_{-1,n} = 0$ *and the coefficients*

$$\begin{aligned}\breve{a}_l &= \frac{2l(l+a+b)}{(2l+a+b-1)(2l+a+b)},\\ \breve{b}_l &= \frac{b^2-a^2}{(2l+a+b)(2l+a+b+2)},\\ \breve{c}_l &= \frac{2(l+a+1)(l+b+1)}{(2l+a+b+2)(2l+a+b+3)},\\ \hat{a}_n &= a_n^{(\alpha,\beta)},\quad \hat{b}_n = b_n^{(\alpha,\beta)},\quad \hat{c}_n = c_n^{(\alpha,\beta)},\end{aligned} \tag{3.2.11}$$

where $\{a_n^{(\alpha,\beta)}, b_n^{(\alpha,\beta)}, c_n^{(\alpha,\beta)}\}$ *are defined in* (2.1.6).

Proof Since $P_0^{(\alpha,\beta)}(x) = 1 = P_0^{(a,b)}(x)$, we have ${}^{(\alpha,\beta)}\mathbf{C}^{(a,b)}_{l0} = \delta_{l0}$, which follows from (3.2.7) and (2.1.3) directly. This leads to (3.2.8).

We now turn to (3.2.9). One verifies that

$$P_1^{(\alpha,\beta)} = \frac{\alpha+\beta+2}{a+b+2}P_1^{(a,b)} + \frac{\alpha(b+1)-\beta(a+1)+b-a}{a+b+2}P_0^{(a,b)}. \tag{3.2.12}$$

Then we can derive (3.2.9) from (3.2.7) and (2.1.6) again.

It remains to prove (3.2.10). By (3.2.7) and the three-term recurrence relation of the Jacobi polynomials, we have

$$\begin{aligned}{}^{(\alpha,\beta)}\mathbf{C}^{(a,b)}_{l,n+1} &= \frac{a_n^{(\alpha,\beta)}}{\gamma_l^{(a,b)}}\int_{-1}^{1} xP_l^{(a,b)}(x)P_n^{(\alpha,\beta)}(x)\omega^{(a,b)}(x)\mathrm{d}x\\ &\quad - b_n^{(\alpha,\beta)}\,{}^{(\alpha,\beta)}\mathbf{C}^{(a,b)}_{l,n} - c_n^{(\alpha,\beta)}\,{}^{(\alpha,\beta)}\mathbf{C}^{(a,b)}_{l,n-1}\\ &= \frac{a_n^{(\alpha,\beta)}}{a_l^{(a,b)}}\frac{\gamma_{l+1}^{(a,b)}}{\gamma_l^{(a,b)}}\,{}^{(\alpha,\beta)}\mathbf{C}^{(a,b)}_{l+1,n} + \frac{a_n^{(\alpha,\beta)}b_l^{(a,b)}}{a_l^{(a,b)}}\,{}^{(\alpha,\beta)}\mathbf{C}^{(a,b)}_{l,n}\\ &\quad + \frac{a_n^{(\alpha,\beta)}c_l^{(a,b)}}{a_l^{(a,b)}}\frac{\gamma_{l-1}^{(a,b)}}{\gamma_l^{(a,b)}}\,{}^{(\alpha,\beta)}\mathbf{C}^{(a,b)}_{l-1,n}\\ &\quad - b_n^{(\alpha,\beta)}\,{}^{(\alpha,\beta)}\mathbf{C}^{(a,b)}_{l,n} - c_n^{(\alpha,\beta)}\,{}^{(\alpha,\beta)}\mathbf{C}^{(a,b)}_{l,n-1}.\end{aligned}$$

Working out the constants, we obtain the desired formula. □

3.2.2 Fractional Pseudo-spectral Differentiation Matrix (F-PSDM)

It is known that the pseudo-spectral differentiation process is the heart of a collocation/pseudo-spectral method for PDEs. Typically, for any $u \in \mathbb{P}_N$, the differentiation $D^k u$ is carried out via (3.2.2) in an exact manner, that is,

$$D^k u(x) = \sum_{j=0}^{N} u(x_j) D^k l_j(x), \quad k \in \mathbb{N}. \tag{3.2.13}$$

It is straightforward to extend this to the fractional pseudo-spectral differentiation. More precisely, for any $u \in \mathbb{P}_N$,

$$(D^\mu u)(x) = \sum_{j=0}^{N} u(x_j) D^\mu l_j(x), \tag{3.2.14}$$

where D^μ is the Riemann–Liouville or Caputo fractional derivative of order $\mu > 0$. However, in distinct contrast to (3.2.13), we have $D^\mu u, D^\mu l_j \notin \mathbb{P}_N$, if $\mu \notin \mathbb{N}$. To have insights into this, we introduce the space:

$$\mathcal{F}_N^{(\nu)} := \big\{(1+x)^\nu \phi \ : \ \forall \phi \in \mathbb{P}_N\big\}, \quad \nu \in \mathbb{R}, \tag{3.2.15}$$

and show the following properties.

Lemma 3.19 *For $\mu \in (k-1, k)$ with $k \in \mathbb{N}$, and for any $u \in \mathbb{P}_N$, we have*

$$ {}_{-1}^{\,\mathrm{C}}D_x^\mu u \in \mathcal{F}_{N-k}^{(k-\mu)}, \quad {}_{-1}^{\mathrm{RL}}D_x^\mu u \in \mathcal{F}_N^{(-\mu)}, \tag{3.2.16}$$

and

$$ {}_{-1}^{\,\mathrm{C}}D_x^\mu u \to 0 \quad \text{as} \quad x \to -1^+. \tag{3.2.17}$$

Proof It is clear that

$$D^k u \in \mathbb{P}_{N-k} = \operatorname{span}\big\{P_n \ : \ 0 \le n \le N-k\big\}.$$

Thus, we obtain

$$ {}_{-1}^{\,\mathrm{C}}D_x^\mu u = {}_{-1}^{\mathrm{RL}}I_x^{k-\mu}(D^k u) = (1+x)^{k-\mu}\phi \quad \text{for some } \phi \in \mathbb{P}_{N-k}. \tag{3.2.18}$$

This implies (3.2.16)–(3.2.17) for the Caputo case.

Similarly, in the Riemann–Liouville case, we deduce from (2.4.15) that ${}_{-1}^{\mathrm{RL}}I_x^{k-\mu} u \in \mathcal{F}_N^{(k-\mu)}$. Then by the definition of the Riemann–Liouville fractional derivative, we obtain from a direct calculation that

$$ {}_{-1}^{\mathrm{RL}}D_x^\mu u = D^k\big({}_{-1}^{\mathrm{RL}}I_x^{k-\mu} u\big) = (1+x)^{-\mu}\psi \quad \text{for some } \psi \in \mathbb{P}_N. \tag{3.2.19}$$

This ends the proof. □

Observe from (3.2.19) that the Riemann–Liouville fractional derivative of a polynomial might tend to infinity as $x \to -1$. This brings about some inconvenience for the computation of the related F-PSDM and implementation of the collocation scheme. This inspires us to multiply both sides of (3.2.14) by the singular factor $(1+x)^{\mu}$, leading to the modified Riemann–Liouville fractional pseudo-spectral differentiation:

$$\big({}^{\rm RL}_{-1}\widehat{D}^{\mu}_{x}u\big)(x) = \sum_{j=0}^{N} u(x_j)\big({}^{\rm RL}_{-1}\widehat{D}^{\mu}_{x}l_j\big)(x), \tag{3.2.20}$$

where

$${}^{\rm RL}_{-1}\widehat{D}^{\mu}_{x} := (1+x)^{\mu}\, {}^{\rm RL}_{-1}D^{\mu}_{x}.$$

With such a modification, we can recover the Riemann–Liouville fractional derivative values at $x_i \neq -1$ by

$$\big({}^{\rm RL}_{-1}D^{\mu}_{x}u\big)(x_i) = (1+x_i)^{-\mu}\big({}^{\rm RL}_{-1}\widehat{D}^{\mu}_{x}u\big)(x_i), \quad 1 \le i \le N. \tag{3.2.21}$$

Correspondingly, we can define the modified fractional integral and state some important properties as follows.

Lemma 3.20 *Let $u \in \mathbb{P}_N$ and $\{l_j\}$ be the Lagrange interpolating basis polynomials at JGL points as before. Define*

$${}^{\rm RL}_{-1}\hat{I}^{\mu}_{x} = (1+x)^{-\mu}\, {}^{\rm RL}_{-1}I^{\mu}_{x}, \quad {}^{\rm RL}_{-1}\widehat{D}^{\mu}_{x} := (1+x)^{\mu}\, {}^{\rm RL}_{-1}D^{\mu}_{x}, \quad \forall \mu \in \mathbb{R}^{+}. \tag{3.2.22}$$

Then we have

$${}^{\rm RL}_{-1}\hat{I}^{\mu}_{x}u,\, {}^{\rm RL}_{-1}\widehat{D}^{\mu}_{x}u \in \mathbb{P}_N, \quad {}_0\mathbb{P}_N = {\rm span}\big\{{}^{\rm RL}_{-1}\hat{I}^{\mu}_{x}h_j \,:\, 1 \le j \le N\big\}, \tag{3.2.23}$$

where ${}_0\mathbb{P}_N = \{\phi \in \mathbb{P}_N \,:\, \phi(-1) = 0\}$.

Proof By (3.2.19), ${}^{\rm RL}_{-1}\widehat{D}^{\mu}_{x}u \in \mathbb{P}_N$. Recall the formula

$${}^{\rm RL}_{-1}\hat{I}^{\mu}_{x}\{P_n(x)\} = \frac{n!}{\Gamma(n+\mu+1)}P_n^{(-\mu,\mu)}(x), \quad \mu \in \mathbb{R}^{+}. \tag{3.2.24}$$

Note that for any real $\mu > 0$, $P_n^{(-\mu,\mu)}(x)$ is a polynomial of degree n (cf. Subsection 2.3). Thus, we have

$$\mathbb{P}_N = {\rm span}\big\{{}^{\rm RL}_{-1}\hat{I}^{\mu}_{x}P_n \,:\, 0 \le n \le N\big\} = {\rm span}\big\{{}^{\rm RL}_{-1}\hat{I}^{\mu}_{x}l_j \,:\, 0 \le j \le N\big\}, \tag{3.2.25}$$

and ${}^{\rm RL}_{-1}\hat{I}^{\mu}_{x}u \in \mathbb{P}_N$.

We now show $({}^{\rm RL}_{-1}\hat{I}^{\mu}_{x}l_j)(-1) = 0$ for $1 \le j \le N$. It is clear that $\big\{(1+x)P_n^{(\mu,1)}\big\}_{n=0}^{N-1}$ forms a basis of ${}_0\mathbb{P}_N$, and by (2.4.13) with $\alpha = \mu$ and $\beta = 1$,

$$ {}_{-1}^{\mathrm{RL}}\hat{I}_x^{\mu}\{(1+x)P_n^{(\mu,1)}(x)\} = \frac{(n+1)!}{\Gamma(n+\mu+2)}(1+x)P_n^{(0,1+\mu)}(x). \quad (3.2.26) $$

Since $l_j \in {}_0\mathbb{P}_N$, the identity (3.2.26) implies $({}_{-1}^{\mathrm{RL}}\hat{I}_x^{\mu} l_j)(-1) = 0$ for $1 \le j \le N$. □

Caputo Fractional Pseudo-spectral Differentiation Matrices (F-PSDM) As before, we use boldface uppercase (resp. lowercase) letters to denote matrices (resp. vectors), and simply denote the entries of a matrix $\boldsymbol{A}$ by $\boldsymbol{A}_{ij}$. Introduce the Caputo F-PSDM of order μ :

$$ {}^{\mathrm{C}}\boldsymbol{D}^{(\mu)} \in \mathbb{R}^{(N+1)\times(N+1)}, \quad {}^{\mathrm{C}}\boldsymbol{D}_{ij}^{(\mu)} = {}_{-1}^{\mathrm{C}}D_x^{\mu}\, l_j(x_i), \quad 0 \le i, j \le N. \quad (3.2.27) $$

In particular, for $\mu = k \in \mathbb{N}$, we denote $\boldsymbol{D}^{(k)} = {}^{\mathrm{C}}\boldsymbol{D}^{(k)}$ and $\boldsymbol{D} = \boldsymbol{D}^{(1)}$.

Remarkably, the higher-order Caputo fractional PSDM at JGL points can be computed by using the following recursive relation.

Theorem 3.21 *Let $\mu \in (0,1)$. Then we have*

$$ {}^{\mathrm{C}}\boldsymbol{D}^{(k+\mu)} = {}^{\mathrm{C}}\boldsymbol{D}^{(\mu)}\boldsymbol{D}^{(k)} = {}^{\mathrm{C}}\boldsymbol{D}^{(\mu)}\boldsymbol{D}^{k}, \quad k \in \mathbb{N}, \quad (3.2.28) $$

where $\boldsymbol{D}^k$ stands for the product of k copies of the first-order PSDM at JGL points.

Proof For any $u \in \mathbb{P}_N$, we have

$$ u'(x) = \sum_{i=0}^{N} u(x_i) l_i'(x). \quad (3.2.29) $$

Taking $u = l_j'$ in (3.2.29), leads to

$$ l_j''(x) = \sum_{i=0}^{N} l_j'(x_l) l_i'(x), \quad (3.2.30) $$

which implies

$$ {}_{-1}^{\mathrm{C}}D_x^{1+\mu} l_j(x) = {}_{-1}^{\mathrm{RL}}I_x^{1-\mu} l_j''(x) = \sum_{i=0}^{N} l_j'(x_i)\, {}_{-1}^{\mathrm{RL}}I_x^{1-\mu} l_i'(x) = \sum_{i=0}^{N} l_j'(x_i)\, {}_{-1}^{\mathrm{C}}D_x^{\mu} l_i(x). \quad (3.2.31) $$

Taking $x = x_i$ in the above, we obtain the matrix identity:

$$ {}^{\mathrm{C}}\boldsymbol{D}^{(1+\mu)} = {}^{\mathrm{C}}\boldsymbol{D}^{(\mu)}\boldsymbol{D}, \quad \mu \in (0,1). \quad (3.2.32) $$

This leads to (3.2.28) with $k = 1$. Taking $u = h_j^{(k)}(x)$ in (3.2.29), we can derive the first identity in (3.2.28) in the same fashion.

Using the property (see Shen et al. (2011a, Thm. 3.10))

$$\boldsymbol{D}^{(k)} = \boldsymbol{D}^k, \quad k \in \mathbb{N}, \tag{3.2.33}$$

we obtain the second identity in (3.2.28). □

It is seen from Theorem 3.21 that the computation of Caputo F-PSDM of any order at JGL points boils down to computing the first-order usual PSDM $\boldsymbol{D}$ (whose explicit formula can be found in e.g., Shen et al. (2011a)), and the Caputo F-PSDM ${}^{C}\boldsymbol{D}^{(\mu)}$ with $\mu \in (0, 1)$. We present the formulas below.

Theorem 3.22 *Let* $\{x_j = x_{N,j}^{(\alpha,\beta)}\}_{j=0}^{N}$ *with* $\alpha, \beta > -1$, $x_0 = -1$ *and* $x_N = 1$ *be the JGL points, and let* $\{\omega_j = \omega_{N,j}^{(\alpha,\beta)}\}_{j=0}^{N}$ *be the corresponding quadrature weights. Then the entries of* ${}^{C}\boldsymbol{D}^{(\mu)}$ *with* $\mu \in (0, 1)$ *can be computed by*

$${}^{C}\boldsymbol{D}_{ij}^{(\mu)} = (1 + x_i)^{1-\mu} \sum_{l=1}^{N} \frac{(l-1)!}{\Gamma(l+1-\mu)} s_{lj}\, P_{l-1}^{(\mu-1,1-\mu)}(x_i) \tag{3.2.34}$$

for $0 \le i, j \le N$, *where*

$$\begin{aligned} s_{lj} &= \frac{1}{2} \sum_{n=l-1}^{N} (n + \alpha + \beta + 1)\, {}^{(\alpha+1,\beta+1)}C_{l-1,n-1}^{(0,0)}\, t_{nj}, \\ t_{nj} &:= \frac{\omega_j}{\tilde{\gamma}_n^{(\alpha,\beta)}} P_n^{(\alpha,\beta)}(x_j), \end{aligned} \tag{3.2.35}$$

$\{{}^{(\alpha+1,\beta+1)}C_{l-1,n-1}^{(0,0)}\}$ *are the Jacobi-to-Legendre connection coefficients, and* $\{\tilde{\gamma}_n^{(\alpha,\beta)}\}$ *are defined in* (3.2.4). *For* $\alpha = \beta = 0$, *we can alternatively compute the coefficients* $\{s_{lj}\}$ *by*

$$s_{lj} = \frac{1}{\gamma_{l-1}} \{\delta_{jN} + (-1)^l \delta_{j0} - \omega_j P_{l-1}'(x_j)\}, \quad \gamma_{l-1} = \frac{2}{2l-1}. \tag{3.2.36}$$

Modified Riemann–Liouville Fractional Pseudo-spectral Differentiation Matrices

Introduce the matrices

$${}^{RL}\widehat{\boldsymbol{D}}^{(\mu)}, \hat{\boldsymbol{I}}^{(\mu)} \in \mathbb{R}^{(N+1)\times(N+1)}, \tag{3.2.37}$$

where

$${}^{RL}\widehat{\boldsymbol{D}}_{ij}^{(\mu)} = \left({}_{-1}^{RL}\widehat{D}_x^{\mu} l_j\right)(x_i), \quad \hat{\boldsymbol{I}}_{ij}^{(\mu)} = \left({}_{-1}^{RL}\hat{I}_x^{\mu} l_j\right)(x_i), \quad 0 \le i, j \le N.$$

We can show the following important property similar to Theorem 3.21.

Theorem 3.23 *Let $\{l_j\}$ be the JGL interpolating basis polynomials. Then for $\mu \in (k-1, k)$ with $k \in \mathbb{N}$, we have*

$$ {}^{\rm RL}\widehat{\boldsymbol{D}}^{(\mu)} = \breve{\boldsymbol{D}}^{(k)} \hat{\boldsymbol{I}}^{(k-\mu)}, \tag{3.2.38} $$

where the entries of $\breve{\boldsymbol{D}}^{(k)}$ are given by

$$ \breve{\boldsymbol{D}}_{ij}^{(k)} = (1+x)^{\mu} D^k\big\{(1+x)^{k-\mu} l_j(x)\big\}\big|_{x=x_i}, \quad 0 \le i, j \le N. \tag{3.2.39} $$

Proof By (3.2.23), we can write that for any $u \in \mathbb{P}_N$,

$$ ({}_{-1}^{\rm RL}\hat{I}_x^{k-\mu} u)(x) = \sum_{i=0}^{N} ({}_{-1}^{\rm RL}\hat{I}_x^{k-\mu} u)(x_i) l_i(x). \tag{3.2.40} $$

Multiplying both sides by $(1+x)^{k-\mu}$, and using (3.2.22), we find

$$ ({}_{-1}^{\rm RL}I_x^{k-\mu} u)(x) = \sum_{i=0}^{N} ({}_{-1}^{\rm RL}\hat{I}_x^{k-\mu} u)(x_i)(1+x)^{k-\mu} l_i(x), \tag{3.2.41} $$

which implies

$$ \begin{aligned} \big({}^{\rm RL}D_{-}^{(\mu)} u\big)(x) &= D^k ({}_{-1}^{\rm RL}I_x^{k-\mu} u)(x) \\ &= \sum_{i=0}^{N} ({}_{-1}^{\rm RL}\hat{I}_x^{k-\mu} u)(x_i) D^k\big\{(1+x)^{k-\mu} l_i(x)\big\} \end{aligned} \tag{3.2.42} $$

for $x \in (-1, 1]$. To remove the singularity at $x = -1$, we multiply both sides of (3.2.42) by $(1+x)^{\mu}$, and reformulate the resulted identity by the modified operator in (3.2.20), leading to

$$ \big({}_{-1}^{\rm RL}\widehat{D}_x^{\mu} u\big)(x) = \sum_{i=0}^{N} ({}_{-1}^{\rm RL}\hat{I}_x^{k-\mu} u)(x_i)\Big\{(1+x)^{\mu} D^k\big\{(1+x)^{k-\mu} l_i(x)\big\}\Big\}. \tag{3.2.43} $$

Taking $u = l_j$ and $x = x_i$ in the above equation yields (3.2.38). □

Observe from (3.2.38) that it suffices to compute the modified fractional integral matrix $\hat{\boldsymbol{I}}^{(k-\mu)}$ with $k - \mu \in (0, 1)$, since $\breve{\boldsymbol{D}}^{(k)}$ can be expressed in terms of the PSDM of integer order, e.g., for $k = 1$,

$$ \breve{\boldsymbol{D}}^{(1)} = (1-\mu)\boldsymbol{I}_{N+1} + \boldsymbol{\Lambda}\boldsymbol{D}, \quad \boldsymbol{\Lambda} = \mathrm{diag}\big((1+x_0), \ldots, (1+x_N)\big), \tag{3.2.44} $$

where $\boldsymbol{I}_{N+1}$ is an identity matrix.

Theorem 3.24 *Let* $\{x_j = x_{N,j}^{(\alpha,\beta)}\}_{j=0}^{N}$ *with* $\alpha, \beta > -1$, $x_0 = -1$ *and* $x_N = 1$ *be the JGL points, and let* $\{\omega_j = \omega_{N,j}^{(\alpha,\beta)}\}_{j=0}^{N}$ *be the corresponding quadrature weights. Then the entries of* $\hat{\boldsymbol{I}}^{(\mu)}$ *with* $\mu \in (0,1)$ *can be computed by*

$$\hat{\boldsymbol{I}}_{ij}^{(\mu)} = \sum_{l=0}^{N} \frac{l!}{\Gamma(l+\mu+1)} \hat{s}_{lj} P_l^{(-\mu,\mu)}(x_i), \quad 0 \le i, j \le N, \tag{3.2.45}$$

where

$$\hat{s}_{lj} = \sum_{n=l}^{N} {}^{(\alpha,\beta)}\boldsymbol{C}_{ln}^{(0,0)} t_{nj}, \quad t_{nj} := \frac{\omega_j}{\tilde{\gamma}_n^{(\alpha,\beta)}} P_n^{(\alpha,\beta)}(x_j), \tag{3.2.46}$$

with $\{{}^{(\alpha,\beta)}\boldsymbol{C}_{ln}^{(0,0)}\}$ *being the Jacobi-to-Legendre connection coefficients, and* $\tilde{\gamma}_n^{(\alpha,\beta)}$ *defined in* (3.2.4). *In particular, if* $\alpha = \beta = 0$, *we have* $\hat{s}_{lj} = t_{lj}$.

Proof We expand the JGL Lagrange interpolating basis polynomials $\{l_j\}$ in terms of Legendre polynomials, and resort to the connection problem to transform between the bases as before. Equating (3.2.3) and the new expansion leads to

$$l_j(x) = \sum_{n=0}^{N} t_{nj} P_n^{(\alpha,\beta)}(x) = \sum_{l=0}^{N} \hat{s}_{lj} P_l(x), \quad 0 \le j \le N, \tag{3.2.47}$$

which defines a connection problem. Thus by (3.2.5),

$$\hat{s}_{lj} = \sum_{n=0}^{N} {}^{(\alpha,\beta)}\boldsymbol{C}_{ln}^{(0,0)} t_{nj} = \sum_{n=l}^{N} {}^{(\alpha,\beta)}\boldsymbol{C}_{ln}^{(0,0)} t_{nj}, \tag{3.2.48}$$

where we used the property ${}^{(\alpha,\beta)}\boldsymbol{C}_{ln}^{(0,0)} = 0$ if $n < l$. Then it follows from (3.2.24) immediately that for $\mu \in (0,1)$,

$$\hat{\boldsymbol{I}}_{ij}^{(\mu)} = \left({}_{-1}^{\mathrm{RL}}\hat{I}_x^{\mu} h_j\right)(x_i) = \sum_{l=0}^{N} \frac{l!\, \hat{s}_{lj}}{\Gamma(l+\mu+1)} P_l^{(-\mu,\mu)}(x_i), \quad 0 \le i, j \le N. \tag{3.2.49}$$

This leads to the desired formulas.

In the Legendre case, it is clear that the expansions in (3.2.47) are identical, so we have $\hat{s}_{lj} = t_{lj}$. □

We conclude this section by providing some numerical study of (discrete) eigenvalues of F-PSDMs. Observe from (3.2.17) that the first row of ${}^{\mathrm{C}}D^{(\mu)}$ is entirely zero, so ${}^{\mathrm{C}}D^{(\mu)}$ is always singular. We therefore remove the "boundary" row/column, and define

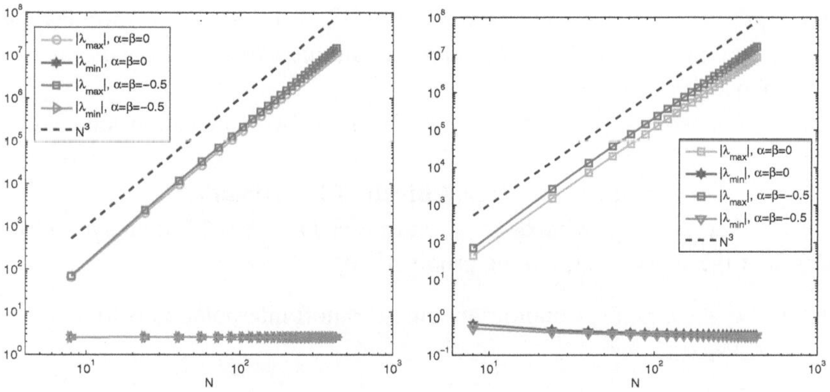

Figure 3.5 Maximum and minimum (in modulus) eigenvalues of F-PSDM with $\mu = 1.5$. Left: Caputo. Right: modified Riemann–Liouville. Source: Reprinted from Jiao et al. (2016) with permission from Elsevier.

$$
{}^{C}\boldsymbol{D}_{\text{in}}^{(\mu)} := \begin{cases} \left({}^{C}D_{ij}^{(\mu)}\right)_{1\le i,j\le N}, & \text{if } \mu \in (0,1), \\ \left({}^{C}D_{ij}^{(\mu)}\right)_{1\le i,j\le N-1}, & \text{if } \mu \in (1,2), \end{cases} \tag{3.2.50}
$$

with ${}^{C}\boldsymbol{D}_{ij}^{(\mu)} = \left({}^{C}D_{-}^{\mu} h_j\right)(x_i)$, which is invertible and allows for incorporating boundary condition(s). Similarly, we define

$$
{}^{RL}\widehat{\boldsymbol{D}}_{\text{in}}^{(\mu)} := \begin{cases} \left({}^{RL}\widehat{\boldsymbol{D}}_{ij}^{(\mu)}\right)_{1\le i,j\le N}, & \text{if } \mu \in (0,1), \\ \left({}^{RL}\widehat{\boldsymbol{D}}_{ij}^{(\mu)}\right)_{1\le i,j\le N-1}, & \text{if } \mu \in (1,2), \end{cases} \tag{3.2.51}
$$

where ${}^{RL}\widehat{\boldsymbol{D}}_{ij}^{(\mu)} = \left({}_{-1}^{RL}\widehat{D}_{x}^{\mu} h_j\right)(x_i)$.

In Figure 3.5, we illustrate the smallest and largest eigenvalues (in modulus) of these matrices. Observe that in both cases, the largest eigenvalue grows like $O(N^{2\mu})$, while the smallest one remains a constant in the Caputo case, and mildly decays with respect to N in the modified RL case.

Explicit Inverse of the Differentiation Matrix

The condition number of the collocation system of an FODE of order μ grows like $O(N^{2\mu})$, so its solution suffers from severe round-off errors, and it also becomes rather prohibitive to solve the linear system by an iterative method. Following the spirit of Costabile and Longo (2010) and Wang et al. (2014), we introduce the Caputo fractional Birkhoff interpolation, which generates a new interpolating polynomial basis with remarkable properties:

(i) It provides a stable way to invert the Caputo F-PSDM in (3.2.50), leading to optimal fractional integration preconditioners for the ill-conditioned collocation schemes.
(ii) It offers a basis for constructing well-conditioned collocation schemes.

Caputo Fractional Birkhoff Interpolation

Let $\{x_j = x_{N,j}^{(\alpha,\beta)}\}_{j=0}^{N}$ (with $x_0 = -1$ and $x_N = 1$) be the JGL points as before. Consider these two interpolating problems:

(i) For $\mu \in (0, 1)$, the Caputo fractional Birkhoff interpolation is to find $p \in \mathbb{P}_N$ such that

$$ {}_{-1}^{\,\mathrm{C}}D_x^{\mu}\, p(x_j) = {}_{-1}^{\,\mathrm{C}}D_x^{\mu} u(x_j), \quad 1 \le j \le N; \quad p(-1) = u(-1), \tag{3.2.52}$$

for any $u \in C(-1, 1)$ satisfying ${}_{-1}^{\,\mathrm{C}}D_x^{\mu} u \in C(-1, 1]$.

(ii) For $\mu \in (1, 2)$, the Caputo fractional Birkhoff interpolation is to find $p \in \mathbb{P}_N$ such that

$$ {}_{-1}^{\,\mathrm{C}}D_x^{\mu}\, p(x_j) = {}_{-1}^{\,\mathrm{C}}D_x^{\mu} u(x_j), \quad 1 \le j \le N-1; \quad p(\pm 1) = u(\pm 1), \tag{3.2.53}$$

for any $u \in C(-1, 1)$ satisfying ${}_{-1}^{\,\mathrm{C}}D_x^{\mu} u \in C(-1, 1)$.

As with the Lagrange interpolation, we search for a nodal basis to represent the interpolating polynomial p. More precisely, we look for $Q_j^{\mu} \in \mathbb{P}_N$ such that

(i) for $\mu \in (0, 1)$,

$$\begin{aligned} &Q_0^{\mu}(-1) = 1, \quad {}_{-1}^{\,\mathrm{C}}D_x^{\mu}\, Q_0^{\mu}(x_i) = 0, \quad 1 \le i \le N, \\ &{}_{-1}^{\,\mathrm{C}}D_x^{\mu}\, Q_j^{\mu}(x_i) = \delta_{ij}, \quad 1 \le i \le N; \quad Q_j^{\mu}(-1) = 0, \quad 1 \le j \le N, \end{aligned} \tag{3.2.54}$$

(ii) for $\mu \in (1, 2)$,

$$\begin{aligned} &Q_0^{\mu}(-1) = 1, \quad {}_{-1}^{\,\mathrm{C}}D_x^{\mu}\, Q_0^{\mu}(x_i) = 0, \quad 1 \le i \le N-1; \quad Q_0^{\mu}(1) = 0, \\ &{}_{-1}^{\,\mathrm{C}}D_x^{\mu}\, Q_j^{\mu}(x_i) = \delta_{ij}, \quad 1 \le i \le N-1; \quad Q_j^{\mu}(\pm 1) = 0, \quad 1 \le j \le N-1, \\ &Q_N^{\mu}(-1) = 0, \quad {}_{-1}^{\,\mathrm{C}}D_x^{\mu}\, Q_N^{\mu}(x_i) = 0, \quad 1 \le i \le N-1; \quad Q_N^{\mu}(1) = 1. \end{aligned} \tag{3.2.55}$$

Remark 3.25 We can show that for $\mu \in (0, 1)$, $Q_0^{\mu}(x) \equiv 1$ in (3.2.54), while for $\mu \in (1, 2)$, $Q_0^{\mu}(x) = (1 - x)/2$ and $Q_N^{\mu}(x) = (1 + x)/2$ in (3.2.55).

Indeed, for $\mu \in (0, 1)$, we find from (3.2.16) that ${}_{-1}^{\,\mathrm{C}}D_x^{\mu} Q_0^{\mu} \in \mathcal{F}_{N-1}^{(1-\mu)}$, i.e., ${}_{-1}^{\,\mathrm{C}}D_x^{\mu} Q_0^{\mu} = (1 + x)^{1-\mu}\psi$ for some $\psi \in \mathbb{P}_{N-1}$. From the interpolation condition: ${}_{-1}^{\,\mathrm{C}}D_x^{\mu}\, Q_0^{\mu}(x_i) = 0$, $1 \le i \le N$, we claim that $\psi(x_i) = 0$, $1 \le i \le$

N, i.e., $\psi(x)$ has N distinct zeros. This implies $\psi(x) \equiv 0$, as $\psi \in \mathbb{P}_{N-1}$. Then solving the fractional equation

$$ {}_{-1}^{\ C}D_x^{\mu}\, Q_0^{\mu}(x) = 0, \quad x \in (-1,1); \quad Q_0^{\mu}(-1) = 1, $$

leads to $Q_0^{\mu}(x) \equiv 1$.

Similarly, for $\mu \in (1,2)$, we deduce from (3.2.16) and (3.2.55) that

$$ \begin{aligned} &{}_{-1}^{\ C}D_x^{\mu}\, Q_0^{\mu}(x) = 0, \quad Q_0^{\mu}(-1) = 1, \;\; Q_0^{\mu}(1) = 0, \\ &{}_{-1}^{\ C}D_x^{\mu}\, Q_N^{\mu}(x) = 0, \quad Q_N^{\mu}(-1) = 0, \;\; Q_N^{\mu}(1) = 1. \end{aligned} $$

Noting that ${}_{-1}^{\ C}D_x^{\mu} = {}_{-1}^{\mathrm{RL}}I_x^{2-\mu} D^2$, we obtain immediately that $Q_0^{\mu}(x) = (1-x)/2$ and $Q_N^{\mu}(x) = (1+x)/2$. □

We can express the Caputo fractional Birkhoff interpolating polynomial p of (3.2.52) and (3.2.53), respectively, as

$$ p(x) = u(-1) + \sum_{j=1}^{N} {}_{-1}^{\ C}D_x^{\mu} u(x_j)\, Q_j^{\mu}(x), \quad \mu \in (0,1), \tag{3.2.56} $$

and for $\mu \in (1,2)$,

$$ p(x) = \frac{1-x}{2} u(-1) + \sum_{j=1}^{N-1} {}_{-1}^{\ C}D_x^{\mu} u(x_j)\, Q_j^{\mu}(x) + \frac{1+x}{2} u(1). \tag{3.2.57} $$

Therefore, $\{Q_j^{\mu}\}$ are dubbed as the Caputo fractional Birkhoff interpolating basis polynomials of order μ.

Introduce the matrices

$$ \boldsymbol{Q}^{(\mu)} = \begin{cases} \big(Q_{lj}^{(\mu)}\big)_{1 \le l,j \le N}, & \text{if } \mu \in (0,1), \\ \big(Q_{lj}^{(\mu)}\big)_{1 \le l,j \le N-1}, & \text{if } \mu \in (1,2), \end{cases} \tag{3.2.58} $$

where $Q_{lj}^{(\mu)} = Q_j^{\mu}(x_l)$. Remarkably, the matrix $\boldsymbol{Q}^{(\mu)}$ is the inverse of ${}^{C}\boldsymbol{D}_{\mathrm{in}}^{(\mu)}$ defined in (3.2.50).

Theorem 3.26 *For $\mu \in (k-1,k)$ with $k = 1,2$, we have*

$$ \boldsymbol{Q}^{(\mu)}\, {}^{C}\boldsymbol{D}_{\mathrm{in}}^{(\mu)} = {}^{C}\boldsymbol{D}_{\mathrm{in}}^{(\mu)}\, \boldsymbol{Q}^{(\mu)} = \boldsymbol{I}_{N+1-k}, \tag{3.2.59} $$

where $\boldsymbol{I}_{N+1-k}$ is the identity matrix of order $N+1-k$.

Proof We just prove (3.2.59) with $\mu \in (0,1)$, as the case $\mu \in (1,2)$ can be shown in a similar fashion. Since $Q_j^{\mu} \in \mathbb{P}_N$ and $Q_j^{\mu}(-1) = 0$ for $1 \le j \le N$, we can write

$$ Q_j^{\mu}(x) = \sum_{l=0}^{N} Q_j^{\mu}(x_l) h_l(x) = \sum_{l=1}^{N} Q_j^{\mu}(x_l) h_l(x), \quad 1 \le j \le N, $$

where $\{h_l\}$ are the Lagrange interpolating basis polynomials associated with JGL points. Thus,

$$ {}_{-1}^{\ C}D_x^{\mu}\, Q_j^{\mu}(x) = \sum_{l=1}^{N} Q_j^{\mu}(x_l)\; {}_{-1}^{\ C}D_x^{\mu}\, h_l(x). $$

Taking $x = x_i$ for $1 \le i \le N$ in the above equation, we obtain (3.2.59) with $\mu \in (0,1)$ from (3.2.54) straightforwardly. □

Computing the New Basis $\{Q_j^{\mu}\}$

The following property plays a crucial role in computing the new basis $\{Q_j^{\mu}\}$, which follows from Lemma 3.19.

Lemma 3.27 *Let $\{x_j\}_{j=0}^{N}$ be the JGL points with $x_0 = -1$ and $x_N = 1$. Then for $\mu \in (k-1,k)$ with $k = 1,2$, we have*

$$ D^k Q_j^{\mu}(x) = {}_{-1}^{\mathrm{RL}}D_x^{k-\mu}\Big\{\Big(\frac{1+x}{1+x_j}\Big)^{k-\mu}\hbar_j(x)\Big\}, \quad 1 \le j \le N+1-k, \tag{3.2.60} $$

where $\{\hbar_j\}_{j=1}^{N+1-k}$ are the Lagrange–Gauss interpolating basis polynomials associated with the JGL points $\{x_j\}_{j=1}^{N+1-k}$, that is,

$$ \hbar_j \in \mathbb{P}_{N-k}, \quad \hbar_j(x_i) = \delta_{ij} \quad \text{for} \quad 1 \le i,j \le N+1-k. \tag{3.2.61} $$

Proof Since $Q_j^{\mu} \in \mathbb{P}_N$, we obtain from Lemma 3.19 that ${}_{-1}^{\ C}D_x^{\mu}\, Q_j^{\mu} \in \mathcal{F}_{N-k}^{(k-\mu)}$. Noting that $\{\hbar_j\}_{j=1}^{N+1-k}$ forms a basis of $\mathbb{P}_{N-k}$, so by (3.2.15),

$$ {}_{-1}^{\ C}D_x^{\mu}\, Q_j^{\mu}(x) = \sum_{l=1}^{N+1-k} c_{lj}\,(1+x)^{k-\mu}\hbar_l(x), \quad 1 \le j \le N+1-k. $$

Letting $x = x_i$ and using the interpolating conditions, we find that $c_{lj} = (1+x_l)^{\mu-k}\delta_{lj}$. Thus, we obtain

$$ {}_{-1}^{\ C}D_x^{\mu}\, Q_j^{\mu}(x) = \Big(\frac{1+x}{1+x_j}\Big)^{k-\mu}\hbar_j(x), \quad 1 \le j \le N+1-k. \tag{3.2.62} $$

We have ${}_{-1}^{\ C}D_x^{\mu} = {}_{-1}^{\mathrm{RL}}I_x^{k-\mu}D^k$, so, we obtain (3.2.60) from (3.2.62) immediately.

□

With the aid of (3.2.60), we are able to derive the explicit formulas for computing the new basis.

Theorem 3.28 *Let $\big\{x_j = x_{N,j}^{(\alpha,\beta)}, \omega_j = \omega_{N,j}^{(\alpha,\beta)}\big\}_{j=0}^{N}$ with $\alpha,\beta > -1$, $x_0 = -1$ and $x_N = 1$ be the JGL quadrature nodes and weights. Then $\{Q_j^{\mu}\}$ can be computed as follows:*

(i) *For* $\mu \in (0,1)$,

$$Q_j^{\mu}(x) = \frac{1}{(1+x_j)^{1-\mu}} \sum_{l=0}^{N-1} \frac{\Gamma(l-\mu+2)}{l!} \breve{\xi}_{lj} \int_{-1}^{x} P_l(x)\mathrm{d}x \tag{3.2.63}$$

for $1 \le j \le N$, *where*

$$\breve{\xi}_{lj} = \sum_{n=l}^{N-1} {}^{(\alpha,\beta)}C_{ln}^{(\mu-1,1-\mu)} \xi_{nj}, \tag{3.2.64}$$

$$\xi_{nj} = \frac{1}{\gamma_n^{(\alpha,\beta)}} \Big\{ - \frac{c_j}{\beta+1} \frac{P_N^{(\alpha,\beta)}(-1)}{P_N^{(\alpha,\beta)}(x_j)} P_n^{(\alpha,\beta)}(-1)\omega_0 + P_n^{(\alpha,\beta)}(x_j)\omega_j \Big\}, \tag{3.2.65}$$

with $c_j = 1$ *for* $1 \le j \le N-1$, *and* $c_N = \alpha + 1$.

(ii) *For* $\mu \in (1,2)$,

$$Q_j^{\mu}(x) = \frac{1}{(1+x_j)^{2-\mu}} \sum_{l=0}^{N-2} \frac{\Gamma(l-\mu+3)}{l!} \breve{\xi}_{lj}\, \Phi_l(x) \tag{3.2.66}$$

for $1 \le j \le N-1$, *where*

$$\breve{\xi}_{lj} = \sum_{n=l}^{N-2} {}^{(\alpha,\beta)}C_{ln}^{(\mu-2,2-\mu)} \xi_{nj}, \tag{3.2.67}$$

$$\begin{aligned} \xi_{nj} = \frac{1}{\gamma_n^{(\alpha,\beta)}} \Big\{ & \frac{(x_j-1)P_N^{(\alpha,\beta)}(-1)}{2(\beta+1)P_N^{(\alpha,\beta)}(x_j)} P_n^{(\alpha,\beta)}(-1)\omega_0 \\ & - \frac{(1+x_j)P_N^{(\alpha,\beta)}(1)}{2(\alpha+1)P_N^{(\alpha,\beta)}(x_j)} P_n^{(\alpha,\beta)}(1)\omega_N + P_n^{(\alpha,\beta)}(x_j)\omega_j \Big\}, \end{aligned} \tag{3.2.68}$$

and

$$\Phi_l(x) := \frac{1+x}{2} \int_{-1}^{1} (t-1)P_l(t)\,dt + \int_{-1}^{x} (x-t)P_l(t)\,dt. \tag{3.2.69}$$

Here, $\{{}^{(\alpha,\beta)}C_{ln}^{(\mu-k,k-\mu)}\}$ *are the connection coefficients as defined in* (3.2.7).

Remark 3.29 Observe from (3.2.64) and (3.2.67) that if we take $(\alpha,\beta) = (\mu-k, k-\mu)$ with $k = 1,2$, we have $\breve{\xi}_{lj} = \xi_{lj}$, so (3.2.63) and (3.2.66) have the simplest form. Thus, it is preferable to choose these special parameters. □

Modified Riemann–Liouville Fractional Birkhoff Interpolation

Like the Caputo case, we consider the modified Riemann–Liouville fractional Birkhoff interpolating problems (i)–(ii) as defined in (3.2.52)–(3.2.53) with

${}_{-1}^{\mathrm{RL}}\widehat{D}_x^{\mu}$ in place of ${}_{-1}^{\mathrm{C}}D_x^{\mu}$. Similarly, we look for the interpolating basis polynomials $\{\widehat{Q}_j^{\mu}\}_{j=0}^{N} \subseteq \mathbb{P}_N$ such that

(i) for $\mu \in (0,1)$,

$$\begin{aligned}
&{}_{-1}^{\mathrm{RL}}\widehat{D}_x^{\mu}\, \widehat{Q}_0^{\mu}(x_i) = 0, \quad 1 \le i \le N; \quad \widehat{Q}_0^{\mu}(-1) = 1,\\
&{}_{-1}^{\mathrm{RL}}\widehat{D}_x^{\mu}\, \widehat{Q}_j^{\mu}(x_i) = \delta_{ij}, \quad 1 \le i \le N, \quad \widehat{Q}_j^{\mu}(-1) = 0, \quad 1 \le j \le N;
\end{aligned} \tag{3.2.70}$$

(ii) for $\mu \in (1,2)$,

$$\begin{aligned}
&{}_{-1}^{\mathrm{RL}}\widehat{D}_x^{\mu}\, \widehat{Q}_0^{\mu}(x_i) = 0, \quad 1 \le i \le N-1; \quad \widehat{Q}_0^{\mu}(-1) = 1, \quad \widehat{Q}_0^{\mu}(1) = 0,\\
&{}_{-1}^{\mathrm{RL}}\widehat{D}_x^{\mu}\, \widehat{Q}_j^{\mu}(x_i) = \delta_{ij}, \quad 1 \le i \le N-1; \quad \widehat{Q}_j^{\mu}(\pm 1) = 0, \quad 1 \le j \le N-1,\\
&{}_{-1}^{\mathrm{RL}}\widehat{D}_x^{\mu}\, \widehat{Q}_N^{\mu}(x_i) = 0, \quad 1 \le i \le N-1; \quad \widehat{Q}_N^{\mu}(-1) = 0, \quad \widehat{Q}_N^{\mu}(1) = 1.
\end{aligned} \tag{3.2.71}$$

Then for any $u \in \mathbb{P}_N$, we can write

$$\begin{aligned}
u(x) &= u(-1)\widehat{Q}_0^{\mu}(x) + \sum_{j=1}^{N} {}_{-1}^{\mathrm{RL}}\widehat{D}_x^{\mu} u(x_j)\, \widehat{Q}_j^{\mu}(x) \qquad (\text{for } \mu \in (0,1))\\
&= u(-1)\widehat{Q}_0^{\mu}(x) + \sum_{j=1}^{N-1} {}_{-1}^{\mathrm{RL}}\widehat{D}_x^{\mu} u(x_j)\, \widehat{Q}_j^{\mu}(x) + u(1)\widehat{Q}_N^{\mu}(x)(\text{for } \mu \in (1,2)).
\end{aligned} \tag{3.2.72}$$

Introduce the matrices generated from the new basis:

$$\widehat{\boldsymbol{Q}}^{(\mu)} = \begin{cases} \big(\widehat{Q}_{lj}^{(\mu)}\big)_{1\le l,j\le N}, & \text{if } \mu \in (0,1),\\ \big(\widehat{Q}_{lj}^{(\mu)}\big)_{1\le l,j\le N-1}, & \text{if } \mu \in (1,2), \end{cases} \tag{3.2.73}$$

where $\widehat{Q}_{lj}^{(\mu)} = \widehat{Q}_j^{\mu}(x_l)$. Like Theorem 3.26, we can claim that $\widehat{\boldsymbol{Q}}^{(\mu)}$ is the inverse of ${}^{\mathrm{RL}}\widehat{\boldsymbol{D}}_{\mathrm{in}}^{(\mu)}$. As the proof of the theorem below is very similar to that of Theorem 3.26, we omit it.

Theorem 3.30 *For $\mu \in (k-1,k)$ with $k = 1,2$, we have*

$$\widehat{\boldsymbol{Q}}^{(\mu)}\,{}^{\mathrm{RL}}\widehat{\boldsymbol{D}}_{\mathrm{in}}^{(\mu)} = {}^{\mathrm{RL}}\widehat{\boldsymbol{D}}_{\mathrm{in}}^{(\mu)}\, \widehat{\boldsymbol{Q}}^{(\mu)} = \boldsymbol{I}_{N+1-k}, \tag{3.2.74}$$

where $\boldsymbol{I}_{N+1-k}$ is the identity matrix of order $N+1-k$.

Computing the New Basis $\{\widehat{Q}_j^{\mu}\}_{j=0}^{N}$

The following lemma is very useful for the computation.

Lemma 3.31 *Let $\mu \in (k-1,k)$ with $k=1,2$. Then for any $f \in {}_0\mathbb{P}_N$, the fractional equation*

$$ {}^{\rm RL}_{-1}\widehat{D}^{\mu}_{x} u(x) = f(x), \quad u(-1)=0, \tag{3.2.75}$$

has a unique solution $u \in {}_0\mathbb{P}_N$ of the form

$$ u(x) = {}^{\rm RL}_{-1}I^{\mu}_{x}\{(1+x)^{-\mu} f(x)\}. \tag{3.2.76}$$

In particular, for any $u \in \mathbb{P}_N$, we have

$$ {}^{\rm RL}_{-1}\widehat{D}^{\mu}_{x} u(-1) = 0 \quad \text{if and only if} \quad u(-1)=0. \tag{3.2.77}$$

For clarity of presentation, we deal with two cases, (i) $\mu \in (0,1)$ and (ii) $\mu \in (1,2)$, separately.

(i). $\{\widehat{Q}^{\mu}_{j}\}_{j=0}^{N}$ with $\mu \in (0,1)$. Using the properties (3.2.23) and (3.2.77), we obtain from the interpolating conditions in (3.2.70) that

$$ \big({}^{\rm RL}_{-1}\widehat{D}^{\mu}_{x}\, \widehat{Q}^{\mu}_{j}\big)(x) = h_j(x), \quad 1 \le j \le N; \quad \big({}^{\rm RL}_{-1}\widehat{D}^{\mu}_{x}\, \widehat{Q}^{\mu}_{0}\big)(x) = \xi h_0(x), \tag{3.2.78}$$

where $\{h_j\}$ are the JGL interpolating basis polynomials defined in (3.2.3), and ξ is a constant to be determined by $\widehat{Q}^{\mu}_{0}(-1)=1$. Note that thanks to (3.2.77), the condition $\widehat{Q}^{\mu}_{j}(-1)=0$ is built-in, as $h_j(-1)=0$ for $1 \le j \le N$. We summarize below the explicit representation of the new basis.

Theorem 3.32 *Let $\{x_j = x^{(\alpha,\beta)}_{N,j}, \omega_j = \omega^{(\alpha,\beta)}_{N,j}\}_{j=0}^{N}$ with $\alpha,\beta > -1, x_0=-1$ and $x_N=1$ be the JGL quadrature points and weights. Then $\{\widehat{Q}^{\mu}_{j}\}_{j=0}^{N}$ with $\mu \in (0,1)$ can be computed by*

$$ \widehat{Q}^{\mu}_{j}(x) = \zeta_j \sum_{l=0}^{N} \frac{\Gamma(l-\mu+1)}{l!}\, \hat{t}_{lj}\, P_l(x) \quad with \quad \hat{t}_{lj} = \sum_{n=l}^{N} {}^{(\alpha,\beta)}C^{(\mu,-\mu)}_{ln}\, t_{nj}, \tag{3.2.79}$$

where $\zeta_0 = 1/\Gamma(1-\mu)$, $\zeta_j = 1$ for $1 \le j \le N$, $\{{}^{(\alpha,\beta)}C^{(\mu,-\mu)}_{ln}\}$ are the connection coefficients defined in Subsection 3.2.1, and

$$ t_{nj} = \frac{\omega_j}{\tilde{\gamma}^{(\alpha,\beta)}_{n}} P^{(\alpha,\beta)}_{n}(x_j), \tag{3.2.80}$$

with $\tilde{\gamma}^{(\alpha,\beta)}_{n}$ being defined in (3.2.4).

Remark 3.33 If $(\alpha,\beta) = (\mu,-\mu)$ with $\mu \in (0,1)$, we have $\hat{t}_{lj} = t_{lj}$, so $\widehat{Q}^{\mu}_{j}$ has the simplest form. □

(ii). $\{\widehat{Q}^{\mu}_{j}\}_{j=0}^{N}$ with $\mu \in (1,2)$. It is essential to derive the identities like (3.2.78). Indeed, using (3.2.23) and (3.2.77), we obtain from the interpolating conditions in (3.2.71) that

$$\left({}^{\mathrm{RL}}_{-1}\widehat{D}^{\mu}_{x}\,\widehat{Q}^{\mu}_{0}\right)(x) = (\tau_0+\kappa_0\,x)\hat{h}_0(x), \quad \widehat{Q}^{\mu}_{0}(-1) = 1, \quad \widehat{Q}^{\mu}_{0}(1) = 0; \tag{3.2.81}$$

$$\left({}^{\mathrm{RL}}_{-1}\widehat{D}^{\mu}_{x}\,\widehat{Q}^{\mu}_{j}\right)(x) = \frac{x+\tau_j}{x_j+\tau_j}\hat{h}_j(x), \quad \widehat{Q}^{\mu}_{j}(1) = 0, \quad 1 \le j \le N-1; \tag{3.2.82}$$

$$\left({}^{\mathrm{RL}}_{-1}\widehat{D}^{\mu}_{x}\,\widehat{Q}^{\mu}_{N}\right)(x) = \tau_N(1+x)\,\hat{h}_0(x), \quad \widehat{Q}^{\mu}_{N}(1) = 1, \tag{3.2.83}$$

where $\{\hat{h}_j\}_{j=0}^{N-1}$ are the Lagrange interpolating basis polynomials at JGL points $\{x_j\}_{j=0}^{N-1}$, that is,

$$\hat{h}_j(x) \in \mathcal{P}_{N-1}, \quad \hat{h}_j(x_i) = \delta_{ij}, \quad 0 \le i,j \le N-1. \tag{3.2.84}$$

In (3.2.81)–(3.2.83), $\{\tau_j\}_{j=0}^{N}$ and κ_0 are constants to be determined by the corresponding conditions at $x = \pm 1$, e.g., $\widehat{Q}^{\mu}_{j}(1) = 0$ in (3.2.82). It is noteworthy that thanks to (3.2.77), the interpolating condition: $\widehat{Q}^{\mu}_{j}(-1) = 0$ is built in $\left({}^{\mathrm{RL}}_{-1}\widehat{D}^{\mu}_{x}\,\widehat{Q}^{\mu}_{j}\right)(-1) = 0$ for $1 \le j \le N$.

In what follows, we shall use the three-term recurrence relation of Jacobi polynomials:

$$x\,P_l^{(\mu,1-\mu)} = a_{l+1}P_{l+1}^{(\mu,1-\mu)} + b_l\,P_l^{(\mu,1-\mu)} + c_{l-1}\,P_{l-1}^{(\mu,1-\mu)}, \quad l \ge 0, \tag{3.2.85}$$

where $c_{-1} = 0$, $\mu \in (1,2)$, and

$$a_{l+1} = \frac{l+2}{2l+3}, \quad b_l = \frac{1-2\mu}{(2l+1)(2l+3)}, \quad c_{l-1} = \frac{(l+\mu)(l-\mu+1)}{(l+1)(2l+1)}. \tag{3.2.86}$$

As before, it is necessary to expand $\{\hat{h}_j\}$ in terms of Jacobi polynomials with different parameters by using the notion of connection problems, so as to use compact and closed-form formulas to compute the new basis. We state below the connections of three expansions.

Lemma 3.34 *Let* $\left\{x_j = x_{N,j}^{(\alpha,\beta)}, \omega_j = \omega_{N,j}^{(\alpha,\beta)}\right\}_{j=0}^{N}$ *with* $\alpha, \beta > -1, x_0 = -1$ *and* $x_N = 1$ *be the JGL quadrature nodes and weights, and let* $\{\hat{h}_j\}_{j=0}^{N-1}$ *be the Lagrange interpolating basis polynomials associated with* $\{x_j\}_{j=0}^{N-1}$ *defined in (3.2.84). Then for* $\mu \in (1,2)$, *we have*

$$
\begin{aligned}
\hat{h}_j(x) &= \sum_{n=0}^{N-1} \rho_{nj}\, P_n^{(\alpha,\beta)}(x) = \sum_{l=0}^{N-1} \tilde{\rho}_{lj}\, P_l^{(\mu,1-\mu)}(x) \\
&= \hat{\rho}_{0j} + \sum_{l=0}^{N-2} \hat{\rho}_{l+1,j}\, (1+x) P_l^{(\mu,1-\mu)}(x), \quad 0 \le j \le N-1,
\end{aligned} \tag{3.2.87}
$$

where $\hat{\rho}_{00} = 1$ *and* $\hat{\rho}_{0j} = 0$ *for* $1 \le j \le N-1$*. Moreover, the coefficients can be computed by*

$$
\rho_{n0} = \frac{1}{\gamma_n^{(\alpha,\beta)}} \left\{ P_n^{(\alpha,\beta)}(-1)\omega_0 - \frac{\beta+1}{\alpha+1} \frac{P_N^{(\alpha,\beta)}(1)}{P_N^{(\alpha,\beta)}(-1)} P_n^{(\alpha,\beta)}(1)\omega_N \right\}, \tag{3.2.88}
$$

$$
\rho_{nj} = \frac{1}{\gamma_n^{(\alpha,\beta)}} \left\{ P_n^{(\alpha,\beta)}(x_j)\omega_j - \frac{1}{\alpha+1} \frac{P_N^{(\alpha,\beta)}(1)}{P_N^{(\alpha,\beta)}(x_j)} P_n^{(\alpha,\beta)}(1)\omega_N \right\}, \tag{3.2.89}
$$

$$
1 \le j \le N-1,
$$

$$
\tilde{\rho}_{lj} = \sum_{n=l}^{N-1} {}^{(\alpha,\beta)}C_{ln}^{(\mu,1-\mu)} \rho_{nj}, \quad 0 \le l, j \le N-1, \tag{3.2.90}
$$

and by the backward recurrence relation:

$$
\begin{aligned}
\hat{\rho}_{ij} &= \frac{1}{a_i}\tilde{\rho}_{ij} - \frac{b_i+1}{a_i}\hat{\rho}_{i+1,j} - \frac{c_i}{a_i}\hat{\rho}_{i+2,j}, \quad i = N-3, N-2, \ldots, 1, \\
\hat{\rho}_{N-1,j} &= \frac{1}{a_{N-1}}\tilde{\rho}_{N-1,j}, \quad \hat{\rho}_{N-2,j} = \frac{1}{a_{N-2}}\tilde{\rho}_{N-2,j} - \frac{b_{N-2}+1}{a_{N-2}}\hat{\rho}_{N-1,j},
\end{aligned} \tag{3.2.91}
$$

where $\{a_i, b_i, c_i\}$ *are given in* (3.2.86).

With the above preparations, we are ready to derive the explicit formulas of the new basis $\{\widehat{Q}_j^{\mu}\}_{j=0}^{N}$ with $\mu \in (1,2)$.

Theorem 3.35 *Let* $\{\tilde{\rho}_{lj}, \hat{\rho}_{lj}\}$ *be the coefficients defined in Lemma 3.34, and denote*

$$
d_l^{\mu} := \frac{\Gamma(l+2-\mu)}{(l+1)!}, \quad \varphi_l(x) := (1+x)P_l^{(0,1)}(x). \tag{3.2.92}
$$

Then $\{\widehat{Q}_j^{\mu}\}_{j=0}^{N}$ *at JGL points with* $\mu \in (1,2)$ *can be computed by the following:*

(i) *for* $j = 0$,

$$\widehat{Q}_0^{\mu}(x) = 1 + \Big(\tau_0 - \frac{1}{\Gamma(1-\mu)}\Big) \sum_{l=0}^{N-1} d_l^{\mu} \tilde{\rho}_{l0}\, \varphi_l(x) \\ + \frac{1}{\Gamma(1-\mu)} \sum_{l=0}^{N-2} d_l^{\mu} \hat{\rho}_{l+1,0}\, \varphi_l(x), \tag{3.2.93}$$

where

$$\tau_0 - \frac{1}{\Gamma(1-\mu)} = -\Big\{\frac{1}{2} + \frac{1}{\Gamma(1-\mu)} \sum_{l=0}^{N-2} d_l^{\mu} \hat{\rho}_{l+1,0}\Big\} \Big/ \sum_{l=0}^{N-1} d_l^{\mu} \tilde{\rho}_{l0};$$

(ii) *for* $j = N$,

$$\widehat{Q}_N^{\mu}(x) = \frac{1}{2} \sum_{l=0}^{N-1} d_l^{\mu} \tilde{\rho}_{l0}\, \varphi_l(x) \Big/ \sum_{l=0}^{N-1} d_l^{\mu} \tilde{\rho}_{l0}\,; \tag{3.2.94}$$

(iii) *for* $1 \le j \le N-1$,

$$\widehat{Q}_j^{\mu}(x) = \frac{\tau_j}{x_j + \tau_j} \sum_{l=0}^{N-2} d_l^{\mu} \hat{\rho}_{l+1,j}\, \varphi_l(x) \\ + \frac{1}{x_j + \tau_j} \times \sum_{l=0}^{N-2} d_l^{\mu} \hat{\rho}_{l+1,j} \Big\{ \frac{l+2-\mu}{l+2} a_{l+1} \varphi_{l+1}(x) \\ + b_l \varphi_l(x) + \frac{l+1}{l+1-\mu} c_{l-1} \varphi_{l-1}(x) \Big\}, \tag{3.2.95}$$

where $\{a_l, b_l, c_l\}$ *(with* $c_{-1} = 0$*) are defined in* (3.2.86), *and*

$$\tau_j = -1 + \mu\, \Gamma(2-\mu) \hat{\rho}_{1j} \Big/ \sum_{l=0}^{N-2} d_l^{\mu} \hat{\rho}_{l+1,j}. \tag{3.2.96}$$

Remark 3.36 We see from (3.2.90) that if $(\alpha, \beta) = (\mu, 1-\mu)$ with $\mu \in (1,2)$, the connections coefficients are not involved, so $\widehat{Q}_j^{\mu}$ has a simpler form. □

In what follows, we apply the tools developed in previous sections to construct well-conditioned collocation schemes for initial-valued or boundary-valued FODEs, and provide ample numerical results to show the accuracy and stability of the methods.

Initial-Value Caputo-Type FODEs

To fix the idea, we first consider the Caputo FODE of order $\mu \in (0,1)$:

$$ {}_{-1}^{\;\,C}D_x^{\mu} u(x) + \lambda(x) u(x) = f(x), \quad x \in (-1,1]; \quad u(-1) = u_-, \tag{3.2.97}$$

where λ, f are given continuous functions, and u_- is a given constant. The collocation scheme is to find $u_N \in \mathbb{P}_N$ such that

$$ {}_{-1}^{\,C}D_x^{\mu} u_N(x_j) + \lambda(x_j) u_N(x_j) = f(x_j), \quad 1 \le j \le N; \quad u_N(-1) = u_-. \tag{3.2.98}$$

The corresponding linear system under the Lagrange basis polynomials $\{h_j\}$ (L-COL) becomes

$$ \big({}^{C}\boldsymbol{D}_{\rm in}^{(\mu)} + \boldsymbol{\Lambda}\big)\boldsymbol{u} = \boldsymbol{f}, \tag{3.2.99}$$

where ${}^{C}\boldsymbol{D}_{\rm in}^{(\mu)}$ is defined as in (3.2.50), $\boldsymbol{\Lambda} = {\rm diag}(\lambda(x_1), \dots, \lambda(x_N))$, and

$$\begin{aligned} \boldsymbol{u} &= \big(u_N(x_1), \dots, u_N(x_N)\big)^T, \\ \boldsymbol{f} &= \big(f(x_1) - u_-\, {}_{-1}^{\,C}D_x^{\mu} h_0(x_1), \dots, f(x_N) - u_-\, {}_{-1}^{\,C}D_x^{\mu} h_0(x_N)\big)^T. \end{aligned} \tag{3.2.100}$$

The collocation system under the Birkhoff interpolation basis polynomials $\{Q_j^{\mu}\}$ in (3.2.54) (B-COL) becomes

$$ \big(\boldsymbol{I}_N + \boldsymbol{\Lambda}\boldsymbol{Q}^{(\mu)}\big)\boldsymbol{v} = \boldsymbol{g}, \tag{3.2.101}$$

where

$$ u_N(x) = u_- + \sum_{i=1}^{N} v_j Q_j^{\mu}(x), \quad \boldsymbol{v} = \big(v_1, \dots, v_N\big)^T, \tag{3.2.102}$$

and

$$ \boldsymbol{g} = \big(f(x_1) - u_-\lambda(x_1), \dots, f(x_N) - u_-\lambda(x_N)\big)^T. $$

It is noteworthy that different from (3.2.99), the unknowns of (3.2.101) are not the approximation of u at the collocation points, but of the Caputo fractional derivative values in view of (3.2.56).

Thanks to Theorem 3.26, we can precondition (3.2.99) to obtain the following preconditioned L-COL (PL-COL) system:

$$ \big(\boldsymbol{I}_N + \boldsymbol{Q}^{(\mu)}\boldsymbol{\Lambda}\big)\boldsymbol{u} = \boldsymbol{Q}^{(\mu)}\boldsymbol{f}. \tag{3.2.103}$$

In the computation, we take $\lambda(x)=2+\sin(25x)$ and $u(x)=E_{\mu,1}(-2(1+x)^{\mu})$ with $\mu = 0.8$ in (3.2.97), where the Mittag-Leffler function is defined by

$$ E_{\alpha,\beta}(z) = \sum_{n=0}^{\infty} \frac{z^n}{\Gamma(n\alpha + \beta)}. \tag{3.2.104}$$

In view of Remark 3.29, we choose the JGL points with $(\alpha, \beta) = (\mu - 1, 1 - \mu) = (-0.2, 0.2)$. We compare the condition numbers, number of iterations

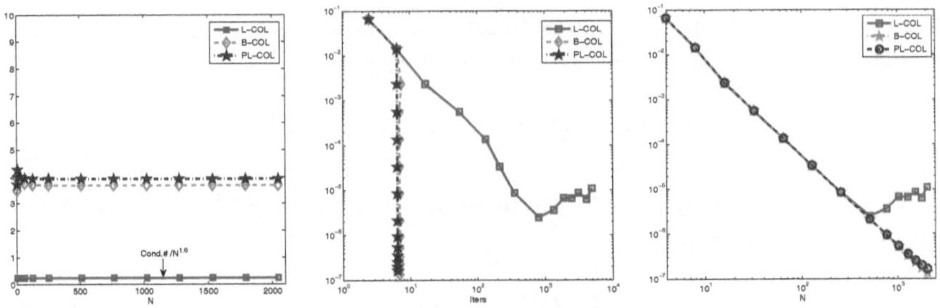

Figure 3.6 Comparison of condition numbers (left), iteration numbers against errors (middle), and errors against N at convergence in log-log scale (right) for (3.2.97). Source: Reprinted from Jiao et al. (2016) with permission from Elsevier.

(using BiCGSTAB in MATLAB) and convergence behavior (in discrete L^2-norm on fine equally-spaced grids) of three schemes (see Figure 3.6). Observe from Figure 3.6 (left) that the condition number of usual L-COL divided by $N^{2\mu}$ behaves like a constant, while that of PL-COL and B-COL remains a constant even for N up to 2000. As a result, the latter two schemes only require about 8 iterations to converge, while the usual L-COL scheme requires much more iterations with a degradation of accuracy as depicted in Figure 3.6 (middle). We record the convergence history of three methods in Figure 3.6 (right), and observe that two new schemes are stable even for very large N.

Boundary-value Caputo FODEs

We now turn to the boundary-value problem:

$$\begin{aligned} &{}_{-1}^{\;C}D_x^{\mu}u(x) + \lambda_1(x)\,{}_{-1}^{\;C}D_x^{\nu}u(x) + \lambda_2(x)u(x) = f(x), \quad x \in (-1,1); \\ &u(-1) = u_-, \quad u(1) = u_+, \quad 0 < \nu < \mu, \quad \mu \in (1,2), \end{aligned} \tag{3.2.105}$$

where λ_1, λ_2 and f are given continuous functions, and $u_\pm$ are given constants.

With a pre-computation of the Caputo fractional differentiation matrices of order μ and ν in Subsection 3.2.2, we can formulate the L-COL scheme as (3.2.99) straightforwardly. The counterpart of (3.2.101) i.e., the B-COL scheme, can be formulated as follows (cf. (3.2.57)): find

$$\begin{aligned} &u_N(x) = u_N^*(x) + \sum_{j=1}^{N-1} v_j\, Q_j^{\mu}(x), \quad \mu \in (1,2); \\ &u_N^*(x) := \frac{1-x}{2}u_- + \frac{1+x}{2}u_+, \end{aligned} \tag{3.2.106}$$

such that

$$\big(\boldsymbol{I}_{N-1} + \boldsymbol{\Lambda}_1 \bar{\boldsymbol{Q}}^{(\nu)} + \boldsymbol{\Lambda}_2 \boldsymbol{Q}^{(\mu)}\big)\boldsymbol{v} = \boldsymbol{g}, \tag{3.2.107}$$

where

$$\boldsymbol{\Lambda}_i = \mathrm{diag}(\lambda_i(x_1), \ldots, \lambda_i(x_{N-1})), \quad i = 1, 2,$$

$$\bar{\boldsymbol{Q}}_{ij}^{(\nu)} = {}_{-1}^{\;\mathrm{C}}D_x^{\nu} Q_j^{\mu}(x_i), \quad 1 \le i, j \le N-1,$$

$$\boldsymbol{g} = \big(f(x_1) - q_*(x_1), \ldots, f(x_{N-1}) - q_*(x_{N-1})\big)^T$$

with

$$q_* = \lambda_1 \, {}_{-1}^{\;\mathrm{C}}D_x^{\nu} u_N^* + \lambda_2 \, u_N^*.$$

Note that the entries of $\bar{\boldsymbol{Q}}^{(\nu)}$ can be evaluated by (3.2.66)–(3.2.69). Here, we omit the details.

Remark 3.37 If $\lambda_1 = 0$ and λ_2 is a constant, we can follow Wang et al. (2014, Proposition 3.5) to justify the coefficient matrix of (3.2.107) is well conditioned. Indeed, thanks to Theorem 3.26, the eigenvalues σ of $\boldsymbol{I}_{N-1} - \lambda_2 \boldsymbol{Q}^{(\mu)}$ satisfy

$$1 + \lambda_2 \lambda_{\max}^{-1} \le \sigma \le 1 + \lambda_2 \lambda_{\min}^{-1},$$

where $\lambda_{\max}$ and $\lambda_{\min}$ are respectively the largest and smallest eigenvalues of ${}^{\mathrm{C}}\boldsymbol{D}_{\mathrm{in}}^{(\mu)}$. Since $\lambda_{\min} = O(1)$ (see Figure 3.5 (right)), the condition number of $\boldsymbol{I}_{N-1} - \lambda_2 \boldsymbol{Q}^{(\mu)}$ is independent of N. □

Like (3.2.103), we can precondition the L-COL scheme by $\boldsymbol{Q}^{(\mu)}$ which leads to the PL-COL system.

In the following comparison, we set $\mu = 1.9, \nu = 0.7$ and $(\alpha, \beta) = (-0.1, 0.1)$ (cf. Remark 3.29), and take

$$\lambda_1(x) = 2 + \sin(4\pi x), \quad \lambda_2(x) = 2 + \cos x, \tag{3.2.108}$$

and

$$u(x) = e^{1+x} + (1+x)^{6+4/7} - 2(1+x)^{5+4/7}, \tag{3.2.109}$$

where we can use the formula

$${}_{-1}^{\;\mathrm{C}}D_x^{\mu} e^{1+x} = (1+x)^{k-\mu} E_{1,k+1-\mu}(1+x), \quad \mu \in (k-1, k), \ \ k = 1, 2,$$

to work out $f(x)$.

Once again, we observe from Figure 3.7 that the new schemes: B-COL and PL-COL are well conditioned, attain the expected convergence order about 10 iterations, and lead to stable computation for large N.

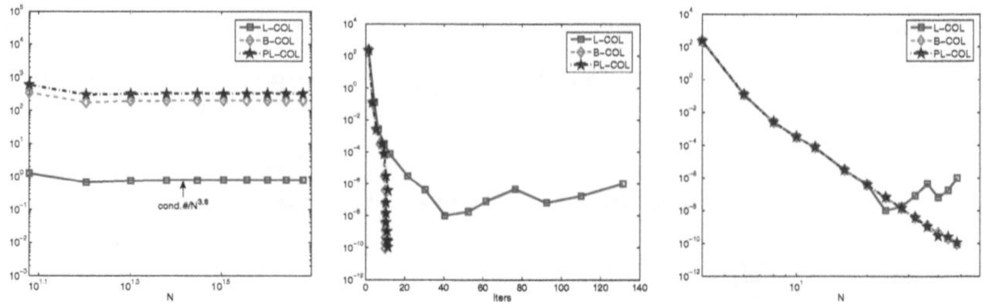

Figure 3.7 Comparison of condition numbers (left), iteration numbers against errors (middle), and errors against N at convergence in log-log scale (right) for (3.2.105). Source: Reprinted from Jiao et al. (2016) with permission from Elsevier.

Riemann–Liouville FODEs

Consider the Riemann–Liouville version of (3.2.105):

$$\begin{aligned} &{}_{-1}^{\mathrm{RL}}D_x^{\mu} u(x) + \lambda_1(x)\, {}_{-1}^{\mathrm{RL}}D_x^{\nu} u(x) + \lambda_2(x) u(x) = f(x), \quad x \in (-1,1); \\ &u(-1) = u_-, \quad u(1) = u_+, \quad 0 < \nu < \mu, \quad \mu \in (1,2), \end{aligned} \tag{3.2.110}$$

where λ_1, λ_2 and f are given continuous functions, and $u_\pm$ are given constants.

For a better treatment of the singularity, we consider the modified Riemann–Liouville fractional collocation scheme: find $u_N \in \mathbb{P}_N$ such that for $1 \le j \le N$,

$$\begin{aligned} &{}_{-1}^{\mathrm{RL}}\widehat{D}_x^{\mu} u_N(x_j) + \hat{\lambda}_1(x_j)\, {}_{-1}^{\mathrm{RL}}\widehat{D}_x^{\nu} u_N(x_j) + \hat{\lambda}_2(x_j) u_N(x_j) = \hat{f}(x_j); \\ &u_N(-1) = u_-, \quad u_N(1) = u_+, \quad 0 < \nu < \mu, \ \mu \in (1,2), \end{aligned} \tag{3.2.111}$$

where

$$\hat{\lambda}_1 = (1+x)^{\mu-\nu}\lambda_1, \quad \hat{\lambda}_2 = (1+x)^{\mu}\lambda_2, \quad \hat{f} = (1+x)^{\mu} f.$$

Here, we just focus on the collocation system using the new basis in (3.2.72), that is,

$$u_N(x) = u_- \widehat{Q}_0^{\mu}(x) + u_+ \widehat{Q}_N^{\mu}(x) + \sum_{j=1}^{N-1} v_j\, \widehat{Q}_j^{\mu}(x), \quad \mu \in (1,2). \tag{3.2.112}$$

Then one can write down the B-COL system in a fashion very similar to (3.2.107) with only a change of basis. Correspondingly, we denote the matrix of the linear system by

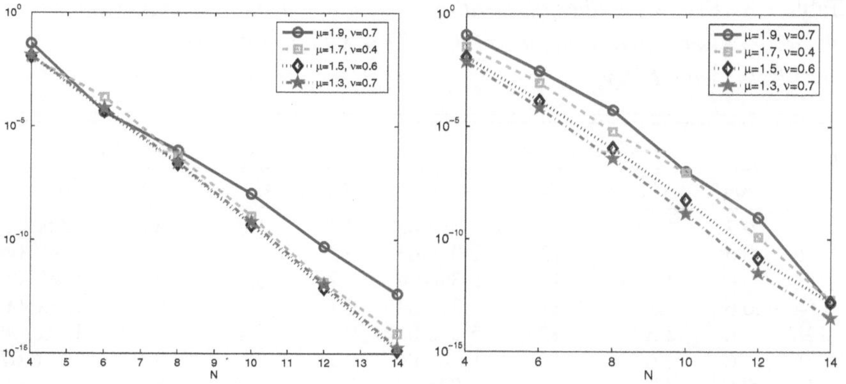

Figure 3.8 Errors against N of the B-COL schemes for Caputo fractional BVP (3.2.105) (left) and Riemann–Liouville fractional BVP (3.2.110) (right) with various μ, ν, where λ_1, λ_2 are given in (3.2.108), and the exact solution u is given in (3.2.113). Source: Reprinted from Jiao et al. (2016) with permission from Elsevier.

$$\boldsymbol{A} := \boldsymbol{I}_{N-1} + \boldsymbol{\Lambda}_1 \widetilde{\boldsymbol{Q}}^{(\nu)} + \boldsymbol{\Lambda}_2 \widehat{\boldsymbol{Q}}^{(\mu)}.$$

We first show that the B-COL scheme enjoys spectral accuracy (i.e., exponential convergence), when the underlying solution is sufficiently smooth. For this purpose, we take

$$u(x) = e^{-(1+x)} - \frac{1-x}{2} - e^{-2}\frac{1+x}{2} \tag{3.2.113}$$

and λ_1, λ_2 to be the same as in (3.2.108). In Figure 3.8, we plot discrete L^2-errors for various pairs of (μ, ν) of the B-COL schemes for both Caputo and Riemann–Liouville fractional boundary-value problems (BVPs) (3.2.105) and (3.2.110) under the same setting. We observe the exponential decay (i.e., $O(e^{-cN})$ for some $c > 0$) of the errors. Both schemes take about 10 iterations to converge, while much more iterations are needed and severe round-off errors are induced if one uses the standard L-COL approach.

We further test the new B-COL method on (3.2.110) with smooth coefficients but large derivative,

$$\lambda_1(x) = 1 + e^{-1000x^2}, \quad \lambda_2(x) = 1 + e^{-1000(x+0.2)^2}, \tag{3.2.114}$$

and with the exact solution having finite regularity in the usual Sobolev space:

$$u(x) = E_{\mu,1}\big(-(1+x)^{\mu}/2\big) + \frac{1 - E_{\mu,1}(-2^{\mu-1})}{2}(1+x) - 1. \tag{3.2.115}$$

Table 3.8. *Errors, number of iterations and concentration of eigenvalues of* $\boldsymbol{A}$. *Source: Reprinted from Jiao et al. (2016). Copyright (2016), with permission from Elsevier.*

N	$\mu = 1.5,\ \nu = 0.6$				$\mu = 1.9,\ \nu = 0.7$			
	$\|\sigma_2\|$	$\|\sigma_{N-2}\|$	Iters	Errors	$\|\sigma_2\|$	$\|\sigma_{N-2}\|$	Iters	Errors
8	0.7	1.0	7	8.58e-03	0.4	1.0	6	2.66e-03
16	0.6	1.1	12	2.03e-03	0.2	1.7	8	3.69e-04
32	0.6	1.2	12	5.39e-04	0.2	3.0	8	5.49e-05
64	0.6	1.3	12	1.31e-04	0.1	5.0	8	7.46e-06
128	0.6	1.5	12	3.24e-05	0.1	7.4	8	1.06e-06
256	0.6	1.6	12	8.08e-06	0.1	9.9	8	1.53e-07
512	0.6	1.7	13	2.02e-06	0.1	12.7	8	2.21e-08
1024	0.6	1.7	13	5.04e-07	0.1	21.1	8	3.69e-09

We tabulate in Table 3.8 the discrete L^2-errors, number of iterations and the second smallest and largest eigenvalues (in modulus). Once again, the scheme converges within a few iterations even for very large N. In fact, as we observed from Figure 3.5 (right), the smallest eigenvalue of ${}^{\rm RL}\widehat{\boldsymbol{D}}_{\rm in}^{(\mu)}$ in (3.2.51) still mildly depends on N. As a result, the condition number of $\boldsymbol{A}$ grows mildly with respect to N. However, it is interesting to find that the eigenvalues in modulus of $\boldsymbol{A}$ (denoted by $\{|\sigma_j|\}_{j=1}^{N-1}$ which are arranged in ascending order) are concentrated in the sense that

$$O(1) = |\sigma_2| \le |\sigma_j| \le |\sigma_{N-2}| = O(1), \quad 2 \le j \le N-2. \tag{3.2.116}$$

Thanks to this remarkable property, the iterative solver for the modified Riemann–Liouville system is actually as fast as the previous Caputo system where the coefficient matrix is well conditioned.

3.2.3 Generalized Jacobi Poly-fractonomial Collocation Methods

Fractional Lagrange Interpolants

To motivate the algorithm development herein, it is necessary to recall the *nodal Jacobi poly-fractonomials* introduced in Zayernouri and Karniadakis (2014a, (3.8)). Let $-1 = x_0 < x_1 < \cdots < x_N = 1$ be a set of generic interpolation points, and define the nodal basis of fractional order $(N+\mu)$ as

$$h_j^{\mu}(x) = \Big(\frac{1+x}{1+x_j}\Big)^{\mu} \prod_{k=0,k\neq j}^{N} \frac{x-x_k}{x_j-x_k} = \Big(\frac{1+x}{1+x_j}\Big)^{\mu} l_j(x), \ \ 1 \le j \le N, \tag{3.2.117}$$

for $\mu \in (0,1)$, where $\{l_j\}_{j=0}^N$ are the Lagrange interpolation basis polynomials in (3.2.3), if $\{x_j\}_{j=0}^N$ are chosen as the JGL points. Note that

$$l_j(x) = \sum_{n=1}^{N+1} s_n^j P_{n-1}^{(-\nu,\nu)}(x),$$

where s_n^j can be computed by (3.2.3) with parameters $\alpha = -\nu$ and $\beta = \nu$. The nodal version (3.2.117) is suggested as a counterpart of the modal Jacobi poly-fractonomials in Zayernouri and Karniadakis (2013), originated from fractional Sturm–Liouville problems. An important special class used in Zayernouri and Karniadakis (2014a) is

$$^{(1)}\mathcal{P}_n^{\mu}(x) = (1+x)^{\mu} P_{n-1}^{(-\mu,\mu)}(x), \quad n \geq 1, \quad \mu \in (0,1). \tag{3.2.118}$$

To have more insights, we consider the model fractional equation of order $\nu \in (0,1)$,

$${}_{-1}^{\mathrm{RL}}D_x^{\nu}\, u(x) = f(x), \quad x \in (-1,1); \quad u(-1) = 0, \tag{3.2.119}$$

where $f(x)$ is continuous in $(-1,1)$. The fractional collocation scheme using nodal Jacobi poly-fractonomials is to find $u_N \in {}_0\mathcal{V}_{N+\nu} = \mathrm{span}\{h_j^{\nu} : 1 \leq j \leq N\}$ such that

$${}_{-1}^{\mathrm{RL}}D_x^{\nu}\, u_N(x_j) = f(x_j), \quad 1 \leq j \leq N, \tag{3.2.120}$$

which leads to the linear system

$$\boldsymbol{D}^{\nu}\boldsymbol{u}_N = \boldsymbol{f} \quad \text{with} \quad \boldsymbol{D}_{ij}^{\nu} = {}_{-1}^{\mathrm{RL}}D_x^{\nu}\, h_j^{\nu}(x_i), \quad 1 \leq i,j \leq N, \tag{3.2.121}$$

and $\boldsymbol{u}_N = (u_N(x_1), \ldots, u_N(x_N))^{\top}$ and $\boldsymbol{f} = (f(x_1), \ldots, f(x_N))^{\top}$. Here, $\boldsymbol{D}^{\nu}$ is the fractional differentiation matrix, whose entries can be evaluated by

$$\boldsymbol{D}_{ij}^{\nu} = \frac{1}{(1+x_j)^{\nu}} \sum_{n=1}^{N+1} \frac{\Gamma(n+\nu)}{\Gamma(n)} s_n^j P_{n-1}(x_i), \quad 1 \leq i,j \leq N. \tag{3.2.122}$$

At this point, it is important to point out that since $l_j(-1) = 0$ for $1 \leq j \leq N$, we have

$$h_j^{\nu}(x) = (1+x)^{1+\nu}\psi_j(x), \quad 1 \leq j \leq N, \tag{3.2.123}$$

where $\psi_j \in \mathbb{P}_{N-1}$ and $\psi_j(-1) \neq 0$. This implies $h_j^{\nu}(-1) = Dh_j^{\nu}(-1) = 0$, so the scheme (3.2.120)–(3.2.121) does not lead to satisfactory approximation if the solution does not meet $u'(-1) = 0$. However, (3.2.119) admits the solution such that $u'(-1) \neq 0$, e.g., $C(1+x)^{\nu}$ if $f(x) = 1$.

Fractional Lagrange Basis Based on Gauss–Radau Interpolation
To overcome the deficiency in (3.2.123), we introduce the following fractional Lagrange basis of order $(N + \mu - 1)$:

$$\bar{h}_j^{\mu}(x) = \left(\frac{1+x}{1+x_j}\right)^{\mu} \bar{l}_j(x), \quad 1 \le j \le N, \quad \mu \in (0,1), \tag{3.2.124}$$

where $\{\bar{l}_j\}_{j=1}^N$ are the Lagrange interpolation polynomial basis associated with the JGL points $\{x_j\}_{j=1}^N$ (exclusive of $x_0 = -1$), satisfying $\bar{l}_j \in \mathbb{P}_{N-1}$ and $\bar{l}_j(x_i) = \delta_{ij}$ (δ_{ij} being the Kronecker delta) for $1 \le i, j \le N$. In fact, one verifies that if $\{x_j\}_{j=0}^N$ are the JGL points (i.e., zeros of $(1-x^2)DP_N^{(\alpha,\beta)}(x)$), then we have

$$h_j^{\mu}(x) = \frac{1+x}{1+x_j}\bar{h}_j^{\mu}(x) = \bar{h}_j^{1+\mu}(x), \quad 1 \le j \le N, \quad \mu \in (0,1). \tag{3.2.125}$$

Indeed, we have from Shen et al. (2011a, Ch. 3) that

$$\begin{aligned} l_j(x) &= \frac{(1-x^2)DP_N^{(\alpha,\beta)}(x)}{(x-x_j)D((1-x^2)DP_N^{(\alpha,\beta)}(x))|_{x=x_j}} \\ &= \frac{1+x}{1+x_j}\frac{(1-x)DP_N^{(\alpha,\beta)}(x)}{(x-x_j)D((1-x)DP_N^{(\alpha,\beta)}(x))|_{x=x_j}} \\ &= \frac{1+x}{1+x_j}\bar{l}_j(x), \quad 1 \le j \le N. \end{aligned} \tag{3.2.126}$$

Thus the relation (3.2.125) follows from (3.2.117) and (3.2.124).

It is noteworthy that the N points $\{x_j\}_{j=1}^N$ of $(N+1)$ JGL points $\{x_j\}_{j=0}^N$ with respect to the Jacobi weight $\omega^{(\alpha,\beta)}$ are also N Jacobi–Gauss–Radau (JGR) points but with respect to the Jacobi weight $\omega^{(\alpha,\beta+1)}$ (cf. Shen et al. (2011a, Ch. 3)).

Remark 3.38 In fact, one can also choose $\{x_j\}_{j=1}^N$ to be the JGR points associated with the Jacobi weight $\omega^{\alpha,\beta}$ (i.e., zeros of $(1-x)P_{N-1}^{(\alpha+1,\beta)}(x)$), and accordingly, we have

$$\bar{l}_j(x) = \frac{(1-x)P_{N-1}^{(\alpha+1,\beta)}(x)}{(x-x_j)D((1-x)P_{N-1}^{(\alpha+1,\beta)}(x))|_{x=x_j}}, \quad 1 \le j \le N, \tag{3.2.127}$$

which is different from (3.2.126) by noting that

$$DP_N^{(\alpha,\beta)}(x) = \frac{1}{2}(N+\alpha+\beta)P_{N-1}^{(\alpha+1,\beta+1)}(x).$$

Observe from the definition (3.2.124) that (i) $\bar{h}_j^{\mu}(1) = 0$ for $1 \le j \le N-1$, which allows for the imposition of boundary conditions at $x = 1$; (ii) the parameter μ in the singular factor $(1+x)^{\mu}$ should be chosen to fit the singularity of the underlying solution, which might be the order ν of the equation; and (iii) in principle, the choice of JGL points (with parameters α, β) can be independent of μ and ν. In the applications, we need to compute the fractional differentiation of order ν on $\{\bar{h}_j^{\mu}\}_{j=1}^{N}$ (with $\mu \in (0,1)$ and $\nu \ge \mu$) at a set of preassigned collocation points.

Lemma 3.39 *For $\alpha, \beta > -1$, let $\{x_j := x_{N,j}^{(\alpha,\beta)}, \omega_j := \omega_{N,j}^{(\alpha,\beta)}\}_{j=0}^{N}$ (with $x_0 = -1$) be the JGL points and corresponding quadrature weights. Then we have*

$$\bar{l}_j(x) = \sum_{n=0}^{N-1} \bar{t}_{nj} P_n^{(\alpha,\beta)}(x) = \sum_{k=0}^{N-1} \bar{s}_{kj} P_k^{(a,b)}(x) \tag{3.2.128}$$

for $a, b > -1$ and $1 \le j \le N$, where

$$\begin{aligned} \bar{t}_{nj} := \bar{t}_{nj}^{(\alpha,\beta)} = \frac{1}{\gamma_n^{(\alpha,\beta)}} \bigg\{ &- \frac{c_j}{\beta+1} \frac{P_N^{(\alpha,\beta)}(-1)}{P_N^{(\alpha,\beta)}(x_j)} P_n^{(\alpha,\beta)}(-1)\omega_0 \\ &+ P_n^{(\alpha,\beta)}(x_j)\omega_j \bigg\}, \end{aligned} \tag{3.2.129}$$

with $c_j = 1$ for $1 \le j \le N-1$, and $c_N = \alpha + 1$, and

$$\bar{s}_{kj} := \bar{s}_{kj}^{(\alpha,\beta,a,b)} = \sum_{n=k}^{N-1} \frac{\bar{t}_{nj}}{\gamma_k^{(a,b)}} \int_{-1}^{1} P_k^{(a,b)}(x) P_n^{(\alpha,\beta)}(x)\, \omega^{(a,b)}(x) dx. \tag{3.2.130}$$

Proof The first expansion in (3.2.128) and the formula (3.2.129) are derived in Jiao et al. (2016, (4.14)), while the formula (3.2.130) follows directly from (3.2.128) and the orthogonality property. □

In the later part, we need a similar expansion for the Lagrange interpolation involving only the interior JGL points $\{x_j\}_{j=1}^{N-1}$ (exclusive of $x_0 = -1$ and $x_N = 1$). Let $\{\hat{l}_j\}_{j=1}^{N-1}$ be the Lagrange interpolation basis satisfying $\hat{l}_j \in \mathbb{P}_{N-2}$ and $\hat{l}_j(x_i) = \delta_{ij}$ for $1 \le i, j \le N-1$. Similar to Lemma 3.39, we have the following useful expressions.

Lemma 3.40 *For $\alpha, \beta > -1$, let $\{x_j := x_{N,j}^{(\alpha,\beta)}, \omega_j := \omega_{N,j}^{(\alpha,\beta)}\}_{j=0}^{N}$ (with $x_0 = -1$ and $x_N = 1$) be the JGL points and corresponding quadrature weights. Then we have*

$$\hat{l}_j(x) = \sum_{n=0}^{N-2} \hat{t}_{nj} P_n^{(\alpha,\beta)}(x) = \sum_{k=0}^{N-2} \hat{s}_{kj} P_k^{(a,b)}(x) \tag{3.2.131}$$

for $a, b > -1$ *and* $1 \le j \le N-1$, *where*

$$\begin{aligned} \hat{t}_{nj} := \hat{t}_{nj}^{(\alpha,\beta)} &= \frac{1}{\gamma_n^{(\alpha,\beta)}} \bigg\{ \frac{(x_j - 1)P_N^{(\alpha,\beta)}(-1)}{2(\beta+1)P_N^{(\alpha,\beta)}(x_j)} P_n^{(\alpha,\beta)}(-1)\omega_0 \\ &- \frac{(1+x_j)P_N^{(\alpha,\beta)}(1)}{2(\alpha+1)P_N^{(\alpha,\beta)}(x_j)} P_n^{(\alpha,\beta)}(1)\omega_N + P_n^{(\alpha,\beta)}(x_j)\omega_j \bigg\} \end{aligned} \tag{3.2.132}$$

and

$$\hat{s}_{kj} := \hat{s}_{kj}^{(\alpha,\beta,a,b)} = \sum_{n=k}^{N-2} \frac{\hat{t}_{nj}}{\gamma_k^{(a,b)}} \int_{-1}^{1} P_k^{(a,b)}(x) P_n^{(\alpha,\beta)}(x)\, \omega^{(a,b)}(x) \mathrm{d}x. \tag{3.2.133}$$

Proof The first expansion in (3.2.131) and the formula (3.2.132) are derived in Jiao et al. (2016, (4.17)), while the formula (3.2.133) follows directly from (3.2.131) and the orthogonality property. □

Fractional Differentiation Matrices

Hereafter, we compute the νth-order fractional differentiation matrix:

$$\bar{\boldsymbol{D}}_\mu^{(\nu)} \in \mathbb{R}^{N\times N} \text{ with entries } \big[\bar{\boldsymbol{D}}_\mu^{(\nu)}\big]_{ij} = {}_{-1}^{\mathrm{RL}}D_x^\nu \, \bar{h}_j^\mu(x_i), \quad 1 \le i, j \le N, \tag{3.2.134}$$

for $\mu \in (0,1)$ and $0 < \nu \le 1+\mu$, where $\{\bar{h}_j^\mu\}$ are defined in (3.2.124).

Theorem 3.41 *For* $\alpha, \beta > -1$, *let* $\big\{x_j := x_{N,j}^{(\alpha,\beta)}, \omega_j := \omega_{N,j}^{(\alpha,\beta)}\big\}_{j=0}^N$ *(with* $x_0 = -1$*) be the JGL points and corresponding quadrature weights. Then the entries of* $\bar{\boldsymbol{D}}_\mu^{(\nu)}$ *with* $\mu \in (0,1)$ *and* $0 < \nu \le 1+\mu$ *can be computed by the following formulas.*

(i) If $0 < \nu < 1+\mu$, *we have*

$$[\bar{\boldsymbol{D}}_\mu^{(\nu)}]_{ij} = \frac{(1+x_i)^{\mu-\nu}}{(1+x_j)^\mu} \sum_{k=0}^{N-1} \tilde{s}_{kj} P_k^{(\nu-\mu,\mu-\nu)}(x_i), \quad 1 \le i, j \le N, \tag{3.2.135}$$

where

$$\tilde{s}_{kj} = \frac{\Gamma(k+\mu+1)}{\Gamma(k+\mu-\nu+1)} \bar{s}_{kj}^{(\alpha,\beta,-\mu,\mu)}, \tag{3.2.136}$$

and $\bar{s}_{kj}^{(\alpha,\beta,-\mu,\mu)}$ *is defined in* (3.2.130).

(ii) If $\nu = 1 + \mu$, *we have*

$$[\bar{\boldsymbol{D}}_{\mu}^{(1+\mu)}]_{ij} = \frac{1}{(1+x_j)^{\mu}} \sum_{k=1}^{N-1} \tilde{s}_{kj}\, P_{k-1}^{(1,1)}(x_i), \quad 1 \le i, j \le N, \tag{3.2.137}$$

where

$$\tilde{s}_{kj} = \frac{(k+1)\Gamma(k+\mu+1)}{2\,\Gamma(k+1)}\, \bar{s}_{kj}^{(\alpha,\beta,-\mu,\mu)}, \tag{3.2.138}$$

and $\bar{s}_{kj}^{(\alpha,\beta,-\mu,\mu)}$ *is defined in* (3.2.130).

Proof By (3.2.124) and (3.2.128),

$$\bar{h}_j^{\mu}(x) = \frac{1}{(x_j+1)^{\mu}} \sum_{k=0}^{N-1} \bar{s}_{kj}^{(\alpha,\beta,-\mu,\mu)}\, (x+1)^{\mu} P_k^{(-\mu,\mu)}(x). \tag{3.2.139}$$

Using the formula (2.4.19) leads to

$$\begin{aligned} {}_{-1}^{\mathrm{RL}}D_x^{\nu}\, \bar{h}_j^{\mu}(x) &= \frac{1}{(x_j+1)^{\mu}} \sum_{k=0}^{N-1} \bar{s}_{kj}^{(\alpha,\beta,-\mu,\mu)}\, {}_{-1}^{\mathrm{RL}}D_x^{\nu}\big\{(x+1)^{\mu} P_k^{(-\mu,\mu)}(x)\big\} \\ &= \frac{1}{(x_j+1)^{\mu}} \sum_{k=0}^{N-1} \bar{s}_{kj}^{(\alpha,\beta,-\mu,\mu)}\, \frac{\Gamma(k+\mu+1)}{\Gamma(k+\mu-\nu+1)} (1+x)^{\mu-\nu} P_k^{(\nu-\mu,\mu-\nu)}(x). \end{aligned} \tag{3.2.140}$$

Taking $x = x_i$ in the above, yields (3.2.135)–(3.2.136).

The formula (3.2.137) can be derived in the same fashion. □

Remark 3.42 Observe from (3.2.130), (3.2.136) and (3.2.138) that if we take $(\alpha, \beta) = (-\mu, \mu)$, then $\bar{s}_{kj} = \bar{t}_{kj}$. In other words, (3.2.135) and (3.2.137) have the simplest from. Thus, it is preferable to choose these special parameters. □

For notational convenience, we allow (in the following theorem and its proof) the nodal basis $\{\bar{h}_j^{\mu}\}$ in (3.2.124) to be defined for $\mu < 0$, and likewise for the differentiation matrix in (3.2.134). We have the following useful rules to compute higher-order fractional differentiation matrices.

Theorem 3.43 *For* $\mu \in (0,1)$ *and* $0 < \nu < 1 + \mu$, *we have*

$$\bar{\boldsymbol{D}}_{\mu}^{(k+\nu)} = \bar{\boldsymbol{D}}_{\mu-\nu}^{(k)}\, \bar{\boldsymbol{D}}_{\mu}^{(\nu)}, \quad k = 0, 1, \ldots, \tag{3.2.141}$$

and in particular, we have

$$\bar{\boldsymbol{D}}_{\mu}^{(k+\mu)} = (\bar{\boldsymbol{D}})^k\, \bar{\boldsymbol{D}}_{\mu}^{(\mu)}, \quad k = 0, 1, \ldots, \tag{3.2.142}$$

where $(\bar{\boldsymbol{D}})^k$ *is a product of k copies of the first-order differentiation matrix* $\bar{\boldsymbol{D}}$ *with entries given by* $[\bar{\boldsymbol{D}}]_{ij} = \bar{l}_j'(x_i)$ *for* $1 \le i, j \le N$.

Proof We find from (3.2.140) that $(1+x)^{\nu-\mu}\, {}^{\rm RL}_{-1}D^{\nu}_x \bar h^{\mu}_j \in \mathbb{P}_{N-1}$, so we can express it as

$$(1+x)^{\nu-\mu}\, {}^{\rm RL}_{-1}D^{\nu}_x \bar h^{\mu}_j(x) = \sum_{p=1}^{N} \big\{(1+x_p)^{\nu-\mu}\, {}^{\rm RL}_{-1}D^{\nu}_x \bar h^{\mu}_j(x_p)\big\}\, \bar l_p(x) \quad (3.2.143)$$

for $1 \le j \le N$. Thus, we have

$$\begin{aligned} {}^{\rm RL}_{-1}D^{k+\nu}_x \bar h^{\mu}_j(x_i) &= D^k\, {}^{\rm RL}_{-1}D^{\nu}_x \bar h^{\mu}_j(x_i) \\ &= \sum_{p=1}^{N} {}^{\rm RL}_{-1}D^{\nu}_x \bar h^{\mu}_j(x_p) D^k \Big\{\frac{(1+x)^{\mu-\nu}}{(1+x_p)^{\mu-\nu}}\bar l_p(x)\Big\}_{x=x_i} \\ &= \sum_{p=1}^{N} {}^{\rm RL}_{-1}D^{\nu}_x \bar h^{\mu}_j(x_p) D^k \bar h^{\mu-\nu}_p(x_i) = \sum_{p=1}^{N} [\bar{\boldsymbol D}^{(k)}_{\mu-\nu}]_{ip}[\bar{\boldsymbol D}^{(\nu)}_{\mu}]_{pj}, \end{aligned} \quad (3.2.144)$$

which yields (3.2.141).

As a special case of (3.2.141), we have $\bar{\boldsymbol D}^{(k+\mu)}_{\mu} = \bar{\boldsymbol D}^{(k)}_0\, \bar{\boldsymbol D}^{(\mu)}_{\mu}$, and by (3.2.144), $[\bar{\boldsymbol D}^{(k)}_0]_{ip} = \bar l^{(k)}_p(x_i)$. According to Shen et al. (2011a, Thm. 3.10), we have $\bar{\boldsymbol D}^{(k)}_0 = (\bar{\boldsymbol D})^k$. This ends the proof. □

We conclude this section by providing some numerical study of (discrete) eigenvalues of $\bar{\boldsymbol D}^{(\nu)}_{\mu}$. Remove the "boundary" row/column, and define

$$\boldsymbol D^{(\nu)}_{\mu,\rm in} := \begin{cases} \big(\bar{\boldsymbol D}^{(\nu)}_{\mu}\big)_{ij}, & 1 \le i,j \le N, \quad \text{if } \mu,\nu \in (0,1), \\ \big(\bar{\boldsymbol D}^{(\nu)}_{\mu}\big)_{ij}, & 1 \le i,j \le N-1, \quad \text{if } \mu \in (0,1),\ 1 < \nu \le 1+\mu, \end{cases} \quad (3.2.145)$$

which is invertible and allows for incorporating boundary condition(s).

In Figure 3.9, we illustrate the condition number of $\boldsymbol D^{(\nu)}_{\mu,\rm in}$ for some μ and ν under Legendre–Gauss–Lobatto points and Chebyshev–Gauss–Lobatto points. One easily observes that in both cases, the condition number of the matrices grows dramatically as the number of collocation points N grows.

Inverse of the Differentiation Matrix

In what follows, we compute the explicit inverse of the fractional differentiation matrix $\bar{\boldsymbol D}^{(\nu)}_{\mu}$ in Theorem 3.41 in a stable manner, through the basis functions of a suitable fractional Birkhoff interpolation problem.

Define the space

$$\mathcal{F}^{(\mu)}_N := \big\{(1+x)^{\mu}\phi(x) : \phi \in \mathbb{P}_{N-1}\big\}, \quad \mu > -1. \quad (3.2.146)$$

We consider the following fractional interpolation problems.

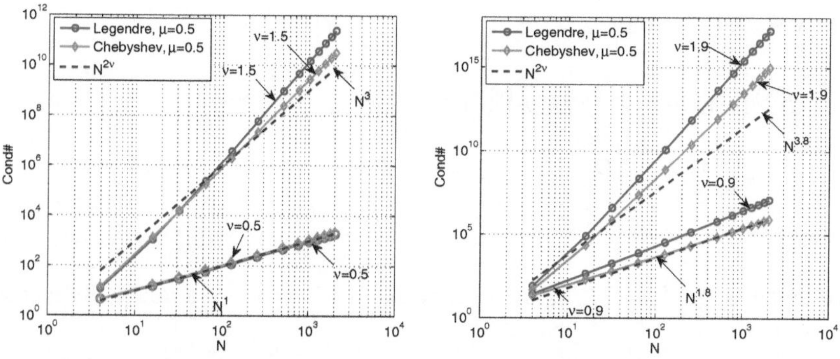

Figure 3.9 Condition number of fractional differentiation matrices $\bar{D}_{0.5}^{0.5}, \bar{D}_{0.5}^{1.5}$ (left) and $\bar{D}_{0.5}^{0.9}, \bar{D}_{0.5}^{1.9}$ (right). Source: Huang et al. (2016). Copyright ©2016 Society for Industrial and Applied Mathematics. Reprinted with permission. All rights reserved.

(i) For $\mu, \nu \in (0,1)$, the RL fractional Birkhoff interpolation is to find $p \in \mathcal{F}_N^{(\mu)}$ such that

$$ {}_{-1}^{\mathrm{RL}}D_x^{\nu}\, p(x_j) = {}_{-1}^{\mathrm{RL}}D_x^{\nu}\, u(x_j), \quad 1 \le j \le N; \quad p(-1) = u(-1), \quad (3.2.147)$$

for any $u \in C(-1,1)$ satisfying $u(-1) = 0$ and ${}_{-1}^{\mathrm{RL}}D_x^{\nu} u \in C(-1,1]$.

(ii) For $\mu \in (0,1)$ and $1 < \nu \le 1+\mu$, the RL fractional Birkhoff interpolation is to find $p \in \mathcal{F}_N^{(\mu)}$ such that

$$\begin{aligned} &{}_{-1}^{\mathrm{RL}}D_x^{\nu}\, p(x_j) = {}_{-1}^{\mathrm{RL}}D_x^{\nu}\, u(x_j), \quad 1 \le j \le N-1; \\ &p(-1) = u(-1), \quad p(1) = u(1), \end{aligned} \quad (3.2.148)$$

for any $u \in C(-1,1)$ satisfying $u(-1) = 0$ and ${}_{-1}^{\mathrm{RL}}D_x^{\nu} u \in C(-1,1)$.

Like the usual Lagrange interpolation, we represent the interpolant (for $\mu \in (0,1)$) as

$$p(x) = \begin{cases} \displaystyle\sum_{j=1}^{N} {}_{-1}^{\mathrm{RL}}D_x^{\nu}\, u(x_j) B_{\mu,j}^{(\nu)}(x), & \nu \in (0,1), \\ \displaystyle\sum_{j=1}^{N-1} {}_{-1}^{\mathrm{RL}}D_x^{\nu}\, u(x_j) B_{\mu,j}^{(\nu)}(x) + u(1) B_{\mu,N}^{(\nu)}(x), & 1 < \nu \le 1+\mu, \end{cases} \quad (3.2.149)$$

where $\{B_{\mu,j}^{(\nu)}\}$ are called the RL fractional Birkhoff interpolation basis in $\mathcal{F}_N^{(\mu)}$, satisfying

$$B_{\mu,j}^{(\nu)}(x_i)(-1) = 0, \quad {}_{-1}^{\mathrm{RL}}D_x^{\nu}\, B_{\mu,j}^{(\nu)}(x_i) = \delta_{ij}, \quad 1 \le i, j \le N, \quad (3.2.150)$$

for $\nu \in (0,1)$; and for $1 < \nu \le 1+\mu$,

$$B^{(\nu)}_{\mu,j}(\pm 1) = 0, \quad {}^{\rm RL}_{-1}D^{\nu}_{x}\, B^{(\nu)}_{\mu,j}(x_i) = \delta_{ij}, \quad 1 \le i,j \le N-1,$$
$$B^{(\nu)}_{\mu,N}(-1) = {}^{\rm RL}_{-1}D^{\nu}_{x}\, B^{(\nu)}_{\mu,N}(x_i) = 0, \quad 1 \le i \le N-1; \quad B^{(\nu)}_{\mu,N}(1) = 1. \tag{3.2.151}$$

Note that as $B^{(\nu)}_{\mu,j} \in \mathcal{F}^{(\mu)}_{N}$, the condition $B^{(\nu)}_{\mu,j}(-1) = 0$ is automatically met if $\mu > 0$.

Remark 3.44 Different from the polynomial-based approach in Jiao et al. (2016), the above fractional Birkhoff interpolation of nodal GJPFs involves different singular indices for fractional derivatives of different orders. Also the derivation of the fractional Birkhoff interpolation basis is much more involved. □

The new basis can be computed by the following explicit formulas in a stable manner.

Theorem 3.45 *For* $\alpha, \beta > -1$, *let* $\{x_j := x^{(\alpha,\beta)}_{N,j}, \omega_j := \omega^{(\alpha,\beta)}_{N,j}\}_{j=0}^{N}$ *(with* $x_0 = -1$*) be the JGL points and corresponding quadrature weights.*

(i) If $0 < \mu, \nu < 1$, *we have*

$$B^{(\nu)}_{\mu,j}(x) = \frac{(1+x)^{\mu}}{(1+x_j)^{\mu-\nu}} \sum_{k=0}^{N-1} \breve{s}_{kj}\, P^{(-\mu,\mu)}_{k}(x), \quad 1 \le j \le N, \tag{3.2.152}$$

where

$$\breve{s}_{kj} = \frac{\Gamma(k+\mu-\nu+1)}{\Gamma(k+\mu+1)}\, \bar{s}^{(\alpha,\beta,\nu-\mu,\mu-\nu)}_{kj}, \tag{3.2.153}$$

and $\bar{s}^{(\alpha,\beta,\nu-\mu,\mu-\nu)}_{kj}$ *is defined in (3.2.130) with* $a = \nu - \mu$ *and* $b = \mu - \nu$.

(ii) If $0 < \mu < 1$ *and* $1 < \nu < 1+\mu$, *we have that for* $1 \le j \le N-1$,

$$B^{(\nu)}_{\mu,j}(x) = \frac{(1+x)^{\mu}}{(1+x_j)^{\mu-\nu}(x_j+\tau_j)} \sum_{k=0}^{N-2} \Big\{(\tau_j - 1)\breve{s}^{(1)}_{kj}\, P^{(-\mu,\mu)}_{k}(x) + \breve{s}^{(2)}_{kj}(1+x) P^{(-\mu,\mu+1)}_{k}(x)\Big\}, \tag{3.2.154}$$

where

$$\breve{s}_{kj}^{(1)} = \frac{\Gamma(k+\mu-\nu+1)}{\Gamma(k+\mu+1)} \hat{s}_{kj}^{(\alpha,\beta,\nu-\mu,\mu-\nu)},$$
$$\breve{s}_{kj}^{(2)} = \frac{\Gamma(k+\mu-\nu+2)}{\Gamma(k+\mu+2)} \hat{s}_{kj}^{(\alpha,\beta,\nu-\mu,\mu-\nu+1)}, \tag{3.2.155}$$
$$\tau_j - 1 = -2\sum_{k=0}^{N-2} \breve{s}_{kj}^{(2)}\, P_k^{(-\mu,\mu+1)}(1) \Big/ \sum_{k=0}^{N-2} \breve{s}_{kj}^{(1)}\, P_k^{(-\mu,\mu)}(1).$$

For $j = N$, we have

$$B_{\mu,N}^{(\nu)}(x) = \tau_N \sum_{k=0}^{N-1} \hat{s}_{kN}^{(\alpha,\beta,\nu-\mu,\mu-\nu)} \frac{\Gamma(k+\mu-\nu+1)}{\Gamma(k+\mu+1)} (1+x)^\mu P_k^{(-\mu,\mu)}(x),$$

where

$$\frac{1}{\tau_N} = 2^\mu \sum_{k=0}^{N-1} \hat{s}_{kN}^{(\alpha,\beta,\nu-\mu,\mu-\nu)} \frac{\Gamma(k+\mu-\nu+1)}{\Gamma(k+\mu+1)} P_k^{(-\mu,\mu)}(1). \tag{3.2.156}$$

In the above, $\{\hat{s}_{kj}^{(\alpha,\beta,\nu-\mu,\mu-\nu)}\}$ and $\{\hat{s}_{kj}^{(\alpha,\beta,\nu-\mu,\mu-\nu+1)}\}$ are defined in (3.2.133).

(iii) If $0 < \mu < 1$ and $\nu = 1+\mu$, we have that for $1 \le j \le N-1$,

$$B_{\mu,j}^{(1+\mu)}(x) = 2(1+x)^\mu \sum_{k=0}^{N-2} \hat{s}_{kj}^{(\alpha,\beta,1,1)} \frac{(k+1)!}{(k+2+\mu)\Gamma(k+\mu)} \times \left\{P_{k+1}^{(-\mu,\mu)}(x) - P_{k+1}^{(-\mu,\mu)}(1)\right\}, \tag{3.2.157}$$

where $\{\hat{s}_{kj}^{(\alpha,\beta,1,1)}\}$ are defined in (3.2.133)*, and for $j = N$,*

$$B_{\mu,N}^{(1+\mu)}(x) = \left(\frac{1+x}{2}\right)^\mu. \tag{3.2.158}$$

Proof We proceed with the proof separately for two cases.

(i) For $\nu \in (0,1)$, since $B_{\mu,j}^{(\nu)} \in \mathcal{F}_N^{(\mu)}$, we can write

$$B_{\mu,j}^{(\nu)}(x) = \sum_{k=0}^{N-1} \breve{b}_{kj}\, (1+x)^\mu P_k^{(-\mu,\mu)}(x), \quad 1 \le j \le N. \tag{3.2.159}$$

We next determine $\{\breve{b}_{kj}\}$. Observe from (3.2.140) that ${}^{\mathrm{RL}}_{-1}D_x^\nu B_{\mu,j}^{(\nu)} \in \mathcal{F}_N^{(\mu-\nu)}$, which, together with (3.2.150), implies

$${}^{\mathrm{RL}}_{-1}D_x^\nu\, B_{\mu,j}^{(\nu)}(x) = \left(\frac{1+x}{1+x_j}\right)^{\mu-\nu} \bar{l}_j(x), \tag{3.2.160}$$

for $x \in (-1,1)$ and $1 \le j \le N$, where $\bar{l}_j(x)$ is defined in (3.2.124). Acting ${}^{\mathrm{RL}}_{-1}D^{\nu}_{x}$ on both sides of (3.2.159), we obtain

$$
\begin{aligned}
{}^{\mathrm{RL}}_{-1}D^{\nu}_{x}\, B^{(\nu)}_{\mu,j}(x) &= \sum_{k=0}^{N-1} \breve{b}_{kj}\, \frac{\Gamma(k+\mu+1)}{\Gamma(k+\mu-\nu+1)}\,(1+x)^{\mu-\nu} P_k^{(\nu-\mu,\mu-\nu)}(x) \\
&= \left(\frac{1+x}{1+x_j}\right)^{\mu-\nu} \sum_{k=0}^{N-1} \bar{s}_{kj}^{(\alpha,\beta,\nu-\mu,\mu-\nu)}\, P_k^{(\nu-\mu,\mu-\nu)}(x),
\end{aligned}
$$

where in the last step, we used (3.2.160) and Lemma 3.39. Comparing the coefficients, we can find $\breve{s}_{kj}$ and obtain (3.2.152)–(3.2.153).

(ii) For $1 < \nu < 1+\mu$, we first derive the formulas for $\{B^{(\nu)}_{\mu,j}\}_{j=1}^{N-1}$. From (3.2.140), we find that ${}^{\mathrm{RL}}_{-1}D^{\nu}_{x} B^{(\nu)}_{\mu,j}(x) \in \mathcal{F}_N^{(\mu-\nu)}$, so by (3.2.151),

$$
{}^{\mathrm{RL}}_{-1}D^{\nu}_{x}\, B^{(\nu)}_{\mu,j}(x) = \frac{x+\tau_j}{x_j+\tau_j}\left(\frac{1+x}{1+x_j}\right)^{\mu-\nu} \hat{l}_j(x), \quad 1 \le j \le N-1, \tag{3.2.161}
$$

where $\{\hat{l}_j\}$ are the Lagrange interpolation basis polynomials associated with the interior JGL points $\{x_j\}_{j=1}^{N-1}$ as in Lemma 3.40, and $\{\tau_j\}$ are constants to be determined by the conditions: $B^{(\nu)}_{\mu,j}(1) = 0$ later. By definition, ${}^{\mathrm{RL}}_{-1}D^{\nu}_{x} = D^2\, {}^{\mathrm{RL}}_{-1}I^{2-\nu}_{x}$, so we have

$$
\begin{aligned}
{}^{\mathrm{RL}}_{-1}I^{2-\nu}_{x}\, B^{(\nu)}_{\mu,j}(x) = C_j^{-1}\, {}^{\mathrm{RL}}_{-1}I^{2}_{x}\{&(1+x)^{\mu-\nu+1}\hat{l}_j(x) \\
&+ (\tau_j-1)(1+x)^{\mu-\nu}\hat{l}_j(x)\} + A_1 + A_2(1+x),
\end{aligned} \tag{3.2.162}
$$

where A_1 and A_2 are arbitrary constants, and $C_j = (x_j+\tau_j)(1+x_j)^{\mu-\nu}$. Acting ${}^{\mathrm{RL}}_{-1}D^{2-\nu}_{x}$ on both sides of the above equation, and using the properties (cf. Diethelm (2010)): ${}^{\mathrm{RL}}_{-1}D^{2-\nu}_{x}\,{}^{\mathrm{RL}}_{-1}I^{2-\nu}_{x} v = v$ and ${}^{\mathrm{RL}}_{-1}D^{2-\nu}_{x}\,{}^{\mathrm{RL}}_{-1}I^{2}_{x} = {}^{\mathrm{RL}}_{-1}I^{\nu}_{x}$, we obtain

$$
\begin{aligned}
B^{(\nu)}_{\mu,j}(x) = C_j^{-1}\, {}^{\mathrm{RL}}_{-1}I^{\nu}_{x}&\{(1+x)^{\mu-\nu+1}\hat{l}_j(x) + (\tau_j-1)(1+x)^{\mu-\nu}\hat{l}_j(x)\} \\
&+ \frac{A_1}{\Gamma(\nu-1)}(1+x)^{\nu-2} + \frac{A_2}{\Gamma(\nu)}(1+x)^{\nu-1},
\end{aligned} \tag{3.2.163}
$$

where we used the basic RL differential formula for 1 and $1+x$. Then we infer that $A_1 = A_2 = 0$, as we require $B^{(\nu)}_{\mu,j} \in \mathcal{F}_N^{(\mu)}$. Imposing $B^{(\nu)}_{\mu,j}(1) = 0$ yields

$$\tau_j = 1 - \frac{{}^{\mathrm{RL}}_{-1}I^{\nu}_{x}\{(1+x)^{\mu-\nu+1}\hat{l}_j(x)\}\big|_{x=1}}{{}^{\mathrm{RL}}_{-1}I^{\nu}_{x}\{(1+x)^{\mu-\nu}\hat{l}_j(x)\}\big|_{x=1}}. \tag{3.2.164}$$

We continue to evaluate the fractional integrals. By Lemma 3.40,

$$\begin{aligned}
&{}^{\mathrm{RL}}_{-1}I^{\nu}_{x}\{(1+x)^{\mu-\nu+1}\hat{l}_j(x)\}\\
&= \sum_{k=0}^{N-2} \hat{s}_{kj}^{(\alpha,\beta,\nu-\mu,\mu-\nu+1)}\, {}^{\mathrm{RL}}_{-1}I^{\nu}_{x}\{(1+x)^{\mu-\nu+1} P_k^{(\nu-\mu,\mu-\nu+1)}(x)\}\\
&= \sum_{k=0}^{N-2} \hat{s}_{kj}^{(\alpha,\beta,\nu-\mu,\mu-\nu+1)} \frac{\Gamma(k+\mu-\nu+2)}{\Gamma(k+\mu+2)} (1+x)^{\mu+1} P_k^{(-\mu,\mu+1)}(x),
\end{aligned} \tag{3.2.165}$$

and

$$\begin{aligned}
&{}^{\mathrm{RL}}_{-1}I^{\nu}_{x}\{(1+x)^{\mu-\nu}\hat{l}_j(x)\}\\
&= \sum_{k=0}^{N-2} \hat{s}_{kj}^{(\alpha,\beta,\nu-\mu,\mu-\nu)}\, {}^{\mathrm{RL}}_{-1}I^{\nu}_{x}\{(1+x)^{\mu-\nu} P_k^{(\nu-\mu,\mu-\nu)}(x)\}\\
&= \sum_{k=0}^{N-2} \hat{s}_{kj}^{(\alpha,\beta,\nu-\mu,\mu-\nu)} \frac{\Gamma(k+\mu-\nu+1)}{\Gamma(k+\mu+1)} (1+x)^{\mu} P_k^{(-\mu,\mu)}(x).
\end{aligned} \tag{3.2.166}$$

With these, we can obtain the formulas in (3.2.154)–(3.2.155).

We now turn to $B^{(\nu)}_{\mu,N}$. Note that by (3.2.151),

$${}^{\mathrm{RL}}_{-1}D^{\nu}_{x}\, B^{(\nu)}_{\mu,N}(x) = \tau_N\, (1+x)^{\mu-\nu}\bar{l}_N(x), \tag{3.2.167}$$

where τ_N will be determined by the boundary condition: $B^{(\nu)}_{\mu,N}(1) = 1$. Following a process similar to the derivation of (3.2.163) and (3.2.166), we obtain

$$\begin{aligned}
B^{(\nu)}_{\mu,N}(x) &= \tau_N\, {}^{\mathrm{RL}}_{-1}I^{\nu}_{x}\{(1+x)^{\mu-\nu}\bar{l}_N(x)\}\\
&= \tau_N \sum_{k=0}^{N-1} \hat{s}_{kj}^{(\alpha,\beta,\nu-\mu,\mu-\nu)}\, {}^{\mathrm{RL}}_{-1}I^{\nu}_{x}\{(1+x)^{\mu-\nu} P_k^{(\nu-\mu,\mu-\nu)}(x)\}\\
&= \tau_N \sum_{k=0}^{N-1} \hat{s}_{kj}^{(\alpha,\beta,\nu-\mu,\mu-\nu)} \frac{\Gamma(k+\mu-\nu+1)}{\Gamma(k+\mu+1)} (1+x)^{\mu} P_k^{(-\mu,\mu)}(x).
\end{aligned}$$

Imposing $B^{(\nu)}_{\mu,N}(1) = 1$, we can determine τ_N and derive the desired formulas.

(iii) For $\nu = 1 + \mu$, one verifies readily that $B^{(1+\mu)}_{\mu,N}(x)$ must be given by (3.2.158), so it suffices for us to derive (3.2.157). We infer that ${}^{\mathrm{RL}}_{-1}D^{1+\mu}_x B^{(1+\mu)}_{\mu,j} \in \mathbb{P}_{N-2}$, as $B^{(1+\mu)}_{\mu,j} \in \mathcal{F}^{(\mu)}_N$. Thus, the interpolating condition (3.2.151) implies that for $1 \le j \le N-1$,

$$D^2\, {}^{\mathrm{RL}}_{-1}I^{1-\mu}_x B^{(1+\mu)}_{\mu,j}(x) = {}^{\mathrm{RL}}_{-1}D^{1+\mu}_x B^{(1+\mu)}_{\mu,j}(x) = \hat{l}_j(x) = \sum_{k=0}^{N-2} \hat{s}^{(\alpha,\beta,1,1)}_{kj} P^{(1,1)}_k(x).$$

Similar to (3.2.162), one has

$${}^{\mathrm{RL}}_{-1}I^{1-\mu}_x B^{(1+\mu)}_{\mu,j}(x) = A_1 + A_2(1+x) + \sum_{k=0}^{N-2} \hat{s}^{(\alpha,\beta,1,1)}_{kj}\, {}^{\mathrm{RL}}_{-1}I^2_x P^{(1,1)}_k(x).$$

Acting ${}^{\mathrm{RL}}_{-1}D^{1-\mu}_x$ on both sides of the above identity leads to

$$\begin{aligned} B^{(1+\mu)}_{\mu,j}(x) = {} & \frac{A_1(1+x)^{\mu-1}}{\Gamma(\mu)} + \frac{A_2(1+x)^{\mu}}{\Gamma(1+\mu)} \\ & + \sum_{k=0}^{N-2} \hat{s}^{(\alpha,\beta,1,1)}_{kj}\, {}^{\mathrm{RL}}_{-1}I^{1+\mu}_x P^{(1,1)}_k(x). \end{aligned} \tag{3.2.168}$$

Requiring $B^{(1+\mu)}_{\mu,j} \in \mathcal{F}^{(\mu)}_N$ leads to $A_1 = 0$. Noting that

$$P^{(1,1)}_k(x) = \frac{2}{k+2} D P_{k+1}(x), \quad k \ge 0,$$

we have from (3.2.168) that

$$\begin{aligned} {}^{\mathrm{RL}}_{-1}I^{1+\mu}_x P^{(1,1)}_k(x) &= \frac{2}{k+2}\, {}^{\mathrm{RL}}_{-1}I^{\mu}_x\, {}^{\mathrm{RL}}_{-1}I^{1}_x D P_{k+1}(x) \\ &= \frac{2}{k+2}\, {}^{\mathrm{RL}}_{-1}I^{\mu}_x \{P_{k+1}(x) - P_{k+1}(-1)\} \\ &= \frac{2(-1)^k}{(k+2)\Gamma(1+\mu)}(1+x)^{\mu} + \frac{2}{k+2}\, {}^{\mathrm{RL}}_{-1}I^{\mu}_x P_{k+1}(x) \\ &= \frac{2(-1)^k}{(k+2)\Gamma(1+\mu)}(1+x)^{\mu} + \frac{2\Gamma(k+2)}{(k+2)\Gamma(k+\mu+2)}(1+x)^{\mu} P^{(-\mu,\mu)}_{k+1}(x). \end{aligned}$$

Therefore, by (3.2.168),

$$B_{\mu,j}^{(1+\mu)}(x) = \frac{A_2(1+x)^{\mu}}{\Gamma(1+\mu)} + (1+x)^{\mu}\sum_{k=0}^{N-2}\frac{2\hat{s}_{kj}^{(\alpha,\beta,1,1)}}{k+2}\left\{\frac{(-1)^k}{\Gamma(1+\mu)} + \frac{\Gamma(k+2)}{\Gamma(k+\mu+2)}P_{k+1}^{(-\mu,\mu)}(x)\right\}.$$

Setting $B_{\mu,j}^{(1+\mu)}(1) = 0$, one obtains

$$\frac{A_2}{\Gamma(1+\mu)} = -\sum_{k=0}^{N-2}\frac{2\hat{s}_{kj}^{(\alpha,\beta,1,1)}}{k+2}\left\{\frac{(-1)^k}{\Gamma(1+\mu)} + \frac{\Gamma(k+2)}{\Gamma(k+\mu+2)}P_{k+1}^{(-\mu,\mu)}(1)\right\}.$$

Then, the formula for $B_{\mu,j}^{(1+\mu)}$ with $1 \le j \le N-1$ in (3.2.157) is derived. □

Introduce the matrices

$$\boldsymbol{B}_{\mu}^{(\nu)} = \begin{cases} \left(\boldsymbol{B}_{\mu}^{(\nu)}\right)_{lj}, & 1 \le l, j \le N, \qquad \mu, \nu \in (0,1), \\ \left(\boldsymbol{B}_{\mu}^{(\nu)}\right)_{lj}, & 1 \le l, j \le N-1, \quad \mu \in (0,1),\ 1 < \nu \le 1+\mu, \end{cases} \tag{3.2.169}$$

where $\left(\boldsymbol{B}_{\mu}^{(\nu)}\right)_{lj} = B_{\mu,j}^{(\nu)}(x_l)$.

Theorem 3.46 *For $\mu \in (0,1)$, we have*

$$\boldsymbol{D}_{\mu,\text{in}}^{(\nu)}\boldsymbol{B}_{\mu}^{(\nu)} = \boldsymbol{B}_{\mu}^{(\nu)}\boldsymbol{D}_{\mu,\text{in}}^{(\nu)} = \boldsymbol{I}, \tag{3.2.170}$$

where $\boldsymbol{I}$ is the identity matrix of order N for $\nu \in (0,1)$, and of order $N-1$ for $1 < \nu \le 1+\mu$.

Proof We only prove the case $\nu \in (0,1)$ because the other case is similar. Since $B_{\mu,j}^{(\nu)} \in \mathcal{F}_N^{(\mu)}$, we can write

$$B_{\mu,j}^{(\nu)}(x) = \sum_{l=1}^{N} B_{\mu,j}^{(\nu)}(x_l)\bar{h}_l^{\mu}(x), \quad 1 \le j \le N,$$

where $\{\bar{h}_l^{\mu}\}$ are defined in (3.2.124). Thus,

$${}_{-1}^{\text{RL}}D_x^{\nu}\, B_{\mu,j}^{(\nu)}(x) = \sum_{l=1}^{N} B_{\mu,j}^{(\nu)}(x_l)\ {}_{-1}^{\text{RL}}D_x^{\nu}\, \bar{h}_l^{\mu}(x).$$

Taking $x = x_i$ for $1 \le i \le N$ in the above equation, we obtain (3.2.170) from (3.2.151) directly. □

3.2.4 Fractional Collocation Methods and Error Analysis

In what follows, we present the fractional collocation schemes for fractional initial-value and fractional boundary-value problems, using the new basis derived previously. We also conduct rigorous error analysis of the schemes at the Chebyshev points.

Fractional Initial-Value Problems of Order $\nu \in (0,1)$

In this section, we consider the fractional initial-value problem (FIVP) of order $\nu \in (0,1)$:

$$ {}_{-1}^{\mathrm{RL}}D_t^{\nu} u(t) + q(t)u(t) = f(t), \quad t \in (-1,1]; \quad u(-1) = 0, \tag{3.2.171}$$

where $q(t)$ and $f(t)$ are continuous functions.

The following decomposition is useful for understanding the singularity of the solution of (3.2.171) so as to compass the design of the collocation scheme. Hereafter, denote $I := (-1,1)$.

Theorem 3.47 *Let u be the solution of* (3.2.171).

- *Assume $q(t) = 0$ and $f(t) \in C^m(I)$. Then*

$$\begin{aligned} u(t) &= {}_{-1}^{\mathrm{RL}}I_t^{\nu} f(t) \\ &= (t+1)^{\nu}\Big\{\sum_{i=0}^{m-1} \frac{f^{(i)}(-1)}{\Gamma(\nu+i+1)}(t+1)^i + \frac{f^{(m)}(\xi_t)}{\Gamma(\nu+m+1)}(t+1)^m\Big\} \end{aligned} \tag{3.2.172}$$

  for some $\xi_t \in (-1,t)$.
- *Assume $q(t) \neq 0$ and $q(t), f(t) \in C^m(I)$. Then, $u(t)$ has the decomposition:*

$$ u(t) = \sum_{(i,j)\in \Upsilon_m^{\nu}} \tilde{\gamma}_{ij}(\nu)(t+1)^{i+\nu(j+1)} + \Phi(\xi_t;\nu)(t+1)^{\nu+m} \tag{3.2.173}$$

  for $t \in (-1,1)$ and $\xi_t \in (-1,t)$, where $\Upsilon_m^{\nu} := \{(i,j) : i + j\nu < m,\ i,j \ge 0\}$, $\Phi(\cdot;\nu) \in C^0(I)$, and $\{\tilde{\gamma}_{ij}(\nu)\}$ are some constants.

Proof We first consider the case with $q(t) = 0$. Since we have $u(t) = {}_{-1}^{\mathrm{RL}}I_t^{\nu} f(t)$, using integration-by-parts repeatedly, leads to

$$\begin{aligned} ({}_{-1}^{\mathrm{RL}}I_t^{\nu} f)(t) &= \frac{1}{\Gamma(\nu)}\int_{-1}^{t} (t-s)^{\nu-1} f(s)\,\mathrm{d}s \\ &= \frac{f(-1)(t+1)^{\nu}}{\Gamma(\nu+1)} + \frac{1}{\Gamma(\nu+1)}\int_{-1}^{t} (t-s)^{\nu} f'(s)\,\mathrm{d}s \end{aligned}$$

$$
\begin{aligned}
&= \frac{f(-1)(t+1)^{\nu}}{\Gamma(\nu+1)} + \frac{f'(-1)(t+1)^{\nu+1}}{\Gamma(\nu+2)} \\
&\quad + \frac{1}{\Gamma(\nu+2)} \int_{-1}^{t} (t-s)^{\nu+1} f''(s)\, \mathrm{d}s \\
&= \cdots \\
&= \sum_{i=0}^{m-1} \frac{f^{(i)}(-1)(t+1)^{\nu+i}}{\Gamma(\nu+i+1)} \\
&\quad + \frac{1}{\Gamma(\nu+m)} \int_{-1}^{t} (t-s)^{\nu+m-1} f^{(m)}(s)\, \mathrm{d}s.
\end{aligned} \tag{3.2.174}
$$

Then (3.2.172) follows from the mean value theorem of integration immediately.

We now turn to the second case. Since ${}^{\mathrm{RL}}_{-1}I_t^{\nu}\ {}^{\mathrm{RL}}_{-1}D_t^{\nu}$ is an identity operator, we introduce $z(t) = {}^{\mathrm{RL}}_{-1}D_t^{\nu}u(t)$, and transform (3.2.171) into

$$
z(t) + \frac{q(t)}{\Gamma(\nu)} \int_{-1}^{t} (t-s)^{\nu-1} z(s)\, \mathrm{d}s = f(t), \quad t \in I, \tag{3.2.175}
$$

which is a Volterra integral equation of the second type with a weakly singular kernel. As (3.2.171) is equivalent to (3.2.175), we obtain from (Brunner, 2017, Theorem 6.1.6) that

$$
z(t) = \sum_{(i,j)\in\Upsilon_m^{\nu}} \gamma_{ij}(\nu)(t+1)^{i+j\nu} + Z_m(t;\nu), \quad t \in I, \tag{3.2.176}
$$

where $Z_m(\cdot;\nu) \in C^m(I)$, and $\{\gamma_{ij}(\nu)\}$ are constants. By (3.2.172) and (3.2.174), we have

$$
\begin{aligned}
{}^{\mathrm{RL}}_{-1}I_t^{\nu} Z_m(t;\nu) &= \sum_{i=0}^{m-1} \frac{Z_m^{(i)}(-1;\nu)}{\Gamma(\nu+i+1)}(t+1)^{i+\nu} \\
&\quad + \frac{Z_m^{(m)}(\xi_t;\nu)}{\Gamma(\nu+m+1)}(t+1)^{\nu+m}.
\end{aligned} \tag{3.2.177}
$$

Recall that (cf. Diethelm (2010))

$$
{}^{\mathrm{RL}}_{-1}I_t^{\nu}\{(t+1)^{i+j\nu}\} = \frac{\Gamma(i+j\nu+1)}{\Gamma(\nu+i+j\nu+1)}(t+1)^{i+(j+1)\nu}. \tag{3.2.178}
$$

Therefore, by (3.2.176)–(3.2.179),

$$\begin{aligned} u(t) &= {}_{-1}^{\mathrm{RL}}I_t^{\nu} z(t) \\ &= \sum_{(i,j)\in \Upsilon_m^{\nu}} \tilde{\gamma}_{ij}(\nu)(1+t)^{i+(j+1)\nu} + \frac{Z_m^{(m)}(\xi_t;\nu)}{\Gamma(\nu+m+1)}(t+1)^{\nu+m}. \end{aligned} \tag{3.2.179}$$

Denoting $\Phi(t;\nu) = \dfrac{Z_m^{(m)}(t;\nu)}{\Gamma(\nu+m+1)}$, we obtain (3.2.173). □

Remark 3.48 We highlight some implications of Theorem 3.47, which can guide the selection of the basis to best fit the singular factor.

- Observe from (3.2.172) that if $q = 0$, $u(t) = (t+1)^{\nu}\Psi(t)$ (where Ψ is smooth), so the use of Jacobi poly-fractonomials, see Chapter 2 and Zayernouri and Karniadakis (2014a), and general Jacobi functions (Chen et al., 2016) leads to spectrally accurate approximation when $f(t)$ is sufficiently smooth.
- If $q(t) = \lambda \neq 0$, the solution of (3.2.171) is given by (cf. Diethelm (2010))

$$u(t) = \int_{-1}^{t} e_{\nu,\nu}(t-\tau;\lambda)\, f(\tau)\,\mathrm{d}\tau = \sum_{k=0}^{\infty}(-\lambda)^k \big({}_{-1}^{\mathrm{RL}}I_t^{(k+1)\nu} f\big)(t), \tag{3.2.180}$$

where the involved Mittag-Leffler (ML) functions of two parameters are given

$$e_{\alpha,\beta}(z;\lambda) := z^{\beta-1}E_{\alpha,\beta}(-\lambda z^{\alpha}), \quad E_{\alpha,\beta}(z) := \sum_{k=0}^{\infty}\frac{z^k}{\Gamma(\alpha k+\beta)}, \tag{3.2.181}$$

for $\alpha, \beta > 0$. We therefore derive from (3.2.173) and (3.2.180) that

$$u(t) = \sum_{k=0}^{\infty}(-\lambda)^k (t+1)^{(k+1)\nu}\Psi_{km}(t;\nu), \tag{3.2.182}$$

where $\Psi_{km}(t;\nu)$ denotes the summation in the brace of (3.2.172). This provide a more explicit formula than (3.2.173), when q is a nonzero constant. □

Collocation Schemes and Error Estimates

Theorem 3.47 implies that an appropriate collocation scheme for (3.2.171) is to find $u_N \in \mathcal{F}_N^{(\nu)}$ (cf. (3.2.146)) such that

$${}_{-1}^{\mathrm{RL}}D_t^{\nu}u_N(t_i) + q(t_i)u_N(t_i) = f(t_i), \quad 1 \le i \le N; \quad u_N(-1) = 0, \tag{3.2.183}$$

where $\{t_i\}$ are the JGL points. Hereafter, we restrict to the Chebyshev–Gauss–Lobatto (CGL) points.

The linear system under *fractional Lagrange interpolation basis* $\{\bar{h}_j^{\nu}\}$ (cf. (3.2.159)) reads

$$(\boldsymbol{D}_{\nu,\text{in}}^{(\nu)} + \boldsymbol{Q})\boldsymbol{u} = \boldsymbol{f}, \tag{3.2.184}$$

where $\boldsymbol{D}_{\nu,\text{in}}^{(\nu)}$ is defined as in (3.2.145), $\boldsymbol{Q} = \text{diag}(q(t_1), q(t_2), \ldots, q(t_N))$ and

$$\boldsymbol{u} = (u_N(t_1), u_N(t_2), \ldots, u(t_N))^T, \quad \boldsymbol{f} = (f(t_1), f(t_2), \ldots, f(t_N))^T.$$

In view of Theorem 3.46, we can precondition (3.2.184) by using $\boldsymbol{B}_{\nu}^{(\nu)}$, and obtain that

$$(\boldsymbol{I}_N + \boldsymbol{B}_{\nu}^{(\nu)}\boldsymbol{Q})\boldsymbol{u} = \boldsymbol{B}_{\nu}^{(\nu)}\boldsymbol{f}. \tag{3.2.185}$$

Alternatively, changing the variable: $\boldsymbol{u} = \boldsymbol{B}_{\nu}^{(\nu)}\boldsymbol{v}$, and using Theorem 3.46, we can rewrite (3.2.184) as

$$(\boldsymbol{I}_N + \boldsymbol{Q}\boldsymbol{B}_{\nu}^{(\nu)})\boldsymbol{v} = \boldsymbol{f}, \tag{3.2.186}$$

which actually is the linear system of the collocation scheme (3.2.183) under the Birkhoff interpolation basis $\{B_{\nu,j}^{(\nu)}\}$ in (3.2.152).

Thanks to the relation between (3.2.171) and (3.2.175), we can use the analysis tools for the integral equations (cf. Huang and Stynes (2016)) to conduct the L^∞-error estimate.

Theorem 3.49 *Let u, u_N and z be defined in* (3.2.171), (3.2.183) *and* (3.2.175) *and $z_N = {}_{-1}^{\text{RL}}D_t^{\nu}u_N(t)$. If $q, f \in C^m(I)$ with $I = (-1, 1)$, then we have*

$$\begin{aligned} \| {}_{-1}^{\text{RL}}D_t^{\nu}u - {}_{-1}^{\text{RL}}D_t^{\nu}u_N\|_\infty &= \|z - z_N\|_\infty \\ &\le C(1 + \log N)\max\{N^{-2\nu}, N^{1-m}\}, \end{aligned} \tag{3.2.187}$$

where the constant C is independent of N.

Proof Note that $z_N(t) = {}_{-1}^{\text{RL}}D_t^{\nu}u_N(t) \in \mathbb{P}_{N-1}$ is a polynomial, which together with the formula (2.4.13), leads to $u_N(t) = {}_{-1}I_t^{\nu}z_N(t)$. For notational simplicity, we denote

$$\mathcal{K}w(t) = \frac{q(t)}{\Gamma(\nu)}\int_{-1}^{t}(t-s)^{\nu-1}w(s)\,ds.$$

The scheme (3.2.183) is equivalent to

$$z_N(t_i) + \mathcal{K}z_N(t_i) = f(t_i), \quad 1 \le i \le N. \tag{3.2.188}$$

We can follow Huang and Stynes (2016) (for the analysis of polynomial approximation to integral equations) to conduct the error analysis. Here, we just sketch the proof. By (3.2.175),

$$z(t_i) + \mathcal{K}z(t_i) = f(t_i), \quad 1 \le i \le N, \tag{3.2.189}$$

where $\{t_i\}_{i=0}^N$ (with $t_0 = -1$) are CGL points. Let $\mathbb{I}_{N-1}^C$ be the Chebyshev–Gauss–Radau interpolation operator at $\{t_i\}_{i=1}^N$. Then we have

$$\mathbb{I}_{N-1}^C z(t) - z_N(t) = \mathbb{I}_{N-1}^C \mathcal{K}e(t), \quad \forall t \in I,$$

where $e = z - z_N$. This leads to

$$e(t) = (I_1 + I_2)(t) + (\mathcal{K}e)(t),$$

where

$$I_1 := z - \mathbb{I}_{N-1}^C z, \quad I_2 := \mathbb{I}_{N-1}^C(\mathcal{K}e) - \mathcal{K}e.$$

It follows from the generalized Gronwall inequality that

$$\|e\|_\infty \le C(\|I_1\|_\infty + \|I_2\|_\infty) \tag{3.2.190}$$

for some constant C independent of N. It is noteworthy that the N-point Lebesgue constant of Chebyshev–Gauss–Radau interpolation is $\mathcal{O}(\log N)$ (cf. Shen et al. (2011a)). With some tedious calculation as in Huang and Stynes (2016, Corollary 1 and Theorem 1) and using (3.2.176), we can show that

$$\begin{aligned} \|I_1\|_\infty &\le C(1+\log N)\max\{N^{-2\nu}, N^{1-m}\}, \\ \|I_2\|_\infty &\le C(1+\log N)N^{-\nu}\|e\|_\infty. \end{aligned} \tag{3.2.191}$$

Substituting (3.2.191) into (3.2.190), and noting that for sufficiently large N, the terms involving $\|e\|_\infty$ can be absorbed into the left-hand side, we obtain the desired estimates. □

Remark 3.50 The approximation of the leading singular factor in (3.2.173) leads to the error term $\mathcal{O}(N^{-2\nu})$, the approximation of the smoother function $\Phi(\xi_t;\nu)(t+1)^{\nu+m}$ contributes the error term $\mathcal{O}(N^{1-m})$ and $\log N$ originates from the Lebesgue constant of the interpolation. □

Remark 3.51 It is worth emphasizing that the approach we use to compute the approximation of $z(t)$ is essentially the iterated method recommended for integral equations in Atkinson (1997, P. 71) since

$$z_N^{it}(t_i) = f(t_i) - q(t_i)u_N(t_i) = f(t_i) - \mathcal{K}z_N(t_i), \quad 1 \le i \le N.$$

This simple iteration can always yield a more accurate approximation because

$$\mathbb{I}_{N-1}^C z_N^{it}(t) = \mathbb{I}_{N-1}^C[f(t) - q(t)u_N(t)] = z_N(t).$$

Note that the approximation space in which $z_N^{it}(t)$ resides is larger than the one for $z_N(t)$. Hence, the convergence rates of the present numerical examples are higher than the one shown in Theorem 3.49, which is also confirmed by the numerical results below. □

Numerical Results

In order to test the efficiency and accuracy of our algorithms, we present below some numerical examples and compare the results obtained by the usual collocation scheme, preconditioned GJF approximation and the method (Z-K method) in Zayernouri and Karniadakis (2014a). In the following tests, we choose $N = 8, 12, 16, 20, 24, 30, 60, 120, 240, 480, 960$. The errors are measured in the norm $\|\cdot\|_\infty$.

Example 3.52 Consider

$$ {}_{-1}^{\mathrm{RL}}D_t^{\mu} u(t) = \sin(10t), \quad \mu \in (0,1),\ t \in (-1,1]; \quad u(-1) = 0. \tag{3.2.192} $$

From Theorem 3.47, we expect to obtain an exponential rate of convergence if we choose $\nu = \mu$. Here, we take $\mu = 0.8$ and use "bicgstab" iterative solver of MATLAB to solve the final linear system. Since the true solution is unknown, we take the approximation when $N = 3840$ as the "true solution" to calculate numerical errors. Indeed, an exponential convergence rate is observed for the present methods, whereas that of the Z-K method is $\mathcal{O}(N^{-1.6})$, see Figure 3.10. Furthermore, we can also observe that the condition number of usual collocation methods grows like $\mathcal{O}(N^{2\mu})$, while the preconditioned scheme remains a constant for all N. Indeed, the new scheme converges within 10 iterations, while the other two methods need more than 4000 iterations to converge to the expected accuracy.

Example 3.53 Consider

$$ {}_{-1}^{\mathrm{RL}}D_t^{\mu} u(t) + \sin(t)u(t) = 1, \quad \mu \in (0,1), \quad t \in (-1,1]; \quad u(-1) = 0. \tag{3.2.193} $$

As in the previous example, we choose $\nu = \mu = 0.8$ in our algorithm and take the approximation when $N = 3840$ as the "true solution". Due to the different nature of singularity, we can only expect an algebraic decay of the errors. From Figure 3.11, we observe that the convergence rate of our methods is faster than that of the Z-K method, which is $\mathcal{O}(N^{-1.6})$. Moreover, as explained in Remark 3.51, the order is higher than the theoretical estimate in Theorem 3.49. On the other hand, to achieve the same numerical error, our preconditioned method only requires several iterations while others require thousands by using "bicgstab" as our linear system solver. We remark that

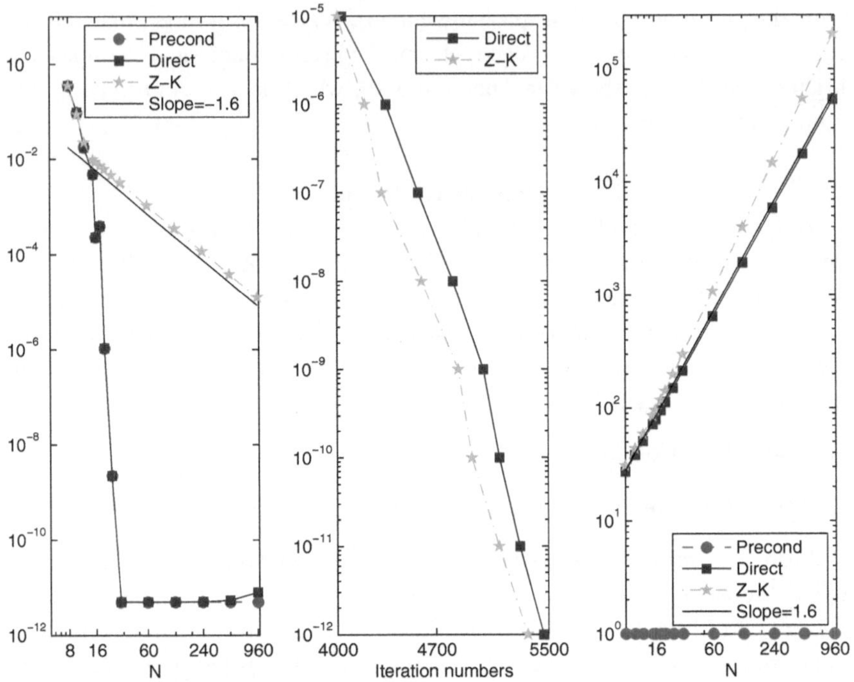

Figure 3.10 (Example 3.52): (Left): L^{∞}-errors in log-log scale of the usual collocation scheme, its preconditioned collocation method and Z-K method (Zayernouri and Karniadakis, 2014a) with respect to $\mu = \nu = 0.8$; (Middle): iteration numbers of usual collocation and Z-K methods against errors for 960 collocation points (note: the preconditioned approach converges within 10 iterations); (Right): plots of condition numbers of obtained linear system of these methods. Source: Huang et al. (2016). Copyright ©2016 Society for Industrial and Applied Mathematics. Reprinted with permission. All rights reserved.

the Z-K method should perform as well as the usual collocation method if $u'(-1) = 0$ as well.

Fractional Boundary-Value Problems of Order $1+\nu \in (1,2)$

Now, we consider the BVP of order $1 + \nu \in (1,2)$:

$$ {}^{\mathrm{RL}}_{-1}D_t^{1+\nu}u(t) + q(t)u(t) = f(t), \quad t \in I = (-1,1), \tag{3.2.194} $$

with $u(-1) = 0$ and $u(1) = u_1$, where $q(t), f(t) \in C^m(I)$.

As with the previous study, we first examine the singular behavior of the solution based on the equivalent fractional integral formulation.

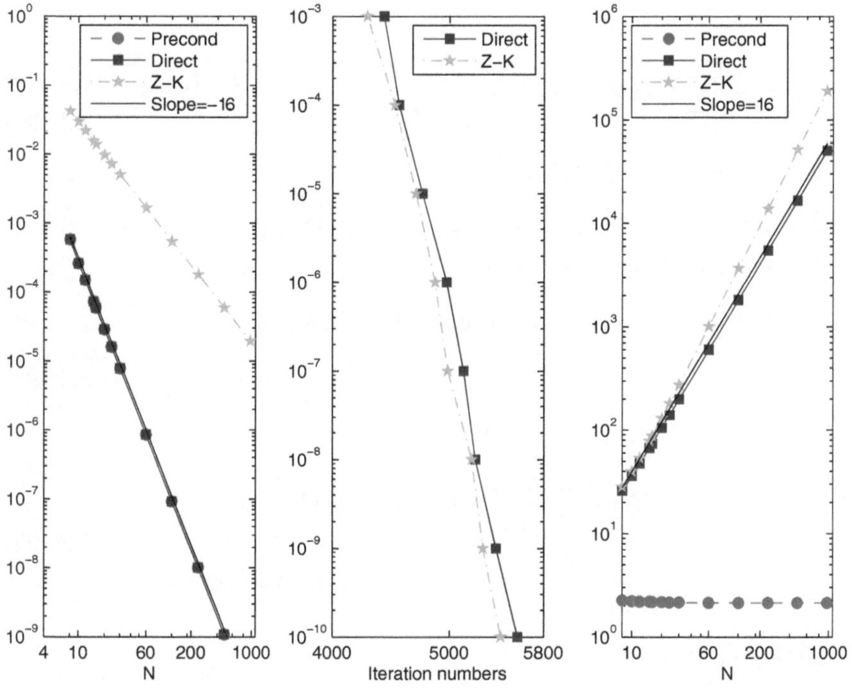

Figure 3.11 (Example 3.53): (Left): L^∞-errors in log-log scale of the usual collocation scheme, its preconditioned collocation method and Z-K method (Zayernouri and Karniadakis, 2014a) with respect to $\mu = \nu = 0.8$; (Middle): iteration numbers of usual collocation and Z-K methods against errors for 960 collocation points (note: the preconditioned approach converges within 10 iterations); (Right): plots of condition numbers of obtained linear system of these methods. Source: Huang et al. (2016). Copyright ©2016 Society for Industrial and Applied Mathematics. Reprinted with permission. All rights reserved.

Theorem 3.54 *Assume that $q(t), f(t) \in C^m(I)$. Then the solution $u(t)$ of* (3.2.194) *has the following decomposition.*

- *If $q(t) = 0$, then*

$$u(t) = \frac{(t+1)^\nu}{2^\nu}\left\{u_1 - {}^{\mathrm{RL}}_{-1}I_t^{\nu+1} f(t)\big|_{t=1}\right\} + (t+1)^{\nu+1}\left\{\sum_{i=0}^{m-1} \frac{f^{(i)}(-1)}{\Gamma(\nu+i+2)}(t+1)^i + \frac{f^{(m)}(\xi_t)}{\Gamma(\nu+m+2)}(t+1)^m\right\} \tag{3.2.195}$$

for some $\xi_t \in (-1, t)$.

- *If $q(t) \neq 0$, then*

$$u(t) = \sum_{i=0}^{m} \breve{\gamma}_i(\nu)(t+1)^{i+\nu} + \sum_{(i,j)\in\Upsilon_m^\nu} \tilde{\gamma}_{ij}(\nu)(1+t)^{i+1+(j+1)\nu} + G(\xi_t,\nu)(t+1)^{\nu+m+1}, \tag{3.2.196}$$

where $\Upsilon_m^\nu := \{(i,j) : i + j\nu < m,\ i,j \geq 0\}$, $G(\cdot;\nu) \in C^0(I)$, and $\{\breve{\gamma}_i(\nu), \tilde{\gamma}_{ij}(\nu)\}$ are some constants.

Proof Let $z(t) = {}_{-1}^{\mathrm{RL}}D_t^\nu u(t)$. Since $u(-1) = 0$, we have

$$u(t) = {}_{-1}I_t^\nu z(t). \tag{3.2.197}$$

Moreover, (3.2.194) can be transformed to

$$z'(t) + \frac{q(t)}{\Gamma(\nu)} \int_{-1}^{t} (t-s)^{\nu-1} z(s) ds = f(t),$$
$$\frac{1}{\Gamma(\nu)} \int_{-1}^{1} (1-s)^{\nu-1} z(s) ds = u(1) = u_1. \tag{3.2.198}$$

If $q(t) = 0$, we have

$$z(t) = z(-1) + \int_{-1}^{t} f(s)\, ds. \tag{3.2.199}$$

Inserting it into the second identity of (3.2.198), and using integration-by-parts, we find

$$z(-1) = \frac{1}{2^\nu}\Big(u_1\Gamma(1+\nu) - \int_{-1}^{1} (1-s)^\nu f(s) ds\Big). \tag{3.2.200}$$

Hence, by (3.2.197), (3.2.199) and a direct calculation,

$$u(t) = {}_{-1}^{\mathrm{RL}}I_t^\nu z(t) = z(-1)\frac{(t+1)^\nu}{\Gamma(\nu+1)} + {}_{-1}^{\mathrm{RL}}I_t^{\nu+1} f(t).$$

Then (3.2.195) is a direct consequence of (3.2.172) and (3.2.200).

We now turn to the second case. Further let $w(t) = z'(t)$, and rewrite (3.2.198) as

$$w(t) + \frac{q(t)}{\Gamma(\nu+1)} \int_{-1}^{t} (t-s)^\nu w(s) ds = f(t) - \frac{q(t) z(-1)}{\Gamma(\nu+1)}(t+1)^\nu, \tag{3.2.201}$$

where like (3.2.200), we have

$$z(-1) = \frac{1}{2^\nu}\Big(u_1\Gamma(1+\nu) - \int_{-1}^{1} (1-s)^\nu w(s) ds\Big). \tag{3.2.202}$$

Note that $z(-1)$ is a well-defined constant by noting that $w(t) = {}^{\rm RL}_{-1}D_t^{1+\nu}u(t) = f(t) - q(t)u(t)$. Then we can derive from Cao et al. (2003) that

$$w(t) = \sum_{(i,j)\in\Upsilon_m^\nu} \gamma_{ij}(\nu)(t+1)^{i+j\nu} + \phi_m(t;\nu), \tag{3.2.203}$$

where $\phi_m(\cdot;\nu) \in C^m(I)$, and $\{\gamma_{ij}(\nu)\}$ are constants. Therefore, $z(t)$ has the decomposition

$$z(t) = \sum_{(i,j)\in\Upsilon_m^\nu} \hat{\gamma}_{ij}(\nu)(t+1)^{i+j\nu+1} + Z_{m+1}(t;\nu), \tag{3.2.204}$$

where $Z_{m+1}(\cdot;\nu) \in C^{m+1}(I)$. Like (3.2.177), we have

$$\begin{aligned} {}^{\rm RL}_{-1}I_t^\nu Z_{m+1}(t;\nu) = &\sum_{i=0}^m \frac{Z_{m+1}^{(i)}(-1;\nu)}{\Gamma(\nu+i+1)}(t+1)^{i+\nu} \\ &+ \frac{Z_{m+1}^{(m+1)}(\xi_t;\nu)}{\Gamma(\nu+m+1)}(t+1)^{\nu+m+1}. \end{aligned} \tag{3.2.205}$$

Thanks to (3.2.197), which we obtain from (3.2.178) and the above, we can apply the same argument for (3.2.173) to derive

$$\begin{aligned} u(t) = {}^{\rm RL}_{-1}I_t^\nu z(t) = &\sum_{(i,j)\in\Upsilon_m^\nu} \tilde{\gamma}_{ij}(\nu)(1+t)^{i+1+(j+1)\nu} \\ &+ \sum_{i=0}^m \frac{Z_{m+1}^{(i)}(-1;\nu)}{\Gamma(\nu+i+1)}(t+1)^{i+\nu} + \frac{Z_{m+1}^{(m+1)}(\xi_t;\nu)}{\Gamma(\nu+m+1)}(t+1)^{\nu+m+1}, \end{aligned} \tag{3.2.206}$$

which implies (3.2.196). □

Collocation Schemes and Error Estimates

From Theorem 3.54, one learns that the appropriate collocation scheme for (3.2.194) is to find $u_N \in \mathcal{F}_N^{(\nu)}$ such that

$${}^{\rm RL}_{-1}D_t^{1+\nu}u_N(t_i) + q(t_i)u_N(t_i) = f(t_i), \quad 1 \le i \le N-1; \quad u_N(1) = u_1, \tag{3.2.207}$$

where $\{t_i\}$ are CGL points, and $u_N(-1) = 0$ is automatically met.

Let $\boldsymbol{D}_{\nu,\rm in}^{(1+\nu)}$ be defined as in (3.2.145), and $\boldsymbol{d} = (d_1, \ldots, d_{N-1})^T$ be the middle $(N-1)$ elements of the last column of (3.2.142). Set

$$\begin{aligned} \boldsymbol{Q} &= \mathrm{diag}(q(t_1), q(t_2), \ldots, q(t_{N-1})), \\ \boldsymbol{u} &= (u_N(t_1), u_N(t_2), \ldots, u(t_{N-1}))^T, \\ \boldsymbol{f} &= (f(t_1) - d_1u_1, f(t_2) - d_2u_1, \ldots, f(t_{N-1}) - d_{N-1}u_1)^T. \end{aligned} \tag{3.2.208}$$

The linear system under the fractional Lagrange interpolation basis $\{\bar{h}_j^\nu\}$ (cf. (3.2.159)) reads

$$\big(\boldsymbol{D}_{\nu,\mathrm{in}}^{(1+\nu)} + \boldsymbol{Q}\big)\boldsymbol{u} = \boldsymbol{f}. \tag{3.2.209}$$

Thanks to Theorem 3.46, we can precondition the above system as

$$\big(\boldsymbol{I}_{N-1} + \boldsymbol{B}_\nu^{(1+\nu)}\boldsymbol{Q}\big)\boldsymbol{u} = \boldsymbol{B}_\nu^{(1+\nu)}\boldsymbol{f}. \tag{3.2.210}$$

Alternatively, changing the variable: $\boldsymbol{u} = \boldsymbol{B}_\nu^{(1+\nu)}\boldsymbol{v}$, and using Theorem 3.46, we can rewrite (3.2.209) as

$$\big(\boldsymbol{I}_{N-1} + \boldsymbol{Q}\boldsymbol{B}_\nu^{(1+\nu)}\big)\boldsymbol{v} = \boldsymbol{f}, \tag{3.2.211}$$

which actually is the linear system of the collocation scheme (3.2.207) under the Birkhoff interpolation basis $\{B_{\nu,j}^{(1+\nu)}\}$ in (3.2.157).

Theorem 3.55 *Let u and u_N be the solutions of* (3.2.194) *and* (3.2.207), *respectively. If $f, q \in C^m(I)$ with $m \geq 1$, we have*

$$\| {}^{\mathrm{RL}}_{-1}D_t^{1+\nu}u - {}^{\mathrm{RL}}_{-1}D_t^{1+\nu}u_N \|_\infty \leq C(1 + \log N)\max\{N^{-2\nu}, N^{1-m}\}, \tag{3.2.212}$$

where u is a positive constant independent of N and any function.

Proof Inserting (3.2.202) into (3.2.201), we can obtain the mixed Volterra–Fredholm equation:

$$\begin{aligned} w(t) &+ \frac{q(t)}{\Gamma(\nu+1)}\int_{-1}^t (t-s)^\nu w(s)ds \\ &- \frac{q(t)(t+1)^\nu}{\Gamma(\nu+1)2^\nu}\int_{-1}^1 (1-s)^\nu w(s)ds \\ &= f(t) - \frac{u_1 q(t)(t+1)^\nu}{2^\nu}. \end{aligned} \tag{3.2.213}$$

For notational convenience, we denote

$$\begin{aligned} \mathcal{K}_1 w &:= \frac{q(t)}{\Gamma(\nu+1)}\int_{-1}^t (t-s)^\nu w(s)ds, \\ \mathcal{K}_2 w &:= \frac{q(t)(t+1)^\nu}{\Gamma(\nu+1)2^\nu}\int_{-1}^1 (1-s)^\nu w(s)ds. \end{aligned} \tag{3.2.214}$$

Both are compact operators from $L^\infty(-1,1)$ to $L^\infty(-1,1)$, so is $\mathcal{K}_2 - \mathcal{K}_1$. Since $u_N \in \mathcal{F}_N^{(\nu)}$, we infer from (2.4.19) that $w_N(t) = {}^{\mathrm{RL}}_{-1}D_t^{1+\nu}u_N \in \mathbb{P}_{N-2}$. Note that for $1 \leq i \leq N-1$,

$$w_N(t_i) + \mathcal{K}_1 w_N(t_i) - \mathcal{K}_2 w_N(t_i) = f(t_i) - \frac{u_1 q(t_i)(t_i+1)^\nu}{2^\nu}.$$

One also verifies

$$w_N(t) + \mathcal{I}_{N-2}^C[\mathcal{K}_1 w_N] - \mathcal{I}_{N-2}^C[\mathcal{K}_2 w_N]$$
$$= \mathcal{I}_{N-2}^C f - \frac{u_1}{2^\nu \Gamma(1+\nu)} \mathcal{I}_{N-2}^C\big[q(t)(t+1)^\nu\big],$$

where $\mathcal{I}_{N-2}^C$ is the interpolation operator associated with the $(N-1)$ "interior" CGL points. Similarly, the true solution satisfies

$$\mathcal{I}_{N-2}^C w + \mathcal{I}_{N-2}^C[\mathcal{K}_1 w] - \mathcal{I}_{N-2}^C[\mathcal{K}_2 w] = \mathcal{I}_{N-2}^C f - \frac{u_1}{2^\nu \Gamma(1+\nu)} \mathcal{I}_{N-2}^C[q(t)(t+1)^\nu].$$

Hence, letting $e_N = w - w_N$, we can obtain

$$\mathcal{I}_{N-2}^C w - w_N + \mathcal{I}_{N-2}^C[\mathcal{K}_1 e_N] - \mathcal{I}_{N-2}^C[\mathcal{K}_2 e_N] = 0.$$

Therefore,

$$\begin{aligned} e_N(t) &= w(t) - \mathcal{I}_{N-2}^C w(t) + \mathcal{I}_{N-2}^C w(t) - w_N(t) \\ &= w(t) - \mathcal{I}_{N-2}^C w(t) - \mathcal{I}_{N-2}^C[\mathcal{K}_1 e_N] + \mathcal{I}_{N-2}^C[\mathcal{K}_2 e_N] \\ &= (\mathcal{K}_2 - \mathcal{K}_1)e_N + \underbrace{w(t) - \mathcal{I}_{N-2}^C w(t)}_{I_1} \\ &\quad - \underbrace{(\mathcal{I}_{N-2}^C \mathcal{K}_1 - \mathcal{K}_1)e_N}_{I_2} + \underbrace{(\mathcal{I}_{N-2}^C \mathcal{K}_2 - \mathcal{K}_2)e_N}_{I_3}. \end{aligned}$$

By the Fredholm alternative, there exists a constant C independent of N and any function such that

$$\|e_N\|_\infty \le C(\|I_1\|_\infty + \|I_2\|_\infty + \|I_3\|_\infty). \tag{3.2.215}$$

From Huang and Stynes (2016, Thm. 2), we obtain

$$\begin{aligned} \|I_1\|_\infty &\le C(1+\log N)\max\{N^{-2\nu}, N^{1-m}\}, \\ \|I_2\|_\infty &\le C(1+\log N)N^{-\nu}\|e_N\|_\infty. \end{aligned} \tag{3.2.216}$$

We now deal with I_3. We first show that $Q(t) = q(t)(t+1)^\nu$ is Holder continuous with index ν in the same way that

$$|Q(t_1) - Q(t_2)| \le C|t_1 - t_2|^\nu, \quad \forall t_1, t_2 \in (-1, 1). \tag{3.2.217}$$

To verify this, we denote $t = t_1$ and $t_2 = t + h$ and assume that $h > 0$. It is evident that

$$\begin{aligned} |Q(t+h) - Q(t)| &\le |q(t+h)[(t+1+h)^\nu - (t+1)^\nu]| \\ &\quad + |(t+1)^\nu[q(t+h) - q(t)]| \\ &\le q_{\max}|W(t)| + 2^\nu|q'(\xi)|h, \end{aligned}$$

where $q_{\max} := \max_{t\in(-1,1)} |q(t)|$ and $W(t) := (t+1+h)^\nu - (t+1)^\nu$. One verifies readily that $W'(t) < 0$ for $t \in (-1,1)$ and $\nu \in (0,1)$, so $W(t) \le W(-1) = h^\nu$. Then we can claim (3.2.217). With this, Jackson's Theorem (Davis, 1975, Thm. 13.3.7) implies there exists a best polynomial $\pi_N Q(t)$ such that

$$\|Q - \pi_N Q\|_\infty \le CN^{-\nu}.$$

Then using the result on the Lebesgue constant, we have

$$\begin{aligned}\|Q(t) - \mathcal{I}_{N-2}^C Q(t)]\|_\infty &\le (1 + \|\mathcal{I}_{N-2}^C\|_\infty)\|Q - \pi_N Q\|_\infty \\ &\le C(1+\log N)N^{-\nu}.\end{aligned} \tag{3.2.218}$$

Therefore,

$$\begin{aligned}\|I_3\|_\infty &= \frac{1}{2^\nu \Gamma(\nu+1)}\|Q(t) - \mathcal{I}_{N-2}^C Q(t)\|_\infty \left|\int_{-1}^{1}(1-s)^\nu e_N(s)\mathrm{d}s\right| \\ &\le C(1+\log N)N^{-\nu}\|e_N\|_\infty.\end{aligned} \tag{3.2.219}$$

Now, substituting (3.2.216) and (3.2.219) into (3.2.215), we find that for N sufficiently large, the terms involving $\|e\|_\infty$ can be absorbed into the left-hand side, so the estimate (3.2.212) follows. □

Numerical Results

We provide some numerical results and also compare with the collocation methods using nodal Jacobi poly-fractonomial approximation (Z-K method) in Zayernouri and Karniadakis (2014a) in (3.2.117) with $\mu = \nu$. It is noteworthy that the numerical results therein were for given exact solutions with singular $f(x)$. Below, we mostly consider the FBVPs with smooth $f(x)$. Note that in the following figures, the samples of N are the same as for the previous FIVPs.

Example 3.56 Consider the fractional BVP:

$${}_{-1}^{\mathrm{RL}}D_t^{1+\nu} u(t) = \cos t,\ \nu \in (0,1),\ \ t \in (-1,1); \quad u(\pm 1) = 0.$$

With understanding of the singularity of the solution in (3.2.195), we expect the collocation scheme (3.2.207) leading to spectrally accurate approximation. Indeed, we observe from Figure 3.12 (leftmost) (where the reference "exact" solution is computed by a large number of points as with the previous two examples) that the errors of (3.2.209) and (3.2.210) decay exponentially, though the former suffers from severe round-off errors as the conditioning of its system behaves like $O(N^{2+2\nu})$ (see Figure 3.12 (rightmost)). Indeed, we also observe that only a few iterations are needed for the system (3.2.210) to converge.

Example 3.57 Consider the fractional BVP:

$${}_{-1}^{\mathrm{RL}}D_t^{1+\nu} u(t) + e^t u(t) = \sin t, \quad \nu \in (0,1),\ \ t \in (-1,1); \quad u(\pm 1) = 0.$$

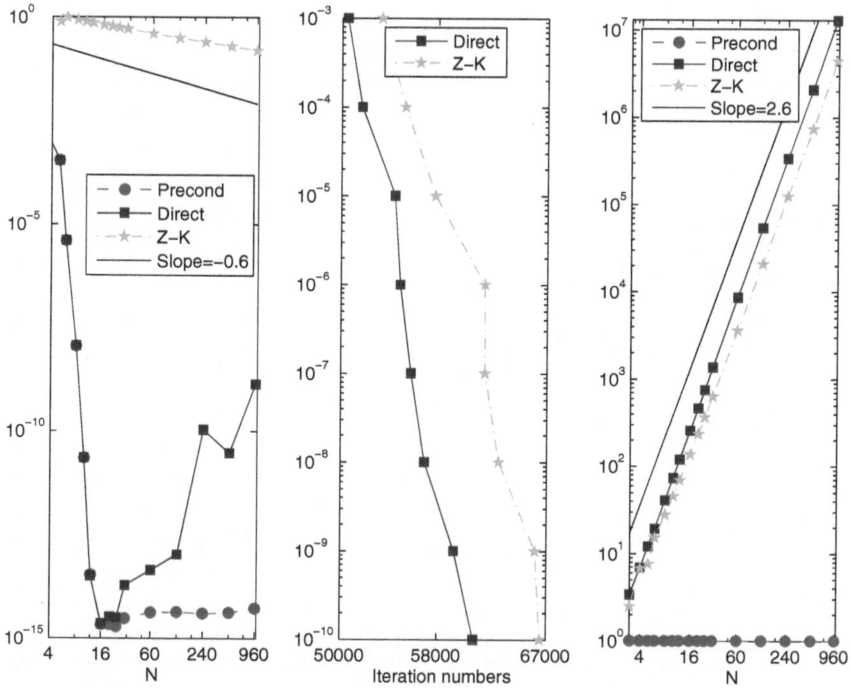

Figure 3.12 (Example 3.56): (Left): L^{∞}-errors in log-log scale of usual collocation method (3.2.209), its preconditioned collocation method (3.2.210) and Z-K method (Zayernouri and Karniadakis, 2014a) with respect to $\nu = 0.3$; (Middle): iteration numbers of usual collocation and Z-K methods against errors for 960 collocation points (note: the preconditioned approach converges within 10 iterations); (Right): plots of condition numbers of the linear systems of these methods. Source: Huang et al. (2016). Copyright ©2016 Society for Industrial and Applied Mathematics. Reprinted with permission. All rights reserved.

We see from Theorems 3.54 and 3.55 that an algebraic convergence is expected for the collocation scheme, which can be clearly observed from Figure 3.13. Once again, the collocation scheme involving the fractional Birkhoff basis in (3.2.210) performs significantly better than its counterparts, so it leads to a stable computation even for large N.

Discussions on the GJPF Collocation Methods

To have more insights into the proposed non-polynomial approach, we compare the approximability of the nodal GJPFs and usual polynomial basis functions, in approximating typical singular solutions.

Let $\omega^a(x) = (1-x^2)^a$, and define the L^2_ω-orthogonal projection:

$$\pi_N^C : L^2_{\omega^{-1/2}}(-1,1) \to \mathbb{P}_N.$$

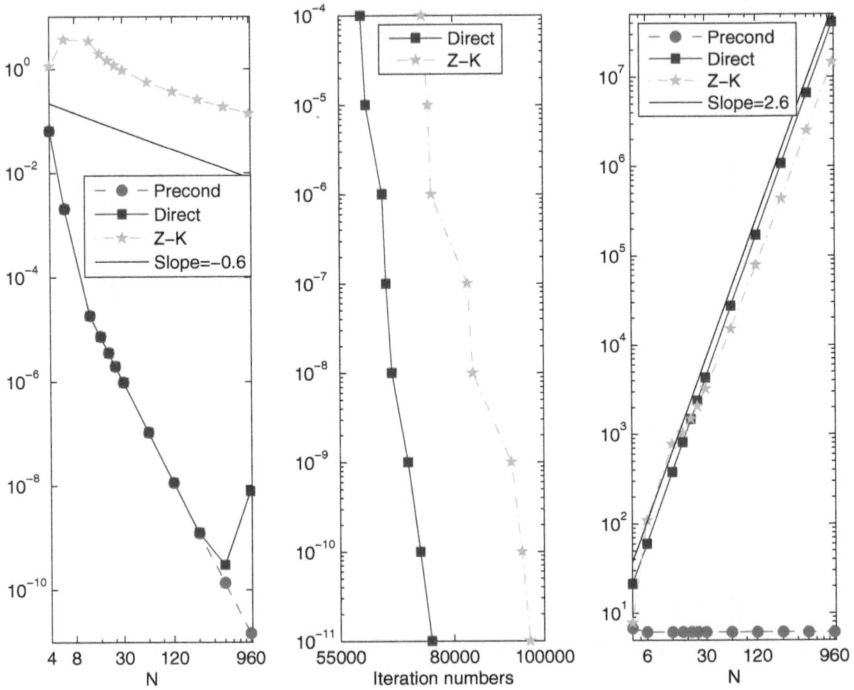

Figure 3.13 (Example 3.57): (Left): L^∞-errors in log-log scale of usual collocation method (3.2.209), its preconditioned collocation method (3.2.210) and Z-K method (Zayernouri and Karniadakis, 2014a) with respect to $\nu = 0.3$; (Middle): iteration numbers of usual collocation and Z-K methods against errors for 960 collocation points (note: the preconditioned approach converges within 10 iterations); (Right): plots of condition numbers of obtained linear system of three methods. Source: Huang et al. (2016). Copyright ©2016 Society for Industrial and Applied Mathematics. Reprinted with permission. All rights reserved.

Then,

$$\left(\pi_N^C u - u, v_N\right)_{\omega^{-1/2}} = 0, \quad \forall v_N \in \mathbb{P}_N, \tag{3.2.220}$$

where $(\cdot,\cdot)_{\omega^{-1/2}}$ is the inner product of $L^2_{\omega^{-1/2}}(-1,1)$. Recall the optimal estimate on Chebyshev polynomial approximation, see, e.g., Shen et al. (2011a, Thm 3.35): if $u^{(l)} \in L^2_{\omega^{l-1/2}}(-1,1)$ for all $0 \le l \le m$, then we have

$$\left\|\pi_N^C u - u\right\|_{\omega^{-1/2}} \le cN^{-m}\left\|u^{(m)}\right\|_{\omega^{m-1/2}}, \tag{3.2.221}$$

where C is a positive constant independent of u and N.

We now consider approximation by non-polynomial basis functions of $\mathcal{F}_N^{(\nu)}$ defined in (3.2.146). Note that

$$\hat{\pi}_N^{\nu} u = (1+x)^{\nu} \pi_{N-1}^{C}\big((1+x)^{-\nu} u\big) \in \mathcal{F}_N^{(\nu)}. \tag{3.2.222}$$

One verifies readily that it is an orthogonal projection in the sense that

$$\big(\hat{\pi}_N^{\nu} u - u, w_N\big)_{\omega^{-1/2}} = 0, \quad \forall\, w_N \in \mathcal{F}_N^{(\nu)}, \tag{3.2.223}$$

since $(1+x)^{-\nu} w_N \in \mathbb{P}_{N-1}$ and by (3.2.220)–(3.2.222),

$$\big(\hat{\pi}_N^{\nu} u - u, w_N\big)_{\omega^{-1/2}} = \big(\pi_{N-1}^{C}((1+x)^{-\nu} u) - (1+x)^{-\nu} u, (1+x)^{-\nu} w_N\big)_{\omega^{-1/2}} = 0.$$

Thus, by (3.2.221)–(3.2.222), we derive that for $\nu > 0$,

$$\begin{aligned} \big\|\hat{\pi}_N^{\nu} u - u\big\|_{\omega^{-1/2}} &\le \big\|(1+x)^{-\nu}(\hat{\pi}_N^{\nu} u - u)\big\|_{\omega^{-1/2}} \\ &= \big\|\pi_{N-1}^{C}((1+x)^{-\nu} u) - (1+x)^{-\nu} u\big\|_{\omega^{-1/2}} \\ &\le c N^{-m} \big\|((1+x)^{-\nu} u)^{(m)}\big\|_{\omega^{m-1/2}}. \end{aligned} \tag{3.2.224}$$

We see that the convergence rate of the non-polynomial approximation of u turns out to be the same as the polynomial approximation of $(1+x)^{-\nu} u$.

As an example, we examine the convergence rates of two approaches in approximating the one-sided singular function:

$$u(x) = (1+x)^{\nu} g_1(x) + (1+x)^{\nu+\rho} g_2(x) + (1+x) g_3(x) \tag{3.2.225}$$

for $0 < \nu < 1$ and $\rho > 0$, where $\{g_i\}_{i=1}^{3}$ are smooth functions. We add the factor $(1+x)$ before g_3 to make u meet the homogeneous initial condition. A direct calculation shows that

- if $m < 2\nu + 1/2$, then $\|u^{(m)}\|_{\omega^{m-1/2}} < \infty$;
- if $m < 2\min\{\rho, 1-\nu\} + 1/2$, then $\|((1+x)^{-\nu} u)^{(m)}\|_{\omega^{m-1/2}} < \infty$.

Thus, it is expected that if $\rho > \nu$ (when $g_2 \neq 0$) and $\nu < 1/2$ (when $g_3 \neq 0$), the non-polynomial approximation leads to a higher convergence rate. In other words, the non-polynomial approach outperforms its counterpart for small ν. Note that if $g_2 = g_3 = 0$, then $(1+x)^{-\nu} u = g_1(x)$, so the non-polynomial approximation can achieve spectral accuracy if g_1 is sufficiently smooth.

We next compare the convergence of two approaches for two FIVPs of small order ν with (i) singular exact solution and singular source term, and (ii) smooth source term (so the solution is singular), respectively.

Example 3.58 Consider the FIVP

$${}_{-1}^{\mathrm{RL}}D_t^{\nu} u(t) + e^t u(t) = f(t), \quad t \in (-1, 1]; \quad u(-1) = 0, \tag{3.2.226}$$

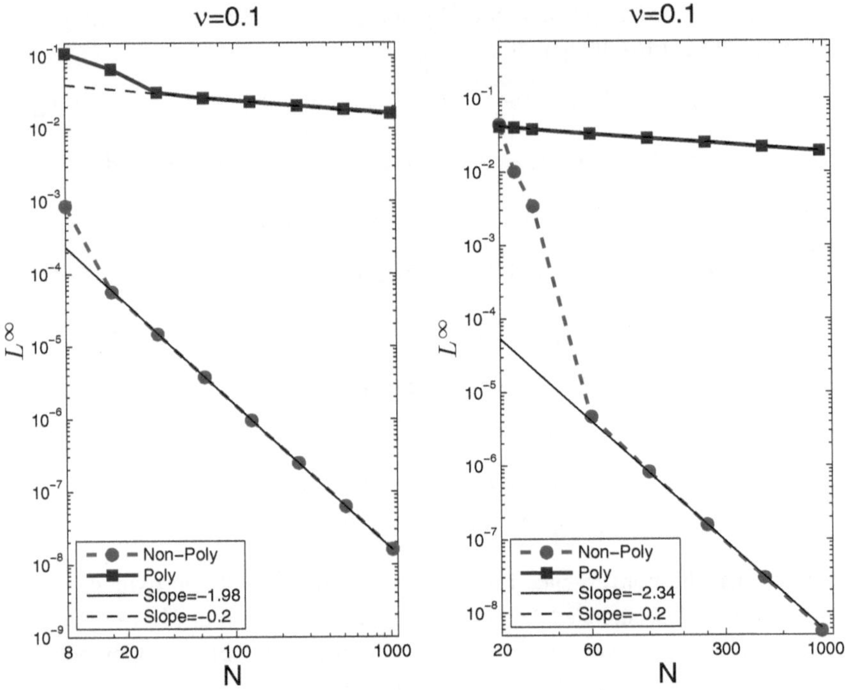

Figure 3.14 Left: Example 3.58 and Right: Example 3.59. L^∞-errors against N in log-log scale obtained by the proposed nodal GJF-collocation scheme, and the polynomial approximation. Source: Huang et al. (2016). Copyright ©2016 Society for Industrial and Applied Mathematics. Reprinted with permission. All rights reserved.

where $f(t)$ is chosen so that the exact solution:

$$u(t) = (1+t)^{\nu}\cos(\pi t) + (1+t)^{3.8} + (1+t) \quad \text{with} \quad \nu = 0.1.$$

In Figure 3.14 (left), we plot the L^∞-errors against various $N = 20, 24, 30, 60, 120, 240, 480, 960$ in log-log scale for both the polynomial and nodal GJF approximations. We observe that for small ν, the proposed GJF-spectral method has a much faster convergence rate than the usual polynomial-based spectral scheme.

Example 3.59 Consider the FIVP:

$$ {}^{\mathrm{RL}}_{-1}D_t^{\nu}u(t) + \sin(\pi t)u(t) = 1, \quad t \in (-1,1]; \quad u(-1) = 0, \qquad (3.2.227)$$

with $\nu = 0.1$.

As before, we consider the numerical solution obtained by the preconditioned method with $N = 3840$ as the "true" solution. In Figure 3.14 (right), we depict the errors for various N as the previous example. Once again, we observe much faster convergence rate of the new approach.

3.2.5 A Tunable Collocation Method for Variable-Order FODEs

The aim of this section is to extend the existing Jacobi–Gauss–Lobatto collocation methods for variable-order FODEs using a singular approximation basis in terms of GJPF of the form $(1 \pm x)^{\mu} P_j^{(a,b)}(x)$, where $\mu > -1$. In order to derive the differentiation matrices of the variable-order fractional derivatives, we introduce the corresponding three-term recurrence relation for both integrals and derivatives of these GJPFs, introduced in Chapter 2, hence extending the three-term recurrence relationship of Jacobi polynomials. The new spectral collocation method is applied to solve fractional ordinary and partial differential equations with end-point singularities. We demonstrate that the singular basis enhances greatly the accuracy of the numerical solution by properly tuning the parameter μ, even for cases where we do not know explicitly the form of singularity in the solution at the boundaries.

We first recall the relevant definitions of the fractional integrals and derivatives with *variable orders* and their related properties.

Definition 3.60 (see Zhuang et al. (2009)) For a function $f(x)$, $x \in [x_L, x_R]$, the left fractional integral (or the left Riemann–Liouville integral) with order $\alpha(x) > 0$ is defined by

$$
{}^{\mathrm{RL}}_{x_L}D_x^{-\alpha(x)} f(x) = {}^{\mathrm{C}}_{x_L}D_x^{-\alpha(x)} f(x) = \frac{1}{\Gamma(\alpha(x))} \int_{x_L}^{x} (x - s)^{\alpha(x)-1} f(s) \mathrm{d}s, \tag{3.2.228}
$$

and the right fractional integral (or the right Riemann–Liouville integral) with order $\alpha(x)$ is defined as

$$
{}^{\mathrm{RL}}_{x}D_{x_R}^{-\alpha(x)} f(x) = {}^{\mathrm{C}}_{x}D_{x_R}^{-\alpha(x)} f(x) = \frac{1}{\Gamma(\alpha(x))} \int_{x}^{x_R} (s - x)^{\alpha(x)-1} f(s) \mathrm{d}s, \tag{3.2.229}
$$

where $\Gamma(\cdot)$ is Euler's Gamma function.

Definition 3.61 (see Zhuang et al. (2009)) The left Riemann–Liouville fractional derivative with order $\alpha(x) > 0, x \in [x_L, x_R]$ of the given function $f(x), x \in [x_L, x_R]$ is defined by

$$
{}^{\mathrm{RL}}_{x_L}D_x^{\alpha(x)} f(x) = \frac{1}{\Gamma(n - \alpha(x))} \left[\frac{\mathrm{d}^n}{\mathrm{d}\xi^n} \int_{x_L}^{\xi} (\xi - s)^{n-\alpha(x)-1} f(s) \mathrm{d}s \right]_{\xi = x}, \tag{3.2.230}
$$

and the right Riemann–Liouville fractional derivative is defined by

$$ {}_{x}^{\mathrm{RL}}D_{x_R}^{\alpha(x)} f(x) = \frac{(-1)^n}{\Gamma(n-\alpha(x))} \left[\frac{\mathrm{d}^n}{\mathrm{d}\xi^n} \int_{\xi}^{x_R} (s-\xi)^{n-\alpha(x)-1} f(s)\mathrm{d}s \right]_{\xi=x}, \tag{3.2.231} $$

where n is a positive integer and $n-1<\alpha(x)<n$.

We now introduce some new definitions relevant to our development here.

Definition 3.62 (Valério and Da Costa, 2011; Zhuang et al., 2009) The left and right Caputo fractional derivatives with order $\alpha(x)>0, x\in[x_L,x_R]$ of the given function $f(x)$, $x\in[x_L,x_R]$ are defined as

$$ {}_{x_L}^{\mathrm{C}}D_{x}^{\alpha(x)} f(x) = \frac{1}{\Gamma(n-\alpha(x))} \int_{x_L}^{x} (x-s)^{n-\alpha(x)-1} f^{(n)}(s)\mathrm{d}s, \tag{3.2.232} $$

$$ {}_{x}^{\mathrm{C}}D_{x_R}^{\alpha(x)} f(x) = \frac{(-1)^n}{\Gamma(n-\alpha(x))} \int_{x}^{x_R} (s-x)^{n-\alpha(x)-1} f^{(n)}(s)\mathrm{d}s, \tag{3.2.233} $$

where n is a positive integer and $n-1<\alpha(x)<n$.

For $n-1<\alpha(x)<n, n\in\mathbb{N}, x\in[x_L,x_R]$, we have the following properties (Zhuang et al., 2009):

$$ \begin{aligned} {}_{x_L}^{\mathrm{RL}}D_{x}^{\alpha(x)} f(x) &= {}_{x_L}^{\mathrm{C}}D_{x}^{\alpha(x)} f(x) + \sum_{k=0}^{n-1} \frac{f^{(k)}(x_L)(x-x_L)^{k-\alpha(x)}}{\Gamma(k-\alpha(x)+1)}, \\ {}_{x}^{\mathrm{RL}}D_{x_R}^{\alpha(x)} f(x) &= {}_{x}^{\mathrm{C}}D_{x_R}^{\alpha(x)} f(x) + \sum_{k=0}^{n-1} \frac{(-1)^k f^{(k)}(x_R)(x_R-x)^{k-\alpha(x)}}{\Gamma(k-\alpha(x)+1)}. \end{aligned} \tag{3.2.234} $$

Now, consider the following FODE of variable order $0<\alpha(x)<1$,

$$ \begin{cases} {}_{x_L}^{\mathrm{C}}D_{x}^{\alpha(x)} u(x) + C(x)u(x) = f(x), & x\in(x_L,x_R), \\ u(x_L)=0. \end{cases} \tag{3.2.235} $$

To approximate the solution of u, we use the following basis

$$ X_N^{\mu} = \left\{ v=(x-x_L)^{\mu} P_n(x) : P_n(x)\in\mathbb{P}_N, n=0,1,\cdots,N \right\}, \quad \mu>-1, $$

where $\mathbb{P}_N$ is a set of algebraic polynomials of degree no more than N. Our collocation method for (3.2.235) is then to find an interpolant $I_N^{\mu,a,b,+}u\in X_N^{\mu}$ such that

$$ \left[{}_{x_L}^{\mathrm{C}}D_{x}^{\alpha(x)} I_N^{\mu,a,b,+} u(x) \right]_{x=x_j} + C(x_j)u_j = f(x_j), \tag{3.2.236} $$

where

$$I_N^{\mu,a,b,+}u(x) = (x - x_L)^{\mu}\, I_N^{a,b}v = (x - x_L)^{\mu} \sum_{j=0}^{N} v_j l_j(x)$$

and $v_j = (x_j - x_L)^{-\mu} u_j$. Here $I_N^{a,b}$ is the Jacobi–Gauss–Lobatto (JGL) interpolation operator associated with the collocation points

$$x_j = \frac{(x_R - x_L)\hat{x}_j + x_L + x_R}{2} \in [x_L, x_R],$$

where $\hat{x}_j$ $(0 \le j \le N)$ are the roots of $(1 - x^2)(P_N^{(a,b)}(x))'$, and $P_N^{(a,b)}(x)$ is the N-order Jacobi polynomial. The Lagrange interpolation basis function $l_j(x)$ is the j-th corresponding interpolation polynomial, i.e., it has value 1 at x_j and value 0 at points x_i $(i \neq j)$.

In its matrix form, (3.2.236) can be expressed as

$$\mathcal{A}(u_1, \ldots, u_N)^T = (f(x_1), f(x_2), \ldots, f(x_N))^T - \mathcal{B}v_0, \tag{3.2.237}$$

where $\mathcal{A} \in \mathbb{R}^{N \times N}$ which satisfies

$$(\mathcal{A})_{j-1,k-1} = \left((x_R - x_L)/2\right)^{-\alpha(x_j)} (\mathcal{D}_L)_{j,k} + C(x_j)\delta_{j,k}, \quad j,k = 1,2,\ldots,N,$$

and $\mathcal{D}_L$ will be given (3.2.280), while $\mathcal{B} \in \mathbb{R}^{N \times 1}$ satisfies

$$(\mathcal{B})_{j-1} = \left((x_R - x_L)/2\right)^{-\alpha(x_j)} (\mathcal{D}_L)_{j,0}, \quad j = 1,2,\ldots,N.$$

With (3.2.236), we actually solve the following equation

$${}_{x_L}^{\mathrm{C}}D_x^{\alpha(x)}((x - x_L)^{\mu}v(x)) + C(x)(x - x_L)^{\mu}v(x) = f(x), \tag{3.2.238}$$

which is equivalent to (3.2.235) with $u = (x - x_L)^{\mu}v$. As Zhang et al. (2015) pointed out, the regularity of v can be much higher than u and thus the polynomial interpolation of v in (3.2.236) is more accurate than the polynomial interpolation of u itself. For example, when $a = b = 0$ (see e.g., Ma and Sun (2001)), we have

$$\begin{aligned}\|I_N^{\mu,0,0,+}u - u\|_{\omega^{0,0}} &= \|(x - x_L)^{\mu}(I_N^{0,0}v - v)\|_{\omega^{0,0}} = \|I_N^{0,0}v - v\|_{\omega^{0,2\mu}} \\ &\le \|I_N^{0,0}v - v\|_{\omega^{-1,-1}} \le CN^{-r}\|\partial_x^r v\|_{\omega^{r-1,r-1}},\end{aligned} \tag{3.2.239}$$

where $\omega^{(a,b)}(x) = (1-x)^a(1+x)^b$ and we have supposed $v \in H^r(-1,1)$, $\mu > -1/2$. If $u(x)$ has very low regularity at the end-point, say $u(x) = (x - x_L)^{0.1}$, then $v = (x - x_L)^{-\mu+0.1}$, which has higher regularity than $u(x)$ when $-1 < \mu < 0$. For FODE (3.2.235) with smooth $f(x)$ and $C(x)$, this is true when α is small, see e.g., Diethelm et al. (2004) and Remark 3.63.

The choice of μ is extremely important in practice. Here we prefer to choose $\mu \le 0$ that will lead to

$$v_0 = v(x_L) = \lim_{x \to x_L^+} (x - x_L)^{-\mu} u(x) = 0 \quad (\mu < 0),$$

or $v_0 = v(x_L) = u(x_L)$ $(\mu = 0)$ whenever $u(x)$ is bounded. For $\mu > 0$, we require to know $v_0 = \lim_{x \to x_L^+} (x - x_L)^{-\mu} u(x)$ that is unknown if the JGL spectral collocation method is applied to (3.2.235), while arbitrary assignment of v_0 may lead to unsatisfactory accuracy; see $\mathcal{B} v_0$ in the right-hand side of (3.2.272), where $v_0 = \lim_{x \to x_L^+} (x - x_L)^{-\mu} u(x)$ is possibly not equal to zero when $\mu > 0$. Moreover, the derivation of the differentiation matrix $\mathcal{D}_L$ in (3.2.272) avoids using $(1 + x_0)^{-\mu}$ for JGL points $(x_0 = -1)$, which is meaningless when $\mu > 0$; see also the numerical results in Tables 3.11–3.12 in Example 3.71, where we also choose $\mu > 0$ and let $v_0 = 0$ in the numerical simulation for comparison.

Remark 3.63 Even for the simple FODE with variable-order as (3.2.235), the regularity of its solution is much more complicated than in the case of the constant-order FODEs. For example, when $f(x) = 1$, $\alpha(x)$ is smooth, and $C(x) = \phi_0 = 0$, the exact solution to (3.2.235) is $u(x) = \frac{(x - x_L)^{\alpha(x)}}{\Gamma(1+\alpha(x))}$. For such a simple case, the regularity of $u(x)$ at $x = x_L$ is low when $\alpha(x)$ is small. In contrast, $v(x) = (x - x_L)^{-\mu} u(x) = \frac{(x - x_L)^{\alpha(x) - \mu}}{\Gamma(1+\alpha(x))}$ can be smoother than u when $-1 < \mu < 0$. Then we have more accurate numerical solutions of $v(x)$ and thus expect more accurate numerical solutions of $u(x)$. See numerical examples for details in Section 3.2.5.

Derivation of the Differentiation Matrices

In this section, we first develop the recurrence formulas to calculate the fractional integrals of the GJPFs $P_{\pm,j}^{a,b,\mu}(x)$, where $P_{\pm,j}^{a,b,\mu}(x)$ is defined by

$$P_{\pm,j}^{a,b,\mu}(x) = (1 \pm x)^{\mu} P_j^{(a,b)}(x), \quad \mu > -1, \quad x \in (-1, 1). \qquad (3.2.240)$$

Then we derive the fractional derivatives of $P_{\pm,j}^{a,b,\mu}(x)$ from the corresponding fractional integrals of $P_{\pm,j}^{a,b,\mu}(x)$, which leads to the desired differentiation matrices associated with these weighted polynomials by a simple matrix-matrix multiplication.

The Jacobi polynomials $\{P_j^{(a,b)}(x)\}$, $a, b > -1, x \in (-1, 1)$ are given by the following three-term recurrence relation (see e.g., Shen et al. (2011a)):

$$\begin{aligned} &P_0^{(a,b)}(x) = 1, \quad P_1^{(a,b)}(x) = \frac{1}{2}(a + b + 2)x + \frac{1}{2}(a - b), \\ &P_{j+1}^{(a,b)}(x) = (A_j x - B_j) P_j^{(a,b)}(x) - C_j P_{j-1}^{(a,b)}(x), \quad j \ge 1, \end{aligned} \qquad (3.2.241)$$

where

$$\begin{aligned} A_j &= \frac{(2j+a+b+1)(2j+a+b+2)}{2(j+1)(j+a+b+1)}, \\ B_j &= \frac{(b^2-a^2)(2j+a+b+1)}{2(j+1)(j+a+b+1)(2j+a+b)}, \\ C_j &= \frac{(j+a)(j+b)(2j+a+b+2)}{(j+1)(j+a+b+1)(2j+a+b)}. \end{aligned} \tag{3.2.242}$$

Let $\omega^{(a,b)}(x) = (1-x)^a(1+x)^b$. Then, one has

$$\int_{-1}^{1} P_m^{(a,b)}(x)P_n^{(a,b)}(x)\omega^{(a,b)}(x)\mathrm{d}x = \delta_{mn}\gamma_n^{(a,b)}, \tag{3.2.243}$$

where δ_{mn} is the Kronecker delta function and

$$\gamma_n^{(a,b)} = \frac{2^{a+b+1}\Gamma(n+a+1)\Gamma(n+b+1)}{(2n+a+b+1)n!\,\Gamma(n+a+b+1)}. \tag{3.2.244}$$

In this section, we also use the following notations:

$$G_{L,j}^{a,b,\alpha,\mu}(x) = {}_{-1}^{\mathrm{RL}}D_x^{-\alpha}\ P_{+,j}^{a,b,\mu}(x), \quad x \in (-1,1), \tag{3.2.245}$$

$$G_{R,j}^{a,b,\alpha,\mu}(x) = {}_{x}^{\mathrm{RL}}D_{-1}^{-\alpha}\ P_{-,j}^{a,b,\mu}(x), \quad x \in (-1,1). \tag{3.2.246}$$

Fractional Integrals of the GJPFs

In this subsection, we develop a recurrence formula to calculate the $\alpha(x)$th-order left (or right) fractional integral of $P_{+,j}^{a,b,\mu}(x)$ (or $P_{-,j}^{a,b,\mu}(x)$).

We can compute $G_{L,j}^{a,b,\alpha,\mu}(x)$ by the following recurrence relation:

Theorem 3.64 *Let $a, b, \mu > -1$ and $\alpha(x) > 0$, $x \in (-1,1)$. Then $G_j = G_{L,j}^{a,b,\alpha,\mu}(x)$ defined by (3.2.245) and $H_j = \alpha(x)G_{L,j}^{a,b,\alpha+1,\mu}(x)$ satisfy, for $j \geq 1$,*

$$G_{j+1} = (A_j x - B_j)G_j - C_j G_{j-1} - A_j H_j, \tag{3.2.247}$$

$$\begin{aligned} H_{j+1} = &-(A_j + B_j)H_j - C_j H_{j-1} \\ &-(\mu+1+\alpha)A_j\left(\widehat{A}_j H_{j-1} + \widehat{B}_j H_j + \widehat{C}_j H_{j+1}\right) \\ &+(1+x)\alpha A_j\left(\widehat{A}_j G_{j-1} + \widehat{B}_j G_j + \widehat{C}_j G_{j+1}\right), \end{aligned} \tag{3.2.248}$$

where

$$\widehat{A}_j = \frac{-2(j+a)(j+b)}{(j+a+b)(2j+a+b)(2j+a+b+1)}, \tag{3.2.249}$$

$$\widehat{B}_j = \frac{2(a-b)}{(2j+a+b)(2j+a+b+2)}, \tag{3.2.250}$$

$$\widehat{C}_j = \frac{2(j+a+b+1)}{(2j+a+b+1)(2j+a+b+2)}. \tag{3.2.251}$$

We can compute $G_{R,j}^{a,b,\alpha,\mu}(x)$ in a similar fashion.

Theorem 3.65 *Let $a, b, \mu > -1$ and $\alpha(x) > 0$, $x \in (-1, 1)$. Then $G_j = G_{R,j}^{a,b,\alpha,\mu}(x)$ defined by (3.2.246) and $H_j = \alpha(x) G_{R,j}^{a,b,\alpha+1,\mu}(x)$ satisfy, for $j \geq 1$,*

$$G_{j+1} = (A_j x - B_j) G_j - C_j G_{j-1} + A_j H_j, \tag{3.2.252}$$

$$\begin{aligned} H_{j+1} = {} & (A_j - B_j) H_j - C_j H_{j-1} \\ & -(\mu + 1 + \alpha) A_j \big(\widehat{A}_j H_{j-1} + \widehat{B}_j H_j + \widehat{C}_j H_{j+1} \big) \\ & +(1 - x) \alpha A_j \big(\widehat{A}_j G_{j-1} + \widehat{B}_j G_j + \widehat{C}_j G_{j+1} \big), \end{aligned} \tag{3.2.253}$$

where $\widehat{A}_j, \widehat{B}_j$, and $\widehat{C}_j$ are defined by (3.2.249), (3.2.250), and (3.2.251), respectively.

Remark 3.66 When $\mu = 0$ and $j \geq 1$, we have (see e.g., Li et al. (2012); Zeng and Li (2014))

$$\begin{aligned} G_{L,j+1}^{a,b,\alpha,0} = {} & \frac{A_j x - B_j - \alpha(x) A_j \widehat{B}_j}{1 + \alpha(x) A_j \widehat{C}_j} G_{L,j}^{a,b,\alpha,0} - \frac{C_j + \alpha(x) A_j \widehat{A}_j}{1 + \alpha(x) A_j \widehat{C}_j} G_{L,j-1}^{a,b,\alpha,0} \\ & + \frac{A_j \big(\widehat{A}_j P_{j-1}^{(a,b)}(-1) + \widehat{B}_j P_j^{(a,b)}(-1) + \widehat{C}_j P_{j+1}^{(a,b)}(-1) \big)}{\Gamma(\alpha(x))(1 + \alpha(x) A_j \widehat{C}_j)} (x+1)^{\alpha(x)} \end{aligned} \tag{3.2.254}$$

and

$$\begin{aligned} G_{R,j+1}^{a,b,\alpha,0} = {} & \frac{A_j x - B_j - \alpha(x) A_j \widehat{B}_j}{1 + \alpha(x) A_j \widehat{C}_j} G_{R,j}^{a,b,\alpha,0} - \frac{C_j + \alpha(x) A_j \widehat{A}_j}{1 + \alpha(x) A_j \widehat{C}_j} G_{R,j-1}^{a,b,\alpha,0} \\ & + \frac{A_j \big(\widehat{A}_j P_{j-1}^{(a,b)}(1) + \widehat{B}_j P_j^{(a,b)}(1) + \widehat{C}_j P_{j+1}^{(a,b)}(1) \big)}{\Gamma(\alpha(x))(1 + \alpha(x) A_j \widehat{C}_j)} (1-x)^{\alpha(x)}. \end{aligned} \tag{3.2.255}$$

For initial values $G_{L,j}^{a,b,\alpha,\mu}$ $(j = 0, 1)$ and $G_{R,j}^{a,b,\alpha,\mu}$ $(j = 0, 1)$, we can easily obtain

$$\begin{aligned} G_{L,0}^{a,b,\alpha,\mu} &= \frac{\Gamma(\mu+1)}{\Gamma(\mu+1+\alpha(x))} (1+x)^{\alpha(x)+\mu}, \\ G_{L,1}^{a,b,\alpha,\mu} &= \frac{(a+b+2)\Gamma(\mu+2)}{2\Gamma(\mu+2+\alpha(x))} (1+x)^{\alpha(x)+\mu+1} - (b+1) G_{L,0}^{a,b,\alpha,\mu}, \\ G_{R,0}^{a,b,\alpha,\mu} &= \frac{\Gamma(\mu+1)}{\Gamma(\mu+1+\alpha(x))} (1-x)^{\alpha(x)+\mu}, \\ G_{R,1}^{a,b,\alpha,\mu} &= -\frac{(a+b+2)\Gamma(\mu+2)}{2\Gamma(\mu+2+\alpha(x))} (1-x)^{\alpha(x)+\mu+1} + (a+1) G_{R,0}^{a,b,\alpha,\mu}. \end{aligned}$$

The initial values of H_j $(j = 0, 1)$ in Theorem 3.64 (or Theorem 3.65) can be obtained from the relation $H_j = \alpha\, G_{L,j}^{a,b,\alpha+1,\mu}$ (or $H_j = \alpha\, G_{R,j}^{a,b,\alpha+1,\mu}$).

When $\mu = b$ in Theorem 3.64 or $\mu = a$ in Theorem 3.65, we have simpler expressions.

Corollary 3.67 *Suppose that $\alpha(x)$, $x \in (-1,1)$ is a real-valued function, $a, b > -1$. Then*

$$
\begin{aligned}
{}^{\mathrm{RL}}_{-1}D_x^{\alpha(x)} P_{+,j}^{a,b,b}(x) &= \frac{\Gamma(j+b+1)}{\Gamma(j+b+1-\alpha(x))}(1+x)^{b-\alpha(x)} P_j^{a+\alpha(x),b-\alpha(x)}(x),\\
{}^{\mathrm{RL}}_{x}D_1^{\alpha(x)} P_{-,j}^{a,b,a}(x) &= \frac{\Gamma(j+a+1)}{\Gamma(j+a+1-\alpha(x))}(1-x)^{a-\alpha(x)} P_j^{a-\alpha(x),b+\alpha(x)}(x).
\end{aligned}
\tag{3.2.256}
$$

Proof Equation (3.2.256) with constant fractional order α can be found in Zayernouri and Karniadakis (2014b); Zhang et al. (2015); Chen et al. (2016), which also holds for the variable-order $\alpha(x)$. For any $\alpha(x)$, (3.2.256) can be proven by using the explicit form of Jacobi polynomials (e.g., see Shen et al. (2011a, p. 71)), the proof of which is omitted here. □

Next, we investigate the stability of the recurrence formulas (3.2.254) and (3.2.255). When $b = 0$, we can easily verify that

$$
\widehat{A}_j P_{j-1}^{(a,b)}(-1) + \widehat{B}_j P_j^{(a,b)}(-1) + \widehat{C}_j P_{j+1}^{(a,b)}(-1) = 0
$$

by using (3.2.249)–(3.2.251). In such a case, (3.2.254) becomes a three-term recurrence formula and defines an orthogonal system with respect to the weight function $\omega^{a+\alpha,0}(x) = (1-x)^{a+\alpha}$ when $a+\alpha, -\alpha > -1$ and α is an constant. Hence, the recurrence formula (3.2.254) is stable for constant-order α when $a+\alpha, -\alpha > -1$. For each $x_0 \in (-1,1)$ and $\alpha_0 = \alpha(x_0)$, we find that $\{G_{L,j}^{a,b,\alpha_0,0}(x)\}$ defines an orthogonal system with respect to the weight function $\omega^{a+\alpha_0,0}(x) = (1-x)^{a+\alpha_0}, x \in (-1,1)$. Thus, (3.2.254) is stable for variable-order $\alpha(x)$ when $a+\alpha(x), -\alpha(x) > -1$. Similarly, (3.2.255) defines an orthogonal system when $a = 0$, α is a constant, and $b+\alpha, -\alpha > -1$, thus is stable for variable-order $\alpha(x)$ when $b+\alpha(x), -\alpha(x) > -1$.

In general cases, i.e., $b \neq 0$ or $a \neq 0$, it is difficult to prove that the three-term recurrence formula (3.2.247)–(3.2.248) or (3.2.252)–(3.2.253) could define an orthogonal system. Thus, we numerically show their performance in stability. Here we only consider (3.2.247)–(3.2.248), since the situation is very similar for (3.2.252)–(3.2.253).

We will test two approaches of computing $G_{L,j}^{a,b,\alpha,\mu}(x)$ in (3.2.245): the recurrence formula (3.2.247)–(3.2.248) and the direct computation using the Gauss quadrature rule given by

$$G_{L,j}^{a,b,\alpha,\mu}(x) = \frac{(x+1)^{\alpha(x)+\mu}}{2^{\alpha(x)+\mu}\Gamma(\alpha(x))} \int_{-1}^{1} \omega^{\alpha(x)-1,\mu}(\hat{s}) P_j^{(a,b)}\left(\frac{(x+1)\hat{s}+x-1}{2}\right) d\hat{s}, \tag{3.2.257}$$

where

$$\omega^{\alpha(x)-1,\mu}(\hat{s}) = (1-\hat{s})^{\alpha(x)-1}(1+\hat{s})^{\mu}.$$

Thus for any given $x \in (-1,1)$, $G_{L,j}^{a,b,\alpha,\mu}(x)$ can be directly calculated by the following exact formula

$$G_{L,j}^{a,b,\alpha,\mu}(x) = \frac{1}{\Gamma(\alpha(x))}\left(\frac{x+1}{2}\right)^{\alpha(x)+\mu} \sum_{k=0}^{M_j} \omega_k P_j^{(a,b)}(s_k) \tag{3.2.258}$$

for $M_j \geq \lceil j/2 \rceil$, where $s_k = \frac{(x+1)\hat{s}_k+x-1}{2}$, $\{\hat{s}_k\}$ are the JGL quadrature points, i.e., $\{\hat{s}_k\}$ are the roots of $(1-\hat{s}^2)\partial_{\hat{s}} P_{M_j}^{\alpha(x)-1,\mu}(\hat{s})$, with the corresponding weights $\{\omega_k\}$. For simplicity, we can choose $M_j = \lceil N/2 \rceil + 2$ in (3.2.258) in the computation.

Now we are ready to test the computational performance of the recurrence formula (3.2.247)–(3.2.248) and the Gauss quadrature (3.2.258). The numerical solutions obtained by the recurrence formula (3.2.247)–(3.2.248) are denoted by ${}^{(1)}G_{L,j}^{a,b,\alpha,\mu}(x)$ and the numerical solutions from the Gauss quadrature (3.2.258) are denoted by ${}^{(2)}G_{L,j}^{a,b,\alpha,\mu}(x)$. Let $e_j^{(m)}(x) = {}^{(m)}G_{L,j}^{a,b,\alpha,\mu}(x) - G_{L,j}^{a,b,\alpha,\mu}(x)$ $(m = 1,2)$, where $G_{L,j}^{a,b,\alpha,\mu}(x)$ is the reference solution from the first equation in (3.2.256) with $\mu = b$. Then the relative L^∞ error is defined by

$$\|e^{(m)}\|_\infty = \frac{\max_{0\leq j\leq N} \max_{0\leq k\leq 256} |e_j^{(m)}(x_k)|}{\max_{0\leq j\leq N} \max_{0\leq k\leq 256} |G_{L,j}^{a,b,\alpha,\mu}(x_k)|}, \quad x_k = -1 + \frac{2k}{256}.$$

In Table 3.9, we observe that the recurrence formula (3.2.247)–(3.2.248) shows higher accuracy while the Gauss quadrature (3.2.258) shows less accurate results, especially when N is large. This is possibly due to the accumulation of the round-off errors in the summation in (3.2.258). Further, the Gauss quadrature (3.2.258) takes much more time than the recurrence formula (3.2.247)–(3.2.248). For fixed x, we need $O(N)$ operations to obtain ${}^{(1)}G_{L,j}^{a,b,\alpha,\mu}(x)(j = 0,1,\ldots,N)$, while we need at least $O(N^2)$ operations to obtain ${}^{(2)}G_{L,j}^{a,b,\alpha,\mu}(x)(j = 0,1,\ldots,N)$. Moreover, we also need to recalculate the Gauss quadrature points $\{s_k\}$ and weights $\{\omega_k\}$ in (3.2.258) when x changes, and thus the computational cost is $O(N^3)$. From Table 3.9, we observe better accuracy of the recurrence formula, which implies the stability of the recurrence formula (3.2.247)–(3.2.248).

Table 3.9. *Comparison between the error* $\|e^{(1)}\|_\infty$ *of the recurrence formula* (3.2.247)–(3.2.248) *and the error* $\|e^{(2)}\|_\infty$ *of the Jacobi–Gauss quadrature* (3.2.258), $a = 0, b = \mu, \alpha(x) = 2|\sin(10x)|$. *Source: Zeng et al. (2015a).*
Copyright ©2015 Society for Industrial and Applied Mathematics. Reprinted with permission. All rights reserved.

N	μ	$\|e^{(1)}\|_\infty$	$\|e^{(2)}\|_\infty$	μ	$\|e^{(1)}\|_\infty$	$\|e^{(2)}\|_\infty$	μ	$\|e^{(1)}\|_\infty$	$\|e^{(2)}\|_\infty$
64		1.59e-15	2.46e-14		1.22e-15	2.76e-14		3.00e-15	2.68e-14
128	0.2	1.59e-15	1.67e-13	0.5	1.22e-15	1.47e-13	0.8	3.34e-15	1.35e-13
256		2.67e-15	7.62e-13		1.22e-15	7.94e-13		3.53e-15	1.53e-12
512		2.67e-15	6.31e-12		1.22e-15	7.63e-12		8.83e-15	8.45e-12
64		1.73e-15	1.41e-14		2.40e-15	1.04e-14		3.16e-15	6.36e-15
128	−0.2	2.25e-15	1.09e-13	−0.5	3.39e-15	5.73e-14	−0.8	5.50e-15	3.66e-14
256		2.32e-15	6.42e-13		4.79e-15	4.10e-13		9.57e-15	1.79e-13
512		5.56e-15	3.88e-12		6.77e-15	4.65e-12		1.66e-14	1.55e-12

Riemann–Liouville Fractional Derivatives of GJPFs

In this subsection, we compute the Riemann–Liouville fractional derivatives of $P^{a,b,\mu}_{+,j}(x)$ and $P^{a,b,u}_{-,j}(x)$. To this end, we adopt the technique used in Li et al. (2012) to get the fractional derivative of $P^{a,b,\mu}_{+,j}(x)$ (or $P^{a,b,\mu}_{-,j}(x)$) from the corresponding fractional integral of $P^{a,b,\mu}_{+,j}(x)$ (or $P^{a,b,\mu}_{-,j}(x)$). For simplicity, we denote

$$\begin{aligned} {}^{\mathrm{RL}}S^{a,b,\alpha,\mu}_{L,j}(x) &= {}^{\mathrm{RL}}_{-1}D^{\alpha(x)}_{x}P^{a,b,\mu}_{+,j}(x),\\ {}^{\mathrm{RL}}S^{a,b,\alpha,\mu}_{R,j}(x) &= {}^{\mathrm{RL}}_{x}D^{\alpha(x)}_{1}P^{a,b,\mu}_{-,j}(x). \end{aligned} \tag{3.2.259}$$

In the rest of this subsection, we use $S_{L,j}(x) = {}^{\mathrm{RL}}S^{a,b,\alpha,\mu}_{L,j}(x)$ and $S_{R,j}(x) = {}^{\mathrm{RL}}S^{a,b,\alpha,\mu}_{R,j}(x)$ for simplicity. When $\alpha(x) = \alpha$ is a constant and $n-1 < \alpha < n, n \in \mathbb{N}$, we have

$$S_{L,j}(x) = \frac{\mathrm{d}^n}{\mathrm{d}x^n}G^{a,b,n-\alpha,\mu}_{L,j}(x), \quad S_{R,j}(x) = (-1)^n\frac{\mathrm{d}^n}{\mathrm{d}x^n}G^{a,b,n-\alpha,\mu}_{R,j}(x).$$

For $0 < \alpha < 1$, let $\tilde{\alpha} = 1 - \alpha$ and $s = (x+1)\hat{s} - 1$, we have

$$\begin{aligned} S_{L,j}(x) = \frac{\mathrm{d}}{\mathrm{d}x}G^{a,b,\tilde{\alpha},\mu}_{L,j}(x) &= \frac{(x+1)^{\tilde{\alpha}+\mu}}{\Gamma(\tilde{\alpha})}\int_0^1(1-\hat{s})^{\tilde{\alpha}-1}\hat{s}^{\mu+1}\frac{\partial P^{(a,b)}_j(s)}{\partial s}\mathrm{d}\hat{s}\\ &+\frac{\tilde{\alpha}+\mu}{x+1}\frac{(x+1)^{\tilde{\alpha}+\mu}}{\Gamma(\tilde{\alpha})}\int_0^1(1-\hat{s})^{\tilde{\alpha}-1}\hat{s}^{\mu}P^{(a,b)}_j(s)\mathrm{d}\hat{s}, \end{aligned}$$

which leads to

$$S_{L,j}(x) = \frac{1}{x+1}\left((\tilde{\alpha}+\mu)G_{L,j}^{a,b,\tilde{\alpha},\mu}(x) + d_{j,1}^{(a,b)} G_{L,j-1}^{a+1,b+1,\tilde{\alpha},\mu+1}(x)\right), \quad (3.2.260)$$

where we used the property

$$\frac{\mathrm{d}^n}{\mathrm{d}x^n} P_j^{(a,b)}(x) = d_{j,n}^{(a,b)} P_{j-n}^{(a+n,b+n)}(x), \quad j \geq n, n \in \mathbb{N}, \quad (3.2.261)$$

with

$$d_{j,n}^{(a,b)} = \frac{\Gamma(j+n+a+b+1)}{2^n \Gamma(j+a+b+1)}. \quad (3.2.262)$$

For $1 < \alpha < 2$, let $\tilde{\alpha} = 2 - \alpha$ and $s = (x+1)\hat{s} - 1$, we have

$$\begin{aligned} S_{L,j}(x) = \frac{\mathrm{d}^2}{\mathrm{d}x^2} G_{L,j}^{a,b,\tilde{\alpha},\mu}(x) &= \frac{(x+1)^{\tilde{\alpha}+\mu}}{\Gamma(\tilde{\alpha})} \int_0^1 (1-\hat{s})^{\tilde{\alpha}-1} \hat{s}^{\mu+2} \frac{\partial^2 P_j^{(a,b)}(s)}{\partial s^2} \mathrm{d}\hat{s} \\ &+ \frac{2(\alpha+\mu)}{x+1} \frac{(x+1)^{\tilde{\alpha}+\mu}}{\Gamma(\tilde{\alpha})} \int_0^1 (1-\hat{s})^{\tilde{\alpha}-1} \hat{s}^{\mu+1} \frac{\partial P_j^{(a,b)}(s)}{\partial s} \mathrm{d}\hat{s} \\ &+ \frac{(\tilde{\alpha}+\mu)(\tilde{\alpha}+\mu-1)}{(x+1)^2} \frac{(x+1)^{\tilde{\alpha}+\mu}}{\Gamma(\tilde{\alpha})} \int_0^1 (1-\hat{s})^{\tilde{\alpha}-1} \hat{s}^{\mu} P_j^{(a,b)}(s) \mathrm{d}\hat{s}. \end{aligned}$$

Applying (3.2.261) and (3.2.262) with $n = 1, 2$, we obtain

$$\begin{aligned} S_{L,j}(x) = \frac{1}{(x+1)^2}\Big[&(\tilde{\alpha}+\mu)(\tilde{\alpha}+\mu-1)G_{L,j}^{a,b,\tilde{\alpha},\mu}(x) \\ &+ 2(\tilde{\alpha}+\mu) d_{j,1}^{(a,b)} G_{L,j-1}^{a+1,b+1,\tilde{\alpha},\mu+1}(x) + d_{j,2}^{(a,b)} G_{L,j-2}^{a+2,b+2,\tilde{\alpha},\mu+2}(x)\Big]. \end{aligned} \quad (3.2.263)$$

When $\alpha(x)$ is a function of $x, x \in (-1, 1]$, we can readily verify that (3.2.260) and (3.2.263) hold from the definitions (3.2.230) and (3.2.231). Also, when $\alpha(x) > 2$, $S_{L,j}(x)$ can be similarly derived, which is omitted here.

For the right fractional derivative operator, we have similar relations. For a given $x \in (-1, 1)$ and $0 < \alpha(x) < 1$, let $\tilde{\alpha}(x) = 1 - \alpha(x)$, we have

$$S_{R,j}(x) = \frac{1}{1-x}\left((\tilde{\alpha}+\mu)G_{R,j}^{a,b,\tilde{\alpha},\mu}(x) - d_{j,1}^{(a,b)} G_{R,j-1}^{a+1,b+1,\tilde{\alpha},\mu+1}(x)\right). \quad (3.2.264)$$

For a given $x \in (-1, 1)$ and $1 < \alpha(x) < 2$, let $\tilde{\alpha}(x) = 2 - \alpha(x)$ and we have

$$\begin{aligned} S_{R,j}(x) = \frac{1}{(1-x)^2}\Big[&(\tilde{\alpha}+\mu)(\tilde{\alpha}+\mu-1)G_{R,j}^{a,b,\tilde{\alpha},\mu}(x) \\ &- 2(\tilde{\alpha}+\mu) d_{j,1}^{(a,b)} G_{R,j-1}^{a+1,b+1,\tilde{\alpha},\mu+1}(x) + d_{j,2}^{(a,b)} G_{R,j-2}^{a+2,b+2,\tilde{\alpha},\mu+2}(x)\Big]. \end{aligned} \quad (3.2.265)$$

Differentiation Matrices of the Riemann–Liouville Fractional Derivatives
Let $u(x) = (1 \pm x)^{\mu} v(x)$ be a function defined on the interval $(-1, 1)$ and N be a positive integer. Denote $x_j (j = 0, 1, \ldots, N)$ as the Jacobi–Gauss–Lobatto (JGL) points defined on the interval $(-1, 1)$. Then $u(x)$ can be approximated by

$$\begin{aligned} I_N^{\mu,a,b,\pm} u(x) &= (1 \pm x)^{\mu} I_N^{a,b} v(x) = (1 \pm x)^{\mu} \sum_{j=0}^{N} v(x_j) l_j(x) \\ &= \sum_{j=0}^{N} \hat{v}_j P_{\pm,j}^{a,b,\mu}(x), \end{aligned} \tag{3.2.266}$$

where $l_j(x)$ is the Lagrange interpolation basis function based on JGL points and $l_j(x)$ can be represented as

$$l_j(x) = \sum_{k=0}^{N} c_{k,j} P_k^{(a,b)}(x). \tag{3.2.267}$$

Then $c_{k,j}$ can be determined by the following relation (Shen et al., 2011a)

$$(\mathcal{C})_{k,j} = c_{k,j} = \frac{P_k^{(a,b)}(x_j)\omega_j}{\delta_k \gamma_k^{(a,b)}}, \quad k = 0, 1, \ldots, N, \tag{3.2.268}$$

in which $\gamma_k^{(a,b)}$ is defined by (3.2.244), $\delta_k = 1\,(0 \leq k < N)$, and $\delta_N = 2 + \frac{a+b+1}{N}$. If $I_N^{a,b}$ in (3.2.266) is the Jacobi–Gauss (JG) or Jacobi–Gauss–Radau (JGR) interpolation, then we have $\delta_k = 1\,(0 \leq k \leq N)$ in (3.2.268). Combining (3.2.267)–(3.2.268) yields

$$\hat{\boldsymbol{v}} = \mathcal{C}\boldsymbol{v}, \tag{3.2.269}$$

where $\hat{\boldsymbol{v}} = \left(\hat{v}_0, \hat{v}_1, \ldots, \hat{v}_N\right)^T$ and $\boldsymbol{v} = (v(x_0), v(x_1), \ldots, v(x_N))^T$.

The Riemann–Liouville derivative ${}_{-1}^{\mathrm{RL}}D_x^{\alpha(x)} I_N^{\mu,a,b,+} u(x)$ $(\alpha(x) > 0, x \in (-1, 1))$ can be calculated by the following relation

$${}_{-1}^{\mathrm{RL}}D_x^{\alpha(x)} I_N^{\mu,a,b,+} u(x) = \sum_{j=0}^{N} \hat{v}_j \, {}^{\mathrm{RL}}S_{L,j}^{a,b,\alpha,\mu}(x), \tag{3.2.270}$$

where ${}^{\mathrm{RL}}S_{L,j}^{a,b,\alpha,\mu}(x)$ is defined in (3.2.259). Combining (3.2.269) and (3.2.270) yields

$${}_{-1}^{\mathrm{RL}}D_x^{\alpha(x)} I_N^{\mu,a,b,+} u(x) = \left({}^{\mathrm{RL}}S_{L,0}^{a,b,\alpha,\mu}(x), \ldots, {}^{\mathrm{RL}}S_{L,N}^{a,b,\alpha,\mu}(x)\right)\mathcal{C}\boldsymbol{v}. \tag{3.2.271}$$

Let

$${}_{-1}^{\mathrm{RL}}D_{x_j}^{\alpha(x_j)} I_N^{\mu,a,b,+} u(x_j) = \left[{}_{-1}^{\mathrm{RL}}D_x^{\alpha(x)} I_N^{\mu,a,b,+} u(x)\right]_{x=x_j}.$$

Then ${}^{\mathrm{RL}}_{-1}D^{\alpha(x)}_{x} I^{\mu,a,b,+}_{N} u$ at the collocation points $x = x_j$ $(j = 0, 1, \ldots, N)$ can be calculated by the following simple formula

$$\left({}^{\mathrm{RL}}_{-1}D^{\alpha(x_0)}_{x_0} I^{\mu,a,b,+}_{N} u(x_0), \ldots, {}^{\mathrm{RL}}_{-1}D^{\alpha(x_N)}_{x_N} I^{\mu,a,b,+}_{N} u(x_N)\right)^T = \mathcal{D}_L \tilde{\boldsymbol{u}}, \tag{3.2.272}$$

where

$$\tilde{\boldsymbol{u}} = (v(x_0), u(x_1), \ldots, u(x_N))^T, \quad v(x_0) = \lim_{x \to x_0^+} (x - x_0)^{-\mu} u(x),$$

the differential matrix

$$\mathcal{D}_L = \mathcal{SC}\,\mathrm{diag}\left(1, (1 + x_1)^{-\mu}, \ldots, (1 + x_N)^{-\mu}\right),$$

the matrix $\mathcal{S}$ satisfies $(\mathcal{S})_{i,j} = {}^{\mathrm{RL}}S^{a,b,\alpha,\mu}_{L,j}(x_i)$ $(i, j = 0, 1, \ldots, N)$, and ${}^{\mathrm{RL}}S^{a,b,\alpha,\mu}_{L,j}(x)$ is defined in (3.2.259).

We can similarly derive the differentiation matrix $\mathcal{D}_R$ such that

$$\left({}^{\mathrm{RL}}_{x_0}D^{\alpha(x_0)}_{1} I^{\mu,a,b,-}_{N} u(x_0), \ldots, {}^{\mathrm{RL}}_{x_N}D^{\alpha(x_N)}_{1} I^{\mu,a,b,-}_{N} u(x_N)\right)^T = \mathcal{D}_R \hat{\boldsymbol{u}}, \tag{3.2.273}$$

where

$$\hat{\boldsymbol{u}} = (u(x_0), \ldots, u(x_{N-1}), v(x_N))^T, \quad v(x_N) = \lim_{x \to x_N^-} (x_N - x)^{-\mu} u(x),$$

and

$$\mathcal{D}_R = \mathcal{SC}\,\mathrm{diag}\left((1 - x_0)^{-\mu}, (1 - x_1)^{-\mu}, \ldots, (1 - x_{N-1})^{-\mu}, 1\right),$$

the matrix $\mathcal{S}$ satisfies $(\mathcal{S})_{i,j} = {}^{\mathrm{RL}}S^{a,b,\alpha,\mu}_{R,j}(x_i)$ $(i, j = 0, 1, \ldots, N)$, and ${}^{\mathrm{RL}}S^{a,b,\alpha,\mu}_{R,j}(x)$ is defined in (3.2.259).

We can similarly derive the differentiation matrices based on the JG and JGR points.

Differentiation Matrices of the Caputo Fractional Derivatives

In this subsection, we focus on the computation of ${}^{\mathrm{C}}S^{a,b,\alpha,\mu}_{L,j}(x) = {}^{\mathrm{C}}_{-1}D^{\alpha(x)}_{x} P^{a,b,\mu}_{j,+}(x)$ and ${}^{\mathrm{C}}S^{a,b,\alpha,\mu}_{R,j}(x) = {}^{\mathrm{C}}_{x}D^{\alpha(x)}_{1} P^{a,b,\mu}_{j,-}(x)$, which leads to the desired differentiation matrices of the Caputo fractional derivatives. In the rest of this subsection, we also use $S_{L,j}(x) = {}^{\mathrm{C}}S^{a,b,\alpha,\mu}_{L,j}(x)$ and $S_{R,j}(x) = {}^{\mathrm{C}}S^{a,b,\alpha,\mu}_{R,j}(x)$ for simplicity.

For $\mu \geq 0$ and $0 < \alpha(x) < 1$ with $\tilde{\alpha} = 1 - \alpha$, we have

$$\begin{aligned} S_{L,j}(x) &= {}^{\mathrm{C}}_{-1}D^{\alpha(x)}_{x} P^{a,b,\mu}_{j,+}(x) \\ &= \frac{1}{\Gamma(\tilde{\alpha})} \int_{-1}^{x} (x - s)^{\tilde{\alpha}-1} \left[\mu(1+s)^{\mu-1} P^{(a,b)}_{j}(s) + d^{(a,b)}_{j,1} (1+s)^{\mu} P^{a+1,b+1}_{j-1}(s)\right] ds, \end{aligned}$$

which leads to

$$S_{L,j}(x) = \mu\, G_{L,j}^{a,b,1-\alpha,\mu-1}(x) + d_{j,1}^{(a,b)}\, G_{L,j-1}^{a+1,b+1,1-\alpha,\mu}(x). \tag{3.2.274}$$

Similarly, we have

$$S_{R,j}(x) = \mu G_{R,j}^{a,b,1-\alpha,\mu-1}(x) - d_{j,1}^{(a,b)} G_{R,j-1}^{a+1,b+1,1-\alpha,\mu}(x). \tag{3.2.275}$$

For $\mu \geq 1$ and $1 < \alpha < 2$, we obtain

$$\begin{aligned} S_{L,j}(x) = \mu(\mu-1)G_{L,j}^{a,b,2-\alpha,\mu-2}(x) + 2\mu d_{j,1}^{(a,b)} G_{L,j-1}^{a+1,b+1,2-\alpha,\mu-1}(x) \\ + d_{j,2}^{(a,b)} G_{L,j-2}^{a+2,b+2,2-\alpha,\mu}(x), \end{aligned} \tag{3.2.276}$$

$$\begin{aligned} S_{R,j}(x) = \mu(\mu-1)G_{R,j}^{a,b,2-\alpha,\mu-2}(x) - 2\mu d_{j,1}^{(a,b)} G_{R,j-1}^{a+1,b+1,2-\alpha,\mu-1}(x) \\ + d_{j,2}^{(a,b)} G_{R,j-2}^{a+2,b+2,2-\alpha,\mu}(x). \end{aligned} \tag{3.2.277}$$

Remark 3.68 According to (3.2.234), we can calculate the Caputo derivative ${}_{-1}^{\ C}D_x^{\alpha(x)} P_{+,j}^{a,b,\mu}(x)$ (or ${}_{x}^{C}D_1^{\alpha(x)} P_{-,j}^{a,b,\mu}(x)$) from ${}_{-1}^{\mathrm{RL}}D_x^{\alpha(x)} P_{+,j}^{a,b,\mu}(x)$ (or ${}_{x}^{\mathrm{RL}}D_1^{\alpha(x)} P_{-,j}^{a,b,\mu}(x)$).

Remark 3.69 When $\mu = 0$, we have (cf. Li et al. (2012); Zeng and Li (2014) where $\alpha(x)$ is a constant)

$${}^{C}S_{L,j}^{a,b,\alpha,0}(x) = {}_{-1}^{\ C}D_x^{\alpha(x)} P_{j,+}^{a,b,0}(x) = d_{j,n}^{(a,b)} G_{L,j-n}^{a+n,b+n,n-\alpha,0}(x), \tag{3.2.278}$$

$${}^{C}S_{R,j}^{a,b,\alpha,0}(x) = {}_{x}^{C}D_1^{\alpha(x)} P_{j,-}^{a,b,0}(x) = (-1)^n d_{j,n}^{(a,b)} G_{R,j-n}^{a+n,b+n,n-\alpha,0}(x), \tag{3.2.279}$$

where $n-1 < \alpha(x) < n, n \in \mathbb{N}$, $d_{j,n}^{(a,b)}$ is defined by (3.2.262), and $G_{L,j}^{a,b,\alpha,\mu}(x)$ and $G_{R,j}^{a,b,\alpha,\mu}(x)$ are defined by (3.2.245) and (3.2.246), respectively.

For the Caputo derivatives, we can similarly derive the following differentiation matrices $\mathcal{D}_L, \mathcal{D}_R \in \mathbb{R}^{(N+1)\times(N+1)}$ such that

$$\left({}_{-1}^{\ C}D_{x_0}^{\alpha(x_0)} I_N^{\mu,a,b,+} u(x_0), \ldots, {}_{-1}^{\ C}D_{x_N}^{\alpha(x_N)} I_N^{\mu,a,b,+} u(x_N)\right)^T = \mathcal{D}_L \boldsymbol{u}, \tag{3.2.280}$$

$$\left({}_{x_0}^{\ C}D_1^{\alpha(x_0)} I_N^{\mu,a,b,-} u(x_0), \ldots, {}_{x_N}^{\ C}D_1^{\alpha(x_N)} I_N^{\mu,a,b,-} u(x_N)\right)^T = \mathcal{D}_R \boldsymbol{u}. \tag{3.2.281}$$

Remark 3.70 The total computational cost of the present algorithm is $O(N^3)$, where $N \times N$ is the size of the fractional differentiation matrices. For the differentiation matrices based on the Chebyshev–Gauss–Lobatto points, the computational cost can be reduced to $O(N^2 \log N)$ due to the fast Fourier transform. In the standard spectral collocation method, i.e., $\alpha = 1, 2$, we have the exact formula to calculate the differentiation matrices with $O(N^2)$ operations, see Shen et al. (2011a).

Applications and Numerical Examples

In this section, we first illustrate the introduced spectral collocation method in this section to solve FODEs and compare the numerical results with existing spectral collocation methods. We present cases, where the end-point singularity index of the solution is known, and we mainly focus on the convergence order of our methods and test the methods for different situations.

Example 3.71 Solve FODE (3.2.235) with $x \in (0,1]$ and the following conditions:

- Case I (Smooth solution): Choose the suitable initial condition and the right-hand side function $f(x)$, such that the analytical solution to (3.2.235) is $u(x) = \sin(2\pi x)$, where $C(x) = \frac{1}{1+x^2}$ and $\alpha(x) = \frac{1}{\sin^2(10x)+1.1}$.
- Case II (Solution with end-point singularity): We choose $C(x) = 0$, $\alpha = \alpha(x)$ is a constant, and $f(x)$ is chosen as $f(x) = x^\sigma \sin(x+1)$, $\sigma \geq 0$.
- Case III (Solution with end-point singularity): Choose the suitable initial condition and the right-hand side function $f(x)$, such that the analytical solution to (3.2.235) is $u(x)$[1] $= x^\sigma + x^{2\sigma}$, $\sigma > 0$, where $C(x) = \frac{1}{1+x^2}$ and $\alpha(x) = |x - \frac{1}{2}| + \frac{1}{3}$.

In this example, the errors are measured in the following sense

$$\|e\|_\infty = \max_{1\leq j\leq K} |u_{\text{ref}}(z_j) - \tilde{u}(z_j)|, \quad z_j = j/K,\ j = 0,1,\ldots,K,\ K = 128,$$

where $\tilde{u}(x) = x^{-\mu}\sum_{j=0}^{N} v_j l_j(x)$, $l_j(x)$ is the Lagrange basis function based on JGL points on the interval $[0,1]$ and u_{ref} is either the exact solution or a numerical solution obtained from the JGL collocation method (3.2.236) with $N = 128$.

In Case I, the solution is smooth and then we use (3.2.237) with $\mu = 0$ (standard spectral collocation method, see e.g., Tian et al. (2014)) to solve it. In Table 3.10, we observe spectral accuracy for the smooth solutions when three types of collocation points are used. For the same N, the three sets of collocation points give the same level of accuracy, which is expected from standard spectral theory, see e.g., Guo (1998).

For Case II, we know from Zhang et al. (2015) that the analytical solution $u(x)$ to (3.2.235) satisfies $u(x) = x^{\sigma+\alpha-\lfloor\sigma+\alpha\rfloor}v(x)$ or $u(x) = x^{\sigma+\alpha-\lceil\sigma+\alpha\rceil}v(x)$

[1] Here we choose such a solution to somewhat mimic the singularity structure of the solution of ${}_0^C D_x^\alpha u + u = 1$, $u(0) = u_0$, $0 < \alpha < 1$, which is of the form $\sum_{i,j=0}^{n} c_{i,j}x^{i\alpha+j}$.

Table 3.10. *Maximum errors for Example 3.71 for Case I, $\mu = 0$. Source: Zeng et al. (2015a). Copyright ©2015 Society for Industrial and Applied Mathematics. Reprinted with permission. All rights reserved.*

	N	(a,b)	Error	(a,b)	Error	(a,b)	Error
	4		2.3866e-01		2.9661e-01		6.9196e-01
	8		1.3750e-03		1.5597e-03		2.8386e-03
Case I	16	(0,0)	7.9171e-11	$(-\frac{1}{2},-\frac{1}{2})$	4.8886e-11	$(-\frac{1}{2},\frac{1}{2})$	3.2089e-10
	32		3.4348e-15		5.1070e-15		6.7724e-15
	64		3.7541e-15		3.8858e-15		7.7091e-15

Table 3.11. *Maximum errors for Example 3.71, Case II, $a = b = 0$. Source: Zeng et al. (2015a). Copyright ©2015 Society for Industrial and Applied Mathematics. Reprinted with permission. All rights reserved.*

σ	α	μ	$N=4$	$N=8$	$N=16$	$N=32$	$N=64$
		-0.6^*	5.8111e-04	1.9957e-09	5.8564e-15	1.0464e-14	4.8850e-15
0.2	0.2	0.4	1.6663e-01	9.7900e-02	4.2005e-02	1.0171e-02	2.9030e-03
		0	1.1811e-01	7.1457e-02	2.5294e-02	3.0128e-03	4.8583e-04
		-0.7^*	2.7865e-04	2.6968e-10	8.2522e-14	6.2637e-14	8.7576e-14
0.8	0.5	0.3^*	4.8338e-05	8.0547e-11	1.4433e-15	1.7764e-15	1.5543e-15
		0	3.0484e-03	6.3265e-04	1.0778e-04	4.0644e-06	3.6246e-07
		-0.5^*	9.9532e-03	1.4705e-07	1.7592e-14	1.4373e-14	2.0317e-14
1.7	0.8	0.5^*	2.0276e-05	1.8398e-11	7.7716e-16	1.9429e-15	1.1102e-15
		0	2.8684e-04	1.9274e-05	6.9531e-07	1.2071e-08	3.1498e-10

when $f(x) = x^\sigma \sin(x+1)$, where $v(x) \in H^\infty([0,1])$. Here, we apply the JGL collocation method (3.2.237) with three choices of μ: $\mu = \sigma + \alpha - \lceil \sigma + \alpha \rceil$ $(-1 < \mu \le 0)$, $\mu = \sigma + \alpha - \lfloor \sigma + \alpha \rfloor$ $(0 \le \mu < 1)$, and $\mu = 0$. As we do not have the exact solution, we use the reference solution u_{ref} obtained by (3.2.237) with $N = 128$ JGL collocation points for each group of parameters $\sigma, \alpha, \mu, a, b$ (see also the values of these parameters in Tables 3.11–3.12). In Table 3.11[2], we show the errors $\|e\|_\infty$ by using the Legendre–Gauss–Lobatto spectral collocation method ($a = b = 0$), and the errors $\|e\|_\infty$ of the Chebyshev–Gauss–Lobatto collocation method ($a = b = -1/2$) are shown in Table 3.12.

[2] In this table and in Tables 3.12 and 3.13, we marked the optimal μ with μ^* for Cases II and III.

Table 3.12. *Maximum errors for Example 3.71, Case II, $a = b = -1/2$. Source: Zeng et al. (2015a).* *Copyright ©2015 Society for Industrial and Applied Mathematics. Reprinted with permission. All rights reserved.*

σ	α	μ	$N = 4$	$N = 8$	$N = 16$	$N = 32$	$N = 64$
		-0.6^*	4.8717e-04	1.3141e-09	1.8874e-15	3.3029e-15	2.4425e-15
0.2	0.2	0.4	1.5900e-01	9.3274e-02	3.2727e-02	1.4969e-02	6.3009e-03
		0	1.1388e-01	6.6410e-02	1.5528e-02	4.2654e-03	1.3560e-03
		-0.7^*	2.3305e-04	1.8148e-10	1.2456e-14	1.9332e-14	2.8945e-14
0.8	0.5	0.3^*	7.2505e-05	1.1160e-10	9.9920e-16	7.7716e-16	1.4433e-15
		0	2.7518e-03	5.1459e-04	7.0073e-05	3.9047e-06	2.0670e-07
		-0.5^*	8.2223e-03	9.2855e-08	1.1102e-15	1.6567e-15	7.0679e-15
1.7	0.8	0.5^*	2.8742e-05	2.5962e-11	7.7716e-16	1.1657e-15	1.9429e-15
		0	1.9856e-04	1.3396e-05	4.3241e-07	5.7959e-09	1.2829e-10

In both tables, we observe spectral accuracy when we choose $\mu = \sigma + \alpha - \lceil \sigma + \alpha \rceil$ or $\mu = \sigma + \alpha - \lfloor \sigma + \alpha \rfloor$ for $\sigma + \alpha \geq 1$. Note that we are actually approximating v instead of u. While we have a smooth v with the chosen parameter μ, it is natural to expect spectral accuracy. However, it is still important to choose $\mu < 0$ if JGL collocation methods are used. When $0 < \sigma + \alpha < 1$, we choose $\mu = \sigma + \alpha - \lceil \sigma + \alpha \rceil$ $(-1 < \mu < 0)$, where we still observe spectral accuracy, see the text before Remark 3.63 for choosing μ. However, if $0 < \sigma + \alpha < 1$ and we choose $\mu = \sigma + \alpha > 0$, then we need to know $v(0)$ to recover spectral accuracy and observe otherwise only slow convergence. When we simply take $v(0)$ any value other than the exact value (here the exact value of $v(0)$ is $\frac{\Gamma(\sigma+1)}{\Gamma(\sigma+1+\alpha)} \sin(1)$), we cannot observe spectral accuracy; see (3.2.237), and Tables 3.11–3.12 where $v(0) = 0$ is used whenever $0 < \mu = \sigma + \alpha < 1$. If $0 < \mu = \sigma + \alpha < 1$ is chosen, then one can use Jacobi–Gauss points to avoid the dilemma instead of JGL points, see Zhang et al. (2015), where the solutions are evaluated only at interior points.

For Case III, we observe in Table 3.13 that higher accuracy is obtained when $\mu < 0$ as we can expect from the fact that $v(x) = x^{-\mu} u(x)$ has better regularity than $u(x)$ $(\mu = 0)$. We observe that the best choice of μ is $\mu = \sigma - \lceil \sigma \rceil < 0$.

Discussion on the Tunable Collocation Methods

In this section, we proposed the use of the spectral collocation methods using the singular approximation basis functions of the GJPFs, i.e., $(1 \pm x)^{\mu} P_j^{(a,b)}(x)$, $P_j^{(a,b)}(x)$ is the Jacobi polynomial and $a, b, \mu > -1$, to solve

Table 3.13. *Maximum errors for Example 3.71, Case III,* $a = -\mu, b = \mu$. *Source: Zeng et al. (2015a). Copyright ©2015 Society for Industrial and Applied Mathematics. Reprinted with permission. All rights reserved.*

σ	μ	$N = 4$	$N = 8$	$N = 16$	$N = 32$	$N = 64$
	−0.7*	5.4992e-04	1.0523e-05	2.5302e-07	5.9610e-09	1.3223e-10
1.3	−0.4	2.5433e-03	3.9183e-04	5.2868e-05	7.0554e-06	8.3719e-07
	0	4.7920e-03	1.2883e-03	2.5965e-04	2.8371e-05	3.5713e-06
	−0.5*	6.2021e-04	5.6469e-06	7.6852e-08	1.0087e-09	1.2946e-11
1.5	−0.55	3.9389e-04	2.4162e-05	3.3420e-06	3.5877e-07	3.3376e-08
	0	4.4015e-03	8.0197e-04	1.2153e-04	9.8811e-06	9.8474e-07
	−0.2*	4.4378e-04	2.0038e-06	1.2603e-08	6.1470e-11	3.2969e-13
1.8	−0.3	2.3472e-04	3.9786e-05	2.7610e-06	1.7865e-07	1.1111e-08
	0	2.1472e-03	1.8291e-04	1.8154e-05	9.4600e-07	6.5811e-08

FODEs. We observed that with proper choices of $\mu > -1$, we have better resolution of solutions to constant-order and variable-order FODEs.

When we know exactly the regularity of the exact solutions, we can choose optimal $\mu > -1$ such that higher accuracy can be obtained. When we do not know the regularity of the analytical solutions but still know that the solution has end-point singularity while the solution is relatively smooth in the interior domain, we always choose suitable $-1 < \mu < 0$ so that higher accuracy can be obtained.

Compared with the existing works, e.g., Tian et al. (2014); Xu and Hesthaven (2014); Zayernouri and Karniadakis (2014a); Zeng and Li (2014), we take the weak regularity at the boundary into account in our spectral collocation method to get better resolution at the boundary and thus the whole domain.

To efficiently evaluate variable-order fractional integrals and derivatives of $(1 \pm x)^{\mu} P_j^{(a,b)}(x)$, we developed the recurrence formulas for these integrals and derivatives, which are shown to be numerically stable. In some special cases, the derived recurrence formulas are reduced to the GJPFs. Comparison with brute force computation of these integrals shows comparable or better accuracy with much improved efficiency.

We will investigate the error estimate of the present spectral collocation methods when the regularity of solutions is known and extend the present methods to solve a wider class of FODEs in the next section yet of a more generalized sense.

3.2.6 A Generalized Tunable Collocation Method for Treating Singularities at End-Points

In this section, we introduce spectral collocation methods for FDEs with variable-order with end-point singularities. Specifically, we present three-term recurrence relations for both integrals and derivatives of the GJPFs of the form $(1+x)^{\mu_1}(1-x)^{\mu_2}P_j^{(a,b)}(x)$ where $a, b, \mu_1, \mu_2 > -1$, leading to the desired differentiation matrices. We apply the new differentiation matrices to construct collocation methods to solve fractional boundary-value problems and fractional partial differential equations with two end-point singularities. We demonstrate that the singular basis enhances greatly the accuracy of the numerical solutions by properly tuning the parameters μ_1 and μ_2, even for cases where we do not know explicitly the form of singularities in the solution at the boundaries. Comparison with other existing methods shows the superior accuracy of the proposed spectral collocation method.

Next, we introduce the *Hadamard finite part integral* that plays a crucial role in the numerical calculation of the fractional derivative operators in this work. There are several ways to define the finite part integral, see e.g., Diethelm (1997, 2010); Monegato (2009); Samko et al. (1993). Here, we introduce the finite part integral defined in the following manner.

Definition 3.72 (Hadamard Finite-Part Integral, see Samko et al. (1993)) Let a function $f(x)$ be integrated on an interval (ϵ, A) for any $A > 0$ and $0 < \epsilon < A$. The function $f(x)$ is said to possess the Hadamard property at the point $x = 0$ if there exist constants a_k, b_0 and $\lambda_k > 0$ such that

$$\int_\epsilon^A f(x)\mathrm{d}x = \sum_{k=1}^{N} a_k \epsilon^{-\lambda_k} + b_0 \ln\frac{1}{\epsilon} + J_0(\epsilon), \tag{3.2.282}$$

where $\lim_{\epsilon\to 0} J_0(\epsilon)$ exists and is finite, which is also denoted by

$$\,\,⨎_0^A f(x)\mathrm{d}x = \lim_{\epsilon\to 0} J_0(\epsilon). \tag{3.2.283}$$

The above limit (3.2.283) is called a finite part of the divergent integral $\int_0^A f(x)\mathrm{d}x$ in the Hadamard sense or simply an integral in the Hadamard sense. One important feature of a_k and b_0 in (3.2.282) is that a_k and b_0 do not depend on A.

Lemma 3.73 (See Samko et al. (1993, p. 112)) *The left Riemann–Liouville fractional derivative* ${}^{\mathrm{RL}}_{x_L}D_x^\alpha f(x)$, $x \in (x_L, x_R), \alpha > 0, \alpha \neq 1, 2, \ldots$, *is equivalent to the following integral in the Hadamard sense, i.e.,*

$${}^{\mathrm{RL}}_{x_L}D_x^\alpha f(x) = \frac{1}{\Gamma(-\alpha)} ⨎_{x_L}^{x} (x-s)^{-\alpha-1} f(s)\,\mathrm{d}s. \tag{3.2.284}$$

For the right Riemann–Liouville fractional derivative operator, we can similarly have

$$ {}^{\text{RL}}_{\ x}D^{\alpha}_{x_R} f(x) = \frac{1}{\Gamma(-\alpha)} \fint_{x}^{x_R} (s-x)^{-\alpha-1} f(s)\mathrm{d}s \tag{3.2.285} $$

for $\alpha > 0$ and $\alpha \neq 1, 2, \ldots$.

In the following, $\fint_0^A$ is reduced to $\int_0^A$ if $\int_0^A f(x)\mathrm{d}x$ exists and is finite, i.e., $\fint_0^A f(x)\mathrm{d}x = \int_0^A f(x)$ when $\int_0^A f(x)$ exists and is finite.

Fractional Integrals/Derivatives of the GJPFs

In this section, we introduce the recurrence formulas to calculate the fractional integrals and derivatives of the GJPFs $P_j^{a,b,\mu_1,\mu_2}(x)$ defined by

$$ P_j^{a,b,\mu_1,\mu_2}(x) = (1+x)^{\mu_1}(1-x)^{\mu_2} P_j^{(a,b)}(x) \tag{3.2.286} $$

for $\mu_1, \mu_2 > -1, x \in (-1, 1)$. Then we derive the corresponding differentiation matrices that will be used to solve FODEs.

We first focus on the calculation of the $\alpha(x)$th-order left (or right) fractional integral and derivative of $P_j^{a,b,\mu_1,\mu_2}(x)$ by the three-term recurrence relation. Accordingly, we define

$$ G_{L,j}^{a,b,\alpha,\mu_1,\mu_2}(x) = {}^{\text{RL}}_{-1}D_x^{-\alpha} P_j^{a,b,\mu_1,\mu_2}(x), \quad x \in (-1,1), \tag{3.2.287} $$

$$ G_{R,j}^{a,b,\alpha,\mu_1,\mu_2}(x) = {}^{\text{RL}}_{\ x}D_1^{-\alpha} P_j^{a,b,\mu_1,\mu_2}(x), \quad x \in (-1,1). \tag{3.2.288} $$

We can compute $G_{L,j}^{a,b,\alpha,\mu_1,\mu_2}(x)$ by the following recurrence relation.

Theorem 3.74 *Let $a, b, \mu_1, \mu_2 > -1$ and $\alpha(x)$ be real, $\alpha(x) \neq -1, -2, \ldots$. Then $G_j = G_{L,j}^{a,b,\alpha,\mu_1,\mu_2}(x)$ defined by* (3.2.287), *$H_j = \alpha(x) G_{L,j}^{a,b,\alpha+1,\mu_1,\mu_2}(x)$, and $R_j = \alpha(x) G_{L,j}^{a,b,\alpha+1,\mu_1,\mu_2-1}(x)$ satisfy, for $j \geq 1$ and $x \in (-1, 1)$,*

$$ G_{j+1} = (A_j x - B_j) G_j - C_j G_{j-1} - A_j H_j, \tag{3.2.289} $$

$$ R_{j+1} = (A_j - B_j) R_j - C_j R_{j-1} - A_j H_j, \tag{3.2.290} $$

$$ \begin{aligned} H_{j+1} = &-(A_j + B_j) H_j - C_j H_{j-1} \\ &- (\mu_1 + \mu_2 + 1 + \alpha) A_j \left(\widehat{A}_j H_{j-1} + \widehat{B}_j H_j + \widehat{C}_j H_{j+1}\right) \\ &+ (1+x)\alpha A_j \left(\widehat{A}_j G_{j-1} + \widehat{B}_j G_j + \widehat{C}_j G_{j+1}\right) \\ &+ 2\mu_2 \left(\widehat{A}_j R_{j-1} + \widehat{B}_j R_j + \widehat{C}_j R_{j+1}\right), \end{aligned} \tag{3.2.291} $$

where A_j, B_j, and C_j are given in (3.2.242), *$\widehat{A}_j, \widehat{B}_j$, and $\widehat{C}_j$ are defined by* (3.2.249), (3.2.250), *and* (3.2.251), *respectively.*

Proof For simplicity, we introduce the notations

$$K_0(x,s) = \frac{1}{\Gamma(\alpha(x))}(x-s)^{\alpha(x)-1}(1+s)^{\mu_1}(1-s)^{\mu_2},$$
$$K_1(x,s) = (x-s)K_0(x,s), \quad K_2(x,s) = (1+s)K_1(x,s),$$
$$K_3(x,s) = K_2(x,s)(1-s)^{-1}.$$

For $j \geq 1$, we can obtain from the three-term recurrence relation (3.2.241) that

$$\begin{aligned} G_{j+1} &= \fint_{-1}^{x} K_0(x,s)P_{j+1}^{(a,b)}(s)\mathrm{d}s \qquad (3.2.292) \\ &= \fint_{-1}^{x} K_0(x,s)\big[(A_j s - B_j)P_j^{(a,b)}(s) - C_j P_{j-1}^{(a,b)}(s)\big]\mathrm{d}s \\ &= (A_j x - B_j)G_j - C_j G_{j-1} - A_j \fint_{-1}^{x} K_1(x,s)P_j^{(a,b)}(s)\mathrm{d}s. \end{aligned}$$

Using $\int_{-1}^{x} K_1(x,s)P_j^{(a,b)}(s)\mathrm{d}s = H_j$ leads to (3.2.289). Using the three-term recurrence relation (3.2.241) again, we have

$$\begin{aligned} H_{j+1} &= \fint_{-1}^{x} K_1(x,s)\left[(A_j(s+1) - (A_j + B_j))P_j^{(a,b)}(s) - C_j P_{j-1}^{(a,b)}(s)\right]\mathrm{d}s \\ &= -(A_j + B_j)H_j - C_j H_{j-1} + A_j \fint_{-1}^{x} K_2(x,s)P_j^{(a,b)}(s)\mathrm{d}s. \end{aligned} \qquad (3.2.293)$$

Next, we consider the integral

$$\int_{A}^{x-\epsilon} K_2(x,s)P_j^{(a,b)}(s)\mathrm{d}s, \quad x - \epsilon > A \geq -1, \quad x \in (-1,1).$$

Let

$$J(x) = \widehat{A}_j P_{j-1}^{(a,b)}(x) + \widehat{B}_j P_j^{(a,b)}(x) + \widehat{C}_j P_{j+1}^{(a,b)}(x).$$

Applying $P_j^{(a,b)}(x) = \frac{\mathrm{d}}{\mathrm{d}x}J(x)$ yields

$$\begin{aligned} \int_{A}^{x-\epsilon} K_2(x,s)P_j^{(a,b)}(s)\mathrm{d}s &= J(s)K_2(x,s)\Big|_{A}^{x-\epsilon} - \int_{A}^{x-\epsilon} J(s)\frac{\partial}{\partial s}K_2(x,s)\mathrm{d}s \\ &= J(x-\epsilon)K_2(x,x-\epsilon) - J(A)K_2(x,A) - \int_{A}^{x-\epsilon} J(s)\frac{\partial}{\partial s}K_2(x,s)\mathrm{d}s. \end{aligned} \qquad (3.2.294)$$

Note that

$$\frac{\partial}{\partial s}K_2(x,s) = (\mu_1+\mu_2+1+\alpha)K_1(x,s) - (1+x)\alpha(x)K_0(x,s) - 2\mu_2 K_3(x,s).$$

Hence, we have

$$\begin{aligned}\int_A^{x-\epsilon} K_2(x,s)\ P_j^{(a,b)}(s)\mathrm{d}s &= J(x-\epsilon)K_2(x,x-\epsilon) - J(A)K_2(x,A)\\ &\quad - (\mu_1+\mu_2+1+\alpha(x))\left(\widehat{A}_j H_{j-1}^{\epsilon,A} + \widehat{B}_j H_j^{\epsilon,A} + \widehat{C}_j H_{j+1}^{\epsilon,A}\right)\\ &\quad + \alpha(x)(1+x)\left(\widehat{A}_j G_{j-1}^{\epsilon,A} + \widehat{B}_j G_j^{\epsilon,A} + \widehat{C}_j G_{j+1}^{\epsilon,A}\right)\\ &\quad + 2\mu_2\left(\widehat{A}_j R_{j-1}^{\epsilon,A} + \widehat{B}_j R_j^{\epsilon,A} + \widehat{C}_j R_{j+1}^{\epsilon,A}\right),\end{aligned} \tag{3.2.295}$$

where

$$G_j^{\epsilon,A} = \int_A^{x-\epsilon} K_0(x,s)P_j^{(a,b)}(s)\mathrm{d}s, \quad H_j^{\epsilon,A} = \int_A^{x-\epsilon} K_1(x,s)P_j^{(a,b)}(s)\mathrm{d}s,$$

and

$$R_j^{\epsilon,A} = \int_A^{x-\epsilon} K_3(x,s)P_j^{(a,b)}(s)\mathrm{d}s.$$

For any $\alpha > 0$, we can let $A = -1$ and obtain

$$\lim_{\epsilon\to 0} G_j^{\epsilon,-1} = G_j, \quad \lim_{\epsilon\to 0} H_j^{\epsilon,-1} = H_j, \quad \lim_{\epsilon\to 0} R_j^{\epsilon,-1} = R_j,$$

$J(-1)K_2(x,-1) = 0$, and $J(x-\epsilon)K_2(x,x-\epsilon) \to 0$ as $\epsilon \to 0$. Therefore, we can combine (3.2.293)–(3.2.295) with $\epsilon \to 0$, $A = -1$ to derive

$$\begin{aligned}H_{j+1} = A_j\Big[&- (\mu_1+\mu_2+1+\alpha(x))\left(\widehat{A}_j H_{j-1} + \widehat{B}_j H_j + \widehat{C}_j H_{j+1}\right)\\ &+ \alpha(x)(1+x)\left(\widehat{A}_j G_{j-1} + \widehat{B}_j G_j + \widehat{C}_j G_{j+1}\right)\\ &+ 2\mu_2\left(\widehat{A}_j R_{j-1} + \widehat{B}_j R_j + \widehat{C}_j R_{j+1}\right)\Big]\\ &- (A_j+B_j)H_j - C_j H_{j-1},\end{aligned} \tag{3.2.296}$$

which leads to (3.2.248). For R_{j+1}, we can easily derive from (3.2.290) via

$$\begin{aligned}R_{j+1} &= \fint_{-1}^{x} K_3(x,s)P_{j+1}^{(a,b)}(s)\mathrm{d}s\\ &= \fint_{-1}^{x} K_3(x,s)\left[(A_j(s-1)+A_j-B_j)P_j^{(a,b)}(s) - C_j P_{j-1}^{(a,b)}(s)\right]\mathrm{d}s\\ &= (A_j-B_j)R_j - C_j R_{j-1} - A_j H_j.\end{aligned} \tag{3.2.297}$$

Next, we turn to the case of $\alpha < 0$ and $\alpha \neq -1, -2, \ldots$. For the term $J(x-\epsilon)K_2(x,x-\epsilon)$ in (3.2.295), we can always find a positive integer m such that $0 < m + \alpha < 1$. Therefore,

$$J(x-\epsilon)K_2(x,x-\epsilon) = J(x-\epsilon)\epsilon^{\alpha}(1+x-\epsilon)^{1+\mu_1}(1-x+\epsilon)^{\mu_2}$$
$$= J(x-\epsilon)\epsilon^{\alpha}\sum_{k=0}^{m-1} c_k \epsilon^k + \widetilde{J}_0(\epsilon), \tag{3.2.298}$$

where $\widetilde{J}_0(\epsilon) = J(x-\epsilon)\epsilon^{\alpha}\big[(1+x-\epsilon)^{1+\mu_1}(1-x+\epsilon)^{\mu_2} - \sum_{k=0}^{m-1} c_k \epsilon^k\big]$ and c_k is the coefficient of the Taylor expansion of $(1+x-\epsilon)^{1+\mu_1}(1-x+\epsilon)^{\mu_2}$ at $\epsilon = 0$. Since α is not a negative integer and $J(s)$ is a polynomial of order no greater than $j+1$, we always have $\widetilde{J}_0(\epsilon) = O(\epsilon^{m+\alpha})$ and $J(x-\epsilon)\epsilon^{\alpha}\sum_{k=0}^{m-1} c_k\epsilon^k = \sum_{k=0}^{j+m} a_k \epsilon^{k+\alpha}$, where a_k is independent of ϵ and A. So (3.2.298) can be rewritten as

$$J(x-\epsilon)K_2(x,x-\epsilon) = \sum_{k=0}^{m-1} a_k \epsilon^{k+\alpha} + J_0(\epsilon), \tag{3.2.299}$$

where

$$J_0(\epsilon) = \sum_{k=m}^{j+m} a_k \epsilon^{k+\alpha} + \widetilde{J}_0(\epsilon) = O(\epsilon^{m+\alpha}) \to 0 \text{ as } \epsilon \to 0.$$

For $\alpha < 0\,(\alpha \neq -1,-2,\ldots)$ and $A = -1$, we have from (3.2.282), (3.2.283), (3.2.295),(3.2.299), and $J(-1)K_2(x,-1) = 0$,

$$\begin{aligned}\fint_{-1}^{x} K_2(x,s)P_j^{(a,b)}(s)\mathrm{d}s = &-(\mu_1+\mu_2+1+\alpha(x))\left(\widehat{A}_jH_{j-1}+\widehat{B}_jH_j+\widehat{C}_jH_{j+1}\right)\\ &+\alpha(x)(1+x)\left(\widehat{A}_jG_{j-1}+\widehat{B}_jG_j+\widehat{C}_jG_{j+1}\right)\\ &+2\mu_2\left(\widehat{A}_jR_{j-1}+\widehat{B}_jR_j+\widehat{C}_jR_{j+1}\right).\end{aligned} \tag{3.2.300}$$

From (3.2.300), we can derive that (3.2.296)–(3.2.297) still hold for $\alpha < 0, \alpha \neq -1,-2,\ldots$. Using the fact that (3.2.292) is true for $\alpha < 0, \alpha \neq -1,-2,\ldots$ completes the proof. □

We can compute $G_{R,j}^{a,b,\alpha,\mu_1,\mu_2}(x)$ in a similar fashion.

Theorem 3.75 *Let $a,b,\mu_1,\mu_2 > -1$ and $\alpha(x)$ be real, $\alpha(x) \neq -1,-2,\ldots$. Then $G_j = G_{R,j}^{a,b,\alpha,\mu_1,\mu_2}(x)$ defined by* (3.2.246), *$H_j = \alpha(x)G_{R,j}^{a,b,\alpha+1,\mu_1,\mu_2}(x)$, and $R_j = \alpha(x)\times G_{R,j}^{a,b,\alpha+1,\mu_1-1,\mu_2}(x)$ satisfy, for $j \geq 1$ and $x \in (-1,1)$,*

$$G_{j+1} = (A_jx - B_j)G_j - C_jG_{j-1} + A_jH_j, \tag{3.2.301}$$
$$R_{j+1} = -(A_j + B_j)R_j - C_jR_{j-1} + A_jH_j, \tag{3.2.302}$$

$$
\begin{aligned}
H_{j+1} = {} & (A_j - B_j)H_j - C_j H_{j-1} \\
& - (\mu_1 + \mu_2 + 1 + \alpha)A_j\left(\widehat{A}_j H_{j-1} + \widehat{B}_j H_j + \widehat{C}_j H_{j+1}\right) \\
& + (1 - x)\alpha A_j\left(\widehat{A}_j G_{j-1} + \widehat{B}_j G_j + \widehat{C}_j G_{j+1}\right) \\
& + 2\mu_1\left(\widehat{A}_j R_{j-1} + \widehat{B}_j R_j + \widehat{C}_j R_{j+1}\right),
\end{aligned} \tag{3.2.303}
$$

where A_j, B_j, *and* C_j *are given in* (3.2.242), $\widehat{A}_j, \widehat{B}_j$, *and* $\widehat{C}_j$ *are defined by* (3.2.249), (3.2.250), *and* (3.2.251), *respectively.*

Proof The proof of Theorem 3.75 is very similar to Theorem 3.74, which is omitted here. □

Next, we discuss how to obtain the initial values $G_{Z,j}^{a,b,\alpha,\mu_1,\mu_2}(x)$ for $Z = R, L$ and $j = 0, 1$. We can easily obtain

$$
\begin{aligned}
G_{L,0}^{a,b,\alpha,\mu_1,\mu_2} &= \frac{1}{\Gamma(\alpha(x))} ⨎_{-1}^{x} (x - s)^{\alpha(x)-1}(1 + s)^{\mu_1}(1 - s)^{\mu_2} \mathrm{d}s \\
&= \frac{2^{\mu_2}(1 + x)^{\alpha(x)+\mu_1}\Gamma(\mu_1 + 1)}{\Gamma(\mu_1 + 1 + \alpha(x))} {}_2F_1(-\mu_2, \mu_1 + 1; \alpha(x) + \mu_1 + 1; (x + 1)/2), \\
G_{R,0}^{a,b,\alpha,\mu_1,\mu_2} &= G_{L,0}^{a,b,\alpha(x),\mu_2,\mu_1}(-x) \\
&= \frac{2^{\mu_1}(1 - x)^{\alpha(x)+\mu_2}\Gamma(\mu_2 + 1)}{\Gamma(\mu_2 + 1 + \alpha(x))} {}_2F_1(-\mu_1, \mu_2 + 1; \alpha(x) + \mu_2 + 1; (1 - x)/2),
\end{aligned}
$$

where ${}_2F_1$ is the hypergeometric function (see, e.g., Becken and Schmelcher (2000)). There are many numerical approaches to calculate the hypergeometric functions (see e.g., Doornik (2015); López and Temme (2013); Pearson et al. (2017)).

The initial values of H_j $(j = 0, 1)$ in Theorem 3.74 (or Theorem 3.75) can be obtained from the relation $H_j = \alpha G_{L,j}^{a,b,\alpha+1,\mu_1,\mu_2}$ (or $H_j = \alpha G_{R,j}^{a,b,\alpha+1,\mu_1,\mu_2}$). The initial values of R_j $(j = 0, 1)$ in Theorem 3.74 (or Theorem 3.75) can be similarly derived by the relation $R_j = \alpha G_{L,j}^{a,b,\alpha+1,\mu_1,\mu_2-1}$ (or $R_j = \alpha G_{R,j}^{a,b,\alpha+1,\mu_1-1,\mu_2}$).

In general cases, i.e., $ab \neq 0$ and $a \neq b$, it is difficult to prove that the three-term recurrence formula (3.2.289)–(3.2.291) or (3.2.301)–(3.2.303) could define an orthogonal system. Thus, we numerically show their performance of stability. Here we consider only (3.2.289)–(3.2.291), since the situation is very similar for (3.2.301)–(3.2.303).

If $x \in [-1, \sigma]$, where $-1 < \sigma < 1$ and σ is far from 1, then $G_{L,j}^{a,b,\alpha,\mu_1,\mu_2}(x)$ can be well approximated by the following approach

$$
\begin{aligned}
&G_{L,j}^{a,b,\alpha,\mu_1,\mu_2}(x)\\
&= \frac{(x+1)^{\alpha(x)+\mu_1}}{2^{\alpha(x)+\mu_1}\Gamma(\alpha(x))} \fint_{-1}^{1} (1-\hat{s})^{\alpha(x)-1}(1+\hat{s})^{\mu_1}(1-s)^{\mu_2} P_j^{(a,b)}(s)\mathrm{d}\hat{s}\\
&\approx \frac{(x+1)^{\alpha(x)+\mu_1}}{2^{\alpha(x)+\mu_1}\Gamma(\alpha(x))} \fint_{-1}^{1} (1-\hat{s})^{\alpha(x)-1}(1+\hat{s})^{\mu_1} \sum_{k=0}^{M} p_{jk} P_k^{(-\frac{1}{2},-\frac{1}{2})}(\hat{s})\mathrm{d}\hat{s}\\
&= \frac{(x+1)^{\alpha(x)+\mu_1}}{2^{\alpha(x)+\mu_1}} \sum_{k=0}^{M} p_{jk} G_{L,k}^{-\frac{1}{2},-\frac{1}{2},\alpha(x),\mu_1}(1), \qquad (3.2.304)
\end{aligned}
$$

where $s = \frac{(x+1)\hat{s}+x-1}{2} \in [-1, x]$, $\hat{s} \in (-1, 1)$, and

$$G_{L,k}^{a,b,\alpha,\mu_1}(x) = \frac{1}{\Gamma(\alpha)} \fint_{-1}^{x} (x-s)^{\alpha-1}(1+s)^{\mu_1} P_k^{(a,b)}(s)\mathrm{d}s$$

can be efficiently calculated by the three-term recurrence formula in Zeng et al. (2015), and p_{jk} satisfies

$$I_M^{-\frac{1}{2},-\frac{1}{2}}\left((1-s)^{\mu_2} P_j^{(a,b)}(s)\right) = \sum_{k=0}^{M} p_{jk} P_k^{(-\frac{1}{2},-\frac{1}{2})}(\hat{s}),$$

which can be efficiently computed using the fast Fourier transform.

Now we are ready to test the computational performance of the recurrence formula (3.2.289)–(3.2.291). We apply (3.2.304) to calculate $G_{L,j}^{a,b,\alpha,\mu_1,\mu_2}(x)$ as the benchmark solutions and the numerical solutions obtained by the recurrence formula (3.2.289)–(3.2.291) are denoted by ${}^{(1)}G_{L,j}^{a,b,\alpha,\mu_1,\mu_2}(x)$. Then the relative L^∞ error $\|e\|_{R,\infty}$ is defined by

$$\|e\|_{\infty,R} = \frac{\max_{0\le j\le N} \max_{1\le k\le 100(1+\sigma)} \left|e_j(x_k)\right|}{\max_{0\le j\le N} \max_{1\le k\le 100(1+\sigma)} \left|G_{L,j}^{a,b,\alpha,\mu_1,\mu_2}(x_k)\right|}, \qquad (3.2.305)$$

where

$$x_k = -1 + \frac{k}{100} \le \sigma, \quad e_j(x) = {}^{(1)}G_{L,j}^{a,b,\alpha,\mu_1,\mu_2}(x) - G_{L,j}^{a,b,\alpha,\mu_1,\mu_2}(x).$$

In Table 3.14, we display the relative maximum error $\|e\|_{R,\infty}$ of the recurrence formula (3.2.289)–(3.2.291), where we used the multiprecision toolbox for MATLAB[1] to compute the benchmark solutions defined by (3.2.304) in order to avoid big round-off errors. Obviously, satisfactory accuracy is observed, which implies good performance of stability of the recurrence formula (3.2.289)–(3.2.291).

[1] Ben Barrowes, Multiple Precision Toolbox for MATLAB, www.mathworks.com/matlabcentral/fileexchange/6446-multiple-precision-toolbox-for-matlab.

Table 3.14. *The relative maximum error* $\|e\|_{R,\infty}$ *(see (3.2.305)) of the recurrence formula* (3.2.289)–(3.2.291), $\alpha(x) = 2\sin(\pi x)$, $N = 256$, $M = 512$ *in* (3.2.304), *and* $\sigma = 0.6$ *in* (3.2.305). *Source: Zeng et al. (2017).* *Copyright ©2017 Society for Industrial and Applied Mathematics. Reprinted with permission. All rights reserved.*

(μ_1, μ_2)	(a, b)	$\|e\|_\infty$	(a, b)	$\|e\|_\infty$
$(0.3, 0.9)$		1.604261e−14		2.244190e−14
$(0.3, -0.9)$		1.486330e−14		2.091652e−14
$(-0.3, 0.9)$	(0,0)	1.642302e−14	$(\frac{1}{2}, -\frac{1}{3})$	2.307503e−14
$(-0.3, -0.9)$		3.404367e−14		3.015168e−14
$(-0.9, -0.3)$		1.803852e−13		1.362458e−13

Differentiation Matrices

Let $u(x)$ be a function defined on the interval $(-1, 1)$. Then we can let $v(x) = (1+x)^{-\mu_1}(1-x)^{-\mu_2}u(x)$ such that $u(x) = (1+x)^{\mu_1}(1-x)^{\mu_2}v(x)$, where $\mu_1, \mu_2 > -1$. Here we can choose suitable μ_1 and μ_2 such that $v(x)$ has the same regularity as $u(x)$ or better regularity than $u(x)$. Let N be a positive integer. Denote $x_j (j = 0, 1, \ldots, N)$ as the Jacobi–Gauss–Lobatto (JGL) points defined on the interval $(-1, 1)$. Then $u(x)$ can be approximated by

$$I_N^{\mu,a,b}u(x) = (1+x)^{\mu_1}(1-x)^{\mu_2}I_N^{a,b}v(x) = \omega^{(\mu_2,\mu_1)}(x)\sum_{j=0}^{N} v(x_j)l_j(x)$$

$$= \omega^{(\mu_2,\mu_1)}(x)\sum_{j=0}^{N} \hat{v}_j P_j^{(a,b)}(x), \tag{3.2.306}$$

where $\mu = (\mu_1, \mu_2)$, $I_N^{a,b}$ is the JGL interpolant, and $l_j(x)$ is the Lagrange interpolation basis function. The basis function $l_j(x)$ can also be represented as

$$l_j(x) = \sum_{k=0}^{N} c_{k,j} P_k^{(a,b)}(x), \tag{3.2.307}$$

where $c_{k,j}$ can be determined by the following relation (Shen et al., 2011a)

$$c_{k,j} = (\mathcal{C})_{k,j} = \frac{P_k^{(a,b)}(x_j)\omega_j}{\delta_k \gamma_k^{(a,b)}}, \quad k = 0, 1, \ldots, N, \tag{3.2.308}$$

in which $\gamma_k^{(a,b)}$ is defined by (3.2.244), $\delta_k = 1\,(0 \le k < N)$, and $\delta_N = 2 + \frac{a+b+1}{N}$. If $I_N^{a,b}$ in (3.2.306) is the Jacobi–Gauss (JG) or Jacobi–Gauss–Radau (JGR) interpolant, then we have $\delta_k = 1\,(0 \le k \le N)$ in (3.2.308). Combining (3.2.306)–(3.2.308) yields

$$\hat{\boldsymbol{v}} = \mathcal{C}\boldsymbol{v}, \tag{3.2.309}$$

where $\hat{\boldsymbol{v}} = (\hat{v}_0, \hat{v}_1, \ldots, \hat{v}_N)^T$ and $\boldsymbol{v} = (v(x_0), v(x_1), \ldots, v(x_N))^T$.

For $x \in (-1, 1)$ and $\alpha(x) \in \mathbb{R}$, ${}_{-1}^{\text{RL}}D_x^{\alpha(x)} I_N^{\mu,a,b} u(x)$ can be calculated by the following relation

$${}_{-1}^{\text{RL}}D_x^{\alpha(x)} I_N^{\mu,a,b} u(x) = \sum_{j=0}^{N} \hat{v}_j \, G_{L,j}^{a,b,-\alpha,\mu_1,\mu_2}(x), \tag{3.2.310}$$

where $G_{L,j}^{a,b,\alpha,\mu_1,\mu_2}(x)$ is defined by (3.2.245) that can be obtained from the three-term recurrence relation in Theorem 3.74. Combining (3.2.309) and (3.2.310) yields

$${}_{-1}^{\text{RL}}D_x^{\alpha(x)} I_N^{\mu,a,b} u(x) = \left(G_{L,0}^{a,b,\alpha,\mu_1,\mu_2}(x), \ldots, G_{L,N}^{a,b,\alpha,\mu_1,\mu_2}(x)\right)\mathcal{C}\boldsymbol{v}. \tag{3.2.311}$$

In what follows, $\tilde{\boldsymbol{u}}$ always denotes

$$\tilde{\boldsymbol{u}} = (v(x_0), u(x_1), u(x_2), \ldots, u(x_{N-1}), v(x_N))^T,$$

where

$$v(x_0) = \lim_{x \to x_0+} \omega^{(-\mu_2,-\mu_1)}(x)u(x), \quad v(x_N) = \lim_{x \to x_N-} \omega^{(-\mu_2,-\mu_1)}(x)u(x),$$

and $v(x_k) = \omega^{(-\mu_2,-\mu_1)}(x_k)u(x_k)$ for $k = 1, 2, \ldots, N-1$.

Let ${}_{-1}^{\text{RL}}D_{x_j}^{\alpha(x_j)} I_N^{\mu,a,b} u(x_j) = \left[{}_{-1}^{\text{RL}}D_x^{\alpha(x)} I_N^{\mu,a,b} u(x)\right]_{x=x_j}$. Then ${}_{-1}^{\text{RL}}D_x^{\alpha(x)}$ $I_N^{\mu,a,b} u(x)$ at the collocation points $x = x_j \; (j = 1, 2, \ldots, N-1)$ can be calculated by the following simple formula

$$\left({}_{-1}^{\text{RL}}D_{x_1}^{\alpha(x_1)} I_N^{\mu,a,b} u(x_1), \ldots, {}_{-1}^{\text{RL}}D_{x_{N-1}}^{\alpha(x_{N-1})} I_N^{\mu,a,b} u(x_{N-1})\right)^T = \mathcal{D}_L \tilde{\boldsymbol{u}}, \tag{3.2.312}$$

where the differentiation matrix $\mathcal{D}_L$ is given by

$$\mathcal{D}_L = \mathcal{S}\mathcal{C}\, \text{diag}\left(1, \omega^{(-\mu_2,-\mu_1)}(x_1), \omega^{(-\mu_2,-\mu_1)}(x_2), \ldots, \omega^{(-\mu_2,-\mu_1)}(x_{N-1}), 1\right),$$

the matrix $\mathcal{S}$ has the entries $(\mathcal{S})_{i,j} = G_{L,j}^{a,b,-\alpha,\mu_1,\mu_2}(x_i)$ $(i = 1, \ldots, N-1, j = 0, \ldots, N)$, and $G_{L,j}^{a,b,-\alpha,\mu_1,\mu_2}(x)$ is defined in (3.2.245), and $\mathcal{C}$ is defined by (3.2.308).

We can similarly obtain the differentiation matrix $\mathcal{D}_R$ such that

$$\left({}_{x_1}^{\text{RL}}D_1^{\alpha(x_1)} I_N^{\mu,a,b} u(x_1), \ldots, {}_{x_{N-1}}^{\text{RL}}D_1^{\alpha(x_{N-1})} I_N^{\mu,a,b} u(x_{N-1})\right)^T = \mathcal{D}_R \tilde{\boldsymbol{u}}. \tag{3.2.313}$$

The differentiation matrices $\mathcal{D}_L$ and $\mathcal{D}_R$ will be used to construct spectral collocation methods for solving FBVPs and FPDEs.

Theoretical Results

In this subsection, we present some error estimates of the interpolation operator $I_N^{\mu,a,b}$. These error estimates can also be used to illustrate high accuracy of the spectral collocation methods in Section 3.2.5.

Theorem 3.76 *Let* $u(x) = (1+x)^{\mu_1}(1-x)^{\mu_2}v(x), \mu_1, \mu_2 > -1, x \in (-1, 1)$. *Then the interpolant* $I_N^{\mu,a,b}$ *defined by* (3.2.306) *satisfies*

$$\|I_N^{\mu,a,b}u - u\|_{\omega^{(a-2\mu_2,b-2\mu_1)}} \le CN^{-r}\|\partial_x^r v\|_{\omega^{(a+r,b+r)}}, \tag{3.2.314}$$

where C is a positive constant independent of N, $v \in B^r_{\omega^{(a,b)}}(I)$, $I = (-1, 1)$, *and* $a - 2\mu_2, b - 2\mu_1 > -1$.

Proof We have

$$\begin{aligned}\|I_N^{\mu,a,b}u - u\|_{\omega^{(a-2\mu_2,b-2\mu_1)}} &= \|(1+x)^{\mu_1}(1-x)^{\mu_2}(I_N^{a,b}v-v)\|_{\omega^{(a-2\mu_2,b-2\mu_1)}}\\ &= \|\omega^{(\mu_2,\mu_1)}(I_N^{a,b}v - v)\|_{\omega^{(a-2\mu_2,b-2\mu_1)}} = \|I_N^{a,b}v - v\|_{\omega^{(a,b)}},\end{aligned}$$

so the desired estimate follows from Shen et al. (2011a, Theorem 3.41). □

If we choose $a = b = -\frac{1}{2}$, then we can obtain (see Eq. (5.5.28) in Canuto et al. (2006))

$$\begin{aligned}\|I_N^{\mu,-\frac{1}{2},-\frac{1}{2}}u - u\|_{L^1} &= \int_{-1}^{1}(1+x)^{\mu_1}(1-x)^{\mu_2}\left|I_N^{-\frac{1}{2},-\frac{1}{2}}v(x) - v(x)\right|\mathrm{d}x\\ &\le \|I_N^{-\frac{1}{2},-\frac{1}{2}}v - v\|_\infty \int_{-1}^{1}(1+x)^{\mu_1}(1-x)^{\mu_2}\mathrm{d}x\\ &\le CN^{\frac{1}{2}-r}\|\partial_x^r v\|_{\omega^{(-\frac{1}{2},-\frac{1}{2})}}.\end{aligned} \tag{3.2.315}$$

Next, we present an error bound of ${}^{\mathrm{RL}}_{-1}D_x^\alpha(I_N^{\mu,a,b}u - u)$ in the following theorem.

Theorem 3.77 *Let m be a positive integer and* $u(x) = (1+x)^b v(x)$, $b > -1, x \in I = (-1, 1)$. *If* $a + \alpha, b - \alpha > -1, \mu = (0, b)$, *then there exists a positive constant C independent of N, such that the interpolant* $I_N^{\mu,a,b}$ *defined by* (3.2.306) *satisfies*

$$\begin{aligned}\|{}^{\mathrm{RL}}_{-1}D_x^\alpha(I_N^{\mu,a,b}u - u)\|_{\omega^{(a+\alpha,\alpha-b)}} &\le CN^{\alpha-m}\|\partial_x^m v\|_{\omega^{(a+m,b+m)}}\\ &= CN^{\alpha-m}\|\partial_x^m(\omega^{(0,-b)}u)\|_{\omega^{(a+m,b+m)}},\end{aligned}$$

where $v = \omega^{(0,-b)}u \in B^m_{\omega^{(a,b)}}(I), m \ge \max\{a+1, b+1\}$.

Proof Define the $L^2_{\omega^{(a,b)}}$-orthogonal projector $\pi_N^{a,b} : L^2_{\omega^{(a,b)}}(I) \to \mathbb{P}_N$ such that (see Shen et al. (2011a))

$$(\pi_N^{a,b} v - v, w)_{\omega^{(a,b)}} = 0, \quad \forall w \in \mathbb{P}_N.$$

The projector $\pi_N^{a,b}$ can also be expressed by

$$(\pi_N^{a,b} v)(x) = \sum_{k=0}^{N} \hat{v}_n^{a,b} P_n^{(a,b)}(x), \quad \hat{v}_n^{a,b} = \frac{(v, P_n^{(a,b)})_{\omega^{(a,b)}}}{\| P_n^{(a,b)} \|^2_{\omega^{(a,b)}}} = \frac{(v, P_n^{(a,b)})_{\omega^{(a,b)}}}{\gamma_n^{(a,b)}}.$$

We also need the following notations and properties:

$$h_{n,k}^{(a,b)} = (d_{n,k}^{(a,b)})^2 \gamma_{n-k}^{(a+k,b+k)}, \quad n \geq k, \tag{3.2.316}$$

$$\frac{h_{n,k}^{(a,b)}}{h_{n,m}^{(a,b)}} = \frac{\Gamma(n+k+a+b+1)\Gamma(n-m+1)}{\Gamma(n+m+a+b+1)\Gamma(n-k+1)} = O(n^{2k-2m}), \tag{3.2.317}$$

$$C_1 \| P_n^{(\tilde{a},\tilde{b})} \|_{\omega^{(\tilde{a},\tilde{b})}} \leq \| P_n^{(a,b)} \|_{\omega^{(a,b)}} \leq C_2 \| P_n^{(\tilde{a},\tilde{b})} \|_{\omega^{(\tilde{a},\tilde{b})}}, \tag{3.2.318}$$

$$\| {}^{\mathrm{RL}}_{-1}D_x^{\alpha}(\omega^{(0,b)}\phi) \|_{\omega^{(a+\alpha,\alpha-b)}} \leq C N^{\alpha} \|\phi\|_{\omega^{(a,b)}}, \quad \phi \in \mathbb{P}_N. \tag{3.2.319}$$

Here C, C_1 and C_2 in (3.2.318) are positive constants independent of n, N, and $a + b = \tilde{a} + \tilde{b}$. Equations (3.2.316) and (3.2.317) can be found in Shen et al. (2011a), equation (3.2.318) can be easily deduced from (3.2.244) and the property

$$\frac{\Gamma(n+a)}{\Gamma(n+b)} = O(n^{a-b}).$$

We have $\phi(x) = \sum_{n=0}^{N} \hat{\phi}_n^{a,b} P_n^{(a,b)}(x)$ for $\phi \in \mathbb{P}_N$. Hence, equation (3.2.319) can be obtained by

$$\begin{aligned} \| {}^{\mathrm{RL}}_{-1}D_t^{\alpha} \omega^{0,b} \phi \|^2_{\omega^{a+\alpha,\alpha-b}} &= \sum_{n=0}^{N} \left| \hat{\phi}_n^{a,b} \right|^2 \frac{\Gamma^2(n+b+1)}{\Gamma^2(n+b+1-\alpha)} \gamma_n^{(a+\alpha,b-\alpha)} \\ &\leq C N^{2\alpha} \sum_{n=0}^{N} \left| \hat{\phi}_n^{a,b} \right|^2 \gamma_n^{(a,b)} = C N^{2\alpha} \|\phi\|^2_{\omega^{(a,b)}}, \end{aligned} \tag{3.2.320}$$

where we used the following relation; see, e.g., Zayernouri and Karniadakis (2014a).

$${}^{\mathrm{RL}}_{-1}D_t^{\alpha}(\omega^{(0,b)} P_n^{(a,b)})(x) = \frac{\Gamma(n+b+1)}{\Gamma(n+b+1-\alpha)} (1+x)^{b-\alpha} P_n^{(a+\alpha,b-\alpha)}.$$

With the help of equations (3.2.316)–(3.2.318), we can obtain

$$
\begin{aligned}
&\| {}_{-1}^{\mathrm{RL}}D_t^{\alpha}(\omega^{(0,b)}(\pi_N^{a,b}v - v))\|^2_{\omega^{a+\alpha,\alpha-b}} \\
&= \sum_{n=N+1}^{\infty} \left|\hat{v}_n^{a,b}\right|^2 \frac{\Gamma^2(n+b+1)}{\Gamma^2(n+b+1-\alpha)} \gamma_n^{(a+\alpha,b-\alpha)} \\
&\le C \sum_{n=N+1}^{\infty} \left|\hat{v}_n^{a,b}\right|^2 n^{2\alpha}\gamma_n^{(a,b)} = C \sum_{n=N+1}^{\infty} \left|\hat{v}_n^{a,b}\right|^2 h_{n,m}^{(a,b)} \frac{n^{2\alpha}h_{n,0}^{(a,b)}}{h_{n,m}^{(a,b)}} \\
&\le C \max_{n\ge N+1} \left\{ \frac{n^{2\alpha}h_{n,0}^{(a,b)}}{h_{n,m}^{(a,b)}} \right\} \sum_{n=N+1}^{\infty} \left|\hat{v}_n^{a,b}\right|^2 h_{n,m}^{(a,b)} \\
&\le C N^{2\alpha-2m} \|\partial_x^m v\|^2_{\omega^{(a+m,b+m)}}.
\end{aligned}
\tag{3.2.321}
$$

Applying the triangle inequality and equations (3.2.319)–(3.2.321) yields

$$
\begin{aligned}
&\| {}_{-1}^{\mathrm{RL}}D_t^{\alpha}(I_N^{a,b,\mu}u - u)\|_{\omega^{(a+\alpha,\alpha-b)}} = \| {}_{-1}^{\mathrm{RL}}D_t^{\alpha}(\omega^{(0,b)}(I_N^{a,b}v - v))\|_{\omega^{(a+\alpha,\alpha-b)}} \\
&\le \| {}_{-1}^{\mathrm{RL}}D_t^{\alpha}(\omega^{(0,b)}(\pi_N^{a,b}v - v))\|_{\omega^{(a+\alpha,\alpha-b)}} + \| {}_{-1}^{\mathrm{RL}}D_t^{\alpha}(\omega^{(0,b)} I_N^{a,b}(\pi_N^{a,b}v - v))\|_{\omega^{(a+\alpha,\alpha-b)}} \\
&\le C N^{\alpha-m} \|\partial_x^m v\|_{\omega^{(a+m,b+m)}} + C N^{\alpha} \|I_N^{a,b}(\pi_N^{a,b}v - v)\|_{\omega^{(a,b)}}.
\end{aligned}
\tag{3.2.322}
$$

Noting that

$$
I_N^{a,b}(\pi_N^{a,b}v - v) = (\pi_N^{a,b}v - v) + (v - I_N^{a,b}v),
$$

we have

$$
\begin{aligned}
\|I_N^{a,b}(\pi_N^{a,b}v - v)\|_{\omega^{(a,b)}} &\le \|\pi_N^{a,b}v - v\|_{\omega^{(a,b)}} + \|I_N^{a,b}v - v\|_{\omega^{(a,b)}} \\
&\le C N^{-m} \|\partial_x^m v\|_{\omega^{(a+m,b+m)}},
\end{aligned}
$$

where we used Theorem 3.35 in Shen et al. (2011a, p. 118) and Theorem 3.43 in Shen et al. (2011a, p. 137). From the above and equation (3.2.322), we obtain the desired result. □

Applications and Numerical Examples

In this section, we present two examples to illustrate the spectral collocation methods using the singular basis to solve a FBVP and a fractional Burgers equation.

Example 3.78 Consider the following FBVP

$$
\begin{cases}
-(p \; {}_{-1}^{\mathrm{RL}}D_x^{\alpha_1} + q \; {}_{x}^{\mathrm{RL}}D_1^{\alpha_2})U(x) + U(x) = f(x), & x \in (-1,1), \\
U(-1) = U(1) = 0,
\end{cases}
\tag{3.2.323}
$$

where $1 < \alpha_1, \alpha_2 \le 2$, p and q are non-negative constants, and $p + q \ne 0$.

The JGL spectral collocation method for (3.2.323) is to find

$$u_N^{\mu} = I_N^{a,b,\mu} u(x) = \omega^{(\mu_2,\mu_1)}(x) I_N^{a,b}(\omega^{(-\mu_2,-\mu_1)} u)(x)$$

such that for $1 \le j \le N-1$,

$$\begin{cases} -\left[p \; {}_{-1}^{\mathrm{RL}}D_x^{\alpha_1} u_N^{\mu}(x) + q \; {}_{x}^{\mathrm{RL}}D_1^{\alpha_2} u_N^{\mu}(x)\right]_{x=x_j} + u_N^{\mu}(x_j) = f(x_j), \\ u_0 = u_N = 0. \end{cases} \tag{3.2.324}$$

- Case I (solution with end-point singularities): Let $p = q = 1$, and choose a suitable source term $f(x)$ and fractional orders $\alpha_1 = \frac{5+4x}{11} + 1$, $\alpha_2 = \frac{7-x}{9} + 1$, such that the analytical solution to (3.2.323) is
$$U(x) = (1+x)^{\sigma_1}(1-x)^{\sigma_2}, \quad \sigma_1, \sigma_2 > 0.$$
- Case II (unknown solution with end-point singularities): Let $\alpha_1 = \alpha_2 = \alpha$ and choose a smooth source term $f(x) = 1 + \sin(5x)$.
- Case III (comparison with the finite element method (FEM)): Consider the model problem (see Wang et al. (2016)) ${}_{0}^{\mathrm{RL}}D_x^{\alpha} U(x) = f(x), x \in (0,1), 1 < \alpha \le 2$, subject to the boundary conditions $U(0) = 0, U(1) = 2$. Choose a suitable source term $f(x)$ such that the analytical solution is $U(x) = x^{\alpha-1} + x^6$.
- Case IV (comparison with the Petrov–Galerkin FEM (PGFEM)): Consider the model problem (see Jin et al. (2016b)) ${}_{0}^{\mathrm{RL}}D_x^{\alpha} U(x) + u'(x) + x(1-x)u(x) = f(x), x \in (0,1), 1 < \alpha \le 2$, subject to the boundary conditions $U(0) = U(1) = 0$.

In this example, the absolute errors are measured in the following sense

$$\|e\|_\infty = \max_{1 \le j \le K-1} \left| u_{\mathrm{ref}}(z_j) - \tilde{u}(z_j) \right|, \quad z_j = -1 + j/K, \quad K = 512,$$

where

$$\tilde{u}(x) = (1+x)^{\mu_1}(1-x)^{\mu_2} \sum_{j=0}^{N} v_j l_j(x), \quad v_j = (1+x_j)^{-\mu_1}(1-x_j)^{-\mu_2} u_j,$$

$l_j(x)$ is the Lagrange basis function based on JGL points (see (3.2.307)) and u_{ref} is either the exact solution or a numerical solution obtained from the JGL collocation method (3.2.324) with $N = 256$.

Next, we apply the collocation method (3.2.324) to solve (3.2.323).

For Case I, we have the exact solution that satisfies $U(x) = (1+x)^{\mu_1}(1-x)^{\mu_2} v(x)$, where $v(x) = (1+x)^{-\mu_1+\sigma_1}(1-x)^{-\mu_2+\sigma_2}$. From Theorem 3.76, we know that $v(x)$ may have much better regularity than $U(x)$, i.e., we can

Table 3.15. *Maximum errors for Example 3.78, Case I, $a = b = 0$, $(\sigma_1, \sigma_2) = (3.5, 3.5)$. Source: Zeng et al. (2017).* *Copyright ©2017 Society for Industrial and Applied Mathematics. Reprinted with permission. All rights reserved.*

μ	$N = 7$	$N = 8$	$N = 16$	$N = 32$	$N = 64$
$(0.5, 0.5)$	1.1102e-14	2.1094e-15	7.7716e-15	6.6613e-15	8.9928e-15
$(0.5, -0.5)$	1.0436e-14	2.2204e-15	9.4369e-15	1.1909e-14	3.9095e-14
$(-0.5, 0.5)$	1.1657e-14	2.8866e-15	1.6104e-14	7.0488e-15	1.3542e-14
$(-0.5, -0.5)$	5.2096e-02	7.3513e-15	3.3401e-14	4.9803e-13	2.0286e-13
$(0, 0)$	1.1899e-03	3.2841e-04	2.3615e-06	1.9635e-08	1.6101e-10

Table 3.16. *Maximum errors for Example 3.78, Case I, $a = b = -1/2$, $(\sigma_1, \sigma_2) = (3.6, 3.4)$. Source: Zeng et al. (2017).* *Copyright ©2017 Society for Industrial and Applied Mathematics. Reprinted with permission. All rights reserved.*

μ	$N = 7$	$N = 8$	$N = 16$	$N = 32$	$N = 64$
$(0.6, 0.4)$	6.2172e-15	2.0095e-14	1.1657e-14	7.8826e-15	9.1038e-15
$(0.6, -0.6)$	7.8826e-15	2.0872e-14	9.1038e-15	1.0614e-14	2.9792e-14
$(-0.6, 0.4)$	4.8850e-15	1.9318e-14	1.2657e-14	5.7732e-15	5.7732e-15
$(-0.6, -0.6)$	2.7801e-02	3.1743e-14	7.8005e-14	2.1217e-13	1.4946e-13
$(0, 0)$	2.1440e-03	2.7482e-04	1.3038e-06	1.1318e-08	1.0634e-10

choose suitable $\mu = (\mu_1, \mu_2)$ such that both $-\mu_1 + \sigma_1$ and $-\mu_2 + \sigma_2$ are non-negative integers. For Case I, we really have several optimal choices of μ, which is given as $\mu = (\sigma_1 - \lfloor \sigma_1 \rfloor, \sigma_2 - \lfloor \sigma_2 \rfloor)$, $(\sigma_1 - \lfloor \sigma_1 \rfloor, \sigma_2 - \lceil \sigma_2 \rceil)$, $(\sigma_1 - \lceil \sigma_1 \rceil, \sigma_2 - \lfloor \sigma_2 \rfloor)$, or $(\sigma_1 - \lceil \sigma_1 \rceil, \sigma_2 - \lceil \sigma_2 \rceil)$. In Tables 3.15–3.16, we display the maximum errors $\|e\|_\infty$ of the standard collocation method (see $\mu = (0, 0)$ in Tables 3.15–3.16) and our collocation method (3.2.324) with four optimal choices of μ. We can see that spectral accuracy is observed for the optimal choices of μ, while spectral accuracy is lost for the standard collocation method by choosing $\mu = (0, 0)$ (this can be explained from Theorem 3.77). We can also observe that very high accuracy is achieved when μ is close to the optimal choices of μ, see Table 3.17. This motivates us to choose suitable μ that is near the optimal choice in real computations when we do not know the regularity of the analytical solution.

For Case II, we do not have analytical solutions and the regularity of the analytical solutions is unknown. The singularity may depend on the fractional orders α_1 and α_2, the coefficients p and q, and the source term $f(x)$.

Table 3.17. *Maximum errors for Example 3.78, Case I,* $a = b = 0$, $(\sigma_1, \sigma_2) = (3.5, 3.5)$, $\mu = (0.5 + \epsilon, 0.5 + \epsilon)$. *Source: Zeng et al. (2017).* *Copyright ©2017 Society for Industrial and Applied Mathematics. Reprinted with permission. All rights reserved.*

ϵ	$N = 6$	$N = 7$	$N = 8$	$N = 16$	$N = 32$	$N = 64$
10^{-1}	5.7345e-04	2.3948e-04	4.5583e-05	3.6062e-07	3.1855e-09	2.6740e-11
10^{-2}	6.3999e-05	2.5580e-05	4.5463e-06	3.6445e-08	3.1788e-10	2.6468e-12
10^{-3}	6.4630e-06	2.5708e-06	4.5380e-07	3.6363e-09	3.1670e-11	2.6398e-13
10^{-4}	6.4692e-07	2.5720e-07	4.5391e-08	3.6354e-10	3.1657e-12	2.6118e-14
10^{-5}	6.4699e-08	2.5721e-08	4.5392e-09	3.6353e-11	3.1613e-13	5.2736e-15
10^{-6}	6.4699e-09	2.5722e-09	4.5392e-10	3.6358e-12	3.1465e-14	1.5821e-14
-10^{-1}	7.1004e-04	2.6537e-04	4.5074e-05	3.3751e-07	2.8811e-09	2.3846e-11
-10^{-2}	6.5389e-05	2.5854e-05	4.5498e-06	3.6231e-08	3.1484e-10	6.8523e-11
-10^{-3}	6.4769e-06	2.5735e-06	4.5404e-07	3.6342e-09	3.1641e-11	2.6354e-13
-10^{-4}	6.4706e-07	2.5723e-07	4.5393e-08	3.6351e-10	3.1654e-12	2.6244e-14
-10^{-5}	6.4700e-08	2.5722e-08	4.5392e-09	3.6351e-11	3.1688e-13	1.2990e-14
-10^{-6}	6.4699e-09	2.5722e-09	4.5392e-10	3.6347e-12	3.1647e-14	5.8287e-15
0	1.2212e-15	1.1102e-14	2.1094e-15	7.7716e-15	6.6613e-15	8.9928e-15

Specifically speaking, we have singularity at the left (or right) end-point when $p \neq 0, q = 0$ (or $p = 0, q \neq 0$) and the regularity is low when α is small, see e.g., Diethelm (2010), or have singularities at both end-points when $pq \neq 0$.

We first choose $p = q = 1$ and relatively small values of fractional order $\alpha = 1.05, 1.1$ in the computation for Case II; numerical solutions are shown in Figure 3.15. We observe that satisfactory numerical solutions are obtained for different choices of μ when $p = q$. This may be explained from the symmetry of the Riesz fractional operator that leads to the partial cancellation of the singularity of the solutions, so that the regularity of the analytical solution is moderately good and leads to satisfactory numerical solutions.

Next, we let p/q be sufficiently large (or small) that leads to stronger singularity at the left (or right) end-point. We still choose a small fractional order $\alpha = 1.05$ in the computation, and numerical solutions are shown in Figure 3.16. We find that wiggles appear with the standard collocation method ($\mu = (0, 0)$), due to the low regularity of the analytical solutions, while wiggles near the left boundary disappear when $\mu_2 = 0, \mu_1$ decreases and p/q is sufficiently large, see Figures 3.16 (a) and (c). A similar phenomenon is observed near the right boundary when $\mu_1 = 0, \mu_2$ decreases and q/p is sufficiently large, see Figures 3.16 (b) and (d). We also observe similar results

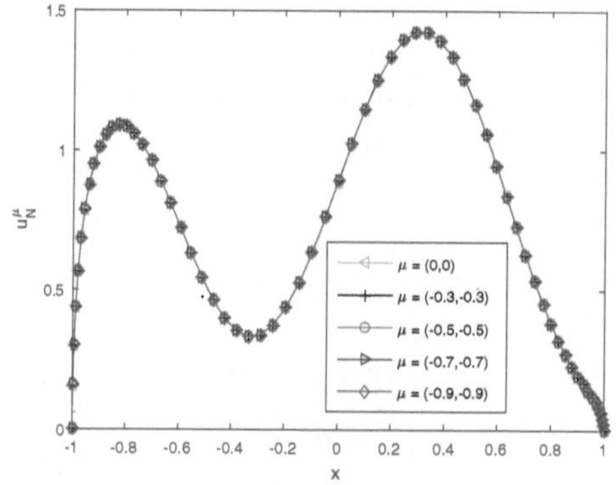

(a) $\alpha = 1.05$.

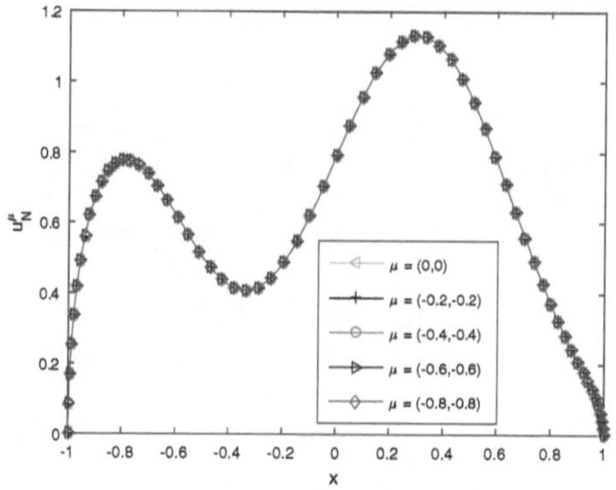

(b) $\alpha = 1.1$.

Figure 3.15 Numerical solutions for Example 3.78, Case II, $a = b = 0$, $p = q = 1$, $N = 128$, $\alpha_1 = \alpha_2 = \alpha$. Source: Zeng et al. (2017). Copyright ©2017 Society for Industrial and Applied Mathematics. Reprinted with permission. All rights reserved.

for $\alpha = 1.1$ in Figure 3.17. Theoretically, we have $U(x) = \sum_{i=0,j=1}^{\infty} c_{i,j}(1 \pm x)^{i+j(\alpha-1)}$ if $q = 0$ (or $p = 0$) and $f(x)$ is smooth, which leads to the optimal choice of $\mu = (\alpha-2, 0)$ (or $\mu = (0, \alpha-2)$), see numerical solutions in Figures 3.16 (a)–(b) and Figures 3.17 (a)–(b).

In Table 3.18, we present the maximum errors for Case II with $\mu = (\mu_1, 0)$, $p = 1, q = 0$. In such a case, we know the regularity of the analytical solutions (see Diethelm (2010)) and we have the optimal choice of $\mu = (\alpha-2, 0)$, which is used to get the benchmark solutions with $N = 256$. We can see that better numerical solutions are obtained as μ_1 approaches $\alpha - 2$.

Next, we compare our collocation method with the indirect finite element method (IFEM) developed in Wang et al. (2016) and the PGFEM developed in Jin et al. (2016b).

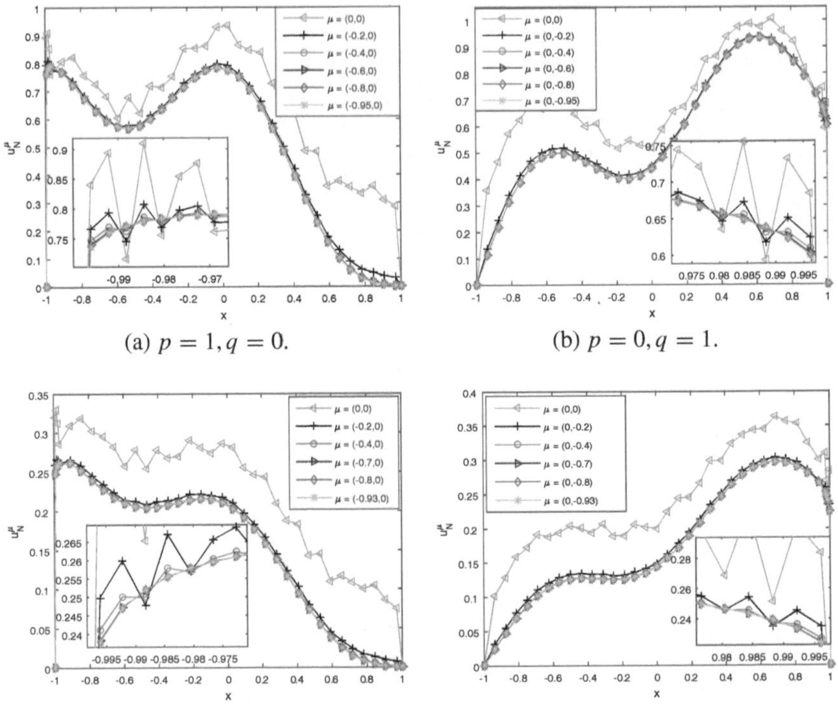

(a) $p = 1, q = 0$.

(b) $p = 0, q = 1$.

(c) $p = 6, q = 1$.

(d) $p = 1, q = 6$.

Figure 3.16 Numerical solutions for Example 3.78, $\alpha_1 = \alpha_2 = \alpha = 1.05$, Case II, $a = b = 0$, $N = 128$. Source: Zeng et al. (2017). Copyright ©2017 Society for Industrial and Applied Mathematics. Reprinted with permission. All rights reserved.

For Case III, we present the L^2 errors of our collocation method and the IFEM in Table 3.19, which shows that the present collocation method displays much better accuracy. Since the analytical solution is known, the optimal $\mu = (\alpha - 2, 0)$ was chosen in the computation that leads to much better results of the collocation method than those from the IFEM in Wang et al. (2016).

For Case IV, we show the L^2 errors of our collocation method and the PGFEM in Tables 3.19–3.20. According to Jin et al. (2016b); Kopteva and Stynes (2017) (see Theorem 4.8 and Section 5 in Jin et al. (2016b), and Eq. (3.1) and Remark in Kopteva and Stynes (2017)), the analytical solution contains the low order term $x^{\alpha-1}$. Hence, we can also choose $\mu = (\alpha - 2, 0)$ in the collocation method, which really shows much better numerical results than those from the PGFEM (Jin et al., 2016b) in this example.

Table 3.18. *Maximum errors for Example 3.78, Case II, $p = 1, q = 0$, $a = b = 0$, $N = 128$. The benchmark solutions are derived with $N = 256$ and $\mu = (\alpha - 2, 0)$. Source: Zeng et al. (2017).* *Copyright ©2017 Society for Industrial and Applied Mathematics. Reprinted with permission. All rights reserved.*

$\alpha = 1.05$		$\alpha = 1.1$	
μ	Error	μ	Error
$(0,0)$	5.4123e-01	$(0,0)$	2.0413e-01
$(-0.20,0)$	5.8088e-02	$(-0.20,0)$	3.6589e-02
$(-0.40,0)$	2.5741e-02	$(-0.40,0)$	1.6761e-02
$(-0.70,0)$	5.8591e-03	$(-0.70,0)$	3.5715e-03
$(-0.80,0)$	2.8077e-03	$(-0.80,0)$	1.4869e-03
$(-0.90,0)$	7.5340e-04	$(-0.89,0)$	1.2835e-04
$(-0.94,0)$	1.3865e-04	$(-0.91,0)$	1.2428e-04

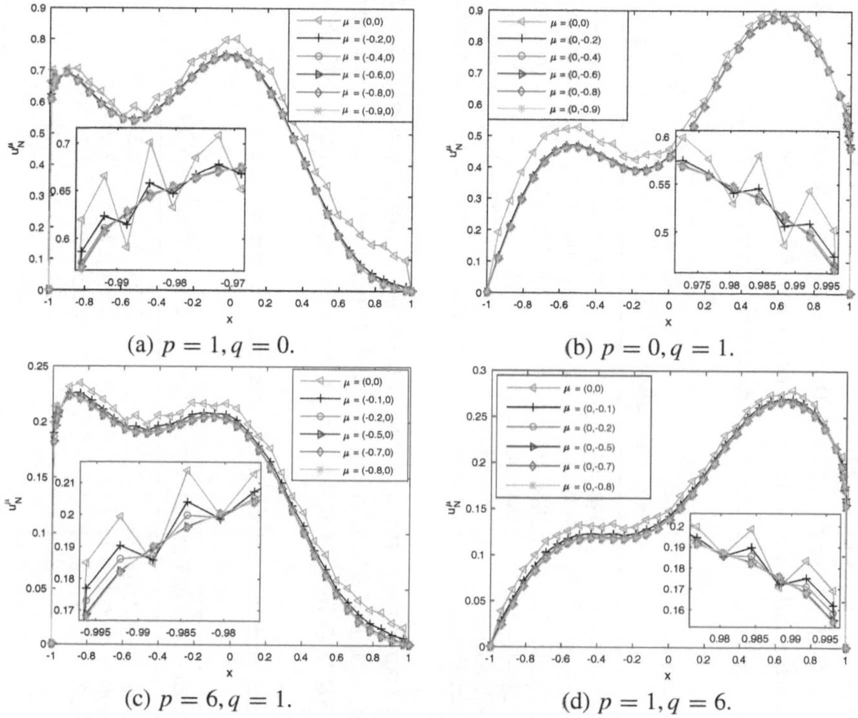

(a) $p = 1, q = 0$. (b) $p = 0, q = 1$.

(c) $p = 6, q = 1$. (d) $p = 1, q = 6$.

Figure 3.17 Numerical solutions for Example 3.78, Case II, $\alpha = 1.1$, $a = b = 0$, $N = 128$. Source: Zeng et al. (2017). Copyright ©2017 Society for Industrial and Applied Mathematics. Reprinted with permission. All rights reserved.

Table 3.19. *Comparison of the L^2 errors between the present collocation method and the IFEM (see $m = 4$ in Tables 1–3 in Wang et al. (2016)) for Case III, $a = b = 0$, $\mu = (\alpha - 2, 0)$. Source: Zeng et al. (2017).* *Copyright ©2017 Society for Industrial and Applied Mathematics. Reprinted with permission. All rights reserved.*

N	$\alpha = 1.1$		$\alpha = 1.5$		$\alpha = 1.9$	
	Collocation	IFEM (Wang et al., 2016)	Collocation	IFEM (Wang et al., 2016)	Collocation	IFEM (Wang et al., 2016)
8	3.621e-07	1.038e-03	2.382e-07	1.369e-04	4.085e-09	1.810e-05
16	8.465e-12	1.308e-04	1.000e-11	1.281e-05	2.026e-13	1.221e-06
32	4.735e-14	1.585e-05	3.774e-14	1.164e-06	5.305e-14	8.201e-08
64	1.222e-13	1.884e-06	9.185e-14	1.043e-07	6.025e-14	5.503e-09
128	3.493e-13	2.219e-07	3.010e-13	9.387e-09	2.733e-13	4.194e-10

Table 3.20. *Comparison of the L^2 errors between our collocation method and the PGFEM (see Table 3 in Jin et al. (2016b)) for Case IV, $f(x) = x$, $a = b = 0$, $\mu = (\alpha - 2, 0)$. The reference solutions are obtained with $N = 640$ and $\mu = (\alpha - 2, 0)$. Source: Zeng et al. (2017).* *Copyright ©2017 Society for Industrial and Applied Mathematics. Reprinted with permission. All rights reserved.*

N	$\alpha = 1.6$		$\alpha = 1.75$		$\alpha = 1.9$	
	Collocation	PGFEM (Jin et al., 2016b)	Collocation	PGFEM (Jin et al., 2016b)	Collocation	PGFEM (Jin et al., 2016b)
10	2.58e-05	2.67e-03	5.75e-06	1.23e-03	8.15e-07	7.49e-04
20	2.75e-06	9.41e-04	5.24e-07	3.69e-04	6.21e-08	1.92e-04
40	3.12e-07	3.89e-04	4.82e-08	1.28e-04	4.64e-09	5.05e-05
80	3.51e-08	1.74e-04	4.39e-09	4.92e-05	3.42e-10	1.40e-05
160	3.74e-09	8.01e-05	3.86e-10	2.00e-05	2.46e-11	4.20e-06

In summary, we have the following conclusion and observation from numerical solutions of (3.78):

(1) Given a smooth input with $q = 0, p \neq 0$ (or $p = 0, q \neq 0$), the analytical solution to (3.2.323) has singularity at the left (or right) boundary, see e.g., (Diethelm, 2010, p. 130–131). Specifically, the solution has very low regularity when α_1, α_2 tend to 1, see also numerical results in Figures 3.16 (a)–(b) and Figures 3.17 (a)–(b).
(2) If $p \gg q$ (or $p \ll q$), then the solution has stronger singularity at the left (or right) boundary than at the right (or left) boundary, especially when the fractional orders α_1, α_2 tend to 1, see Figures 3.16 (c)–(d) and Figures 3.17 (c)–(d).
(3) When $p = q \neq 0$, the solution has singularity at both end-points. However, the singularity is not very strong due to the symmetry of the Riesz fractional operator when $\alpha_1 = \alpha_2$, see Figure 3.15 and also related numerical results in Mao and Shen (2016).

Discussion on the Generalized Tunable Collocation Methods

We presented spectral collocation methods using the singular basis functions of the GJPFs $(1+x)^{\mu_1}(1-x)^{\mu_2}P_j^{(a,b)}(x)$ to solve FBVPs and FPDEs with singularities at both end-points.

In order to construct the spectral collocation method, we develop the three-term recurrence formula to efficiently calculate the left and right fractional derivatives of the GJPFs, which leads to the corresponding differentiation matrices. In some special cases, the three-term recurrence formula can be reduced to the orthogonal Jacobi polynomials (or functions), see e.g., Mao et al. (2016); Zayernouri and Karniadakis (2014a); Zeng et al. (2015). The present three-term recurrence formula greatly simplifies the calculation of the fractional derivative of the GJPFs in Li et al. (2012); Zeng et al. (2015). We also provide some theoretical results (see Subsection 3.2.6) to explain why the present collocation methods can achieve high accuracy.

We presented extensive numerical experiments to illustrate the accuracy of our collocation method, which shows that the present collocation method achieves high accurate numerical solutions for nonsmooth solutions, even if the regularity of the analytical solutions is low and unknown. Furthermore, we compare the present collocation methods with the known methods such as the FEM (see Wang et al. (2016)) and the Petrov–Galerkin FEM (see Jin et al. (2016b)), and find that the present method achieves much better numerical solutions, see Tables 3.19–3.21.

Table 3.21. *Comparison of the L^2 errors between the collocation method and the PGFEM (see Table 6 in Jin et al. (2016b)) for Case IV, $f(x) = 1$, $a = b = 0$, $\mu = (\alpha - 2, 0)$. The reference solutions are obtained with $N = 640$ and $\mu = (\alpha - 2, 0)$. Source: Zeng et al. (2017). Copyright ©2017 Society for Industrial and Applied Mathematics. Reprinted with permission. All rights reserved.*

N	$\alpha = 1.6$		$\alpha = 1.75$		$\alpha = 1.9$	
	Collocation	PGFEM (Jin et al., 2016b)	Collocation	PGFEM (Jin et al., 2016b)	Collocation	PGFEM (Jin et al., 2016b)
10	8.61e-05	6.87e-03	1.94e-05	2.93e-03	2.80e-06	1.37e-03
20	9.05e-06	2.89e-03	1.74e-06	1.05e-03	2.10e-07	3.82e-04
40	1.02e-06	1.28e-03	1.59e-07	4.04e-04	1.56e-08	1.12e-04
80	1.15e-07	5.78e-04	1.45e-08	1.62e-04	1.15e-09	3.54e-05
160	1.23e-08	2.65e-04	1.28e-09	6.65e-05	8.30e-11	1.19e-05

3.3 Petrov–Galerkin Spectral Element Methods for FODEs

We present a C^0-continuous Petrov–Galerkin spectral element method for one-dimensional fractional elliptic problems of the form ${}^{\mathrm{RL}}_{0}D^{\alpha}_{x}u(x) - \lambda u(x) = f(x)$, $\alpha \in (1,2]$, subject to homogeneous boundary conditions. We employ the standard (modal) spectral element basis and the Jacobi poly-fractonomials as the test functions (Zayernouri and Karniadakis, 2013). We formulate a new procedure for assembling the global linear system from elemental (local) mass and stiffness matrices. The Petrov–Galerkin formulation requires performing elemental (local) construction of mass and stiffness matrices in the standard domain only once. Moreover, we efficiently obtain the nonlocal (history) stiffness matrices, in which the nonlocality is presented analytically for uniform grids. We also investigate two distinct choices of basis/test functions: i) local basis/test functions, and ii) local basis with global test functions. We show that the former choice leads to a better-conditioned system and accuracy, while the latter results in ill-conditioned linear systems, and therefore, that is not an efficient and a proper choice of test function. We consider smooth and singular solutions, where the singularity can occur at boundary points as well as in the interior domain. We also construct two non-uniform grids over the whole computational domain in order to capture singular solutions. Finally, we perform a systematic numerical study of nonlocal effects via full and partial history fading in order to further enhance the efficiency of the scheme.

Problem Definition

We study the following fractional Helmholtz problem of order $\alpha = 1 + \mu$, $\mu \in (0, 1]$:

$$
{}^{\mathrm{RL}}_{0}D^{\alpha}_{x}u(x) - \lambda u(x) = f(x), \quad \forall x \in \Omega, \tag{3.3.1}
$$

$$
u(0) = u(L) = 0, \tag{3.3.2}
$$

where $\Omega = [0, L]$. We multiply both sides of (3.3.1) by some proper test function $v(x)$ and transfer the fractional portion, μ, of the derivative onto the test function by taking the fractional integration-by-parts, following the proof of Lemma 2.4 in Kharazmi and Zayernouri (2018). Then we obtain the following bilinear form:

$$
a(u, v) = l(v), \tag{3.3.3}
$$

in which

$$
a(u, v) = \left(\frac{\mathrm{d}u}{\mathrm{d}x}, {}^{\mathrm{RL}}_{x}D^{\mu}_{L}\, v\right)_{\Omega} - \lambda(u\,,\, v)_{\Omega}, \tag{3.3.4}
$$

$$
l(v) = (f\,,\, v)_{\Omega}, \tag{3.3.5}
$$

where $(\cdot\,,\,\cdot)_{\Omega}$ denotes the usual L^2 inner product.

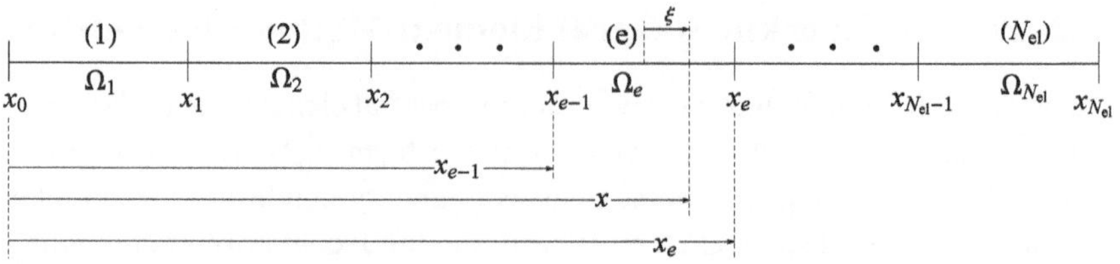

Figure 3.18 Domain partitioning. Source: Reprinted from Kharazmi et al. (2017) with permission from Elsevier.

Local Basis Functions

We partition the computational domain into N_{el} non-overlapping elements $\Omega_e = [x_{e-1}, x_e]$ such that $\Omega = \cup_{e=1}^{Nel} \Omega_e$, see Figure 3.18. Therefore, the bilinear form (3.3.4) can be written as

$$a(u, v) \approx a(u^\delta, v^\delta) = \sum_{e=1}^{N_{el}} \left(\frac{du_N^{(e)}}{dx}, {}^{\mathrm{RL}}_{\ x}D_L^{\mu} v^\delta \right)_{\Omega_e} - \lambda \sum_{e=1}^{N_{el}} \left(u_N^{(e)}, v^\delta \right)_{\Omega_e}, \tag{3.3.6}$$

where $u_N^{(e)}$ is the approximation solution in each element, which is given by

$$u_N^{(e)}(x) = \sum_{p=0}^{P} \hat{u}_p^{(e)} \psi_p(x), \quad x \in \Omega_e, \tag{3.3.7}$$

using the basis functions $\psi_p(x)$ in the element. Thus the approximated solution over the whole domain can be written as

$$u \approx u^\delta(x) = \sum_{e=1}^{N_{el}} \sum_{p=0}^{P} \hat{u}_p^{(e)} \psi_p(x). \tag{3.3.8}$$

The modal bases $\{\psi_p(x)\}$ are defined in the standard (reference) domain $(-1,1)$ as:

$$\psi_p(x(\zeta)) = \begin{cases} \dfrac{1-\zeta}{2}, & p = 0, \\ \left(\dfrac{1-\zeta}{2}\right)\left(\dfrac{1+\zeta}{2}\right) P_{p-1}^{(1,1)}(\zeta), & p = 1, 2, \ldots, P-1, \\ \dfrac{1+\zeta}{2}, & p = P. \end{cases} \tag{3.3.9}$$

The choice of basis functions is the same as in standard spectral element methods for integer-order PDEs (see e.g., Karniadakis et al. (2005)).

Test Functions: Local vs. Global

We choose two types of test functions v^{δ}: i) *local* test functions, and ii) *global* test functions, given for $\varepsilon = 1, 2, \ldots, N_{el}$ as follows:

$$v_k^{local}(x) = v_k^{\varepsilon}(x) = \begin{cases} {}^{(2)}\mathcal{P}_{k+1}^{\mu}(x^{\varepsilon}), & \forall x \in \Omega_{\varepsilon}, \\ 0, & \text{otherwise} \end{cases} \tag{3.3.10}$$

for $k = 0, 1, \ldots, P$, in which ${}^{(2)}\mathcal{P}_{k+1}^{\mu}(x^{\varepsilon})$ represents the Jacobi *poly-fractonomial* of the second kind, defined in the corresponding intervals $\Omega_{\varepsilon} = [x_{\varepsilon-1}, x_{\varepsilon}]$; and

$$v_k^{global}(x) = v_k^{\varepsilon}(x) = \begin{cases} {}^{(2)}\mathcal{P}_{k+1}^{\mu}(x^{1\sim\varepsilon}), & \forall x \in [0, x_{\varepsilon}], \\ 0, & \text{otherwise} \end{cases} \tag{3.3.11}$$

for $k = 0, 1, \ldots, P$, where ${}^{(2)}\mathcal{P}_{k+1}^{\mu}(x^{1\sim\varepsilon})$ represents the Jacobi *poly-fractonomial* of the second kind, defined in the corresponding intervals $[0, x_{\varepsilon}]$. The structure of Jacobi poly-fractonomials is given in Chapter 2. It should be noted that for each element ε, the corresponding local test function has nonzero value only in the element and vanishes elsewhere, unlike the corresponding global test function, which vanishes only where $x > x_{\varepsilon}$.

3.3.1 Petrov–Galerkin Method with Local Test Functions

We develop the Petrov–Galerkin scheme by substituting (3.3.7) and (3.3.10) into (3.3.6) to obtain

$$\sum_{e=1}^{N_{el}} \Big(\sum_{p=0}^{P} \hat{u}_p^{(e)} \frac{d\psi_p(x)}{dx}, {}^{\mathrm{RL}}_{\ x}D_L^{\mu}\, v_k^{\varepsilon}(x) \Big)_{\Omega_e} - \lambda \sum_{e=1}^{N_{el}} \Big(\sum_{p=0}^{P} \hat{u}_p^{(e)}\, \psi_p(x),\ v_k^{\varepsilon}(x) \Big)_{\Omega_e}$$
$$= \Big(f,\ v_k^{\varepsilon}(x) \Big)_{\Omega}, \quad \varepsilon = 1, 2, \ldots, N_{el}, \quad k = 0, 1, \ldots, P. \tag{3.3.12}$$

Since the local test function vanishes $\forall x \in \Omega_e \neq \Omega_{\varepsilon}$, we have

$$\lambda \sum_{e=1}^{N_{el}} \Big(\sum_{p=0}^{P} \hat{u}_p^{(e)} \psi_p(x),\ v_k^{\varepsilon}(x) \Big)_{\Omega_e} = \lambda \Big(\sum_{p=0}^{P} \hat{u}_p^{(\varepsilon)} \psi_p(x),\ v_k^{\varepsilon}(x) \Big)_{\Omega_{\varepsilon}},$$
$$\Big(f,\ v_k^{\varepsilon}(x) \Big)_{\Omega} = \Big(f,\ v_k^{\varepsilon}(x) \Big)_{\Omega_{\varepsilon}}.$$

Moreover, for every ε, the right-sided fractional derivative,

$${}^{\mathrm{RL}}_{\ x}D_L^{\mu}\, v_k^{\varepsilon}(x) = \frac{-1}{\Gamma(1-\mu)} \frac{d}{dx} \int_x^L \frac{v_k^{\varepsilon}(s)}{(s-x)^{\mu}} ds, \quad x \in \Omega_e,$$

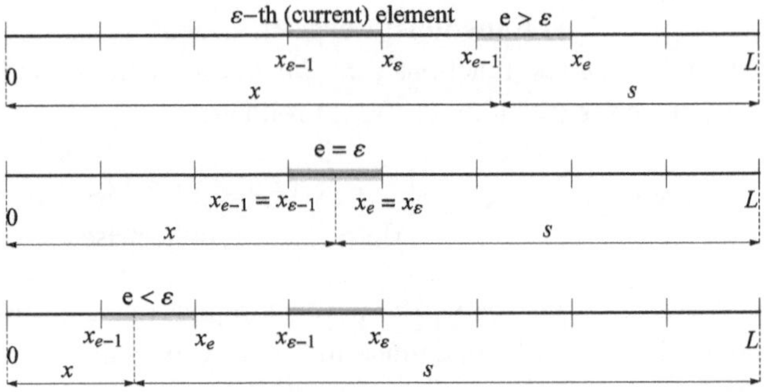

Figure 3.19 Location of the (dummy) element number, e, with respect to the current element, ε. If $e > \varepsilon$, (top), then ${}^{\text{RL}}_{\;x}D^{\mu}_{L}\, v^{\varepsilon}_{k}(x) = 0$. If $e = \varepsilon$, (middle), then ${}^{\text{RL}}_{\;x}D^{\mu}_{L}\, v^{\varepsilon}_{k}(x) = {}^{\text{RL}}_{\;x}D^{\mu}_{x_\varepsilon}\left[{}^{(2)}\mathcal{P}^{\mu}_{k+1}(x)\right]$. If $e < \varepsilon$, (bottom), then ${}^{\text{RL}}_{\;x}D^{\mu}_{L}\, v^{\varepsilon}_{k}(x) = H^{(\varepsilon)}_{k}(x)$. Source: Reprinted from Kharazmi et al. (2017) with permission from Elsevier.

is taken from $x \in \Omega_e$ to $x = L$, where $e = 1, 2, \ldots, N_{el}$ through the summation over the elements and s varies from $x \in \Omega_e$ to L. The local test function vanishes $\forall x \in \Omega_e \neq \Omega_\varepsilon$, thus if $e > \varepsilon$ ($x > x_\varepsilon$, see Figure 3.19 top), then

$$ {}^{\text{RL}}_{\;x}D^{\mu}_{L}\, v^{\varepsilon}_{k}(x) = \frac{-1}{\Gamma(1-\mu)} \frac{\mathrm{d}}{\mathrm{d}x} \int_{x}^{L} \frac{0}{(s-x)^{\mu}} \mathrm{d}s = 0, \tag{3.3.13} $$

and if $e < \varepsilon$ ($x < x_{\varepsilon-1}$, see Figure 3.19 bottom), then

$$ {}^{\text{RL}}_{\;x}D^{\mu}_{L}\, v^{\varepsilon}_{k}(x) = \frac{-1}{\Gamma(1-\mu)} \frac{\mathrm{d}}{\mathrm{d}x} \int_{x_{\varepsilon-1}}^{x_\varepsilon} \frac{{}^{(2)}\mathcal{P}^{\mu}_{k+1}(s)}{(s-x)^{\mu}} \mathrm{d}s \equiv H^{(\varepsilon)}_{k}(x), \tag{3.3.14} $$

and if $e = \varepsilon$, ($x_{\varepsilon-1} < x < x_\varepsilon$, see Figure 3.19 middle), then

$$ {}^{\text{RL}}_{\;x}D^{\mu}_{L}\, v^{\varepsilon}_{k}(x) = \frac{-1}{\Gamma(1-\mu)} \frac{\mathrm{d}}{\mathrm{d}x} \int_{x}^{x_\varepsilon} \frac{{}^{(2)}\mathcal{P}^{\mu}_{k+1}(s)}{(s-x)^{\mu}} \mathrm{d}s = {}^{\text{RL}}_{\;x}D^{\mu}_{x_\varepsilon}\left[{}^{(2)}\mathcal{P}^{\mu}_{k+1}(x)\right]. \tag{3.3.15} $$

Hence, for $\varepsilon = 1, 2, \ldots, N_{el}$ and $k = 0, 1, \ldots, P$,

$$ {}^{\text{RL}}_{\;x}D^{\mu}_{L}\, v^{\varepsilon}_{k}(x) = \begin{cases} 0, & \forall x \in \Omega_e,\ e > \varepsilon, \\ {}^{\text{RL}}_{\;x}D^{\mu}_{x_\varepsilon}\left[{}^{(2)}\mathcal{P}^{\mu}_{k+1}(x)\right], & \forall x \in \Omega_e,\ e = \varepsilon, \\ H^{(\varepsilon)}_{k}(x), & \forall x \in \Omega_e,\ e < \varepsilon. \end{cases} \tag{3.3.16} $$

Therefore, the bilinear form (3.3.12) can be written as

$$\sum_{e=1}^{\varepsilon-1}\sum_{p=0}^{P}\hat{u}_p^{(e)}\left(\frac{d\psi_p(x)}{dx},H_k^{(\varepsilon)}(x)\right)_{\Omega_e}+\sum_{p=0}^{P}\hat{u}_p^{(\varepsilon)}\left(\frac{d\psi_p(x)}{dx},{}^{\mathrm{RL}}_{\ x}\mathcal{D}^{\mu}_{x_\varepsilon}\left[{}^{(2)}\mathcal{P}^{\mu}_{k+1}(x)\right]\right)_{\Omega_\varepsilon}$$

$$-\lambda\sum_{p=0}^{P}\hat{u}_p^{(\varepsilon)}\left(\psi_p(x),\,{}^{(2)}\mathcal{P}^{\mu}_{k+1}(x)\right)_{\Omega_\varepsilon}=\left(f,\,{}^{(2)}\mathcal{P}^{\mu}_{k+1}(x)\right)_{\Omega_\varepsilon},\tag{3.3.17}$$

and the weak form is obtained as

$$\sum_{e=1}^{\varepsilon-1}\sum_{p=0}^{P}\hat{u}_p^{(e)}\,\hat{\mathbf{S}}_{kp}^{(e,\varepsilon)}+\sum_{p=0}^{P}\hat{u}_p^{(\varepsilon)}\left[\,\mathbf{S}_{kp}^{(\varepsilon)}-\lambda\,\mathbf{M}_{kp}^{(\varepsilon)}\right]=\mathbf{f}_k^{(\varepsilon)},\tag{3.3.18}$$

for $\varepsilon=1,2,\ldots,N_{el}$ and $k=0,1,\ldots,P$, in which

$$\begin{aligned}\hat{\mathbf{S}}_{kp}^{(e,\varepsilon)}&=\left(\frac{d\psi_p}{dx},\,H_k^{(\varepsilon)}(x)\right)_{\Omega_e},\quad e=1,2,\ldots,\varepsilon-1,\\ \mathbf{S}_{kp}^{(\varepsilon)}&=\left(\frac{d\psi_p}{dx},\,{}^{\mathrm{RL}}_{\ x}\mathcal{D}^{\mu}_{x_\varepsilon}\left[{}^{(2)}\mathcal{P}^{\mu}_{k+1}(x)\right]\right)_{\Omega_\varepsilon},\\ \mathbf{M}_{kp}^{(\varepsilon)}&=\left(\psi_p(x),\,{}^{(2)}\mathcal{P}^{\mu}_{k+1}(x)\right)_{\Omega_\varepsilon},\\ \mathbf{f}_k^{(\varepsilon)}&=\left(f,\,{}^{(2)}\mathcal{P}^{\mu}_{k+1}(x)\right)_{\Omega_\varepsilon}\end{aligned}\tag{3.3.19}$$

are respectively the *history*, local stiffness, local mass matrices, and local force vector.

Elemental (Local) Operations: The Construction of Local Matrices $\mathbf{S}^{(\varepsilon)}$ and $\mathbf{M}^{(\varepsilon)}$, and Vector $\mathbf{f}^{(\varepsilon)}$

Here, we provide the analytically obtained expressions of the local stiffness matrix as well as the proper quadrature rules to construct the local mass matrix and force vector for all elements $\varepsilon=1,2,\ldots,N_{el}$.

Elemental (Local) Stiffness Matrix $\mathbf{S}^{(\varepsilon)}$: given the structure of the basis functions, we first obtain the first ($p=0$) and last column ($p=P$) of the local stiffness matrix $\mathbf{S}^{(\varepsilon)}$, and then, the rest of the entries corresponding to the interior modes. Hence,

$$\begin{aligned}\mathbf{S}_{k0}^{(\varepsilon)}&=\int_{x_{\varepsilon-1}}^{x_\varepsilon}\frac{d\psi_0}{dx}\,{}^{\mathrm{RL}}_{\ x}\mathcal{D}^{\mu}_{x_\varepsilon}\left[{}^{(2)}\mathcal{P}^{\mu}_{k+1}(x)\right]dx\\ &=Jac(\varepsilon,\mu)\int_{-1}^{1}\left(\frac{-1}{2}\right)\left(\frac{d\zeta}{dx}\right)\frac{\Gamma[1+k+\mu]}{\Gamma[1+k]}P_k(\zeta)\left(\frac{dx}{d\zeta}\right)d\zeta\end{aligned}\tag{3.3.20}$$

$$= -Jac(\varepsilon,\mu)\frac{\Gamma(1+k+\mu)}{2\,\Gamma(1+k)}\int_{-1}^{1} P_k(\zeta)\mathrm{d}\zeta$$
$$= -Jac(\varepsilon,\mu)\frac{\Gamma(1+k+\mu)}{\Gamma(1+k)}\delta_{k,0}, \quad \text{(by the orthogonality)}$$

in which the Jacobian constant, associated with the element ε and the fractional order μ, is $Jac(\varepsilon,\mu) = (\frac{2}{x_\varepsilon - x_{\varepsilon-1}})^\mu$. Thus, the first column of the local stiffness matrix is obtained as

$$\mathbf{S}_{k0}^{(\varepsilon)} = -\Big(\frac{2}{x_\varepsilon - x_{\varepsilon-1}}\Big)^\mu \frac{\Gamma(1+k+\mu)}{\Gamma(1+k)}\delta_{k,0}, \quad k = 0,1,\ldots,P. \tag{3.3.21}$$

Similarly, we can obtain the last column of the local stiffness matrix $\mathbf{S}_{kP}^{(\varepsilon)}$ as

$$\mathbf{S}_{kP}^{(\varepsilon)} = \Big(\frac{2}{x_\varepsilon - x_{\varepsilon-1}}\Big)^\mu \frac{\Gamma(1+k+\mu)}{\Gamma(1+k)}\delta_{k,0} = -\mathbf{S}_{k0}^{(\varepsilon)}, \quad k = 0,1,\ldots,P. \tag{3.3.22}$$

In order to obtain the rest of the entries of $\mathbf{S}_{kp}^{(\varepsilon)}$ ($k = 0,1,\ldots,P$ and $p = 1,2,\ldots,P-1$), we carry out the integration-by-parts and transfer another derivative onto the test function, taking into account that the interior modes vanish at the boundary points x_ε and $x_{\varepsilon-1}$. Therefore,

$$\mathbf{S}_{kp}^{(\varepsilon)} = \int_{x_{\varepsilon-1}}^{x_\varepsilon} \frac{\mathrm{d}\psi_p}{\mathrm{d}x}\, {}^{\mathrm{RL}}_{\ x}D^\mu_{x_\varepsilon}\Big[{}^{(2)}\mathcal{P}^\mu_{k+1}(x)\Big]\mathrm{d}x \tag{3.3.23}$$
$$= -\int_{x_{\varepsilon-1}}^{x_\varepsilon} \psi_p(x)\,\frac{\mathrm{d}}{\mathrm{d}x}\, {}^{\mathrm{RL}}_{\ x}D^\mu_{x_\varepsilon}\Big[{}^{(2)}\mathcal{P}^\mu_{k+1}(x)\Big]\mathrm{d}x$$
$$= -\int_{-1}^{1} \psi_p(\zeta)\,\frac{\mathrm{d}}{\mathrm{d}\zeta}\frac{\mathrm{d}\zeta}{\mathrm{d}x}\, Jac(\varepsilon,\mu)\, {}^{\mathrm{RL}}_{\ \zeta}D^\mu_{1}\Big[{}^{(2)}\mathcal{P}^\mu_{k+1}(\zeta)\Big]\frac{\mathrm{d}x}{\mathrm{d}\zeta}\mathrm{d}\zeta$$
$$= -Jac(\varepsilon,\mu)\frac{\Gamma(1+k+\mu)}{4\,\Gamma(1+k)}\int_{-1}^{1}(1-\zeta)(1+\zeta)P_{p-1}^{(1,1)}(\zeta)\,\frac{\mathrm{d}}{\mathrm{d}\zeta}\Big[P_k(\zeta)\Big]\mathrm{d}\zeta$$
$$= -Jac(\varepsilon,\mu)\frac{\Gamma(1+k+\mu)}{4\,\Gamma(1+k)}\frac{k+1}{2}\int_{-1}^{1}(1-\zeta)(1+\zeta)P_{p-1}^{(1,1)}(\zeta)\,P_{k-1}^{(1,1)}(\zeta)\mathrm{d}\zeta.$$

Hence, for $\varepsilon = 1,2,\ldots,N_{el}$, $k = 0,\ldots,P$, and $p = 1,\ldots,P-1$,

$$\mathbf{S}_{kp}^{(\varepsilon)} = -\Big(\frac{2}{x_\varepsilon - x_{\varepsilon-1}}\Big)^\mu \frac{\Gamma(1+k+\mu)(k+1)}{8\,\Gamma(1+k)}C_{k-1}^{1,1}\delta_{k,p},$$

where $C_{k-1}^{1,1}$ represents the corresponding orthogonality constant of Jacobi polynomials of order $k-1$ with parameters $\alpha = \beta = 1$. We note that the entries of $\mathbf{S}_{kp}^{(\varepsilon)}$ are obtained analytically, using the orthogonality of the Jacobi polynomial. Also, the interior modes lead to a *diagonal* matrix due to $\delta_{k,p}$. Figure 3.20 shows the sparsity of the local stiffness matrix.

Elemental (Local) Mass Matrix $\mathbf{M}^{(\varepsilon)}$: given the structure of basis functions and definition (3.3.19), we first obtain the corresponding first ($p = 0$) and last

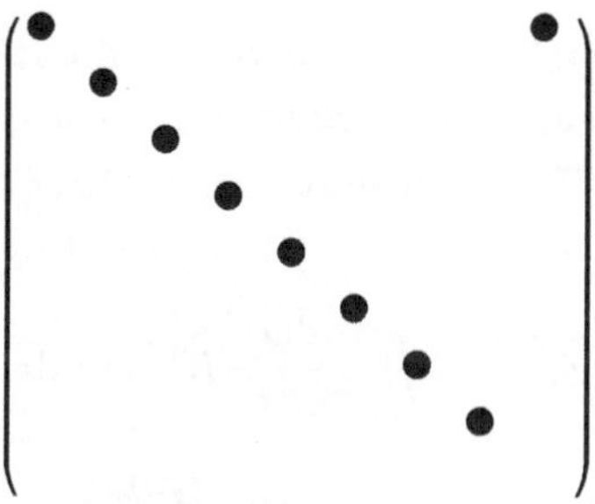

Figure 3.20 Sparsity of local stiffness matrix. Source: Reprinted from Kharazmi et al. (2017) with permission from Elsevier.

column ($p = P$) of the local mass matrix $\mathbf{M}^{(\varepsilon)}$, and then, we compute the rest of the entries associated with the interior modes, using proper quadrature rules. Therefore,

$$\mathbf{M}_{k0}^{(\varepsilon)} = \frac{x_\varepsilon - x_{\varepsilon-1}}{4} \sum_{q=1}^{Q} w_q^{1+\mu,0} \; P_k^{(\mu,-\mu)}(z_q^{1+\mu,0}), \tag{3.3.24}$$

$$\mathbf{M}_{kP}^{(\varepsilon)} = \frac{x_\varepsilon - x_{\varepsilon-1}}{4} \sum_{q=1}^{Q} w_q^{\mu,1} \; P_k^{(\mu,-\mu)}(z_q^{\mu,1}), \tag{3.3.25}$$

$$\mathbf{M}_{kp}^{(\varepsilon)} = \frac{x_\varepsilon - x_{\varepsilon-1}}{8} \sum_{q=1}^{Q} w_q^{1+\mu,1} \; P_k^{(\mu,-\mu)}(z_q^{1+\mu,1}), \tag{3.3.26}$$

where $k = 0, 1, \ldots, P$ and $\{w_q^{\alpha,\beta}, z_q^{\alpha,\beta}\}_{q=1}^{Q}$ are the Gauss–Lobatto–Jacobi weights and points corresponding to the parameters α and β.

Elemental (Local) Load Vector $\mathbf{f}^{(\varepsilon)}$: the local load vector is obtained as:

$$\begin{aligned} \mathbf{f}_k^{(\varepsilon)} &= \int_{x_{\varepsilon-1}}^{x^\varepsilon} f(x) \; {}^{(2)}\mathcal{P}_{k+1}^{\mu}(x)dx \\ &= \frac{x_\varepsilon - x_{\varepsilon-1}}{2} \int_{-1}^{1} (1-\zeta)^\mu \; f(x(\zeta)) \; P_k^{(\mu,-\mu)}(\zeta) \; d\zeta. \end{aligned}$$

Hence, for $\varepsilon = 1, 2, \ldots, N_{el}$,

$$\mathbf{f}_k^{(\varepsilon)} = \frac{x_\varepsilon - x_{\varepsilon-1}}{2} \sum_{q=1}^{Q} w_q^{\mu,0} \; f(x^\varepsilon(\zeta_q)) \, P_k^{(\mu,-\mu)}(z_q^{\mu,0}), \quad k = 0, 1, \ldots, P, \tag{3.3.27}$$

where $\{w_q^{\mu,0}, z_q^{\mu,0}\}_{q=1}^{Q}$ are the Gauss–Lobatto–Jacobi weights and points corresponding to the parameters $\alpha = \mu$ and $\beta = 0$.

Nonlocal Operation: The Construction of the History Matrix $\hat{\mathbf{S}}^{(e,\varepsilon)}$
The most challenging part of constructing the linear system is to compute the global history matrix $\hat{\mathbf{S}}^{(e,\varepsilon)}$. The history matrix relates the current element $\varepsilon = 1, 2, \ldots, N_{el}$ to its past elements $e = 1, 2, \ldots, \varepsilon - 1$ by

$$\hat{\mathbf{S}}_{kp}^{(e,\varepsilon)} = \int_{x_{e-1}}^{x_e} \frac{\mathrm{d}\psi_p}{\mathrm{d}x} H_k^{(\varepsilon)}(x)\, \mathrm{d}x \tag{3.3.28}$$

for $k = 0, \ldots, P$, $p = 1, \ldots, P-1$, where $H_k^{(\varepsilon)}(x)$ is given in (3.3.14) as

$$\begin{aligned} H_k^{(\varepsilon)}(x) &= \frac{-1}{\Gamma(1-\mu)} \frac{\mathrm{d}}{\mathrm{d}x} \int_{x_{\varepsilon-1}}^{x^{\varepsilon}} \frac{{}^{(2)}\mathcal{P}_{k+1}^{\mu}(s)}{(s-x)^{\mu}} \mathrm{d}s \\ &= \frac{-\mu}{\Gamma(1-\mu)} \int_{x_{\varepsilon-1}}^{x^{\varepsilon}} \frac{{}^{(2)}\mathcal{P}_{k+1}^{\mu}(s)}{(s-x)^{1+\mu}} \mathrm{d}s, \end{aligned}$$

in which $x \in \Omega_e = [x_{e-1}, x_e]$ and $s \in \Omega_\varepsilon = [x_{\varepsilon-1}, x_\varepsilon]$. By performing the following affine mappings

$$s = \frac{x_\varepsilon + x_{\varepsilon-1}}{2} + \frac{x_\varepsilon - x_{\varepsilon-1}}{2}\zeta \quad x = \frac{x_e + x_{e-1}}{2} + \frac{x_e - x_{e-1}}{2}\xi,$$

from Ω_e and Ω_ε to the standard domain $[-1, 1]$, the history function $H_k^{(\varepsilon,e)}(\xi) = H_k^{(\varepsilon)}(x)$ is obtained as

$$\begin{aligned} H_k^{(\varepsilon,e)}(\xi) = &\frac{x_\varepsilon - x_{\varepsilon-1}}{2} \frac{-\mu}{\Gamma(1-\mu)} \\ &\times \int_{-1}^{1} \frac{{}^{(2)}\mathcal{P}_{k+1}^{\mu}(\zeta)}{\left[\frac{(x_\varepsilon + x_{\varepsilon-1}) - (x_e + x_{e-1})}{2} + \frac{x_\varepsilon - x_{\varepsilon-1}}{2}\zeta - \frac{x_e - x_{e-1}}{2}\xi\right]^{1+\mu}} \mathrm{d}\zeta. \end{aligned} \tag{3.3.29}$$

If the mesh is "uniform", then

$$\begin{aligned} &\frac{x_\varepsilon + x_{\varepsilon-1}}{2} = \frac{2\varepsilon - 1}{2}\Delta x, \quad \frac{x_e + x_{e-1}}{2} = \frac{2e-1}{2}\Delta x, \\ &\frac{x_\varepsilon - x_{\varepsilon-1}}{2} = \frac{x_e - x_{e-1}}{2} = \frac{\Delta x}{2}, \end{aligned} \tag{3.3.30}$$

and thus,

$$\begin{aligned} H_k^{(\varepsilon,e)}(\xi) &= \frac{\Delta x}{2} \frac{-\mu}{\Gamma(1-\mu)} \int_{-1}^{1} \frac{{}^{(2)}\mathcal{P}_{k+1}^{\mu}(\zeta)}{\left[\frac{(2\varepsilon-1)-(2e-1)}{2}\Delta x + \frac{\Delta x}{2}(\zeta - \xi)\right]^{1+\mu}} \mathrm{d}\zeta \\ &= \frac{-\mu}{\Gamma(1-\mu)} \left(\frac{2}{\Delta x}\right)^{\mu} \int_{-1}^{1} \frac{{}^{(2)}\mathcal{P}_{k+1}^{\mu}(\zeta)}{[2(\varepsilon - e) + \zeta - \xi]^{1+\mu}} \mathrm{d}\zeta \\ &= \frac{-\mu}{\Gamma(1-\mu)} \left(\frac{2}{\Delta x}\right)^{\mu} \int_{-1}^{1} \frac{{}^{(2)}\mathcal{P}_{k+1}^{\mu}(\zeta)}{[2\,\Delta\varepsilon + \zeta - \xi]^{1+\mu}} \mathrm{d}\zeta, \end{aligned} \tag{3.3.31}$$

where $\Delta\varepsilon = \varepsilon - e > 0$ denotes the element difference between the current element ε and the e-th element. Next, we expand the poly-fractonomials ${}^{(2)}\mathcal{P}^{\mu}_{k+1}(\zeta)$ in terms of fractonomials $(1-\zeta)^{\mu+m}$ as

$$ {}^{(2)}\mathcal{P}^{\mu}_{k+1}(\zeta) = (1-\zeta)^{\mu} P_k^{(\mu,-\mu)}(\zeta) = \sum_{m=0}^{k} C_{km}(1-\zeta)^{\mu+m}, \quad (3.3.32) $$

in which $C_{km} = \binom{k+m}{m}\binom{k+\mu}{k-m}(-\frac{1}{2})^m$ is a lower-triangle matrix. Therefore, (3.3.31) can be written as

$$ H_k^{(\varepsilon,e)}(\xi) = \frac{-\mu}{\Gamma(1-\mu)}\left(\frac{2}{\Delta x}\right)^{\mu} \sum_{m=0}^{k} C_{km} h_m^{(\varepsilon,e)}(\xi), \quad (3.3.33) $$

where we call

$$ h_m^{(\varepsilon,e)}(\xi) \equiv \int_{-1}^{1} \frac{(1-\zeta)^{\mu+m}}{[2\,\Delta\varepsilon + \zeta - \xi]^{1+\mu}} \mathrm{d}\zeta, \quad m = 0, 1, \dots, k, \quad (3.3.34) $$

the *(modal) memory mode*. Also, $h_m^{(\varepsilon,e)}(\xi)$ can be obtained analytically as

$$ h_m^{(\varepsilon,e)}(\xi) = \frac{2^{-\mu}(\Delta\varepsilon - \xi/2)^{-\mu}}{1+m+\mu}\left[h_{m,I}(\xi,\Delta\varepsilon) + h_{m,II}(\xi,\Delta\varepsilon) + h_{m,III}(\xi,\Delta\varepsilon)\right], $$

in which

$$ \begin{aligned} h_{m,I}(\xi,\Delta\varepsilon) &= -Z_I(\xi,\Delta\varepsilon)\ {}_2F_1\Big(1\,,\,1+m\,,\,2+m+\mu\,,\,Z_I(\xi,\Delta\varepsilon)\Big), \\ h_{m,II}(\xi,\Delta\varepsilon) &= \left(\frac{1-2\Delta\varepsilon+\xi}{-2\Delta\varepsilon+\xi}\right)^{-\mu}(2^{m+\mu})\ Z_{II}(\xi,\Delta\varepsilon) \\ &\quad \times {}_2F_1\Big(1\,,\,1+m\,,\,2+m+\mu\,,\,Z_{II}(\xi,\Delta\varepsilon)\Big), \quad (3.3.35) \\ h_{m,III}(\xi,\Delta\varepsilon) &= -Z_{III}(\xi,\Delta\varepsilon)\,{}_2F_1\Big(1\,,\,1+m\,,\,2+m+\mu\,,\,Z_{III}(\xi,\Delta\varepsilon)\Big), \end{aligned} $$

and the group variables are $Z_I(\xi,\Delta\varepsilon) = \frac{1}{1+2\Delta\varepsilon-\xi}$, $Z_{II}(\xi,\Delta\varepsilon) = \frac{-2}{-1-2\Delta\varepsilon+\xi}$, and $Z_{III}(\xi,\Delta\varepsilon) = \frac{1}{-2\Delta\varepsilon+\xi}$. Therefore, by (3.3.33) and obtaining the derivative of the basis function in the standard domain, the entries of the history matrix can be efficiently computed, using a Gauss quadrature. Hence,

$$ \hat{\mathbf{S}}_{kp}^{(\varepsilon,e)} \equiv \hat{\mathbf{S}}_{kp}^{(\Delta\varepsilon)} = \int_{-1}^{1} \frac{\mathrm{d}\psi_p}{\mathrm{d}\xi} H_k(\xi,\Delta\varepsilon)\mathrm{d}\xi, \quad k, p = 0, 1, \dots, P. \quad (3.3.36) $$

Remark 3.79 We note that when a uniform mesh is employed, the history function $H_k^{(\varepsilon,e)}(\xi) \equiv H_k(\xi,\Delta\varepsilon)$, defined in the standard domain, only depends on the "element difference", $\Delta\varepsilon = \varepsilon - e$. This is significant since one only needs to construct $N_{el} - 1$ history functions, and thus, history matrices $\hat{S}^{(e,\varepsilon)}$.

3.3.2 Assembling the Global System with Local Test Functions

We generalize the notion of global linear system assembly by taking into account the presence of the history stiffness matrices and recalling that the corresponding local mass matrix $\mathbf{M}^{(\varepsilon)}$ or the local load vector $\mathbf{f}^{(\varepsilon)}$ do not contribute to any history calculations. We impose the C^0-*continuity* by employing the "mapping arrays", map[e][p], defined as

$$map[e][p] = P(e-1)+p, \quad p = 1,2,\ldots,P, \quad e = 1,2,\ldots,N_{el}, \tag{3.3.37}$$

as for instance in Mathematica, the first entry of a vector is labelled by 1 rather than 0 as in C++. Then, the corresponding $(P+1) \times (P+1)$ "local" linear system, which is associated with the element Ω_ε, is obtained as

$$\mathcal{M}^{(\varepsilon)} = \mathbf{S}^{(\varepsilon)} - \lambda \mathbf{M}^{(\varepsilon)}. \tag{3.3.38}$$

We assemble the corresponding global linear matrix $\mathbb{M}_G$ and the global load vector $\mathbb{F}_G$ as follows:

$do\ \varepsilon = 1, N_{el}$

$\quad do\ k = 1, P+1$

$\quad\quad \mathbb{F}_G\Big[\, map[\varepsilon][k] \,\Big] = \mathbf{f}^{(\varepsilon)}[k]$

$\quad\quad do\ p = 1, P+1$

$\quad\quad \mathbb{M}_G\Big[\, map[\varepsilon][k] \Big]\Big[\, map[\varepsilon][p] \Big] = \mathbb{M}_G\Big[\, map[\varepsilon][k] \Big]\Big[\, map[\varepsilon][p] \Big] + \mathcal{M}^{(\varepsilon)}[k][p]$

$\quad\quad\quad do\ e = 1, \varepsilon - 1$

$\quad\quad \mathbb{M}_G\Big[\, map[\varepsilon][k] \Big]\Big[\, map[e][p] \Big] = \mathbb{M}_G\Big[\, map[\varepsilon][k] \Big]\Big[\, map[e][p] \Big] + \hat{\mathbf{S}}^{(\Delta\varepsilon)}[k][p]$

End

This global operation leads to the following linear system:

$$\mathbb{M}_G\ \hat{u}_G = \mathbb{F}_G, \tag{3.3.39}$$

in which $\hat{u}_G$ denotes the global degrees of freedom. The homogeneous Dirichlet boundary conditions require the first and last entries of the global degree of freedom to be zero, i.e., $\hat{u}_0^1 = \hat{u}_P^{N_{el}} = 0$. This is enforced by removing the first and last rows as well as the first and last columns of the global matrix, in addition to removing the first and last entries of the load vector. We also note that the C^0-continuity and decomposition of basis functions into boundary and interior modes lead to the standard scattering process from the global to local degrees of freedom (see e.g., Karniadakis et al. (2005)).

Off-Line Computation of History Matrices and History Retrieval

As mentioned in Remark 3.79 (on uniform grid generation), the history matrices solely depend on the element difference, $\Delta\varepsilon = \varepsilon - e$. Thus, for all

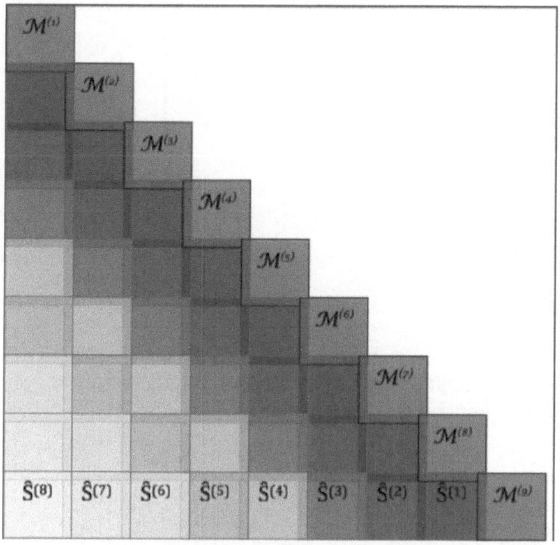

Figure 3.21 The assembled global matrix corresponding to a uniform grid with $N_{el} = 9$. In this global matrix, $\mathcal{M}^{(\varepsilon)} = \mathbf{S}^{(\varepsilon)} - \lambda\mathbf{M}^{(\varepsilon)}$, $\varepsilon = 1, 2, \ldots, N_{el}$, represents the local matrix, associated with the element Ω_ε. To fill the lower-triangular block matrices, we construct only $(N_{el} - 1)$ history matrices $\hat{\mathbf{S}}^{(\Delta\varepsilon)}$, where $\Delta\varepsilon = 1, 2, .., N_{el} - 1$, rather than $\frac{N_{el}(N_{el}-1)}{2}$ matrices. Source: Reprinted from Kharazmi et al. (2017) with permission from Elsevier.

local elements ε, where $\varepsilon = 1, 2, \ldots, N_{el}$, the history matrices corresponding to the past element e with similar element difference, are the same. See Figure 3.21, where similarly-shaded blocks represent the same history matrix and one can see that, for example, all the history matrices adjacent to the local stiffness matrices have the same element difference, $\Delta\varepsilon = 1$, and thus are in the same shade. Therefore, given number of elements N_{el}, we only need to construct the total number of $N_{el} - 1$ history matrices.

For a maximum number of elements, $N_{el}|_{max}$, and a maximum number of modes, $P|_{max}$, we can compute off-line and store the total $N_{el}|_{max} - 1$ history matrices of size $(P|_{max} + 1) \times (P|_{max} + 1)$, which we can fetch later for any specific $N_{el} \leq N_{el}|_{max}$ and $P \leq P|_{max}$.

3.3.3 Non-uniform Kernel-Based Grids

We present a non-uniform grid generation based on the power-law kernel in the definition of fractional derivative. There are different sources of singularity in the proposed problem that can be caused mainly due to the force function

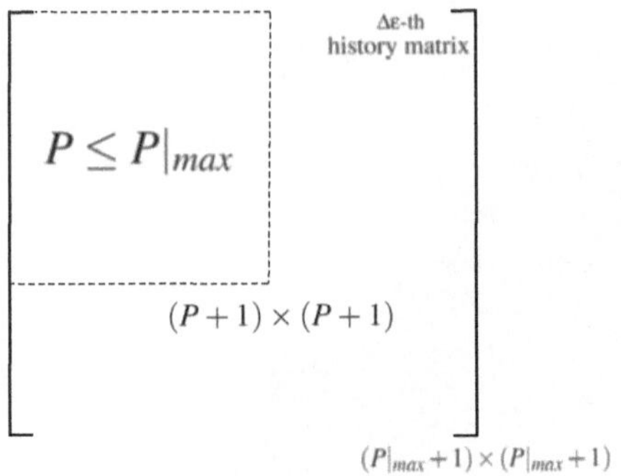

Figure 3.22 History computation and retrieval. Source: Reprinted from Kharazmi et al. (2017) with permission from Elsevier.

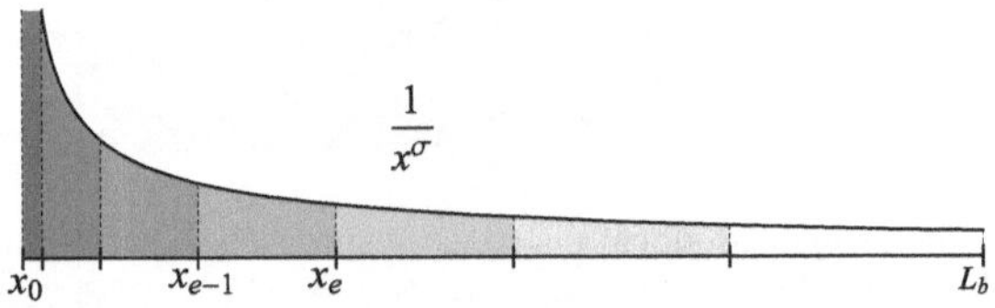

Figure 3.23 Kernel-based non-uniform grid in the boundary layer; L_b and N_b are the length of and the number of elements in the boundary layer, respectively. Source: Reprinted from Kharazmi et al. (2017) with permission from Elsevier.

$f(x)$. However, even if the force term is smooth the underlying kernel of a fractional derivative leads to formation of singularities at the boundaries. Herein, we propose a new kernel-based grid generation method that considers a sufficiently small boundary layer at the vicinity of a singular point and partitions that particular region non-uniformly. In this approach, we treat the kernel of the form $\frac{1}{x^\sigma}$ as a density function and then, we construct the grid such that the integral of the kernel function over each element $\Omega_e \in [x_{e-1}, x_e]$ (in the boundary layer) is constant. Since the operator is a left-sided fractional derivative, we represent the non-uniform grid refinement at the left boundary. Let L_b be the length of the boundary layer and $\int_0^{L_b} \frac{1}{x^\sigma}\,\mathrm{d}x = \frac{L_b^{1-\sigma}}{1-\sigma} = A$. Then, the integral over each element is

$$\frac{1}{A}\int_{x_{e-1}}^{x_e} \frac{1}{x^\sigma}\,\mathrm{d}x = \frac{1}{L_b^{1-\sigma}}\left[(x_{e-1}+\Delta x_e)^{1-\sigma} - x_{e-1}^{1-\sigma}\right] = C,$$

where $\Delta x_e = x_e - x_{e-1}$ and C is a constant. Thus,

$$\Delta x_e = \left[x_{e-1}^{1-\sigma} + C\,L_b^{1-\sigma}\right]^{\frac{1}{1-\sigma}} - x_{e-1}.$$

Starting from $x_0 = 0$ and calculating the rest of the grid locations successively, we obtain

$$x_e = \delta \, e^{\frac{1}{1-\sigma}}, \quad e = 1, 2, \ldots, N_b, \quad \text{element numbers}, \tag{3.3.40}$$

in which $\delta = L_b \, C^{\frac{1}{1-\sigma}}$ and N_b is the number of elements in the boundary layer. The constant C is obtained by the constraint $\sum_{e=1}^{N_b} \Delta x_e = L_b$ and hence,

$$C = \left(\sum_{e=1}^{N_b} \left[e^{\frac{1}{1-\sigma}} - (e-1)^{\frac{1}{1-\sigma}} \right] \right)^{\sigma - 1}.$$

We consider $\sigma = 1 - \mu$ and thus when $\mu = 1$, we recover the uniform grid $x_e = \frac{L_b}{N_b} e$, where the kernel is 1, $C = \frac{1}{N_b}$, $\delta = \frac{L_b}{N_b}$. We note that in the boundary layer, where the grid is non-uniform, equations (3.3.30)–(3.3.34) no longer hold. Thus, using (3.3.40), we obtain

$$H_k^{(\varepsilon,e)}(\xi) = \frac{-\mu}{\Gamma(1-\mu)} \frac{\delta}{2} \left(\varepsilon^{\frac{1}{\mu}} - (\varepsilon - 1)^{\frac{1}{\mu}} \right) \int_{-1}^{1} \frac{{}^{(2)}\mathcal{P}_{k+1}^{\mu}(\zeta) d\zeta}{\mathcal{Z}}, \tag{3.3.41}$$

in which

$$\mathcal{Z} = \left(\frac{\delta}{2} \right)^{1+\mu} \Big[\left(\varepsilon^{\frac{1}{\mu}} + (\varepsilon - 1)^{\frac{1}{\mu}} \right) - \left(e^{\frac{1}{\mu}} + (e-1)^{\frac{1}{\mu}} \right)$$
$$+ \left(\varepsilon^{\frac{1}{\mu}} - (\varepsilon - 1)^{\frac{1}{\mu}} \right) \zeta - \left(e^{\frac{1}{\mu}} - (e-1)^{\frac{1}{\mu}} \right) \xi \Big]^{1+\mu}.$$

Therefore, by (3.3.28), the entries of the history matrix for the boundary layer elements, where $\varepsilon = 1, 2, \ldots, N_b$ and $e = 1, 2, \ldots, \varepsilon - 1$, can be numerically obtained as

$$\hat{\mathbf{S}}_{kp}^{(\varepsilon,e)} = \int_{-1}^{1} \frac{d\psi_p}{d\xi} H_k^{(\varepsilon,e)}(\xi) \, d\xi, \quad k, p = 0, 1, \ldots, P. \tag{3.3.42}$$

These matrices are the small squares in the upper left corner of Figure 3.24 (interaction of boundary layer elements e and ϵ). For the interior elements, $\varepsilon = N_b + 1, N_b + 2, \ldots, N_{el}$, when $N_b + 1 \le e \le \epsilon - 1$, the grid is uniform and therefore, we use (3.3.36) to obtain the history matrices. These matrices are shown as the big squares in Figure 3.24 (interaction of interior elements e and ϵ). However, when $1 \le e \le N_b$, the grid is non-uniform and we use (3.3.42) to obtain the history matrices. These matrices are shown as skinny rectangles in Figure 3.24 (interaction of interior elements with boundary layer elements).

In uniform grid generation, the history function (3.3.31) only depends on element difference $\Delta\epsilon$, which leads to a fast and efficient construction of history matrices (see Remark 3.79). However, in non-uniform kernel-based

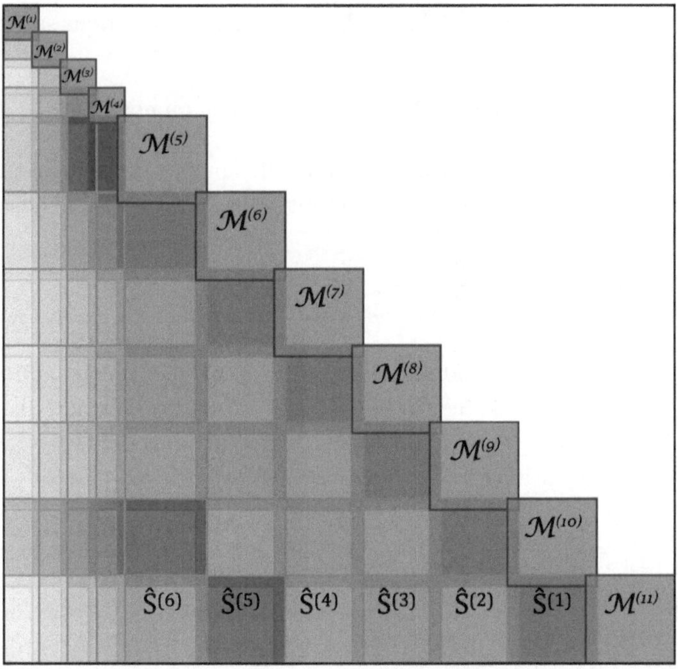

Figure 3.24 The assembled global matrix corresponding to $N_{el} = 11$ with $N_b = 4$ non-uniform boundary elements and 7 uniform interior elements. In this global matrix, $\mathcal{M}^{(\varepsilon)} = \mathbf{S}^{(\varepsilon)} - \lambda\mathbf{M}^{(\varepsilon)}$, $\varepsilon = 1, 2, \ldots, N_{el}$, represents the local matrix, associated with the element Ω_ε. The lower-triangle consists of three parts: 1) The small square $\frac{N_b(N_b-1)}{2}$ history matrices (interaction of boundary elements, $\varepsilon = 1, 2, \ldots, N_b$). 2) The big square history matrices (interaction of interior elements, $\varepsilon = N_b + 1, \ldots, N_{el}$). 3) The skinny rectangular $(N_{el} - N_b)N_b$ history matrices (interaction of boundary elements with interior elements). Source: Reprinted from Kharazmi et al. (2017) with permission from Elsevier.

grid generation, this is not the case anymore and construction of history matrices is computationally expensive. Improving the history construction on non-uniform grids requires further investigations, to be done in our future works.

Non-uniform Geometrically Progressive Grids

In addition to the non-uniform grid generation based on the kernel of the fractional derivative, we consider a non-uniform grid using geometrically progressive series (Ainsworth and Oden, 1997). In this case, the length of elements are increased by a constant factor r (see Figure 3.25). By considering

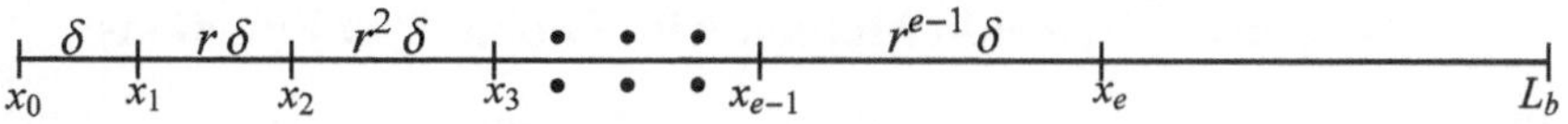

Figure 3.25 Non-uniform geometrically progressive grid. Source: Reprinted from Kharazmi et al. (2017) with permission from Elsevier.

the length of the first element to be δ, we construct the grid as $x_0 = 0,\ x_1 = \delta,\ x_2 = \delta(1+r),\ x_3 = \delta(1+r+r^2)$ and so on. Hence,

$$x_e = \delta \sum_{i=0}^{e-1} r^i = \delta \frac{r^e - 1}{r - 1}, \quad e = 1, 2, \ldots, N_b. \tag{3.3.43}$$

Choosing r and N_b, the constant δ is obtained by the constraint $x_{N_b} = L_b$, which gives $\delta = L_b \frac{r-1}{r^{N_b}-1}$. Since the grid is non-uniform, equations (3.3.30)–(3.3.34) do not hold anymore. Thus, using (3.3.43), we obtain

$$H_k^{(\varepsilon,e)}(\xi) = \frac{-\mu}{\Gamma(1-\mu)} \left(\frac{2}{\delta}\right)^{\mu} r^{\Delta\varepsilon - \mu(e-1)} \int_{-1}^{1} \frac{{}^{(2)}\mathcal{P}_{k+1}^{\mu}(\zeta)\,d\zeta}{\left[\frac{r+1}{r-1}(r^{\Delta\varepsilon} - 1) + (\zeta - r^{\Delta\varepsilon}\xi)\right]^{1+\mu}}, \tag{3.3.44}$$

where $\Delta\varepsilon = \varepsilon - e > 0$, denotes the *element difference* between the current element ε and the e-th element. Using the same expansion as in (3.3.32), we can write (3.3.44) as

$$H_k^{(\varepsilon,e)}(\xi) = \frac{-\mu}{\Gamma(1-\mu)} \left(\frac{2}{\delta}\right)^{\mu} r^{\Delta\varepsilon - \mu(e-1)} \sum_{m=0}^{k} C_{km} \tilde{h}_m^{(\varepsilon,e)}(\xi), \tag{3.3.45}$$

where the (modal) memory mode

$$\tilde{h}_m^{(\varepsilon,e)}(\xi) = \int_{-1}^{1} \frac{(1-\zeta)^{\mu+m}}{[\frac{r+1}{r-1}(r^{\Delta\varepsilon} - 1) + (\zeta - r^{\Delta\varepsilon}\xi)]^{1+\mu}} d\zeta, \quad m = 0, 1, \ldots, k, \tag{3.3.46}$$

can be obtained analytically using hypergeometric functions. Therefore, by (3.3.45), the entries of the history matrix can be efficiently computed using the Gauss quadrature in (3.3.36). The construction of the assembled global linear system is the same as the kernel-based grid generation approach. We note that similar to the uniform grid, in the non-uniform grid generation using the geometrical progression, the history functions depend on the element difference $\Delta\varepsilon = \varepsilon - e$, leading to a fast and efficient construction of history matrices.

3.3.4 Petrov–Galerkin Method with Global Test Functions

In this section, similar to the case of local test functions, we develop the Petrov–Galerkin scheme by substituting (3.3.7) into (3.3.6) and considering the global test function given in (3.3.11), to obtain:

$$\sum_{e=1}^{N_{el}}\Big(\sum_{p=0}^{P}\hat{u}_p^{(e)}\ \frac{d\psi_p(x)}{dx},\ {}^{\mathrm{RL}}_{\ x}D_L^{\mu}\, v_k^{\varepsilon}(x)\Big)_{\Omega_e}-\lambda\sum_{e=1}^{N_{el}}\Big(\sum_{p=0}^{P}\hat{u}_p^{(e)}\ \psi_p(x),\ v_k^{\varepsilon}(x)\Big)_{\Omega_e}$$
$$=\sum_{e=1}^{N_{el}}\Big(f,\ v_k^{\varepsilon}(x)\Big)_{\Omega_e},\quad \varepsilon=1,2,\dots,N_{el},\quad k=0,1,\dots,P. \tag{3.3.47}$$

Since the test function vanishes only $\forall x\in\Omega_e\neq\Omega_\varepsilon$ and $e>\varepsilon$, (3.3.47) reduces to

$$\sum_{e=1}^{\varepsilon}\sum_{p=0}^{P}\hat{u}_p^{(e)}\Big(\frac{d\psi_p}{dx},\ {}^{RL}_{\ x}\mathcal{D}_{x_\varepsilon}^{\mu}\, v_k^{\varepsilon}(x)\Big)_{\Omega_e}-\lambda\sum_{e=1}^{\varepsilon}\sum_{p=0}^{P}\hat{u}_p^{(e)}\Big(\psi_p(x),\ v_k^{\varepsilon}(x)\Big)_{\Omega_e}$$
$$=\sum_{e=1}^{\varepsilon}\Big(f,\ v_k^{\varepsilon}(x)\Big)_{\Omega_e}.$$

By substituting (3.3.11), we obtain

$$\sum_{e=1}^{\varepsilon}\sum_{p=0}^{P}\hat{u}_p^{(e)}\Big(\frac{d\psi_p}{dx},\ {}^{RL}_{\ x}\mathcal{D}_{x_\varepsilon}^{\mu}\ {}^{(2)}\mathcal{P}_{k+1}^{\mu}(x^{1\sim\varepsilon})\Big)_{\Omega_e}$$
$$-\lambda\sum_{e=1}^{\varepsilon}\sum_{p=0}^{P}\hat{u}_p^{(e)}\Big(\psi_p(x),\ {}^{(2)}\mathcal{P}_{k+1}^{\mu}(x^{1\sim\varepsilon})\Big)_{\Omega_e}=\int_0^{x_\varepsilon}f(x)\ {}^{(2)}\mathcal{P}_{k+1}^{\mu}(x^{1\sim\varepsilon})dx \tag{3.3.48}$$

for $\varepsilon=1,2,\dots,N_{el},\quad k=0,1,\dots,P$, which can be written in the matrix form as

$$\sum_{e=1}^{\varepsilon}\sum_{p=0}^{P}\hat{u}_p^{(e)}\left[\hat{\mathbf{S}}_{kp}^{(\varepsilon,e)}-\lambda\,\hat{\mathbf{M}}_{kp}^{(\varepsilon,e)}\right]=\mathbf{f}_k^{(\varepsilon)} \tag{3.3.49}$$

for $\varepsilon=1,2,\dots,N_{el},\quad k=0,1,\dots,P$, where

$$\hat{\mathbf{S}}_{kp}^{(\varepsilon,e)}=\Big(\frac{d\psi_p^{\varepsilon}}{dx},\ {}^{RL}_{\ x}\mathcal{D}_{x_\varepsilon}^{\mu}\Big[{}^{(2)}\mathcal{P}_{k+1}^{\mu}(x^{1\sim\varepsilon})\Big]\Big)_{\Omega_\varepsilon}, \tag{3.3.50}$$

$$\hat{\mathbf{M}}_{kp}^{(\varepsilon,e)}=\Big(\psi_p^{\varepsilon}(x),\ {}^{(2)}\mathcal{P}_{k+1}^{\mu}(x^{1\sim\varepsilon})\Big)_{\Omega_\varepsilon}, \tag{3.3.51}$$

$$\mathbf{f}_k^{(\varepsilon)}=\int_0^{x_\varepsilon}f\ {}^{(2)}\mathcal{P}_{k+1}^{\mu}(x^{1\sim\varepsilon})dx. \tag{3.3.52}$$

Remark 3.80 The benefit of choosing such global test functions is now clear since we can analytically evaluate ${}^{\mathrm{RL}}_{\ x}D^{\mu}_{x_\varepsilon}\ {}^{(2)}\mathcal{P}^{\mu}_{k+1}(x^{1\sim\varepsilon})$. However, we note that this choice of test functions introduces "extra" work associated with the construction of the "history mass matrix" $\hat{\mathbf{M}}^{(\varepsilon,e)}$, $\forall e = 1, 2, \ldots, \varepsilon - 1$, when $\lambda \neq 0$.

Remark 3.81 The choice of global test functions leads to extra cost of quadrature carried out over the increasing-in-length domains of integration in (3.3.52). Depending on the behavior of the force-term $f(x)$, this approach might require adaptive/multi-element quadrature rules to obtain the corresponding entries of the desired precision.

Elemental (Local) Operations: The Construction of $\mathbf{f}^{(\varepsilon)}$

Here, the construction of the load vector is the only operation that could be regarded as "local operations". Hence,

$$\begin{aligned}\mathbf{f}^{(\varepsilon)}_k &= \int_0^{x_\varepsilon} f(x)\ {}^{(2)}\mathcal{P}^{\mu}_{k+1}(x^{1\sim\varepsilon})\mathrm{d}x\\ &= \left(\frac{x_\varepsilon}{2}\right)\int_{-1}^{1}(1-\zeta)^{\mu} f(x^{1\sim\varepsilon}(\zeta))\ P^{(\mu,-\mu)}_k(\zeta)\mathrm{d}\zeta,\end{aligned}$$

and thus,

$$\mathbf{f}^{(\varepsilon)}_k = \left(\frac{x_\varepsilon}{2}\right)\sum_{q=1}^{Q} w^{\mu,0}_q\ f(x^{1\sim\varepsilon}(\zeta_q))\ P^{(\mu,-\mu)}_k(z^{\mu,0}_q), \tag{3.3.53}$$

where $\{w^{\mu,0}_q, z^{\mu,0}_q\}_{q=1}^{Q}$ are the Gauss–Lobatto–Jacobi weights and points corresponding to the parameters $\alpha = \mu$ and $\beta = 0$.

Global Operations: The Construction of $\hat{\mathbf{S}}^{(\varepsilon,e)}$ and $\hat{\mathbf{M}}^{(\varepsilon,e)}$

The corresponding stiffness and mass matrices are global in nature and we obtain their entries using proper Gauss quadrature rules.

3.3.5 Assembling the Global System with Global Test Functions

We extend the notion of global linear system assembly by taking into account the presence of the history stiffness and mass matrices. We similarly impose the C^0-*continuity* by employing the same "mapping arrays", $map[e][p]$, defined in (3.3.37). Let us define the $(P+1)\times(P+1)$ matrix

$$\hat{\mathcal{M}}^{(\varepsilon,e)} = \hat{\mathbf{S}}^{(\varepsilon,e)} - \lambda\hat{\mathbf{M}}^{(\varepsilon,e)}, \tag{3.3.54}$$

$\forall\varepsilon$, e fixed. Then, we assemble the corresponding global linear matrix $\mathbb{M}_G$ and the global load vector $\mathbb{F}_G$ as follows:

$do\ \varepsilon = 1, N_{el}$
$\quad do\ k = 1, P+1$
$\qquad \mathbb{F}_G\Big[\ map[\varepsilon][k]\ \Big] = \mathbf{f}^{(\varepsilon)}[k]$
$\quad do\ p = 1, P+1$
$\qquad do\ e = 1, \varepsilon$
$\qquad \mathbb{M}_G\Big[map[\varepsilon][k]\Big]\Big[map[e][p]\Big] = \mathbb{M}_G\Big[map[\varepsilon][k]\Big]\Big[map[e][p]\Big] + \hat{\mathcal{M}}^{(\varepsilon,e)}[k][p]$
End

This leads to a linear system similar to that in (3.3.39), shown in Figure 3.26, where the homogeneous Dirichlet boundary conditions are enforced in a similar fashion as before. We note that the scattering operation from global to local degrees of freedom is similar to the standard scattering process.

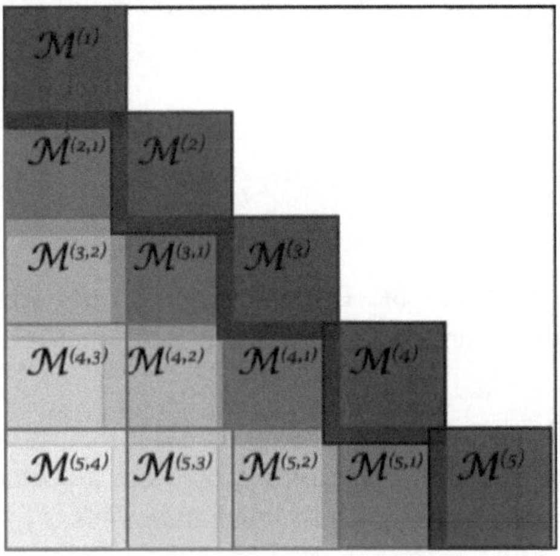

Figure 3.26 The assembled global matrix corresponding to $N_{el} = 5$ elements when global test functions are employed. In this global matrix, $\mathcal{M}^{(\varepsilon)} = \hat{\mathbf{S}}^{(\varepsilon)} - \lambda\hat{\mathbf{M}}^{(\varepsilon)}$, $\varepsilon = 1, 2, \ldots, N_{el}$, represents the local matrix, associated with the element Ω_ε. To fill the lower-triangular block matrices, we must construct $\frac{N_{el}(N_{el}-1)}{2}$ history matrices $\hat{\mathbf{S}}^{(\varepsilon,e)}$. Source: Reprinted from Kharazmi et al. (2017) with permission from Elsevier.

Numerical Examples for the PG SEM

We consider numerical examples of the two PG schemes we have proposed. We provide examples of smooth and singular solutions with singularities at boundary points and in the interior domain, where we show the efficiency of the developed schemes in capturing the singularities. We also perform the off-line computation of history matrices and show the improvement of computational cost. Moreover, we construct non-uniform kernel-based and geometrically progressive grids and present the success of the two approaches in capturing singular solutions. Furthermore, we investigate the nonlocal effects for different cases of history fading. In this section, we consider the computational domain $L = 1$.

Smooth Problems

In the proposed schemes, the choice of basis functions are polynomials, enabling the scheme to accurately and efficiently approximate the smooth solutions over the whole domain. We consider two smooth solutions of the form $u^{ext} = x^7 - x^6$ and $u^{ext} = x^6 \sin(2\pi x)$. The corresponding force functions are obtained by substituting the exact solutions into (3.3.1) (with $\lambda = 0$). By employing PG SEM, using local basis/test functions and local basis with global test functions (developed in Section 3.3.1 and Section 3.3.4, respectively), we observe that the former leads to a better approximability and condition number. Figure 3.27 presents the L_2-norm error of the PG SEM, employing local basis/test functions, where we show the exponential convergence of the scheme in approximating the two smooth solutions. The condition numbers of the resulting assembled global matrix, using the two developed schemes, are also presented in Table 3.22. We show that the choice

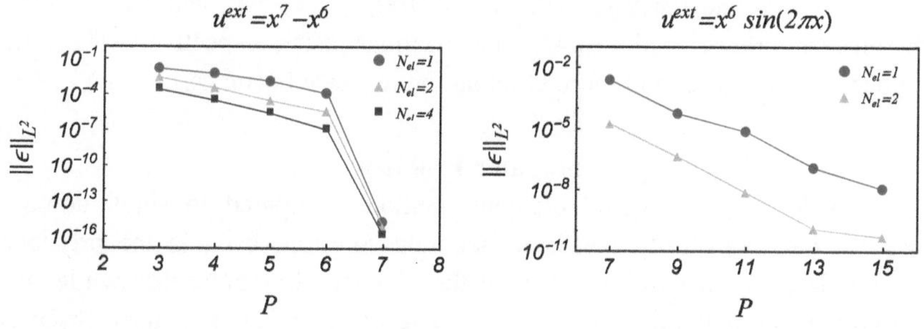

Figure 3.27 PG SEM with local basis/test functions. Plotted is the error with respect to the polynomial degree of each element (spectral order). Source: Reprinted from Kharazmi et al. (2017) with permission from Elsevier.

Table 3.22. *Condition number of the resulting assembled global matrix for the two choices of local bases/test functions (left) and local bases with global test functions (right) for different number of elements and modes. Source: Reprinted from Kharazmi et al. (2017) with permission from Elsevier.*

(Local Test Functions)		
P	$N_{el} = 2$	$N_{el} = 10$
3	7.13	86.13
5	13.21	153.86
10	35.39	420.24

(Global Test Function)		
P	$N_{el} = 2$	$N_{el} = 10$
3	3.46×10^4	1.84×10^{16}
5	4.3×10^7	7.2×10^{16}
10	2.73×10^{15}	5.1×10^{17}

of local bases/test functions leads to a better conditioning for different number of elements and modes.

History Retrieval

As discussed in Section 3.3.2, a large number of history matrices can be computed off-line, stored, and retrieved for later use. The retrieval process, compared to on-line construction of the history matrices, leads to higher computational efficiency. In this section, by considering 1000 elements, we compute and store 999 history matrices for different number of modes, $P = 2, 3$, and 4 (here $\mu = \frac{1}{2}$). Then, for different number of elements, we compute the CPU time required for constructing and solving the linear system, obtained by retrieving the stored history matrices from hard drive. We also compute the CPU time required for constructing and solving the linear system, obtained by on-line computation of the history matrices. Table 3.23 shows that in the case of $p = 4$ and for $N_{el} = 10$, $N_{el} = 100$, $N_{el} = 500$, and $N_{el} = 1000$, the retrieval process is almost 4, 5, and 10 times faster, respectively. Thus, the higher p is, the faster and more efficient the retrieval becomes.

Singular Problems

The developed PG spectral element method, compared to single-domain spectral methods, further leads to accurate solutions even in the presence of singularities via *hp*-refinements at the vicinity of singularities, while still employing smooth polynomial bases. The error in the boundary layer is controlled by considering sufficient number of modes in the boundary layer elements. The error in the interior domain is then improved by performing *p*-refinement in those elements. In order to investigate the performance of the

Table 3.23. *CPU time of constructing and solving the linear system based on off-line retrieval and on-line calculation of history matrices. Source: Reprinted from Kharazmi et al. (2017) with permission from Elsevier.*

	CPU Time							
	$N_{el} = 10$		$N_{el} = 100$		$N_{el} = 500$		$N_{el} = 1000$	
P	Off-line retrieval	On-line computation	Off-line retrieval	On-line computation	Off-line retrieval	On-line computation	Off-line retrieval	On-line computation
2	2.6520	7.2540	24.7105	83.5229	141.3525	429.3147	370.6895	790.3478
3	4.7580	18.9073	46.0826	161.8042	266.0441	1308.8327	746.4959	4423.7671
4	8.8140	32.2922	84.8645	499.9988	485.7715	5599.4062	1392.8705	14709.4902

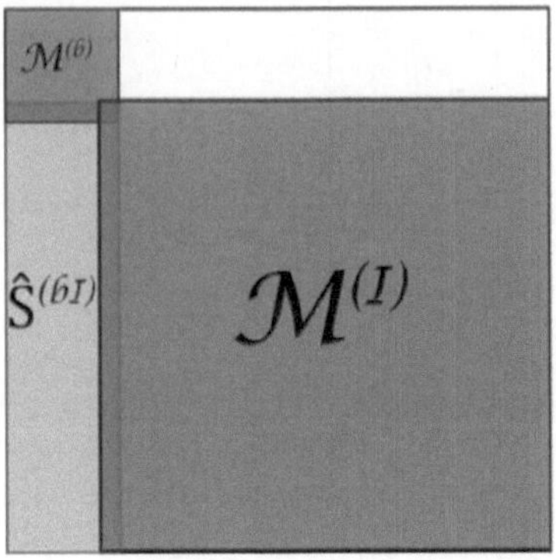

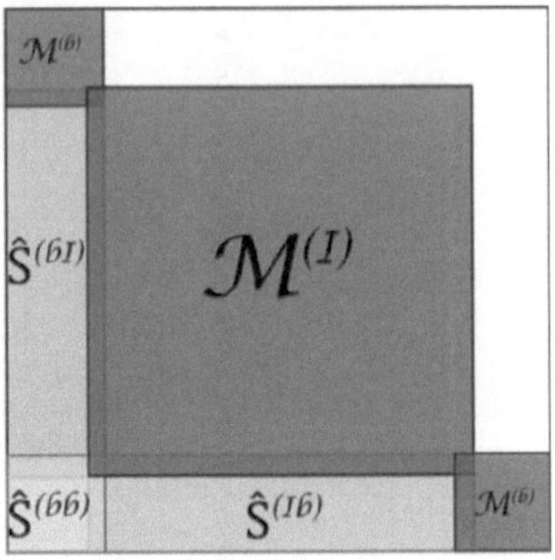

Figure 3.28 Schematic of global matrices corresponding to the case of singular solutions. (left): left boundary singularity, (right): left and right boundary singularities. $\hat{S}^{(bI)}$, $\hat{S}^{(Ib)}$, and $\hat{S}^{(bb)}$ denote the interaction of boundary/interior, interior/boundary and boundary/boundary elements, respectively. Source: Reprinted from Kharazmi et al. (2017) with permission from Elsevier.

scheme in capturing a singularity, we consider three types of singularities, including: i) single-boundary singularity, ii) full-boundary singularity, and iii) interior singularity (when discontinuous force functions are applied).

I) Single-Boundary Singularity: we consider two singular solutions of the form $u^{ext} = (1-x)x^{2+\mu}$ and $u^{ext} = (1-x)x^{5+\mu}$ with left boundary singularity. We partition the domain into two non-overlapping elements, including one boundary element of length L_b at the vicinity of the singular point in addition to an interior element for the rest of the computational domain. The schematic of the corresponding global system is shown in Figure 3.28 (left). Table 3.24 shows the exponential convergence of L_2-norm error in the interior domain. The error in the boundary layer element is then controlled by choosing sufficient number of modes in the boundary element. The results are obtained for the two cases of $L_b = 10^{-2}L$ and $L_b = 10^{-4}L$.

II) End-Point Singularities: we consider the solution of the form $u^{ext} = (1-x)^{3+\mu_1}x^{3+\mu_2}$ with singular points at two ends, i.e., $x = 0$ and $x = 1$. Herein, we partition the domain into three non-overlapping elements including two boundary elements of length L_b in the vicinity of the singular points, and one interior element for the rest of the domain. The schematic of the corresponding global system is shown in Figure 3.28 (right). Similar to the previous example, the PG SEM can accurately capture the singularities at both ends, where increasing the number of modes in the interior element results in exponential convergence. Table 3.25 shows the L_2-norm error in the

Table 3.24. *Single-Boundary Singularity: L_2-norm error in the boundary and interior elements using PG SEM with local basis/test functions. Here, L_b represents the size of the left boundary element, P_b and P_I denote the number of modes in the boundary and interior elements respectively. Source: Reprinted from Kharazmi et al. (2017) with permission from Elsevier.*

	$u^{ext} = (1-x)x^{2+\mu},\quad \mu = 1/2$		
	Boundary Element Error		
P_b	$L_b = 10^{-1}L$	$L_b = 10^{-2}L$	$L_b = 10^{-4}L$
6	1.29387×10^{-7}	1.29634×10^{-10}	1.19525×10^{-16}
10	1.46601×10^{-8}	1.4193×10^{-11}	4.07955×10^{-18}
	Interior Element Error, $P_b = 10$		
P_I	$L_b = 10^{-1}L$	$L_b = 10^{-2}L$	$L_b = 10^{-4}L$
6	5.49133×10^{-6}	2.6893×10^{-5}	3.38957×10^{-5}
10	9.39045×10^{-8}	1.08594×10^{-6}	1.91087×10^{-6}
14	8.27224×10^{-8}	1.13249×10^{-7}	3.02065×10^{-7}
	$u^{ext} = (1-x)x^{5+\mu},\quad \mu = 1/2$		
	Boundary Element Error		
P_b	$L_b = 10^{-1}L$	$L_b = 10^{-2}L$	$L_b = 10^{-4}L$
6	3.94221×10^{-11}	2.96862×10^{-17}	4.8243×10^{-29}
10	7.07024×10^{-13}	2.54089×10^{-18}	2.26939×10^{-29}
	Interior Element Error, $P_b = 10$		
P_I	$L_b = 10^{-1}L$	$L_b = 10^{-2}L$	$L_b = 10^{-4}L$
6	1.73622×10^{-5}	3.80264×10^{-5}	4.13249×10^{-5}
10	1.3122×10^{-9}	8.76951×10^{-9}	1.10139×10^{-8}
14	4.39611×10^{-12}	1.07775×10^{-10}	1.66044×10^{-10}

boundary layers and interior elements with two choices of $P_b = 6, 10$ and $L_b = 10^{-2}L,\ 10^{-4}L$.

III) Interior Singularities (Discontinuous Force Function): we consider the solutions with singularity in the middle of the domain. The force function, obtained by substituting the solution into (3.3.1), is considered to be discontinuous at the point of singularity. Figure 3.29 shows the two exact solutions of

Table 3.25. *End-Point Singularities: L_2-norm error in the boundary element (BE) and interior element (IE) by PG SEM with local basis/test functions. Here, $u^{ext} = (1-x)^{3+\mu_1} x^{3+\mu_2}$ with $\mu_1 = \frac{1}{4}$, $\mu_2 = \frac{2}{3}$, L_b represents the size of the left and right boundary elements, P_b and P_I denote the number of modes in the boundary and interior elements respectively. Source: Reprinted from Kharazmi et al. (2017) with permission from Elsevier.*

	$L_b = 10^{-2} L$				$L_b = 10^{-4} L$		
	$P_b = 6$				$P_b = 6$		
P_I	**Left BE Error**	**IE Error**	**Right BE Error**	P_I	**Left BE Error**	**IE Error**	**Right BE Error**
6	2.73893×10^{-7}	6.52605×10^{-5}	3.51075×10^{-6}	6	3.61679×10^{-10}	5.85397×10^{-5}	4.60538×10^{-8}
10	2.46964×10^{-11}	1.52215×10^{-7}	2.2902×10^{-9}	10	1.2676×10^{-10}	2.43295×10^{-7}	1.62151×10^{-10}
14	3.08719×10^{-12}	9.30483×10^{-9}	2.69541×10^{-10}	14	1.53993×10^{-13}	2.09933×10^{-8}	1.97677×10^{-11}
	$P_b = 10$				$P_b = 10$		
P_I	**Left BE Error**	**IE Error**	**Right BE Error**	P_I	**Left BE Error**	**IE Error**	**Right BE Error**
6	2.73892×10^{-7}	6.52605×10^{-5}	3.51075×10^{-6}	6	3.61679×10^{-10}	5.85397×10^{-5}	4.60538×10^{-8}
10	2.48058×10^{-11}	1.52215×10^{-7}	2.29003×10^{-9}	10	1.2676×10^{-12}	2.43295×10^{-7}	1.62151×10^{-10}
14	3.19684×10^{-12}	9.30511×10^{-9}	2.69506×10^{-10}	14	1.53993×10^{-13}	2.09933×10^{-8}	1.97677×10^{-11}

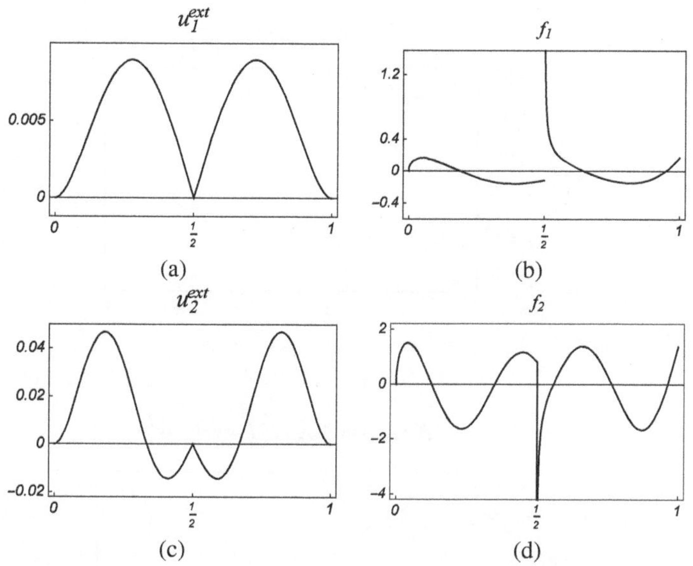

Figure 3.29 Interior singularities. (left): exact solutions, (right): the corresponding force functions. Source: Reprinted from Kharazmi et al. (2017) with permission from Elsevier.

the form $u_1^{ext} = x^2\,(1-x)^2\,|x-\frac{1}{2}|$ (top) and $u_2^{ext} = sin(3\pi x)\,x\,(1-x)\,|x-\frac{1}{2}|$ (bottom) and their corresponding force functions. We partition the domain at the vicinity of the singular point using two non-overlapping interior elements, in which the solution is smooth. The PG scheme with local basis/test functions is shown to be able to accurately capture the singularity in the middle of the domain. In the case of u_1^{ext}, we approximate the solution in the range of machine precision with $P = 5$ within each element. We also show the exponential rate of convergence in the case of u_2^{ext} by increasing the number of modes, P, in each element. The results are shown in Figure 3.30.

Non-uniform Grids

We consider a singular solution of the form $u^{ext} = (1-x)x^{1+\mu}$ (here $\mu = \frac{1}{10}$ and $\lambda = 0$) with singularity at the left boundary. In order to solve the problem, we consider three grid generation approaches with similar degrees of freedom, including one uniform and two non-uniform grids over the computational domain. The non-uniform grids are generated based on the power-law kernel in the definition of fractional derivative and the geometric progression series (discussed in Section 3.3.3 and page 274, respectively). Here, we choose $L_b = L$. Table 3.26 shows the L_2-norm error considering the uniform and

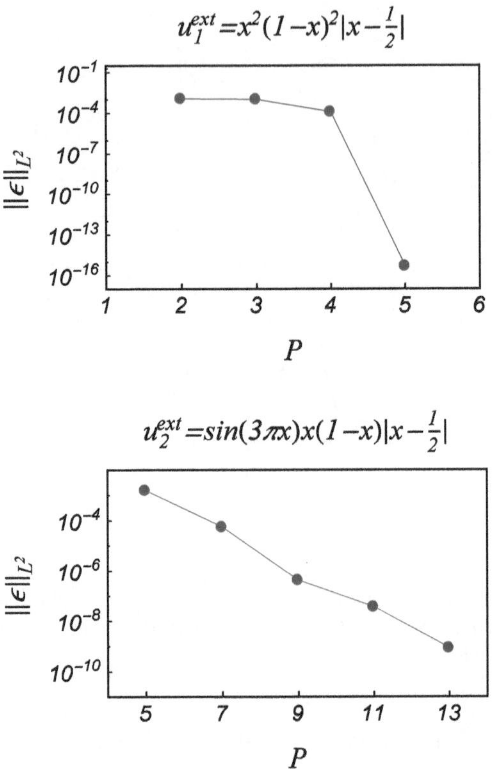

Figure 3.30 Interior singularities: PG SEM with local basis/test functions. Plotted is the error with respect to spectral order in each element. Source: Reprinted from Kharazmi et al. (2017) with permission from Elsevier.

non-uniform grids. We keep the total degrees of freedom fixed, but we increase the polynomial order P in each simulation. The success of the non-uniform grid in providing more accurate results is observed, where fewer number of elements are used, while higher-order polynomials are employed. We recall that the size of the boundary layer has been set to its maximum possible length, i.e., $L_b = L$. Clearly, one can obtain even more accurate results when L_b is set to much smaller length (e.g., $10^{-1}L$, $10^{-3}L$, etc.).

A Systematic Memory Fading Analysis

In order to investigate the effect of truncating the history matrices, we perform a systematic memory fading analysis.

Table 3.26. *L_2-norm error, using uniform and non-uniform grids. The exact singular solution is $u^{ext} = (1-x)x^{1+\mu}$ with $\mu = 1/10$. Source: Reprinted from Kharazmi et al. (2017) with permission from Elsevier.*

	Uniform Grid	Kernel-Based Non-Uniform Grid	Geometrically Progressive Non-Uniform Grid
$N_{el} = 50,\ P = 2$	5.83943×10^{-4}	2.33461×10^{-5}	3.93956×10^{-4}
$N_{el} = 25,\ P = 4$	3.04739×10^{-5}	1.77458×10^{-7}	1.38755×10^{-6}
$N_{el} = 10,\ P = 10$	1.39586×10^{-5}	2.10813×10^{-9}	1.45695×10^{-9}

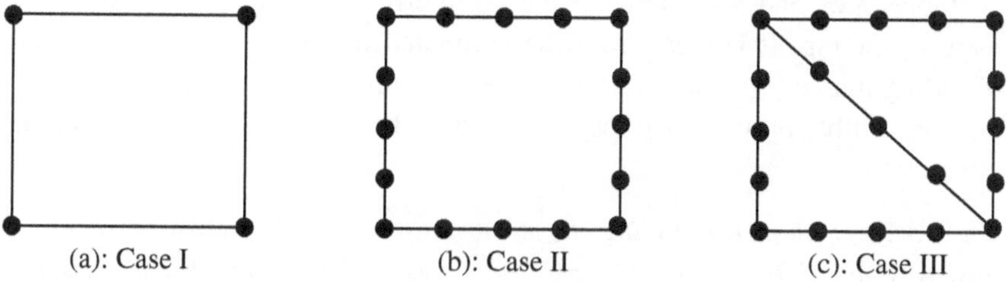

Figure 3.31 Memory fading: (a) B-B interaction, the corner entries (b) B-B and B-I interaction, the boundary entries (c) B-B, B-I and S-I interaction, boundary and diagonal entries. Source: Reprinted from Kharazmi et al. (2017) with permission from Elsevier.

In *full* memory fading, we fade the memory by truncating the history matrices, i.e., we consider the full history matrices up to some specific number and then truncate the rest of the history. For instance, we consider up to the first 4 history matrices for each element and thus compute $\hat{\mathbf{S}}^1$, $\hat{\mathbf{S}}^2$, $\hat{\mathbf{S}}^3$ and $\hat{\mathbf{S}}^4$, and truncate the rest $N_{el} - 1 - 4$ matrices; see Figure 3.21 for better visualization.

In *partial* memory fading, we fade the memory by partially computing the history matrices. Similar to the *full* memory fading, we consider the full history matrices up to some specific number; however, for the rest of the history matrices we partially compute the entries of the matrices. In partial memory fading, we consider three different cases as follows.

- **Case I:** Boundary-Boundary (B-B) interaction. In this case, we only consider the interactions of boundary mode and boundary test functions, i.e., $p = 0, P$ and $k = 0, P$, and thus, only compute the corner entries (see Figure 3.31a).
- **Case II:** Boundary-Boundary (B-B) and Boundary-Interior (B-I) interaction. In addition to the corner entries, here we also consider the

interaction of boundary mode/test functions with the interior test/mode functions, i.e.,

$$\begin{cases} k=0, \quad p=0,1,\dots,P, \quad \text{and} \quad k=P, \quad p=0,1,\dots,P, \\ p=0, \quad k=0,1,\dots,P, \quad \text{and} \quad p=P, \quad k=0,1,\dots,P, \end{cases}$$

and thus, we compute the boundary entries (see Figure 3.31b).

- **Case III:** Boundary-Boundary (B-B), Boundary-Interior (B-I), Self-Interior (S-I) interaction. In addition to the last two cases, we consider the interaction of each mode with its corresponding test function and thus, we compute the boundaries as well as the diagonal entries (see Figure 3.31c).

Tables 3.27 and 3.28 show the L_2-norm error for cases of full and partial memory fading. It is clear from the computed norms that even in the case of fading memory, we can still accurately obtain the approximation solution, however with a proportional loss of accuracy depending on the lack of modal interaction.

Table 3.27. *Full history fading: L_2-norm error using PG SEM with local basis/test functions, where $u^{ext} = x^7 - x^6$, $N_{el} = 19$, $P = 6$. The first column in the table shows the number of fully faded history matrices. Source: Reprinted from Kharazmi et al. (2017) with permission from Elsevier.*

	Full fading		
# faded history matrices	$\mu = 1/10$	$\mu = 1/2$	$\mu = 9/10$
0	9.26034×10^{-12}	2.31391×10^{-11}	4.24903×10^{-9}
2	7.8905×10^{-11}	1.26365×10^{-10}	4.25456×10^{-9}
5	1.42423×10^{-8}	6.39474×10^{-8}	1.95976×10^{-8}
8	2.69431×10^{-7}	2.47423×10^{-6}	8.45001×10^{-7}
11	2.09737×10^{-6}	3.19995×10^{-5}	1.37959×10^{-5}
14	9.07427×10^{-6}	2.44911×10^{-4}	1.40684×10^{-4}
17	2.94001×10^{-5}	1.39043×10^{-3}	1.6001×10^{-3}

Table 3.28. *Partial history fading: L_2-norm error using PG SEM with local basis/test functions, where $u^{ext} = x^7 - x^6$, $N_{el} = 19$, $P = 6$. The first column in the tables shows the number of partially faded history matrices. Source: Reprinted from Kharazmi et al. (2017) with permission from Elsevier.*

Partial fading case I			
# faded history matrices	$\mu = 1/10$	$\mu = 1/2$	$\mu = 9/10$
0	9.26034×10^{-12}	2.31391×10^{-11}	4.24903×10^{-9}
2	7.8905×10^{-11}	1.26365×10^{-10}	4.25456×10^{-9}
5	1.42423×10^{-8}	6.39474×10^{-8}	1.95976×10^{-8}
8	2.69431×10^{-7}	2.47423×10^{-6}	8.45001×10^{-7}
11	2.09737×10^{-6}	3.19995×10^{-5}	1.37959×10^{-5}
14	9.07427×10^{-6}	2.44911×10^{-4}	1.40684×10^{-4}
17	2.94001×10^{-5}	1.39043×10^{-3}	1.6001×10^{-3}

Partial fading case II			
# faded history matrices	$\mu = 1/10$	$\mu = 1/2$	$\mu = 9/10$
0	9.26034×10^{-12}	2.31391×10^{-11}	4.24903×10^{-9}
2	9.27241×10^{-12}	2.34361×10^{-11}	4.2491×10^{-9}
5	3.37716×10^{-11}	6.6476×10^{-10}	4.44832×10^{-9}
8	3.8092×10^{-10}	1.99961×10^{-8}	1.36941×10^{-8}
11	1.47228×10^{-9}	2.2715×10^{-7}	1.47786×10^{-7}
14	1.15821×10^{-8}	1.64103×10^{-6}	1.47098×10^{-6}
17	5.06103×10^{-7}	7.87929×10^{-6}	1.85274×10^{-5}

Partial fading case III			
# faded history matrices	$\mu = 1/10$	$\mu = 1/2$	$\mu = 9/10$
0	9.26034×10^{-12}	2.31391×10^{-11}	4.24903×10^{-9}
2	9.26023×10^{-12}	2.3113×10^{-11}	4.24903×10^{-9}
5	1.18462×10^{-11}	7.7854×10^{-11}	4.23683×10^{-9}
8	1.60656×10^{-10}	3.38055×10^{-9}	3.65689×10^{-9}
11	1.32413×10^{-9}	4.35421×10^{-8}	8.84638×10^{-9}
14	7.10271×10^{-9}	3.87096×10^{-7}	1.7226×10^{-7}
17	2.87023×10^{-8}	3.71057×10^{-6}	5.12104×10^{-6}

4
Global Schemes for Fractional PDEs (FPDEs)

We selectively present global methods for efficiently solving fractional PDEs (FPDEs), employing the basis functions introduced in Chapters 2 and 3. Here in this chapter, we adopt the term "global" in the sense of Section 1.4 and often in the context of *space-time*, considering time as another (space-like) spectral direction. We will examine a number of typical FPDEs, which we introduced and probabilistically interpreted in Chapter 1, including the subdiffusion equation, the tempered fractional diffusion on the half/whole line, and also the generalized and unified $(1+d)$-dimensional sub-to-superdiffusion FPDE model, for $d \geq 1$, where a single FPDE form can model a range of physical processes by just varying the corresponding temporal/spatial fractional derivatives in the model, hence, rendering the FPDE elliptic, parabolic, and/or hyperbolic on the $(1 + d)$-dimensional space-time hypercube. In this chapter, we employ one-sided, two-sided, constant/variable-order and fully distributed-order fractional operators, introduced in Chapters 1 and 2.

In Section 4.1, we formulate a generalized Jacobi poly-fractonomial (GJPF) space-time Petrov–Galerkin spectral method for solving the subdiffusion equation, being modeled by a time-fractional diffusion problem on a bounded domain, where the left- and right-sided GJPFs are employed as the basis (trial) and test functions, respectively. We study the well-posedness of the problem and introduce multiple algorithms, including: eigen-decomposition, partial QZ decomposition, and full decomposition, for the fast solution of the resulting linear system. Next, we present the corresponding numerical analysis followed with enriching the approximation space in the temporal direction. We examine different linear and nonlinear FPDEs including the time-fractional Allen–Cahn (phase-field) equation.

In Section 4.2, we present a unified GJPF space-time Petrov–Galerkin spectral method for general FPDEs on hypercubes. We describe the corresponding stochastic interpretation of the strong form of the proposed FPDE.

Interestingly, the generalized form of the model can code sub-to-superdiffusion in any $(1+d)$-dimensional space-time domain for any spatial dimension $d \geq 1$, where the FPDE can model a range of physical processes by just varying the corresponding temporal/spatial fractional derivatives rendering the underlying model elliptic, parabolic, and/or hyperbolic. We perform the stability and convergence analysis of the proposed model and develop a unified fast FPDE solver, following with numerical tests for hyperbolic/parabolic/elliptic FPDEs.

Subsequently, we construct a logrithmic orthogonal function (LOF) space-time Galerkin spectral method for the subdiffusion problem in Section 4.3. Due to their importance and rather distinguished structures, we provide a quick review of LOFs and their properties, followed by the relevant approximations to Mittag-Leffler functions also constructing the shifted generalized log-orthogonal functions (GLOFs). Harmoniously with the earlier section, we provide the corresponding well-posedness and error analyses of the method, followed with a formulation of a fast solver and numerical examples to echo the efficiency of the proposed scheme in the context of subdiffusion with constant and time-dependent coefficients, and the nonlinear time-fractional Allen–Cahn equation.

We devote Section 4.4 to tempered FPDEs on the whole/half line. We first provide a quick review of the notations used in this chapter and the definitions of tempered fractional integrals and derivatives, in addition to introducing the Laguerre polynomials along with some useful formulas, needed for the construction of our method. With this preparation, we present the generalized Laguerre functions (GLFs) and propose a well-posed Galerkin spectral method for the tempered fractional diffusion on the half line and a two-domain (i.e., two-element) spectral element method for solving the tempered fractional diffusion on the whole line.

In Section 4.5, we further generalize the method developed in Section 4.2 in two ways: i) we employ two-sided Riemann–Liouville fractional derivatives in all the spatial dimensions and ii) we render all the temporal and spatial derivatives *distributed-order* with respect to some distributional weight functions.

In the last section and employing the spectral collocation methods presented in Section 3.2.6, we introduce a GJPF tunable collocation method for variable-order linear and nonlinear FPDEs. Of particular interest, we solve space-variable order FPDEs including the superdiffusion model and the space-fractional Burgers equation, where we demonstrate the accuracy, stability, and flexibility of the proposed spectral collocation method with examining a variety of different variable derivative order fields.

4.1 GJPF Space-Time Method for the Subdiffusion Equation

We consider the following time fractional diffusion equation (TFDE):

$$ {}_{0}^{\mathrm{C}}D_t^{\alpha} v(\boldsymbol{x},t) + \mathcal{L}v(\boldsymbol{x},t) + \mathcal{N}(v(\boldsymbol{x},t),t) = 0 \tag{4.1.1} $$

for $(\boldsymbol{x},t) \in D := \Omega \times (0,T]$, with the initial condition

$$ v(\boldsymbol{x},0) = v_0(\boldsymbol{x}), \quad \forall \boldsymbol{x} \in \Omega, \tag{4.1.2} $$

and the homogeneous boundary conditions. In the above, $\alpha \in (0,1)$, Ω is a bounded domain in $\mathbb{R}^d$, $\mathcal{L}$ is a linear self-adjoint elliptic operator and $\mathcal{N}$ is a lower-order nonlinear operator, and ${}_{0}^{\mathrm{C}}D_t^{\alpha}$ $(0 < \alpha < 1)$ is the left-sided Caputo fractional derivative of order α.

Some Functional Spaces and Their Properties

We now introduce some functional spaces to be used later.

Let $I = (0,T)$. Let $C_0^{\infty}(I)$ be the space of smooth functions with compact support in I, and let $H_0^s(I)$ denote the closure of $C_0^{\infty}(I)$ with respect to norm $\|\cdot\|_{s,I}$.

For any real $s > 0$, we introduce some useful spaces.

- Let $\widetilde{H}^s(I)$ be the set of functions in $H^s(I)$ whose extension by zero to $\mathbb{R}$ is in $H^s(\mathbb{R})$.
- Let $\widetilde{H}_l^s(I)$ be the set of functions u whose extension by zero, denoted by $\tilde{u}$ is in $H^s(-\infty,T)$.
- Let $\widetilde{H}_r^s(I)$ be the set of functions u whose extension by zero, denoted by $\tilde{u}$ is in $H^s(0,\infty)$.

Therefore, a direct consequence is as follows, see, e.g., Jin et al. (2015b).

Lemma 4.1 *The operators* ${}_{0}^{\mathrm{RL}}D_t^{s}$ *and* ${}_{t}^{\mathrm{RL}}D_T^{s}$ *extend continuously to bounded operators from* $\widetilde{H}_l^s(I)$ *and* $\widetilde{H}_r^s(I)$*, respectively, to* $L^2(I)$.

We define the following space in the x-direction:

$$ H_{\mathcal{L}}(\Omega) = \{v \,:\, v \in L^2(\Omega), \langle \mathcal{L}v, v\rangle < \infty, \; v|_{\partial\Omega} = 0\}, \tag{4.1.3} $$

equipped with the norm

$$ \|v\|_{\mathcal{L}} = \left(\int_{\Omega} v^2(\boldsymbol{x}) \mathrm{d}\boldsymbol{x} + \int_{\Omega} \mathcal{L}v(\boldsymbol{x}) v(\boldsymbol{x}) \mathrm{d}\boldsymbol{x} \right)^{1/2}. \tag{4.1.4} $$

The notation $\langle\cdot,\cdot\rangle_{L^2(\Omega)}$ denotes the duality pairing between $H_{\mathcal{L}}(\Omega)$ and its dual $H_{\mathcal{L}}^{-1}(\Omega)$, also the inner product in $L^2(\Omega)$. For $u, v \in L^2(D)$, for each $t \in I$, $u(t)$, $v(t) \in H_{\mathcal{L}}(\Omega)$, we denote

$$b(u, v) := \langle \mathcal{L}u, v\rangle_{L^2(D)} = \int_0^T \int_\Omega \mathcal{L}u(\boldsymbol{x}, t)\, v(\boldsymbol{x}, t)\mathrm{d}\boldsymbol{x}\mathrm{d}t, \tag{4.1.5}$$

and the Bochner spaces on D:

$$\begin{aligned} &\mathbb{V} := L^2(I; H_{\mathcal{L}}(\Omega)) \quad \text{with norm} \quad \|v\|_{\mathbb{V}}^2 = b(v, v), \\ &\mathbb{V}^* := L^2(I; H_{\mathcal{L}}^{-1}(\Omega)) \quad \text{with norm} \quad \|v\|_{\mathbb{V}^*} = \sup_{0\neq\phi\in\mathbb{V}} \frac{\langle v, \phi\rangle_{L^2(D)}}{\|\phi\|_{\mathbb{V}}}. \end{aligned} \tag{4.1.6}$$

Below we will also use $\langle\cdot,\cdot\rangle_{L^2(D)}$ for the duality pairing between $\mathbb{V}$ and $\mathbb{V}^*$. For any $s \in (0, 1)$, we also define space

$$B^s(D) := \widetilde{H}_l^s(I, H_{\mathcal{L}}^{-1}(\Omega)) \cap L^2(I, H_{\mathcal{L}}(\Omega)),$$

equipped with the norm

$$\|v\|_{B^s(D)} := \left(\| {}^{\mathrm{RL}}_{\ 0}D_t^s v\|^2_{L^2(I, H_{\mathcal{L}}^{-1}(\Omega))} + \|v\|^2_{L^2(I, H_{\mathcal{L}}(\Omega))}\right)^{1/2}.$$

The following Lemma is useful, see, e.g., Duan et al. (2018).

Lemma 4.2 *For all* $0 < \alpha < 1$, *if* $u \in B^\alpha(D)$, *then*

$$\|u\|_{\widetilde{H}_l^\alpha(I; H_{\mathcal{L}}^{-1}(\Omega))} \sim \| {}^{\mathrm{RL}}_{\ 0}D_t^\alpha u\|_{L^2(I; H_{\mathcal{L}}^{-1}(\Omega))}. \tag{4.1.7}$$

Basis (Trial) and Test Functions in Time

We introduce below the trial and test functions that we will use for the time variable. For $\alpha, \beta > -1$, let $P_n^{(\alpha,\beta)}(x)$, $x \in \Lambda$ be the standard Jacobi polynomial of degree n, and denote the weight function $\chi^{(\alpha,\beta)}(x) = (1 - x)^\alpha(1+x)^\beta$. The set of Jacobi polynomials is a complete $L^2_{\chi^{(\alpha,\beta)}}(\Lambda)$-orthogonal system, i.e.,

$$\int_{-1}^1 P_l^{(\alpha,\beta)}(x)P_m^{(\alpha,\beta)}(x)\chi^{(\alpha,\beta)}(x)\mathrm{d}x = \gamma_l^{(\alpha,\beta)}\delta_{l,m}, \tag{4.1.8}$$

where $\delta_{l,m}$ is the Kronecker delta, and

$$\gamma_l^{(\alpha,\beta)} = \frac{2^{\alpha+\beta+1}}{(2l + \alpha + \beta + 1)} \frac{\Gamma(l + \alpha + 1)\Gamma(l + \beta + 1)}{l!\,\Gamma(l + \alpha + \beta + 1)}.$$

In particular, $P_0^{(\alpha,\beta)}(x) = 1$.

The shifted Jacobi polynomial of degree n is defined by

$$\widetilde{P}_n^{(\alpha,\beta)}(t) = P_n^{(\alpha,\beta)}\Big(\frac{2t - T}{T}\Big), \quad t \in I, \quad n \geq 0. \tag{4.1.9}$$

Clearly, the set of $\{\widetilde{P}_n^{(\alpha,\beta)}(t)\}_{n\geq 0}$ is a complete $L^2_{\omega^{(\alpha,\beta)}}(I)$-orthogonal system with the weight function $\omega^{(\alpha,\beta)}(t) = (T - t)^\alpha t^\beta$. From (4.1.8) and (4.1.9) we get that

$$\int_I \widetilde{P}_l^{(\alpha,\beta)}(t)\widetilde{P}_m^{(\alpha,\beta)}(t)\omega^{(\alpha,\beta)}(t)\mathrm{d}t = \left(\frac{T}{2}\right)^{\alpha+\beta+1}\gamma_l^{(\alpha,\beta)}\delta_{l,m}. \tag{4.1.10}$$

For any α, $\beta > -1$, the shifted generalized Jacobi function on I is defined by Shen et al. (2015)

$$J_n^{(\alpha,\beta)}(t) = t^\beta \widetilde{P}_n^{(\alpha,\beta)}(t), \quad t \in I, \quad n \geq 0, \tag{4.1.11}$$

and our approximation space on I is defined by

$$\begin{aligned}\mathcal{F}_N^{(\alpha)}(I) &:= \{t^\alpha\psi(t) : \psi(t) \in \mathcal{P}_N(I)\}\\ &= \mathrm{span}\{J_n^{(-\alpha,\alpha)}(t) = t^\alpha \widetilde{P}_n^{(-\alpha,\alpha)}(t) : 0 \leq n \leq N\},\end{aligned} \tag{4.1.12}$$

which incorporates the homogeneous boundary conditions at $t = 0$.

A particular case is the shifted Legendre polynomial $L_n(t)$, $t \in I$ defined by

$$L_n(t) = P_n^{(0,0)}\left(\frac{2t}{T} - 1\right), \quad n = 0, 1, 2, \ldots. \tag{4.1.13}$$

The set of $L_n(t)$ is a complete $L^2(I)$-orthogonal system, namely,

$$\int_I L_l(t)L_m(t)\mathrm{d}t = \frac{T}{2l+1}\delta_{l,m}. \tag{4.1.14}$$

Clearly, we derive from (4.1.11)–(4.1.13) and a direct calculation that

$${}_0^{\mathrm{RL}}D_t^\alpha J_n^{(-\alpha,\alpha)}(t) = \frac{\Gamma(n+\alpha+1)}{n!}L_n(t). \tag{4.1.15}$$

4.1.1 A Petrov–Galerkin Space-Time Method

We start with a special case of (4.1.1) with $\mathcal{N}v = \beta v - g$:

$${}_0^{\mathrm{C}}D_t^\alpha v(\boldsymbol{x},t) + \mathcal{L}v(\boldsymbol{x},t) + \beta v(\boldsymbol{x},t) = g(\boldsymbol{x},t), \quad \forall(\boldsymbol{x},t) \in D, \tag{4.1.16}$$

with the initial and boundary conditions (4.1.2) and the constant $\beta \geq 0$. Note that the above linear equation will be used as a preconditioner for nonlinear equation (4.1.1) later.

GJPF Petrov–Galerkin Method for (4.1.16)

We assume that $v_0 \in H^2(\Omega) \cap H_0^1(\Omega)$. For the case of non-homogeneous initial conditions $v(\boldsymbol{x},0) = v_0(\boldsymbol{x})$, we first decompose the solution $v(\boldsymbol{x},t)$ into two parts as

$$v(\boldsymbol{x},t) = u(\boldsymbol{x},t) + v_0(\boldsymbol{x}), \tag{4.1.17}$$

with $u(\boldsymbol{x},0) = 0$. Hence, the equation (4.1.16) is equivalent to the following equation with Riemann–Liouville fractional derivative:

$$ {}^{\mathrm{RL}}_{0}D^{\alpha}_{t}u(\boldsymbol{x},t) + \mathcal{L}u(\boldsymbol{x},t) + \beta u(\boldsymbol{x},t) = f(\boldsymbol{x},t), \quad \forall(\boldsymbol{x},t) \in D, \qquad (4.1.18) $$

where

$$ f(\boldsymbol{x},t) = g(\boldsymbol{x},t) - \mathcal{L}v_0(\boldsymbol{x}) - \beta v_0(\boldsymbol{x}), $$

with the homogeneous initial condition

$$ u(\boldsymbol{x},0) = 0, \quad \forall \boldsymbol{x} \in \Omega, \qquad (4.1.19) $$

and the corresponding boundary conditions.

A weak formulation of (4.1.16) is to find $u \in B^{\alpha}(D)$ such that for all $v \in \mathbb{V}$,

$$ \mathcal{A}(u,v) := \langle {}^{\mathrm{C}}_{0}D^{\alpha}_{t}u, v\rangle_{L^2(D)} + b(u,v) + \beta(u,v)_{L^2(D)} = (g,v)_{L^2(D)}, \quad (4.1.20) $$

where $b(\cdot,\cdot)$ and $\mathbb{V}$ are defined in (4.1.5) and (4.1.6), respectively. Without loss of generality, we assume that the bilinear form $b(\cdot,\cdot)\colon \mathbb{V}\times\mathbb{V} \to \mathbb{R}$ satisfies

- Continuity: $\exists\, C_1 > 0$ such that $|b(u,v)| \le C_1\|u\|_{\mathbb{V}}\|v\|_{\mathbb{V}}$,
- Coercivity: $\exists\, C_2 > 0$ such that $|b(u,u)| \ge C_2\|u\|^2_{\mathbb{V}}$.

A direct consequence is the following.

Theorem 4.3 *For any $f \in \mathbb{V}^*$, problem* (4.1.20) *has a unique solution $u \in B^{\alpha}(D)$, and it satisfies*

$$ \|u\|_{B^{\alpha}(D)} \le c\|f\|_{\mathbb{V}^*}. $$

Proof By using an argument similar to the proof of Lemma 2.4 in Duan et al. (2018), we deduce that

$$ \begin{aligned} |\mathcal{A}(u,v)| &\le \left|\langle {}^{\mathrm{RL}}_{0}D^{\alpha}_{t}u, v\rangle_{L^2(D)}\right| + |b(u,v)| + \beta|(u,v)_{L^2(D)}| \\ &\le c_{\beta}\|u\|_{B^{\alpha}(D)}\|v\|_{\mathbb{V}}, \\ \sup_{v\in\mathbb{V}} \frac{\mathcal{A}(u,v)}{\|v\|_{\mathbb{V}}} &\ge \|u\|_{B^{\alpha}(D)}, \quad \forall u \in B^{\alpha}(D). \end{aligned} \qquad (4.1.21) $$

Hence, the existence, uniqueness and stability follow directly from the continuity and inf-sup condition of the bilinear form $\mathcal{A}(\cdot,\cdot)$. □

Remark 4.4 The Galerkin weak formulation (see, e.g., Li and Xu (2009)) is to find $u \in \tilde{B}^{\frac{\alpha}{2}} := H^{\frac{\alpha}{2}}(I; L^2(\Omega)) \cap L^2(I; H^1_0(\Omega))$ such that

$$ a(u,v) = \left\langle {}^{\mathrm{RL}}_{0}D^{\frac{\alpha}{2}}_{t}u, {}^{\mathrm{RL}}_{0}D^{\frac{\alpha}{2}}_{T}v\right\rangle_{L^2(D)} + (\nabla u, \nabla v)_{L^2(D)} = \langle f, v\rangle_{L^2(D)}, \quad \forall v \in \tilde{B}^{\frac{\alpha}{2}}. $$

Then, the bilinear form $a(\cdot,\cdot)$ is continuous and coercive on $\tilde{B}^{\frac{\alpha}{2}}$, and the variational problem is well-posed.

Let V_h be a finite-dimensional approximation space of $V := H_{\mathcal{L}}(\Omega)$ defined by

$$V_h = \text{span}\{\phi_1, \phi_2, \ldots, \phi_M\}. \tag{4.1.22}$$

For the time variable, we shall use the fractional polynomial space $\mathcal{F}_N^{(\alpha)}(I)$ defined in (4.1.12). Then, the Petrov–Galerkin method for (4.1.18) is to seek $u_L(\boldsymbol{x},t) := u_{MN} \in V_h \otimes \mathcal{F}_N^{(\alpha)} \subset B^\alpha(D)$ such that

$$\begin{aligned}\mathcal{A}(u_L, v) &:= \langle {}^{\mathrm{RL}}_{\ 0}D_t^\alpha u_L, v\rangle_{L^2(D)} + b(u_L, v) + \beta(u_L, v)_{L^2(D)} \\ &= (f, v)_{L^2(D)}, \quad \forall v \in V_h \otimes \mathcal{P}_N \subset \mathbb{V}.\end{aligned} \tag{4.1.23}$$

Next, we shall construct suitable basis functions of $\mathcal{F}_N^{(\alpha)}(I)$ so that the above system can be solved efficiently.

For the time variable, we use the generalized Jacobi functions $J_n^{(-\alpha,\alpha)}(t)$ defined in the last section,

$$\mathcal{F}_N^{(\alpha)}(I) = \text{span}\{J_n^{(-\alpha,\alpha)}(t) : 0 \le n \le N\}, \tag{4.1.24}$$

and for $\mathcal{P}_N(I)$ in the test space, we simply use the scaled Legendre polynomials, namely,

$$\mathcal{P}_N(I) = \text{span}\left\{L_n^{(\alpha)}(t) := \kappa_{n,\alpha} L_n(t) : 0 \le n \le N\right\}, \tag{4.1.25}$$

with the constant

$$\kappa_{n,\alpha} = \frac{n!\,(2n+1)}{T \cdot \Gamma(n+\alpha+1)}.$$

We now describe the numerical implementations for (4.1.23) under this set of basis functions. We write

$$u_L(\boldsymbol{x}, t) = \sum_{m=1}^{M} \sum_{n=0}^{N} \widetilde{u}_{mn} \phi_m(\boldsymbol{x}) J_n^{(-\alpha,\alpha)}(t). \tag{4.1.26}$$

Denote

$$\begin{aligned}&f_{mn} = (f, \phi_m L_n^{(\alpha)})_{L^2(D)}, \quad F = (f_{mn})_{1\le m\le M, 0\le n\le N}, \\ &s^t_{pq} = \int_I {}^{\mathrm{RL}}_{\ 0}D_t^\alpha J_q^{(-\alpha,\alpha)}(t) L_p^{(\alpha)}(t)\mathrm{d}t, \\ &m^t_{pq} = \int_I J_q^{(-\alpha,\alpha)}(t) L_p^{(\alpha)}(t)\mathrm{d}t, \\ &\boldsymbol{S}^t = (s^t_{pq})_{0\le p,q\le N}, \quad \boldsymbol{M}^t = (m^t_{pq})_{0\le p,q\le N}, \\ &U = (\widetilde{u}_{mn})_{1\le m\le M, 0\le n\le N}.\end{aligned} \tag{4.1.27}$$

It can be verified readily from (4.1.14), (4.1.15) and (4.1.25) that $\boldsymbol{S}^t = \boldsymbol{I}$ with $\boldsymbol{I}$ being the identity matrix. On the other hand,

$$m^t_{pq} = \int_I J_p^{(-\alpha,\alpha)}(t) L_q^{(\alpha)}(t)\mathrm{d}t = \int_I \widetilde{P}_p^{(-\alpha,\alpha)}(t) L_q^{(\alpha)}(t) t^{\alpha}\mathrm{d}t. \tag{4.1.28}$$

Note that $\boldsymbol{M}^t$ is not sparse but can be accurately computed by Jacobi–Gauss quadrature with index $(0, \alpha)$.

Then, we find that (4.1.23) is equivalent to the following linear system:

$$\boldsymbol{M}^h U(\boldsymbol{S}^t)^T + \boldsymbol{S}^h U(\boldsymbol{M}^t)^T + \beta \boldsymbol{M}^h U(\boldsymbol{M}^t)^T = F, \tag{4.1.29}$$

where $\boldsymbol{M}^h$ and $\boldsymbol{S}^h$ are the mass matrix and stiffness matrix in the space variable, namely,

$$\begin{aligned} &s^h_{pq} = \int_\Omega \mathcal{L}\phi_q(\boldsymbol{x})\phi_p(\boldsymbol{x})\mathrm{d}\boldsymbol{x}, \quad m^h_{pq} = \int_\Omega \phi_q(\boldsymbol{x})\phi_p(\boldsymbol{x})\mathrm{d}\boldsymbol{x}, \\ &\boldsymbol{S}^h = (s^h_{pq})_{1\le p,q\le M}, \quad \boldsymbol{M}^h = (m^h_{pq})_{1\le p,q\le M}. \end{aligned} \tag{4.1.30}$$

It is noteworthy that it is not computationally efficient to use the matrix decomposition method in the space direction for non-tensor product domains, because the size of $\boldsymbol{M}^h$ and $\boldsymbol{S}^h$ is always very large in (4.1.29). In this paper, we devote to doing matrix decomposition in the time direction. Notice that an advantage of using matrix decomposition in the time direction is that we only need to solve N times an elliptic equation in the space direction. Moreover, it is straightforward to extend this approach to the nonlinear case (4.1.1); we omitted all the details in this paper.

Several Algorithms for Solving (4.1.29)

This subsection focuses on the efficient algorithms for solving the nonsymmetric linear system.

Algorithm I: Eigen-decomposition. We consider the generalized eigenvalue problem: let $\boldsymbol{E} := (\bar{e}_0, \ldots, \bar{e}_N) = (e_{pq})_{p,q=0,\ldots,N}$ be the matrix formed by the orthonormal eigenvectors of the generalized eigenvalue problem $\boldsymbol{M}^t \bar{e}_j = \lambda_j \boldsymbol{S}^t \bar{e}_j$ and $\Lambda = \mathrm{diag}(\lambda_0, \ldots, \lambda_N)$, i.e.,

$$\boldsymbol{M}^t \boldsymbol{E} = \boldsymbol{S}^t \boldsymbol{E}\Lambda. \tag{4.1.31}$$

Setting $U = V\boldsymbol{E}^T$, and multiplying (4.1.29) (or (4.1.56) respectively) both sides by $(\boldsymbol{S}^t \boldsymbol{E})^{-T}$, we arrive at

$$\boldsymbol{M}^h V + \boldsymbol{S}^h V\Lambda + \beta \boldsymbol{M}^h V\Lambda = G := F(\boldsymbol{S}^t \boldsymbol{E})^{-T}. \tag{4.1.32}$$

Hence, the m-th column of the above matrix equation becomes

$$(\lambda_n \boldsymbol{S}^h + (1+\beta\lambda_n)\boldsymbol{M}^h)\boldsymbol{v}_n = \boldsymbol{g}_n, \quad 0 \le n \le N, \tag{4.1.33}$$

where

$$V = (\widetilde{v}_{m,n})_{1\le m\le M, 0\le n\le N}, \quad \boldsymbol{v}_n = (\widetilde{v}_{1,n}, \widetilde{v}_{2,n}, \ldots, \widetilde{v}_{M,n})^T,$$
$$G = (g_{m,n})_{1\le m\le M, 0\le n\le N}, \quad \boldsymbol{g}_n = (g_{1,n}, g_{2,n}, \ldots, g_{M,n})^T.$$

The coefficient U can be computed by $U = V\boldsymbol{E}^T$. Finally, we can use (4.1.26) and (4.1.17) to get the solution u_L. This approach is very efficient because we only need to solve $N+1$ times an elliptic equation in space, and the total cost is $O(NT_M)$, where T_M is the cost of the underlying algorithm in the space direction. However, since $\boldsymbol{M}^t$ is nonsymmetric, this approach suffers from large round-off errors. In Table 4.1, we tabulate the maximum error of $(\boldsymbol{S}^t\boldsymbol{E})^{-1}\boldsymbol{M}^t\boldsymbol{E} - \Lambda$ for the stiffness/mass matrix in (4.1.27) with various N, which indicates that the error of eigen-decomposition is grown as N increases. We then conclude that the eigen-decomposition algorithm for the linear system with the full nonsymmetric matrix is not stable.

Algorithm II: Partial QZ Decomposition. We consider the following QZ decomposition:

$$\begin{aligned} &\boldsymbol{Q}\,(\boldsymbol{S}^t)^T\boldsymbol{Z} = \boldsymbol{A},\ \boldsymbol{Q}\,(\boldsymbol{M}^t)^T\boldsymbol{Z} = \boldsymbol{B} \\ &\quad \Leftrightarrow \boldsymbol{Q}^T\boldsymbol{A}\boldsymbol{Z}^T = (\boldsymbol{S}^t)^T,\ \boldsymbol{Q}^T\boldsymbol{B}\boldsymbol{Z}^T = (\boldsymbol{M}^t)^T, \end{aligned} \tag{4.1.34}$$

where $\boldsymbol{Q}$, $\boldsymbol{Z}$ are the unitary matrices, and $\boldsymbol{A}$, $\boldsymbol{B}$ are upper triangular matrices, namely,

$$\boldsymbol{A} = \begin{pmatrix} a_{0,0} & a_{0,1} & \cdots & a_{0,N} \\ & a_{1,1} & \cdots & a_{1,N} \\ & & \ddots & \vdots \\ & & & a_{N,N} \end{pmatrix}, \quad \boldsymbol{B} = \begin{pmatrix} b_{0,0} & b_{0,1} & \cdots & b_{0,N} \\ & b_{1,1} & \cdots & b_{1,N} \\ & & \ddots & \vdots \\ & & & b_{N,N} \end{pmatrix}.$$

Setting $U = V\boldsymbol{Q}$, and multiplying (4.1.29) both sides by $\boldsymbol{Z}$, we arrive at

$$\boldsymbol{M}^h\,V\,\boldsymbol{A} + \boldsymbol{S}^h\,V\,\boldsymbol{B} + \beta\boldsymbol{M}^h\,V\,\boldsymbol{B} = G := F\boldsymbol{Z}. \tag{4.1.35}$$

We can obtain the column of coefficients $\boldsymbol{v}_n = (\widetilde{v}_{1,n}, \widetilde{v}_{2,n}, \ldots, \widetilde{v}_{M,n})^T$, one by one, namely,

$$a_{n,n}\boldsymbol{M}^h\,\boldsymbol{v}_n + b_{n,n}\boldsymbol{S}^h\,\boldsymbol{v}_n + \beta b_{n,n}\boldsymbol{M}^h\,\boldsymbol{v}_n = \boldsymbol{g}_n - \boldsymbol{h}_{n-1}, \quad 0 \le n \le N,$$

where

$$\boldsymbol{h}_{n-1} = \sum_{k=0}^{n-1}\big(a_{k,n}\boldsymbol{M}^h\,\boldsymbol{v}_k + b_{k,n}\boldsymbol{S}^h\,\boldsymbol{v}_k + \beta b_{k,n}\boldsymbol{M}^h\,\boldsymbol{v}_k\big). \tag{4.1.36}$$

Briefly, we obtain the unknown coefficients $\boldsymbol{v}_n$ in terms of previous computed coefficients $\{\boldsymbol{v}_k\}_{k=0}^{n-1}$, as stated in **Algorithm II** below.

Table 4.1. *The error of eigen-decomposition. Source: Shen and Sheng (2019).*

	Max error of $(\boldsymbol{S}^t\boldsymbol{E})^{-1}\boldsymbol{M}^t\boldsymbol{E}-\Lambda$ in (4.1.27) with $\alpha = 0.3$								
N	4	8	12	16	20	24	28	32	100
error	7.98e-15	6.36e-13	1.33e-10	1.73e-08	2.20e-06	3.70e-04	2.81e-02	8.79e-02	1.58e-02
	Max error of $(\boldsymbol{S}^t\boldsymbol{E})^{-1}\boldsymbol{M}^t\boldsymbol{E}-\Lambda$ in (4.1.27) with $\alpha = 0.7$								
N	4	8	12	16	20	24	28	32	100
error	3.64e-15	1.91e-13	7.36e-11	3.85e-09	3.72e-07	2.22e-04	2.93e-03	2.11e-03	1.11e-02

Algorithm II for (4.1.35)

For $n = 0, 1, \ldots, N$,
  If $n = 0$,
  Solve $\boldsymbol{v}_0^h$ by: $a_{0,0}\boldsymbol{M}^h\, \boldsymbol{v}_0 + b_{0,0}\boldsymbol{S}^h\, \boldsymbol{v}_0 + \beta b_{0,0}\boldsymbol{M}^h\, \boldsymbol{v}_0 = \boldsymbol{g}_0$;
  Else
    Compute the history part $\boldsymbol{h}_{n-1}$ by (4.1.36);
    Solve $\boldsymbol{v}_n^h$ by: $a_{n,n}\boldsymbol{M}^h\, \boldsymbol{v}_n + b_{n,n}\boldsymbol{S}^h\, \boldsymbol{v}_n + \beta b_{n,n}\boldsymbol{M}^h\, \boldsymbol{v}_n = \boldsymbol{g}_n - \boldsymbol{h}_{n-1}$;
  End
End

Finally, the coefficient matrix U can be computed by $U = V\boldsymbol{Q}$, and we can use (4.1.26) and (4.1.17) to get the value of solution u_L. In this algorithm, we only need to solve $N+1$ times an elliptic equation, so this new approach is also very efficient. Let T_M be the cost of solving one elliptic equation in space, then the total cost is $O(N^2M) + (N+1)T_M$ (usually $N \ll M$). In Table 4.2, we tabulate the maximum error of $\max\{|\boldsymbol{Q}^T\boldsymbol{A}\boldsymbol{Z}^T - \boldsymbol{S}^t|, |\boldsymbol{Q}^T\boldsymbol{B}\boldsymbol{Z}^T - \boldsymbol{M}^t|\}$ for the stiffness/mass in (4.1.27) with various N, which indicates that the error of eigen-decomposition is always stable. We find that QZ decomposition is much more stable than eigen-decomposition for a full nonsymmetric matrix.

Algorithm III: QZ Decomposition. In cases where it is feasible to diagonalize also in the space variables, e.g., when Ω is a simple geometric (such as rectangular domains), we can simplify the algorithm even further.

We consider the following eigen-decomposition in space and QZ decomposition in time:

$$\boldsymbol{M}^h\boldsymbol{E}^h = \boldsymbol{S}^h\boldsymbol{E}^h\Lambda^h, \quad \boldsymbol{Q}^t\,(\boldsymbol{S}^t)^T\boldsymbol{Z}^t = \boldsymbol{A}^t, \quad \boldsymbol{Q}^t\,(\boldsymbol{M}^t)^T\boldsymbol{Z}^t = \boldsymbol{B}^t,$$

where $\boldsymbol{E}^h$ and Λ^h are matrix form eigenvectors and eigenvalues, $\boldsymbol{Q}^t$, $\boldsymbol{Z}^t$ are the unitary matrices, and $\boldsymbol{A}^t$, $\boldsymbol{B}^t$ are upper triangular matrices.

Setting $U = \boldsymbol{E}\,V\boldsymbol{Q}^t$, and multiplying the left (resp. right) of equation (4.1.29) by $(\boldsymbol{S}^h\boldsymbol{E}^h)^{-1}$ (resp. $\boldsymbol{Z}^t$), we arrive at

$$\Lambda^h\, V\boldsymbol{A}^t + V\boldsymbol{B}^t + \beta\Lambda^h\, V\boldsymbol{B}^t = G := (\boldsymbol{S}^h\boldsymbol{E}^h)^{-1}F\boldsymbol{Z}^t. \tag{4.1.37}$$

We can solve the coefficients $\{\widetilde{v}_{mn}\}_{0\le n\le N}$, one by one, namely,

$$\widetilde{v}_{mn} = \frac{g_{mn} - h_{mn}}{\lambda_{mm}^h a_{nn}^t + b_{nn}^t + \beta\lambda_{mm}^h b_{nn}^t}, \tag{4.1.38}$$

where for $1 \le m \le M$,

$$h_{mn} = \sum_{k=0}^{n-1}\left(\lambda_{m,m}^h\widetilde{v}_{mk}a_{kn}^t + \widetilde{v}_{mk}b_{kn}^t + \beta\lambda_{m,m}^h\widetilde{v}_{mk}b_{kn}^t\right). \tag{4.1.39}$$

Table 4.2. *The error of QZ-decomposition. Source: Shen and Sheng (2019).*

	Max error of $\max\{\lvert \boldsymbol{Q}^T \boldsymbol{A}\boldsymbol{Z}^T - \boldsymbol{S}^t \rvert, \lvert \boldsymbol{Q}^T \boldsymbol{B}\boldsymbol{Z}^T - \boldsymbol{M}^t \rvert\}$ in (4.1.27) with $\alpha = 0.3$							
N	4	8	12	16	20	24	28	32	100
error	2.22e-15	2.22e-15	2.25e-15	1.65e-15	1.82e-15	1.57e-15	2.46e-15	3.10e-15	3.02e-15
	Max error of $\max\{\lvert \boldsymbol{Q}^T \boldsymbol{A}\boldsymbol{Z}^T - \boldsymbol{S}^t \rvert, \lvert \boldsymbol{Q}^T \boldsymbol{B}\boldsymbol{Z}^T - \boldsymbol{M}^t \rvert\}$ in (4.1.27) with $\alpha = 0.7$								
N	4	8	12	16	20	24	28	32	100
error	1.31e-15	1.36e-15	1.59e-15	1.56e-15	2.00e-15	2.30e-15	2.10e-15	2.16e-15	1.24e-14

Briefly, we obtain the unknown column coefficients $\{\widetilde{v}_{mn}\}_{0\le n\le N}$, in terms of previous computed coefficients $\{\widetilde{v}_{mj}\}_{j=0,1,\ldots,n-1}$, as stated in **Algorithm III** below.

Algorithm III for (4.1.37)

For $n = 0,\ldots,N$,

  If $n = 0$,

  Solve $\widetilde{v}_{m0}$ by: $g_{m0}/\left(\lambda^h_{mm}a^t_{00} + b^t_{00} + \beta\lambda^h_{mm}b^t_{00}\right)$ with $1 \le m \le M$;

  Else

    Compute the history part h_{mn} by (4.1.39);

    Solve $\widetilde{v}_{mn}$ by: $\widetilde{v}_{mn} = \dfrac{g_{mn} - h_{mn}}{\lambda^h_{mm}a^t_{nn} + b^t_{nn} + \beta\lambda^h_{mm}b^t_{nn}}$ with $1 \le m \le M$;

  End

End

Finally, the coefficient matrix U can be computed by $U = \boldsymbol{E}\,V\,\boldsymbol{Q}^t$, and we can use (4.1.17) to get the value of solution u_L. Notice that the full decomposition algorithm procedure is simpler to implement than the partial QZ decomposition algorithm, when Ω is a simple geometric.

We summarize the three algorithms below: (i) all the three algorithms are only needed to solve $N+1$ times systems of order M; (ii) Algorithms II and III are very stable for solving nonsymmetric linear systems, while Algorithm I is less stable; (iii) Algorithms II and III provide the same numerical results when Ω is a simple geometric, and Algorithm III is simpler to implement than Algorithm II; (iv) it is important to point out that we can use QZ decomposition in both time and space directions when the spatial discretization is nonsymmetric too.

4.1.2 Error Estimates for GJPF Space-Time Spectral Methods

In this subsection, we conduct error analysis for the Petrov–Galerkin method described in the previous section. Here we shall consider the analysis for the linear case of (4.1.16) with $\mathcal{L} = -\Delta$, that is $\widetilde{\mathcal{N}}(u) = \beta(u + u_0)$.

In order to characterize the regularity of u in t, we define the following non-uniformly weighted space involving fractional derivatives

$$\mathcal{B}^s_{\alpha,\beta}(I) := \left\{v \in L^2_{\omega^{(\alpha,-\beta)}}(I) : {}^{\mathrm{RL}}_{0}D^{\beta+r}_t v \in L^2_{\omega^{(\alpha+\beta+r,r)}}(I),\ 0 \le r \le s\right\},\quad s \in \mathbb{N}_0.$$

The corresponding projection operator in time is $\pi^{(-\alpha,\alpha)}_N :\ L^2_{\omega^{(-\alpha,-\alpha)}}(I) \to \mathcal{F}^{(\alpha)}_N(I)$ defined by

$$\left(\pi_N^{(-\alpha,\alpha)} v - v, \psi\right)_{\omega^{(-\alpha,-\alpha)}} = 0, \quad \forall \psi \in \mathcal{F}_N^{(\alpha)}(I). \tag{4.1.40}$$

It is shown in Chen et al. (2016) that

$$({}^{\mathrm{RL}}_{0}D_t^{\alpha}(\pi_N^{(-\alpha,\alpha)} v - v), p) = 0, \quad \forall p \in \mathcal{P}_N(I). \tag{4.1.41}$$

We recall from Shen et al. (2015) the following.

Lemma 4.5 *Let $\alpha \in (0,1)$, for any $v \in \mathcal{B}^s_{-\alpha,\alpha}(I)$, with integer $0 \le s \le N$. Then*

$$\|\pi_N^{(-\alpha,\alpha)} v - v\|_{\omega^{(-\alpha,-\alpha)}} \lesssim N^{-(\alpha+s)} \| {}^{\mathrm{RL}}_{0}D_t^{\alpha+s} v\|_{\omega^{(s,s)}} \tag{4.1.42}$$

and

$$\| {}^{\mathrm{RL}}_{0}D_t^{\alpha}(\pi_N^{(-\alpha,\alpha)} v - v)\|_I \lesssim N^{-s} \| {}^{\mathrm{RL}}_{0}D_t^{\alpha+s} v\|_{\omega^{(s,s)}}. \tag{4.1.43}$$

Let π^h be the H_0^1-orthogonal projection in the space direction

$$\langle \pi^h u - u, v\rangle_{H_{\mathcal{L}}(\Omega)} = (\nabla(\pi^h u - u), \nabla v)_{L^2(\Omega)} = 0, \quad \forall v \in V_h. \tag{4.1.44}$$

Using Lemma 4.5, we obtain the following result.

Theorem 4.6 *Let u be the solution of* (4.1.18) *and u_L be the numerical solution* (4.1.23). *If $u \in B^{\alpha}(D)$ and ${}^{\mathrm{RL}}_{0}D_t^{\alpha+s} u \in L^2_{\omega^{(s,s)}}(I, H_{\mathcal{L}}(\Omega))$, then we have*

$$\|u - u_L\|_{B^{\alpha}(D)} \lesssim N^{-s} \| {}^{\mathrm{RL}}_{0}D_t^{\alpha+s} u\|_{L^2_{\omega^{(s,s)}}(I, H_{\mathcal{L}}(\Omega))} + \|u - \pi^h u\|_{B^{\alpha}(D)}. \tag{4.1.45}$$

Proof The discrete inf-sup condition can be derived as follows. There is a constant $c_\alpha > 0$, independent of N, such that

$$\sup_{v \in V_h \otimes \mathcal{P}_N} \frac{\mathcal{A}(u_L, v)}{\|v\|_{\mathbb{V}}} \ge c_\alpha \|u_L\|_{B^{\alpha}(D)}, \quad \forall u_L \in V_h \otimes \mathcal{F}_N^{(\alpha)}. \tag{4.1.46}$$

We can derive from continuity and Theorem 4.3 and (4.1.46) that (4.1.23) admits a unique solution $u_L \in B^{\alpha}(D)$ satisfying

$$\|u_L\|_{B^{\alpha}(D)} \le \frac{1}{c_\alpha} \|f\|_{V^*}.$$

Moreover, we have

$$\|u - u_L\|_{B^{\alpha}(D)} \le C \inf_{v \in V_h \otimes \mathcal{F}_N^{(\alpha)}} \|u - v\|_{B^{\alpha}(D)}. \tag{4.1.47}$$

Taking $v = \pi_N^{(-\alpha,\alpha)}\pi^h u = \pi^h \pi_N^{(-\alpha,\alpha)} u$ into (4.1.47), we obtain from the above equation that

$$\begin{aligned}
\|u - u_L\|^2_{B^\alpha(D)} &\le C\|u - \pi_N^{(-\alpha,\alpha)} \circ \pi^h u\|^2_{B^\alpha(D)} \\
&\lesssim \| {}^{\mathrm{RL}}_{0}D_t^\alpha (u - \pi_N^{(-\alpha,\alpha)} \circ \pi^h u)\|^2_{L^2(D)} + \|u - \pi_N^{(-\alpha,\alpha)} \circ \pi^h u\|^2_{L^2(I,H_{\mathcal{L}}(\Omega))} \\
&:= I_1 + I_2.
\end{aligned} \tag{4.1.48}$$

In view of this, taking $s = 0$ in (4.1.43), we conclude that

$$\begin{aligned}
I_1 &= \| {}^{\mathrm{RL}}_{0}D_t^\alpha (u - \pi_N^{(-\alpha,\alpha)} \circ \pi^h u)\|^2_{L^2(D)} \\
&\lesssim \| {}^{\mathrm{RL}}_{0}D_t^\alpha \pi_N^{(-\alpha,\alpha)} (u - \pi^h u)\|^2_{L^2(D)} + \| {}^{\mathrm{RL}}_{0}D_t^\alpha (u - \pi_N^{(-\alpha,\alpha)} u)\|^2_{L^2(D)} \\
&\lesssim \| {}^{\mathrm{RL}}_{0}D_t^\alpha (u - \pi^h u)\|^2_{L^2(D)} + \| {}^{\mathrm{RL}}_{0}D_t^\alpha (u - \pi_N^{(-\alpha,\alpha)} u)\|^2_{L^2(D)} \\
&\lesssim N^{-2s} \| {}^{\mathrm{RL}}_{0}D_t^{\alpha+s} u\|^2_{L^2_{\omega^{(s,s)}}(I,L^2(\Omega))} + \| {}^{\mathrm{RL}}_{0}D_t^\alpha (u - \pi^h u)\|^2_{L^2(D)}.
\end{aligned}$$

By (4.1.44), we treat the second term I_2 as

$$\begin{aligned}
I_2 &= \|u - \pi_N^{(-\alpha,\alpha)} \circ \pi^h u\|^2_{L^2(I,H_{\mathcal{L}}(\Omega))} \\
&\lesssim \|\pi^h (u - \pi_N^{(-\alpha,\alpha)} u)\|^2_{L^2(I,H_{\mathcal{L}}(\Omega))} + \|u - \pi^h u\|^2_{L^2(I,H_{\mathcal{L}}(\Omega))} \\
&\lesssim \|u - \pi_N^{(-\alpha,\alpha)} u\|^2_{L^2(I,H_{\mathcal{L}}(\Omega))} + \|u - \pi^h u\|^2_{L^2(I,H_{\mathcal{L}}(\Omega))} \\
&\lesssim N^{-2(\alpha+s)} \| {}^{\mathrm{RL}}_{0}D_t^{\alpha+s} u\|^2_{L^2_{\omega^{(s,s)}}(I,H_{\mathcal{L}}(\Omega))} + \|u - \pi^h u\|^2_{L^2(I,H_{\mathcal{L}}(\Omega))}.
\end{aligned}$$

A combination of the above estimates and Lemma 4.1 leads to the desired estimate. □

Numerical Examples

We employ the Petrov–Galerkin method to the linear TFDEs to demonstrate the spatial and temporal convergence rates of the proposed algorithms.

Example 4.7 We first consider the following linear TFDEs in rectangular domains

$${}^{\mathrm{C}}_{0}D_t^\alpha u(\boldsymbol{x},t) - \mathrm{div}(a(\boldsymbol{x})\nabla u(\boldsymbol{x},t)) + \beta u(\boldsymbol{x},t) = f(\boldsymbol{x},t) \tag{4.1.49}$$

for $(\boldsymbol{x},t) \in (-1,1)^2 := \Omega \times (0,T]$, with $a(x,y) = 1$, the exact solution

$$u(x,y,t) = \sin(\pi x)\sin(\pi y)(t^\alpha \sin(\pi t) + 1),$$

and the initial conditions $u_0(x,y) = \sin(\pi x)\sin(\pi y)$.

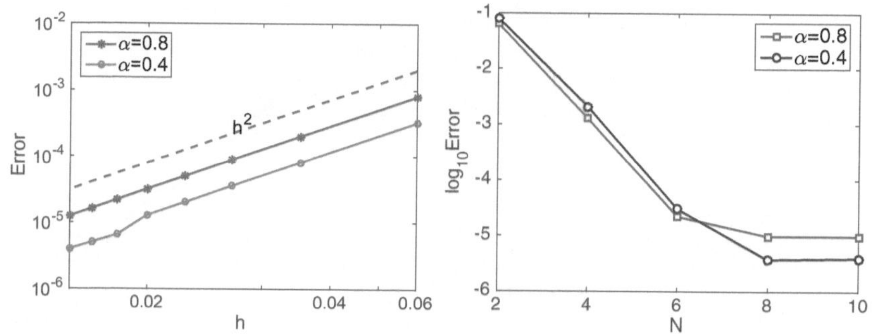

Figure 4.1 Left: the maximum error in log-log scale against various h and α for Example 4.7 with $M = 20$; Right: the $L^2(\Omega)$-error in semi-log scale against various N and α for Example 4.7 with $h = 1/90$. Source: Shen and Sheng (2019).

Without loss of generality, we use piecewise linear FEM for spatial discretization and GJPFs Petrov–Galerkin method for temporal discretization. Let $\mathcal{T}_h = \{K\}$ be a uniformly regular family rectangular in $(-1, 1)^2$, and define the mesh size h. In Figure 4.1, we list the maximum errors of (4.1.49), in log-log scale (or semi-log scale) at $T = 1$. We see from Figure 4.1 (left) the proposed scheme has a second-order convergence rate in time. In Figure 4.1 (right), we observe an exponential decrease of the numerical errors with increasing N and a level-off of the error curves beyond $N = 8$, due to the saturation of spatial errors.

Example 4.8 In this example, we employ GJPFs Petrov–Galerkin method in temporal discretization and Legendre spectral method in spatial discretization to resolve Example 4.7 numerically.

In Figure 4.2 (left), we compare the $B^\alpha(D)$-errors of Algorithm I with Algorithm III, for which we take $M = 40$. Notice that Algorithms II and III give the same numerical result, so we prefer to use Algorithm III in this example. We observe that for $N \le 10$ the numerical results from the two algorithms are the same; but for $N > 10$, Algorithm III (or Algorithm II) provides more accurate numerical results than Algorithm I. This is because the eigen-decomposition is not stable for the nonsymmetric linear system. In Figure 4.2 (right), we also list the $B^\alpha(D)$-errors, in semi-log scale, with $N = 40$. Although the exact solution has singularity at $t = 0$, but in the weighted Sobolev spaces $\mathcal{B}^s_{-\alpha,\alpha}(I)$ which involve fractional derivatives, we have $u|_t \in \mathcal{B}^s_{-\alpha,\alpha}(I)$ for any large s. We observe that the numerical errors decay exponentially as M/N increases. We note in particular that very high accuracy is achieved with quite small M/N for this special problem.

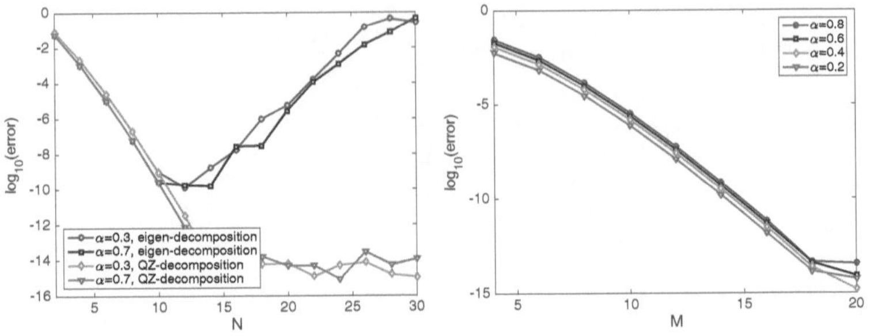

Figure 4.2 Left: the $B^\alpha(D)$-error in semi-log scale against various N and α for Example 4.8 with $M = 20$; Right: the $B^\alpha(D)$-error in semi-log scale against various M and α for Example 4.8 with $N = 20$. Source: Shen and Sheng (2019).

Example 4.9 We also consider equation (4.1.49) in rectangular domains with the exact solution $u(x,y,t) = \sin(\pi x)\sin(\pi y)\cdot\sin(\pi t^\alpha)$, $a(x,y) = \frac{x^2+y^2}{2}$, and the homogeneous initial conditions $u_0(\boldsymbol{x}) = 0$.

Clearly, the singularity of the exact solution in time is more complicated than Example 4.7, since we have

$$u|_t = \sin(\pi t^\alpha) = t^\alpha - \frac{1}{3!}t^{3\alpha} + O(t^{5\alpha}). \tag{4.1.50}$$

The GJPFs can match the first leading singularity of the solution, then the convergence rate only depends on the second leading singularity of the solution. Here we take $\alpha = 0.8, 0.6, 0.4, 0.2$. Then, we can derive from the theoretical estimate and (4.1.50) that the corresponding convergence rate is $N^{-4}, N^{-3}, N^{-2}, N^{-1}$. In Figure 4.3 (left) we plot the B^α-errors in log-log scale against various N and α with $M = 20$. We observe that the numerical results are consistent with theoretical results. We also list the B^α-errors in semi-log scale against various M and α with $N = 60$ in Figure 4.3 (right). They indicate that the numerical errors decay exponentially as M increases, and numerical errors decay algebraically as N increases.

Example 4.10 We consider TFDEs (4.1.49) in unit disk $\Omega = \{x^2 + y^2 < 1\}$ with the exact solution $u(x,y,t) = (x^2+y^2-1)(\cos(8(x+y)) + \sin(8(x+y))) \cdot t^\alpha \sin(\pi t)$, $a(x,y) = 1$, and the initial condition: $u_0(x,y) = 0$.

In Figure 4.4, we plot the $B^\alpha(D)$-errors, in semi-log scale. We observe that the numerical errors decay exponentially as M/N increases. We note in particular that very high accuracy is achieved with quite small M/N for this special problem.

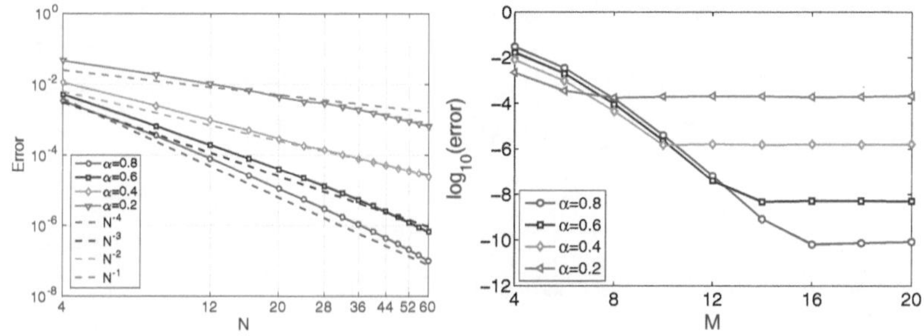

Figure 4.3 Left: the $B^\alpha(D)$-error in log-log scale against various N and α for Example 4.9 with $M = 20$; Right: the $B^\alpha(D)$-error in semi-log scale against various M and α for Example 4.9 with $N = 60$. Source: Shen and Sheng (2019).

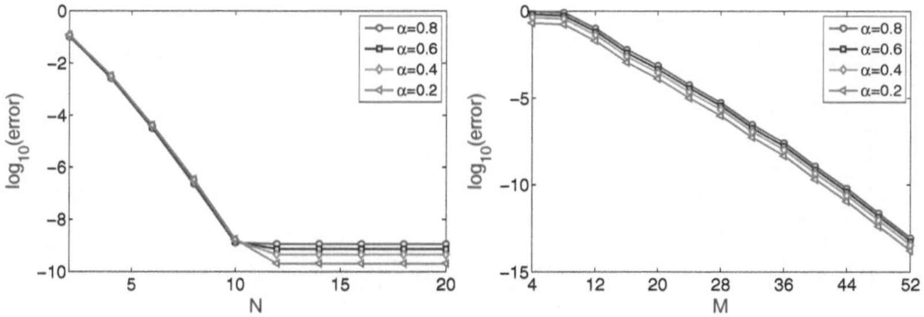

Figure 4.4 Left: the $B^\alpha(D)$-error in semi-log scale against various N and α for Example 4.10 with $M = 40$; Right: the $B^\alpha(D)$-error in semi-log scale against various M and α for Example 4.10 with $N = 20$. Source: Shen and Sheng (2019).

Example 4.11 We also consider equation (4.1.49) in unit disk $\Omega = \{x^2 + y^2 < 1\}$ but with a different exact solution $u(x, y, t) = (x^2 + y^2 - 1)(\cos(8(x + y)) + \sin(8(x + y))) \cdot \sin(\pi t^\alpha)$, and the same $a(x, y) = 1$, and the initial condition: $u_0(x, y) = 0$.

Figure 4.5 (left) shows the $B^\alpha(D)$-errors in log-log scale against various N and α with $M = 20$. We also list the $B^\alpha(D)$-errors in semi-log scale against various M and α with $N = 60$ in Figure 4.5 (right). We see that the numerical errors decay exponentially as M increases, and numerical errors decay algebraically as N increases.

Example 4.12 To eliminate the error of spatial direction, we also consider equation (4.1.49) in unit disk $\Omega = \{x^2 + y^2 < 1\}$ with the given right-hand side term $f(x, y, t) = x^2 + y^2 - 1$, $a(x, y) = 1$, and the homogeneous initial condition. Since no exact solution is available, we use the numerical solution

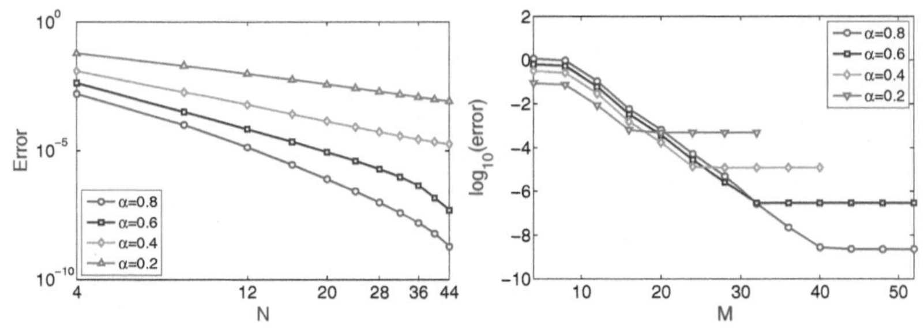

Figure 4.5 Left: the B^{α}-error in log-log scale against various N and α for Example 4.11 with $M = 40$; Right: the B^{α}-error in semi-log scale against various M and α for Example 4.11 with $N = 60$. Source: Shen and Sheng (2019).

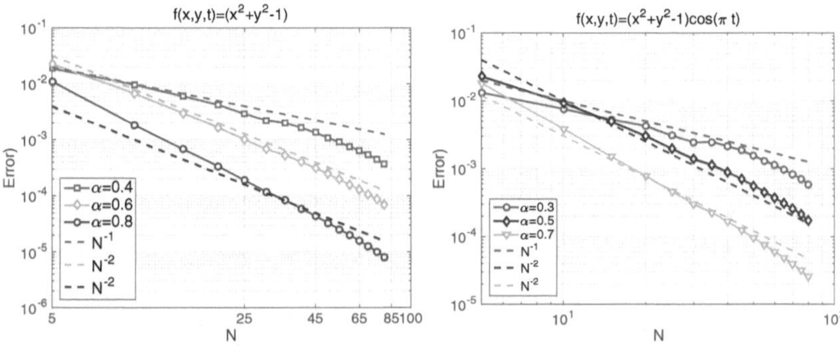

Figure 4.6 Left: the B^{α}-error in log-log scale against various N and α for Example 4.12 with $M = 40$; Right: the B^{α}-error in semi-log scale against various N and α for Example 4.12 with $M = 40$. Source: Shen and Sheng (2019).

with $M = 50$, $N = 100$ and add three singular functions as the reference solution.

In Figure 4.6 (left), we plot the $B^{\alpha}(D)$-errors in log-log scale against various N and α with $M = 40$. We observe that the numerical errors decay algebraically as M increases. Since the GJPF Petrov–Galerkin method only matches the first leading singularity of the solution, then the second leading singularities are $t^{\nu} = t^{0.8}, t^{1.2}, t^{1.6}$ with respect to $\alpha = 0.4, 0.6, 0.8$. We can derive from Theorem 4.6 that the convergence rates are N^{-1}, N^{-2}, N^{-2} respectively. Similarly, we plot the $B^{\alpha}(D)$-errors in log-log scale against various N and α with $M = 40$ with $f(x, y, t) = (x^2 + y^2 - 1) \cdot \cos(\pi t)$ in Figure 4.6 (right).

4.1.3 Improved Algorithm with Enriched Approximation in Time

This section is devoted to the improved algorithm with enriched approximation space in time for TFDEs as well as the error estimate for the proposed method. Some numerical examples to illustrate the efficiency of our method are presented.

Algorithm

In order to introduce the enriched Petrov–Galerkin method for (4.1.18), we first need the following basis function space for time variable

$$\begin{aligned}\mathcal{F}_N^{(k,\alpha)}(I) := \operatorname{span}\big\{J_n^{(-\alpha,\alpha)}(t) = t^\alpha \widetilde{P}_n^{(-\alpha,\alpha)}(t) : 0 \le n \le N\big\}\\ \oplus \big\{\psi_i = t^{\nu_i},\ i = 1, \ldots, k\big\},\end{aligned} \tag{4.1.51}$$

and in the test space, we take

$$\mathcal{P}_{N+k}(I) = \operatorname{span}\big\{L_n^{(\alpha)}(t) := \kappa_{n,\alpha} L_n(t) : 0 \le n \le N+k\big\}, \tag{4.1.52}$$

where ψ_i, $i = 1, \ldots, k$ are some known singular functions.

Remark 4.1 Although we can use the basis in (4.1.51) directly, due to the added basis $\{\psi_i\}_{i=1}^k$ are not orthogonal to the original one $\{J_n^{-\alpha,\alpha}(t)\}_{n=0}^N$, which leads to ill-conditioned matrix. Moreover, the ill-conditioning will affect the accuracy of solving the linear system.

In order to make the implementation more efficient, it appears necessary to use the following modified Gram–Schmidt process:

$$\varphi_n := J_n^{(-\alpha,\alpha)}(t) = t^\alpha \widetilde{P}_n^{(-\alpha,\alpha)}(t), \quad 0 \le n \le N,$$

and φ_{N+i}, $i = 1, \ldots, k$ can be computed as

$$\begin{aligned}\varphi_{N+i}^{(0)} &= t^{\nu_i} - \operatorname{proj}_{\varphi_0}(t^{\nu_i}),\\ \varphi_{N+i}^{(1)} &= \varphi_{N+i}^{(0)} - \operatorname{proj}_{\varphi_1}(\varphi_{N+i}^{(0)}),\\ &\ \vdots\\ \varphi_{N+i}^{(N+i-2)} &= \varphi_{N+i}^{(N+i-3)} - \operatorname{proj}_{\varphi_{N+i-2}}(\varphi_{N+i}^{(N+i-3)}),\\ \varphi_{N+i} &= \varphi_{N+i}^{(N+i-2)} - \operatorname{proj}_{\varphi_{N+i-1}}(\varphi_{N+i}^{(N+i-2)}),\end{aligned} \tag{4.1.53}$$

where

$$\operatorname{proj}_u(v) = \frac{(u, v)_{\omega^{(-\alpha,-\alpha)}}}{(u, u)_{\omega^{(-\alpha,-\alpha)}}} u.$$

Then, the enriched Petrov–Galerkin method for (4.1.18) is to seek $u_L^k(\boldsymbol{x},t) := u_{MN}^k \in V_h \otimes \mathcal{F}_N^{(k,\alpha)}$ such that

$$({}_0^{\rm RL}D_t^\alpha u_L^k, v)_{L^2(D)} + b(u_L^k, v) + \beta(u_L^k, v)_{L^2(D)} = (f, v)_{L^2(D)} \tag{4.1.54}$$

for all $v \in V_h \otimes \mathcal{P}_{N+k}$.

We now describe the numerical implementations for (4.1.54) under this set of basis functions. We write

$$u_L^k(\boldsymbol{x},t) = \sum_{m=1}^{M}\sum_{n=0}^{N+k} \widetilde{u}_{mn}\phi_m(\boldsymbol{x})\varphi_n(t). \tag{4.1.55}$$

Denote

$$\begin{aligned}
f_{mn}^{en} &= (f, \phi_m(\boldsymbol{x})L_n^{(\alpha)}(t))_D, \qquad & F^{en} &= (f_{mn}^{en})_{1\le m\le M, 0\le n\le N+k},\\
s_{pq}^{en} &= \int_I {}_0^{\rm RL}D_t^\alpha \varphi_q(t)L_p^{(\alpha)}(t)dt, \qquad & m_{pq}^{en} &= \int_I \varphi_q(t)L_p^{(\alpha)}(t)dt,\\
\boldsymbol{S}^{en} &= (s_{pq}^{en})_{0\le p,q\le N+k}, \qquad & \boldsymbol{M}^{en} &= (m_{pq}^{en})_{0\le p,q\le N+k},\\
U &= (\widetilde{u}_{mn})_{1\le m\le M, 0\le n\le N+k}.
\end{aligned}$$

Then, from (4.1.1), we find that (4.1.54) is equivalent to the following linear system:

$$\boldsymbol{M}^h\, U(\boldsymbol{S}^{en})^T + \boldsymbol{S}^h\, U(\boldsymbol{M}^{en})^T + \beta\boldsymbol{M}^h\, U(\boldsymbol{M}^{en})^T = F^{en}. \tag{4.1.56}$$

Hence we can use Algorithm II or Algorithm III to resolve linear system (4.1.56).

In Table 4.3, we tabulate the maximum error of $(\boldsymbol{S}^{en}\boldsymbol{E})^{-1}\boldsymbol{M}^{en}\boldsymbol{E} - \Lambda$ (resp. $\max\{|\boldsymbol{Q}^T \boldsymbol{A}\boldsymbol{Z}^T - \boldsymbol{S}^{en}|, |\boldsymbol{Q}^T \boldsymbol{B}\boldsymbol{Z}^T - \boldsymbol{M}^{en}|\}$) for eigen-decomposition (resp. QZ decomposition) in (4.1.56) with various N, which indicates that the error of eigen-decomposition grows as N increases. We observe that the QZ decomposition is much more stable than eigen-decomposition for the nonsymmetric enriched matrices, similar to the behaviors for the examples in the previous section.

Error Analysis

In this subsection, we shall restrict our attention to the linear case (4.1.18) with $\mathcal{L} = -\Delta$. To this end, we begin with the representation of the solution of (4.1.18). Let us consider the following eigenvalue problem

$$\mathcal{L}\varphi = \lambda\varphi \quad \text{in} \quad \Omega,$$

where φ satisfies the homogeneous boundary condition, and it admits a non-decreasing sequence $\{\lambda_j\}_{j=1}^\infty$ of positive eigenvalues and the eigenfunctions

Table 4.3. *A comparison of decomposition error between eigen and QZ decomposition. Source: Shen and Sheng (2019).*

	matrices in (4.1.56) with $\alpha = 0.3$ and $k = 3$		matrices in (4.1.56) with $\alpha = 0.7$ and $k = 3$	
N	Eigen	QZ	Eigen	QZ
4	1.37e-13	1.33e-15	3.58e-12	8.09e-16
8	2.73e-12	1.99e-15	3.33e-12	1.79e-15
12	3.17e-10	1.77e-15	8.77e-11	2.33e-15
16	9.55e-08	2.22e-15	3.39e-08	1.57e-15
20	6.45e-06	2.22e-15	7.29e-07	1.66e-15
24	1.81e-04	3.11e-15	3.60e-05	1.79e-15
28	6.10e-03	1.88e-15	2.41e-04	2.67e-15
32	2.34e-02	2.22e-15	1.35e-03	3.58e-15
80	6.86e-03	4.88e-15	3.61e-01	7.32e-15

$\{\varphi_j\}_{j=1}^{\infty}$ form an orthonormal basis in $L^2(\Omega)$. Then, the solution of (4.1.18) can be expressed as (see, e.g., Sakamoto and Yamamoto (2011))

$$u(x,t) = \sum_{n=1}^{\infty} \int_0^t s^{\alpha-1} E_{\alpha,\alpha}(-(\lambda_n + \beta)s^{\alpha})(f(\cdot, t-s), \varphi_n)_{L^2(\Omega)} \varphi_n(x) ds, \tag{4.1.57}$$

where $E_{a,b}(z)$ denotes the two-parameter Mittag-Leffler function:

$$E_{a,b}(z) = \sum_{j=0}^{\infty} \frac{z^j}{\Gamma(aj+b)}, \quad z \in \mathbb{C},\ a > 0,\ b \in \mathbb{R}.$$

Denote by $k \in \mathbb{N}$ the number of elements contained in the set

$$I(k;\alpha) := \{(i,j) : i, j \in \mathbb{N}_0, i + j\alpha < \bar{k}\}, \quad \bar{k} \in \mathbb{N}. \tag{4.1.58}$$

For any given right-hand side $f \in C^{\bar{k}}(I; L^2(\Omega))$ in (4.1.18), we can easily get the singularity of the solution in t-direction. More precisely, for any k, the solution of (4.1.57) in t-direction can be rewritten as

$$u|_t = \sum_{I(k;\alpha)} \gamma_{ij}^{\alpha} t^{i+j\alpha} + \tilde{u}^k(t), \quad t \in I,$$

where the coefficients $\gamma_{i,j}^{\alpha}$ are dependent on i, j and α. Let the leading singularity of $\tilde{u}^k(t)$ be $t^{\nu+\bar{k}-1}$, with $\nu \in (0,1]$. Clearly, the singularity of our approximation space is $A(k;\alpha) = \{i \in \mathbb{N}_0, i + \alpha < \bar{k}\}$, $\bar{k} \in \mathbb{N}$, so the remaining part of the singularity is given by

$$R(k;\alpha) := I(k;\alpha)/A(k;\alpha) = \{(i,j) : i, j \in \mathbb{N}_0, i + j\alpha < \bar{k}, j \neq 1\}, \tag{4.1.59}$$

where $k \in \mathbb{N}$ denotes the number of elements contained in $R(k;\alpha)$ or $A(k;\alpha)$.

Following the same procedure in the Petrov–Galerkin spectral method, yields

$$\|u-u_L^k\|_{B^\alpha(D)} \lesssim \inf_{v|_t\in\mathcal{F}_N^{(k,\alpha)}} \|v-u\|_{B^\alpha(D)} + \inf_{v|_x\in V_h} \|v-u\|_{B^\alpha(D)}.$$

To this end, we assume that

$$u(\cdot,t) = \widetilde{u}^k(\cdot,t) + \sum_{I(k;\alpha)} \gamma_{ij}^\alpha t^{i+j\alpha} = \widetilde{u}^k(\cdot,t) + \sum_{i=1}^{k} u_i^s \psi_i(t). \tag{4.1.60}$$

Then we take

$$v|_t = \pi_N^{-\alpha,\alpha}\widetilde{u}^k(\cdot,t) + \sum_{i=1}^{k} u_i^s\psi_i(t),$$

and by applying Theorem 4.6, we have

$$\|v-u\|_{B^\alpha(D)} \lesssim N^{-s}\| {}^{\mathrm{RL}}_{0}D_t^{\alpha+s}\widetilde{u}^k\|_{L^2_{\omega^{(s,s)}}(I,H_{\mathcal{L}}(\Omega))} + \inf_{v|_x\in V_h}\|u-v\|_{B^\alpha(D)}.$$

Theorem 4.2 *Let u be the solution of* (4.1.18) *and u_L^k be the numerical solution* (4.1.54)*. Assume* (4.1.60) *holds and the leading singularity of $\widetilde{u}^k|_t$ near $t=0$ is $t^{\nu+\overline{k}-1}$ with $\nu\in(0,1]$ and $\overline{k}$ is defined in* (4.1.58)*. If $u\in B^\alpha(D)$ and ${}^{\mathrm{RL}}_{0}D_t^{\alpha+m}\widetilde{u}^k \in L^2_{\omega^{(m,m)}}(I,H_{\mathcal{L}}(\Omega))$, then we have*

$$\begin{aligned}\|u-u_L^k\|_{B^\alpha(D)} &\lesssim N^{-m}\| {}^{\mathrm{RL}}_{0}D_t^{\alpha+m}\,\widetilde{u}^k\|_{L^2_{\omega^{(m,m)}}(I,H_{\mathcal{L}}(\Omega))}\\ &\quad + \inf_{v_L|_x\in V_h}\|u-v_L\|_{B^\alpha(D)},\end{aligned} \tag{4.1.61}$$

where the convergence index $m\in\mathbb{N}$ satisfies

- *for $\nu-\alpha\in(-1,-\frac{1}{2}]$, we have $m=2\overline{k}-3$ with $\overline{k}\geq 2$;*
- *for $\nu-\alpha\in(-\frac{1}{2},0]$, we have $m=2\overline{k}-2$ with $\overline{k}\geq 2$;*
- *for $\nu-\alpha\in(0,\frac{1}{2}]$, we have $m=2\overline{k}-1$ with $\overline{k}\geq 1$;*
- *for $\nu-\alpha\in(\frac{1}{2},1]$, we have $m=2\overline{k}$ with $\overline{k}\geq 1$.*

Proof Since the leading singularity of $\widetilde{u}^k|_t$ is $t^{\nu+\overline{k}-1}$, $\nu\in(0,1]$, which implies $\nu-\alpha\in(-\alpha,1-\alpha]$. According to Lemma 4.5 and Theorem 4.6, we only need to consider

$$\begin{aligned}\| {}^{\mathrm{RL}}_{0}D_t^{\alpha+m}t^{\nu+\overline{k}-1}\|^2_{L^2_{\omega^{(m,m)}}(I)} &\approx \int_I \left(t^{\nu-\alpha+\overline{k}-m-1}\right)^2 t^m\,\mathrm{d}t < \infty,\\ \| {}^{\mathrm{RL}}_{0}D_t^{\alpha+m+1}t^{\nu+\overline{k}-1}\|^2_{L^2_{\omega^{(m+1,m+1)}}(I)} &\approx \int_I \left(t^{\nu-\alpha+\overline{k}-m-2}\right)^2 t^{m+1}\,\mathrm{d}t = \infty,\end{aligned} \tag{4.1.62}$$

or equivalently,

$$\begin{aligned} &2\nu - 2\alpha + 2\bar{k} - m - 2 > -1 \;\Rightarrow\; 2\nu - 2\alpha + 2\bar{k} - m > 1, \\ &2\nu - 2\alpha + 2\bar{k} - m - 2 \le -1 \;\Rightarrow\; 2\nu - 2\alpha + 2\bar{k} - m \le 2. \end{aligned} \tag{4.1.63}$$

To determine the value of $m \ge 0$, by the fact that $\bar{k} \ge 1$ and a direct computation from (4.1.63), gives:

- For $\nu - \alpha \in (-1, -\frac{1}{2}]$ (or $2(\nu - \alpha) \in (-2, -1]$), we obtain that $m = 2\bar{k} - 3$, namely,

$$\text{If } \bar{k} = 2, 3, 4, 5, \ldots, \quad \text{we have } m = 1, 3, 5, 7 \ldots.$$

- For $\nu - \alpha \in (-\frac{1}{2}, 0]$ (or $2(\nu - \alpha) \in (-1, 0]$), we obtain that $m = 2\bar{k} - 2$, namely,

$$\text{If } \bar{k} = 1, 2, 3, 4, \ldots, \quad \text{we have } m = 0, 2, 4, 6, \ldots.$$

- For $\nu - \alpha \in (0, \frac{1}{2}]$ (or $2(\nu - \alpha) \in (0, 1]$), we obtain that $m = 2\bar{k} - 1$, namely,

$$\text{If } \bar{k} = 1, 2, 3, 4, \ldots, \quad \text{we have } m = 1, 3, 5, 7 \ldots.$$

- For $\nu - \alpha \in (\frac{1}{2}, 1]$ (or $2(\nu - \alpha) \in (1, 2]$), we obtain that $m = 2\bar{k}$, namely,

$$\text{If } \bar{k} = 1, 2, 3, 4, \ldots, \quad \text{we have } m = 2, 4, 6, 8, \ldots.$$

Note that in the case $\nu - \alpha \in (-\frac{1}{2}, 0]$: when $m = 0$, we got $\alpha > \nu + \bar{k} - 1$ in (4.1.62), but this is not true. Hence, we have $m \ge 2$ in the case $\nu - \alpha \in (-\frac{1}{2}, 0]$. A combination of the above estimates leads to the desired result. □

Numerical Results

In order to ignore the error in spatial direction, we consider TFDEs (4.1.49) in 1D with a given source function $f(x,t) = \sin(\pi x) \cdot \cos(\pi t)$, $a(x) = \frac{x^2}{2}$ and homogeneous initial conditions. In this example, we use an improved algorithm to reslove (4.1.49) numerically. More specifically, we can take the first k-th singular term t^{ν_j} in $R(k; \alpha) = \{t^{i+j\alpha},\ i, j \in \mathbb{N}_0, i + j\alpha < \bar{k}, j \ne 1\}$. The exact solution is unknown; we compute a numerical solution with $M = 40$, $N = 100$ and $k = 4$ as the exact solution. In Figure 4.7, we list the max-errors in log-log scale against various N, k and α with $M = 30$. They indicate that the enriched Petrov–Galerkin method is better than the Petrov–Galerkin method, and the numerical errors decay algebraically.

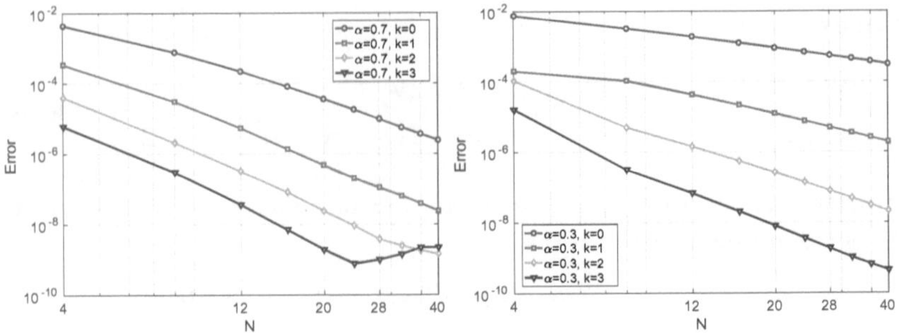

Figure 4.7 Left: the max-error in log-log scale against various N and α for Example 4.7 with $M = 30$; Right: the max-error in semi-log scale against various M and α for Example 4.7 with $M = 30$. Source: Shen and Sheng (2019).

Numerical Results for Nonlinear Problems

In this section, we shall present some numerical results obtained by the proposed methods introduced in this section for nonlinear problems.

Example 4.3 Consider the following time-fractional Burgers equation in 1D:

$$ {}_{0}^{C}D_t^{\alpha} u(x,t) - \gamma \Delta u(x,t) + \partial_x \Big(\frac{1}{2} u^2(x,t) \Big) = 0, \quad \forall (x,t) \in D, \tag{4.1.64} $$

with the initial data $u_0(x) = -\sin(\pi x)$.

We initially solved this problem by using the preconditioned Jacobian-free Newton–Krylov method. To accelerate the convergence, we employed a two-grid approach. Namely, we use $\widetilde{N} \approx N/2$ and $M \approx M/2$ as the initial data for the iteration of the coarse approximation solution with N for the fine approximation solution with N and M. This two-grid approach significantly improved the convergence, only less than 10 iterations are needed to achieve 10^{-7}-digit accuracy.

We plot in Figure 4.8 (left) the numerical solution with $M = 100$, $N = 80$, $\alpha = 0.7$, $\gamma = 0.01$ at $t = 0, 0.25, 0.5, 0.75, 1$. We observe that, similar to the usual Burgers equation, the profile of the solution becomes steeper as t increases to about $t = 0.25$, and the solution starts to relax towards zero for $t > 0.25$. To investigate the influence of α on the solution, we plot the numerical solution with $\gamma = 0.01, N = 64, M = 100$ at $t = 1$ for various α in Figure 4.8 (right). As a comparison, we also plot the solution with $\alpha = 1$, i.e., the usual Burgers equation. We observe that, as α increases, the profile of the solution becomes steeper and dissipates faster.

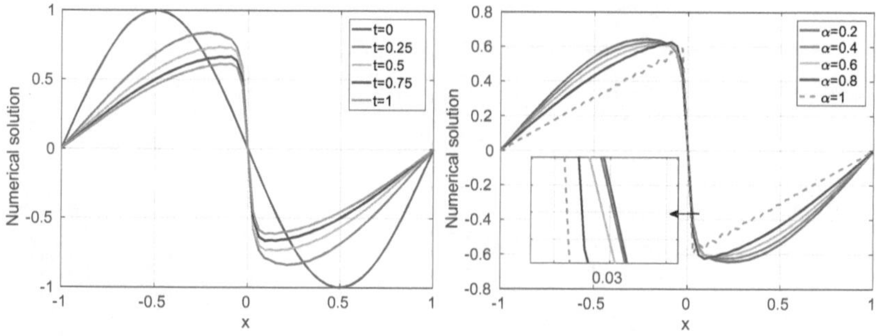

Figure 4.8 Left: the numerical solution with $\alpha = 0.7$, $\gamma = 0.01$, $N = 80$, $M = 100$ against various t; Right: the numerical solution with $\gamma = 0.01$, $N = 80$, $M = 100$ at $t = 1$ against various α. Source: Shen and Sheng (2019).

Example 4.4 Consider the following time-fractional Allen–Cahn equation in 1D:

$$ {}_0^C D_t^\alpha u(x,t) - \epsilon^2 \Delta u(x,t) + f(u(x,t)) = 0, \quad \forall (x,t) \in D, \tag{4.1.65} $$

with the initial data

$$ u_0(x) = \begin{cases} 1, & 0 \le x \le 1, \\ -1, & -1 \le x < 0, \end{cases} \tag{4.1.66} $$

and homogeneous Neumann boundary condition: $\partial u/\partial n|_{\partial\Omega} = 0$. Here, we take $f(u) = (u^2 - 1)u = F'(u)$ to be the usual double-well potential, Ω is the bounded domain and n is the outward normal. The parameter we use in the following presentation is $\epsilon = 0.1$.

We also employed a two-grid approach (see Example 4.3) to resolve (4.1.65). Note that in this example, only less than 12 iterations are needed to achieve 10^{-7}-digit accuracy. We plot in Figure 4.9 (left) the numerical solution with $M = 250$, $N = 80$, $\alpha = 0.7$, $\epsilon = 0.1$ at $t = 0,\ 0.5,\ 1,\ 1.5,\ 2$. We observe that the relaxation process as t increases is similar to the usual Allen–Cahn equation.

To further investigate the influence of α on the solution, we plot the numerical solution with $\epsilon = 0.1$, $N = 80$, $M = 250$ at $t = 1$ for various α in Figure 4.8 (right). As a comparison, we also plot the solution with $\alpha = 1$, i.e., the usual Allen–Cahn equation. We observe that, as α decreases, the speed of profile relaxation also decreases.

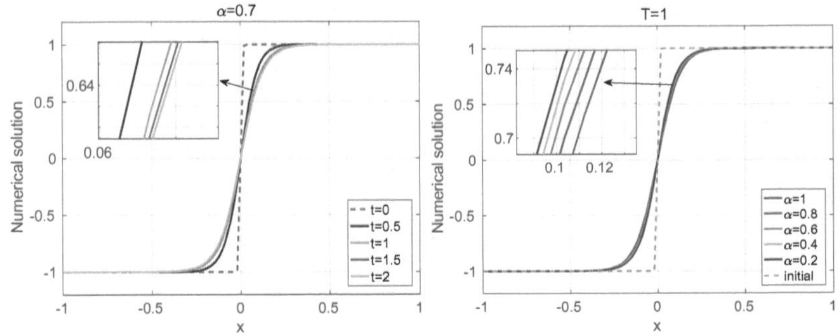

Figure 4.9 Left: the numerical solution with $\alpha = 0.7$, $\epsilon = 0.1$, $N = 80$, $M = 250$ against various t; Right: the numerical solution with $\epsilon = 0.1$, $N = 80$, $M = 250$ at $t = 1$ against various α. Source: Shen and Sheng (2019).

4.2 Unified GJPF Space-Time Method on Hypercubes

In Chapter 2, we introduced the family of *Jacobi poly-fractonomials* and their generalization as the eigenfunctions of fractional Sturm–Liouville problems, explicitly given by

$$ {}^{(1)}\mathcal{P}_n^{\alpha,\beta,\mu}(\xi) = (1+\xi)^{-\beta+\mu-1} P_{n-1}^{(\alpha-\mu+1,-\beta+\mu-1)}(\xi), \quad \xi \in (-1,1), \quad (4.2.1) $$

with $\mu \in (0,1)$, $-1 \le \alpha < 2-\mu$, and $-1 \le \beta < \mu-1$, which are representing the eigenfunctions of the *singular* FSLP of the first kind (SFSLP-I), and

$$ {}^{(2)}\mathcal{P}_n^{\alpha,\beta,\mu}(\xi) = (1-\xi)^{-\alpha+\mu-1} P_{n-1}^{(-\alpha+\mu-1,\beta-\mu+1)}(\xi), \quad \xi \in (-1,1), \quad (4.2.2) $$

where $-1 < \alpha < \mu - 1$ and $-1 < \beta < 2 - \mu$, and $\mu \in (0,1)$, denoting the eigenfunctions of the *singular* FSLP of the *second* kind (SFSLP-II). In this chapter, we develop a unified and spectrally accurate Petrov–Galerkin (PG) spectral method for the general FPDEs of the following weak form

$$ \left({}^{\mathrm{RL}}_{0}D_t^{\tau} u, {}^{\mathrm{RL}}_{t}D_T^{\tau} v\right)_\Omega + \sum_{j=1}^{d} c_j \left({}^{\mathrm{RL}}_{a_j}D_{x_j}^{\mu_j} u, {}^{\mathrm{RL}}_{x_j}D_{b_j}^{\mu_j} v\right)_\Omega + \gamma\,(u,v)_\Omega = (f,v)_\Omega, \quad (4.2.3) $$

where 2τ, $\mu_j \in (0,1)$, in a $(1+d)$-dimensional *space-time* domain subject to Dirichlet initial and boundary conditions. Such a weak form is equivalent to the *strong form*

$$ {}^{\mathrm{RL}}_{x_j}D_{b_j}^{\mu_j} u + \sum_{j=1}^{d} c_j\, {}^{\mathrm{RL}}_{a_j}D_{x_j}^{2\mu_j} u + \gamma\, u = f, \quad (4.2.4) $$

when u possesses enough smoothness. This method applies equally well to the entire family of linear fractional *hyperbolic*, *parabolic* and *elliptic* equations with the same ease.

Stochastic Interpretation of the Strong Form

Following Baeumer et al. (2018b), we provide a brief stochastic interpretation of the FPDEs in (4.2.4) that sheds light on the well-posedness of the problem from the perspective of probability theory. Suppose that in (4.2.4), $f \equiv 0$ and $\gamma = 0$ and $0 < 2\tau < 1$ and that $a_i = -\infty$ and $b_i = +\infty$ for $i = 1, 2, \ldots$. Then (4.2.4) governs (see Baeumer et al. (2018b)) a time-changed Lévy process $X(E_t)$ on $\mathbb{R}^d$ whose Fourier transform is $\mathbb{E}[e^{-ik\cdot X(t)}] = e^{t\psi(k)}$ with the Fourier symbol

$$\begin{aligned}\psi(k) = &-\sum_{n=1}^{d}[c_{l_n}(\mathrm{i}k_n)^{2\mu_n} + c_{r_n}(-\mathrm{i}k_n)^{2\mu_n}] \\ &+\sum_{m=1}^{d}[\kappa_{l_m}(-\mathrm{i}k_m)^{2\nu_m} + \kappa_{r_m}(\mathrm{i}k_m)^{2\nu_m}].\end{aligned} \tag{4.2.5}$$

Recalling that in one dimension the Lévy process $Y(t)$ with Fourier transform $\mathbb{E}[e^{-ikY(t)}] = e^{t\psi_0(k)}$ where $\psi_0(k) = pD(\mathrm{i}k)^{\alpha} + qD(-\mathrm{i}k)^{\alpha}$ for $D > 0$ and $1 < \alpha \leq 2$, $p \geq 0$, $q \geq 0$, and $p + q = 1$ is a stable Lévy process with index α and skewness $p - q$ (see Baeumer et al. (2018b)).

In brief, the fractional advection–dispersion equation on unbounded domain is represented by a solution that involves an inverse stable subordinator time-changed, resulting in a non-Markovian process. You can find complete details in Meerschaert and Sikorskii (2011).

Regarding a computational domain, Chen et al. (2012) developed a solution for the case of equation (4.2.4) where $f \equiv 0$ and $\gamma = 0$ and $0 < 2\tau < 1$ and all $a_m = a_n > -\infty$ and $b_m = b_n < \infty$, with zero Dirichlet boundary conditions. It follows from Baeumer et al. (2018b) that

$$Lu(x) = cu'(x) + \kappa_l \, {}^{\mathrm{RL}}_{a}D^{2\nu}_{x} u(x) + \kappa_r \, {}^{\mathrm{RL}}_{x}D^{2\nu}_{b} u(x) \tag{4.2.6}$$

is the generator of the killed semigroup on the bounded domain $\Omega = (a, b)$ which is also the point source to (4.2.4). In other words, starting with the point source initial condition $u(x, 0) = \delta(x)$, the solution to (4.2.4) with the restrictions discussed in Meerschaert and Sikorskii (2011); Chen et al. (2012) is the PDF of a *killed non-Markovian* process.

Remark 4.5 Here in this section, we consider the FPDEs with only left-sided derivatives in each dimension. In the next section however, not only will the case of two-sided fractional derivatives be employed, but they will also be generalized to the fully distributed-order case in each dimension with rather a more complete well-posedness, stability, and error analysis.

4.2.1 Mathematical Formulation of the Unified Space-Time Method

The main feature of the proposed unified Petrov–Galerkin spectral methods in this chapter is the *global* discretization of the temporal term, in addition to the spatial derivatives, rather than utilizing traditional low-order time-integration methods. Regarding the strong form, we note that there is a much larger and richer set of fractional derivative operators in $d > 1$ dimensions (see, e.g., p.157 in Meerschaert and Sikorskii (2011)) to be considered in our future work.

We essentially develop our method based on the fractional Sturm–Liouville problems (FSLPs), given in Chapter 2 and Zayernouri and Karniadakis (2013). Specifically, we employ the eigenfunctions of the FSLP of the *first* kind (FSLP-I), called *Jacobi poly-fractonomials*, as temporal/spatial bases. Next, we construct a different space for test functions from poly-fractonomial eigenfunctions of the FSLP of the *second* kind (FSLP-II). We show that this choice of basis and test functions leads to a stable bilinear form; moreover, we perform the corresponding error analysis. In the present method, all the aforementioned matrices are constructed *exactly* and *efficiently*. Moreover, we formulate a new general fast linear solver based on the eigenpairs of the corresponding temporal and spatial mass matrices with respect to the stiffness matrices, which significantly reduces the computational cost in higher-dimensional problems.

Let $u\colon \mathbb{R}^{d+1} \to \mathbb{R}$ for some positive integer d. For $u \in U$ (see e.g., (4.2.10)), we consider the following general weak form in $\Omega = [0,T] \times [a_1,b_1] \times \cdots \times [a_d,b_d]$ as

$$({}^{\text{RL}}_{0}D^{\tau}_{t}\, u, {}^{\text{RL}}_{t}D^{\tau}_{T}\, v)_\Omega + \sum_{j=1}^{d} c_j ({}^{\text{RL}}_{a_j}D^{\mu_j}_{x_j}\, u, {}^{\text{RL}}_{x_j}D^{\mu_j}_{b_j}\, v)_\Omega + \gamma(u,v)_\Omega = (f,v)_\Omega \tag{4.2.7}$$

for all $v \in V$, where γ, c_j are constant, 2τ, $\mu_j \in (0,1)$, $j = 1,2,\ldots,d$, subject to the following homogeneous Dirichlet initial and boundary conditions:

$$\begin{aligned} u|_{t=0} &= 0, \quad \tau \in (0,1/2), \\ u|_{x_j=a_j} &= 0, \quad \mu_j \in (0,1/2), \quad j = 1,2,\ldots,d, \\ u|_{x_j=a_j} = u|_{x_j=b_j} &= 0, \quad \mu_j \in (1/2,1), \quad j = 1,2,\ldots,d. \end{aligned} \tag{4.2.8}$$

We note that the variational form (4.2.7) is equivalent to the following linear FPDE of order 2τ in time and $2\mu_j$ in the j-th spatial dimension, $j = 1, 2, \ldots, d$,

$$ {}^{\mathrm{RL}}_{x_j}D^{\mu_j}_{b_j} u(t, x_1, x_2, \ldots, x_d) + \sum_{j=1}^{d} c_j [\, {}^{\mathrm{RL}}_{a_j}D^{2\mu_j}_{x_j} u(t, x_1, x_2, \ldots, x_d)\,] \qquad (4.2.9)$$
$$ + \gamma\, u(t, x_1, x_2, \ldots, x_d) = f(t, x_1, x_2, \ldots, x_d), $$

when solution u is smooth enough. We define the solution space U as

$$ U := \left\{ u \,:\, u \in C(\Omega),\ \|u\|_U < \infty,\ s.t.\ u|_{t=0} = u|_{x_j=a_j} = 0 \right\} $$

if $\mu_j \in (0, 1/2)$ and

$$ U := \left\{ u \,:\, u \in C(\Omega),\ \|u\|_U < \infty,\ s.t.\ u|_{t=0} = u|_{x_j=a_j} = u|_{x_j=b_j} = 0 \right\} \qquad (4.2.10)$$

when $\mu_j \in (1/2, 1)$, in which

$$ \|u\|_U = \left\{ \| {}^{\mathrm{RL}}_{0}D^{\tau}_{t} u \|^2 + \sum_{j=1}^{d} \| {}^{\mathrm{RL}}_{a_j}D^{\mu_j}_{x_j} u \|^2 + \|u\|^2 \right\}^{1/2}. \qquad (4.2.11)$$

Correspondingly, we define the test space V as

$$ V := \left\{ v \,:\, \|v\|_V < \infty,\ s.t.\ v|_{t=T} = v|_{x_j=b_j} = 0 \right\} $$

when $\mu_j \in (0, 1)$, in which

$$ \|v\|_V = \left\{ \| {}^{\mathrm{RL}}_{t}D^{\tau}_{T} v \|^2 + \sum_{j=1}^{d} \| {}^{\mathrm{RL}}_{x_j}D^{\mu_j}_{b_j} v \|^2 + \|v\|^2 \right\}^{1/2}, $$

where by Fix and Roof (2004); Roop (2004), we can show that U and V are Hilbert spaces; moreover, the associated norms $\|\cdot\|_U$ and $\|\cdot\|_V$ are equivalent. Now, let $a\colon U \times V \to \mathbb{R}$ be a bilinear form, defined as

$$ a(u, v) = \left({}^{\mathrm{RL}}_{0}D^{\tau}_{t}\, u, {}^{\mathrm{RL}}_{t}D^{\tau}_{T}\, v \right)_{\Omega} + \sum_{j=1}^{d} c_j \left({}^{\mathrm{RL}}_{a_j}D^{\mu_j}_{x_j}\, u,\ {}^{\mathrm{RL}}_{x_j}D^{\mu_j}_{b_j}\, v \right)_{\Omega} + \gamma (u, v)_{\Omega}. $$

Moreover, let $L \in V^*$, the dual space of V, be a continuous linear functional defined as

$$ L(v) = (f, v), \quad \forall v \in V. $$

Now, the problem is to find $u \in U$ such that

$$ a(u, v) = L(v), \quad \forall v \in V. $$

Next, we define $U_N \subset U$ and $V_N \subset V$ to be finite-dimensional subspaces of U and V with $\dim(U_N) = \dim(V_N) = N$. Now, our PG spectral method reads as: find $u_N \in U_N$ such that

$$a(u_N, v_N) = L(v_N), \quad \forall v_N \in V_N. \tag{4.2.12}$$

By representing u_N as a linear combination of points/elements in U_N i.e., the corresponding $(1+d)$-dimensional space-time basis functions, the finite-dimensional problem (4.2.12) leads to a linear system known as the *Lyapunov* matrix equation. For instance, if $d = 1$, i.e., 1-D in time and 1-D space, we obtain the corresponding Lyapunov equation in the space-time domain $[0, T] \times [a_1, b_1]$ as

$$S_\tau \mathcal{U} M_{\mu_1}^T + M_\tau \mathcal{U} S_{\mu_1}^T = F, \tag{4.2.13}$$

in which $\mathcal{U}$ is the matrix of unknown coefficients, S_τ and M_τ denote, respectively, the temporal stiffness and mass matrices; similarly, S_{μ_1} and M_{μ_1} represent the spatial stiffness and mass matrices, and F is the corresponding load matrix. In general, numerical solutions to such a linear system, originating from a fractional differential operator, become expensive since the corresponding mass and stiffness matrices usually turn out to be *full* and *nonsymmetric*. Moreover, we note that the size of the above linear system grows as the product of the degrees of freedom in each dimension. To address this problem in this chapter, we present a new class of basis and test functions yielding stiffness matrices, which are either *diagonal* or *tridiagonal*. Similarly, by introducing proper quadrature rules, we compute exactly the corresponding mass matrices, which are *symmetric*. Such useful properties allow us to subsequently develop a general fast linear solver for (4.2.13) with a substantially reduced computational cost. To this end, we first introduce the corresponding finite-dimensional spaces of basis U_N and test functions V_N in our PG framework.

Space of Basis Functions for U_N

We develop a unified Petrov–Galerkin spectral method for (4.2.9), subject to homogeneous Dirichlet initial and boundary conditions. We construct the basis function space as the space of some temporal and spatial functions to globally treat the time-dimension in addition to the spatial-dimensions. To this end, the eigensolutions, introduced in Chapter 2 and Zayernouri and Karniadakis (2013), yield the sets of *basis* and *test* functions, properly suited for our Petrov–Galerkin framework. We represent the solution in the entire space-time computational domain Ω in terms of specially chosen basis functions, constructed as the tensor product of the eigenfunctions in the following manner. Let

$$^{(1)}\mathcal{P}_n^{\mu}(\xi) = (1+\xi)^{\mu}\, P_{n-1}^{(-\mu,\mu)}(\xi), \quad n = 1,2,\ldots, \quad x \in (-1,1),$$

denote the eigenfunctions of the *regular* FSLP of the first kind (RFSLP-I), which are corresponding to the case where $\alpha = \beta = -1$. We construct our basis for the spatial discretization using the univariate poly-fractonomials defined by

$$\phi_m^{\mu}(\xi) = \sigma_m \begin{cases} ^{(1)}\mathcal{P}_m^{\mu}(\xi), & m = 1,2,\ldots, \quad \mu \in (0,1/2], \\ ^{(1)}\mathcal{P}_m^{\mu}(\xi) - \epsilon_m^{\mu}\ {}^{(1)}\mathcal{P}_{m-1}^{\mu}(\xi), & m = 2,3,\ldots, \quad \mu \in (1/2,1), \end{cases} \tag{4.2.14}$$

where $\sigma_m = 2 + (-1)^m$ and the μ-dependent coefficient $\epsilon_{m_j}^{\mu} = (m - 1 - \mu)/(m-1)$. The definition reflects the fact that if $\mu \leq 1/2$ then only one boundary condition needs to be presented, whereas if $\mu > 1/2$ then two end-point conditions are prescribed. Naturally, for the temporal basis functions only initial conditions are prescribed and as a consequence the basis functions for the temporal discretization are constructed using the univariate poly-fractonomials

$$\psi_n^{\tau}(\eta) = \sigma_n\ {}^{(1)}\mathcal{P}_n^{\tau}(\eta), \quad \tau \in (0,1), \tag{4.2.15}$$

for $n \geq \lceil 2\tau \rceil$. With these notations established, we define the space-time trial space to be

$$U_N = \text{span}\Big\{\Big(\psi_n^{\tau} \circ \eta\Big)(t) \prod_{j=1}^{d} \Big(\phi_{m_j}^{\mu_{x_j}} \circ \xi_j\Big)(x_j) : n = 1,\ldots,\mathcal{N},\ m_j = \lceil \sigma_j \rceil,\ldots,\mathcal{M}_j\Big\}, \tag{4.2.16}$$

where $\eta(t) = 2t/T - 1$ and $\xi_j(s) = 2\frac{s-a_j}{b_j-a_j} - 1$. The construction of the univariate functions ensures that $U_N \subset U$, since $\phi_m^{\mu}(-1) = 0$ for all $\mu \in (0,1)$, also $\phi_m^{\mu}(1) = 0$ for all $\mu \in (1/2,1)$. Then, we shall approximate the solution to (4.2.9) in terms of a linear combination of elements in U_N, whose bases satisfy exactly the homogeneous initial and boundary condition in Ω.

Space of Test Functions for V_N

Let the poly-fractonomials

$$^{(2)}\mathcal{P}_k^{\mu}(\xi) = (1-\xi)^{\mu}\, P_{k-1}^{(\mu,-\mu)}(\xi), \quad k = 1,2,\ldots, \quad \xi \in (-1,1),$$

denote the eigenfunctions of the *regular* FSLP of the second kind (RFSLP-II), corresponding to the case $\alpha = \beta = -1$ in (4.2.2). Next, we construct our spatial test functions using the univariate poly-fractonomials defined by

$$\Phi_k^\mu(\xi) = \widetilde{\sigma}_k \begin{cases} {}^{(2)}\mathcal{P}_k^{\mu}(\xi), & k = 1,2,\ldots, \quad \mu \in (0,1/2], \\ {}^{(2)}\mathcal{P}_k^{\mu}(\xi) \; + \; \epsilon_k^{\mu} \; {}^{(2)}\mathcal{P}_{k-1}^{\mu}(\xi), & k = 2,3,\ldots, \quad \mu \in (1/2,1), \end{cases} \tag{4.2.17}$$

where $\widetilde{\sigma}_k = 2\,(-1)^k + 1$. Next, we define the temporal test functions using the univariate poly-fractonomials

$$\Psi_r^{\tau}(\eta) = \widetilde{\sigma}_r \; {}^{(2)}\mathcal{P}_r^{\tau}(\eta), \quad \tau \in (0,1), \tag{4.2.18}$$

for all $r \geq \lceil 2\tau \rceil$. With these notations established, we define the space-time test space to be

$$V_N = \text{span}\Big\{ \Big(\Psi_r^{\tau} \circ \eta\Big)(t) \prod_{j=1}^{d} \Big(\Phi_{k_j}^{\mu_j} \circ \xi_j\Big)(x_j) \; : r = 1,\ldots,\mathcal{N},\, k_j = \lceil \sigma_j \rceil,\ldots,\mathcal{M}_j \Big\}. \tag{4.2.19}$$

Having defined the space of trial and test functions, we can now define the corresponding temporal/spatial stiffness and mass matrices.

Remark 4.6 We show later that the choice of σ_m in (4.2.14) and (4.2.15), also $\widetilde{\sigma}_k$ in (4.2.17) and (4.2.18), leads to the construction of *symmetric* spatial/temporal mass and stiffness matrices. We will exploit this property to formulate a general fast linear solver for the resulting linear system.

4.2.2 Stability and Convergence Analysis

The following theorems provide the stability analysis of the scheme when the pair of $U_N \subset U$ and $V_N \subset V$ are given as in (4.2.16) and (4.2.19), respectively. We first consider the discrete stability of the method for one-dimensional case.

Theorem 4.7 *The Petrov–Galerkin spectral method for the problem*

$$\begin{aligned} {}^{\text{RL}}_{-1}D_x^{2\mu} u(x) &= f(x), \quad \forall x \in (-1,1), \\ u(-1) &= 0, \quad \textit{if } 0 < \mu < 1/2, \\ u(\pm 1) &= 0, \quad \textit{if } 1/2 < \mu < 1, \end{aligned} \tag{4.2.20}$$

is stable, i.e., the discrete inf-sup *condition*

$$\sup_{v_N \in V_N} \frac{a(u_N, v_N)}{\|v_N\|_V} \geq \beta \|u_N\|_U, \quad \forall u_N \in U_N \subset U,$$

holds with $\beta = 1$.

Proof We note that in the absence of the time-derivative and since $\gamma = 0$, the corresponding norm defined on U (see equation 4.2.11) just reduces to

$\|u\|_U = \|\,{}^{\mathrm{RL}}_{-1}D^{\mu}_{x} u\|$. While $\|\,{}^{\mathrm{RL}}_{-1}D^{\mu}_{x} u\|$ has been traditionally treated as a seminorm in the literature (see, e.g., Ervin and Roop (2006); Li and Xu (2009)), one can easily show that it satisfies all the properties of a norm since the Riemann–Liouville fractional derivative of a constant is nonzero. Correspondingly, in this one-dimensional setting, $\|v\|_V = \|\,{}^{\mathrm{RL}}_{x}D^{\mu}_{1} v\|$.

Case I) $0 < \mu < 1/2$. We represent u_N as

$$u_N(x) = \sum_{n=1}^{N} \hat{u}_n (1+x)^{\mu} P^{(-\mu,\mu)}_{n-1}(x), \tag{4.2.21}$$

and choose v_N to be the following linear combination of elements in V_N as

$$v_N(x) = \sum_{k=1}^{N} \hat{u}_k (1-x)^{\mu} P^{(\mu,-\mu)}_{k-1}(x), \tag{4.2.22}$$

in which we employ the same coefficients $\hat{u}_k$ as in (4.2.21). Hence, we obtain

$$\begin{aligned}
a(u_N, v_N) &= \int_{-1}^{1} {}^{\mathrm{RL}}_{-1}D^{\mu}_{x} u_N \; {}^{\mathrm{RL}}_{x}D^{\mu}_{1} v_N \,\mathrm{d}x \\
&= \sum_{n=1}^{N} \hat{u}_n \sum_{k=1}^{N} \hat{u}_k \int_{-1}^{1} {}^{\mathrm{RL}}_{-1}D^{\mu}_{x}[(1+x)^{\mu} P^{(-\mu,\mu)}_{n-1}(x)],\, {}^{\mathrm{RL}}_{x}D^{\mu}_{1}[(1-x)^{\mu} P^{(\mu,-\mu)}_{k-1}(x)]\,\mathrm{d}x \\
&= \sum_{n=1}^{N} \hat{u}_n \frac{\Gamma(n+\mu)}{\Gamma(n)} \sum_{k=1}^{N} \hat{u}_k \frac{\Gamma(k+\mu)}{\Gamma(k)} \int_{-1}^{1} P_{n-1}(x) P_{k-1}(x)\,\mathrm{d}x \\
&= \sum_{n=1}^{N} \hat{u}_n^2 \Big(\frac{\Gamma(n+\mu)}{\Gamma(n)}\Big)^2 \frac{2n+1}{2} \\
&= \|\,{}^{\mathrm{RL}}_{x}D^{\mu}_{1} v_N\|^2_{L^2(-1,1)} = \|v_N\|^2_V .
\end{aligned}$$

Thus we have

$$\sup_{v_N \in V_N} \frac{a(u_N, v_N)}{\|v_N\|_V} = \|u_N\|_U, \quad \forall u_N \in U_N,$$

which means that the stability is guaranteed for $\beta = 1$.

Case II) $1/2 < \mu < 1$. We expand u_N this time as

$$u_N(x) = \sum_{n=1}^{N} \hat{u}_n \big[(1+x)^{\mu} P^{(-\mu,\mu)}_{n-1}(x) - \epsilon^{\mu}_{n} (1+x)^{\mu} P^{(-\mu,\mu)}_{n-2}(x)\big], \tag{4.2.23}$$

and choose v_N to be the following linear combination of elements in V_N as

$$v_N(x) = \sum_{k=1}^{N} \hat{u}_k \big[(1-x)^{\mu} P^{(\mu,-\mu)}_{k-1}(x) + \epsilon^{\mu}_{k} (1-x)^{\mu} P^{(\mu,-\mu)}_{k-2}(x)\big],$$

where the coefficients $\hat{u}_k$ are the same as the ones in (4.2.23). Hence, it is easy to again show that

$$a(u_N, v_N) = \| {}^{\mathrm{RL}}_{\ x}D_1^{\mu} v_N \|^2_{L^2(-1,1)} = \| v_N \|^2_V.$$

Similarly, the stability is guaranteed for $\beta = 1$. □

Remark 4.8 We performed the discrete stability analysis for the 1-D case. The multi-D case is more involved and we will address it in a separate paper in future.

Theorem 4.9 (Projection Error) *In (4.2.12), let* $\| {}^{\mathrm{RL}}_{\ 0}D_t^{r+\tau} u \|_{L^2(\Omega)} < \infty$ *and* $\| {}^{\mathrm{RL}}_{-1}D_{x_j}^{r+\mu_j} u \|_{L^2(\Omega)} < \infty$ *for all* $j = 1, 2, \ldots, d$, *for some integer* $r \geq 1$. *Moreover, let* u_N *denote the projection of the exact solution* u. *Then,*

$$\|u - u_N\|^2_U \leq C\,N^{-2r} \Big(\| {}^{\mathrm{RL}}_{\ 0}D_t^{r+\tau} u \|^2_{L^2(\Omega)} + \sum_{j=1}^{d} \| {}^{\mathrm{RL}}_{-1}D_{x_j}^{r+\mu_j} u \|^2_{L^2(\Omega)} \Big).$$

Proof We first consider the one-dimensional problem (4.2.20) with $\Omega = (-1, 1)$. We expand the exact solution u, when $2\mu \in (0, 1)$, in terms of the following infinite series of Jacobi poly-fractonomials

$$u(x) = \sum_{n=1}^{\infty} \hat{u}_n (1 + x)^{\mu} P_{n-1}^{(-\mu,\mu)}(x).$$

Here, we would like to bound $\|u - u_N\|_U$ in terms of higher-order derivative. We first note that

$${}^{\mathrm{RL}}_{-1}D_x^{r+\mu} u(x) = \frac{\mathrm{d}^r}{\mathrm{d}x^r}[{}^{\mathrm{RL}}_{-1}D_x^{\mu} u(x)] = \sum_{n=1}^{\infty} \hat{u}_n \frac{\Gamma(n+\mu)}{\Gamma(n)} \frac{\mathrm{d}^r}{\mathrm{d}x^r}[P_{n-1}(x)],$$

where

$$\frac{\mathrm{d}^r}{\mathrm{d}x^r}[P_{n-1}(x)] = \begin{cases} \frac{(n-1+r)!}{2r(n-1)!} P_{n-1-r}^{(r,r)}(x), & r \leq n, \\ 0, & r > n. \end{cases}$$

Hence,

$${}^{\mathrm{RL}}_{-1}D_x^{r+\mu} u(x) = \sum_{n=r}^{\infty} \hat{u}_n \frac{\Gamma(n+\mu)}{\Gamma(n)} \frac{(n-1+r)!}{2r(n-1)!} P_{n-1-r}^{(r,r)}(x).$$

Therefore,

$$
\begin{aligned}
&\|(1-x^2)^{r/2}\,{}^{\mathrm{RL}}_{-1}D_x^{r+\mu}u(x)\|^2\\
&=\int_{-1}^{1}(1-x^2)^r\Big(\sum_{n=r}^{\infty}\hat{u}_n\frac{\Gamma(n+\mu)}{\Gamma(n)}\frac{(n-1+r)!}{2r(n-1)!}P_{n-1-r}^{(r,r)}(x)\Big)^2\mathrm{d}x\\
&=\sum_{n=r}^{\infty}\Big(\hat{u}_n\frac{\Gamma(n+\mu)}{\Gamma(n)}\frac{(n-1+r)!}{2r(n-1)!}\Big)^2\gamma_{n-1-r}^{(r,r)}\\
&=\sum_{n=r}^{\infty}\Big[\hat{u}_n\frac{\Gamma(n+\mu)}{\Gamma(n)}\frac{(n-1+r)!}{2r(n-1)!}\Big]^2\frac{2^{2r+1}((n-1)!)^2}{(n-1-r)!(n-1+r)!}\\
&=\sum_{n=r}^{\infty}\Big[\hat{u}_n\frac{\Gamma(n+\mu)}{\Gamma(n)}\Big]^2\frac{2}{2n+1}\frac{(n-1+r)!}{(n-1-r)!}.
\end{aligned}
$$

We also note that $\frac{(n-1+r)!}{(n-1-r)!}$ is minimized when $n=N+1$. Hence,

$$
\begin{aligned}
\|u-u_N\|_U^2&=\sum_{n=N+1}^{\infty}\Big[\hat{u}_n\frac{\Gamma(n+\mu)}{\Gamma(n)}\Big]^2\\
&\le\sum_{n=N+1}^{\infty}\Big[\hat{u}_n\frac{\Gamma(n+\mu)}{\Gamma(n)}\Big]^2\frac{(n-1+r)!}{(n-1-r)!}\frac{(N-r)!}{(N+r)!}\\
&=\frac{(N-r)!}{(N+r)!}\sum_{n=N+1}^{\infty}\Big[\hat{u}_n\frac{\Gamma(n+\mu)}{\Gamma(n)}\Big]^2\frac{(n-1+r)!}{(n-1-r)!},\\
&=\frac{(N-r)!}{(N+r)!}\|(1-x)^{r/2}(1+x)^{r/2}\,{}^{\mathrm{RL}}_{-1}D_x^{r+\mu}u(x)\|^2_{L^2(\Omega)}\\
&\le\frac{(N-r)!}{(N+r)!}\|\,{}^{\mathrm{RL}}_{-1}D_x^{r+\mu}u(x)\|^2_{L^2(\Omega)}\\
&\le c\,N^{-2r}\|\,{}^{\mathrm{RL}}_{-1}D_x^{r+\mu}u(x)\|^2_{L^2(\Omega)},
\end{aligned}
\tag{4.2.24}
$$

where $r\ge 1$ and $2\mu\in(0,1)$. Similar steps are done for the case $2\mu\in(1,2)$ to obtain (4.2.24) noting that in either case, μ remains between 0 and 1.

Next, we consider the following *two-dimensional* problem in $\Omega=(-1,1)^2$:

$$
\begin{aligned}
&{}^{\mathrm{RL}}_{-1}D_x^{2\mu_x}u(x,y)+{}^{\mathrm{RL}}_{-1}D_y^{2\mu_y}u(x,y)=f(x,y),\quad\forall(x,y)\in\Omega;\\
&u(-1,y)=u(x,-1)=0
\end{aligned}
$$

for $0<\mu_x,\mu_y<1/2$, whose corresponding weak form is given by

$$
\big({}^{\mathrm{RL}}_{-1}D_x^{\mu_x}u\,,\,{}^{\mathrm{RL}}_{x}D_1^{\mu_x}v\big)_\Omega+\big({}^{\mathrm{RL}}_{-1}D_y^{\mu_y}u\,,\,{}^{\mathrm{RL}}_{y}D_1^{\mu_y}v\big)_\Omega=(f,v)_\Omega.
$$

We represent the exact solution u when $2\mu_x, 2\mu_y \in (0,1)$, in terms of the following infinite series of tensor product Jacobi poly-fractonomials as

$$u(x,y) = \sum_{n=1}^{\infty}\sum_{m=1}^{\infty} \hat{u}_{nm} \mathcal{P}_n^{\mu_x}(x)\mathcal{P}_m^{\mu_y}(y),$$

where we denote

$$\mathcal{P}_k^{\mu}(z) = (1+z)^{\mu} P_{k-1}^{(-\mu,\mu)}(z).$$

Hence,

$$ {}^{\mathrm{RL}}_{-1}D_x^{r+\mu_x} u = \sum_{n=1}^{\infty}\sum_{m=1}^{\infty} \hat{u}_{nm} \mathbb{C}_n^{\mu_x,r} P_{n-1-r}^{(r,r)}(x)\, \mathcal{P}_m^{\mu_y}(y),$$

$$ {}^{\mathrm{RL}}_{-1}D_y^{r+\mu_y} u = \sum_{n=1}^{\infty}\sum_{m=1}^{\infty} \hat{u}_{nm} \mathbb{C}_m^{\mu_y,r} \mathcal{P}_n^{\mu_x}(x) P_{m-1-r}^{(r,r)}(y),$$

where for simplicity, we denote

$$\mathbb{C}_k^{\mu,r} = \frac{\Gamma(k+\mu)}{\Gamma(k)} \frac{(k-1+r)!}{2r(k-1)!},$$

and

$$w_1(z) = (1-z^2)^{r/2}, \quad w_2^{\mu}(z) = (1-z^2)^{-\mu/2}.$$

Then we have

$$\begin{aligned}
&\| w_1(x)\, w_2^{\mu_y}(y)\, {}^{\mathrm{RL}}_{-1}D_x^{r+\mu_x} u \|^2_{L^2(\Omega)} \\
&= \sum_{n=1}^{\infty}\sum_{m=1}^{\infty} [\hat{u}_{nm} \mathbb{C}_n^{\mu_x,r}]^2 \int_{-1}^{1} (1-x^2)^r [P_{n-1-r}^{(r,r)}(x)]^2 \mathrm{d}x \\
&\quad \times \int_{-1}^{1} (1-y)^{-\mu_y}(1+y)^{\mu_y} [P_{m-1}^{(-\mu_y,\mu_y)}(y)]^2 \mathrm{d}y \\
&= \sum_{n=1}^{\infty}\sum_{m=1}^{\infty} [\hat{u}_{nm} \mathbb{C}_n^{\mu_x,r}]^2 \gamma_{n-1-r}^{(r,r)} \gamma_{m-1}^{(-\mu_y,\mu_y)}.
\end{aligned}$$

Similarly, we can show

$$\| w_2^{\mu_x}(x)\, w_1(y)\, {}^{\mathrm{RL}}_{-1}D_y^{r+\mu_y} u \|^2_{L^2(\Omega)} = \sum_{n=1}^{\infty}\sum_{m=1}^{\infty} [\hat{u}_{nm} \mathbb{C}_m^{\mu_y,r}]^2 \gamma_{n-1}^{(-\mu_x,\mu_x)} \gamma_{m-1-r}^{(r,r)}.$$

We note that $u(x, y)$ can be decomposed into four contributions as

$$u = \Big(\sum_{n=1}^{N}\sum_{m=1}^{N} + \sum_{n=1}^{N}\sum_{m=N+1}^{\infty} + \sum_{n=N+1}^{\infty}\sum_{m=1}^{N} + \sum_{n=N+1}^{\infty}\sum_{m=N+1}^{\infty}\Big)\hat{u}_{nm}\mathcal{P}_n^{\mu_x}(x)\mathcal{P}_m^{\mu_y}(y),$$

or equivalently,

$$u - u_N = \Big(\sum_{n=1}^{N}\sum_{m=N+1}^{\infty} + \sum_{n=N+1}^{\infty}\sum_{m=1}^{N} + \sum_{n=N+1}^{\infty}\sum_{m=N+1}^{\infty}\Big)\hat{u}_{nm}\mathcal{P}_n^{\mu_x}(x)\mathcal{P}_m^{\mu_y}(y).$$

Next, we aim to bound $\|u - u_N\|_U$ in terms of higher-order derivative as

$$\begin{aligned}\|u - u_N\|_U^2 \le \|\sum_{n=1}^{N}\sum_{m=N+1}^{\infty} &\hat{u}_{nm}\mathcal{P}_n^{\mu_x}(x)\mathcal{P}_m^{\mu_y}(y)\|_U^2\\ &+ \|\sum_{n=N+1}^{\infty}\sum_{m=1}^{N}\hat{u}_{nm}\mathcal{P}_n^{\mu_x}(x)\mathcal{P}_m^{\mu_y}(y)\|_U^2\\ &+ \|\sum_{n=N+1}^{\infty}\sum_{m=N+1}^{\infty}\hat{u}_{nm}\mathcal{P}_n^{\mu_x}(x)\mathcal{P}_m^{\mu_y}(y)\|_U^2,\end{aligned} \tag{4.2.25}$$

in which we note the symmetry between the first two terms on the right-hand side. Let us consider the second term first:

$$\begin{aligned}&\|\sum_{n=N+1}^{\infty}\sum_{m=1}^{N}\hat{u}_{nm}\mathcal{P}_n^{\mu_x}(x)\mathcal{P}_m^{\mu_y}(y)\|_U^2\\ &\quad= \|{}^{\mathrm{RL}}_{-1}D_x^{\mu_x}\Big(\sum_{n=N+1}^{\infty}\sum_{m=1}^{N}\hat{u}_{nm}\mathcal{P}_n^{\mu_x}(x)\mathcal{P}_m^{\mu_y}(y)\Big)\|_{L^2(\Omega)}^2\\ &\qquad+ \|{}^{\mathrm{RL}}_{-1}D_y^{\mu_y}\Big(\sum_{n=N+1}^{\infty}\sum_{m=1}^{N}\hat{u}_{nm}\mathcal{P}_n^{\mu_x}(x)\mathcal{P}_m^{\mu_y}(y)\Big)\|_{L^2(\Omega)}^2,\end{aligned}$$

where

$$\begin{aligned}&\|{}^{\mathrm{RL}}_{-1}D_x^{\mu_x}\Big(\sum_{n=N+1}^{\infty}\sum_{m=1}^{N}\hat{u}_{nm}\mathcal{P}_n^{\mu_x}(x)\mathcal{P}_m^{\mu_y}(y)\Big)\|_{L^2(\Omega)}^2\\ &\quad\le \int_{-1}^{1}\int_{-1}^{1} w_2^{\mu_y}(y)\Big(\sum_{n=N+1}^{\infty}\sum_{m=1}^{N}\hat{u}_{nm}\frac{\Gamma(n+\mu_x)}{\Gamma(n)}P_{n-1}(x)\mathcal{P}_m^{\mu_y}(y)\Big)^2 \mathrm{d}x\mathrm{d}y\\ &\quad\le \sum_{n=N+1}^{\infty}\sum_{m=1}^{N}\Big[\hat{u}_{nm}\frac{\Gamma(n+\mu_x)}{\Gamma(n)}\Big]^2\frac{2}{2n+1}\frac{2}{2m-1}\frac{\Gamma(m-\mu_y)\Gamma(m+\mu_y)}{(m-1)!\,\Gamma(m)}\end{aligned}$$

$$\leq \frac{(N-r)!}{(N+r)!} \| {}^{\mathrm{RL}}_{-1}D_x^{r+\mu_x} u \|^2_{L^2(\Omega)}$$
$$\leq cN^{-2r} \| {}^{\mathrm{RL}}_{-1}D_x^{r+\mu_x} u \|^2_{L^2(\Omega)}.$$

Following similar steps, we obtain

$$\| {}^{\mathrm{RL}}_{-1}D_y^{\mu_y} \Big(\sum_{n=N+1}^{\infty} \sum_{m=1}^{N} \hat{u}_{nm} \mathcal{P}_n^{\mu_x}(x) \mathcal{P}_m^{\mu_y}(y) \Big) \|^2_{L^2(\Omega)} \leq cN^{-2r} \| {}^{\mathrm{RL}}_{-1}D_y^{r+\mu_y} u \|^2_{L^2(\Omega)}.$$

Therefore, we obtain the following estimate for the second term on the right-hand side of (4.2.25):

$$\begin{aligned} &\| \sum_{n=N+1}^{\infty} \sum_{m=1}^{N} \hat{u}_{nm} \mathcal{P}_n^{\mu_x}(x) \mathcal{P}_m^{\mu_y}(y) \|^2_U \\ &\quad \leq \tilde{c} N^{-2r} \big(\| {}^{\mathrm{RL}}_{-1}D_x^{r+\mu_x} u \|^2_{L^2(\Omega)} + \| {}^{\mathrm{RL}}_{-1}D_y^{r+\mu_y} u \|^2_{L^2(\Omega)} \big). \end{aligned} \tag{4.2.26}$$

Moreover, by symmetry, we have the following results for the first term on the right-hand side of (4.2.25):

$$\begin{aligned} &\| \sum_{n=1}^{N} \sum_{m=N+1}^{\infty} \hat{u}_{nm} \mathcal{P}_n^{\mu_x}(x) \mathcal{P}_m^{\mu_y}(y) \|^2_U \\ &\quad \leq \tilde{C} N^{-2r} \big(\| {}^{\mathrm{RL}}_{-1}D_x^{r+\mu_x} u \|^2_{L^2(\Omega)} + \| {}^{\mathrm{RL}}_{-1}D_y^{r+\mu_y} u \|^2_{L^2(\Omega)} \big). \end{aligned} \tag{4.2.27}$$

It is easy to check that $\| \sum_{n=N+1}^{\infty} \sum_{m=N+1}^{\infty} \hat{u}_{nm} \mathcal{P}_n^{\mu_x}(x) \mathcal{P}_m^{\mu_y}(y) \|^2_U$ can be bounded by the first two terms on the right-hand side; hence, by substituting (4.2.26) and (4.2.27) into (4.2.25), we finally obtain the following estimate

$$\|u - u_N\|^2_U \leq CN^{-2r} \big(\| {}^{\mathrm{RL}}_{-1}D_x^{r+\mu_x} u \|^2_{L^2(\Omega)} + \| {}^{\mathrm{RL}}_{-1}D_y^{r+\mu_y} u \|^2_{L^2(\Omega)} \big). \tag{4.2.28}$$

Such an error estimate can be isotropically tensor producted up for higher-dimensional problems to get the following estimate

$$\|u - u_N\|^2_U \leq C\, N^{-2r} \Big(\| {}^{\mathrm{RL}}_{0}D_t^{r+\tau} u \|^2_{L^2(\Omega)} + \sum_{j=1}^{d} \| {}^{\mathrm{RL}}_{-1}D_{x_j}^{r+\mu_j} u \|^2_{L^2(\Omega)} \Big), \tag{4.2.29}$$

in which N denotes the number of terms in the expansion in all $(d+1)$ dimensions. □

Remark 4.10 Since the inf-sup condition holds (see Theorem 4.7), by the Banach–Necas–Babuska theorem in Ern and Guermond (2013), the error in the numerical scheme is less than or equal to a constant times the projection error. Choosing the projection u_N in Theorem 4.9, we conclude the spectral accuracy of the scheme.

4.2.3 Implementation of Petrov–Galerkin Spectral Method

We now seek the solution to (4.2.9) in terms of a linear combination of elements in the space U_N of the form

$$u_N(x,t) = \sum_{n=\lceil 2\tau \rceil}^{\mathcal{N}} \sum_{m_1=\lceil 2\mu_1 \rceil}^{\mathcal{M}_1} \cdots \sum_{m_d=\lceil 2\mu_d \rceil}^{\mathcal{M}_d} \hat{u}_{n,m_1,\ldots,m_d} \Big[\psi_n^{\tau}(t) \prod_{j=1}^{d} \phi_{m_j}^{\mu_j}(x_j) \Big] \tag{4.2.30}$$

in Ω. Next, we require the corresponding residual

$$R_N(t,x_1,\ldots,x_d) = {}_{x_j}^{\mathrm{RL}}D_{b_j}^{\mu_j}(u_N) + \sum_{j=1}^{d} c_j \; {}_{a_j}^{\mathrm{RL}}D_{x_j}^{2\mu_j}(u_N) + \gamma\, u_N - f$$

to be L^2-orthogonal to the elements in $v_N \in V_N$, which leads to the finite-dimensional variational form given in (4.2.12). Specifically, by choosing $v_N = \Psi_r^{\mu_t}(t) \prod_{j=1}^{d} \Phi_{k_j}^{\mu_{x_j}}(x_j)$, when $r = \lceil 2\tau \rceil, \ldots, \mathcal{N}$ and $k_j = \lceil 2\mu_j \rceil, \ldots, \mathcal{M}_j$, we obtain

$$\begin{aligned}
\sum_{n=\lceil 2\tau \rceil}^{\mathcal{N}} \sum_{m_1=\lceil 2\mu_1 \rceil}^{\mathcal{M}_1} \cdots \sum_{m_d=\lceil 2\mu_d \rceil}^{\mathcal{M}_d} \hat{u}_{n,m_1,\ldots,m_d} \Big(&\{S_\tau\}_{r,n} \, \{M_{\mu_1}\}_{k_1,m_1} \cdots \{M_{\mu_d}\}_{k_d,m_d} \\
&+ \sum_{j=1}^{d} c_j \, [\, \{M_\tau\}_{r,n} \, \{M_{\mu_1}\}_{k_1,m_1} \cdots \{S_{\mu_j}\}_{k_j,m_j} \cdots \{M_{\mu_d}\}_{k_d,m_d} \,] \\
&+ \gamma \, \{M_\tau\}_{r,n} \, \{M_{\mu_1}\}_{k_1,m_1} \cdots \{M_{\mu_d}\}_{k_d,m_d} \Big) = F_{r,k_1,\ldots,k_d},
\end{aligned} \tag{4.2.31}$$

where S_τ and M_τ denote, respectively, the temporal stiffness and mass matrices, whose entries are defined as

$$\{S_\tau\}_{r,n} = \int_0^T {}_{0}^{\mathrm{RL}}D_t^{\tau}\Big(\psi_n^{\tau} \circ \eta\Big)(t) \; {}_{t}^{\mathrm{RL}}D_T^{\tau}\Big(\Psi_r^{\tau} \circ \eta\Big)(t)\, dt$$

and

$$\{M_\tau\}_{r,n} = \int_0^T \Big(\Psi_r^{\tau} \circ \eta\Big)(t) \Big(\psi_n^{\tau} \circ \eta\Big)(t)\, dt.$$

Moreover, S_{μ_j} and M_{μ_j}, $j = 1,2,\ldots,d$, are the corresponding spatial stiffness and mass matrices

$$\{S_{\mu_j}\}_{k_j,m_j} = \int_{a_j}^{b_j} {}_{a_j}^{\mathrm{RL}}D_{x_j}^{\mu_j}\Big(\phi_{m_j}^{\mu_j} \circ \xi_j\Big)(x_j) \; {}_{x_j}^{\mathrm{RL}}D_{b_j}^{\mu_j}\Big(\Phi_{k_j}^{\mu_j} \circ \xi_j\Big)(x_j)\, dx_j$$

and

$$\{M_{\mu_j}\}_{k_j,m_j} = \int_{a_j}^{b_j} \left(\Phi_{k_j}^{\mu_j} \circ \xi_j\right)(x_j) \left(\phi_{m_j}^{\mu_j} \circ \xi_j\right)(x_j)\, \mathrm{d}x_j,$$

respectively, to be exactly computed. Moreover, $F_{r,k_1,\ldots,k_d}$ is

$$\int_{\Omega} f(t, x_1, \ldots, x_d)\left(\Psi_r^{\tau} \circ \eta\right)(t) \prod_{j=1}^{d} \left(\Phi_{k_j}^{\mu_j} \circ \xi_j\right)(x_j)\, \mathrm{d}\Omega. \tag{4.2.32}$$

Assuming that all the aforementioned stiffness and mass matrices are symmetric, we can render the linear system (4.2.31) as the following general Lyapunov equation

$$\begin{aligned}\Big(& S_\tau \otimes M_{\mu_1} \otimes M_{\mu_2} \cdots \otimes M_{\mu_d} \\ & + \sum_{j=1}^{d} c_j \; [\; M_\tau \otimes M_{\mu_1} \otimes \cdots \otimes M_{\mu_{j-1}} \otimes S_{\mu_j} \otimes M_{\mu_{j+1}} \cdots \otimes M_{\mu_d} \;] \\ & + \gamma\; M_\tau \otimes M_{\mu_1} \otimes M_{\mu_2} \cdots \otimes M_{\mu_d} \Big)\mathcal{U} = F,\end{aligned} \tag{4.2.33}$$

in which $\otimes$ represents the Kronecker product, F denotes the multi-dimensional load matrix whose entries are given in (4.2.32), and $\mathcal{U}$ is the corresponding multi-dimensional matrix of unknown coefficients whose entries are $\hat{u}_{n,m_1,\cdots,m_d}$.

4.2.4 Calculation of the Matrices

We investigate the properties of the aforementioned matrices in addition to presenting efficient ways of deriving the stiffness matrices explicitly and computing the mass matrices exactly through proper quadrature rules.

Theorem 4.11 *The temporal stiffness matrix S_τ corresponding to the time-fractional order $\tau \in (0, 1)$ is a diagonal $\mathcal{N} \times \mathcal{N}$ matrix, whose entries are explicitly given as*

$$\{S_\tau\}_{n,n} = \widetilde{\sigma}_n\, \sigma_n \left[\frac{\Gamma(n+\tau)}{\Gamma(n)}\right]^2 \left(\frac{2}{T}\right)^{2\tau-1} \frac{2}{2n-1}, \quad n = 1, 2, \ldots, \mathcal{N}.$$

Proof By the PG projection, the (r,n)-th entry of the stiffness matrix, r, $n = 1, 2, \ldots, \mathcal{N}$, is defined as

$$\{S_\tau\}_{r,n} = \int_0^T {}^{\mathrm{RL}}_{\ t}D_T^{\tau}\left(\Psi_r^{\tau} \circ \eta\right)(t)\; {}^{\mathrm{RL}}_{\ 0}D_t^{\tau}\left(\psi_n^{\tau} \circ \eta\right)(t)\, \mathrm{d}t. \tag{4.2.34}$$

We now obtain the Riemann–Liouville left-sided time-fractional derivative of the temporal basis, following Chapter 2, as

$$ {}^{\mathrm{RL}}_{0}D^{\tau}_{t}\big(\psi^{\tau}_{n}\circ\eta\big)(t) = \sigma_n\, \zeta^{\tau}_{n} \left(\frac{2}{T}\right)^{\tau} P_{n-1}(2t/T-1), \tag{4.2.35} $$

where $P_{n-1}(2t/T-1)$ represents the $(n-1)$-th order *Legendre polynomial* in $t \in [0,T]$, and we denote

$$ \zeta^{\tau}_{n} := \frac{\Gamma(n+\tau)}{\Gamma(n)}. $$

Also, we obtain the right-sided time-fractional derivative of the temporal basis as

$$ {}^{\mathrm{RL}}_{t}D^{\tau}_{T}\Big(\Psi^{\tau}_{r}\circ\eta\Big)(t) = \widetilde{\sigma}_r\, \zeta^{\tau}_{r} \left(\frac{2}{T}\right)^{\tau} P_{r-1}(2t/T-1). \tag{4.2.36} $$

Now, by plugging (4.2.35) and (4.2.36) into (4.2.34), we obtain

$$ \begin{aligned} \{S_\tau\}_{r,n} &= \widetilde{\sigma}_r\, \sigma_n\, \zeta^{\tau}_{r}\, \zeta^{\tau}_{n} \left(\frac{2}{T}\right)^{2\tau} \int_0^T P_{r-1}(x(t))\, P_{n-1}(x(t))\, \mathrm{d}t \\ &= \widetilde{\sigma}_r\, \sigma_n\, \sigma_n\, \zeta^{\tau}_{r}\, \zeta^{\tau}_{n} \left(\frac{2}{T}\right)^{2\tau-1} \int_{-1}^{1} P_{r-1}(x)\, P_{n-1}(x)\, \mathrm{d}x \\ &= \widetilde{\sigma}_r\, \sigma_n\, \sigma_n\, \zeta^{\tau}_{r}\, \zeta^{\tau}_{n} \left(\frac{2}{T}\right)^{2\tau-1} \frac{2}{2n-1}\delta_{rn} \end{aligned} \tag{4.2.37} $$

by the orthogonality of the Legendre polynomials, where δ_{rn} is the Kronecker delta function. □

Theorem 4.12 *(I) If $\mu_j \in (0,1/2]$, the spatial stiffness matrix S_{μ_j} is a diagonal $\mathcal{M}_j \times \mathcal{M}_j$ matrix, whose entries are explicitly given as*

$$ \{S_{\mu_j}\}_{k,k} = \widetilde{\sigma}_k\, \sigma_k \left[\frac{\Gamma(k+\mu_j)}{\Gamma(k)}\right]^2 \left(\frac{2}{L_j}\right)^{2\mu_j-1} \frac{2}{2k-1}, \quad k = 1,2,\dots,\mathcal{M}_j. $$

(II) If $\mu_j \in (1/2,1)$, S_{μ_j} is a symmetric tridiagonal $(\mathcal{M}_j-1)\times(\mathcal{M}_j-1)$ matrix with entries, explicitly given as

$$ \begin{aligned} \{S_{\mu_j}\}_{k,m} &= b_k\, a_m \left[\frac{\Gamma(k+\mu_j)}{\Gamma(k)}\right]^2 \left(\frac{2}{L_j}\right)^{2\mu_j-1} \frac{2}{2k-1}\big(\delta_{k,m} - \epsilon^{\mu_j}_{m}\, \delta_{k,m-1}\big) \\ &+ \epsilon^{\mu_j}_{k}\, b_k\, a_m \left[\frac{\Gamma(k-1+\mu_j)}{\Gamma(k-1)}\right]^2 \left(\frac{2}{L_j}\right)^{2\mu_j-1} \frac{2}{2k-3}\big(\delta_{k-1,m} - \epsilon^{\mu_j}_{m}\, \delta_{k-1,m-1}\big), \end{aligned} $$

$k,m = 2,3,\dots,\mathcal{M}_j$ and $L_j = b_j - a_j$.

Proof The first part, when $\mu_j \in (0,1/2]$ is similar to the proof in Theorem 4.11, however, carried out on the interval $[a_j,b_j]$ rather than $[0,T]$. Here, the μ_j-th order left-sided Riemann–Liouville fractional derivative of $\left(\phi_{m_j}^{\mu_j} \circ \xi_j\right)(x_j)$ is given as

$$ {}^{\mathrm{RL}}_{a_j}D^{\mu_j}_{x_j}\left(\phi^{\mu_j}_m \circ \xi_j\right)(x_j) = \begin{cases} B^{\mu_j}_m \; P_{m-1}\left(\xi(x_j)\right), & \mu_j \in (0, 1/2], \\ \\ B^{\mu_j}_m \; P_{m-1}\left(\xi(x_j)\right) \; - C^{\mu_j}_m \; P_{m-2}\left(\xi(x_j)\right), & \mu_j \in (1/2, 1) \end{cases} \tag{4.2.38} $$

for $m = \lceil 2\mu_j \rceil, \ldots, \mathcal{M}_j$ in the j-th spatial dimension, where the coefficients

$$ B^{\mu_j}_{m_j} = \sigma_m \, (2/L_j)^{2\mu_j} \, \Gamma(m + \mu_j)/\Gamma(m) $$

and

$$ C^{\mu_j}_m = \sigma_m \, (2/L_j)^{2\mu_j} \, \epsilon^{\mu_j}_m \, \Gamma(m - 1 + \mu_j)/\Gamma(m - 1); $$

in addition, the μ_j-th order left-sided Riemann–Liouville fractional derivative of $\left(\Phi^{\mu_j}_m \circ \xi_j\right)(x_j)$ is obtained as

$$ {}^{\mathrm{RL}}_{x_j}D^{\mu_j}_{b_j}\left(\Phi^{\mu_j}_m \circ \xi_j\right)(x_j) = \begin{cases} \mathcal{B}^{\mu_j}_k \; P_{k-1}\left(\xi(x_j)\right), & \mu_j \in (0, 1/2], \\ \\ \mathcal{B}^{\mu_j}_k \; P_{k-1}\left(\xi(x_j)\right) - \mathcal{C}^{\mu_j}_k \; P_{k-2}\left(\xi(x_j)\right), & \mu_j \in (1/2, 1), \end{cases} \tag{4.2.39} $$

for $k = \lceil 2\mu_j \rceil, \ldots, \mathcal{M}_j$ in the j-th spatial dimension, in which we set the coefficients

$$ \mathcal{B}^{\mu_j}_k = \widetilde{\sigma}_k \, (2/L_j)^{2\mu_j} \, \Gamma(k + \mu_j)/\Gamma(k) $$

and

$$ \mathcal{C}^{\mu_j}_k = \widetilde{\sigma}_k \, (2/L_j)^{2\mu_j} \, \epsilon^{\mu_j}_k \, \Gamma(k - 1 + \mu_j)/\Gamma(k - 1). $$

For the second part, when $\mu_j \in (1/2, 1)$, the (k, m)-th entry of S_{μ_j} is

$$ \{S_{\mu_j}\}_{k,m} = \int_{a_j}^{b_j} {}^{\mathrm{RL}}_{x_j}D^{\mu_j}_{b_j}\left(\Phi^{\mu_j}_k \circ \xi_j\right)(x_j) \, {}^{\mathrm{RL}}_{a_j}D^{\mu_j}_{x_j}\left(\phi^{\mu_j}_m \circ \xi_j\right)(x_j) \mathrm{d}x_j. $$

Next, by virtue of (4.2.38) and (4.2.39), also by an affine mapping from $[a_j, b_j]$ to the standard interval $[-1, 1]$, we obtain

$$ \begin{aligned} \{S_{\mu_j}\}_{k,m} = \frac{L_j}{2} \int_{-1}^{1} & \left[\, B^{\mu_j}_m \; P_{m-1}(\xi_j) \; - C^{\mu_j}_m \; P_{m-2}(\xi_j) \right] \\ & \left[\, \mathcal{B}^{\mu_j}_k \; P_{k-1}(\xi_j) \; - \mathcal{C}^{\mu_j}_k \; P_{k-2}(\xi_j) \right] \mathrm{d}\xi_j. \end{aligned} $$

Hence, by the orthogonality of the Legendre polynomials we obtain

$$\{S_{\mu_j}\}_{k,m} = \frac{L_j}{2}\Big[B_m^{\mu_j}\ \mathcal{B}_k^{\mu_j}\ \frac{2}{2k-1}\delta_{k,m} - C_m^{\mu_j}\ \mathcal{B}_k^{\mu_j}\ \frac{2}{2k-1}\delta_{k,m-1}$$
$$+ B_m^{\mu_j}\ \mathcal{C}_k^{\mu_j}\ \frac{2}{2k-3}\ \delta_{k-1,m} - C_m^{\mu_j}\ \mathcal{C}_k^{\mu_j}\ \frac{2}{2k-3}\ \delta_{k-1,m-1}\Big],$$

which completes the proof, while the symmetry of the stiffness matrix can be easily checked. □

Theorem 4.13 *The temporal and the spatial mass matrices M_τ as well as M_{μ_j} are symmetric. Moreover, their entries can be computed exactly by employing a Gauss–Lobatto–Jacobi (GLJ) rule with respect to the weight function $(1-\xi^2)^\alpha$, $\xi \in (-1,1)$, where $\alpha = \tau/2$ in the temporal and $\alpha = \mu_j$ for the spatial case.*

Proof The entries of M_τ in our PG spectral method are defined as

$$\{M_\tau\}_{r,n} = \int_0^T \big(\Psi_r^\tau \circ \eta\big)(t)\big(\psi_m^{\mu_t} \circ \eta\big)(t)\mathrm{d}t,$$

which can be computed exactly as

$$\{M_\tau\}_{r,n} = \widetilde{\sigma}_r\ \sigma_n\ \left(\frac{2}{T}\right)^\tau \int_0^T t^\tau (T-t)^\tau P_{r-1}^{(\tau,-\tau)}(\eta(t))\ P_{n-1}^{(-\tau,\tau)}(\eta(t))\,\mathrm{d}t$$
$$= \widetilde{\sigma}_r\ \sigma_n\ \frac{T}{2}\int_{-1}^1 (1-\eta)^\tau(1+\eta)^\tau P_{r-1}^{(\tau,-\tau)}(\eta)\ P_{n-1}^{(-\tau,\tau)}(\eta)\mathrm{d}\eta$$
$$= \widetilde{\sigma}_r\ \sigma_n\ \frac{T}{2}\sum_{q=1}^{\mathcal{Q}} w_q\ P_{r-1}^{(\tau,-\tau)}(\eta_q)P_{n-1}^{(-\tau,\tau)}(\eta_q),$$

in which $\mathcal{Q} \geq \mathcal{N}+2$ represents the minimum number of GLJ quadrature points $\{\eta_q\}_{q=1}^{\mathcal{Q}}$, associated with the weight function $(1-\eta)^\tau(1+\eta)^\tau$, for *exact* quadrature, and $\{w_q\}_{q=1}^{\mathcal{Q}}$ are the corresponding quadrature weights. From the exact discrete rule, recalling the definition of σ_n and $\widetilde{\sigma}_r$, employing the property of the Jacobi polynomials where $P_n^{(\alpha,\beta)}(-x) = (-1)^n P_n^{(\beta,\alpha)}(x)$, moreover, noting that $\{\eta_q\}_{q=1}^{\mathcal{Q}}$ and $\{w_q\}_{q=1}^{\mathcal{Q}}$ are symmetric with respect to the reference point, it is easy to show that $\{M_\tau\}_{r,n} = \{M_\tau\}_{n,r}$.

The spatial mass matrix M_{μ_j}, when $\mu_j \in (0, 1/2]$, is also $\mathcal{M}_j \times \mathcal{M}_j$, whose entries are computed similarly as

$$\{M_{\mu_j}\}_{k,m} = \widetilde{\sigma}_k\ \sigma_m\ \frac{L_j}{2}\sum_{q=1}^{\mathcal{Q}} w_q\ P_{k-1}^{(\mu_j,-\mu_j)}(\xi_q)P_{m-1}^{(-\mu_j,\mu_j)}(\xi_q),$$

in which $\mathcal{Q} \geq \mathcal{M}_j + 2$ represents the minimum number of GLJ quadrature points $\{\xi_q\}_{q=1}^{\mathcal{Q}}$, associated with the weight function $(1-\xi)^{\mu_j}(1+\xi)^{\mu_j}$, for *exact* quadrature. We can also show that M_{μ_j} is symmetric and that the GJL rule is exact when $\mathcal{Q} \geq \mathcal{M}_j + 2$.

Finally, when $\mu_j \in (1/2, 1)$, the spatial mass matrix M_{μ_j}, becomes $(\mathcal{M}_j - 1) \times (\mathcal{M}_j - 1)$, whose entries are computed exactly as

$$
\begin{aligned}
\{M_{\mu_j}\}_{k,m} &= \int_{a_j}^{b_j} \big(\Phi_k^{\mu_j/2} \circ \xi\big)(x_j)\big(\phi_m^{\mu_j/2} \circ \xi\big)(x_j) \mathrm{d}x_j \\
&= \widetilde{\sigma}_{k_j}\, \sigma_{m_j} \Big[\int_{-1}^{1} {}^{(2)}\mathcal{P}_{k_j}^{\mu_j}(\xi(x_j))\; {}^{(1)}\mathcal{P}_{m}^{\mu_x}(\xi(x_j))\, \mathrm{d}x_j \\
&\quad - \epsilon_m^{\mu_x} \int_{-1}^{1} {}^{(2)}\mathcal{P}_{k_j}^{\mu_j}(\xi(x_j))\; {}^{(1)}\mathcal{P}_{m_j-1}^{\mu_j}(\xi(x_j))\, \mathrm{d}x_j \\
&\quad + \epsilon_k^{\mu_x} \int_{-1}^{1} {}^{(2)}\mathcal{P}_{k_j-1}^{\mu_j}(\xi(x_j))\; {}^{(1)}\mathcal{P}_{m_j}^{\mu_j}(\xi(x_j))\, \mathrm{d}x_j \\
&\quad - \epsilon_{k_j}^{\mu_j}\, \epsilon_{m_j}^{\mu_j} \int_{-1}^{1} {}^{(2)}\mathcal{P}_{k_j-1}^{\mu_j}(\xi(x_j))\; {}^{(1)}\mathcal{P}_{m_j-1}^{\mu_j}(\xi(x_j)) \mathrm{d}x_j \Big],
\end{aligned}
$$

where we note that all the above integrations share the same weight function by construction. Hence, we obtain

$$
\begin{aligned}
\{M_{\mu_j}\}_{k,m} = \widetilde{\sigma}_{k_j}\, \sigma_{m_j} \frac{L_j}{2} \int_{-1}^{1} (1-\xi^2)^{\mu_j} \Big[& P_{k-1}^{(\mu_j,-\mu_j)}(\xi)\, P_{m-1}^{(-\mu_j,\mu_j)}(\xi) \\
& - \epsilon_m^{\mu_x} P_{k-1}^{(\mu_j,-\mu_j)}(\xi)\, P_{m-2}^{(-\mu_j,\mu_j)}(\xi) + \epsilon_k^{\mu_x} P_{k-2}^{(\mu_j,-\mu_j)}(\xi)\, P_{m-1}^{(-\mu_j,\mu_j)}(\xi) \\
& - \epsilon_{k_j}^{\mu_j}\, \epsilon_{m_j}^{\mu_j} P_{k-2}^{(\mu_j,-\mu_j)}(\xi)\, P_{m-2}^{(-\mu_j,\mu_j)}(\xi) \Big] \mathrm{d}\xi,
\end{aligned}
$$

which leads to the following exact GLJ rule

$$
\begin{aligned}
\{M_{\mu_j}\}_{k,m} = \widetilde{\sigma}_{k_j}\, \sigma_{m_j} \frac{L_j}{2} \sum_{q=1}^{Q} w_q \Big[& P_{k-1}^{(\mu_j,-\mu_j)}(\xi_q)\, P_{m-1}^{(-\mu_j,\mu_j)}(\xi_q) \\
& - \epsilon_m^{\mu_x} P_{k-1}^{(\mu_j,-\mu_j)}(\xi_q)\, P_{m-2}^{(-\mu_j,\mu_j)}(\xi_q) \\
& + \epsilon_k^{\mu_x} P_{k-2}^{(\mu_j,-\mu_j)}(\xi_q)\, P_{m-1}^{(-\mu_j,\mu_j)}(\xi_q) \\
& - \epsilon_{k_j}^{\mu_j}\, \epsilon_{m_j}^{\mu_j} P_{k-2}^{(\mu_j,-\mu_j)}(\xi_q)\, P_{m-2}^{(-\mu_j,\mu_j)}(\xi_q) \Big],
\end{aligned}
$$

which is also exact when $\mathcal{Q} \geq \mathcal{M}_j + 2$, and the same argument on the symmetry of the matrix applies here. □

4.2.5 Unified Fast Fractional PDE Solver

So far, we have formulated a suitable Petrov–Galerkin variational framework for the general $(1+d)$-dimensional FPDE, given in (4.2.9), by choosing proper basis and test functions. The main advantage of such framework is that we can explicitly obtain the corresponding stiffness matrices to be symmetric diagonal/tridiagonal, and moreover, to exactly compute the mass matrices, which we showed to be symmetric. The following result better highlights the benefit of this scheme, where we formulate a closed-form solution for the Lyapunov system (4.2.33) in terms of the generalized eigensolutions that can be computed very efficiently.

Theorem 4.14 *Let $\{\vec{e}_{m_j}^{\,\mu_j}, \lambda_{m_j}^{\mu_j}\}_{m_j=\lceil 2\mu_j \rceil}^{\mathcal{M}_j}$ be the set of general eigensolutions of the spatial mass matrix M_{μ_j} with respect to the stiffness matrix S_{μ_j}. Moreover, let us assume that $\{\vec{e}_n^{\,\tau}, \lambda_n^{\tau}\}_{n=\lceil 2\tau \rceil}^{\mathcal{N}}$ are the set of general eigensolutions of the temporal mass matrix M_τ with respect to the stiffness matrix S_τ. (I) if $d > 1$, then the multi-dimensional matrix of unknown solution $\mathcal{U}$ is explicitly obtained as*

$$\mathcal{U} = \sum_{n=\lceil 2\tau \rceil}^{\mathcal{N}} \sum_{m_1=\lceil 2\mu_1 \rceil}^{\mathcal{M}_1} \cdots \sum_{m_d=\lceil 2\mu_d \rceil}^{\mathcal{M}_d} \kappa_{n,m_1,\ldots,m_d}\; \vec{e}_n^{\,\tau} \otimes \vec{e}_{m_1}^{\,\mu_1} \otimes \cdots \otimes \vec{e}_{m_d}^{\,\mu_d}, \tag{4.2.40}$$

where the unknown $\kappa_{n,m_1,\ldots,m_d}$ are given by

$$\kappa_{n,m_1,\ldots,m_d} = \frac{(\vec{e}_n^{\,\tau}\vec{e}_{m_1}^{\,\mu_1}\cdots\vec{e}_{m_d}^{\,\mu_d})F}{\Big[(\vec{e}_n^{\,\tau^T} S_\tau \vec{e}_n^{\,\tau}) \prod_{j=1}^{d} (\vec{e}_{m_j}^{\,\mu_j^T} S_{\mu_j} \vec{e}_{m_j}^{\,\mu_j})\Big] \Lambda_{n,m_1,\ldots,m_d}}, \tag{4.2.41}$$

in which the numerator represents the standard multi-dimensional inner product, and $\Lambda_{n,m_1,\ldots,m_d}$ are obtained in terms of the eigenvalues of all mass matrices as

$$\Lambda_{n,m_1,\ldots,m_d} = \Big[(1+\gamma\,\lambda_n^{\tau}) \prod_{j=1}^{d} \lambda_{m_j}^{\mu_j} + \lambda_n^{\tau} \sum_{j=1}^{d} c_j \,\Big(\prod_{s=1,s\neq j}^{d} \lambda_{m_s}^{\mu_s}\Big)\Big].$$

(II) If $d = 1$, then the two-dimensional matrix of the unknown solution $\mathcal{U}$ is obtained as

$$\mathcal{U} = \sum_{n=\lceil 2\tau \rceil}^{\mathcal{N}} \sum_{m_1=\lceil 2\mu_1 \rceil}^{\mathcal{M}_1} \kappa_{n,m_1}\; \vec{e}_n^{\,\tau}\, \vec{e}_{m_1}^{\,\mu_1^T},$$

where κ_{n,m_1} is explicitly obtained as

$$\kappa_{n,m_1} = \frac{\vec{e}_n^{\,\tau^T} F\, \vec{e}_{m_1}^{\,\mu_1}}{(\vec{e}_n^{\,\tau^T} S_\tau \vec{e}_n^{\,\tau})(\vec{e}_{m_1}^{\,\mu_1^T} S_{\mu_1} \vec{e}_{m_1}^{\,\mu_1})\Big[\,\lambda_{m_1-1}^{\mu_1} + c_1\,\lambda_n^{\tau} + \gamma\,\lambda_n^{\tau}\lambda_{m_1}^{\mu_1}\Big]}.$$

Proof Let us consider the following generalized eigenvalue problems

$$M_{\mu_j}\vec{e}_{m_j}^{\,\mu_j} = \lambda_{m_j}^{\mu_j} S_{\mu_j}\vec{e}_{m_j}^{\,\mu_j}, \quad m_j = \lceil 2\mu_j \rceil, \ldots, \mathcal{M}_j, \quad j = 1,2,\ldots,d, \tag{4.2.42}$$

$$M_\tau \vec{e}_n^{\,\tau} = \lambda_n^{\tau} S_\tau \vec{e}_n^{\,\tau}, \quad n = \lceil 2\tau \rceil, 2, \ldots, \mathcal{N}. \tag{4.2.43}$$

We aim to represent the unknown coefficient matrix $\mathcal{U}$ in the expansion (4.2.30) in terms of the spatial and temporal eigenvectors as

$$\mathcal{U} = \sum_{n=\lceil 2\tau \rceil}^{\mathcal{N}} \sum_{m_1=\lceil 2\mu_1 \rceil}^{\mathcal{M}_1} \cdots \sum_{m_d=\lceil 2\mu_d \rceil}^{\mathcal{M}_d} \kappa_{n,m_1,\ldots,m_d} \, \vec{e}_n^{\,\tau} \otimes \vec{e}_{m_1}^{\,\mu_1} \otimes \cdots \otimes \vec{e}_{m_d}^{\,\mu_d}, \tag{4.2.44}$$

where the unknown $\kappa_{n,m_1,\ldots,m_d}$ are obtained as follows. We first take the multi-dimensional inner product of $\vec{e}_q^{\,\tau} \vec{e}_{p_1}^{\,\mu_1} \cdots \vec{e}_{p_d}^{\,\mu_d}$ on both sides of the Lyapunov equation (4.2.33) to obtain

$$\begin{aligned}
(\vec{e}_q^{\,\tau} \vec{e}_{p_1}^{\,\mu_1} \vec{e}_{p_2}^{\,\mu_2} \cdots \vec{e}_{p_d}^{\,\mu_d}) \Big[& S_\tau \otimes M_{\mu_1} \otimes \cdots \otimes M_{\mu_d} \\
& + \sum_{j=1}^{d} c_j \, [M_\tau \otimes M_{\mu_1} \otimes \cdots \otimes M_{\mu_{j-1}} \otimes S_{\mu_j} \otimes M_{\mu_{j+1}} \cdots \otimes M_{\mu_d}] \\
& + \gamma \, M_\tau \otimes M_{\mu_1} \otimes \cdots \otimes M_{\mu_d} \Big] \mathcal{U} = (\vec{e}_q^{\,\tau} \vec{e}_{p_1}^{\,\mu_1} \cdots \vec{e}_{p_d}^{\,\mu_d}) F.
\end{aligned}$$

Next, we substitute (4.2.44) and rearrange the terms to obtain

$$\begin{aligned}
& \sum_{n=\lceil 2\tau \rceil}^{\mathcal{N}} \sum_{m_1=\lceil 2\mu_1 \rceil}^{\mathcal{M}_1} \cdots \sum_{m_d=\lceil 2\mu_d \rceil}^{\mathcal{M}_d} \kappa_{n,m_1,\ldots,m_d} \\
& \times \Big(\vec{e}_q^{\,\tau^T} S_\tau \vec{e}_n^{\,\tau} \vec{e}_{p_1}^{\,\mu_j^T} M_{\mu_1} \vec{e}_{m_1}^{\,\mu_j} \cdots \vec{e}_{p_d}^{\,\mu_j^T} M_{\mu_d} \vec{e}_{m_d}^{\,\mu_j} \\
& + \sum_{j=1}^{d} c_j \, \vec{e}_q^{\,\tau^T} M_\tau \vec{e}_n^{\,\tau}, \vec{e}_{p_1}^{\,\mu_j^T} M_{\mu_1} \vec{e}_{m_1}^{\,\mu_j} \cdots \vec{e}_{p_j}^{\,\mu_j^T} S_{\mu_j} \vec{e}_{m_j}^{\,\mu_j} \vec{e}_{p_{j+1}}^{\,\mu_{j+1}^T} M_{\mu_{j+1}} \vec{e}_{m_{j+1}}^{\,\mu_{j+1}} \vec{e}_{p_d}^{\,\mu_j^T} M_{\mu_d} \vec{e}_{m_d}^{\,\mu_j} \\
& + \gamma \, \vec{e}_q^{\,\tau^T} M_\tau \vec{e}_n^{\,\tau} \quad \vec{e}_{p_1}^{\,\mu_j^T} M_{\mu_1} \vec{e}_{m_1}^{\,\mu_j} \quad \vec{e}_{p_2}^{\,\mu_j^T} M_{\mu_2} \vec{e}_{m_2}^{\,\mu_j} \quad \cdots \quad \vec{e}_{p_d}^{\,\mu_j^T} M_{\mu_d} \vec{e}_{m_d}^{\,\mu_j} \Big) \\
& = (\vec{e}_q^{\,\tau} \vec{e}_{p_1}^{\,\mu_1} \vec{e}_{p_2}^{\,\mu_2} \cdots \vec{e}_{p_d}^{\,\mu_d}) F,
\end{aligned}$$

where we recall that $M_{\mu_j} \vec{e}_{m_j}^{\,\mu_j} = (\lambda_{m_j}^{\mu_j} S_{\mu_j} \vec{e}_{m_j}^{\,\mu_j})$ and $M_\tau \, \vec{e}_n^{\,\tau} = (\lambda_n^\tau \, S_\tau \, \vec{e}_n^{\,\tau})$. Hence,

$$\begin{aligned}
& \sum_{n=\lceil 2\tau \rceil}^{\mathcal{N}} \sum_{m_1=\lceil 2\mu_1 \rceil}^{\mathcal{M}_1} \cdots \sum_{m_d=\lceil 2\mu_d \rceil}^{\mathcal{M}_d} \kappa_{n,m_1,\ldots,m_d} \\
& \times \Big(\vec{e}_q^{\,\tau^T} S_\tau \vec{e}_n^{\,\tau} \quad \vec{e}_{p_1}^{\,\mu_j^T} (\lambda_{m_1}^{\mu_1} S_{\mu_1} \vec{e}_{m_1}^{\,\mu_1}) \quad \vec{e}_{p_2}^{\,\mu_j^T} (\lambda_{m_2}^{\mu_2} S_{\mu_2} \vec{e}_{m_2}^{\,\mu_2}) \quad \cdots \quad (\lambda_{m_d}^{\mu_d} S_{\mu_d} \vec{e}_{m_d}^{\,\mu_d}) \\
& + \sum_{j=1}^{d} c_j \, \vec{e}_q^{\,\tau^T} (\lambda_n^\tau \, S_\tau \, \vec{e}_n^{\,\tau}) \quad \vec{e}_{p_1}^{\,\mu_j^T} (\lambda_{m_1}^{\mu_1} S_{\mu_1} \vec{e}_{m_1}^{\,\mu_1}) \\
& \cdots \vec{e}_{p_j}^{\,\mu_j^T} S_{\mu_j} \vec{e}_{m_j}^{\,\mu_j} \quad \vec{e}_{p_{j+1}}^{\,\mu_{j+1}^T} (\lambda_{m_{j+1}}^{\mu_{j+1}} S_{\mu_{j+1}} \vec{e}_{m_{j+1}}^{\,\mu_{j+1}}) \cdots \vec{e}_{p_d}^{\,\mu_j^T} (\lambda_{m_d}^{\mu_d} S_{\mu_d} \vec{e}_{m_d}^{\,\mu_d}) \\
& + \gamma \, \vec{e}_q^{\,\tau^T} (\lambda_n^\tau \, S_\tau \, \vec{e}_n^{\,\tau}) \quad \vec{e}_{p_1}^{\,\mu_j^T} (\lambda_{m_1}^{\mu_1} S_{\mu_1} \vec{e}_{m_1}^{\,\mu_1}) \quad \vec{e}_{p_2}^{\,\mu_j^T} (\lambda_{m_2}^{\mu_2} S_{\mu_2} \vec{e}_{m_2}^{\,\mu_2}) \quad \cdots \quad (\lambda_{m_d}^{\mu_d} S_{\mu_d} \vec{e}_{m_d}^{\,\mu_d}) \Big) \\
& = (\vec{e}_q^{\,\tau} \vec{e}_{p_1}^{\,\mu_1} \vec{e}_{p_2}^{\,\mu_2} \cdots \vec{e}_{p_d}^{\,\mu_d}) F.
\end{aligned}$$

or alternatively,

$$\sum_{n=\lceil 2\tau\rceil}^{\mathcal{N}} \sum_{m_1=\lceil 2\mu_1\rceil}^{\mathcal{M}_1} \cdots \sum_{m_d=\lceil 2\mu_d\rceil}^{\mathcal{M}_d} \kappa_{n,m_1,\ldots,m_d}\, (\vec{e}_q^{\,\tau^T} S_\tau \vec{e}_n^{\,\tau})(\vec{e}_{p_1}^{\,\mu_j^T} S_{\mu_1}\vec{e}_{m_1}^{\,\mu_1}) \cdots (\vec{e}_{p_d}^{\,\mu_d^T} S_{\mu_d}\vec{e}_{m_d}^{\,\mu_d})$$

$$\times\Big[(1+\gamma\,\lambda_n^\tau)\prod_{j=1}^{d}\lambda_{m_j}^{\mu_j} + \lambda_n^\tau \sum_{j=1}^{d} c_j\,(\prod_{s=1,s\neq j}^{d} \lambda_{m_s}^{\mu_s}\,)\Big] = (\vec{e}_q^{\,\tau}\,\vec{e}_{p_1}^{\,\mu_1}\,\vec{e}_{p_2}^{\,\mu_2}\cdots\vec{e}_{p_d}^{\,\mu_d})F,$$

and since the spatial and temporal stiffness matrices S_{μ_j} and S_τ are diagonal, then $(\vec{e}_q^{\,\tau^T} S_\tau \vec{e}_n^{\,\tau}) = 0$ if $q \neq n$, also $(\vec{e}_{p_j}^{\,\mu_j^T} S_{\mu_j}\vec{e}_{m_j}^{\,\mu_j}) = 0$, if $p_j \neq m_j$, which completes the proof for the case $d > 1$. Following similar steps for the two-dimensional problem in the t-μ_1 domain, it is easy to see that if $d = 1$, the relationship for κ can be derived as

$$\kappa_{q,p_1} = \frac{\vec{e}_q^{\,\tau^T} F\, \vec{e}_{p_1}^{\,\mu_1}}{(\vec{e}_q^{\,\tau^T} S_\tau \vec{e}_q^{\,\tau})(\vec{e}_{p_1}^{\,\mu_1^T} S_{\mu_1}\vec{e}_{p_1}^{\,\mu_1})\Big[\,\lambda_p^{\mu_1} \;+\; c_{\mu_1}\,\lambda_q^\tau + \gamma\,\lambda_q^\tau\lambda_p^{\mu_1}\Big]}. \tag{4.2.45}$$

□

Remark 4.15 If $\mu_j = \mu \neq \tau$ and $\mathcal{M}_j = \mathcal{M}$, $j = 1, 2, \ldots, d$, then the complexity of the calculations of (4.2.42) and (4.2.43) reduces to two linear generalized eigenproblems for space and time. Moreover, if $\mu = \tau$ and $\mathcal{M} = \mathcal{N}$, then we only need to solve a single one-dimensional eigenproblem $M_\mu \vec{e}_q = \lambda_q S_\mu \vec{e}_q$ once.

Remark 4.16 For time-independent (steady-state) problems, where the time-fractional derivative vanishes in (4.2.9), the same general framework holds. For such problems, the time-dependent basis and test functions in U_N and V_N consequently vanish, and we construct the d-dimensional basis space $\mathcal{U}_N$ in $\Omega = [a_1, b_1] \times \cdots \times [a_d, b_d]$ as

$$\mathcal{U}_N = \text{span}\Big\{\prod_{j=1}^{d}\Big(\phi_{m_j}^{\mu_j}\circ\xi_j\Big)(x_j)\;:\; m_j = \lceil 2\mu_j\rceil, \ldots, \mathcal{M}_j\Big\},$$

where we seek the solution in terms of elements in the space $\mathcal{U}_N$ of the form

$$u_N(\mu_1, \ldots, \mu_d) = \sum_{m_1=\lceil 2\mu_1\rceil}^{\mathcal{M}_1} \cdots \sum_{m_d=\lceil 2\mu_d\rceil}^{\mathcal{M}_d} \hat{u}_{m_1,\ldots,m_d} \prod_{j=1}^{d}\Big(\phi_{m_j}^{\mu_j}\circ\xi_j\Big)(x_j) \tag{4.2.46}$$

and test the problem against the elements in

$$\mathcal{V}_N = \text{span}\Big\{\prod_{j=1}^{d}\Big(\Phi_{k_j}^{\mu_j}\circ\xi_j\Big)(x_j)\;:\; k_j = \lceil 2\mu_j\rceil, \ldots, \mathcal{M}_j\Big\}.$$

Subsequently, we obtain a similar Lyapunov equation as in (4.2.33) where M_τ no longer appears; however, M_{μ_j} and S_{μ_j} possess all the properties presented in the end of this section.

Lemma 4.17 *If $d > 1$, in the absence of the fractional time-derivative in (4.2.9), i.e., when S_τ vanishes, we obtain the matrix of unknown solution $\mathcal{U}$ in (4.2.46) as*

$$\mathcal{U} = \sum_{m_1=\lceil 2\mu_1 \rceil}^{\mathcal{M}_1} \cdots \sum_{m_d=\lceil 2\mu_d \rceil}^{\mathcal{M}_d} \kappa_{m_1,\dots,m_d}\, \vec{e}_{m_1}^{\,\mu_1} \otimes \cdots \otimes \vec{e}_{m_d}^{\,\mu_d},$$

where the unknown $\kappa_{m_1,\dots,m_d}$ is given by

$$\kappa_{m_1,\dots,m_d} = \frac{(\vec{e}_{m_1}^{\,\mu_1} \cdots \vec{e}_{m_d}^{\,\mu_d})F}{\left[\prod_{j=1}^{d} (\vec{e}_{m_j}^{\,\mu_j^T} S_{\mu_j} \vec{e}_{m_j}^{\,\mu_j})\right]\left[\gamma \prod_{j=1}^{d} \lambda_{m_j}^{\mu_j} + \sum_{j=1}^{d} c_j \left(\prod_{s=1, s\neq j}^{d} \lambda_{m_s}^{\mu_s}\right)\right]}.$$

Proof It follows the proof in Theorem 4.14. □

Computational Considerations

In Theorem 4.14, we assume that the eigenvectors and eigenvalues of each mass matrix with respect to the corresponding stiffness matrices are known. Therefore, employing the PG spectral method in a $(1+d)$-dimensional problem when $(1+d) \geq 2$ leads to efficient computations. Otherwise, the computational cost of the eigensolver, which is $\mathcal{O}(N^3)$ in practice, becomes dominant. As we shall demonstrate, this approach appears to be even more beneficial as $(1 + d)$ increases. In fact, the cost of the fast FPDE solver is associated with the following two steps: (i) the computation of $\kappa_{n,m_1,\dots,m_d}$ in (4.2.41), and (ii) the cost of representing $\mathcal{U}$ in (4.2.40). In what follows, we show that the computational complexity of mathematical operations in our PG spectral method is $\mathcal{O}(N^{2+d})$, the dimension of the space-time domain Ω, and if we assume $\mathcal{N} = \mathcal{M}_1 = \cdots = \mathcal{M}_d$.

Step (i): In order to compute the $(1 + d)$-dimensional array κ in (4.2.41), we need to first calculate the numerator

$$(\vec{e}_q^{\,\tau} \vec{e}_{p_1}^{\,\mu_1} \cdots \vec{e}_{p_d}^{\,\mu_d})F = \sum_{i=\lceil 2\tau \rceil}^{\mathcal{N}} \sum_{s_1=\lceil 2\mu_1 \rceil}^{\mathcal{M}_1} \cdots \sum_{s_d=\lceil 2\mu_d \rceil}^{\mathcal{M}_d} \{\vec{e}_q^{\,\tau}\}_i \{\vec{e}_{p_1}^{\,\mu_1}\}_{s_1} \cdots \{\vec{e}_{p_d}^{\,\mu_d}\}_{s_d} \{F\}_{i,s_1,\dots,s_d},$$

for which naive computations for all the entries leads to a computational complexity $\mathcal{O}(N^{2(1+d)})$ that can be intractable when d increases. Alternatively, by performing *sum-factorization*, the operation counts can be reduced to

$\mathcal{O}(N^{2+d})$, including the *time-dimension* in our calculations. Following this technique we re-write the inner product as

$$(\vec{e}_q^{\,\tau} \vec{e}_{p_1}^{\,\mu_1} \cdots \vec{e}_{p_d}^{\,\mu_d})F = \sum_{i=\lceil 2\tau \rceil}^{\mathcal{N}} \{\vec{e}_q^{\,\tau}\}_i \sum_{s_1=\lceil 2\mu_1 \rceil}^{\mathcal{M}_1} \{\vec{e}_{p_1}^{\,\mu_1}\}_{s_1}$$
$$\cdots \sum_{s_{d-1}=\lceil 2\mu_{d-1} \rceil}^{\mathcal{M}_{d-1}} \{\vec{e}_{p_{d-1}}^{\,\mu_{d-1}}\}_{s_{d-1}} \sum_{s_d=\lceil 2\mu_d \rceil}^{\mathcal{M}_d} \{\vec{e}_{p_d}^{\,\mu_d}\}_{s_d} \{F\}_{i,s_1,\dots,s_d}, \quad (4.2.47)$$

in which we separately obtain the inner-most sum as

$$\mathcal{F}^d_{i,s_1,\dots,s_{d-1},p_d} = \sum_{s_d=\lceil 2\mu_d \rceil}^{\mathcal{M}_d} \{\vec{e}_{p_d}^{\,\mu_d}\}_{s_d} \{F\}_{i,s_1,\dots,s_d},$$

and similarly we write the second inner-most sum as

$$\mathcal{F}^{d-1}_{i,s_1,\dots,s_{d-2},p_{d-1},p_d} = \sum_{s_{d-1}=\lceil 2\mu_{d-1} \rceil}^{\mathcal{M}_{d-1}} \{\vec{e}_{p_{d-1}}^{\,\mu_{d-1}}\}_{s_{d-1}} \mathcal{F}^d_{i,s_1,\dots,s_{d-1},p_d}.$$

Finally, we recursively obtain

$$\mathcal{F}^1_{i,p_1,\dots,p_d} = \sum_{s_1=\lceil 2\mu_1 \rceil}^{\mathcal{M}_1} \{\vec{e}_{p_1}^{\,\mu_1}\}_{s_1} \mathcal{F}^2_{i,p_1,p_2,\dots,p_d}. \quad (4.2.48)$$

We note that the operation count in computing the entries of $\mathcal{F}^j_{i,s_1,\dots,s_{j-1},p_j,\dots,p_d}$ in each recursion is $\mathcal{O}(N^{2+d})$. Now, by substituting (4.2.48) back into (4.2.47), we obtain the whole inner product as

$$(\vec{e}_q^{\,\tau} \vec{e}_{p_1}^{\,\mu_1} \cdots \vec{e}_{p_d}^{\,\mu_d})F = \sum_{i=\lceil 2\tau \rceil}^{\mathcal{N}} \{\vec{e}_q^{\,\tau}\}_i \mathcal{F}^1_{i,p_1,\dots,p_d}, \quad (4.2.49)$$

which is again of complexity $\mathcal{O}(N^{2+d})$. We observe that the total computational complexity of evaluating the inner product is $\mathcal{O}(N^{2+d})$. Moreover, the operation count for computing the denumerator in (4.2.41) and for each entry of $\kappa_{n,m_1,\dots,m_d}$ is $\mathcal{O}(N)$. This is true since the stiffness matrix is either *diagonal* or *tridiagonal* due to the choice of our poly-fractonomial bases. Hence, the total complexity for computing the denumerator is again $\mathcal{O}(N^{2+d})$. We recall that we have already included the time-dimension into account, i.e., the space-time domain $\Omega \subset \mathbb{R}^{1+d}$. Hence, κ in (4.2.41) is obtained with cost $\mathcal{O}(N^{2+d})$.

Step (ii): In the computation of (4.2.40), we observe that the sum-factorization technique helps to reduce the complexity to $\mathcal{O}(N^{2+d})$.

4.2.6 Numerical Tests for Hyperbolic/Parabolic/Elliptic FPDEs

In Section 4.2.5, we introduced general $(1+d)$-dimensional linear FPDEs, for which we developed a general Petrov–Galerkin spectral method in addition to the general fast solver. Here, we reduce this general framework to the special well-known (i) *hyperbolic* FPDEs such as the fractional advection equation, (ii) *parabolic* FPDEs such as the fractional subdiffusion problems, and (iii) *elliptic* FPDEs such as the fractional Helmholtz/Poisson equations. In the following numerical examples, we carry out the spatial/temporal p-refinement test via fixing correspondingly the temporal/spatial expansion order fixed at 15.

Hyperbolic FPDEs

We consider the hyperbolic FPDE

$$ {}^{\text{RL}}_{x_j}D^{\mu_j}_{b_j} u(t,x) + c_x \, [{}^{\text{RL}}_{-1}D^{2\mu}_{x} u(t,x)\,] = f(t,x) \tag{4.2.50} $$

for $(t,x) \in (0,T] \times (-1,1)$, subject to $u(x,0) = 0$ and $u(-1,t) = 0$ when $\tau, \mu \in (0,1/2]$. In this case, the FPDE (4.2.50) appears as a time- and space-fractional advection (TSFA) equation, where we set $c_x = 1$. We then seek the solution to (4.2.50) in terms of a linear combination of elements in U_N, now consisting of only two dimensions, i.e., time t and space x, of the form

$$ u_N(t,x) = \sum_{n=\lceil 2\tau \rceil}^{\mathcal{N}} \sum_{m=\lceil 2\mu \rceil}^{\mathcal{M}} \hat{u}_{n,m} \big(\psi^{\tau}_n \circ \eta\big)(t)\, \big(\phi^{\mu}_m \circ \xi\big)(x). \tag{4.2.51} $$

Next, we obtain the corresponding linear system of the Lyapunov equation after carrying out the Kronecker product as

$$ S_\tau \mathcal{U} M_\mu + M_\tau \mathcal{U} S_\mu = F, \tag{4.2.52} $$

where we represent the unknown coefficient matrix $\mathcal{U}$ in terms of the spatial and temporal eigenvectors as

$$ \mathcal{U} = \sum_{q=\lceil 2\tau \rceil}^{\mathcal{N}} \sum_{p=\lceil 2\mu \rceil}^{\mathcal{M}} \kappa_{q,p}\, \vec{e}^{\,\tau}_q \vec{e}^{\,\mu^T}_p, $$

for which $\kappa_{q,p}$ is followed by (4.2.45) setting $\gamma = 0$ as

$$ \kappa_{q,p} = \frac{\vec{e}^{\,\tau^T}_q \, F \, \vec{e}^{\,\mu}_p}{(\vec{e}^{\,\tau^T}_q S_\tau \vec{e}^{\,\tau}_q) \cdot (\vec{e}^{\,\mu^T}_p S_\mu \vec{e}^{\,\mu}_p) \cdot (c_1 \, \lambda^\tau_q \; + \; \lambda^\mu_p)}. $$

In Figure 4.10, we examine the TSFA problem (4.2.50) and study the p-refinement in both the temporal (left) and the spatial (right) dimensions.

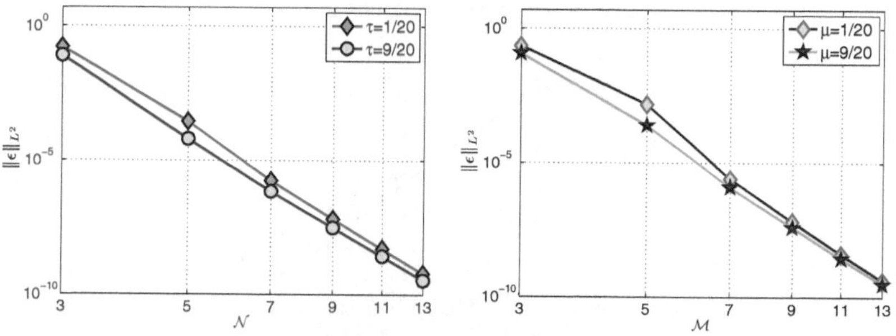

Figure 4.10 TSFA, temporal/spatial *p-refinement*: log-log L^2-error versus temporal and spatial expansion orders $\mathcal{N}$, $\mathcal{M}$. In the temporal p-refinement $\tau = 1/20$ and $9/20$ while $\mu = 1/4$, also in the spatial p-refinement, the spatial orders $\mu = 1/20$ and $9/20$ while $\tau = 1/4$. Here, the exact solution is $u^{ext}(x,t) = t^{6+2/7}\,(1+x)^{6+3/4}$. Source: Reprinted from Zayernouri et al. (2015) with permission from Elsevier.

To demonstrate the spectral convergence of the PG spectral method, we plot the log-log L^2-error versus temporal and spatial expansion orders $\mathcal{N}$, $\mathcal{M}$. In the temporal p-refinement $\tau = 1/20$ and $9/20$ while $\mu = 1/4$; also in the spatial p-refinement, the spatial orders $\mu = 1/20$ and $9/20$ while $\tau = 1/4$. In this test, we set the simulation time to $T = 1$, while the exact solution is $u^{ext}(x,t) = t^{6+2/7}\,(1+x)^{6+3/4}$.

Parabolic FPDEs

First, we consider the following parabolic time- and space-fractional diffusion (TSFD) equation

$$
\begin{aligned}
{}^{\text{RL}}_{x_j}D^{\mu_j}_{b_j}\, u(x,t) &= K\ {}^{\text{RL}}_{-1}D^{2\mu}_{x} u(x,t) + f(x,t), \qquad (4.2.53)\\
u(x,0) &= 0, \quad u(\pm 1, t) = 0,
\end{aligned}
$$

for $(t,x) \in (0,T] \times (-1,1)$, where $\tau \in (0,1/2]$, $\mu \in (1/2,1)$, $K > 0$, which is a well-known model for anomalous *subdiffusion* processes. In this case, we seek the solution also of the form (4.2.51), and obtain a similar linear system as in (4.2.52), and hence we obtain

$$
\kappa_{q,p} = \frac{\vec{e}_q^{\,\tau^T}\ F\, \vec{e}_p^{\,\mu}}{(\vec{e}_q^{\,\tau^T} S_\tau\, \vec{e}_q^{\,\tau}) \cdot (\vec{e}_p^{\,\mu^T} S_\mu\, \vec{e}_p^{\,x}) \cdot (\lambda_p^{\mu}\ -\ K\ \lambda_q^{\tau})}
$$

for $p = \lceil 2\tau \rceil, \ldots, \mathcal{N}$ and $q = \lceil 2\mu \rceil, \ldots, \mathcal{M}$. In Figure 4.11, we solve the TSFD problem (4.2.53) and study the p-refinement in the spacial dimension. We plot the log-log L^2-error versus the spatial expansion order $\mathcal{M}$. Similarly, in the spatial p-refinement, the spatial orders $\mu = 11/20$ and $19/20$ while

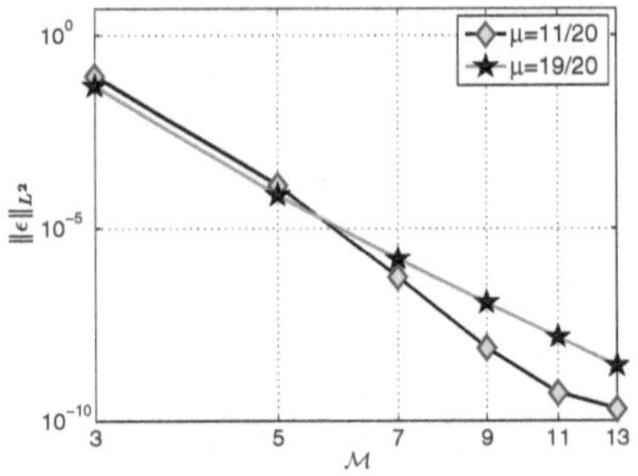

Figure 4.11 TSFD, spatial *p-refinement*: log-log L^2-error versus spatial expansion orders M. Here, the spatial orders $\mu = 11/20$ and $19/20$ while $\tau = 1/4$, and the exact solution is $u^{ext}(x,t) = t^{6+2/7}\,[(1+x)^{6+3/4} - 2^{38/35}(1+x)^{5+1/2}]$. Source: Reprinted from Zayernouri et al. (2015) with permission from Elsevier.

$\tau = 1/4$. In both cases, the spectral convergence of the solution is achieved. Since, the exact solution is fixed in each case, the convergence rate corresponding to $\mu = 11/20$ appears to be larger than the case where $\mu = 19/20$ as expected. It is naturally due to the higher regularity requirement in the weak form corresponding to the second test-case. In the temporal p-refinement similar results were observed as in Figure 4.10 (left).

Elliptic FPDEs

We examine the well-known *elliptic Helmholtz/Poisson* equations, rendered fractional in two-dimensional (in space) domains. We choose the spatial computational domain as $\Omega = (-1,1)^2$, and consider the following problem

$$\begin{aligned} {}^{\text{RL}}_{-1}D^{2\mu_1}_{x_1} u(x_1,x_2) + {}^{\text{RL}}_{-1}D^{2\mu_2}_{x_2} u(x_1,x_2) + \gamma\, u(x_1,x_2) = f(x_1,x_2) \ \text{ in } \ \Omega, \\ u(x_1,x_2) = 0 \ \text{ on } \ \partial\Omega, \quad (4.2.54) \end{aligned}$$

where $\gamma > 0$, $\mu_1, \mu_2 \in (1/2, 1)$, which reduces to the space-fractional Poisson equation when $\gamma = 0$. Here, we present a general scheme in addition to a linear fast solver for both problems.

We then seek the solution to (4.2.54) in terms of a linear combination of elements in U_N in absence of the time-basis, consisting of only two dimensions of the form

$$u_N(x_1,x_2) = \sum_{m_1=2}^{\mathcal{M}_1} \sum_{m_2=2}^{\mathcal{M}_2} \hat{u}_{m_1,m_2} \big(\phi^{\mu_1}_{m_1} \circ \xi_1\big)(x_1) \big(\phi^{\mu_2}_{m_2} \circ \xi_2\big)(x_2), \quad (4.2.55)$$

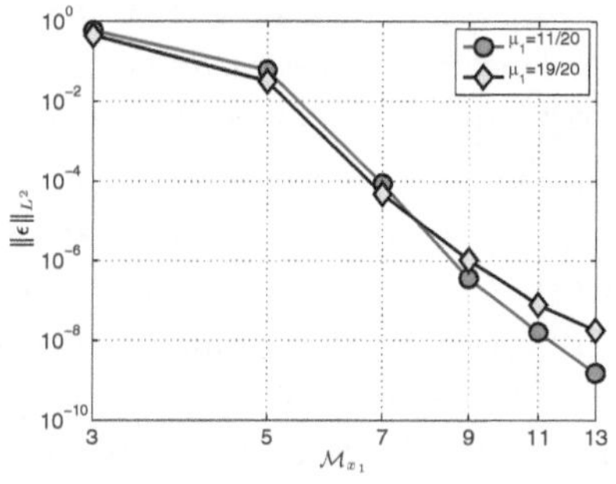

Figure 4.12 Space-fractional Helmholtz problem with $\gamma = 1$, spatial *p-refinement* in x-dimension: log-log L^2-error versus spatial expansion orders $\mathcal{M}_{x_1}$. Here, the spatial orders are $\mu_1 = 11/20$ and $19/20$ while $\mu_2 = 15/20$, is kept constant. The exact solution is $u^{ext}(x_1, x_2) = [\,(1+x_1)^{6+3/4} - 2^{5/4}\,(1+x_1)^{5+1/2}][\,(1+x_2)^{6+4/9} - 2^{73/63}\,(1+x_2)^{5+2/7}]$. A similar convergence curve is achieved in the p-refinement performed in the y-dimension, also for the case of $\gamma = 0$. Source: Reprinted from Zayernouri et al. (2015) with permission from Elsevier.

for which we represent the unknown coefficient matrix $\mathcal{U}$ in terms of the spatial eigenvectors as

$$\mathcal{U} = \sum_{p_1=2}^{\mathcal{M}_1} \sum_{p_2=2}^{\mathcal{M}_2} \kappa_{p_1,p_2}\, \vec{e}_{p_1}^{\,\mu_1}\, \vec{e}_{p_2}^{\,\mu_2\,T},$$

where

$$\kappa_{p_1,p_2} = \frac{\vec{e}_{p_1}^{\,\mu_1\,T}\, G\, \vec{e}_{p_2}^{\,\mu_2}}{(\vec{e}_{p_1}^{\,\mu_1\,T} S_{\mu_1} \vec{e}_{p_1}^{\,\mu_1}) \cdot (\vec{e}_{p_2}^{\,\mu_2\,T} S_{\mu_2} \vec{e}_{p_2}^{\,\mu_2}) \cdot (\lambda_{p_1}^{\mu_1} + \lambda_{p_2}^{\mu_2} + \gamma\, \lambda_{p_1}^{\mu_1} \lambda_{p_2}^{x_2})}.$$

In Figure 4.12, we solve the fractional Helmholtz problem (4.2.54) and study the p-refinement in the spacial x_1-dimension. To demonstrate the spectral convergence of the fast FPDE solver, we plot the log-log L^2-error versus the spatial expansion order $\mathcal{M}_{x_1}$. The spatial orders $\mu_1 = 11/20$ and $19/20$ while $\mu_2 = 15/20$. Similar to the spatial p-convergence in Figure 4.11, the convergence rate corresponding to $\mu_1 = 11/20$ appears to be larger than the case where $\mu_2 = 19/20$. We observe a similar p-refinement in the x_2-dimension as well.

Higher-Dimensional FPDEs

Next, we employ our PG method and the fast solver in even higher dimensional problems to exhibit the generality and efficiency of the scheme. In

Table 4.4. *Convergence study and CPU time of the unified PG spectral method employed in the time- and space-fractional advection equation (TSFA)* ${}^{RL}_{x_j}D^{\mu_j}_{b_j} u + \sum_{j=1}^{d} [{}^{RL}_{-1}D^{2\mu_j}_{x_j} u] = f$, *where* $2\tau = 2\mu_j = 1/2$, $j = 1, 2, \dots, d$, *subject to homogeneous Dirichlet boundary conditions in four-dimensional (4-D), six-dimensional (6-D), and ten-dimensional (10-D) space-time hypercube domains, where* $D = 1 + d$. *The error is measured by the essential norm* $\|\epsilon\|_{L^\infty} = \|u - u^{ext}\|_{L^\infty}/\|u^{ext}\|_{L^\infty}$, *which is normalized by the essential norm of the exact solution* $u^{ext}(t, \vec{x}) = [t \prod_{j=1}^{d}(1 + x_j)]^{2+2/5}$, *where* $t \in [0, 1]$ *and* $x \in [-1, 1]^d$. *The CPU time (seconds) is obtained on an Intel (Xeon X5550) 2.67GHz processor. In each step, we uniformly increase the bases order by one in all dimensions. Source: Reprinted from Zayernouri et al. (2015) with permission from Elsevier.*

4-D TSFA

$\mathcal{N} = \mathcal{M}_1 = \cdots = \mathcal{M}_d$	$\|\epsilon\|_{L^\infty}$	CPU Time (seconds)
2	0.576869	0.006333
3	0.034706	0.015997
4	0.003990	0.041994
5	0.0009071	0.105984

6-D TSFA

$\mathcal{N} = \mathcal{M}_1 = \cdots = \mathcal{M}_d$	$\|\epsilon\|_{L^\infty}$	CPU Time (seconds)
2	0.741056	0.014748
3	0.055171	0.134313
4	0.006578	0.821208
5	0.001525	3.546791

10-D TSFA

$\mathcal{N} = \mathcal{M}_1 = \cdots = \mathcal{M}_d$	$\|\epsilon\|_{L^\infty}$	CPU Time (seconds)
2	0.903357	0.288956
3	0.095305	18.45320
4	0.0119229	370.1 ($\approx$ 6 min)
5	0.0028102	3332.4 ($\approx$ 55 min)

Table 4.4, the convergence results and CPU time of the unified PG spectral method in higher-dimensional problems are examined. Particularly, we employ this scheme to solve the time- and space-fractional advection equation (TSFA)

$$ {}^{\mathrm{RL}}_{x_j}D^{\mu_j}_{b_j} u + {}^{\mathrm{RL}}_{-1}D^{2\mu_1}_{x_1} u + {}^{\mathrm{RL}}_{-1}D^{2\mu_2}_{x_2} u + \cdots + {}^{\mathrm{RL}}_{-1}D^{2\mu_d}_{x_d} u = f, $$

where $2\tau = 2\mu_j = 1/2$, subject to homogeneous Dirichlet boundary conditions in four-dimensional (4-D), six-dimensional (6-D), and ten-dimensional (10-D) *space-time* hypercube domains. The error is measured by the essential norm $\|\epsilon\|_{L^\infty} = \|u - u^{ext}\|_{L^\infty}/\|u^{ext}\|_{L^\infty}$, which is stronger than the L^2-norm and is normalized by the essential norm of the exact solution $u^{ext}(t, \vec{x}) = [t \prod_{j=1}^{d}(1 + x_j)]^{2+2/5}$ for the sake of consistency. The CPU time (seconds) is measured on a single-core Intel (Xeon X5550) 2.67GHz processor. In each step of the p-refinement, we uniformly increase the bases order by one in all dimensions. All the computations are performed in Mathematica 8. These simulations highlight that the unified PG spectral method is efficient even for a 10-D problem run on a PC in less than an hour!

Time-Integration when $2\tau = 1$

We recall that our unified PG spectral method works equally well when the temporal time-derivative order $2\tau = 1$. In general, a first-order in time PDE/FPDE reads

$$ \frac{\partial u}{\partial t} = F(u; t, x_1, \ldots, x_d), \tag{4.2.56} $$

where the operator $F(u; t, x_1, \ldots, x_d)$ is given as

$$ F(u; t, x_1, \ldots, x_d) = f(t, x_1, \ldots, x_d) - \sum_{j=1}^{d} c_j [{}^{\mathrm{RL}}_{a_j}D^{2\mu_j}_{x_j} u] + \gamma\, u, $$

in view of (4.2.9). Here, we regard the PG method as an alternative scheme for *spectrally* accurate time-integration for a general $F(u; t, x_1, \ldots, x_d)$, rather than utilizing existing *algebraically* accurate methods, including *multi-step* methods such as the Adams family and stiffly-stable schemes, also *multi-stage* approaches such as the Runge–Kutta method.

The idea of employing the PG spectral method when $2\tau = 1$ is simply based on a useful property by which a full first-order derivative d/dt can be decomposed into a product of the sequential $(\frac{1}{2})$-th order derivatives ${}^{\mathrm{RL}}_{0}D^{1/2}_{t}\ {}^{\mathrm{RL}}_{0}D^{1/2}_{t}$, a result that is not valid in the standard (integer-order) calculus. Hence, by virtue of the *fractional* integration-by-parts, such a decomposition artificially induces *nonlocality* to the temporal term in the corresponding weak form. Therefore, it provides an appropriate framework for *global* (spectral) treatment of the temporal term using our unified PG spectral method. To this end, we carry out the time-integration when $2\tau = 1$ in the following FPDE

Table 4.5. *Time-Integration when* $2\tau = 1$*:* $\partial u/\partial t + \sum_{j=1}^{3} [{}^{\text{RL}}_{-1}D_{x_j}^{2\mu_j} u] = f$ *in* $\Omega \subset \mathbb{R}^{1+3}$*, where* $t \in [0, 1]$ *and* $x_j \in [-1, 1]$*,* $j = 1, 2, 3$*. Here, we set* $\mu_j = 1/2$ *to fully recover the standard time-dependent advection equation in three-dimensional spatial domain. However, in general* $\mu_j \in (0, 1)$*. The error is measured by the essential norm* $\|\epsilon\|_{L^\infty} = \|u - u^{ext}\|_{L^\infty}/\|u^{ext}\|_{L^\infty}$*, which is normalized by the essential norm of the exact solution* $u^{ext}(t, \vec{x}) = [t \prod_{j=1}^{3}(1 + x_j)]^{6+2/5}$*. The CPU time (seconds) is obtained on an Intel (Xeon X5550) 2.67GHz processor. In each step, we uniformly increase the bases order by one in all dimensions. Source: Reprinted from Zayernouri et al. (2015) with permission from Elsevier.*

Integer-Order Time-Integration

$\mathcal{N} = \mathcal{M}_1 = \mathcal{M}_2 = \mathcal{M}_3$	$\|\epsilon\|_{L^\infty}$	CPU Time (seconds)
3	0.6225970	0.051992
5	0.0336570	0.352947
7	1.34×10^{-5}	1.9737
9	1.06×10^{-7}	3.7894
11	3.52×10^{-9}	9.7365
13	2.54×10^{-10}	21.472

$$\partial u/\partial t + \sum_{j=1}^{3} {}^{\text{RL}}_{-1}D_{x_j}^{2\mu_j} u = f$$

in $\Omega \subset \mathbb{R}^{1+3}$, where in general $\mu_j \in (0, 1)$. Here, we set $\mu_j = 1/2$ for simplicity, which recovers the standard time-dependent advection equation in three-dimensional spatial domain.

In Table 4.5, we again measure the error by the normalized essential norm, where the exact solution is $u^{ext}(t, \vec{x}) = [t \prod_{j=1}^{3}(1+x_j)]^{6+2/5}$, where $t \in [0, 1]$ and $x_j \in [-1, 1]$, $j = 1, 2, 3$. Similar to the previous case, the CPU time (seconds) is obtained on a single-core Intel (Xeon X5550) 2.67GHz processor, where we uniformly increase the bases order by one in all dimensions in each step. In these simulations, we globally treat the time-axis in addition to other spatial dimensions. The CPU time and the spectral convergence strongly highlights the efficiency of our approach, where a 4-D problem (i.e., 1-D in time and 3-D in space) can be highly accurately solved in a fraction of minute!

Discussion

In practice, the enforcement of *periodic* boundary conditions to FPDEs is not possible since it is not clear how to define history (memory) for a periodic function. Moreover, we note that Riemann–Liouville fractional derivatives in time/space only allow us to impose homogeneous initial/boundary conditions to the corresponding FPDEs to be well-posed. However, we note that our PG spectral method is also applicable in such problems in the following manner. When inhomogeneous Dirichlet conditions are enforced, the corresponding derivatives are usually replaced by *Caputo* fractional derivatives. We illustrate such a treatment in the following model problem posed subject to an inhomogeneous initial condition:

$$\begin{aligned} {}^{\mathrm{C}}_{0}D_t^{2\tau} u &= {}^{\mathrm{RL}}_{-1}D_x^{2\mu} u + f(x,t), \\ u(x,0) &= g(x), \quad u(\pm 1,t) = 0, \end{aligned} \tag{4.2.57}$$

in which $2\mu \in (1,2)$, $g(x) \in C^0([-1,1])$, and ${}^{\mathrm{C}}_{0}D_t^{2\tau}(\cdot)$ denotes the Caputo fractional derivative of order $2\tau \in (0,1)$, which is defined via interchanging the order of differentiation and integration. Now, we define $U(x,t) = u(x,t) - g(x)$, and take into account that ${}^{\mathrm{C}}_{0}D_t^{2\tau} g(x) \equiv 0$. Then, by substituting $u = U + g$ into (4.2.57) and noting that ${}^{\mathrm{C}}_{0}D_t^{2\tau} U = {}^{\mathrm{RL}}_{x_j}D_{b_j}^{\mu_j} U$ due to the homogeneity of $U(x,0)$, we obtain the transformed problem as

$$\begin{aligned} {}^{\mathrm{RL}}_{x_j}D_{b_j}^{\mu_j} U &= {}^{\mathrm{RL}}_{-1}D_x^{2\mu} U + \widetilde{f}(x,t), \\ U(x,0) &= 0, \quad U(\pm 1,t) = 0, \end{aligned}$$

in which

$$\widetilde{f}(x,t) = f(x,t) + {}^{\mathrm{RL}}_{-1}D_x^{2\mu} g(x).$$

Therefore, we can treat such inhomogeneous conditions by our unified PG spectral method through homogenizing the problem and modifying the forcing term on the right-hand side. The same approach applies to inhomogeneous boundary conditions.

4.3 GLOF Space-Time Method for the Subdiffusion Equation

Let $\Omega \subset \mathbb{R}^d$ $(d = 1,2,3)$ be a bounded domain with a convex polygonal boundary $\partial\Omega$. Consider the following time-fractional evolution problem for the function $u(x,t)$ with $\nu \in (0,1)$:

$$\begin{cases} {}^{\mathrm{C}}_{0}D_t^{\nu} u(x,t) = Lu(x,t) + f(x,t), & x \in \Omega,\ t \in \Lambda := (0,T), \\ u(x,t) = 0, & (x,t) \in \partial\Omega \times (0,T], \\ u(x,0) = u_0(x), & x \in \Omega, \end{cases} \tag{4.3.1}$$

where $T > 0$ is a fixed final time, f and u_0 are given source term and initial data, respectively, and ${}_0^{C}D_t^{\gamma}$ denotes the Caputo fractional derivative with respect to t. Here $Lu = \nabla \cdot (a(x)\nabla u) - b(x)u$ and $a(x)$ is a symmetric $d \times d$ matrix-valued measurable function on the domain Ω with smooth entries and $b(x) \geq 0$ is an $L_\infty(\Omega)$-function. We assume that

$$c_0|\xi|^2 \leq \xi^T a(x)\xi \leq c_1|\xi|^2, \quad \forall \xi \in \mathbb{R}^d,\ x \in \Omega, \tag{4.3.2}$$

where $c_0, c_1 > 0$ are constants. Then $-L$ is a symmetric and positive definite operator.

In Jin et al. (2015b, Theorems 2.1 and 3.1), it has been proved that for any $f \in H_0^{\beta}(\Lambda)$ with $\beta \in (0,1)$, there exist $c_{\beta,1}$ and $c_{\beta,2}$ such that

$$c_{\beta,1}\| {}_0^{RL}D_t^{\beta} f\|_{L^2(\Lambda)} \leq |f|_{H_0^{\beta}(\Lambda)} \leq c_{\beta,2}\| {}_0^{RL}D_t^{\beta} f\|_{L^2(\Lambda)}. \tag{4.3.3}$$

Moreover, for any $f, g \in H_0^{\beta}(\Lambda)$ with $0 \leq \beta < 1/2$, it has been proved in Li and Xu (2010, Lemma 2.8) that

$$({}_0^{RL}D_t^{2\beta} f, g)_\Lambda = ({}_0^{RL}D_t^{\beta} f, {}_t^{RL}D_T^{\beta} g)_\Lambda, \tag{4.3.4}$$

where $(\cdot,\cdot)_\Lambda$ denotes the inner product of $L^2(\Lambda)$ or the duality between $H_0^s(\Lambda)$ and its dual space $(H_0^s(\Lambda))^* = H^{-s}(\Lambda)$ with $s \in [0,1]$. Note that the fractional operator ${}_0^{RL}D_t^{\beta} f$ is defined in the distribution sense as in Li and Xu (2010), or ${}_0^{RL}D_t^{2\beta} f$ is not well defined for $f \in H_0^{\beta}(\Lambda)$. In addition, given $f \in H_0^{\beta}(\Lambda)$, the following coercivity is valid

$$({}_0^{RL}D_t^{\beta} f, {}_t^{RL}D_T^{\beta} f)_\Lambda \geq c_\beta |f|^2_{H_0^{\beta}(\Lambda)}. \tag{4.3.5}$$

We first reformulate (4.3.1) into an equivalent form. In the case that $u_0 \in H^2(\Omega) \cap H_0^1(\Omega)$, we let $w = u - u_0$, and observe that w satisfies

$$\begin{cases} {}_0^{C}D_t^{\gamma} w(x,t) = Lw(x,t) + F(x,t), & x \in \Omega,\ t \in \Lambda := (0,T), \\ w(x,t) = 0, & (x,t) \in \partial\Omega \times (0,T], \\ w(x,0) = 0, & x \in \Omega, \end{cases}$$

with $F(x,t) = f(x,t) + Lu_0$. Since $w(0) = 0$, we have ${}_0^{C}D_t^{\gamma} w(t) = {}_0^{RL}D_t^{\gamma} w(t),\quad 0 < \gamma < 1$. Then without loss of generality, we only consider the following subdiffusion problem with trivial initial data:

$$\begin{cases} {}_0^{RL}D_t^{\gamma} u(x,t) = Lu(x,t) + f(x,t), & x \in \Omega,\ t \in \Lambda := (0,T), \\ u(x,t) = 0, & (x,t) \in \partial\Omega \times (0,T], \\ u(x,0) = 0, & x \in \Omega. \end{cases} \tag{4.3.6}$$

Solutions to Subdiffusion Equations

In this part, we introduce a representation of the solution to (4.3.1) by spectral decomposition, which will be intensively used in the error analysis. To this end, we consider the eigenvalue problem

$$-L\varphi = \lambda\varphi \text{ in } \Omega; \quad \varphi = 0 \text{ on } \partial\Omega.$$

Since $-L$ is a symmetric uniformly elliptic operator, it admits a nondecreasing sequence $\{\lambda_j\}_{j=1}^{\infty}$ of positive eigenvalues, which tend to ∞ with $j \to \infty$, and a corresponding sequence $\{\varphi_j\}_{j=1}^{\infty}$ of eigenfunctions, $\varphi_j \in \mathrm{Dom}(L) = H_0^1(\Omega) \cap H^2(\Omega)$, forms an orthonormal basis in $L^2(\Omega)$, whose inner product is denoted by $(\cdot,\cdot)_\Omega$. Further, $\|v\|_{\dot{H}^0(\Omega)} = \|v\|_{L^2(\Omega)} = (v,v)_\Omega^{1/2}$ is the norm in $L^2(\Omega)$. Besides, it is easy to verify that $\|v\|_{\dot{H}^1(\Omega)} = \|\nabla v\|_{L^2(\Omega)}$ is also the norm in $H_0^1(\Omega)$ and $\|v\|_{\dot{H}^2(\Omega)} = \|\Delta v\|_{L^2(\Omega)}$ is equivalent to the norm in $H^2(\Omega) \cap H_0^1(\Omega)$ (see Thomée (2006, Lemma 3.1)).

Next, we represent the solution to problem (4.3.6) using the eigenpairs $\{(\lambda_j, \varphi_j)\}_{j=1}^{\infty}$. Define an operator $E(t)$ for $v \in L^2(\Omega)$ by

$$E(t)v(x) = \sum_{j=1}^{\infty} t^{\nu-1} E_{\nu,\nu}(-\lambda_j t^\nu)\,(v,\varphi_j)_\Omega\,\varphi_j(x),$$

where $E_{a,b}(z)$ denotes the two-parameter Mittag-Leffler function:

$$E_{a,b}(z) = \sum_{j=0}^{\infty} \frac{z^j}{\Gamma(aj+b)}, \qquad z \in \mathbb{C},\ a > 0,\ b \in \mathbb{R}.$$

The Mittag-Leffler function plays an crucial role in solving fractional differential equations. For clarity, here we list some results which will be used in the subsequent parts below.

Lemma 4.18 *Let* $0 < a < 2$, $b \in \mathbb{R}$ *and* $a\pi/2 < \mu < \min(\pi, a\pi)$. *There exists a constant* $C = C(a,b,\mu) > 0$ *such that*

$$|E_{a,b}(z)| \le \frac{C}{1+|z|}, \qquad \mu \le |arg(z)| \le \pi.$$

Lemma 4.19 *For* $\lambda > 0$, $a > 0$ *and* $b \in \mathbb{R}$ *and fixed positive integer* k, *we have*

$$\partial_t^k\{t^{b-1}E_{a,b}(-\lambda t^a)\} = t^{b-k-1}E_{a,b-k}(-\lambda t^a), \quad t > 0. \tag{4.3.7}$$

In particular, for $b = 1$ *and* $b = a$, *there exists*

$$\partial_t^k E_{a,1}(-\lambda t^a) = -\lambda t^{a-k} E_{a,a-k+1}(-\lambda t^a),$$
$$\partial_t^k\{t^{a-1}E_{a,a}(-\lambda t^a)\} = t^{a-k-1}E_{a,a-k}(-\lambda t^a),$$

where $\partial_t^k := \dfrac{\mathrm{d}^k}{\mathrm{d}t^k}$ *is the k-fold derivative with respect to t.*

Proof The proof can be ended by the following relation

$$\partial_t^k\{t^{b-1}E_{a,b}(-\lambda t^a)\} = \partial_t^k \sum_{j=0}^{\infty} \frac{(-\lambda)^j t^{ja+b-1}}{\Gamma(ja+b)} = t^{b-k-1}E_{a,b-k}(-\lambda t^a)$$

for any $t > 0$. □

Then the solution to (4.3.6) could be expressed as (see Sakamoto and Yamamoto (2011, Theorem 2.1))

$$\begin{aligned} u(x,t) &= \int_0^t E(t-s)f(s)\,\mathrm{d}s \\ &:= \sum_{n=1}^{\infty} \int_0^t \tau^{\nu-1}E_{\nu,\nu}(-\lambda_n\tau^\nu)\big(f(\cdot,t-\tau),\varphi_n\big)_\Omega \varphi_n(x)\mathrm{d}\tau. \end{aligned} \tag{4.3.8}$$

Remark 4.20 From the series expansion of the Mittag-Leffler function, it is easy to observe that the solution (4.3.8) is weakly singular near $t = 0$. This leads to low accuracy of many popular time-stepping methods. For example, in the case that $f(x,t) \equiv f(x)$, the solution can be written as

$$\begin{aligned} u(x,t) &= \sum_{n=1}^{\infty} \int_0^t \tau^{\nu-1}E_{\nu,\nu}(-\lambda_n\tau^\nu)\mathrm{d}\tau\,\big(f,\varphi_n\big)_\Omega\, \varphi_n(x) \\ &= \sum_{n=1}^{\infty} \frac{1}{\lambda_n} E_{\nu,1}(-\lambda_n\tau^\nu)\big(f,\varphi_n\big)_\Omega\, \varphi_n(x). \end{aligned}$$

In fact, Lemma 4.19 indicates that, for $m \geq 1$, the m-fold derivative $\partial_t^m E_{\nu,1}(-\lambda t^\nu) \notin L^1(0,T)$. As a result, the solution of the subdiffusion model fails to meet the regularity assumptions of many existing algorithms. In order to get rid of this dilemma, we introduce the following log orthogonal functions which can approximate the Mittag-Leffler functions with exponential convergence rate.

4.3.1 A Review of Log Orthogonal Functions (LOFs)

In order to design an efficient spectral method in the time direction of the subdiffusion equation, we shall use the following generalized log orthogonal functions (LOFs) introduced in Chapter 2:

$$\mathcal{S}_n(t) := \mathcal{S}_n^{0,\beta,0}(t;\beta) = t^{\beta/2}\mathscr{L}_n(-(\beta+1)\log t), \quad t \in (0,1),$$

where $\mathscr{L}_n(x) = \mathscr{L}_n(-(\beta+1)\log t)$ is the classical Laguerre polynomial of the variable $x \in (0,\infty)$. The mapping parameter β is designed for handling functions with distinct singular behavior when t is close to 0 (see Chen and Shen (2022)).

Review of Basic Properties

As introduced in Chapter 2, we review a rather shorter list of useful properties below:

- **Orthogonality**: Owing to the orthogonality of Laguerre polynomials, there exists

$$\int_0^1 \mathcal{S}_n(t)\,\mathcal{S}_m(t)\,\mathrm{d}t = (\beta+1)^{-1}\delta_{mn}, \quad t \in I := (0,1).$$

- **Gauss-LOFs Quadrature**: Let $\{x_j,\ \omega_j\}_{j=0}^N$ be the Gauss nodes and weights of Laguerre polynomial $\mathscr{L}_{N+1}(x)$. Denote

$$\{t_j = e^{-x_j/(\beta+1)}, \quad \chi_j = \omega_j t_j^{-\beta}/(\beta+1)\}_{j=0}^N.$$

  Then, for any $p \in \mathcal{P}_{2N+1}^{\beta,\log t}$, there exists

$$\int_0^1 p(t)\mathrm{d}x = \sum_{j=0}^N p(t_j)\,\chi_j, \tag{4.3.9}$$

  where the approximation space

$$\mathcal{P}_K^{\beta,\log t} := \mathrm{span}\{t^\beta,\ t^\beta\log t,\ t^\beta(\log t)^2, \ldots, t^\beta(\log t)^K\}.$$

- **Closed Form**: The closed form can be read

$$\mathcal{S}_n(t) = \sum_{k=0}^n \frac{(-1)^k}{k!}\binom{n}{n-k} t^{\frac{\beta}{2}}[-(\beta+1)\log t]^k.$$

  In particular, $\mathcal{S}_n(1) = 1$.

- **Generalized Derivative Relation**: Define the generalized derivative

$$\partial_{\gamma,t} u := t^{\gamma+1}\partial_t\{t^{-\gamma}u\} = t\partial_t u - \gamma u.$$

  For parameter $\gamma = \beta/2$, it holds that

$$\partial_{\gamma,t}\,\mathcal{S}_n(t) = \partial_{\frac{\beta}{2},t}\,\mathcal{S}_n(t) = (\beta+1)\sum_{l=0}^{n-1}\mathcal{S}_l(t).$$

Then combining the above equalities, it holds that

$$\partial_t \mathcal{S}_n(t) = t^{-1}\Big(\frac{\beta}{2}\mathcal{S}_n(t) + (\beta+1)\sum_{l=0}^{n-1}\mathcal{S}_l(t)\Big).$$

Approximation Properties of LOFs

Here we shall introduce the approximation properties of the LOFs, which will be intensively used in the subsequent section. To this end, we denote the L^2-projection $\Pi_N u$ from $L^2(I)$ to $\mathcal{P}_N^{\frac{\beta}{2},\log t}$ by satisfying

$$\int_0^1 (u - \Pi_N u)v \,\mathrm{d}t = 0, \quad \forall v \in \mathcal{P}_N^{\frac{\beta}{2},\log t}. \tag{4.3.10}$$

Owing to the orthogonality of the LOFs, it holds that

$$\Pi_N u(t) = \sum_{n=0}^{N} \hat{u}_n \,\mathcal{S}_n(t), \quad \text{with} \quad \hat{u}_n = (\beta+1)\int_0^1 u(t)\,\mathcal{S}_n(t)\,\mathrm{d}t.$$

Moreover, for describing the approximability of the projection $\Pi_N u$, we define the following non-uniformly weighted Sobolev space

$$A_\gamma^m(I) := \{v \in L^2(I) : \partial_{\gamma,t}^k v \in L^2_{\chi^k}(I),\ k = 1,2,\dots,m\} \text{ with } \chi(t) := |\log t|$$

equipped with semi-norm and norm

$$|v|_{A_\gamma^m(I)} := \|\partial_{\gamma,t}^m v\|_{L^2_{\chi^m}(I)}, \quad \|v\|_{A_a^m(I)} := \Big(\sum_{k=0}^{m} |v|^2_{A_a^k(I)}\Big)^{1/2}.$$

Here $L^2_w(I)$ denotes the weighted $L^2(I)$ space with norm $\|u\|^2_{L^2_w(I)} = \int_I |u(t)|^2 w(t)\,\mathrm{d}t$.

Lemma 4.21 (See Theorem 2.1 in Chen and Shen (2022)) *Let $m, N, k \in \mathbb{N}$ and $\beta > -1$. For any $u \in A^m_{\beta/2}(I)$ and $0 \le k \le \tilde{m}$, $\tilde{m} = \min\{m, N+1\}$, we have*

$$|u - \Pi_N u|_{A^k_{\beta/2}(I)} \le \sqrt{\frac{(\beta+1)^{k-\tilde{m}}(N-\tilde{m}+1)!}{(N-k+1)!}}\, |u|_{A^{\tilde{m}}_{\beta/2}(I)}.$$

In particular, for fixed $m < N$, there exists

$$|u - \Pi_N u|_{A^k_{\beta/2}(I)} \le cN^{(k-m)/2}\, |u|_{A^m_{\beta/2}(I)}, \tag{4.3.11}$$

where the constant c depends on β, k and m.

4.3.2 Approximation to Mittag-Leffler Functions

Recalling the solution representation given in (4.3.8) and Remark 4.20, the main part in the time direction consists of Mittag-Leffler functions $E_{\nu,1}(-\lambda t^\nu)$, which are weakly singular near $t = 0$. This fact leads to the ineffectiveness of many numerical methods for solving the subdiffusion model.

However, for any fixed integer m, it is easy to observe that for any $\beta > 0$ and $\nu \in (0, 1)$,

$$\partial^m_{\frac{\beta}{2},t} E_{\nu,1}(-\lambda t^\nu) = \sum_{j=0}^{\infty} \frac{(\nu j - \frac{\beta}{2})^m}{\Gamma(\nu j + 1)} (-\lambda t^\nu)^j,$$

which is still in $C[0, 1]$. Therefore, $E_{\nu,1}(-\lambda t^\nu) \in A^m_{\beta/2}(I)$ for any $\beta > 0$, and the bound of the semi-norm is independent of λ. This observation together with the approximation properties given in Lemma 4.21 indicates that the Mittag-Leffler functions $E_{\nu,1}(-\lambda t^\nu)$ can be approximated efficiently by using LOFs, and hence motivates us to use the LOFs spectral method to solve the subdiffusion equation (4.3.6).

To see this, we test the approximation to the Mittag-Leffler function $E_{\nu,1}(-\lambda t^\nu)$ on the unit interval $I = (0, 1)$, using the LOFs $\{S_n(t; \beta)\}_{n=0}^N$ as basis functions. We evaluate the L^2-error

$$e_N = \|(\Pi_N - I)E_{\nu,1}(-\lambda t^\nu)\|_{L^2(I)}.$$

In Figure 4.13 (left), we plot the numerical error curves for different ν, where we fixed $\beta = 6$ and $\lambda = 5$. The numerical results demonstrate that the numerical approximation of LOFs exponentially converges to the Mittag-Leffler function. Moreover, since the solution (4.3.8) consists of Mittag-Leffler

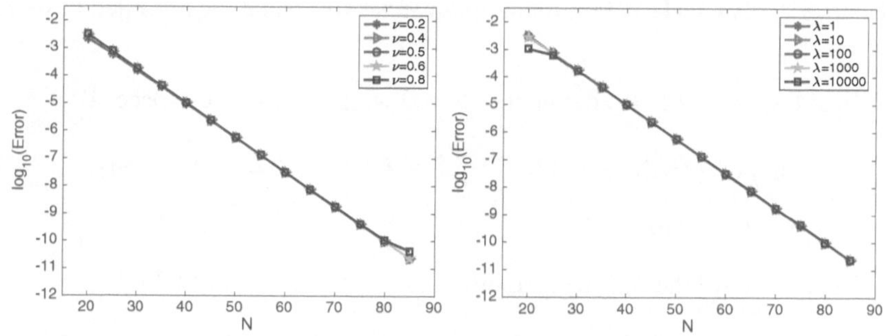

Figure 4.13 Left: Plot of e_N for $E_{\nu,1}(-5t^\nu)$ with different ν. Right: Plot of e_N for $E_{0.7,1}(-\lambda t^{0.7})$ with different λ. Source: Chen et al. (2020). Copyright ©2020 Society for Industrial and Applied Mathematics. Reprinted with permission. All rights reserved.

functions with different eigenvalues $\lambda_1 < \lambda_2 < \ldots < \lambda_n < \ldots$, we also check the error e_N with different λ, in order to verify that LOFs are efficient for approximating Mittag-Leffler funtions with a large eigenvalue. In Figure 4.13 (right), we draw the numerical error curves, with fixed $\beta = 6$ and $\nu = 0.7$, for different λ. Our numerical results show that the value of λ does not significantly affect the projection error which decays exponentially. Those numerical results indicate that the LOFs are suitable for approximating Mittag-Leffler functions with different parameters, and hence also proper for solving fractional subdiffusion equations (4.3.6).

4.3.3 Shifted Generalized Log Orthogonal Functions

In the preceding section, we discuss the approximation to Mittag-Leffler functions on the unit interval by using the LOFs. Next, we shall consider the general interval $\Lambda = (0, T)$ and the solution of subdiffusion equation (4.3.6). To this end, with a slight modification, we define a new class of log orthogonal functions as

$$\widehat{\mathcal{S}}_n^{\nu}(t) := (t/T)^{\frac{\nu}{2}} \mathcal{S}_n(t/T; \beta), \quad t \in (0, T), \tag{4.3.12}$$

and an L^2-projection $\widehat{\Pi}_N$ from $L^2(\Lambda)$ onto the space spanned by $\{\mathcal{S}_n(t/T)\}_{n=0}^{N}$ by

$$\widehat{\Pi}_N u(t) := \Pi_N u(T\tau), \quad \text{with} \quad \tau \in (0, 1).$$

Remark 4.22 The modified basis functions $\widehat{\mathcal{S}}_n^{\nu}(t)$ coincide with a special case of the shifted generalized LOFs in Chen and Shen (2022). The power functions $(t/T)^{\frac{\nu}{2}}$ multiplying the shifted LOFs $\mathcal{S}_n(t/T; \beta)$ are for the convenience of the error analysis. Hereafter, for simplicity, we still name the modified basis as LOFs.

Besides, we shall also define the shifted weighted Sobolev space $A^m_{\gamma,T}(\Lambda)$

$$A^m_{\gamma,T}(\Lambda) := \{v \in L^2(\Lambda) : \partial^k_{\gamma,t} v \in L^2_{\chi_T^k}(\Lambda),\ k = 1, 2, \ldots, m\},$$

$$\chi_T(t) := |\log(t/T)|.$$

Then we have the following approximation properties of SGLOFs.

Lemma 4.23 *Let* $0 < \nu < 1$, $\beta > 0$ *and* $\Lambda = (0, T)$. *For any* $u \in H_0^{\frac{\nu}{2}}(\Lambda)$ *and* ${}^{\mathrm{RL}}_{\ 0}D_t^{\frac{\nu}{2}} u \in A^m_{\frac{\beta}{2},T}(\Lambda)$, *the global-in-time projection*

$$\Pi^t_N u(t) := {}^{\mathrm{RL}}_{\ 0}I_t^{\frac{\nu}{2}}\, \widehat{\Pi}_N\, {}^{\mathrm{RL}}_{\ 0}D_t^{\frac{\nu}{2}} u$$

belongs to the linear vector space spanned by $\{\widehat{S}_n^{\nu}\}_{n=0}^{N}$ *and satisfies*

$$\begin{aligned} \|\Pi_N^t u - u\|_{H^{\frac{\nu}{2}}(\Lambda)} &\le c\| {}^{\mathrm{RL}}_{\ 0}D_t^{\frac{\nu}{2}}(\Pi_N^t u - u)\|_{L^2(\Lambda)} \\ &\le cN^{-m/2}\, | {}^{\mathrm{RL}}_{\ 0}D_t^{\frac{\nu}{2}} u|_{A^m_{\frac{\beta}{2},T}(\Lambda)}, \end{aligned} \tag{4.3.13}$$

where the constant c depends on β, k *and* m.

Proof First of all, we note that for any $\nu \in (0,1)$ and $u \in H_0^{\frac{\nu}{2}}(\Lambda)$ it holds that ${}^{\mathrm{RL}}_{\ 0}D_t^{\frac{\nu}{2}} u \in L^2(\Lambda)$, and hence $\widehat{\Pi}_N({}^{\mathrm{RL}}_{\ 0}D_t^{\frac{\nu}{2}} u)$ is well defined. Meanwhile, using the fact that $\nu \in (0,1)$, we know the fractional integral operator ${}^{\mathrm{RL}}_{\ 0}I_t^{\frac{\nu}{2}}$ is bounded from $L^2(\Lambda)$ to $H_0^{\frac{\nu}{2}}(\Lambda)$ in Jin et al. (2015b, Theorems 2.1), and hence we have $\Pi_N^t u \in H^{\frac{\nu}{2}}(\Lambda)$. Then the first inequality is valid owing to the Poincaré inequality and the equivalence relation (4.3.3)

$$\|\Pi_N^t u - u\|_{H^{\frac{\nu}{2}}(\Lambda)} \le c|\Pi_N^t u - u|_{H_0^{\frac{\nu}{2}}(\Lambda)} \le c\| {}^{\mathrm{RL}}_{\ 0}D_t^{\frac{\nu}{2}}(\Pi_N^t u - u)\|_{L^2(\Lambda)}.$$

Next, for any v defined on Λ, we define $\widehat{v}(\tau) = v(T\tau)$ with $\tau \in I = (0,1)$. Then, owing to the approximation result given in (4.3.11), it holds that

$$\|(\widehat{\Pi}_N - I)v\|_{L^2(\Lambda)} = \sqrt{T}\|(\Pi_N - I)\widehat{v}\|_{L^2(I)} \le c\sqrt{T}N^{-m/2}\, |\widehat{v}|_{A^m_{\beta/2}(I)}.$$

Hence, we arrive at

$$\|(\widehat{\Pi}_N - I)v\|_{L^2(\Lambda)} \le cN^{-m/2}\, \|v\|_{A^m_{\beta/2,T}(\Lambda)}. \tag{4.3.14}$$

Therefore, the claim (4.3.13) is valid. Then we shall prove that $\Pi_N^t u = {}^{\mathrm{RL}}_{\ 0}I_t^{\frac{\nu}{2}}\widehat{\Pi}_N\, {}^{\mathrm{RL}}_{\ 0}D_t^{\frac{\nu}{2}} u$ can be expanded by SGLOFs $\{\widehat{S}_n^{\nu}(t)\}_{n=1}^{N}$. In fact, due to the facts that

$$\mathrm{span}\{\widehat{S}_n^{\nu}(t)\}_{n=0}^{N} = \mathrm{span}\{t^{\frac{\nu+\beta}{2}}(\log t)^k\}_{k=0}^{N}$$

and

$$\widehat{\Pi}_N\, {}^{\mathrm{RL}}_{\ 0}D_t^{\frac{\nu}{2}} u = \sum_{n=0}^{N} a_n \widehat{S}_n^0 = \sum_{k=0}^{N} a_n' t^{\frac{\beta}{2}}(\log t)^k,$$

it suffices to show that ${}^{\mathrm{RL}}_{0}I_t^{\frac{\gamma}{2}}\{t^{\frac{\beta}{2}}(\log t)^k\}_{k=0}^N$ belongs to $\mathrm{span}\{t^{\frac{\gamma+\beta}{2}}(\log t)^k\}_{k=0}^N$. Now we observe that

$$
\begin{aligned}
{}^{\mathrm{RL}}_{0}I_t^{\frac{\gamma}{2}}\{t^{\frac{\beta}{2}}(\log t)^k\} =& \frac{1}{\Gamma(\frac{\gamma}{2})}\int_0^t (t-\tau)^{\frac{\gamma}{2}-1}\tau^{\frac{\beta}{2}}(\log \tau)^k \mathrm{d}\tau \\
\overset{\tau=ts}{=}& \frac{1}{\Gamma(\frac{\gamma}{2})}\int_0^1 t^{\frac{\gamma+\beta}{2}}(1-s)^{\frac{\gamma}{2}-1}s^{\frac{\beta}{2}}(\log t+\log s)^k \mathrm{d}s \\
=& \sum_{j=0}^{k} t^{\frac{\gamma+\beta}{2}}(\log t)^j \frac{1}{\Gamma(\frac{\gamma}{2})}\int_0^1 (1-s)^{\frac{\gamma}{2}-1}s^{\frac{\beta}{2}}\binom{k}{j}(\log s)^k \mathrm{d}s.
\end{aligned}
$$

Finally, the estimate (4.3.13) can be derived by using (4.3.14). □

4.3.4 Well-Posedness of the Weak Problem

In order to solve the subdiffusion equation, we follow the standard space-time Galerkin framework. To this end, we define a space-time Hilbert space

$$
H_0^{s,1}(\mathcal{O}) := H_0^s\big(\Lambda; L^2(\Omega)\big) \cap L^2\big(\Lambda; H_0^1(\Omega)\big), \quad \mathcal{O} := \Lambda\times\Omega
$$

endowed with the norm

$$
\|v\|_{H_0^{s,1}(\mathcal{O})} := \left(\|v\|^2_{H^s(\Lambda;L^2(\Omega))} + \|v\|^2_{L^2(\Lambda;H^1(\Omega))}\right)^{1/2}.
$$

Recalling the assumptions on the elliptic operator L in (4.3.2), it induces a coercive bilinear form

$$
\mathcal{A}(u,v) = -(a\nabla u, \nabla v)_\Omega + (bu, v)_\Omega, \quad \forall u, v \in H_0^1(\Omega),
$$

which satisfies

$$
\mathcal{A}(u,u) \ge c_1\|u\|^2_{H^1(\Omega)} \text{ and } \mathcal{A}(u,v) \le c_2\|u\|_{H^1(\Omega)}\|v\|_{H^1(\Omega)}. \tag{4.3.15}
$$

Then a weak formulation of the subdiffusion equation (4.3.6) reads: find $u \in H_0^{\frac{\gamma}{2},1}(\mathcal{O})$ such that

$$
\mathcal{B}(u,v) = \mathcal{F}(v), \quad \forall v \in H_0^{\frac{\gamma}{2},1}(\mathcal{O}), \tag{4.3.16}
$$

where the bilinear form $\mathcal{B}(\cdot,\cdot)$ and the funtional $\mathcal{F}(\cdot)$ are respectively defined by

$$
\mathcal{B}(u,v) := \int_0^T \left({}^{\mathrm{RL}}_{0}D_t^{\frac{\gamma}{2}}u, {}^{\mathrm{RL}}_{t}D_T^{\frac{\gamma}{2}}v\right)_\Omega + \mathcal{A}\big(u,v\big)\,\mathrm{d}t, \quad \mathcal{F}(v) := \int_0^T (f,v)_\Omega\,\mathrm{d}t.
$$

Lemma 4.24 *For any* $u, v \in H_0^{\frac{\nu}{2},1}(\mathcal{O})$, *there exist constants* c_1, c_2 *such that*

$$\mathcal{B}(u,u) \geq c_1 \|u\|_{H_0^{\frac{\nu}{2},1}(\mathcal{O})}, \quad \mathcal{B}(u,v) \leq c_2 \|u\|_{H_0^{\frac{\nu}{2},1}(\mathcal{O})} \|v\|_{H_0^{\frac{\nu}{2},1}(\mathcal{O})}. \tag{4.3.17}$$

Proof The coercivity and continuity of the bilinear form $\mathcal{B}(\cdot,\cdot)$ are the straightforward results from the elliptic conditions (4.3.15), the Cauchy–Schwarz inequality, the properties (4.3.3) and (4.3.5), and the fractional Poincaré inequality. □

With the elliptic conditions (4.3.17) in hand, we can claim the well-posedness of the weak formulation of the subdiffusion equation by the Lax–Milgram lemma as follows.

Theorem 4.25 *Given a function* f *belonging to* $\left(H_0^{\frac{\nu}{2},1}(\mathcal{O})\right)'$, *the dual space of* $H_0^{\frac{\nu}{2},1}(\mathcal{O})$. *For any* $\nu \in (0,1)$, *the weak problem* (4.3.16) *admits a unique solution* u *satisfying*

$$\|u\|_{H_0^{\frac{\nu}{2},1}(\mathcal{O})} \leq c\|f\|_{\left(H_0^{\frac{\nu}{2},1}(\mathcal{O})\right)'}.$$

4.3.5 Galerkin Spectral Method and Error Estimation

In this part, we shall develop and study a semi-discrete spectral-Galerkin method using the basis $\widehat{\mathcal{S}}_m^{\nu}(t)$ defined in (4.3.12). Let $X_N^t(\Lambda)$ be a finite-dimensional subspace of $H_0^{\frac{\nu}{2}}(\Lambda)$

$$X_N^t = \text{span}\{\widehat{\mathcal{S}}_n^{\nu}(t)\}_{n=0}^N, \tag{4.3.18}$$

which is a finite-dimensional subspace of $H_0^{\frac{\nu}{2}}(\Lambda)$. Then the semi-discrete-in-time scheme reads: find $u_N \in V_N = X_N^t \otimes H_0^1(\Omega)$ such that

$$\mathcal{B}(u_N, v) = \mathcal{F}(v), \quad \forall v \in V_N. \tag{4.3.19}$$

By the Lax–Milgram lemma, for any given function f belonging to $\left(H_0^{\frac{\nu}{2},1}(\mathcal{O})\right)'$, the semi-discrete problem admits a unique solution u_N satisfying

$$\|u_N\|_{H_0^{\frac{\nu}{2},1}(\mathcal{O})} \leq c\|f\|_{\left(H_0^{\frac{\nu}{2},1}(\mathcal{O})\right)'}. \tag{4.3.20}$$

Next, we shall derive the error estimate for the semi-discrete solution. The error $u - u_N$ satisfies the Galerkin orthogonality

$$\mathcal{B}(u_N - u, v) = 0, \quad \forall v \in V_N.$$

Recalling that $\Pi_N^t u \in V_N$ (by Lemma 4.23), the standard argument leads to

$$\begin{aligned}\|u_N - u\|_{H_0^{\frac{\gamma}{2},1}(\mathcal{O})} &\le c\|\Pi_N^t u - u\|_{H_0^{\frac{\gamma}{2},1}(\mathcal{O})} \\ &\le c\|{}^{\mathrm{RL}}_{0}D_t^{\frac{\gamma}{2}}(\Pi_N^t u - u)\|_{L^2(\Lambda;H^1(\Omega))}.\end{aligned} \tag{4.3.21}$$

To derive the error estimate, we shall use the following regularity estimate.

Theorem 4.26 *Let u be the solution of the subdiffusion equation* (4.3.6). *For $k = 0, 1, \ldots, m$, if the source term f satisfies*

$$\partial_t^j(t^j f) \in L^2_{\chi_T^k}(\Lambda; H_0^1(\Omega)) \quad \textit{for all } 0 \le j \le k,$$

then the solution u satisfies

$$\partial_{\gamma,t}^k({}^{\mathrm{RL}}_{0}D_t^{\frac{\gamma}{2}} u) \in L^2_{\chi_T^k}(\Lambda; H_0^1(\Omega)), \quad k = 0, 1, \ldots, m,$$

where $\partial_{\gamma,t} := t\partial_t - \gamma$ with $\gamma > 0$ is a generalized derivative.

Proof For a given suitable function $g(t)$, we have

$$\begin{aligned}&{}^{\mathrm{RL}}_{0}D_t^{\frac{\gamma}{2}} \int_0^t (t-\tau)^{\nu-1} E_{\nu,\nu}(-\lambda_n(t-\tau)^\nu) g(\tau) d\tau \\ &= \frac{1}{\Gamma(1-\frac{\gamma}{2})}\partial_t \int_0^t (t-s)^{-\frac{\gamma}{2}} \int_0^s (s-\tau)^{\nu-1} E_{\nu,\nu}(-\lambda_n(s-\tau)^\nu) g(\tau)\, d\tau\, ds \\ &= \frac{1}{\Gamma(1-\frac{\gamma}{2})}\partial_t \int_0^t g(\tau) \int_\tau^t (t-s)^{-\frac{\gamma}{2}}(s-\tau)^{\nu-1} E_{\nu,\nu}(-\lambda_n(s-\tau)^\nu)\, ds\, d\tau \\ &= \partial_t \int_0^t (t-\tau)^{\frac{\gamma}{2}} E_{\nu,1+\frac{\gamma}{2}}(-\lambda_n(t-\tau)^\nu)\, g(\tau)\, d\tau,\end{aligned} \tag{4.3.22}$$

where the validation of the last equality owes to

$$\begin{aligned}&\int_\tau^t (t-s)^{-\frac{\gamma}{2}}(s-\tau)^{\nu-1} E_{\nu,\nu}(-\lambda_n(s-\tau)^\nu)\, ds \\ &\quad = (t-\tau)^{\frac{\gamma}{2}} \int_0^1 (1-\theta)^{-\frac{\gamma}{2}} \theta^{\nu-1} \sum_{j=0}^\infty \frac{(-\lambda_n)^j \theta^{j\nu}(t-\tau)^{j\nu}}{\Gamma(j\nu+\nu)}\, d\theta \\ &\quad = (t-\tau)^{\frac{\gamma}{2}} \sum_{j=0}^\infty \frac{(-\lambda_n)^j (t-\tau)^{j\nu}}{\Gamma(j\nu+\nu)} \int_0^1 (1-\theta)^{-\frac{\gamma}{2}} \theta^{j\nu+\nu-1}\, d\theta \\ &\quad = \Gamma(1-\frac{\gamma}{2})(t-\tau)^{\frac{\gamma}{2}} E_{\nu,1+\frac{\gamma}{2}}(-\lambda_n(t-\tau)^\nu).\end{aligned} \tag{4.3.23}$$

Using the solution representation (4.3.8) and the derivative relation (4.3.7), we have that

$$\begin{aligned}
&{}^{\mathrm{RL}}_{0}D_t^{\frac{\gamma}{2}} u(x,t) \\
&= \sum_{n=1}^{\infty} \varphi_n(x)\, \partial_t \int_0^t (t-\tau)^{\frac{\gamma}{2}} E_{\nu,1+\frac{\gamma}{2}}(-\lambda_n (t-\tau)^{\nu}) \big(f(\cdot,\tau), \varphi_n\big)_{\Omega} d\tau \\
&= \sum_{n=1}^{\infty} \varphi_n(x) \int_0^t (t-\tau)^{\frac{\gamma}{2}-1} E_{\nu,\frac{\gamma}{2}}(-\lambda_n (t-\tau)^{\nu}) \big(f(\cdot,\tau), \varphi_n\big)_{\Omega} d\tau.
\end{aligned} \tag{4.3.24}$$

Then we can rewrite ${}^{\mathrm{RL}}_{0}D_t^{\frac{\gamma}{2}} u$ as

$$ {}^{\mathrm{RL}}_{0}D_t^{\frac{\gamma}{2}} u(t) = \int_0^t \bar{E}(t-\tau) f(\tau)\, d\tau, $$

where

$$ \bar{E}(t)v = \sum_{n=1}^{\infty} t^{\frac{\gamma}{2}-1} E_{\nu,\frac{\gamma}{2}}(-\lambda_n t^{\nu}) \big(v, \varphi_n\big)_{\Omega} \varphi_n. $$

Now we are ready to prove that $\partial_{\gamma,t}^k [{}^{\mathrm{RL}}_{0}D_t^{\frac{\gamma}{2}} (\Delta^{\frac{1}{2}} u)] \in L^2_{\chi_T^k}(\Lambda; L^2(\Omega))$. Thanks to the definition of the generalized derivative $\partial_{\gamma,t} = t\partial_t - \gamma$, we only need to check $\partial_t^k \big(t^k [{}^{\mathrm{RL}}_{0}D_t^{\frac{\gamma}{2}} (\Delta^{\frac{1}{2}} u)]\big) \in L^2_{\chi_T^k}(\Lambda; L^2(\Omega))$ for $k = 0, 1, \ldots, m$. For this purpose, we observe that

$$\begin{aligned}
\partial_t^k \big(t^k [{}^{\mathrm{RL}}_{0}D_t^{\frac{\gamma}{2}} (\Delta^{\frac{1}{2}} u)]\big)(t) &= \partial_t^k \Big(t^k \int_0^t \bar{E}(t-\tau) \Delta^{\frac{1}{2}} f(\tau)\, d\tau\Big) \\
&= \sum_{j=0}^{k} \binom{k}{j} \partial_t^k \Big(\int_0^t (t-\tau)^{k-j} \bar{E}(t-\tau)\, \tau^j f(\tau) d\tau\Big) \\
&= \sum_{j=0}^{k} \binom{k}{j} \int_0^t \big(\partial_t^{k-j}[(t-\tau)^{k-j} \bar{E}(t-\tau)]\big) \big(\partial_\tau^j [\tau^j \Delta^{\frac{1}{2}} f(\tau)]\big) d\tau.
\end{aligned}$$

The last equality holds due to the facts that

$$ \lim_{t\to 0} t^{k+1} \partial_t^k f(t) = 0 \quad \text{and} \quad \lim_{t\to 0} t^{k+1} \partial_t^k \bar{E}(t) v = 0, \quad \forall v \in L^2(\Omega). $$

Thanks to Lemma 4.27, we have

$$\|\partial_t^k\big(t^k[{}^{\mathrm{RL}}_{\ 0}D_t^{\frac{\gamma}{2}}(\Delta^{\frac{1}{2}}u)]\big)\|_{L^2(\Omega)} \le c\sum_{j=0}^{k}\int_0^t (t-\tau)^{\frac{\gamma}{2}-1}\|\partial_\tau^j[\tau^j\Delta^{\frac{1}{2}}f(\tau)]\|_{L^2(\Omega)}\,\mathrm{d}\tau$$
$$=: c\sum_{j=0}^{k}K_j(t).$$

By Young's convolution inequality, we have

$$\begin{aligned}
&\int_0^T |K_j(t)|^2|\log(t/T)|^k\,\mathrm{d}t\\
&= \int_0^T \Big|\int_0^t (t-\tau)^{\frac{\gamma}{2}-1}\|\partial_\tau^j[\tau^j\Delta^{\frac{1}{2}}f(\tau)]\|_{L^2(\Omega)}\,\mathrm{d}\tau\Big|^2 |\log(t/T)|^k\,\mathrm{d}t\\
&\le \int_0^T \Big|\int_0^t (t-\tau)^{\frac{\gamma}{2}-1}\Big(\|\partial_\tau^j[\tau^j\Delta^{\frac{1}{2}}f(\tau)]\|_{L^2(\Omega)}|\log(\tau/T)|^{\frac{k}{2}}\Big)\,\mathrm{d}\tau\Big|^2\,\mathrm{d}t\\
&\le \Big(\int_0^T t^{\frac{\gamma}{2}-1}\,\mathrm{d}t\Big)^2\int_0^T \|\partial_\tau^j[\tau^j\Delta^{\frac{1}{2}}f(\tau)]\|_{L^2(\Omega)}^2|\log(\tau/T)|^k\,\mathrm{d}t\\
&\le c\int_0^T \|\partial_\tau^j[\tau^j f(\tau)]\|_{H^1(\Omega)}^2|\log(\tau/T)|^k\,\mathrm{d}t \le c.
\end{aligned}$$

Therefore $\partial_t^k\big(t^k[{}^{\mathrm{RL}}_{\ 0}D_t^{\frac{\gamma}{2}}u]\big) \in L^2_{\chi_T^k}(\Lambda; H_0^1(\Omega))$ for $k = 0, 1, \ldots, m$, and so does $\partial_{\gamma,t}^k[{}^{\mathrm{RL}}_{\ 0}D_t^{\frac{\gamma}{2}}u]$. This completes the proof of this theorem. □

Lemma 4.27 *Let $\bar{E}(t)$ be the operator defined in* (4.3.5). *Then it holds that*

$$\|\partial_t^k(t^k\bar{E}(t))v\|_{L^2(\Omega)} \le ct^{\frac{\gamma}{2}-1}\|v\|_{L^2(\Omega)}, \quad \forall v \in L^2(\Omega),$$

where $\partial_t^k := \dfrac{\mathrm{d}^k}{\mathrm{d}t^k}$ is the k-fold derivative with respect to t for all $k \ge 0$.

Proof Using the identity (4.3.7), we have

$$\begin{aligned}
\partial_t^k(t^k\bar{E}(t))v &= \sum_{j=0}^{k}\binom{k}{j}\sum_{n=1}^{\infty}(\partial_t^{k-j}t^k)\partial_t^j\Big(t^{\frac{\gamma}{2}-1}E_{\gamma,\frac{\gamma}{2}}(-\lambda_n t^\gamma)\Big)\big(v,\varphi_n\big)_\Omega\varphi_n\\
&= \sum_{j=0}^{k}\frac{k!}{j!}\binom{k}{j}\sum_{n=1}^{\infty}t^{\frac{\gamma}{2}-1}E_{\gamma,\frac{\gamma}{2}-j}(-\lambda_n t^\gamma)\big(v,\varphi_n\big)_\Omega\varphi_n.
\end{aligned}$$

This together with Lemma 4.18 leads to

$$\|\partial_t^k(t^k \bar{E}(t))v\|_{L^2(\Omega)}^2 \le c \sum_{j=0}^{k} \sum_{n=1}^{\infty} \left| t^{\frac{\nu}{2}-1} E_{\nu,\frac{\nu}{2}-j}(-\lambda_n t^\nu) \right|^2 (v, \varphi_n)_\Omega^2$$
$$\le c t^{\nu-2} \|v\|_{L^2(\Omega)}^2,$$

which completes the proof. □

Combining (4.3.21) and Theorem 4.26 with the approximation properties stated in Lemma 4.23, we immediately obtain the following error estimate.

Theorem 4.28 *Let u be the solution of the subdiffusion equation* (4.3.6) *and u_N be the semi-discrete solution of the spectral-Galerkin scheme* (4.3.19). *If the source term f satisfies*

$$\partial_t^j(t^j f) \in L^2_{\chi_T^k}(\Lambda; H_0^1(\Omega)), \quad \textit{for all } 0 \le j \le k \textit{ and } k = 0, 1, \ldots, m,$$

then it holds that

$$\|u_N - u\|_{H_0^{\frac{\nu}{2},1}(\mathcal{O})} \le c N^{-m/2}.$$

Here the constant is independent of N and u, but may depend on ν, β, m, T and f.

Remark 4.29 In the case that the initial condition of the subdiffusion problem is not zero, we may derive the same result by assuming that $u_0 \in \dot{H}^3(\Omega)$. Recalling that the initial value $u_0(x)$ of the subdiffusion equation (4.3.1) transforms to Lu_0 arising in the source term in (4.3.6), it is easy to verify that $\partial_t^n(t^n L u_0) \in C^\infty([0,T]; H_0^1(\Omega))$ for any positive integer n.

Remark 4.30 Jin et al. (2016a, 2017b) developed time-stepping schemes using convolution quadrature generated by high-order BDF methods, motivated by the pioneer work in Lubich (1988); Cuesta et al. (2006). An initial correction strategy was proposed to restore high-order convergence. See also Yan et al. (2018) for the analysis of the $L1$ scheme with initial correction and Fischer (2019); Banjai and López-Fernández (2019) for the convolution quadrature Runge–Kutta schemes. One requirement of those time-stepping schemes in aforementioned works is that the source term f needs to be smooth in the time direction. As an example, the k-th order BDF method requires $f \in W^{k,1}((0,T); L^2(\Omega)) \cap f \in W^{k,\infty}((\epsilon,T); L^2(\Omega))$, and the high-order convergence will deteriorate if f is not regular enough (see Jin et al. (2017b, Section 4)). Therefore, it is nontrivial to extend the strategy and analysis to nonlinear problems (see Jin et al. (2018)) or problems with time-dependent diffusion coefficients (see Jin et al. (2020, 2019)). Compared with those schemes, the spectral-Galerkin method (4.3.19) allows singularity of the source term near $t = 0$, and hence also performs well for solving linear subdiffusion equations with time-dependent diffusion coefficients as well as nonlinear subdiffusion problems. See Section 5 for numerical results and more discussion. On the other hand, we note that the spectral-Galerkin method asks for smooth initial condition and the error estimate was derived in the space-time energy norm, while the time-stepping schemes allow optimal pointwise-in-time error estimate at a fixed time (even for nonsmooth initial data $u_0 \in L^2(\Omega)$). The pointwise-in-time error estimate of the spectral-Galerkin method (4.3.19) is still unclear and deserves further investigation.

4.3.6 Fully Discrete Scheme and Error Estimation

In this section, we shall discuss the fully discrete scheme based on our preceding results. As an example, we apply the $P1$ conforming finite element in space. Here we introduce a quasi-uniform shape regular partition of the domain Ω into simplicial elements of maximal diameter h, which is denoted by $\mathcal{T}_h$. We consider the space of continuous piecewise linear functions on $\mathcal{T}_h$ with $N \in \mathbb{N}$ being the number of degrees of freedom. Let $\{\varphi_i\}_{i=1}^M \subset H_0^1(\Omega)$ be the nodal basis functions and denote

$$X_h^x := \text{span}\{\varphi_i\}_{i=1}^M. \tag{4.3.25}$$

The L^2-projection $P_h\colon L^2(\Omega) \to X_h^x$ defined by

$$(\phi - P_h\phi, v)_\Omega = 0, \quad \forall v \in X_h^x.$$

It is well known that it satisfies the following error estimate (see Thomée (2006)):

$$\|P_h\phi - \phi\|_{L^2(\Omega)} + h\|\nabla(P_h\phi - \phi)\|_{L^2(\Omega)} \le ch^q \|\phi\|_{H^q(\Omega)} \tag{4.3.26}$$

for $\phi \in H_0^1(\Omega) \cap H^q(\Omega)$, $q = 1, 2$. Then the spatially semi-discrete scheme reads: find $u_h \in W_h = H_0^{\frac{\gamma}{2}}(0,T) \otimes X_h^x$ such that

$$\mathcal{B}(u_h, v) = \mathcal{F}(v), \quad \forall v \in W_h. \tag{4.3.27}$$

The well-posedness follows from the coercivity of the bilinear form and the Lax–Milgram lemma.

Upon introducing the discrete Laplace operator $L_h\colon X_h^x \to X_h^x$ defined by $-(L_h\psi, \chi) = a(\psi, \chi)$ for all $\psi, \chi \in X_h$, let $\{\lambda_j^h, \varphi_j^h\}_{j=1}^N$ be the eigenpairs of the discrete Laplacian $-L_h$. Now we introduce the discrete analogue E_h of the operator E defined in (4.3.8) for $t > 0$ and $v_h \in X_h$:

$$E_h(t)v_h = \sum_{j=1}^{N} t^{\nu-1} E_{\nu,\nu}(-\lambda_j^h t^\nu)\,(v_h, \varphi_j^h)_\Omega\, \varphi_j^h. \tag{4.3.28}$$

Then the solution $u_h(t)$ of the spatially semi-discrete problem (4.3.27) can be represented by Jin et al. (2015a, eq (3.4)):

$$u_h(t) = \int_0^t E_h(t-s) P_h f(s)\, ds. \tag{4.3.29}$$

Lemma 4.31 *For $f \in L^2(\Lambda; L^2(\Omega))$, the spatially semi-discrete solution u_h in* (4.3.27) *satisfies*

$$\|u_h - u\|_{H_0^{\frac{\nu}{2},1}(\mathcal{O})} \le ch\|f\|_{L^2(\Lambda;L^2(\Omega))}.$$

Proof By the approximation property (4.3.26), we have for $\rho = P_h u - u$,

$$\begin{aligned}\|\rho\|^2_{H_0^{\frac{\nu}{2},1}(\mathcal{O})} &= \|\rho\|^2_{H_0^{\frac{\nu}{2}}(\Lambda;L^2(\Omega))} + \|\nabla\rho\|^2_{L^2(\Lambda;L^2(\Omega))}\\ &\le ch^2\Big(\|\nabla u\|^2_{H_0^{\frac{\nu}{2}}(\Lambda;L^2(\Omega))} + \|u\|^2_{L^2(\Lambda;H^2(\Omega))}\Big).\end{aligned}$$

Then by applying (4.3.3), (4.3.4) and regularity results in Jin et al. (2015a, Theorem 2.1), we obtain

$$\begin{aligned}\|\nabla u\|^2_{H_0^{\frac{\nu}{2}}(\Lambda;L^2(\Omega))} &\le c\int_0^T \left({}^{\mathrm{RL}}_{\ \ 0}D_t^{\frac{\nu}{2}}\nabla u\,,\,{}^{\mathrm{RL}}_{\ \ t}D_T^{\frac{\nu}{2}}\nabla u\right)_\Omega dt\\ &= c\int_0^T \left({}^{\mathrm{RL}}_{\ \ 0}D_t^{\nu} u\,, -\Delta u\right)_\Omega dt\\ &\le c\|{}^{\mathrm{RL}}_{\ \ 0}D_t^{\nu} u\|^2_{L^2(\Lambda;L^2(\Omega))} + c\|\Delta u\|^2_{L^2(\Lambda;L^2(\Omega))}\\ &\le c\|f\|^2_{L^2(\Lambda;L^2(\Omega))}.\end{aligned}$$

Therefore, we arrive at

$$\|\rho\|_{H_0^{\frac{\nu}{2},1}(\mathcal{O})} \le ch\|f\|_{L^2(\Lambda;L^2(\Omega))}. \tag{4.3.30}$$

The function $\vartheta := u_h - P_h u$ satisfies $\vartheta(0) = 0$ and ${}_0\partial_t^\alpha \vartheta - L_h\vartheta = L_h(P_h u - R_h u) = L_h R_h \rho$ (with R_h being the Ritz projection), see Jin et al. (2015a, (3.9)). By the regularity results, we have

$$\|\vartheta\|_{H^\nu(\Lambda;H^{-1}(\Omega))} + \|\vartheta\|_{L^2(\Lambda;H_0^1(\Omega))} \le c\|L_h R_h \rho\|_{L^2(0,T;H^{-1}(\Omega))}.$$

Then the interpolation between $H^{\nu}(\Lambda; H^{-1}(\Omega))$ and $L^2(\Lambda; H_0^1(\Omega))$ yields (note that $\frac{\nu}{2} < \frac{1}{2}$)

$$\|\vartheta\|_{H_0^{\frac{\nu}{2}}(\Lambda;L^2(\Omega))} + \|\vartheta\|_{L^2(\Lambda;H_0^1(\Omega))} \le c\|L_h R_h \rho\|_{L^2(\Lambda;H^{-1}(\Omega))}.$$

This together with the estimate (4.3.30) leads to

$$\begin{aligned}\|\vartheta\|_{H_0^{\frac{\nu}{2},1}(\mathcal{O})} &\le c\|L_h R_h \rho\|_{L^2(\Lambda;H^{-1}(\Omega))} \\ &\le c\|\nabla R_h \rho\|_{L^2(\Lambda;L^2(\Omega))} \le ch\|f\|_{L^2(\Lambda;L^2(\Omega))}.\end{aligned} \tag{4.3.31}$$

As a result, (4.3.30), (4.3.31) and the triangle inequality complete the proof. □

The next lemma, which is an analogue to Theorem 4.26, provides regularity results of u_h in the time direction. The proof, relying on the solution representation (4.3.29) and the property of Mittag-Leffler functions, is similar to the proof of Theorem 4.26.

Lemma 4.32 *Let u_h be the spatially semi-discrete solution in* (4.3.27). *For $k = 0, 1, \ldots, m$, if the source term f satisfies*

$$\partial_t^j(t^j f) \in L^2_{\chi_T^k}(\Lambda; H_0^1(\Omega)), \quad \textit{for all } 0 \le j \le k,$$

then the solution u satisfies

$$\partial_{\gamma,t}^k \, {}^{\mathrm{RL}}_{\;\;0}D_t^{\frac{\nu}{2}} u_h \in L^2_{\chi_T^k}(\Lambda; H_0^1(\Omega)), \quad \textit{for } k = 0, 1, \ldots, m,$$

where $\partial_{\gamma,t} := t\partial_t - \gamma$ with $\gamma > 0$ is a generalized derivative.

Now we are ready to develop a fully discrete scheme. Let $X_N^t(\Lambda)$ and $X_h^x(\Omega)$ be finite-dimensional spaces defined in (4.3.18) and (4.3.25), respectively. Then the fully discrete scheme reads: find $u_{hN} \in X_{hN} := X_N^t \otimes X_h^x$ such that

$$\mathcal{B}(u_{hN}, v) = (f, v), \quad \forall v \in X_{hN}. \tag{4.3.32}$$

Similarly, the well-posedness of the above numerical scheme can be guaranteed by the coercivity of linear form $\mathcal{B}(\cdot,\cdot)$ and the Lax–Milgram lemma.

The following theorem provides an estimate on the difference between the fully discrete solution u_{hN} and the spatially semi-discrete solution u_h.

Theorem 4.33 *Let u_h be the solution of the spatially semi-discrete scheme* (4.3.27) *and u_{hN} be the fully discrete solution satisfying* (4.3.32). *If the resource term f satisfies*

$$\partial_t^j(t^j f) \in L^2_{\chi_T^k}(\Lambda; H_0^1(\Omega)), \quad \forall 0 \le j \le k, \quad k = 0, 1, \ldots, m,$$

then it holds that

$$\|u_{hN} - u_h\|_{H_0^{\frac{\nu}{2},1}(\mathcal{O})} \le c_2 N^{-m/2}.$$

Here the constant is independent of h, N and u, may depend on ν, β, m, T *and* f.

To estimate the numerical error $u_{hN} - u$, we split it into two components

$$u_{hN} - u = (u_{hN} - u_h) + (u_h - u).$$

Note that the bound of $u_h - u$ and $u_{hN} - u_h$ has already been established in Theorem 4.28 and Theorem 4.33, respectively.

Corollary 4.34 *Let u be the solution of the subdiffusion equation* (4.3.6) *and* u_{hN} *be the fully discrete solution satisfying* (4.3.32). *If the resource term* $f \in L^2(\Lambda; L^2(\Omega))$ *also satisfies*

$$\partial_t^j(t^j f) \in L^2_{\chi_T^k}(\Lambda; H_0^1(\Omega)), \quad \forall 0 \le j \le k, \quad k = 0, 1, \ldots, m,$$

then it holds that

$$\|u_{hN} - u\|_{H_0^{\frac{\nu}{2},1}(\mathcal{O})} \le c_1 h + c_2 N^{-m/2}.$$

Here the constants are independent of h, N and u, but c_1 *may depend on* ν, f, T, *and* c_2 *may depend on* ν, β, m, T *and* f.

4.3.7 Fast Solver

In this part, we intend to develop a fast algorithm to solve the space-time linear system (4.3.32). By substituting $u_{hN} = \sum_{m=1}^{M}\sum_{n=0}^{N} \tilde{u}_{mn}\phi_m(x)\widehat{S}_n^\nu(t)$ and $v = \phi_p(x)\widehat{S}_q^\nu(t)$, $p = 1, \ldots, M, q = 0, 1, \ldots, N$, the fully discrete scheme (4.3.32) is equivalent to solving the following matrix system

$$\boldsymbol{S}^x\, \boldsymbol{U}\, (\boldsymbol{M}^t)^T + \boldsymbol{M}^x\, \boldsymbol{U}\, (\boldsymbol{S}^t)^T = \boldsymbol{F}, \tag{4.3.33}$$

where $\boldsymbol{M}^x$ and $\boldsymbol{S}^x$ are the mass and stiffness matrices in x-direction(s), and $\boldsymbol{M}^t$ and $\boldsymbol{S}^t$ are the mass and stiffness matrices in t-direction below

$$\begin{aligned}
\boldsymbol{S}^x &= (s_{pm}^x), & s_{pm}^x &= \big(L\phi_m, \phi_p\big)_\Omega,\\
\boldsymbol{M}^x &= (m_{pm}^x), & m_{ml}^x &= \big(\phi_m, \phi_p\big)_\Omega,\\
\boldsymbol{S}^t &= (s_{qn}^t), & s_{qn}^t &= \big({}^{\mathrm{RL}}_{\ 0}D_t^{\frac{\nu}{2}}\widehat{S}_n^\nu, {}^{\mathrm{RL}}_{\ t}D_T^{\frac{\nu}{2}}\widehat{S}_q^\nu\big)_\Lambda,\\
\boldsymbol{M}^t &= (m_{qn}^t), & m_{pn}^t &= \big(\widehat{S}_n^\nu, \widehat{S}_q^\nu\big)_\Lambda,\\
\boldsymbol{U} &= (u_{mn}), & u_{mn} &= \tilde{u}_{mn},\\
\boldsymbol{F} &= (f_{pq}), & f_{pq} &= \big(f, \phi_p\widehat{S}_q^\nu\big).
\end{aligned}$$

Notice that both the matrices $\boldsymbol{M}^t$ and $\boldsymbol{S}^t$ are nonsymmetric, so the classical matrix diagonalization method in Shen (1994) cannot be applied directly. To overcome this difficulty, we shall apply an efficient QZ decomposition method proposed by Shen and Sheng (2019) recently.

The key point of the new method is the following QZ decomposition,

$$\boldsymbol{Q}(\boldsymbol{S}^t)^T\boldsymbol{Z} = \boldsymbol{A}, \quad \boldsymbol{Q}(\boldsymbol{M}^t)^T\boldsymbol{Z} = \boldsymbol{B},$$

where $\boldsymbol{Q}$, $\boldsymbol{Z}$ are unitary matrices satisfying $\boldsymbol{Q}\boldsymbol{Q}^T = \boldsymbol{Z}\boldsymbol{Z}^T = \boldsymbol{I}$, and $\boldsymbol{A}$, $\boldsymbol{B}$ are upper triangular matrices, namely,

$$\boldsymbol{A} = \begin{pmatrix} a_{11} & a_{12} & \dots & a_{1N} \\ & a_{22} & \dots & a_{2N} \\ & & \ddots & \vdots \\ & & & a_{NN} \end{pmatrix}, \quad \boldsymbol{B} = \begin{pmatrix} b_{11} & b_{12} & \dots & b_{1N} \\ & b_{22} & \dots & b_{2N} \\ & & \ddots & \vdots \\ & & & b_{NN} \end{pmatrix}.$$

Then, by setting $\boldsymbol{U} = \boldsymbol{V}\boldsymbol{Q}$ and multiplying (4.3.33) both sides by $\boldsymbol{Z}$, we have an equivalent form

$$\boldsymbol{M}^x\,\boldsymbol{V}\,\boldsymbol{A} + \boldsymbol{S}^x\,\boldsymbol{V}\,\boldsymbol{B} = \boldsymbol{G} := \boldsymbol{F}\boldsymbol{Z}. \tag{4.3.34}$$

Denote by $\boldsymbol{v}_n = (v_{1n}, v_{2n}, \dots, v_{Mn})^T$ the n-th column of the matrix $\boldsymbol{V}$, i.e.,

$$\boldsymbol{V} = [\boldsymbol{v}_1, \boldsymbol{v}_2, \dots, \boldsymbol{v}_N] = \begin{pmatrix} v_{11} & v_{12} & \dots & v_{1N} \\ v_{21} & v_{22} & \dots & v_{2N} \\ \vdots & \vdots & \ddots & \vdots \\ v_{M1} & v_{M2} & \dots & v_{MN} \end{pmatrix}.$$

The matrix system (4.3.34) can be solved by the following fast algorithm:

$$(a_{nn}\boldsymbol{M}^x + b_{nn}\boldsymbol{S}^x)\,\boldsymbol{v}_n = \boldsymbol{g}_n - \boldsymbol{h}_{n-1}, \quad n = 1, 2, \dots, N, \tag{4.3.35}$$

where $\boldsymbol{g}_n$ is the n-th column of the matrix $\boldsymbol{G}$ and

$$\boldsymbol{h}_0 = \boldsymbol{0}, \quad \boldsymbol{h}_{n-1} = \boldsymbol{M}^x\Big(\sum_{k=1}^{n-1} a_{kn}\,\boldsymbol{v}_k\Big) + \boldsymbol{S}^x\Big(\sum_{k=1}^{n-1} b_{kn}\,\boldsymbol{v}_k\Big).$$

Note that the new algorithm (4.3.35) is indeed equivalent to solving N times elliptical problems below

$$-a_{nn}Lv(x) + b_{nn}v(x) = g_h(x), \quad n = 1, 2, \dots N.$$

As is shown in Theorem 4.33, our scheme enjoys spectral convergence in time, so in general only small N is needed to achieve the desired accuracy. On the other hand, the above elliptic problems can be solved efficiently by the usual multigrid or other fast solvers. Therefore, our space-time scheme is very efficient.

Numerical Experiments, Extensions and Discussions

In this section, we present some numerical examples to illustrate our theoretical results proposed in previous sections, and in addition, apply the proposed techniques to problems with time-dependent diffusion coefficients and to a non-linear time-fractional Allen–Cahn equation. In our experiments, we computed the numerical solutions of the fully discrete scheme (4.3.32), with the spectral-Galerkin method in time and the Galerkin finite element method in space. Since the spatially semi-discrete solution has been verified in Jin et al. (2015a), we focus on the temporal discretization error below. All the computations are carried out in MATLAB R2015a on a personal laptop.

Example (a). Subdiffusion Problems with Time-Independent Coefficients

In the first example, we present the numerical results for the following two-dimension subdiffusion problem on the unit square domain $\Omega = (0,1)^2$. Let $0 < \nu < 1$ and $x := (x_1, x_2) \in \Omega$; we look for the function u satisfying

$$\begin{cases} {}_0^C D_t^\nu u(x,t) - \Delta u(x,t) = f(x,t), & (x,t) \in \Omega \times (0,T], \\ u(x,0) = 0, & x \in \Omega, \\ u(x,t)\big|_{\partial\Omega} = 0, & t \in (0,T). \end{cases} \tag{4.3.36}$$

Here we choose the source term

$$f(x,t) = t^{0.3}(1 - |2x_1 - 1|)(1 - |2x_2 - 1|). \tag{4.3.37}$$

In our computation, the domain $\Omega = (0,1)^2$ is first divided into M^2 small equal squares, and we obtain a symmetric triangulation by connecting the diagonal of each small square. To verify the temporal error, we fixed $M = 100$, $\beta = 5$ and change the number of basis functions of the spectral method in the time direction. Since the closed form of the exact solution to problem (4.3.36) is unavailable, we compute a reference solution $u_h \approx u_{hN_{ref}}$ with $N_{ref} = 70$.

In this case, the source term f satisfies the smoothness requirement in Theorems 4.28 and 4.33,

$$\partial_t^j (t^j f) \in L^2_{\chi_T^k}(\Lambda; H_0^1(\Omega)), \quad \forall 0 \le j \le k, \quad k = 0, 1, \ldots, m, \tag{4.3.38}$$

with an arbitrary positive integer m. Therefore, all the theoretical results hold valid and we expect an exponential convergence in the time direction. The numerical results are presented in Figure 4.14, where the temporal error E_L in L^2 space and error E_H in space-time energy space are defined by

$$E_L = \frac{\|u_h - u_{hN}\|_{L^2(\Lambda;L^2(\Omega))}}{\|u_h\|_{L^2(\Lambda;L^2(\Omega))}} \quad \text{and} \quad E_H = \frac{\|u_h - u_{hN}\|_{H^{\frac{\nu}{2},1}(\mathcal{O})}}{\|u_h\|_{H^{\frac{\nu}{2},1}(\mathcal{O})}}, \tag{4.3.39}$$

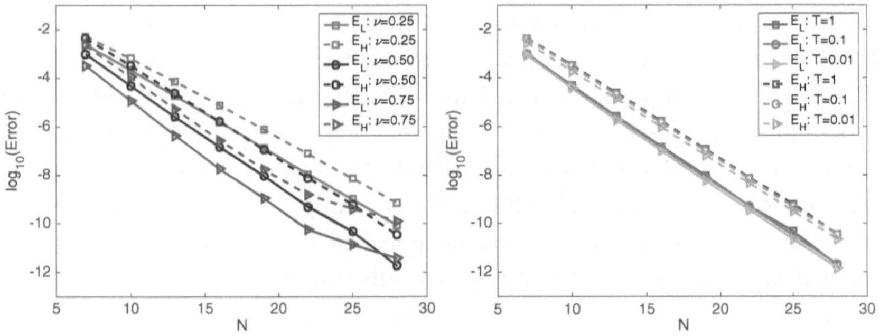

Figure 4.14 Example (a) with the source term (4.3.37). Left: plot of E_L and E_H with $T = 1$ and fractional orders $\nu = 0.25, 0.5, 0.75$. Right: plot of E_L and E_H with $\nu = 0.5$ and terminal times $T = 0.01, 0.1, 1$. Source: Chen et al. (2020). Copyright ©2020 Society for Industrial and Applied Mathematics. Reprinted with permission. All rights reserved.

respectively. In Figure 4.14, we draw the error curves against N for distinct parameters ν and final time T. All the experiments show that the proposed spectral-Galerkin method is highly efficient for subdiffusion problems, and the numerical solutions u_{hN} exponentially converge to u_h.

Next, we test the problem (4.3.36) with another source term

$$f(x,t) = (1 + t^{0.5})x_1(1 - x_1)x_2(1 - x_2), \tag{4.3.40}$$

and compare the spectral-Galerkin method with some high-order BDF schemes developed in Jin et al. (2017b). Note that the source is nonsmooth in the time direction, and hence the BDF schemes fail to achieve the desired high accuracy. In particular, it can be verified that

$$f \in W^{\frac{3}{2}-\epsilon,1}(0, T/2; L^2(\Omega)) \cap W^{m,\infty}(T/2, T; L^2(\Omega))$$

for any small $\epsilon > 0$ and positive integer $m > 0$. The proof of Jin et al. (2017b, Theorem 2.2) indicates a convergence rate $O(\tau^{\frac{3}{2}-\epsilon})$ for this example, where τ denotes the step size in time. This prediction is consistent to the numerical results plotted in Figure 4.15 (right), where we plot $L^2(\Omega)$-norm of the numerical error at $T = 1$. On the other hand, the source term still satisfies the smoothness requirement (4.3.38). Therefore, we can observe the high accuracy of the spectral method in Figure 4.15 (left). Those experiments indicate that the spectral-Galerkin method (4.3.32) performs much better than the time-stepping approach in this special case.

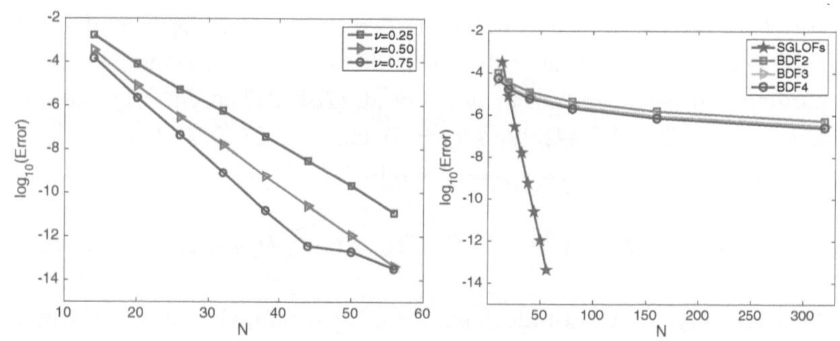

Figure 4.15 Example (a) with the source term (4.3.40). Left: plot of E_L with $T = 1$ and fractional orders $\nu = 0.25, 0.5, 0.75$. Right: plot of $L^2(\Omega)$-norm of the error at $T = 1$ with $\nu = 0.5$, for spectral-Galerkin method (4.3.32) and the k-th order BDF method proposed in Jin et al. (2017b). Source: Chen et al. (2020). Copyright ©2020 Society for Industrial and Applied Mathematics. Reprinted with permission. All rights reserved.

Example (b). Subdiffusion Problems with Time-Dependent Coefficients

Consider the subdiffusion problem with time-dependent coefficient:

$$\begin{cases} {}^{\mathrm{RL}}_{0}D_t^{\nu}u(x,t) - \nabla \cdot (a(x,t)\nabla u(x,t)) = f(x,t), & x \in \Omega,\ t \in \Lambda := (0,T], \\ u(x,t) = 0, & (x,t) \in \partial\Omega \times (0,T], \\ u(x,0) = 0, & x \in \Omega. \end{cases} \tag{4.3.41}$$

Here we assume that the diffusion coefficient $a(x,t)\colon \Omega \times (0,T) \to \mathbb{R}^{d\times d}$ has the following regularity for some real number $\lambda \geq 1$ and for any positive integer m:

$$\lambda^{-1}|\xi|^2 \leq a(x,t)\xi \cdot \xi \leq \lambda|\xi|^2, \tag{4.3.42}$$

$$|\tfrac{\partial}{\partial t}a(x,t)| + |\nabla_x a(x,t)| + |\nabla_x \tfrac{\partial}{\partial t}a(x,t)| \leq c, \tag{4.3.43}$$

$$|\nabla_x \tfrac{\partial^j}{\partial t^j}a(x,t)| + |\nabla_x \tfrac{\partial^{j+1}}{\partial t^{j+1}}a(x,t)| \leq c \tag{4.3.44}$$

for $\xi \in \mathbb{R}^d$, $(x,t) \in \Omega \times (0,T]$, and $j = 1,\ldots,m$.

For standard parabolic problems with time-dependent coefficients, there are a few relevant works. Unfortunately, the fractional derivative does not satisfy the well-known Leibnitz rule, and hence some traditional techniques working for the heat equation cannot be directly applied. Mustapha (2018) analyzed the spatially semi-discrete Galerkin FEM approximation of the subdiffusion problem involving time-dependent coefficients, but without the source term, by using a novel energy argument. A perturbation argument was developed in Jin et al. (2019) to derive regularity results, analyze the spatially semi-discrete Galerkin scheme, as well as some first-order time-stepping schemes. In particular, it has been proved in Jin et al. (2019, Theorem 2.1) that under conditions (4.3.42)–(4.3.44), with $u_0 = 0$ and $f \in L^p(0,T;L^2(\Omega))$, $1/\alpha < p < \infty$, problem (4.3.6) has a unique solution

$$u \in C([0,T];L^2(\Omega)) \cap L^p(0,T;\dot{H}^2(\Omega)) \text{ s.t. } {}^{\mathrm{RL}}_{\ 0}D_t^{\gamma} u \in L^p(0,T;L^2(\Omega)).$$

Very recently, a second-order time-stepping scheme, based on convolution quadrature generated by the second-order BDF scheme and initial correction technique, was developed and analyzed in Jin et al. (2020). To the best of our knowledge, there is no other high-order numerical scheme for such models with rigorous analysis in the literature. The main difficulty is caused by the initial singularity of the solution, which will make trouble in the estimation of the perturbation term. However, under the assumptions (4.3.42)–(4.3.44), it has been proved in Jin et al. (2020, Theorem 3.2) that the solution u satisfies the regularity results for all $t \in (0,T]$ and $k \in \mathbb{N}$:

$$\left\| \partial_t^k (t^k u(t)) \right\|_{\dot{H}^3(\Omega)} \le c \sum_{j=0}^{k} t^j \| f^{(j)}(0) \|_{\dot{H}^1(\Omega)} + c t^k \int_0^t \| f^{(k)}(s) \|_{\dot{H}^1(\Omega)} ds. \tag{4.3.45}$$

Therefore, using this estimate, we have the following result.

Theorem 4.35 *Assume that* $f \in W^{m+1,1}(\Lambda;\dot{H}^1(\Omega))$. *Then the solution u of the subdiffusion problem* (4.3.41) *satisfies*

$$\partial_{\gamma,t}^k ({}^{\mathrm{RL}}_{\ 0}D_t^{\frac{\gamma}{2}} u) \in L^2_{\chi_T^k}(\Lambda; H_0^1(\Omega)), \quad \text{for } k = 0,1,\ldots,m,$$

where $\partial_{\gamma,t} = t\partial_t - \gamma$ *with* $\gamma > 0$.

Proof It suffices to show that $\partial_t^k (t^k \, {}^{\mathrm{RL}}_{\ 0}D_t^{\frac{\gamma}{2}} u) \in L^2_{\chi_T^k}(\Lambda; H_0^1(\Omega))$ for all $k = 0,1,\ldots,m$. Appealing to the subdiffusion equation (4.3.41), we have

$${}^{\mathrm{RL}}_{\ 0}D_t^{\frac{\gamma}{2}} u(t) = \int_0^t \frac{(t-\tau)^{\nu/2-1}}{\Gamma(\nu/2)} L(\tau) u(\tau)\, d\tau + \int_0^t \frac{(t-\tau)^{\nu/2-1}}{\Gamma(\nu/2)} f(s)\, ds.$$

Using the a priori estimate (4.3.45), we derive for all $k = 0,1,\ldots,m-1$,

$$\lim_{t\to 0} \partial_t^k (t^{k+1} u(t)) = 0 \quad \text{and} \quad \lim_{t\to 0} \partial_t^k (t^{k+1} f(t)) = 0,$$

and hence

$$\partial_t^k (t^k \, {}^{\mathrm{RL}}_{0}D_t^{\frac{\nu}{2}} u(t)) =$$
$$\frac{1}{\Gamma(\nu/2)} \sum_{j=0}^{k} \binom{k}{j} \int_0^t \partial_t^j (t-\tau)^{\nu/2+j-1} \partial_\tau^{k-j} \big[\tau^{(k-j)} (L(\tau)u(\tau) + f(\tau))\big] \, \mathrm{d}\tau.$$

Then taking the $\dot{H}^1(\Omega)$ norm, we obtain

$$\|\partial_t^k (t^k \, {}^{\mathrm{RL}}_{0}D_t^{\frac{\nu}{2}} u(t))\|_{\dot{H}^1(\Omega)} \le$$
$$c \sum_{j=0}^{k} \int_0^t (t-\tau)^{\nu/2-1} \big[\|\partial_\tau^j (\tau^j u(\tau))\|_{\dot{H}^3(\Omega)} + \|\partial_\tau^j (\tau^j f(\tau))\|_{\dot{H}^1(\Omega)}\big] \, \mathrm{d}\tau \le c.$$

This immediately implies that $\partial_t^k (t^k \, {}^{\mathrm{RL}}_{0}D_t^{\frac{\nu}{2}} u(t)) \in L^2_{\chi_T^k}(\Lambda; H_0^1(\Omega))$. □

The above result indicates that ${}^{\mathrm{RL}}_{0}D_t^{\frac{\nu}{2}} u(x,t)$ belongs to the non-uniformly weighted Sobolev space $A^m_{\frac{\beta}{2},T}(\Lambda; H_0^1(\Omega))$, provided that the source term is smooth enough in the time direction. This regularity result motivates us to develop a spectral-Galerkin method using the LOFs. In fact, it is possible to prove the above regularity result in the case that f satisfies the smoothness requirement (4.3.38). But the proof requires some technical argument and is out of the scope of the current book.

Let $X_N^t(\Lambda)$ and $X_h^x(\Omega)$ be finite-dimensional spaces defined in (4.3.18) and (4.3.25), respectively. Then our fully discrete scheme for (4.3.41) reads: find $u_{hN} \in X_{hN} := X_N^t \otimes X_h^x$ such that for all $v \in X_N^t \otimes X_h^x$,

$$\int_0^T \big({}^{\mathrm{RL}}_{0}D_t^{\frac{\nu}{2}} u_{hN}, {}^{\mathrm{RL}}_{t}D_T^{\frac{\nu}{2}} v\big)_\Omega + \big(a(t)\nabla u_{hN}, \nabla v\big) \, dt = \int_0^T (f, v)_\Omega \, dt. \quad (4.3.46)$$

In the first experiment, we let $a(x,t) := 2 + \cos(t)$, $\Omega = (0,1)^2$, $T = 1$, and test

$$f(x,t) = \cos(t) x_1 (1 - x_1) x_2 (1 - x_2), \quad (4.3.47)$$

which is smooth in the time direction. Therefore, by Theorem 4.35 and the approximation property in Lemma 4.23, we expect that the numerical solution u_{hN} converges to u_h exponentially. This estimate is fully supported by the error curves (against N) plotted in Figure 4.16 (left).

Further, we test the numerical results in the case of nonsmooth (in time) source term

$$f(x,t) = (1 + t^{0.1}) x_1 (1 - x_1) x_2 (1 - x_2). \quad (4.3.48)$$

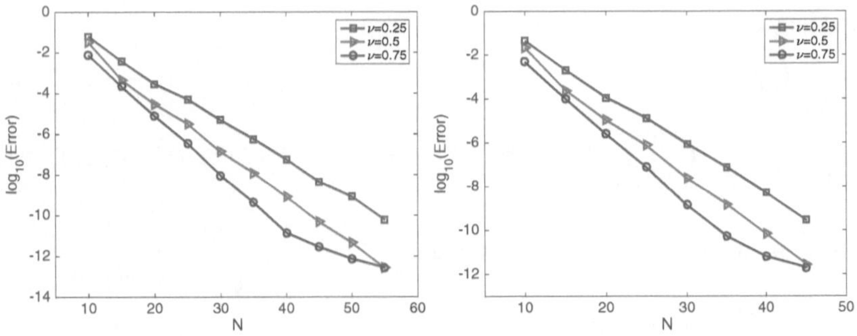

Figure 4.16 Example (b). Left: plot of E_L with $T = 1$, with the smooth-in-time source term (4.3.47) and fractional orders $\nu = 0.25, 0.5, 0.75$. Right: plot of E_L with $T = 1$, with the nonsmooth-in-time source term (4.3.47) and fractional orders $\nu = 0.25, 0.5, 0.75$. Source: Chen et al. (2020). Copyright ©2020 Society for Industrial and Applied Mathematics. Reprinted with permission. All rights reserved.

Although it does not satisfy the smoothness condition in Theorem 4.35, we still observe the exponential convergence (e.g., Figure 4.16 (right)). The theoretical confirmation awaits studies in the future. The numerical results verify the high efficiency of the spectral-Galerkin method for solving the subdiffusion problem with time-dependent coefficients.

Example (c). Time-Fractional Allen–Cahn Equation.

Finally, we consider the following one-dimensional time-fractional Allen–Cahn problem:

$$\begin{cases} {}_0^C D_t^\nu u - \epsilon^2 \Delta u = u - u^3, & x \in \Omega := (-1,1),\ t > 0, \\ u(-1,t) = u(1,t) = 0, & t \geq 0, \\ u(x,0) = u_0(x), & x \in \Omega, \end{cases} \tag{4.3.49}$$

where ε is a small parameter which describes the inter-facial width.

The first rigorous studies of the semilinear subdiffusion problem were given in Jin et al. (2018), where a general framework was proposed for mathematical and numerical analysis of the semilinear subdiffusion equation with a globally Lipschitz continuous potential $f(u)$. A time-stepping scheme based on backward Euler convolution quadrature scheme was studied and a convergence rate of order $O(\tau^\alpha)$ was proved, where τ denotes the step size in time. Then the analysis was extended to the time-fractional Allen–Cahn equation in Du et al. (2020), where several α-th order accurate time-stepping schemes satisfying a weighted energy dissipation law were developed and analyzed. See also Al-Maskari and Karaa (2019) for the argument in the case of nonsmooth initial data. As far as we know, high-order time-stepping schemes by using convolution quadrature or the collocation method for the nonlinear problem (4.3.49) are still missing in the literature.

In order to develop a spectral-Galerkin scheme for solving the time-fractional Allen–Cahn equation (4.3.49), we define an auxiliary function $w(x,t) = u(x,t) - u_0(x)$, and note that the function $w(x,t)$ satisfies the following initial-boundary value problem

$$\begin{cases} F(w) := {}_0^{\mathrm{C}}D_t^{\nu} w - \varepsilon^2 \Delta w + (w+u_0)^3 - w - (\varepsilon^2 \Delta u_0 + u_0) = 0, & x \in \Omega,\ t > 0, \\ w(-1,t) = w(1,t) = 0, & t \geq 0, \\ w(x,0) = 0, & x \in \bar{\Omega}. \end{cases}$$

Using the Newton iterative method, we are led to solve, at each iteration,

$$\begin{cases} {}_0^{\mathrm{C}}D_t^{\nu} E_k - \varepsilon^2 \Delta E_k + (3(w_k+u_0)^2 - 1) E_k = -F(w_k), & x \in \Omega,\ t > 0, \\ E_k(-1,t) = E_k(1,t) = 0, & t \geq 0, \\ E_k(x,0) = 0, & x \in \Omega, \\ w_{k+1} = E_k + w_k, & x \in \Omega,\ t \geq 0. \end{cases}$$

In the computation, we choose the initial guess $w_0 = 0$, and derive the approximation $w_{k+1} = w_k + E_k$ by the above iteration process. For each iteration step, we apply the fully discrete method (4.3.32).

In our computation, we test the time-fractional Allen–Cahn equation (4.3.49) with $\varepsilon = 0.05$, $h = 0.01$, $\beta = 7$ and smooth initial distribution $u_0(x) = \sin \pi x$. The profile of the numerical solution at $T = 10$ is plotted in Figure 4.17 (left). Furthermore, in order to verify the high efficiency of the spectral-Galerkin method in time, we plot the error curve in Figure 4.17 (right). The exponential decay of the error shows again that our new method is very efficient for solving the semilinear time-fractional problems (4.3.49).

4.4 Laguerre Spectral Methods for Tempered FPDEs

The anomalous diffusion equation takes the form

$$\partial_t^{\nu} p(x,t) = \partial_x^{\mu} p(x,t),$$

where $0 < \nu \leq 1$ and $0 < \mu < 2$ (see Metzler and Klafter (2000) for a review on this subject), whose solution exhibits heavy tails, i.e., power law decays

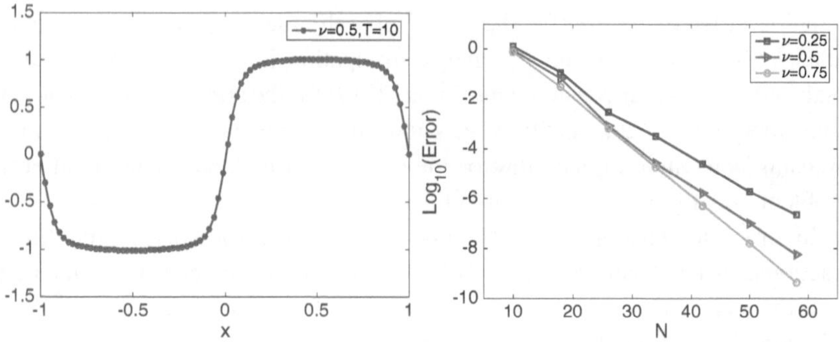

Figure 4.17 Example (c). Left: profile of numerical solutions at $T = 10$. Right: plot of E_L with $T = 10$ and $\nu = 0.25, 0.5, 0.75$. Source: Chen et al. (2020). Copyright ©2020 Society for Industrial and Applied Mathematics. Reprinted with permission. All rights reserved.

at infinity. In order to "temper" the power law decay, Sabzikar et al. (2015) incorporated an exponential factor $e^{-\lambda|x|}$ into the particle jump density, and showed that the Fourier transform of the tempered probability density function $p(x,t)$ takes the form

$$\mathscr{F}[p](\omega,t) = e^{-[pA_+^{\mu,\lambda}(\omega)+qA_-^{\mu,\lambda}(\omega)]Dt}, \quad 0 < \mu < 2,$$

where $0 \le p \le 1$, $q = 1 - p$, D is a constant and

$$A_\pm^{\mu,\lambda}(\omega) := \begin{cases} (\lambda \pm i\omega)^\mu - \lambda^\mu, & 0 < \mu < 1, \\ (\lambda \pm i\omega)^\mu - \lambda^\mu - \pm i\omega\mu\lambda^{\mu-1}, & 1 < \mu < 2. \end{cases} \tag{4.4.1}$$

Moreover, they defined tempered fractional derivative operators $\partial_{\pm,x}^{\mu,\lambda}$ through Fourier transform, $\mathscr{F}[\partial_{\pm,x}^{\mu,\lambda}u](\omega) = A_\pm^{\mu,\lambda}(\omega)\mathscr{F}[u](\omega)$, and derived the tempered fractional diffusion equation:

$$\partial_t u(x,t) = (-1)^k C_T\{p\ \partial_{+,x}^{\mu,\lambda} + q\ \partial_{-,x}^{\mu,\lambda}\}u(x,t), \quad \mu \in (k-1,k),\ k = 1,2. \tag{4.4.2}$$

It is believed that tempered anomalous diffusion models have advantages over the normal diffusion models in some applications in geophysics (see, e.g., Meerschaert et al. (2008); Zhang and Meerschaert (2011)) and finance (see, e.g., Carr et al. (2002)).

4.4.1 Tempered Fractional Integrals and Derivatives on $\mathbb{R}$

Recently, Sabzikar et al. (2015, (19)–(23)) introduced the tempered fractional integrals and derivatives on the whole line.

Definition 4.36 (Tempered Fractional Integrals) For $\lambda \in \mathbb{R}_0^+$, the left tempered fractional integral of a suitable function $u(x)$ of order $\mu \in \mathbb{R}^+$ is defined by

$$ {}_{-\infty}I_x^{\mu,\lambda} u(x) = \frac{1}{\Gamma(\mu)} \int_{-\infty}^{x} \frac{e^{-\lambda(x-y)}}{(x-y)^{1-\mu}} u(y)\, dy, \quad x \in \mathbb{R}, \tag{4.4.3} $$

and the right tempered fractional integral of order $\mu \in \mathbb{R}^+$ is defined by

$$ {}_xI_\infty^{\mu,\lambda} u(x) = \frac{1}{\Gamma(\mu)} \int_{x}^{\infty} \frac{e^{-\lambda(y-x)}}{(y-x)^{1-\mu}} u(y)\, dy, \quad x \in \mathbb{R}. \tag{4.4.4} $$

It is evident that for $\lambda = 0$, they reduce to the usual (non-tempered) fractional integrals:

$$ {}_{-\infty}^{\ \mathrm{RL}}I_x^{\mu} = {}_{-\infty}I_x^{\mu,0}, \quad {}_{x}^{\mathrm{RL}}I_\infty^{\mu} = {}_xI_\infty^{\mu,0}. \tag{4.4.5} $$

Moreover, one verifies readily from the above definitions that

$$ \begin{aligned} {}_{-\infty}I_x^{\mu,\lambda} u(x) &= e^{-\lambda x}\, {}_{-\infty}^{\ \mathrm{RL}}I_x^{\mu} \{e^{\lambda x} u(x)\}, \\ {}_xI_\infty^{\mu,\lambda} u(x) &= e^{\lambda x}\, {}_{x}^{\mathrm{RL}}I_\infty^{\mu} \{e^{-\lambda x} u(x)\}. \end{aligned} \tag{4.4.6} $$

As shown in Sabzikar et al. (2015), the tempered fractional derivative can be characterized by its Fourier transform. Recall that, for any $u \in L^2(\mathbb{R})$, its Fourier transform and inverse Fourier transform are defined by

$$ \begin{aligned} &\mathscr{F}[u](\omega) = \int_{-\infty}^{\infty} u(x) e^{-\mathrm{i}\omega x}\, dx; \\ &u(x) = \mathscr{F}^{-1}\big[\mathscr{F}[u](\omega)\big](x) = \frac{1}{2\pi} \int_{-\infty}^{\infty} \mathscr{F}[u](\omega) e^{\mathrm{i}\omega x}\, d\omega. \end{aligned} \tag{4.4.7} $$

There holds the well-known Parseval's identity,

$$ \int_{-\infty}^{\infty} u(x)\, \overline{v(x)}\, dx = \frac{1}{2\pi} \int_{-\infty}^{\infty} \mathscr{F}[u](\omega)\, \overline{\mathscr{F}[v](\omega)}\, d\omega, \tag{4.4.8} $$

where $\bar{v}$ is the complex conjugate of v. Let $H(x)$ be the Heaviside function, i.e., $H(x) = 1$ for $x \geq 0$, and vanishing for all $x < 0$. Then we can reformulate the left tempered fractional integral as

$$\begin{aligned}{}_{-\infty}I_x^{\mu,\lambda}u(x) &= \frac{1}{\Gamma(\mu)}\int_0^{\infty} y^{\mu-1}e^{-\lambda y}u(x-y)\,\mathrm{d}y \\ &= \frac{1}{\Gamma(\mu)}\int_{-\infty}^{\infty} y^{\mu-1}e^{-\lambda y}H(y)\,u(x-y)\,\mathrm{d}y \\ &= \big(K*u\big)(x), \quad \text{where} \quad K(x) := x^{\mu-1}e^{-\lambda x}H(x)/\Gamma(\mu).\end{aligned} \tag{4.4.9}$$

Note that $K(x)$ is related to the particle jump density (see Sabzikar et al. (2015, (8))). Using the formula $\mathscr{F}[K](\omega) = (\lambda + \mathrm{i}\omega)^{-\mu}$, and the convolution property of the Fourier transform, we can derive

$$\begin{aligned}\mathscr{F}[{}_{-\infty}I_x^{\mu,\lambda}u](\omega) &= \mathscr{F}[K*u](\omega) = \mathscr{F}[K](\omega)\,\mathscr{F}[u](\omega) \\ &= (\lambda+\mathrm{i}\omega)^{-\mu}\mathscr{F}[u](\omega).\end{aligned} \tag{4.4.10}$$

Similarly, the Fourier transform of the right tempered fractional integral is

$$\mathscr{F}[{}_xI_\infty^{\mu,\lambda}u](\omega) = (\lambda - \mathrm{i}\omega)^{-\mu}\mathscr{F}[u](\omega). \tag{4.4.11}$$

In view of (4.4.10)–(4.4.11), Sabzikar et al. (2015) then introduced the left and right tempered fractional derivatives as follows.

Definition 4.37 (Tempered Fractional Derivatives) For $\lambda \in \mathbb{R}_0^+$, the left and right tempered fractional derivatives of order $\mu \in \mathbb{R}^+$ of a suitable function $u(x)$, are defined by

$$\begin{aligned}\mathscr{F}\big[{}_{-\infty}D_x^{\mu,\lambda}u\big](\omega) &= (\lambda+\mathrm{i}\omega)^{\mu}\mathscr{F}[u](\omega), \\ \mathscr{F}\big[{}_xD_\infty^{\mu,\lambda}u\big](\omega) &= (\lambda-\mathrm{i}\omega)^{\mu}\mathscr{F}[u](\omega),\end{aligned} \tag{4.4.12}$$

that is, for any $x \in \mathbb{R}$,

$$\begin{aligned}{}_{-\infty}D_x^{\mu,\lambda}u(x) &= \mathscr{F}^{-1}\big[(\lambda+\mathrm{i}\omega)^{\mu}\mathscr{F}[u](\omega)\big](x), \\ {}_xD_\infty^{\mu,\lambda}u(x) &= \mathscr{F}^{-1}\big[(\lambda-\mathrm{i}\omega)^{\mu}\mathscr{F}[u](\omega)\big](x).\end{aligned} \tag{4.4.13}$$

Introduce the space

$$W_\lambda^{\mu,2}(\mathbb{R}) := \Big\{u \in L^2(\mathbb{R}) : \int_{\mathbb{R}}(\lambda^2+\omega^2)^{\mu}\,\big|\mathscr{F}[u](\omega)\big|^2\mathrm{d}\omega < \infty\Big\} \tag{4.4.14}$$

for $\mu, \lambda \in \mathbb{R}^+$. Thanks to Parseval's identity (4.4.8), the above tempered fractional derivatives are well defined for any $u \in W_\lambda^{\mu,2}(\mathbb{R})$. Moreover, one verifies from (4.4.10)–(4.4.13) that

$$\begin{aligned}&{}_{-\infty}I_x^{\mu,\lambda}\,{}_{-\infty}D_x^{\mu,\lambda}u(x) = u(x), \quad {}_xI_\infty^{\mu,\lambda}\,{}_xD_\infty^{\mu,\lambda}u(x) = u(x), \quad \forall\, u \in W_\lambda^{\mu,2}(\mathbb{R}); \\ &{}_{-\infty}D_x^{\mu,\lambda}\,{}_{-\infty}I_x^{\mu,\lambda}u(x) = u(x), \quad {}_xD_\infty^{\mu,\lambda}\,{}_xI_\infty^{\mu,\lambda}u(x) = u(x), \quad \forall\, u \in L^2(\mathbb{R}).\end{aligned} \tag{4.4.15}$$

Similar to (4.4.6), we have the following explicit representations (see Li and Deng (2016, Lemma 1 and Remark 2)).

Proposition 4.38 *For any $u \in W_\lambda^{\mu,2}(\mathbb{R})$, with $\lambda \in \mathbb{R}_0^+$, the left and right tempered fractional derivatives of order $\mu \in [k-1,k)$ with $k \in \mathbb{N}$, have the explicit representations:*

$$\begin{aligned} {}_{-\infty}D_x^{\mu,\lambda}u(x) &= e^{-\lambda x}\, {}_{-\infty}^{\;\mathrm{RL}}D_x^{\mu}\{e^{\lambda x}u(x)\}, \\ {}_xD_\infty^{\mu,\lambda}u(x) &= e^{\lambda x}\, {}_{x}^{\mathrm{RL}}D_\infty^{\mu}\{e^{-\lambda x}u(x)\}, \end{aligned} \tag{4.4.16}$$

where ${}_{-\infty}^{\;\mathrm{RL}}D_x^{\mu}$ and ${}_{x}^{\mathrm{RL}}D_\infty^{\mu}$ are the non-tempered Riemann–Liouville fractional derivative operators. Alternatively, we have

$$\begin{aligned} {}_{-\infty}D_x^{\mu,\lambda}u(x) &= (\lambda+D)^k\left\{{}_{-\infty}^{\;\mathrm{RL}}I_x^{k-\mu}u(x)\right\} \\ &= (\lambda+D)^k\left\{e^{-\lambda x}\,{}_{-\infty}^{\;\mathrm{RL}}I_x^{k-\mu}\{e^{\lambda x}u(x)\}\right\}; \\ {}_xD_\infty^{\mu,\lambda}u(x) &= (\lambda-D)^k\left\{{}_{x}^{\mathrm{RL}}I_\infty^{k-\mu}u(x)\right\} \\ &= (\lambda-D)^k\left\{e^{\lambda x}\,{}_{x}^{\mathrm{RL}}I_\infty^{k-\mu}\{e^{-\lambda x}u(x)\}\right\}. \end{aligned} \tag{4.4.17}$$

We collect below some useful properties (see Sabzikar et al. (2015)).

Lemma 4.39 *Given $\lambda > 0$ and $\mu \in [k-1,k)$, $k \in \mathbb{N}$, the tempered fractional derivative*

$$\begin{aligned} {}_{-\infty}D_x^{\mu,\lambda}u(x) &= {}_{-\infty}D_x^{k,\lambda}\,{}_{-\infty}I_x^{k-\mu,\lambda}u(x), \\ {}_xD_\infty^{\mu,\lambda}u(x) &= {}_xD_\infty^{k,\lambda}\,{}_xI_\infty^{k-\mu,\lambda}u(x). \end{aligned} \tag{4.4.18}$$

In addition, we have

$$ {}_{-\infty}I_x^{\mu+\nu,\lambda}u = {}_{-\infty}I_x^{\mu,\lambda}\,{}_{-\infty}I_x^{\nu,\lambda}u, \quad {}_xI_\infty^{\mu+\nu,\lambda}u = {}_xI_\infty^{\mu,\lambda}\,{}_xI_\infty^{\nu,\lambda}u, \tag{4.4.19}$$

$$ {}_{-\infty}D_x^{\mu+\nu,\lambda}u = {}_{-\infty}D_x^{\mu,\lambda}\,{}_{-\infty}D_x^{\nu,\lambda}u, \quad {}_xD_\infty^{\mu+\nu,\lambda}u(x) = {}_xD_\infty^{\mu,\lambda}\,{}_xD_\infty^{q\nu,\lambda}u, \tag{4.4.20}$$

$$ ({}_{-\infty}D_x^{\mu,\lambda}u, v) = (u, {}_xD_\infty^{\mu,\lambda}v), \quad ({}_xD_\infty^{\mu,\lambda}u, v) = (u, {}_{-\infty}D_x^{\mu,\lambda}v), \tag{4.4.21}$$

where $\mu, \nu \geq 0$.

Remark 4.40 For a suitable function $f(x)$, $x \in \mathbb{R}^+$, its reflection $g(y) = f(-y)$, $y \in \mathbb{R}^-$ satisfies

$$\begin{aligned} {}_{-\infty}I_y^{\mu,\lambda}g(y) &= \frac{e^{-\lambda y}}{\Gamma(\mu)}\int_{-\infty}^{y} e^{\lambda\tau}(y-\tau)^{\mu-1}f(-\tau)\mathrm{d}\tau \\ &\overset{t=-\tau}{=} \frac{e^{-\lambda y}}{\Gamma(\mu)}\int_{-y}^{\infty} e^{-\lambda t}(y+t)^{\mu-1}f(t)\mathrm{d}t \\ &\overset{x=-y}{=} \frac{e^{\lambda x}}{\Gamma(\mu)}\int_{x}^{\infty} e^{-\lambda t}(t-x)^{\mu-1}f(t)\mathrm{d}t = {}_xI_\infty^{\mu,\lambda}f(x). \end{aligned} \tag{4.4.22}$$

Hence, we can use (4.4.18) and derivative relation $\frac{\mathrm{d}^k}{\mathrm{d}y^k} = (-1)^k\frac{\mathrm{d}^k}{\mathrm{d}x^k}$ to obtain the tempered derivative relation

$$ {}_{-\infty}D_y^{\mu,\lambda}f(-y) = {}_xD_\infty^{\mu,\lambda}f(x), \quad y = -x,\ x \in \mathbb{R}^+. \tag{4.4.23}$$

Recall the properties

$$
{}^{\mathrm{RL}}_{a}I^{\rho}_{t}v(\lambda t) = \lambda^{-\rho}\,{}^{\mathrm{RL}}_{\lambda a}I^{\rho}_{x}v(x), \quad {}^{\mathrm{RL}}_{a}D^{s}_{t}v(\lambda t) = \lambda^{s}\,{}^{\mathrm{RL}}_{\lambda a}D^{s}_{x}v(x) \qquad (4.4.24)
$$

and

$$
{}^{\mathrm{RL}}_{t}I^{\rho}_{b}v(\lambda t) = \lambda^{-\rho}\,{}^{\mathrm{RL}}_{x}I^{\rho}_{\lambda b}v(x), \quad {}^{\mathrm{RL}}_{t}D^{s}_{b}v(\lambda t) = \lambda^{s}\,{}^{\mathrm{RL}}_{x}D^{s}_{\lambda b}v(x) \qquad (4.4.25)
$$

for $s, \rho, \lambda > 0$.

4.4.2 Laguerre Polynomials and Some Useful Formulas

For any $a \in \mathbb{R}$ and $j \in \mathbb{N}_0$, we recall that the rising factorial in the Pochhammer symbol and the Gamma function have the relation:

$$
(a)_0 = 1; \quad (a)_j := a(a+1)\cdots(a+j-1) = \frac{\Gamma(a+j)}{\Gamma(a)}, \quad \text{for } j \geq 1. \qquad (4.4.26)
$$

Recall the hypergeometric function (see Abramowitz and Stegun (1972)):

$$
{}_1F_1(a;b;x) = \sum_{j=0}^{\infty} \frac{(a)_j}{(b)_j}\frac{x^j}{j!}, \quad a, b, x \in \mathbb{R}^+, \quad -b \notin \mathbb{N}_0. \qquad (4.4.27)
$$

If $b - a > 0$, then ${}_1F_1(a;b;x)$ is absolutely convergent for all $x \in \mathbb{R}$. If a is a negative integer, then it reduces to a polynomial.

The Laguerre polynomial with parameter $\alpha > -1$ is defined as in Szegő (1975, (5.3.3)):

$$
L_n^{(\alpha)}(x) = \frac{(\alpha+1)_n}{n!}\,{}_1F_1\big(-n; \alpha+1; x\big), \quad n \geq 1, \quad x \in \mathbb{R}^+, \qquad (4.4.28)
$$

and $L_0^{(\alpha)}(x) \equiv 1$. Note that

$$
L_n^{(\alpha)}(0) = \frac{(\alpha+1)_n}{n!}, \qquad (4.4.29)
$$

and the Laguerre polynomials (with $\alpha > -1$) are orthogonal with respect to the weight function $x^\alpha e^{-x}$, namely,

$$
\int_0^\infty L_n^{(\alpha)}(x)\, L_m^{(\alpha)}(x)\, x^\alpha e^{-x}\,\mathrm{d}x = \gamma_n^\alpha\, \delta_{mn}, \quad \gamma_n^\alpha = \frac{\Gamma(n+\alpha+1)}{\Gamma(n+1)}. \qquad (4.4.30)
$$

They are eigenfunctions of the Sturm–Liouville problem:

$$
x^{-\alpha}e^x D\big(x^{\alpha+1}e^{-x} DL_n^{(\alpha)}(x)\big) + \lambda_n L_n^{(\alpha)}(x) = 0, \quad \lambda_n = n. \qquad (4.4.31)
$$

We have the following relations:

$$
L_n^{(\alpha)}(x) = DL_n^{(\alpha)}(x) - DL_{n+1}^{(\alpha)}(x), \qquad (4.4.32)
$$

$$xDL_n^{(\alpha)}(x) = nL_n^{(\alpha)}(x) - (n+\alpha)L_{n-1}^{(\alpha)}(x), \tag{4.4.33}$$

$$DL_n^{(\alpha)}(x) = -L_{n-1}^{(\alpha+1)}(x) = -\sum_{k=0}^{n-1} L_k^{(\alpha)}(x). \tag{4.4.34}$$

In particular, for $\alpha = -k,\ k = 1, 2, \ldots$, (see Szegő (1975, (5.2.1))),

$$L_n^{(-k)}(x) = (-1)^k \frac{\Gamma(n-k+1)}{\Gamma(n+1)} x^k L_{n-k}^{(k)}(x), \quad n \geq k.$$

For notational convenience, we denote

$$h_n^{a,b} := \frac{\Gamma(n+1+a)}{\Gamma(n+1+a-b)}. \tag{4.4.35}$$

We present below some formulas related to Laguerre polynomials and fractional integrals and derivatives, which play an important role in the algorithm development and analysis later. We provide their derivations.

Lemma 4.41 *For $\mu \in \mathbb{R}^+$, we have*

$${}^{\mathrm{RL}}_{\ 0}I^{\mu}_{x}\{x^{\alpha} L_n^{(\alpha)}(x)\} = h_n^{\alpha,-\mu}\, x^{\alpha+\mu} L_n^{(\alpha+\mu)}(x), \quad \alpha > -1; \tag{4.4.36}$$

$${}^{\mathrm{RL}}_{\ 0}D^{\mu}_{x}\{x^{\alpha} L_n^{(\alpha)}(x)\} = h_n^{\alpha,\mu}\, x^{\alpha-\mu} L_n^{(\alpha-\mu)}(x), \quad \alpha > \mu - 1, \tag{4.4.37}$$

and

$${}^{\mathrm{RL}}_{\ x}I^{\mu}_{\infty}\{e^{-x} L_n^{(\alpha)}(x)\} = e^{-x} L_n^{(\alpha-\mu)}(x), \quad \alpha > \mu - 1; \tag{4.4.38}$$

$${}^{\mathrm{RL}}_{\ x}D^{\mu}_{\infty}\{e^{-x} L_n^{(\alpha)}(x)\} = e^{-x} L_n^{(\alpha+\mu)}(x), \quad \alpha > -1. \tag{4.4.39}$$

Moreover, we have that for $k \in \mathbb{N}$ and $\alpha > k - 1$,

$$D^k\left\{x^{\alpha} e^{-x} L_n^{(\alpha)}(x)\right\} = \frac{\Gamma(n+k+1)}{\Gamma(n+1)} x^{\alpha-k} L_{n+k}^{(\alpha-k)}(x) e^{-x}. \tag{4.4.40}$$

Proof We first prove (4.4.36)–(4.4.37). Recall the fractional integral formula of hypergeometric functions (see Andrews et al. (1999, P. 287)): for real $b, \mu \geq 0$ and $x > 0$,

$$x^{b+\mu-1}\,{}_1F_1(a; b+\mu; x) = \frac{\Gamma(b+\mu)}{\Gamma(b)\Gamma(\mu)} \int_0^x (x-t)^{\mu-1} t^{b-1}\,{}_1F_1(a; b; t)\, dt. \tag{4.4.41}$$

Taking $a = -n,\ b = \alpha + 1$ and using the hypergeometric representation (4.4.28) of the Laguerre polynomials, we obtain

$$x^{\alpha+\mu} L_n^{(\alpha+\mu)}(x) = \frac{\Gamma(n+\alpha+\mu+1)}{\Gamma(n+\alpha+1)\Gamma(\mu)} \int_0^x (x-t)^{\mu-1} t^{\alpha} L_n^{(\alpha)}(t)\, dt,$$

which yields (4.4.36), i.e.,

$$ {}^{\mathrm{RL}}_{\ 0}I^{\mu}_{x}\{x^{\alpha}L^{(\alpha)}_{n}(x)\} = h^{\alpha,-\mu}_{n}\, x^{\alpha+\mu}\, L^{(\alpha+\mu)}_{n}(x). $$

Then, performing ${}^{\mathrm{RL}}_{\ 0}D^{\mu}_{x}$ on both sides and taking $\alpha+\mu\to\alpha$, we derive for $\alpha-\mu>-1$,

$$ \begin{aligned} {}^{\mathrm{RL}}_{\ 0}D^{\mu}_{x}\{x^{\alpha}L^{(\alpha)}_{n}(x)\} &= \frac{1}{h^{\alpha-\mu,-\mu}_{n}} x^{\alpha-\mu}L^{(\alpha-\mu)}_{n}(x) \\ &= \frac{\Gamma(n+\alpha+1)}{\Gamma(n+\alpha-\mu+1)} x^{\alpha-\mu}L^{(\alpha-\mu)}_{n}(x). \end{aligned} $$

This leads to (4.4.37).

We now turn to (4.4.38)–(4.4.39). According to Podlubny (1999, (6.146), P. 191), we have

$$ {}^{\mathrm{RL}}_{\ x}I^{\mu}_{\infty}\{e^{-x}L^{(\alpha+\mu)}_{n}(x)\} = e^{-x}L^{(\alpha)}_{n}(x), \quad \alpha>-1,\ \mu>0. $$

Similarly, from the property ${}^{\mathrm{RL}}_{\ x}D^{\mu}_{\infty}\ {}^{\mathrm{RL}}_{\ x}I^{\mu}_{\infty}\, u(x) = u(x)$, we derive

$$ {}^{\mathrm{RL}}_{\ x}D^{\mu}_{\infty}\{e^{-x}L^{(\alpha)}_{n}(x)\} = e^{-x}L^{(\alpha+\mu)}_{n}(x). $$

Finally, we prove (4.4.40). Noting that

$$ {}_1F_1(a;c;x) = e^{x}\ {}_1F_1(c-a;c;-x) \tag{4.4.42} $$

(see Andrews et al. (1999, P. 191)), we derive from (4.4.28) and (4.4.42) that

$$ \begin{aligned} x^{\alpha}L^{(\alpha)}_{n}(x)e^{-x} &= x^{\alpha}\frac{(\alpha+1)_n}{n!}\ {}_1F_1\big(-n;\alpha+1;x\big)e^{-x} \\ &= \frac{(\alpha+1)_n}{n!}x^{\alpha}\ {}_1F_1\big(n+\alpha+1;\alpha+1;-x\big) \\ &= \frac{(\alpha+1)_n}{n!}x^{\alpha}\ {}_1F_1\big(n+\alpha+1;\alpha+1;-x\big) \\ &= \frac{(\alpha+1)_n}{n!}\sum_{j=0}^{\infty}\frac{(n+\alpha+1)_j(-1)^j}{(\alpha+1)_j}\frac{x^{j+\alpha}}{j!}. \end{aligned} \tag{4.4.43} $$

Then acting the derivative D^k on (4.4.43) and using the identities (4.4.28), (4.4.42) again, we obtain

$$ \begin{aligned} D^{k}\big\{x^{\alpha}L^{(\alpha)}_{n}(x)e^{-x}\big\} &= \frac{(\alpha+1)_n}{n!}\sum_{j=0}^{\infty}\frac{(n+\alpha+1)_j(-1)^j}{(\alpha+1)_j}\frac{D^k x^{j+\alpha}}{j!} \\ &= \frac{(\alpha+1)_n}{n!}\sum_{j=0}^{\infty}\frac{(n+\alpha+1)_j(-1)^j}{(\alpha+1)_j}\frac{\Gamma(j+\alpha+1)}{\Gamma(j+\alpha-k+1)}\frac{x^{j+\alpha-k}}{j!} \end{aligned} $$

$$
\begin{aligned}
&= x^{\alpha-k}\frac{(\alpha+1)_n}{n!}\frac{\Gamma(\alpha+1)}{\Gamma(\alpha-k+1)}\sum_{j=0}^{\infty}\frac{(n+\alpha+1)_j}{(\alpha-k+1)_j}\frac{(-x)^j}{j!}\\
&= x^{\alpha-k}\frac{(\alpha-k+1)_{n+k}}{n!}\,{}_1F_1\big(n+\alpha+1;\alpha-k+1;-x\big)\\
&= \frac{(n+k)!}{n!}x^{\alpha-k}\frac{(\alpha-k+1)_{n+k}}{(n+k)!}\,{}_1F_1\big(-n-k;\alpha-k+1;x\big)e^{-x}\\
&= \frac{\Gamma(n+k+1)}{\Gamma(n+1)}x^{\alpha-k}L_{n+k}^{(\alpha-k)}(x)e^{-x}.
\end{aligned}
$$

This completes the proof. □

4.4.3 Generalized Laguerre Functions

We introduce the generalized Laguerre functions (GLFs), and study their approximation properties. In what follows, the operators ${}_0I_x^{\mu,\lambda}, {}_0D_x^{\mu,\lambda}$ on the half line should be understood as 0 in place of $-\infty$ in (4.4.3) and (4.4.16)–(4.4.17).

Definition and Properties

We first introduce the GLFs and their associated properties related to tempered fractional integrals/derivatives.

Definition 4.42 (GLFs) For real $\alpha \in \mathbb{R}$ and $\lambda > 0$, we define the GLFs as

$$
\mathcal{L}_n^{(\alpha,\lambda)}(x) := \begin{cases} x^{-\alpha}e^{-\lambda x}\,L_n^{(-\alpha)}(2\lambda x), & \alpha < 0,\\ e^{-\lambda x}\,L_n^{(\alpha)}(2\lambda x), & \alpha \ge 0, \end{cases} \tag{4.4.44}
$$

for all $x \in \mathbb{R}^+$ and $n \in \mathbb{N}_0$.

Remark 4.43 It is noteworthy that Zhang and Guo (2012) introduced the GLFs

$$
\tilde{\mathscr{L}}_l^{(\alpha,\beta)}(x) = \begin{cases} x^{-\alpha}e^{-\frac{\beta}{2}x}L_l^{(-\alpha)}(\beta x), & \alpha \le -1,\ \ l \ge \bar{l}_\alpha = [-\alpha],\\ e^{-\frac{\beta}{2}x}L_l^{(\alpha)}(x), & \alpha > -1,\ \ l \ge \bar{l}_\alpha = 0, \end{cases} \tag{4.4.45}
$$

where the scaling factor $\beta > 0$. It is seen that we modified the definition in the range of $0 < \alpha < 1$ (with $\beta = 2\lambda$). This turns out to be essential for the numerical solution of FDEs of order $\mu \in (0, 1)$, as we shall see in the subsequent sections.

We next present the basic properties of GLFs. Firstly, one verifies readily from the orthogonality (4.4.30) and Definition 4.42 that for $\alpha \in \mathbb{R}$ and $\lambda > 0$,

$$\int_0^\infty \mathcal{L}_n^{(\alpha,\lambda)}(x)\mathcal{L}_m^{(\alpha,\lambda)}(x)\, x^\alpha \mathrm{d}x = \gamma_n^{|\alpha|,\lambda}\delta_{nm}, \quad \gamma_n^{|\alpha|,\lambda} = \frac{\gamma_n^{|\alpha|}}{(2\lambda)^{|\alpha|+1}}, \tag{4.4.46}$$

where $\gamma_n^{|\alpha|}$ is defined in (4.4.30).

We have the following important (left) "tempered" fractional integral and derivative rules.

Lemma 4.44 *For $\mu, \nu, \lambda, x \in \mathbb{R}_0^+$, we have*

$${}_0I_x^{\mu,\lambda}\mathcal{L}_n^{(-\nu,\lambda)}(x) = h_n^{\nu,-\mu}\mathcal{L}_n^{(-\nu-\mu,\lambda)}(x), \tag{4.4.47}$$

$${}_0D_x^{\mu,\lambda}\mathcal{L}_n^{(-\nu,\lambda)}(x) = h_n^{\nu,\mu}\mathcal{L}_n^{(\mu-\nu,\lambda)}(x), \quad \nu \ge \mu, \tag{4.4.48}$$

and

$${}_0D_x^{\mu+k,\lambda}\mathcal{L}_n^{(-\mu,\lambda)}(x) = (-2\lambda)^k h_n^{\mu,\mu}\mathcal{L}_{n-k}^{(k,\lambda)}(x), \quad n \ge k \in \mathbb{N}_0, \tag{4.4.49}$$

where $h_n^{a,b}$ is defined in (4.4.35).

Proof We obtain from (4.4.3) and (4.4.16)–(4.4.17) (with replacing $-\infty$ by 0) that

$${}_0I_x^{\mu,\lambda}\mathcal{L}_n^{(-\nu,\lambda)}(x) = e^{-\lambda x}\,{}_0^{\mathrm{RL}}I_x^{\mu}\{e^{\lambda x}\mathcal{L}_n^{(-\nu,\lambda)}(x)\} = e^{-\lambda x}\,{}_0^{\mathrm{RL}}I_x^{\mu}\{x^\nu L_n^{(\nu)}(2\lambda x)\},$$

and

$${}_0D_x^{\mu,\lambda}\mathcal{L}_n^{(-\nu,\lambda)}(x) = e^{-\lambda x}\,{}_0^{\mathrm{RL}}D_x^{\mu}\{e^{\lambda x}\mathcal{L}_n^{(-\nu,\lambda)}(x)\} = e^{-\lambda x}\,{}_0^{\mathrm{RL}}D_x^{\mu}\{x^\nu L_n^{(\nu)}(2\lambda x)\}.$$

Thus using (4.4.24) and Lemma 4.41, we obtain (4.4.47)–(4.4.48).

Using (4.4.48) and the derivative relation (4.4.34) (with $\alpha = \mu$), we obtain

$$\begin{aligned}{}_0D_x^{\mu+k,\lambda}\mathcal{L}_n^{(-\mu,\lambda)}(x) &= {}_0D_x^{k,\lambda}\,{}_0D_x^{\mu,\lambda}\mathcal{L}_n^{(-\mu,\lambda)}(x) = e^{-\lambda x}D^k\big\{e^{\lambda x}h_n^{\mu,\mu}\mathcal{L}_n^{(0,\lambda)}(x)\big\}\\ &= h_n^{\mu,\mu}\,e^{-\lambda x}D^k\big\{L_n^{(0)}(2\lambda x)\big\} = (-2\lambda)^k h_n^{\mu,\mu}\,L_{n-k}^{(k)}(2\lambda x)\,e^{-\lambda x}.\end{aligned}$$

This leads to (4.4.49). □

Similarly, we have the following rules of the (right) "tempered" fractional integrals and derivatives.

Lemma 4.45 *For $\mu, \nu, \lambda, x \in \mathbb{R}_0^+$, we have*

$${}_xI_\infty^{\mu,\lambda}\mathcal{L}_n^{(\nu,\lambda)}(x) = (2\lambda)^{-\mu}\mathcal{L}_n^{(\nu-\mu,\lambda)}(x), \quad \nu \ge \mu, \tag{4.4.50}$$

$${}_xD_\infty^{\mu,\lambda}\mathcal{L}_n^{(\nu,\lambda)}(x) = (2\lambda)^{\mu}\mathcal{L}_n^{(\mu+\nu,\lambda)}(x). \tag{4.4.51}$$

Proof Identities (4.4.50) and (4.4.51) can be easily derived from (4.4.24), (4.4.25) and Lemma 4.41. □

We highlight the fractional derivative formulas, which play an important role in the forthcoming algorithm and analysis.

Theorem 4.46 *Let $k \in \mathbb{N}$ and $k - \nu \le 0$,*

$$ {}_{-\infty}D_x^{k,\lambda}\{\mathcal{L}_n^{(-\nu,\lambda)}(x)\} = \frac{\Gamma(n+\nu+1)}{\Gamma(n+\nu-k+1)}\,\mathcal{L}_n^{(k-\nu,\lambda)}(x), \tag{4.4.52}$$

and

$$ {}_xD_\infty^{k,\lambda}\{\mathcal{L}_n^{(-\nu,\lambda)}(x)\} = (-1)^k\frac{\Gamma(n+k+1)}{\Gamma(n+1)}\mathcal{L}_{n+k}^{(k-\nu,\lambda)}(x). \tag{4.4.53}$$

Proof From Lemma 4.41 and relations

$$ {}_{-\infty}D_x^{k,\lambda}u = e^{-\lambda x}D^k\{e^{\lambda x}u\}, \quad {}_xD_\infty^{k,\lambda}u = e^{\lambda x}(-1)^kD^k\{e^{-\lambda x}u\}, \tag{4.4.54}$$

we obtain that for $k - \nu \le 0$,

$$\begin{aligned} {}_{-\infty}D_x^{k,\lambda}\{x^\nu\mathcal{L}_n^{(\nu,\lambda)}(x)\} &= e^{-\lambda x}D^k\{(2\lambda)^{-\nu}(2\lambda x)^\nu L_n^{(\nu)}(2\lambda x)\} \\ &\overset{(4.4.37)}{=} \frac{\Gamma(n+1+\nu)}{\Gamma(n+\nu-k+1)}x^{\nu-k}L_n^{(\nu-k)}(2\lambda x)e^{-\lambda x} \\ &= \frac{\Gamma(n+1+\nu)}{\Gamma(n+\nu-k+1)}\,\mathcal{L}_n^{(k-\nu,\lambda)}(x), \end{aligned}$$

and

$$\begin{aligned} {}_xD_\infty^{k,\lambda}\{x^\nu\mathcal{L}_n^{(\nu,\lambda)}(x)\} &= e^{\lambda x}(-1)^kD^k\{(2\lambda)^{-\nu}(2\lambda x)^\nu L_n^{(\nu)}(2\lambda x)e^{-2\lambda x}\} \\ &\overset{(4.4.40)}{=} (-1)^k\frac{\Gamma(n+k+1)}{\Gamma(n+1)}x^{\nu-k}L_{n+k}^{(\nu-k)}(2\lambda x)e^{-\lambda x} \\ &= (-1)^k\frac{\Gamma(n+k+1)}{\Gamma(n+1)}\,\mathcal{L}_n^{(k-\nu,\lambda)}(x). \end{aligned} \tag{4.4.55}$$

This ends the proof. □

Another attractive property of GLFs is that they are eigenfunctions of Sturm–Liouville problem.

Theorem 4.47 *Let $s, \nu, x \in \mathbb{R}_0^+$ and $n \in \mathbb{N}_0$. Then,*

$$ x^\nu{}_xD_\infty^{s,\lambda}\{x^{s-\nu}{}_0D_x^{s,\lambda}\,\mathcal{L}_n^{(-\nu,\lambda)}(x)\} = \lambda_{n,-}^{s,\nu}\,\mathcal{L}_n^{(-\nu,\lambda)}(x), \quad \nu - s \ge 0, \tag{4.4.56}$$

$$ x^{-\nu}{}_0D_x^{s,\lambda}\{x^{s+\nu}{}_xD_\infty^{s,\lambda}\,\mathcal{L}_n^{(\nu,\lambda)}(x)\} = \lambda_{n,+}^{s,\nu}\,\mathcal{L}_n^{(\nu,\lambda)}(x), \tag{4.4.57}$$

where the corresponding eigenvalues $\lambda_{n,-}^{s,\nu} = (2\lambda)^s h_n^{\nu,s}$ and $\lambda_{n,+}^{s,\nu} = (2\lambda)^s h_n^{\nu+s,s}$.

Proof Due to (4.4.48) and (4.4.51),

$$ {}_0D_x^{s,\lambda}\mathcal{L}_n^{(-\nu,\lambda)}(x) = h_n^{\nu,s}\mathcal{L}_n^{(s-\nu,\lambda)}(x), \quad {}_xD_\infty^{s,\lambda}\mathcal{L}_n^{(\nu-s,\lambda)}(x) = (2\lambda)^s\mathcal{L}_n^{(\nu,\lambda)}(x). $$

It is straightforward to obtain that

$$ \begin{aligned} x^\nu{}_xD_\infty^{s,\lambda}\{x^{s-\nu}{}_0D_x^{s,\lambda}\,\mathcal{L}_n^{(-\nu,\lambda)}(x)\} &= h_n^{\nu,s}x^\nu{}_xD_\infty^{s,\lambda}\{x^{s-\nu}\mathcal{L}_n^{(s-\nu,\lambda)}(x)\} \\ &= h_n^{\nu,s}x^\nu{}_xD_\infty^{s,\lambda}\mathcal{L}_n^{(\nu-s,\lambda)}(x) = (2\lambda)^s h_n^{\nu,s}\,\mathcal{L}_n^{(-\nu,\lambda)}(x). \end{aligned} $$

Similarly, we have

$$ \begin{aligned} x^{-\nu}{}_0D_x^{s,\lambda}\{x^{s+\nu}{}_xD_\infty^{s,\lambda}\,\mathcal{L}_n^{(\nu,\lambda)}(x)\} &= (2\lambda)^s x^{-\nu}{}_0D_x^{s,\lambda}\{x^{s+\nu}\mathcal{L}_n^{(s+\nu,\lambda)}(x)\} \\ &= (2\lambda)^s x^{-\nu}{}_0D_x^{s,\lambda}\mathcal{L}_n^{(-\nu-s,\lambda)}(x) = (2\lambda)^s h_n^{\nu+s,s}\,\mathcal{L}_n^{(\nu,\lambda)}(x). \end{aligned} $$

This ends the derivation. □

Remark 4.48 The above identities can be viewed as an extension of the standard Sturm–Liouville problem of generalized Laguerre functions (cf. (4.4.31)) to the tempered fractional derivative. We derive immediately from (4.4.56), (4.4.57) and Stirling's formula (see (4.4.66)) that for fixed s and ν,

$$ \lambda_{n,-}^{s,\nu} = \lambda_{n,+}^{s,\nu} = O\big((2\lambda n)^s\big), \quad n \gg 1. $$

When $s \to 1$ and $\lambda = 1/2$, it recovers the $O(n)$ growth of eigenvalues of the standard Sturm–Liouville problem.

4.4.4 Approximation by $\big\{\mathcal{L}_n^{(-\nu,\lambda)} : \nu > 0\big\}_{n=0}^{\infty}$

Denote by $\mathbb{P}_N$ the set of all polynomials of degree at most N, and define the finite-dimensional space

$$ \mathcal{F}_N^{\nu,\lambda}(\mathbb{R}^+) := \big\{x^\nu e^{-\lambda x}p(x) : p \in \mathbb{P}_N\big\}, \quad N \in \mathbb{N}_0. \tag{4.4.58} $$

Define the $L^2_\omega(\mathbb{R}^+)$ with the inner product and norm:

$$ (f,g)_\omega := \int_{\mathbb{R}^+} f\,\bar{g}\,\omega\,\mathrm{d}x, \quad \|f\|_\omega^2 = (f,f)_\omega, \tag{4.4.59} $$

where $\omega(x)$ is a generic weight function and $\bar{g}$ is the conjugate of the function g. In particular, we omit ω when $\omega \equiv 1$.

To characterize the approximation errors, we define the non-uniformly weighted Sobolev space

$$ A_{\nu,\lambda}^m(\mathbb{R}^+) := \Big\{u \in L^2_{\omega^{-\nu}}(\mathbb{R}^+) : {}_0D_x^{\nu+k,\lambda}u \in L^2_{\omega^k}(\mathbb{R}^+),\ k = 0,\dots,m\Big\} \tag{4.4.60} $$

for $m \in \mathbb{N}_0$, equipped with the norm and semi-norm

$$\|u\|_{A^m_{\nu,\lambda}} := \Big(\|u\|^2_{\omega^{-\nu}} + \sum_{k=0}^{m} \|{}_0D_x^{\nu+k,\lambda} u\|^2_{\omega^k}\Big)^{1/2}, \quad |u|_{A^m_{\nu,\lambda}} := \|{}_0D_x^{\nu+m,\lambda} u\|_{\omega^m}, \tag{4.4.61}$$

where the weight function $\omega^a(x) = x^a$.

Consider the orthogonal projection $\pi_N^{-\nu,\lambda}:\ L^2_{\omega^{-\nu}}(\mathbb{R}^+) \to \mathcal{F}_N^{\nu,\lambda}(\mathbb{R}^+)$ defined by

$$(\pi_N^{-\nu,\lambda} u - u,\ \phi)_{\omega^{-\nu}} = 0, \quad \forall \phi \in \mathcal{F}_N^{\nu,\lambda}(\mathbb{R}^+). \tag{4.4.62}$$

Then, by the orthogonality (4.4.46), u and its L^2-orthogonal projection can be expanded as

$$u(x) = \sum_{n=0}^{\infty} \hat{u}_n \mathcal{L}_n^{(-\nu,\lambda)}(x), \quad (\pi_N^{-\nu,\lambda} u)(x) = \sum_{n=0}^{N} \hat{u}_n \mathcal{L}_n^{(-\nu,\lambda)}(x), \tag{4.4.63}$$

where

$$\hat{u}_n = \big(u,\ \mathcal{L}_n^{(-\nu,\lambda)}\big)_{\omega^{-\nu}} / \gamma_n^{\nu,\lambda}.$$

Theorem 4.49 *For $\lambda, \nu > 0$, we have that for any $u \in A^m_{\nu,\lambda}(\mathbb{R}^+)$ with $m \le N+1$,*

$$\|\pi_N^{-\nu,\lambda} u - u\|_{\omega^{-\nu}} \le c\,(2\lambda N)^{-\frac{\nu+m}{2}}\ \|{}_0D_x^{\nu+m,\lambda} u\|_{\omega^m}, \tag{4.4.64}$$

and for any $k \le m$,

$$\big\|{}_0D_x^{\nu+k,\lambda}\big(\pi_N^{-\nu,\lambda} u - u\big)\big\|_{\omega^k} \le c\,(2\lambda N)^{\frac{k-m}{2}}\ \|{}_0D_x^{\nu+m,\lambda} u\|_{\omega^m}, \tag{4.4.65}$$

where $c \approx 1$ for large N.

Proof By (4.4.63), we have

$$(u - \pi_N^{-\nu,\lambda} u)(x) = \sum_{n=N+1}^{\infty} \hat{u}_n\, \mathcal{L}_n^{(-\nu,\lambda)}(x).$$

By the orthogonality (4.4.46) and (4.4.49),

$$\begin{aligned}\big\|{}_0D_x^{\nu+k,\lambda} \mathcal{L}_n^{(-\nu,\lambda)}\big\|^2_{\omega^k} &= (-2\lambda)^{2k}\,(h_n^{\nu,\nu})^2 \int_0^\infty \big(L_{n-k}^{(k)}(2\lambda x)\big)^2 e^{-2\lambda x}\,\omega^k(x)dx\\ &= (d_{n,k}^{\nu,\lambda})^2 \gamma_{n-k}^{k,\lambda},\end{aligned}$$

where we denoted $d_{n,k}^{\nu,\lambda} := (2\lambda)^k\, h_n^{\nu,\nu}$ and used the fact:

$$\int_0^\infty L_{n-m}^{(m)}(2\lambda x) L_{n-k}^{(k)}(2\lambda x)\, e^{-2\lambda x}\,\omega^k dx = \frac{\gamma_{n-k}^k}{(2\lambda)^{k+1}}\delta_{km} = \gamma_{n-k}^{k,\lambda}\delta_{km}.$$

Thus we can obtain

$$\|\pi_N^{-\nu,\lambda}u - u\|^2_{\omega^{-\nu}} = \sum_{n=N+1}^{\infty} (\hat{u}_n)^2 \gamma_n^{\nu,\lambda},$$

$$\left|\pi_N^{-\nu,\lambda}u - u\right|^2_{A^k_{\nu,\lambda}} = \sum_{n=N+1}^{\infty} (\hat{u}_n d^{\nu,\lambda}_{n,k})^2 \gamma^{k,\lambda}_{n-k},$$

$$\left|u\right|^2_{A^m_{\nu,\lambda}} = \sum_{n=m}^{\infty} (\hat{u}_n d^{\nu,\lambda}_{n,m})^2 \gamma^{m,\lambda}_{n-m}.$$

Then one verifies readily that

$$\|\pi_N^{-\nu,\lambda}u - u\|^2_{\omega^{-\nu}} \leq \frac{\gamma^{\nu,\lambda}_{N+1}}{(d^{\nu,\lambda}_{N+1,m})^2 \gamma^{m,\lambda}_{N+1-m}} \left|u\right|^2_{A^m_{\nu,\lambda}},$$

$$\left|\pi_N^{-\nu,\lambda}u - u\right|^2_{A^k_{\nu,\lambda}} \leq \left(\frac{d^{\nu,\lambda}_{N+1,k}}{d^{\nu,\lambda}_{N+1,m}}\right)^2 \frac{\gamma^{k,\lambda}_{N+1-k}}{\gamma^{m,\lambda}_{N+1-m}} \left|u\right|^2_{A^m_{\nu,\lambda}}.$$

Recall the property of the Gamma function (see Abramowitz and Stegun (1972, (6.1.38))):

$$\Gamma(x+1) = \sqrt{2\pi} x^{x+1/2} \exp\left(-x + \frac{\theta}{12x}\right), \quad \forall x > 0, \ 0 < \theta < 1. \tag{4.4.66}$$

One can then obtain that for any constants a, b, and for $n \geq 1$, $n + a > 1$ and $n + b > 1$,

$$\frac{\Gamma(n+a)}{\Gamma(n+b)} \leq \nu_n^{a,b} n^{a-b}, \tag{4.4.67}$$

where

$$\nu_n^{a,b} = \exp\left(\frac{a-b}{2(n+b-1)} + \frac{1}{12(n+a-1)} + \frac{(a-b)^2}{n}\right). \tag{4.4.68}$$

Therefore,

$$\begin{aligned}
\frac{\gamma^{\nu,\lambda}_{N+1}}{(d^{\nu,\lambda}_{N+1,m})^2 \gamma^{m,\lambda}_{N+1-m}} &= \frac{\Gamma(N+2-m)}{(2\lambda)^{\nu+m}\Gamma(N+2+\nu)} \\
&\leq (2\lambda)^{-\nu-m} \nu_n^{2-m,2+\nu} N^{-\nu-m}, \\
\left(\frac{d^{\nu,\lambda}_{N+1,k}}{d^{\nu,\lambda}_{N+1,m}}\right)^2 \frac{\gamma^{k,\lambda}_{N+1-k}}{\gamma^{m,\lambda}_{N+1-m}} &= \frac{(2\lambda)^k \Gamma(N+2-m)}{(2\lambda)^m \Gamma(N+2-k)} \\
&\leq (2\lambda)^{k-m} \nu_n^{2-m,2-k} N^{k-m},
\end{aligned} \tag{4.4.69}$$

where $\nu_n^{2-m,2+\nu} \approx 1$ and $\nu_n^{2-m,2-k} \approx 1$ for fixed m and $n \geq N \gg 1$. Then (4.4.64)–(4.4.65) follow. □

Approximation by $\{\mathcal{L}_n^{(\nu,\lambda)}(x) : \nu \geq 0\}_{n=0}^{\infty}$

Introduce the non-uniformly weighted Sobolev space:

$$B^r_{\nu,\lambda}(\mathbb{R}^+) := \left\{u \in L^2_{\omega^\nu}(\mathbb{R}^+) : {}_x D^{s,\lambda}_{\infty} u \in L^2_{\omega^{\nu+s}}(\mathbb{R}^+),\ 0 \leq s \leq r\right\}, \quad (4.4.70)$$

endowed with the norm and semi-norm

$$\|u\|_{B^r_{\nu,\lambda}} := \left(\|u\|^2_{\omega^\nu} + |u|^2_{B^r_{\nu,\lambda}}\right)^{1/2}, \qquad |u|_{B^r_\nu} := \|{}_x D^{\nu+r,\lambda}_{\infty} u\|_{\omega^{\nu+r}}. \quad (4.4.71)$$

Consider the orthogonal projection $\Pi_N^{\nu,\lambda}$: $L^2_{\omega^\nu}(\mathbb{R}^+) \to \mathcal{F}_N^{0,\lambda}(\mathbb{R}^+)$, defined by

$$\left(\Pi_N^{\nu,\lambda} u - u,\ \phi\right)_{\omega^\nu} = 0, \quad \forall \phi \in \mathcal{F}_N^{0,\lambda}(\mathbb{R}^+), \quad \nu > -1. \quad (4.4.72)$$

Theorem 4.50 *Let $\lambda, r, \nu > 0$. For any $u \in B^r_{\nu,\lambda}(\mathbb{R}^+)$ with $0 \leq s \leq r \leq N$, we have*

$$\left\|{}_x D^{s,\lambda}_{\infty}\{\Pi_N^{\nu,\lambda} u - u\}\right\|_{\omega^{\nu+s}} \leq c\,(2\lambda N)^{\frac{s-r}{2}}\ \|{}_x D^{r,\lambda}_{\infty} u\|_{\omega^{\nu+r}}, \quad (4.4.73)$$

where $c \approx 1$ for large N.

Proof Note that by definition,

$$u - \Pi_N^{\nu,\lambda} u = \sum_{n=N+1}^{\infty} \hat{u}_n \mathcal{L}_n^{(\nu,\lambda)}(x), \quad \hat{u}_n = \left(u,\ \mathcal{L}_n^{(\nu,\lambda)}\right)_{\omega^\nu} / \gamma_n^{\nu,\lambda}.$$

Then by (4.4.51), and the orthogonality

$$\|{}_x D^{s,\lambda}_{\infty} \mathcal{L}_n^{(\nu,\lambda)}\|^2_{\omega^{\nu+s}} = \|(2\lambda)^s \mathcal{L}_n^{(\nu+s,\lambda)}\|^2_{\omega^{\nu+s}} = (2\lambda)^{2s} \gamma_n^{\nu+s,\lambda},$$

we can derive

$$\begin{aligned}
\left|\Pi_N^{\nu,\lambda} u - u\right|^2_{B^s_{\nu,\lambda}} &= \left\| \sum_{n=N+1}^{\infty} \hat{u}_n (2\lambda)^s \mathcal{L}_n^{(\nu+s,\lambda)} \right\|^2_{\omega^{\nu+s}} \\
&= \sum_{n=N+1}^{\infty} (\hat{u}_n)^2 (2\lambda)^{2s} \gamma_n^{\nu+s,\lambda}, \\
|u|^2_{B^r_{\nu+r,\lambda}} &= \left\| \sum_{n=0}^{\infty} \hat{u}_n (2\lambda)^r \mathcal{L}_n^{(\nu+r,\lambda)} \right\|^2_{\omega^{\nu+r}} = \sum_{n=0}^{\infty} (\hat{u}_n)^2 (2\lambda)^{2r} \gamma_n^{\nu+r,\lambda}.
\end{aligned}$$

Then,

$$\begin{aligned}\left|\Pi_N^{\nu,\lambda}u-u\right|^2_{B^s_{\nu,\lambda}} &= \sum_{n=N+1}^{\infty}\hat{u}_n^2(2\lambda)^{2s}\gamma_n^{\nu+s,\lambda}\\ &\le (2\lambda)^{2s-2r}\frac{\gamma_{N+1}^{\nu+s,\lambda}}{\gamma_{N+1}^{\nu+r,\lambda}}\sum_{n=N+1}^{\infty}\hat{u}_n^2(2\lambda)^{2r}\gamma_n^{\nu+r,\lambda},\end{aligned}$$

where by (4.4.67)–(4.4.68) and an argument similar to (4.4.69), we obtain

$$\frac{\gamma_{N+1}^{\nu+s,\lambda}}{\gamma_{N+1}^{\nu+r,\lambda}}=\frac{(2\lambda)^r\Gamma(N+\nu+s+2)}{(2\lambda)^s\Gamma(N+\nu+r+2)}\le c(2\lambda)^{r-s}N^{s-r}.$$

Consequently, we have

$$\|{}_xD_{\infty}^{s,\lambda}\{\Pi_N^{\nu,\lambda}u-u\}\|_{\omega^{\nu+s}}\le\ c\,(2\lambda N)^{\frac{s-r}{2}}\,|u|_{B^r_{\nu,\lambda}}.$$

This ends the proof. □

A Model Problem and Numerical Results

In what follows, we consider the GLF approximation to a model tempered fractional equation of order $s\in[k-1,k)$ with $k\in\mathbb{N}$:

$$ {}_0D_x^{s,\lambda}u(x)=f(x),\ \ x\in\mathbb{R}^+,\ \ \lambda>0;\ \ u^{(j)}(0)=0,\quad j=0,1,\ldots,k-1, \tag{4.4.74}$$

where $f\in L^2(\mathbb{R}^+)$ is a given function. Using the fractional derivative relation, one can find

$$u(x)={}_0I_x^{s,\lambda}f(x)+\sum_{i=1}^{k}c_i\,x^{s-i}e^{-\lambda x},$$

where $\{c_i\}$ can be determined by the conditions at $x=0$. In fact, we have all $c_i=0$, and

$$\begin{aligned}u(x)={}_0I_x^{s,\lambda}f(x)&=\frac{e^{-\lambda x}}{\Gamma(s)}\int_0^x(x-\tau)^{s-1}e^{\lambda\tau}f(\tau)\mathrm{d}\tau\\ &=\frac{x^s}{\Gamma(s)}\int_0^1(1-t)^{s-1}e^{-\lambda(1-t)x}f(xt)\mathrm{d}t.\end{aligned}\tag{4.4.75}$$

We see that if $f(x)$ is smooth, then $u(x)=x^sF(x)$, where $F(x)$ is smoother than $f(x)$. With this understanding, we construct the GLF Petrov–Galerkin approximation as: find $u_N\in\mathcal{F}_N^{s,\lambda}(\mathbb{R}^+)$ (defined in (4.4.58)) such that

$$({}_0D_x^{s,\lambda}u_N,v_N)=(f,v_N),\quad\forall\,v_N\in\mathcal{F}_N^{0,\lambda}(\mathbb{R}^+).\tag{4.4.76}$$

We expand f and u_N as

$$f(x) = \sum_{n=0}^{\infty} \hat{f}_n \mathcal{L}_n^{(0,\lambda)}(x), \quad u_N(x) = \sum_{n=0}^{N} \hat{u}_n \mathcal{L}_n^{(s,\lambda)}(x). \tag{4.4.77}$$

Using the derivative relation (4.4.48), we find immediately that $\hat{u}_n = \hat{f}_n / h_n^{s,s}$ for $n = 0, 1, \ldots, N$, which also implies ${}_0D_x^{s,\lambda} u_N = \pi_N^{0,\lambda} f$.

Moreover, we can show that the numerical solution u_N is precisely the orthogonal projection in the following sense:

$$(u_N - u\,,\, w_N)_{\omega^{-s}} = 0, \quad \forall\, w_N \in \mathcal{F}_N^{s,\lambda}(\mathbb{R}^+). \tag{4.4.78}$$

To this end, we first show

$$(u_N - u\,,\, {}_xD_\infty^{s,\lambda} v_N) = ({}_0D_x^{s,\lambda} u_N - {}_0D_x^{s,\lambda} u, v_N) = 0, \quad \forall\, v_N \in \mathcal{F}_N^{0,\lambda}(\mathbb{R}^+). \tag{4.4.79}$$

Indeed, thanks to $u^{(j)}(0) = 0$ for $j = 0, \ldots, k-1$, we have

$${}_0D_x^{s,\lambda}\{u_N - u\} = e^{-\lambda x}\, {}^{\mathrm{RL}}_{\ 0}D_x^s \{e^{\lambda x}(u_N - u)\} = e^{-\lambda x}\, {}_0I_x^{k-s}\, {}^{\mathrm{RL}}_{\ 0}D_x^k \{e^{\lambda x}(u_N - u)\}.$$

Then,

$$\begin{aligned}\left({}_0D_x^{s,\lambda} u_N - {}_0D_x^{s,\lambda} u\,,\, v_N\right) &= \left({}_0I_x^{k-s}\, {}^{\mathrm{RL}}_{\ 0}D_x^k \{e^{\lambda x}(u_N - u)\}, e^{-\lambda x} v_N\right)\\ &= \left(e^{\lambda x}(u_N - u)\,,\, (-1)^k \mathrm{D}^k{}_x I_\infty^{k-s,\lambda} v_N\right) = \left(u_N - u\,,\, {}_xD_\infty^{s,\lambda} v_N\right),\end{aligned}$$

so (4.4.79) is valid. In addition, thanks to Lemma 4.41 and (4.4.25), we have

$$\begin{aligned}{}_xD_\infty^{s,\lambda} \mathcal{L}_n^{(0,\lambda)}(x) &= e^{\lambda x}\, {}^{\mathrm{RL}}_{\ x}D_\infty^s \left\{e^{-2\lambda x} L_n^{(0)}(2\lambda x)\right\}\\ &= (2\lambda)^s e^{-\lambda x} L_n^{(s)}(2\lambda x) = (2\lambda)^s x^{-s} \mathcal{L}_n^{(-s,\lambda)}(x).\end{aligned}$$

Hence, (4.4.78) is valid.

Thanks to (4.4.78), we derive from Theorem 4.49 the following estimate where the convergence rate only depends on the regularity of the source term.

Theorem 4.51 *Let u and u_N be respectively the solutions of* (4.4.74) *and* (4.4.76). *Then for ${}_0D_x^{m,\lambda} f \in L^2_{\omega^m}(I)$ with $m \in \mathbb{N}_0$, we have*

$$\|u - u_N\|_{\omega^{-s}} \le c\,(2\lambda N)^{-\frac{s+m}{2}} \|{}_0D_x^{s+m,\lambda} u\|_{\omega^m} = c\,(2\lambda N)^{-\frac{s+m}{2}} \|{}_0D_x^{m,\lambda} f\|_{\omega^m},$$

where $c \approx 1$ for large N.

We provide some numerical results to illustrate the convergence behavior. We take $f(x) = e^{-x}\sin x$ and then evaluate the exact solution by (4.4.75). Note that as ${}_0D_x^{m,\lambda} f = e^{-\lambda x} D^m\{e^{\lambda x} f\}$, a direct calculation leads to

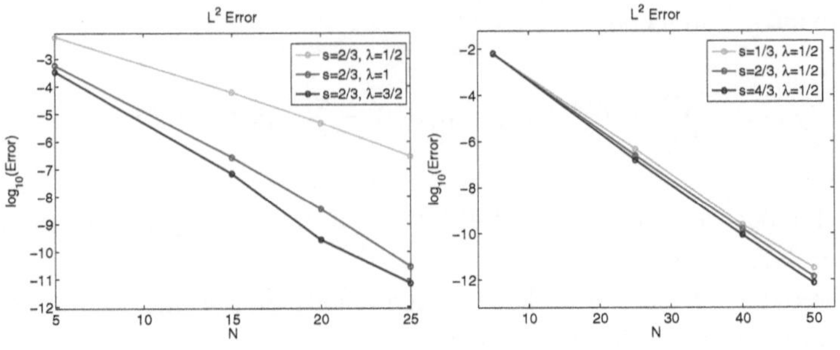

Figure 4.18 Convergence of the GLF approximation to (4.4.74) with $f(x) = e^{-x}\sin x$. Source: Chen et al. (2018).

$$e^{-\lambda x} D^m \{e^{\lambda x} f\} = e^{-\lambda x} \sum_{k=0}^{m} \binom{m}{k} \lambda^{m-k} e^{\lambda x} D^k f = \sum_{k=0}^{m} \binom{m}{k} \lambda^{m-k} D^k f.$$

We infer from Theorem 4.51 that the spectral accuracy can be achieved by the GLF approximation. Indeed, we observe from Figure 4.18 such a convergence behavior.

Assume that $f \in L^2_{\omega^\alpha}(\mathbb{R}^+)$ with $\alpha > s-1$. Then we have from (4.4.46) that

$$f(x) = \sum_{n=0}^{\infty} \hat{f}_n \breve{\mathcal{L}}_n^{(\alpha,\lambda)}(x), \quad \|f\|^2_{\omega^\alpha} = \sum_{n=0}^{\infty} |\hat{f}_n|^2 \gamma_n^{\alpha,\lambda}.$$

Thanks to (4.4.50), we find that for any $0 \le \nu \le s$,

$$\begin{aligned} {}_x D_\infty^{\nu,\lambda} u(x) &= {}_x I_\infty^{s-\nu,\lambda} f(x) = \sum_{n=0}^{\infty} \hat{f}_n \, {}_x I_\infty^{s-\nu,\lambda} \breve{\mathcal{L}}_n^{(\alpha,\lambda)}(x) \\ &= \sum_{n=0}^{\infty} (2\lambda)^{\nu-s} \hat{f}_n \, \breve{\mathcal{L}}_n^{(\alpha-s+\nu,\lambda)}(x). \end{aligned} \tag{4.4.80}$$

Then,

$$\begin{aligned} \|{}_x D_\infty^{\nu,\lambda} u\|^2_{\omega^{\alpha-s+\nu}} &= \sum_{n=0}^{\infty} |(2\lambda)^{\nu-s} \hat{f}_n|^2 \gamma_n^{\alpha-s+\nu,\lambda} \\ &= \sum_{n=0}^{\infty} \frac{\Gamma(n+\alpha-s+\nu+1)}{(2\lambda)^{s-\nu+\alpha+1}\Gamma(n+1)} |\hat{f}_n|^2 \\ &\le \frac{\Gamma(\alpha-s+\nu+1)}{(2\lambda)^{s-\nu}\Gamma(\alpha+1)} \|f\|^2_{\omega^\alpha} < \infty, \end{aligned} \tag{4.4.81}$$

which implies that ${}_xD_\infty^{\nu,\lambda}u \in L^2_{\omega^{\alpha-s+\nu}}(\mathbb{R}^+)$, $0 \le \nu \le s$, namely, $u \in B^s_{\alpha-s,\lambda}(\mathbb{R}^+)$.

The GLF-II Galerkin method for (4.4.74) is to find $u_N \in \mathcal{F}_N^\lambda(\mathbb{R}^+)$ such that

$$({}_xD_\infty^{s,\lambda}u_N, v_N) = (f, v_N), \quad \forall v_N \in \mathcal{F}_N^\lambda(\mathbb{R}^+). \tag{4.4.82}$$

Furthermore, thanks to the orthogonality (4.4.46), we deduce that the solution of (4.4.82) is given by

$$u_N = \sum_{n=0}^{N}(2\lambda)^{-s}\hat{f}_n\breve{\mathcal{L}}_n^{(\alpha-s,\lambda)}(x),$$

which is exactly the orthogonal projection from $L^2_{\omega^{\alpha-s}}(\mathbb{R}^+)$ to $\mathcal{F}_N^\lambda(\mathbb{R}^+)$. Thus, the following estimate is an immediate consequence of Theorem 4.50.

Theorem 4.52 *Let u and u_N be the solutions of (4.4.74) and (4.4.82), respectively. If $f \in L^2_{\omega^\alpha}(\mathbb{R}^+)$ and ${}_xD_\infty^{r-s,\lambda}f \in L^2_{\omega^{\alpha-s+r}}(I)$ with $\alpha > s-1$, then we have*

$$\|u-u_N\|_{\omega^{\alpha-s}} \le c\,(2\lambda N)^{-\frac{r}{2}}\|{}_xD_\infty^{r,\lambda}u\|_{\omega^{\alpha-s+r}} = c\,(2\lambda N)^{-\frac{r}{2}}\|{}_xD_\infty^{r-s,\lambda}f\|_{\omega^{\alpha-s+r}},$$

where $c \approx 1$ for large N.

4.4.5 Tempered Fractional Diffusion on the Half Line

In this section, we apply the GLFs to solve a tempered fractional diffusion equation on the half line. Consider the tempered fractional diffusion equation of order $\mu \in (0,1)$ on the half line:

$$\begin{cases} \partial_t u(x,t) + {}_0D_x^{\mu,\lambda}u(x,t) - \lambda^\mu u(x,t) = f(x,t), & (x,t) \in \mathbb{R}^+ \times (0,T], \\ u(0,t) = 0, \quad \lim\limits_{x\to\infty} u(x,t) = 0, & 0 < t \le T, \\ u(x,0) = u_0(x), & x \in \mathbb{R}^+. \end{cases} \tag{4.4.83}$$

This equation models the particles jumping on the half line $\mathbb{R}^+$ with the probability density function:

$$f_\varepsilon(x) = C_\varepsilon^{-1}x^{-\mu-1}e^{-\lambda x}\mathbf{1}_{(\varepsilon,\infty)}(x), \quad 0 < \mu < 1.$$

Remark 4.53 Note that (4.4.83) can be viewed as the TFDE (4.4.2) on the half line with

$$\partial_{+,x}^{\mu,\lambda}u = {}_0D_x^{\mu,\lambda}u - \lambda^\mu u, \quad 0 < \mu < 1.$$

Indeed, we can show that for $\mu \in (0,1)$ and real $\lambda > 0$,

$$ {}_0D_x^{\mu,\lambda}u = e^{-\lambda x}\, {}^{\mathrm{RL}}_{\ 0}D_x^{\mu}\{e^{\lambda x}u(x)\} = e^{-\lambda x}\, {}^{\ \mathrm{RL}}_{-\infty}D_x^{\mu}\{e^{\lambda x}\tilde{u}(x)\}, \quad x \in \mathbb{R}^+, $$

where $\tilde{u} = u$ for $x \in \mathbb{R}^+$ and $\tilde{u} = 0$ for $x \in (-\infty, 0)$. Moreover, we have

$$ \begin{aligned} &\mathscr{F}\big[e^{-\lambda x}\, {}^{\ \mathrm{RL}}_{-\infty}D_x^{\mu}\{e^{\lambda x}\tilde{u}(x)\}\big](\omega) \\ &= \int_{\mathbb{R}} D\, {}^{\ \mathrm{RL}}_{-\infty}I_x^{1-\mu}\{e^{\lambda x}\tilde{u}(x)\}\, e^{-(\lambda+\mathrm{i}\omega)x}\mathrm{d}x \\ &= (\lambda+\mathrm{i}\omega)\int_{\mathbb{R}} {}^{\ \mathrm{RL}}_{-\infty}I_x^{1-\mu}\{e^{\lambda x}\tilde{u}(x)\}\, e^{-(\lambda+\mathrm{i}\omega)x}\mathrm{d}x \\ &= (\lambda+\mathrm{i}\omega)\int_{\mathbb{R}} e^{\lambda x}\tilde{u}(x)\, {}^{\mathrm{RL}}_{\ x}I_{\infty}^{1-\mu}e^{-(\lambda+\mathrm{i}\omega)x}\mathrm{d}x \\ &= (\lambda+\mathrm{i}\omega)^{\mu}\mathscr{F}[\tilde{u}](\omega) \overset{(4.4.12)}{=} \mathscr{F}\big[{}_{-\infty}D_x^{\mu,\lambda}\tilde{u}(x)\big](\omega). \end{aligned} $$

This implies $\tilde{u} \in W_{\lambda}^{\mu,2}(\mathbb{R})$ and the extended tempered fractional derivative ${}_0D_x^{\mu,\lambda}u$ can be understood in the sense of the original definition in Sabzikar et al. (2015).

Galerkin Spectral Scheme

Observe from Remark 4.53 that the identities (4.4.21) are also valid on $\mathbb{R}^+$. A weak form of the problem (4.4.83) is to find $u(\cdot,t) \in W_{\lambda}^{\mu/2,2}(\mathbb{R}^+)$ such that for all $v \in W_{\lambda}^{\mu/2,2}(\mathbb{R}^+)$ and $t \in (0,T]$,

$$ \begin{cases} \big(\partial_t u(\cdot,t), v\big) + a_{\mu,\lambda}\big(u(\cdot,t), v\big) = \big(f(\cdot,t), v\big), \\ \big(u(\cdot,0), v\big) = (u_0, v), \end{cases} \tag{4.4.84} $$

where the space $W_{\lambda}^{\mu/2,2}(\mathbb{R}^+)$ consists of all functions whose zero extensions are in $W_{\lambda}^{\mu/2,2}(\mathbb{R})$ (cf. (4.4.14)), and the bilinear form reads

$$ a_{\mu,\lambda}(u,v) := \big({}_0D_x^{\mu/2,\lambda}u, {}_xD_{\infty}^{\mu/2,\lambda}v\big) - \lambda^{\mu}(u,v). \tag{4.4.85} $$

The semi-discrete Galerkin approximation scheme is to find $u_N(\cdot,t) \in \mathcal{F}_N^{\nu,\lambda}(\mathbb{R}^+)$ such that

$$ \big(\partial_t u_N(\cdot,t), v\big) + a_{\mu,\lambda}\big(u_N(\cdot,t), v\big) = \big(f(\cdot,t), v\big), \quad \forall v \in \mathcal{F}_N^{\nu,\lambda}(\mathbb{R}^+), \tag{4.4.86} $$

with

$$ u_N(x,0) = \pi_N^{-\nu,\lambda}u_0(x) = \sum_{n=0}^{N} c_{0,n}\, \mathcal{L}_n^{(-\nu,\lambda)}(x), $$

where the projection operator $\pi_N^{-\nu,\lambda}$ is defined in (4.4.62) with $\max\{0, \mu - 1/2\} < \nu \le 1$. Note that the boundary condition $u(0,t) = 0$ is automatically met, and we choose the parameter ν to better fit the singularity behavior of the solution near $x = 0$.

Remark 4.54 We show in the next section (see Theorem 4.55 and Remark 4.56) the positivity of the bilinear form, that is, for any $0 \neq v \in W_\lambda^{\mu/2,2}(\mathbb{R}^+)$, $a_{\mu,\lambda}(v, v) > 0$. Then we can show the stability of the solutions of (4.4.84) and (4.4.86) as in (4.4.98) (with $\mathbb{R}^+$ in place of $\mathbb{R}$). Moreover, we can conduct the error analysis of the semi-discrete scheme by using the approximation results in Section 4.4.3, and following a standard argument for usual diffusion equations.

Numerical Algorithm

Now, set

$$u_N(x,t) = \sum_{n=0}^{N} c_n(t)\, \varphi_n(x), \quad \varphi_n(x) := \mathcal{L}_n^{(-\nu,\lambda)}(x). \tag{4.4.87}$$

We derive from the scheme (4.4.86) that

$$M\frac{\mathrm{d}}{\mathrm{d}t}\vec{c}(t) + A\vec{c}(t) = \vec{f}(t), \quad \vec{c}(0) = \vec{c}_0, \tag{4.4.88}$$

where for fixed $t > 0$, vectors

$$\begin{aligned} &\vec{c}(t) = \big(c_0(t), c_1(t), \ldots, c_N(t)\big)^T, \quad \vec{c}_0 = \big(c_{0,0}(t), c_{0,1}(t), \ldots, c_{0,N}(t)\big)^T, \\ &\vec{f}(t) = \big(f_0(t), f_1(t), \ldots, f_N(t)\big)^T, \quad f_n(t) = (f, \varphi_n), \quad 0 \le n \le N. \end{aligned}$$

Note that for any $u, v \in \mathcal{F}_N^{\nu,\lambda}(\mathbb{R}^+)$, there exists

$$({}_0D_x^{\mu/2,\lambda}u, {}_xD_\infty^{\mu/2,\lambda}v) = ({}_0D_x^{\mu,\lambda}u, v).$$

The mass matrix M and the stiffness matrix S can be computed by the Laguerre–Gauss quadrature formula. In fact, by using the tempered fractional derivative relations (4.4.48), it is straightforward to obtain that for $m, n = 0, 1, 2, \ldots, N$,

$$\begin{aligned} M_{mn} &= (\varphi_n, \varphi_m) = \big(\mathcal{L}_n^{(-\nu,\lambda)}, \mathcal{L}_m^{(-\nu,\lambda)}\big), \\ A_{mn} &= a_\mu(\varphi_n, \varphi_m) = h_n^{\nu,\mu}\big(\mathcal{L}_n^{(\mu-\nu,\lambda)}, \mathcal{L}_m^{(-\nu,\lambda)}\big) - \lambda^\mu M_{mn}, \end{aligned} \tag{4.4.89}$$

where $h_n^{\nu,\mu} = \Gamma(n+1+\nu)/\Gamma(n+1+\nu-\mu)$ was defined in (4.4.35).

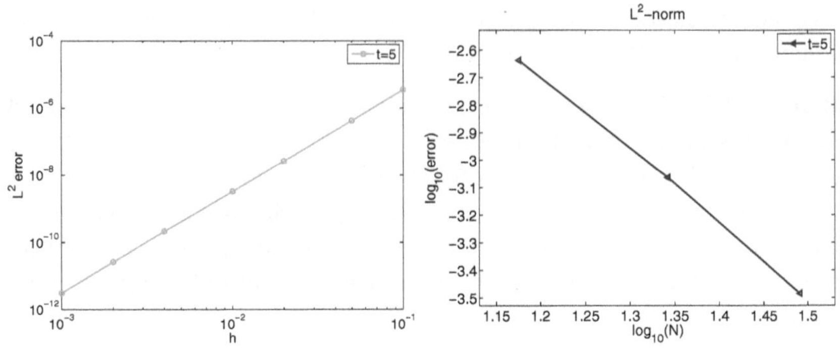

Figure 4.19 Left: $u = x\exp(-\lambda x)\cos(t)$. Right: $f = \cos(x)\exp(-x)\sin(t)$. Source: Chen et al. (2018).

Numerical Results

Typically, we test three cases as follows.

(i). Choose the exact solution to be $u(x,t) = xe^{-\lambda x}\cos(t)$. By a direct calculation, the source term is given by

$$f(x,t) = -xe^{-\lambda x}\sin(t) + \Big(\frac{\Gamma(2)}{\Gamma(2-\mu)}x^{1-\mu} - \lambda^{\mu}x\Big)e^{-\lambda x}\cos(t).$$

Figure 4.19 (left) illustrates that the error decays to zero rapidly for the spectral method built upon the GLF basis with $\nu = -1$ and $N = 50$, and the third-order explicit Runge–Kutta method in time with $\lambda = \mu = 2/3$ and time-stepping size $h \in (10^{-3}, 10^{-1})$.

(ii). Set $f(x,t) = \cos(x)e^{-x}\sin(t)$, and choose λ, μ as above. Figure 4.19 (right) verifies that the solution is singular even though $f(x,t)$ is a smooth function. Here, we compare the error with a reference "exact" solution computed with $N = 100$.

(iii). Consider $f(x,t) \equiv 0$, and let $\mu = 2/3$, $\lambda = 2/3$ in (4.4.83). Figure 4.20 (left) exhibits the evolution of the tempered fractional diffusion model with the initial distribution $u_0(x) = xe^{-x}$. Figure 4.20 (right) shows the convergence rate of the scheme, where the error is compared with the reference solution $u_{100}(x,t)$ with different ν at $t = 10$.

4.4.6 Tempered Fractional Diffusion Equation on the Whole Line

In this section, we present a multi-domain spectral element method for the tempered fractional diffusion equation on the whole line. Consider the

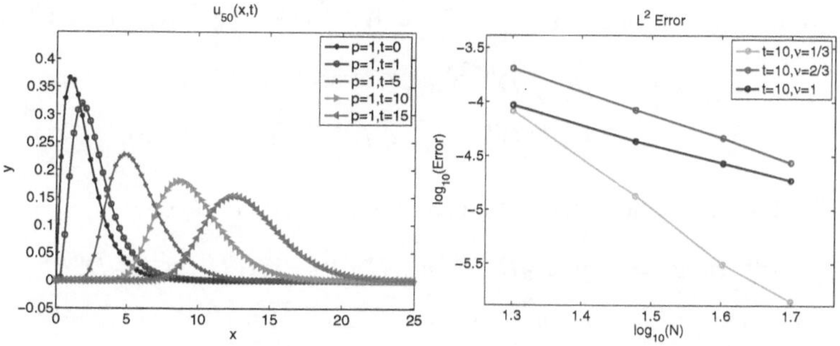

Figure 4.20 TFDE with $f \equiv 0, \lambda = 2/3, \ \mu = 2/3$. Left: Profiles of the solutions at different time. Right: Convergence behavior for different ν at $t = 10$. Source: Chen et al. (2018).

tempered fractional diffusion equation of order $\mu \in (k-1,k), \ k = 1,2$ on the whole line:

$$\begin{cases} \partial_t u(x,t) + \mathcal{A}^{\mu,\lambda}_{p,q} u(x,t) = f(x,t), \quad x \in \mathbb{R}, \ \ 0 < t \le T, \\ u(x,0) = u_0(x), \ \ x \in \mathbb{R}, \quad \lim\limits_{|x|\to\infty} u(x,t) = 0, \quad 0 \le t \le T, \end{cases} \tag{4.4.90}$$

where p, q are non-negative constants such that $p+q = 1$, and f, u_0 are given functions. Here, we denote

$$\mathcal{A}^{\mu,\lambda}_{p,q} u = (-1)^{k-1} \{ p\partial^{\mu,\lambda}_{+,x} + q\partial^{\mu,\lambda}_{-,x} \} u, \tag{4.4.91}$$

where the involved fractional operators are

(i) for $0 < \mu < 1$,

$$\partial^{\mu,\lambda}_{+,x} u = {}_{-\infty}D^{\mu,\lambda}_{x} u - \lambda^\mu u, \quad \partial^{\mu,\lambda}_{-,x} u = {}_{x}D^{\mu,\lambda}_{\infty} u - \lambda^\mu u; \tag{4.4.92}$$

(ii) for $1 < \mu < 2$,

$$\begin{aligned} \partial^{\mu,\lambda}_{+,x} u &= {}_{-\infty}D^{\mu,\lambda}_{x} u - \mu\lambda^{\mu-1}\partial_x u - \lambda^\mu u, \\ \partial^{\mu,\lambda}_{-,x} u &= {}_{x}D^{\mu,\lambda}_{\infty} u + \mu\lambda^{\mu-1}\partial_x u - \lambda^\mu u. \end{aligned} \tag{4.4.93}$$

We refer to Definition 4.37 for the tempered derivative operators.

A weak form of (4.4.90) is to find $u(\cdot,t) \in V^{\mu}_{\lambda}(\mathbb{R})$ for $0 < t \le T$ such that

$$\begin{cases} \partial_t (u(\cdot,t), v) + a^{\mu,\lambda}_{p,q}(u(\cdot,t), v) = (f(\cdot,t), v), \quad \forall\, v \in V^{\mu}_{\lambda}(\mathbb{R}), \\ (u(\cdot,0), w) = (u_0, w), \quad \forall\, w \in V^{\mu}_{\lambda}(\mathbb{R}), \end{cases} \tag{4.4.94}$$

where $(\cdot,\cdot)$ is the inner product of $L^2(\mathbb{R})$ as before. The space $V^{\mu}_{\lambda}(\mathbb{R})$ and the bilinear form $a^{\mu,\lambda}_{p,q}(\cdot,\cdot)$ are defined as

(i) for $0 < \mu < 1$, and $u, v \in V_\lambda^\mu(\mathbb{R}) = W_\lambda^{\mu/2,2}(\mathbb{R}) \cap L^2(\mathbb{R})$ (cf. (4.4.14)),

$$\begin{aligned} a_{p,q}^{\mu,\lambda}(u,v) &= p\big({}_{-\infty}D_x^{\mu/2,\lambda}u, {}_xD_\infty^{\mu/2,\lambda}v\big) \\ &\quad + q\big({}_xD_\infty^{\mu/2,\lambda}u, {}_{-\infty}D_x^{\mu/2,\lambda}v\big) - \lambda^\mu\,(u,v), \end{aligned} \tag{4.4.95}$$

(ii) for $1 < \mu < 2$, and $u, v \in V_\lambda^\mu(\mathbb{R}) = W_\lambda^{\mu-1,2}(\mathbb{R}) \cap H^1(\mathbb{R})$,

$$\begin{aligned} a_{p,q}^{\mu,\lambda}(u,v) = &- p({}_{-\infty}D_x^{\mu-1,\lambda}u, {}_xD_\infty^{1,\lambda}v) - q({}_xD_\infty^{1,\lambda}u, {}_{-\infty}D_x^{\mu-1,\lambda}v) \\ &+ \lambda^\mu(u,v) + (p-q)\mu\lambda^{\mu-1}(\partial_x u, v). \end{aligned} \tag{4.4.96}$$

Note that in (4.4.96), ${}_xD_\infty^{1,\lambda}u = \lambda u - \partial_x u$ (cf. (4.4.17)).

Importantly, we can show that the involved bilinear form is strictly positive, so the well-posedness of (4.4.94) follows.

Theorem 4.55 *For any $0 \neq v \in V_\lambda^\mu(\mathbb{R})$ with $\mu \in (0,1) \cup (1,2)$ and $\lambda > 0$, we have*

$$a_{p,q}^{\mu,\lambda}(v,v) > 0. \tag{4.4.97}$$

If $u_0 \in L^2(\mathbb{R})$ and $f \in L^2(\mathbb{R} \times (0,T))$, then the problem (4.4.94) has a unique solution $u \in L^\infty(0,T; V_\lambda^\mu(\mathbb{R}))$ such that

$$\|u(\cdot,t)\|_{L^2(\mathbb{R})}^2 \le e^t\Big(\|u_0\|_{L^2(\mathbb{R})}^2 + \int_0^t \|f(\cdot,s)\|_{L^2(\mathbb{R})}^2 \mathrm{d}s\Big), \quad \forall\, 0 \le t \le T. \tag{4.4.98}$$

Proof We first consider $0 < \mu < 1$. Using (4.4.12) and Parseval's identity (4.4.8) leads to

$$\begin{aligned} \big({}_{-\infty}D_x^{\mu/2,\lambda}v, {}_xD_\infty^{\mu/2,\lambda}v\big) &= \int_{\mathbb{R}} (\lambda + \mathrm{i}\omega)^\mu \big|\mathscr{F}[v](\omega)\big|^2 \mathrm{d}\omega \\ &= \int_{\mathbb{R}} (\lambda^2 + \omega^2)^{\frac{\mu}{2}} e^{\mathrm{i}\mu\Theta(\omega)} \big|\mathscr{F}[v](\omega)\big|^2 \mathrm{d}\omega, \end{aligned} \tag{4.4.99}$$

where $\Theta(\omega)$ is the argument of $\lambda + \mathrm{i}\omega$, i.e.,

$$\begin{aligned} \Theta(\omega) &= \arccos\big(\lambda/\sqrt{\lambda^2+\omega^2}\big), \quad \text{if } \omega > 0, \\ \Theta(\omega) &= -\arccos\big(\lambda/\sqrt{\lambda^2+\omega^2}\big), \quad \text{if } \omega < 0. \end{aligned} \tag{4.4.100}$$

It is evident that $\Theta(\omega)$ is odd in ω. In fact, for real function v, $\big|\mathscr{F}[v](\omega)\big|^2$ is an even function in ω, thanks to the property $\mathscr{F}[v](\omega) = \overline{\mathscr{F}[v](-\omega)}$, which can be derived from the definition (4.4.7) straightforwardly. Thus, we derive from (4.4.99) that

$$\left({}_{-\infty}D_x^{\mu/2,\lambda}v, {}_xD_\infty^{\mu/2,\lambda}v\right) = 2\int_0^\infty (\lambda^2+\omega^2)^{\frac{\mu}{2}}\cos(\mu\Theta(\omega))\left|\mathscr{F}[v](\omega)\right|^2 \mathrm{d}\omega. \tag{4.4.101}$$

For notational convenience, we denote

$$\mathcal{K}_\mu(\omega) := (\lambda^2+\omega^2)^{\frac{\mu}{2}}\cos(\mu\Theta(\omega)). \tag{4.4.102}$$

Noting that

$$\Theta'(\omega) = \frac{\lambda}{\lambda^2+\omega^2}, \quad \cos\Theta = \frac{\lambda}{\sqrt{\lambda^2+\omega^2}}, \quad \sin\Theta = \frac{\omega}{\sqrt{\lambda^2+\omega^2}},$$

we find

$$\begin{aligned}\mathcal{K}'_\mu(\omega) &= \mu(\lambda^2+\omega^2)^{\frac{\mu-1}{2}}\left(\frac{\omega}{\sqrt{\lambda^2+\omega^2}}\cos(\mu\Theta) - \frac{\lambda}{\sqrt{\lambda^2+\omega^2}}\sin(\mu\Theta)\right)\\ &= \mu(\lambda^2+\omega^2)^{\frac{\mu-1}{2}}\sin\big((1-\mu)\Theta\big).\end{aligned} \tag{4.4.103}$$

As $\Theta \in (0,\pi/2)$, $\mathcal{K}_\mu(\omega)$ is ascending with respect to ω, when $\mu \in (0,1)$. Consequently, for $\mu \in (0,1)$,

$$\mathcal{K}_\mu(\omega) > \mathcal{K}_\mu(0) = \lambda^\mu, \quad \forall\,\omega > 0. \tag{4.4.104}$$

Combining (4.4.101) and (4.4.104) leads to

$$\begin{aligned}\left({}_{-\infty}D_x^{\mu/2,\lambda}v, {}_xD_\infty^{\mu/2,\lambda}v\right) &= 2\int_0^\infty \mathcal{K}_\mu(\omega)\left|\mathscr{F}[v](\omega)\right|^2 \mathrm{d}\omega\\ &> 2\lambda^\mu \int_0^\infty \left|\mathscr{F}[v](\omega)\right|^2 \mathrm{d}\omega = \lambda^\mu \|v\|^2_{L^2(\mathbb{R})},\end{aligned}$$

where in the last, we used the property that $|\mathscr{F}[v](\omega)|^2$ is even in ω. As $p+q=1$, we obtain from (4.4.95) and the above that

$$a_{p,q}^{\mu,\lambda}(v,v) = \left({}_{-\infty}D_x^{\mu/2,\lambda}v, {}_xD_\infty^{\mu/2,\lambda}v\right) - \lambda^\mu\,\|v\|^2_{L^2(\mathbb{R})} > 0. \tag{4.4.105}$$

For $1 < \mu < 2$, we can follow the same derivation and show that

$$({}_{-\infty}D_x^{\mu-1,\lambda}v, {}_xD_\infty^{1,\lambda}v) = 2\int_0^\infty \mathcal{K}_\mu(\omega)\left|\mathscr{F}[v](\omega)\right|^2 \mathrm{d}\omega, \tag{4.4.106}$$

but by (4.4.103), we have $\mathcal{K}'_\mu(\omega) < 0$, so

$$\mathcal{K}_\mu(\omega) < \mathcal{K}_\mu(0) = \lambda^\mu, \quad \forall\,\omega > 0. \tag{4.4.107}$$

Observe that

$$(v', v) = \frac{1}{2}\int_{\mathbb{R}} (v^2(x))'\,\mathrm{d}x = 0. \tag{4.4.108}$$

From (4.4.96) and (4.4.106)–(4.4.108), we obtain

$$a_{p,q}^{\mu,\lambda}(v,v) = -({}_{-\infty}D_x^{\mu-1,\lambda}v, {}_xD_{\infty}^{1,\lambda}v) + \lambda^{\mu}\,\|v\|_{L^2(\mathbb{R})}^2 > 0. \tag{4.4.109}$$

This ends the proof of (4.4.97).

Next, taking $v = u$ in (4.4.94), we obtain from the Cauchy–Schwarz inequality that

$$\begin{aligned}\frac{1}{2}\frac{\mathrm{d}}{\mathrm{d}t}\|u(\cdot,t)\|_{L^2(\mathbb{R})}^2 + a_{p,q}^{\mu,\lambda}(u(\cdot,t),u(\cdot,t)) &= (f(\cdot,t),u(\cdot,t)) \\ &\le \|f(\cdot,t)\|_{L^2(\mathbb{R}^+)}\|u(\cdot,t)\|_{L^2(\mathbb{R})},\end{aligned} \tag{4.4.110}$$

which, together with (4.4.97), implies

$$\frac{\mathrm{d}}{\mathrm{d}t}\|u(\cdot,t)\|_{L^2(\mathbb{R})}^2 \le \|u(\cdot,t)\|_{L^2(\mathbb{R})}^2 + \|f(\cdot,t)\|_{L^2(\mathbb{R}^+)}^2, \quad \forall\, 0 \le t \le T. \tag{4.4.111}$$

We immediately obtain (4.4.98), which implies the uniqueness of the solution. The existence follows from the equivalence of uniqueness and existence for linear problems. □

Remark 4.56 We see from the proof that the same result is valid for $\mu \in (0,1)$ and $u(x,0) = 0$ on the half line $\mathbb{R}^+$ through zero extension. Therefore, we can show the stability of the model in the previous section (see Remark 4.54).

A Two-Domain Spectral Element Method

An interesting observation of the model in Sabzikar et al. (2015) (i.e., (4.4.90)) is its solution might have a limited regularity across $x = 0$. This motivates us to use the Laguerre polynomial approximations on $(-\infty,0)$ and $(0,\infty)$, respectively. Thus, we decompose the whole line as

$$\mathbb{R} = \Lambda_1 \cup \Lambda_2, \quad \Lambda_1 = (-\infty,0), \quad \Lambda_2 = [0,\infty),$$

and denote $u_{\Lambda_j}(x,t) := u(x,t)\big|_{\Lambda_j}$, $j = 1,2$. Introduce the approximation space:

$$V_N^{\lambda}(\mathbb{R}) := \big\{\phi \in C(\mathbb{R}) : \phi(x) = e^{-\lambda|x|}p, \ \ p|_{\Lambda_i} \in \mathcal{P}_{N_i-1}(\Lambda_i)\big\}, \tag{4.4.112}$$

and define

$$\begin{aligned}\phi^*(x) &= e^{-\lambda|x|}, \quad \phi_{n_1}^-(x) = \begin{cases}\mathcal{L}_{n_1}^{(-1,\lambda)}(-x), & x \le 0,\\ 0, & x > 0,\end{cases}\\ \phi_{n_2}^+(x) &= \begin{cases}0, & x \le 0,\\ \mathcal{L}_{n_2}^{(-1,\lambda)}(x), & x > 0,\end{cases}\end{aligned} \tag{4.4.113}$$

where $\mathcal{L}_n^{(-1,\lambda)}(x) = e^{-\lambda x}xL_n^{(1)}(2\lambda x)$. One verifies readily that

$$V_N^{\lambda}(\mathbb{R}) = \mathrm{span}\big\{\phi^*(x);\ \phi_{n_1}^-(x),\ 0 \le n_1 \le N_1 - 1;\ \phi_{n_2}^+(x),\ 0 \le n_2 \le N_2 - 1\big\}. \tag{4.4.114}$$

Then, our semi-discrete spectral-Galerkin method is to find $u_N(\cdot,t) \in V_N^\lambda(\mathbb{R})$ such that

$$\begin{cases} \big(\partial_t u_N(\cdot,t), v\big) + a_{p,q}^{\mu,\lambda}\big(u_N(\cdot,t), v\big) = \big(f(\cdot,t), v\big), & \forall v \in V_N^\lambda(\mathbb{R}), \\ \big(u_N(\cdot,0), v\big) = (u_0, v), & \forall v \in V_N^\lambda(\mathbb{R}). \end{cases} \tag{4.4.115}$$

We provide below some details of the algorithm.

$$\begin{aligned} u_N(x,t) &= c^*(t)\phi^*(x) + \sum_{n_1=0}^{N_1-1} c_{n_1}^-(t)\phi_{n_1}^-(x) + \sum_{n_2=0}^{N_2-1} c_{n_2}^+(t)\phi_{n_2}^+(x), \\ u_N(x,0) &= c_0^*\phi^*(x) + \sum_{n_1=0}^{N_1-1} c_{0,n_1}^-\phi_{n_1}^-(x) + \sum_{n_2=0}^{N_2-1} c_{0,n_2}^+\phi_{n_2}^+(x). \end{aligned} \tag{4.4.116}$$

Let $H(x)$ be the Heaviside function as before. Thanks to the tempered fractional derivative and integral relations with GLFs, and a reflected mapping from positive half line $\mathbb{R}^+$ to negative half line $\mathbb{R}^-$, we can derive the following identities:

$$\begin{aligned} &{}_xD_\infty^{1,\lambda}\phi^*(x) = -2\lambda e^{-\lambda x}H(x), \\ &{}_xD_\infty^{1,\lambda}\phi_{n_1}^-(x) = (n_1+1)\mathcal{L}_{n_1}^{(0,\lambda)}(-x)H(-x), \\ &{}_{-\infty}D_x^{s,\lambda}\phi^*(x) = \begin{cases} (2\lambda)^s e^{\lambda x}, & x \le 0, \\ \dfrac{2\lambda e^{\lambda x}}{\Gamma(1-s)}\displaystyle\int_x^\infty \frac{e^{-2\lambda t}}{t^s}\mathrm{d}t, & x > 0, \end{cases} \\ &{}_{-\infty}D_x^{s,\lambda}\phi_{n_1}^-(x) = \begin{cases} -(2\lambda)^{s-1}(n_1+1)L_{n_1+1}^{(s-1)}(-2\lambda x)e^{\lambda x}, & x \le 0, \\ -e^{\lambda x}\dfrac{n_1+1}{\Gamma(1-s)}\displaystyle\int_x^\infty \frac{L_{n_1+1}^{(0)}(2\lambda(t-x))e^{-2\lambda t}}{t^s}\mathrm{d}t, & x > 0, \end{cases} \\ &{}_xD_\infty^{1,\lambda}\phi_{n_2}^+(x) = -(n_2+1)\mathcal{L}_{n_2+1}^{(0,\lambda)}(x)H(x), \\ &{}_{-\infty}D_x^{s,\lambda}\phi_{n_2}^+(x) = \frac{\Gamma(n_2+2)}{\Gamma(n_2+2-s)}x^{1-s}\mathcal{L}_{n_2}^{(1-s,\lambda)}(x)H(x). \end{aligned} \tag{4.4.117}$$

Derivation of (4.4.117).

- For $x \in \mathbb{R}^-$, $0 < s < 1$,

$$\begin{aligned} {}_{-\infty}D_x^{s,\lambda}\phi^*(x) &= {}_{-\infty}I_x^{1-s,\lambda}\,{}_{-\infty}D_x^{1,\lambda}\phi^*(x) = \frac{e^{-\lambda x}}{\Gamma(1-s)}\int_{-\infty}^x \frac{e^{\lambda\tau}(2\lambda)e^{\lambda\tau}}{(x-\tau)^s}\mathrm{d}\tau \\ &\overset{t=x-\tau}{=} \frac{2\lambda e^{-\lambda x}}{\Gamma(1-s)}\int_0^\infty \frac{e^{2\lambda(x-t)}}{t^s}dt \end{aligned} \tag{4.4.118}$$

$$
\begin{aligned}
&= \frac{(2\lambda)^s e^{\lambda x}}{\Gamma(1-s)} \int_0^{\infty} e^{-2\lambda t} (2\lambda t)^{-s} \mathrm{d}(2\lambda t) = (2\lambda)^s e^{\lambda x}, \\
{}_{-\infty}D_x^{s,\lambda} \phi_{n_1}^{-}(x) &= {}_{-\infty}I_x^{1-s,\lambda}\, {}_{-\infty}D_x^{1,\lambda} \phi_{n_1}^{-}(x) \\
&\overset{(4.4.53)}{=} {}_{-\infty}I_x^{1-s,\lambda} \{-(n_1+1)\mathcal{L}_{n_1+1}^{(0,\lambda)}(-x)\} \\
&\overset{(4.4.38)}{=} -(n_1+1)(2\lambda)^{s-1} L_{n_1+1}^{(s-1)}(-2\lambda x) e^{-\lambda x}, \\
{}_{-\infty}D_x^{s,\lambda} \phi_{n_2}^{+}(x) &= 0.
\end{aligned} \tag{4.4.119}
$$

- For $x \in \mathbb{R}^+, 0 < s < 1,$

$$
\begin{aligned}
{}_{-\infty}D_x^{s,\lambda} \phi^*(x) &= {}_{-\infty}I_x^{1-s,\lambda}\, {}_{-\infty}D_x^{1,\lambda} \phi^*(x) = \frac{e^{-\lambda x}}{\Gamma(1-s)} \int_{-\infty}^{0} \frac{2\lambda e^{2\lambda \tau}}{(x-\tau)^s} \mathrm{d}\tau \\
&\overset{\tau = x - t}{=} \frac{2\lambda e^{\lambda x}}{\Gamma(1-s)} \int_x^{\infty} \frac{e^{-2\lambda t}}{t^s} \mathrm{d}t, \\
{}_{-\infty}D_x^{s,\lambda} \phi_{n_1}^{-}(x) &= {}_{-\infty}I_x^{1-s,\lambda}\, {}_{-\infty}D_x^{1,\lambda} \phi_{n_1}^{-}(x) \\
&\overset{(4.4.53)}{=} \frac{e^{-\lambda x}}{\Gamma(1-s)} \int_{-\infty}^{0} \frac{-(n_1+1) L_{n_1+1}^{(0)}(-2\lambda\tau) e^{2\lambda\tau}}{(x-\tau)^s} \mathrm{d}\tau \\
&\overset{\tau = x - t}{=} -e^{\lambda x} \frac{n_1+1}{\Gamma(1-s)} \int_x^{\infty} \frac{L_{n_1+1}^{(0)}(2\lambda(t-x)) e^{-2\lambda t}}{t^s} \mathrm{d}t, \\
{}_{-\infty}D_x^{s,\lambda} \phi_{n_2}^{+}(x) &= {}_{-\infty}I_x^{1-s,\lambda}\, {}_{-\infty}D_x^{1,\lambda} \phi_{n_2}^{+}(x) \\
&\overset{(4.4.52)}{=} \frac{e^{-\lambda x}}{\Gamma(1-s)} \int_0^{x} \frac{(n_2+1) L_{n_2}^{(0)}(x)}{(x-\tau)^s} \mathrm{d}\tau \\
&= \frac{\Gamma(n_2+2)}{\Gamma(n_2+2-s)} x^{1-s} \mathcal{L}_{n_2}^{(1-s,\lambda)}(x).
\end{aligned} \tag{4.4.120}
$$

Then (4.4.115) leads to the linear system of ordinary differential equations:

$$
\boldsymbol{M} \frac{\mathrm{d}}{\mathrm{d}t} \vec{\boldsymbol{C}}(t) + \boldsymbol{A}\vec{\boldsymbol{C}}(t) = \vec{\boldsymbol{F}}(t), \tag{4.4.121}
$$

where

$$
\begin{aligned}
&\vec{\boldsymbol{C}}(t) = \left(c^*(t), \vec{\boldsymbol{C}}^-(t), \vec{\boldsymbol{C}}^+(t)\right)^T, \qquad \vec{\boldsymbol{F}}(t) = \left(f^*(t), \vec{\boldsymbol{F}}^-(t), \vec{\boldsymbol{F}}^+(t)\right)^T, \\
&\vec{\boldsymbol{C}}^-(t) = \left(c_0^-(t), c_1^-(t), \ldots, c_{N_1-1}^-(t)\right)^T, \ \vec{\boldsymbol{C}}^+(t) = \left(c_0^+(t), c_1^+(t), \ldots, c_{N_2-1}^+(t)\right)^T, \\
&\vec{\boldsymbol{F}}^-(t) = \left(f_0^-(t), f_1^-(t), \ldots, f_{N_1-1}^-(t)\right)^T, \ \vec{\boldsymbol{F}}^+(t) = \left(f_0^+(t), f_1^+(t), \ldots, f_{N_2-1}^+(t)\right)^T, \\
&f^*(t) = (f, \phi^*), \ \ f_{n_1}^-(t) = (f, \phi_{n_1}^-), \ \ f_{n_2}^+(t) = (f, \phi_{n_2}^+)
\end{aligned}
$$

for $0 \le n_i \le N_i - 1, \ i = 1, 2,$ and the matrices

$$\boldsymbol{M} = \begin{pmatrix} \boldsymbol{M}^{(*,*)}_{1\times 1} & \boldsymbol{M}^{(*,-)}_{1\times N_1} & \boldsymbol{M}^{(*,+)}_{1\times N_2} \\ \boldsymbol{M}^{(-,*)}_{N_1\times 1} & \boldsymbol{M}^{(-,-)}_{N_1\times N_1} & \boldsymbol{M}^{(-,+)}_{N_1\times N_2} \\ \boldsymbol{M}^{(+,*)}_{N_2\times 1} & \boldsymbol{M}^{(+,-)}_{N_2\times N_1} & \boldsymbol{M}^{(+,+)}_{N_2\times N_2} \end{pmatrix},$$

$$\boldsymbol{A} = \begin{pmatrix} \boldsymbol{A}^{(*,*)}_{1\times 1} & \boldsymbol{A}^{(*,-)}_{1\times N_1} & \boldsymbol{A}^{(*,+)}_{1\times N_2} \\ \boldsymbol{A}^{(-,*)}_{N_1\times 1} & \boldsymbol{A}^{(-,-)}_{N_1\times N_1} & \boldsymbol{A}^{(-,+)}_{N_1\times N_2} \\ \boldsymbol{A}^{(+,*)}_{N_2\times 1} & \boldsymbol{A}^{(+,-)}_{N_2\times N_1} & \boldsymbol{A}^{(+,+)}_{N_2\times N_2} \end{pmatrix}, \tag{4.4.122}$$

with the entries

$$\boldsymbol{M}^{(a,b)}_{c\times d}(i+1,j+1) = (\phi^b_j, \phi^a_i), \quad \boldsymbol{A}^{(a,b)}_{c\times d}(i+1,j+1) = a^{\mu,\lambda}_{p,q}(\phi^b_j, \phi^a_i),$$
$$a,b = *,-,+, \quad c,d = 1, N_1, N_2, \qquad 0 \le i \le c-1, \quad 0 \le j \le d-1,$$

and $\vec{\boldsymbol{C}}(0)$ is determined by the initial data. The derivation of the tempered derivative relation (4.4.117), and of the entries of the matrix $\boldsymbol{A}$ can be found below.

The Entries of Matrix A with $1 < \mu = 1 + s < 2$.

$$\begin{aligned} \left({}_{-\infty}D^{s,\lambda}_x \phi^*, {}_x D^{1,\lambda}_\infty \phi^*\right) &= \frac{-(2\lambda)^2}{\Gamma(1-s)} \int_0^\infty \int_x^\infty \frac{e^{-2\lambda t}}{t^s} \mathrm{d}t\mathrm{d}x \\ &= \frac{-(2\lambda)^2}{\Gamma(1-s)} \int_0^\infty \frac{e^{-2\lambda t}}{t^s} \int_0^t 1\mathrm{d}x\mathrm{d}t \\ &= \frac{-(2\lambda)^2}{\Gamma(1-s)} \int_0^\infty t^{1-s} e^{-2\lambda t}\mathrm{d}t \\ &\overset{\tau=2\lambda t}{=} \frac{-(2\lambda)^s}{\Gamma(1-s)} \int_0^\infty \tau^{1-s} e^{-\tau}\mathrm{d}\tau = (s-1)(2\lambda)^s. \end{aligned} \tag{4.4.123}$$

Since

$$D\left\{(2\lambda x) L^{(1)}_{n_2+1}(2\lambda x)\right\} = 2\lambda(n_2+2) L^{(0)}_{n_2+1}(2\lambda x),$$

i.e.,

$$\int_0^t L^{(0)}_{n_2+1}(2\lambda x)\mathrm{d}x = \frac{1}{n_2+2} t L^{(1)}_{n_2+1}(2\lambda t),$$

then

$$\begin{aligned}\left({}_{-\infty}D_x^{s,\lambda}\phi^*, {}_xD_\infty^{1,\lambda}\phi_{n_2}^+\right) &= \frac{-2\lambda(n_2+1)}{\Gamma(1-s)}\int_0^\infty\int_x^\infty \frac{e^{-2\lambda t}}{t^s}\mathrm{d}t\, L_{n_2+1}^{(0)}(2\lambda x)\mathrm{d}x \\ &= \frac{-2\lambda(n_2+1)}{\Gamma(1-s)}\int_0^\infty \frac{e^{-2\lambda t}}{t^s}\int_0^t L_{n_2+1}^{(0)}(2\lambda x)\mathrm{d}x\,\mathrm{d}t \\ &= \frac{-2\lambda(n_2+1)}{(n_2+2)\Gamma(1-s)}\int_0^\infty t^{1-s}L_{n_2+1}^{(1)}(2\lambda t)e^{-2\lambda t}\mathrm{d}t.\end{aligned}$$

Similarly, we have

$$\begin{aligned}&\left({}_{-\infty}D_x^{s,\lambda}\phi_{n_1}^-, {}_xD_\infty^{1,\lambda}\phi_{n_2}^+\right) \\ &= \frac{(n_1+1)(n_2+1)}{\Gamma(1-s)}\int_0^\infty\int_x^\infty \frac{L_{n_1+1}^{(0)}(2\lambda(t-x))e^{-2\lambda t}}{t^s}\mathrm{d}t\, L_{n_2+1}^{(0)}(2\lambda x)\mathrm{d}x \\ &= \frac{(n_1+1)(n_2+1)}{\Gamma(1-s)}\int_0^\infty t^{-s}e^{-2\lambda t}\int_0^t L_{n_2+1}^{(0)}(2\lambda x)L_{n_1+1}^{(0)}(2\lambda(t-x))\mathrm{d}x\mathrm{d}t \\ &\overset{x=t\xi}{=} \frac{(n_1+1)(n_2+1)}{\Gamma(1-s)}\int_0^\infty t^{1-s}e^{-2\lambda t}\int_0^1 L_{n_2+1}^{(0)}(2\lambda t\xi)L_{n_1+1}^{(0)}(2\lambda t(1-\xi))\mathrm{d}\xi\mathrm{d}t.\end{aligned}$$

The Entries of Matrix A with $0 < \mu = s < 1$.

$$\begin{aligned}\left({}_{-\infty}D_x^{s,\lambda}\phi^*, \phi^*\right) &= (2\lambda)^s\int_{-\infty}^0 e^{2\lambda x}\mathrm{d}x + \frac{2\lambda}{\Gamma(1-s)}\int_0^\infty\int_x^\infty \frac{e^{-2\lambda t}}{t^s}\mathrm{d}t\mathrm{d}x \\ &= (2\lambda)^{s-1} + \frac{2\lambda}{\Gamma(1-s)}\int_0^\infty \frac{e^{-2\lambda t}}{t^s}\int_0^t 1\mathrm{d}x\mathrm{d}t \\ &= (2\lambda)^{s-1} + \frac{2\lambda}{\Gamma(1-s)}\int_0^\infty t^{1-s}e^{-2\lambda t}\mathrm{d}t \\ &\overset{\tau=2\lambda t}{=} (2\lambda)^{s-1} + \frac{(2\lambda)^{s-1}}{\Gamma(1-s)}\int_0^\infty \tau^{1-s}e^{-\tau}\mathrm{d}\tau = (2-s)(2\lambda)^{s-1}.\end{aligned}$$

Owing to

$$D\left\{(2\lambda x)^2 L_{n_2}^{(2)}(2\lambda x)\right\} = (2\lambda)^2(n_2+2)xL_{n_2}^{(1)}(2\lambda x),$$

i.e.,

$$\int_0^t xL_{n_2}^{(1)}(2\lambda x)\mathrm{d}x = \frac{1}{n_2+2}t^2L_{n_2}^{(2)}(2\lambda t),$$

we obtain that

$$
\begin{aligned}
({}_{-\infty}D_x^{s,\lambda}\phi^*, \phi_{n_2}^+) &= \int_0^\infty \frac{e^{-\lambda x}}{\Gamma(1-s)} \int_{-\infty}^0 \frac{2\lambda e^{2\lambda\tau}}{(x-\tau)^s} \mathrm{d}\tau \, x L_{n_2}^{(1)}(2\lambda x) e^{-\lambda x} \mathrm{d}x \\
&\overset{\tau=x-t}{=} \frac{2\lambda}{\Gamma(1-s)} \int_0^\infty \int_x^\infty \frac{e^{-2\lambda t}}{t^s} \mathrm{d}t \, x L_{n_2}^{(1)}(2\lambda x) \mathrm{d}x \\
&= \frac{2\lambda}{\Gamma(1-s)} \int_0^\infty \frac{e^{-2\lambda t}}{t^s} \int_0^t x L_{n_2}^{(1)}(2\lambda x) \mathrm{d}x \, \mathrm{d}t \\
&= \frac{2\lambda}{(n_2+2)\Gamma(1-s)} \int_0^\infty t^{2-s} L_{n_2}^{(2)}(2\lambda t) e^{-2\lambda t} \mathrm{d}t.
\end{aligned}
$$

Similarly, we have

$$
\begin{aligned}
&({}_{-\infty}D_x^{s,\lambda}\phi_{n_1}^-, \phi_{n_2}^+) \\
&= -\frac{n_1+1}{\Gamma(1-s)} \int_0^\infty \int_x^\infty \frac{L_{n_1+1}^{(0)}(2\lambda(t-x)) e^{-2\lambda t}}{t^s} \mathrm{d}t \, x L_{n_2}^{(1)}(2\lambda x) \mathrm{d}x \\
&= -\frac{n_1+1}{\Gamma(1-s)} \int_0^\infty t^{-s} e^{-2\lambda t} \int_0^t x L_{n_2}^{(1)}(2\lambda x) L_{n_1+1}^{(0)}(2\lambda(t-x)) \mathrm{d}x \mathrm{d}t \\
&\overset{x=t\xi}{=} -\frac{n_1+1}{\Gamma(1-s)} \int_0^\infty t^{2-s} e^{-2\lambda t} \int_0^1 \xi L_{n_2}^{(1)}(2\lambda t\xi) L_{n_1+1}^{(0)}(2\lambda t(1-\xi)) \mathrm{d}\xi \mathrm{d}t.
\end{aligned}
$$

The above equations are enough to calculate out the matrix $\boldsymbol{A}$ due to some symmetric properties of the entries.

Base on the semi-discrete scheme (4.4.121), we further use the third-order explicit Runge–Kutta method in time direction with step size $h = 10^{-3}$ to numerically solve the problem.

Numerical Results

We solve (4.4.90) with $C_T = 1$ and $u_0 = 10e^{-5|x|}$ as the initial distribution by using the proposed method. We first test its accuracy. In Figure 4.21, we plot the convergence rate of the spectral method at $T = 5$ with fixed time-step $h = 10^{-3}$. We choose a slow decay $f(x,t) = (1+x^2)^{-1}$, and an exponential decay $f(x,t) = \cos t \; e^{-x^2}$, respectively. Observe from Figure 4.21 an exponential convergence for the latter, but an algebraic convergence for the former. In fact, we expect the solution with $f(x,t) = \cos t \; e^{-x^2}$ decays exponentially in space. Indeed, as the Fourier transform of the Gaussian e^{-x^2} is invariant, so we can use the model by using Fourier transform and then solve the resulted equation. However, the solution with $f(x,t) = (1+x^2)^{-1}$ should decay very slowly. As a result, we observe a very different convergence behavior.

Next, we examine behaviors of the solution under various situations. In Figure 4.22, we plot the snapshots at different times of the tempered fractional

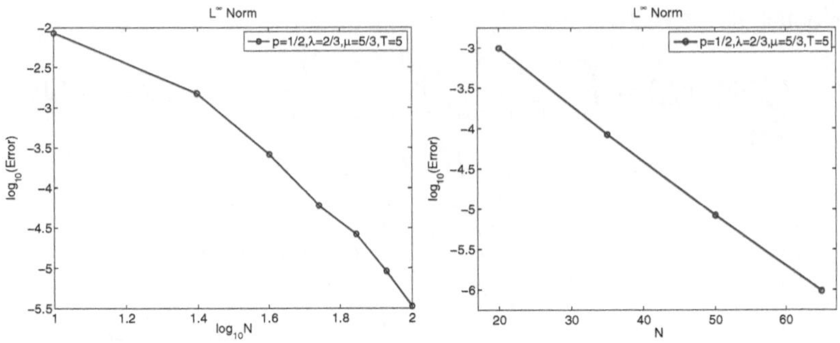

Figure 4.21 Left: $f(x,t) = (1+x^2)^{-1}$. Right: $f(x,t) = \cos t\ e^{-x^2}$. Source: Chen et al. (2018).

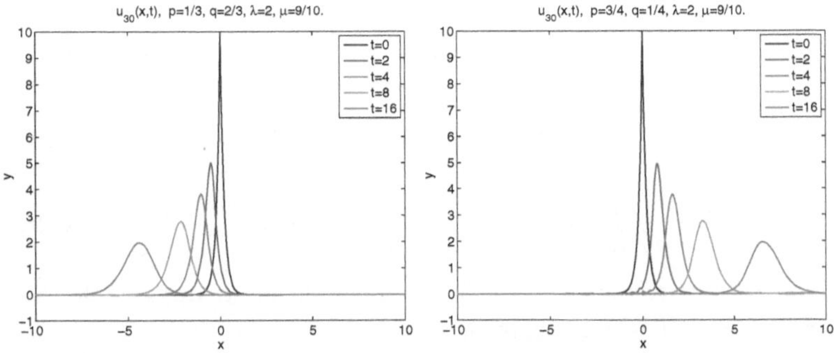

Figure 4.22 Left: $p = 1/3$, $q = 2/3$. Right: $p = 3/4$, $q = 1/4$. Source: Chen et al. (2018).

diffusion with $p = 1/3$, $q = 2/3$ and $p = 3/4$, $q = 1/4$, respectively. The case with $p = q = 1/2$ is plotted in Figure 4.23.

- The parameters p and q reflect the directional preference of the particle jumping. More precisely, if $p > q$, the particles tend to jump to the right, and if $p < q$, the particles tend to jump to the left, see Figure 4.22. In particular, $p = q$ produces a symmetric profile in the case of $f(x,t) = 0$, see the left in Figure 4.23.
- The parameter λ determines the probability of the jump distance of the particles. A larger λ indicates a shorter jump distance, see the right of Figure 4.23.
- To compare with the usual fractional diffusion equation, i.e., $\lambda = 0$, we plot in Figure 4.24 the particle distributions of the usual fractional diffusion and the tempered fractional diffusion with initial distribution $u_0(x) = 10e^{-4x^2}$

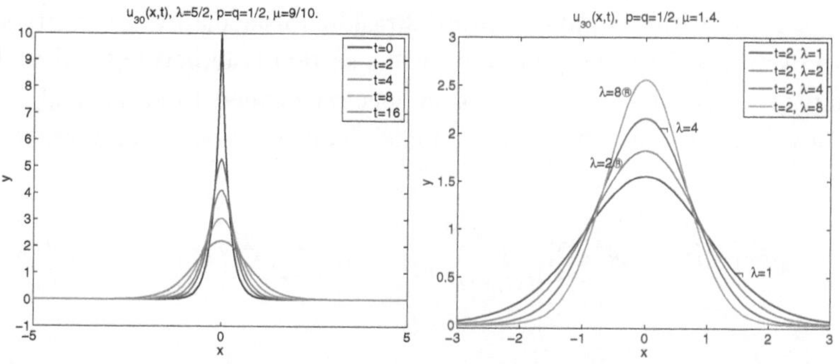

Figure 4.23 Left: $p = q = 1/2$, $\lambda = 5/2$. Right: $p = q = 1/2$, $t = 2$. Source: Chen et al. (2018).

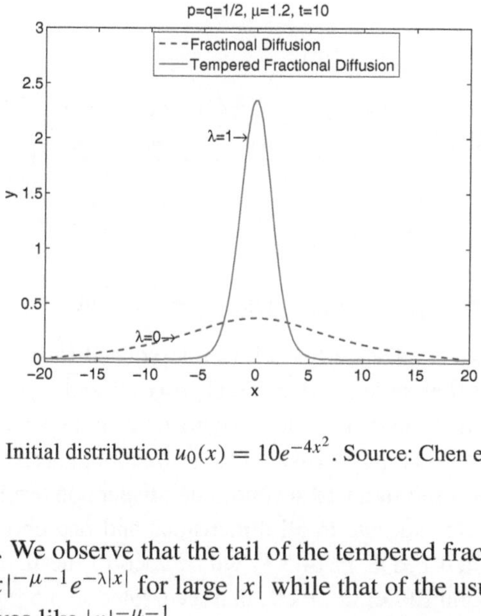

Figure 4.24 Initial distribution $u_0(x) = 10e^{-4x^2}$. Source: Chen et al. (2018).

at time $t = 10$. We observe that the tail of the tempered fractional diffusion behaves like $|x|^{-\mu-1}e^{-\lambda|x|}$ for large $|x|$ while that of the usual fractional diffusion behaves like $|x|^{-\mu-1}$.

4.5 Unified GJPF Space-Time Method for Distributed FPDEs

In complex multi-fractal anomalous transport phenomena, distributed-order FPDEs appear as tractable mathematical models, where the underlying derivative orders are distributed over a range of values, hence taking into account a wide range of multi-physics from ultraslow-to-standard-to-superdiffusion/wave dynamics. The main purpose of this section is to develop and analyze a unified distributed Petrov–Galerkin spectral method to solve a $(1 + d)$-dimensional fully distributed-order FPDE with two-sided derivatives of the form

$$\int_{\tau^{min}}^{\tau^{max}} \varphi(\tau)\ {}_{0}^{C}D_{t}^{2\tau} u \, d\tau + \sum_{i=1}^{d} \int_{\mu_i^{min}}^{\mu_i^{max}} \rho_i(\mu_i)\, [c_{l_i}\ {}_{a_j}^{RL}D_{x_j}^{2\mu_j} u + c_{r_i}\ {}_{x_j}^{RL}D_{b_j}^{2\mu_j} u]\, d\mu_i$$

$$= \sum_{j=1}^{d} \int_{\nu_j^{min}}^{\nu_j^{max}} \rho_j(\nu_j)\, [\kappa_{l_j}\ {}_{a_j}^{RL}D_{x_j}^{2\nu_j} u + \kappa_{r_j}\ {}_{x_j}^{RL}D_{b_j}^{2\nu_j} u]\, d\nu_j - \gamma\, u + f, \quad (4.5.1)$$

subject to homogeneous Dirichlet boundary conditions and zero initial condition, where for $i, j = 1, 2, ..., d$,

$$\begin{aligned}
&t \in [0, T], \quad x_j \in [a_j, b_j],\\
&2\tau^{min} < 2\tau^{max} \in (0, 2],\ 2\tau^{min} \neq 1,\ 2\tau^{max} \neq 1,\\
&2\mu_i^{min} < 2\mu_i^{max} \in (0, 1),\ 2\nu_j^{min} < 2\nu_j^{max} \in (1, 2],\\
&0 < \varphi(\tau) \in L^1((\tau^{min}, \tau^{max})),\ 0 < \rho_i(\mu_i) \in L^1((\mu_i^{min}, \mu_i^{max})),\\
&0 < \rho_j(\nu_j) \in L^1((\nu_j^{min}, \nu_j^{max})),
\end{aligned}$$

and the coefficients c_{l_i}, c_{r_i}, κ_{l_i}, κ_{r_i}, and γ are constant.

Remark 4.57 We emphasize that (4.5.1) is reduced to fractional advection-dispersion-reaction equations when $\varphi(\tau)$, $\rho_i(\mu_i)$, and $\rho_j(\nu_j)$ are chosen to be a Dirac delta function/distribution, where e.g., for the special case of $d = 1$ and $\rho_1(\nu_1) = \delta(\nu_1 - 1)$, the two-sided Riesz fractional derivatives with proper coefficients recover the standard second order dispersion terms. When only the left-sides derivatives survive in all dimensions and one chooses $\rho_i(\nu_i) = 0$, the fully distributed FPDE in (4.5.1) will reduce to the fixed-order $(1 + d)$-dimensional general FPDE in (4.2.9) that we treated in the previous section.

Short Review of Definitions

For the reader's convenience, we provide a short review of important definitions here. We denote by ${}_{a}^{RL}D_{x}^{\sigma} g(x)$ and ${}_{x}^{RL}D_{b}^{\sigma} g(x)$ the left-sided and the right-sided Riemann–Liouville fractional derivatives of order $\sigma > 0$,

$${}_{a}^{RL}D_{x}^{\sigma} g(x) = \frac{1}{\Gamma(n-\sigma)} \frac{d^n}{dx^n} \int_{a}^{x} \frac{g(s)}{(x-s)^{\sigma+1-n}}\, ds, \quad x \in [a, b],$$

$${}_{x}^{RL}D_{b}^{\sigma} g(x) = \frac{(-1)^n}{\Gamma(n-\sigma)} \frac{d^n}{dx^n} \int_{x}^{b} \frac{g(s)}{(s-x)^{\sigma+1-n}}\, ds, \quad x \in [a, b],$$

in which $g(x) \in L^1[a,b]$ and $\int_a^x \frac{g(s)}{(x-s)^{\sigma+1-n}}\, ds$, $\int_x^b \frac{g(s)}{(s-x)^{\sigma+1-n}}\, ds \in C^n[a,b]$, respectively, where $n = \lceil\sigma\rceil$. Besides, ${}^{\mathrm{C}}_{a}D^{\sigma}_{x} g(x)$ and ${}^{\mathrm{C}}_{x}D^{\sigma}_{b} g(x)$ represent the left-sided and the right-sided Caputo fractional derivatives, where

$$
\begin{aligned}
{}^{\mathrm{C}}_{a}D^{\sigma}_{x} f(x) &= \frac{1}{\Gamma(n-\sigma)} \int_a^x \frac{g^{(n)}(s)}{(x-s)^{\sigma+1-n}}\, ds, \quad x \in [a,b],\\
{}^{\mathrm{C}}_{x}D^{\sigma}_{b} f(x) &= \frac{(-1)^n}{\Gamma(n-\sigma)} \int_x^b \frac{g^{(n)}(s)}{(s-x)^{\sigma+1-n}}\, ds, \quad x \in [a,b].
\end{aligned}
$$

The relationship between the RL and the Caputo fractional derivatives is given by

$$
\begin{aligned}
{}^{\mathrm{RL}}_{a}D^{\nu}_{x} f(x) &= \frac{f(a)}{\Gamma(1-\nu)(x-a)^{\nu}} + {}^{\mathrm{C}}_{a}D^{\nu}_{x} f(x),\\
{}^{\mathrm{RL}}_{x}D^{\nu}_{b} f(x) &= \frac{f(b)}{\Gamma(1-\nu)(b-x)^{\nu}} + {}^{\mathrm{C}}_{x}D^{\nu}_{b} f(x),
\end{aligned}
$$

when $\lceil\nu\rceil = 1$. In the case of homogeneous boundary conditions, we obtain ${}^{\mathrm{RL}}_{a}D^{\nu}_{x} f(x) = {}^{\mathrm{C}}_{a}D^{\nu}_{x} f(x) := {}_{a}D^{\nu}_{x} f(x)$ and ${}^{\mathrm{RL}}_{x}D^{\nu}_{b} f(x) = {}^{\mathrm{C}}_{x}D^{\nu}_{b} f(x) := {}_{x}D^{\nu}_{b} f(x)$. The Riemann–Liouville fractional integrals of Jacobi *poly-fractonomials* are analytically obtained in the standard domain $\xi \in [-1,1]$ as

$$
{}^{\mathrm{RL}}_{-1}I^{\sigma}_{\xi}\{(1+\xi)^{\beta} P_n^{(\alpha,\beta)}(\xi)\} = \frac{\Gamma(n+\beta+1)}{\Gamma(n+\beta+\sigma+1)} (1+\xi)^{\beta+\sigma} P_n^{(\alpha-\sigma,\beta+\sigma)}(\xi),
$$

and

$$
{}^{\mathrm{RL}}_{\xi}I^{\sigma}_{1}\{(1-\xi)^{\alpha} P_n^{(\alpha,\beta)}(\xi)\} = \frac{\Gamma(n+\alpha+1)}{\Gamma(n+\alpha+\sigma+1)} (1-\xi)^{\alpha+\sigma} P_n^{(\alpha+\sigma,\beta-\sigma)}(\xi),
$$

where $0 < \sigma < 1$, $\alpha > -1$, $\beta > -1$, and $P_n^{(\alpha,\beta)}(\xi)$ denotes the standard Jacobi polynomials of order n and parameters α and β, introduced in Chapter 2. Accordingly,

$$
{}^{\mathrm{RL}}_{-1}I^{\sigma}_{\xi} P_n(\xi) = \frac{\Gamma(n+1)}{\Gamma(n-\sigma+1)} P_n^{(\sigma,-\sigma)}(\xi)\,(1+\xi)^{-\sigma}
$$

and

$$
{}^{\mathrm{RL}}_{\xi}I^{\sigma}_{1} P_n(\xi) = \frac{\Gamma(n+1)}{\Gamma(n-\sigma+1)} P_n^{(-\sigma,\sigma)}(\xi)\,(1-\xi)^{-\sigma},
$$

where $P_n(\xi) := P_n^{(0,0)}(\xi)$ represents Legendre polynomial of degree n.

Let us define the distributed-order derivative as

$$
{}^{\mathrm{D}}D^{\phi}_{t} f(t,x) := \int_{\tau_{min}}^{\tau_{max}} \phi(\tau)\, {}^{\mathrm{RL}}_{0}D^{\tau}_{t} f(t,x) d\tau,
$$

where $\alpha \to \phi(\alpha)$ is a continuous mapping in $[\alpha^{min}, \alpha^{max}]$ (see Kharazmi and Zayernouri (2018)) and $t > 0$. We note that by choosing the distribution function in the distributed-order derivatives to be the Dirac delta function $\delta(\tau - \tau_0)$, we recover a single (fixed) term fractional derivative, i.e.,

$$\int_{\tau^{min}}^{\tau^{max}} \delta(\tau - \tau_0)\, {}^{\mathrm{RL}}_{0}D_t^{\tau} f(t,x) d\tau = {}^{\mathrm{RL}}_{0}D_t^{\tau_0} f(t,x),$$

where $\tau_0 \in (\tau^{min}, \tau^{max})$.

4.5.1 Mathematical Formulation

We introduce the underlying solution and test spaces along with their proper norms, and also provide some useful lemmas to derive the corresponding bilinear form and thus, prove the well-posedness of problem.

Let $C_0^{\infty}(\Lambda)$ represent the space of smooth functions with compact support in $\Lambda = (a,b)$. Recalling the definition of Sobolev space for real $\sigma \geq 0$ from Kharazmi and Zayernouri (2018); Li and Xu (2010), the usual Sobolev space denoted by $H^{\sigma}(\Lambda)$ is the closure of $C_0^{\infty}(\Lambda)$ on the finite interval Λ, which is associated with the norm $\|\cdot\|_{H^{\sigma}(\Lambda)}$. For the real index $\sigma \geq 0$ and $\sigma \neq n - \frac{1}{2}$ on the bounded interval Λ the following norms are equivalent (see Li and Xu (2010)):

$$\|\cdot\|_{H^{\sigma}(\Lambda)} \cong \|\cdot\|_{{}^{l}H^{\sigma}(\Lambda)} \cong \|\cdot\|_{{}^{r}H^{\sigma}(\Lambda)} \cong |\cdot|^{*}_{H^{\sigma}(\Lambda)}, \tag{4.5.2}$$

where "$\cong$" denotes equivalence relation,

$$\|\cdot\|_{{}^{l}H^{\sigma}(\Lambda)} = \left(\|{}^{\mathrm{RL}}_{a}D_x^{\sigma}(\cdot)\|^2_{L^2(\Lambda)} + \|\cdot\|^2_{L^2(\Lambda)}\right)^{\frac{1}{2}},$$

$$\|\cdot\|_{{}^{r}H^{\sigma}(\Lambda)} = \left(\|{}^{\mathrm{RL}}_{x}D_b^{\sigma}(\cdot)\|^2_{L^2(\Lambda)} + \|\cdot\|^2_{L^2(\Lambda)}\right)^{\frac{1}{2}},$$

and

$$|\cdot|^{*}_{H^{\sigma}(\Lambda)} = |({}^{\mathrm{RL}}_{a}D_x^{\sigma}(\cdot), {}^{\mathrm{RL}}_{x}D_b^{\sigma}(\cdot))_{\Lambda}|^{\frac{1}{2}}.$$

Moreover, from Ervin and Roop (2006, Lemma 5.2), we have

$$|\cdot|^{*}_{{}^{l}H^{\sigma}(\Lambda)} \cong |\cdot|^{\frac{1}{2}}_{{}^{l}H^{\sigma}(\Lambda)}\, |\cdot|^{\frac{1}{2}}_{{}^{r}H^{\sigma}(\Lambda)} = \|{}^{\mathrm{RL}}_{a}D_x^{\sigma}(\cdot)\|^{\frac{1}{2}}_{L^2(\Lambda)}\, \|{}^{\mathrm{RL}}_{x}D_b^{\sigma}(\cdot)\|^{\frac{1}{2}}_{L^2(\Lambda)},$$

where $|\cdot|_{{}^{l}H^{\sigma}(\Lambda)} = \|{}^{\mathrm{RL}}_{a}D_x^{\sigma}(\cdot)\|_{L^2(\Lambda)}$, and $|\cdot|_{{}^{r}H^{\sigma}(\Lambda)} = \|{}^{\mathrm{RL}}_{x}D_b^{\sigma}(\cdot)\|_{L^2(\Lambda)}$. According to Lemma 3.1 in Samiee et al. (2019), the norms $\|\cdot\|_{{}^{l}H^{\sigma}(\Lambda)}$ and $\|\cdot\|_{{}^{r}H^{\sigma}(\Lambda)}$ are equivalent to $\|\cdot\|_{{}^{c}H^{\sigma}(\Lambda)}$ in space $C_0^{\infty}(\Lambda)$, where

$$\|\cdot\|_{{}^{c}H^{\sigma}(\Lambda)} = \left(\|{}^{\mathrm{RL}}_{x}D_b^{\sigma}(\cdot)\|^2_{L^2(\Lambda)} + \|{}^{\mathrm{RL}}_{a}D_x^{\sigma}(\cdot)\|^2_{L^2(\Lambda)} + \|\cdot\|^2_{L^2(\Lambda)}\right)^{\frac{1}{2}}.$$

In the usual Sobolev space, for $u \in H^\sigma(\Lambda)$, we define

$$|u|^*_{H^\sigma(\Lambda)} = |({}^{\text{RL}}_{a}D^{\sigma}_{x}\, u, {}^{\text{RL}}_{x}D^{\sigma}_{b}\, v)_\Lambda|^{\frac{1}{2}} + |({}^{\text{RL}}_{x}D^{\sigma}_{b}\, u, {}^{\text{RL}}_{a}D^{\sigma}_{x}\, v)_\Lambda|^{\frac{1}{2}}, \quad \forall v \in H^\sigma(\Lambda),$$

where we assume

$$\sup_{u \in H^\sigma(\Lambda)} |({}^{\text{RL}}_{a}D^{\sigma}_{x}\, u, {}^{\text{RL}}_{x}D^{\sigma}_{b}\, v)_\Lambda|^{\frac{1}{2}} + |({}^{\text{RL}}_{x}D^{\sigma}_{b}\, u, {}^{\text{RL}}_{a}D^{\sigma}_{x}\, v)_\Lambda|^{\frac{1}{2}} > 0, \quad \forall v \in H^\sigma(\Lambda),$$

which excludes the solutions to $|u|^*_{H^\sigma(\Lambda)} = 0$. Denoted by ${}^{l}H^{\sigma}_0(\Lambda)$ and ${}^{r}H^{\sigma}_0(\Lambda)$ are the closure of $C^\infty_0(\Lambda)$ with respect to the norms $\|\cdot\|_{{}^{l}H^\sigma(\Lambda)}$ and $\|\cdot\|_{{}^{r}H^\sigma(\Lambda)}$in Λ, respectively.

Recalling from Chapter 3, ${}^{\mathfrak{D}}H^{\varphi}(\mathbb{R})$ represents the *distributed Sobolev* space on $\mathbb{R}$, which is associated with the norm

$$\|\cdot\|_{{}^{\mathfrak{D}}H^{\varphi}(\mathbb{R})} = \left(\int_{\tau^{min}}^{\tau^{max}} \varphi(\tau)\ \|(1+|\omega|^2)^{\frac{\tau}{2}} \mathcal{F}(\cdot)(\omega)\|^2_{L^2(\mathbb{R})}\ d\tau \right)^{\frac{1}{2}},$$

where $0 < \varphi(\tau) \in L^1([\tau^{min}, \tau^{max}])$, $0 < \tau^{min} < \tau^{max} < 1$ $(1 < \tau^{min} < \tau^{max} \leq 2)$. Subsequently, we denote by ${}^{\mathfrak{D}}H^{\varphi}(I)$ the *distributed Sobolev* space on the bounded open interval $I = (0, T)$, which is defined as

$${}^{\mathfrak{D}}H^{\varphi}(I) = \{v \in L^2(I)|\ \exists \tilde{v} \in {}^{\mathfrak{D}}H^{\varphi}(\mathbb{R})\ \ s.t.\ \ \tilde{v}|_I = v\}$$

with the equivalent norms $\|\cdot\|_{l,{}^{\mathfrak{D}}H^{\varphi}(I)}$ and $\|\cdot\|_{r,{}^{\mathfrak{D}}H^{\varphi}(I)}$ in Kharazmi and Zayernouri (2018), where

$$\|\cdot\|_{l,{}^{\mathfrak{D}}H^{\varphi}(I)} = \left(\|\cdot\|^2_{L^2(I)} + \int_{\tau^{min}}^{\tau^{max}} \varphi(\tau)\ \|\ {}^{\text{RL}}_{0}D^{\tau}_{t}(\cdot)\|^2_{L^2(I)}\ d\tau \right)^{\frac{1}{2}}$$

and

$$\|\cdot\|_{r,{}^{\mathfrak{D}}H^{\varphi}(I)} = \left(\|\cdot\|^2_{L^2(I)} + \int_{\tau^{min}}^{\tau^{max}} \varphi(\tau)\ \|\ {}^{\text{RL}}_{t}D^{\tau}_{T}(\cdot)\|^2_{L^2(I)}\ d\tau \right)^{\frac{1}{2}}.$$

In each realization of a physical process (e.g., sub- or superdiffusion), the distribution function $\varphi(\tau)$ can be obtained from experimental observations, while the theoretical setting of the problem remains invariant. More importantly, choice of *distributed Sobolev* space and the associated norms provides a sharper estimate for the accuracy of the proposed PG method.

Let $\Lambda_1 = (a_1, b_1)$, $\Lambda_i = (a_i, b_i) \times \Lambda_{i-1}$ for $i = 2, \ldots, d$. We define $\mathcal{X}_1 = {}^{\mathfrak{D}}H^{\rho_1}(\Lambda_1)$ with the associated norm $\|\cdot\|_{{}^{\mathfrak{D}}H^{\rho_1}(\Lambda_1)}$, where

$$\|\cdot\|_{{}^{\mathfrak{D}}H^{\rho_1}(\Lambda_1)} = \tag{4.5.3}$$

$$\left\{ \|\cdot\|^2_{L^2(I)} + \int_{\nu_1^{min}}^{\nu_1^{max}} \rho_1(\nu_1) \left(\|\ {}^{\text{RL}}_{a_1}D^{\nu_1}_{x_1}(\cdot)\|^2_{L^2(\Lambda_1)} + \|\ {}^{\text{RL}}_{x_1}D^{\nu_1}_{b_1}(\cdot)\|^2_{L^2(\Lambda_1)} \right\} d\nu_1 \right)^{\frac{1}{2}}.$$

Subsequently, we construct $\mathcal{X}_d$ such that

$$\mathcal{X}_2 = {}^{\mathfrak{D}}H^{\rho_2}\big((a_2,b_2);L^2(\Lambda_1)\big) \cap L^2((a_2,b_2);\mathcal{X}_1),$$
$$\vdots$$
$$\mathcal{X}_d = {}^{\mathfrak{D}}H^{\rho_d}\big((a_d,b_d);L^2(\Lambda_{d-1})\big) \cap L^2((a_d,b_d);\mathcal{X}_{d-1}),$$

associated with the norm

$$\|\cdot\|_{\mathcal{X}_d} = \left\{ \|\cdot\|^2_{{}^{\mathfrak{D}}H^{\rho_d}((a_d,b_d);L^2(\Lambda_{d-1}))} + \|\cdot\|^2_{L^2((a_d,b_d);\mathcal{X}_{d-1})} \right\}^{\frac{1}{2}}.$$

Lemma 4.58 *Let* $\nu_i > 0$ *and* $\nu_i \neq n - \frac{1}{2}$ *for* $i = 1, \dots, d$. *Then*

$$\|\cdot\|_{\mathcal{X}_d} \cong \left\{ \sum_{i=1}^{d} \int_{\nu_i^{min}}^{\nu_i^{max}} \rho_i(\nu_i)\left(\| {}^{RL}_{x_i}D^{\nu_i}_{b_i}(\cdot)\|^2_{L^2(\Lambda_d)} + \| {}^{RL}_{a_i}D^{\nu_i}_{x_i}(\cdot)\|^2_{L^2(\Lambda_d)}\right)d\nu_i + \|\cdot\|^2_{L^2(\Lambda_d)} \right\}^{\frac{1}{2}}. \tag{4.5.4}$$

Proof Considering (4.5.3), $\mathcal{X}_1$ is endowed with $\|\cdot\|_{\mathcal{X}_1} \cong \|\cdot\|_{{}^{\mathfrak{D}}H^{\rho_1}(\Lambda_1)}$. $\mathcal{X}_2$ is associated with

$$\|\cdot\|_{\mathcal{X}_2} = \left\{ \|\cdot\|^2_{{}^{c}H^{\rho_2}((a_2,b_2);L^2(\Lambda_1))} + \|\cdot\|^2_{L^2((a_2,b_2);\mathcal{X}_1)} \right\}^{\frac{1}{2}},$$

which is proved to be

$$\begin{aligned}
&\|u\|^2_{{}^{\mathfrak{D}}H^{\rho_2}((a_2,b_2);L^2(\Lambda_1))} \\
&= \int_{\nu_2^{min}}^{\nu_2^{max}} \rho_2(\nu_2) \int_{a_1}^{b_1} \Big(\int_{a_2}^{b_2} | {}^{RL}_{a_2}D^{\nu_2}_{x_2} u|^2 \, dx_2 \\
&\quad + \int_{a_2}^{b_2} | {}^{RL}_{x_2}D^{\nu_2}_{b_2} u|^2 \, dx_2 + \int_{a_2}^{b_2} |u|^2 \, dx_2 \Big) \, dx_1 \, d\nu_2 \\
&= \int_{\nu_2^{min}}^{\nu_2^{max}} \rho_2(\nu_2)\left(\| {}^{RL}_{x_2}D^{\nu_2}_{b_2}(u)\|^2_{L^2(\Lambda_2)} + \| {}^{RL}_{a_2}D^{\nu_2}_{x_2}(u)\|^2_{L^2(\Lambda_2)}\right)d\nu_2 + \|u\|^2_{L^2(\Lambda_2)}
\end{aligned} \tag{4.5.5}$$

and

$$\begin{aligned}
&\|u\|^2_{L^2((a_2,b_2);\mathcal{X}_1)} \\
&= \int_{\nu_1^{min}}^{\nu_1^{max}} \rho_1(\nu_1) \int_{a_2}^{b_2} \Big(\int_{a_1}^{b_1} | {}^{RL}_{a_1}D^{\nu_1}_{x_1} u|^2 \, dx_1 \\
&\qquad + \int_{a_1}^{b_1} | {}^{RL}_{x_1}D^{\nu_1}_{b_1} u|^2 \, dx_1 + \int_{a_1}^{b_1} |u|^2 \, dx_1 \Big) \, dx_2 \, d\nu_1 \\
&= \int_{\nu_1^{min}}^{\nu_1^{max}} \rho_1(\nu_1)\left(\| {}^{RL}_{x_1}D^{\nu_1}_{b_1} u\|^2_{L^2(\Lambda_2)} + \| {}^{RL}_{a_1}D^{\nu_1}_{x_1} u\|^2_{L^2(\Lambda_2)}\right)d\nu_1 + \|u\|^2_{L^2(\Lambda_2)}.
\end{aligned} \tag{4.5.6}$$

Next, providing that

$$\|\cdot\|_{\mathcal{X}_{d-1}} \cong \left\{ \sum_{i=1}^{d-1} \int_{\nu_i^{min}}^{\nu_i^{max}} \rho_i(\nu_i) \Big(\| {}^{\mathrm{RL}}_{x_i}D_{b_i}^{\nu_i} (\cdot) \|^2_{L^2(\Lambda_{d-1})} + \| {}^{\mathrm{RL}}_{a_i}D_{x_i}^{\nu_i} (\cdot) \|^2_{L^2(\Lambda_{d-1})} \Big) d\nu_i + \| \cdot \|^2_{L^2(\Lambda_{d-1})} \right\}^{\frac{1}{2}},$$

then the inductive step is attained according to

$$\begin{aligned} &\|u\|^2_{{}^{\mathfrak{D}}H^{\rho_d}((a_d,b_d);L^2(\Lambda_{d-1}))} \qquad (4.5.7) \\ &= \int_{\Lambda_{d-1}} \Big(\int_{a_d}^{b_d} |u|^2 \, dx_d \\ &\quad + \int_{a_d}^{b_d} \int_{\nu_d^{min}}^{\nu_d^{max}} \rho_d(\nu_d) (| {}^{\mathrm{RL}}_{a_d}D_{x_d}^{\nu_D} u|^2 + | {}^{\mathrm{RL}}_{x_d}D_{b_d}^{\nu_D} u|^2) d\nu_d \, dx_d \Big) d\Lambda_{d-1} \\ &= \int_{\nu_d^{min}}^{\nu_d^{max}} \rho_d(\nu_d) \Big(\| {}^{\mathrm{RL}}_{x_d}D_{b_d}^{\nu_D} (u) \|^2_{L^2(\Lambda_d)} + \| {}^{\mathrm{RL}}_{a_d}D_{x_d}^{\nu_D} (u) \|^2_{L^2(\Lambda_d)} \Big) d\nu_d + \|u\|^2_{L^2(\Lambda_d)}, \end{aligned}$$

and

$$\begin{aligned} &\|u\|^2_{L^2((a_d,b_d);\mathcal{X}_{d-1})} \qquad (4.5.8) \\ &= \int_{a_d}^{b_d} \int_{\Lambda_{d-1}} \Big(\sum_{i=1}^{d-1} \int_{\nu_i^{min}}^{\nu_i^{max}} \rho_i(\nu_i) \, (| {}^{\mathrm{RL}}_{a_i}D_{x_i}^{\nu_i} u|^2 + | {}^{\mathrm{RL}}_{x_i}D_{b_i}^{\nu_i} u|^2) d\nu_i \Big) d\Lambda_{d-1} \, dx_d \\ &\quad + \int_{a_d}^{b_d} \int_{\Lambda_{d-1}} |u|^2 d\Lambda_{d-1} \, dx_d \\ &= \sum_{i=1}^{d-1} \int_{\nu_i^{min}}^{\nu_i^{max}} \rho_i(\nu_i) \, (\| {}^{\mathrm{RL}}_{x_i}D_{b_i}^{\nu_i} \, u \|^2_{L^2(\Lambda_d)} + \| {}^{\mathrm{RL}}_{a_i}D_{x_i}^{\nu_i} \, u \|^2_{L^2(\Lambda_d)}) d\nu_i + \|u\|^2_{L^2(\Lambda_d)}. \end{aligned}$$

Therefore, (4.5.4) arises from (4.5.5), (4.5.6), (4.5.7), and (4.5.8) by induction and the proof is complete. □

The following assumptions allow us to prove the uniqueness of the bilinear form by excluding the solutions to $|({}^{\mathrm{RL}}_{0}D_{t}^{\tau}u, {}^{\mathrm{RL}}_{t}D_{T}^{\tau}v)_{\Omega}| = 0$ and

$$|\big({}^{\mathrm{RL}}_{a_i}D_{x_i}^{\nu_i} u, {}^{\mathrm{RL}}_{x_i}D_{b_i}^{\nu_i} v \big)_{\Lambda_d}| + |\big({}^{\mathrm{RL}}_{x_i}D_{b_i}^{\nu_i} u, {}^{\mathrm{RL}}_{a_i}D_{x_i}^{\nu_i} v \big)_{\Lambda_d}| = 0 \quad \text{for } i = 1, \ldots, d.$$

Assumption 4.59 *For* $u \in \mathcal{X}_d$,

$$\sup_{u \in \mathcal{X}_d} \int_{\nu_i^{min}}^{\nu_i^{max}} \rho_i(\nu_i) \Big(|\big({}^{\mathrm{RL}}_{a_i}D_{x_i}^{\nu_i} u, {}^{\mathrm{RL}}_{x_i}D_{b_i}^{\nu_i} v \big)_{\Lambda_d}| + |\big({}^{\mathrm{RL}}_{x_i}D_{b_i}^{\nu_i} u, {}^{\mathrm{RL}}_{a_i}D_{x_i}^{\nu_i} v \big)_{\Lambda_d}| \Big) d\nu_i > 0$$

for all $v \in \mathcal{X}_d$, *when* $i = 1, \dots, d$, *and* $\Lambda_d^i = \prod_{\substack{j=1 \\ j \neq i}}^{d} (a_j, b_j)$.

Assumption 4.60 *For* $u \in {}^{l,\mathfrak{D}}H^{\varphi}(I; L^2(\Lambda_d))$,

$$\sup_{0 \neq u \in {}^{l,\mathfrak{D}}H^{\varphi}(I;L^2(\Lambda_d))} \int_{\tau^{min}}^{\tau^{max}} \varphi(\tau)\, |({}^{\mathrm{RL}}_{0}D_t^{\tau} u, {}^{\mathrm{RL}}_{t}D_T^{\tau} v)_{\Omega}| d\tau > 0$$

for all $v \in {}^{r,\mathfrak{D}}H^{\varphi}(I; L^2(\Lambda_d))$.

In Lemma 3.3 in Samiee et al. (2019), it is shown that if $1 < 2\nu_i < 2$ for $i = 1, \dots, d$ and $u, v \in \mathcal{X}_d$, then $\left({}^{\mathrm{RL}}_{x_i}D_{b_i}^{2\nu_i} u, v\right)_{\Lambda_d} = \left({}^{\mathrm{RL}}_{x_i}D_{b_i}^{\nu_i} u, {}^{\mathrm{RL}}_{a_i}D_{x_i}^{\nu_i} v\right)_{\Lambda_d}$, and $\left({}^{\mathrm{RL}}_{a_i}D_{x_i}^{2\nu_i} u, v\right)_{\Lambda_d} = \left({}^{\mathrm{RL}}_{a_i}D_{x_i}^{\nu_i} u, {}^{\mathrm{RL}}_{x_i}D_{b_i}^{\nu_i} v\right)_{\Lambda_d}$. Consequently, we derive

$$\int_{\nu_i^{min}}^{\nu_i^{max}} \rho_i(\nu_i) \left({}^{\mathrm{RL}}_{x_i}D_{b_i}^{2\nu_i} u, v\right)_{\Lambda_d} d\nu_i = \int_{\nu_i^{min}}^{\nu_i^{max}} \rho_i(\nu_i) \left({}^{\mathrm{RL}}_{x_i}D_{b_i}^{\nu_i} u, {}^{\mathrm{RL}}_{a_i}D_{x_i}^{\nu_i} v\right)_{\Lambda_d} d\nu_i \tag{4.5.9}$$

and

$$\int_{\nu_i^{min}}^{\nu_i^{max}} \rho_i(\nu_i) \left({}^{\mathrm{RL}}_{a_i}D_{x_i}^{2\nu_i} u, v\right)_{\Lambda_d} d\nu_i = \int_{\nu_i^{min}}^{\nu_i^{max}} \rho_i(\nu_i) \left({}^{\mathrm{RL}}_{a_i}D_{x_i}^{\nu_i} u, {}^{\mathrm{RL}}_{x_i}D_{b_i}^{\nu_i} v\right)_{\Lambda_d} d\nu_i. \tag{4.5.10}$$

Additionally, in the light of Lemma 3.2 in Samiee et al. (2019), we have

$$\begin{aligned} &\int_{\nu_i^{min}}^{\nu_i^{max}} \rho_i(\nu_i) \Big(|\left({}^{\mathrm{RL}}_{a_i}D_{x_i}^{\nu_i} u, {}^{\mathrm{RL}}_{x_i}D_{b_i}^{\nu_i} v\right)_{\Lambda_d}| + |\left({}^{\mathrm{RL}}_{x_i}D_{b_i}^{\nu_i} u, {}^{\mathrm{RL}}_{a_i}D_{x_i}^{\nu_i} v\right)_{\Lambda_d}|\Big) d\nu_i \qquad (4.5.11) \\ &\cong |u|_{{}^{\mathfrak{D}}H^{\rho_i}\left((a_i,b_i);L^2(\Lambda_d^i)\right)} |v|_{{}^{\mathfrak{D}}H^{\rho_i}\left((a_i,b_i);L^2(\Lambda_d^i)\right)} \end{aligned}$$

for $i = 1, \dots, d$, where Assumption 4.59 holds.

Next, we study the property of the fractional time-derivative in the following lemmas.

Lemma 4.61 *If* $0 < 2\tau^{min} < 2\tau^{max} < 1$ $(1 < 2\tau^{min} < 2\tau^{max} < 2)$ *and* $u, v \in {}^{l,\mathfrak{D}}H^{\varphi}(I)$, *when* $u|_{t=0} (= \frac{du}{dt}|_{t=0}) = 0$, *then*

$$\int_{\tau^{min}}^{\tau^{max}} \varphi(\tau) \left({}^{\mathrm{RL}}_{x_j}D_{b_j}^{\mu_j} u, v\right)_I d\tau = \int_{\tau^{min}}^{\tau^{max}} \varphi(\tau) \left({}^{\mathrm{RL}}_{0}D_t^{\tau} u, {}^{\mathrm{RL}}_{t}D_T^{\tau} v\right)_I d\tau, \tag{4.5.12}$$

where $I = (0, T)$, $0 < \varphi(\tau) \in L^1\Big([\tau^{min}, \tau^{max}]\Big)$.

Proof It follows from Kharazmi and Zayernouri (2018) that for $u, v \in H^{\tau}(I)$, when $u|_{t=0} (= \frac{du}{dt}|_{t=0}) = 0$ and $v|_{t=T} (= \frac{dv}{dt}|_{t=T}) = 0$, we have

$$\left({}^{\mathrm{RL}}_{x_j}D_{b_j}^{\mu_j} u, v\right)_I = \left({}^{\mathrm{RL}}_{0}D_t^{\tau} u, {}^{\mathrm{RL}}_{t}D_T^{\tau} v\right)_I. \tag{4.5.13}$$

Then (4.5.12) arises from (4.5.13). □

Let $0 < 2\tau^{min} < 2\tau^{max} < 1$ $(1 < 2\tau^{min} < 2\tau^{max} \leq 2)$, and $\Omega = I \times \Lambda_d$, where $I = (0, T)$ and $\Lambda_d = \prod_{i=1}^{d}(a_i, b_i)$. We define

$$
\begin{aligned}
{}^{l,\mathfrak{D}}H^{\varphi}\Big(I; L^2(\Lambda_d)\Big) := \Big\{u \mid \|u(t,\cdot)\|_{L^2(\Lambda_d)} \in {}^{l,\mathfrak{D}}H^{\varphi}(I),& \\
u|_{t=0} = (\frac{\mathrm{d}u}{\mathrm{d}t})|_{t=0} = u|_{x_i=a_i} = u|_{x_i=b_i} = 0,\ i = 1, \ldots, d\Big\},&
\end{aligned}
\tag{4.5.14}
$$

which is endowed with the norm $\|\cdot\|_{{}^{l,\mathfrak{D}}H^{\varphi}\left(I;L^2(\Lambda_d)\right)}$, where we have

$$
\begin{aligned}
\|u\|_{{}^{l,\mathfrak{D}}H^{\varphi}(I;L^2(\Lambda_d))} &= \Big\| \|u(t,\cdot)\|_{L^2(\Lambda_d)} \Big\|_{{}^{l,\mathfrak{D}}H^{\varphi}(I)} \\
&= \Big(\int_{\tau^{min}}^{\tau^{max}} \varphi(\tau) \| {}^{\mathrm{RL}}_{0}D_t^{\tau}(u)\|^2_{L^2(\Omega)} \,\mathrm{d}\tau + \|u\|^2_{L^2(\Omega)}\Big)^{\frac{1}{2}}.
\end{aligned}
\tag{4.5.15}
$$

Similarly, we define

$$
\begin{aligned}
{}^{r,\mathfrak{D}}H^{\varphi}\Big(I; L^2(\Lambda_d)\Big) := \Big\{v \mid \|v(t,\cdot)\|_{L^2(\Lambda_d)} \in {}^{r,\mathfrak{D}}H^{\varphi}(I),& \\
v|_{t=T}(= \frac{\mathrm{d}v}{\mathrm{d}t}|_{t=0}) = v|_{x_i=a_i} = v|_{x_i=b_i} = 0,\ i = 1, \ldots, d\Big\},&
\end{aligned}
\tag{4.5.16}
$$

which is equipped with the norm $\|\cdot\|_{{}^{r,\mathfrak{D}}H^{\varphi}(I;L^2(\Lambda_d))}$. Following (4.5.15),

$$
\begin{aligned}
\|v\|_{{}^{r,\mathfrak{D}}H^{\varphi}(I;L^2(\Lambda_d))} &= \Big\| \|v(t,\cdot)\|_{L^2(\Lambda_d)} \Big\|_{{}^{r,\mathfrak{D}}H^{\varphi}(I)} \\
&= \Big(\int_{\tau^{min}}^{\tau^{max}} \varphi(\tau) \| {}^{\mathrm{RL}}_{t}D_T^{\tau}(v)\|^2_{L^2(\Omega)} \mathrm{d}\tau + \|v\|^2_{L^2(\Omega)}\Big)^{\frac{1}{2}}.
\end{aligned}
$$

Lemma 4.62 *For $u \in {}^{r,\mathfrak{D}}H^{\varphi}(I; L^2(\Lambda_d))$ and $0 < 2\tau^{min} < 2\tau^{max} < 1$ $(1 < 2\tau^{min} < 2\tau^{max} < 2)$, we have*

$$
\int_{\tau^{min}}^{\tau^{max}} \varphi(\tau)\, |({}^{\mathrm{RL}}_{0}D_t^{\tau}u, {}^{\mathrm{RL}}_{t}D_T^{\tau}v)_{\Omega}| \,\mathrm{d}\tau \leq \|u\|_{{}^{l,\mathfrak{D}}H^{\varphi}(I;L^2(\Lambda_d))} \|v\|_{{}^{r,\mathfrak{D}}H^{\varphi}(I;L^2(\Lambda_d))}
$$

for all $v \in {}^{r,\mathfrak{D}}H^{\varphi}(I; L^2(\Lambda_d))$.

Proof From the previous section, we have

$$
|({}^{\mathrm{RL}}_{0}D_t^{\tau}u, {}^{\mathrm{RL}}_{t}D_T^{\tau}v)_{\Omega}| \leq \Big(\| {}^{\mathrm{RL}}_{0}D_t^{\tau}u\|^2_{L^2(\Omega)} + \|u\|^2_{L^2(\Omega)}\Big)^{\frac{1}{2}} \Big(\| {}^{\mathrm{RL}}_{t}D_T^{\tau}v\|^2_{L^2(\Omega)} + \|v\|^2_{L^2(\Omega)}\Big)^{\frac{1}{2}}.
$$

Then by the Hölder inequality,

$$
\begin{aligned}
&\int_{\tau^{min}}^{\tau^{max}} \varphi(\tau)\,|({}^{\mathrm{RL}}_{0}D_t^{\tau}u, {}^{\mathrm{RL}}_{t}D_T^{\tau}v)_{\Omega}|\mathrm{d}\tau \\
&= \int_{\tau^{min}}^{\tau^{max}} \varphi(\tau) \int_{\Lambda_d}\int_0^T |{}^{\mathrm{RL}}_{0}D_t^{\tau}u|\,|{}^{\mathrm{RL}}_{t}D_T^{\tau}v|\mathrm{d}t\,\mathrm{d}\Lambda_d\,\mathrm{d}\tau \\
&\le \Big(\int_{\tau^{min}}^{\tau^{max}} \int_{\Lambda_d}\int_0^T \varphi(\tau)\,|{}^{\mathrm{RL}}_{0}D_t^{\tau}u|^2\,\mathrm{d}t\mathrm{d}\Lambda_d\Big)^{\frac{1}{2}} \\
&\quad\times\Big(\int_{\tau^{min}}^{\tau^{max}} \int_{\Lambda_d}\int_0^T \varphi(\tau)\,|{}^{\mathrm{RL}}_{t}D_T^{\tau}v|^2\,\mathrm{d}t\mathrm{d}\Lambda_d\Big)^{\frac{1}{2}} \\
&= \Big(\int_{\tau^{min}}^{\tau^{max}} \varphi(\tau)\,\|{}^{\mathrm{RL}}_{0}D_t^{\tau}u\|^2_{L^2(\Omega)}\mathrm{d}\tau + \|u\|^2_{L^2(\Omega)}\Big)^{\frac{1}{2}} \\
&\quad\times\Big(\int_{\tau^{min}}^{\tau^{max}} \varphi(\tau)\,\|{}^{\mathrm{RL}}_{t}D_T^{\tau}v\|^2_{L^2(\Omega)}\mathrm{d}\tau + \|v\|^2_{L^2(\Omega)}\Big)^{\frac{1}{2}} \\
&= \|u\|_{{}^{r,\mathfrak{D}}H^{\varphi}(I;L^2(\Lambda_d))}\,\|v\|_{{}^{r,\mathfrak{D}}H^{\varphi}(I;L^2(\Lambda_d))}.
\end{aligned}
$$

This ends the proof. □

Lemma 4.63 *For any $u \in {}^{l,\mathfrak{D}}H^{\varphi}(I;L^2(\Lambda_d))$ and $0 < 2\tau^{min} < 2\tau^{max} < 1$ $(1 < 2\tau^{min} < 2\tau^{max} \le 2)$, there exists a constant $c > 0$ and independent of u such that*

$$
\sup_{0\neq v\in {}^{r,\mathfrak{D}}H^{\varphi}(I;L^2(\Lambda_d))} \frac{\int_{\tau^{min}}^{\tau^{max}} \varphi(\tau)\,|({}^{\mathrm{RL}}_{0}D_t^{\tau}u, {}^{\mathrm{RL}}_{t}D_T^{\tau}v)_{\Omega}|\mathrm{d}\tau}{|v|_{{}^{r,\mathfrak{D}}H^{\varphi}(I;L^2(\Lambda_d))}} \ge c|u|_{{}^{l,\mathfrak{D}}H^{\varphi}(I;L^2(\Lambda_d))}, \tag{4.5.17}
$$

under Assumption 4.60.

Proof Following Lemma 2.4 in Duan et al. (2017) and Lemma 3.7 in Samiee et al. (2019), for any $u \in {}^{l,\mathfrak{D}}H^{\varphi}(I;L^2(\Lambda_d))$ let $\mathcal{V}_u = H(t-T)\big(u - u|_{t=T}\big)$ assuming that $\int_{\tau^{min}}^{\tau^{max}} \varphi(\tau)\,|({}^{\mathrm{RL}}_{0}D_t^{\tau}u, {}^{\mathrm{RL}}_{t}D_T^{\tau}u|_{t=T})_{\Omega}|\mathrm{d}\tau > 0$, where $H(t)$ is the Heaviside function. Evidently, $\mathcal{V}_u \in {}^{r,\mathfrak{D}}H^{\varphi}(I;L^2(\Lambda_d))$. From the Hölder inequality, we obtain

$$
\begin{aligned}
&\|\mathcal{V}_u\|^2_{{}^{r,\mathfrak{D}}H^{\varphi}(I;L^2(\Lambda_d))} \\
&= \int_{\tau^{min}}^{\tau^{max}} \varphi(\tau)\,\|{}^{\mathrm{RL}}_{t}D_T^{\tau}(H(t-T)\left(u-u|_{t=T}\right))\|^2_{L^2(\Omega)}\mathrm{d}\tau \\
&= \int_{\tau^{min}}^{\tau^{max}} \varphi(\tau)\,\|{}^{\mathrm{RL}}_{t}I_T^{1-\tau}\frac{\mathrm{d}}{\mathrm{d}t}(H(t-T)\left(u-u|_{t=T}\right))\|^2_{L^2(\Omega)}\mathrm{d}\tau
\end{aligned}
$$

$$
\begin{aligned}
&= \int_{\tau^{min}}^{\tau^{max}} \varphi(\tau) \left\| {}^{\mathrm{RL}}_{\ t}I_T^{1-\tau} \left(\frac{\mathrm{d}\, H(t-T)}{\mathrm{d}t} \left(u - u|_{t=T}\right) + H(t-T) \frac{d\left(u - u|_{t=T}\right)}{dt} \right) \right\|_{L^2(\Omega)}^2 \mathrm{d}\tau \\
&= \int_{\tau^{min}}^{\tau^{max}} \varphi(\tau) \left\| {}^{\mathrm{RL}}_{\ t}I_T^{1-\tau} \left(H(t-T) \frac{\mathrm{d}\left(u - u|_{t=T}\right)}{\mathrm{d}t} \right) \right\|_{L^2(\Omega)}^2 \mathrm{d}\tau \\
&= \int_{\tau^{min}}^{\tau^{max}} \varphi(\tau) \, \| {}^{\mathrm{RL}}_{\ t}D_T^{\tau} u \|_{L^2(\Omega)}^2 \mathrm{d}\tau. \qquad (4.5.18)
\end{aligned}
$$

Using Samiee et al. (2019, (12)), we have

$$
\|\mathcal{V}_u\|^2_{{}^{r,\mathfrak{D}} H^{\varphi}(I;L^2(\Lambda_d))} \cong \int_{\tau^{min}}^{\tau^{max}} \varphi(\tau) \, \| {}^{\mathrm{RL}}_{\ 0}D_t^{\tau} u \|_{L^2(\Omega)}^2 \mathrm{d}\tau = \|u\|^2_{{}^{l,\mathfrak{D}} H^{\varphi}(I;L^2(\Lambda_d))}.
$$

Hence, $\| {}^{\mathrm{RL}}_{\ t}D_T^{\tau} \mathcal{V}_u \|^2_{L^2(\Omega)} \cong \| {}^{\mathrm{RL}}_{\ 0}D_t^{\tau} u \|^2_{L^2(\Omega)}$. Therefore,

$$
\begin{aligned}
\int_{\tau^{min}}^{\tau^{max}} \varphi(\tau) \, |({}^{\mathrm{RL}}_{\ 0}D_t^{\tau} u, {}^{\mathrm{RL}}_{\ t}D_T^{\tau} \mathcal{V}_u)_{\Omega}| \, \mathrm{d}\tau &= \int_{\tau^{min}}^{\tau^{max}} \varphi(\tau) \int_{\Lambda_d} \int_0^T | {}^{\mathrm{RL}}_{\ 0}D_t^{\tau} u| \, | {}^{\mathrm{RL}}_{\ t}D_T^{\tau} \mathcal{V}_u | dt \, \mathrm{d}\Lambda_d \, \mathrm{d}\tau \\
&\geq \tilde{\beta} \int_{\tau^{min}}^{\tau^{max}} \varphi(\tau) \int_{\Lambda_d} \int_0^T | {}^{\mathrm{RL}}_{\ 0}D_t^{\tau} u|^2 dt \, \mathrm{d}\Lambda_d \, \mathrm{d}\tau \\
&= |u|^2_{{}^{l,\mathfrak{D}} H^{\varphi}(I;L^2(\Lambda_d))}, \qquad (4.5.19)
\end{aligned}
$$

where $\tilde{\beta} > 0$ and independent of u. Considering (4.5.18) and (4.5.19), we obtain

$$
\begin{aligned}
&\sup_{0 \neq v \in {}^{r,\mathfrak{D}} H^{\varphi}(I;L^2(\Lambda_d))} \frac{\int_{\tau^{min}}^{\tau^{max}} \varphi(\tau) \, |({}^{\mathrm{RL}}_{\ 0}D_t^{\tau} u, {}^{\mathrm{RL}}_{\ t}D_T^{\tau} v)_{\Omega}| \mathrm{d}\tau}{|v|_{{}^{r,\mathfrak{D}} H^{\varphi}(I;L^2(\Lambda_d))}} \\
&\geq \frac{\int_{\tau^{min}}^{\tau^{max}} \varphi(\tau) \, |({}^{\mathrm{RL}}_{\ 0}D_t^{\tau} u, {}^{\mathrm{RL}}_{\ t}D_T^{\tau} \mathcal{V}_u)_{\Omega}| \mathrm{d}\tau}{|\mathcal{V}_u|_{{}^{r,\mathfrak{D}} H^{\varphi}(I;L^2(\Lambda_d))}} \\
&\geq \tilde{\beta} \, |u|_{{}^{l,\mathfrak{D}} H^{\varphi}(I;L^2(\Lambda_d))}. \qquad (4.5.20)
\end{aligned}
$$

This ends the proof. □

Lemma 4.64 *If* $0 < 2\tau^{min} < 2\tau^{max} < 1$ $(1 < 2\tau^{min} < 2\tau^{max} \leq 2)$ *and* $u, v \in {}^{l,\mathfrak{D}} H^{\varphi}(I; L^2(\Lambda_d))$, *then*

$$
\int_{\tau^{min}}^{\tau^{max}} \varphi(\tau) \left({}^{\mathrm{RL}}_{\ x_j}D_{b_j}^{\mu_j} u, v \right)_{\Omega} \mathrm{d}\tau = \int_{\tau^{min}}^{\tau^{max}} \varphi(\tau) \left({}^{\mathrm{RL}}_{\ 0}D_t^{\tau} u, {}^{\mathrm{RL}}_{\ t}D_T^{\tau} v \right)_{\Omega} \mathrm{d}\tau, \qquad (4.5.21)
$$

where $0 < \varphi(\tau) \in L^1\Big([\tau^{min}, \tau^{max}]\Big)$.

Proof By Lemma 4.61,

$$\int_{\tau^{min}}^{\tau^{max}} \varphi(\tau)\left({}^{\mathrm{RL}}_{x_j}D^{\mu_j}_{b_j}u, v\right)_{\Omega} d\tau = \int_{\tau^{min}}^{\tau^{max}} \varphi(\tau) \int_{\Lambda_d} \int_0^T |\,{}^{\mathrm{RL}}_{x_j}D^{\mu_j}_{b_j}u|\,|v|\, dt\, d\Lambda_d\, d\tau$$

$$= \int_{\Lambda_d} \int_{\tau^{min}}^{\tau^{max}} \varphi(\tau) \int_0^T |\,{}^{\mathrm{RL}}_{0}D^{\tau}_{t}u|\,|\,{}^{\mathrm{RL}}_{t}D^{\tau}_{T}v|\, dt\, d\tau\, d\Lambda_d$$

$$= \int_{\tau^{min}}^{\tau^{max}} \varphi(\tau)\left({}^{\mathrm{RL}}_{0}D^{\tau}_{t}u, {}^{\mathrm{RL}}_{t}D^{\tau}_{T}v\right)_{\Omega} d\tau.$$

This ends the derivation. □

4.5.2 Solution and Test Function Spaces

Take $0 < 2\tau^{min} < 2\tau^{max} < 1$ ($1 < 2\tau^{min} < 2\tau^{max} \leq 2$) and $1 < 2\nu_i^{min} < 2\nu_i^{max} \leq 2$ for $i = 1, \dots, d$. We define the solution space

$$\mathcal{B}^{\varphi,\rho_1,\dots,\rho_d}(\Omega) := {}^{l,\mathfrak{D}}H^{\varphi}\Big(I; L^2(\Lambda_d)\Big) \cap L^2(I; \mathcal{X}_d),$$

associated with the norm

$$\|u\|_{\mathcal{B}^{\varphi,\rho_1,\dots,\rho_d}(\Omega)} = \left\{\|u\|^2_{{}^{l,\mathfrak{D}}H^{\varphi}(I;L^2(\Lambda_d))} + \|u\|^2_{L^2(I;\mathcal{X}_d)}\right\}^{\frac{1}{2}}.$$

Considering Lemma 4.58,

$$\|u\|_{L^2(I;\mathcal{X}_d)} = \Big\|\, \|u(t,.)\|_{\mathcal{X}_d} \Big\|_{L^2(I)} \tag{4.5.22}$$
$$= \left\{\sum_{i=1}^{d} \int_{\nu_i^{min}}^{\nu_i^{max}} \rho_i(\nu_i)\left(\|\,{}^{\mathrm{RL}}_{x_i}D^{\nu_i}_{b_i}(u)\|^2_{L^2(\Omega)} + \|\,{}^{\mathrm{RL}}_{a_i}D^{\nu_i}_{x_i}(u)\|^2_{L^2(\Omega)}\right) d\nu_i + \|u\|^2_{L^2(\Lambda_d)}\right\}^{\frac{1}{2}}.$$

Therefore, from (4.5.15) and (4.5.22),

$$\|u\|_{\mathcal{B}^{\varphi,\rho_1,\dots,\rho_d}(\Omega)} = \left\{\|u\|^2_{L^2(\Omega)} + \int_{\tau^{min}}^{\tau^{max}} \varphi(\tau)\,\|\,{}^{\mathrm{RL}}_{0}D^{\tau}_{t}(u)\|^2_{L^2(\Omega)} d\tau\right.$$
$$\left. + \sum_{i=1}^{d} \int_{\nu_i^{min}}^{\nu_i^{max}} \rho_i(\nu_i)\left(\|\,{}^{\mathrm{RL}}_{x_i}D^{\nu_i}_{b_i}(u)\|^2_{L^2(\Omega)} + \|\,{}^{\mathrm{RL}}_{a_i}D^{\nu_i}_{x_i}(u)\|^2_{L^2(\Omega)}\right) d\nu_i\right\}^{\frac{1}{2}}. \tag{4.5.23}$$

Similarly, we define the test space

$$\mathfrak{B}^{\varphi,\rho_1,\dots,\rho_d}(\Omega) := {}^{r,\mathfrak{D}}H^{\varphi}\Big(I; L^2(\Lambda_d)\Big) \cap L^2(I; \mathcal{X}_d),$$

equipped with the norm

$$\begin{aligned}\|v\|_{\mathfrak{B}^{\varphi,\rho_1,\ldots,\rho_d}(\Omega)} &= \left\{\|v\|^2_{{}^rH^{\varphi}(I;L^2(\Lambda_d))} + \|v\|^2_{L^2(I;\mathcal{X}_d)}\right\}^{\frac{1}{2}} \\ &= \left\{\|v\|^2_{L^2(\Omega)} + \int_{\tau^{min}}^{\tau^{max}} \varphi(\tau)\, \| {}^{RL}_{t}D^{\tau}_{T}\,(v)\|^2_{L^2(\Omega)} d\tau \right. \\ &\left. + \sum_{i=1}^{d}\int_{\nu_i^{min}}^{\nu_i^{max}} \rho_i(\nu_i)\left(\| {}^{RL}_{x_i}D^{\nu_i}_{b_i}\,(v)\|^2_{L^2(\Omega)} + \| {}^{RL}_{a_i}D^{\nu_i}_{x_i}\,(v)\|^2_{L^2(\Omega)}\right) d\nu_i\right\}^{\frac{1}{2}}\end{aligned}$$

by Lemma (4.58) and (4.5.15). Take $\Omega = I \times \Lambda_d$ for a positive integer d. The Petrov–Galerkin spectral method reads as: find $u \in \mathcal{B}^{\varphi,\rho_1,\ldots,\rho_d}(\Omega)$ such that

$$a(u, v) = l(v), \quad \forall v \in \mathfrak{B}^{\varphi,\rho_1,\ldots,\rho_d}(\Omega), \tag{4.5.24}$$

where the functional $l(v) = (f, v)_\Omega$ and

$$\begin{aligned} a(u, v) = &\int_{\tau^{min}}^{\tau^{max}} \varphi(\tau)\, ({}^{RL}_{0}D^{\tau}_{t}\,u, {}^{RL}_{t}D^{\tau}_{T}\,v)_\Omega d\tau \\ &+ \sum_{i=1}^{d}\int_{\mu_i^{min}}^{\mu_i^{max}} \rho_i(\mu_i)\left(c_{l_i}({}^{RL}_{a_i}D^{\mu_i}_{x_i}\,u, {}^{RL}_{x_i}D^{\mu_i}_{b_i}\,v)_\Omega + c_{r_i}({}^{RL}_{a_i}D^{\mu_i}_{x_i}\,v, {}^{RL}_{x_i}D^{\mu_i}_{b_i}\,u)_\Omega\right) d\mu_i \\ &- \sum_{j=1}^{d}\int_{\nu_j^{min}}^{\nu_j^{max}} \rho_j(\nu_j)\left(k_{l_j}({}^{RL}_{a_j}D^{\nu_j}_{x_j}\,u, {}^{RL}_{x_j}D^{\nu_j}_{b_j}\,v)_\Omega + k_{r_j}({}^{RL}_{a_j}D^{\nu_j}_{x_j}\,v, {}^{RL}_{x_j}D^{\nu_j}_{b_j}\,u)_\Omega\right) d\nu_j \\ &+ \gamma(u, v)_\Omega, \end{aligned} \tag{4.5.25}$$

following (4.5.9), (4.5.10) and Lemma 4.64 and $\gamma, c_{l_i}, c_{r_i}, \kappa_{l_i}$, and κ_{r_i} are all constant. Besides, $0 < 2\tau^{min} < 2\tau^{max} < 1$ $(1 < 2\tau^{min} < 2\tau^{max} \le 2)$, $0 < 2\mu_i^{min} < 2\mu_i^{max} < 1$ and $1 < 2\nu_j^{min} < 2\nu_j^{max} \le 2$ for $i, j = 1, 2, \ldots, d$.

Remark 4.65 In the case $\tau < \frac{1}{2}$, additional regularity assumptions are required to ensure equivalence between the weak and strong formulations, see Jin et al. (2017a) for more details.

U_N and V_N are chosen as the finite-dimensional subspaces of $\mathcal{B}^{\varphi,\rho_1,\ldots,\rho_d}(\Omega)$ and $\mathfrak{B}^{\varphi,\rho_1,\ldots,\rho_d}(\Omega)$, respectively. Then, the Petrov–Galerkin scheme reads as: find $u_N \in U_N$ such that

$$a(u_N, v_N) = l(v_N), \quad \forall v \in V_N, \tag{4.5.26}$$

where

$$
\begin{aligned}
a(u_N, v_N) &= \int_{\tau^{min}}^{\tau^{max}} \varphi(\tau)\, ({}^{\text{RL}}_{0}D_t^{\tau}\, u_N, {}^{\text{RL}}_{t}D_T^{\tau}\, v_N)_\Omega d\tau \\
&+ \sum_{i=1}^{d} \int_{\mu_i^{min}}^{\mu_i^{max}} \rho_i(\mu_i) \Big[c_{l_i} ({}^{\text{RL}}_{a_i}D_{x_i}^{\mu_i}\, u_N, {}^{\text{RL}}_{x_i}D_{b_i}^{\mu_i}\, v_N)_\Omega + c_{r_i} ({}^{\text{RL}}_{a_i}D_{x_i}^{\mu_i}\, u_N, {}^{\text{RL}}_{x_i}D_{b_i}^{\mu_i}\, v_N)_\Omega \Big] d\mu_i \\
&- \sum_{j=1}^{d} \int_{\nu_j^{min}}^{\nu_j^{max}} \rho_j(\nu_j) \Big[k_{l_j} ({}^{\text{RL}}_{a_j}D_{x_j}^{\nu_j}\, u_N, {}^{\text{RL}}_{x_j}D_{b_j}^{\nu_j}\, v_N)_\Omega + k_{r_j} ({}^{\text{RL}}_{a_j}D_{x_j}^{\nu_j}\, u_N, {}^{\text{RL}}_{x_j}D_{b_j}^{\nu_j}\, v_N)_\Omega \Big] d\nu_j \\
&+ \gamma (u_N, v_N)_\Omega. \qquad (4.5.27)
\end{aligned}
$$

Representing u_N as a linear combination of elements in U_N, the finite-dimensional problem (4.5.27) leads to a linear system, known as a Lyapunov system, introduced in Section 4.5.4.

4.5.3 Well-Posedness Analysis

The following assumption permits us to prove uniqueness of the weak form in (4.5.24) in Theorem 4.69.

Assumption 4.66 *For all* $v \in \mathfrak{B}^{\varphi,\rho_1,\dots,\rho_d}(\Omega)$,

$$
\begin{aligned}
&\sup_{u \in \mathcal{B}^{\varphi,\rho_1,\dots,\rho_d}(\Omega)} \int_{\tau^{min}}^{\tau^{max}} \varphi(\tau)\, |({}^{\text{RL}}_{0}D_t^{\tau}\, u, {}^{\text{RL}}_{t}D_T^{\tau}\, v)_\Omega| d\tau > 0, \\
&\sup_{u \in \mathcal{B}^{\varphi,\rho_1,\dots,\rho_d}(\Omega)} \int_{\nu_j^{min}}^{\nu_j^{max}} \rho_j(\nu_j) \Big(|({}^{\text{RL}}_{a_j}D_{x_j}^{\nu_j}\, u, {}^{\text{RL}}_{x_j}D_{b_j}^{\nu_j}\, v)_\Omega| + |({}^{\text{RL}}_{x_j}D_{b_j}^{\nu_j}\, u, {}^{\text{RL}}_{a_j}D_{x_j}^{\nu_j}\, v)_\Omega| \Big) d\nu_j > 0, \\
&\sup_{u \in \mathcal{B}^{\varphi,\rho_1,\dots,\rho_d}(\Omega)} |(u, v)_\Omega| > 0,
\end{aligned}
$$

when $j = 1, \dots, d$.

Lemma 4.67 *(Continuity) Let Assumption 4.66 hold. The bilinear form in* (4.5.25) *is continuous, i.e., for* $u \in \mathcal{B}^{\varphi,\rho_1,\dots,\rho_d}(\Omega)$,

$$
\exists \beta > 0, \;\; |a(u, v)| \le \beta\, \|u\|_{\mathcal{B}^{\varphi,\rho_1,\dots,\rho_d}(\Omega)} \|v\|_{\mathfrak{B}^{\varphi,\rho_1,\dots,\rho_d}(\Omega)} \;\; \forall v \in \mathfrak{B}^{\varphi,\rho_1,\dots,\rho_d}(\Omega).
$$

Proof It follows from (4.5.11) and Lemma 4.62. □

Theorem 4.68 *Let Assumption 4.66 hold. The inf-sup condition of the bilinear form* (4.5.25) *for any* $d \ge 1$ *holds with* $\beta > 0$*, i.e.,*

$$
\inf_{0 \neq u \in \mathcal{B}^{\varphi,\rho_1,\dots,\rho_d}(\Omega)} \; \sup_{0 \neq v \in \mathfrak{B}^{\varphi,\rho_1,\dots,\rho_d}(\Omega)} \frac{|a(u, v)|}{\|v\|_{\mathfrak{B}^{\varphi,\rho_1,\dots,\rho_d}(\Omega)} \|u\|_{\mathcal{B}^{\varphi,\rho_1,\dots,\rho_d}(\Omega)}} \ge \beta > 0,
$$

where $\Omega = I \times \Lambda_d$.

Proof For $u \in \mathcal{B}^{\varphi,\rho_1,\ldots,\rho_d}(\Omega)$ and $v \in \mathfrak{B}^{\varphi,\rho_1,\ldots,\rho_d}(\Omega)$ under Assumption 4.66,

$$
\begin{aligned}
|a(u,v)| \cong |(u,v)_\Omega| &+ \int_{\tau^{min}}^{\tau^{max}} \varphi(\tau)\, |({}^{\mathrm{RL}}_{0}D_t^{\tau}\, u, {}^{\mathrm{RL}}_{t}D_T^{\tau}\, v)_\Omega| d\tau \\
&+ \sum_{i=1}^{d} \int_{\mu_i^{min}}^{\mu_i^{max}} \rho_i(\mu_i) \Big(|({}^{\mathrm{RL}}_{a_i}D_{x_i}^{\mu_i}\, u, {}^{\mathrm{RL}}_{x_i}D_{b_i}^{\mu_i}\, v)_\Omega| + |({}^{\mathrm{RL}}_{x_i}D_{a_i}^{\mu_i}\, u, {}^{\mathrm{RL}}_{b_i}D_{x_i}^{\mu_i}\, v)_\Omega| \Big) d\mu_i \\
&+ \sum_{j=1}^{d} \int_{\nu_j^{min}}^{\nu_j^{max}} \rho_j(\nu_j) \Big(|({}^{\mathrm{RL}}_{a_j}D_{x_j}^{\nu_j}\, u, {}^{\mathrm{RL}}_{x_j}D_{b_j}^{\nu_j}\, v)_\Omega| + |({}^{\mathrm{RL}}_{x_j}D_{b_j}^{\nu_j}\, u, {}^{\mathrm{RL}}_{a_j}D_{x_j}^{\nu_j}\, v)_\Omega| \Big) d\nu_j.
\end{aligned}
\tag{4.5.28}
$$

Following (4.5.11) and Theorem 4.3 in Samiee et al. (2019),

$$
\begin{aligned}
&\sum_{i=1}^{d} \int_{\nu_i^{min}}^{\nu_i^{max}} \rho_i(\nu_i) \Big(|({}^{\mathrm{RL}}_{a_i}D_{x_i}^{\nu_i}\,(u), {}^{\mathrm{RL}}_{x_i}D_{b_i}^{\nu_i}\,(v))_\Omega| + |({}^{\mathrm{RL}}_{x_i}D_{b_i}^{\nu_i}\,(u), {}^{\mathrm{RL}}_{a_i}D_{x_i}^{\nu_i}\,(v))_\Omega| \Big) d\nu_i \\
&\geq \tilde{C}_1 \sum_{i=1}^{d} \Big[\int_{\nu_i^{min}}^{\nu_i^{max}} \rho_i(\nu_i) \Big(\| {}^{\mathrm{RL}}_{a_i}D_{x_i}^{\nu_i}\,(u) \|_{L^2(\Omega)} \Big) d\nu_i \int_{\nu_i^{min}}^{\nu_i^{max}} \rho_i(\nu_i) \Big(\| {}^{\mathrm{RL}}_{x_i}D_{b_i}^{\nu_i}\,(v) \|_{L^2(\Omega)} \Big) d\nu_i \\
&\quad + \int_{\nu_i^{min}}^{\nu_i^{max}} \rho_i(\nu_i) \Big(\| {}^{\mathrm{RL}}_{x_i}D_{b_i}^{\nu_i}\,(u) \|_{L^2(\Omega)} \Big) d\nu_i \int_{\nu_i^{min}}^{\nu_i^{max}} \rho_i(\nu_i) \Big(\| {}^{\mathrm{RL}}_{a_i}D_{x_i}^{\nu_i}\,(v) \|_{L^2(\Omega)} \Big) d\nu_i \Big].
\end{aligned}
$$

Thus,

$$
\begin{aligned}
&\sum_{i=1}^{d} \int_{\nu_i^{min}}^{\nu_i^{max}} \rho_i(\nu_i) \Big(|({}^{\mathrm{RL}}_{a_i}D_{x_i}^{\nu_i}\,(u), {}^{\mathrm{RL}}_{x_i}D_{b_i}^{\nu_i}\,(v))_\Omega| + |({}^{\mathrm{RL}}_{x_i}D_{b_i}^{\nu_i}\,(u), {}^{\mathrm{RL}}_{a_i}D_{x_i}^{\nu_i}\,(v))_\Omega| \Big) d\nu_i \\
&\geq \tilde{C}_1 \sum_{i=1}^{d} \int_{\nu_i^{min}}^{\nu_i^{max}} \rho_i(\nu_i) \Big(\| {}^{\mathrm{RL}}_{a_i}D_{x_i}^{\nu_i}\,(u) \|_{L^2(\Omega)} + \| {}^{\mathrm{RL}}_{x_i}D_{b_i}^{\nu_i}\,(u) \|_{L^2(\Omega)} \Big) d\nu_i \\
&\quad \times \sum_{j=1}^{d} \int_{\nu_j^{min}}^{\nu_j^{max}} \rho_j(\nu_j) \Big(\| {}^{\mathrm{RL}}_{x_j}D_{b_j}^{\nu_j}\,(v) \|_{L^2(\Omega)} + \| {}^{\mathrm{RL}}_{a_j}D_{x_j}^{\nu_j}\,(v) \|_{L^2(\Omega)} \Big) d\nu_j \\
&= \tilde{C}_1 |u|_{L^2(I;\mathcal{X}_d)}\, |v|_{L^2(I;\mathcal{X}_d)},
\end{aligned}
\tag{4.5.29}
$$

where $\tilde{C}_1$ is a positive constant and independent of u. Considering Lemma 4.63, there exists a positive constant $\tilde{C}_2 > 0$ and independent of u such that

$$
\sup_{0 \neq v \in \mathfrak{B}^{\varphi,\rho_1,\ldots,\rho_d}(\Omega)} \frac{\int_{\tau^{min}}^{\tau^{max}} \varphi(\tau)\, |({}^{\mathrm{RL}}_{0}D_t^{\tau}(u), {}^{\mathrm{RL}}_{t}D_T^{\tau}(v))_\Omega| d\tau}{|v|_{r,\mathfrak{D}}\, {}_{H^{\varphi}(I;L^2(\Lambda_d))}} \geq \tilde{C}_2 |u|_{l,\mathfrak{D}}\, {}_{H^{\varphi}(I;L^2(\Lambda_d))}.
\tag{4.5.30}
$$

Furthermore, for $u \in \mathcal{B}^{\varphi,\rho_1,\ldots,\rho_d}(\Omega)$,

$$
\begin{aligned}
&\sup_{0 \neq v \in \mathfrak{B}^{\varphi,\ldots,\rho_d}(\Omega)} \frac{\int_{\tau^{min}}^{\tau^{max}} \varphi(\tau)\, |({}^{\mathrm{RL}}_{0}D_t^{\tau}(u), {}^{\mathrm{RL}}_{t}D_T^{\tau}(v))_\Omega| d\tau}{|v|_{r,\mathfrak{D}}\, {}_{H^{\varphi}(I;L^2(\Lambda_d))}} \\
&\cong \sup_{0 \neq v \in \mathfrak{B}^{\varphi,\ldots,\rho_d}(\Omega)} \frac{\int_{\tau^{min}}^{\tau^{max}} \varphi(\tau)\, |({}^{\mathrm{RL}}_{0}D_t^{\tau}(u), {}^{\mathrm{RL}}_{t}D_T^{\tau}(v))_\Omega| d\tau}{|v|_{\mathcal{B}^{\varphi,\rho_1,\ldots,\rho_d}(\Omega)}}
\end{aligned}
\tag{4.5.31}
$$

and

$$\sup_{0\neq v\in \mathfrak{B}^{\varphi,\rho_1,\dots,\rho_d}(\Omega)} \frac{\sum_{j=1}^{d}\int_{\nu_j^{min}}^{\nu_j^{max}} \rho_j(\nu_j)\Big(|({}^{RL}_{a_j}D_{x_j}^{\nu_j}\,u,\, {}^{RL}_{x_j}D_{b_j}^{\nu_j}\,v)_\Omega| + |({}^{RL}_{x_j}D_{b_j}^{\nu_j}\,u,\, {}^{RL}_{a_j}D_{x_j}^{\nu_j}\,v)_\Omega|\Big)d\nu_j}{\|v\|_{L^2(I;\mathcal{X}_d)}}$$

$$\cong \sup_{0\neq v\in \mathfrak{B}^{\varphi,\rho_1,\dots,\rho_d}(\Omega)} \frac{\sum_{j=1}^{d}\int_{\nu_j^{min}}^{\nu_j^{max}} \rho_j(\nu_j)\Big(|({}^{RL}_{a_j}D_{x_j}^{\nu_j}\,u,\, {}^{RL}_{x_j}D_{b_j}^{\nu_j}\,v)_\Omega| + |({}^{RL}_{x_j}D_{b_j}^{\nu_j}\,u,\, {}^{RL}_{a_j}D_{x_j}^{\nu_j}\,v)_\Omega|\Big)d\nu_j}{\|v\|_{\mathfrak{B}^{\varphi,\rho_1,\dots,\rho_d}(\Omega)}}. \tag{4.5.32}$$

Therefore, from (4.5.29), (4.5.30), (4.5.31), and (4.5.32), we have

$$\sup_{0\neq v\in \mathfrak{B}^{\varphi,\rho_1,\dots,\rho_d}(\Omega)} \frac{|a(u,v)|}{\|v\|_{\mathfrak{B}^{\varphi,\rho_1,\dots,\rho_d}(\Omega)}} \tag{4.5.33}$$

$$\geq \bar{\beta} \sup_{0\neq v\in \mathfrak{B}^{\varphi,\rho_1,\dots,\rho_d}(\Omega)} \frac{|(u,v)_\Omega| + \int_{\tau^{min}}^{\tau^{max}} \varphi(\tau)\,|({}^{RL}_{0}D_{t}^{\tau}\,u,\, {}^{RL}_{t}D_{T}^{\tau}\,v)_\Omega|d\tau}{\|v\|_{\mathfrak{B}^{\varphi,\rho_1,\dots,\rho_d}(\Omega)}}$$

$$+\frac{\sum_{j=1}^{d}\int_{\nu_j^{min}}^{\nu_j^{max}} \rho_j(\nu_j)\Big(|({}^{RL}_{a_j}D_{x_j}^{\nu_j}\,u,\, {}^{RL}_{x_j}D_{b_j}^{\nu_j}\,v)_\Omega| + |({}^{RL}_{x_j}D_{b_j}^{\nu_j}\,u,\, {}^{RL}_{a_j}D_{x_j}^{\nu_j}\,v)_\Omega|\Big)d\nu_j}{\|v\|_{\mathfrak{B}^{\varphi,\rho_1,\dots,\rho_d}(\Omega)}}$$

$$\geq \bar{\beta}\,\bar{C}\,\Big(\|u\|_{L^2(\Omega)} + |u|_{I,\mathfrak{D}\,H^{\varphi}(I;L^2(\Lambda_d))} + |u|_{L^2(I;\mathcal{X}_d)}\Big), \tag{4.5.34}$$

where $\bar{C} = \min\{\tilde{C}_2,\, \tilde{C}_1\}$. Accordingly,

$$\inf_{0\neq u\in \mathcal{B}^{\varphi,\rho_1,\dots,\rho_d}(\Omega)}\ \sup_{0\neq v\in \mathfrak{B}^{\varphi,\rho_1,\dots,\rho_d}(\Omega)} \frac{|a(u,v)|}{\|v\|_{\mathfrak{B}^{\varphi,\rho_1,\dots,\rho_d}(\Omega)}} \geq \beta\, \|u\|_{\mathcal{B}^{\varphi,\rho_1,\dots,\rho_d}(\Omega)}, \tag{4.5.35}$$

where $\beta = \bar{\beta}\,\bar{C}$ is a positive constant and independent. □

Theorem 4.69 (Well-Posedness) *For $0 < 2\tau^{min} < 2\tau^{max} < 1$ $(1 < 2\tau^{min} < 2\tau^{max} \leq 2)$, $1 < 2\nu_i^{min} < 2\nu_i^{max} \leq 2$, and $i = 1,\dots,d$, there exists a unique solution to* (4.5.26), *which is continuously dependent on $f \in \big(\mathcal{B}^{\tau,\nu_1,\dots,\nu_d}\big)^{\star}(\Omega)$, where $\big(\mathcal{B}^{\tau,\nu_1,\dots,\nu_d}\big)^{\star}(\Omega)$ is the dual space of $\mathcal{B}^{\tau,\nu_1,\dots,\nu_d}(\Omega)$.*

Proof In virtue of the generalized Babuska–Lax–Milgram theorem (see Shen et al. (2011a)), the well-posedness of the weak form in (4.5.24) in $(1+d)$ dimensions is guaranteed by the continuity and the *inf-sup* condition, which are proven in Lemma 4.67 and Theorem 4.68, respectively. □

4.5.4 Construction of the Unified Spectral Method

To construct a Petrov–Galerkin spectral method for the finite-dimensional weak form problem in (4.5.26), we first define the proper finite-dimensional basis/test spaces and then implement the numerical scheme.

Space of Basis (U_N) and Test (V_N) Functions

As discussed in the previous section, we take the spatial basis, given in the standard domain $\xi \in [-1,1]$ as $\phi_m(\xi) = \sigma_m\big(P_{m+1}(\xi) - P_{m-1}(\xi)\big)$, $m = 1,2,\ldots$, where $P_m(\xi)$ is the Legendre polynomials of order m and $\sigma_m = 2+(-1)^m$. Besides, employing Jacobi *poly-fractonomials* of the first kind (see, e.g., Zayernouri and Karniadakis (2013); Zayernouri et al. (2015)), the temporal basis functions are given in the standard domain $\eta \in [-1,1]$ as $\psi_n^\tau(\eta) = \sigma_n(1+\eta)^\tau P_{n-1}^{(-\tau,\tau)}(\eta)$, $n = 1,2,\ldots$.

We also let $\eta(t) = 2t/T - 1$ and $\xi_j(s) = 2\frac{s-a_j}{b_j-a_j} - 1$ be temporal and spatial affine mappings from $t \in [0,T]$ and $x_j \in [a_j,b_j]$ to the standard domain $[-1,1]$, respectively. Therefore,

$$U_N = \mathrm{span}\Big\{\Big(\psi_n^\tau\circ\eta\Big)(t)\prod_{j=1}^{d}\Big(\phi_{m_j}\circ\xi_j\Big)(x_j) : n = 1,2,\ldots,\mathcal{N},\ m_j = 1,2,\ldots,\mathcal{M}_j\Big\}.$$

Similarly, we employ Legendre polynomials and Jacobi *poly-fractonomials* of the second kind in the standard domain to construct the finite-dimensional test space as

$$V_N = \mathrm{span}\Big\{\Big(\Psi_r^\tau\circ\eta\Big)(t)\prod_{j=1}^{d}\Big(\Phi_{k_j}\circ\xi_j\Big)(x_j) : r = 1,2,\ldots,\mathcal{N},\ k_j = 1,2,\ldots,\mathcal{M}_j\Big\},$$

where $\Psi_r^\tau(\eta) = \widetilde{\sigma}_r(1-\eta)^\tau P_{r-1}^{(\tau,-\tau)}(\eta)$, $r = 1,2,\ldots$ and $\Phi_k(\xi) = \widetilde{\sigma}_k\big(P_{k+1}(\xi) - P_{k-1}(\xi)\big)$, $k = 1,2,\ldots$. The coefficient $\widetilde{\sigma}_k$ is defined as $\widetilde{\sigma}_k = 2\,(-1)^k + 1$.

Since the univariate basis/test functions belong to the fractional Sobolev spaces and $0 < \varphi(\tau) \in L^1((\tau^{min},\tau^{max}))$, $0 < \rho_j(\nu_j) \in L^1((\nu_j^{min},\nu_j^{max}))$ for $j = 1,\ldots,d$, then $U_N \subset \mathcal{B}^{\varphi,\rho_1,\ldots,\rho_d}(\Omega)$ and $V_N \subset \mathfrak{B}^{\varphi,\rho_1,\ldots,\rho_d}(\Omega)$. Accordingly, we approximate the solution in terms of a linear combination of elements in U_N, which satisfies initial and boundary conditions.

Implementation of the PG Spectral Method

The solution u_N of (4.5.26) can be represented as

$$u_N(x,t) = \sum_{n=1}^{\mathcal{N}}\sum_{m_1=1}^{\mathcal{M}_1}\cdots\sum_{m_d=1}^{\mathcal{M}_d}\hat{u}_{n,m_1,\ldots,m_d}\Big[\psi_n^\tau(t)\prod_{j=1}^{d}\phi_{m_j}(x_j)\Big]$$

in Ω and also we take $v_N = \Psi_r^\tau(t)\prod_{j=1}^{d}\Phi_{k_j}(x_j)$, $r = 1,2,\ldots,\mathcal{N}$, $k_j = 1,2,\ldots,\mathcal{M}_j$. Accordingly, by replacing u_N and v_N in (4.5.26), we obtain the following Lyapunov system:

$$\Big(S_\tau^{\varphi} \otimes M_1 \otimes M_2 \cdots \otimes M_d + \sum_{j=1}^{d} [M_\tau \otimes M_1 \otimes \cdots \otimes M_{j-1} \otimes S_j^{Tot} \otimes M_{j+1} \cdots \otimes M_d]$$
$$+ \gamma M_\tau \otimes M_1 \otimes M_2 \cdots \otimes M_d \Big) \mathcal{U} = F, \tag{4.5.36}$$

in which $\otimes$ represents the Kronecker product, F denotes the multi-dimensional load matrix whose entries are given as

$$F_{r,k_1,\dots,k_d} = \int_\Omega f(t, x_1, \dots, x_d) \Big(\Psi_r^\tau \circ \eta\Big)(t) \prod_{j=1}^{d} \Big(\Phi_{k_j} \circ \xi_j\Big)(x_j)\, d\Omega, \tag{4.5.37}$$

and $S_j^{Tot} = c_{l_j} S_l^{\rho_j} + c_{r_j} S_r^{\rho_j} - \kappa_{l_j} S_l^{\rho_j} - \kappa_{r_j} S_r^{\rho_j}$. The matrices S_τ^{φ} and M_τ denote the temporal stiffness and mass matrices, respectively; $S_l^{\rho_j}$, $S_r^{\rho_j}$, $S_l^{\rho_j}$, $S_r^{\rho_j}$, and M_j denote the spatial stiffness and mass matrices. The entries of the spatial mass matrix M_j are computed analytically, while we employ proper quadrature rules to accurately compute the entries of the temporal mass matrix M_τ as discussed in the previous section. The entries of S_τ^{φ} are also computed based on the spectrally/exponentially accurate quadrature rule in α-dimension in Chapter 2. Likewise, we present the computation of S_j^{Tot}. Here, we provide the computation of entries of the spatial stiffness matrix by performing an affine mapping ϑ from the standard domain $\mu_j^{stn} \in [-1, 1]$ to $\mu_j \in [\mu_j^{max}, \mu_j^{min}]$.

Lemma 4.70 *The total spatial stiffness matrix S_j^{Tot} is symmetric and its entries can be exactly computed as:*

$$S_j^{Tot} = c_{l_j} \times S_l^{\rho_j} + c_{r_j} \times S_r^{\rho_j} - \kappa_{l_j} \times S_l^{\rho_j} - \kappa_{r_j} \times S_r^{\rho_j},$$

where $j = 1, 2, \dots, d$.

Proof Regarding the definition of the stiffness matrix, we have

$$\begin{aligned}
\{S_l^{\rho_j}\}_{r,n} &= \int_{-1}^{1} \int_{\mu_j^{min}}^{\mu_j^{max}} \rho_j(\mu_j)\, {}^{\mathrm{RL}}_{-1}D_{\xi_j}^{\mu_j}\Big(\phi_n(x_j)\Big)\, {}^{\mathrm{RL}}_{\xi_j}D_{1}^{\mu_j}\Big(\Phi_r(x_j)\Big)\, d\mu_j dx_j,\\
&= \beta_1 \int_{-1}^{1} \int_{-1}^{1} \rho_j(\vartheta(\mu_j^{stn}))\, {}^{\mathrm{RL}}_{-1}D_{\xi_j}^{\mu_j^{stn}}\Big(P_{n+1}(\xi_j) - P_{n-1}(\xi_j)\Big)\\
&\qquad \times {}^{\mathrm{RL}}_{\xi_j}D_{1}^{\mu_j^{stn}}\Big(P_{k+1}(\xi_j) - P_{k-1}(\xi_j)\Big)\, d\xi_j d\mu_j^{\mathrm{stn}},\\
&= \beta_1 \Big[\widetilde{S}_{r+1,n+1}^{\rho_j} - \widetilde{S}_{r+1,n-1}^{\rho_j} - \widetilde{S}_{r-1,n+1}^{\rho_j} + \widetilde{S}_{r-1,n-1}^{\rho_j}\Big],
\end{aligned}$$

where $\beta_1 = \widetilde{\sigma}_r \, \sigma_n \left(\frac{\mu_j^{max} - \mu_j^{min}}{2} \right)$ and

$$
\begin{aligned}
\widetilde{S}^{\rho_j}_{r,n} &= \int_{-1}^{1} \int_{-1}^{1} \rho_j(\vartheta(\mu_j^{stn}))\, {}^{\mathrm{RL}}_{-1}D^{\mu_j^{stn}}_{\xi_j} \left(P_n(\xi_j) \right) {}^{\mathrm{RL}}_{\xi_j}D^{\mu_j^{stn}}_{1} \left(P_r(\xi_j) \right) \mathrm{d}\xi_j \, \mathrm{d}\mu_j^{stn} \\
&= \int_{-1}^{1} \rho_j(\vartheta(\mu_j^{stn})) \, \frac{\Gamma(r+1)}{\Gamma(r-\mu_j^{stn}+1)} \, \frac{\Gamma(n+1)}{\Gamma(n-\mu_j^{stn}+1)} \\
&\quad \times \int_{-1}^{1} (1-\xi_j^2)^{-\mu_j^{stn}} \, P_r^{(-\mu_j^{stn},\mu_j^{stn})} \, P_n^{(\mu_j^{stn},-\mu_j^{stn})} \mathrm{d}\xi_j \, \mathrm{d}\mu_j^{stn}.
\end{aligned}
$$

$\widetilde{S}^{\rho_j}_{r,n}$ can be computed accurately using Gauss–Legendre (GL) quadrature rules as

$$
\begin{aligned}
\widetilde{S}^{\rho_j^{stn}}_{r,n} &= \sum_{q=1}^{Q} \frac{\Gamma(r+1)}{\Gamma(r-\mu_j^{stn}|_q+1)} \, \frac{\Gamma(n+1)}{\Gamma(n-\mu_j^{stn}|_q+1)} \rho_j|_q \, w_q \\
&\quad \times \int_{-1}^{1} (1-\xi_j^2)^{-\mu_j^{stn}|_q} \, P_r^{(-\mu_j^{stn}|_q,\mu_j^{stn}|_q)}(\xi_j) \, P_n^{(\mu_j^{stn}|_q,-\mu_j^{stn}|_q)}(\xi_j) \mathrm{d}\xi_j,
\end{aligned}
$$

in which $\mathcal{Q} \geq \mathcal{M}_j + 2$ represents the minimum number of GL quadrature points $\{\mu_j^{stn}|_q\}_{q=1}^{\mathcal{Q}}$ for *exact* quadrature, and $\{w_q\}_{q=1}^{\mathcal{Q}}$ are the corresponding quadrature weights. Exploiting the property of the Jacobi polynomials where $P_n^{(\alpha,\beta)}(-\xi_j) = (-1)^n P_n^{(\beta,\alpha)}(\xi_j)$, we have $\widetilde{S}^{\rho_j^{stn}}_{r,n} = (-1)^{(r+n)} \, \widetilde{S}^{\rho_j^{stn}}_{n,r}$. Following Samiee et al. (2019), $\widetilde{\sigma}_r$ and σ_n are chosen such that $(-1)^{(n+r)}$ is canceled. Accordingly, $\{S_l^{\rho_j}\}_{n,r} = \{S_l^{\rho_j}\}_{r,n} = \{S_r^{\rho_j}\}_{r,n} = \{S_r^{\rho_j}\}_{r,n}$ due to the symmetry of $S_l^{\rho_j}$ and $S_r^{\rho_j}$. Similarly, we get $\{S_l^{\rho_j}\}_{n,r} = \{S_l^{\rho_j}\}_{r,n} = \{S_r^{\rho_j}\}_{n,r} = \{S_r^{\rho_j}\}_{r,n}$. Eventually, we conclude that the stiffness matrix $S_l^{\rho_j}$, $S_r^{\rho_j}$, $S_l^{\rho_j}$, $S_r^{\rho_j}$, and thereby $\{S_j^{Tot}\}_{n,r}$ as the sum of symmetric matrices is symmetric. □

Remark 4.71 The choices of coefficients in the construction of finite-dimensional basis/test functions lead to symmetric mass/stiffness matrices, which help formulate the following fast solver.

4.5.5 Unified Distributed-Order Fast Solver

In order to formulate a closed-form solution to the Lyapunov system (4.5.36), we follow the previous section and develop a fast solver in terms of the further generalized eigensolutions.

Theorem 4.72 *See the previous section. Take* $\{\vec{e}^{\,j}_{m_j}, \lambda^j_{m_j}\}_{m_j=1}^{\mathcal{M}_j}$ *as the set of general eigensolutions of the spatial stiffness matrix* S_j^{Tot} *with respect to the mass matrix* M_j*. Besides, let* $\{\vec{e}^{\,\tau}_n, \lambda^\tau_n\}_{n=1}^{\mathcal{N}}$ *be the set of general eigensolutions* *of the temporal mass matrix M_τ with respect to the stiffness matrix S^φ_τ. Then the unknown coefficients matrix $\mathcal{U}$ is obtained as*

$$\mathcal{U} = \sum_{n=1}^{\mathcal{N}} \sum_{m_1=1}^{\mathcal{M}_1} \cdots \sum_{m_d=1}^{\mathcal{M}_d} \kappa_{n,m_1,\dots,m_d}\ \vec{e}^{\,\tau}_n \otimes \vec{e}^{\,1}_{m_1} \otimes \cdots \otimes \vec{e}^{\,d}_{m_d}, \tag{4.5.38}$$

where

$$\kappa_{n,m_1,\dots,m_d} = \frac{(\vec{e}^{\,\tau}_n \vec{e}^{\,1}_{m_1} \cdots \vec{e}^{\,d}_{m_d})F}{\left[(\vec{e}^{\,\tau^T}_n S^\varphi_\tau \vec{e}^{\,\tau}_n) \prod_{j=1}^{d} (\vec{e}^{\,j^T}_{m_j} M_j \vec{e}^{\,j}_{m_j})\right] \Lambda_{n,m_1,\dots,m_d}}, \tag{4.5.39}$$

and

$$\Lambda_{n,m_1,\dots,m_d} = \left[(1 + \gamma\, \lambda^\tau_n) + \lambda^\tau_n \sum_{j=1}^{d} (\lambda^j_{m_j})\right].$$

Remark 4.73 The naive computation of all entries in (4.5.39) leads to a computational complexity of $O(\mathcal{N}^{2+2d})$, including construction of stiffness and mass matrices. By performing *sum-factorization*, the operator counts can be reduced to $O(\mathcal{N}^{2+d})$.

4.5.6 Stability and Error Analysis

The following theorems provide the finite-dimensional stability and error analysis of the proposed scheme, based on the well-posedness analysis from Section 4.5.3.

Theorem 4.74 *Let Assumption 4.66 hold. The Petrov–Galerkin spectral method for* (4.5.27) *is stable, i.e.,*

$$\inf_{0 \neq u_N \in U_N} \sup_{0 \neq v \in V_N} \frac{|a(u_N, v_N)|}{\|v_N\|_{\mathfrak{B}^{\varphi,\rho_1,\dots,\rho_d}(\Omega)} \|u_N\|_{\mathcal{B}^{\varphi,\rho_1,\dots,\rho_d}(\Omega)}} \geq \beta > 0 \tag{4.5.40}$$

holds with $\beta > 0$ and independent of N.

Proof Regarding $U_N \subset \mathcal{B}^{\varphi,\rho_1,\dots,\rho_d}(\Omega)$ and $V_N \subset \mathfrak{B}^{\varphi,\rho_1,\dots,\rho_d}(\Omega)$, (4.5.40) follows directly from Theorem 4.68. □

Remark 4.75 The bilinear form (4.5.27) can be expanded in terms of the basis and test functions to obtain the lower limit of β, see the previous section.

Error Analysis

Denoting by $P_{\mathcal{M}}(\Lambda)$ the space of all polynomials of degree $\leq \mathcal{M}$ on $\Lambda \subset \mathbb{R}$, $P^\varphi_{\mathcal{M}}(\Lambda) := P_{\mathcal{M}}(\Lambda) \cap {}^{\mathfrak{D}}H^\varphi(\Lambda)$, where $0 < \varphi(\tau) \in L^1((\tau^{min}, \tau^{max}))$ and ${}^{\mathfrak{D}}H^\varphi(\Lambda)$ is the *distributed Sobolev* space associated with the norm $\|\cdot\|_{{}^{\mathfrak{D}}H^\varphi(\Lambda)}$.

In this section, we take $I_0 = (0,T)$, $I_i = (a_i,b_i)$ for $i = 1,...,d$, $\Lambda_i = I_i \times \Lambda_{i-1}$, and $\Lambda_i^j = \prod_{\substack{k=1\\k\neq j}}^{i} I_k$. Besides, $0 < 2\tau^{min} < 2\tau^{max} < 1$ $(1 < 2\tau^{min} < 2\tau^{max} \leq 2)$, $1 < 2\nu_i^{min} < 2\nu_i^{max} \leq 2$ for $i = 1,\dots,d$. Where there is no confusion, the symbols I_i, Λ_i, and Λ_i^j and the intervals of (τ^{min},τ^{max}) and $(\nu_i^{min},\nu_i^{max})$ will be dropped from the notations.

Theorem 4.76 (Maday, 1990) *Let r_1 be a real number, where $r_1 \neq \mathcal{M}_1 + \frac{1}{2}$, and $1 \leq r_1$. There exists a projection operator $\Pi^{\nu_1}_{r_1,\mathcal{M}_1}$ from $H^{r_1}(\Lambda_1) \cap H_0^{\nu_1}(\Lambda_1)$ to $P^{\nu_1}_{\mathcal{M}_1}(\Lambda_1)$ such that for any $u \in H^{r_1}(\Lambda_1) \cap H_0^{\nu_1}(\Lambda_1)$, we have*

$$\|u - \Pi^{\nu_1}_{r_1,\mathcal{M}_1} u\|_{{}^cH^{\nu_1}(\Lambda_1)} \leq c_1 \mathcal{M}_1^{\nu_1 - r_1} \|u\|_{H^{r_1}(\Lambda_1)},$$

where c_1 is a positive constant.

Theorem 4.77 (Kharazmi and Zayernouri, 2018) *Let $r_0 \geq 1$, $r_0 \neq \mathcal{N} + \frac{1}{2}$. There exists an operator $\Pi^{\varphi}_{r_0,\mathcal{N}}$ from $H^{r_0}(I) \cap {}^{l,\mathfrak{D}}H^{\varphi}(I)$ to $P^{\varphi}_{\mathcal{N}}(\Lambda_1)$ such that for any $u \in H^{r_0}(I) \cap {}^{l,\mathfrak{D}}H^{\varphi}(I)$, we have*

$$\|u - \Pi^{\varphi}_{r_0,\mathcal{N}} u\|^2_{{}^lH^{\varphi}(I)} \leq c_0 \mathcal{N}^{-2r_0} \int_{\tau^{min}}^{\tau^{max}} \varphi(\tau)\, \mathcal{N}^{2\tau} \|u\|_{H^{r_0}(I)} d\tau,$$

where c_0 is a positive constant and $0 < \varphi(\tau) \in L^1((\tau^{min},\tau^{max}))$.

In the following, employing Theorems 4.76 and 4.77 and also Theorem 5.3 from Samiee et al. (2019), we study the properties of higher-dimensional approximation operators in the following lemmas.

Theorem 4.78 *Let $r_1 \geq 1$, $r_1 \neq \mathcal{M}_1 + \frac{1}{2}$. There exists a projection operator $\Pi^{\rho_1}_{r_1,\mathcal{M}_1}(I_1)$ from $H^{r_1}(I_1) \cap {}^{l,\mathfrak{D}}H^{\rho_1}(I_1)$ to $P^{\rho_1}_{\mathcal{M}_1}(I_1)$ such that for any $u \in H^{r_1}(I_1) \cap {}^{l,\mathfrak{D}}H^{\rho_1}(I_1)$, we have*

$$\|u - \Pi^{\rho_1}_{r_1,\mathcal{M}_1} u\|^2_{{}^{l,\mathfrak{D}}H^{\rho_1}(I_1)} \leq \mathcal{M}_1^{-2r_1} \int_{\nu_1^{min}}^{\nu_1^{max}} \rho_1(\nu_1)\, \mathcal{M}_1^{2\nu_1} \|u\|_{H^{r_1}(I_1)} d\nu_1,$$

where $0 < \rho_1(\nu_1) \in L^1((\nu_1^{min},\nu_1^{max}))$.

Proof From Theorem 4.76 for $u \in H^{r_1} \cap {}^{c}H^{\nu_1}$, we have $\|u - \Pi^{\nu_1}_{r_1,\mathcal{M}_1} u\|_{^{c}H^{\nu_1}(\Lambda_1)} \leq \mathcal{M}_1^{\nu_1 - r_1} \|u\|_{H^{r_1}(\Lambda_1)}$. Therefore, for $u \in H^{r_1}(I_1) \cap {}^{l}H^{\rho_1}(I_1)$, we have

$$\begin{aligned}
\|u - \Pi^{\rho_1}_{r_1,\mathcal{M}_1} u\|^2_{^{l,\mathfrak{D}}H^{\rho_1}(I_1)} &= \int_{\nu_1^{min}}^{\nu_1^{max}} \rho_1(\nu_1)\, \|u - \Pi^{\nu_1}_{r_1,\mathcal{M}_1} u\|^2_{^{c}H^{\nu_1}(\Lambda_1)} d\nu_1 \\
&\leq \int_{\nu_1^{min}}^{\nu_1^{max}} \rho_1(\nu_1)\, \mathcal{M}_1^{2\nu_1 - 2r_1} \|u\|^2_{H^{r_1}(\Lambda_1)} d\nu_1 \qquad (4.5.41) \\
&= \mathcal{M}_1^{-2r_1} \int_{\nu_1^{min}}^{\nu_1^{max}} \rho_1(\nu_1)\, \mathcal{M}_1^{2\nu_1} \|u\|_{H^{r_1}(I_1)} d\nu_1.
\end{aligned}$$

□

Lemma 4.79 *Let the real-valued* $1 \leq r_1$, r_2 *and* $\Omega = I_1 \times I_2$. *If*

$$u \in {}^{l,\mathfrak{D}}H^{\rho_2}(I_2, H^{r_1}(I_1)) \cap H^{r_2}(I_2, {}^{l,\mathfrak{D}}H_0^{\rho_1}(I_1)),$$

then

$$\begin{aligned}
&\|u - \Pi^{\rho_1}_{r_1,\mathcal{M}_1} \Pi^{\rho_2}_{r_2,\mathcal{M}_2} u\|^2_{\mathcal{B}^{\rho_1,\rho_2}(\Omega)} \\
&\leq \mathcal{M}_2^{-2r_2} \int_{\nu_2^{min}}^{\nu_2^{max}} \rho_2(\nu_2) \left(\mathcal{M}_2^{2\nu_2} \|u\|_{H^{r_2}(I_2,L^2(I_1))} + \mathcal{M}_2^{2\nu_2} \mathcal{M}_1^{-2r_1} \|u\|_{H^{r_2}(I_2,H^{r_1}(I_1))}\right) d\nu_2 \\
&+ \mathcal{M}_1^{-2r_1} \int_{\nu_1^{min}}^{\nu_1^{max}} \rho_1(\nu_1) \left(\mathcal{M}_1^{2\nu_1} \|u\|_{H^{r_1}(I_1,L^2(I_2))} + \mathcal{M}_1^{2\nu_1} \mathcal{M}_2^{-2r_2} \|u\|_{H^{r_1}(I_1,H^{r_2}(I_2))}\right) d\nu_1 \\
&+ \mathcal{M}_2^{-2r_2} \|u\|_{\mathfrak{D}H^{\rho_1}(I_1,H^{r_2}(I_2))} + \mathcal{M}_1^{-2r_1} \|u\|_{\mathfrak{D}H^{\rho_2}(I_2,H^{r_1}(I_1))}, \qquad (4.5.42)
\end{aligned}$$

where $\|\cdot\|_{\mathcal{B}^{\rho_1,\rho_2}(\Omega)} = \left\{\|\cdot\|^2_{H^{\rho_1}(I_1,L^2(I_2))} + \|\cdot\|^2_{H^{\rho_2}(I_1,L^2(I_1))}\right\}^{\frac{1}{2}}$, $0 < \rho_1(\nu_1) \in L^1([\nu_1^{min}, \nu_1^{max}])$, *and* $0 < \rho_2(\nu_2) \in L^1([\nu_2^{min}, \nu_2^{max}])$.

Proof For $u \in {}^{l,\mathfrak{D}}H^{\rho_2}(I_2, H^{r_1}(I_1)) \cap H^{r_2}(I_2, H^{\rho_1}(I_1))$, evidently $u \in H^{r_2}(I_2, H^{r_1}(I_1))$, $u \in H^{r_2}(I_2, L^2(I_1))$, and $u \in H^{r_1}(I_1, L^2(I_2))$. Besides, from the definition of $\|\cdot\|_{\mathcal{B}^{\rho_1,\rho_2}(\Omega)}$, we have

$$\begin{aligned}
&\|u - \Pi^{\rho_1}_{r_1,\mathcal{M}_1} \Pi^{\rho_2}_{r_2,\mathcal{M}_2} u\|_{\mathcal{B}^{\rho_1,\rho_2}(\Omega)} \\
&= \left\{\|u - \Pi^{\rho_1}_{r_1,\mathcal{M}_1} \Pi^{\rho_2}_{r_2,\mathcal{M}_2} u\|^2_{\mathfrak{D}H^{\rho_1}(I_1,L^2(I_2))} + \|u - \Pi^{\rho_1}_{r_1,\mathcal{M}_1} \Pi^{\rho_2}_{r_2,\mathcal{M}_2} u\|^2_{\mathfrak{D}H^{\rho_2}(I_2,L^2(I_1))}\right\}^{\frac{1}{2}}.
\end{aligned}$$

Following Lemma 5.3 in Samiee et al. (2019) and Theorem 4.78, we have

$$
\begin{aligned}
&\|u - \Pi^{\rho_1}_{r_1,\mathcal{M}_1}\Pi^{\rho_2}_{r_2,\mathcal{M}_2}u\|^2_{\mathfrak{D}H^{\rho_2}(I_2,L^2(I_1))}\\
&= \|u - \Pi^{\rho_2}_{r_2,\mathcal{M}_2}u + \Pi^{\rho_2}_{r_2,\mathcal{M}_2}u - \Pi^{\rho_1}_{r_1,\mathcal{M}_1}\Pi^{\rho_2}_{r_2,\mathcal{M}_2}u\|^2_{\mathfrak{D}H^{\rho_2}(I_2,L^2(I_1))}\\
&\leq \|u - \Pi^{\rho_2}_{r_2,\mathcal{M}_2}u\|^2_{\mathfrak{D}H^{\rho_2}(I_2,L^2(I_1))} + \|\Pi^{\rho_2}_{r_2,\mathcal{M}_2}u - \Pi^{\rho_1}_{r_1,\mathcal{M}_1}\Pi^{\rho_2}_{r_2,\mathcal{M}_2}u\|^2_{\mathfrak{D}H^{\rho_2}(I_2,L^2(I_1))}\\
&\leq \mathcal{M}_2^{-2r_2}\int_{\nu_2^{min}}^{\nu_2^{max}} \rho_2(\nu_2)\,\mathcal{M}_2^{2\nu_2}\|u\|^2_{H^{r_2}(I_2,L^2(I_1))}d\nu_2\\
&\quad + \|(\Pi^{\rho_2}_{r_2,\mathcal{M}_2} - \mathcal{I})(u - \Pi^{\rho_1}_{r_1,\mathcal{M}_1}u)\|^2_{\mathfrak{D}H^{\rho_2}(I_2,L^2(I_1))}\\
&\quad + \|u - \Pi^{\rho_1}_{r_1,\mathcal{M}_1}u\|^2_{\mathfrak{D}H^{\rho_2}(I_2,L^2(I_1))}\\
&\leq \mathcal{M}_2^{-2r_2}\int_{\nu_2^{min}}^{\nu_2^{max}} \rho_2(\nu_2)\,\mathcal{M}_2^{2\nu_2}\|u\|^2_{H^{r_2}(I_2,L^2(I_1))}d\nu_2\\
&\quad + \mathcal{M}_2^{-2r_2}\mathcal{M}_1^{-2r_1}\int_{\nu_2^{min}}^{\nu_2^{max}} \rho_2(\nu_2)\,\mathcal{M}_2^{2\nu_2}\|u\|^2_{H^{r_2}(I_2,H^{r_1}(I_1))}d\nu_2\\
&\quad + \mathcal{M}_1^{-2r_1}\|u\|^2_{\mathfrak{D}H^{\rho_2}(I_2,H^{r_1}(I_1))},
\end{aligned}
\tag{4.5.43}
$$

where $\mathcal{I}$ is the identity operator. Furthermore,

$$
\begin{aligned}
&\|u - \Pi^{\rho_1}_{r_1,\mathcal{M}_1}\Pi^{\rho_2}_{r_2,\mathcal{M}_2}u\|^2_{L^2(I_2,H^{\rho_1}(I_1))}\\
&= \|u - \Pi^{\rho_1}_{r_1,\mathcal{M}_1}u + \Pi^{\rho_1}_{r_1,\mathcal{M}_1}u - \Pi^{\rho_1}_{r_1,\mathcal{M}_1}\Pi^{\rho_2}_{r_2,\mathcal{M}_2}u\|^2_{\mathfrak{D}H^{\rho_1}(I_1,L^2(I_2))}\\
&\leq \|u - \Pi^{\rho_1}_{r_1,\mathcal{M}_1}u\|^2_{\mathfrak{D}H^{\rho_1}(I_1,L^2(I_2))} + \|\Pi^{\rho_1}_{r_1,\mathcal{M}_1}u - \Pi^{\rho_1}_{r_1,\mathcal{M}_1}\Pi^{\rho_2}_{r_2,\mathcal{M}_2}u\|^2_{\mathfrak{D}H^{\rho_1}(I_1,L^2(I_2))}\\
&\leq \mathcal{M}_1^{-2r_1}\int_{\nu_1^{min}}^{\nu_1^{max}} \rho_1(\nu_1)\,\mathcal{M}_1^{2\nu_1}\|u\|^2_{H^{r_1}(I_1,L^2(I_2))}d\nu_1\\
&\quad + \|(\Pi^{\rho_1}_{r_1,\mathcal{M}_1} - \mathcal{I})(u - \Pi^{\rho_2}_{r_2,\mathcal{M}_2}u)\|^2_{\mathfrak{D}H^{\rho_1}(I_1,L^2(I_2))}\\
&\quad + \|u - \Pi^{\rho_2}_{r_2,\mathcal{M}_2}u\|^2_{\mathfrak{D}H^{\rho_1}(I_1,L^2(I_2))}\\
&\leq \mathcal{M}_1^{-2r_1}\int_{\nu_1^{min}}^{\nu_1^{max}} \rho_1(\nu_1)\,\mathcal{M}_1^{2\nu_1}\|u\|^2_{H^{r_1}(I_1,L^2(I_2))}d\nu_1\\
&\quad + \mathcal{M}_2^{-2r_2}\mathcal{M}_1^{-2r_1}\int_{\nu_1^{min}}^{\nu_1^{max}} \rho_1(\nu_1)\,\mathcal{M}_1^{2\nu_1}\|u\|^2_{H^{r_1}(I_1,H^{r_2}(I_2))}d\nu_1\\
&\quad + \mathcal{M}_2^{-2r_2}\|u\|^2_{\mathfrak{D}H^{\rho_1}(I_1,H^{r_2}(I_2))}.
\end{aligned}
\tag{4.5.44}
$$

Therefore, (4.5.42) can be derived immediately from (4.5.43) and (4.5.44). □

Likewise, Lemma 4.78 can be easily extended to the d-dimensional approximation operator as

$$
\begin{aligned}
&\|u - \Pi^h_d u\|^2_{\mathfrak{D}H^{\rho_i}(I_i,L^2(\Lambda^i_d))}\\
&\quad \leq \mathcal{M}_i^{-2r_i}\int_{\nu_i^{min}}^{\nu_i^{max}} \rho_i(\nu_i)\,\mathcal{M}_i^{2\nu_i}\|u\|^2_{H^{r_i}(I_i,L^2(\Lambda^i_d))}d\nu_i
\end{aligned}
$$

$$
+\sum_{\substack{j=1\\ j\neq i}}^{d}\mathcal{M}_j^{-2r_j}\,\|u\|^2_{\mathfrak{D}H^{\rho_i}(I_i,H^{r_j}(I_j,L^2(\Lambda_d^{i,j})))}
$$

$$
+\mathcal{M}_i^{-2r_i}\int_{\nu_i^{min}}^{\nu_i^{max}}\rho_i(\nu_i)\,\mathcal{M}_i^{2\nu_i}\sum_{\substack{j=1\\ j\neq i}}^{d}\mathcal{M}_j^{-2r_j}\,\|u\|^2_{H^{r_i}(I_i,H^{r_j}(I_j,L^2(\Lambda_d^{i,j})))}d\nu_i
$$

$$
+\sum_{\substack{k=1\\ k\neq i}}^{d}\sum_{\substack{j=1\\ j\neq i,k}}^{d}\mathcal{M}_j^{-2r_j}\mathcal{M}_k^{-2r_k}\,\|u\|^2_{\mathfrak{D}H^{\rho_i}(I_i,H^{r_k,r_j}(I_k\times I_j,L^2(\Lambda_d^{i,j,k})))}
$$

$$
+\cdots+\mathcal{M}_i^{-2r_i}\int_{\nu_i^{min}}^{\nu_i^{max}}\rho_i(\nu_i)\,\mathcal{M}_i^{2\nu_i}\prod_{\substack{j=1\\ j\neq i}}^{d}\mathcal{M}_j^{-r_j}\,\|u\|^2_{{}^cH^{\nu_i}(I_i,H^{r_1,\dots,r_d}(\Lambda_d^{i}))}d\nu_i,
\tag{4.5.45}
$$

where $\Pi_d^h=\Pi_{r_1,\mathcal{M}_1}^{\rho_1}\cdots\Pi_{r_d,\mathcal{M}_d}^{\rho_d}$.

Theorem 4.80 *Let* $1\leq r_i$, $I_0=(0,T)$, $I_i=(a_i,b_i)$, $\Omega=I_0\times\left(\prod_{i=1}^d I_i\right)$, $\Lambda_k=\prod_{i=1}^k I_i$, $\Lambda_k^j=\prod_{\substack{i=1\\ i\neq j}}^k I_i$ *and* $\frac{1}{2}<\nu_i^{min}<\nu_i^{max}\leq 1$ *for* $i=1,\dots,d$. *If*

$$
u\in\Big(\bigcap_{i=1}^{d}H^{r_0}(I_0,{}^{\mathfrak{D}}H^{\rho_i}(I_i,H^{r_1,\dots,r_{i-1},r_{i+1},\dots,r_d}(\Lambda_d^i))\Big)\cap{}^{l,\mathfrak{D}}H^{\varphi}(I_0,H^{r_1,\dots,r_d}(\Lambda_d)),
$$

then,

$$
\begin{aligned}
&\|u-\Pi_{r_0,\mathcal{N}}^{\varphi}\Pi_d^h u\|^2_{\mathcal{B}^{\tau,\nu_1,\dots,\nu_d}(\Omega)}\\
&\leq\mathcal{N}^{-2r_0}\int_{\tau^{min}}^{\tau^{max}}\varphi(\tau)\,\mathcal{N}^{2\tau}\|u\|_{H^{r_0}(I_0,L^2(\Lambda_d))}d\tau\\
&\quad+\mathcal{N}^{-2r_0}\int_{\tau^{min}}^{\tau^{max}}\varphi(\tau)\,\mathcal{N}^{2\tau}\sum_{j=1}^{d}\mathcal{M}_j^{-2r_j}\,\|u\|^2_{H^{r_0}(I_0,H^{r_j}(I_j,L^2(\Lambda_d^j)))}d\tau+\cdots\\
&\quad+\mathcal{N}^{-2r_0}\int_{\tau^{min}}^{\tau^{max}}\varphi(\tau)\,\mathcal{N}^{2\tau}\Big(\prod_{j=1}^{d}\mathcal{M}_j^{-2r_j}\Big)\|u\|_{H^{r_0}(I_0,H^{r_1,\dots,r_d}(\Lambda_d)))}d\tau\\
&\quad+\sum_{i=1}^{d}\int_{\nu_i^{min}}^{\nu_i^{max}}\rho_i(\nu_i)\Big\{\mathcal{M}_i^{2\nu_i-2r_i}\|u\|_{H^{r_i}(I_i,L^2(\Lambda_d^i\times I_0))}+\cdots\\
&\quad+\mathcal{M}_i^{2\nu_i-2r_i}\Big(\prod_{\substack{j=1\\ j\neq i,k}}^{d}\mathcal{M}_j^{-2r_j}\Big)\|u\|_{H^{r_i}(I_i,H^{r_1,\dots,r_d}(\Lambda_d^i,L^2(I_0)))}\Big\}d\nu_i,
\end{aligned}
\tag{4.5.46}
$$

where $\Pi_d^h=\Pi_{r_1,\mathcal{M}_1}^{\rho_1}\cdots\Pi_{r_d,\mathcal{M}_d}^{\rho_d}$ *and* β *is a real positive constant.*

Proof Directly from (4.5.23) we conclude that

$$\|u\|^2_{\mathcal{B}^{\tau,\nu_1,\dots,\nu_d}(\Omega)} \leq \|u\|^2_{{}^lH^{\tau}(I_0,L^2(\Lambda_d))} + \sum_{i=1}^{d} \|u\|^2_{L^2(I_0,{}^{\mathfrak{D}}H^{\rho_i}(I_i,L^2(\Lambda_d^i)))}.$$

Next, it follows from Theorem 4.77 that

$$\begin{aligned}
&\|u - \Pi^{\varphi}_{r_0,\mathcal{N}}\Pi^h_d u\|^2_{l,\mathfrak{D}H^{\varphi}(I_0,L^2(\Lambda_d))} \\
&\leq \mathcal{N}^{-2r_0}\int_{\tau_{min}}^{\tau_{max}} \varphi(\tau)\,\mathcal{N}^{2\tau}\Bigg[\|u\|^2_{H^{r_0}(I_0,L^2(\Lambda_d))} + \sum_{j=1}^{d} \mathcal{M}_j^{-r_j}\|u\|^2_{H^{r_0}(I_0,H^{r_j}(I_j,L^2(\Lambda_d)))} + \cdots \\
&\qquad + \Big(\prod_{j=1}^{d}\mathcal{M}_j^{-r_j}\Big)\|u\|_{H^{r_0}(I_0,H^{r_1,\cdots,r_d}(\Lambda_d)))}\Bigg]\,d\tau. \qquad (4.5.47)
\end{aligned}$$

Therefore, (4.5.46) is obtained immediately from (4.5.45) and (4.5.47). □

Remark 4.81 Since the inf-sup condition holds (see Theorem 4.74), by Lemma 4.67, the error in the numerical scheme is less than or equal to a constant times the projection error. Hence the results above imply the spectral accuracy of the scheme.

4.5.7 Numerical Tests

We provide several numerical examples to investigate the performance of the proposed scheme. We consider a $(1+d)$-dimensional fully distributed diffusion problem with left-sided derivative by letting $c_{l_i} = c_{r_i} = \kappa_{r_i} = 0$, $\kappa_{l_i} = 1$, $0 < 2\tau^{min} < 2\tau^{max} < 1$ and $1 < 2\nu_i^{min} < 2\nu_i^{max} \leq 2$ in (4.5.26) for $i = 1, \dots, d$, where the computational domain is $\Omega = (0, 2) \times \prod_{i=1}^{i=d}(-1, 1)$. We report the measured L^{∞} error, $\|e\|_{L^{\infty}} = \|u_N - u^{ext}\|_{L^{\infty}}$ as the maximum bound of $\|e\|_{L^2}$.

In each of the following test cases, we use the method of fabricated solutions to construct the load vector, given an exact solution u^{ext}. Here, we assume $u^{ext} = u_t \times \prod_{i=1}^{d} u_{x_i}$. We project the spatial part in each dimension, u_{x_i}, on the spatial bases, and then, construct the load vector by plugging the projected exact solution into the weak form of the problem. This helps us take the fractional derivative of the exact solution more efficiently, while by truncating the projection with a sufficient number of terms, we make sure that the corresponding projection error does not dominantly propagate into the convergence analysis of the numerical scheme.

Case I: We consider a smooth solution in space with finite regularity in time as

$$u^{ext} = t^{p_1+\alpha} \times \left((1 + x_1)^{p_2}(1 - x_1)^{p_3}\right) \tag{4.5.48}$$

to investigate the spatial/temporal p-refinement. We allow the singularity to take order of $\alpha = 10^{-4}$, while p_1, p_2, and p_3 take some integer values. We show the L^∞-error for different test cases in Figure 4.25, where by tuning the fractional parameter of the temporal basis, we can accurately capture the

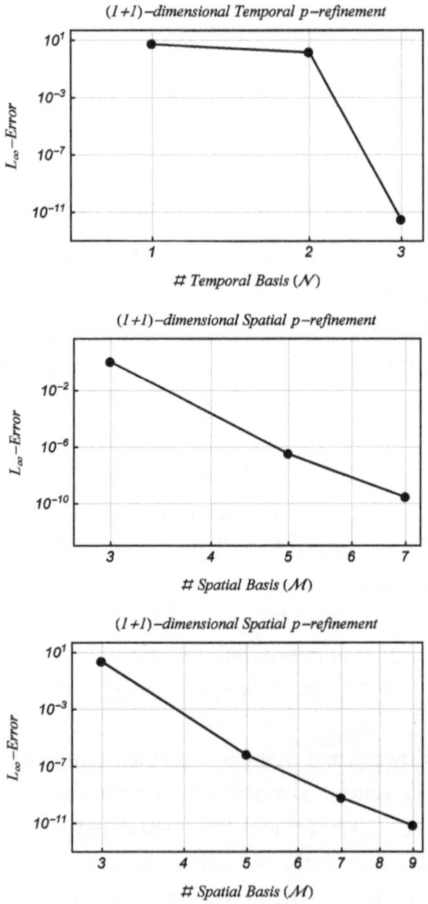

Figure 4.25 Temporal/spatial *p-refinement* for case I with singularity of order $\alpha = 10^{-4}$. (Top): $p_1 = 3$, $p_2 = p_3 = 2$, and expansion order of $\mathcal{N} \times 9$. (Middle): $p_1 = 2$, $p_2 = p_3 = 2$, and expansion order of $3 \times \mathcal{M}$. (Bottom): $p_1 = 3$, $p_2 = p_3 = 2$, and expansion order of $4 \times \mathcal{M}$. Source: Samiee et al. (2021).

singularity of the exact solution, when the approximate solution converges as we increase the expansion order. In each case of spatial/temporal p-refinement, we choose sufficient number of bases in other directions to make sure their corresponding error is of machine precision order. Considering $\alpha = 10^{-4}$, $p_1 = 2$, $p_2 = p_3 = 2$ in (4.5.48), and the temporal order of expansion being fixed ($\mathcal{N} = 4$) in the spatial p-refinement, we get the rate of convergence as a function of the minimum regularity in the spatial direction. From Theorem 4.80, the rate of convergence is bounded by the spatial approximation error, i.e., $\|e\|_{L^2(\Omega)} \leq \|e\|_{L^\infty(\Omega)} \leq \mathcal{M}_1^{-2r_1} \int_{\nu_1^{min}}^{\nu_1^{max}} \rho_1(\nu_1)\, \mathcal{M}_1^{2\nu_1} \|u\|_{H^{r_1}(I_1,L^2(I_0))} d\nu_1$, where $r_1 = p_2 + \frac{1}{2} - \epsilon$ is the minimum regularity of the exact solution in the spatial direction for $\epsilon < \frac{1}{2}$. Conforming to Theorem 4.80, the practical rate of convergence $\bar{r}_1 = 16.05$ in $\|e\|_{L^\infty(\Omega)}$ is greater than $r_1 \approx 2.50$.

Case II: We consider $u^{ext} = t^{p_1+\alpha} \sin(2\pi x_1)$, where $p_1 = 3$, and let $\alpha = 0.1$ and $\alpha = 0.9$. We set the number of temporal basis functions, $\mathcal{N} = 4$, and show the convergence of the approximate solution by increasing the number of spatial bases, $\mathcal{M}$ in Figure 4.26. The main difficulty in this case is the construction of the load vector. To accurately compute the integrals in the load vector, we project the spatial part of the forcing function, $\sin(2\pi x_1)$, on the spatial bases. To make sure that the corresponding error is of machine-precision order and thus, not dominant, we truncate the projection at 25 terms, where there error is of order 10^{-16}. Therefore, the quadrature rule over derivative order should be performed for 25 terms rather than only a single $\sin(2\pi x_1)$ term. This will increase the computational cost.

Case III: (High-dimensional p-refinement) We consider the exact solution of the form

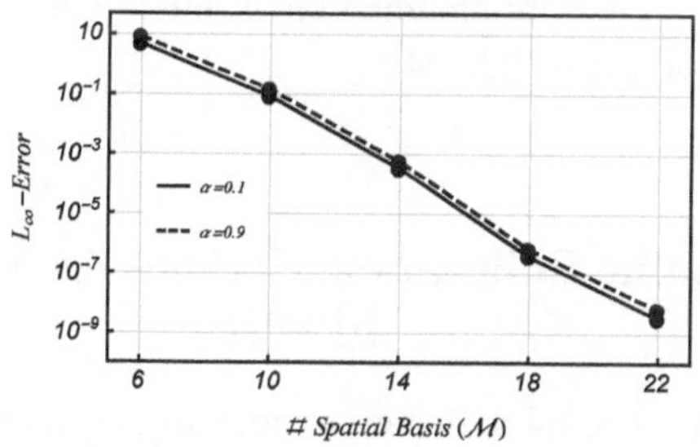

Figure 4.26 Spatial p-refinement for case II, $p_1 = 3$, $\alpha = 0.1$, and $\alpha = 0.9$. Source: Samiee et al. (2021).

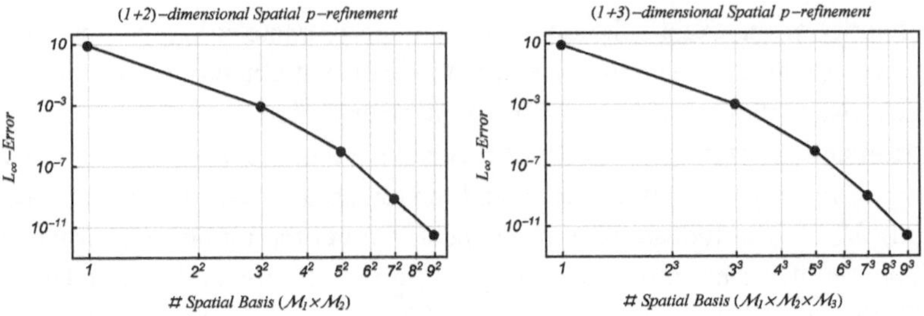

Figure 4.27 Spatial p-refinement for case III with singularity of order $\alpha = 10^{-4}$. (Left): $(1+2)$-dimensional, $p_1 = 3$, $p_{2i} = p_{2i+1} = 1$, where the expansion order is $\mathcal{N}\times\mathcal{M}_1\times\mathcal{M}_2$. (Right): $(1+3)$-dimensional, $p_1 = 3$, $p_{2i} = p_{2i+1} = 1$, where the expansion order is $\mathcal{N}\times\mathcal{M}_1\times\mathcal{M}_2\times\mathcal{M}_3$. Source: Samiee et al. (2021).

$$u^{ext} = t^{p_1+\alpha} \times \prod_{i=1}^{3}(1+x_i)^{p_{2i}}(1-x_i)^{p_{2i+1}} \tag{4.5.49}$$

with singularity of order $\alpha = 10^{-4}$, where $p_1 = 3$, and $p_{2i} = p_{2i+1} = 1$. Similar to previous cases, we set the number of temporal bases, $\mathcal{N} = 4$, and study convergence by uniformly increasing the number of spatial bases in all dimensions. Figure 4.27 shows the results for $(1+2)$-dimensional and $(1+3)$-dimensional problems with expansion order of $\mathcal{N}\times\mathcal{M}_1\times\mathcal{M}_2$, and $\mathcal{N}\times\mathcal{M}_1\times\mathcal{M}_2\times\mathcal{M}_3$, respectively. Following Case I, the computed rate of convergence $\bar{r}_1 = \bar{r}_2 = \bar{r}_3 = 16.13$ in (4.5.49) for $\alpha = 10^{-4}$ is greater than the minimum regularity of the exact solution $r \approx 2.05$, which is in agreement with Theorem 4.80.

In addition to the convergence study, we examine the efficiency of the developed method and fast solver by comparing the CPU times for $(1+1)$-, $(1+2)$-, and $(1+3)$-dimensional space-time hypercube domains in case III. The computed CPU times are obtained on an Intel (Xeon E52670) processor of 2.5 GHZ, and reported in Table 4.6.

4.6 GJPF Tunable Collocation Methods for Variable-Order FPDEs

We now employ the spectral collocation methods presented in Chapter 3 and specifically in Section 3.2.6 to solve space-time FPDEs and test these methods when solutions are known and smooth.

Table 4.6. *CPU time, PG spectral method for fully distributed (1+d)-dimensional diffusion problems.* $u^{ext} = t^{p_1+\alpha} \times \prod_{i=1}^{3}(1+x_i)^{p_{2i}}(1-x_i)^{p_{2i+1}}$, *where* $\alpha = 10^{-4}$, $p_1 = 3$, *and the expansion order is* 4×11^d. *Source: Samiee et al. (2021).*

	$p_{2i} = p_{2i+1} = 2$			$p_{2i} = p_{2i+1} = 3$		
	d=1	d=2	d=3	d=1	d=2	d=3
CPU Time [Sec]	1546.81	1735.03	2358.67	1596.16	1786.61	2407.22
$\|e\|_{L^\infty(\Omega)}$	6.84×10^{-12}	4.45×10^{-12}	3.27×10^{-12}	6.27×10^{-12}	3.86×10^{-12}	2.71×10^{-12}

Example 4.82 Consider the following space-time-fractional advection–diffusion equation over $(-1,1)\times(0,1]$ (see, e.g., Zayernouri and Karniadakis (2014a))

$$\begin{cases} {}_0^{\mathrm{C}}D_t^{\gamma(x,t)}u(x,t) + {}_{-1}^{\mathrm{RL}}D_x^{\beta(x,t)}u(x,t) = {}_{-1}^{\mathrm{RL}}D_x^{\alpha(x,t)}u(x,t) + f(x,t), \\ u(x,0) = 0, \quad x \in [-1,1], \\ u(-1,t) = u(1,t) = 0, \quad t \in (0,1], \end{cases} \tag{4.6.1}$$

where $0 < \gamma(x,t), \beta(x,t) < 1$ and $1 < \alpha(x,t) < 2$.

We represent the solution $u(x,t)$ to (4.6.1) as $u(x,t) = t^{\mu}(x-x_L)^{\sigma}v(x,t)$. We interpolate $v(x,t)$ on the JGL points (x_j,t_k), where

$$x_j = \frac{x_R - x_L}{2}\hat{x}_j + \frac{x_L + x_R}{2}, \quad t_k = \frac{T(\hat{t}_k + 1)}{2}. \tag{4.6.2}$$

Here $\{\hat{x}_j\}$ and $\{\hat{t}_k\}$ are the roots of $(1-x^2)(P_N^{(a_1,b_1)}(x))'$ and $(1-t^2)(P_M^{(a_2,b_2)}(t))'$, respectively. Hence $u(x,t)$ can be approximated by

$$\begin{aligned} I_{M,N}^{\mu,\sigma,a_1,b_1,a_2,b_2,+}u(x,t) &= t^{\mu}(x-x_L)^{\sigma} I_{M,N}^{a_1,b_1,a_2,b_2} v(x,t) \\ &= t^{\mu}(x-x_L)^{\sigma} \sum_{j=0}^{N}\sum_{k=0}^{M} v_{j,k} l_j(x) h_k(t), \end{aligned}$$

where $l_j(x)$ and $h_k(t)$ are Lagrange interpolation polynomials, see Section 3.2.6.

Replacing $u(x,t)$ in (4.6.1) with $I_{M,N}^{\mu,\sigma,a_1,b_1,a_2,b_2,+}u$ and letting $(x,t) = (x_j,t_k)$ yields

$$\begin{cases} \left[{}_0^{\mathrm{C}}D_t^{\gamma(x_j,t)} I_{M,N}^{\mu,\sigma,a_1,b_1,a_2,b_2,+}u(x_j,t)\right]_{t=t_k} = \left[{}_{x_L}^{\mathrm{RL}}D_x^{\alpha(x,t_k)} I_{M,N}^{\mu,\sigma,a_1,b_1,a_2,b_2,+}u(x,t_k)\right]_{x=x_j} \\ \quad + \left[{}_{x_L}^{\mathrm{RL}}D_x^{\beta(x,t_k)} I_{M,N}^{\mu,\sigma,a_1,b_1,a_2,b_2,+}u(x,t_k)\right]_{x=x_j} + f(x_j,t_k), \\ \quad j = 1,2,\ldots,N-1, k = 1,2,\ldots,M, \\ u_{j,0} = 0, \quad j = 0,1,\ldots,N, \\ u_{0,k} = 0, \quad u_{N,k} = \psi(t_k), \quad k = 0,1,\ldots,M. \end{cases} \tag{4.6.3}$$

In our computation, we take $x_L = -1, x_R = 1$, $T = 1$, and choose proper $f(x,t)$ such that equation (4.6.1) has the exact solution $u(x,t) = t^{6+2/3}\sin(\pi x)$ with different variable-orders given by:

- Case I: $\gamma = \beta = \frac{5+4x}{10}\frac{1+4t}{10}$, $\alpha = 1 + \frac{1+4t}{10}$;
- Case II: $\gamma = \frac{1}{|\sin(\pi(x-t))|+1.2}$, $\beta = \frac{|xt|+1}{4}$, $\alpha = \exp(-|xt| - 0.1) + 1$.

The relative L^∞ error is defined by

$$\|e\|_\infty = \frac{\max_{0\le j\le N} \max_{0\le k\le M} |u(x_j,t_k) - u_{i,k}|}{\max_{0\le j\le N} \max_{0\le k\le M} |u_{i,k}|},$$

where (x_j, t_k) are the LGL points from (4.6.2), i.e., the LGL spectral collocation method is applied in the computation ($a_1 = b_1 = a_2 = b_2 = 0$ in (4.6.3)). Since the analytical solution is smooth in space, we choose $\sigma = 0$ in (4.6.3). In Figures 4.28 (a)–(b), we observe machine accuracy when $M \ge 6, \mu = 2/3$ or $M > 6, \mu = -1/3$. This is because the analytical solution $u(x,t)$ to (4.6.1) satisfies $u(x,t) = t^{2/3}v(x,t)$ or $u(x,t) = t^{-1/3}v(x,t)$, where $v(x,t) = t^6 \sin(\pi x)$ or $t^7 \sin(\pi x)$. When we choose $\mu = 2/3$ or $\mu = -1/3$, we interpolate $v(x,t)$ instead of $u(x,t)$, which yields much higher accuracy due to the smoothness of $v(x,t)$, thus higher accurate numerical solutions of $u(x,t)$.

Conservation laws that involve fractional Laplacians have been widely applied in several fields, see, e.g., Chen and Holm (2004); Constantin and Vicol (2012). We illustrate the spectral collocation methods to solve FPDEs. In the next example, we consider a fractional Burgers equation where the end-point singularity index of the solution is not explicitly known. Here, we apply our generalized and tunable collocation method to solve a suitable benchmark problem, namely the space-fractional Burgers equation, where the solution is not known explicitly. Numerical results show the superiority of choosing μ close to -1.

Example 4.83 Consider the space-fractional Burgers equation

$$\partial_t U(x,t) + U(x,t)\partial_x U(x,t) = \left(p \;\; {}^{\text{RL}}_{-1}D_x^{\alpha(x,t)} + q \;\; {}^{\text{RL}}_{x}D_1^{\alpha(x,t)}\right) U(x,t), \tag{4.6.4}$$

subject to the homogeneous boundary conditions and initial condition $u(x,0) = \varphi(x)$, $p,q \ge 0$ are constants, $p + q \ne 0$, $1 < \alpha(x,t) \le 2$, and $(x,t) \in (-1,1)\times(0,T]$, and $T > 0$.

We first use the standard Jacobi–Gauss–Lobatto (JGL) spectral collocation method in space and the first-order backward difference (backward Euler)

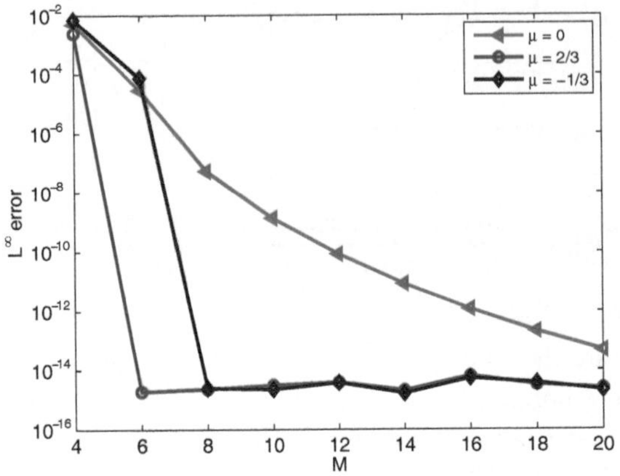

(a) Case I.

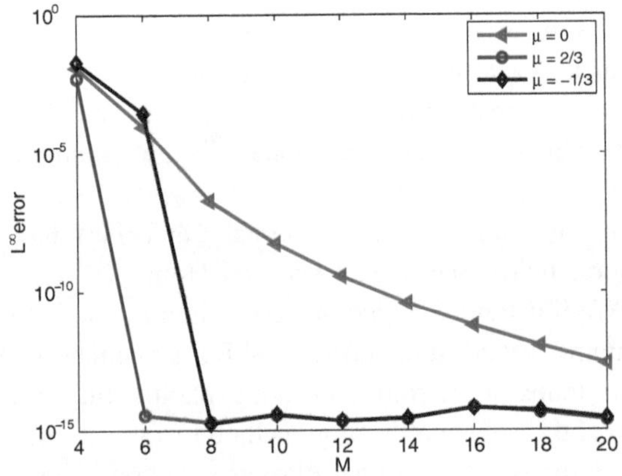

(b) Case II.

Figure 4.28 The relative L^∞ errors for Example 4.82, $N = 24$, $a_1 = b_1 = a_2 = b_2 = 0$. Source: Zeng et al. (2015a). Copyright ©2015 Society for Industrial and Applied Mathematics. Reprinted with permission. All rights reserved.

method in time to solve (4.6.4) (when $p = \epsilon$ and $q = 0$). The method is given by: find $I_N^{\mu,a,b,+}u^n(x) = (x - x_L)^\mu \sum_{j=0}^N v_j^n l_j(x)$ $(n \geq 1)$ such that

$$\begin{cases} \frac{1}{\tau}\left(u_j^n - u_j^{n-1}\right) + u_j^n \left[\partial_x I_N^{\mu,a,b,+} u^n(x)\right]_{x=x_j} = \epsilon \left[{}_{x_L}^{\mathrm{RL}}D_x^{\alpha(x,t_n)} I_N^{\mu,a,b,+} u^n(x)\right]_{x=x_j} \\ \qquad j = 1, 2, ..., N-1, \\ u_j^0 = \varphi(x_j), \quad j = 0, 1, ..., N, \\ u_0^n = u_N^n = 0, \quad n = 0, 1, ..., n_T, \end{cases} \tag{4.6.5}$$

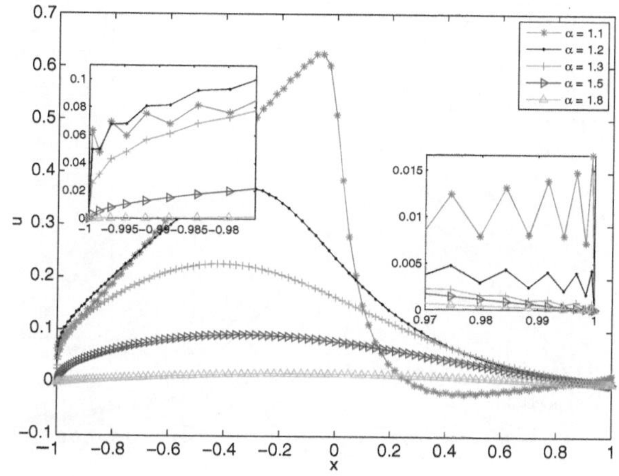

Figure 4.29 Numerical solutions of Example 4.83 (Case 1) at $t = 1$ by the method (4.6.5), $\epsilon = 1$, $a = b = 0$, $N = 128$, $\tau = 10^{-3}$, and $\mu = 0$. The two inset plots show the behavior of the numerical solution near the boundaries. Source: Zeng et al. (2015a). Copyright ©2015 Society for Industrial and Applied Mathematics. Reprinted with permission. All rights reserved.

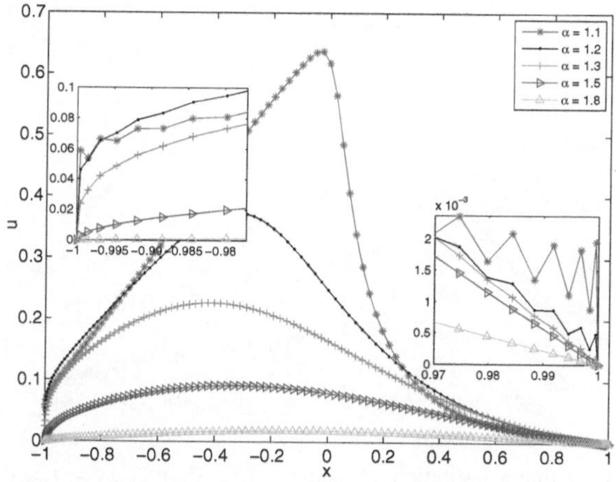

Figure 4.30 Numerical solutions of Example 4.83 (Case 1) at $t = 1$ by the method (4.6.5), $\epsilon = 1$, $a = b = 0$, $N = 128$, $\tau = 10^{-3}$, and $\mu = -0.2$. Source: Zeng et al. (2015a). Copyright ©2015 Society for Industrial and Applied Mathematics. Reprinted with permission. All rights reserved.

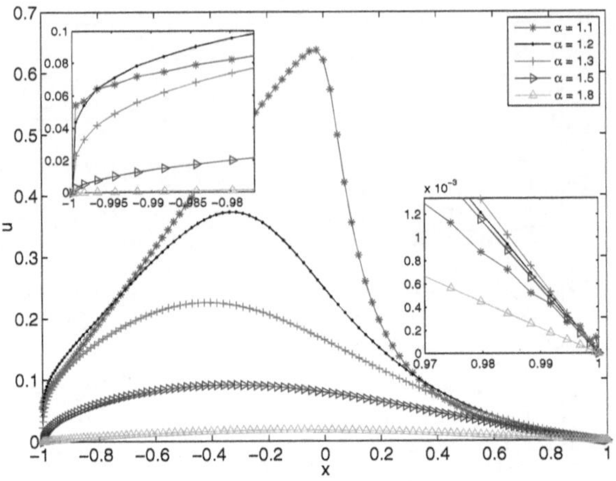

Figure 4.31 Numerical solutions of Example 4.83 (Case 1) at $t = 1$ by the method (4.6.5), $\epsilon = 1$, $a = b = 0$, $N = 128$, $\tau = 10^{-3}$, and $\mu = -0.45$. Source: Zeng et al. (2015a). Copyright ©2015 Society for Industrial and Applied Mathematics. Reprinted with permission. All rights reserved.

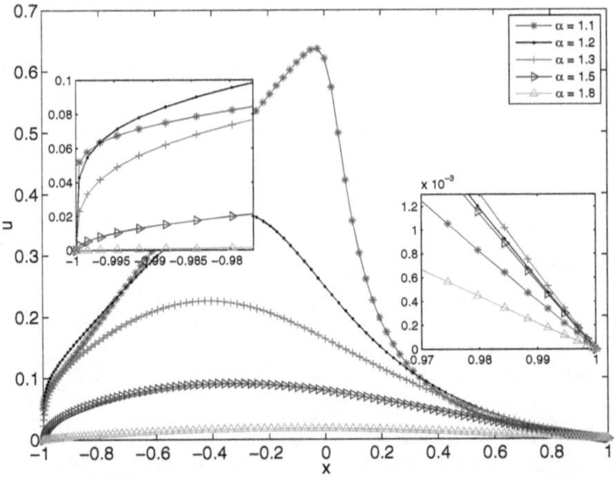

Figure 4.32 Numerical solutions of Example 4.83 (Case 1) at $t = 1$ by the method (4.6.5), $\epsilon = 1$, $a = b = 0$, $N = 128$, $\tau = 10^{-3}$, and $\mu = -0.7$; Zeng et al. (2015a). Copyright ©2016 Society for Industrial and Applied Mathematics. Reprinted with permission. All rights reserved. Source: Zeng et al. (2015a). Copyright ©2015 Society for Industrial and Applied Mathematics. Reprinted with permission. All rights reserved.

where τ is the time-step size with $\tau = T/n_T, n_T \in \mathbb{N}, t_n = n\tau$, $\{x_j\}$ are the roots of $(1-x^2)(P_N^{(a,b)}(x))'$. As (4.6.5) is implicit, we employ the fixed-point iterations to solve the resulted nonlinear system. As in the previous two examples, we observe better accuracy if $-1 < \mu < 0$ rather than $\mu \geq 0$.

Next, we apply the new general tunable collocation method in space and second-order semi-implicit time discretization to solve (4.6.4). The fully discrete spectral collocation method for (4.6.4) is given by: find $u_N^{\mu,n+1} = I_N^{\mu,a,b} u^{n+1}(x)$ (see (3.2.266)) for $n = 1, 2, ..., n_T - 1$, such that

$$\begin{cases} \dfrac{3u_N^{\mu,n+1}(x_j) - 4u_N^{\mu,n}(x_j) + u_N^{\mu,n-1}(x_j)}{2\tau} + \left((2u_N^{\mu,n} - u_N^{\mu,n-1})\partial_x u_N^{\mu,n+1}\right)(x_j) \\ = \left[\left(p\ {}_{-1}^{\mathrm{RL}}D_x^{\alpha(x,t_{n+1})} + q\ {}_{x}^{\mathrm{RL}}D_1^{\alpha(x,t_{n+1})}\right) u_N^{\mu,n+1}(x)\right]_{x=x_j}, \quad 1 \leq j \leq N-1, \\ u_j^0 = \varphi(x_j), \quad j = 0, 1, ..., N, \\ u_j^1 = u_j^0 + \tau\partial_t U(x_j, 0), \quad j = 0, 1, ..., N, \\ u_0^k = u_N^k = 0, \quad k = 0, 1, ..., n_T, \end{cases} \tag{4.6.6}$$

where $\tau = T/n_T$ is the time-step size, n_T is a positive integer, and $t_n = n\tau$.

- Case I: (constant fractional orders) $\alpha(x,t) = 1.05, 1.1, 1.2, 1.3, 1.4$;
- Case II: (monotonic fractional order) $\alpha(x,t) = \frac{10-9x}{20} + 1$;
- Case III: (non-monotonic fractional order) $\alpha(x,t) = 0.9\exp(-10tx^2) + 1.05$.

In the numerical simulations, we always choose the initial condition $\varphi(x) = \sin(\pi x)$ and $a = b = 0$ (Legendre collocation) in the method (4.6.6).

We first consider Case I using the standard collocation method ($\mu = (0,0)$ in (4.6.6)) with $N = 128$; numerical solutions at $t = 0.4$ are shown in Figure 4.33 for different fractional orders $\alpha = 1.05, 1.1, 1.2, 1.3, 1.4$. We observe wiggles at both left and right boundaries for $\alpha = 1.05$ in Figure 4.33 (a), where we choose $p = q = 0.5$. In Figure 4.33 (b), we observe wiggles for $\alpha = 1.05, 1.1$ at the left boundary due to $p/q = 1.5 > 1$. This coincides with the numerical results in the previous example. Wiggles do not appear due to the relatively good regularity of the analytical solutions when the fractional order α is relatively large. In the following, we are mainly concerned with smaller values of the fractional order α that may cause stronger singularity of solutions to (4.6.4), which is not well solved numerically by the known methods.

Next, we show the effectiveness of negative μ_1 and/or μ_2 on the improvement of accuracy of numerical solutions when the regularity of analytical solutions to the considered FDEs is low, i.e., α is small in (4.6.4). For Case

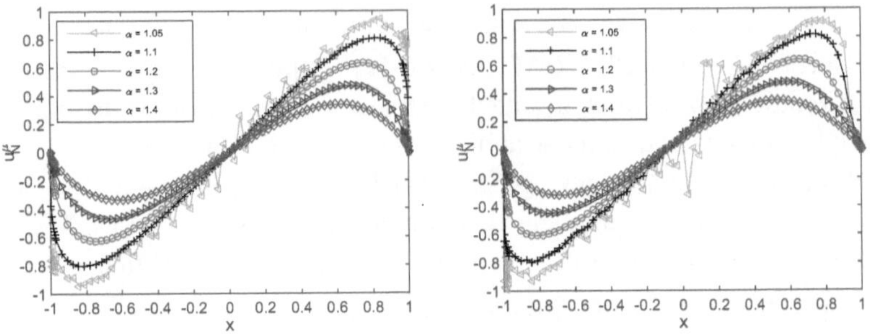

(a) $p = q = 0.5$. (b) $p = 0.6, q = 0.4$.

Figure 4.33 Numerical solutions at $t = 0.4$ for Example 4.83 by the collocation method (4.6.6), Case I, $\mu = (0,0)$, $a = b = 0$, and $N = 128$. Source: Zeng et al. (2017). Copyright ©2017 Society for Industrial and Applied Mathematics. Reprinted with permission. All rights reserved.

I with $p = q = 0.5$ and $\alpha = 1.05$, the singularity of the analytical solution also depends on time t as shown in Figure 4.33 (a) and becomes stronger at both end-points as time evolves. We still choose $N = 128$ and $\mu_1 = \mu_2 = 0, -0.1, -0.2, -0.4, -0.6, -0.85$ in the computation; the numerical solutions at $t = 0.2, 0.4, 0.6, 0.8$ are displayed in Figure 4.34. Obviously, much better numerical solutions are obtained as $\mu_k (k = 1,2)$ tends to its optimal choice. For the regular collocation method ($\mu_1 = \mu_2 = 0$), the numerical solutions at $t = 0.6, 0.8$ oscillate violently, hence they are not completely displayed in Figures 4.34 (c) and (d). In fact, the magnitude of the numerical solutions of the regular collocation method is about 15 for $t = 0.6$ and 800 for $t = 0.8$. The singly-singular basis in Zeng et al. (2015) also failed to obtain accurate numerical solutions when t is sufficiently large due to the two end-point singularities of the present problem, and the results are not presented here. Since we have much stronger singularity at the left boundary for $(p,q) = (0.6,0.4)$ or $(0.7,0.3)$ and $\alpha = 1.05$, we set $N = 512$ and $\mu = (\mu_1, 0)$ in the computation with $\mu_1 = 0, -0.1, -0.2, -0.4, -0.6, -0.85$; numerical results at $t = 0.4$ are displayed in Figure 4.35, which shows better numerical solutions as μ_1 decreases.

In Figure 4.36, we show numerical solutions at $t = 0.2, 0.4, 0.8, 1$ for Case II and find that the solution has stronger singularity at the right boundary for $p = 0.3, q = 0.7$ as t evolves. Hence, we choose $\mu = (0, \mu_2)$ and $N = 128$ in the computation. Wiggles appear using the standard collocation method due to the low regularity of the analytical solution at the right end-point, see numerical solutions for $\mu = (0,0)$ in Figures 4.36 (a)–(d). However, wiggles

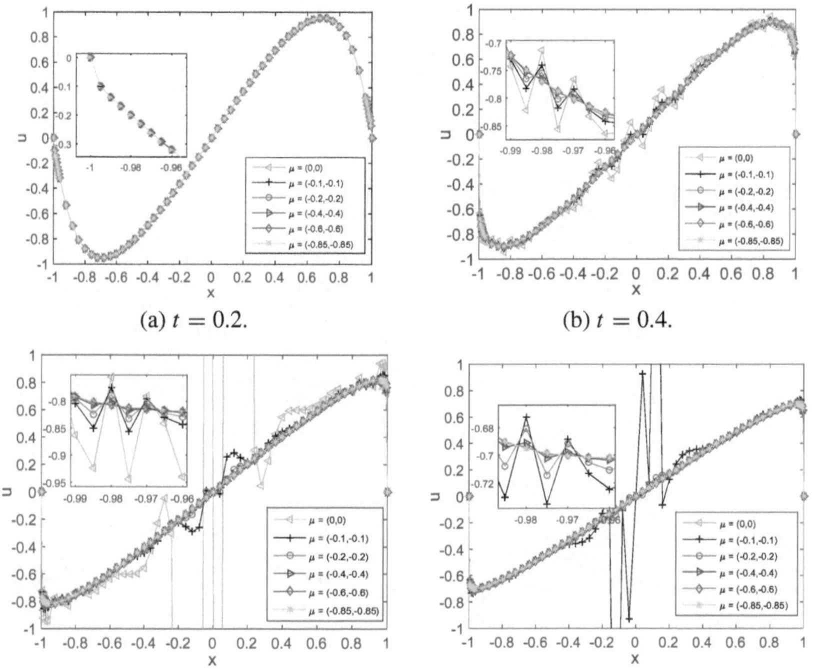

(a) $t = 0.2$. (b) $t = 0.4$.

(c) $t = 0.6$. (d) $t = 0.8$.

Figure 4.34 Numerical solutions for Example 4.83 by the collocation method (4.6.6), Case I, $p = q = 0.5$, $\alpha = 1.05$, $a = b = 0$, and $N = 128$. Source: Zeng et al. (2017). Copyright ©2017 Society for Industrial and Applied Mathematics. Reprinted with permission. All rights reserved.

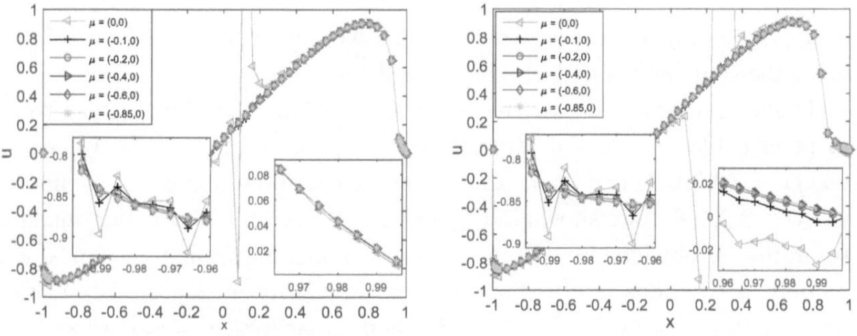

(a) $p = 0.6, q = 0.4$. (b) $p = 0.7, q = 0.3$.

Figure 4.35 Numerical solutions for Example 4.83 at $t = 0.4$ by the collocation method (4.6.6), Case I, $\alpha = 1.05$, $a = b = 0$, and $N = 512$. Source: Zeng et al. (2017). Copyright ©2017 Society for Industrial and Applied Mathematics. Reprinted with permission. All rights reserved.

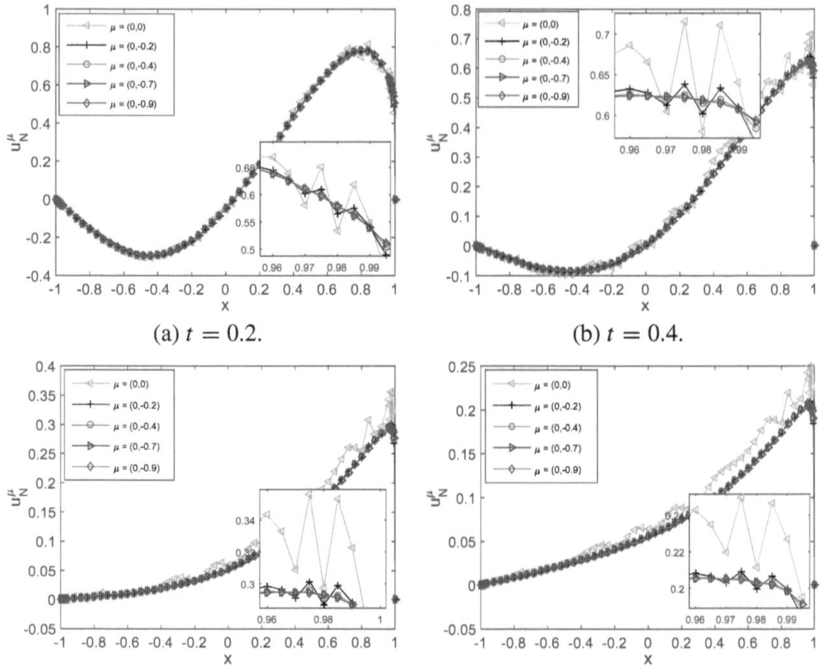

(a) $t = 0.2$. (b) $t = 0.4$.

(c) $t = 0.8$. (d) $t = 1$.

Figure 4.36 Numerical solutions for Example 4.83 by the collocation method (4.6.6), Case II, $p = 0.3, q = 0.7$, $\alpha(x,t) = \frac{10-9x}{20} + 1$, $a = b = 0$, and $N = 128$. Source: Zeng et al. (2017). Copyright ©2017 Society for Industrial and Applied Mathematics. Reprinted with permission. All rights reserved.

disappear as μ_2 decreases, which means we get better numerical solutions using the singular basis functions $(1-x)^{\mu_2} P_j^{(a,b)}(x)$.

In the last example, we choose in Case III the fractional order $\alpha(x,t)$ to behave like a Gaussian distribution in space when $t > 0$. We choose $(p,q) = (0.7, 0.3)$ and $N = 128$ in the computation; numerical solutions at $t = 0.2, 0.5, 0.8, 1$ are shown in Figure 4.37. We observe that better numerical solutions are obtained by tuning $\mu = (\mu_1, 0)$, i.e., decreasing μ_1 yields better numerical solutions due to the stronger singularity at the left boundary as time t evolves. If $(p,q) = (0.3, 0.7)$, then the analytical solution has strong singularity near the right boundary as time t evolves, and numerical solutions become better when tuning $\mu = (0, \mu_2)$ with μ_2 decreased; the results are not provided here.

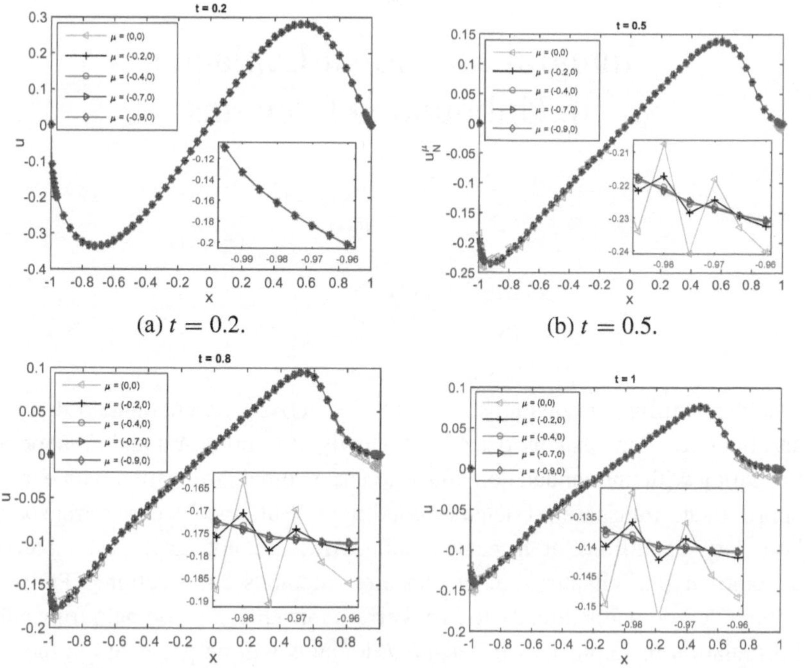

(a) $t = 0.2$. (b) $t = 0.5$.

(c) $t = 0.8$. (d) $t = 1$.

Figure 4.37 Numerical solutions for Example 4.83 by the collocation method (4.6.6), Case III, $p = 0.7, q = 0.3$, $\alpha = 0.9\exp(-10tx^2) + 1.05$, $a = b = 0$, and $N = 128$. Source: Zeng et al. (2017). Copyright ©2017 Society for Industrial and Applied Mathematics. Reprinted with permission. All rights reserved.

5
Integral Fractional Laplacian in Unbounded Domains

Fractional diffusion equations are naturally derived on unbounded domains, and their solutions usually decay very slowly at infinity. A standard approach to dealing with unbounded domains is to use a domain truncation with exact or approximate transparent boundary conditions. But since accurate transparent boundary conditions at truncated boundaries are not easily available, we develop in this chapter efficient spectral methods for fractional PDEs on unbounded domains directly to avoid errors introduced by domain truncation. Formulation of Laplacians in *bounded* domains will be presented in the next chapter.

5.1 Hermite Spectral-Galerkin Methods

The integral fractional Laplacian is a nonlocal operator involving a singular kernel function in the physical space. Thus, it is very difficult and costly to deal with the integral fractional Laplacian in the physical unbounded space. However, an alternative definition of the integral fractional Laplacian is through Fourier transform, and the fact that Hermite functions are eigenfunctions of Fourier transforms makes it possible approximate the integral fractional Laplacian in the phase space.

We first introduce some definitions and notation which will be used hereafter. We shall use bold letters such as $\boldsymbol{x} = (x_1, \ldots, x_d)$ and $\boldsymbol{j} = (j_1, \ldots, j_d)$ to denote multi-variables and multi-indices. For any function $u(\boldsymbol{x}) \in L^2(\mathbb{R}^d)$, we denote its Fourier transform by $\widehat{u}(\xi)$ with $\xi \in \mathbb{R}^d$. We denote by $|\xi|_1$, $|\xi|_2$ and $|\xi|_\infty$ the l^1, l^2 and l^∞ norm of ξ in $\mathbb{R}^d$, respectively. Let c be a generic positive constant independent of any functions and of any discretization parameters. We use the expression $A \lesssim B$ (respectively $A \gtrsim B$) to mean that $A \leqslant cB$ (respectively $A \geqslant cB$), and use the expression $A \cong B$ to mean that $A \lesssim B \lesssim A$.

Definition 5.1 Given $s > 0$, the fractional Laplacian operator $(-\Delta)^s$ is defined in the Fourier transformed domain by

$$\widehat{(-\Delta)^s u}(\xi) = |\xi|_2^{2s}\,\widehat{u}(\xi). \tag{5.1.1}$$

Let $\omega(\boldsymbol{x}) > 0, \boldsymbol{x} \in \Omega$, be a weight function, and denote by $L^2_\omega(\Omega)$ the usual weighted Hilbert space with the inner product and norm defined by

$$(u, v)_{\Omega,\omega} = \int_\Omega u(\boldsymbol{x})v(\boldsymbol{x})\omega(\boldsymbol{x})\,\mathrm{d}\boldsymbol{x}, \quad \|u\|_{\Omega,\omega} = (u, u)^{\frac{1}{2}}_{\Omega,\omega}, \tag{5.1.2}$$

for any $u, v \in L^2_\omega(\Omega)$. When $\omega \equiv 1$, we drop ω from the above notation. The Plancherel theorem states that

$$\|u\|_\Omega = \|\widehat{u}\|_\Omega. \tag{5.1.3}$$

We denote by $H^\mu(\Omega)$ (with $\mu \geq 0$) the usual Hilbert spaces with the semi-norm

$$|u|_{\mu,\Omega} = \||\xi|_2^\mu\,\widehat{u}\|_\Omega, \tag{5.1.4}$$

and the norm

$$\|u\|_{\mu,\Omega} = (\|u\|^2_\Omega + |u|^2_{\mu,\Omega})^{1/2} = (\|\widehat{u}\|^2_\Omega + \||\xi|_2^\mu\widehat{u}\|^2_\Omega)^{1/2}. \tag{5.1.5}$$

We recall that the orthonormal Hermite polynomials $\{H_n(x)\}$ are defined by the three-term recurrence relation:

$$\begin{aligned} &H_{n+1}(x) = x\sqrt{\frac{2}{n+1}}H_n(x) - \sqrt{\frac{n}{n+1}}H_{n-1}(x), \quad n \geq 1, \\ &H_0(x) = \pi^{-1/4}, \quad H_1(x) = \sqrt{2}\pi^{-1/4}x, \quad x \in \mathbb{R}. \end{aligned} \tag{5.1.6}$$

They are mutually orthogonal with respect to the weight function $\omega(x) = e^{-x^2}$, i.e.,

$$\int_{-\infty}^{\infty} H_m(x)H_n(x)\omega(x)dx = \delta_{mn}, \tag{5.1.7}$$

and satisfy

$$H_n'(x) = \sqrt{2n}H_{n-1}(x), \; n \geq 1. \tag{5.1.8}$$

Moreover, $H_n(x)$ is an odd (resp. even) function for n odd (resp. even), that is,

$$H_n(-x) = (-1)^n H_n(x). \tag{5.1.9}$$

In practice, it is advised to use a problem dependent scaling factor λ to redistribute the Hermite–Gauss collocation points (cf. Shen et al. (2011b)). We define a sequence of Hermite functions with a scaling parameter λ:

$$\psi_n(x;\lambda) = \sqrt{\lambda}e^{-(\lambda x)^2/2}H_n(\lambda x), \tag{5.1.10}$$

and denote in particular $\psi_n(x) = \psi_n(x;1)$. It follows from (5.1.7) that the Hermite functions form an orthonormal basis in $L^2(\mathbb{R})$, i.e.,

$$\int_{-\infty}^{\infty} \psi_m(x;\lambda)\psi_n(x;\lambda)\mathrm{d}x = \delta_{mn}. \tag{5.1.11}$$

We can derive from (5.1.6) the following three-term recurrence relation:

$$\begin{aligned}
&\psi_{n+1}(x;\lambda) = \lambda x\sqrt{\frac{2}{n+1}}\psi_n(x;\lambda) - \sqrt{\frac{n}{n+1}}\psi_{n-1}(x;\lambda),\ n \ge 1,\\
&\psi_0(x;\lambda) = \sqrt{\lambda}\pi^{-1/4}e^{-(\lambda x)^2/2}, \quad \psi_1(x;\lambda) = \lambda^{\frac{3}{2}}\sqrt{2}\pi^{-1/4}xe^{-(\lambda x)^2/2}.
\end{aligned} \tag{5.1.12}$$

Obviously, $\psi_n(x;\lambda)$ is an also odd (resp. even) function for n odd (resp. even).

The following result (see e.g., Duoandikoetxea (2001)) plays a key role in the algorithm development.

Lemma 5.2 *The Hermite functions* $\{\psi_n(x)\}$ *are the eigenfunctions of the Fourier transform operator with eigenvalues* $(-\mathrm{i})^n$, $n = 0, 1, \ldots$, *i.e.,*

$$(\widehat{\psi_n(x)})(\xi) = \frac{1}{\sqrt{2\pi}}\int_{-\infty}^{\infty} \psi_n(x;1)e^{-\mathrm{i}\xi x}\mathrm{d}x = (-\mathrm{i})^n\psi_n(\xi), \tag{5.1.13}$$

where $\mathrm{i} = \sqrt{-1}$.

We can easily derive from the above and from

$$(\widehat{f(ax)})(\xi) = \frac{1}{\sqrt{2\pi}}\int_{-\infty}^{\infty} f(ax)e^{-\mathrm{i}\xi x}\mathrm{d}x = \frac{1}{|a|}\widehat{f}\Big(\frac{\xi}{a}\Big) \tag{5.1.14}$$

the following lemma.

Lemma 5.3

$$(\widehat{\psi_n(x;\lambda)})(\xi) = (-\mathrm{i})^n\psi_n\Big(\xi;\frac{1}{\lambda}\Big). \tag{5.1.15}$$

To simplify the presentation, we shall omit the scaling factor λ in the analysis, but we will use different scaling factors in our numerical experiments.

As before, let $\mathbb{P}_N$ be the space of polynomials of degree less than or equal to N, and denote $X_N = \{v : v = e^{-x^2/2}w,\ \forall\, w \in \mathbb{P}_N\}$. Let $\{\eta_j\}_{0\le j\le N}$ be the Hermite–Gauss points, i.e., $H_{N+1}(\eta_j) = 0,\ 0 \le j \le N$, and $\{\omega_j\}_{0\le j\le N}$ be the corresponding weights such that the following Hermite–Gauss quadrature holds (cf. Shen et al. (2011b)):

$$\int_{\mathbb{R}} p(\eta)\mathrm{d}\eta = \sum_{j=0}^{N} p(\eta_j)\omega_j, \quad \forall p \in X_{2N+1}. \tag{5.1.16}$$

In the d-dimensional case, we set

$$\boldsymbol{j} = (j_1, \ldots, j_d), \quad \eta_{\boldsymbol{j}} = (\eta_{j_1} \ldots, \eta_{j_d}), \quad \hat{\omega}_{\boldsymbol{j}} = \Pi_{k=1}^d \omega_{j_k},$$
$$\psi_{\boldsymbol{j}}(\beta) = \Pi_{k=1}^d \psi_{j_k}(\eta_k), \quad \mathbb{X}_N := X_N^d.$$

Then, the d-dimensional Hermite–Gauss quadrature reads

$$\int_{\mathbb{R}^d} p(\beta) d\beta = \sum_{|\boldsymbol{j}|_\infty = 0}^{N} p(\beta_{\boldsymbol{j}}) \hat{\omega}_{\boldsymbol{j}}, \quad \forall p \in \mathbb{X}_{2N+1}. \tag{5.1.17}$$

We denote by I_N the Hermite–Gauss interpolation operator from $C(\mathbb{R}^d)$ to $\mathbb{X}_N$ based on the Hermite–Gauss points. We shall use $x_{\boldsymbol{j}} = \eta_{\boldsymbol{j}}$ and $\xi_{\boldsymbol{j}} = \eta_{\boldsymbol{j}}$ to denote the Hermite–Gauss points in the physical space and phase space, respectively. For the sake of simplicity, we use $\mathbb{X}_N$ to denote the subspaces in both the (real) physical space and (complex) phase space.

5.1.1 Multivariate Hermite Approximations

We denote $\mathbf{1} = \{1, 1, \ldots, 1\} \in \mathbb{N}^d$, and use the following conventions:

$$\alpha \geq \boldsymbol{k} \Leftrightarrow \alpha_j \geq k_j, \ \forall 1 \leq j \leq d. \tag{5.1.18}$$

For a given multivariate function $u(\boldsymbol{x})$, we denote the $|\boldsymbol{k}|_1$-th (mixed) partial derivative by

$$\partial_{\boldsymbol{x}}^{\boldsymbol{k}} = \frac{\partial^{|\boldsymbol{k}|_1} u}{\partial_{x_1}^{k_1} \cdots \partial_{x_d}^{k_d}} = \partial_{x_1}^{k_1} \cdots \partial_{x_d}^{k_d} u. \tag{5.1.19}$$

In particular, we denote $\partial_{\boldsymbol{x}}^{s} u := \partial_{\boldsymbol{x}}^{(s,s,\ldots,s)} u$.

Using (5.1.8) repeatedly, we have

$$\partial_x^k H_n(x) = d_{n,k} H_{n-k}(x), \quad n \geq k \geq 0, \tag{5.1.20}$$

where

$$d_{n,k} = \sqrt{\frac{2^k n!}{(n-k)!}}, \quad n \geq k \geq 0. \tag{5.1.21}$$

For the sake of notational convenience, we extend the definition of $d_{n,k}$ to $0 \leq n < k$ by defining

$$d_{n,k} = 0, \quad 0 \leq n < k. \tag{5.1.22}$$

Then (5.1.20) implies that $\{\partial_x^k H_n\}$ are orthogonal with respect to the weighted inner product. More precisely,

$$\|\partial_x^k H_n\|_\omega^2 = h_{n,k} \ \text{ with } \ h_{n,k} = d_{n,k}^2 \gamma_{n-k}, \tag{5.1.23}$$

where γ_{n-k} is given in (5.1.7).

Defining the d-dimensional tensorial Hermite polynomial and Hermite weight function as

$$\boldsymbol{H}_{\boldsymbol{n}}(\boldsymbol{x}) = \prod_{j=1}^{d} H_{n_j}(x_j), \quad \omega(\boldsymbol{x}) = \prod_{j=1}^{d} \omega(x_j), \quad \boldsymbol{x} \in \mathbb{R}^d, \tag{5.1.24}$$

we derive from (5.1.20) and (5.1.23) that

$$\partial_{\boldsymbol{x}}^{\boldsymbol{k}} \boldsymbol{H}_{\boldsymbol{n}}(\boldsymbol{x}) = \boldsymbol{d}_{\boldsymbol{n},\boldsymbol{k}} \boldsymbol{H}_{\boldsymbol{n}-\boldsymbol{k}}(\boldsymbol{x}) \text{ with } \boldsymbol{d}_{\boldsymbol{n},\boldsymbol{k}} = \prod_{j=1}^{d} d_{n_j,k_j}, \tag{5.1.25}$$

and

$$\int_{\mathbb{R}^d} \partial_{\boldsymbol{x}}^{\boldsymbol{k}} \boldsymbol{H}_{\boldsymbol{n}}(\boldsymbol{x}) \partial_{\boldsymbol{x}}^{\boldsymbol{k}} \boldsymbol{H}_{\boldsymbol{m}}(\boldsymbol{x}) \omega(\boldsymbol{x}) \mathrm{d}\boldsymbol{x} = \boldsymbol{h}_{\boldsymbol{n},\boldsymbol{k}} \delta_{\boldsymbol{n}\boldsymbol{m}}, \tag{5.1.26}$$

where $\boldsymbol{n}, \boldsymbol{k} \geq \boldsymbol{0}$, $\boldsymbol{h}_{\boldsymbol{n},\boldsymbol{k}} = \prod_{j=1}^{d} h_{n_j,k_j}$ and $\delta_{\boldsymbol{n}\boldsymbol{m}} = \prod_{j=1}^{d} \delta_{n_j m_j}$.

For $u(\boldsymbol{x}) \in L_{\omega}^2(\mathbb{R}^d)$, we write

$$u(\boldsymbol{x}) = \sum_{\boldsymbol{n} \geq 0} \hat{u}_{\boldsymbol{n}} \boldsymbol{H}_{\boldsymbol{n}}(\boldsymbol{x}) \text{ with } \hat{u}_{\boldsymbol{n}} = \int_{\mathbb{R}^d} u(\boldsymbol{x}) \boldsymbol{H}_{\boldsymbol{n}}(\boldsymbol{x}) \omega(\boldsymbol{x}) \mathrm{d}\boldsymbol{x}. \tag{5.1.27}$$

Formally, we have

$$\partial_{\boldsymbol{x}}^{\boldsymbol{k}} u = \sum_{\boldsymbol{n} \geq \boldsymbol{k}} \hat{u}_{\boldsymbol{n}} \partial_{\boldsymbol{x}}^{\boldsymbol{k}} \boldsymbol{H}_{\boldsymbol{n}}(\boldsymbol{x}),$$

and by the orthogonality (5.1.26), we get

$$\|\partial_{\boldsymbol{x}}^{\boldsymbol{k}} u\|_{\omega}^2 = \sum_{\boldsymbol{n} \geq \boldsymbol{k}} \boldsymbol{h}_{\boldsymbol{n},\boldsymbol{k}} |\hat{u}_{\boldsymbol{n}}|^2 = \sum_{\boldsymbol{n} \in \mathbb{N}^d} \boldsymbol{h}_{\boldsymbol{n},\boldsymbol{k}} |\hat{u}_{\boldsymbol{n}}|^2. \tag{5.1.28}$$

Note that the second equality is a consequence of (5.1.22).

Let $\mathbb{P}_N^d$ be the d dimension tensor of P_N. Consider the orthogonal projection $\boldsymbol{\Pi}_N : L_{\omega}^2(\mathbb{R}^d) \to \mathbb{P}_N^d$, defined by

$$\int_{\mathbb{R}^d} (\boldsymbol{\Pi}_N u - u) v_N \omega(\boldsymbol{x}) \mathrm{d}\boldsymbol{x} = 0, \quad \forall v_N \in \mathbb{P}_N^d. \tag{5.1.29}$$

We define the d-dimensional weighted Sobolev spaces

$$B_{\omega}^m(\mathbb{R}^d) := \{u : \partial_{\boldsymbol{x}}^{\boldsymbol{k}} u \in L_{\omega}^2(\mathbb{R}^d),\ 0 \leq |\boldsymbol{k}|_1 \leq m\}, \quad \forall m \in \mathbb{N}, \tag{5.1.30}$$

equipped with the norm and semi-norm

$$\|u\|_{B_{\omega}^m(\mathbb{R}^d)} = \Big(\sum_{0 \leq |\boldsymbol{k}|_1 \leq m} \|\partial_{\boldsymbol{x}}^{\boldsymbol{k}} u\|_{\omega}^2 \Big)^{\frac{1}{2}}, \quad |u|_{B_{\omega}^m(\mathbb{R}^d)} = \Big(\sum_{j=1}^{d} \|\partial_{x_j}^m u\|_{\omega}^2 \Big)^{\frac{1}{2}}. \tag{5.1.31}$$

In particular, we denote $H^m(\mathbb{R}^d) = B_{\omega}^m(\mathbb{R}^d)$ with $\omega = 0$ and $\|\cdot\| = \|\cdot\|_{L^2(\mathbb{R}^d)}$.

The following results can be found in Shen et al. (2011b).

Theorem 5.4 *For any $u \in B^m_\omega(\mathbb{R}^d)$ and $0 \le l \le m$, we have*

$$\|\mathbf{\Pi}_N u - u\|_{B^l_\omega(\mathbb{R}^d)} \lesssim N^{(l-m)/2}|u|_{B^m_\omega(\mathbb{R}^d)}; \tag{5.1.32}$$

and if, in addition, $m \ge d$, then

$$\|I_N u - u\|_{B^l_\omega(\mathbb{R}^d)} \lesssim N^{d/6+(l-m)/2}|u|_{B^m_\omega(\mathbb{R}^d)}. \tag{5.1.33}$$

Next, we consider approximations by multivariate Hermite functions. Note that for any $u \in L^2(\mathbb{R}^d)$, we have $u\omega^{-1/2} \in L^2_\omega(\mathbb{R}^d)$. Define

$$\hat{\mathbf{\Pi}}_N u := \omega^{1/2}\mathbf{\Pi}_N(u\omega^{-1/2}) \in \mathbb{X}_N. \tag{5.1.34}$$

Then for $u \in L^2(\mathbb{R}^d)$, we derive immediately from (5.1.29) that

$$\int_{\mathbb{R}^d} (\hat{\mathbf{\Pi}}_N u - u)v_N \, d\boldsymbol{x} = 0, \quad \forall\, v_N \in \mathbb{X}_N. \tag{5.1.35}$$

Introduce the operator $\hat{\partial}_{x_j} = \partial_{x_j} + x_j$ which also satisfies

$$\omega^{-1/2}(x_j)\hat{\partial}_{x_j}u(x_j) = \partial_{x_j}\big[\omega^{-1/2}(x_j)u(x_j)\big],$$

and denote $\hat{\partial}_{\boldsymbol{x}} := \prod_{j=1}^d \hat{\partial}_{x_j}$, $\hat{\boldsymbol{\partial}}_{\boldsymbol{x}}^{\boldsymbol{k}} := \prod_{j=1}^d \hat{\partial}_{x_j}^{k_j}$, and define

$$\hat{B}^m(\mathbb{R}^d) := \big\{u : \hat{\partial}_{\boldsymbol{x}}^{\boldsymbol{k}} u \in L^2(\mathbb{R}^d),\ 0 \le |\boldsymbol{k}|_1 \le m\big\},\ \forall m \in \mathbb{N}, \tag{5.1.36}$$

equipped with the norm and semi-norm

$$\|u\|_{\hat{B}^m(\mathbb{R}^d)} = \Big(\sum_{0 \le |\boldsymbol{k}|_1 \le m} \|\hat{\partial}_{\boldsymbol{x}}^{\boldsymbol{k}} u\|^2\Big)^{\frac{1}{2}}, \quad |u|_{\hat{B}^m(\mathbb{R}^d)} = \Big(\sum_{j=1}^d \|\hat{\partial}_{x_j}^m u\|^2\Big)^{\frac{1}{2}}. \tag{5.1.37}$$

We derive below approximation results for the errors measured in the usual Hilbert space.

Theorem 5.5 *For any $u \in \hat{B}^m(\mathbb{R}^d)$ with $m \ge 1$, we have*

$$\|\hat{\mathbf{\Pi}}_N u - u\|_{H^\mu(\mathbb{R}^d)} \lesssim N^{(\mu-m)/2}|u|_{\hat{B}^m(\mathbb{R}^d)}, \quad 0 \le \mu \le m; \tag{5.1.38}$$

and if additionally $m \ge d$, we also have

$$\|I_N u - u\|_{H^\mu(\mathbb{R}^d)} \lesssim N^{d/6+(\mu-m)/2}|u|_{\hat{B}^m(\mathbb{R}^d)}, \quad 0 \le \mu \le 1. \tag{5.1.39}$$

Proof We start with the proof of (5.1.38).

The case $\mu = 0$ is a direct consequence of (5.1.32) with $l = 0$.

For any $1 \le j \le d$,

$$\begin{aligned}\partial_{x_j}(\hat{\mathbf{\Pi}}_N u - u) &= \omega^{1/2}\partial_{x_j}\big(\mathbf{\Pi}_N(u\omega^{-1/2}) - (u\omega^{-1/2})\big)\\ &\quad - x_j\omega^{1/2}\big(\mathbf{\Pi}_N(u\omega^{-1/2}) - (u\omega^{-1/2})\big).\end{aligned}$$

We recall (cf. Shen et al. (2011a, Lemma B.6)) that

$$\|x_j u\|_\omega^2 \le \|u\|_\omega^2 + \|\partial_{x_j} u\|_\omega^2, \quad 1 \le j \le d, \quad \forall u \in H^1(\mathbb{R}^d). \tag{5.1.40}$$

Therefore, thanks to (5.1.40), we obtain

$$\begin{aligned}
\|\partial_{x_j}(\hat{\mathbf{\Pi}}_N u - u)\|^2 &\le 2\|\partial_{x_j}\big(\mathbf{\Pi}_N(u\omega^{-1/2}) - (u\omega^{-1/2})\big)\|_\omega^2 \\
&\quad + 2\|x_j\big(\mathbf{\Pi}_N(u\omega^{-1/2}) - (u\omega^{-1/2})\big)\|_\omega^2 \\
&\le 2\|\mathbf{\Pi}_N(u\omega^{-1/2}) - (u\omega^{-1/2})\|_\omega^2 \\
&\quad + 4\|\partial_{x_j}\big(\mathbf{\Pi}_N(u\omega^{-1/2}) - (u\omega^{-1/2})\big)\|_\omega^2 \\
&\lesssim N^{1-m}|u\omega^{-1/2}|^2_{B^m_\omega(\mathbb{R}^d)} \lesssim N^{1-m}|u|^2_{H^m(\mathbb{R}^d)}.
\end{aligned}$$

Summing up the above for $1 \le j \le d$, we obtain (5.1.38) with $\mu = 1$. By recursively applying the above argument, we can prove (5.1.38) for all positive integer $\mu \le m$. Finally, (5.1.38) for $0 \le \mu \le m$ can be obtained by a standard space interpolation (cf. Adams (1975)).

Similarly, we can prove (5.1.39) from (5.1.33). □

A Model Problem

In the next two sections, we shall consider, as a model problem, the following FPDE:

$$\begin{aligned}
&(-\Delta)^{\alpha/2}u(\boldsymbol{x}) + \rho u(\boldsymbol{x}) = f(\boldsymbol{x}) \quad \text{in} \quad \Omega = \mathbb{R}^d; \\
&\lim_{|\boldsymbol{x}|\to\infty} u(\boldsymbol{x}) = 0, \quad \alpha,\ \rho \ge 0.
\end{aligned} \tag{5.1.41}$$

We start by deriving a suitable weak formulation for (5.1.41).

Lemma 5.6 *Let $\mu > 0$, then for any $u,\ v \in H^\mu(\Omega)$, we have*

$$\big((-\Delta)^\mu u, v\big) = \big((-\Delta)^{\frac{\mu}{2}}u, (-\Delta)^{\frac{\mu}{2}}v\big). \tag{5.1.42}$$

Proof First, let us recall a useful property of Fourier transform, namely, for $\phi,\ \varphi \in L^2(\Omega)$, we have

$$\int_\Omega \phi\,\overline{\varphi}\,\mathrm{d}\boldsymbol{x} = \int_\Omega \widehat{\phi}\,\overline{\widehat{\varphi}}\,\mathrm{d}\xi, \tag{5.1.43}$$

where $\overline{\varphi}$ is the complex conjugate of φ. Then

$$\begin{aligned}
\big((-\Delta)^\mu u, v\big) &= \big(\widehat{(-\Delta)^\mu u}, \widehat{v}\big) = \big(|\xi|_2^{2\mu}\widehat{u}, \widehat{v}\big) = \big(|\xi|_2^{\mu}\widehat{u}, |\xi|_2^{\mu}\widehat{v}\big) \\
&= \big(\widehat{(-\Delta)^{\frac{\mu}{2}}u}, \widehat{(-\Delta)^{\frac{\mu}{2}}v}\big) = \big((-\Delta)^{\frac{\mu}{2}}u, (-\Delta)^{\frac{\mu}{2}}v\big).
\end{aligned}$$

This completes the proof. □

Here we only focus on real-valued functions. To simplify the notation, we denote $s = \alpha/2$, and define

$$A(u, v) = \rho(u, v) + \big((-\Delta)^{\frac{s}{2}} u, (-\Delta)^{\frac{s}{2}} v\big). \tag{5.1.44}$$

By the definition in (5.1.4), we immediately obtain that $A(\cdot, \cdot)$ is continuous and coercive in $H^s(\Omega) \times H^s(\Omega)$.

Then by Lemma 5.6, a weak formulation for (5.1.41) is: find $u \in H^s(\Omega)$ such that

$$A(u, v) = (f, v), \quad \forall v \in H^s(\Omega). \tag{5.1.45}$$

Thanks to the Lax–Milgram lemma, the above problem admits a unique solution $u \in H^s(\Omega)$ satisfying

$$\|u\|_{H^s} \lesssim \|f\|_{(H^s)'}, \tag{5.1.46}$$

where $(H^s)'$ is the dual space of H^s.

5.1.2 Hermite Collocation Method

In this section, we present a Hermite collocation method for (5.1.41) and derive corresponding error estimates.

Applying the Fourier transform on both sides of problem (5.1.41), we derive an equivalent formulation in frequency space:

$$\rho\widehat{u}(\xi) + |\xi|_2^{\alpha} \widehat{u}(\xi) = \widehat{f}(\xi), \tag{5.1.47}$$

which implies that

$$\widehat{u}(\xi) = \frac{\widehat{f}(\xi)}{\rho + |\xi|_2^{\alpha}}. \tag{5.1.48}$$

Hence, the Fourier transform of the solution can be expressed directly by using the Fourier transform of f. However, we need a numerical procedure to compute the approximation of the (continuous) Fourier transform, and to compute function values in physical space from its approximation of the (continuous) Fourier transform.

We propose the following Hermite collocation method: find $u_N \in \mathbb{X}_N$ such that

$$\widehat{u_N}(\beta_j) = \frac{\widehat{I_N f}(\beta_j)}{\rho + |\beta_j|_2^{\alpha}}, \quad \forall 0 \le |j|_1 \le N, \tag{5.1.49}$$

where $I_N f$ is the interpolation in $\mathbb{X}_N$ of f at the Hermite–Gauss points.

Denote $\psi_n(x) = \Pi_{n=n_1}^{n_d} \psi_n(x_n)$; we describe below the detailed algorithm for (5.1.49).

Step 1. From the values of $f(\beta_j)$ with $0 \le |j|_1 \le N$, perform a forward discrete Hermite transform to obtain $\tilde{f}_k$ such that

$$I_N f(x) = \sum_{0\le|k|_1\le N} \tilde{f}_k \psi_k(x).$$

Step 2. Taking the Fourier transform of the above; thanks to (5.1.13), we obtain

$$\widehat{I_N f}(\xi) = \sum_{0\le|k|_1\le N} \tilde{f}_k (-\mathrm{i})^{|k|_1} \psi_k(\xi).$$

Then, compute $\widehat{I_N f}(\beta_j)$ ($0 \le |j|_1 \le N$) from the above with a backward discrete Hermite transform.

Step 3. Compute $\widehat{u_N}(\beta_j)$ from (5.1.49), and perform a forward discrete Hermite transform to obtain $\tilde{v}_k$ such that

$$\widehat{u_N}(\xi) = \sum_{0\le|k|_1\le N} \tilde{v}_k \psi_k(\xi).$$

Step 4. Take the inverse Fourier transform of the above to obtain

$$u_N(x) = \sum_{0\le|k|_1\le N} \tilde{v}_k \mathrm{i}^{|k|_1} \psi_k(x).$$

Finally, we obtain $u_N(\beta_j)$ for $0 \le |j|_1 \le N$ from the above by performing a backward Hermite transform.

Remark 5.7 We observe that the main cost of the above algorithm is one discrete Hermite transform at each step, the cost of which is $O(N^{d+1})$ for the d-dimensional problem.

By using the exactness of the Hermite–Gauss quadrature, we rewrite the scheme (5.1.49) in the following variational form:

$$\rho(\widehat{u_N}, \widehat{v}_N) + (|\xi|_2^\alpha \widehat{u_N}, \widehat{v}_N)_N = (\widehat{I_N f}, \widehat{v}_N) \quad \forall \widehat{v}_N \in \mathbb{X}_N, \tag{5.1.50}$$

where

$$(u, v)_N \equiv \sum_{0\le|j|_\infty\le N} u(\beta_j) v(\beta_j) \hat{\omega}_j \tag{5.1.51}$$

is the multi-dimensional discrete inner product defined by the Hermite–Gauss quadrature in (5.1.17).

We also derive from (5.1.47) that

$$\rho(\widehat{u}, \widehat{v}_N) + (|\xi|_2^\alpha \widehat{u}, \widehat{v}_N)_N = (\widehat{f}, \widehat{v}_N) + (I_N(|\xi|_2^\alpha \widehat{u}) - |\xi|_2^\alpha \widehat{u}, \widehat{v}_N) \tag{5.1.52}$$

for all $\widehat{v}_N \in \mathbb{X}_N$. Let us define the bilinear form

$$a_N(\widehat{u}, \widehat{v}) = \rho(\widehat{u}, \widehat{v}) + (|\xi|_2^\alpha \widehat{u}, \widehat{v})_N \quad \forall \widehat{u}, \widehat{v} \in \mathbb{X}_N, \tag{5.1.53}$$

and the discrete norm

$$\|\widehat{v}\|_{N,\alpha} \equiv \sqrt{a_N(\widehat{v},\widehat{v})}. \tag{5.1.54}$$

By definition, we have

$$\|\widehat{v}\|^2 \lesssim a_N(\widehat{v},\widehat{v}) = \|\widehat{v}\|_{N,\alpha}^2, \quad a_N(\widehat{u},\widehat{v}) \lesssim \|\widehat{u}\|_{N,\alpha}\|\widehat{v}\|_{N,\alpha}. \tag{5.1.55}$$

Hence, applying the first Strang lemma to (5.1.52) and (5.1.50), we obtain immediately

$$\|\widehat{u}-\widehat{u_N}\|_{N,\alpha} \lesssim \inf_{\widehat{v}_N\in\mathbb{X}_N} \|\widehat{u}-\widehat{v}_N\|_{N,\alpha} + \|\widehat{f}-\widehat{I_N f}\| + \|I_N(|\xi|_2^\alpha\widehat{u}) - |\xi|_2^\alpha\widehat{u}\|. \tag{5.1.56}$$

Since

$$\inf_{\widehat{v}_N\in\mathbb{X}_N} \|\widehat{u}-\widehat{v}_N\|_{N,\alpha} \le \|\widehat{u}-I_N\widehat{u}\|_{N,\alpha} = \sqrt{\rho}\|\widehat{u}-I_N\widehat{u}\|,$$

$$\|\widehat{f}-\widehat{I_N f}\| = \|f-I_N f\|,$$

and

$$\begin{aligned} I_N(|\xi|_2^\alpha\widehat{u}) - |\xi|_2^\alpha\widehat{u} &= I_N(\rho\widehat{u}+|\xi|_2^\alpha\widehat{u}) - (\rho\widehat{u}+|\xi|_2^\alpha\widehat{u}) + \rho(\widehat{u}-I_N\widehat{u}) \\ &= I_N\widehat{f} - \widehat{f} + \rho(\widehat{u}-I_N\widehat{u}), \end{aligned}$$

we derive from the above and the error estimate (5.1.39) the following theorem.

Theorem 5.8 *Let $\widehat{u}$ and $\widehat{u_N}$ be the solutions to* (5.1.48) *and* (5.1.49), *respectively. Assuming that $\widehat{u} \in \hat{B}^{m_1}(\mathbb{R}^d)$, $f \in \hat{B}^{m_2}(\mathbb{R}^d)$, and $\widehat{f} \in \hat{B}^{m_3}(\mathbb{R}^d)$ with $m_1, m_2, m_3 \ge d$, we have*

$$\begin{aligned} \|u-u_N\| \lesssim \|\widehat{u}-\widehat{u_N}\|_{N,\alpha} &\lesssim N^{d/6-m_1/2}|\widehat{u}|_{\hat{B}^{m_1}(\mathbb{R}^d)} \\ &+ N^{d/6-m_2/2}|f|_{\hat{B}^{m_2}(\mathbb{R}^d)} + N^{d/6-m_3/2}|\widehat{f}|_{\hat{B}^{m_3}(\mathbb{R}^d)}. \end{aligned}$$

Remark 5.9 We are unable to obtain an optimal error estimate in the energy norm $\|\cdot\|_{H^{\alpha/2}}$ due to the unboundedness of the coefficient $|\xi|_2^\alpha$.

5.1.3 Hermite–Galerkin Method

While the Hermite-collocation method presented above is very easy to implement, its error estimate is not quite optimal as the discrete norm in Theorem 5.8 is only an approximation to the energy norm $\|\cdot\|_{H^{\alpha/2}}$. Below, we present a Hermite–Galerkin method which enjoys optimal error-estimates in the energy norm.

The Method and Error Estimates

The Hermite–Galerkin approximation for (5.1.45) is: find $u_N \in \mathbb{X}_N$ such that

$$A(u_N, v_N) = (I_N f, v_N), \quad \forall\, v_N \in \mathbb{X}_N. \tag{5.1.57}$$

Since $\mathbb{X}_N \subset H^s(\Omega)$ for any $s \geq 0$, it is clear that the problem (5.1.57) admits a unique solution. We derive from (5.1.45) and (5.1.57) that

$$A(u - u_N, v_N) = (f - I_N f, v_N), \quad \forall\, v_N \in \mathbb{X}_N.$$

We then derive from a standard argument that

$$\|u - u_N\|_{H^s} \lesssim \inf_{v_N \in \mathbb{X}_N} \|u - v_N\|_{H^s} + \|f - I_N f\|. \tag{5.1.58}$$

Hence, taking $v_N = \hat{\mathbf{\Pi}}_N u$ in the above and using Theorem 5.5, we obtain the following result.

Theorem 5.10 *Let u and u_N be the solutions to* (5.1.45) *and* (5.1.57) *respectively. Assuming that $u \in B^{\hat{m}_1}(\mathbb{R}^d)$ and $f \in B^{\hat{m}_2}(\mathbb{R}^d)$ with $m_1 \geq s$, $m_2 \geq d$, we have*

$$\|u - u_N\|_{H^s} \lesssim N^{(s-m_1)/2} |u|_{\hat{B}^{m_1}(\mathbb{R}^d)} + N^{d/6 - m_2/2} |f|_{\hat{B}^{m_2}(\mathbb{R}^d)}.$$

Remark 5.11 The error estimate for the Hermite–Galerkin method is better than that for the Hermite-collocation method in the following two aspects: (i) the error estimate in the Galerkin case is in energy norm while it only implies an estimate in L^2-norm in the collocation case; (ii) it only depends on the smoothness of u and f, while in the collocation case it depends on the smoothness of $\widehat{u}$, f and $\widehat{f}$.

Remark 5.12 Since the usual duality argument cannot be applied to the fractional differential equations due to the lack of a regularity result, it is not clear how to derive an improved error estimate in L^2 norm.

Implementation in a 1-D Domain

To implement (5.1.57), we need to compute the mass matrix $M_{kj} = (\phi_j, \phi_k)$ and stiffness matrix $S_{kj} = \big((-\Delta)^{\frac{s}{2}} \phi_j, (-\Delta)^{\frac{s}{2}} \phi_k\big)$, where $\{\phi_k\}$ are the basis functions of $\mathbb{X}_N$.

Let us begin with the one-dimensional case. Since $\mathbb{X}_N = \mathrm{span}\{\psi_k(x) : 0 \leq k \leq N\}$, we have $M_{kj} = (\psi_j, \psi_k) = \delta_{kj}$, so we only need to compute S_{kj}. By Lemma 5.6 and (5.1.13), we have

$$S_{kj} = \big((-\Delta)^{\frac{s}{2}} \psi_j, (-\Delta)^{\frac{s}{2}} \psi_k\big) = (-\mathrm{i})^j \cdot \mathrm{i}^k \int_{-\infty}^{\infty} |\xi|^{\alpha} \psi_j(\xi) \psi_k(\xi)\, d\xi. \tag{5.1.59}$$

To simplify the presentation, we denote $S_{kj} := (-\mathrm{i})^j \cdot \mathrm{i}^k s_{kj}$, and will compute s_{kj}. We have from (5.1.11) that

$$s_{kj} = \int_{-\infty}^{\infty} |\xi|^\alpha \exp(-\xi^2) H_j(\xi) H_k(\xi)\, d\xi.$$

For $0 \le j,k \le N$, the above integral can be computed by deriving an analytical recurrence formula for the quantities. Let $\{\tilde{H}_j\}_{j=0}^{\infty}$ denote the generalized Hermite polynomial family orthonormal with respect to the weight function $|\xi|^\alpha \exp(-\xi^2)$ satisfying the following three-term recurrence, see Gautschi (2004, Table 1.1):

$$\sqrt{b_{k+1}}\tilde{H}_{k+1}(x) = x\tilde{H}_k(x) + \sqrt{b_k}\tilde{H}_{k-1}(x),\ k \ge 1,$$
$$\tilde{H}_0(x) = 1/\Gamma\left(\frac{\alpha+1}{2}\right),\ \tilde{H}_1(x) = x/\Gamma\left(\frac{\alpha+1}{2}\right),$$

where $b_k = k/2$ for k is even while $b_k = k/2 + \alpha/2$ for k is odd. Then there are connection coefficients $c_{n,j}$, so that

$$H_n(x) = \sum_{j=0}^{n} c_{n,j}\tilde{H}_j(x).$$

Therefore, there is an explicit recurrence formula that translates these connection coefficients into $c_{n,j}$ (see Maroni and da Rocha (2008, (12) and Theorem 1)). Collect these coefficients into the $(N+1)\times(N+1)$ matrix C with $(C)_{j,k} = c_{j,k}$ (using 0-based indexing, $0 \le j,k \le N$). Then the desired matrix is $\hat{S} = CC^T$, where $\hat{S} = \{s_{kj}\}_{k,j=0}^{N}$.

Implementation in Multi-Dimensional Domains

In the one-dimensional case, we developed above an efficient algorithm to compute the stiffness matrix. However, in the multi-dimensional case, the term $|\xi|_2^\alpha = (\xi_1^2 + \cdots + \xi_d^2)^{\frac{\alpha}{2}}$ in (5.1.47) is non-separable, making it very expensive to compute the stiffness matrix explicitly and to solve the resulting linear system by a direct method. Hence, we shall use an iterative approach for (5.1.57) with a suitable separable problem as a preconditioner. To simplify the presentation, we shall only consider the two-dimensional case, although the method can be directly applied to multi-dimensional case.

Unlike in the integer PDE case, the multi-dimensional fractional PDE (5.1.41) is no longer separable. This can be easily seen from its equivalent formulation in frequency space (5.1.47), which, in the two-dimensional case, can be written as

$$\rho\widehat{u}(\xi) + (|\xi_1|^2 + |\xi_2|^2)^{\alpha/2}\widehat{u}(\xi) = \widehat{f}(\xi), \tag{5.1.60}$$

whose Hermite–Galerkin approximation, an equivalent formulation to (5.1.60), is to find $u_N \in \mathbb{X}_N$ such that

$$\rho(\widehat{u_N}, \widehat{v_N}) + ((|\xi_1|^2 + |\xi_2|^2)^{\alpha/2}\widehat{u_N}, \widehat{v_N}) = (\widehat{I_N f}, \widehat{v_N}), \quad \forall \widehat{v_N} \in \mathbb{X}_N. \quad (5.1.61)$$

However, it is convenient to use the separable form

$$\rho\widehat{u}(\xi) + |\xi_1|^\alpha \widehat{u}(\xi) + |\xi_2|^\alpha \widehat{u}(\xi) = \widehat{f}(\xi) \quad (5.1.62)$$

to build a preconditioner. Indeed, a Hermite–Galerkin method for (5.1.62) is to find $u_N \in \mathbb{X}_N$ such that

$$\rho(\widehat{u_N}, \widehat{v_N}) + (|\xi_1|^\alpha \widehat{u_N}, \widehat{v_N}) + (|\xi_2|^\alpha \widehat{u_N}, \widehat{v_N}) = (\widehat{I_N f}, \widehat{v_N}), \quad \forall \widehat{v_N} \in \mathbb{X}_N. \quad (5.1.63)$$

Let $u_N = \sum_{k,j=0}^N \tilde{u}_{kj} \psi_k(x_1;\lambda)\psi_j(x_2;\lambda) \in X_N$ be the solution of (5.1.63), and set $\bar{U} = (\tilde{u}_{kj})_{0\le k,j\le N}$ and $\bar{F} = (\tilde{f}_{kj})_{0\le k,j\le N}$ with $\tilde{f}_{kj} = (I_N f, \psi_k(x_1;\lambda)\psi_j(x_2;\lambda))$; the problem (5.1.63) is reduced to

$$\rho\bar{U} + S\bar{U} + \bar{U}S = \bar{F}, \quad (5.1.64)$$

where S is the one-dimensional stiffness matrix computed above.

Thus, the above linear system can be efficiently solved by using the matrix diagonalization method (cf. Lynch et al. (1964); Shen (1994)) in a small multiple of N^3 operations (a small multiple of N^{d+1} operations for d-dimensional problems).

The next lemma shows that (5.1.63) provides an optimal preconditioner for (5.1.61).

Lemma 5.13 *For* $0 < \alpha \le 2$, *we have*

$$d^{\alpha/2-1}\sum_{j=1}^d |\xi_j|^\alpha \le \left(\sum_{j=1}^d |\xi_j|^2\right)^{\alpha/2} \le \sum_{j=1}^d |\xi_j|^\alpha, \quad \forall \xi_1, \ldots, \xi_d \in \mathbb{R}.$$

For $\alpha > 2$, *we have*

$$\sum_{j=1}^d |\xi_j|^\alpha \le \left(\sum_{j=1}^d |\xi_j|^2\right)^{\alpha/2} \le d^{1-2/\alpha}\sum_{j=1}^d |\xi_j|^\alpha, \quad \forall \xi_1, \ldots, \xi_d \in \mathbb{R}.$$

Proof It is obvious that the result holds for $\alpha = 2$.

Since $f(x) = -x^{\alpha/2}$ is convex for $x > 0$ with $\alpha \in (0,2)$, we have

$$f\left(\frac{\sum_{j=1}^d |\xi_j|^2}{d}\right) \le \frac{1}{d}\sum_{j=1}^d f(|\xi_j|^2),$$

which implies

$$\Big(\sum_{j=1}^{d}|\xi_j|^2\Big)^{\alpha/2}\geq d^{\alpha/2-1}\sum_{j=1}^{d}|\xi_j|^{\alpha}.$$

On the other hand, we have

$$\sum_{j=1}^{d}a_j\leq\Big(\sum_{j=1}^{d}a_j^{\nu}\Big)^{1/\nu},\quad \forall a_j\geq 0,\ \ 0\leq\nu\leq 1.$$

Then let $a_j=|\xi_j|^2$ and $\nu=\alpha/2$ for $\alpha\in(0,2)$ in the above; we obtain

$$\sum_{j=1}^{d}|\xi_j|^2\leq\Big(\sum_{j=1}^{d}|\xi_j|^{\alpha}\Big)^{2/\alpha}.$$

We obtain immediately from the above that

$$\Big(\sum_{j=1}^{d}|\xi_j|^2\Big)^{\alpha/2}\leq\sum_{j=1}^{d}|\xi_j|^{\alpha}.$$

The case with $\alpha>2$ can be obtained by using a similar argument. □

Let the linear systems for the original Hermite–Galerkin approximation (5.1.61) and for the preconditioner (5.1.63) be $A\bar{u}=\bar{f}$ and $B\bar{u}=\bar{f}$, respectively. Then, the above lemma shows that the condition number of $B^{-1}A$ is uniformly bounded. Since the preconditioner system $B\bar{u}=\bar{f}$ can be efficiently solved, and the matrix-vector product $A\bar{u}$ can be efficiently performed in the frequency space, the linear system $A\bar{u}=\bar{f}$ can then be solved efficiently by using the preconditioned conjugate gradient method.

We now present a numerical example to show the effectiveness of this preconditioner. We consider the problem (5.1.41) with a random function f and let $\rho\equiv 1$. We set the threshold $\epsilon=10^{-8}$ and list in Table 5.1 the iteration numbers of the CG method for solving (5.1.57) with preconditioner (5.1.62). We observe that the iteration numbers are bounded independent of N.

5.1.4 Numerical Results and Applications

Accuracy Tests

We present a few numerical results below to test the accuracy and validate our algorithm. In the first two examples, we shall use the Hermite–Galerkin method, and in Example 5.16, we shall compare the accuracy of the Hermite–Galerkin method and Hermite-collocation method.

Table 5.1. *Number of iterations for solving* (5.1.60). *Source: Mao and Shen (2017). Copyright ©2017 Society for Industrial and Applied Mathematics. Reprinted with permission. All rights reserved.*

	# of iteration					
	$\alpha = 1.2$		$\alpha = 1.5$		$\alpha = 1.8$	
N	$\lambda = 1$	$\lambda = 0.7$	$\lambda = 1$	$\lambda = 0.7$	$\lambda = 1$	$\lambda = 0.7$
60	7	8	6	7	5	6
80	7	8	6	7	5	6
100	7	8	6	7	5	6
120	7	8	6	7	5	6
140	7	8	6	7	5	6
160	7	8	6	7	5	6

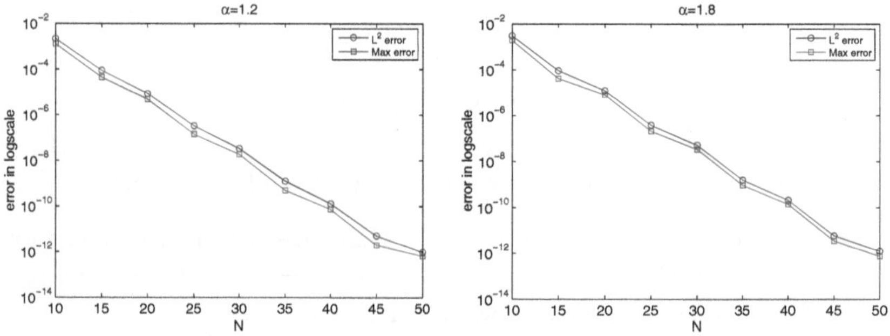

Figure 5.1 Convergent rates for $u(x) = e^{-x^2}$, left: $\alpha = 1.2$, right: $\alpha = 1.8$. Source: Mao and Shen (2017). Copyright ©2017 Society for Industrial and Applied Mathematics. Reprinted with permission. All rights reserved.

Example 5.14 (Problem with exponential decay solution) Let $\rho \equiv 1$, and the exact solution be $u(x) = e^{-x^2}$. Here we take the scaling parameter $\lambda = 1$.

The L^2 and max errors in semi-log scale are shown in Figure 5.1 for $\alpha = 1.2, 1.8$. We observe an exponential convergence as expected.

Example 5.15 (Problem with smooth and exponential decay forcing function) Let $\rho \equiv 1$, we take $f(x) = e^{-x^2/2}(1+x)$. Since no exact solution is available, we use the numerical solution with $N = 512$ as the reference solution. We choose $\lambda = 0.4$.

According to Theorem 5.10, the convergence rate will essentially depend on the decay rate of the solution at infinity. Hence, we plot in Figure 5.2

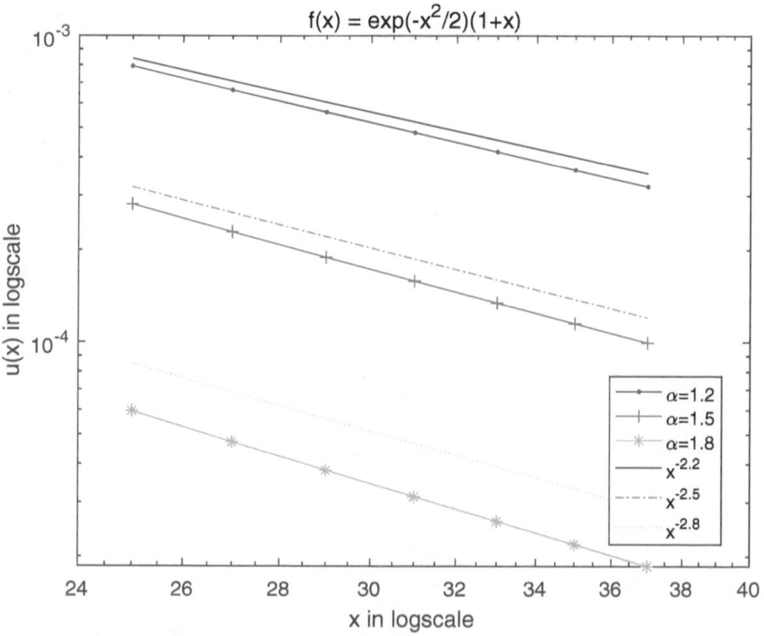

Figure 5.2 Solution $u(x), x \in (25,37)$ for different values of α with $f(x) = e^{-x^2/2}(1+x)$. Source: Mao and Shen (2017). Copyright ©2017 Society for Industrial and Applied Mathematics. Reprinted with permission. All rights reserved.

the numerical solution in the interval $x \in (25,37)$ in log-log scale in order to study the asymptotic behavior of the solution when $|x| \to \infty$. We observe that the solutions have algebraic decay for all non-integer values of α. The asymptotic decay rates for different values of α are also shown in Figure 5.2, which indicate that the solution behaves like $|x|^{-\alpha-1}$ for large $|x|$. Note that when $\alpha=2$, the solution does converge exponentially at infinity.

The convergent results in log-log scale are presented in Figure 5.3. We observe that both L^2-error and $H^{\frac{\alpha}{2}}$-error have algebraic convergence. As expected, the convergence rate improves as α increases.

Example 5.16 In this test, we compare the accuracy of the Hermite–Galerkin method with the Hermite-collocation method. Let $\rho \equiv 1$; we consider two cases: (i) $f(x) = e^{-x^2/2}(1+x)$ and (ii) $f(\boldsymbol{x}) = e^{-x_1^2/2-x_2^2/2}(1+x_1)(1+x_2)$, and examine the maximum errors by the collocation and Galerkin methods for different values of α. The scaling parameter λ is fixed to be 0.4, and the numerical solution with $N = 512$ is used as reference solution.

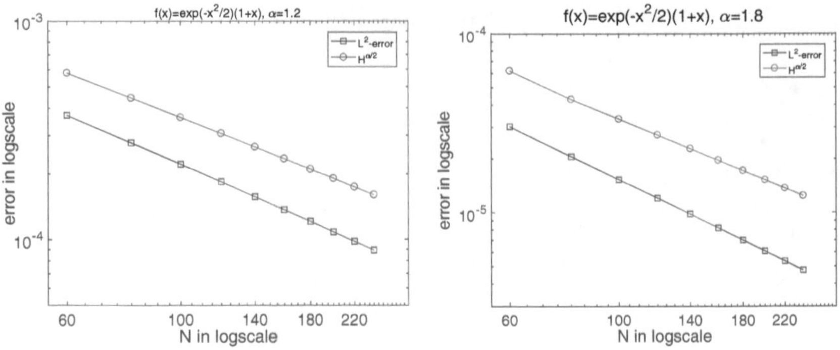

Figure 5.3 Convergent rates for $f(x) = e^{-x^2/2}(1+x)$, Left: $\alpha = 1.2$, Right: $\alpha = 1.8$. Source: Mao and Shen (2017). Copyright ©2017 Society for Industrial and Applied Mathematics. Reprinted with permission. All rights reserved.

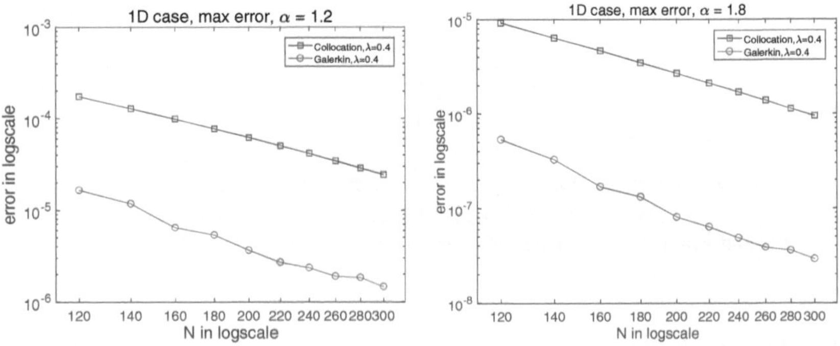

Figure 5.4 Max error with spectral collocation and spectral-Galerkin method for 1D case, $f(x) = e^{-x^2/2}(1+x)$, Left: $\alpha = 1.2$, Right: $\alpha = 1.8$. Source: Mao and Shen (2017). Copyright ©2017 Society for Industrial and Applied Mathematics. Reprinted with permission. All rights reserved.

From Figures 5.4–5.5, we observe that the Galerkin method leads to a higher accuracy than the collocation method. However, the Galerkin spectral method is more expensive than the collocation method, particularly in the multi-dimensional case. We list the CPU time in Table 5.2 with the Galerkin and the collocation methods. For the Galerkin method, two different values of threshold $\epsilon = 10^{-8}$ and 10^{-12} are used. The table lists the total CPU time for solving the two-dimensional problem 10 times and using MATLAB 2016 on a Lenovo Thinkpad laptop with Intel Core i7-6600 CPU. We observe that the collocation method is much faster than the Galerkin method with the same number of unknowns, while the accuracy improvement of the Galerkin method

Table 5.2. *Total CPU time (in seconds) for solving the 2D problem 10 times with Galerkin or collocation method (G – Galerkin, C – collocation), $\lambda = 0.4$.*

	$\alpha = 1.2$			$\alpha = 1.8$		
N	G, $\epsilon = 10^{-8}$	G, $\epsilon = 10^{-12}$	C	G, $\epsilon = 10^{-8}$	G, $\epsilon = 10^{-12}$	C
200	5.08	7.02	0.241	3.94	5.13	0.225
300	23.1	31.0	0.663	17.8	23.5	0.660

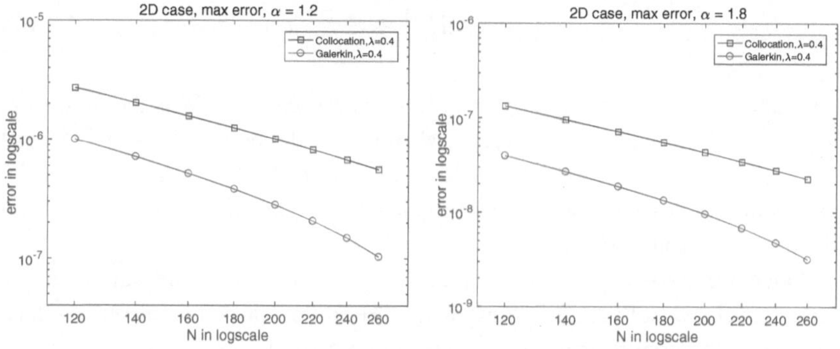

Figure 5.5 Max error with spectral collocation and spectral-Galerkin method for 2D case, $f(\boldsymbol{x}) = e^{-x_1^2/2 - x_2^2/2}(1 + x_1)(1 + x_2)$, Left: $\alpha = 1.2$, Right: $\alpha = 1.8$. Source: Mao and Shen (2017). Copyright ©2017 Society for Industrial and Applied Mathematics. Reprinted with permission. All rights reserved.

is problem dependent. It appears that for most problems, the collocation method would be more efficient than the Galerkin method. However, for more general problems with variable coefficients, the efficiency of the collocation method will be lost.

Application to a Fractional Nonlinear Schrödinger Equation

We consider the following two-dimensional fractional nonlinear Schrödinger equation (fNLS) (see Klein et al. (2014)):

$$\begin{aligned} &\mathrm{i}\partial_t u(\boldsymbol{x},t) = \frac{1}{2}(-\Delta)^{\frac{\alpha}{2}} u(\boldsymbol{x},t) + \gamma |u|^{2p} u(\boldsymbol{x},t), \quad \boldsymbol{x} \in \mathbb{R},\ t \in (0,T], \\ &u(\boldsymbol{x},0) = u_0(\boldsymbol{x}), \end{aligned} \tag{5.1.65}$$

where $\mathrm{i}^2 = -1$ and $p > 0$, $u(\boldsymbol{x},t)$ is a complex-valued wave function. Let $f(u) = \gamma |u|^{2p} u$. Applying the Fourier transform on both sides of (5.1.65), we get

$$
\begin{aligned}
&\mathrm{i}\partial_t\widehat{u}(\xi,t) = \frac{1}{2}|\xi|^\alpha\widehat{u}(\xi,t) + \widehat{f}(u)(\xi,t), \quad \xi \in \mathbb{R}, \ t \in (0,T], \\
&\widehat{u}(\xi,0) = \widehat{u}_0(\xi).
\end{aligned} \tag{5.1.66}
$$

In order to solve the above equation, we use the Hermite collocation method for the space discretization and use a second order backward Euler method for time discretization. More precisely, the full discretization scheme in the phase space reads:

$$
\mathrm{i}\frac{3\widehat{u}_{kl}^{n+1} - 4\widehat{u}_{kl}^{n} + \widehat{u}_{kl}^{n-1}}{2\Delta t} = \frac{1}{2}|\xi_{kl}|^\alpha\widehat{u}_{kl}^{n+1} + (2\widehat{f}_{kl}(u^n) - \widehat{f}_{kl}(u^{n-1})), \tag{5.1.67}
$$

with $\widehat{u}_{kl}^0 = \widehat{u}_0(\xi_{kl})$, for $k,l = 0,1,\ldots,N$, where $\Delta t = T/M$ is the time-step, $\widehat{u}_{kl}^n, k,l = 0,1,\ldots,N, n = 0,1,\ldots,M$ represent the point values of $\widehat{u}(\xi,t)$ at $\xi = \xi_{kl} := ((\xi_1)_k, (\xi_2)_l)$ and time level $t_n = n\Delta t$, $\widehat{f}_{kl}(u^n)$ represent the point values of $\widehat{f}(u^n)(\xi)$ at $\xi = \xi_{kl}$ and time level t_n. Again, $\widehat{u}_{kl}^1$ are computed with the backward Euler method.

Once we have $\widehat{u}_{kl}^n, k,l = 0,1,\ldots,N$, we can obtain the values at the Hermite–Gauss collocation points with a Hermite–Gauss transform. At each time-step, the nonlinear terms $\widehat{f}_{kl}(u^n)$ are computed with a standard pseudo-spectral approach.

We take the initial condition to be

$$
u_0(\boldsymbol{x}) = \eta\,\mathrm{sech}(x_1)\mathrm{sech}(x_2), \tag{5.1.68}
$$

where η is a positive constant, and focus on the quintic case, i.e., $p = 2$. The other parameters are set to be: $\eta = 1.0,\ N = 200,\ \Delta t = 10^{-3},\ \lambda = 0.6$.

We first consider the defocusing case with $\gamma = 1$. The modulus squared of the numerical solution at times $T = 1$ and $T = 2$ for different values of fractional order $\alpha = 1.1,\ 1.6$ are given in Figures 5.6 and 5.7. We observe that the solution diffuses as expected.

Next we consider the focusing case with $\gamma = -1$. In this case, the solution will eventually blow up. We plot the modulus squared of the numerical solution at time $T = 1$ for different values of fractional order $\alpha = 1.1,\ 1.6$ in Figure 5.8. The numerical results indicate that the maximum of the solution increases faster with smaller α, which is consistent with the observation in Klein et al. (2014) where the one-dimensional fNLS is solved with up to $N = 2^{17}$ points using a Fourier spectral method. We also show the L^∞ norm of the modulus squared of the numerical solution in Figure 5.9. The modulus squared of the numerical solution quickly increases as time increases and eventually blows up.

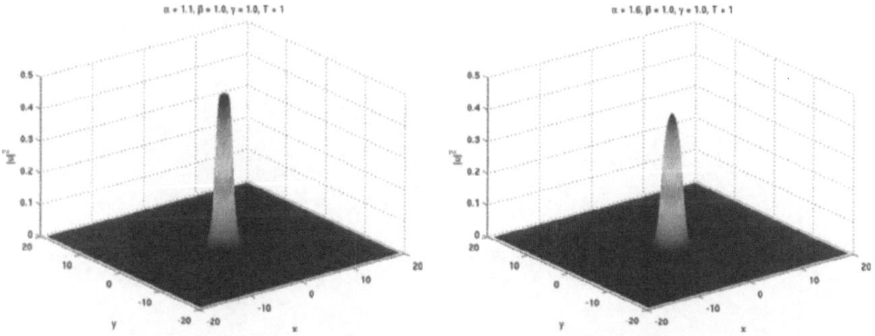

Figure 5.6 Defocusing case: Modulus squared of the solution to the fNLS at time $T = 1$ with initial condition: $\mathrm{sech}(x_1)\mathrm{sech}(x_2)$. Left: $\alpha = 1.1$, Right: $\alpha = 1.6$. Source: Mao and Shen (2017). Copyright ©2017 Society for Industrial and Applied Mathematics. Reprinted with permission. All rights reserved.

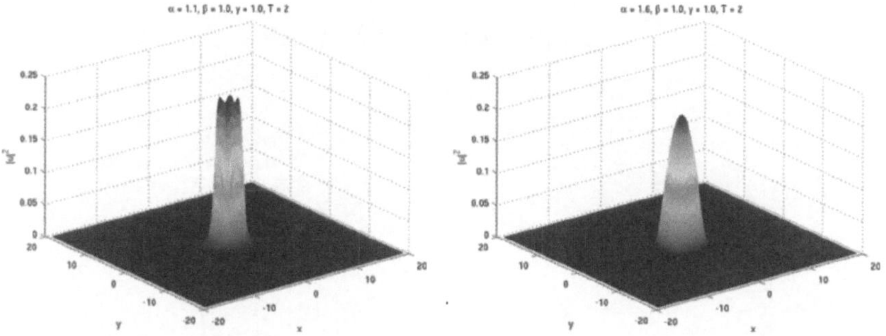

Figure 5.7 Modulus squared of the solution to the fNLS at time $T = 2$ with initial condition: $\mathrm{sech}(x_1)\mathrm{sech}(x_2)$, Left: $\alpha = 1.1$, Right: $\alpha = 1.6$. Source: Mao and Shen (2017). Copyright ©2017 Society for Industrial and Applied Mathematics. Reprinted with permission. All rights reserved.

5.2 Non-tensorial Hermite Spectral Methods

In this section, we introduce two families of non-tensorial generalized Hermite polynomials/functions (GHPs/GHFs) in arbitrary dimensions that can lead to more efficient and accurate spectral methods for solving PDEs with integral fractional Laplacian (IFL) in $\mathbb{R}^d$.

5.2.1 Notation and Preliminaries

Let $\mathbb{R} = (-\infty, \infty)$, $\mathbb{N} = \{1, 2, \cdots\}$, and $\mathbb{N}_0 := \{0\} \cup \mathbb{N}$. For $d \in \mathbb{N}$, we denote by $\mathbb{R}^d$ the d-dimensional Euclidean space with the inner product and norm

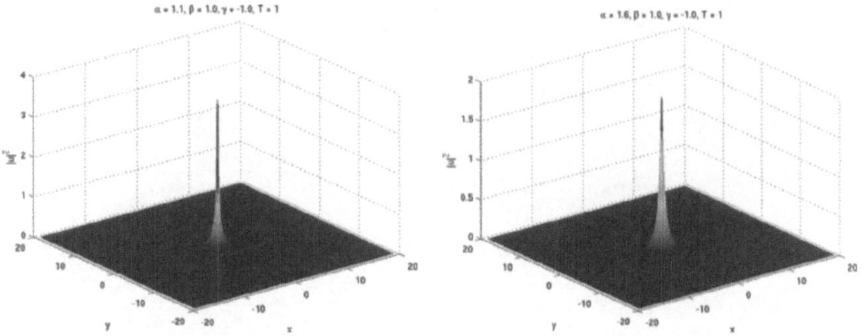

Figure 5.8 Focusing case: Modulus squared of the solution to the fNLS at time $T = 1$ with initial condition: $\operatorname{sech}(x_1)\operatorname{sech}(x_2)$, Left: $\alpha = 1.1$, Right: $\alpha = 1.6$. Source: Mao and Shen (2017). Copyright ©2017 Society for Industrial and Applied Mathematics. Reprinted with permission. All rights reserved.

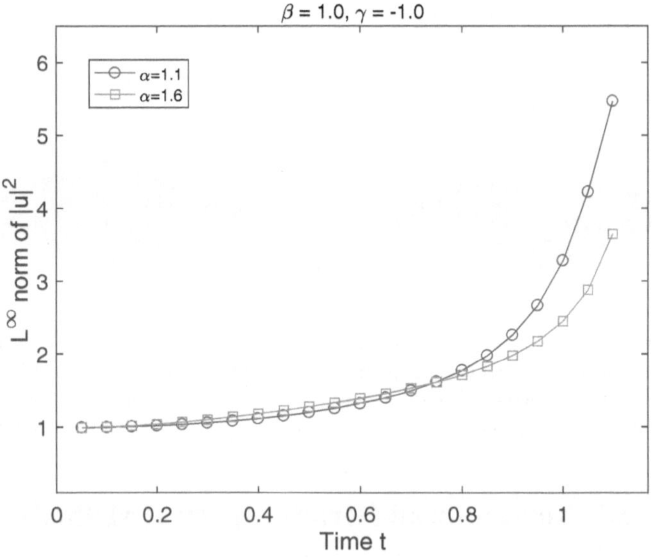

Figure 5.9 Evolution of the L^∞ norm of modulus squared of the solution to the fNLS for the focusing case, initial condition: $\operatorname{sech}(x_1)\operatorname{sech}(x_2)$. Source: Mao and Shen (2017). Copyright ©2017 Society for Industrial and Applied Mathematics. Reprinted with permission. All rights reserved.

$\langle \boldsymbol{x}, \boldsymbol{y}\rangle := \sum_{i=1}^d x_i y_i$, and $r = |\boldsymbol{x}| := \sqrt{\langle \boldsymbol{x}, \boldsymbol{x}\rangle}$, respectively, for any $\boldsymbol{x}, \boldsymbol{y} \in \mathbb{R}^d$. Denote the unit vector along any nonzero vector $\boldsymbol{x}$ by $\hat{\boldsymbol{x}} = \boldsymbol{x}/|\boldsymbol{x}|$. Recall the d-dimensional spherical coordinates

$$\begin{aligned} x_1 &= r\cos\theta_1;\ x_2 = r\sin\theta_1\cos\theta_2;\ \cdots\cdots; \\ x_{d-1} &= r\sin\theta_1\cdots\sin\theta_{d-2}\cos\theta_{d-1}; \\ x_d &= r\sin\theta_1\cdots\sin\theta_{d-2}\sin\theta_{d-1} \end{aligned} \tag{5.2.1}$$

for $\theta_1,\ldots,\theta_{d-2}\in[0,\pi]$ and $\theta_{d-1}\in[0,2\pi]$, with the spherical volume element

$$\begin{aligned} \mathrm{d}\boldsymbol{x} &= r^{d-1}\sin^{d-2}(\theta_1)\sin^{d-3}(\theta_2)\cdots\sin(\theta_{d-2})\,\mathrm{d}r\,\mathrm{d}\theta_1\,\mathrm{d}\theta_2\cdots\mathrm{d}\theta_{d-1} \\ &:= r^{d-1}\mathrm{d}r\,\mathrm{d}\sigma(\hat{\boldsymbol{x}}). \end{aligned} \tag{5.2.2}$$

In spherical coordinates, the d-dimensional Laplacian takes the form

$$\Delta = \partial_r^2 + \frac{d-1}{r}\partial_r + \frac{1}{r^2}\Delta_{\mathbb{S}^{d-1}}, \tag{5.2.3}$$

where $\Delta_{\mathbb{S}^{d-1}}$ is the Laplace–Beltrami operator on the unit sphere $\mathbb{S}^{d-1} := \{\boldsymbol{x}\in\mathbb{R}^d : |\boldsymbol{x}| = 1\}$. Define the inner product of $L^2(\mathbb{S}^{d-1})$ as

$$\langle f, g\rangle_{\mathbb{S}^{d-1}} := \int_{\mathbb{S}^{d-1}} f(\hat{\boldsymbol{x}})g(\hat{\boldsymbol{x}})\,\mathrm{d}\sigma(\hat{\boldsymbol{x}}).$$

We next introduce the d-dimensional spherical harmonics as in Dai and Xu (2013). Let $\mathcal{P}_n^d$ be the space of all real d-dimensional homogeneous polynomials of degree n as follows:

$$\mathcal{P}_n^d = \mathrm{span}\big\{\boldsymbol{x}^{\boldsymbol{k}} = x_1^{k_1}x_2^{k_2}\ldots x_d^{k_d} : k_1+k_2+\cdots+k_d = n\big\}. \tag{5.2.4}$$

As an important subspace of $\mathcal{P}_n^d$, the space of all real harmonic polynomials of degree n is defined as

$$\mathcal{H}_n^d := \big\{P\in\mathcal{P}_n^d : \Delta P(\boldsymbol{x}) = 0\big\}. \tag{5.2.5}$$

It is known that the dimensionality

$$\begin{aligned} \dim(\mathcal{P}_n^d) &= \binom{n+d-1}{n}, \\ \dim(\mathcal{H}_n^d) &= \binom{n+d-1}{n} - \binom{n+d-3}{n-2} := a_n^d, \end{aligned} \tag{5.2.6}$$

where it is understood that for $n = 0, 1$, the value of the second binomial coefficient is zero (cf. Dai and Xu (2013, (1.1.5))). In fact, for $d = 1$, all harmonic polynomials are spanned by $\{1, x\}$.

The d-dimensional spherical harmonics are the restrictions of harmonic polynomials in $\mathcal{H}_n^d$ to $\mathbb{S}^{d-1}$, denoted by $\mathcal{H}_n^d|_{\mathbb{S}^{d-1}}$. It is important to remark the correspondence between a harmonic polynomial and the related spherical harmonic function (cf. Dai and Xu (2013, Ch. 1)): for any $Y(\boldsymbol{x})\in\mathcal{H}_n^d$,

$$Y(\boldsymbol{x}) = |\boldsymbol{x}|^n Y(\boldsymbol{x}/|\boldsymbol{x}|) = r^n Y(\hat{\boldsymbol{x}}), \tag{5.2.7}$$

with $Y(\hat{x}) \in \mathcal{H}_n^d|_{\mathbb{S}^{d-1}}$. It is noteworthy that $Y(x)$ is a homogeneous polynomial in $\mathbb{R}^d$, while $Y(\hat{x})$ is a non-polynomial function on the unit sphere. For $n \in \mathbb{N}_0$, let $\{Y_\ell^n : 1 \le \ell \le a_n^d\}$ be the real (orthogonal) spherical harmonic basis of $\mathcal{H}_n^d|_{\mathbb{S}^{d-1}}$, and note that the spherical harmonics of different degree are mutually orthogonal (cf. Dai and Xu (2013, Thm. 1.1.2)), i.e., $\mathcal{H}_n^d|_{\mathbb{S}^{d-1}} \perp \mathcal{H}_m^d|_{\mathbb{S}^{d-1}}$ for $m \neq n$. Thus, we have

$$\langle Y_\ell^n, Y_\iota^m \rangle_{\mathbb{S}^{d-1}} = \int_{\mathbb{S}^{d-1}} Y_\ell^n(\hat{x}) Y_\iota^m(\hat{m})\, \mathrm{d}\sigma(\hat{x}) = \delta_{nm}\delta_{\ell\iota} \tag{5.2.8}$$

for all $(\ell, n), (\iota, m) \in \Upsilon_\infty^d$, where we introduce the two related index sets

$$\begin{aligned} \Upsilon_\infty^d &= \{(\ell, n) : 1 \le \ell \le a_n^d,\ \ 0 \le n < \infty,\ \ \ell, n \in \mathbb{N}_0\}, \\ \Upsilon_N^d &= \{(\ell, n) : 1 \le \ell \le a_n^d,\ \ 0 \le n \le N,\ \ \ell, n \in \mathbb{N}_0\}. \end{aligned} \tag{5.2.9}$$

The spherical harmonic basis functions are eigenfunctions of the Laplace–Beltrami problem

$$\Delta_{\mathbb{S}^{d-1}} Y_\ell^n(\hat{x}) = -n(n+d-2) Y_\ell^n(\hat{x}). \tag{5.2.10}$$

The second building block of the GHPs/GHFs is the generalized Laguerre polynomials, denoted by $L_k^{(\alpha)}(z)$ for $z \in (0, \infty)$ and $\alpha > -1$. They are orthogonal with respect to the weight function $z^\alpha \mathrm{e}^{-z}$, that is,

$$\int_0^\infty L_k^{(\alpha)}(z) L_j^{(\alpha)}(z)\, z^\alpha\, \mathrm{e}^{-z}\, \mathrm{d}z = \frac{\Gamma(k+\alpha+1)}{k!}\delta_{kj}, \quad k, j \in \mathbb{N}_0, \tag{5.2.11}$$

where we refer to Szegő (1975) and Shen et al. (2011b, Ch. 7) for more properties.

5.2.2 Generalized Hermite Polynomials/Functions in $\mathbb{R}^d$

Definition 5.17 For $\mu > -\frac{1}{2}$, $k \in \mathbb{N}_0$ and $(\ell, n) \in \Upsilon_\infty^d$, we define the d-dimensional generalized Hermite polynomials as

$$\begin{aligned} H_{k,\ell}^{\mu,n}(x) :&= H_{k,\ell}^{\mu,n}(x; d) = L_k^{(n+\frac{d-2}{2}+\mu)}(|x|^2) Y_\ell^n(x) \\ &= r^n L_k^{(n+\frac{d-2}{2}+\mu)}(r^2) Y_\ell^n(\hat{x}), \quad x = r\,\hat{x}, \end{aligned} \tag{5.2.12}$$

and the d-dimensional generalized Hermite functions as

$$\widehat{H}_{k,\ell}^{\mu,n}(x) = \sqrt{1/\gamma_{k,n}^{\mu,d}}\, \mathrm{e}^{-\frac{|x|^2}{2}} H_{k,\ell}^{\mu,n}(x), \tag{5.2.13}$$

where

$$\gamma_{k,n}^{\mu,d} := \frac{\Gamma(k+n+\frac{d}{2}+\mu)}{2\,k!}.$$

Remark 5.18 As we shall see later, the one-dimensional GHPs (up to a constant multiple) coincide with the one-dimensional generalization first introduced in Szegő (1975), after which we name the above new families. □

Before we consider the applications of GHPs and GHFs, we first present some attractive properties. By construction, they enjoy the following important orthogonality.

Theorem 5.19 *For $\mu > -\frac{1}{2}$, $k, j \in \mathbb{N}_0$ and $(\ell, n), (\iota, m) \in \Upsilon_\infty^d$, the GHPs are mutually orthogonal with respect to the weight function $|\boldsymbol{x}|^{2\mu} \mathrm{e}^{-|\boldsymbol{x}|^2}$, namely,*

$$\int_{\mathbb{R}^d} H_{k,\ell}^{\mu,n}(\boldsymbol{x}) H_{j,\iota}^{\mu,m}(\boldsymbol{x})\, |\boldsymbol{x}|^{2\mu}\, \mathrm{e}^{-|\boldsymbol{x}|^2}\, \mathrm{d}\boldsymbol{x} = \gamma_{k,n}^{\mu,d}\, \delta_{mn}\delta_{kj}\delta_{\ell\iota}, \tag{5.2.14}$$

and the GHFs are orthonormal, viz.,

$$\int_{\mathbb{R}^d} \widehat{H}_{k,\ell}^{\mu,n}(\boldsymbol{x}) \widehat{H}_{j,\iota}^{\mu,m}(\boldsymbol{x})\, |\boldsymbol{x}|^{2\mu}\, \mathrm{d}\boldsymbol{x} = \delta_{mn}\delta_{kj}\delta_{\ell\iota}. \tag{5.2.15}$$

Proof The orthogonality (5.2.15) is a direct consequence of (5.2.13) and (5.2.14), so we only need to show (5.2.14).

In view of the definition (5.2.12), we use the spherical coordinates transformation (5.2.1)–(5.2.2), and find from the orthogonality (5.2.8) and (5.2.11) that

$$\begin{aligned}
&\int_{\mathbb{R}^d} H_{k,\ell}^{\mu,n}(\boldsymbol{x}) H_{j,\iota}^{\mu,m}(\boldsymbol{x})\, |\boldsymbol{x}|^{2\mu}\, \mathrm{e}^{-|\boldsymbol{x}|^2}\, \mathrm{d}\boldsymbol{x} \\
&= \int_0^\infty L_k^{(n+\frac{d-2}{2}+\mu)}(r^2)\, L_j^{(m+\frac{d-2}{2}+\mu)}(r^2)\, r^{2\mu+2n+d-1} \mathrm{e}^{-r^2} \mathrm{d}r \\
&\quad \int_{\mathbb{S}^{d-1}} Y_\ell^n(\hat{\boldsymbol{x}}) Y_\iota^m(\hat{\boldsymbol{x}})\, \mathrm{d}\sigma(\hat{\boldsymbol{x}}) \\
&= \delta_{mn}\delta_{\ell\iota} \int_0^\infty L_k^{(n+\frac{d-2}{2}+\mu)}(r^2)\, L_j^{(m+\frac{d-2}{2}+\mu)}(r^2)\, r^{2\mu+2n+d-1} \mathrm{e}^{-r^2}\, \mathrm{d}r \\
&= \frac{1}{2}\delta_{mn}\delta_{\ell\iota} \int_0^\infty L_k^{(n+\frac{d-2}{2}+\mu)}(\rho)\, L_j^{(n+\frac{d-2}{2}+\mu)}(\rho)\, \rho^{n+\frac{d-2}{2}+\mu} \mathrm{e}^{-\rho}\, \mathrm{d}\rho \\
&= \frac{\Gamma(k+n+\frac{d}{2}+\mu)}{2k!} \delta_{mn}\delta_{kj}\delta_{\ell\iota} = \gamma_{k,n}^{\mu,d}\, \delta_{mn}\delta_{kj}\delta_{\ell\iota},
\end{aligned}$$

which yields (5.2.14) and ends the proof. □

The d-dimensional GHPs/GHFs satisfy the recurrence relations.

Proposition 5.20 *For $\mu > -\frac{1}{2}$ and fixed $(\ell, n) \in \Upsilon_\infty^d$, we have the following recurrence relations in k :*

$$(k+1)H^{\mu,n}_{k+1,\ell}(\boldsymbol{x}) = \Big(2k+n+\frac{d}{2}+\mu-|\boldsymbol{x}|^2\Big)H^{\mu,n}_{k,\ell}(\boldsymbol{x}) \\ -\Big(k+n+\frac{d}{2}-1+\mu\Big)H^{\mu,n}_{k-1,\ell}(\boldsymbol{x}), \tag{5.2.16}$$

and for the GHFs,

$$a_k\,\widehat{H}^{\mu,n}_{k+1,\ell}(\boldsymbol{x}) = (b_k-|\boldsymbol{x}|^2)\widehat{H}^{\mu,n}_{k,\ell}(\boldsymbol{x}) - c_k\widehat{H}^{\mu,n}_{k-1,\ell}(\boldsymbol{x}), \tag{5.2.17}$$

where

$$a_k = \sqrt{(k+1)(k+n+d/2+\mu)},\quad b_k = 2k+n+d/2+\mu,$$
$$c_k = \sqrt{k(k-1+n+d/2+\mu)}.$$

Proof Recall the three-term recurrence relation of the Laguerre polynomials (cf. Szegő (1975)):

$$(k+1)L^{(\alpha)}_{k+1}(z) = (2k+\alpha+1-z)L^{(\alpha)}_k(z) - (k+\alpha)L^{(\alpha)}_{k-1}(z). \tag{5.2.18}$$

Then the relation (5.2.16) is a direct consequence of (5.2.12) and (5.2.18). From (5.2.13), we have

$$H^{\mu,n}_{k,\ell}(\boldsymbol{x}) = \sqrt{\gamma^{\mu,d}_{k,n}}\,\mathrm{e}^{\frac{|\boldsymbol{x}|^2}{2}}\,\widehat{H}^{\mu,n}_{k,\ell}(\boldsymbol{x}). \tag{5.2.19}$$

Substituting it into (5.2.16) and working out the constants, we obtain (5.2.17). □

The GHFs with different parameters are connected through the following identity, which is found very useful in the algorithm development.

Proposition 5.21 *For $\mu,\nu > -\frac{1}{2}$ and $(\ell,n)\in\Upsilon^d_\infty$, there holds*

$$\widehat{H}^{\mu,n}_{k,\ell}(\boldsymbol{x}) = \sum_{j=0}^{k}{}^{\mu}_{\nu}\mathbb{C}^k_j\,\widehat{H}^{\nu,n}_{j,\ell}(\boldsymbol{x}),\quad \boldsymbol{x}\in\mathbb{R}^d,\ \ k\in\mathbb{N}_0, \tag{5.2.20}$$

where the connection coefficients are given by

$${}^{\mu}_{\nu}\mathbb{C}^k_j = \frac{\Gamma(k-j+\mu-\nu)}{\Gamma(\mu-\nu)\,(k-j)!}\sqrt{\frac{k!\,\Gamma(j+n+\frac{d}{2}+\nu)}{j!\,\Gamma(k+n+\frac{d}{2}+\mu)}}. \tag{5.2.21}$$

Proof Recall the property of the generalized Laguerre polynomials (cf. Askey (1975, (7.4))):

$$L^{(\mu+\beta+1)}_k(z) = \sum_{j=0}^{k}\frac{\Gamma(k-j+\beta+1)}{\Gamma(\beta+1)(k-j)!}L^{(\mu)}_j(z), \tag{5.2.22}$$

so we can derive the identity from Definition 5.17 and direct calculation. □

Remark 5.22 As $\Gamma(0)=\infty$, we can find that in the limiting sense: ${}^{\mu}_{\mu}\mathbb{C}^k_j=\delta_{kj}$. □

Remarkably, for $\mu = 0$, the GHFs are the eigenfunctions of the harmonic oscillator: $-\Delta+|\boldsymbol{x}|^2$, are essential for the error analysis to be conducted in the forthcoming section.

Lemma 5.23 *For $k \in \mathbb{N}_0$, $(\ell, n) \in \Upsilon^d_\infty$, the GHFs with $\mu = 0$ satisfy*

$$\big(-\Delta+|\boldsymbol{x}|^2\big)\widehat{H}^{0,n}_{k,\ell}(\boldsymbol{x}) = (4k+2n+d)\,\widehat{H}^{0,n}_{k,\ell}(\boldsymbol{x}). \tag{5.2.23}$$

Proof According to Ma et al. (2019, Lemma 2.1) with $\alpha = n + d/2 - 1$ and $\beta = \alpha + 1 - d/2$, we have

$$\begin{aligned}&\Big[\partial_r^2+\frac{d-1}{r}\partial_r-\frac{n(n+d-2)}{r^2}-r^2+4k+2n+d\Big]\\ &\qquad\times\big[r^n L_k^{(n+d/2-1)}(r^2)\,\mathrm{e}^{-\frac{r^2}{2}}\big]=0.\end{aligned} \tag{5.2.24}$$

In view of $Y(\boldsymbol{x}) = r^n Y(\hat{\boldsymbol{x}})$, we obtain from (5.2.3), (5.2.10), (5.2.12), (5.2.13) and (5.2.24) that

$$\begin{aligned}&-\Delta\widehat{H}^{0,n}_{k,\ell}(\boldsymbol{x})\\ &=-\sqrt{1/\gamma^{0,d}_{k,n}}\Big[\partial_r^2+\frac{d-1}{r}\partial_r-\frac{n(n+d-2)}{r^2}\Big]\big[r^n\,L_k^{(n+d/2-1)}(r^2)\,\mathrm{e}^{-\frac{r^2}{2}}\big]\,Y(\hat{\boldsymbol{x}})\\ &=\sqrt{1/\gamma^{0,d}_{k,n}}\big[-r^2+4k+2n+d\big]\big[r^n L_k^{(n+d/2-1)}(r^2)\,\mathrm{e}^{-\frac{r^2}{2}}\big]\,Y(\hat{\boldsymbol{x}})\\ &=(-r^2+4k+2n+d)\,\widehat{H}^{0,n}_{k,\ell}(\boldsymbol{x}),\end{aligned}$$

which leads to (5.2.23). □

5.2.3 Adjoint Generalized Hermite Functions in $\mathbb{R}^d$

Definition 5.24 For $\mu > -\frac{1}{2}$ and $(\ell, n) \in \Upsilon^d_\infty$, the d-dimensional adjoint GHFs are defined by

$$\check{H}^{\mu,n}_{k,\ell}(\boldsymbol{x}) = \sum_{j=0}^{k}(-1)^{k-j}\,{}^{\mu}_{0}\mathbb{C}^k_j\,\widehat{H}^{0,n}_{j,\ell}(\boldsymbol{x}), \quad \boldsymbol{x}\in\mathbb{R}^d,\ \ k\in\mathbb{N}_0, \tag{5.2.25}$$

where the coefficients $\{{}^{\mu}_{0}\mathbb{C}^k_j\}$ are given by (5.2.21).

Remark 5.25 In light of the connection relation in Proposition 5.21, it is evident that $\check{H}^{\mu,n}_{k,\ell}(\boldsymbol{x})$ can be expressed as a linear combination of the counterparts $\big\{\widehat{H}^{\mu,n}_{j,\ell}(\boldsymbol{x})\big\}_{j=0}^{k}$. □

It is seen from (5.2.20) (with $\nu = 0$) that the GHF can be represented as

$$\widehat{H}^{\mu,n}_{k,\ell}(\boldsymbol{x}) = \sum_{j=0}^{k} {}^{\mu}_{0}\mathbb{C}^{k}_{j}\, \widehat{H}^{0,n}_{j,\ell}(\boldsymbol{x}), \tag{5.2.26}$$

which differs from its adjoint in (5.2.25) by the sign of the coefficients. Notably, such a subtlety results in an intimate relation between this adjoint pair through the Fourier transform:

$$\begin{aligned} \hat{u}(\xi) &:= \mathscr{F}[u](\xi) = \frac{1}{(2\pi)^{\frac{d}{2}}} \int_{\mathbb{R}^d} u(\boldsymbol{x})\, \mathrm{e}^{-\mathrm{i}\langle \xi, \boldsymbol{x}\rangle}\, \mathrm{d}\boldsymbol{x}, \\ \mathscr{F}^{-1}[\widehat{u}](\boldsymbol{x}) &= \frac{1}{(2\pi)^{\frac{d}{2}}} \int_{\mathbb{R}^d} \widehat{u}(\xi)\, \mathrm{e}^{\mathrm{i}\langle \xi, \boldsymbol{x}\rangle}\, \mathrm{d}\xi. \end{aligned} \tag{5.2.27}$$

Moreover, the use of A-GHFs as basis functions in a spectral-Galerkin framework can diagonalize the nonlocal integral fractional Laplacian $(-\Delta)^s$ for $s > 0$. Recall that for $s > 0$, the fractional Laplacian of $u \in \mathscr{S}(\mathbb{R}^d)$ (functions of Schwartz class) can be naturally defined via the Fourier transform:

$$(-\Delta)^s u(\boldsymbol{x}) = \mathscr{F}^{-1}\big[|\xi|^{2s} \mathscr{F}[u](\xi)\big](\boldsymbol{x}), \quad \boldsymbol{x} \in \mathbb{R}^d. \tag{5.2.28}$$

Recall that for $0 < s < 1$, the fractional Laplacian can be equivalently defined by the point-wise formula:

$$(-\Delta)^s u(\boldsymbol{x}) = C_{d,s}\, \text{p.v.} \int_{\mathbb{R}^d} \frac{u(\boldsymbol{x}) - u(\boldsymbol{y})}{|\boldsymbol{x} - \boldsymbol{y}|^{d+2s}}\, \mathrm{d}\boldsymbol{y}, \tag{5.2.29}$$

where

$$C_{d,s} := \frac{2^{2s} s \Gamma(s + d/2)}{\pi^{d/2} \Gamma(1 - s)},$$

and "p.v." stands for the principal value.

Theorem 5.26 *For $\mu > -\frac{1}{2}$, $(\ell, n) \in \Upsilon^d_\infty$ and $k \in \mathbb{N}_0$, we have*

$$\mathscr{F}[\check{H}^{\mu,n}_{k,\ell}](\xi) = (-\mathrm{i})^{n+2k} \widehat{H}^{\mu,n}_{k,\ell}(\xi), \quad \mathscr{F}^{-1}[\widehat{H}^{\mu,n}_{k,\ell}](\boldsymbol{x}) = \mathrm{i}^{n+2k} \check{H}^{\mu,n}_{k,\ell}(\boldsymbol{x}), \tag{5.2.30}$$

and for $s > 0$,

$$\mathscr{F}[(-\Delta)^s \check{H}^{\mu,n}_{k,\ell}](\xi) = (-\mathrm{i})^{n+2k} |\xi|^{2s} \widehat{H}^{\mu,n}_{k,\ell}(\xi). \tag{5.2.31}$$

Moreover, the adjoint GHFs are orthonormal in the sense that for $s > 0$,

$$\big((-\Delta)^{\frac{s}{2}} \check{H}^{s,n}_{k,\ell}, (-\Delta)^{\frac{s}{2}} \check{H}^{s,m}_{j,\iota}\big)_{\mathbb{R}^d} = \delta_{jk}\delta_{mn}\delta_{\ell\iota}. \tag{5.2.32}$$

Proof We first show that the GHFs with $\mu = 0$ are eigenfunctions of the Fourier transform, namely,

$$\mathscr{F}[\widehat{H}_{k,\ell}^{0,n}](\xi) = (-\mathrm{i})^{n+2k}\,\widehat{H}_{k,\ell}^{0,n}(\xi). \tag{5.2.33}$$

According to Andrews et al. (1999, Lemma 9.10.2), we have that for $\omega > 0$,

$$\int_{\mathbb{S}^{d-1}} Y_\ell^n(\hat{\boldsymbol{x}})\,\mathrm{e}^{-\mathrm{i}\omega\langle\hat{\xi},\hat{\boldsymbol{x}}\rangle}\mathrm{d}\sigma(\hat{\boldsymbol{x}}) = \frac{(-\mathrm{i})^n(2\pi)^{\frac{d}{2}}}{\omega^{\frac{d-2}{2}}} J_{n+\frac{d-2}{2}}(\omega)\,Y_\ell^n(\hat{\xi}) \tag{5.2.34}$$

for $\hat{\xi} \in \mathbb{S}^{d-1}$, where $J_\nu(z)$ are the Bessel functions of the first kind of order ν. Then using Definition 5.17 with $\mu = 0$, and (5.2.34) with $\omega = \rho r$ and $\rho = |\xi|$, leads to

$$\begin{aligned}
\mathscr{F}[\widehat{H}_{k,\ell}^{0,n}](\xi) &= \frac{1}{(2\pi)^{\frac{d}{2}}}\int_{\mathbb{R}^d} \widehat{H}_{k,\ell}^{0,n}(\boldsymbol{x})\mathrm{e}^{-\mathrm{i}\langle\xi,\boldsymbol{x}\rangle}\mathrm{d}\boldsymbol{x} \\
&= \frac{1}{\sqrt{\gamma_{k,n}^{0,d}}}\frac{1}{(2\pi)^{\frac{d}{2}}}\int_0^\infty r^n L_k^{(n+\frac{d-2}{2})}(r^2)\mathrm{e}^{-\frac{r^2}{2}}\Big\{\int_{\mathbb{S}^{d-1}} Y_\ell^n(\hat{\boldsymbol{x}})\mathrm{e}^{-\mathrm{i}\rho r\langle\hat{\xi},\hat{\boldsymbol{x}}\rangle}\mathrm{d}\sigma(\hat{\boldsymbol{x}})\Big\} r^{d-1}\mathrm{d}r \\
&= \frac{1}{\sqrt{\gamma_{k,n}^{0,d}}}\frac{(-\mathrm{i})^n}{\rho^{\frac{d-2}{2}}}\Big\{\int_0^\infty r^{n+\frac{d}{2}} L_k^{(n+\frac{d-2}{2})}(r^2)\,\mathrm{e}^{-\frac{r^2}{2}}\,J_{n+\frac{d-2}{2}}(\rho r)\,\mathrm{d}r\Big\}Y_\ell^n(\hat{\xi}), \quad \rho > 0.
\end{aligned} \tag{5.2.35}$$

Recall the integral identity of the generalized Laguerre polynomials (cf. Gradshteyn and Ryzhik (2015, P. 820)): for $\alpha > -1$,

$$\int_0^\infty r^{\alpha+1} L_k^{(\alpha)}(r^2)\,\mathrm{e}^{-\frac{r^2}{2}}\,J_\alpha(\rho r)\,\mathrm{d}r = (-1)^k\rho^\alpha L_k^{(\alpha)}(\rho^2)\,\mathrm{e}^{-\frac{\rho^2}{2}}, \quad \rho > 0. \tag{5.2.36}$$

Thus, taking $\alpha = n + \frac{d-2}{2}$ in (5.2.36), we can work out the integral in (5.2.35) and then obtain from (5.2.13) with $\mu = 0$ that

$$\begin{aligned}
\mathscr{F}[\widehat{H}_{k,\ell}^{0,n}](\xi) &= \frac{(-1)^k}{\sqrt{\gamma_{k,n}^{0,d}}}\frac{(-\mathrm{i})^n}{\rho^{\frac{d-2}{2}}}\rho^{n+\frac{d-2}{2}} L_k^{(n+\frac{d-2}{2})}(\rho^2)\mathrm{e}^{-\frac{\rho^2}{2}}\,Y_\ell^n(\hat{\xi}) \\
&= (-\mathrm{i})^{n+2k}\,\widehat{H}_{k,\ell}^{0,n}(\xi).
\end{aligned} \tag{5.2.37}$$

This yields (5.2.33).

From Definition 5.24 and the property (5.2.37), we obtain

$$\begin{aligned}\mathscr{F}[\check{H}_{k,\ell}^{\mu,n}](\xi) &= \sum_{j=0}^{k}(-1)^{k-j}\,{}_0^{\mu}\mathbb{C}_j^k\,\mathscr{F}[\widehat{H}_{j,\ell}^{0,n}](\xi)\\ &= \sum_{j=0}^{k}(-1)^{k-j}\,(-\mathrm{i})^{n+2j}\,{}_0^{\mu}\mathbb{C}_j^k\,\widehat{H}_{j,\ell}^{0,n}(\xi)\\ &= (-\mathrm{i})^{n+2k}\sum_{j=0}^{k}{}_0^{\mu}\mathbb{C}_j^k\,\widehat{H}_{j,\ell}^{0,n}(\xi) = (-\mathrm{i})^{n+2k}\,\widehat{H}_{k,\ell}^{\mu,n}(\xi),\end{aligned} \tag{5.2.38}$$

where in the last step, we used (5.2.26). This gives the first identity in (5.2.30), and the second is its immediate consequence. The property (5.2.31) follows directly from (5.2.30) and the definition of fractional Laplacian (5.2.28).

Finally, using Parseval's identity and (5.2.31), we derive from the orthogonality (5.2.15) that

$$\begin{aligned}\big((-\Delta)^{\frac{s}{2}}\check{H}_{k,\ell}^{s,n},(-\Delta)^{\frac{s}{2}}\check{H}_{j,\iota}^{s,m}\big)_{\mathbb{R}^d} &= \big(\mathscr{F}[(-\Delta)^{\frac{s}{2}}\check{H}_{k,\ell}^{s,n}],\mathscr{F}[(-\Delta)^{\frac{s}{2}}\check{H}_{j,\iota}^{s,m}]\big)_{\mathbb{R}^d}\\ &= (-\mathrm{i})^{n-m+2k-2j}\big(|\xi|^{2s}\,\widehat{H}_{k,\ell}^{s,n},\widehat{H}_{j,\iota}^{s,m}\big)_{\mathbb{R}^d} = \delta_{mn}\delta_{kj}\delta_{\ell\iota}.\end{aligned}$$

This yields (5.2.32) and ends the proof. □

Note that the GHFs are real-valued, so we infer from (5.2.27) readily that

$$\mathscr{F}[\widehat{H}_{k,\ell}^{\mu,n}(\boldsymbol{x})](\xi) = \big\{\mathscr{F}^{-1}[\widehat{H}_{k,\ell}^{\mu,n}(\boldsymbol{x})](\xi)\big\}^{*}.$$

Thus, we find from (5.2.30) immediately the following "inverse" form of (5.2.30).

Corollary 5.27 *For $\mu > -\frac{1}{2}$, $(\ell,n) \in \Upsilon_\infty^d$ and $k \in \mathbb{N}_0$, we have*

$$\mathscr{F}[\widehat{H}_{k,\ell}^{\mu,n}](\xi) = \mathrm{i}^{n+2k}\,\check{H}_{k,\ell}^{\mu,n}(\xi), \quad \mathscr{F}^{-1}[\check{H}_{k,\ell}^{\mu,n}](\boldsymbol{x}) = (-\mathrm{i})^{n+2k}\,\widehat{H}_{k,\ell}^{\mu,n}(\boldsymbol{x}). \tag{5.2.39}$$

Remark 5.28 The fractional Sobolev orthogonality (5.2.32) has profound implications even for the integral-order Laplacian $(-\Delta)^m$ with $m \in \mathbb{N}$. For example, we find from (5.2.25) with $s = 1$ that the A-GHFs read

$$\check{H}_{k,\ell}^{1,n}(\boldsymbol{x}) = \sqrt{\frac{k!}{\Gamma(k+n+\frac{d}{2}+1)}}\sum_{j=0}^{k}\sqrt{\frac{\Gamma(j+n+\frac{d}{2})}{j!}}\,\widehat{H}_{j,\ell}^{0,n}(\boldsymbol{x}),$$

which are orthogonal with respect to $(\nabla\cdot,\nabla\cdot)_{\mathbb{R}^d}$. However, this attractive property is not valid for the usual Hermite-based methods based on tensorial Hermite functions $\prod_{j=1}^{d}\widehat{H}_{n_j}(x_j)$. Thus, it is advantageous to use the A-GHFs for usual Laplacian and bi-harmonic Laplacian (using the A-GHFs with $s = 2$) in $\mathbb{R}^d$. □

5.2.4 One-Dimensional GHPs/GHFs by Szegő

Note from (5.2.6) that for $d = 1$, $a_0^1 = a_1^1 = 1$ and $a_n^1 = 0$ for $n \geq 2$, so there are only two orthonormal harmonic polynomials: $Y_1^0(x) = \frac{1}{\sqrt{2}}$ and $Y_1^1(x) = \frac{x}{\sqrt{2}}$. Accordingly, the GHPs in Definition 5.17 (with a constant multiple) reduce to

$$\begin{aligned} H_{2k}^{(\mu)}(x) &:= (-1)^k\, 2^{2k+\frac{1}{2}}\, k!\ H_{k,1}^{\mu,0}(x) = (-1)^k\, 2^{2k}\, k!\ L_k^{(\mu-\frac{1}{2})}(x^2), \\ H_{2k+1}^{(\mu)}(x) &:= (-1)^k\, 2^{2k+\frac{3}{2}}\, k!\ H_{k,1}^{\mu,1}(x) = (-1)^k\, 2^{2k+1}\, k!\ x\, L_k^{(\mu+\frac{1}{2})}(x^2), \end{aligned} \tag{5.2.40}$$

which are mutually orthogonal with respect to the weight function $|x|^{2\mu}\, \mathrm{e}^{-x^2}$ on $\mathbb{R}$. In fact, this family of GHPs is first introduced by Szegő (1975) as an exercised problem and promoted by Chihara (1955, 1978). According to Szegő (1975, Prob. 25), the GHPs with $\mu > -\frac{1}{2}$ satisfy the differential equation:

$$xy'' + 2(\mu - x^2)y' + (2nx - \theta_n x^{-1})y = 0, \quad \theta_n = \begin{cases} 0, & n \text{ even}, \\ 2\mu, & n \text{ odd}, \end{cases} \tag{5.2.41}$$

where $y = H_n^{(\mu)}(x)$. Some other properties of $\{H_n^{(\mu)}\}$ can be found in Chihara (1955, 1978). We also refer to some limited works on the analytic studies or further generalizations (see, e.g., Rösler (1998); Masjed-Jamei and Koepf (2012)). With the normalization in (5.2.40), the orthonormal GHFs (in (5.2.13) with $d = 1$) take the form

$$\widehat{H}_n^{(\mu)}(x) := \sqrt{1/\gamma_n^{(\mu)}}\ \mathrm{e}^{-\frac{x^2}{2}}\, H_n^{(\mu)}(x), \tag{5.2.42}$$

where

$$\gamma_n^{(\mu)} := 2^{2n} \left[\frac{n}{2}\right]!\ \Gamma\Big(\Big[\frac{n+1}{2}\Big] + \mu + \frac{1}{2}\Big).$$

In particular, for $\mu = 0$, they reduce to the usual Hermite polynomials/functions. For distinction, we denote them by $H_n(x)$ and $\widehat{H}_n(x)$, respectively.

From (5.2.20) and (5.2.40), we have the following transformation between GHFs with different parameters

$$\widehat{H}_n^{(\mu)}(x) = \sum_{j=0;j+n \text{ even}}^{n} {}^{\mu}_{\nu}\widehat{C}^n_j \, \widehat{H}_j^{(\nu)}(x), \tag{5.2.43}$$

where for even $j+n$, the connection coefficients are given by

$${}^{\mu}_{\nu}\widehat{C}^n_j = \frac{(-1)^{\frac{n-j}{2}} \sqrt{\Gamma([\frac{n}{2}]+1)\Gamma([\frac{j+1}{2}]+\nu+\frac{1}{2})}\ \Gamma(\frac{n-j}{2}+\mu-\nu)}{\Gamma(\mu-\nu)\sqrt{\Gamma([\frac{n+1}{2}]+\mu+\frac{1}{2})\Gamma([\frac{j}{2}]+1)}\ \Gamma(\frac{n-j}{2}+1)}. \tag{5.2.44}$$

It is known that the usual Hermite functions $\{\widehat{H}_n\}$ are the eigenfunctions of the Fourier transform. However, this property cannot carry over to the GHFs with $\mu \neq 0$. In Masjed-Jamei and Koepf (2012, (2.34)), the Fourier transform of $\widehat{H}_n^{(\mu)}(x)$ is expressed in terms of the Kummer hypergeometric function ${}_1F_1(\cdot)$. Here we find from Corollary 5.27 with $d = 1$ the following more informative representation

$$\frac{1}{\sqrt{2\pi}} \int_{\mathbb{R}} \widehat{H}_n^{(\mu)}(x) \mathrm{e}^{-\mathrm{i}\xi x} dx = \mathrm{i}^n \check{H}_n^{(\mu)}(\xi), \quad \xi \in \mathbb{R}, \tag{5.2.45}$$

where the one-dimensional adjoint GHFs are given by

$$\check{H}_n^{(\mu)}(x) = \sum_{\substack{j=0 \\ j+n \text{ even}}}^{n} (-1)^{\frac{n-j}{2}} \, {}^{\mu}_{0}\widehat{C}^n_j \, \widehat{H}_j(x), \quad x \in \mathbb{R}. \tag{5.2.46}$$

We remark that the formulation (5.2.46) requires some simple calculation from (5.2.25) and (5.2.40).

5.2.5 GHF Spectral Method for a Fractional Model Problem

As an illustrative example, we consider

$$(-\Delta)^s u(\boldsymbol{x}) + \gamma u(\boldsymbol{x}) = f(\boldsymbol{x}) \quad \text{in } \mathbb{R}^d; \quad u(\boldsymbol{x}) \to 0 \quad \text{as } |\boldsymbol{x}| \to \infty, \tag{5.2.47}$$

where $s \in (0,1), \gamma > 0$, $f \in H^{-s}(\mathbb{R}^d)$, and the fractional Laplacian operator is defined in (5.2.28)–(5.2.29). Here, the fractional Sobolev space $H^s(\mathbb{R}^d)$ with real s is defined as in Nezza et al. (2012).

A weak formulation of (5.2.47) is to find $u \in H^s(\mathbb{R}^d)$ such that

$$\mathcal{A}_s(u,v) = \big((-\Delta)^{\frac{s}{2}} u, (-\Delta)^{\frac{s}{2}} v\big)_{\mathbb{R}^d} + \gamma(u,v)_{\mathbb{R}^d} = (f,v)_{\mathbb{R}^d} \tag{5.2.48}$$

for all $v \in H^s(\mathbb{R}^d)$. From (5.2.28), we find readily the continuity and coercivity of the bilinear form $\mathcal{A}_s(\cdot,\cdot)$. Then we conclude from the standard Lax–Milgram lemma that the problem (5.2.48) admits a unique solution satisfying

$$\|u\|_{H^s(\mathbb{R}^d)} \le c\|f\|_{H^{-s}(\mathbb{R}^d)}.$$

We choose the finite-dimensional approximation space spanned by the d-dimensional GHFs in Definition 5.17 or equivalently by the A-GHFs in Definition 5.24. However, in view of (5.2.32), it is advantageous to use the latter as the basis functions, so we define

$$\begin{aligned} \mathcal{V}_N^d := \operatorname{span}\big\{ \check{H}_{k,\ell}^{s,n}(\boldsymbol{x}) : 0 \le n \le N,\ 1 \le \ell \le a_n^d, \\ 0 \le 2k \le N-n,\ k,\ell,n \in \mathbb{N}_0 \big\}. \end{aligned} \tag{5.2.49}$$

Then, the spectral-Galerkin approximation to (5.2.48) is to find $u_N \in \mathcal{V}_N^d$ such that

$$\mathcal{A}_s(u_N, v_N) = (f, v_N)_{\mathbb{R}^d}, \quad \forall v_N \in \mathcal{V}_N^d. \tag{5.2.50}$$

As with the continuous problem (5.2.48), it has a unique solution $u_N \in \mathcal{V}_N^d$.

In real implementation, we write

$$u_N(\boldsymbol{x}) = \sum_{n=0}^{N} \sum_{\ell=1}^{a_n^d} \sum_{k=0}^{[\frac{N-n}{2}]} \tilde{u}_{k,\ell}^n \check{H}_{k,\ell}^{s,n}(\boldsymbol{x}), \tag{5.2.51}$$

and arrange the unknown coefficients in the order

$$\begin{aligned} \boldsymbol{u} &= \big(\tilde{\boldsymbol{u}}_1^0, \tilde{\boldsymbol{u}}_2^0, \ldots, \tilde{\boldsymbol{u}}_{a_0^d}^0, \tilde{\boldsymbol{u}}_1^1, \tilde{\boldsymbol{u}}_2^1, \ldots, \tilde{\boldsymbol{u}}_{a_1^d}^1, \ldots, \tilde{\boldsymbol{u}}_1^N, \tilde{\boldsymbol{u}}_2^N, \ldots, \tilde{\boldsymbol{u}}_{a_N^d}^N\big)^T, \\ \tilde{\boldsymbol{u}}_\ell^n &= \big(\tilde{u}_{0,\ell}^n, \tilde{u}_{1,\ell}^n, \ldots, \tilde{u}_{[\frac{N-n}{2}],\ell}^n\big)^T, \end{aligned} \tag{5.2.52}$$

and likewise for $\boldsymbol{f}$, but with the components $\tilde{f}_{k,\ell}^n = (f, \check{H}_{k,\ell}^{s,n})_{\mathbb{R}^d}$. The orthogonality (5.2.32) implies that the stiffness matrix is an identity matrix. Moreover, in view of the orthogonality of the spherical harmonic basis (cf. (5.2.8)), the corresponding mass matrix is block diagonal as follows

$$\begin{aligned} \boldsymbol{M} = \operatorname{diag}\big\{ &\boldsymbol{M}_1^0, \boldsymbol{M}_2^0, \ldots, \boldsymbol{M}_{a_0^d}^0, \\ &\boldsymbol{M}_1^1, \boldsymbol{M}_2^1, \ldots, \boldsymbol{M}_{a_1^d}^1, \ldots, \boldsymbol{M}_1^N, \boldsymbol{M}_2^N, \ldots, \boldsymbol{M}_{a_N^d}^N \big\}, \end{aligned} \tag{5.2.53}$$

where the entries of each diagonal block can be computed by

$$\begin{aligned} (\boldsymbol{M}_\ell^n)_{kj} &= \big(\check{H}_{k,\ell}^{s,n}, \check{H}_{j,\ell}^{s,n}\big)_{\mathbb{R}^d} \\ &= \sum_{p=0}^{k} (-1)^{k-p}\, {}_0^s\mathbb{C}_p^k \sum_{q=0}^{j} (-1)^{j-q}\, {}_0^s\mathbb{C}_q^j \big(\widehat{H}_{p,\ell}^{0,n}, \widehat{H}_{q,\ell}^{0,n}\big)_{\mathbb{R}^d} \\ &= (-1)^{k+j} \sum_{p=0}^{\min\{j,k\}} {}_0^s\mathbb{C}_p^k\, {}_0^s\mathbb{C}_p^j. \end{aligned} \tag{5.2.54}$$

Thus the linear system of (5.2.50) can be written as

$$(\boldsymbol{I} + \gamma \boldsymbol{M})\boldsymbol{u} = \boldsymbol{f}. \tag{5.2.55}$$

Remark 5.29 With the new basis at our disposal, the above method has advantages over the usual Hermite approaches described in the previous section. Although the usual one-dimensional Hermite functions are eigenfunctions of the Fourier transform, we observe from (5.2.28) that the factor $|\xi|^{2s}$ is nonseparable and singular, so the use of tensorial Hermite functions leads to a dense stiffness matrix whose entries are difficult to evaluate due to the involved singularity for $d \geq 2$. □

Error Analysis

Applying the first Strang lemma for the standard Galerkin framework (i.e., (5.2.48) and (5.2.50)), we obtain immediately that

$$\|u - u_N\|_{H^s(\mathbb{R}^d)} \leq c \inf_{v_N \in \mathcal{V}_N^d} \|u - v_N\|_{H^s(\mathbb{R}^d)}. \tag{5.2.56}$$

To obtain optimal error estimates, we have to resort to some intermediate approximation results related to certain orthogonal projection. To this end, we consider the L^2-orthogonal projection $\pi_N^d : L^2(\mathbb{R}^d) \to \mathcal{V}_N^d$ such that

$$(\pi_N^d u - u, v)_{\mathbb{R}^d} = 0, \quad \forall v \in \mathcal{V}_N^d. \tag{5.2.57}$$

From Definition 5.24 and with a change of basis functions, we find readily that

$$\mathcal{V}_N^d := \operatorname{span}\big\{\widehat{H}_{k,\ell}^{0,n}(\boldsymbol{x}) : 0 \leq n \leq N,\ 1 \leq \ell \leq a_n^d,\ 0 \leq 2k \leq N-n,\ k, \ell, n \in \mathbb{N}_0\big\}. \tag{5.2.58}$$

Thus, we can equivalently write

$$\pi_N^d u(\boldsymbol{x}) = \sum_{n=0}^{N} \sum_{\ell=1}^{a_n^d} \sum_{k=0}^{[\frac{N-n}{2}]} \widehat{u}_{k,\ell}^n \widehat{H}_{k,\ell}^{0,n}(\boldsymbol{x}). \tag{5.2.59}$$

Based on (5.2.23), we introduce the function space $\mathcal{B}^r(\mathbb{R}^d)$ equipped with the norm

$$\|u\|_{\mathcal{B}^r(\mathbb{R}^d)}^2 = \begin{cases} \|(-\Delta + |\boldsymbol{x}|^2)^m u\|_{L^2(\mathbb{R}^d)}^2, & r = 2m, \\ \dfrac{1}{2}\big(\|(\boldsymbol{x} + \nabla)(-\Delta + |\boldsymbol{x}|^2)^m u\|_{L^2(\mathbb{R}^d)}^2 & \\ \quad + \|(\boldsymbol{x} - \nabla)(-\Delta + |\boldsymbol{x}|^2)^m u\|_{L^2(\mathbb{R}^d)}^2\big), & r = 2m+1, \end{cases} \tag{5.2.60}$$

where integer $r \geq 0$, and $\boldsymbol{x} + \nabla$ and $\boldsymbol{x} - \nabla$ are the lowering and raising operators, respectively.

The main approximation result is stated below.

Theorem 5.30 *Let $s \in (0,1)$. For any $u \in \mathcal{B}^r(\mathbb{R}^d)$ with integer $r \geq 1$, we have*

$$\|\pi_N^d u - u\|_{H^s(\mathbb{R}^d)} \leq (2N+d+2)^{(s-r)/2} \|u\|_{\mathcal{B}^r(\mathbb{R}^d)}. \tag{5.2.61}$$

Proof (i). We first estimate the L^2-error. For $r = 2m+1$, a direct calculation gives

$$\begin{aligned} \|u\|^2_{\mathcal{B}^r(\mathbb{R}^d)} &= \frac{1}{2}\Big(\|(\boldsymbol{x}+\nabla)(-\Delta+|\boldsymbol{x}|^2)^m u\|^2_{L^2(\mathbb{R}^d)} \\ &\quad + \|(\boldsymbol{x}-\nabla)(-\Delta+|\boldsymbol{x}|^2)^m u\|^2_{L^2(\mathbb{R}^d)}\Big) \\ &= \big((-\Delta+|\boldsymbol{x}|^2)^{m+1}u, (-\Delta+|\boldsymbol{x}|^2)^m u\big)_{\mathbb{R}^d}. \end{aligned} \tag{5.2.62}$$

Thanks to the orthogonality (5.2.15), (5.2.23)–(5.2.60) and (5.2.62), we have that for any $r \geq 0$,

$$\|u\|^2_{\mathcal{B}^r(\mathbb{R}^d)} = \sum_{n=0}^{\infty}\sum_{\ell=1}^{a_n^d}\sum_{k=0}^{\infty} h_{k,d}^{n,r}|\hat{u}_{k,\ell}^n|^2, \quad h_{k,d}^{n,r} = (4k+2n+d)^r. \tag{5.2.63}$$

Then, we derive from (5.2.59) and (5.2.63) that

$$\begin{aligned} \|\pi_N^d u - u\|^2_{L^2(\mathbb{R}^d)} &= \sum_{n=0}^{\infty}\sum_{\ell=1}^{a_n^d}\sum_{k=\lceil\frac{N+1-n}{2}\rceil}^{\infty} h_{k,d}^{n,0}|\hat{u}_{k,\ell}^n|^2 \\ &\leq \max_{2k+n\geq N+1}\left\{\frac{h_{k,d}^{n,0}}{h_{k,d}^{n,r}}\right\}\sum_{n=0}^{\infty}\sum_{\ell=1}^{a_n^d}\sum_{k=\lceil\frac{N+1-n}{2}\rceil}^{\infty} h_{k,d}^{n,r}|\hat{u}_{k,\ell}^n|^2 \\ &\leq (2N+2+d)^{-r}\|u\|^2_{\mathcal{B}^r(\mathbb{R}^d)}. \end{aligned} \tag{5.2.64}$$

If $r = 2m$, we find from (5.2.60) that (5.2.62) simply becomes

$$\|u\|^2_{\mathcal{B}^r(\mathbb{R}^d)} = \big((-\Delta+|\boldsymbol{x}|^2)^m u, (-\Delta+|\boldsymbol{x}|^2)^m u\big)_{\mathbb{R}^d}, \tag{5.2.65}$$

so we can follow the same lines as above to derive the L^2-estimate.

(ii). We next estimate the H^1-error. Using the triangle inequality and (5.2.62), we obtain that

$$\begin{aligned} &\|\nabla(\pi_N^d u - u)\|^2_{L^2(\mathbb{R}^d)} \\ &\leq \frac{1}{2}\big(\|(\boldsymbol{x}+\nabla)(\pi_N^d u - u)\|^2_{L^2(\mathbb{R}^d)} + \|(\boldsymbol{x}-\nabla)(\pi_N^d u - u)\|^2_{L^2(\mathbb{R}^d)}\big) \\ &= \big((-\Delta+|\boldsymbol{x}|^2)(\pi_N^d u - u), (\pi_N^d u - u)\big)_{\mathbb{R}^d} \\ &\leq \sum_{n=0}^{\infty}\sum_{\ell=1}^{a_n^d}\sum_{k=\lceil\frac{N+1-n}{2}\rceil}^{\infty} h_{k,d}^{n,1}|\hat{u}_{k,\ell}^n|^2 \end{aligned} \tag{5.2.66}$$

$$\leq \max_{2k+n\geq N+1}\left\{\frac{h_{k,d}^{n,1}}{h_{k,d}^{n,r}}\right\}\sum_{n=0}^{\infty}\sum_{\ell=1}^{a_n^d}\sum_{k=\lceil\frac{N+1-n}{2}\rceil}^{\infty} h_{k,d}^{n,r}|\hat{u}_{k,\ell}^{n}|^2$$

$$\leq (2N+2+d)^{1-r}\|u\|^2_{\mathcal{B}^r(\mathbb{R}^d)}.$$

Finally, the desired results can be obtained by the L^2- and H^1-bounds derived above and the following space interpolation inequality (cf. Agranovich (2015, Ch. 1)):

$$\|u\|_{H^s(\mathbb{R}^d)} \leq \|u\|^{1-s}_{L^2(\mathbb{R}^d)}\,\|u\|^{s}_{H^1(\mathbb{R}^d)}, \quad s\in[0,1]. \tag{5.2.67}$$

This ends the proof. □

Taking $v_N = \pi_N^d u$ in (5.2.56) and using Theorem 5.30, we immediately obtain the following error estimate.

Theorem 5.31 *Let u and u_N be the solutions to* (5.2.48) *and* (5.2.50), *respectively. If $u \in \mathcal{B}^r(\mathbb{R}^d)$ with integer $r \geq 1$, then we have*

$$\|u-u_N\|_{H^s(\mathbb{R}^d)} \leq c(2N+d+2)^{(s-r)/2}\|u\|_{\mathcal{B}^r(\mathbb{R}^d)}, \quad s\in[0,1], \tag{5.2.68}$$

where c is a positive constant independent of N and u.

Useful Analytic Formulas

We first derive analytical formulas for the fractional Laplacian of some functions with typical exponential or algebraic decay, upon which we construct the exact solutions to test the accuracy of the proposed method. Moreover, we reveal that the fractional Laplacian has a very different property from the usual Laplacian. For example, the image of an exponential function decays algebraically as opposite to the usual one.

We have the following exact formulas for the Gaussian function and rational function.

Proposition 5.32 *For real $s > 0$ and integer $d \geq 1$, we have that*

$$(-\Delta)^s\{e^{-|\boldsymbol{x}|^2}\} = \frac{2^{2s}\Gamma(s+d/2)}{\Gamma(d/2)}\,{}_1F_1\Big(s+\frac{d}{2};\frac{d}{2};-|\boldsymbol{x}|^2\Big). \tag{5.2.69}$$

Moreover for non-integer $s > 0$ and $|x| \to \infty$, we have the asymptotic behavior

$$(-\Delta)^s\{e^{-|\boldsymbol{x}|^2}\} = -\frac{2^{2s}\sin(\pi s)}{\pi}\frac{\Gamma(s+d/2)\Gamma(1+s)}{|\boldsymbol{x}|^{d+2s}}\{1+O(|\boldsymbol{x}|^{-2})\}. \tag{5.2.70}$$

Proof The results with $d = 1$ were derived in Tang et al. (2018), so it suffices to prove them for integer $d \geq 2$. Note that

$$\mathscr{F}\{e^{-|\boldsymbol{x}|^2}\}(\xi) = \frac{1}{(2\pi)^{d/2}} \int_{\mathbb{R}^d} e^{-|\boldsymbol{x}|^2} e^{-\mathrm{i}\boldsymbol{x}\cdot\xi} \mathrm{d}\boldsymbol{x}$$
$$= \frac{1}{(2\pi)^{d/2}} \int_{\mathbb{R}} e^{-x_1^2} e^{-\mathrm{i}x_1\xi_1} \mathrm{d}x_1 \cdots \int_{\mathbb{R}} e^{-x_d^2} e^{-\mathrm{i}x_d\xi_d} \mathrm{d}x_d = \frac{1}{2^{d/2}} e^{-\frac{|\xi|^2}{4}},$$

where we used the identity (cf. Gradshteyn and Ryzhik (2015, P. 339)):

$$\int_{\mathbb{R}} e^{-x^2} e^{-\mathrm{i}x\xi} \mathrm{d}x = \sqrt{\pi} e^{-\frac{\xi^2}{4}}.$$

Thus from the definition, we obtain

$$\begin{aligned}(-\Delta)^s\{e^{-|\boldsymbol{x}|^2}\}(\boldsymbol{x}) &= \mathscr{F}^{-1}\Big\{|\xi|^{2s}\mathscr{F}\{e^{-|\boldsymbol{x}|^2}\}(\xi)\Big\} = \frac{1}{2^{d/2}(2\pi)^{d/2}} \int_{\mathbb{R}^d} |\xi|^{2s} e^{-\frac{|\xi|^2}{4}} e^{\mathrm{i}\boldsymbol{x}\cdot\xi} \mathrm{d}\xi \\ &= \frac{2^d}{2^{d/2}(2\pi)^{d/2}} \int_{\mathbb{R}_+^d} |\xi|^{2s} e^{-\frac{|\xi|^2}{4}} \cos(x_1\xi_1)\cos(x_2\xi_2)\cdots\cos(x_d\xi_d)\mathrm{d}\xi.\end{aligned} \tag{5.2.71}$$

We proceed with the calculation by using the d-dimensional spherical coordinates:

$$\begin{aligned}\xi_1 &= r\cos\theta_1;\ \xi_2 = r\sin\theta_1\cos\theta_2;\ \cdots\cdots; \\ \xi_{d-1} &= r\sin\theta_1\cdots\sin\theta_{d-2}\cos\theta_{d-1}; \\ \xi_d &= r\sin\theta_1\cdots\sin\theta_{d-2}\sin\theta_{d-1}, \quad r = |\xi|,\end{aligned} \tag{5.2.72}$$

so we can write

$$(-\Delta)^s\{e^{-|\boldsymbol{x}|^2}\}(\boldsymbol{x}) = \frac{1}{\pi^{d/2}} \int_0^\infty r^{2s+d-1} e^{-\frac{r^2}{4}} \mathcal{I}(r;\boldsymbol{x})\mathrm{d}r, \tag{5.2.73}$$

where

$$\begin{aligned}\mathcal{I}(r;\boldsymbol{x}) = &\int_{[0,\frac{\pi}{2}]^{d-1}} \cos\big(rx_1\cos\theta_1\big)\cos\big(rx_2\sin\theta_1\cos\theta_2\big)\cdots \\ &\cos\big(rx_{d-1}\sin\theta_1\cdots\sin\theta_{d-2}\cos\theta_{d-1}\big) \\ &\cos\big(rx_d\sin\theta_1\cdots\sin\theta_{d-2}\sin\theta_{d-1}\big)(\sin\theta_1)^{d-2}(\sin\theta_2)^{d-3}\cdots \\ &(\sin\theta_{d-2})\,\mathrm{d}\theta_1\mathrm{d}\theta_2\cdots\mathrm{d}\theta_{d-1}.\end{aligned}$$

We first integrate $\mathcal{I}(r;\boldsymbol{x})$ with respect to θ_{d-1}. To do this, we recall the integral formula involving the Bessel functions (cf. Gradshteyn and Ryzhik (2015, P. 732)): for real μ, $\nu > -1$ and $a, b > 0$,

$$\int_0^{\frac{\pi}{2}} J_\nu(a\sin\theta)J_\mu(b\cos\theta)\sin^{\nu+1}\theta\cos^{\mu+1}\theta\,d\theta \\ = \frac{a^\nu b^\mu J_{\nu+\mu+1}(\sqrt{a^2+b^2})}{(a^2+b^2)^{(\nu+\mu+1)/2}}. \tag{5.2.74}$$

Then using the identity $\cos z = \sqrt{\pi z/2}J_{-1/2}(z)$ and (5.2.74) (with $a = rx_{d-1}\sin\theta_1\cdots\sin\theta_{d-2}$, $b = rx_d\sin\theta_1\cdots\sin\theta_{d-2}$ and $\mu = \nu = -1/2$), we derive

$$\int_0^{\frac{\pi}{2}} \cos\big(rx_{d-1}\sin\theta_1\cdots\sin\theta_{d-2}\cos\theta_{d-1}\big) \\ \cos\big(rx_d\sin\theta_1\cdots\sin\theta_{d-2}\sin\theta_{d-1}\big)\,d\theta_{d-1} \\ = \frac{\pi}{2}J_0\Big(r\sin\theta_1\cdots\sin\theta_{d-2}\sqrt{x_{d-1}^2+x_d^2}\Big).$$

Substituting the above into $\mathcal{I}(r,x)$, and applying the same argument to $\theta_{d-2},\theta_{d-3},\ldots,\theta_1$ repeatedly $d-2$ times, we obtain

$$\mathcal{I}(r;\boldsymbol{x}) = \Big(\frac{\pi}{2}\Big)^{\frac{d}{2}}(r|\boldsymbol{x}|)^{1-\frac{d}{2}}J_{\frac{d}{2}-1}(r|\boldsymbol{x}|). \tag{5.2.75}$$

We proceed with the integral identity (cf. Gradshteyn and Ryzhik (2015, P. 713)): for real $\mu+\nu > -1$ and $p > 0$,

$$\int_{\mathbb{R}^+} J_\mu(bt)e^{-p^2t^2}t^{\nu-1}dt \\ = \frac{b^\mu\Gamma((\mu+\nu)/2)}{2^{\mu+1}p^{\nu+\mu}\Gamma(\mu+1)}\,{}_1F_1\Big(\frac{\mu+\nu}{2};\mu+1;-\frac{b^2}{4p^2}\Big). \tag{5.2.76}$$

Then, substituting (5.2.75) into (5.2.73) and using (5.2.76) (with $\mu = d/2-1$ and $\nu = 2s+d/2+1$), we derive

$$(-\Delta)^s\big\{e^{-|\boldsymbol{x}|^2}\big\} = \frac{|\boldsymbol{x}|^{1-\frac{d}{2}}}{2^{d/2}}\int_0^\infty r^{2s+\frac{d}{2}}e^{-\frac{r^2}{4}}J_{\frac{d}{2}-1}(r|\boldsymbol{x}|)dr \\ = \frac{2^{2s}\Gamma(s+d/2)}{\Gamma(d/2)}\,{}_1F_1\Big(s+\frac{d}{2};\frac{d}{2};-|\boldsymbol{x}|^2\Big).$$

This yields (5.2.69). The asymptotic behavior (5.2.70) follows from the property (cf. Bateman (1953, P. 278)):

$${}_1F_1(a;b;z) = \frac{\Gamma(b)}{\Gamma(b-a)}(-z)^{-a}\big\{1+O(|z|^{-1})\big\}. \tag{5.2.77}$$

Then (5.2.70) follows. This completes the proof. □

Proposition 5.33 *For real $s, r > 0$, and integer $d \geq 1$, we have*

$$\begin{aligned}(-\Delta)^s&\Big\{\frac{1}{(1+|\boldsymbol{x}|^2)^r}\Big\}\\&=\frac{2^{2s}\Gamma(s+\gamma)\Gamma(s+d/2)}{\Gamma(\gamma)\Gamma(d/2)}\,{}_2F_1\Big(s+r,s+\frac{d}{2};\frac{d}{2};-|\boldsymbol{x}|^2\Big).\end{aligned}\tag{5.2.78}$$

Moreover for non-integer $s > 0$ and $|x| \to \infty$, we have the asymptotic properties: for $r \neq d/2$,

$$(-\Delta)^s\Big\{\frac{1}{\big(1+|\boldsymbol{x}|^2\big)^r}\Big\}\sim\frac{1}{(1+|\boldsymbol{x}|^2)^{s+\mu}},\quad \mu=\min\{r,d/2\},\tag{5.2.79}$$

while for $r = d/2$,

$$(-\Delta)^s\Big\{\frac{1}{(1+|\boldsymbol{x}|^2)^r}\Big\}\sim\frac{\ln(1+|\boldsymbol{x}|^2)}{(1+|\boldsymbol{x}|^2)^{s+d/2}}.\tag{5.2.80}$$

Proof The identity with $d = 1$ can be found in Tang et al. (2018), so we assume that $d \geq 2$. Using the d-spherical coordinate system in (5.2.72), we obtain from (5.2.75) that

$$\begin{aligned}\mathscr{F}\Big\{\frac{1}{(1+|\boldsymbol{x}|^2)^\gamma}\Big\}(\xi)&=\frac{1}{(2\pi)^{d/2}}\int_{\mathbb{R}^d}\frac{e^{-\mathrm{i}\boldsymbol{x}\cdot\xi}}{\big(1+|\boldsymbol{x}|^2\big)^\gamma}\mathrm{d}\boldsymbol{x}\\&=\frac{2^d}{(2\pi)^{d/2}}\int_{\mathbb{R}^d_+}\frac{\cos(x_1\xi_1)\cos(x_2\xi_2)\cdots\cos(x_d\xi_d)}{\big(1+|\boldsymbol{x}|^2\big)^\gamma}\mathrm{d}\boldsymbol{x}\\&=\Big(\frac{2}{\pi}\Big)^{\frac{d}{2}}\int_0^\infty\frac{r^{d-1}}{\big(1+r^2\big)^\gamma}\mathcal{I}(r;\xi)\,\mathrm{d}r=|\xi|^{1-\frac{d}{2}}\int_0^\infty\frac{r^{\frac{d}{2}}}{\big(1+r^2\big)^\gamma}J_{\frac{d}{2}-1}(r|\xi|)\mathrm{d}r.\end{aligned}$$

Recall the integral formula (cf. Gradshteyn and Ryzhik (2015, P. 686)): for $-1 < \nu < 2\mu + \frac{3}{2}$ and $a, b > 0$,

$$\int_0^\infty\frac{x^{\nu+1}}{(x^2+a^2)^{\mu+1}}J_\nu(bx)\mathrm{d}x=\frac{a^{\nu-\mu}b^\mu}{2^\mu\Gamma(\mu+1)}K_{\nu-\mu}(ab),\tag{5.2.81}$$

where $K_\nu(x)$ are the modified Bessel functions of the second kind. Note that $K_{-\nu}(x) = K_\nu(x)$. Then letting $\mu = \gamma - 1$ and $\nu = d/2 - 1$ in (5.2.81), we obtain

$$\mathscr{F}\Big\{\frac{1}{(1+|\boldsymbol{x}|^2)^\gamma}\Big\}(\xi)=\frac{|\xi|^{\gamma-\frac{d}{2}}}{2^{\gamma-1}\Gamma(\gamma)}K_{\gamma-\frac{d}{2}}(|\xi|).$$

We also use the integral formula (cf. Gradshteyn and Ryzhik (2015, P. 692)): for real $a > 0$, real b, and $\nu - \lambda + 1 > |\mu|$,

$$\begin{aligned}\int_0^\infty & x^{-\lambda} K_\mu(ax) J_\nu(bx)\mathrm{d}x \\ &= \frac{b^\nu \Gamma\big((\nu-\lambda+\mu+1)/2\big)\Gamma\big((\nu-\lambda-\mu+1)/2\big)}{2^{\lambda+1}a^{\nu-\lambda+1}\Gamma(\nu+1)} \\ &\quad \times {}_2F_1\Big(\frac{\nu-\lambda+\mu+1}{2}, \frac{\nu-\lambda-\mu+1}{2}; \nu+1; -\frac{b^2}{a^2}\Big).\end{aligned} \tag{5.2.82}$$

Once again, using the d-spherical coordinate system (5.2.72), (5.2.75) and (5.2.82) (with $\lambda = -2s - \gamma$, $\mu = \gamma - d/2$ and $\nu = d/2 - 1$), we have

$$\begin{aligned}(-\Delta)^s\Big\{\frac{1}{(1+|\boldsymbol{x}|^2)^\gamma}\Big\} &= \frac{1}{(2\pi)^{\frac{d}{2}}2^{\gamma-1}\Gamma(\gamma)}\int_{\mathbb{R}^d} e^{\mathrm{i}\boldsymbol{x}\cdot\xi}|\xi|^{2s+\gamma-\frac{d}{2}} K_{\gamma-\frac{d}{2}}(|\xi|)\,\mathrm{d}\xi \\ &= \frac{2^d}{(2\pi)^{\frac{d}{2}}2^{\gamma-1}\Gamma(\gamma)}\int_{\mathbb{R}^d_+} \cos(x_1\xi_1)\cos(x_2\xi_2)\cdots\cos(x_d\xi_d)|\xi|^{2s+\gamma-\frac{d}{2}} K_{\gamma-\frac{d}{2}}(|\xi|)\,\mathrm{d}\xi \\ &= \frac{2^{\frac{d}{2}-\gamma+1}}{\pi^{\frac{d}{2}}\Gamma(\gamma)}\int_0^\infty r^{2s+\gamma+\frac{d}{2}-1} K_{\gamma-\frac{d}{2}}(r)\mathcal{I}(r,\boldsymbol{x})\,\mathrm{d}r \\ &= \frac{2^{-\gamma+1}}{\Gamma(\gamma)}|\boldsymbol{x}|^{1-\frac{d}{2}}\int_0^\infty r^{2s+\gamma}K_{\gamma-\frac{d}{2}}(r)J_{\frac{d}{2}-1}(r|\boldsymbol{x}|)\mathrm{d}r \\ &= \frac{2^{2s}\Gamma(s+\gamma)\Gamma(s+\frac{d}{2})}{\Gamma(\gamma)\Gamma(\frac{d}{2})}\,{}_2F_1\Big(s+\gamma, s+\frac{d}{2}; \frac{d}{2}; -|\boldsymbol{x}|^2\Big).\end{aligned}$$

This completes the derivation of (5.2.78).

We know from Bateman (1953, P. 76) that the asymptotic behavior of the hypergeometric function for large $|x|$ (unless $a-b$ is an integer) is

$$ {}_2F_1(a,b;c;x) = \lambda_1|x|^{-a} + \lambda_2|x|^{-b} + O(|x|^{-a-1}) + O(|x|^{-b-1}), \tag{5.2.83}$$

where λ_1 and λ_2 are constants; if $a-b$ is an integer, z^{-a} or z^{-b} has to be multiplied by a factor $\ln(x)$. Thus we have the asymptotic behaviors in (5.2.79)–(5.2.80). □

Numerical Results

For convenience of implementation, we fix the degree of the numerical solution in both radial and angular direction in (5.2.51), so the numerical solution takes the form

$$u_{N,K}(\boldsymbol{x}) = \sum_{n=0}^{N}\sum_{\ell=1}^{a_n^d}\sum_{k=0}^{K}\hat{u}_{k,\ell}^n\,\check{H}_{k,\ell}^{s,n}(\boldsymbol{x}). \tag{5.2.84}$$

Here, we focus on $d = 2, 3$.

Example 5.34 (Problem (5.2.47) with exact solution) We first consider (5.2.47) with the following exact solutions:

$$u_e(\boldsymbol{x}) = \mathrm{e}^{-|\boldsymbol{x}|^2}, \quad u_a(\boldsymbol{x}) = (1 + |\boldsymbol{x}|^2)^{-r}, \quad r > 0, \ \ \boldsymbol{x} \in \mathbb{R}^d. \tag{5.2.85}$$

By Propositions 5.32–5.33, the source terms $f_e(\boldsymbol{x})$ and $f_a(\boldsymbol{x})$ are respectively given by

$$f_e(\boldsymbol{x}) = \gamma \mathrm{e}^{-|\boldsymbol{x}|^2} + \frac{2^{2s}\Gamma(s+d/2)}{\Gamma(d/2)} {}_1F_1\Big(s+\frac{d}{2}; \frac{d}{2}; -|\boldsymbol{x}|^2\Big),$$

$$f_a(\boldsymbol{x}) = \gamma(1+|\boldsymbol{x}|^2)^{-r} + \frac{2^{2s}\Gamma(s+r)\Gamma(s+d/2)}{\Gamma(r)\Gamma(d/2)} {}_2F_1\Big(s+r, s+\frac{d}{2}; \frac{d}{2}; -|\boldsymbol{x}|^2\Big).$$

For $d = 2, 3$, we take $s = 0.3,\ 0.5,\ 0.7$ and the degree in angular direction is fixed $N \equiv 10$ (see (5.2.84)). In Figure 5.10, we plot the maximum errors, in semi-log scale and log-log scale, for u_e and u_a with $d = 2, 3$ against

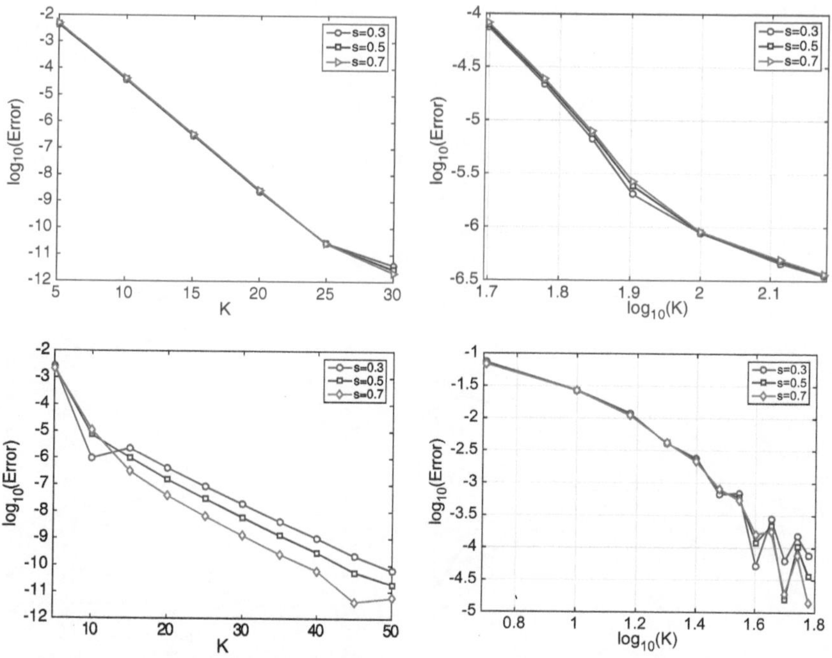

Figure 5.10 The maximum errors of the GHF-spectral-Galerkin scheme with $\gamma = 1$ for Example 5.34 with exact solutions in (5.2.85) and $s = 0.3,\ 0.5,\ 0.7$. Top: (left) $d = 2$ and $u_e(\boldsymbol{x}) = e^{-|\boldsymbol{x}|^2}$ and (right) $d = 2$ and $u_a(\boldsymbol{x}) = (1 + |\boldsymbol{x}|^2)^{-2}$. Bottom: (left) $d = 3$ and $u_e(\boldsymbol{x}) = e^{-|\boldsymbol{x}|^2}$ and (right) $d = 3$ and $u_a(\boldsymbol{x}) = (1 + |\boldsymbol{x}|^2)^{-2}$. Source: Sheng et al. (2021).

various K, respectively. As expected, we observe the exponential and algebraic convergence for u_e and u_a, respectively.

Example 5.35 (Problem (5.2.47) with a source term) We next consider (5.2.47) with the following source functions:

$$f_e(\boldsymbol{x}) = \sin(|\boldsymbol{x}|)e^{-|\boldsymbol{x}|^2}, \quad f_a(\boldsymbol{x}) = \cos(|\boldsymbol{x}|)(1+|\boldsymbol{x}|^2)^{-r}, \tag{5.2.86}$$

for $r > 0$ and $\boldsymbol{x} \in \mathbb{R}^d$.

The exact solutions are unknown, and we use the numerical solution with $K = 80$, $N = 20$ as the reference solution. For $d = 2, 3$, we plot the maximum errors, in log-log scale, for (5.2.47) against various K in Figure 5.11, in which we take $s = 0.3,\ 0.5,\ 0.7$ and fix $N \equiv 10$. Note that the solution of (5.2.47) decays algebraically, even for exponentially decaying source terms. Indeed, we observe an algebraic order of convergence.

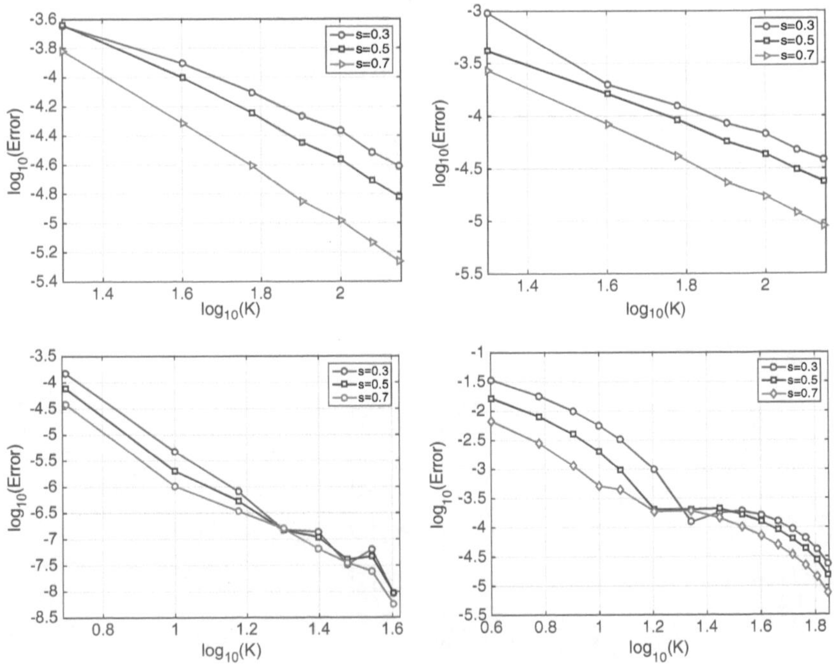

Figure 5.11 The maximum errors of the GHF-spectral-Galerkin scheme with $\gamma = 1$ for Example 5.35 with given source functions in (5.2.86) and $s = 0.3,\ 0.5,\ 0.7, r = 2$. Top: (left) $d = 2$ with given source term $f_e(\boldsymbol{x})$ and (right) $d = 2$ with given source term $f_a(\boldsymbol{x})$. Bottom: (left) $d = 3$ with given source term $f_e(\boldsymbol{x})$ and (right) $d = 3$ with given source term $f_a(\boldsymbol{x})$. Source: Sheng et al. (2021).

6
Fractional Laplacian in Bounded Domains

The fractional Laplacian in $\mathbb{R}^d$, which we write as $(-\Delta)^{\alpha/2}$ with $\alpha \in (0, 2)$, has multiple equivalent characterizations. Moreover, in bounded domains, boundary conditions must be incorporated in these characterizations in mathematically distinct ways, and there is currently no consensus in the literature as to which definition of the fractional Laplacian in bounded domains is most appropriate for a given application. Although the study of the fractional Laplacian is far from complete, this chapter can serve as a proper educational/research starting point for students/researchers in order to employ these operators to model complex anomalous systems.

The Riesz (or integral) definition, for example, admits a nonlocal boundary condition, where the value of a function must be prescribed on the entire exterior of the domain in order to compute its fractional Laplacian. By contrast, the spectral definition requires only the standard local boundary condition. We are going to provide a rather comprehensive and comparative answer to the question: *What is the fractional Laplacian?*

Beginning from first principles, we compare several commonly used definitions of the fractional Laplacian theoretically, through their stochastic interpretations as well as their analytical properties. Then, we present quantitative comparisons using a sample of state-of-the-art methods. We discuss recent advances on nonzero boundary conditions and present new methods to discretize such boundary-value problems: radial basis function collocation for the Riesz fractional Laplacian and nonharmonic lifting for the spectral fractional Laplacian. To this end, here is a guide to the notation and terminology, employed throughout this chapter:

- We define the fractional Laplacian to be $(-\Delta)^{\alpha/2}$, where $\Delta = \partial^2/\partial x_1^2 + \cdots + \partial^2/\partial x_d^2$. We take the fractional power of $(-\Delta)$ to obtain a positive operator. As a result, our definition of the fractional Laplacian $(-\Delta)^{\alpha/2}$ is the *negative* generator of the standard isotropic α-stable Lévy process, and reduces to $-\Delta = -\partial^2/\partial x_1^2 - \cdots - \partial^2/\partial x_d^2$ when $\alpha = 2$.

- In this chapter, the symbol α is always used as the fractional order. In particular, the fractional Laplacian is represented as $(-\Delta)^{\alpha/2}$ and $\alpha \in (0, 2)$.
- The symbol s is always used in the representation of a real-ordered (fractional) Sobolev space, H^s, and is often used in this work to describe the regularity of source function f of a fractional Poisson equation. The symbol s should not be confused with the fractional order of the Laplacian, as often appears in the book.
- We do not make a notational distinction between the definitions of fractional Laplacian, as the definition should be clear from the context or can be understood from the subsection heading.
- *Homogeneous* fractional Laplacians are defined in the context of zero boundary conditions, and *inhomogeneous* fractional Laplacians are defined with nonzero boundary conditions. The type of boundary condition (in this work, Dirichlet or Neumann) is specified in the text.
- In the following sections, multiple definitions appear together for the purpose of comparison, so we use the following notation:

  $(-\Delta_{\text{Riesz}})^{\alpha/2}$ represents the *Riesz* (or *integral*) definition;
  $(-\Delta_{\text{spectral}})^{\alpha/2}$ represents the *spectral* definition;
  $(-\Delta_{\text{M}})^{\alpha/2}$ represents the *directional* definition.

Over the past few decades, scientists have been exploring fractional calculus as a tool for developing more sophisticated mathematical models that can accurately describe complex anomalous systems (Pozrikidis, 2016; Bucur and Valdinoci, 2016; Meerschaert and Sikorskii, 2019; Vázquez, 2017). In particular, the fractional Laplacian has been used in place of the integer-order Laplacian in many applications, including, for example, the fractional models listed in Table 6.1, in a range of applications, e.g, diffusion-reaction (Yamamoto, 2012), quasi-geostrophic flows (Constantin and Wu, 1999), the Cahn–Hilliard equation (Akagi et al., 2016; Ainsworth and Mao, 2017a,b), porous medium (Akagi et al., 2016; de Pablo et al., 2011), quantum mechanics (Laskin, 2000), and ultrasound wave mechanics (Treeby and Cox, 2010; Chen and Holm, 2004). The fractional Laplacian can be defined in $\mathbb{R}^d$ in several equivalent ways (Kwaśnicki, 2017); however, when these definitions are restricted to *bounded domains*, the associated *boundary conditions* lead to distinct operators.

Table 6.1. *Important equations involving the fractional Laplacian. Source: Reprinted from Lischke et al. (2020) with permission from Elsevier.*

	Fractional Equation	Domain
Diffusion-Reaction	$\partial_t u + (-\Delta)^{\alpha/2} u + c(t,x)u = 0$	$(0,+\infty) \times \mathbb{R}^d$
Quasi-geostrophic	$\partial_t \theta + u \cdot \nabla\theta + \kappa(-\Delta)^{\alpha/2}\theta = f$	$(0,T] \times \mathbb{R}^2$
Cahn–Hilliard	$\partial_t u + (-\Delta)^{\alpha/2}(-\varepsilon^2 \Delta u + f(u)) = 0$	$(0,T] \times (0,2\pi)^2$
Porous Medium	$\partial_t u + (-\Delta)^{\alpha/2}(\lvert u\rvert^{m-1} \text{sign} u) = 0$	$(0,+\infty) \times \mathbb{R}^d$
Schrödinger	$i\hbar\partial_t \psi = D_\alpha(-\hbar^2\Delta)^{\alpha/2}\psi + V(r,t)\psi$	$\mathbb{R}^3 \times (0,+\infty)$
Ultrasound	$\frac{1}{c_0^2}\partial_t^2 p = \nabla^2 p - \{\tau\partial_t(-\Delta)^{\alpha/2} + \eta(-\Delta)^{(\alpha+1)/2}\}p$	$(-\infty,+\infty) \times \mathbb{R}^d$

The purpose of this chapter is two-fold: (i) to give a comprehensive report of the commonly used definitions of the fractional Laplacian and examine their differences in bounded domains, and (ii) to quantitatively explore these differences in benchmark problems using a sample of state-of-the-art numerical methods that are described in a detailed and reasonably self-contained way.

As research on numerical methods for approximating the fractional Laplacian is extensive and ongoing, we do not attempt to include a comprehensive description of all available numerical methods. Of significance in this chapter is the emphasis on boundary-value problems with nonzero boundary conditions, including stochastic methods, and the inclusion of new methods for discretizing such problems. This work will be of use to practitioners looking to gain insight into which fractional Laplacian definition and associated numerical methods may be appropriate for their application. A number of articles which include comparisons of the different fractional Laplacians on bounded domains have appeared recently, such as those of Bonito et al. (2018) and Duo et al. (2019).

6.1 A Motivating Example

We consider some one-dimensional benchmark problems involving different definitions of the fractional Laplacian, which are later defined in detail. We will return to this problem in higher dimensions and in different domains. Consider the one-dimensional fractional Poisson problem on an interval $\Omega = (-L, L)$:

$$(-\Delta)^{\alpha/2} u(x) = f(x), \qquad x \in \Omega, \tag{6.1.1}$$

with zero Dirichlet boundary conditions and $\alpha \in (0,2)$. Importantly, we consider two cases for the operator $(-\Delta)^{\alpha/2}$: (i) Riesz fractional Laplacian and (ii) spectral fractional Laplacian.

The formulation of the zero Dirichlet boundary conditions depends on the definition of the fractional Laplacian. For the Riesz fractional Laplacian, the boundary condition is formulated as $u(x) = 0$ in $\mathbb{R} \setminus (-L, L)$, and for the spectral fractional Laplacian, the boundary condition is $u(\pm L) = 0$. The reasons for these formulations will be discussed. The benchmark problems are posed with source functions $f = 1$ and $f = \sin(\pi x)$ on Ω. Below, we observe that the spectral and Riesz solutions evolve in different ways as the fractional order α is changed, and that these evolutions are dependent on the size of the interval Ω. Additionally, we discuss the differing behaviors of the solutions to the benchmark problems near the end-points of the interval Ω.

Case 1: $f(x) = 1$. To discretize the spectral definition, we use the discrete eigenfunction method described in Section 6.5.1, and the Riesz fractional Poisson equation is solved numerically using the one-dimensional spectral method in Mao et al. (2016). We plot numerical solutions of the fractional Poisson equation for both the spectral and Riesz definitions in Figure 6.1 with various values of the fractional order α. We observe from Figure 6.1 that the maximum value of the Riesz solution at $x = 0$ does not vary in a monotone fashion as α ranges from 1.99 to 0.01. Indeed, from $\alpha = 1.99$ to $\alpha = 0.5$, the maximum value increases, and from $\alpha = 0.5$ to $\alpha = 0.1$, the maximum value decreases. In contrast, the maximum values of the solutions corresponding with the spectral definition increase in a monotone fashion with α.

If we instead defined (6.1.1) on $\mathbb{R}$, we would expect the solutions for different fractional Laplacian definitions to be the same, as these definitions are equivalent on $\mathbb{R}$. This observation leads to the following question: is this monotonicity property of the solutions to (6.1.1) affected by changing the size of the computational domain? To investigate this, we solved the same fractional Poisson problems (using both the spectral and Riesz definitions) but changed the length of the interval Ω.

In this example, we denote the solutions to (6.1.1) by u_L, to make the dependence on L explicit. Figure 6.2 includes plots of the trajectories $M_L(\alpha) = \max(u_L)$ of the maximum values of the solutions u_L for $\alpha \in (0, 2]$ and for different lengths L of the interval Ω (each curve corresponds to a different value of L). Using a change of variables, one can show that the solution $u_L(x)$, for both the spectral and the Riesz fractional Laplacian in (6.1.1), has the property $u_L(x) = L^\alpha u_1(x/L)$. However, as u_1 itself depends on α, the interaction between the factors L^α and $u_1(x/L)$ leads to a switch in the trajectories $M_L(\alpha)$ from monotonically decreasing behavior, to non-monotonic behavior, and finally to monotonically increasing behavior as L increases. As is clear in Figure 6.2, this occurs at different values of α for

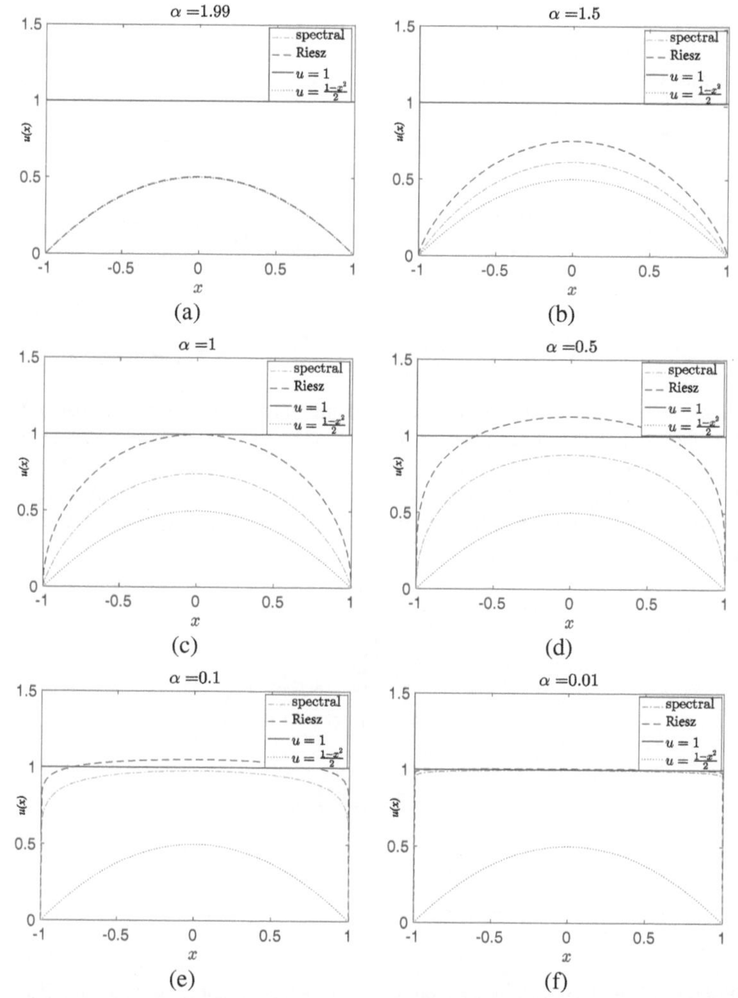

Figure 6.1 One-dimensional study: the profiles of the numerical solutions of (6.1.1) for different fractional orders α: (a)–(e) for $\alpha = 1.99, 1.5, 1.0, 0.5, 0.1, 0.01$. The pink dotted curve corresponds to the solution in the case $\alpha = 2$, and the solid line represents the (discontinuous) solution for $\alpha = 0$, and are included for reference. Note that the Riesz solution has greater amplitude than the spectral solution, and this amplitude increases above $u = 1$ as α goes to zero before decreasing to one in the plot for $\alpha = 0.01$. Source: Reprinted from Lischke et al. (2020) with permission from Elsevier.

the different fractional Laplacians (Riesz and spectral). Thus, we observe that the size of the computational domain affects the behavior of the solutions in relation to the fractional order α, and it does so in different ways for the different fractional Laplacian definitions.

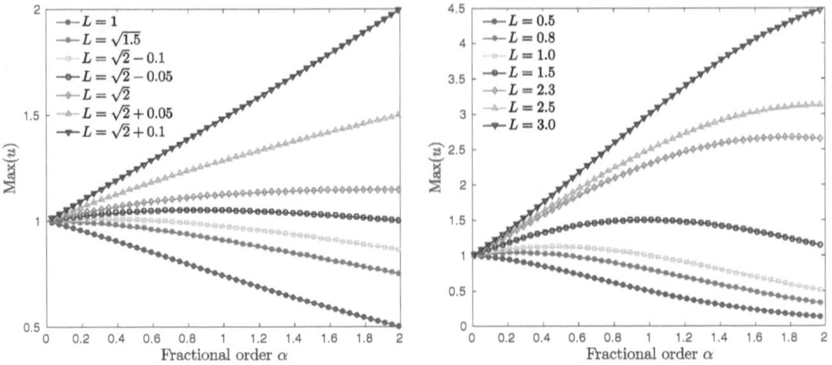

Figure 6.2 Monotonicity study: The maximum value of the numerical solutions for the spectral fractional Poisson equation (6.1.1) in the interval $(-L, L)$ for $\alpha \in (0, 2)$ with $f(x) = 1$ (left). The maximum value of the numerical solutions for the Riesz fractional Poisson equation in the interval $(-L, L)$ for $\alpha \in (0, 2)$ with $f = 1$ (right). Source: Reprinted from Lischke et al. (2020) with permission from Elsevier.

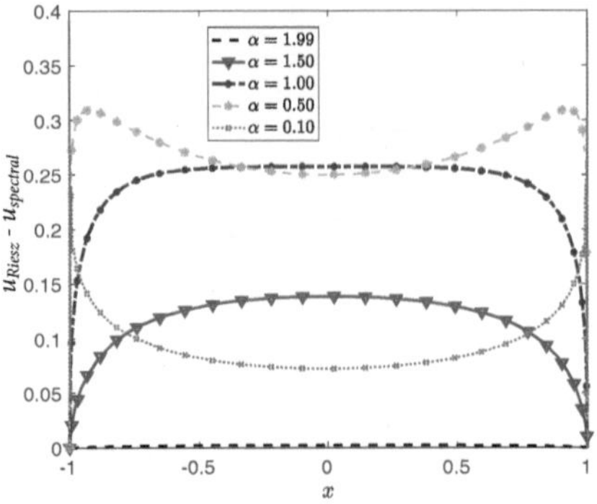

Figure 6.3 Plot of differences between the spectral solutions and the Riesz solutions of (6.1.1) for the right-hand side $f = 1$ on the interval $(-1, 1)$. For values of α between 1 and 2, the difference is greatest in the interior of the interval. When $\alpha < 1$, a boundary layer forms and sharpens as $\alpha \to 0$. Source: Reprinted from Lischke et al. (2020) with permission from Elsevier.

In Figure 6.3, we plot the differences $u_{\text{Riesz}} - u_{\text{spectral}}$ of solutions to (6.1.1) on $\Omega = (-1, 1)$ with $f = 1$ for different values of α. We observe that boundary layers in the differences start to form as α drops below 1, becoming particularly sharp and developing singularities in their derivatives as α approaches zero. This behavior can be understood by examining boundary regularity of solutions arising from the two different fractional Laplacian definitions, which we discuss in detail in Section 6.5.2. Futhermore, we notice that the differences are non-negative in all of Ω, indicating that the Riesz solutions, for any value of $\alpha \in (0, 2)$, lie above the spectral solutions. This is consistent with the theoretical result that the inverse Riesz fractional Laplacian minus the inverse spectral fractional Laplacian (for zero Dirichlet boundary conditions) is positivity preserving (Musina and Nazarov, 2016). When the problem has nonzero boundary conditions, this property need not hold, as discussed later in this chapter. We also observe this property in the two-dimensional zero Dirichlet boundary condition comparisons.

Case 2: $f(x) = \sin(\pi x)$. The solution to the Riesz fractional Poisson equation is computed using a spectral method (Mao et al., 2016), and the solution to the spectral Poisson equation is computed using the discrete eigenfunction method discussed in Section 6.5.1. The solutions are shown in Figure 6.4. The interesting feature to note is the boundary layer in the Riesz solutions that sharpens as $\alpha \to 0$ in comparison with the smooth behavior of the spectral solutions at the boundaries. Since the source function in this example, $f = \sin(\pi x)$, is an eigenfunction of the spectral Laplacian, the spectral solution is analytic in Ω and no boundary layer forms. For the Riesz solution, however, the boundary regularity decreases with α, resulting in the boundary singularities observed in Figure 6.4. In fact, for smooth source functions that satisfy the zero boundary conditions, we can always expect this difference in behaviors near the boundary. On the other hand, it is possible to achieve a

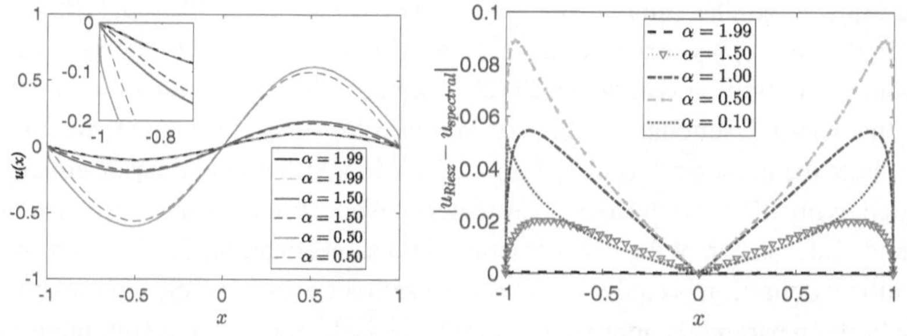

Figure 6.4 One-dimensional study: The numerical solutions of both the spectral and Riesz fractional Poisson equations with $f = \sin(\pi x)$. The solid curves are solutions of the spectral Poisson equation computed with the spectral element method where $\mathcal{N} = 64$. The dashed curves represent the solutions of the Riesz fractional Poisson equation which are computed using a finite element approximation (left). The numerical solutions between the spectral solutions u_{spectral} and the Riesz solutions u_{Riesz} of their respective fractional Poisson equations for the smooth right-hand side $f = \sin(\pi x)$ on the interval $(-1, 1)$ (right). Source: Reprinted from Lischke et al. (2020) with permission from Elsevier.

singular (at the boundaries) solution using the spectral definition if the source function itself is sufficiently singular. This is a useful observation for modeling anomalous diffusion systems, since the Riesz definition may be a better choice to model data that exhibits such a boundary layer, given a smooth forcing function.

Scope of Numerical Methods for the Fractional Laplacian

We aim to compare numerical solutions for the fractional Poisson problem using different definitions of the fractional Laplacian. To this end, we report a series of numerical methods to discretize each definition. All the methods that we are going to introduce here are to compute the solutions of the benchmark problems as described in this section. To discretize the Riesz fractional Laplacian, we use the adaptive finite element method (AFEM) in Ainsworth and Glusa (2018b) and the walk-on-spheres (WOS) method in Kyprianou et al. (2018). We discretize the spectral fractional Laplacian directly using the spectral element method (SEM) of Song et al. (2017), and the heat semigroup approach (Stinga, 2010; Cusimano et al., 2018; Stinga and Torrea, 2010), and elliptic extension approaches (Caffarelli and Silvestre, 2007; Stinga and Torrea, 2010; Stinga, 2010; Galé et al., 2013; Caffarelli and Stinga, 2016) are also discussed for completeness. We demonstrate a recent approximation method for the directional definition using a radial basis function (RBF) collocation method, which also makes use of the vector Grünwald scheme of Meerschaert et al. (2004). This is also the first work in which numerical results have been produced using the vector Grünwald scheme, as no other work (to our knowledge) has implemented the method of Meerschaert et al. (2004).

Furthermore, we examine another nonlocal operator, the "horizon-based nonlocal" definition (Du et al., 2012), which can be seen as an approximation to the Riesz fractional Laplacian. We develop a finite-volume method to compute the numerical results of one-dimensional fractional Poisson equations posed with different horizon parameters. We compare these results to the numerical solution of the Riesz fractional Poisson equation. The horizon-based nonlocal definition is equivalent to the Riesz fractional Laplacian in the limit as the horizon parameter approaches infinity, as is demonstrated by our numerical examples. We can also see in these examples approximately how large the horizon parameter should be to result in a reasonably close approximation of the solution of the Riesz fractional Poisson equation.

6.2 Sobolev Spaces and the Trace Theorem

In this section, we assemble the commonly-used Sobolev spaces for the fractional Laplacian. These spaces are discussed in Grubb (2015, 2016), but we have more closely followed the exposition and notation of Antil et al. (2017).

Definition 6.1 (See Antil et al. (2017)) For $0 < s < 1$, we define the fractional Sobolev space

$$H^s(\Omega) := \left\{ u \in L^2(\Omega) : \int_\Omega \int_\Omega \frac{|u(x) - u(y)|^2}{|x - y|^{n+2s}} \mathrm{d}x\mathrm{d}y < \infty \right\}, \tag{6.2.1}$$

with the semi-norm and norm

$$|u|^2_{H^s(\Omega)} := \int_\Omega \int_\Omega \frac{|u(x) - u(y)|^2}{|x - y|^{n+2s}} \mathrm{d}x\mathrm{d}y, \tag{6.2.2}$$

$$\|u\|_{H^s(\Omega)} := \left(\|u\|^2_{L^2(\Omega)} + |u|^2_{H^s(\Omega)} \right)^{1/2}. \tag{6.2.3}$$

Definition 6.2 (See Ding (1996)) For any $u \in C^\infty(\Omega)$, define the trace operator $\gamma|_{\partial\Omega}$ by

$$\gamma|_{\partial\Omega} u(x) = u(x), \quad x \in \partial\Omega. \tag{6.2.4}$$

Theorem 6.3 (Trace theorem (Ding, 1996)) *Let Ω be a bounded, simply connected Lipschitz domain and $1/2 < s < 3/2$. Then the trace operator $\gamma|_{\partial\Omega}$ is a bounded linear operator from $H^s(\Omega)$ to $H^{s-\frac{1}{2}}(\partial\Omega)$.*

This range for s, $1/2 < s < 3/2$, is sufficient for the discussion of traces in this section. Now we can define the subspace $H^s_0(\Omega)$ of $H^s(\Omega)$.

Definition 6.4 For $s > 1/2$,

$$H^s_0(\Omega) := \left\{ u \in H^s(\Omega) : \gamma|_{\partial\Omega} u(x) = 0 \right\}. \tag{6.2.5}$$

In the case $s = 1/2$, we must define another fractional Sobolev space, denoted $H^{1/2}_{00}(\Omega)$.

Definition 6.5 (See Antil et al. (2017))

$$H^{1/2}_{00}(\Omega) := \left\{ u \in H^{1/2}(\Omega) : \int_\Omega \frac{u^2(x)}{\mathrm{dist}(x, \partial\Omega)} \mathrm{d}x < \infty \right\}. \tag{6.2.6}$$

The associated norm is defined

$$\|u\|_{H^{1/2}_{00}(\Omega)} := \left(\|u\|^2_{H^{1/2}(\Omega)} + \int_\Omega \frac{u^2(x)}{\mathrm{dist}(x, \partial\Omega)} \mathrm{d}x \right)^{1/2}. \tag{6.2.7}$$

We define the space $\mathbb{H}^s(\Omega)$, which is used to characterize the regularity properties of the spectral fractional Laplacian.

Definition 6.6 (See Antil et al. (2017)) For any $s \geq 0$,

$$\mathbb{H}^s(\Omega) := \left\{ u = \sum_{k=1}^{\infty} u_k \phi_k \in L^2(\Omega) : \|u\|^2_{\mathbb{H}^s(\Omega)} := \sum_{k=1}^{\infty} \lambda_k^s u_k^2 < \infty \right\}, \tag{6.2.8}$$

where (λ_k, ϕ_k) are the eigenpairs of the integer Laplacian $-\Delta$ on Ω with homogeneous Dirichlet boundary conditions.

As noted in Antil et al. (2017), the following relationship exists between the fractional Sobolev spaces presented in Definitions 6.1, 6.4, 6.5, and 6.6:

$$\mathbb{H}^s(\Omega) = \begin{cases} H^s(\Omega) = H_0^s(\Omega), & \text{if } 0 < s < 1/2, \\ H_{00}^{1/2}(\Omega), & \text{if } s = 1/2, \\ H_0^s(\Omega), & \text{if } 1/2 < s < 1. \end{cases} \tag{6.2.9}$$

Definition 6.7 The space $\mathbb{H}^{-s}(\Omega)$, for $s \geq 0$, is defined to be the dual space of $\mathbb{H}^s(\Omega)$.

Definition 6.8 $H_{\text{loc}}^s(\Omega) = \{u \in H^s(K) \text{ for all compact sets } K \subseteq \bar{\Omega}\}$.

6.3 Definitions and Properties of Fractional Laplacians

To describe and compare the different definitions of the fractional Laplacian considered in this work from a theoretical perspective, we review the derivations, regularity properties, and stochastic connections for the Riesz, spectral, directional, and regional definitions of the fractional Laplacian. We first discuss these different characterizations in $\mathbb{R}^d$, including their derivations, equivalence, and relations to Lévy processes. Next, we describe the breakdown in the equivalence when these characterizations are restricted to a bounded domain with either zero or nonzero boundary conditions. We provide a summary of the stochastic processes for which each fractional Laplacian (with the associated implicit boundary conditions) is the infinitesimal generator. This information is useful for developing Feynman–Kac type formulas for solving fractional elliptic/parabolic problems, as in the walk-on-spheres algorithm discussed in Section 6.4.2. Furthermore, we describe extension methods that have been used to reformulate the spectral fractional Laplacian, which make more conventional computational techniques useful in discretizing the differential equations involving these operators. We also summarize well-posedness and regularity properties of each operator.

6.3.1 The Fractional Laplacian on $\mathbb{R}^d$

Although the fractional Laplacian on $\mathbb{R}^d$ is not considered in the numerical experiments contained in this work, the characterizations discussed in this section will lead to the definitions of the most commonly used fractional

Laplacians (Riesz, directional, and spectral) in a *bounded* domain. Thus, this section serves as a basis for the entire section. For proofs that the characterizations we consider in this section, as well as other characterizations, are equivalent in $\mathbb{R}^d$ see Kwaśnicki (2017).

Spectral/Fourier Definition

We wish to construct a fractional power of the Laplacian, $(-\Delta)^{\alpha/2}$, for $0 < \alpha < 2$. A general approach to define the positive real powers L^ρ, with $\rho \in [-1, 1]$, of a positive self-adjoint linear operator L, such as $L = -\Delta$, is facilitated by the spectral theorem. This result states that for a self-adjoint, densely defined linear operator L (not necessarily bounded) on a Hilbert space $\mathcal{H}$,

$$L : \mathcal{D}(L) \to \mathcal{H}, \quad \mathcal{D}(L) \text{ a dense subspace of } \mathcal{H}, \tag{6.3.1}$$

there is a projection-valued measure E_λ such that

$$L = \int_{\lambda \in \sigma(L)} \lambda \, \mathrm{d}E_\lambda \quad \text{on } \mathcal{D}(L) \subset \mathcal{H}. \tag{6.3.2}$$

Here, E_λ is the unique operator-valued spectral measure (resolution of the identity) associated to L, and $\sigma(L) \subset \mathbb{R}$ is the spectrum of L, which is the support of E_λ. See Reed (2012, P. 263) or Rudin (1991, P. 368) for a full discussion of the spectral theorem for self-adjoint operators. The dense domain $\mathcal{D}(L)$ is often referred to as a *core* for the operator L (Ethier and Kurtz, 2009), and may be chosen to be a smaller and more convenient space than a "maximal" domain of definition of L. Using the above spectral representation, powers of the operator L of order $-1 \leq \alpha/2 \leq 1$ can be defined as the self-adjoint operator

$$L^{\alpha/2} = \int_{\lambda \in \sigma(L)} \lambda^{\alpha/2} \mathrm{d}E_\lambda. \tag{6.3.3}$$

The domain of the operator $L^{\alpha/2}$ can then be extended to $\mathcal{H}$ by continuity.

We wish to consider $L = -\Delta$ on $\mathbb{R}^d$ on the Sobolev space $\mathcal{H} = H^2(\mathbb{R}^d)$, with domain the dense subspace $\mathcal{D}(-\Delta) = C_0^\infty(\mathbb{R}^d) \subset \mathcal{H}$ taken with the $\mathcal{H}$-norm. This gives $-\Delta = \int_{\sigma(-\Delta)} \lambda \mathrm{d}E_\lambda$ from (6.3.1) and (6.3.2). For a regular domain Ω, the spectrum $\sigma(-\Delta)$ is a point spectrum, i.e., consisting entirely of eigenvalues (Reed, 2012), so one can think of $\mathrm{d}E_\lambda$ as being, for each λ, a projection operator onto the eigenspace of λ. Then, from (6.3.3), the fractional Laplacian on $\mathbb{R}^d$ is defined by

$$(-\Delta)^{\alpha/2} := \int_{\sigma(-\Delta)} \lambda^{\alpha/2} \mathrm{d}E_\lambda. \tag{6.3.4}$$

Let us briefly discuss the use of $-\Delta$ instead of Δ above. Since $\Delta = \frac{\partial^2}{\partial x_1^2} + \cdots + \frac{\partial^2}{\partial x_d^2}$ has negative eigenvalues, if we were to push such an operator through the above machinery, the resulting fractional operator would have complex eigenvalues. For this reason, the spectral theorem is usually applied to positive-definite operators.

To make the definition (6.3.4) more explicit, the spectrum $\sigma(-\Delta)$ must be known exactly. On $\mathbb{R}^d$, this spectrum consists of eigenvalues $|\xi|^2$, where $\xi \in \mathbb{R}^d$, with corresponding generalized eigenfunctions $e^{-\mathrm{i}\xi\cdot x}$. Thus, the projection valued measure is given on $\mathcal{D}(-\Delta)$ by

$$\mathrm{d}E = \frac{1}{(2\pi)^d}(\cdot\,, e^{-\mathrm{i}\xi\cdot x})e^{\mathrm{i}\xi\cdot x}\mathrm{d}\xi, \tag{6.3.5}$$

where $(u, v) = \int uv\mathrm{d}x$ denotes the L^2 inner product on $\mathbb{R}^d$. The scale factor $1/(2\pi)^d$ is required so that $\int \mathrm{d}E_\lambda = I$ (where I is the identity). The fractional Laplacian in $\mathbb{R}^d$ can therefore be written as[1]

$$\begin{aligned}(-\Delta)^{\alpha/2}u(x) &= \frac{1}{(2\pi)^d}\int_{\mathbb{R}^d} |\xi|^\alpha (u, e^{-\mathrm{i}\xi\cdot x})e^{\mathrm{i}\xi\cdot x}\mathrm{d}\xi \\ &= \mathcal{F}^{-1}\left\{|\xi|^\alpha \mathcal{F}\{u\}(\xi)\right\}(x),\end{aligned} \tag{6.3.6}$$

where $\mathcal{F}$ and $\mathcal{F}^{-1}$ denote the Fourier and inverse Fourier transforms, respectively. Here, we use the convention

$$\mathcal{F}\{u\}(\xi) = \frac{1}{(2\pi)^{d/2}}\int_{\mathbb{R}^d} u(x)e^{-\mathrm{i}\xi\cdot x}\mathrm{d}x,\ \mathcal{F}^{-1}\{\hat{u}\}(x) = \frac{1}{(2\pi)^{d/2}}\int_{\mathbb{R}^d} \hat{u}(\xi)e^{\mathrm{i}\xi\cdot x}\mathrm{d}\xi.$$

Thus, we see that $(-\Delta)^{\alpha/2}$, defined by (6.3.4), is a *Fourier multiplier operator* with symbol $|\xi|^\alpha$, i.e.,

$$\mathcal{F}\big\{(-\Delta)^{\alpha/2}u\big\}(\xi) = |\xi|^\alpha \mathcal{F}\{u\}(\xi), \tag{6.3.7}$$

which generalizes the well-known Fourier multiplier property of $-\Delta$. Many authors use this relation to define the fractional Laplacian as a pseudodifferential operator (Stein, 1970). The drawback of taking this as a starting point is that the Fourier transform is no longer available for bounded domains, although the functional calculus approach (6.3.4) using the spectral theorem is applicable for the case of zero boundary conditions. Of course, in that setting, the Hilbert space $\mathcal{H}$, spectrum $\sigma(-\Delta)$, and measure $\mathrm{d}E_\lambda$ must be taken accordingly, as discussed in Section 6.3.5.

[1] A alternate statement of the spectral theorem can be made which involves representations of operators as multiplication operators. From that point of view, this result is not surprising, as $\mathcal{F}$ is precisely the unitary transformation $\mathcal{H} \to L^2$ specified in that theorem that diagonalizes the Laplacian, turning it into a multiplication operator (Reed, 2012, P. 260).

Singular Integral Representation

The fractional Laplacian can be expressed directly as a singular integral in real space $\mathbb{R}^d$, rather than as a $2d$-integral in both real and frequency space, as in equation (6.3.6). This section focuses on positive powers $0 \leq \alpha \leq 2$, but the negative α case bears mentioning in connection with this goal.

For $-d < \alpha < 0$, i.e., for fractional *inverse* Laplacians, the multiplier $|\xi|^\alpha$ is decaying and has a Fourier inverse in the sense of distributions: $\mathcal{F}^{-1}\{|\xi|^\alpha\} = C(d,\alpha)|x|^{-d-\alpha}$ (see Stein (1970) or Landkof (1972)). The constant $C(d,\alpha)$ is given by

$$C(d,\alpha) = \frac{2^\alpha \Gamma\left(\frac{\alpha}{2} + \frac{d}{2}\right)}{\pi^{d/2}|\Gamma\left(-\frac{\alpha}{2}\right)|}. \qquad (6.3.8)$$

Then, (6.3.7) and the convolution property of the Fourier transform imply that the fractional inverse Laplacian is given, for $-2 < \alpha < 0$, by

$$\begin{aligned}(-\Delta)^{\alpha/2}u(x) &= C(d,\alpha)|x|^{-d-\alpha} * u(x)\\ &= C(d,\alpha)\int_{\mathbb{R}^d} \frac{u(y)}{|x-y|^{d+\alpha}}\mathrm{d}y =: I_{-\alpha}u(x).\end{aligned} \qquad (6.3.9)$$

This results in a well-defined function if $u(x)$ is, say, a smooth function with sufficient decay (Friedlander et al., 1998; Reed, 2012). This operator $I_{-\alpha}$ is known as the Riesz potential, the properties of which (such as L^p boundedness) are discussed at length in Stein (1970). The Riesz potential is an important tool in harmonic analysis and the analysis of linear PDEs (Hormander, 2003).

For $0 < \alpha < 2$ (the fractional Laplacians in which we are interested), the above derivation fails because the inverse Fourier transform of the symbol $|\xi|^\alpha$ no longer exists, even as a distribution. In addition, the representation (6.3.9) does not continue for $\alpha > 0$, since the singularity would no longer be integrable. However, starting from the negative α case (6.3.9), a nice argument that can be found in Landkof (1972, P. 45) based on analytic continuation in α of $(-\Delta)^{\alpha/2}u(x)$, for fixed u and x, yields the following real-space formula for the fractional Laplacian:

$$(-\Delta)^{\alpha/2}u = C(d,\alpha)\,\text{p.v.}\int_{\mathbb{R}^d} \frac{u(x)-u(y)}{|x-y|^{d+\alpha}}\mathrm{d}y. \qquad (6.3.10)$$

The constant $C(d,\alpha)$ is the same as in (6.3.8), and "p.v." denotes the principal value of the integral:

$$\text{p.v.}\int_{\mathbb{R}^d} \frac{u(x)-u(y)}{|x-y|^{d+\alpha}}\mathrm{d}y = \lim_{\epsilon\to 0}\int_{\mathbb{R}^d\setminus B_\epsilon(x)} \frac{u(x)-u(y)}{|x-y|^{d+\alpha}}\mathrm{d}y,$$

where $B_\epsilon(x)$ is a ball of radius ϵ centered at x. The difference $u(x) - u(y)$ in the numerator of (6.3.10), which vanishes at the singularity, provides a regularization, which together with averaging of positive and negative parts allows the principal value to exist, e.g., for smooth u with sufficient decay.

The relation between the Riesz potential and the fractional Laplacian is discussed in detail in Samko et al. (1993, Sections 5.25 and 5.26), where it is shown that

$$(-\Delta)^{\alpha/2} I_\alpha u = u. \tag{6.3.11}$$

This identity leads to the representation of the fractional Laplacian directly in terms of the Riesz potential:

$$(-\Delta)^{\alpha/2} u(x) = -\Delta\, I_{2-\alpha} u(x). \tag{6.3.12}$$

Elliptic Extension, Heat Semigroup, and Balakrishnan Formula

Next, we describe three representations of the fractional Laplacian $(-\Delta)^{\alpha/2}$ of a function $u(x)$ on $\mathbb{R}^d$ that require the solution of equations involving the standard Laplacian, albeit in the $(d+1)$-dimensional half-plane. The first is the *extension method*, or the *Dirichlet-to-Neumann map*, which requires the solution of a degenerate *elliptic* equation in the half-plane using $u(x)$ as the Dirichlet boundary data, followed by a type of normal derivative of the solution. The second is the *heat semigroup method*, which requires the solution of a *parabolic* equation – the simple heat equation – in the half-plane with $u(x)$ as the intial condition, followed by long-time integration. The third is the Balakrishnan formula, which expresses the fractional Laplacian in terms of the resolvent $(sI - \Delta)^{-1}$.

The *extension method* is based on the following result. Given a function $u(x)$ on $\mathbb{R}^d$, consider the extension $\tilde{u}(x,y)$ on $\mathbb{R}^d \times [0,\infty)$ that solves

$$\begin{aligned} \Delta_x \tilde{u} + \frac{1-\alpha}{y}\partial_y \tilde{u} + \partial_y^2 \tilde{u} &= 0, \\ \tilde{u}(x,0) &= u(x). \end{aligned} \tag{6.3.13}$$

Then

$$(-\Delta)^{\alpha/2} u(x) = c \lim_{y\to 0} \frac{u(x,y) - u(x,0)}{y^\alpha} \tag{6.3.14}$$

for a certain constant c that depends on d and α.

The above statement is taken directly from Caffarelli and Silvestre (2007), which is the most widely read source for the extension method. The extension method was reported as early as 1968 by Molchanov and Ostrovskii, in their studies of symmetric stable processes (Molchanov and Ostrovskii, 1968).

The result was also used by other authors (e.g., DeBlassie (1990)), before it was systematically addressed by Caffarelli and Silvestre.

The *heat semigroup representation* of the fractional Laplacian uses the solution of the heat equation in $\mathbb{R}^d \times [0, \infty)$:

$$(-\Delta)^{\alpha/2}u(x) = \frac{1}{\Gamma(-\alpha)} \int_0^{\infty} \left(e^{t\Delta}u(x) - u(x)\right) \frac{\mathrm{d}t}{t^{1+\alpha/2}}. \tag{6.3.15}$$

Here, $e^{t\Delta}$ is the propagator of the heat equation, i.e., $w(x,t) = e^{t\Delta}u(x)$ is the solution of the problem

$$\begin{aligned} \partial_t w - \Delta w &= 0, \quad \text{on } \mathbb{R}^d \times [0, \infty), \\ w(x, t = 0) &= u. \end{aligned} \tag{6.3.16}$$

The family $\{e^{t\Delta}\}$ is called the heat semigroup. See Stinga (2010) and Stinga and Torrea (2010) for a full discussion. An implementation of the formula (6.3.15) on $\mathbb{R}^d$ was studied in Ciaurri et al. (2018).

The *Balakrishnan formula*, introduced in Balakrishnan (1960), is a result from spectral theory and the theory of semigroups. This formula for the fractional Laplacian is

$$(-\Delta)^{\alpha/2}u(x) = \frac{\sin(\alpha\pi/2)}{\pi} \int_0^{\infty} \Delta(sI - \Delta)^{-1}u(x)s^{\alpha/2-1}\mathrm{d}s. \tag{6.3.17}$$

For a further discussion of these characterizations in $\mathbb{R}^d$, we refer to Kwaśnicki (2017) and references therein. Although fractional Laplacians in bounded domains will be introduced in the next subsection, we preface that discussion by pointing out that analogues of these methods hold in bounded domains, depending on the fractional Laplacian being considered. In the case of the spectral fractional Laplacian for functions that satisfy zero Dirichlet boundary conditions, the characterization via an elliptic extension holds with the half-plane being replaced by a cylinder over the original domain (Stinga, 2010; Stinga and Torrea, 2010). The Balakrishnan formula and the closely related Dunford-Taylor formula (for the inverse spectral fractional Laplacian) are valid and have efficient numerical implementations (Bonito and Pasciak, 2015; Bonito et al., 2019). The heat kernel formula has been found to be valid even for nonzero boundary conditions (Cusimano et al., 2018). See Sections 6.3.5 and 6.3.6 for further discussion and references to proofs and implementations of these methods.

Relation to Lévy Processes

While we had a proper discussion of the subject in Chapter 1, we find it useful to provide a timely overview on Lévy processes and their interconnection to the fractional Laplacian in here. The fractional Laplacian is connected to *anomalous* diffusion, which accounts for much of the interest in modeling with fractional equations. Just as the Laplacian is the negative generator of Brownian motion (scaled by $\sqrt{2}$), the fractional Laplacian is the infinitesimal generator of a standard isotropic α-stable Lévy motion X_t^α, which can be expressed by

$$-(-\Delta)^{\alpha/2} f(x) = \lim_{h \to 0} \frac{\mathbb{E}\left[f(x - X_h^\alpha) - f(x)\right]}{h}. \tag{6.3.18}$$

This connection is explained more fully in the last two paragraphs of this section. This process X_t^α is a Lévy process (Meerschaert and Sikorskii (2019, P. 100)) in which the increments are drawn from a spherically-symmetric α-stable distribution (Meerschaert and Sikorskii, 2019, Ex. 6.24). This process can be viewed as the long-time scaling limit of a random walk with power-law jumps (Meerschaert and Sikorskii, 2019, Thm. 6.17). For $\alpha = 2$, X_t^α reduces to scaled Brownian motion $X_t^2 = \sqrt{2} B_t$, with the 2-stable distribution being the normal distribution $\mathcal{N}(0, \sigma^2 = 2)$. For $\alpha < 2$, the α-stable distribution exhibits heavy tails and infinite variance. Moreover, the mean is finite if and only if $\alpha > 1$. The Lévy process X_t^α has superdiffusive scaling $X_{ct}^\alpha \sim c^{1/\alpha} X_t^\alpha$, and exhibits long, infinite-variance jumps when $\alpha < 2$. As an example, a superdiffusing cloud of particles would spread in space like $t^{1/\alpha}$. In many ways, α-stable Lévy flights are the simplest generalization of Brownian motion. As a result of (6.3.18), the fractional Laplacian naturally appears in macroscopic governing equations of systems of particles undergoing α-stable Lévy motion, making it a powerful and useful generalization (Meerschaert and Sikorskii, 2019). In Figure 6.5, we include examples of 2D isotropic stable Lévy motion and standard Brownian motion.

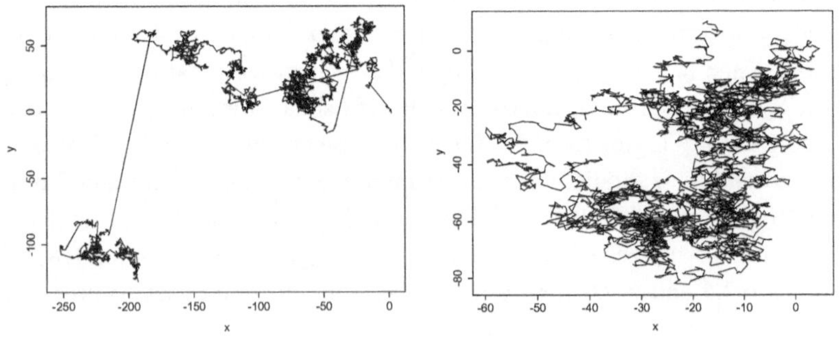

Figure 6.5 Comparison of Lévy motion and Brownian motion. Left: A 2D standard isotropic stable Lévy motion with $\alpha = 1.8$. Right: A 2D standard isotropic stable Lévy motion with $\alpha = 2.0$ (Brownian motion). Source: Reprinted from Lischke et al. (2020) with permission from Elsevier.

The heavy-tailed behavior of α-stable random variables is very much in demand for modeling, although the infinite variance property may sometimes be undesirable, for physical or numerical reasons. Recently, *tempered fractional calculus* has been developed to avoid this issue (Meerschaert et al., 2015). Tempered fractional diffusion equations model particles that undergo *tempered α-stable Lévy motion*, which is based on an exponentially tempered α-stable density (Koponen, 1995). Roughly, this density exhibits the power-law decay up to a certain argument, then decays exponentially. The generator of tempered α-stable Lévy motion, i.e., the tempered fractional Laplacian, was discussed in Zhang et al. (2018), where a Riesz basis Galerkin method was proposed to solve the Poisson problem with this operator.

This characterization as an infinitesimal generator of X_t^α also lends itself to the stochastic (Monte Carlo) solution of boundary-value problems involving the various fractional Laplacians in bounded domains. These formulas are typically referred to as Feynman–Kac (Dirichlet boundary conditions, (Kac, 1949)) or Brosamler (Neumann boundary conditions, after Brosamler (1967); Lions and Sznitman (1984); Hsu (1984)) formulas. One such solution method is discussed in Section 6.3.3, and its implementation is described in Section 6.4.2.

We now give a brief outline of the probabilistic theory behind the relation of the fractional Laplacian to Lévy processes. This will also result in the *generator form* of the fractional Laplacian, which is related to the directional representation in the next Section 6.3.1. Following Meerschaert and Sikorskii (2019), we review the classical Lévy–Khintchine formula for generators of Lévy processes and sketch how it results in the fractional Laplacian for the special case of isotropic α-stable processes. Given a Lévy process $\{Z_t : t > 0\}$, define the family of linear operators

$$T_t f(x) = \mathbb{E}[f(x - Z_t)], \quad t \geq 0, \tag{6.3.19}$$

for suitable functions f. Suppose that the characteristic function of the random variable Z_1 is $\mathbb{E}[e^{ik \cdot Z_1}] = e^{\psi(k)}$, where

$$\psi(k) = ik \cdot a - \frac{1}{2} k \cdot Bk + \int \left(e^{ik \cdot y} - 1 - \frac{ik \cdot y}{1 + |y|^2} \right) \phi(dy). \tag{6.3.20}$$

Then $T_t f(x)$ defines a C_0-semigroup on $C_0(\mathbb{R}^d)$ with generator (Meerschaert and Sikorskii, 2019, Thm. 6.26):

$$\begin{aligned} Lf(x) = &-a \cdot \nabla f(x) + \frac{1}{2} \nabla \cdot Q \nabla f(x) \\ &+ \int \left(f(x - y) - f(x) + \frac{y \cdot \nabla f(x)}{1 + |y|^2} \right) \phi(dy), \end{aligned} \tag{6.3.21}$$

where $\phi(\mathrm{d}y)$ is some Lévy measure (Meerschaert and Sikorskii, 2019, (6.20)). The domain of L contains $\{f : f, f', f'' \in C_0(\mathbb{R}^d)\}$ (Meerschaert and Sikorskii, 2019, Thm. 6.26). This is the Lévy–Khintchine formula. The first order differential term in this operator can be understood as the "drift" term, the second order differential term can be understood as the standard diffusion, and the Lévy measure describes the jumps of the Lévy process (Meerschaert and Sikorskii, 2019, P. 159).

As we will now see, if we choose Z_t to be the standard isotropic α-stable Lévy motion, then (6.3.21) reduces to the negative of the fractional Laplacian. The details of the following discussion can be found in Meerschaert and Sikorskii (2019, Examples 6.28 and 6.29). If $0 < \alpha < 1$ and Z_t is a Lévy process such that Z_1 has the characteristic function

$$\mathbb{E}[e^{ik\cdot Z_1}] = \exp\left[-C\Gamma(1-\alpha)\int_{|\boldsymbol{\theta}|=1}(-ik\cdot\boldsymbol{\theta})^{\alpha}M(\mathrm{d}\boldsymbol{\theta})\right], \tag{6.3.22}$$

then the generator of the corresponding stable semigroup can be written in the form

$$Lf(x) = \int (f(x-y) - f(x))\phi(\mathrm{d}y), \tag{6.3.23}$$

where

$$\phi(\mathrm{d}y) = \alpha C r^{-\alpha-1}\mathrm{d}r M(\mathrm{d}\boldsymbol{\theta}) \text{ with } y = r\boldsymbol{\theta},\ r > 0, \text{ and } |\boldsymbol{\theta}| = 1. \tag{6.3.24}$$

If $M(\mathrm{d}\boldsymbol{\theta})$ is a uniform measure on the unit sphere, and if

$$C^{-1} = B\Gamma(1-\alpha), \text{ and } B = \cos(\pi\alpha/2)\int_{|\boldsymbol{\theta}|=1}|\theta_1|^{\alpha}M(\mathrm{d}\boldsymbol{\theta}), \tag{6.3.25}$$

then the right-hand side of (6.3.22) reduces to $\exp[-|k|^{\alpha}]$, so that $Z_t = X_t^{\alpha}$, the standard isotropic α-stable Lévy motion[2], and $L = -(-\Delta)^{\alpha/2}$, the negative of the fractional Laplacian (Meerschaert and Sikorskii, 2019, Example 6.24). Indeed, (6.3.23) reduces to (6.3.10) after a simple change of variables (see Defterli et al. (2015, Sec. 4)). Now if $1 < \alpha < 2$, and Z_1 has the characteristic function

$$\mathbb{E}[e^{ik\cdot Z_1}] = \exp\left[C\frac{\Gamma(2-\alpha)}{\alpha-1}\int_{|\boldsymbol{\theta}|=1}(-ik\cdot\boldsymbol{\theta})^{\alpha}M(\mathrm{d}\boldsymbol{\theta})\right], \tag{6.3.26}$$

[2] The general isotropic α-stable random variable has characteristic function $\exp(-c^{\alpha}|k|^{\alpha})$, where c is the scale parameter. The standard α-stable random variable has scale parameter $c = 1$. Note that the standard isotropic 2-stable random variable corresponds to a normal random variable with characteristic function $\exp(-\sigma^2|k|^2/2)$ and variance $\sigma^2 = 2$.

then the generator of the corresponding stable semigroup can be written in the form

$$Lf(x) = \int (f(x-y) - f(x) + y \cdot \nabla f(x))\phi(dy), \tag{6.3.27}$$

where ϕ is defined as in (6.3.24), and the constants B and C are defined as in (6.3.25). In this case, if $M(d\theta)$ is a uniform measure on the unit sphere, then $L = (-\Delta)^{\alpha/2}$, which has a different sign than in the case $0 < \alpha < 1$. The sign change is due to the fact that $B > 0$ for $0 < \alpha < 1$, and $B < 0$ for $1 < \alpha < 2$. Equations (6.3.23) and (6.3.27) yield alternate forms of the fractional Laplacian (6.3.10), known as the generator forms, without using the principal value. The generator form (6.3.21) of the fractional Laplacian is useful for probabilists working in this area.

Directional Representation

Another integral characterization of the fractional Laplacian, which we refer to as the directional representation, can be found in Samko et al. (1993, (26.24)). The operator is written as

$$(-\Delta)^{\alpha/2} u(x) = C_{\alpha,d} \int_{|\theta|=1} D_\theta^\alpha u(x) d\theta, \quad x, \theta \in \mathbb{R}^d, \ \alpha \in (0,1) \cup (1,2], \tag{6.3.28}$$

where the scaling constant before the integral is (Samko et al., 1993)

$$C_{\alpha,d} = \frac{\Gamma(\frac{1-\alpha}{2})\Gamma(\frac{d+\alpha}{2})}{2\pi^{\frac{1+d}{2}}}. \tag{6.3.29}$$

The Fourier transform of (6.3.28) corresponds to multiplication of the Fourier transform of u by (6.3.22) when $\alpha < 1$ and by (6.3.26) when $\alpha > 1$. Here, $D_\theta^\alpha(\cdot)$ is the Riemann–Liouville fractional directional derivative (Meerschaert et al., 2006) given by

$$\begin{aligned} D_\theta^\alpha(\cdot) &= (\theta \cdot \nabla)\, I_\theta^{1-\alpha}(\cdot), \quad \text{for } 0 < \alpha < 1; \\ D_\theta^\alpha(\cdot) &= (\theta \cdot \nabla)^2 I_\theta^{2-\alpha}(\cdot), \quad \text{for } 1 < \alpha < 2, \end{aligned} \tag{6.3.30}$$

where $\theta \cdot \nabla = \sum_{i=1}^d \theta_i \frac{\partial}{\partial x_i}$ is the directional derivative, thus

$$(\theta \cdot \nabla)^2 = \sum_{j=1}^d \sum_{i=1}^d \theta_i \theta_j \frac{\partial^2}{\partial x_j \partial x_i},$$

and the fractional directional integral $I_\theta^\beta(\cdot)$ is defined by (for $\beta \in (0,1)$)

$$I_\theta^\beta u(x) = \frac{1}{\Gamma(1-\beta)} \int_0^{+\infty} \varsigma^{-\beta} u(x - \varsigma\theta) d\varsigma. \tag{6.3.31}$$

Note that (6.3.28) excludes the case $\alpha = 1$. When $\alpha = 1$, the directional representation is more complicated; for the one-dimensional case, see Kelly et al. (2018).

We now explicitly write out the directional representation (6.3.28) for one and two dimensions; these formulas will be used later in this section. In one dimension, using the identity $\Gamma(x)\Gamma(1-x) = \pi/\sin(\pi x)$, we have

$$\begin{aligned}\Gamma\Big(\frac{1-\alpha}{2}\Big)\Gamma\Big(\frac{1+\alpha}{2}\Big) &= \Gamma\Big(\frac{1-\alpha}{2}\Big)\Gamma\Big(1-\frac{1-\alpha}{2}\Big)\\ &= \frac{\pi}{\sin\left(\pi\frac{1-\alpha}{2}\right)} = \frac{\pi}{\cos\left(\frac{\pi\alpha}{2}\right)},\end{aligned}$$

and so

$$\begin{aligned}(-\Delta)^{\alpha/2}u(x) &= \frac{1}{2\cos\left(\frac{\pi\alpha}{2}\right)}\left[D^{\alpha}_{\theta=-1}u(x) + D^{\alpha}_{\theta=1}u(x)\right]\\ &= \frac{1}{2\cos\left(\frac{\pi\alpha}{2}\right)}\left[{}^{RL}_{-\infty}D^{\alpha}_{x}u(x) + {}^{RL}_{x}D^{\alpha}_{\infty}u(x)\right],\end{aligned} \tag{6.3.32}$$

where ${}^{RL}_{-\infty}D^{\alpha}_{x}$ and ${}^{RL}_{x}D^{\alpha}_{\infty}$ are the well-known one-dimensional left- and right-sided Riemann–Liouville fractional derivatives, respectively (Yang et al., 2010). Note that, using the Fourier transforms of these operators,

$$\mathcal{F}[(-\Delta)^{\alpha/2}u] = \frac{1}{\cos(\alpha\pi/2)}\left[\frac{1}{2}(-\mathrm{i}\xi)^{\alpha} + \frac{1}{2}(\mathrm{i}\xi)^{\alpha}\right]\hat{u} = |\xi|^{\alpha}\hat{u}, \quad \text{for} \quad \alpha \neq 1.$$

In two dimensions and for $\alpha \neq 1$, the fractional Laplacian can be written in polar form as

$$(-\Delta)^{\alpha/2}u(x) = C_{\alpha,2}\int_0^{2\pi} D^{\alpha}_{\theta}u(x)\mathrm{d}\theta, \tag{6.3.33}$$

where D^{α}_{θ} is a derivative along the direction $[\cos(\theta), \sin(\theta)]$.

Meerschaert et al. (1999) extended the above operator (6.3.28) to an anisotropic version, which they called the general (asymmetric) fractional derivative operator ∇^{α}_{M},

$$\nabla^{\alpha}_{M}u(x) = \int_{|\theta|=1} D^{\alpha}_{\theta}u(x)M(\mathrm{d}\theta), \quad x, \theta \in \mathbb{R}^d, \alpha \in (0,1) \cup (1,2], \tag{6.3.34}$$

where $M(\mathrm{d}\theta)$ is an arbitrary probability measure on the unit sphere $\{|\theta| = 1\}$. The motivation for the definition extension is to generate "the full range of Lévy-stable motions." For more details, see equation (6.3.22) and the corresponding discussion. The study of this more general operator is beyond the scope of the current section. However, it is an interesting and pertinent topic for future extensions of many of the results and methods discussed in this work. It should be noted that the Riesz fractional Laplacian is recovered if the measure $M(d\theta)$ is uniform, namely, $M(d\theta) = d\theta/S_d$ where $S_d = 2\pi^{d/2}/\Gamma(d/2)$ is the surface area of the unit sphere in $\mathbb{R}^d$. In this case, the two types of operators are related by

$$(-\Delta)^{\alpha/2}u(x) = \frac{\Gamma\left(\frac{d+\alpha}{2}\right)\Gamma\left(\frac{1-\alpha}{2}\right)}{\Gamma\left(\frac{d}{2}\right)\sqrt{\pi}}\nabla_M^{\alpha}u(x). \tag{6.3.35}$$

6.3.2 Fractional Laplacians on Bounded Domains

We have discussed several characterizations and formulas for the fractional Lapacian in $\mathbb{R}^d$, which are all equivalent in that setting. If we apply these different formulas on a bounded domain $\Omega \subset \mathbb{R}^d$, certain equivalences break down, which leads to the definition of several distinct fractional Laplacians in a bounded domain. Some researchers have defined other fractional Laplacians on a bounded domain directly, instead of restricting the definition for $\mathbb{R}^d$ (Chen and Pang, 2016). Here, we survey the *Riesz* fractional Laplacian (sometimes called the *integral* fractional Laplacian), the *directional* fractional Laplacian, and the *spectral* fractional Laplacian. We focus on the following topics.

- Appropriate boundary conditions for the Poisson problem

$$(-\Delta)^{\alpha/2}u = f \text{ in } \Omega. \tag{6.3.36}$$

  The spectral fractional Laplacian admits local boundary conditions on $\partial\Omega$; the regional fractional Laplacian has also been considered with local boundary conditions. On the other hand, the Riesz (and directional) fractional Laplacians require *exterior* boundary conditions on $\mathbb{R}^d \setminus \Omega$.
- Connection of each fractional Laplacian with given boundary conditions to a stochastic process. The standard negative Laplacian $-\Delta$ in a bounded domain is the negative generator of *stopped* (Kolokoltsov, 2011) Brownian motion if taken with Dirichlet BCs (Privault, 2008; Freidlin, 1985), and *reflected* Brownian motion if taken with Neumann BCs (Brosamler, 1967; Hsu, 1984; Bencherif-Madani and Pardoux, 2009). Similarly, the fractional Laplacians are negative generators of certain stopped[3] or reflected Lévy

[3] Many results in the probability literature (e.g., Song and Vondracek (2003)) are stated and proved for *killed* processes rather than *stopped* processes. Without going into details here, we note that the resulting semigroups are equivalent when applied to functions f that vanish outside the domain, in the sense that $\mathbb{E}_x\left[f(X_t)\right]$ is the same whether X_t is the killed or the stopped process. Here, we are assuming that $f(\partial) = 0$ where ∂ is the cemetery point specified for the killed process.

motions. This is useful for physical interpretation as well as stochastic solution methods.

- Well-posedness and regularity properties of the fractional Poisson problem (6.3.36) with appropriate boundary conditions.

We do not consider the Laplacian with periodic boundary conditions (i.e., on the torus $\mathbb{T}^d$) in this section. The construction of such an operator is unambiguous, and follows the construction of the fractional Laplacian in $\mathbb{R}^d$ in Section 6.3.1 except that the spectrum is discrete, so the operator is given by a Fourier series. For a basic discussion, see, e.g., Hermann (2014), and for discussions of extension characterizations as well as well-posedness and regularity, see Roncal and Stinga (2016, 2014).

6.3.3 Riesz Fractional Laplacian

Definition

One approach to defining the fractional Laplacian on a bounded domain Ω is to apply the real space formula (6.3.10) to functions on Ω. This leads to the *Riesz* fractional Laplacian in Ω. Let us first consider Dirichlet boundary conditions. Since formula (6.3.10) requires values of u on all of $\mathbb{R}^d$, an *exterior* boundary condition

$$u = g \quad \text{in} \quad \mathbb{R}^d \setminus \Omega \tag{6.3.37}$$

is required, even to define the Riesz Laplacian within Ω. For functions u that satisfy (6.3.37), the *Riesz* fractional Laplacian is defined for $x \in \Omega$ by

$$\begin{aligned}(-\Delta)^{\alpha/2}u(x) &= C(d,\alpha)\,\text{p.v.}\int_{\mathbb{R}^d}\frac{u(x)-u(y)}{|x-y|^{d+\alpha}}\mathrm{d}y \\ &= C(d,\alpha)\left[\text{p.v.}\int_{\Omega}\frac{u(x)-u(y)}{|x-y|^{d+\alpha}}\mathrm{d}y + \int_{\mathbb{R}^d\setminus\Omega}\frac{u(x)-g(y)}{|x-y|^{d+\alpha}}\mathrm{d}y\right].\end{aligned} \tag{6.3.38}$$

The Riesz fractional Laplacian depends directly on Ω and the exterior boundary values g. It also referred to in the literature as the *integral* fractional Laplacian (Bonito et al., 2018) or as the *restricted* (Dirichlet) fractional Laplacian (Vázquez, 2014; Grubb, 2016); for consistency, we do not use these terms in this section.

Boundary Conditions: Dirichlet vs. Neumann

The Neumann condition for the Riesz Laplacian is, at the time of this writing, an area of active development. A type of exterior fractional normal derivative is needed to specify u on $\mathbb{R}^d \setminus \Omega$, but there is no widely studied or accepted definition. Recently, a fractional Neumann operator was proposed in Du et al. (2012), and the properties were studied in detail in Dipierro et al. (2017). Another approach to defining reflecting boundary conditions based on mass conservation for diffusion in one dimension was explored in Baeumer et al. (2018a); Kelly et al. (2019).

The Poisson problem with Dirichlet boundary conditions

$$\begin{aligned}(-\Delta)^{\alpha/2}u &= f \quad \text{in } \Omega\,,\\ u &= g \quad \text{in } \mathbb{R}^d \setminus \Omega,\end{aligned} \tag{6.3.39}$$

has been extensively studied in the $g \equiv 0$ case (Acosta and Borthagaray, 2017; Ros-Oton and Serra, 2014; Ainsworth and Glusa, 2018b, 2017; Grubb, 2015), but literature based on the nonzero case is more recent and limited (Grubb, 2015; Kyprianou et al., 2018; Borthagaray et al., 2018; Antil et al., 2019). For the case of zero exterior Dirichlet condition $g \equiv 0$ in (6.3.39), finite element algorithms have been developed in Acosta and Borthagaray (2017), Ainsworth and Glusa (2018b), and in particular the adaptive finite element scheme of Ainsworth and Glusa (2017) has been used for the computations of this paper. For the case of nonzero exterior Dirichlet condition, in Kyprianou et al. (2018) a Monte Carlo method based on the Feynman–Kac formula for the problem (6.3.39) was developed and studied. This method is described in more detail in Section 6.4.2. The radial basis function collocation method introduced in Section 6.6.1 (based on discretizing the directional representation of the Riesz fractional Laplacian) readily admits nonzero exterior Dirichlet condition as well. In Acosta et al. (2018), a finite element method for nonzero exterior Dirichlet conditions was introduced in which the exterior condition was enforced through a Lagrange multiplier formulation. Another approach for the case of nonzero exterior condition, with application to exterior control problems, is that of Antil et al. (2019), which implemented nonzero exterior Robin conditions and used this to approximate the solution to the nonzero exterior Dirichlet problem.

Stopped/Killed Lévy Motion

The Riesz fractional Laplacian with Dirichlet boundary conditions is the generator of the stopped α-stable Lévy motion. The stopped Lévy motion is defined as follows (Kolokoltsov, 2011; Privault, 2008; Freidlin, 1985). Given a stable Lévy motion X_t^α and a bounded domain Ω, we define the stopping time (or exit time)

$$\sigma_\Omega = \inf\{t : X_t^\alpha \notin \Omega\}. \tag{6.3.40}$$

Then, the stopped Lévy process is defined as $X^{\alpha}_{t\times\sigma_\Omega}$. Unlike stopped Brownian motion, which has almost surely continuous paths that are stopped at the boundary, the paths of α-stable Lévy motion are discontinuous and almost surely exit the domain by a jump. Thus, the paths of stopped α-stable Lévy motion pass into the exterior $\mathbb{R}^d \setminus \Omega$ where they are stopped immediately.

In Kyprianou et al. (2018), this stochastic connection was exploited to prove a Feynman–Kac for the Poisson problem.

$$\begin{aligned} (-\Delta)^{\alpha/2}u(x) &= f(x), \quad x \in \Omega \subset \mathbb{R}^d, \\ u(x) &= g(x), \quad x \in \mathbb{R}^d \setminus \Omega. \end{aligned} \tag{6.3.41}$$

Under mild conditions on f and g, the following Feynman–Kac formula was proven:

$$u(x) = \mathbb{E}_x\left[g(X^{\alpha}_{\sigma_\Omega})\right] + \mathbb{E}_x\left[\int_0^{\sigma_\Omega} f(X^{\alpha}_s)\mathrm{d}s\right], \quad x \in \Omega, \tag{6.3.42}$$

where X^{α}_t is the symmetric α-stable process, $\mathbb{E}_x(\cdot)$ denotes the expectation with respect to all the sample paths with the initial location x, and $\sigma_\Omega = \inf\{t > 0 : X_t \notin \Omega\}$ is the first exit time of the sample path.

In addition to a proof, in Kyprianou et al. (2018) a walk-on-spheres (WOS) method for much faster computation of (6.3.42) was introduced and analyzed. This speeds up the implementation of (6.3.42) by replacing the simulation of exact stopped paths in Ω by a series of maximal inscribed spheres. Conceptually, the WOS method is based on the isotropy of the problem (6.3.41) as well as the analytic formula for mean-exit time of X^{α}_t on a sphere (Getoor, 1961). In the Brownian motion case, the center of each sphere is chosen by sampling a uniform distribution on the boundary of the previous sphere, and the procedure terminates when the process comes within some tolerance ε of the boundary of Ω. In the α-stable case, the center of each sphere is chosen by sampling a distribution (which is described in detail in Section 6.4.2) on the exterior of the previous sphere, and the procedure terminates when the process jumps outside the domain Ω. This is because, in the α-stable case, the sample path may exit the sphere by a jump, instead of by passing through the boundary; convergence in finite steps was proven in Kyprianou et al. (2018).

Well-Posedness and Regularity

In this section, we consider the problem

$$\begin{aligned} (-\Delta)^{\alpha/2}u &= f \quad \text{in } \Omega, \\ u &= 0 \quad \text{in } \mathbb{R}^d \setminus \Omega. \end{aligned} \tag{6.3.43}$$

Using the Lax–Milgram lemma, for a bounded Lipschitz domain Ω, it is easy to show that this problem is well-posed if $f(x) \in \mathbb{H}^{-\alpha/2}(\Omega)$ and the resulting solution $u(x) \in \mathbb{H}^{\alpha/2}(\Omega)$. See, for example, Ainsworth and Glusa (2018b); Bonito et al. (2018); Acosta and Borthagaray (2017). For the definitions of the Sobolev space $\mathbb{H}^s$, which will also feature in the following regularity results, see Section 6.2.

As for regularity, our intention is not to give a complete survey of all the known regularity results for the fractional Laplacian. Instead, we mention two results that give an indication of the regularity for the Riesz fractional Laplacian, for the case of the zero boundary condition. These properties should be compared with the regularity of the spectral fractional Laplacian (see Section 6.3.6). The regularity up to the boundary is a key difference between the two definitions.

The Hölder regularity for the solution of (6.3.43) was studied in Ros-Oton and Serra (2014). The authors were motivated by the exact solution for $f \equiv 1$ in the ball $B_r(x_0)$ centered at x_0 of radius r:

$$\begin{aligned} (-\Delta)^{\alpha/2} u &= 1 \quad \text{in } B_r(x_0), \\ u &= 0 \quad \text{in } \mathbb{R}^d \setminus B_r(x_0), \end{aligned} \tag{6.3.44}$$

which has solution

$$u(x) = \frac{2^{-\alpha}\Gamma(n/2)}{\Gamma\left(\frac{n+\alpha}{2}\right)\Gamma(1+\alpha/2)} \left(r^2 - |x - x_0|^2\right)^{\alpha/2} \quad \text{in } B_r(x_0).$$

This solution exhibits strict $C^{\alpha/2}$ regularity up to the boundary. The authors proved the following result. Let Ω be a bounded $C^{1,1}$ domain and $\delta(x) = d(x, \partial\Omega)$ denote the distance to the boundary, Then $g \in L^\infty(\Omega)$ implies that $u/\delta^{\alpha/2}$ can be extended to a $C^r(\overline{\Omega})$ function for some $r < \min(\alpha/2, 1 - \alpha/2)$, and

$$\|u/\delta^{\alpha/2}\|_{C^r(\overline{\Omega})} \leq C\|g\|_{L^\infty(\Omega)}.$$

This result is sharp for the problem (6.3.44). Boundary-value problems in which weighted boundary conditions (on the trace of $u/\delta^{\alpha/2}$ rather than u) are specified on $\partial\Omega$ are studied in Grubb (2015).

The following theorem relates to the Sobolev regularity of (6.3.43). This form is presented nicely in Ainsworth and Glusa (2017) and is based on the results of Grubb (2015) and Acosta et al. (2017). We mention that although the formulation in Ainsworth and Glusa (2017) considers $s \geq -\alpha/2$, Grubb (2015) establishes the first case below for $-1/2 < \alpha/2 + s < 1/2$. Regularity for the inhomogeneous problem (6.3.39) is also studied in Grubb (2015), although we restrict ourselves to the case below.

Theorem 6.9 (See Ainsworth and Glusa (2017); Grubb (2015); Acosta et al. (2017)) *Let* $\partial\Omega \in C^{\infty}$, $f \in H^s(\Omega)$ *for* $s \geq -\alpha/2$ *and* $u \in \mathbb{H}^{\alpha/2}(\Omega)$ *be the solution of the fractional Poisson problem* (6.3.43). *Then*

$$u \in \begin{cases} H^{\alpha+s}(\Omega) & \text{if } 0 < \alpha/2 + s < 1/2, \\ H^{\alpha/2+1/2-\varepsilon}(\Omega), \ \forall\varepsilon > 0, & \text{if } 1/2 \leq \alpha/2 + s. \end{cases} \tag{6.3.45}$$

An interesting feature of the Sobolev regularity for the Riesz fractional Laplacian is that if $\alpha/2 + s \geq 1/2$, the global regularity of u need not improve if s, the regularity of f, is increased. Rather, s merely improves the interior regularity of u. This was proven in Grubb (2015), where it was shown that for a smooth boundary $\partial\Omega$, $f \in H^s(\Omega)$ implies that $u \in H^{\alpha+s}_{\text{loc}}(\Omega)$. Therefore, the global regularity (i.e., regularity up to the boundary) for the Riesz fractional Laplacian is in contrast to the global regularity for the spectral fractional Laplacian (see Section 6.3.6). In the case when $f \in L^2(\Omega)$ for the spectral Laplacian, $u \in \mathbb{H}^{\alpha}(\Omega)$ for all $\alpha \in (0, 2)$, while this is only true if $\alpha < 1$ for the Riesz fractional Laplacian. In the case when $f \in H^s(\Omega)$ and $s > 1/2$, then $u \in \mathbb{H}^{s+\alpha}(\Omega)$ for the spectral fractional Laplacian provided $f \equiv 0$ on the boundary $\partial\Omega$, while for the Riesz definition, $u \in H^{\alpha/2+1/2-\varepsilon}(\Omega)$, independent of s. Even if $f \not\equiv 0$ on $\partial\Omega$, the regularity of u for the spectral definition improves as $\alpha + 1/2 - \varepsilon$, while in the Riesz case, it improves as $\alpha/2 + 1/2 - \varepsilon$.

6.3.4 Directional Fractional Laplacian

We can also apply the directional fractional Laplacian $(-\Delta)^{\alpha/2}_M$ to functions on Ω with an exterior boundary condition, which is applied in the same way as for the Riesz definition. Although the directional characterization was motivated in Meerschaert et al. (1999) by the desire to capture anisotropic anomalous diffusion, in this work, we always choose $M(\mathrm{d}\theta) = \mathrm{d}\theta/S_d$ where $S_d = 2\pi^{d/2}/\Gamma(d/2)$ is the surface area of the unit sphere in $\mathbb{R}^d$, so that our computational results for this definition can be compared to those of the Riesz definition. With this choice of measure, the associated Lévy process is symmetric stable motion that is stopped upon exiting the domain Ω, which is the same process as the one associated with the Riesz fractional Laplacian. A nice proof of this equivalence can be found in Chen et al. (2012), Lemma 4.1. Hence the discussions of Section 6.3.3 also apply to the directional fractional Laplacian with our choice of integration measure.

6.3.5 Spectral Fractional Laplacian

In this section, we have split the discussion of the spectral fractional Laplacian into two: the case of zero boundary condition and the case of nonzero boundary conditions. This is due to the vast body of work that considers only the homogeneous spectral fractional Laplacian (i.e., the zero boundary condition case), as only recently was the inhomogeneous version considered. In deference to the fact that many results in the literature apply only to the homogeneous case, we have taken this approach to avoid misleading the reader. The results reported in Sections 6.3.5 and 6.3.6 on the homogeneous spectral fractional Laplacian should not be taken to apply in the case of nonzero boundary conditions without modification.

The inhomogeneous spectral fractional Laplacian has been considered by Antil et al. (2017) (see Section 6.3.6) and Cusimano et al. (2018) (see Section 6.3.6). In Section 6.3.6, we point out that the definitions considered in Antil et al. (2017) and Cusimano et al. (2018) are the same, and are essentially a harmonic lifting to the homogeneous spectral fractional Laplacian. Moreover, we show that this amounts to another natural definition in terms of the inverse spectral fractional Laplacian.

Zero Boundary Condition

The spectral approach to defining the fractional Laplacian in Ω is to start with the standard negative Laplacian $-\Delta$ on that domain and take the spectral power defined by (6.3.4). Let us first consider the case of zero boundary conditions. We will take $\mathcal{D}(-\Delta)$ to be the subspace of $H^2(\Omega)$ consisting of smooth functions with zero Dirichlet or zero Neumann boundary values[4], depending on the choice of boundary condition.

The spectrum $\sigma(-\Delta)$, which is now discrete, depends on the domain Ω and on whether Dirichlet or Neumann conditions are used. The spectrum of the Laplacian in the Dirichlet case is defined by the problem,

$$\begin{aligned} -\Delta e_k &= \lambda_k e_k \quad \text{in } \Omega, \\ e_k &= 0 \quad \text{on } \partial\Omega, \end{aligned} \tag{6.3.46}$$

and in the Neumann case by

$$\begin{aligned} -\Delta e_k &= \lambda_k e_k \quad \text{in } \Omega, \\ \frac{\partial e_k}{\partial n} &= 0 \quad \text{on } \partial\Omega. \end{aligned} \tag{6.3.47}$$

The λ_k are the eigenvalues, and e_k the eigenfunctions, of $-\Delta$ with zero Dirichlet or Neumann boundary condition, respectively. The spectral decomposition (6.3.2) then reads

[4] We remind the reader that, for using the spectral theorem as in Section 6.3.1, $\mathcal{D}(-\Delta)$ need not be all of $\mathcal{H}$, or even a "maximal" domain of definition of $-\Delta$ within $\mathcal{H}$, but rather a more convenient, dense subspace of $\mathcal{H}$ (a *core* for $-\Delta$) .

$$-\Delta u(x) = \sum_{k=1}^{\infty} \lambda_k (u, e_k)_\Omega e_k(x). \tag{6.3.48}$$

We remark that this identity is valid on $\mathcal{D}(\Omega)$, and can be extended by continuity to $H_0^2(\Omega)$ in the Dirichlet case, or

$$H^2_{\partial u/\partial n=0}(\Omega) = \left\{ u \in H^2(\Omega) \ : \ \frac{\partial u}{\partial n}\bigg|_{\partial\Omega} = 0 \right\}$$

in the Neumann case. It is not in general true for functions u with nonzero boundary values (Dirichlet or Neumann). For example, in one dimension with $-\Delta = -\frac{\partial^2}{\partial x^2}$, and (λ_k, e_k) from (6.3.46) on $[0, 2\pi]$ and $u = \cos(x)$, the equation (6.3.48) does not hold. In fact, the series on the right-hand side diverges.

Thus, on $\mathcal{D}(-\Delta)$, in accordance with (6.3.4), the *spectral fractional Laplacian* is defined by

$$(-\Delta)^{\alpha/2} u(x) := \sum_{k=1}^{\infty} \lambda_k^{\alpha/2} (u, e_k)_\Omega e_k(x). \tag{6.3.49}$$

As usual, by continuity, this can be extended to an operator on $\mathbb{H}^\alpha(\Omega)$ (see Section 6.2). The spectral fractional Laplacian is nonlocal on the interior of Ω for non-integer $\alpha \in (0, 2)$. We see that to compute the inner product $(u, e_k)_{L^2_\Omega}$, it suffices for u to be defined on the interior of Ω. Unlike the Riesz fractional Laplacian, the spectral fractional Laplacian requires no information about u on the exterior $\mathbb{R}^d \setminus \Omega$. Thus, from a conceptual viewpoint, in boundary-value problems the spectral fractional Laplacian admits the same type of boundary conditions as the standard, local Laplacian $-\Delta$. Precise conditions and references for rigorous proofs for well-posedness of boundary-value problems for the spectral fractional Laplacian with such local boundary conditions (Dirichlet and Neumann) are discussed in Section 6.3.6.

Other Representations

In this section, we merely point out the analogues of the characterizations in Section 6.3.1 which have been proven (and implemented) for the spectral fractional Laplacian. Unless otherwise stated, these are not applicable without modification in the case of nonzero boundary conditions. We do not use these characterizations in our numerical comparisons, but each has its advantages and disadvantages.

The analogue of the extension method (6.3.13) for the homogeneous spectral fractional Laplacian on a bounded domain was derived in Stinga and Torrea (2010). In this characterization, a degenerate elliptic equation is solved in the extruded domain ("cylinder") over Ω, followed by a similar Neumann-type trace. Further studies include Galé et al. (2013); Caffarelli and Stinga (2016); Stinga (2010). As this higher-dimensional formulation involves only integer-order operators, standard discretization approaches may be applied, as in Nochetto et al. (2015); Gatto and Hesthaven (2015); Ainsworth and Glusa (2018a).

The heat semigroup formula on a bounded domain Ω was studied and implemented in Cusimano et al. (2018), where the heat semigroup was used to define the spectral fractional Laplacian. This approach is robust for nonzero boundary conditions, and is discussed at length in Section 6.3.6.

The Balakrishnan formula (6.3.17) was implemented in, e.g., Bonito and Pasciak (2015). More recently, in Bonito et al. (2019), a rapidly convergent sinc quadrature was developed for this formula, which is used together with a finite element approximation to the resolvent $(sI - \Delta)^{-1}$.

Yet another approach in Vabishchevich (2015) considered a reformulation of the spectral fractional Poisson equation with zero Robin boundary conditions as an integer-order (time-dependent) pseudo-parabolic equation that could be solved with standard FEMs in space and finite difference methods in time. The solution of the pseudo-parabolic equation at time $t = 1$ turns out to be the inverse spectral fractional Laplacian applied to the source function, which yields the solution to the fractional Poisson problem.

Subordinate Stopped/Reflected Brownian Motion

It is well known that the Laplacian in a bounded domain is the generator of stopped Brownian motion for Dirichlet boundary conditions, and reflected Brownian motion for Neumann boundary conditions (Yosida, 1980). The construction of stopped Brownian motion uses the same procedure as in (6.3.40), while the construction of reflected Brownian motion involves the solution of the Skorokhod problem (Hsu, 1984). The corresponding processes for the spectral fractional Laplacian can be obtained by means of subordination. Subordination of a process X_t results in a process $X_{T(t)}$, in which time t is replaced by "operational time" $T(t)$, itself a stochastic process – more specifically, an increasing Lévy process. In $\mathbb{R}^d$, an isotropic α-stable Lévy motion can be constructed by subordinating the isotropic Brownian motion with the stable subordinator (see Cont and Tankov (2004, Section 4.4)). The same time change provides a way of converting Brownian paths that are stopped at the boundary into α-stable Lévy paths that are stopped at the boundary.

The spectral fractional Laplacian with Dirichlet boundary conditions is the generator of subordinate stopped Brownian motion (Song and Vondracek, 2003), i.e., stopped Brownian motion that is then subordinated by the standard stable subordinator. The spectral fractional Laplacian with Neumann boundary conditions is the generator of subordinate reflected Brownian motion. These are both results from the semigroup theory of Markov processes (see Yosida (1980), Chapter IX; a detailed discussion may be found in Gulian (2019); Gulian and Pang (2018)). Thus, one imposes the boundary condition on the process first, before subordinating and turning the stopped/reflected Brownian motion into a Lévy motion. It is important to note the order of the modifications; the reverse order corresponds to the *Riesz* fractional Laplacian in Ω, as discussed in Section 6.3.3.

6.3.6 Inhomogeneous Spectral Fractional Laplacian

The construction of the spectral power (6.3.49) cannot simply be repeated for the case of nonzero boundary conditions. This is because subsets of $C^\infty(\Omega)$ that satisfy a fixed nonzero boundary condition are no longer linear spaces, which prohibits the use of the spectral theorem. However, in the case of the standard Laplacian $-\Delta$ on a bounded domain, using a lifting technique, a spectral series can be derived (see (6.3.53)). The generalization of this approach can be used to derive a suitable series representation for the inhomogeneous spectral fractional Laplacian, namely

$$(-\Delta_{\Omega,g})^{\alpha/2}u = \sum_{i=1}^{\infty}\left(\lambda_i^{\alpha/2}(u,e_i)_\Omega - \lambda_i^{\alpha/2-1}\left(u,\frac{\partial e_i}{\partial n}\right)_{\partial\Omega}\right)e_i, \qquad (6.3.50)$$

where g is the nonzero Dirichlet boundary condition for u. This approach was taken by Antil et al. (2017) and is discussed in Section 6.3.6. The same authors show that under suitable regularity conditions, the inhomogeneous spectral fractional Laplacian can be written as

$$(-\Delta_{\Omega,g})^{\alpha/2}u = (-\Delta_{\Omega,0})^{\alpha/2}[u-v], \qquad (6.3.51)$$

where $-\Delta v = 0$ in the weak sense, and $v\big|_{\partial\Omega} = g$. This essentially reduces the inhomogeneous operator to the well-studied homogeneous spectral fractional Laplacian. The details of the harmonic lifting are discussed in Section 6.3.6. Nonharmonic lifting functions may also be used; this is discussed in Section 6.3.6. We note here a consequence of (6.3.51) that may be surprising at first: as $\alpha \to 0$, $(-\Delta_{\Omega,g})^{\alpha/2}u \to u - v$; in other words, for $g \not\equiv 0$, $(-\Delta_{\Omega,g})^{\alpha/2}$ does not reduce to the identity operator as $\alpha \to 0$.

A different approach was taken by Cusimano et al. (2018) where the heat semigroup was used to define the inhomogeneous spectral fractional Laplacian, for which they used the notation $\mathcal{L}_{\Omega,g}^{\alpha/2}$:

$$\mathcal{L}_{\Omega,g}^{\alpha/2} u = -\frac{1}{\Gamma(-\alpha/2)} \int_0^\infty \left(e^{t\Delta_{\Omega,g}} u(x) - u(x)\right) \frac{dt}{t^{1+\alpha/2}}. \tag{6.3.52}$$

This approach is discussed in Section 6.3.6. A lifting characterization equivalent to (6.3.51) was obtained for this operator $\mathcal{L}_{\Omega,g}^{\alpha/2}$; thus both the operator $(-\Delta_{\Omega,g})^{\alpha/2}$ defined by Antil et al. (2017) and the operator $\mathcal{L}_{\Omega,g}^{\alpha/2}$ defined by Cusimano et al. (2018) are the same. Hence, we use the symbol $(-\Delta_{\Omega,g})^{\alpha/2}$ in the remainder of this work. We compare numerical methods that can be applied to both approaches in Section 6.9.1.

Series Representation

We motivate the approach in Antil et al. (2017) by discussing the spectral representation of the standard Laplacian $-\Delta u = -\frac{\partial^2 u}{\partial x_1^2} - \cdots - \frac{\partial^2 u}{\partial x_d^2}$ on a bounded domain Ω for arguments u with nonzero boundary values. Given a function u such that $u|_{\partial\Omega} = g$, where g need not be zero, subtract from u a harmonic function v inside Ω with the same boundary values:

$$-\Delta v = 0, \quad v|_{\partial\Omega} = g.$$

We refer to this function v as a *harmonic lifting* of g. Then

$$-\Delta u = -\Delta(u - v), \quad (u - v)|_{\partial\Omega} = 0.$$

In other words, harmonic lifting of the boundary values does not change the Laplacian u, and since the boundary value for $u - v$ is zero, we can use it in formula (6.3.48):

$$\begin{aligned} -\Delta u = -\Delta(u - v) &= \sum \lambda_k (u - v, e_k)_\Omega \, e_k \\ &= \sum \lambda_k (u, e_k)_\Omega \, e_k - \sum \lambda_k (v, e_k)_\Omega \, e_k. \end{aligned}$$

Let us rewrite the second inner product in the sum, which involves the harmonic function v:

$$\begin{aligned} (v, e_k)_\Omega &= (v, -\Delta(-\Delta)^{-1} e_k)_\Omega \\ &= (v, -\Delta\lambda_k^{-1} e_k)_\Omega \qquad \text{(by the eigenfunction property)} \\ &= \lambda_k^{-1} \int_\Omega -\Delta v e_k + \lambda_k^{-1} \int_{\partial\Omega} v \frac{\partial e_k}{\partial n} - \lambda_k^{-1} \int_{\partial\Omega} e_k \frac{\partial v}{\partial n} \\ &= \lambda_k^{-1} \int_{\partial\Omega} v \frac{\partial e_k}{\partial n} \qquad \text{(since } -\Delta v = 0 \text{ on } \Omega \text{ and } e_k = 0 \text{ on } \partial\Omega) \\ &= \lambda_k^{-1} \int_{\partial\Omega} u \frac{\partial e_k}{\partial n} \qquad \text{(since } u - v = 0 \text{ on } \partial\Omega). \end{aligned}$$

Here we used Green's second identity to derive the identity in the third line. This gives us the spectral expansion of $-\Delta$ which is now valid for *any* smooth function u on Ω, regardless of boundary values:

$$-\Delta u = \sum \left(\lambda_k \int_\Omega u e_k - \int_{\partial\Omega} u \frac{\partial e_k}{\partial n} \right) e_k. \tag{6.3.53}$$

This is a key result for defining the inhomogeneous spectral fractional Laplacian. However, before we can proceed, the operator $(-\Delta)^{-1}$ that appeared in the derivation above requires clarification. *A priori*, it is a multiple-valued operator, since $(-\Delta)^{-1} f$ is only specified up to a harmonic function in Ω. However, owing to the uniqueness of the (standard) Poisson problem, this arbitrariness may be removed by requiring $(-\Delta)^{-1} f$ to have fixed boundary value $\tilde{g}$ on $\partial\Omega$. Thus, we define $u = (-\Delta_{\tilde{g}})^{-1} f$ as the function such that $-\Delta u = f$ and $u = \tilde{g}$ on $\partial\Omega$. Regardless of the fixed boundary condition, this will result in a single-valued operator $(-\Delta_{\tilde{g}})^{-1}$ such that

$$-\Delta(-\Delta_{\tilde{g}})^{-1} = \mathrm{Id}_{L^2}, \tag{6.3.54}$$

and

$$(-\Delta_{\tilde{g}})^{-1}(-\Delta) = \mathrm{Id}_{\{u \in H^2 \text{ such that } u|_{\partial\Omega} = \tilde{g}\}}. \tag{6.3.55}$$

For definiteness, and to obtain an identity operator on a linear subset of H^2 in (6.3.55), the zero boundary value $\tilde{g} \equiv 0$ is chosen. We sometimes refer to this as the *zero gauge* inverse Laplacian, and simply write $(-\Delta)^{-1}$ rather than $(-\Delta_0)^{-1}$. By the spectral theorem, the series expansion of the inverse Laplacian is then the spectral power

$$(-\Delta)^{-1} f = \sum_{i=1}^{\infty} \lambda_i^{-1} (f, e_i)_\Omega \, e_i,$$

which defines an operator from $H^{-2}(\Omega)$ to $L^2(\Omega)$. We remark that, even with the choice of a zero gauge, (6.3.54) is valid on functions with arbitrary boundary values.

Using the result (6.3.53), a natural approach to defining the spectral fractional Laplacian with nonzero boundary conditions is to write

$$(-\Delta)^{\alpha/2} u := (-\Delta)^{\alpha/2-1}(-\Delta) u,$$

and to combine a definition of $(-\Delta)^{\alpha/2-1}$ with the series (6.3.53) above. At first glance, this merely shifts the problem to defining the *inverse* spectral fractional Laplacian with nonzero boundary conditions. However, defining this inverse inhomogeneous operator is considerably easier. By the above discussion of $(-\Delta)^{-1}$, the spectral power

$$(-\Delta)^{\alpha/2-1} := \sum_{i=1}^{\infty} \lambda_i^{\alpha/2-1} (\cdot, e_i)_\Omega \, e_i$$

is a convergent series defining a single-valued operator on $H^{2(\alpha/2-1)}(\Omega)$, regardless of boundary conditions. For $\alpha = 0$, we understand the operator $(-\Delta)^{-1}u$ for $u \in L^2$ to be the projection onto the zero-boundary value functions $H_0^2(\Omega)$ of all functions v such that $-\Delta v = u$. Thus, the zero-boundary condition is used to eliminate multi-valuedness of $(-\Delta)^{-1}$, but the operator $(-\Delta)^{\alpha/2-1}$ may take as argument a function with arbitrary boundary values, in particular $(-\Delta)u$.

As a result of these definitions, using the spectral expansion (6.3.53) of the inhomogeneous (integer-order) Laplacian, we obtain

$$\begin{aligned}
(-\Delta)^{\alpha/2}u &= (-\Delta)^{\alpha/2-1}(-\Delta)u \\
&= \sum_{i=1}^{\infty} \lambda_i^{\alpha/2-1} (-\Delta u, e_i)_\Omega \, e_i \\
&= \sum_{i=1}^{\infty} \lambda_i^{\alpha/2-1} \left(\sum_{j=1}^{\infty} \lambda_j (u, e_j)_\Omega \, e_j - \sum_{j=1}^{\infty} \left(\left(u, \frac{\partial e_j}{\partial n} \right)_\Omega e_j, e_i \right) e_i \right) \\
&= \sum_{i=1}^{\infty} \lambda_i^{\alpha/2-1} \left(\lambda_i (u, e_i)_\Omega - \left(u, \frac{\partial e_i}{\partial n} \right)_\Omega \right) e_i \\
&= \sum_{i=1}^{\infty} \lambda_i^{\alpha/2} (u, e_i)_\Omega \, e_i - \sum_{i=1}^{\infty} \lambda_i^{\alpha/2-1} \left(u, \frac{\partial e_i}{\partial n} \right)_\Omega e_i .
\end{aligned} \tag{6.3.56}$$

This is the definition in Antil et al. (2017) of a general inhomogeneous spectral fractional Laplacian. This can be considered the fractional analogue of (6.3.53). In their paper, Antil et al. also proved a similar formula for nonzero Neumann boundary conditions. Furthermore, they developed an integration-by-parts formula for these formulations, proved regularity results, introduced finite element discretizations, derived the associated error estimates, and included numerical experiments.

Harmonic Lifting

Taking the above series representation of the operator $(-\Delta_{\Omega,g})^{\alpha/2}$ as a definition, Antil et al. (2017) considered the inhomogeneous spectral fractional Poisson problem:

$$\begin{aligned}
(-\Delta_{\Omega,g})^{\alpha/2}u(x) &= f(x), \quad x \in \Omega, \quad \alpha \in (0,2), \\
u(x) &= g(x), \quad x \in \partial\Omega.
\end{aligned} \tag{6.3.57}$$

By linearity, the solution u can be written as

$$u(x) = a(x) + b(x), \tag{6.3.58}$$

where a solves

$$\begin{aligned} (-\Delta_{\Omega,g})^{\alpha/2} a = (-\Delta_{\Omega,0})^{\alpha/2} a = f \quad &\text{in } \Omega, \\ a|_{\partial\Omega} = 0 \quad &\text{on } \partial\Omega, \end{aligned} \tag{6.3.59}$$

and the component b solves the equation

$$\begin{aligned} (-\Delta_{\Omega,g})^{\alpha/2} b = 0 \quad &\text{in } \Omega, \\ b|_{\partial\Omega} = g \quad &\text{on } \partial\Omega. \end{aligned} \tag{6.3.60}$$

However, this "fractional harmonic" function b, under certain conditions on the regularity of g, is simply the solution to the standard Laplace equation

$$\begin{aligned} -\Delta b = 0 \quad &\text{in } \Omega, \\ b|_{\partial\Omega} = g \quad &\text{on } \partial\Omega. \end{aligned} \tag{6.3.61}$$

The simple intuition behind this equivalence involves the characterization of the operator $(-\Delta_{\Omega,g})^{\alpha/2}$ as $(-\Delta)^{\alpha/2-1}(-\Delta)$ and by our choice of the inverse Laplacian with $(-\Delta)^{\alpha/2-1}0 = 0$, which yield

$$-\Delta b = 0 \implies (-\Delta_{\Omega,g})^{\alpha/2} b = 0. \tag{6.3.62}$$

Indeed, the authors of Antil et al. (2017) show that equation (6.3.60) is, for boundary functions $g \in L^2(\partial\Omega)$, equivalent to the *very weak variational form* of equation (6.3.61):

$$\int_\Omega b(-\Delta)\phi = \int_{\partial\Omega} g \frac{\partial \phi}{\partial n}, \quad \forall\, \phi \in H_0^1(\Omega) \cap H^2(\Omega). \tag{6.3.63}$$

The phrase "very weak" refers to the transfer of all derivatives of v to the test function ϕ via fractional integration-by-parts, a result of the same work (Antil et al. (2017)). Of course, if the boundary data g is sufficiently regular, then b may be sought as a solution to the weak (rather than very weak) variational form of equation (6.3.61):

$$\int_\Omega \nabla b \cdot \nabla \phi = \int_{\partial\Omega} g \frac{\partial \phi}{\partial n}, \quad \forall\, \phi \in H_0^1(\Omega). \tag{6.3.64}$$

For the precise regularity estimate for the problem (6.3.63) in terms of the boundary function g, we refer to Lemma 4.1 and the surrounding discussion in the article by Antil et al. (2017). We point out the simplest case, in which $g \in H^{1/2}(\partial\Omega)$ implies that $b \in H^1(\Omega)$ and b satisfies (6.3.61) in the weak sense. This approach is implemented in Section 6.9.1.

Nonharmonic Lifting

While Antil et al. (2017) used a fractional harmonic lifting in their approach, it is possible to obtain a variational form with an arbitrary (i.e., nonharmonic) lifting function. In this section, we describe a lifting approach in which the lifting function $v \in H^1(\Omega)$ need only satisfy the boundary condition $v|_{\partial\Omega} = u|_{\partial\Omega}$. Again, we wish to solve the spectral fractional Poisson problem

$$\begin{aligned} (-\Delta)^{\alpha/2} u &= f \quad \text{in } \Omega, \\ u\big|_{\partial\Omega} &= g \quad \text{on } \partial\Omega. \end{aligned} \tag{6.3.65}$$

The unknown function u can be decomposed as $u = w - v$, where v is the lifting function. This function v need not be unique, and the fractional harmonic function v described above is also admissible.

To derive the variational form of (6.3.65), we need the following integration-by-parts formula for the inverse spectral fractional Laplacian $(-\Delta)^{\alpha/2-1}$.

Theorem 6.10 *Let* $0 \le \alpha \le 2$ *and* $f, \phi \in L^2(\Omega)$. *Then*

$$((-\Delta)^{\alpha/2-1} f, \phi) = (f, (-\Delta)^{\alpha/2-1} \phi). \tag{6.3.66}$$

Proof We use the notation $\hat{f}_i = (f, e_i)$ and $\hat{\phi}_i = (\phi, e_i)$.

$$\begin{aligned} ((-\Delta)^{\alpha/2-1} f, \phi) &= \Big(\sum_{i=1}^{\infty} \lambda_i^{\alpha/2-1} \hat{f}_i e_i, \sum_{j=1}^{\infty} \hat{\phi}_j e_j \Big) \\ &= \sum_{i=1}^{\infty} \sum_{j=1}^{\infty} \lambda_i^{\alpha/2-1} \hat{f}_i \hat{\phi}_j (e_i, e_j) = \sum_{i=1}^{\infty} \lambda_i^{\alpha/2-1} \hat{f}_i \hat{\phi}_i \\ &= \sum_{i=1}^{\infty} \sum_{j=1}^{\infty} \lambda_j^{\alpha/2-1} \hat{f}_i \hat{\phi}_j (e_i, e_j) \\ &= \Big(\sum_{i=1}^{\infty} \hat{f}_i e_i, \sum_{j=1}^{\infty} \lambda_j^{\alpha/2-1} \hat{\phi}_j e_j \Big) = (f, (-\Delta)^{\alpha/2-1} \phi). \end{aligned} \tag{6.3.67}$$

This ends the proof. □

This formula is valid regardless of the boundary values of the functions f and ϕ. This is due to our choice of zero gauge for the definition of $(-\Delta)^{\alpha/2-1}$. Then, the variational form of the problem is written as follows: find the function $w \in \mathbb{H}^\alpha$ such that for any $\phi \in L^2(\Omega)$,

$$((-\Delta)^{\alpha/2}(w + v), \phi) = (f, \phi). \tag{6.3.68}$$

Now we can define the inhomogeneous spectral fractional Laplacian of v in weak form by integrating-by-parts in the v term on the left-hand side.

$$((-\Delta)^{\alpha/2}v,\phi) = (-\Delta v,(-\Delta)^{\alpha/2-1}\phi) \tag{6.3.69}$$

$$= (\nabla v,\nabla(-\Delta)^{\alpha/2-1}\phi), \tag{6.3.70}$$

which is our (weak sense) definition of the inhomogeneous spectral fractional Laplacian.

Finally, we can solve for w by solving the variational equation

$$((-\Delta)^{\alpha/2}w,\phi) = (f,\phi) - (\nabla v,\nabla((-\Delta)^{\alpha/2-1}\phi)), \tag{6.3.71}$$

where all the fractional operators that appear are now applied to functions satisfying zero boundary conditions. Note that this (standard) weak form (6.3.71) requires that $v \in H^1(\Omega)$, which by $v|_{\partial\Omega} = g$ and the trace Theorem 6.3, requires the boundary data $g \in H^{1/2}(\Omega)$. While other weak variational forms leading to less regular solution spaces may be studied to treat rougher boundary data, for the purposes of this section, (6.3.71) will suffice. We recover the solution to equation (6.3.65) using the relation $u = w + v$. These lifting approaches are compared using numerical examples in Section 6.9.1.

Heat Semigroup

Cusimano et al. (2018) proposed an extension of the heat semigroup method to define an inhomogeneous spectral fractional Laplacian. In their work, only discretizations of their newly-defined operator are considered, not the solution of the Poisson problem using this operator. We compare these discretizations with the approach of Antil et al. (2017) numerically in Section 6.9.1.

Cusimano et al. define their inhomogeneous spectral fractional Laplacian as in equation (6.3.52), in which $e^{t\Delta_{\Omega,g}}$ is the propagator of the heat equation. Next, the authors use a splitting for $w(x,t)$, which allows one to relate $\mathcal{L}^{\alpha/2}_{\Omega,g}$ to $(-\Delta)^{\alpha/2}_{\Omega,0}$, the homogeneous spectral fractional Laplacian. The unknown function $w(x,t)$ can be expressed as

$$w(x,t) = v(x,t) + z(x),$$

where $z(x)$ is a harmonic function, i.e.,

$$-\Delta z = 0 \quad \text{in } \Omega; \qquad z = g \quad \text{on } \partial\Omega, \tag{6.3.72}$$

and $v(x,t)$ solves a zero boundary-value heat equation:

$$\begin{aligned} \partial_t v - \Delta v = 0 \quad &\text{in } \Omega \times [0,+\infty), \\ v(x,t=0) = u(x) - z(x)&, \\ v(x,t) = 0 \quad &\text{on } \partial\Omega \times [0,+\infty). \end{aligned} \tag{6.3.73}$$

We notice that

$$w(x,0) = v(x,0) + z(x) = u(x) - z(x) + z(x) = u(x).$$

Then, using the definition of $w(x,t)$ and equation (6.3.52), we see that

$$\begin{aligned}
\mathcal{L}^{\alpha/2}_{\Omega,g} u &= \frac{1}{\Gamma(-\alpha/2)} \int_0^\infty (w(x,t) - w(x,0)) \frac{dt}{t^{1+\alpha/2}} \\
&= \frac{1}{\Gamma(-\alpha/2)} \int_0^\infty ([v(x,t) + z(x)] - [v(x,0) + z(x)] \frac{dt}{t^{1+\alpha/2}} \\
&= \frac{1}{\Gamma(-\alpha/2)} \int_0^\infty (v(x,t) - v(x,0)) \frac{dt}{t^{1+\alpha/2}} \qquad (6.3.74) \\
&= \frac{1}{\Gamma(-\alpha/2)} \int_0^\infty (e^{t\Delta_{\Omega,g}}[u-z](x) - [u-z](x)) \frac{dt}{t^{1+\alpha/2}} \\
&= (-\Delta_{\Omega,0})^{\alpha/2}[u-z](x).
\end{aligned}$$

Given this information, we can see that the two formulations of the spectral fractional Laplacians in Antil et al. (2017) and Cusimano et al. (2018) are equivalent. We know that the heat semigroup approach leads to the relation

$$\mathcal{L}^{\alpha/2}_{\Omega,g} u = (-\Delta_{\Omega,0})^{\alpha/2}[u-z], \qquad (6.3.75)$$

where z is the harmonic function given by equation (6.3.72). We can also obtain a similar formula for the operator of Antil et al. From equations (6.3.60) and (6.3.59), we know that

$$(-\Delta_{\Omega,g})^{\alpha/2} u = (-\Delta_{\Omega,0})^{\alpha/2} a. \qquad (6.3.76)$$

Inserting $a = u - b$ from the splitting (6.3.58), we have

$$(-\Delta_{\Omega,g})^{\alpha/2} u = (-\Delta_{\Omega,0})^{\alpha/2}[u-b]. \qquad (6.3.77)$$

This shows that the inhomogeneous spectral fractional Laplacian $(-\Delta_{\Omega,g})^{\alpha/2}$ is just the standard homogeneous operator applied to the lifting $u - b$. Since b is also a harmonic function in Ω with $b\big|_{\partial\Omega} = g$, by uniqueness, $b = z$. Therefore

$$(-\Delta_{\Omega,g})^{\alpha/2} u = (-\Delta_{\Omega,0})^{\alpha/2}[u-z] = \mathcal{L}^{\alpha/2}_{\Omega,g} u. \qquad (6.3.78)$$

This fact is demonstrated numerically in Section 6.9.1.

Well-Posedness and Regularity

We begin by discussing well-posedness (existence and uniqueness) of the spectral fractional Poisson problem. We consider the general case of nonzero boundary conditions. Thus, the fractional Laplacian that is used in the two problems below is the inhomogeneous operator that has been reviewed in this section. We have transcribed the existence and uniqueness theorems proven in Antil et al. (2017). These results have been abbreviated by considering $\alpha/2 \geq 1/2$, resulting in minimal regularity of boundary data $g \in L^2(\partial\Omega)$ in the Dirichlet case, and $g \in H^{-1}(\partial\Omega)$ for the Neumann case; the full theorems in Antil et al. (2017) actually allow for $\alpha/2 < 1/2$ and rougher boundary data. The Sobolev space notations used in this section are defined in Section 6.2.

For the Dirichlet boundary condition, let Ω be a bounded quasi-convex domain. If $f \in \mathbb{H}^{-\frac{\alpha}{2}}(\Omega)$ and $g \in H^{\frac{\alpha}{2}-\frac{1}{2}}(\partial\Omega)$ for $\alpha/2 \geq 1/2$, then there exists a unique solution $u \in H^{\frac{\alpha}{2}}(\Omega)$ to

$$(-\Delta)^{\alpha/2} u = f \quad \text{in} \quad \Omega; \quad u|_{\partial\Omega} = g, \tag{6.3.79}$$

which satisfies

$$\|u\|_{H^{\frac{\alpha}{2}}(\Omega)} \leq C\Big(\|f\|_{\mathbb{H}^{-\frac{\alpha}{2}}(\Omega)} + \|g\|_{H^{\frac{\alpha}{2}-\frac{1}{2}}(\partial\Omega)}\Big).$$

For $\alpha/2 = 1/2$, the convention is $H^0(\partial\Omega) = L^2(\partial\Omega)$. This statement is a special case of Theorem 4.5 in Antil et al. (2017).

For the Neumann boundary condition, again let Ω be a bounded quasi-convex domain. Let $1/2 \leq \alpha/2 \leq 1$. If $f \in H^{\frac{\alpha}{2}}(\Omega)^*$ (the dual of $H^{\frac{\alpha}{2}}(\Omega)$; for $\alpha/2 > 1/2$ this is distinct from $\mathbb{H}^{-\frac{\alpha}{2}}(\Omega)$) and $g \in H^{\frac{\alpha}{2}-\frac{3}{2}}(\partial\Omega)$ satisfy the compatibility condition

$$\int_\Omega f + \int_{\partial\Omega} g = 0, \tag{6.3.80}$$

then there exists a unique solution $u \in H^{\frac{\alpha}{2}}(\Omega)$ such that $\int_\Omega u = 0$ to

$$(-\Delta)^{\alpha/2} u = f \quad \text{in } \Omega; \quad \partial u/\partial n\big|_{\partial\Omega} = g, \tag{6.3.81}$$

which satisfies

$$|u|_{H^{\frac{\alpha}{2}}(\Omega)} \leq C\Big(\|f\|_{H^{\frac{\alpha}{2}}(\Omega)^*} + \|g\|_{\mathbb{H}^{\frac{\alpha}{2}-\frac{3}{2}}(\partial\Omega)}\Big). \tag{6.3.82}$$

This statement is a special case of Theorem 5.4 in Antil et al. (2017).

Next, we discuss regularity for the spectral fractional Poisson equation. Unlike the well-posedness results discussed above, we now focus exclusively on the case of *zero* boundary conditions. This is because almost all regularity results that have been published at the time of this writing have been presented in this form. Various extensions to nonzero boundary conditions can be made by considering these results in combination with the harmonic lifting property discussed in Section 6.3.6.

For the standard homogeneous spectral fractional Laplacian, we state only the simplest case of the Sobolev regularity results in Grubb (2016), and direct the reader to that article for a full discussion. First, we consider the Dirichlet problem, using the zero Dirichlet boundary-value fractional Sobolev space $\mathbb{H}^{\alpha/2}(\Omega)$. Let Ω be a bounded, C^∞-smooth subset of $\mathbb{R}^d$. Let u satisfy the equation

$$(-\Delta)^{\alpha/2}u = f \quad \text{in } \Omega; \quad u|_{\partial\Omega} = 0, \tag{6.3.83}$$

for $0 < \alpha < 2$. Then

(i) If $s < 1/2$, then $f \in H^s(\Omega)$ implies $u \in \mathbb{H}^{s+\alpha}(\Omega)$.
(ii) If $s = 1/2$, then $f \in H^s_{00}(\Omega)$ implies $u \in \mathbb{H}^{s+\alpha}(\Omega)$.
(iii) If $1/2 < s < 2 + 1/2$, then $f \in H^s(\Omega)$ implies only that $u \in \mathbb{H}^{1/2-\epsilon+\alpha}(\Omega)$, and we have the stronger result $u \in \mathbb{H}^{s+\alpha}(\Omega)$ if and only if $f = 0$ on $\partial\Omega$.

We illustrate that this last result is sharp with a simple example. Note that the Sobolev norm in $H^s(I)$ of a Fourier sine series is given by

$$\left\| \sum a_k \sin(kx) \right\|_{H^s(I)} = \sum (1 + k^2)^s |a_k^2|.$$

Consider the interval $[0, 2\pi]$, where the eigenvalues of the (Dirichlet) Laplacian are $\lambda_k = k^2$, with corresponding eigenfunctions $e_{\lambda_k} = \sin(kx)$. Let us begin by constructing a low regularity $f \in L^2$:

$$f = \sum_{k=1}^{\infty} \frac{1}{\sqrt{k}\log(k+1)} \sin(kx). \tag{6.3.84}$$

Then, $f \in L^2$ since

$$\begin{aligned}
\sum_{k=1}^{\infty} \frac{1}{k\log^2(k+1)} &\leq \int_1^\infty \frac{dx}{x\log^2(x+1)} \\
&= \int_1^\infty \frac{dx}{(x+1)\left(\log^2(x+1) - \frac{\log^2(x+1)}{x+1}\right)} \\
&= \int_{\log 2}^\infty \frac{dx}{x^2(1-e^{-x})} \leq 2.
\end{aligned}$$

But $f \notin H^s$ for any $s > 0$ since

$$\|f\|_{H^s} \sim \sum_{k=1}^{\infty} \frac{(1+k^2)^s}{k(\log^2(k+1))} \geq \sum_{k=1}^{\infty} \frac{k^{2s}}{k\log^2(k+1)} \geq \sum_{k=1}^{\infty} \frac{1}{k} = \infty.$$

Now, the solution to the fractional Poisson problem with zero Dirichlet boundary conditions and f given by (6.3.84) is

$$u = \sum_{k=1}^{\infty} \frac{k^{-\alpha}}{\sqrt{k}\log(k+1)} \sin(kx). \tag{6.3.85}$$

We see $\|u\|_{H^{s+\alpha}} \sim \|f\|_{H^s}$, so u has strict H^α regularity: $u \in H^\alpha$ and $u \notin H^{\alpha+\epsilon}$. *Thus, the α-gain in regularity as stated in the result above is sharp.*

Next, we consider the Neumann problem, and for $s > 3/2$ we define the zero Neumann boundary-value fractional Sobolev space

$$H^s_{\partial u/\partial n=0}(\Omega) = \left\{ u \in H^s(\Omega) \text{ such that } \frac{\partial u}{\partial n}\Big|_{\partial\Omega} = 0 \right\}. \tag{6.3.86}$$

If $s < 3/2$, we define $H^s_{\partial u/\partial n=0}(\Omega) = H^s(\Omega)$. Let Ω be a bounded, C^∞-smooth subset of $\mathbb{R}^d$. Let u satisfy the equation

$$(-\Delta)^{\alpha/2} u = f \text{ in } \Omega; \quad \partial u/\partial n|_{\partial\Omega} = 0. \tag{6.3.87}$$

Then,

(i) If $s < 3/2$, then $f \in H^s(\Omega)$ implies $u \in H^{s+\alpha}_{\partial u/\partial n=0}(\Omega)$.
(ii) If $3/2 < s < 7/2$, then $f \in H^s(\Omega)$ implies $u \in H^{s+\alpha}_{\partial u/\partial n=0}(\Omega)$ if and only if $\partial f/\partial n = 0$ on $\partial\Omega$.

Among additional results, Grubb (2016) discusses the extension by induction of the above points to higher $H^s(\Omega)$ regularity of f.

Regularity in the Hölder spaces $C^{k,r}$ for the spectral fractional Poisson problem was studied extensively in Caffarelli and Stinga (2016). In that work, an array of results was obtained, for both interior and boundary regularity of the solution u in the Dirichlet problem (6.3.83) and in the Neumann problem (6.3.87), with conditions of the form $f \in C^{0,r}(\Omega)$ or of the form $f \in L^p(\Omega)$, under fairly general conditions on the domain Ω. The results also allow for powers of more general, variable-coefficient elliptic operators. We transcribe just one of these results for the fractional Laplacian, namely, the interior regularity for $f \in C^{0,r}(\Omega)$ of the Poisson problem, which is the same for both zero Dirichlet or zero Neumann boundary condition:

Assume that Ω is a bounded Lipschitz domain and that $f \in C^{0,r}(\Omega)$, for some $0 < r < 1$. Let u be a solution to (6.3.83) or (6.3.87).

- If $0 < r + \alpha < 1$, then $u \in C^{0,r+\alpha}(\Omega)$ and

$$[u]_{C^{0,r+\alpha}(\Omega)} \le C\big(\|u\|_{L^2(\Omega)} + [u]_{H^{\alpha/2}(\Omega)} + \|f\|_{C^{0,r}(\Omega)}\big).$$

- If $1 < r + \alpha < 2$, then $u \in C^{1,r+\alpha-1}(\Omega)$ and

$$[u]_{C^{1,r+\alpha-1}(\Omega)} \le C\big(\|u\|_{L^2(\Omega)} + [u]_{H^{\alpha/2}(\Omega)} + \|f\|_{C^{0,r}(\Omega)}\big).$$

The constants C depend only on d, Ω, r, and α. For the proof of this and other regularity results, see Caffarelli and Stinga (2016). Similar results have also been obtained for the equation $(-\epsilon\Delta)^{1/2}u + u = f$ with zero Neumann boundary condition in Stinga and Volzone (2015).

In Table 6.2, we have summarized the fundamental properties and stochastic interpretations of the Riesz and spectral fractional Laplacians, as they were discussed in this section. From the stochastic perspective, these operators differ in that the Riesz Laplacian is associated to processes that leave the closure of the domain, while the spectral Laplacian is associated to processes that are confined to the closure of the domain.

Remark 6.11 The Riesz and the spectral fractional Laplacians are merely the two most commonly used definitions. Another fractional Laplacian that has been studied is the *regional* definition:

$$(-\Delta_{\text{regional}})^{\alpha/2}u(x) = C(d,\alpha)\ \text{p.v.} \int_{\Omega} \frac{u(x)-u(y)}{|x-y|^{d+\alpha}} dy, \tag{6.3.88}$$

where $C(d,\alpha)$ is given by (6.3.8). Note that the domain of this operator consists of functions defined on Ω, rather than $\mathbb{R}^d$. The regional definition differs from the Riesz definition, even if $u(x) \equiv 0$ for $x \in \mathbb{R}^d \setminus \Omega$:

$$\begin{aligned}(-\Delta_{\text{Riesz}})^{\alpha/2}u(x) &= C(d,\alpha)\ \text{p.v.} \int_{\mathbb{R}^d} \frac{u(x)-u(y)}{|x-y|^{d+\alpha}}\, dy \\ &= C(d,\alpha)\ \text{p.v.} \int_{\Omega} \frac{u(x)-u(y)}{|x-y|^{d+\alpha}}\, dy \\ &\quad - u(x) \int_{\mathbb{R}^d\setminus\Omega} \frac{C(d,\alpha)}{|x-y|^{d+\alpha}}\, dy \\ &= (-\Delta_{\text{regional}})^{\alpha/2}u(x) - u(x) \int_{\mathbb{R}^d\setminus\Omega} \frac{C(d,\alpha)}{|x-y|^{d+\alpha}}\, dy.\end{aligned} \tag{6.3.89}$$

The well-posedness of the fractional Poisson problem involving the regional Laplacian was studied using the Feynman–Kac formula (Guan and Ma, 2005). For a further discussion of the regional Laplacian and the relation to reflected and censored α-stable processes, see Guan and Ma (2006). Neumann and Robin boundary conditions for the regional Laplacian have been discussed in Warma (2015).

Other notions of fractional Laplacians that arise from related processes are discussed in Kim et al. (2020). In general, the probabilistic literature on stable-type processes in bounded domains and related notions of fractional Laplacians is very rich (Bogdan et al., 2009). For example, estimates for eigenvalues of the spectral Laplacian were derived using probabilistic techniques in Chen and Song (2005).

Table 6.2. *Comparison of the Riesz and spectral fractional Laplacians in a bounded domain* $\Omega \subset \mathbb{R}^d$. *Reprinted from the Journal of Computational Physics, 404, Lischke, Anna, Pang, Guofei, Gulian, Mamikon, Song, Fangying, Glusa, Christian, Zheng, Xiaoning, Mao, Zhiping, Cai, Wei, Meerschaert, Mark M, Ainsworth, Mark, Karniadakis, George EM, What is the fractional Laplacian? A comparative review with new results, 109009, Copyright (2020), with permission from Elsevier.*

$(-\Delta)^{\alpha/2}u$	Definition	Domain	BC type	Stopped process for Dirichlet BC	Reflected process for zero Neumann BC
Spectral	$\sum \lambda^{\alpha/2}(u,e_\lambda)e_\lambda$, where (λ, e_λ) is the spectrum of $-\Delta$ on Ω.	Functions on Ω	Boundary ($\partial\Omega$)	Subordinate stopped Brownian motion (Yosida, 1980; Song and Vondracek, 2003; Gulian and Pang, 2018)	Subordinate reflected Brownian motion (Yosida, 1980; Gulian, 2019)
Riesz	C p.v. $\int_{\mathbb{R}^d} \frac{u(x)-u(y)}{\lvert x-y\rvert^{d+\alpha}} \mathrm{d}y$, for $C = \frac{2^\alpha \Gamma\left(\frac{\alpha}{2}+\frac{d}{2}\right)}{\pi^{d/2}\lvert\Gamma\left(-\frac{\alpha}{2}\right)\rvert}$.	Functions on $\mathbb{R}^d$.	Exterior ($\mathbb{R}^d \setminus \Omega$)	Stopped α-stable motion (Kyprianou et al., 2018).	Various conditions/ processes proposed (Du et al., 2012; Dipierro et al., 2017)

6.4 Riesz Definition

Using the Riesz definition, we consider the fractional Poisson equation (6.1.1) with zero Dirichlet boundary conditions $u(x) = 0$ for $x \in \mathbb{R}^d \setminus \Omega$, where $\Omega \subset \mathbb{R}^d$ is a bounded Lipschitz domain, and we define the Riesz fractional Laplacian to be the *restriction* of the operator to functions with compact support in Ω. The boundary condition for this definition is considered "nonlocal" and is also called a "volume constraint", as it is defined on the exterior of the domain Ω.

6.4.1 Adaptive Finite Element Method (AFEM)

The Riesz fractional Poisson problem takes the variational form (Ainsworth and Glusa, 2018b)

$$\text{Find} \quad u \in \mathbb{H}^{\alpha/2}(\Omega): \quad a(u, v) = (f, v) \quad \forall v \in \mathbb{H}^{\alpha/2}(\Omega), \tag{6.4.1}$$

where

$$a(u, v) = \frac{C(d, \alpha)}{2} \int_{\Omega} \mathrm{d}x \int_{\Omega} \mathrm{d}y \frac{(u(x) - u(y))(v(x) - v(y))}{|x - y|^{d+\alpha}} \tag{6.4.2}$$

$$+ \frac{C(d, \alpha)}{\alpha} \int_{\Omega} \mathrm{d}x \int_{\partial\Omega} \mathrm{d}y \frac{u(x)\, v(x)\, n_y \cdot (x - y)}{|x - y|^{d+\alpha}}, \tag{6.4.3}$$

and where n_y is the *inward* normal to $\partial\Omega$ at y. The space $\mathbb{H}^{\alpha/2}(\Omega)$ is defined in Section 6.2, Definition 6.6.

A straightforward finite element discretization of (6.4.1) encounters several difficulties:

(i) The element contributions for adjacent or identical element pairs of (6.4.3) are given by singular integrals. Special Gaussian quadrature methods are given in Ainsworth and Glusa (2018b, 2017).

(ii) The resulting linear algebraic system is dense. In Ainsworth and Glusa (2018b), the panel clustering technique known from the boundary element literature is adapted for the fractional Laplacian. This reduced the complexity of the matrix-vector product from $\mathcal{O}(n^2)$ to $\mathcal{O}(n \log^{2d} n)$, where $d = 1$ or $d = 2$ is the spatial dimension.

(iii) As shown in Grubb (2015), solutions to the Riesz fractional Laplacian display generally low regularity close to the boundary of the domain. This behavior is noticeably different from the classical integer-order Laplacian for which higher regularity of the domain and right-hand side imply higher regularity of the solution. Therefore, globally quasi-uniform meshes are ill-suited for the discretization of the integral fractional Laplacian. In Ainsworth and Glusa (2017), posteriori error estimates of residual type and a Dörfler marking strategy are employed to adaptively refine the discretization, resulting in meshes that are highly refined close to the boundary. It was shown that, using piecewise linear finite elements, optimal rates of convergence are obtained:

	$\mathbb{H}^{\alpha/2}$	L^2
$d = 1$	$n^{\alpha/2-2}$	n^{-2}
$d = 2$	$n^{-1/2}$	$n^{-1/2-\alpha/4}$

Again, a clustering approach leads to $\mathcal{O}\left(n \log^{2d} n\right)$ complexity for the computation of the error indicators.

(iv) The efficient solution of the arising linear systems of equations can be achieved by a standard multigrid solver (Ainsworth and Glusa, 2018b, 2017). Quasi-optimal complexity is shown to be obtained.

Remark 6.12 Notice that in the variational form (6.4.1), we only require $u \in \mathbb{H}^{\alpha/2}(\Omega)$, and in the case where $0 < \alpha < 1$, this variational form accepts functions u that may not admit a trace (see Theorem 6.3). Although the weak solution of the Riesz fractional Poisson problem has the regularity described in Theorem 6.9, and therefore admits a trace if $\alpha + s > 1/2$, the AFEM approximation space is $\mathbb{H}^{\alpha/2}(\Omega)$, which is only contained in the trace space $\mathbb{H}^{1/2+\varepsilon}(\Omega)$ in the case $\alpha > 1$. This is the reason that the examples in Section 6.8 (namely, Figures 6.16, 6.21, and 6.22) where $\alpha < 1$ show solutions to the homogeneous Riesz fractional Poisson equation in which the zero boundary condition is not strongly enforced on the approximation, despite the solution belonging to $H^{1/2+\varepsilon}(\Omega)$ and possessing a zero trace. As the finite element mesh is refined and the approximant converges to the true weak solution, the oscillations at the boundary are diminished.

6.4.2 Walk-on-Spheres (WOS) Method

The *walk-on-spheres* method is a type of Monte Carlo method for simulating solutions to the Dirichlet fractional Poisson problem with both zero and nonzero boundary conditions. It was originally proposed by Muller (1956) for solving Laplace equations with Dirichlet boundary conditions (see also Yan et al. (2013)), and has been used for Neumann boundary conditions (Zhou et al., 2017), and Robin boundary conditions (Zhou and Cai, 2016) as well. This approach has recently been extended Kyprianou et al. (2018) to the following Riesz fractional Poisson problem:

$$
\begin{aligned}
(-\Delta_{\text{Riesz}})^{\alpha/2} u(x) &= f(x), \quad \text{in } \Omega, \\
u(x) &= g(x), \quad \text{in } \mathbb{R}^d \setminus \Omega,
\end{aligned}
\tag{6.4.4}
$$

where g and f are suitably regular functions, $f\colon \Omega \to \mathbb{R}$, and $g\colon \mathbb{R} \setminus \Omega \to \mathbb{R}^d$. In order to formulate the walk-on-spheres method, one must first identify the stochastic process that is generated by the operator $(-\Delta_{\text{Riesz}})^{\alpha/2}$. In this case, the process is a killed isotropic α-stable process with $\alpha \in (0, 2)$ (Kyprianou et al., 2018). One significant difference in this setting is that, in contrast with the Brownian motion setting when $\alpha = 2$, the stable process exits Ω by a jump rather than by passing through the boundary. A consequence of this is that disconnected domains can be considered, and the walk-on-spheres algorithm will terminate in finite steps, whereas the walk-on-spheres algorithm for Brownian motion will not terminate and must be truncated.

Kyprianou et al. (2018) proved the following Feynman–Kac formula for the solution of equation (6.4.4) (see Theorem 6.1). Given a Borel set S, define the space $L^1_\alpha(S)$ to be all real-valued, measurable functions ϕ that satisfy

$$
\int_S \frac{|\phi(x)|}{1 + |x|^{\alpha+d}}\, \mathrm{d}x < \infty.
$$

Let $g \in C(\Omega) \cap L^1_\alpha(\mathbb{R}^d \setminus \Omega)$, $f \in C^{\alpha+\varepsilon}(\overline{\Omega})$ with some $\varepsilon > 0$. Then there exists a continuous solution $u(x) \in L^1_\alpha(\mathbb{R}^d)$ to equation (6.4.4), where

$$
u(x) = \mathbb{E}_x \left[g(X_{\sigma_\Omega}) \right] + \mathbb{E}_x \left[\int_0^{\sigma_\Omega} f(X_s)\mathrm{d}s \right], \tag{6.4.5}
$$

and $\sigma_\Omega = \inf\{t > 0 : X_t \notin \Omega\}$ is the first exit time of the process X_t from Ω. This equation is known as a Feynman–Kac formula and allows for stochastic solution of the associated boundary-value problem; in principle, the derivation of such formulas can make possible stochastic solution methods for other definitions of fractional Laplacian and other types of boundary conditions as well (Gulian and Pang, 2018).

In order to use the Feynman–Kac formula (6.4.5), we must simulate a very large number of paths of the α-stable process X_t, beginning from the point $x \in \Omega$ at which we want to compute the solution. This approach is embarrassingly parallel, as one could assign each point x in the domain to a different compute node, and no information needs to be shared between processors. Further, while one could generate an exact simulation of each path, the idea of the walk-on-spheres approach is to "speed up" these simulations by only computing a few points along each path until the process jumps out of the domain. This is especially effective when the right-hand side f in the problem (6.4.4) is zero, since the Feynman–Kac formula (6.4.5) reduces to an expectation over the distribution of exit points of the process. We will describe the walk-on-spheres algorithm, first presented in Kyprianou et al. (2018), below.

First, we discuss some of the key ingredients of the paper by Kyprianou et al. (2018) that are used to formulate the walk-on-spheres algorithm. The following result gives the distribution of a stable process that begins from the origin when it first exits the unit sphere:

Theorem 6.13 (See Kyprianou et al. (2018)) *Suppose that $B(0,1)$ is the unit ball centered at the origin, and write $\sigma_{B(0,1)} = \inf\{t > 0 : X_t \notin B(0,1)\}$. Then, for $|y| > 1$,*

$$\mathbb{P}_0(X_{\sigma_{B(0,1)}} \in dy) = \pi^{-(d/2+1)}\Gamma(d/2)\sin(\pi\alpha/2)|1-|y|^2|^{-\alpha/2}|y|^{-d}dy. \tag{6.4.6}$$

Using this theorem, we can construct a series of points along the sample paths of stable processes. One must choose a tolerance ε and an initial point $x = X_0$. Then x is circumscribed by a sphere of radius ε. Let E_1 represent a sampling from the distribution of (6.4.6), which gives the exit from a ball of radius one centered at the origin. Using a scaling property and the fact that X has stationary and independent increments, $x + \varepsilon E_1$ gives the exit position from the ball $B(X_0 = x, \varepsilon)$. Then, we define $X_1 = x + \varepsilon E_1$ and proceed inductively, generating X_{n+1} as the exit point of the ball centered at X_n with radius ε, i.e., $X_{n+1} = X_n + \varepsilon E_{n+1}$, where E_{n+1} is an i.i.d. copy of E_1.

Now we describe the walk-on-spheres solution method for the fractional Poisson equation (Kyprianou et al., 2018). Suppose that Ω is a convex domain in $\mathbb{R}^d$, $d \geq 2$. Ω can be bounded or unbounded, as long as the measure of $\mathbb{R}^d \setminus \Omega$ is not zero. Given a starting position $x \in \Omega$, we inscribe the largest sphere that fits inside Ω and is centered at $\rho_0 := x$. The radius of this sphere is denoted by r_1. We continue inductively to generate the "walk-on-spheres" $(\rho_n,\ n \geq 0)$. Given ρ_{n-1}, we select the distribution of ρ_n according to an independent copy of $X_{\sigma_{B_n}}$ under $\mathbb{P}_{\rho_n - 1}$ (the shifted version of $\mathbb{P}_0$ from (6.4.6)), where

$$B_n = \{x \in \mathbb{R}^d : |x - \rho_{n-1}| < r_n\} \quad \text{and} \quad \sigma_{B_n} = \inf\{t > 0 : X_t \notin B_n\}. \tag{6.4.7}$$

The algorithm terminates at the random index $N = \min\{n \geq 0 : \rho_n \notin \Omega\}$, i.e., when the walk-on-spheres jumps out of the domain Ω. We refer the reader to Kyprianou et al. (2018) for details of the implementation of this procedure.

For two different classes of domain Ω, Kyprianou et al. also proved that for all $x \in \Omega$, the index N will always be finite and at most geometrically distributed. The first class is convex (possibly unbounded) domains, and the second class is non-convex bounded domains that satisfy the uniform exterior-cone condition. See Kyprianou et al. (2018) for proofs in either case and a discussion of how convexity relates to the proof. Furthermore, the convergence of the walk-on-spheres method is proved, and the code for the numerical examples implemented in Kyprianou et al. (2018) can be found at

https://bitbucket.org/wos_paper/wos_repo.

Using the sequences of spheres, the Feynman–Kac formula for the solution u can be replaced by an expectation over the boundary condition g evaluated at the terminal centers ρ_N, and the right-hand side f integrated over the expected occupation measure of the stable process prior to exiting each sphere. The final representation obtained by Kyprianou et al. (2018) is, for $x \in D$, $g \in L^1_\alpha(D^c)$ and $f \in C^{\alpha+\varepsilon}(\overline{D})$,

$$u(x) = \mathbb{E}_x[g(\rho_N)] + \mathbb{E}_x\left[\sum_{n=0}^{N-1} r_n^\alpha V_1\big(0, f(\rho_n + r_n\cdot)\big)\right],$$

where $V_1(x, f(\cdot)) = \int_{|y-x|<1} f(y)\, V_1(x, \mathrm{d}y)$. The expected occupation measure $V_1(x, \mathrm{d}y)$ of the stable process prior to exiting a unit ball centered at the origin is given by Boggio (1905); Blumenthal et al. (1961), for $|y| < 1$, by

$$V_1(0, \mathrm{d}y) = 2^{-\alpha}\pi^{-d/2}\frac{\Gamma(d/2)}{\Gamma(\alpha/2)^2}|y|^{\alpha-d}\left(\int_0^{|y|^{-2}-1}(u+1)^{-d/2}u^{\alpha/2-1}\mathrm{d}u\right)\mathrm{d}y.$$

The implementation of this formula, using Monte Carlo integration for the integral $V_1\big(0, f(\rho_n + r_n\cdot)\big)$, is discussed in Kyprianou et al. (2018). We use the WOS algorithm in Figure 6.11 and throughout Sections 6.8 and 6.9.

6.5 Spectral Definition

Several approaches have been proposed for discretizing the spectral definition, which is defined in equation (6.3.49). Stinga and Torrea (2010) showed that the fractional Poisson problem

$$\begin{cases} \big(-\Delta_{\text{spectral}}\big)^{\alpha/2}u(x) = f(x), & x \in \Omega, \\ u(x) = 0, & x \in \partial\Omega, \end{cases} \tag{6.5.1}$$

can be recast as a problem over the extruded domain $\mathcal{C} = \Omega \times [0, \infty)$:

$$\begin{cases} -\nabla \cdot y^\beta \nabla U(x, y) = 0, & (x, y) \in \mathcal{C}, \\ U(x, y) = 0, & (x, y) \in \partial_L\mathcal{C} := \partial\Omega \times [0, \infty), \\ \frac{\partial U}{\partial n^\beta}(x) = d_\alpha f(x), & x \in \Omega, \end{cases} \tag{6.5.2}$$

where $\beta = 1 - \alpha/2$, $d_\alpha = 2^{1-\alpha}\frac{\Gamma(1-\alpha/2)}{\Gamma(\alpha/2)}$, and

$$\frac{\partial U}{\partial n^\beta}(x) = - \lim_{y \to 0^+} y^\beta \frac{\partial U}{\partial y}(x, y).$$

The solution to (6.5.1) can then be recovered by taking the trace of U on Ω, i.e., $u = \mathrm{tr}_\Omega U$. As this higher-dimensional formulation involves only integer-order operators, standard discretization approaches may be applied, as in Nochetto et al. (2015); Gatto and Hesthaven (2015); Ainsworth and Glusa (2018a).

The work of Nochetto et al. (2015) used the Dirichlet-to-Neumann map for a singular elliptic problem posed on a semi-infinite cylinder in order to study solution techniques for problems on bounded domains with Dirichlet, Neumann, and Robin boundary conditions. A truncation was proposed based on the rapid decay of the solution to the problem on the cylinder, and a priori error estimates were derived in weighted Sobolev spaces. Also along these lines, a hybrid finite element-spectral method was recently introduced by Ainsworth and Glusa (2018a), where the discretization along the direction of the problem domain Ω was done using a finite element method, and the direction along the cylinder was discretized with a spectral method. We emphasize that these methods are all for the fractional Poisson problem with zero Dirichlet boundary conditions. For the computations in this section, we prefer to use the spectral element method of Song et al. (2017), described below, due to its accuracy and ease of implementation for the considered examples.

6.5.1 Discrete Eigenfunction and Spectral Element Methods

We consider the following eigenvalue problem (EVP) for the spectral Laplacian operator:

$$-\Delta\phi - \lambda\phi = 0, \quad x \in \Omega; \quad \phi|_{\partial\Omega} = 0, \tag{6.5.3}$$

where $\Omega \subset \mathbb{R}^d$, with $d = 1, 2, 3$, is a bounded domain. When $d = 1$, we use a Galerkin expansion to discretize (6.5.3). For a non-negative integer N, let $\mathbb{P}_N(\Omega)$ be the space of polynomials on Ω up to order N. The Galerkin basis functions $\{p_n(x)\}_{n=1}^N$ are chosen from the space $\mathcal{S}_N(\Omega) := \mathbb{P}_N(\Omega) \cap H_0^1(\Omega)$, so that the eigenfunctions ϕ can be approximated as

$$\phi \approx \sum_{n=1}^{N} \hat{\phi}_n p_n(x), \tag{6.5.4}$$

where $\hat{\phi}_n = (\phi, p_n)$ and $(\cdot, \cdot)$ represents the $L^2(\Omega)$-inner product. The weak form of equation (6.5.3) is then written as

$$(-\Delta\phi, p_k) - \lambda(\phi, p_k) = (\nabla\phi, \nabla p_k) - \lambda(\phi, p_k) = 0. \tag{6.5.5}$$

The inner product of the gradients is discretized according to

$$(\nabla\phi, \nabla p_k) = \sum_{n=1}^{N} \hat{\phi}_n (\nabla p_n, \nabla p_k) = A_N \phi, \tag{6.5.6}$$

where $(A_N)_{kn} = (\nabla p_n, \nabla p_k)$ and $\phi = (\hat{\phi}_1, \hat{\phi}_2, \ldots, \hat{\phi}_N)$. Again using the expansion (6.5.4) of ϕ in (ϕ, p_k) yields

$$(\phi, p_k) = \sum_{n=1}^{N} \hat{\phi}_n (p_n, p_k) = M_N \, \phi, \tag{6.5.7}$$

where $(M_N)_{kn} = (p_n, p_k)$. These discretizations result in the discrete eigenproblem

$$A_N \phi - \lambda M_N \phi = 0. \tag{6.5.8}$$

We left-multiply equation (6.5.8) by M_N^{-1}, so that the discrete eigenproblem becomes

$$(M_N^{-1} A_N - \lambda)\phi = 0. \tag{6.5.9}$$

We define the matrix $K_N := M_N^{-1} A_N$ and compute its discrete eigenpairs $\{(\lambda_i, \phi_i)\}_{i=1}^{N}$ using the QR algorithm, following Song et al. (2017). The eigenpairs are used to approximate the fractional Laplacian of a function u according to

$$(-\Delta_{\text{spectral}})^{\alpha/2} u \approx \sum_{i=1}^{N} \lambda_i^{\alpha/2} (u, \phi_i)_{L^2(\Omega)} \phi_i. \tag{6.5.10}$$

Furthermore, the solution of the spectral fractional Poisson equation with zero Dirichlet boundary conditions may be approximated using the formula

$$u \approx \sum_{k=1}^{N} u_k \phi_k, \quad \text{where} \quad u_k := \frac{(f, \phi_k)}{\lambda_k^{\alpha/2}}. \tag{6.5.11}$$

We call this method the *discrete eigenfunction method*. When $d > 1$, we use a spectral element method (SEM) to discretize (6.5.3). First, a grid with ℓ elements is generated to discretize the domain Ω. The SEM is developed using nodal Lagrangian polynomials of degree N on each element. The number of degrees of freedom for the SEM is denoted by $\mathcal{N} := \ell \cdot N$. $A_{\mathcal{N}}$ represents the $\mathcal{N} \times \mathcal{N}$ stiffness matrix associated with the integer Laplacian, where each element of $A_{\mathcal{N}}$ is the L^2-inner product of the gradients of the Lagrangian polynomials. Similarly, the mass matrix $M_{\mathcal{N}}$ of size $\mathcal{N} \times \mathcal{N}$ is computed as the

L^2-inner product of the Lagrangian polynomials. The eigenpairs $\{\lambda_i, \phi_i\}_{i=1}^{\mathcal{N}}$ are computed by solving the eigenproblem (6.5.8). Then, a weighted Gram–Schmidt procedure is applied to transform the basis $\{\phi_i\}$ into the orthonormal basis $\{\tilde{\phi}_i\}_{i=1}^{\mathcal{N}}$ (see Song et al. (2017) for the details of this procedure). Finally, we use the equation

$$\sum_{i=1}^{\mathcal{N}} \lambda_i^{\alpha/2}(u, \tilde{\phi}_i)\tilde{\phi}_i \approx \sum_{i=1}^{\infty}(f, \tilde{\phi}_i)\tilde{\phi}_i \approx \sum_{i=1}^{\mathcal{N}}(f, \tilde{\phi}_i)\tilde{\phi}_i, \tag{6.5.12}$$

from which we infer that $\hat{u}_i = (u, \tilde{\phi}_i) = (f, \tilde{\phi}_i)\lambda_i^{-\alpha/2}$. Then, the solution is written as $u \approx \sum_{i=1}^{\mathcal{N}} \hat{u}_i \tilde{\phi}_i$. Further details of this method can be found in Song et al. (2017), where it is demonstrated in numerical examples that the method is stable and accurate for the same domains that we consider in the following sections. Rigorous proofs of these properties are under development and have not been published at the time of this writing.

6.5.2 Boundary Regularity of Solutions Using the Discrete Eigenfunction Method

Now we address an important issue of regularity near the boundary and the enforcement of boundary conditions of solutions to the spectral fractional Poisson equation using the discrete eigenfunction method. Recall the regularity results of Grubb (2016) for the Dirichlet problem for the spectral fractional Laplacian for $\alpha < 1$, reviewed in Section 6.3.6: if $(-\Delta)^{\alpha/2}u = f$, and if $f \in H^s$, then $u \in H^{s+\alpha}$. This means that if $s + \alpha \leq 1/2$, then $u \notin H^{1/2+\varepsilon}$. However, numerical solutions obtained by the discrete eigenfunction method even with $s + \alpha \leq 1/2$ will always satisfy the zero boundary condition exactly, even when the true solution, by the regularity result just stated, does not admit a trace. This is an artifact of the discrete eigenfunction method, in which the approximant is always a finite-dimensional projection of u onto the eigenfunctions, which are zero on the boundary. As a simple example in one dimension, one can plot the partial sums of the solution (6.3.85) to the homogeneous spectral fractional Poisson with the strictly L^2 right-hand side (6.3.84); for $\alpha \leq 0.5$, the lack of trace of the true solution will manifest as Gibbs oscillations as the number of discrete eigenfunctions is increased.

6.6 Directional Representation

The directional fractional Laplacian ∇_M^α in $\mathbb{R}^d$ (6.3.34) was reviewed in Section 6.3.1. It was pointed out that for a uniform measure M, the directional fractional Laplacian reduces to the standard fractional Laplacian $(-\Delta)^{\alpha/2}$ that is the focus; otherwise, it represents the generator of general multivariate α-stable Levy motions. Thus, when applied to functions in a bounded domain satisfying an exterior boundary condition, this (uniform) directional fractional Laplacian agrees with the Riesz fractional Laplacian, and is an advantageous representation in certain geometries.

In this section, we present a new radial basis function collocation method based on the directional representation and a modified vector Grünwald–Letnikov formula to discretize the Poisson equation in bounded domains for the Riesz fractional Laplacian. The method has a clear extension to directional fractional Laplacians (with non-uniform measures), but to maintain the scope of the section, we focus only on the uniform case resulting in the Riesz fractional Laplacian. The use of this method for numerical studies of general directional fractional Laplacians in bounded domains is an interesting topic for future work.

To introduce the method and illustrate the main idea, we split the fractional directional integral (6.3.31) appearing in the directional representation (6.3.34) of the fractional Laplacian as the sum of two integrals corresponding to integration of u inside and outside the bounded domain Ω:

$$\begin{aligned} I_\theta^\beta u(x) = \frac{1}{\Gamma(1-\beta)} \bigg(& \int_0^{\delta(x,\theta,\Omega)} \varsigma^{-\beta} u(x - \varsigma\theta) d\varsigma \\ & + \int_{\delta(x,\theta,\Omega)}^{+\infty} \varsigma^{-\beta} g(x - \varsigma\theta) d\varsigma \bigg). \end{aligned} \tag{6.6.1}$$

Here, $\delta(x, \theta, \Omega)$ is the distance from x to the domain boundary $\partial\Omega$ along the direction $-\theta$. In Pang et al. (2013, 2015), $\delta(\cdot, \cdot, \cdot)$ is termed the *backward distance*. The nonlocal exterior Dirichlet condition $u(x) = g(x)$ for $x \in \mathbb{R}^d \setminus \Omega$ is assumed. Note that for $g \equiv 0$, the second integral vanishes. For general (non-zero) g, we introduce a finite difference (Grünwald–Letnikov type) scheme to approximate the fractional directional derivative $D_\theta^\alpha u(x)$ following the above representation for $I_\theta^\beta u(x)$, which is then used in a collocation method.

6.6.1 A Radial Basis Function Collocation Method for the Riesz Fractional Laplacian

The radial basis function (RBF) collocation method has been used to solve equations involving directional fractional Laplacians (Pang et al., 2015). Here, we focus on the fractional Poisson problem

$$\begin{aligned} (-\Delta_{\text{Riesz}})^{\alpha/2} u(x) &= f(x), \quad x \in \Omega \subset \mathbb{R}^d, \\ u(x) &= g(x), \quad x \in \mathbb{R}^d \setminus \Omega. \end{aligned} \tag{6.6.2}$$

The first step of the RBF collocation method is to approximate $u(x)$ inside Ω as the weighted sum of RBFs:

$$u(x) \approx \sum_{j=1}^{M+N} \lambda_j \phi(|x - x_j|), \quad x \in \Omega, \tag{6.6.3}$$

where λ_j's are unknown coefficients. The RBF $\phi(\cdot)$ is taken as the multiquadric function with the shape parameter c,

$$\phi(r) = \sqrt{r^2 + c^2}.$$

It is straightforward to consider other RBFs (Chen et al., 2014), but here we restrict our attention to the multiquadric RBF. The collection of points $\{x_j\}_{j=1}^{j=M+N}$ are called collocation points. The first M collocation points are located inside the domain Ω, and the last N collocation points are located on the boundary $\partial\Omega$ (to obtain higher accuracy near $\partial\Omega$). Enforcing the approximate solution (6.6.3) to satisfy both the equation inside Ω and the exterior condition on $\partial\Omega$ from (6.6.2) produces the following equations:

$$\begin{aligned} &C_{\alpha,d} \int_{|\theta|=1} D_\theta^\alpha u(x_i) \mathrm{d}\theta = f(x_i), \quad x_i \in \Omega,\ i = 1, 2, \ldots, M, \\ &\sum_{j=1}^{M+N} \lambda_j \phi(|x_i - x_j|) = g(x_i), \quad x_i \in \partial\Omega,\ i = M+1, M+2, \ldots, M+N. \end{aligned} \tag{6.6.4}$$

The full exterior condition $u = g$ on $\mathbb{R}^d \setminus \Omega$ in (6.6.2) is enforced directly through the discretization of the operator D_θ^α in the first equation above as alluded to by (6.6.1). We discretize D_θ^α through a finite difference scheme known as the vector Grünwald–Letnikov (GL) formula that was proposed in Meerschaert et al. (2004), splitting the terms into interior and exterior contributions. The original vector GL formula takes the form

$$D_\theta^\alpha u(x_i) = h^{-\alpha} \sum_{k=0}^{\infty} (-1)^k \binom{\alpha}{k} u(x_i - kh\theta) + O(h). \tag{6.6.5}$$

This formula has first-order accuracy with respect to the spatial step h. The coefficients $c_k = (-1)^k \binom{\alpha}{k}$ can be alternatively calculated by the iterative formula $c_0 = 1, c_k = \left(1 - \frac{\alpha+1}{k}\right) c_{k-1}, k \geq 1$. For our scheme, we truncate the vector GL formula and combine it with the RBF approximation (6.6.3):

$$
\begin{aligned}
D_\theta^\alpha u(x_i) &\approx h^{-\alpha}\sum_{k=0}^{K_1} c_k u(x_i - kh\theta) + h^{-\alpha}\sum_{k=K_1+1}^{K_2} c_k g(x_i - kh\theta) \\
&= h^{-\alpha}\sum_{k=0}^{K_1} c_k \sum_{j=1}^{M+N} \lambda_j \phi(|x_i - kh\theta - x_j|) \\
&\quad + h^{-\alpha}\sum_{k=K_1+1}^{K_2} c_k g(x_i - kh\theta).
\end{aligned}
\tag{6.6.6}
$$

The integer K_1 depends on x_i, θ and Ω – it is determined by the value of the backward distance $\delta(x, \theta, \Omega)$, and is an integer taken such that $x_i - K_1 h\theta \in \Omega$ and $x_i - (K_1 + 1)h\theta \in \mathbb{R}^d \setminus \Omega$. It is easy to see that $\delta(x_i, \theta, \Omega) \approx K_1 h$ by noting that $|\theta| = 1$. We refer to formula (6.6.6) as the modified GL formula.

Before discussing how (6.6.6) is used in (6.6.4) to obtain a numerical scheme, we discuss in more detail the truncation parameter K_2. Unlike K_1, the integer K_2 is user-selected and affects the accuracy of the modified GL formula (6.6.6), introducing the truncation error

$$
R(x_i) = h^{-\alpha}\sum_{k=K_2+1}^{\infty} c_k g(x_i - kh\theta) \tag{6.6.7}
$$

into the calculation of $D_\theta^\alpha u(x_i)$. In the case that g has compact support, choosing $K_2 h \geq \operatorname{diam}(\operatorname{supp} g)$ ensures no truncation error. In the case that g does not have compact support, but decays at infinity, the following are two strategies to truncate:

(i) Replace g in the problem (6.6.2) by $g\chi_{\mathcal{D}}$, where $\chi_{\mathcal{D}}$ denotes the indicator function of an appropriate truncation domain $\mathcal{D}$ outside of which g is sufficiently small:

$$
\begin{aligned}
(-\Delta_{\text{Riesz}})^{\alpha/2}\tilde{u} &= f(x), \quad x \in \Omega \subset \mathbb{R}^d, \\
\tilde{u}(x) &= g(x)\chi_{\mathcal{D}}(x), \quad x \in \mathbb{R}^d \setminus \Omega.
\end{aligned}
\tag{6.6.8}
$$

One can then use the modified GL formula (6.6.6) with $g\chi_{\mathcal{D}}$ as the exterior condition and $K_2 h \geq \operatorname{diam}(\mathcal{D})$. Theoretically, the total error in solving (6.6.2) is controlled by the numerical error in solving the truncated problem (6.6.8) plus the truncation error, i.e., the difference of exact solutions $u - \tilde{u}$. This truncation error has been studied by Acosta et al. (2018), where the same truncation strategy was used to implement a finite element solver for the Riesz fractional Poisson problem with nonzero exterior condition. It was shown that $g \in H^s(\mathbb{R}^d \setminus \Omega)$ implies that as $\text{diam}(\mathcal{D}) \to \infty$, $\|u - \tilde{u}\|_{H^{\alpha/2}(\Omega)} \to 0$. Practically, this strategy corresponds to choosing $K_2 h \approx \delta(x_i, \theta, \mathcal{D})$.

(ii) A more direct way to exploit the decay of the coefficients in (6.6.5) is to choose K_2 such that $R(x_i)$ is below a threshold value. In the simplest case where K_2 is chosen to be a constant independent of x_i, theoretically this corresponds to truncating the integral (6.6.1) or equivalently (6.3.31) in the definition of the operator. In effect, this introduces a truncating parameter or *horizon* into the directional fractional Laplacian (6.3.28) (more generally (6.3.34)). Truncated forms of the Riesz fractional Laplacian are discussed in Section 6.7, where references are given that establish convergence of the solutions of the associated Poisson problem to those of (6.6.2) in the case of zero exterior condition. However, we are not aware of such convergence results for the inhomogeneous problem, or for works that establish a connection to truncating the directional representations (6.3.34).

We adopt the second strategy; in practice, regardless of the strategy, the choice of K_2 should take into account the rate of decay of the exterior condition g away from Ω. The truncation error should be studied both at the level of approximating $D_\theta^\alpha u(x)$ (i.e., by computing (6.6.7)) at various points x and at the level of the final numerical solution using the RBF collocation scheme outlined below; the numerical solution should be converged with respect to K_2 for a fixed set of collocation points.

We demonstrate this study for the case of $\Omega = \{(x_1, x_2) \in \mathbb{R}^2 \mid x_1^2 + x_2^2 < 1\}$ and $g(x) = \exp(-|x|^2)$ for $x \in \mathbb{R}^2 \setminus \Omega$. In Figure 6.6, the remainder $|R(x_i)|$ in (6.6.7) is compared for $K_2 = 2000$ and $K_2 = 6000$ with the spatial step fixed to be $h = 0.001$ at two “extreme cases” of collocation points $x_1 = (0, 0)$ and $x_2 = (0.98, 0)$. The summation in $R(x_i)$ is taken from $K_2 + 1$ to $20,000$. Comparing the subplots indicates that for the collocation point $x_1 = (0, 0)$ at the center of the disk, the truncation error is isotropic with respect to θ due to the symmetry of $g(x) = \exp(-|x|^2)$ around x_1, but for the point $x_2 = (0.98, 0)$ near the boundary, the truncation error is direction-dependent. However, the choice of $K_2 = 6000$ yields $|R(x_i)|$ sufficiently close to machine precision for use in the RBF collocation scheme.

We also verify that the solution obtained by solving the RBF collocation scheme below is converged at this value of K_2 in Figure 6.7. In that figure, we show the relative L^2-error of the final numerical solution from the reference solution with respect to K_2 (for full details, see the discussion of (6.6.18) below). Not surprisingly, the final numerical solution u is less sensitive to

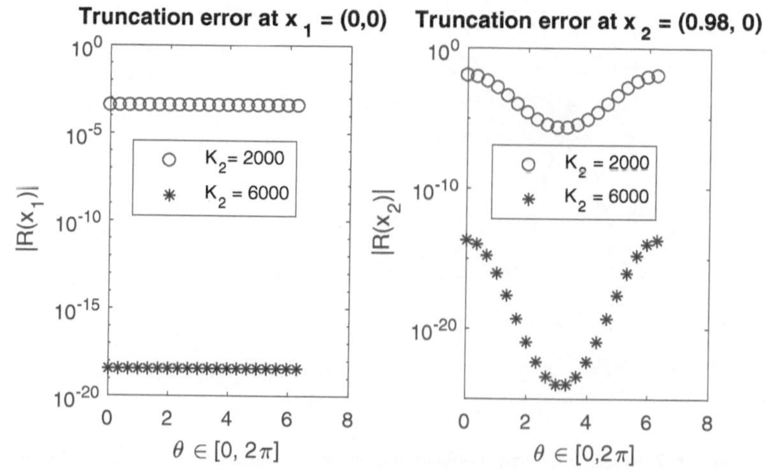

(a) $\alpha = 0.5$

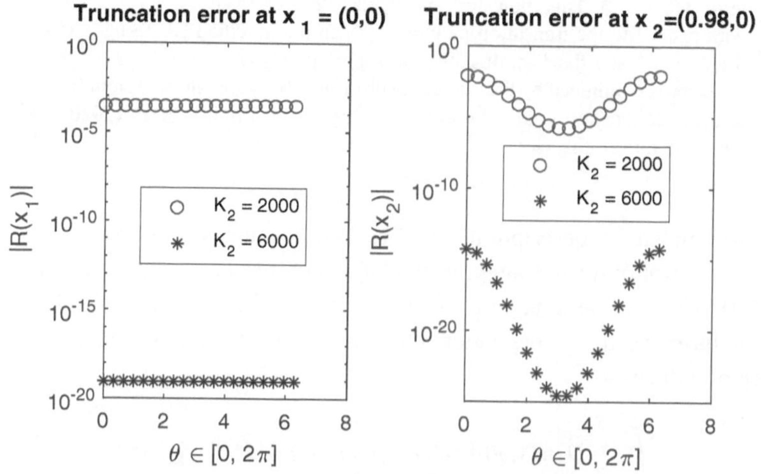

(b) $\alpha = 1.5$

Figure 6.6 Absolute values of truncation errors $|R(x_i)|$ (6.6.7) for the modified GL formula (6.6.6) evaluated at the collocation points $x_1 = (0,0)$ and $x_2 = (0.98, 0)$. Two truncation parameters $K_2 = 2000$ and $K_2 = 6000$ are considered. Subplots (a) and (b) correspond to $\alpha = 0.5$ and $\alpha = 1.5$, respectively. Source: Reprinted from Lischke et al. (2020) with permission from Elsevier.

K_2 than is $D_\theta^\alpha u$, and is actually converged near $K_2 = 2000$ (at which point the error becomes dominated by the numerical error of solving the RBF collocation system below). This justifies the use of $K_2 = 6000$ with $h = 0.001$ for this choice of Ω and g, which are used as test cases below.

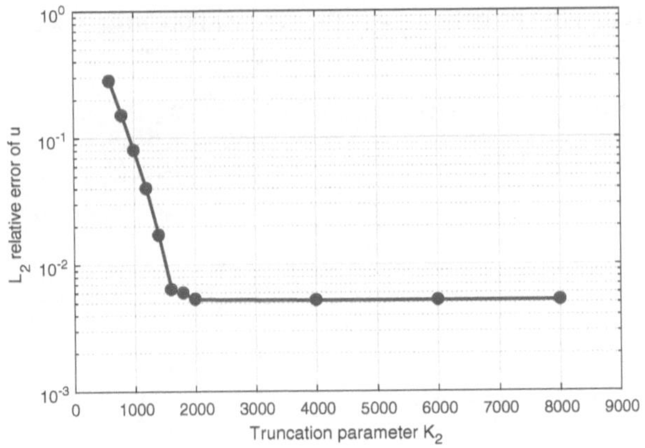

Figure 6.7 Relative L^2 error between the reference solution and the RBF solution to the problem (6.6.2) on the unit disk with $g(x) = \exp(-|x|^2)$ and zero source term $f(x) = 0$. This indicates convergence to a consistent numerical solution with respect to the truncation parameter K_2 in the modified GL formula (6.6.6) for $\alpha = 1.5$ at a fixed set of 2500 collocation points ($I = 39$); for $K_2 > 2000$, the error is dominated by the numerical error in solving the linear system (6.6.11) at these collocation points. Source: Reprinted from Lischke et al. (2020) with permission from Elsevier.

To complete the description of the RBF collocation scheme, next we discuss the discretization of the integral of $D_\theta^\alpha(\cdot)$ with respect to θ, as called for by (6.6.4). This can be done, e.g., with the trapezoidal rule or Gauss–Legendre quadrature; we adopt the latter. Of course, in one dimension, the desired integral reduces to

$$\int_{|\theta|=1} D_\theta^\alpha u(x_i) d\theta = D_{\theta=(1,0)}^\alpha u(x_i) + D_{\theta=(-1,0)}^\alpha u(x_i)$$

and no quadrature rule is needed. In two dimensions, the integral over θ is approximated by

$$\begin{aligned} \int_{|\theta|=1} D_\theta^\alpha u(x_i) d\theta &= \int_0^{2\pi} D_{\theta=(\cos\theta,\sin\theta)}^\alpha u(x_i) d\theta \\ &\approx \sum_{l=1}^{P} w_l D_{\theta_l=(\cos\theta_l,\sin\theta_l)}^\alpha u(x_i). \end{aligned} \tag{6.6.9}$$

w_l and θ_l are the Gauss–Legendre weights and points, respectively. In the three dimensions case, Gauss–Legendre quadrature yields

$$\int_{|\theta|=1} D_\theta^\alpha u(x_i)d\theta = \int_0^{2\pi}\int_0^\pi D^\alpha_{\theta=(\sin\phi\cos\theta,\sin\phi\sin\theta,\cos\phi)} u(x_i)\sin\phi d\phi d\theta$$
$$\approx \sum_{l=1}^{P}\sum_{m=1}^{Q} w_l v_m \sin\phi_m D^\alpha_{\theta_l=(\sin\phi_m\cos\theta_l,\sin\phi_m\sin\theta_l,\cos\phi_m)} u(x_i),$$

where (ϕ_m, v_m) and (θ_l, w_l) are Gauss–Legendre quadrature point and weight pairs, respectively. The $\sin\phi$ appearing in the integrand comes from the determinant of the Jacobian matrix in the transformation from Cartesian coordinates to spherical coordinates. Once the integral of $D_\theta^\alpha u$ has been discretized as above, it can be combined with the modified GL formula (6.6.6) to obtain a linear system for the coefficients λ_j in the RBF expansion (6.6.3).

For simplicity, we consider two dimensions. According to the modified GL formula (6.6.6) and the quadrature rule (6.6.9), we can rearrange the discretized system (6.6.4) as a system of $M+N$ equations in $M+N$ unknown coefficients:

$$C_{\alpha,d}h^{-\alpha}\sum_{j=1}^{M+N}\sum_{l=1}^{P} w_l \sum_{k=0}^{K_1} c_k\phi(|x_i - kh\theta_l - x_j|)\lambda_j =$$

$$f(x_i) - C_{\alpha,d}h^{-\alpha}\sum_{l=1}^{P} w_l \sum_{k=K_1+1}^{K_2} c_k g(x_i - kh\theta_l), \quad x_i \in \Omega, \quad i = 1,2,\ldots,M;$$

$$\sum_{j=1}^{M+N} \phi(|x_i - x_j|)\lambda_j = g(x_i), \quad x_i \in \partial\Omega, \quad i = M+1, M+2,\ldots,M+N.$$

$$(6.6.10)$$

This linear system can also be abbreviated in matrix-vector form as

$$\begin{bmatrix} (-\Delta)^{\alpha/2}\boldsymbol{\Phi}_{M\times(M+N)} \\ \boldsymbol{\Phi}_{N\times(M+N)} \end{bmatrix}\begin{bmatrix} \lambda_{M\times 1} \\ \lambda_{N\times 1}\end{bmatrix} = \begin{bmatrix} \boldsymbol{F} - \boldsymbol{R}(K_2) \\ \boldsymbol{G}\end{bmatrix}. \quad (6.6.11)$$

Validation Using Reference Solutions and Walk-on-Spheres

Next, we demonstrate the convergence of the numerical solutions produced by the RBF collocation method by considering a circular domain $\{(x_1, x_2) \in \mathbb{R}^2 \mid x_1^2 + x_2^2 < 1\}$. In principle, the collocation points $\{x_i\}$ can have arbitrary distribution in Ω provided they do not overlap. We choose a total of I^2+1 collocation points in $\Omega \cup \partial\Omega$ on a uniform grid in polar coordinates:

$$x_i = \big(i_1\Delta r\cos(i_2\Delta\theta), i_1\Delta r\sin(i_2\Delta\theta)\big), \quad i = (i_1-1)\times I + i_2, \quad i_1, i_2 = 1,2,\ldots I, \quad (6.6.12)$$

where $\Delta r = 1/I$, and $\Delta\theta = 2\pi/I$. The subscript pair (i_1, i_2) labeling a collocation point in two dimensions is used to define the single subscript i to index the vector of collocation points/weights. There are $M = (I-1)\times I+1$ collocation points inside Ω, consisting of $I-1$ concentric rings of I collocation points together with the center point $(0,0)$, and $N = I$ collocation points on $\partial\Omega$. The number of collocation points varies for convergence studies, but the numerical errors are evaluated at a fixed collection of test points $\mathbf{x}_{\text{test}}$ which are generated via (6.6.12) using $I = 39$ throughout. The shape parameter c in the RBF $\phi(\cdot)$ is selected as $c = 3/I$. For the Gauss–Legendre quadrature (6.6.9), 16 quadrature points are taken, i.e, $P = 16$. The spatial step in the modified GL formula is $h = 0.001$. Standard LU decomposition in MATLAB is used to solve the linear system (6.6.11).

Below, we test both the case of zero exterior condition, in which case the term $\sum_{k=K_1+1}^{K2} c_k g(x_i - kh\theta)$ in the modified GL formula (6.6.6) is simply zero, and the nonzero exterior condition case, in which case we take $K_2 = 6000$, i.e., $K_2 h = 6$, three times the diameter of the unit disk.

Convergence of the GL Formula. Before demonstrating the convergence of RBF solutions, we first demonstrate the convergence of the vector GL formula (6.6.6) with respect to the parameter h. In two dimensions, we consider the following two functions:

$$\text{(i) } u_1(x) = (1-|x|)_+^{\alpha/2} \text{ and (ii) } u_2(x) = (1-|x|)_+^{1+\alpha/2}, \tag{6.6.13}$$

where $(x)_+ = x$ if $x \geq 0$ and $(x)_+ = 0$ if $x < 0$. The domain is taken to be a unit disc: $\Omega = \{(x_1, x_2) \mid x_1^2 + x_2^2 < 1\}$. The Riesz fractional Laplacians of these functions are (Dyda, 2012)

$$f_1(x) = (-\Delta_{\text{Riesz}})^{\alpha/2} u_1(x) = 2^\alpha \Gamma\left(\frac{\alpha}{2}+1\right)^2, \tag{6.6.14}$$

and

$$\begin{aligned} f_2(x) &= (-\Delta_{\text{Riesz}})^{\alpha/2} u_2(x) \\ &= 2^\alpha \Gamma\left(\frac{\alpha}{2}+2\right)\Gamma\left(\frac{\alpha}{2}+1\right)\left(1-\left(1+\frac{\alpha}{2}|x|^2\right)\right), \end{aligned} \tag{6.6.15}$$

respectively.

To measure the error, we use the L^∞-error

$$\begin{aligned} \epsilon_1 &= \|(-\Delta)^{\alpha/2} u_1(x_{\text{test}}) - f_1(x_{\text{test}})\|_{L^\infty}, \\ \epsilon_2 &= \|(-\Delta)^{\alpha/2} u_2(x_{\text{test}}) - f_2(x_{\text{test}})\|_{L^\infty}, \end{aligned} \tag{6.6.16}$$

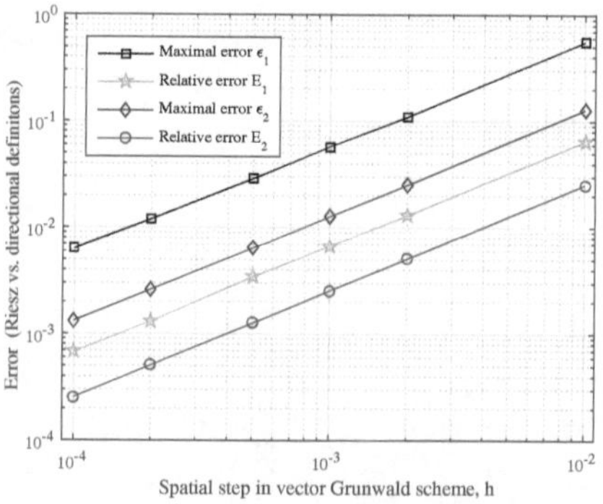

Figure 6.8 *Convergence of the modified GL formula with respect to the step size h for the 2D problem with* $g(x) = 0$ *on the unit disk.* We plot the error between the modified vector GL formula (6.6.6) approximation of the fractional Laplacian of $u_1(x)$ and $u_2(x)$ and the source functions $f_1(x)$ and $f_2(x)$, respectively. The slope of each line is 1, demonstrating that the convergence rate of the vector Grünwald scheme is $\mathcal{O}(h)$. Source: Reprinted from Lischke et al. (2020) with permission from Elsevier.

as well as the relative L^2-error

$$\begin{aligned} E_1 &= \frac{\|(-\Delta)^{\alpha/2} u_1(x_{\text{test}}) - f_1(x_{\text{test}})\|_{L^2}}{\|f_1(x_{\text{test}})\|_{L^2}}, \\ E_2 &= \frac{\|(-\Delta)^{\alpha/2} u_2(x_{\text{test}}) - f_2(x_{\text{test}})\|_{L^2}}{\|f_2(x_{\text{test}})\|_{L^2}}, \end{aligned} \tag{6.6.17}$$

where $(-\Delta)^{\alpha/2} u(x_{\text{test}})$ and $f_i(x_{\text{test}})$ are the vectors formed by the directional Laplacian using (6.6.6) and the functions (6.6.14), (6.6.15) evaluated at a collection of test points, respectively. Figure 6.8 demonstrates first-order convergence of the errors with respect to h, which agrees with the theoretical convergence rate given in Meerschaert et al. (2004).

Convergence of the RBF Solution for $g(x) = 0$. We demonstrate the convergence of the RBF collocation method with respect to the number of collocation points (namely, $M + N$ in RBF expansion (6.6.3)) for solving the fractional Poisson problem with a zero Dirichlet boundary condition $g(x) = 0$ for $x \in \mathbb{R}^d \backslash \Omega$. We plot in Figure 6.9 the convergence of the RBF method for two fabricated solutions u_1 and u_2 and four different values of $\alpha = 0.1, 0.5, 1.5$ and 1.9. The convergence is measured in the relative L_2 error. It is observed from Figure 6.9 that the RBF method is convergent, although the convergence rate is reduced at larger numbers of collocation points due to the rapidly increasing condition number of the collocation matrix in the linear system

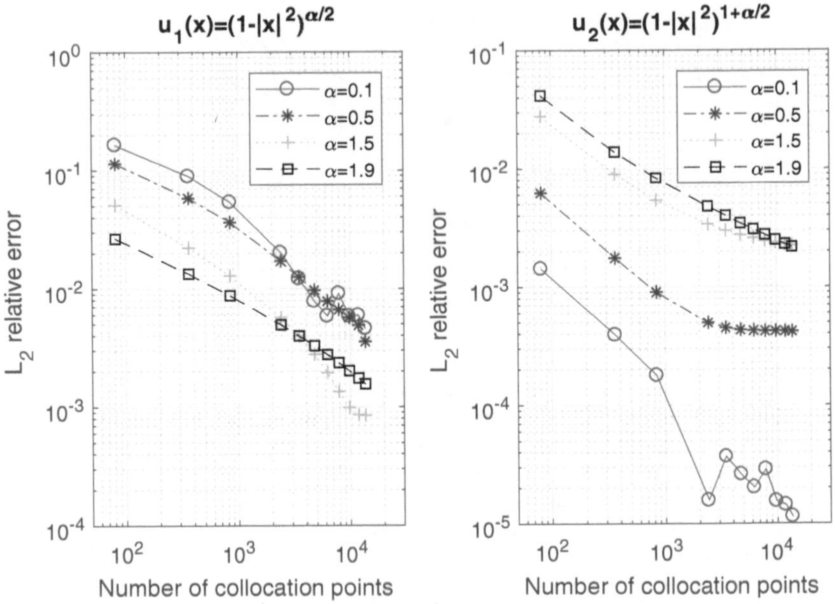

Figure 6.9 Convergence of the RBF method for the 2D fractional Poisson problem with zero exterior values on a unit disk. Left: fabricated solution u_1 with $\alpha = 0.1, 0.5, 1.5$, and 1.9. Right: fabricated solution u_2 with $\alpha = 0.1, 0.5, 1.5$, and 1.9. Source: Reprinted from Lischke et al. (2020) with permission from Elsevier.

(6.6.10); see Chen et al. (2014) for a discussion of this well-known issue. Also, we see that the RBF method achieves lower accuracy for u_1 compared to that for u_2; this can be expected since u_1 has much higher gradients than u_2 near the boundary. We study the local behavior of the error in Figure 6.10, in which exact and RBF solutions are compared. We see that the error is larger near the boundary because the collocation points are sparser near the boundary than around the domain center.

Convergence of the RBF Solution for $g(x) \neq 0$ and Comparison with Walk-on-Spheres.

From Theorem 2.11 of Bucur (2016), a solution formula for (6.6.2) with $f \equiv 0$ in the case that Ω is a ball B_r of radius r is

$$u(x) = \begin{cases} \displaystyle\int_{\mathbb{R}^d \setminus \Omega} P_r(y,x) g(y) \mathrm{d}y, & x \in B_r, \\ g(x), & x \in \mathbb{R}^d \setminus B_r. \end{cases} \tag{6.6.18}$$

The fractional Poisson kernel $P_r(y,x)$ is given by

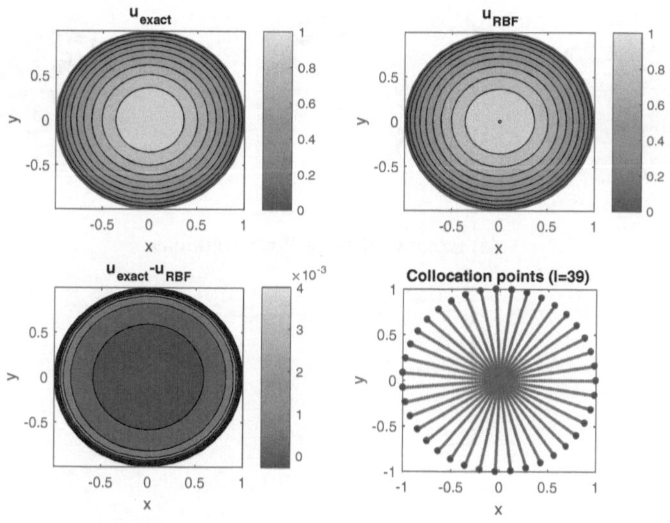

Figure 6.10 Comparison of reference and RBF solutions for fractional Poisson problem with zero exterior values on a unit disk. Upper left: Reference solution u_1 with $\alpha = 1.5$. Upper right: RBF solution with roughly 10,000 collocation points (namely $I = 99$). Lower left: Pointwise error between exact and RBF solutions. Lower right: Illustration of collocation points with $I = 39$ where darker gray and lighter gray points are boundary and domain collocation points, respectively. Source: Reprinted from Lischke et al. (2020) with permission from Elsevier.

$$P_r(y,x) = \frac{\Gamma(\frac{d}{2}) \sin \frac{\pi\alpha}{2}}{\pi^{\frac{d}{2}+1}} \left(\frac{r^2 - |x|^2}{|y|^2 - r^2} \right)^{\alpha/2} \frac{1}{|x-y|^d}. \tag{6.6.19}$$

Using the RBF collocation method, we solve this problem with $g(x) = \exp(-|x|^2)$. To obtain the reference solution, the integral in (6.6.18) is computed using a costly direct Monte Carlo (MC) integration with 5×10^7 samples. To obtain the RBF solution, as discussed in Section 6.6.1, we use the truncation parameter $K_2 = 6000$ and step size $h = 0.001$ in (6.6.6). Recall that we verified in Figure 6.6 that these values yielded sufficiently high accuracy of the discretization (6.6.6), and at the same time in Figure 6.7 that the final RBF solution was converged with respect to these truncation parameters. Using these parameters, we compute and plot the pointwise errors of RBF and WOS solutions with respect to the reference solution in Figure 6.11(a); we see that both RBF and WOS methods can achieve accurate results.

Next we compare errors and compute (wall) times of the RBF collocation and WOS methods in Figure 6.11(b) and Figure 6.11(c), respectively. Both methods are parallelized over 24 cores. We observe that the error of the

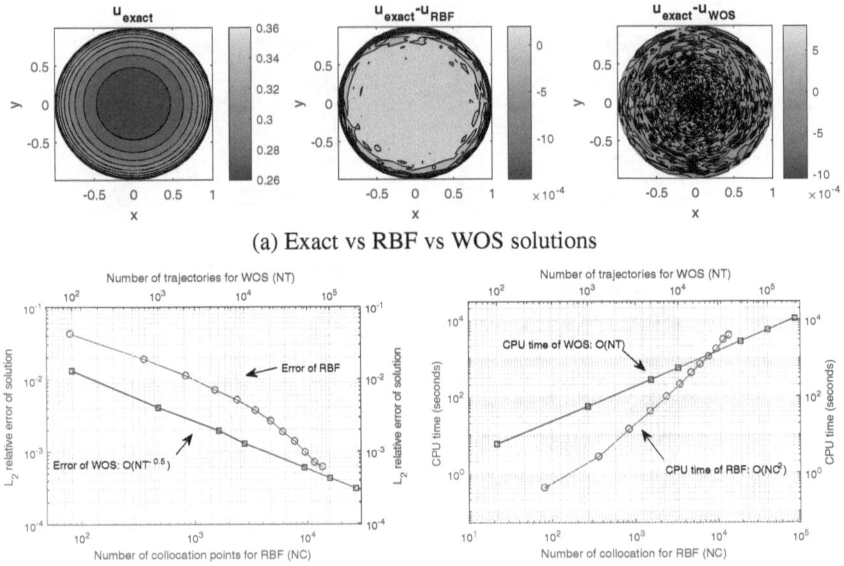

(a) Exact vs RBF vs WOS solutions

(b) Convergence

(c) CPU time

Figure 6.11 Comparison of the RBF and the WOS methods in terms of solution error and CPU time for the problem (6.6.2) on the unit disk with nonzero exterior condition $g(x) = \exp(-|x|^2)$ and zero source term $f(x) = 0$. (a) Pointwise errors of RBF and WOS methods with respect to the reference solution (b) Comparison of L_2 relative errors of RBF and WOS methods. (c) Comparison of CPU times of RBF and WOS methods; both methods are parallelized over 24 CPU cores. Source: Reprinted from Lischke et al. (2020) with permission from Elsevier.

WOS solution decays with the rate $O(NT^{-0.5})$ where NT is the number of trajectories, as expected for a Monte Carlo method. The error of the RBF method does not exhibit consistent convergence rate. Clearly, the convergence of the RBF method is stymied due to the increasing condition number of the collocation matrix. In terms of the computational time, WOS (which is embarassingly parallel) has time complexity $O(NT)$, whereas the time complexity of the RBF method is $O(NC^2)$ where NC is the number of RBF collocation points. The assembly time of the collocation matrix (6.6.11) is $O(NC^2)$, which dominates the solution time and accounts for the quadratic time complexity here. However, if we solve (6.6.11) by direct methods, cubic complexity will be observed as we increase the number of collocation points. For this reason, iterative methods should be adopted. However, the increasing condition number of the collocation matrix prevents successful application of iterative methods, and preconditioning is required. It is still an open problem to develop preconditioners for the fractional Laplacian collocation matrix.

For these simulation parameters, the two methods are roughly comparable and the WOS method has a slight edge over the RBF method. For example, to achieve a L_2 relative error of 10^{-3}, the WOS and the RBF methods take around 900 and 1100 seconds, respectively. Nevertheless, the RBF method has the potential to be more flexible and converge faster than the WOS method. The WOS method requires establishing the Feynman–Kac formula for the associated boundary-value problem, but for many fractional PDEs and boundary conditions, the corresponding Feynman–Kac formulas have not yet been found. Moreover, efficient sampling algorithms for WOS require characterization of exit distributions of the associated stochastic process.

The RBF collocation method introduced here for the Riesz fractional Laplacian takes advantage of the directional representation (6.3.28) and extends readily to general nonisotropic directional fractional Laplacians (6.3.34). We established two problems for further work: (1) Development of effective preconditioners for the RBF collocation matrices. This will improve convergence rates as the number of collocation points increases. (2) Study of *truncated* fractional directional Laplacians and error bounds for the difference between solutions of the associated boundary-value problems and those of (6.6.2). While in our example we drove the truncation error in (6.6.6) close to machine precision by choosing K_2 sufficiently large, theoretical justification for truncation in the modified GL formula should be studied. To have full impact, these questions should be studied in the context of the most general directional fractional Laplacians (6.3.34) connected to the class of multivariate stable processes, which goes beyond the scope of the current section.

6.7 Horizon-Based Nonlocal Definition

Even though we consider fractional equations involving bounded domains, the region of dependence for the Riesz and directional fractional Poisson equations is still $\mathbb{R}^d$. This is in contrast with the spectral fractional Laplacian, which admits only the usual locally-defined boundary conditions, and does not utilize information about the solution outside the bounded domain. However, in the Riesz or directional fractional Poisson problems, it can be noticed through computational experiments that the dependence of the solution on points far away from the domain decreases with distance. Du et al. (2012) proposed a type of nonlocal problem admitting a "horizon parameter", which is used to describe an "interaction domain", a proper subset of $\mathbb{R}^d$, in which all computational considerations take place. In other words, data from outside the interaction domain is ignored. Thus, the operator used in the formulation of these equations is distinct from operators we have presented above, although the nonlocal operator does approximate the Riesz fractional Laplacian as the interaction domain increases in size. Indeed, taking the horizon parameter to $+\infty$, we recover the Riesz definition. In this section, we describe this nonlocal formulation and demonstrate how the value of the horizon parameter affects the behavior of the solution to the following problem. A similar study of the nonlocal diffusion operator was carried out by D'Elia and Gunzburger (2013). Another noteworthy paper by Duo et al. (2019) compared numerical results of some one-dimensional fractional diffusion problems for different definitions of the fractional Laplacian and the horizon-based nonlocal operator, which they call the "peridynamic" operator.

6.7.1 Nonlocal Problem

Let $\Omega := (-1, 1) \subset \mathbb{R}$, $L = |\Omega| = 2$, and $B_\delta(x) = \{y \in \mathbb{R} : |y - x| < \delta\}$ denotes the interval centered at x having length 2δ, where δ is the horizon parameter. Define the interaction domain as:

$$\Omega_{\mathcal{I}} = \{y \in \mathbb{R} \setminus \Omega : |y - x| < \delta \text{ for } x \in \Omega\} = (-1-\delta, -1] \cup [1, 1+\delta),$$

i.e., $\Omega_{\mathcal{I}}$ consists of those points outside of Ω that interact with points in Ω. Consider the one-dimensional nonlocal problem with a Dirichlet constraint,

$$\begin{aligned} -\mathcal{L}_\delta u(x) &= f(x) \quad \text{in } \Omega, \\ u &= 0 \quad \text{in } \Omega_{\mathcal{I}}, \end{aligned} \tag{6.7.1}$$

where

$$-\mathcal{L}_\delta u = C_\alpha \, \text{p.v.} \int_{B_\delta(x)} \frac{u(x) - u(y)}{|x - y|^{1+\alpha}} \mathrm{d}y, \quad 0 < \alpha < 2, \tag{6.7.2}$$

with

$$C_\alpha = \frac{2^\alpha \alpha \Gamma((\alpha+1)/2)}{2\pi^{1/2}\Gamma(1 - \alpha/2)}.$$

6.7.2 Finite-Volume Discretization

In order to solve the nonlocal problem (6.7.1), we consider a finite volume discretization. We divide the domain $\bar{\Omega} = [-1, 1]$ into N sub-domains, $h = 2/N$ is the space step size, and we divide the interaction domain $\Omega_{\mathcal{I}}$ into $2K$ sub-domains where $K = \delta/h$. Then, we have a partition $x_k = -1 + kh$, $k = -K, \ldots, 0, 1, \ldots, N, N+1, \ldots, N+K$, and we denote the partition by $I_j = [x_{j-1}, x_j]$, $j = 1, \ldots, N$. The finite volume formulation of (6.7.1) is written as:

$$\frac{C_\alpha}{h} \text{p.v.} \int_{I_j} \int_{B_\delta(x)} \frac{u(x) - u(y)}{|y - x|^{1+\alpha}} \mathrm{d}y\mathrm{d}x = \frac{1}{h} \int_{I_j} f \mathrm{d}x, \quad j = 1, \ldots, N. \tag{6.7.3}$$

Let $\bar{v}_j := \frac{1}{h}\int_{I_j} v(x)\mathrm{d}x$ be the mean value of $v(x)$ in the interval I_j. Note that

$$\begin{aligned}\frac{C_\alpha}{h}\int_{I_j}\int_{B_\delta(x)}\frac{u(x)-u(y)}{|y-x|^{1+\alpha}}\mathrm{d}y\mathrm{d}x &= \frac{C_\alpha}{h}\int_{I_j}\int_{B_\delta(x)}\frac{u(x)-u(x+z)}{|z|^{1+\alpha}}\mathrm{d}z\mathrm{d}x\\ &= \frac{C_\alpha}{h}\int_{I_j}\int_{-\delta}^{\delta}\frac{u(x)-u(x+z)}{|z|^{1+\alpha}}\mathrm{d}z\mathrm{d}x.\end{aligned} \tag{6.7.4}$$

Using the right rectangle rule to discretize the inner integral of the above equation, the above equation is equal to

$$\frac{C_\alpha}{h}\int_{I_j}\int_{0<|z|<Kh}\frac{u(x)-u(x+z)}{|z|^{1+\alpha}}\mathrm{d}z\mathrm{d}x \approx C_\alpha\sum_{0<k\le K}\frac{-\bar{u}_{j+k}-\bar{u}_{j-k}+2\bar{u}_j}{|kh|^{1+\alpha}}.$$

Thus, (6.7.3) becomes

$$C_\alpha\sum_{0<k\le K}\frac{-\bar{u}_{j+k}-\bar{u}_{j-k}+2\bar{u}_j}{|kh|^{1+\alpha}} = \bar{f}_j, \quad j=1,\ldots,N. \tag{6.7.5}$$

Writing the discretized equation in matrix form, we have

$$\boldsymbol{S}\boldsymbol{U} = \boldsymbol{F}, \tag{6.7.6}$$

where $\boldsymbol{U} = (\bar{u}_1,\bar{u}_1,\ldots,\bar{u}_N)^T$ and $\boldsymbol{F} = (\bar{f}_1,\bar{f}_1,\ldots,\bar{f}_N)^T$. $\boldsymbol{S}$ is the stiffness matrix with its elements:

$$S_{jj} = \sum_{k=1}^{K}\frac{2}{k^{1+\alpha}}, \quad S_{j,j+|k|} = -\frac{1}{|k|^{1+\alpha}},$$
$$\text{for } |k| = 1,2,\ldots,\min(K,N-|k|), \ j=1,\ldots,N.$$

If $K<N$, then $S_{j,j+|k|}=0$ for $|k|>K$, from which, we can see that $\boldsymbol{S}$ is a symmetric, diagonally dominant Toeplitz matrix.

The diagonal dominance property ensures that the system (6.7.6) admits a *unique* solution. The third property of $\boldsymbol{S}$ suggests that we can efficiently solve the linear system (6.7.6) in $O(N\log N)$ operations and $O(N)$ storage by using a preconditioned iterative method, such as GMRES.

Numerical Examples

We first test the convergence of the finite volume scheme.

Example 6.14 Let $f(x)=1$. Recall that $L=2$ is the length of the domain Ω. We test two different values of horizon parameter $\delta = 2L, 64L$ for different values of fractional order $\alpha = 0.5, 1.5$. Since there is no exact solution, we use the numerical solution with $N=2^{15}$ as the reference solution. The convergence results of the L^2- and L^∞-error are shown in Figure 6.12.

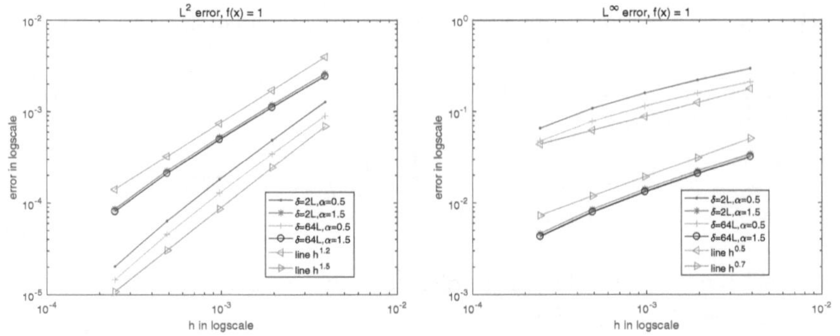

Figure 6.12 Convergence of the errors of the finite volume discretization, left: L^2-error, right: L^∞-error. Source: Reprinted from Lischke et al. (2020) with permission from Elsevier.

We observe in Figure 6.12 that the convergence rates of the L^2-error for $\alpha = 0.5$ and $\alpha = 1.5$ are about $O(h^{1.2})$ and $O(h^{1.5})$, respectively, while the convergence rates of the L^∞-error for $\alpha = 0.5$ and $\alpha = 1.5$ are about $O(h^{0.5})$ and $O(h^{0.7})$, respectively. We point out here that theoretical estimates for the convergence rate for finite-volume schemes are not known at the time of this writing. However, we mention that L^2 convergence rates for finite element schemes for such problems have been established in Borthagaray et al. (2018); Burkovska and Gunzburger (2018).

We now study how the value of the horizon parameter affects the behavior of the solution of the nonlocal problem. We also compare it with the solution of the nonlocal diffusion problem

$$\begin{aligned} C_\alpha \,\text{p.v.} \int_{\mathbb{R}} \frac{u(x) - u(y)}{|x - y|^{(1+\alpha)}} \mathrm{d}y &= f(x) \quad \text{in} \quad \Omega, \\ u &= 0 \quad \text{in} \quad \Omega^c := \mathbb{R} \setminus \Omega. \end{aligned} \tag{6.7.7}$$

The above problem is equivalent to the Riesz fractional diffusion problem

$$(-\Delta)^{\alpha/2} u(x) = f(x) \quad \text{in} \quad \Omega; \quad u = 0 \quad \text{in} \quad \Omega^c. \tag{6.7.8}$$

The above fractional diffusion problem can be solved by using the Petrov–Galerkin method proposed in Mao et al. (2016) (see also Mao and Karniadakis (2018)).

Example 6.15 Let $N = 2^{14}$. We vary the value of horizon parameter δ for $f(x) = 1$ and $f(x) = \sin(\pi x)$, and consider different values of the fractional order $\alpha = 0.5, 1.5$. The results are shown in Figure 6.13.

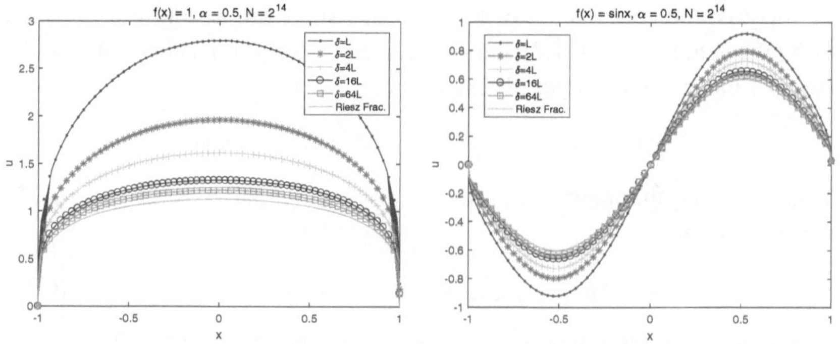

Figure 6.13 Numerical solutions with different values of horizon parameter, $N = 2^{14}$, $\alpha = 0.5$. Left: $f = 1$; Right: $f = \sin(\pi x)$. Source: Reprinted from Lischke et al. (2020) with permission from Elsevier.

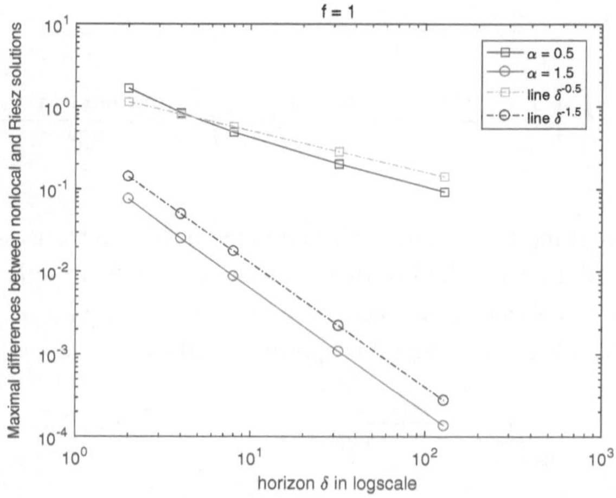

Figure 6.14 Convergence of $\|u - u_\delta\|_\infty$ with respect to the horizon δ with different values of fractional order α, $L = 2$, and $f(x) = 1$. Source: Reprinted from Lischke et al. (2020) with permission from Elsevier.

We observe that the numerical solution converges to the solution of the fractional diffusion problem (6.7.7) as the horizon parameter becomes infinite. This phenomenon has also been observed by D'Elia and Gunzburger (2013). Moreover, we study the convergence rate in L^∞ in the sense of the asymptotic behavior; in particular, we plot $\|u - u_\delta\|_\infty$ against the horizon parameter δ in log-log scale in Figure 6.14. We observe that the convergence rate of $\|u-u_\delta\|_\infty$ is 0.5 for $\alpha = 0.5$ while it is 1.5 for $\alpha = 1.5$. The convergence rate can also be analyzed theoretically for $\alpha > 1$. We now show this as follows: Let u and u_δ be the solutions of (6.7.7) and (6.7.1), respectively; it has been shown that (see D'Elia and Gunzburger (2013, Theorem 3.1))

$$\|u - u_\delta\|_{H^{\alpha/2}} \le C_1 \delta^{-\alpha}, \quad \|u - u_\delta\|_{L^2} \le C_2 \delta^{-\alpha}.$$

Thus, by using the interpolation theory and noting that $\|v\|_\infty \le c\|v\|_{H^s}$ for $s > 1/2$, we have

$$\|u - u_\delta\|_\infty \le C\delta^{-\alpha}, \quad \text{for } \alpha > 1. \tag{6.7.9}$$

We point out in the following that, for solving the equation (6.7.7) using the FVM, the corresponding integral (6.7.4) can be partitioned as

$$\begin{aligned}
\frac{C_\alpha}{h}\int_{I_j}\int_{\mathbb{R}} \frac{u(x)-u(y)}{|y-x|^{1+\alpha}}\mathrm{d}y\mathrm{d}x &= \frac{C_\alpha}{h}\int_{I_j}\int_{\mathbb{R}} \frac{u(x)-u(x+z)}{|z|^{1+\alpha}}\mathrm{d}z\mathrm{d}x \\
= \frac{C_\alpha}{h}\int_{I_j}\int_{-L}^{L} \frac{u(x)-u(x+z)}{|z|^{1+\alpha}}\mathrm{d}z& \\
+\frac{C_\alpha}{h}\int_{I_j}\left\{\int_{-\infty}^{-L} \frac{u(x)-u(x+z)}{|z|^{1+\alpha}}\mathrm{d}z + \int_{L}^{\infty} \frac{u(x)-u(x+z)}{|z|^{1+\alpha}}\mathrm{d}z\right\}\mathrm{d}x.&
\end{aligned} \tag{6.7.10}$$

The first term of the above integral is computed as before by using a rectangle rule discretization, while the last two terms are computed exactly. Thus, using the fact that $u = 0$ outside Ω and $x + z$ lies outside Ω for $x \in I_j \subset \bar{\Omega}$ and $|z| \ge L = |\Omega|$, the above integral is approximated by

$$C_\alpha \sum_{0<k\le L/h} \frac{-\bar{u}_{j+k} - \bar{u}_{j-k} + 2\bar{u}_j}{(kh)^{1+\alpha}} + \frac{2C_\alpha}{\alpha} L^{-\alpha}\bar{u}_j.$$

6.8 Numerical Solutions: Comparisons

In the preceding sections, we have thoroughly discussed the theoretical properties of the Riesz and spectral fractional Laplacians, and have established significant differences in the fractional Poisson problem associated to each operator. In this section, we leverage some of the numerical methods described in Section 6.1 to demonstrate the practical differences between these different operators in benchmark problems. As expected from the preliminary numerical studies at the start of this chapter and the theoretical discussion in Section 6.3, these differences in the solutions of the respective Poisson problems are highly dependent on the fractional order α and the properties of the forcing term f. We focus on numerical methods that are generalizable to higher dimensions, and we use these methods to compute benchmark solutions in two dimensions on the square, L-shaped, and unit disk domains. We solve fractional Poisson problems on each of these domains using the spectral, Riesz, and directional definitions with $\alpha = 0.5$ and $\alpha = 1.5$ with zero Dirichlet boundary conditions. We make observations about the differences between these numerical solutions, including regularity issues near the boundary, equivalence of the directional and Riesz definitions, and boundary oscillations in the case $\alpha = 0.5$. We also provide plots of the spectral and Riesz solutions compared along one-dimensional slices of the two-dimensional domains. Finally, we summarize the computational advantages and disadvantages of each method so that practitioners are better able to choose which method is best suited for a given equation.

We consider four fractional Poisson problems in three 2-dimensional domains of different shapes. In particular, we compare the solutions of equation (6.1.1) in a bounded domain $\Omega \subset \mathbb{R}^2$ with both the Riesz and spectral fractional Laplacians and zero Dirichlet boundary conditions, where the choices of Ω, f, and α are described in Sections 6.8.1–6.8.3 and outlined in Table 6.3.

To solve the Riesz fractional Poisson equation, we use the AFEM (see Section 6.4.1), the WOS method (see Section 6.4.2), and the RBF collocation method (see Section 6.6.1). The shape parameters of the RBF collocation method for each example are reported in the figure captions for the corresponding grid figures in below. Sixteen-point Gauss–Legendre quadrature is used to approximate the integral of the directional derivative in (6.6.9). We use multiple methods to verify the accuracy of the solutions and to ensure that the observed differences in the numerical solutions reflect differences in the fractional operators instead of inaccuracies in the numerical results. The meshes or collocation points used in each example below are displayed in below.

6.8.1 Square Domain

In this section, we show the solutions to the benchmark problems on the domain $[-1, 1]^2$, see Figure 6.15. Moreover in Figure 6.16, we show cases 1 and 2, where $f = 1$ and $\alpha = 0.5$ and 1.5, respectively. In each case, the spectral solution u_{spectral} lies below the Riesz solution u_{Riesz}, just as for the one-dimensional comparisons in Figure 6.1. Since $f \geq 0$ in the domain, this is a special case of the theoretical results of Musina and Nazarov (2016). We remark that the apparent failure in the plotted solution u_{Riesz} at enforcing

Table 6.3. *Guide to benchmark problems formulated with zero Dirichlet boundary conditions. The test cases were used to make comparisons on the square, disk, and L-shaped domains using the Riesz and spectral definitions as discussed in Sections 6.8.1–6.8.3. Reprinted from the Journal of Computational Physics, 404, Lischke, Anna, Pang, Guofei, Gulian, Mamikon, Song, Fangying, Glusa, Christian, Zheng, Xiaoning, Mao, Zhiping, Cai, Wei, Meerschaert, Mark M, Ainsworth, Mark, Karniadakis, George EM, What is the fractional Laplacian? A comparative review with new results, 109009, Copyright (2020), with permission from Elsevier.*

Benchmark Problems with Zero Dirichlet BCs		
Section 6.8.1, Square: $[-1,1]^2$	Order α	Source function f
Case 1 & 2	0.5 & 1.5	$f = 1$
Case 3 & 4	0.5 & 1.5	$f = \sin(\pi x)\sin(\pi y)$
Section 6.8.2, Disk: $\{(x, y) : x^2 + y^2 \leq 1\}$	Order α	Source function f
Case 1 & 2	0.5 & 1.5	$f = 1$
Case 3 & 4	0.5 & 1.5	$f = \sin(\pi r^2)$
Section 6.8.3, L-shape: $[-1,1]^2 \setminus [0,1)^2$	Order α	Source function f
Case 1 & 2	0.5 & 1.5	$f = 1$
Case 3 & 4	0.5 & 1.5	$f = \sin(\pi x)\sin(\pi y)$

the zero boundary condition for $\alpha = 0.5$ is an artifact of the AFEM method (Ainsworth and Glusa, 2018b); the true solution does in fact possess a zero trace. This is discussed in detail in Remark 6.12. The solution u_{spectral} for $\alpha = 0.5$ also possesses a trace (as discussed in Section 6.5.2), and in contrast, the numerical method used to compute it (SEM) is able to enforce the zero boundary condition.

In Figure 6.17, we consider Cases 3 and 4, where $f = \sin(\pi x)\sin(\pi y)$, and $\alpha = 0.5$ and 1.5, respectively. In this case, the issue of satisfying the boundary condition in the Riesz solution when $\alpha = 0.5$ is not as pronounced, but the oscillations are visible in the plot of $u_{\text{Riesz}} - u_{\text{spectral}}$.

Figure 6.18 displays the solutions plotted along the lines $y = 0$ for $f = 1$ and $f = \sin(\pi x)\sin(\pi y)$, respectively. These figures highlight the difference between the boundary layers for the spectral and Riesz solutions, as we discussed in relation to Figure 6.1. Indeed, the one- and two-dimensional profiles are qualitatively similar for all four cases, and the boundary layer in the

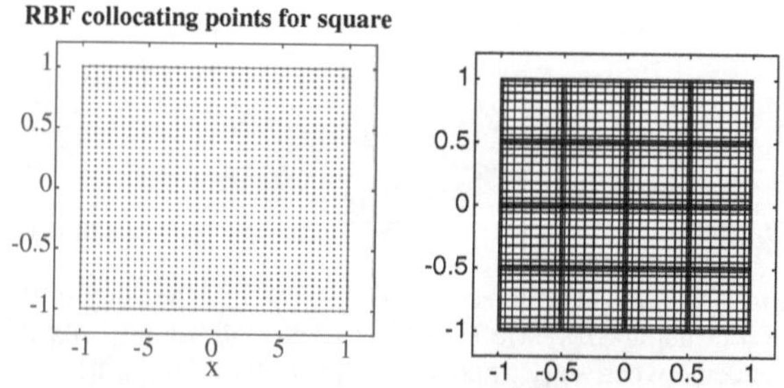

(a) 1600 RBF collocation points with shape parameter 0.3 used for RBF collocation method (left) and mesh used for SEM method (right).

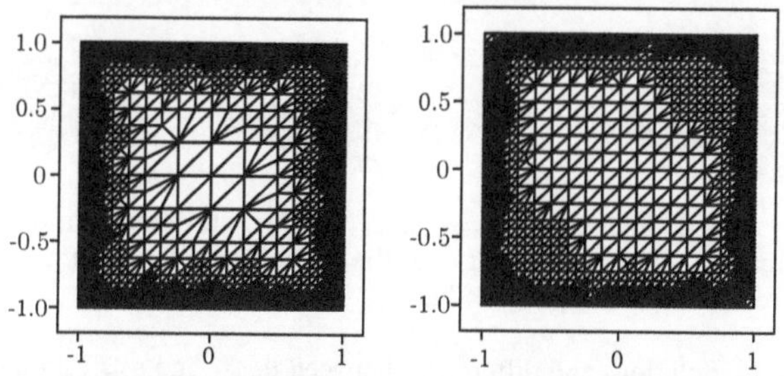

(b) Adaptively refined meshes for finite element solution to Riesz fractional Poisson equation corresponding to the constant source term $f = 1$ for $\alpha = 0.5$ (left) and for $\alpha = 1.5$ (right).

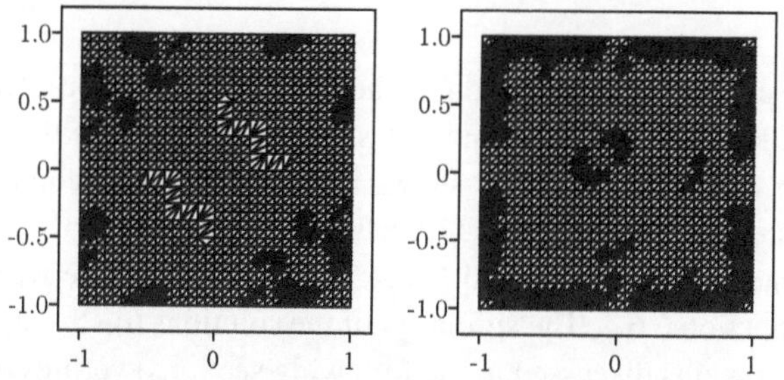

(c) Adaptively refined meshes for finite element solution to Riesz fractional Poisson equation corresponding to the source term $f = \sin(\pi x)\sin(\pi y)$ for $\alpha = 0.5$ (left) and for $\alpha = 1.5$ (right).

Figure 6.15 Comparisons on the square: Meshes and collocation points used for each numerical method in computing the solutions of the directional (top left), spectral (top right), and Riesz (center and bottom) fractional Poisson equations. Source: Reprinted from Lischke et al. (2020) with permission from Elsevier.

Riesz solution for Case 4 (see Figure 6.18) due to the singularity in the Riesz definition (6.3.9) is particularly noteworthy, as the spectral solution always will have smooth behavior near the boundary given smooth f.

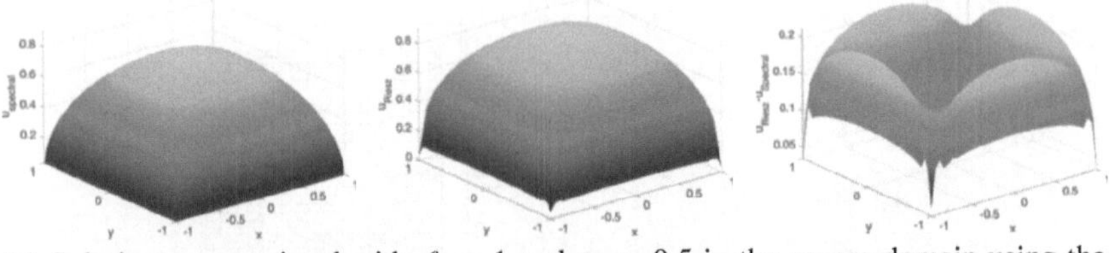

(a) Solutions u associated with $f = 1$ and $\alpha = 0.5$ in the square domain using the spectral definition (using SEM) (left) and the Riesz definition (using AFEM) (center), and the difference between u_{Riesz} and u_{spectral} for this case (right).

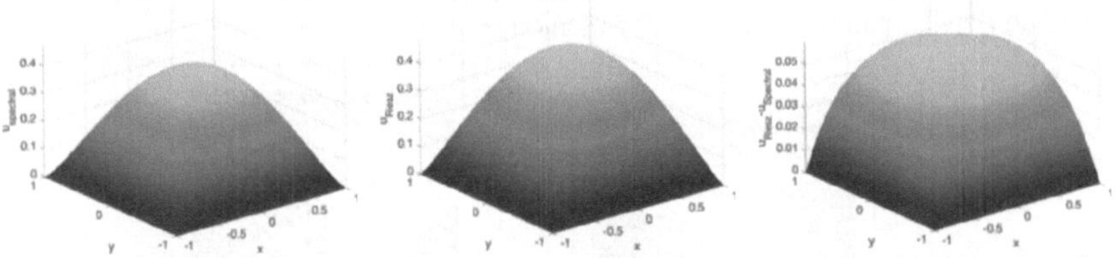

(b) Solutions u associated with $f = 1$ and $\alpha = 1.5$ in the square domain using the spectral definition (using SEM) (left) and the Riesz definition (using AFEM) (center), and the difference between u_{Riesz} and u_{spectral} for this case (right).

Figure 6.16 Solutions and differences between u_{Riesz} and u_{spectral} on the square domain for $\alpha = 0.5$ and 1.5. Source: Reprinted from Lischke et al. (2020) with permission from Elsevier.

We also computed the solution to the four benchmark problems using the directional definition (6.3.28), and this solution is plotted in the top left panel of Figure 6.19. This figure also shows the Riesz solution computed in three ways: (i) using the AFEM method (Ainsworth and Glusa, 2018b), (ii) using the walk-on-spheres method of Section 6.4.2, and (iii) using the RBF collocation method of Section 6.6.1. The similarity of the solutions (on the square domains and on the domains discussed in the following sections) verifies our numerical results, and demonstrates that the directional solution is equivalent to the Riesz solution, up to some numerical error. The spectral solution in the top right panel is significantly different from the Riesz and directional solutions, as indicated by their maximal values reported in the titles of the plots.

6.8.2 Disk Domain

In this section, we solve the benchmark problems in Table 6.3 on the disk domain: $\Omega := \{(x, y)|\sqrt{x^2 + y^2} \leq 1\}$. We use the same numerical methods as in Section 6.8.1, where we solved the benchmark problems on the square. The mesh for the SEM used to compute the spectral solution was generated

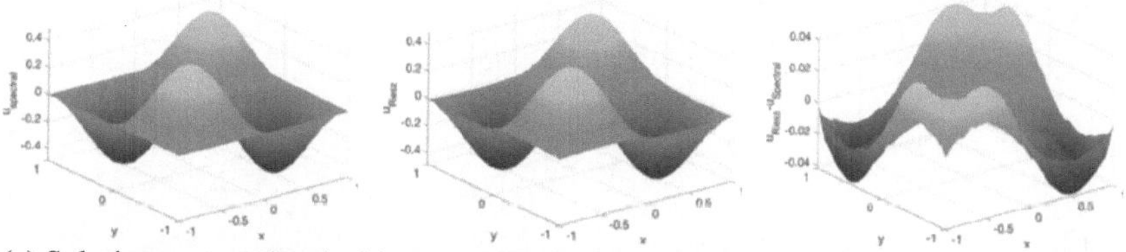

(a) Solutions u associated with $f = \sin(\pi x)\sin(\pi y)$ and $\alpha = 0.5$ in the square domain using the spectral definition (using SEM) (left) and the Riesz definition (using AFEM) (center), and the difference between u_{Riesz} and u_{spectral} for this case (right).

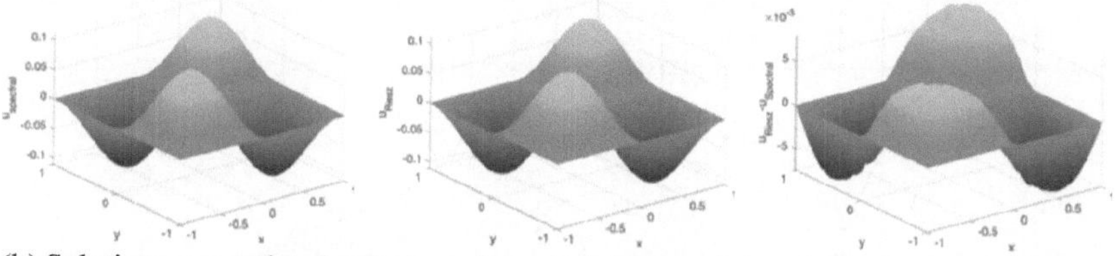

(b) Solutions u associated with $f = \sin(\pi x)\sin(\pi y)$ and $\alpha = 1.5$ in the square domain using the spectral definition (using SEM) (left) and the Riesz definition (using AFEM) (center), and the difference between u_{Riesz} and u_{spectral} for this case (right).

Figure 6.17 Solutions and differences between u_{Riesz} and u_{spectral} on the square domain for $\alpha = 0.5$ and 1.5. Source: Reprinted from Lischke et al. (2020) with permission from Elsevier.

by transforming the square mesh onto the circular domain. The details of this transformation are included in Song et al. (2017). The collocation points used for the RBF method and the adaptively refined meshes used to compute the Riesz solutions are included in Figure 6.20.

In Figure 6.21, we again see that the Riesz solutions have oscillations near the boundary in the case with $\alpha = 0.5$, where the spectral solutions satisfy the boundary conditions. For Cases 3 and 4 with $f = \sin(\pi r^2)$ (see Table 6.3), our observations are very similar to those of Figure 6.17, as shown in Figure 6.23. As for the case $f \equiv 1$, the property $f \geq 0$ leads to the Riesz solution lying above the spectral solution (Musina and Nazarov, 2016). We also include one-dimensional slice plots of the Riesz and spectral solutions and their differences along the line $y = 0$ in Figure 6.22; the profiles are similar to those we saw in the square examples, where the Riesz solutions exhibit sharper boundary layers than the spectral solutions.

The comparisons with the directional definition solution using the RBF collocation method are similar to the square domain examples, and so we include the figure containing these plots in Figure 6.23.

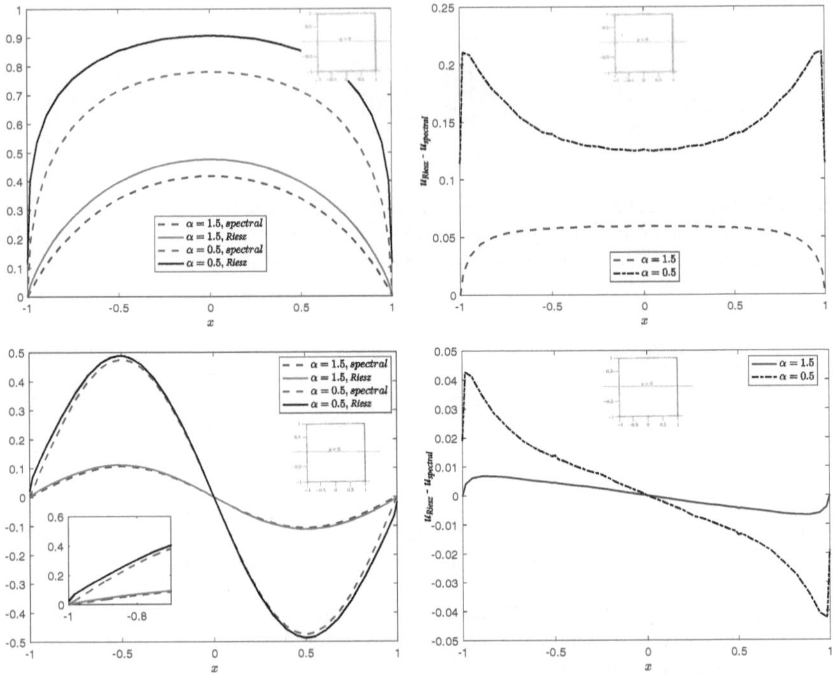

Figure 6.18 Slices along the line $y = 0$ in the square domain with both $\alpha = 0.5$ and $\alpha = 1.5$. (top left) Plots of the solutions for the Riesz and spectral formulations with $f = 1$ and (top right) plot of the differences $u_{\text{Riesz}} - u_{\text{spectral}}$ along the line $y = 0$ with $f = 1$. (bottom left) Plots of the solutions for the Riesz and spectral formulations with $f = \sin(\pi x)\sin(\pi y)$ and (bottom right) plot of the differences $u_{\text{Riesz}} - u_{\text{spectral}}$ along the line $y = 0$ with $f = \sin(\pi x)\sin(\pi y)$. Some small wiggles that appear in the difference plots are due to the computation of the differences on two different meshes, even though the solutions are sufficiently converged and stable. Source: Reprinted from Lischke et al. (2020) with permission from Elsevier.

6.8.3 L-shaped Domain

In this section, we solve the four benchmark problems of Table 6.3 on the L-shaped domain: $\Omega := \{[-1,1]^2 \setminus [0,1)^2\}$, i.e., the square $[-1,1]^2$ with the upper right corner removed. We display the solutions from two viewing angles: from the points $(-1,-1)$ and $(1,1)$, so that all features are clearly visible.

We use the same numerical methods as in Sections 6.8.1 and 6.8.2, where we solved the benchmark problems on the square and disk domains. The mesh for the SEM used to compute the spectral solution includes additional refinement near the boundary to ensure a converged numerical solution, and

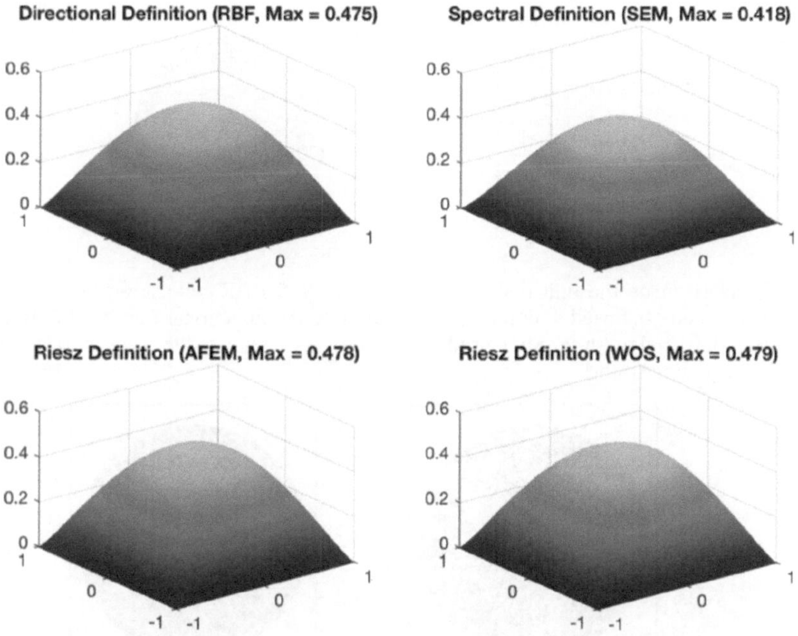

Figure 6.19 Square, $f(x) = 1$, and $g(x) = 0$: Comparison, for $\alpha = 1.5$, of u_{Riesz} and u_{spectral}, using three methods to compute the Riesz solution and one method to compute the spectral solution. Top left: The Riesz solution obtained using the RBF collocation method based on the directional representation. Top right: The spectral solution obtained using the SEM. Bottom left: The Riesz solution obtained using the AFEM. Bottom right: The Riesz solution obtained using the WOS method. The only solution with a significant difference is the spectral solution (top right); all other solutions are equivalent up to numerical error. Source: Reprinted from Lischke et al. (2020) with permission from Elsevier.

is plotted in Figure 6.24. The collocation points used for the RBF method and the adaptively refined meshes used to compute the Riesz solutions are also included in Figure 6.24.

Figure 6.25 shows the Riesz and spectral solutions for Cases 1 and 2, where $\alpha = 0.5$ and 1.5, respectively, and $f = 1$ at the two viewing angles, and Figure 6.26 shows the solutions for Cases 3 and 4. We observe that the difference plots, particularly for the cases with $\alpha = 0.5$, exhibit a relatively minor spike near the inside corner of the domain. Note that this spike does not occur (or is at least much less pronounced) for the cases with $\alpha = 1.5$.

We also compute the directional definition solution on the L-shaped domain using the RBF collocation method. We show the comparison in Figure

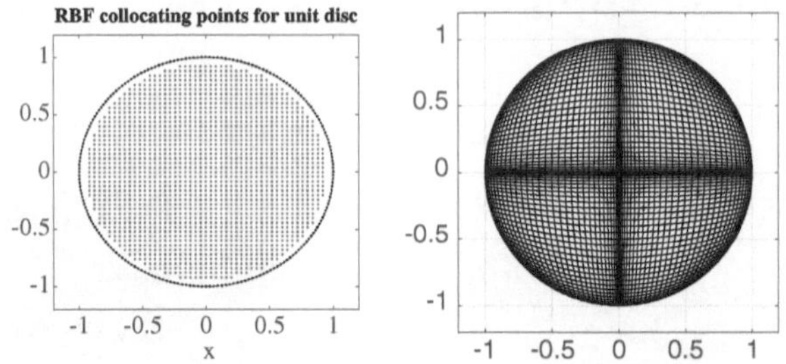

(a) Comparison on the unit disk: distribution of 1965 RBF collocation points with shape parameter 0.3 used to compute the solution to the directional fractional Poisson equation (left) and mesh for SEM used to compute the solution to the spectral fractional Poisson equation (right).

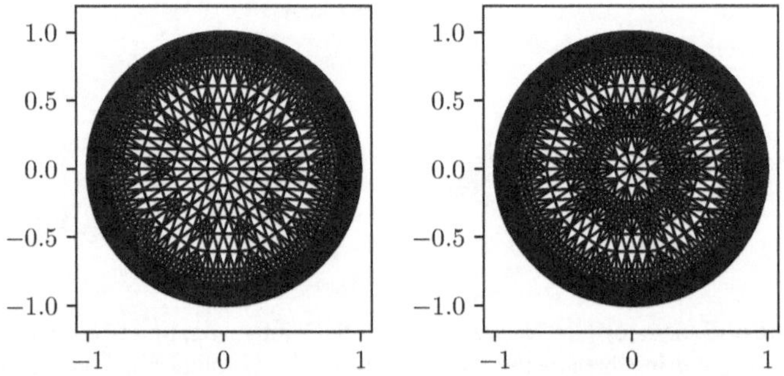

(b) Adaptively refined meshes for finite element solution to Riesz fractional Poisson equation corresponding to the constant source term $f = 1$ for $\alpha = 0.5$ (left) and for $\alpha = 1.5$ (right).

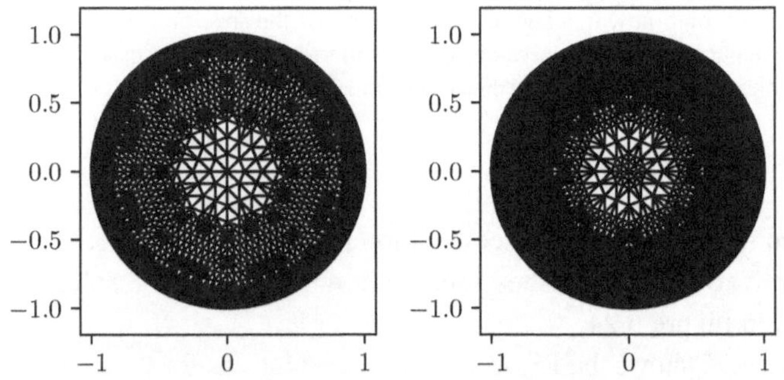

(c) Adaptively refined meshes for finite element solution to Riesz fractional Poisson equation corresponding to the source term $f = \sin(\pi x)\sin(\pi y)$ for $\alpha = 0.5$ (left) and for $\alpha = 1.5$ (right).

Figure 6.20 Comparisons on the disk: Meshes and collocation points used for each numerical method in computing the solutions of the directional (top left), spectral (top right), and Riesz (center and bottom) fractional Poisson equations. Source: Reprinted from Lischke et al. (2020) with permission from Elsevier.

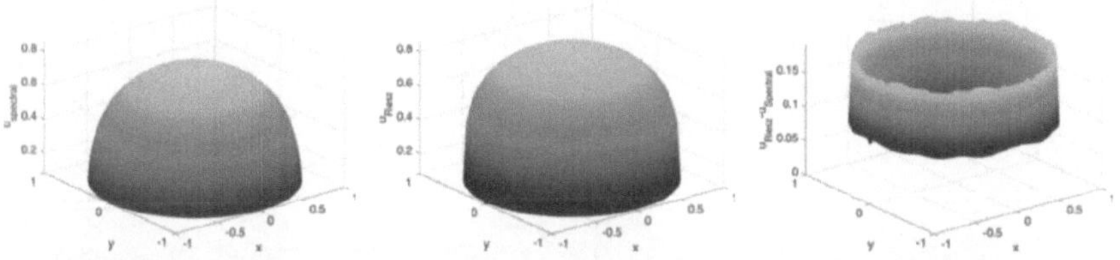

(a) Solutions u associated with $f = 1$ and $\alpha = 0.5$ in the disk domain using the spectral definition (using SEM) (left) and the Riesz definition (using AFEM) (center), and the difference between u_{Riesz} and u_{spectral} for this case (right).

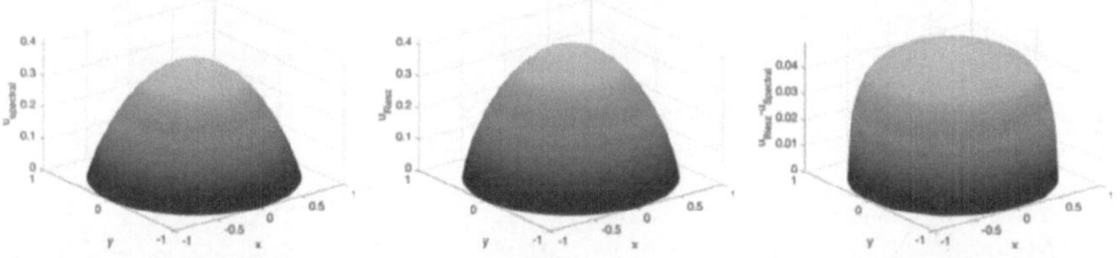

(b) Solutions u associated with $f = 1$ and $\alpha = 1.5$ in the disk domain using the spectral definition (using SEM) (left) and the Riesz definition (using AFEM) (center), and the difference between u_{Riesz} and u_{spectral} for this case (right).

Figure 6.21 Solutions and differences between u_{Riesz} and u_{spectral} on the disk domain for $f = 1$ and $\alpha = 0.5$ and 1.5. Source: Reprinted from Lischke et al. (2020) with permission from Elsevier.

6.27 using the view from $(1, 1)$, and the view from $(-1, -1)$ is included subsequently.

In this set of examples, we also plot the solutions and differences along the slices defined by the lines $y = x$ and $y = 1 - x$ in Figure 6.28 to further observe the behaviors near the corner. Again, the fact that the Riesz solution lies above the spectral solution is consistent with $f \geq 0$ in the domain and the theoretical result of Musina and Nazarov (2016). The spike near the corner is intensified in the cases where $\alpha = 0.5$.

6.8.4 Computational Considerations

Many of the numerical methods used earlier in this work have only recently been reported, and some require further development. Therefore, it is not reasonable to directly compare the computational complexities of these methods, since they are all at differing stages of development. Instead, in this section, we present an overview of the computational advantages and disadvantages of each method.

The AFEM of Ainsworth and Glusa (2018b, 2017) for solving Riesz (or integral) fractional Poisson equations has been well developed, and its

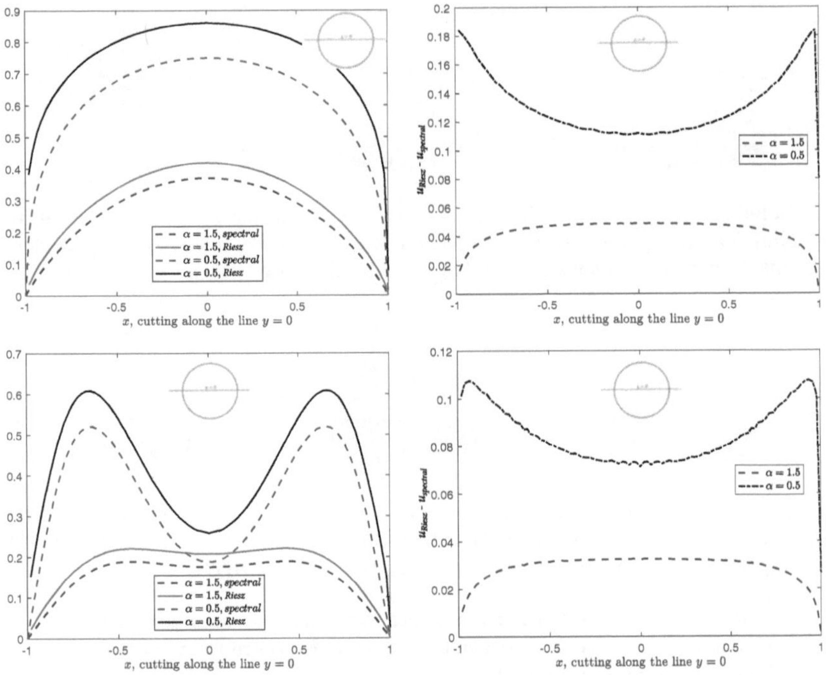

Figure 6.22 Slices along the line $y = 0$ in the unit disk domain with both $\alpha = 0.5$ and $\alpha = 1.5$. (top left) Plots of the solutions for the Riesz and spectral formulations with $f = 1$ and (top right) plot of the differences $u_{\text{Riesz}} - u_{\text{spectral}}$ along the line $y = 0$ and $f = 1$. (bottom left) Plots of the solutions for the Riesz and spectral formulations with $f = \sin(\pi r^2)$ and (bottom right) plot of the differences $u_{\text{Riesz}} - u_{\text{spectral}}$ along the line $y = 0$ and $f = \sin(\pi r^2)$. Some small wiggles that appear in the difference plots are due to the computation of the differences on two different meshes, even though the solutions are sufficiently converged and stable. Source: Reprinted from Lischke et al. (2020) with permission from Elsevier.

overall computational complexity is $\mathcal{O}(n(\log n)^{2d})$ operations, including the assembly, computation of error indicators for the adaptive refinement, and solving the linear system. A detailed study of the complexity was included in Ainsworth and Glusa (2017). At this time, the AFEM has been applied only to zero Dirichlet boundary conditions, although it may be possible to extend this method to nonzero Dirichlet conditions using a similar clustering approach as was mentioned in Section 3.1.1. As can be seen in the adaptively refined meshes, the AFEM captures the boundary singularity with a high degree of accuracy, and is a robust method for discretizing the Riesz fractional Laplacian.

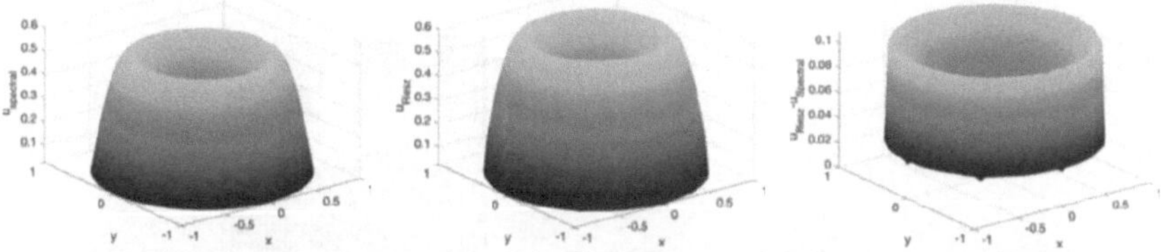

(a) Solutions u associated with $f = \sin(\pi r^2)$ and $\alpha = 0.5$ in the disk domain using the spectral definition (using SEM) (left) and the Riesz definition (using AFEM) (center), and the difference between u_{Riesz} and $u_{spectral}$ for this case (right).

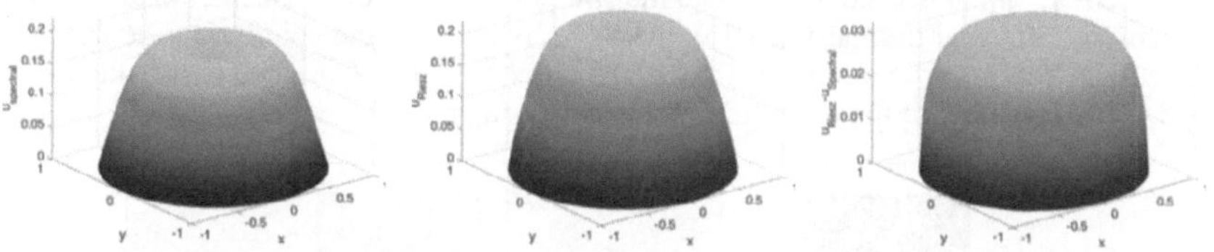

(b) Solutions u associated with $f = \sin(\pi r^2)$ and $\alpha = 1.5$ in the disk domain using the spectral definition (using SEM) (left) and the Riesz definition (using AFEM) (center), and the difference between u_{Riesz} and $u_{spectral}$ for this case (right).

Figure 6.23 Solutions and differences between u_{Riesz} and $u_{spectral}$ on the disk domain for $\alpha = 1.5$. Source: Reprinted from Lischke et al. (2020) with permission from Elsevier.

Another approach we used to discretize the Riesz definition was the WOS algorithm (see Section 6.4.2), which can be applied to quasi-convex domains and zero or nonzero Dirichlet boundary conditions. In this method, the computational complexity is dominated by the Monte Carlo simulation of many sample paths for each point in the bounded domain Ω, although the complexity of our implementation scaled as the product of the number of sample paths multiplied by the number of points in the computational domain. Furthermore, the WOS computation of the solution at each point in Ω is independent of the computations for any other point, which means all of the walks-on-spheres can be done simultaneously. Therefore, WOS is an embarrassingly parallel algorithm. The WOS algorithm also has the advantage that no modification needs to be made when the Dirichlet boundary condition is nonzero other than inserting the appropriate function $g(x)$ into the Feynman–Kac formula (6.3.42). In Kyprianou et al. (2018), the convergence of the WOS algorithm is proved, and sample code for solving the fractional Poisson equation is provided.

To discretize the directional definition (6.3.28), we used the RBF collocation method of Section 6.6.1. An advantage of this approach is that it

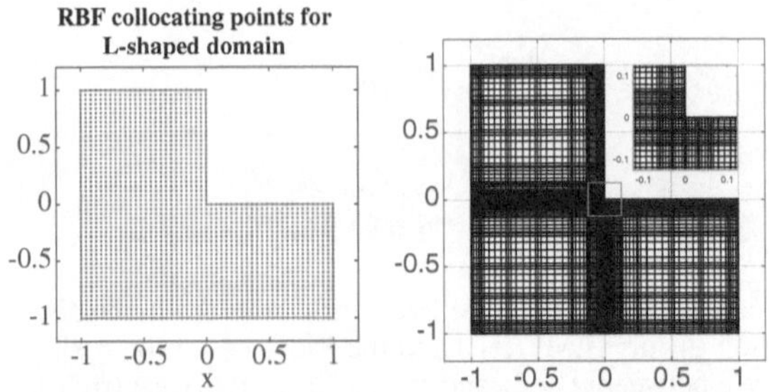

(a) Distribution of 1976 RBF collocating points using shape parameter 0.2 for the RBF collocation, which is used to compute the solution to the directional fractional Poisson equation (left), and the mesh used in the SEM to compute the solution to the spectral fractional Poisson equation for both $\alpha = 0.5$ and 1.5 (right). The same points and mesh are used regardless of the definition of the source function f.

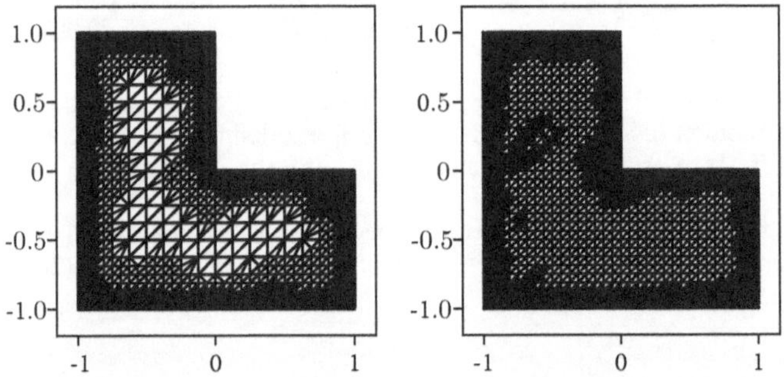

(b) Adaptively refined meshes for finite element solution to Riesz fractional Poisson equation corresponding to the constant source term $f = 1$ for $\alpha = 0.5$ (left) and for $\alpha = 1.5$ (right).

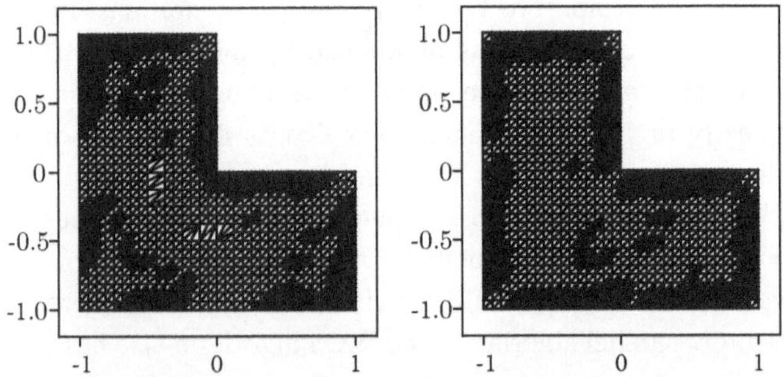

(c) Adaptively refined meshes for finite element solution to Riesz fractional Poisson equation corresponding to the source term $f = \sin(\pi x)\sin(\pi y)$ for $\alpha = 0.5$ (left) and for $\alpha = 1.5$ (right).

Figure 6.24 Comparisons on the L-shaped domain: Meshes and collocation points used for each numerical method in computing the solutions of the directional (top left), spectral (top right), and Riesz (center and bottom) fractional Poisson equations. Source: Reprinted from Lischke et al. (2020) with permission from Elsevier.

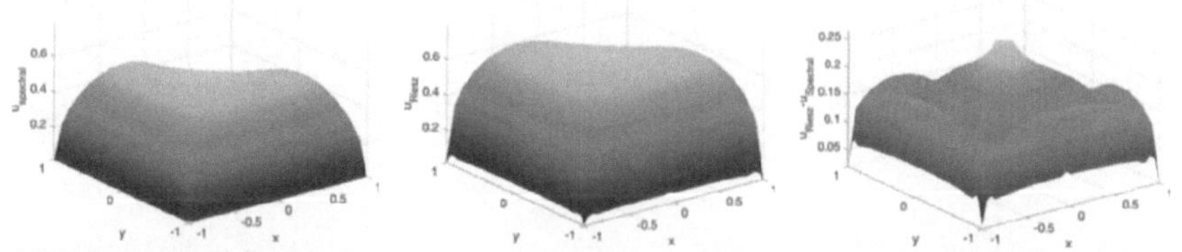

(a) Solutions u associated with $f = 1$ and $\alpha = 0.5$ in the L-shaped domain using the spectral definition (using SEM) (left) and the Riesz definition (using AFEM) (center), and the difference between u_{Riesz} and u_{spectral} for this case (right).

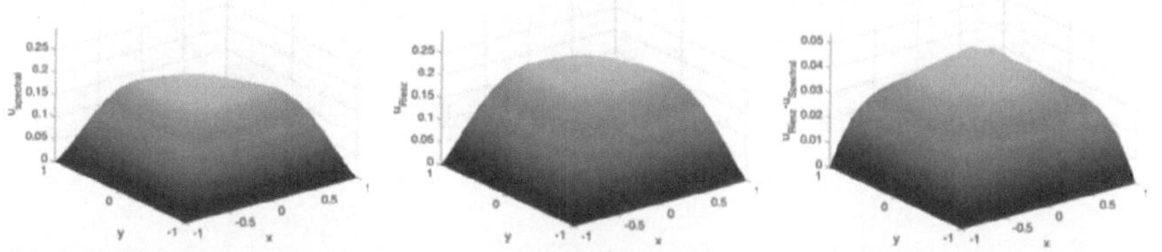

(b) Solutions u associated with $f = 1$ and $\alpha = 1.5$ in the L-shaped domain using the spectral definition (using SEM) (left) and the Riesz definition (using AFEM) (center), and the difference between u_{Riesz} and u_{spectral} for this case (right).

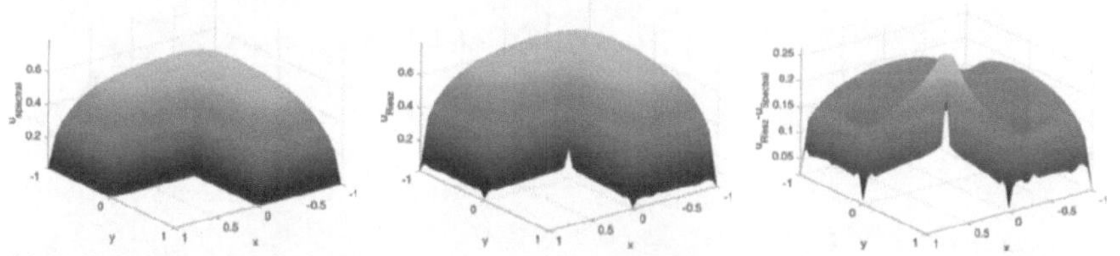

(c) Solutions u associated with $f = 1$ and $\alpha = 0.5$ in the L-shaped domain using the spectral definition (using SEM) (left) and the Riesz definition (using AFEM) (center), and the difference between u_{Riesz} and u_{spectral} for this case (right).

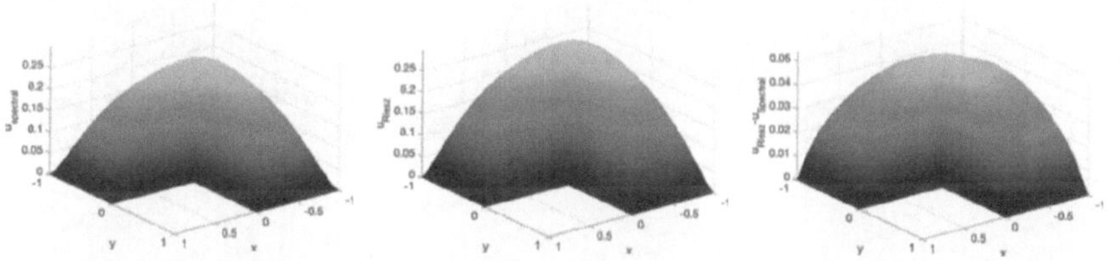

(d) Solutions u associated with $f = 1$ and $\alpha = 1.5$ in the L-shaped domain using the spectral definition (using SEM) (left) and the Riesz definition (using AFEM) (center), and the difference between u_{Riesz} and u_{spectral} for this case (right).

Figure 6.25 Solutions and differences between u_{Riesz} and u_{spectral} on the L-shaped domain for $\alpha = 0.5$ and 1.5. Source: Reprinted from Lischke et al. (2020) with permission from Elsevier.

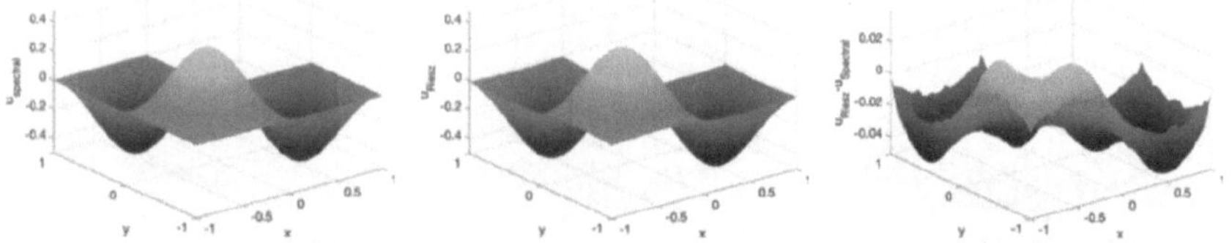

(a) Solutions u associated with $f = \sin(\pi x)\sin(\pi y)$ and $\alpha = 0.5$ in the L-shaped domain using the spectral definition (using SEM) (left) and the Riesz definition (using AFEM) (center), and the difference between u_{Riesz} and u_{spectral} for this case (right).

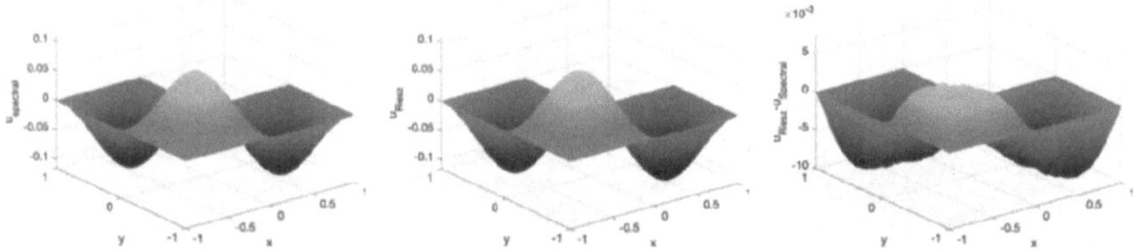

(b) Solutions u associated with $f = \sin(\pi x)\sin(\pi y)$ and $\alpha = 1.5$ in the L-shaped domain using the spectral definition (using SEM) (left) and the Riesz definition (using AFEM) (center), and the difference between u_{Riesz} and u_{spectral} for this case (right).

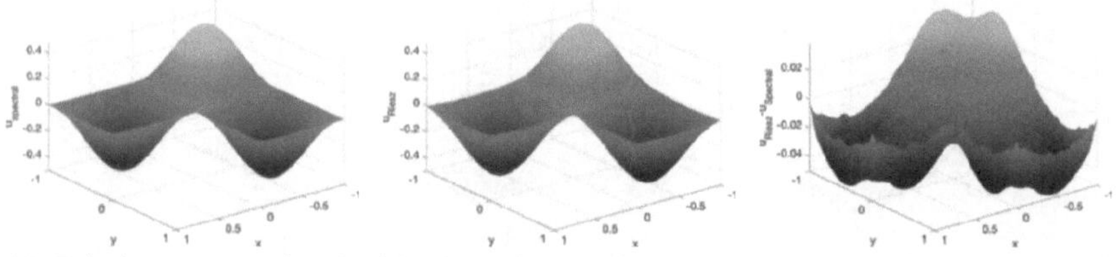

(c) Solutions u associated with $f = \sin(\pi x)\sin(\pi y)$ and $\alpha = 0.5$ in the L-shaped domain using the spectral definition (using SEM) (left) and the Riesz definition (using AFEM) (center), and the difference between u_{Riesz} and u_{spectral} for this case (right).

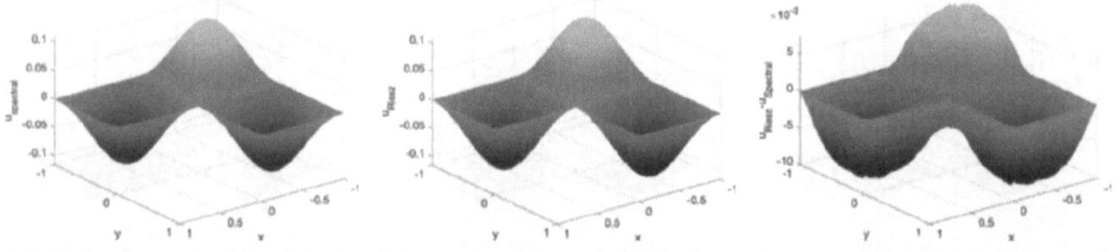

(d) Solutions u associated with $f = \sin(\pi x)\sin(\pi y)$ and $\alpha = 1.5$ in the L-shaped domain using the spectral definition (using SEM) (left) and the Riesz definition (using AFEM) (center), and the difference between u_{Riesz} and u_{spectral} for this case (right).

Figure 6.26 Solutions and differences between u_{Riesz} and u_{spectral} on the L-shaped domain for $\alpha = 0.5$ and 1.5. Source: Reprinted from Lischke et al. (2020) with permission from Elsevier.

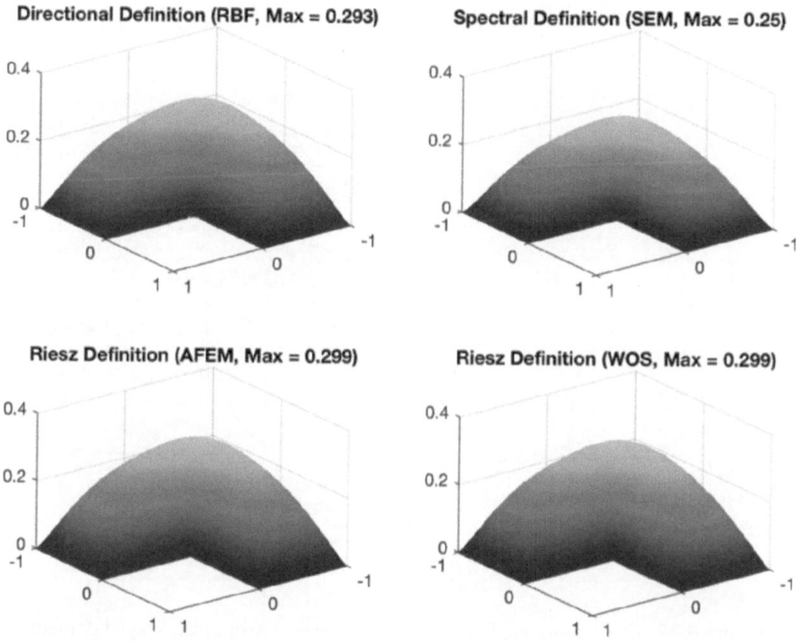

Figure 6.27 L-shaped domain, $f(x) = 1$, and $g(x) = 0$: Comparison, for $\alpha = 1.5$, of u_{Riesz} and u_{spectral}, using three methods to compute the Riesz solution and one method to compute the spectral solution. Top left: The Riesz solution obtained using the RBF collocation method based on the directional representation. Top right: The spectral solution obtained using the SEM. Bottom left: The Riesz solution obtained using the AFEM. Bottom right: The Riesz solution obtained using the WOS method. The only solution with a significant difference is the spectral solution (top right); all other solutions are equivalent up to numerical error. Source: Reprinted from Lischke et al. (2020) with permission from Elsevier.

can be easily modified to handle the case where the measure $M(\mathrm{d}\boldsymbol{\theta})$ is nonuniform, as was done using another version of this method in Pang et al. (2015). The RBF collocation method is applicable to problems with nonzero Dirichlet boundary conditions as long as the exterior boundary value decays at infinity. In addition to these advantages, there are some aspects in which further development is needed. When collocation points are too close together, the method becomes unstable, which limits the level of accuracy this method can achieve (Chen et al., 2014). Furthermore, for direct and iterative solvers, the linear system resulting from the RBF approach requires cubic and squared complexity to solve, respectively. Currently, we lack good preconditioners due to the complicated matrix structure formed by the Grünwald–Letnikov scheme as well as Gaussian quadrature for the integral with respect to θ.

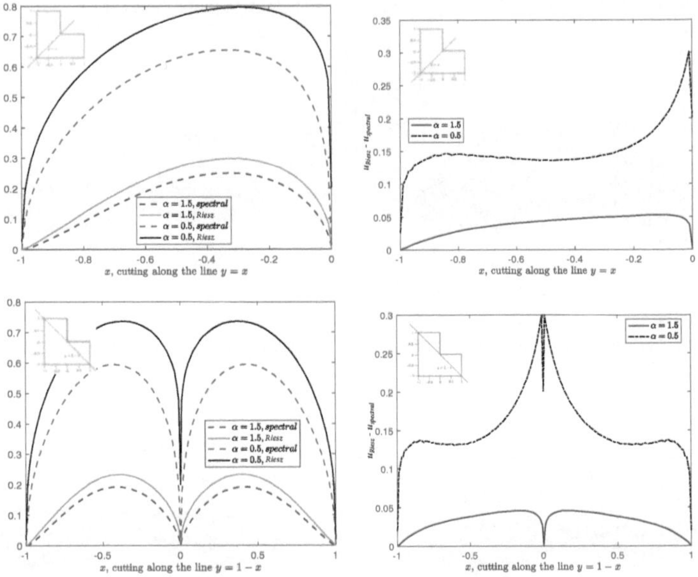

Figure 6.28 Slices along the line $y = x$ and $y = 1 - x$ in the L-shaped domain with $f = 1$ and both $\alpha = 0.5$ and $\alpha = 1.5$. (top left) Plots of the solutions for the Riesz and spectral formulations along $y = x$ and (top right) plot of the differences $u_{\text{Riesz}} - u_{\text{spectral}}$ along the line $y = x$. (bottom left) Plots of the solutions for the Riesz and spectral formulations along $y = 1 - x$ and (bottom right) plot of the differences $u_{\text{Riesz}} - u_{\text{spectral}}$ along the line $y = 1 - x$. Source: Reprinted from Lischke et al. (2020) with permission from Elsevier.

For the spectral definition, there are many choices of numerical methods. In the comparisons above, we reported results computed using the SEM described in Section 6.5.1. Given a mesh of elements on a bounded domain Ω, this method requires the computation of a large number of eigenvalues and eigenfunctions of the standard Laplacian. This is computationally expensive, and for larger eigenvalues, the computation of these eigenpairs becomes inaccurate. Furthermore, the resulting linear system is of cubic complexity. The advantage of this approach is its flexibility with respect to different types of boundary conditions and complex domains. Furthermore, if the SEM is applied to a time-dependent problem, e.g., the space-fractional heat equation, the eigenpairs need only be computed once at the beginning, and can be reused during each time-step. To understand why this is the case, recall that the discretized form of the fractional Poisson equation is

$$AU_{\mathcal{N}} = \hat{f}, \tag{6.8.1}$$

where $\hat{f}$ is the load vector, $A_{\mathcal{N}}$ is the discrete counterpart of the spectral fractional Laplacian:

$$(A_{ij}) = \sum_{n=1}^{\mathcal{N}} \lambda_n^{\alpha/2}(\tilde{\phi}_i, \tilde{\phi}_j), \tag{6.8.2}$$

and $\{\tilde{\phi}_i\}$ are the orthonormal Lagrange basis functions generated by the WGS procedure. Hence, A is a diagonal matrix, so solving the linear system is simply a computation of the matrix-vector product $A^{-1}\hat{f}$. So this method is inexpensive in the context of time-dependent problems. For more details of the implementation and computational complexity for the SEM, see Song et al. (2017).

6.9 Nonzero Boundary Conditions

Depending on the fractional Laplacian definition, nonzero boundary conditions can lead to a significant increase in the computational cost of solving our benchmark equations. Most importantly, modifications to the definition itself may be necessary, particularly in the case of the spectral fractional Laplacian, as discussed in Section 6.3.5. We demonstrate how our numerical methods may be adapted for this case. We also present computationally feasible approaches to solving the Riesz fractional Poisson equation with nonzero Dirichlet boundary conditions, and we make comparisons between the solutions to some inhomogeneous benchmark problems. We note some qualitative differences in the solutions for the inhomogeneous case from our observations of the solutions to the homogeneous problems.

We are unable to compute solutions to the fractional Neumann problem involving the Riesz definition for two reasons. Firstly, because the formulation of the nonlocal Neumann condition is a subject of some controversy, as there are multiple formulations in the literature with no consensus (Dipierro et al., 2017; Song and Vondracek, 2003; Bogdan et al., 2003; Guan and Ma, 2006; Barles et al., 2014). Secondly, numerical approaches have not yet been developed for these fractional Neumann problems. It is more straightforward to define Neumann boundary conditions for the spectral fractional Laplacian, since this definition only requires local boundary conditions, as discussed above. However, the points we make in the following section can be understood fully using only nonzero Dirichlet boundary conditions.

6.9.1 Spectral Definition

In this section, we follow up on the discussion of the different (equivalent) formulations of the inhomogeneous spectral fractional Laplacian in Section 6.3.5 by presenting some numerical comparisons involving nonzero Dirichlet boundary conditions. We demonstrate numerically that the approach of Antil et al. (2017), the heat semigroup formulation of Cusimano et al. (2018), and our nonharmonic lifting approach are all equivalent ways of describing the inhomogeneous spectral fractional Laplacian. As these formulations require distinct numerical approaches, we also compare these approaches in Section 6.9.1.

Numerical Comparison of Nonharmonic Lifting and Harmonic Lifting Methods

In Section 6.3.6, we described the nonharmonic lifting approach for discretizing the inhomogeneous spectral fractional Laplacian. In this section, we compare the results of the nonharmonic lifting method with the harmonic lifting method of Antil et al. (2017), described in Section 6.3.6. In our figures, we frequently refer to the latter as the "APR" method. Recall that this method allows for lifting the boundary data by functions v that solve $-\Delta v = 0$ in a *very weak* variational form (6.3.63); however, in all of the examples we consider, the boundary data is smooth enough that we can lift by standard harmonic functions.

For the harmonic lifting method, as explained in Section 6.3.6, any method used to discretize the homogeneous spectral fractional Laplacian can be applied after the harmonic lifting. In our examples, we use the spectral element method described in Section 6.5.1 in order to discretize $(-\Delta)^{\alpha/2}u(x)$. For the nonharmonic lifting method, we also apply the spectral element method (or the discrete eigenfunction method in the case $d = 1$) of Section 6.5.1, where the numerical solution of the fractional Poisson problem is approximated by a function in the space $\mathbb{P}_N(\Omega) \cap H_0^1(\Omega)$. We consider the fractional Poisson equation (6.1.1) with nonzero Dirichlet boundary $u(x)\big|_{\partial\Omega} = g(x)$. Let $u = w + v$, where w is an unknown function and $w\big|_{\partial\Omega} = 0$, and v is a function chosen to satisfy $v\big|_{\partial\Omega} = g$. Hence this function v is not unique. Assuming that $\partial\Omega$ is smooth, g must belong to the space $H^{1/2}(\partial\Omega)$, since the solution $u \in H_0^1(\Omega)$ and has trace equal to g. Then, following the approach of Section 6.3.6, the weak form of the inhomogeneous fractional Poisson problem can be written as follows: find $w \in \mathbb{P}_N(\Omega) \cap H_0^1(\Omega)$ such that

$$((-\Delta)^{\alpha/2}w, \phi) = (f, \phi) - \left(\nabla v, \nabla((-\Delta)^{\alpha/2-1}\phi)\right) \tag{6.9.1}$$

for all $\phi \in \mathbb{P}_N(\Omega) \cap H_0^1(\Omega)$.

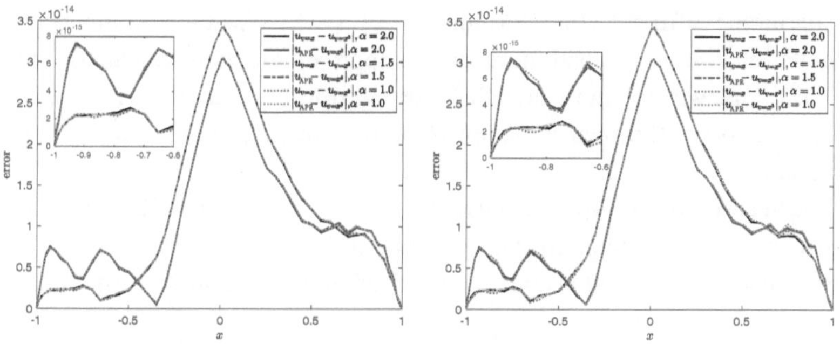

Figure 6.29 Differences between the APR harmonic lifting method (Antil et al., 2017) and nonharmonic lifting method using (left) $f = x$ and (right) $f = -x$ with different fractional orders α. The solid black lines, light gray dotted-dashed lines, and dark gray dotted lines represent the differences between the two versions of the lifting method with $v = x$ and $v = x^3$ for different values of α. These differences turn out to be very similar regardless of the values of α. We also plot the differences (for different α's) between the very weak solution, u_A, and the lifting solution with $v = x^3$ denoted $u_{v=x^3}$. Source: Reprinted from Lischke et al. (2020) with permission from Elsevier.

We solve this problem in the domain $\Omega = (-1, 1)$, where we test two cases with different forcing terms, $f = x$ and $f = -x$ with boundary conditions $u(-1) = -1$ and $u(1) = +1$. We have exact solutions in the case $\alpha = 2$: $u = \frac{1}{6}x^3 + \frac{5}{6}x$ with the forcing term $f = -x$ (*Case I*), and $u = -\frac{1}{6}x^3 + \frac{7}{6}x$ with the forcing term $f = x$ (*Case II*). We choose the lifting functions $v = x$ and $v = x^3$ for our numerical tests. We plot the differences of the numerical solutions between the nonharmonic lifting method and the harmonic lifting method of Antil et al. (2017) in Figure 6.29 for different values of α and different lifting functions $v(x)$. We can see from Figure 6.29 that the differences are on the order of machine precision, demonstrating that these approaches yield equivalent solutions.

Numerical Comparison of the Harmonic Lifting and Heat Semigroup Approaches

In this section, we compute the spectral fractional Laplacian of $u = \cos(\pi x)\sin(\pi y)$ on the domain $\Omega = [0, 1]^2$ using both the harmonic lifting method (Antil et al., 2017) as well as the heat semigroup formulation of Cusimano et al. (2018), which we abbreviate as "Heat SG" in our figures. The idea of this section is to give an indication of the computational cost of computing $(-\Delta)^{\alpha/2}u(x)$ using the different (equivalent) formulations. Once again, we use the SEM described in Section 6.5.1 in order to discretize the harmonic lifting (APR) formulation of $(-\Delta)^{\alpha/2}u(x)$. For the heat SG formulation, we use a quadrature rule to approximate the (truncated) integral (6.3.52), and we use the same SEM of Section 6.5.1 along with the Euler method to solve the heat equation (6.3.16). We give more details of these computations below.

The heat semigroup approach uses the formulations of (6.3.52) and (6.3.16), as discussed in Section 6.3.6. The integration (6.3.52) over $(0, \infty)$ is truncated to the interval $(0, T)$ and is approximated using the quadrature formula

$$\begin{aligned}\mathcal{L}_{\Omega,0}^{\alpha/2}u(x_i) &:= \sum_{j=1}^{N_t}(e^{t_j\Delta_{\mathcal{B}}}u(x_i) - u(x_i))\beta_j \\ &= \sum_{j=1}^{N_t}(w_h(x_i,t_j) - w_h(x_i,0))\beta_j,\end{aligned} \tag{6.9.2}$$

where

$$\beta_j := \frac{1}{\Gamma(-\alpha/2)}\int_{t_j-\Delta t/2}^{t_j+\Delta t/2}\frac{dt}{t^{1+\alpha/2}}.$$

The w_h represents the SEM approximation of the solution $w(x,t)$ to (6.3.16). Equation (6.3.16) is discretized in space using the SEM of Section 6.5.1, and the time discretization is done using a first order Euler method. The solution $w_h(x_i,t_j)$ is computed by evaluating the polynomial expansion produced by the SEM at x_i at time-step j. For the case with nonzero boundary condition $u|_{\partial\Omega} = g$, we use a lifting function v that satisfies the integer order equation

$$\begin{aligned}-\Delta v &= 0, \quad x \in \Omega, \\ v &= g, \quad x \in \partial\Omega.\end{aligned} \tag{6.9.3}$$

Then we subtract the harmonic lifting function v from u and proceed as above to calculate $\mathcal{L}_{\Omega,0}^{\alpha/2}(u - v)$.

In Figure 6.30, we plot $(-\Delta)^{\alpha/2}u(x)$ with $\alpha = 0.5$ and $\alpha = 0.8$ for both the harmonic lifting and Heat SG formulations, both in the square $[0, 1]^2$ and along the slice $y = 0.46$. We also plot the differences between these approximations along the same slice, $y = 0.46$. In these examples, we use the SEM with a single element and expansion order $N = 20$, both in the harmonic lifting discretization and in the space-discretization of the heat equation (6.3.16) associated with the Heat SG formulation. For the time-discretization of the heat equation (6.3.16), we used the forward Euler method with time-step size 6e-5 and 200,000 iterations. The integral (6.3.52) was approximated with 30 quadrature points and the integration was truncated to the interval $(0, T = 1)$.

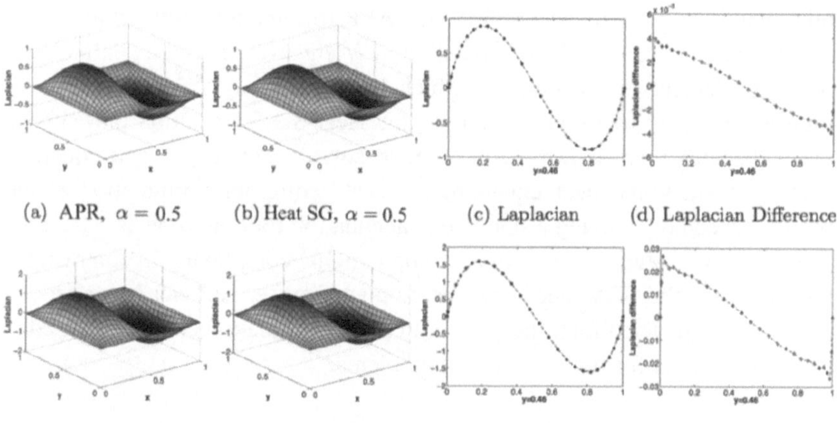

(a) APR, $\alpha = 0.5$ (b) Heat SG, $\alpha = 0.5$ (c) Laplacian (d) Laplacian Difference

(e) APR, $\alpha = 0.8$ (f) Heat SG, $\alpha = 0.8$ (g) Laplacian (h) Laplacian Difference

Figure 6.30 $\alpha = \mathbf{0.5}$: Plots of $(-\Delta)^{\alpha/2}u(x)$ for $u(x) = \cos(\pi x)\sin(\pi y)$ using (a) the harmonic lifting (APR) formulation (Antil et al., 2017) and (b) the Heat SG formulation (Cusimano et al., 2018) on the square domain $[0, 1]^2$. (c) Plots of the solutions along the slice $y = 0.46$, where the discretizations appear on top of each other. (d) Difference of the two discretizations plotted along the slice $y = 0.46$. $\alpha = \mathbf{0.8}$: Plots of $(-\Delta)^{\alpha/2}u(x)$ for $u(x) = \cos(\pi x)\sin(\pi y)$ using (e) the harmonic lifting (APR) formulation and (f) the Heat SG formulation on the square domain $[0, 1]^2$. (g) Plots of the solutions along the slice $y = 0.46$, where the discretizations appear on top of each other. (h) Difference of the two discretizations plotted along the slice $y = 0.46$. In subplots (d) and (h), the difference between the two numerical solutions arrives at a maximum near the boundaries, but is zero exactly at the boundaries because the boundary condition is strongly enforced. This difference will decrease and converge to zero as the accuracy of each method is increased, by increasing the integration time and decreasing the time-step size in the Heat SG method and increasing the number of eigenfunctions and fineness of the mesh in the SEM discretization used in the APR method. Source: Reprinted from Lischke et al. (2020) with permission from Elsevier.

Although theoretical estimates for the convergence of the SEM in Section 6.5.1 are currently under development, it is still possible to have a heuristic discussion comparing the efficiency of using the SEM together with the APR and/or harmonic lifting formulation of Antil et al. (2017) vs. using the heat semigroup formulation of Cusimano et al. (2018) to discretize the inhomogeneous spectral fractional Laplacian. We remark for the heat semigroup formulation, one must solve the (standard) heat equation for many time-steps and use the solution to compute the integral (6.3.52) over $t \in (0, \infty)$. The advantage of this approach is that many robust methods exist to discretize the heat equation with arbitrary boundary conditions in high dimensions. On the other hand, discretizing according to the APR formulation (Antil et al., 2017) using the SEM requires the computation of N eigenvalues and eigenfunctions on each element of the domain, which has complexity N^3. Therefore, for time-*independent* fractional equations, the Heat SG formulation may offer a faster discretization. However, in order to solve a time-*dependent* equation, such as a fractional heat equation, the heat semigroup formulation would require repeatedly solving a standard heat equation for long time *at each time-step*, leading to significantly higher computational complexity. In contrast, the eigenpairs of the SEM need only be computed once, and can be re-used at each time-step. Therefore, the APR with SEM formulation can be expected to be more efficient for time-dependent problems.

6.9.2 Comparison of Spectral, Directional, and Riesz Solutions

Now we can compare the solutions of the inhomogeneous spectral, directional, and Riesz fractional Laplacians. Consider the following equation with nonzero Dirichlet boundary conditions.

$$\begin{aligned} (-\Delta)^{\alpha/2} u = 1, \quad & x \in \Omega := (-1,1) \times (-1,1), \\ u(x) = g(x) = \exp(-|x|^2), \quad & x \in \partial\Omega \;\text{ or }\; \mathbb{R}^d \setminus \Omega, \end{aligned} \tag{6.9.4}$$

where α is chosen to be 1.5. The boundary condition is posed on $\partial\Omega$ for the spectral fractional Laplacian, and it is posed on $\mathbb{R}^d \setminus \Omega$ for the Riesz and directional definitions.

We use the method of Antil et al. (2017) to solve (6.9.4) using the spectral fractional Laplacian, the WOS method for the Riesz definition, and the RBF collocation method for the directional definition. We again find that the directional and Riesz solutions are equivalent up to numerical error, as shown in Figure 6.31. We also observe that the spectral solution is of greater magnitude than the other solutions in this example, contrary to the observations we made in the zero boundary condition case. This indicates that the relative magnitude of the solutions for different definitions relies on the boundary condition. In Figure 6.32, we plot the solutions and differences along the line $y = 0$, where this property can be more easily observed.

The influence of the truncation parameter on the solution accuracy for the RBF collocation method is shown in Table 6.4. The solution in this case is converged when the truncation parameter $K_2 = 6000$, due to the rapid decay of the exterior condition $g(x) = \exp(-|x|^2)$.

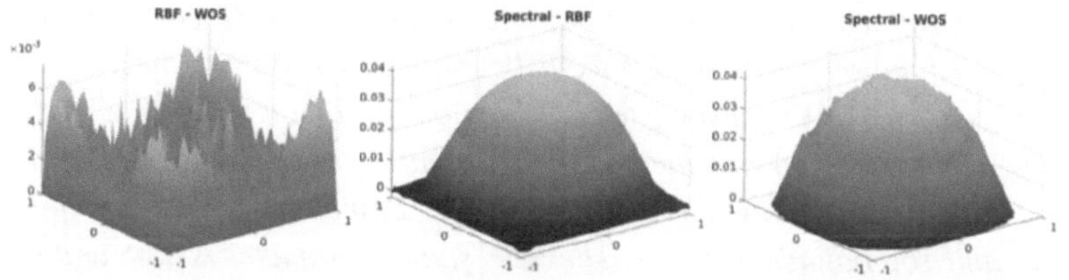

Figure 6.31 Differences of the solutions to the inhomogeneous fractional Poisson equation: (left) the difference between the RBF and WOS solution demonstrates the close similarity between these solutions, as the directional and Riesz definitions are equivalent; (center) the difference between the spectral and RBF solutions; (right) the difference between the spectral and WOS solutions. In these examples, the source function is $f = 1$ and the BC is $g = \exp(-|x|^2)$, and $\alpha = 1.5$. Furthermore, the spectral solution has greater magnitude than the directional and Riesz solutions, which is in contrast with the results for zero boundary conditions. Source: Reprinted from Lischke et al. (2020) with permission from Elsevier.

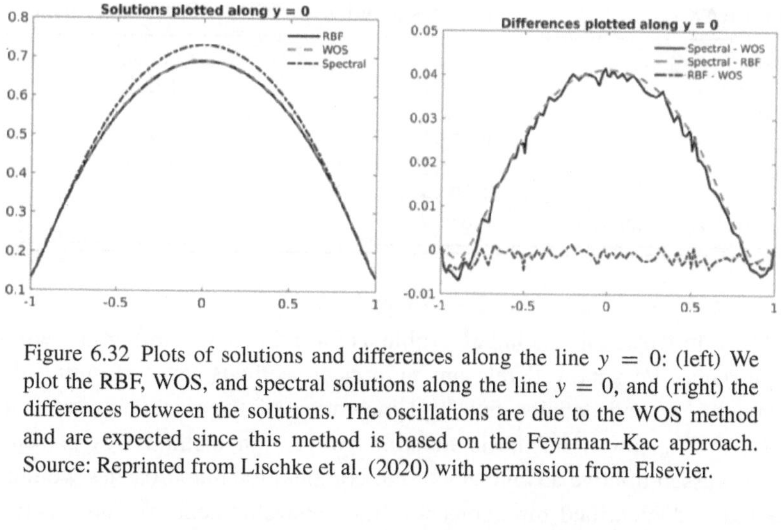

Figure 6.32 Plots of solutions and differences along the line $y = 0$: (left) We plot the RBF, WOS, and spectral solutions along the line $y = 0$, and (right) the differences between the solutions. The oscillations are due to the WOS method and are expected since this method is based on the Feynman–Kac approach. Source: Reprinted from Lischke et al. (2020) with permission from Elsevier.

6.10 Summary and Discussion

In this chapter, we followed a joint theoretical and computational approach to examining the different characteristics of different fractional Laplacians and solutions of related fractional Poisson equations formulated on bounded domains. This included the spectral and horizon-based nonlocal definitions of the fractional Laplacian as well as various formulations of the Riesz fractional Laplacian. We made numerical comparisons using different methods and high levels of refinement in order to compare solutions to both one-

Table 6.4. *Influence of truncation parameter K_2 on the RBF collocation numerical solution for inhomogeneous boundary condition on the unit square* $[-1,1] \times [-1,1]$. *We consider the 2D fractional Poisson problem with the forcing term* $f(x) = 1$ *and the boundary condition* $g(x) = \exp(-|x|^2)$. *The collocation points are on a 41 × 41 regular grid, including 1521 domain points and 160 boundary points. The RBF shape parameter is 0.05 and the spatial step size h is set to be 0.001. We increase the truncation parameter from* $K_2 = 1000$ *to* $K_2 = 6000$ *until the numerical solution becomes sufficiently consistent with respect to* K_2. *Reprinted from the Journal of Computational Physics, 404, Lischke, Anna, Pang, Guofei, Gulian, Mamikon, Song, Fangying, Glusa, Christian, Zheng, Xiaoning, Mao, Zhiping, Cai, Wei, Meerschaert, Mark M, Ainsworth, Mark, Karniadakis, George EM, What is the fractional Laplacian? A comparative review with new results, 109009, Copyright (2020), with permission from Elsevier.*

Truncation Parameter K_2	L_2 Difference from the Previous Solution	L_2 Difference from the Reference Solution
1000	N/A	1.06e+0
1200	5.41e-1	5.19e-1
1600	4.05e-1	1.14e-1
2000	9.27e-2	2.15e-2
3000	2.13e-2	1.50e-4
4000	1.50e-4	2.63e-7
6000 (reference solution)	2.63e-7	0

and two-dimensional benchmark problems formulated with different fractional Laplacians. We surveyed relevant numerical methods for performing these computations and detailed the implementation of the methods used in this work, and a new radial basis function collocation method was presented. We discussed relative advantages of the computational approaches we implemented and identified directions for future development. We also outlined the theoretical derivation of each fractional Laplacian definition in $\mathbb{R}^d$ and contrasted the different approaches to restricting these definitions to bounded domains. Connections were made between fractional Laplacians with different types of boundary conditions with their associated stochastic processes. We surveyed relevant regularity results for the fractional Poisson problems, and discussed the well-posedness of the considered equations.

The nonlocality of the fractional Poisson problems presented significant computational challenges even when posed with zero Dirichlet conditions, and nonzero boundary conditions further compounded these difficulties.

We discussed recently-proposed approaches for approximating the inhomogeneous fractional Laplacians (Antil et al., 2017; Cusimano et al., 2018) and compared these approaches to nonharmonic lifting of the equation to a homogeneous reformulation. We found that all considered methods resulted in equivalent solutions to the inhomogeneous benchmark problems, and we proved this equivalence analytically for the first time.

The numerical methods used in this work are not representative of the breadth of research on this topic, and the types of methods considered reflect the desire of the authors to maintain a focus on fundamental questions (such as nonzero boundary conditions) related to the fractional Laplacian rather than a comprehensive survey of numerical methods. For example, finite-difference approaches to fractional Laplacians and fractional diffusion, such as Huang and Oberman (2014); Li and Chen (2018); Baeumer et al. (2018a); Kelly et al. (2018), have not been discussed at length despite being of classical importance and an area of active development. However, the discussion of the varied types of methods implemented and developed for this work, which we believe to be state-of-the art at the time of this writing, has significant value for aiding researchers simulating fractional models in choosing an appropriate numerical method.

Using these numerical methods, we were able to make several interesting observations. We found that the size of the domain has a significant effect on the evolution of the numerical solutions as the fractional order α was changed, as discussed in the beginning of the chapter. We also emphasized the singular behavior of the Riesz fractional Poisson equation solutions near the boundary, which sharply contrasted with the smooth behavior of the spectral solutions in the same locations. We observed in our examples that given a source function f with no boundary singularity, we can expect that the solution to the Riesz fractional Poisson equation will have a boundary singularity, and the solution to the spectral fractional Poisson equation will be smooth at the boundaries. This observation should be considered when choosing which definition to use to fit data most appropriately. The analytical perspective on this boundary regularity issue was discussed in detail in Sections 6.3 and 6.5.2.

Although the study of the fractional Laplacian is far from complete, this work can serve as a starting point for researchers using these operators to model systems exhibiting anomalous transport phenomena.

In this work, we followed a computational approach to examining the different characteristics of solutions of fractional Poisson equations formulated on bounded domains with the spectral, directional, Riesz, and horizon-based nonlocal definitions of the fractional Laplacian. We made numerical comparisons using high levels of refinement in order to identify differing features in the solutions to both one- and two-dimensional benchmark problems. We surveyed relevant numerical methods for performing these computations and detailed the implementation of the methods used in this work. In addition to the derivation of the methods, we included a discussion of both their advantageous aspects and facets that require further development. We also outlined the theoretical background of each fractional Laplacian definition in $\mathbb{R}^d$ and contrasted the different approaches to restricting these definitions to bounded domains. The regularity of solutions to the fractional Poisson equations with different fractional Laplacians was discussed, in addition to each definitions' stochastic connections.

The nonlocal benchmark equations presented computational challenges when posed with zero Dirichlet boundary conditions, and nonzero boundary conditions further compounded these difficulties. We discussed recently-proposed approaches for approximating the inhomogeneous fractional Laplacians (Antil et al., 2017; Cusimano et al., 2018) and compared these techniques to nonharmonic lifting of the equation to a homogeneous reformulation. We found that these approaches resulted in equivalent solutions to the inhomogeneous benchmark problems, and we proved this equivalence analytically.

The second goal of this chapter was to provide a concise summary of the current state-of-the-art in numerical methods for discretizing each definition of the fractional Laplacian. We demonstrated those methods for both the spectral and Riesz fractional Laplacians. Note that the directional definition is a generalization of the Riesz fractional Laplacian when the probability measure $M(\mathrm{d}\boldsymbol{\theta})$ is chosen to be non-uniform, so the method developed for the directional definition is grouped with the methods for the Riesz definition. To our knowledge, no methods are currently published for discretizing the Riesz fractional Laplacian with Neumann or Robin boundary conditions, and in fact, the formulation of the Neumann Riesz fractional Laplacian is the subject of debate at this time.

While we did present numerical results using previously reported numerical methods, some of the numerical methods used had not been implemented before. We also demonstrated the implementation of the vector Grünwald method (Meerschaert et al., 2004), which was used as an element of the RBF collocation method. Using these methods, we made several interesting observations as discussed in the relevant sections above. We observed that for homogeneous zero Dirichlet conditions in the considered domains, the spectral solutions always had smaller magnitude than their Riesz and directional counterparts. This behavior was reversed in the inhomogeneous examples we tested, leading us to infer that the relative difference in magnitudes of the solution is dependent on the boundary values defined in the problem formulation. We found that the size of the domain has a significant effect on the evolution of the numerical solutions as the fractional order α was changed, as discussed in the beginning of the chapter. We also emphasized the singular behavior of the Riesz and directional solutions near the boundary, which sharply contrasted with the smooth behavior of the spectral solutions in the same locations. We observed in our examples that given a source function f with no boundary singularity, we can expect that the solution to the Riesz fractional Poisson equation will have a boundary singularity, and the solution to the spectral fractional Poisson equation will be smooth at the boundaries. This observation should be considered when choosing which definition to use to fit data most appropriately. The analytical perspective on this boundary regularity issue was discussed in detail in Sections 6.3 and 6.5.2.

7

Time-Integration of Fractional Models

We present efficient time-stepping schemes for accurate and long-time integration of time-fractional models. While we devoted almost all of the preceding chapters to formulating *global* methods (i.e., spectral and spectral element methods) for time-/space-fractional PDEs, we now direct our attention to introducing *local* multi-step finite-difference methods for time-fractional models as described in Section 1.4.

We introduce the fractional Adams family of schemes in Section 7.1, which seamlessly generalize the classical explicit Adams–Bashforth and implicit Adams–Moulton schemes, yet, with a new fractional-order dependent set of coefficients, analytically obtained for each case in addition to a new *history load* vector, arising from the memory-dependent structure of the time-fractional derivatives. Subsequently in Section 7.2, we combine the aforementioned fractional Adams implicit-explicit (IMEX) schemes for stable and long-time integrations along with employing new *correction terms*, which enrich the underlying approximation space, especially in the context of nonlinear fractional ODEs (FODEs). We also investigate the linear stability of the fractional IMEX methods along their fast implementations. To this end, we present a fast approximate inversion scheme and fast computation of hypergeometric functions, which makes the IMEX algorithms amenable for accurate long-time integration of FODEs. In order to reduce the number of correction terms, hence reducing the cost of the fractional IMEX schemes, we formulate a self-singularity-capturing scheme in Section 7.3, which automatically captures the singular structure of the unknown solution (with even several random singularities without any prior knowledge), employing a two-stage time-integration algorithm. We will test the ease and efficiency of the method in the context of challenging cases e.g., long-integration of singular-oscillatory solutions and nonlinear FODEs.

Background and Preliminaries

Let $C(\Xi)$ be the Banach space of continuous functions in $\Xi = [a,b]$ with the norm

$$\|u\|_{C(\Xi)} = \max_{t\in\Xi}|u(t)|. \tag{7.0.1}$$

Also let $0 < \gamma \leq \alpha < 1$ and define $C_\gamma(\Xi)$ to be the following weighted space of continuous functions:

$$C_\gamma(\Xi) = \{f(t) \,:\, (t-a)^\gamma f(t) \in C(\Xi)\}, \tag{7.0.2}$$

with

$$\|f\|_{C_\gamma} = \|(t-a)^\gamma f(t)\|_{C(\Xi)}. \tag{7.0.3}$$

Let I be an open set in $\mathbb{R}$, and consider a function $f(t,u(t))\colon (a,b] \times I \to \mathbb{R}$ such that $f(t,u(t)) \in C_\gamma(\Xi)$ and is Lipschitz continuous with respect to any $u \in I$. Let the fractional Cauchy problem of interest given by:

$$ {}^{\mathrm{C}}_{a}D^{\alpha}_{t}u(t) = g(t,u), \quad \forall t \in \Xi, \tag{7.0.4}$$

$$u(a) = u_0. \tag{7.0.5}$$

Given the aforementioned conditions, there exists a unique solution for (7.0.4) in the following space of functions (see e.g., Kilbas et al. (2006)):

$$C^{\alpha}_{\gamma}(\Xi) = \{u(t) \in C(\Xi) \big| \, {}^{\mathrm{C}}_{a}D^{\alpha}_{t}u(t) \in C_\gamma(\Xi)\}. \tag{7.0.6}$$

A series of numerical methods for FODEs were developed in the 1980s, with several significant contributions summarized in Figure 7.1. In such schemes it is fundamental to incorporate the history effects arising from the fractional operators. The pioneering works are attributed to Lubich (1983, 1986) on *discretized fractional calculus* in the sense of fractional multi-step and finite-difference (FD) schemes. Later on, Tang (1993) developed a super-linear convergent FD scheme, followed by a numerical quadrature approach for fractional derivatives introduced by Diethelm (1997). In the 2000s, Diethelm developed a predictor-corrector approach in addition to a fractional Adams method (Diethelm and Ford, 2002; Diethelm et al., 2004). Later on, Lin and Xu (2007) developed an FD discretization with order $2-\alpha$, which was applied to the time-fractional diffusion equation. More recently, Garrappa (2015) developed trapezoidal methods for fractional multi-step approaches, Zeng et al. (2015) developed a second-order scheme for time-fractional diffusion equations, while Li and Cai (2017) developed a high-order scheme with order $3-\alpha$.

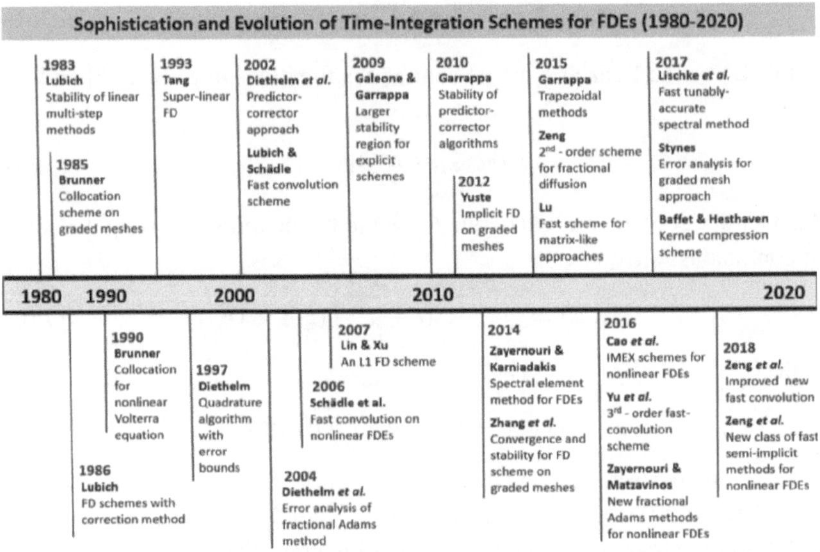

Figure 7.1 Research timeline on significant and diverse numerical schemes for time-fractional differential equations. Source: Reprinted from Zhou et al. (2020b) with permission from Elsevier.

It is known that time-fractional operators possess power-law kernels with a singularity near the initial time, which produces nonsmooth solutions that deteriorate the accuracy of many existing numerical schemes. In order to handle such problems, Lubich (1986) introduced the *so-called* correction method, which was later applied to a series of direct/multi-step schemes for linear/nonlinear FODEs (Cao et al., 2016; Zeng et al., 2017, 2018a,b), and also employed in a *self-singularity-capturing* approach by Suzuki and Zayernouri (2021). In the aforementioned works, the correct determination of singularity powers leads to global high accuracy of the numerical schemes. The idea of *graded meshes* was also introduced with the same objectives by Brunner (1986), who developed a spline collocation scheme for Volterra integro-differential equations, where the graded meshes correspond to non-uniform time-grids which are simple to incorporate in existing FD schemes. Graded meshes were later applied to nonlinear Volterra integral equations (Brunner, 1990). An implicit FD approach was developed in the context of graded meshes by Yuste and Quintana-Murillo (2012). More recently, stability issues of existing/new FD approaches were addressed by Zhang et al. (2014) for the time-fractional diffusion equation, and also by Stynes et al. (2017) for a reaction-diffusion problem, where, in the latter, an optimal mesh grading parameter was obtained. For a comparison between the performance of Lubich's corrections and graded meshes, we refer the reader to (Zeng et al., 2018a).

The main computational challenge of direct FD schemes for time-integration of FODEs is the evaluation of the history term, which usually leads to a computational complexity of $\mathcal{O}(N^2)$ and memory storage of $\mathcal{O}(N)$, where N represents the total number of time-steps. To address such issues, fast schemes were developed, starting with the first-order *fast convolution method* by Lubich and Schädle (2002), which reduced the computational complexity to $\mathcal{O}(N \log N)$, and memory requirements to $\mathcal{O}(\log N)$. The main idea of the scheme is to approximate the power-law kernel via numerical inverse Laplace transforms and split the integral operator (but not the time-grid) into exponentially increasing time-steps. Later on, Schädle et al. (2006) extended the developed fast convolution for nonlinear FODEs. A third-order extension was developed by Yu et al. (2016) and applied to the time-integration for 3D simulation of a class of time-fractional PDEs. Zeng et al. (2018a) developed an improved version of the fast-convolution approach, which considers real-valued integration contours with the order $3 - \alpha$. Of particular interest, fast matrix-based schemes were also developed, such as the kernel compression method by Baffet and Hesthaven (2017) and the KPS preconditioner by Zhou et al. (2020c,a). The main idea of such approaches is to represent the time-stepping equation in a global linear system and exploit the resulting Toeplitz-like structure through fast Fourier transforms (FFTs).

In addition to the aforementioned challenges, dealing with stiff/nonlinear problems further deteriorates the accuracy and might pose stability issues to existing numerical schemes. Fractional linear multi-step approaches become interesting alternatives to handle such issues. Diethelm and Ford (2002) developed a predictor-corrector scheme and later analyzed the error of a family of fractional Adams–Bashforth/Moulton schemes (Diethelm et al., 2004). Galeone and Garrappa (2006, 2009) studied the stability of implicit and explicit fractional multi-step methods and proposed new explicit schemes with larger stability regions. In addition, the stability analysis of fractional predictor-corrector schemes was studied by Garrappa (2010). Also in the context of fractional multi-step schemes, Zayernouri and Matzavinos (2016) developed a family of fractional Adams schemes for high-order explicit/implicit treatment of nonlinear problems, where a particular time-splitting preserved the original structure of integer-order Adams schemes. Larger stability regions can be obtained through semi-implicit schemes, where for instance, Cao et al. (2016) developed two IMEX schemes for nonlinear FODEs, utilizing two distinct force extrapolation formulas and also analyzed the stability of the developed schemes. Recently, Zeng et al. (2018b) developed a new class of fast, second-order semi-implicit methods for nonlinear FODEs through new fast convolutions. Zhou et al. (2020b) also developed and analyzed the convergence and stability of one-leg approaches.

7.1 Fractional Adams Family of Schemes

The classical Adams family of schemes are linear multi-step methods for time-integration of integer-order ODEs. The explicit Adams schemes are known as Adams–Bashforth methods, and the corresponding implicit Adams schemes are known as Adams–Moulton methods. Here, following Zayernouri and Matzavinos (2016), we aim to generalize the existing schemes to fractional ODEs as they can be effectively employed in solving nonlinear problems through mixed implicit-explicit (IMEX) splitting of fractional operators, while maintaining a higher accuracy compared to L^1 formulation without resorting to any iterative solver at each time-step.

We again consider the fractional ODE in (7.0.4) and (7.0.5). By the definition of Caputo fractional derivative, we have

$$ {}^{\mathrm{C}}_{0}D^{\tau}_{t}u = \frac{1}{\Gamma(1-\tau)}\int_0^t \frac{u'(s)}{(t-s)^{\tau}}\mathrm{d}s = H^k(t) + {}^{\mathrm{C}}_{t_k}D^{\tau}_{t}u, \tag{7.1.1} $$

where

$$ H^k(t) = \frac{1}{\Gamma(1-\tau)}\int_0^{t_k} \frac{u'(s)}{(t-s)^{\tau}}\,\mathrm{d}s. $$

Moreover,

$$ {}^{\mathrm{C}}_{t_k}D^{\tau}_{t}u = {}^{\mathrm{C}}_{t_k}D^{\tau}_{t}(u - u_k + u_k) = {}^{\mathrm{C}}_{t_k}D^{\tau}_{t}(u-u_k) = {}^{\mathrm{RL}}_{t_k}D^{\tau}_{t}(u-u_k), \tag{7.1.2} $$

since ${}^{\mathrm{RL}}_{t_k}D^{\tau}_{t}u_k \equiv 0$, and $(u - u_k)$ vanishes at $t = t_k$, hence, replacing the Caputo derivative with its Riemann–Liouville counterpart. Next, by substituting (7.1.2) into (7.1.1), then, plugging (7.1.1) into (7.0.4), we obtain

$$ {}^{\mathrm{RL}}_{t_k}D^{\tau}_{t}(u-u_k) = g(t;u) - H^k(t), \quad t \in (0,T]. \tag{7.1.3} $$

Next, by applying the fractional integration ${}^{\mathrm{RL}}_{t_k}I^{\tau}_{t}(\cdot)$ to (7.1.3) and evaluating both sides at $t = t_{k+1}$, we obtain

$$ u_{k+1} - u_k = \left[\, {}^{\mathrm{RL}}_{g}I^{t}_{(};u) - \mathcal{H}^k \right]\Big|_{t=t_{k+1}}, \tag{7.1.4} $$

noting that

$$ {}^{\mathrm{RL}}_{t_k}I^{\tau}_{t}[{}^{\mathrm{RL}}_{t_k}D^{\tau}_{t}(u-u_k)] = u - u_k $$

by Podlubny (1999, (2.113)), where $f(t) = u - u_k$ is a continuous function vanishing at t_k; moreover, $\mathcal{H}^k = {}^{\mathrm{RL}}_{t_k}I^{\tau}_t(H^k)$ will be referred to as the *history load* term in what follows.

In the next section, we introduce fractional generalizations of the Adams–Bashforth (A-B) and Adams–Moulton (A-M) methods through an explicit and an implicit extrapolation of the term ${}^{\mathrm{RL}}_{t_k}I^{\tau}_t g(t; u)$, respectively.

7.1.1 Fractional Adams–Bashforth Method

In this method, we apply a time-*extrapolation* of the right-hand side $g(t; u)$ to the interval $[t_k, t_{k+1}]$ using $(J + 1)$ past grid points as

$$g(t; u) \approx \sum_{j=0}^{J} g_{k-j} L_j(t), \tag{7.1.5}$$

where the coefficients $g_{k-j} = g(t_{k-j}; u(t_{k-j}, x))$, for $j = 0, 1, \ldots, J$, and the Lagrange interpolant $L_j(t)$ is given by

$$L_j(t) = \prod_{\substack{i=0\\ i\neq j}}^{N} \frac{t - t_{k-i}}{t_{k-j} - t_{k-i}}, \qquad j = 0, 1, \ldots, J. \tag{7.1.6}$$

Hence, we obtain

$$\begin{aligned} {}^{\mathrm{RL}}_{t_k}I^{\tau}_t g(t; u) &\approx \frac{1}{\Gamma(\tau)} \int_{t_k}^{t_{k+1}} \frac{\sum_{j=0}^{J} g_{k-j} L_j(s)}{(t_{k+1} - s)^{1-\tau}} ds \\ &= (\Delta t)^{\tau} \sum_{j=0}^{J} \beta_j^{AB} g_{k-j}, \end{aligned}$$

where β_j^{AB}, $j = 0, 1, \ldots, J$, are the τ-dependent coefficients analytically obtained and given in Table 7.1. Finally, we represent the fractional A-B method as

$$\frac{u_{k+1} - u_k}{(\Delta t)^{\tau}} = \sum_{j=0}^{J} \beta_j^{AB} g_{k-j} - \frac{1}{(\Delta t)^{\tau}} \mathcal{H}^k, \tag{7.1.7}$$

where we later show how to obtain the history load $\mathcal{H}^k$, $k = 1, 2, \ldots, N$.

Table 7.1. *τ-dependent coefficients of the generalized (fractional) Adams–Bashforth method. Reprinted from the Journal of Computational Physics, 317, Zayernouri, Mohsen, and Matzavinos, Anastasios, Fractional Adams–Bashforth/Moulton methods: an application to the fractional Keller–Segel chemotaxis system, 1–14, Copyright (2016), with permission from Elsevier.*

β_j	$J=0$	$J=1$	$J=2$
β_0^{AB}	$\frac{1}{\Gamma(1+\tau)}$	$\frac{1}{\Gamma(1+\tau)}+\frac{1}{\Gamma(2+\tau)}$	$\frac{1}{\Gamma(1+\tau)}+\frac{3/2}{\Gamma(2+\tau)}+\frac{1}{\Gamma(3+\tau)}$
β_1^{AB}	0	$\frac{-1}{\Gamma(2+\tau)}$	$\frac{-2}{\Gamma(2+\tau)}+\frac{-2}{\Gamma(3+\tau)}$
β_2^{AB}	0	0	$\frac{1/2}{\Gamma(2+\tau)}+\frac{1}{\Gamma(3+\tau)}$

7.1.2 Fractional Adams–Moulton Method

We can alternatively *interpolate* $g(t;u)$ between t_k and t_{k+1} as

$$g(t;u) \approx \sum_{j=0}^{J} g_{k+1-j}\mathbb{L}_j(t), \tag{7.1.8}$$

where $g_{k+1-j} = g(t_{k+1-j}; u(t_{k+1-j}, x))$, and the interpolant Lagrange bases are constructed as

$$\mathbb{L}_j(t) = \prod_{\substack{i=0 \\ i\neq j}}^{N} \frac{t - t_{k+1-i}}{t_{k+1-j} - t_{k+1-i}}, \quad j = 0, 1, \ldots, J. \tag{7.1.9}$$

Therefore, we obtain

$$\begin{aligned} {}^{\mathrm{RL}}_{t_k}I^{\tau}_{t} g(t;u) &\approx \frac{1}{\Gamma(\tau)} \int_{t_k}^{t_{k+1}} \frac{\sum_{j=0}^{J} g_{k+1-j}\mathbb{L}_j(s)}{(t_{k+1}-s)^{1-\tau}} ds \\ &= (\Delta t)^{\tau} \sum_{j=0}^{J} \beta_j^{AM} g_{k+1-j}, \end{aligned}$$

where the coefficients β_j^{AM} are τ-dependent and are given in Table 7.2. Hence, we obtain the generalized A-M method as

$$\frac{u_{k+1} - u_k}{(\Delta t)^{\tau}} = \sum_{j=0}^{J} \beta_j^{AM} g_{k+1-j} - \frac{1}{(\Delta t)^{\tau}} \mathcal{H}^k. \tag{7.1.10}$$

Table 7.2. *τ-dependent coefficients of the generalized (fractional) Adams–Moulton method. Reprinted from the Journal of Computational Physics, 317, Zayernouri, Mohsen, and Matzavinos, Anastasios, Fractional Adams–Bashforth/Moulton methods: an application to the fractional Keller–Segel chemotaxis system, 1–14, Copyright (2016), with permission from Elsevier.*

β_j	$J=0$	$J=1$	$J=2$
β_0^{AM}	$\frac{1}{\Gamma(1+\tau)}$	$\frac{1}{\Gamma(2+\tau)}$	$\frac{1}{2\Gamma(2+\tau)} + \frac{1}{\Gamma(3+\tau)}$
β_1^{AM}	0	$\frac{1}{\Gamma(1+\tau)} - \frac{1}{\Gamma(2+\tau)}$	$\frac{1}{\Gamma(1+\tau)} + \frac{-2}{\Gamma(3+\tau)}$
β_2^{AM}	0	0	$\frac{-1}{2\Gamma(2+\tau)} + \frac{1}{\Gamma(3+\tau)}$

Remark 7.1 We note that the generalized (fractional) A-B and A-M methods, given in (7.1.7) and (7.1.10), respectively, share the same history load $\mathcal{H}^k$. Moreover, $\mathcal{H}^k = 0$ when the temporal fractional order $\tau = 1$; hence, we recover the classical A-B and A-M methods. Moreover when $J = 0$, we observe that the generalized A-B and A-M methods, given in (7.1.7) and (7.1.10), represent the fractional forward (explicit) Euler method and backward (implicit) Euler method, respectively.

7.1.3 Computation of the History Load $\mathcal{H}^k$

We recall that the history load term is given by $\mathcal{H}^k = {}^{\text{RL}}_{t_k}I^{\tau}_t(H^k)$, where

$$H^k(t) = \frac{1}{\Gamma(1-\tau)} \int_0^{t_k} \frac{u'(s)}{(t-s)^{\tau}} ds = \frac{1}{\Gamma(1-\tau)} \sum_{j=0}^{k-1} \int_{t_j}^{t_{j+1}} \frac{u'(s)}{(t-s)^{\tau}} ds.$$

In order to approximate $\frac{du}{ds}$ in $[t_j, t_{j+1}]$, we employ two interpolation methods to approximate the solution $u(t)$.

Method I: Linear Interpolation of $u(t)$

The linear interpolation of u leads to a locally second-order approximation of the solution in $[t_j, t_{j+1}]$, which yields

$$H^k(t) \approx H_I^k(t) = \frac{1}{\Gamma(1-\tau)} \sum_{j=0}^{k-1} (\frac{u_{j+1}-u_j}{\Delta t}) \int_{t_j}^{t_{j+1}} \frac{1}{(t-s)^{\tau}} \mathrm{d}s$$
$$= \frac{1}{\Gamma(2-\tau)} \sum_{j=0}^{k-1} (\frac{u_{j+1}-u_j}{\Delta t}) G_j^I(t),$$

where

$$G_j^I(t) = (t-t_j)^{1-\tau} - (t-t_{j+1})^{1-\tau}.$$

Hence,

$$\mathcal{H}^k(t) \approx \mathcal{H}_I^k(t) = \frac{1}{\Gamma(2-\tau)} \sum_{j=0}^{k-1} (\frac{u_{j+1}-u_j}{\Delta t}) \, {}^{\mathrm{RL}}_{t_k}\mathcal{I}_t^{\tau}\Big(G_j^I(t)\Big).$$

By evaluating both sides at t_{k+1}, we obtain the corresponding history load as

$$\mathcal{H}^k \approx \mathcal{H}_I^k = \frac{1}{\Gamma(\tau)\Gamma(2-\tau)} \sum_{j=0}^{k-1} (\frac{u_{j+1}-u_j}{\Delta t}) \Big[\int_{t_k}^{t_{k+1}} (t_{k+1}-s)^{\tau-1}(s-t_j)^{1-\tau}\mathrm{d}s$$
$$- \int_{t_k}^{t_{k+1}} (t_{k+1}-s)^{\tau-1}(s-t_{j+1})^{1-\tau}\mathrm{d}s \Big],$$

which can be written as

$$\boxed{\mathcal{H}^k \approx \mathcal{H}_I^k = \frac{1}{\Gamma(\tau)\Gamma(2-\tau)} \sum_{j=0}^{k-1} (\frac{u_{j+1}-u_j}{\Delta t}) \Big[\mathcal{A}_{k,j} - \mathcal{A}_{k,j+1} \Big],} \quad (7.1.11)$$

where

$$\mathcal{A}_{kj} = \int_{t_k}^{t_{k+1}} (t_{k+1}-s)^{\tau-1}(s-t_j)^{1-\tau}\mathrm{d}s.$$

The latter is obtained analytically in terms of the following hypergeometric functions (e.g., see Aomoto et al. (2011)):

$$\mathcal{A}_{kj} = \begin{cases} -\frac{(\Delta t)^{\tau}}{\tau}(t_k-t_j)^{1-\tau}\big[-\tau + (\tau-1)\,{}_2F_1(1,1,1+\tau,\frac{\Delta t}{t_{k+1}-t_j})\big], & 0 \le j < k, \\ -(\Delta t)(\tau-1)\pi\csc(\pi\tau), & j = k, \end{cases} \quad (7.1.12)$$

where we recall from Chapter 2 that the hypergeometric function is defined for $|z| < 1$ by the power series

$${}_2F_1(a,b,c,z) = \sum_{n=0}^{\infty} \frac{(a)_n\,(b)_n}{(c)_n} \frac{z^n}{n!},$$

in which $(q)_n$ is the (rising) Pochhammer symbol, defined by:

$$(q)_n = \begin{cases} 1, & n = 0, \\ q(q+1)\cdots(q+n-1), & n > 0. \end{cases}$$

Therefore, the history load $\mathcal{H}_I^k$ is computed exactly in terms of $\mathcal{A}_{kj}$ up to the discretization accuracy of the first derivative of the solution, i.e., $\frac{u_{j+1}-u_j}{\Delta t}$.

Method II: Quadratic Interpolation of $u(t)$

A third-order approximation of $u(t)$ in $[t_j, t_{j+1}]$ yields

$$u(t) \approx u_{j+1}\left(\frac{t-t_j}{t_{j+1}-t_j}\right)\left(\frac{t-t_{j-1}}{t_{j+1}-t_{j-1}}\right) + u_j\left(\frac{t-t_{j+1}}{t_j-t_{j+1}}\right)\left(\frac{t-t_{j-1}}{t_j-t_{j-1}}\right) + u_{j-1}\left(\frac{t-t_{j+1}}{t_{j-1}-t_{j+1}}\right)\left(\frac{t-t_j}{t_{j-1}-t_j}\right),$$

where by taking the first derivative with respect to time, we obtain

$$\frac{du}{dt} \approx \left.\frac{du}{dt}\right|_{t=t_{j+1/2}} + \left.\frac{d^2u}{dt^2}\right|_{t=t_{j+1/2}} (t - t_{j+1/2}).$$

The last expression yields a higher-order approximation of $\frac{du}{dt}$ to be employed in the calculation of history load. Therefore, $H^k(t) \approx H_{II}^k(t)$, where

$$H_{II}^k(t) = \frac{1}{\Gamma(1-\tau)} \sum_{j=1}^{k-1} \int_{t_j}^{t_{j+1}} \frac{u'(s)|_{s=t_{j+1/2}} + u''(s)|_{s=t_{j+1/2}}(s - t_{j+1/2})}{(t-s)^\tau} ds + \frac{1}{\Gamma(1-\tau)} \int_0^{t_1} \frac{u'(s)|_{s=t_{1/2}}}{(t-s)^\tau} ds,$$

where we employed the previously obtained lower-order approximation in $[0, t_1]$ corresponding to $j = 0$. Hence,

$$\begin{aligned} H_{II}^k(t) &= \frac{1}{\Gamma(1-\tau)} \sum_{j=0}^{k-1} \left(\frac{u_{j+1}-u_j}{\Delta t}\right) \int_{t_j}^{t_{j+1}} \frac{1}{(t-s)^\tau} ds \\ &\quad + \frac{1}{\Gamma(1-\tau)} \sum_{j=1}^{k-1} \frac{d^2u}{ds^2}|_{s=t_{j+1/2}} \int_{t_j}^{t_{j+1}} \frac{(s-t_{j+1/2})}{(t-s)^\tau} ds \\ &= H_I^k(t) + \frac{1}{\Gamma(\tau)\,\Gamma(2-\tau)} \sum_{j=1}^{k-1} \left(\frac{u_{j+1}-2u_j+u_{j-1}}{(\Delta t)^2}\right) G_j^{II}(t), \end{aligned} \tag{7.1.13}$$

in which the first term is associated with the linear representation of $u(t)$, and the second term in (7.1.13) denotes the correction for obtaining a higher-order accuracy. Moreover, $G_j^{II}(t)$ is given by

$$\begin{aligned} G_j^{II}(t) &= -\frac{\Delta t}{2}\left[(t-t_j)^{1-\tau} + (t-t_{j+1})^{1-\tau}\right] \\ &\quad + \frac{1}{2-\tau}\left[(t-t_j)^{2-\tau} - (t-t_{j+1})^{2-\tau}\right]. \end{aligned} \tag{7.1.14}$$

Accordingly, the corresponding load vector is obtained as

$$\begin{aligned} \mathcal{H}^k \approx \mathcal{H}^k_{II} &= {}^{\mathrm{RL}}_{t_k}I^{\tau}_{t_k+1} H^k_{II}(t) = {}^{\mathrm{RL}}_{t_k}I^{\tau}_{t_k+1} H^k_I(t) \\ &\quad + \frac{1}{\Gamma(\tau)\Gamma(2-\tau)} \sum_{j=1}^{k-1} \left(\frac{u_{j+1} - 2u_j + u_{j-1}}{(\Delta t)^2}\right) {}^{\mathrm{RL}}_{t_k}I^{\tau}_{t_k+1} G_j^{II}(t), \end{aligned}$$

which can be written as

$$\begin{aligned} \mathcal{H}^k \approx \mathcal{H}^k_{II} &= \mathcal{H}^k_I + \frac{1}{\Gamma(\tau)\Gamma(2-\tau)} \\ &\quad \times \sum_{j=1}^{k-1} \left(\frac{u_{j+1} - 2u_j + u_{j-1}}{(\Delta t)^2}\right)\left[-\frac{\Delta t(\mathcal{A}_{k,j} + \mathcal{A}_{k,j+1})}{2} + \frac{\mathcal{B}_{k,j} - \mathcal{B}_{k,j+1}}{2-\tau}\right], \end{aligned}$$

where

$$\mathcal{B}_{kj} = \int_{t_k}^{t_{k+1}} (t_{k+1} - s)^{\tau-1}(s - t_j)^{2-\tau} ds.$$

The integral can be computed exactly in terms of the following hypergeometric function:

$$\mathcal{B}_{kj} = \begin{cases} -\frac{(\Delta t)^{\tau}}{\tau}\frac{(t_k - t_j)^{3-\tau}}{(t_{k+1} - t_j)}\left[{}_2F_1(1, 3, 1+\tau, \frac{\Delta t}{t_{k+1} - t_j})\right], & 0 \le j < k, \\ \\ -\frac{(\Delta t)^{\tau}\pi}{2}(-2+\tau)(\tau-1)\csc(\pi\tau), & j = k. \end{cases} \tag{7.1.15}$$

Finally, we note that $\mathcal{B}_{kj}$ is constructed only once.

In Table 7.3, we examine the performance of the fractional A-B scheme at different values of J = 0, 1, and 2, and for different temporal orders τ. The equation ${}^{\mathrm{C}}_{0}D^{\tau}_{t}u(t) = f(t;u)$ is solved subject to homogeneous initial conditions. In the upper table, we present the fractional A-B method with $J = 0$ and $\mathcal{H}^k \approx \mathcal{H}^k_I$. The exact solution in this case is given by $u^{ext} = t^{\tau}$, in which the right-hand side $f(t) = \frac{\Gamma(1+\tau+J)}{\Gamma(1+\tau+J-\tau)} t^J$, where J =0, 1, and 2, respectively. In the middle part of Table 7.3, we present the fractional A-B method with $J = 1$ and $\mathcal{H}^k \approx \mathcal{H}^k_{II}$, considering $u^{ext} = t^{1+\tau}$. In the lower part of Table 7.3, we demonstrate the performance of the fractional A-B method when $J = 2$, $\mathcal{H}^k \approx \mathcal{H}^k_I$, and $u^{ext} = t^{2+\tau}$. The simulation time is

Table 7.3. *Fractional Adams–Bashforth time-integration of* ${}^{C}_{0}D^{\tau}_{t}u(t) = f(t;u)$ *subject to homogeneous initial conditions. (upper table): fractional A-B (J=0), where the exact solution is* $u^{ext} = t^{\tau}$*, (middle table): fractional A-B (J=1), where* $u^{ext} = t^{1+\tau}$*, and (lower table): fractional A-B (J = 2), where* $u^{ext} = t^{2+\tau}$*. Here, the simulation time is* $T = 1$ *and* $f(t) = \frac{\Gamma(1+\tau+J)}{\Gamma(1+\tau+J-\tau)} t^{J}$*, where* $J = 0, 1,$ *and 2, respectively. Reprinted from the Journal of Computational Physics, 317, Zayernouri, Mohsen, and Matzavinos, Anastasios, Fractional Adams–Bashforth/Moulton methods: an application to the fractional Keller–Segel chemotaxis system, 1–14, Copyright (2016), with permission from Elsevier.*

Fractional A-B, J=0, ($\mathcal{H}^k \approx \mathcal{H}^k_I$)

Δt	$\tau = 1/10$	Order	$\tau = 1/2$	Order	$\tau = 9/10$	Order
1/8	0.823385	0.43	0.388225	0.88	0.261049	1.00
1/16	0.613304	0.55	0.211394	0.92	0.129804	1.00
1/32	0.416202	0.70	0.111384	0.96	0.064746	1.00
1/64	0.255626	0.82	0.057415	0.98	0.032357	0.999
1/128	0.144731	0.90	0.029210	0.99	0.016188	0.998

Fractional A-B, J=1, ($\mathcal{H}^k \approx \mathcal{H}^k_{II}$)

Δt	$\tau = 1/10$	Order	$\tau = 1/2$	Order	$\tau = 9/10$	Order
1/8	0.229075	1.65	0.0517457	1.91	0.029365	1.975
1/16	0.073136	1.83	0.013736	1.96	0.000747	1.993
1/32	0.020521	1.90	0.003527	1.978	0.001876	1.998
1/64	0.005516	1.91	0.000895	1.984	0.000469	1.999
1/128	0.001473	1.93	0.000226	1.998	0.000117	2.00

Fractional A-B, J=2, ($\mathcal{H}^k \approx \mathcal{H}^k_{II}$)

Δt	$\tau = 1/10$	Order	$\tau = 1/2$	Order	$\tau = 9/10$	Order
1/8	0.088115	2.72	0.017732	2.92	0.010759	2.92
1/16	0.013366	2.76	0.002345	2.99	0.001423	2.98
1/32	0.001969	2.78	0.000294	3.05	0.000180	3.02
1/64	0.000286	2.90	0.000035	3.12	0.000022	3.09
1/128	0.000038	2.91	4.06×10^{-6}	3.20	2.60×10^{-6}	3.22

$T = 1$ in all cases, and all errors are computed using the L^{∞} norm at the final simulation time. Interestingly, we observe a first-, second-, and third-order of accuracy when $J = 0$, 1, and 2 respectively. Here in this context when the order of accuracy is $p = 1, 2, 3$, that means by doubling the number of time-steps the corresponding time-integration error in L^∞ norm would reduce 2^p times. In other words, the numerical evidence suggests that the fractional A-B method not only generalizes the standard A-B method, but it also preserves the traditional order of accuracy of the A-B methods originally developed for integer-order systems.

In a similar fashion, we examine the performance of the fractional A-M scheme with different values of J and for different choices of τ in Table 7.4. We keep the setting as in Table 7.3 to experimentally observe the order accuracy associated with different J's. Again, we observe a τ-independent first- and second-order of accuracy when $J = 0$, and 1. However, our results show a τ-dependent accuracy of less than third order when $J = 2$, as opposed to the fractional A-B scheme. That simply highlights the need for a higher-order approximation of the history term, which turns out to be inefficient in terms of computation cost. We note that the numerical test cases are chosen such that the J-th derivative of the exact solution, where $J = 0, 1, 2$, vanishes at $t = 0$ in each corresponding method.

While fractional Adams schemes show more flexibility in treatment of nonlinear problems, there are still two remaining challenges associated with such family of FDMS, which are capturing singularities and more efficient/faster evaluation of history terms for long-time integration of fractional ODEs. The following implicit FDM will address them properly.

7.2 Adams Implicit-Explicit Schemes for Long-Time Integration

Efficient long-time integration of nonlinear fractional differential equations is significantly challenging due to the integro-differential nature of the fractional operators. In addition, the inherent nonsmoothness introduced by the inverse power-law kernels deteriorates the accuracy and efficiency of many existing numerical methods. We present two efficient first- and second-order implicit-explicit (IMEX) methods for accurate time-integration of stiff/nonlinear fractional differential equations with fractional order $\alpha \in (0, 1]$ and prove their convergence and linear stability properties. These methods are based on a linear multi-step fractional Adams–Moulton method (FAMM), followed by the extrapolation of the nonlinear force terms. In order to handle the singularities near the initial time, we employ Lubich-like corrections to the resulting fractional operators. The obtained linear stability regions of the developed IMEX methods are larger than existing IMEX methods in the literature. Furthermore, the size of the stability regions increases with the decrease of

Table 7.4. *Fractional Adams–Moulton time-integration of* ${}^{C}_{0}D^{\tau}_{t}u(t) = f(t;u)$ *subject to homogeneous initial conditions. (upper table): fractional A-M (J=0), where the exact is solution* $u^{ext} = t^{\tau}$, *(middle table): fractional A-M (J=1), where* $u^{ext} = t^{1+\tau}$, *and (lower table): fractional A-M (J=2), where* $u^{ext} = t^{2+\tau}$. *Here, the simulation time is* $T = 1$ *and* $f(t) = \frac{\Gamma(1+\tau+J)}{\Gamma(1+\tau+J-\tau)} t^J$, *where* $J = 0$, *1, and 2, respectively. Reprinted from the Journal of Computational Physics, 317, Zayernouri, Mohsen, and Matzavinos, Anastasios, Fractional Adams–Bashforth/Moulton methods: an application to the fractional Keller–Segel chemotaxis system, 1–14, Copyright (2016), with permission from Elsevier.*

Fractional A-M, J=0, $(\mathcal{H}^k \approx \mathcal{H}^k_I)$

Δt	$\tau = 1/10$	Order	$\tau = 1/2$	Order	$\tau = 9/10$	Order
1/8	0.566053	1.82	0.186798	1.32	0.146284	1.19
1/16	0.1602418	1.51	0.074632	1.15	0.063814	1.09
1/32	0.056163	1.28	0.033644	1.07	0.029979	1.04
1/64	0.022971	1.15	0.016002	1.03	0.014544	1.02
1/128	0.010352	1.07	0.007710	1.02	0.007161	1.01

Fractional A-M, J=1, $(\mathcal{H}^k \approx \mathcal{H}^k_{II})$

Δt	$\tau = 1/10$	Order	$\tau = 1/2$	Order	$\tau = 9/10$	Order
1/8	0.051728	2.09	0.037242	2.06	0.031385	2.05
1/16	0.012254	2.02	0.008879	2.03	0.007565	2.02
1/32	0.003004	1.97	0.002171	2.01	0.001859	2.01
1/64	0.000765	1.93	0.000537	2.006	0.000465	2.007
1/128	0.000200	1.94	0.000134	2.00	0.000114	2.00

Fractional A-M, J=2, $(\mathcal{H}^k \approx \mathcal{H}^k_{II})$

Δt	$\tau = 1/10$	Order	$\tau = 1/2$	Order	$\tau = 9/10$	Order
1/8	0.012274	3.00	0.010341	2.94	0.009544	2.92
1/16	0.001528	2.95	0.001351	2.88	0.001260	2.82
1/32	0.000197	2.93	0.000184	2.83	0.000177	2.70
1/64	0.000026	2.91	0.000026	2.78	0.000027	2.57
1/128	3.45×10^{-6}	2.90	3.74×10^{-6}	2.74	4.57×10^{-6}	2.43

fractional order values, which is suitable for stiff problems. We also rewrite the resulting IMEX methods in the language of nonlinear Toeplitz systems, where we employ a fast inversion scheme to achieve a computational complexity of $\mathcal{O}(N \log N)$, where N denotes the number of time-steps. Our computational results demonstrate that the developed schemes can achieve global first- and second-order accuracy for highly-oscillatory stiff/nonlinear problems with singularities.

Although there are significant developments in this context, they usually address the aforementioned singularity/performance/stability issues for stiff/nonlinear problems separately. In this regard, there is still a need for numerical schemes in the context of stiff/nonlinear FODEs that **I)** efficiently handle the numerical solution with low-regularity for both the solution $u(t)$ and nonlinear term $f(t, u(t))$; **II)** present linear complexity with respect to the number of time-steps N; **III)** have larger stability regions compared to the existing numerical schemes; **IV)** mimic and generalize the structure of existing integer-order IMEX schemes, widely employed by scientists and engineers to its fractional-order counterparts. The main contribution of the present work is to develop a class of IMEX methods for accurate time-integration of stiff/nonlinear FODEs. Specifically:

- We start with the linear multi-step FAMMs developed in the previous section in the sense of linear/nonlinear fractional Cauchy equations. For the linear problem, two sets of Lubich-like correction terms (Lubich, 1986) are employed.
- We develop a class of new first- and second-order IMEX methods with the combination of the work in Zayernouri and Matzavinos (2016) FAMMs with two extrapolation methods for the nonlinear term. The obtained methods are denoted by IMEX(p), which is first-order accurate when $p = 0$, and second-order accurate when $p = 1$.
- The convergence and linear stability of the developed IMEX methods are proved and the corresponding regions of stability are shown to be larger for smaller values of the fractional order α.
- The convolution nature of the fractional operators allows us to represent the corresponding IMEX methods in the language of global-in-time Toeplitz-like nonlinear systems, and employ the fast approximate inversion approach by Lu et al. (2015). Since the Toeplitz system is nonlinear, we utilize a Picard iteration scheme which takes N^p iterations until convergence with respect to a specified tolerance ϵ^p. Under suitable conditions, N^p does not significantly increase with N.
- The corresponding *history load* for the developed IMEX schemes is given by hypergeometric functions, which are efficiently evaluated through a Gauss–Jacobi quadrature with a fixed number of Q integration points.
- The asymptotic computational complexity of the scheme is $\mathcal{O}(N \log N)$, with memory storage of order $\mathcal{O}(N)$.

7.2.1 Linear FODEs as Initial-Value Problems

Consider the numerical solutions of the following linear FODE:

$$ {}_0^{\mathrm{C}}D_t^{\alpha}u(t) = \lambda u(t), \quad \alpha \in (0,1], \ t \in (0,T]; \quad u(0) = u_0, \tag{7.2.1} $$

where $\lambda \in \mathbb{C}$, $u_0 \in \mathbb{R}^d$. Now, we adopt the FAMMs developed in Zayernouri and Matzavinos (2016) to solve (7.2.1). Let $h > 0$ be the time-step size with $h = T/N$ ($N \in \mathbb{N}$) and $t_k = kh$ ($0 \le k \le N$). By the definition of the Caputo fractional derivative, we can split the fractional operator in history and local parts:

$$ \begin{aligned} {}_0^{\mathrm{C}}D_t^{\alpha}u(t) =& \frac{1}{\Gamma(1-\alpha)} \int_0^{t_k} \frac{u'(v)}{(t-v)^{\alpha}} \mathrm{d}v + \frac{1}{\Gamma(1-\alpha)} \int_{t_k}^{t} \frac{u'(v)}{(t-v)^{\alpha}} \mathrm{d}v \\ :=& H^k(t) + {}_{t_k}^{\mathrm{C}}D_t^{\alpha}u(t). \end{aligned} \tag{7.2.2} $$

Moreover, we have

$$ \begin{aligned} {}_{t_k}^{\mathrm{C}}D_t^{\alpha}u(t) &= {}_{t_k}^{\mathrm{C}}D_t^{\alpha}[u(t) - u(t_k) + u(t_k)] \\ &= {}_{t_k}^{\mathrm{C}}D_t^{\alpha}[u(t) - u(t_k)] = {}_{t_k}^{\mathrm{RL}}D_t^{\alpha}[u(t) - u(t_k)]. \end{aligned} \tag{7.2.3} $$

Then, by substituting (7.2.2) and (7.2.3) into (7.2.1), it holds that

$$ {}_{t_k}^{\mathrm{RL}}D_t^{\alpha}[u(t) - u(t_k)] = \lambda u(t) - H^k(t), \quad \alpha \in (0,1], \ t \in (0,T]. \tag{7.2.4} $$

Applying the inverse operator ${}_{t_k}I_t^{\alpha}$ on both sides of (7.2.4) and evaluating at $t = t_{k+1}$, we obtain

$$ u(t_{k+1}) - u(t_k) = \left[\lambda\, {}_{t_k}^{\mathrm{RL}}I_t^{\alpha}u(t) - \mathcal{H}^k(t)\right]\Big|_{t=t_{k+1}}, \tag{7.2.5} $$

where $\mathcal{H}^k(t)$ denotes the *history load term*, which is given by

$$ \begin{aligned} \mathcal{H}^k(t) &= {}_{t_k}^{\mathrm{RL}}I_t^{\alpha}(H^k(t)) \\ &= \frac{1}{\Gamma(\alpha)\Gamma(1-\alpha)} \int_{t_k}^{t} \frac{1}{(t_{k+1}-v)^{1-\alpha}} \int_0^{t_k} \frac{u'(s)}{(v-s)^{\alpha}} \mathrm{d}s\, \mathrm{d}v. \end{aligned} \tag{7.2.6} $$

We follow the FAMMs from Zayernouri and Matzavinos (2016) and interpolate $u(t)$ from ${}_{t_k}^{\mathrm{RL}}I_t^{\alpha}u(t)\Big|_{t=t_{k+1}}$ in an implicit fashion

$$ {}_{t_k}^{\mathrm{RL}}I_t^{\alpha}u(t)\Big|_{t=t_{k+1}} \approx h^{\alpha} \sum_{j=0}^{p} \beta_j^{(p)} u(t_{k+1-j}), \quad p = 0, 1: \tag{7.2.7} $$

with the following fractional Adams–Moulton coefficients, respectively, for $p = 0$ and $p = 1$:

$$ \beta_0^{(0)} = \frac{1}{\Gamma(\alpha+1)}, \quad \beta_0^{(1)} = \frac{1}{\Gamma(\alpha+2)}, \quad \beta_1^{(1)} = \frac{\alpha}{\Gamma(\alpha+2)}. $$

Moreover, these coefficients recover the standard Adams–Moulton method's coefficients in the limit case when $\alpha = 1$. To compute the history load term $\mathcal{H}^k(t_{k+1})$, on each small interval $[t_j, t_{j+1}]$ $(0 \le j \le k-1)$, we linearly interpolate $u(t)$ when $p = 0$,

$$\Pi_{1,j}u(t) = \frac{t - t_{j+1}}{t_j - t_{j+1}} u(t_j) + \frac{t - t_j}{t_{j+1} - t_j} u(t_{j+1});$$

therefore, we obtain the following form for the history load:

$$\begin{aligned}\mathcal{H}^k(t_{k+1}) &\approx \mathcal{H}_0^k(t_{k+1}) \\ &= \frac{1}{\Gamma(\alpha)\Gamma(2-\alpha)} \sum_{j=0}^{k-1} \frac{u(t_{j+1}) - u(t_j)}{h} \big(\mathcal{A}_{k,j} - \mathcal{A}_{k,j+1}\big),\end{aligned} \tag{7.2.8}$$

where

$$\begin{aligned}\mathcal{A}_{k,j} &= \int_{t_k}^{t_{k+1}} (t_{k+1} - v)^{\alpha-1}(v - t_j)^{1-\alpha} dv \\ &= h \int_0^1 (1-\theta)^{\alpha-1}(k - j + \theta)^{1-\alpha} d\theta \\ &= h \begin{cases} \frac{(k-j)^{1-\alpha}}{\alpha} {}_2F_1\big(\alpha - 1, 1; \alpha + 1; \frac{1}{j-k}\big), & 0 \le j < k, \\ \Gamma(\alpha)\Gamma(2-\alpha), & j = k, \end{cases}\end{aligned} \tag{7.2.9}$$

in which ${}_2F_1(a, b; c; z)$ denotes the hypergeometric function. Also, when $p = 1$, we utilize quadratic interpolation function $\Pi_{2,j}u(t)$ to approximate $u(t)$ on the interval $[t_j, t_{j+1}]$ $(0 \le j \le k-1)$ as:

$$\begin{aligned}\Pi_{2,j}u(t) = &\frac{(t - t_j)(t - t_{j+1})}{(t_{j-1} - t_j)(t_{j-1} - t_{j+1})} u(t_{j-1}) \\ &+ \frac{(t - t_{j-1})(t - t_{j+1})}{(t_j - t_{j-1})(t_j - t_{j+1})} u(t_j) \\ &+ \frac{(t - t_{j-1})(t - t_j)}{(t_{j+1} - t_{j-1})(t_{j+1} - t_j)} u(t_{j+1}),\end{aligned}$$

and therefore, the history load for the choice of $p = 1$ is given by

$$\begin{aligned}\mathcal{H}^k(t_{k+1}) \approx \mathcal{H}_1^k(t_{k+1}) = \mathcal{H}_0^k(t_{k+1}) + \frac{1}{\Gamma(\alpha)\Gamma(2-\alpha)} \\ \times \sum_{j=1}^{k-1} \frac{u(t_{j+1}) - 2u(t_j) + u(t_{j-1})}{h^2} \Big[- \frac{h(\mathcal{A}_{k,j} + \mathcal{A}_{k,j+1})}{2} + \frac{\mathcal{B}_{k,j} - \mathcal{B}_{k,j+1}}{2-\alpha} \Big],\end{aligned} \tag{7.2.10}$$

where

$$
\begin{aligned}
\mathcal{B}_{k,j} &= \int_{t_k}^{t_{k+1}} (t_{k+1}-v)^{\alpha-1}(v-t_j)^{2-\alpha}dv \\
&= h^2 \int_0^1 (1-\theta)^{\alpha-1}(k-j+\theta)^{2-\alpha}d\theta \\
&= h^2 \begin{cases} \frac{(k-j)^{2-\alpha}}{\alpha} {}_2F_1\left(\alpha-2, 1; \alpha+1; \frac{1}{j-k}\right), & 0 \le j < k, \\ \frac{\Gamma(\alpha)\Gamma(3-\alpha)}{2}, & j = k. \end{cases}
\end{aligned} \tag{7.2.11}
$$

Let u_k be the approximate solution of $u(t_k)$ $(0 \le k \le N)$, and denote

$$
\mathcal{H}_p^k = \frac{1}{\Gamma(\alpha)\Gamma(2-\alpha)} \sum_{j=0}^{k} \gamma_{k,j}^{(p)} u_j, \quad p = 0, 1, \tag{7.2.12}
$$

where the coefficients $\gamma_{k,j}^{(p)}$ are presented below. Then, we get the FAMMs for (7.2.5) which have the following discrete form:

$$
\frac{u_{k+1}-u_k}{h^\alpha} = \lambda \sum_{j=0}^{p} \beta_j^{(p)} u_{k+1-j} - \frac{1}{h^\alpha}\mathcal{H}_p^k, \quad p = 0, 1. \tag{7.2.13}
$$

In order to lay the analytical basis for the convergence and linear stability analysis of the methods, we introduce several preparatory results through a series of lemmas.

Lemma 7.2 *For $0 \le j \le k$, $0 \le k \le N$ and $p = 0, 1$, the coefficients $\{\gamma_{k,j}^{(p)}\}$ in (7.2.12) have the following properties:*

- $\gamma_{k,j}^{(0)} < 0$ $(0 \le j \le k-1)$, $0 < \gamma_{k,k}^{(0)} \le C_1$;
- $\gamma_{k,j}^{(1)} < 0$ $(0 \le j \le k-3)$, $|\gamma_{k,j}^{(1)}| \le C_1$ *$(j = k-2,\ k-1$ or $k)$;*
- $\sum_{j=0}^{k} \gamma_{k,j}^{(0)} = 0$, $\sum_{j=0}^{k} \gamma_{k,j}^{(1)} = 0$,

where $C_1 > 0$ is a constant.

Lemma 7.3 *Let $u(t) = t^\sigma$ $(\sigma \ge 0)$. Then there exists a constant $C_2 > 0$ independent of h such that*

$$
\left| {}_{t_k}^{\mathrm{RL}}I_t^\alpha u(t)\big|_{t=t_{k+1}} - h^\alpha \sum_{j=0}^{p} \beta_j^{(p)} u(t_{k+1-j}) \right| \le C_2 h^{\alpha+p+1} t_{k+1}^{\sigma-p-1}.
$$

Lemma 7.4 *Let $u(t) = t^\sigma$ ($\sigma \geq 0$). Then there exists a constant $C_3 > 0$ independent of h such that*

$$\left|\mathcal{H}^k(t_{k+1}) - \mathcal{H}_p^k(t_{k+1})\right| \leq C_3\big(h^{\sigma+\alpha+1}t_{k+1}^{-\alpha-1} + h^{\alpha+p+1}t_{k+1}^{\sigma-\alpha-p-1} + h^{p+1}\big).$$

7.2.2 Correction Terms

It is well known that the solutions of (7.2.1) usually exhibit weak singularity at the initial time and it can be written as the summation of regular and singular parts (see e.g., Diethelm et al. (2004, Theorem 2.1)):

$$u(t) = \psi(t) + \sum_{v=1}^{\hat{v}} \hat{c}_v t^{v\alpha} + \sum_{v=1}^{\tilde{v}} \tilde{c}_v t^{1+v\alpha}, \tag{7.2.14}$$

where

$$\hat{v} = \lceil 2/\alpha \rceil - 1, \quad \tilde{v} = \lceil 1/\alpha \rceil - 1, \quad \psi \in C^2[0,T],$$

and $\hat{c}_v, \tilde{c}_v \in \mathbb{R}$. Hence, the optimal convergence rates of the above discussed numerical methods cannot be achieved (see Lemmas 7.3 and 7.4). To improve the accuracy near the initial time, we follow Lubich's idea (cf. Lubich (1986)) by adding correction terms to the resulting fractional operators of (7.2.5). From (7.2.14), we assume that the solution $u(t)$ of (7.2.1) has the form (see, e.g., Cao et al. (2016); Diethelm et al. (2004); Zeng et al. (2017) for more discussions on the regularity of FODEs):

$$u(t) - u(0) = \sum_{r=1}^{m+1} d_r t^{\sigma_r} + \zeta(t)t^{\sigma_{m+2}}, \quad 0 < \sigma_r < \sigma_{r+1}, \tag{7.2.15}$$

where $d_1, d_2, \ldots, d_{m+1} \in \mathbb{R}$ are some constants, m is a positive integer and $\zeta(t)$ is a uniformly continuous function for $t \in [0,T]$. The term $\{\sigma_r\}$ represents positive correction powers which is usually uniquely determined by the considered equation. We will now introduce the correction approach for each term of the right-hand side of (7.2.5).

I) We start with the term ${}^{\mathrm{RL}}_{t_k}I_t^{\alpha}u(t)\big|_{t=t_{k+1}}$. It follows from Lemma 7.3 that

$$\begin{aligned}
{}^{\mathrm{RL}}_{t_k}I_t^{\alpha}u(t)\big|_{t=t_{k+1}} &= {}^{\mathrm{RL}}_{t_k}I_t^{\alpha}\left(u(t)-u(0)\right)\big|_{t=t_{k+1}} + {}^{\mathrm{RL}}_{t_k}I_t^{\alpha}u(0)\big|_{t=t_{k+1}} \\
&= h^{\alpha}\sum_{j=0}^{p}\beta_j^{(p)}\left(u(t_{k+1-j}) - u_0\right) + h^{\alpha}\sum_{j=1}^{m_u} W_{k,j}^{(\alpha,\sigma,p)}\left(u(t_j) - u_0\right)
\end{aligned}$$

$$
\begin{aligned}
&+\frac{h^{\alpha}}{\Gamma(\alpha+1)}u_0+R^u_{k+1}\\
&=h^{\alpha}\sum_{j=0}^{p}\beta_j^{(p)}u(t_{k+1-j})+h^{\alpha}\sum_{j=1}^{m_u}W_{k,j}^{(\alpha,\sigma,p)}\left(u(t_j)-u_0\right)+R^u_{k+1}\\
&:=h^{\alpha}I_{m_u}^{\alpha,\sigma,p}u(t_{k+1})+R^u_{k+1},
\end{aligned}
\tag{7.2.16}
$$

with $R^u_{k+1}=\mathcal{O}(h^{\alpha+p+1}t_{k+1}^{\sigma_{m_u+1}-p-1})$. The correction weights $\{W_{k,j}^{(\alpha,\sigma,p)}\}$ are chosen such that (7.2.16) is exact for $u(t)=t^{\sigma_r}$ $(1\le r\le m_u)$, and therefore are obtained through the following linear system of order $m_u\times m_u$ to be solved for $k=0,1,\ldots,N-1$ time-steps:

$$
\sum_{j=1}^{m_u}W_{0,j}^{(\alpha,\sigma,p)}j^{\sigma_r}=\frac{\Gamma(\sigma_r+1)}{\Gamma(\sigma_r+\alpha+1)}-\sum_{j=0}^{p}\beta_j^{(p)}(1-j)^{\sigma_r},\quad 1\le r\le m_u,
\tag{7.2.17}
$$

$$
\sum_{j=1}^{m_u}W_{k,j}^{(\alpha,\sigma,p)}j^{\sigma_r}=\frac{k^{\sigma_r}}{\Gamma(\alpha+1)}{}_2F_1\left(-\sigma_r,1;\alpha+1;-\frac{1}{k}\right)-\sum_{j=0}^{p}\beta_j^{(p)}(k+1-j)^{\sigma_r},
$$
$$
1\le k\le N-1,\quad 1\le r\le m_u. \tag{7.2.18}
$$

II) For the history load $\mathcal{H}^k(t_{k+1})$, we introduce the correction terms as follows. Substituting (7.2.15) into (7.2.6) and using Lemma 7.4 yields

$$
\begin{aligned}
\mathcal{H}^k(t_{k+1})&=\frac{1}{\Gamma(\alpha)\Gamma(1-\alpha)}\int_{t_k}^{t_{k+1}}\frac{1}{(t_{k+1}-v)^{1-\alpha}}\int_0^{t_k}\frac{[u(s)-u(0)]'}{(v-s)^{\alpha}}ds\,dv\\
&+\frac{1}{\Gamma(\alpha)\Gamma(1-\alpha)}\int_{t_k}^{t_{k+1}}\frac{1}{(t_{k+1}-v)^{1-\alpha}}\int_0^{t_k}\frac{u'(0)}{(v-s)^{\alpha}}ds\,dv\\
&=\frac{1}{\Gamma(\alpha)\Gamma(2-\alpha)}\sum_{j=0}^{k}\gamma_{k,j}^{(p)}u(t_j)+\sum_{j=1}^{\tilde{m}_u}\tilde{W}_{k,j}^{(\alpha,\sigma,p)}\left(u(t_j)-u_0\right)+\tilde{R}^u_{k+1}\\
&:=\mathcal{H}_{\tilde{m}_u}^{\alpha,\sigma,p}u(t_{k+1})+\tilde{R}^u_{k+1},
\end{aligned}
\tag{7.2.19}
$$

with

$$
\tilde{R}^u_{k+1}=\mathcal{O}(h^{\sigma_{\tilde{m}_u+1}+\alpha+1}t_{k+1}^{-\alpha-1})+\mathcal{O}(h^{\alpha+p+1}t_{k+1}^{\sigma_{\tilde{m}_u+1}-\alpha-p-1})+\mathcal{O}(h^{p+1}).
$$

The history load correction weights $\{\tilde{W}_{k,j}^{(\alpha,\sigma,p)}\}$ are chosen such that (7.2.19) is exact for $u(t)=t^{\sigma_r}$ $(1\le r\le m_{\tilde{u}})$. However, we remark that it is very difficult to obtain the analytical solution of $\mathcal{H}^k(t_{k+1})$, given $u(t) = t^{\sigma_r}$. Fortunately, we know from Deng (2007) that (7.2.5) is also equivalent to

$$u(t_{k+1}) = u(t_k) + \frac{\lambda}{\Gamma(\alpha)} \int_{t_k}^{t_{k+1}} (t_{k+1} - v)^{\alpha-1} u(v) dv \qquad (7.2.20)$$
$$+ \frac{\lambda}{\Gamma(\alpha)} \int_0^{t_k} \left[(t_{k+1} - v)^{\alpha-1} - (t_k - v)^{\alpha-1} \right] u(v) dv.$$

Comparing (7.2.5) with (7.2.20) and using (7.2.1), we can obtain

$$\mathcal{H}^k(t_{k+1}) = -\frac{\lambda}{\Gamma(\alpha)} \int_0^{t_k} \left[(t_{k+1} - v)^{\alpha-1} - (t_k - v)^{\alpha-1} \right] u(v) dv$$
$$= -\frac{1}{\Gamma(\alpha)} \int_0^{t_k} \left[(t_{k+1} - v)^{\alpha-1} - (t_k - v)^{\alpha-1} \right] {}_0^C D_t^\alpha u(v) dv.$$

Therefore, we have the following linear system of size $\tilde{m}_u \times \tilde{m}_u$ to be solved for $k = 1, 2, \ldots, N-1$ time-steps:

$$\sum_{j=1}^{\tilde{m}_u} \tilde{W}_{k,j}^{(\alpha,\sigma,p)} j^{\sigma_r} = k^{\sigma_r} - \frac{\Gamma(\sigma_r + 1)}{\Gamma(\alpha)\Gamma(\sigma_r - \alpha + 1)} (k+1)^{\sigma_r} B\left(\frac{k}{k+1}; \sigma_r - \alpha + 1, \alpha \right)$$
$$- \frac{1}{\Gamma(\alpha)\Gamma(2-\alpha)} \sum_{j=0}^{k} \gamma_{k,j}^{(p)} j^{\sigma_r}, \quad 1 \le k \le N-1, \quad 1 \le r \le \tilde{m}_u, \qquad (7.2.21)$$

where $B(z; a, b)$ denotes the incomplete beta function, which is defined by

$$B(z; a, b) = \int_0^z v^{a-1} (1 - v)^{b-1} dv.$$

Substituting (7.2.16) and (7.2.19) into (7.2.5) yields

$$u(t_{k+1}) = u(t_k) + \lambda h^\alpha I_{m_u}^{\alpha,\sigma,p} u(t_{k+1}) - \mathcal{H}_{\tilde{m}_u}^{\alpha,\sigma,p} u(t_{k+1}) + R_{k+1}^u + \tilde{R}_{k+1}^u \qquad (7.2.22)$$

for $p = 0, 1$. Dropping the truncation errors R_{k+1}^u and $\tilde{R}_{k+1}^u$ in (7.2.22) and replacing $u(t_k)$ with approximate solution u_k, we obtain the following FAMMs with correction terms for solving (7.2.1):

$$\boxed{\frac{u_{k+1} - u_k}{h^\alpha} = \lambda I_{m_u}^{\alpha,\sigma,p} u_{k+1} - \frac{1}{h^\alpha} \mathcal{H}_{\tilde{m}_u}^{\alpha,\sigma,p} u_{k+1}, \quad p = 0, 1,} \qquad (7.2.23)$$

with $I_{m_u}^{\alpha,\sigma,p} u_{k+1}$ and $\mathcal{H}_{\tilde{m}_u}^{\alpha,\sigma,p} u_{k+1}$ given, respectively, by

$$I_{m_u}^{\alpha,\sigma,p} u_{k+1} = \sum_{j=0}^{p} \beta_j^{(p)} u_{k+1-j} + \sum_{j=1}^{m_u} W_{k,j}^{(\alpha,\sigma,p)} (u_j - u_0) \qquad (7.2.24)$$

and

$$\mathcal{H}_{\tilde{m}_u}^{\alpha,\sigma,p} u_{k+1} = \frac{1}{\Gamma(\alpha)\Gamma(2-\alpha)} \sum_{j=0}^{k} \gamma_{k,j}^{(p)} u_j + \sum_{j=1}^{\tilde{m}_u} \tilde{W}_{k,j}^{(\alpha,\sigma,p)} \left(u_j - u_0\right). \tag{7.2.25}$$

7.2.3 Nonlinear FODEs as Initial-Value Problems

Having defined the discretization and corrections for the linear case, we now consider the numerical solutions of the following nonlinear FODE:

$${}_0^C D_t^\alpha u(t) = \lambda u(t) + f(t,u), \quad \alpha \in (0,1],\ t \in (0,T]; \quad u(0) = u_0, \tag{7.2.26}$$

where the nonlinear function $f\colon (0,T] \times \mathbb{R}^d \to \mathbb{R}^d$ satisfies the Lipschitz condition with constant $L > 0$:

$$\left\| f(t,u) - f(t,\hat{u}) \right\|_\infty \le L \left\| u - \hat{u} \right\|_\infty, \quad t \in (0,T], \quad u, \hat{u} \in \mathbb{R}^d, \tag{7.2.27}$$

where $\| \cdot \|_\infty$ denotes the usual maximum norm on $\mathbb{R}^d$. Under these assumptions, it has been proved by Diethelm and Ford (2002, Theorem 2.1 and 2.2) that problem (7.2.26) has a unique solution on the interval $(0,T]$.

By the same token, we adopt the FAMMs developed in Zayernouri and Matzavinos (2016) to solve (7.2.26), and in a similar fashion to Section 7.2.1, but with the addition of the nonlinear term $f(t,u)$, we have:

$$u(t_{k+1}) - u(t_k) = \left[\lambda\, {}_{t_k}I_t^\alpha u(t) + {}_{t_k}I_t^\alpha f(t,u) - \mathcal{H}^k(t)\right]\Big|_{t=t_{k+1}}. \tag{7.2.28}$$

Therefore, the FAMMs for (7.2.28) are given by

$$\frac{u_{k+1} - u_k}{h^\alpha} = \sum_{j=0}^{p} \beta_j^{(p)} \left(\lambda u_{k+1-j} + f_{k+1-j}\right) - \frac{1}{h^\alpha}\mathcal{H}_p^k, \quad p = 0, 1, \tag{7.2.29}$$

where $f_{k+1-j} = f(t_{k+1-j}, u_{k+1-j})$.

Correction Terms for $f(t,u)$

The regularity of $f(t,u)$ is related to the regularity of $u(t)$. If $u(t)$ satisfies (7.2.15), we know from (7.2.26) that

$$f(t,u) - f(0,u(0)) = -\lambda \sum_{r=1}^{m+1} d_r t^{\sigma_r} + \sum_{r=1}^{m+1} d_r \frac{\Gamma(\sigma_r + 1)}{\Gamma(\sigma_r - \alpha + 1)} t^{\sigma_r - \alpha} + \cdots$$

$$:= \sum_{r=1}^{l+1} h_r t^{\delta_r} + \hat{\zeta}(t) t^{\delta_{l+2}}, \quad \delta_r < \delta_{r+1}, \tag{7.2.30}$$

where $\hat{\zeta}(t)$ is a uniformly continuous function for $t \in [0, T]$ and $\delta_r \in \{\sigma_l\} \cup \{\sigma_l - \alpha\}$. Similar to (7.2.16) and by using Lemma 7.3, we have

$$
\begin{aligned}
{}_{t_k}I_t^{\alpha} f(t,u)\big|_{t=t_{k+1}} &= h^{\alpha} \sum_{j=0}^{p} \beta_j^{(p)} f(t_{k+1-j}, u(t_{k+1-j})) && (7.2.31)\\
&\quad + h^{\alpha} \sum_{j=1}^{m_f} W_{k,j}^{(\alpha,\delta,p)} \big(f(t_j,u(t_j)) - f(0,u_0)\big) + R_{k+1}^{f} \\
&:= h^{\alpha} I_{m_f}^{\alpha,\delta,p} f(t_{k+1}, u(t_{k+1})) + R_{k+1}^{f}, && (7.2.32)
\end{aligned}
$$

where $R_{k+1}^{f} = \mathcal{O}\big(h^{\alpha+p+1} t_{k+1}^{\delta_{m_f+1}-p-1}\big)$, and $\{W_{k,j}^{(\alpha,\delta,p)}\}$ with $k = 0, 1, \ldots, N-1$ and $j = 1, 2, \ldots, m_f$ are given by

$$
\sum_{j=1}^{m_f} W_{0,j}^{(\alpha,\delta,p)} j^{\delta_r} = \frac{\Gamma(\delta_r+1)}{\Gamma(\delta_r+\alpha+1)} - \sum_{j=0}^{p} \beta_j^{(p)} (1-j)^{\delta_r}, \quad 1 \le r \le m_f, \tag{7.2.33}
$$

$$
\sum_{j=1}^{m_f} W_{k,j}^{(\alpha,\delta,p)} j^{\delta_r} = \frac{k^{\delta_r}}{\Gamma(\alpha+1)} {}_2F_1\Big(-\delta_r, 1; \alpha+1; -\frac{1}{k}\Big) - \sum_{j=0}^{p} \beta_j^{(p)} (k+1-j)^{\delta_r},
$$
$$
1 \le k \le N-1, \quad 1 \le r \le m_f. \tag{7.2.34}
$$

Inserting (7.2.16), (7.2.19) and (7.2.31) into (7.2.28) yields

$$
\begin{aligned}
u(t_{k+1}) &= u(t_k) + \lambda h^{\alpha} I_{m_u}^{\alpha,\sigma,p} u(t_{k+1}) + h^{\alpha} I_{m_f}^{\alpha,\delta,p} f(t_{k+1}, u(t_{k+1})) && (7.2.35)\\
&\quad - \mathcal{H}_{\tilde{m}_u}^{\alpha,\sigma,p} u(t_{k+1}) + R_{k+1}^{u} + R_{k+1}^{f} + \tilde{R}_{k+1}^{u}, \quad p = 0, 1.
\end{aligned}
$$

Therefore, we obtain the following FAMMs with correction terms for solving (7.2.26):

$$
\boxed{\frac{u_{k+1} - u_k}{h^{\alpha}} = \lambda I_{m_u}^{\alpha,\sigma,p} u_{k+1} + I_{m_f}^{\alpha,\delta,p} f_{k+1} - \frac{1}{h^{\alpha}} \mathcal{H}_{\tilde{m}_u}^{\alpha,\sigma,p} u_{k+1}, \quad p = 0, 1,} \tag{7.2.36}
$$

with $I_{m_u}^{\alpha,\sigma,p} u_{k+1}$, $\mathcal{H}_{\tilde{m}_u}^{\alpha,\sigma,p} u_{k+1}$ and $I_{m_f}^{\alpha,\delta,p} f_{k+1}$ given, respectively, by (7.2.24), (7.2.25) and

$$
I_{m_f}^{\alpha,\delta,p} f_{k+1} = \sum_{j=0}^{p} \beta_j^{(p)} f_{k+1-j} + \sum_{j=1}^{m_f} W_{k,j}^{(\alpha,\delta,p)} (f_j - f_0). \tag{7.2.37}
$$

In order to obtain the IMEX methods, we follow the idea from Cao et al. (2016), and employ an extrapolation to linearize the nonlinear force term $f(t_{k+1}, u(t_{k+1}))$ in (7.2.35), which is given by

$$f(t_{k+1}, u(t_{k+1})) = E_k^p f(t_k, u(t_k)) + \sum_{j=1}^{\tilde{m}_f} W_{k,j}^{(\delta,p)} (f(t_j, u(t_j)) - f(0, u_0)) + \tilde{R}_{k+1}^f, \tag{7.2.38}$$

where

$$E_k^p f(t_k, u(t_k)) = \begin{cases} f(t_k, u(t_k)), & p = 0, \\ 2f(t_k, u(t_k)) - f(t_{k-1}, u(t_{k-1})), & p = 1, \end{cases} \tag{7.2.39}$$

and $\tilde{R}_k^f = \mathcal{O}(h^{p+1} t_{k+1}^{\delta_{\tilde{m}_f+1}-p-1})$. In addition, the $\tilde{m}_f \times \tilde{m}_f$ linear system for correction weights is given by

$$\sum_{j=1}^{\tilde{m}_f} W_{k,j}^{(\delta,0)} j^{\delta_r} = (k+1)^{\delta_r} - k^{\delta_r}, \quad 1 \le r \le \tilde{m}_f, \tag{7.2.40}$$

and

$$\sum_{j=1}^{\tilde{m}_f} W_{k,j}^{(\delta,1)} j^{\delta_r} = (k+1)^{\delta_r} - 2k^{\delta_r} + (k-1)^{\delta_r}, \quad 1 \le r \le \tilde{m}_f. \tag{7.2.41}$$

Inserting (7.2.38) into (7.2.35) yields

$$\begin{aligned} u(t_{k+1}) &= u(t_k) + \lambda h^\alpha I_{m_u}^{\alpha,\sigma,p} u(t_{k+1}) + h^\alpha I_{m_f}^{\alpha,\delta,p} f(t_{k+1}) \\ &\quad + h^\alpha \beta_0^{(p)} \Big[E_k^p f(t_k) - f(t_{k+1}) + \sum_{j=1}^{\tilde{m}_f} W_{k,j}^{(\delta,p)} (f(t_j) - f(t_0)) \Big] \\ &\quad - \mathcal{H}_{\tilde{m}_u}^{\alpha,\sigma,p} u(t_{k+1}) + R_{k+1}^u + R_{k+1}^f + \tilde{R}_{k+1}^u + h^\alpha \beta_0^{(p)} \tilde{R}_{k+1}^f, \quad p = 0, 1, \end{aligned} \tag{7.2.42}$$

where we denoted $f(t_i) := f(t_i, u(t_i))$. Finally, we obtain the IMEX(p) methods in the following form:

$$\boxed{\begin{aligned} \frac{u_{k+1} - u_k}{h^\alpha} &= \lambda I_{m_u}^{\alpha,\sigma,p} u_{k+1} + I_{m_f}^{\alpha,\delta,p} f_{k+1} \\ &\quad + \beta_0^{(p)} \Big[- f_{k+1} + E_k^p f_k + \sum_{j=1}^{\tilde{m}_f} W_{k,j}^{(\delta,p)} (f_j - f_0) \Big] \\ &\quad - \frac{1}{h^\alpha} \mathcal{H}_{\tilde{m}_u}^{\alpha,\sigma,p} u_{k+1}, \quad p = 0, 1, \end{aligned}} \tag{7.2.43}$$

with $I_{m_u}^{\alpha,\sigma,p}u_{k+1}$, $\mathcal{H}_{\tilde{m}_u}^{\alpha,\sigma,p}u_{k+1}$, $I_{m_f}^{\alpha,\delta,p}f_{k+1}$ and $E_k^p f_k$ given, respectively, by (7.2.24), (7.2.25), (7.2.37) and

$$E_k^p f_k = \begin{cases} f_k, & p = 0, \\ 2f_k - f_{k-1}, & p = 1. \end{cases} \qquad (7.2.44)$$

Remark 7.5 If $\alpha = 1$, the history load term for (7.2.26) is zero and we don't need to add the correction terms to solve (7.2.26). Moreover, (7.2.43) recovers the standard IMEX methods.

In the following, we will present the convergence results for the IMEX(p) methods (7.2.43). For this purpose, we first introduce some preparatory results. Both of the proofs can be found in Zhou et al. (2019).

Lemma 7.6 *The correction weights* $W_{k,j}^{(\alpha,\sigma,p)}$, $\tilde{W}_{k,j}^{(\alpha,\sigma,p)}$, $W_{k,j}^{(\alpha,\delta,p)}$ *and* $W_{k,j}^{(\delta,p)}$ *in* (7.2.16), (7.2.19), (7.2.31) *and* (7.2.38), *respectively, satisfy*

$$\begin{aligned} \left|W_{k,j}^{(\alpha,\sigma,p)}\right| &= \mathcal{O}((k+1)^{\sigma_{m_u}-p-1}), \\ \left|W_{k,j}^{(\alpha,\delta,p)}\right| &= \mathcal{O}((k+1)^{\delta_{m_f}-p-1}), \\ \left|\tilde{W}_{k,j}^{(\alpha,\sigma,p)}\right| &= \mathcal{O}((k+1)^{-\alpha-1}) + \mathcal{O}((k+1)^{\sigma_{\tilde{m}_u}-p-1}), \\ \left|W_{k,j}^{(\delta,p)}\right| &= \mathcal{O}((k+1)^{\delta_{\tilde{m}_f}-p-1}). \end{aligned}$$

Theorem 7.7 *Suppose that the Lipschitz condition* (7.2.27) *holds. If* $\sigma_{\tilde{m}_u} \le p+1$, $\sigma_{m_u}, \delta_{m_f}, \delta_{\tilde{m}_f} \le p + \alpha + 1$, *then, for the IMEX(p) methods* (7.2.43), *there exists a constant* $C_4 > 0$ *independent of h such that*

$$\max_{0 \le k \le N} \|u(t_k) - u_k\|_\infty \le C_4\Big(\sum_{j=1}^{M} \|u(t_j) - u_j\|_\infty + h^q\Big), \qquad (7.2.45)$$

where $M = \max\{m_u, m_f, \tilde{m}_u, \tilde{m}_f\}$ *and*

$$q = \min\{p+1, \sigma_{\tilde{m}_u+1} + p, \sigma_{m_u+1} + \alpha + p, \delta_{m_f+1} + \alpha + p, \delta_{\tilde{m}_f+1} + \alpha + p\}.$$

7.2.4 Linear Stability of IMEX(p) Methods

In this section, we investigate the linear stability of the proposed IMEX(p) methods (7.2.43) by considering the following usual scalar test equation

$${}_0^C D_t^\alpha u(t) = \lambda u(t) + \rho u(t), \quad \alpha \in (0,1], \; t \ge 0, \; \lambda, \rho \in \mathbb{C}; \quad u(0) = u_0. \qquad (7.2.46)$$

For this, the following result from Lubich (1986) is useful to determine the stability regions of the obtained numerical schemes.

Lemma 7.8 (See Lubich (1986)) *Assume that the sequence $\{g_k\}$ is convergent and that the quadrature weights $w_k (k \geq 1)$ satisfy*

$$w_k = \frac{k^{\alpha-1}}{\Gamma(\alpha+1)} + v_k, \text{ where } \sum_{k=1}^{\infty} |v_k| < \infty.$$

Then the stability region of the convolution quadrature

$$u_k = g_k + \hat{h} \sum_{j=0}^{k} w_{k-j} u_j$$

is

$$S = \left\{ \hat{h} \in \mathbb{C} : 1 - \hat{h} w^{\alpha}(\xi) \neq 0, \ |\xi| \leq 1 \right\},$$

where $w^{\alpha}(\xi) = \sum_{j=0}^{\infty} w_j \xi^j$, and $\hat{h} = h^{\alpha}(\lambda + \rho)$ or $\hat{h}$ is some function of $h^{\alpha}(\lambda + \rho)$.

We first consider the linear stability of the IMEX(0) for the test equation (7.2.46). It holds that

$$\begin{aligned} u_{k+1} = u_k &+ \frac{\lambda h^{\alpha}}{\Gamma(\alpha+1)} u_{k+1} + h^{\alpha} \left(\lambda \sum_{j=1}^{m_u} W_{k,j}^{(\alpha,\sigma,0)} + \rho \sum_{j=1}^{m_f} W_{k,j}^{(\alpha,\delta,0)} \right) (u_j - u_0) \\ &+ \frac{\rho h^{\alpha}}{\Gamma(\alpha+1)} \left[u_k + \sum_{j=1}^{\tilde{m}_f} W_{k,j}^{(\delta,0)} (u_j - u_0) \right] \qquad (7.2.47) \\ &- \frac{1}{\Gamma(\alpha)\Gamma(2-\alpha)} \sum_{j=0}^{k} \gamma_{k,j}^{(0)} u_j - \sum_{j=1}^{\tilde{m}_u} \tilde{W}_{k,j}^{(\alpha,\sigma,0)} (u_j - u_0) \\ = u_k &+ \frac{h^{\alpha}}{\Gamma(\alpha+1)} (\lambda u_{k+1} + \rho u_k) - \frac{1}{\Gamma(\alpha)\Gamma(2-\alpha)} \sum_{j=0}^{k} \gamma_{k,j}^{(0)} u_j + \sum_{j=1}^{M} W_{k,j} (u_j - u_0), \end{aligned}$$

where

$$\begin{aligned} \sum_{j=1}^{M} W_{k,j} (u_j - u_0) = \lambda h^{\alpha} \sum_{j=1}^{m_u} W_{k,j}^{(\alpha,\sigma,0)} (u_j - u_0) + \rho h^{\alpha} \sum_{j=1}^{m_f} W_{k,j}^{(\alpha,\delta,0)} (u_j - u_0) \\ + \frac{\rho h^{\alpha}}{\Gamma(\alpha+1)} \sum_{j=1}^{\tilde{m}_f} W_{k,j}^{(\delta,0)} (u_j - u_0) - \sum_{j=1}^{\tilde{m}_u} \tilde{W}_{k,j}^{(\alpha,\sigma,0)} (u_j - u_0). \end{aligned}$$

Since $\sum_{j=1}^{M} W_{k,j} (u_j - u_0)$ does not affect the stability analysis, we don't give the exact expression of $W_{k,j}$. Denote $U(\xi) = \sum_{k=0}^{\infty} u_k \xi^k, \ |\xi| \leq 1$. Then it follows from (7.2.47) that

$$\sum_{k=0}^{\infty} u_{k+1}\xi^k = \sum_{k=0}^{\infty} u_k\xi^k + \frac{\lambda h^\alpha}{\Gamma(\alpha+1)}\sum_{k=0}^{\infty} u_{k+1}\xi^k + \frac{\rho h^\alpha}{\Gamma(\alpha+1)}\sum_{k=0}^{\infty} u_k\xi^k$$
$$-\frac{1}{\Gamma(\alpha)\Gamma(2-\alpha)}\sum_{k=0}^{\infty}\sum_{j=0}^{k}\gamma_{k,j}^{(0)}u_j\xi^k + \sum_{k=0}^{\infty}\sum_{j=1}^{M} W_{k,j}(u_j-u_0)\xi^k,$$

which leads to

$$\frac{1}{\xi}(U(\xi)-u_0) = U(\xi) + \frac{\lambda h^\alpha}{\Gamma(\alpha+1)\xi}(U(\xi)-u_0) + \frac{\rho h^\alpha}{\Gamma(\alpha+1)}U(\xi)$$
$$-\frac{1}{\Gamma(\alpha)\Gamma(2-\alpha)}\gamma_k^{(0)}(\xi)U(\xi) + \sum_{k=1}^{\infty}\sum_{j=1}^{M} W_{k,j}(u_j-u_0)\xi^k, \tag{7.2.48}$$

where $\gamma_k^{(0)}(\xi) = \sum_{j=0}^{\infty}\gamma_{k,j}^{(0)}\xi^j$. We simplify (7.2.48) as

$$\left[1-\xi-\frac{\lambda h^\alpha}{\Gamma(\alpha+1)}-\frac{\rho h^\alpha}{\Gamma(\alpha+1)}\xi+\frac{1}{\Gamma(\alpha)\Gamma(2-\alpha)}\gamma_k^{(0)}(\xi)\xi\right]U(\xi) = H(\xi),$$

where

$$H(\xi) = \sum_{k=0}^{\infty} H_k\xi^k = \left[1-\frac{\lambda h^\alpha}{\Gamma(\alpha+1)}\right]u_0 + \sum_{k=1}^{\infty}\sum_{j=1}^{m} W_{k,j}(u_j-u_0)\xi^{k+1}.$$

By using Lemma 7.6, when

$$\sigma_{m_u}, \delta_{m_f}, \delta_{\tilde{m}_f} < 1, \quad \sigma_{\tilde{m}_u} < \alpha+1,$$

we can obtain that $\{H_k\}$ is a convergent sequence. Moreover, we know from Lemma 7.2 that $\sum_{j=0}^{\infty}\left|\gamma_{k,j}^{(0)}\right| < \infty$. Then it follows from Lemma 7.8 that method IMEX(0) is stable if

$$1-\xi-\frac{\lambda h^\alpha}{\Gamma(\alpha+1)}-\frac{\rho h^\alpha}{\Gamma(\alpha+1)}\xi+\frac{1}{\Gamma(\alpha)\Gamma(2-\alpha)}\gamma_k^{(0)}(\xi)\xi \neq 0, \quad \forall|\xi|\leq 1.$$

Similarly, we can obtain the stability region of the method IMEX(1). Then we have the following theorem.

Theorem 7.9 *Let* $\rho = \kappa\lambda$ *and* $\hat{h} = \lambda h^\alpha$. *Then for* $\sigma_{m_u}, \delta_{m_f}, \delta_{\tilde{m}_f} < p+1$, $\sigma_{\tilde{m}_u} < \alpha+p+1$, *we have the stability region of IMEX(0):*

$$S_0 = \mathbb{C}\setminus\left\{\hat{h} : \hat{h} = \frac{\Gamma(\alpha+1)}{1+\kappa\xi}\eta,\ |\xi|\leq 1\right\},$$

and the stability region of IMEX(1):

$$S_1 = \mathbb{C} \setminus \left\{ \hat{h} : \hat{h} = \frac{\Gamma(\alpha+2)\eta}{(1+\kappa)(\alpha\xi+1) - \kappa(\xi-1)^2},\ |\xi| \le 1 \right\},$$

where

$$\eta := 1 - \xi + \frac{\gamma_k^{(0)}(\xi)\xi}{\Gamma(\alpha)\Gamma(2-\alpha)}, \quad \gamma_k^{(1)}(\xi) = \sum_{j=0}^{\infty} \gamma_{k,j}^{(1)} \xi^j.$$

In Figure 7.2 (top left, top right, bottom left), we plot the stability regions of the method IMEX(0) with $\alpha = 0.2, 0.5, 0.8$ and $\rho = 0.5\lambda$, respectively. We also plot the stability regions of the method IMEX(1) with $\alpha = 0.2, 0.5, 0.8$ and $\rho = 0.5\lambda$ in Figure 7.2 (bottom right). As the functions $\gamma_k^{(p)}(\xi) = \sum_{j=0}^{\infty} \gamma_{k,j}^{(p)} \xi^j$ are not explicitly known, in all of these figures we take $k = 10^5$.

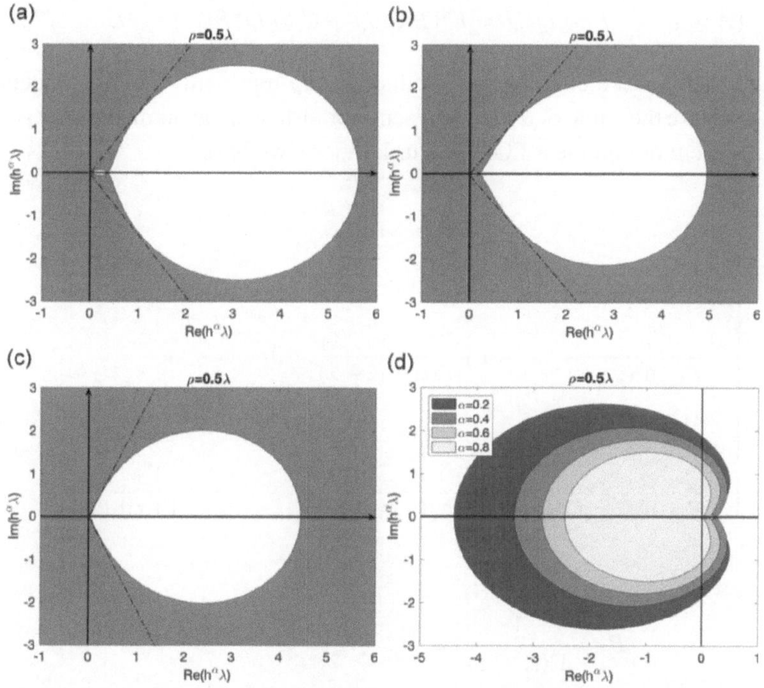

Figure 7.2 (a) Stability region of IMEX(0) with $\alpha = 0.2$ and $\rho = 0.5\lambda$; (b) Stability region of IMEX(0) with $\alpha = 0.5$ and $\rho = 0.5\lambda$; (c) Stability region of IMEX(0) with $\alpha = 0.8$ and $\rho = 0.5\lambda$; (d) Stability region of the IMEX(1) with different α and $\rho = 0.5\lambda$. Source: Reprinted from Zhou et al. (2020b) with permission from Elsevier.

7.2.5 The Fast Implementation of IMEX(p) Methods

The step-by-step numerical solution of (7.2.43) for N time-steps requires $\mathcal{O}(N^2)$ evaluations of the time-dependent coefficients given by the hypergeometric functions, making the scheme expensive. Hence, we rewrite (7.2.43) as the matrices form, where the corresponding convolution matrices of coefficients have the Toeplitz structure and thus we leverage the use of FFTs to obtain the solution of the problem with complexity $\mathcal{O}(N \log N)$. For simplicity and objectivity, we demonstrate the procedure only for the IMEX(0) scheme, for which we start by introducing the notations

$$U = (u_1, u_2, \dots, u_N)^T, \quad F(U) = (f_1, f_2, \dots, f_N)^T,$$

where U denotes the unknown solution vector. Then IMEX(0) can be written in a compact form:

$$(A \otimes I_d - \lambda B \otimes I_d)U = h^\alpha (B \otimes I_d + C \otimes I_d) F(U) + D, \tag{7.2.49}$$

where $\otimes$ denotes the Kronecker product and I_d represents the $d \times d$ identity matrix. Here the value of d can represent, for instance, the number of equations for a system of nonlinear FODEs. Furthermore, we have

$$A = \begin{bmatrix} 1 & & & & \\ -1+\dfrac{\gamma_{1,1}^{(0)}}{\Gamma(\alpha)\Gamma(2-\alpha)} & 1 & & & \\ \dfrac{\gamma_{2,1}^{(0)}}{\Gamma(\alpha)\Gamma(2-\alpha)} & -1+\dfrac{\gamma_{2,2}^{(0)}}{\Gamma(\alpha)\Gamma(2-\alpha)} & 1 & & \\ \vdots & \vdots & \ddots & \ddots & \\ \dfrac{\gamma_{N-1,1}^{(0)}}{\Gamma(\alpha)\Gamma(2-\alpha)} & \dfrac{\gamma_{N-1,2}^{(0)}}{\Gamma(\alpha)\Gamma(2-\alpha)} & \cdots & -1+\dfrac{\gamma_{N-1,N-1}^{(0)}}{\Gamma(\alpha)\Gamma(2-\alpha)} & 1 \end{bmatrix}, \tag{7.2.50}$$

$$B = \frac{1}{\Gamma(\alpha+1)} I_N, \tag{7.2.51}$$

$$C = \frac{1}{\Gamma(\alpha+1)} \begin{bmatrix} -1 & & & & \\ 1 & -1 & & & \\ 0 & 1 & -1 & & \\ \vdots & \ddots & \ddots & \ddots & \\ 0 & \cdots & 0 & 1 & -1 \end{bmatrix},$$

in which A, B and $C \in \mathbb{R}^{N\times N}$ and D is a vector related to the corrections and solutions for the initial steps, in the following way:

$$\begin{aligned} D = &\lambda h^{\alpha}(W_{m_u} \otimes I_d)U_{m_u} + h^{\alpha}(W_{m_f} \otimes I_d)F(U_{m_f}) - (W_{\tilde{m}_u} \otimes I_d)U_{\tilde{m}_u} \\ &+ \frac{h^{\alpha}}{\Gamma(\alpha+1)}(W_{\tilde{m}_f} \otimes I_d)F(U_{\tilde{m}_f}) + (a_0 \otimes I_d)u_0 + h^{\alpha}(b_0 \otimes I_d) f_0, \end{aligned} \tag{7.2.52}$$

where the correction weights $W_{\hat{M}}$ ($\hat{M} = m_u, m_f, \tilde{m}_u, \tilde{m}_f$) are denoted by the $N \times \hat{M}$ matrices with the element $W_{k,j}^{(\alpha,\sigma,0)}$, $W_{k,j}^{(\alpha,\delta,0)}$, $\tilde{W}_{k,j}^{(\alpha,\sigma,0)}$ and $\tilde{W}_{k,j}^{(\delta,0)}$, respectively, for $k = 0, 1, \dots, N-1$ and $j = 1, 2, \dots, \hat{M}$. We also have

$$a_0 = \left(1, \frac{\gamma_{1,0}^{(0)}}{\Gamma(\alpha)\Gamma(2-\alpha)}, \frac{\gamma_{2,0}^{(0)}}{\Gamma(\alpha)\Gamma(2-\alpha)}, \dots, \frac{\gamma_{N-1,0}^{(0)}}{\Gamma(\alpha)\Gamma(2-\alpha)}\right)^T \in \mathbb{R}^N,$$

$$b_0 = \left(\frac{1}{\Gamma(\alpha+1)}, 0, \dots, 0\right)^T \in \mathbb{R}^N, \quad U_{\hat{M}} = \left(u_1 - u_0, u_2 - u_0, \dots, u_{\hat{M}} - u_0\right)^T.$$

From (7.2.9), we observe that $\mathcal{A}_{k+1,j+1} = \mathcal{A}_{k,j}$, and therefore A is a lower-triangular Toeplitz matrix.

In what follows, we will analyze the unique solvability of the IMEX(0) scheme. For this purpose, we introduce the mapping $\Phi_h\colon \mathbb{R}^{Nd} \to \mathbb{R}^{Nd}$ as follows:

$$\Phi_h(X) := h^{\alpha}\left[(A-\lambda B)^{-1}(B+C) \otimes I_d\right] F(X) + \left[(A-\lambda B)^{-1} \otimes I_d\right] D,$$

and therefore, we have the following result.

Theorem 7.10 *Suppose that Lipschitz condition* (7.2.27) *holds and*

$$h^{\alpha} L \|(A-\lambda B)^{-1}(B+C)\|_{\infty} < 1. \tag{7.2.53}$$

Then the method IMEX(0) has a unique solution $U \in \mathbb{R}^{Nd}$.

Proof Let $X = (x_1^T, x_2^T, \dots, x_N^T)^T$, $\hat{X} = (\hat{x}_1^T, \hat{x}_2^T, \dots, \hat{x}_N^T)^T$ be two arbitrary vectors in $\mathbb{R}^{Nd}$. It follows from the Lipschitz condition (7.2.27) that

$$\begin{aligned} \left\|\Phi_h(X) - \Phi_h(\hat{X})\right\|_{\infty} &\le h^{\alpha} \left\|(A-\lambda B)^{-1}(B+C)\right\|_{\infty} \left\|F(X) - F(\hat{X})\right\|_{\infty} \\ &\le h^{\alpha} L \left\|(A-\lambda B)^{-1}(B+C)\right\|_{\infty} \left\|X - \hat{X}\right\|_{\infty}. \end{aligned}$$

If condition (7.2.53) holds, we know that $\Phi_h(X)$ is a contraction mapping with contraction factor $h^{\alpha} L\|(A-\lambda B)^{-1}(B+C)\|_{\infty}$. Moreover, it is well known that space $\mathbb{R}^{Nd}$ with norm $\|\cdot\|_{\infty}$ is complete. Hence, according to the Banach contraction mapping principle (see e.g., Agarwal et al. (2001)), mapping $\Phi_h(X)$ has a unique fixed point in $\mathbb{R}^{Nd}$. Namely, the method IMEX(0) has a unique solution $U \in \mathbb{R}^{Nd}$. □

Remark 7.11 It should be pointed out that the coefficient matrices A for IMEX(1) are not the Toeplitz matrices, but we can choose the first column a_1 of A as

$$\hat{a}_1 = \left(1, -1+\frac{\gamma_{2,2}^{(1)}}{\Gamma(\alpha)\Gamma(2-\alpha)}, \frac{\gamma_{3,2}^{(1)}}{\Gamma(\alpha)\Gamma(2-\alpha)}, \ldots, \frac{\gamma_{N,2}^{(1)}}{\Gamma(\alpha)\Gamma(2-\alpha)}\right)^T \in \mathbb{R}^N.$$

Then A will be a Toeplitz matrix. Moreover, if we do this, the corresponding vector D for IMEX(1) will be changed by adding the term $\left[(\hat{a}_1 - a_1) \otimes I_d\right] U$.

7.2.6 Fast Approximate Inversion Scheme

In order to obtain a fast solution to the Toeplitz system (7.2.49), we employ a scheme which approximates the lower-triangular Toeplitz matrix K^{-1}. In particular, we have $K = (A \otimes I_d - \lambda B \otimes I_d)$ for the method IMEX(0) in (7.2.49). The first step involves approximating the matrix K by the following block ϵ-circulant matrix:

$$K_\epsilon = \begin{bmatrix} K_0 & \epsilon K_{N-1} & \ldots & \epsilon K_2 & \epsilon K_1 \\ K_1 & K_0 & \epsilon K_{N-1} & \ldots & \epsilon K_2 \\ \vdots & K_1 & K_0 & \ddots & \vdots \\ K_{N-2} & \ldots & \ddots & \ddots & \epsilon K_{N-1} \\ K_{N-1} & K_{N-2} & \ldots & K_1 & K_0 \end{bmatrix}, \tag{7.2.54}$$

with $\epsilon > 0$. The accuracy of the fast inversion is $\mathcal{O}(\epsilon)$, where mathematically ϵ can be taken arbitrarily small. However, for double-precision arithmetic, ϵ cannot be set too small due to rounding errors. Numerical experiments demonstrated the smallest practical value to be $\epsilon = 5 \times 10^{-9}$. Moreover, K_ϵ^{-1} is also a block ϵ-circulant matrix, and therefore the solution to system (7.2.49) can be written as

$$U \approx K_\epsilon^{-1} R,$$

where R denotes the right-hand side of (7.2.49).

Let $D_\rho = \mathrm{diag}(1, \rho, \ldots, \rho^{N-1})$, $\rho = \epsilon^{1/N}$ be a diagonal matrix and F_N be a $N \times N$ Fourier matrix. We then have the following spectral decomposition:

$$K_\epsilon^{-1} = \left[(D_\rho^{-1} F_N^*) \otimes I_d\right] \mathrm{diag}\left(\Lambda_0^{-1}, \Lambda_1^{-1}, \ldots, \Lambda_{N-1}^{-1}\right) \left[(F_N D_\rho) \otimes I_d\right],$$

with

$$\begin{bmatrix} \Lambda_0 \\ \Lambda_1 \\ \vdots \\ \Lambda_{N-1} \end{bmatrix} = \left[(\sqrt{N} F_N D_\rho) \otimes I_d \right] \begin{bmatrix} K_0 \\ K_1 \\ \vdots \\ K_{N-1} \end{bmatrix}. \tag{7.2.55}$$

Finally, the approximate solution for U becomes

$$U \approx \left[(D_\rho^{-1} F_N^*) \otimes I_d \right] \operatorname{diag}\left(\Lambda_0^{-1}, \Lambda_1^{-1}, \ldots, \Lambda_{N-1}^{-1} \right) \left[(F_N D_\rho) \otimes I_d \right] R, \tag{7.2.56}$$

where in practical implementations, we replace the Fourier matrices F_N in (7.2.55) and (7.2.56) with FFT operations in order to achieve a computational complexity of $\mathcal{O}(N \log N)$. The other operations to form R do not require FFTs, since matrices B and C are sparse, lower-Toeplitz nature, and the vector D is formed through the multiplication of tall matrices with small vectors.

Remark 7.12 Note that (7.2.56) is a nonlinear system, therefore an iterative solver should be applied to solve this problem. As is known, the Newton iteration method may be the most popular solver for a general system of nonlinear equations $G(x) = 0$. However, the disadvantage of the Newton iteration method is that, at each iteration step, it requires the explicit form of the $N \times N$ Jacobian matrix $G'(x^{(n)})$, where $x^{(n)}$ denotes the n-th approximation to x. So the computation of the Newton iteration method could be much more expensive. In order to overcome this disadvantage, the Picard iteration method has been used to solve the system (7.2.56), where, for a given iteration n, we have

$$U^{(n+1)} = K_\epsilon^{-1} R(U^{(n)}), \tag{7.2.57}$$

until $||U^{(n+1)} - U^{(n)}|| > \epsilon^p$, where $U^{(n)}$ denotes the n-th approximation to U and ϵ^p represents the tolerance of the Picard iteration scheme.

7.2.7 Fast Computation of Hypergeometric Functions

Accurate and efficient computation of the Gauss hypergeometric function ${}_2F_1(a, b; c; z)$ is also fundamental to the developed scheme. We know that there is no simple answer for this problem, and different methods are optimal for different parameter regimes. When $\Re(c) > \Re(a) > 0$ or $\Re(c) > \Re(b) > 0$, the Gauss–Jacobi quadrature method is effective. As stated in Abramowitz and Stegun (1972), when $|\arg(1 - z)| < \pi$, we have

$${}_2F_1(a, b; c; z) = \frac{\Gamma(c)}{\Gamma(b)\Gamma(c - b)} \int_0^1 (1 - zt)^{-a} w_{b,c}(t) dt, \tag{7.2.58}$$

where $w_{b,c}(t) = (1-t)^{c-b-1}t^{b-1}$ and $\Re(c) > \Re(b) > 0$. The parameters a and b in (7.2.58) can be interchanged due to the basic power series definition of the hypergeometric function. Transforming $t \to \frac{\tilde{t}+1}{2}$, we can obtain that

$$\begin{aligned}\int_0^1 (1-zt)^{-a} w_{b,c}(t)\mathrm{d}t =& \frac{1}{2^{c-1}} \int_{-1}^{1} \left(1-\frac{z}{2}-\frac{z}{2}\tilde{t}\right)^{-a} (1-\tilde{t})^{c-b-1} (1+\tilde{t})^{b-1} \mathrm{d}\tilde{t} \\ =& \sum_{j=1}^{Q} w_j^{GJ} \left(1-\frac{z}{2}-\frac{z}{2}t_j^{GJ}\right)^{-a} + E_Q(a,b,z),\end{aligned}$$

where t_j^{GJ} and w_j^{GJ} are the Gauss–Jacobi nodes and weights on $[-1,1]$, and Q is the number of mesh points. Error bounds for this method are discussed in Gautschi (2002).

7.2.8 IMEX(p) Algorithms for Accurate Long-Time Integration

We present the main stages of the developed fast IMEX(p) methods for efficient time-integration of nonlinear FODEs in Algorithm I. The algorithm particularly describes the IMEX(0) approach, but the main steps remain the same for IMEX(1), with slight modifications regarding the number of sets of correction weights and forms for the nonlinear system matrices. The operators $\mathcal{F}(\cdot)$ and $\mathcal{F}^{-1}(\cdot)$ represent, respectively, the forward and inverse fast Fourier transforms.

Remark 7.13 For the FAMMs (7.2.13), (7.2.29) and the FAMMs with correction terms (7.2.23), (7.2.36), we also can use the fast implementation proposed in this section to construct the corresponding fast methods.

Numerical Tests

We present several numerical examples to verify our theoretical analysis presented in the previous sections. In all presented numerical examples, we utilize a numerical tolerance $\epsilon = 5 \times 10^{-9}$ for the fast inversion step. For all hypergeometric functions involved in the evaluation of correction weights and history load term, we utilize $Q = 200$ Gauss–Jacobi quadrature points. One exception is the incomplete beta function evaluated for the history load correction in (7.2.21). For this case, the argument $\frac{k}{k+1}$ approaches 1 as N increases, and a numerical quadrature becomes a poor choice due to singularities. In that sense we evaluate the incomplete beta function using the

Algorithm I Fast IMEX(0) Integration Scheme for Nonlinear FODEs.

Known database:
Initial parameters h, N, ϵ, ϵ^p, λ, number of correction terms and powers σ, δ.
Compute the first column a_1 of A defined in (7.2.50). Compute B and C defined in (7.2.51) with sparse allocation.
Compute corrections using (7.2.17)–(7.2.18), (7.2.21), (7.2.33)–(7.2.34) and (7.2.40) (or (7.2.41)) for all N time-steps.
Compute D_ρ. Assemble the first column $K_{\epsilon,1}$ of K_ϵ in (7.2.54).
Compute $\Lambda = \mathcal{F}((D_\rho \otimes I_d)K_{\epsilon,1})$ to obtain (7.2.55).
Picard Iteration: Initial guess $U^{(0)}$.
while $e > \epsilon^p$ **do**
  Compute $D(U^{(n)})$ using (7.2.52). Compute $F(U^{(n)})$.
  Compute $R(U^{(n)})$ using (7.2.49).
  Compute $r_\epsilon = \mathcal{F}\big((D_\rho \otimes I_d)R(U^{(n)})\big)$.
  Solve $\tilde{r}_\epsilon = \Lambda^{-1} r_\epsilon$.
  Compute the updated solution vector $U^{(n+1)} = D_\rho^{-1}\mathcal{F}^{-1}(\tilde{r}_\epsilon)$.
  Compute $e = ||U^{(n+1)} - U^{(n)}||$.
  $n = n + 1$.
end while
return $U^{(n+1)}$

native MATLAB implementation. Furthermore, given $\Omega = (0, T]$, we utilize the quantities

$$\mathrm{err}_N(h) = \frac{\|u(t_N) - u_N\|_\infty}{\|u(t_N)\|_\infty}, \quad \mathrm{err}(h) = \frac{\max\limits_{0\le k\le N} \|u(t_k) - u_k\|_\infty}{\max\limits_{0\le k\le N} \|u(t_k)\|_\infty},$$

$$\mathrm{Order}_1 = \log_2\left[\frac{\mathrm{err}_N(h)}{\mathrm{err}_N(h/2)}\right], \quad \mathrm{Order}_2 = \log_2\left[\frac{\mathrm{err}(h)}{\mathrm{err}(h/2)}\right]$$

to denote the error at the end-point T, the global error on the solution interval Ω and convergence order of the used method at the end-point T and on the solution interval Ω, respectively. The developed framework was implemented in MATLAB R2019a and was run in a desktop computer with Intel Core i7-6700 CPU with 3.40 GHz, 16 GB RAM and Ubuntu 18.04.2 LTS operating system.

Example 7.14 Linear FODE (see e.g., Zayernouri and Matzavinos (2016)):

$$ {}_0^C D_t^\alpha u(t) = f(t), \quad \alpha \in (0, 1], \; t \in (0, 1]; \quad u(0) = 0. \tag{7.2.59}$$

The exact solution of (7.2.59) is $u(t) = t^{p+\alpha}$ for $p = 1, 2$. The corresponding force term is $f(t) = \frac{\Gamma(\alpha+p+1)}{\Gamma(p+1)} t^p$. Recalling Remark 7.13, we can employ the fast inversion scheme directly to the FAMM (7.2.13) in order to obtain a fast FAMM. Therefore, in this example we compare the performance between the fast and original FAMMs (7.2.13), where we verify the computational complexity and accuracy of both original and fast schemes. Table 7.5 presents the obtained results for the implemented FAMMs at the end-point $T = 1$ and no correction terms. Similar to the results in Zayernouri and Matzavinos (2016), we observe that the convergence order is independent of the fractional order α, preserving the accuracy of the integer-order methods. For the case of $\alpha = 0.9$ and $p = 0$, the scheme required a smaller time-step size $h = 2^{-13}$ to achieve a linear convergence rate with $\text{Order}_1 = 0.9967$ and error $\text{err}_N(h) = 1.9520\text{e}{-06}$. The computational times for the original and fast FAMMs are illustrated in Figure 7.3. We observe the computational complexity of $\mathcal{O}(N \log N)$ for the developed fast FAMMs. A break-even point at $N = 4$ is observed between both methods, which makes the fast method more computationally efficient when $N > 4$ time-steps.

Example 7.15 Stiff FODEs (see e.g., Cao et al. (2016)):

$$\begin{cases} {}^{C}_{0}D^{\alpha}_{t} u(t) = Pu(t) + Su(t) + g(t), \quad \alpha \in (0, 1], \quad t \in (0, 10], \\ u(0) = [1,\ 1,\ 1]^T, \end{cases} \tag{7.2.60}$$

where

$$P = \begin{bmatrix} -1 & 0 & 0.001 \\ -0.0005 & -0.0008 & -0.0002 \\ 0.001 & 0 & -0.001 \end{bmatrix}, \quad S = \begin{bmatrix} -0.006 & 0 & 0.002 \\ -0.001 & -0.002 & 0 \\ 0 & -0.005 & -0.008 \end{bmatrix},$$

$$g(t) = \begin{pmatrix} a_1\Gamma_1 t^{\sigma_1-\alpha} + a_2\Gamma_2 t^{\sigma_2-\alpha} \\ a_3\Gamma_3 t^{\sigma_3-\alpha} + a_4\Gamma_4 t^{\sigma_4-\alpha} \\ a_5\Gamma_5 t^{\sigma_5-\alpha} + a_6\Gamma_6 t^{\sigma_6-\alpha} \end{pmatrix} - (P+S) \begin{pmatrix} a_1 t^{\sigma_1} + a_2 t^{\sigma_2} + 1 \\ a_3 t^{\sigma_3} + a_4 t^{\sigma_4} + 1 \\ a_5 t^{\sigma_5} + a_6 t^{\sigma_6} + 1 \end{pmatrix},$$

with $\Gamma_k = \frac{\Gamma(\sigma_k+1)}{\Gamma(\sigma_k+1-\beta)}$, $1 \le k \le 6$. Therefore, the exact solution for the stiff FODE (7.2.60) is given by

$$u(t) = \left(a_1 t^{\sigma_1} + a_2 t^{\sigma_2} + 1,\ a_3 t^{\sigma_3} + a_4 t^{\sigma_4} + 1,\ a_5 t^{\sigma_5} + a_6 t^{\sigma_6} + 1\right)^T,$$

where, as in Cao et al. (2016), we consider $\sigma_1 = \alpha$, $\sigma_2 = 2\alpha$, $\sigma_3 = 1 + \alpha$, $\sigma_4 = 5\alpha$, $\sigma_5 = 2$, $\sigma_6 = 2 + \alpha$, and $a_1 = 0.5$, $a_2 = 0.8$, and $a_3 = a_4 = a_5 = a_6 = 1$. For the numerical solution of (7.2.60), we take $f(t, u) = Su(t) + g(t)$ and employ the IMEX(p) scheme with $\alpha = 0.3$, utilizing a Picard iteration

Table 7.5. *The errors at the end-point and convergence orders of the FAMMs for* (7.2.59) *with* $p = 0$ *(upper table) and* $p = 1$ *(lower table). Reprinted from the Journal of Applied Numerical Mathematics, 156, Zhou, Yongtao, Suzuki, Jorge L, Zhang, Chengjian, and Zayernouri, Mohsen, Implicit-explicit time integration of nonlinear fractional differential equations, 555–583, Copyright (2020), with permission from Elsevier and IMACS.*

The fast FAMM with $p = 0$.

	$\alpha = 0.1$		$\alpha = 0.5$		$\alpha = 0.9$	
h	$\text{err}_N(h)$	Order_1	$\text{err}_N(h)$	Order_1	$\text{err}_N(h)$	Order_1
2^{-3}	5.2795e–03	–	9.7878e–03	–	1.0461e–03	–
2^{-4}	2.6031e–03	1.0202	5.0535e–03	0.9537	6.0576e–04	0.7883
2^{-5}	1.2912e–03	1.0115	2.5658e–03	0.9779	3.3733e–04	0.8446
2^{-6}	6.4272e–04	1.0064	1.2925e–03	0.9892	1.8357e–04	0.8778
2^{-7}	3.2058e–04	1.0035	6.4866e–04	0.9947	9.8414e–05	0.8994

The fast FAMM with $p = 1$.

	$\alpha = 0.1$		$\alpha = 0.5$		$\alpha = 0.9$	
h	$\text{err}_N(h)$	Order_1	$\text{err}_N(h)$	Order_1	$\text{err}_N(h)$	Order_1
2^{-3}	2.1934e–05	–	4.0975e–04	–	4.9840e–04	–
2^{-4}	5.5334e–06	1.9869	9.6888e–05	2.0804	1.2210e–04	2.0292
2^{-5}	1.4102e–06	1.9723	2.3574e–05	2.0391	2.9761e–05	2.0366
2^{-6}	3.5857e–07	1.9756	5.8150e–06	2.0193	7.2452e–06	2.0383
2^{-7}	8.9586e–08	2.0009	1.4452e–06	2.0085	1.7663e–06	2.0363

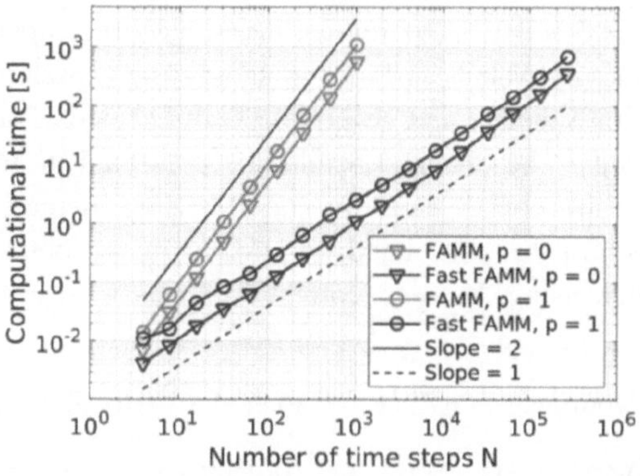

Figure 7.3 Computational time versus number of time-steps N for the original (lighter gray curves) and the developed, fast (darker gray curves) FAMMs. Source: Reprinted from Zhou et al. (2020) with permission from Elsevier.

Table 7.6. *The global errors and convergence orders of the methods IMEX(p) for* (7.2.60) *with* $p = 0$ *(upper table) and* $p = 1$ *(lower table) and* $\alpha = 0.3$. *Reprinted from the Journal of Applied Numerical Mathematics, 156, Zhou, Yongtao, Suzuki, Jorge L, Zhang, Chengjian, and Zayernouri, Mohsen, Implicit-explicit time integration of nonlinear fractional differential equations, 555–583, Copyright (2020), with permission from Elsevier and IMACS.*

IMEX(0)

	$M = 0$		$M = 1$	
h	err(h)	Order_2	err(h)	Order_2
2^{-3}	1.9340e–02	–	1.9271e–02	–
2^{-4}	9.7036e–03	0.9950	9.6793e–03	0.9935
2^{-5}	4.8614e–03	0.9972	4.8518e–03	0.9964
2^{-6}	2.4334e–03	0.9984	2.4294e–03	0.9979
2^{-7}	1.2175e–03	0.9991	1.2157e–03	0.9988

IMEX(1)

	$M = 0$		$M = 1$		$M = 2$		$M = 3$	
h	err(h)	Order_2	err(h)	Order_2	err(h)	Order_2	err(h)	Order_2
2^{-3}	6.5717e–04	–	4.3931e–04	–	2.4945e–04	–	2.1176e–04	–
2^{-4}	3.8296e–04	0.7791	1.5083e–04	1.5424	6.2702e–05	1.9922	5.6167e–05	1.9146
2^{-5}	2.3992e–04	0.6746	5.5600e–05	1.4397	1.5730e–05	1.9950	1.4606e–05	1.9431
2^{-6}	1.5407e–04	0.6390	2.1541e–05	1.3680	4.0658e–06	1.9519	3.7593e–06	1.9581
2^{-7}	9.9312e–05	0.6335	8.5481e–06	1.3334	1.7074e–06	1.2517	9.6316e–07	1.9646

tolerance of $\epsilon^p = 5 \times 10^{-7}$. We remark that the coefficients of P and S are taken as small enough values in order to satisfy the Lipschitz condition for the Picard iteration scheme, and nevertheless, the choice of such values still makes (7.2.60) stiff. The obtained results are presented in Table 7.6, where we obtain first-order convergence for the IMEX(0) method without using correction terms. On the other hand, for IMEX(1), we obtain second-order convergence when using $M = 3$ correction terms with correction powers $\sigma = \{\alpha, 2\alpha, 1 + \alpha\}$ and $\delta = \{2\alpha, 1 + \alpha, 5\alpha\}$.

Example 7.16 Nonlinear FODEs (cf. Cao et al. (2016); Zeng et al. (2018b)):

$$ {}_0^C D_t^{\alpha} u(t) = \lambda u(t) + f(t, u(t)), \quad \alpha \in (0, 1], \; t \in \Omega; \quad u(0) = u_0. \qquad (7.2.61)$$

We consider the following cases for (7.2.61):

Table 7.7. *The global errors and convergence orders of the methods IMEX(p) for solving Case I with $\alpha = 0.4$, varying h and correction terms M with corresponding powers $\sigma_k = \delta_k = k\alpha$. Reprinted from the Journal of Applied Numerical Mathematics, 156, Zhou, Yongtao, Suzuki, Jorge L, Zhang, Chengjian, and Zayernouri, Mohsen, Implicit-explicit time integration of nonlinear fractional differential equations, 555–583, Copyright (2020), with permission from Elsevier and IMACS.*

h	M	IMEX(0)			M	IMEX(1)		
		err(h)	Order$_2$	CPU		err(h)	Order$_2$	CPU
2^{-2}		7.6111e–03	–	0.21		5.9001e–03	–	0.37
2^{-3}		5.7608e–03	0.4019	0.37		4.6835e–03	0.3352	0.68
2^{-4}	0	4.3450e–03	0.4069	0.75	0	3.6810e–03	0.3475	1.37
2^{-5}		3.2724e–03	0.4090	1.35		2.8698e–03	0.3591	2.65
2^{-6}		2.4640e–03	0.4093	2.65		2.2231e–03	0.3684	5.32
2^{-2}		1.9406e–03	–	0.22		2.1673e–05	–	1.06
2^{-3}		1.2144e–03	0.6762	0.35		1.0252e–05	1.0800	2.13
2^{-4}	1	7.4541e–04	0.7042	0.73	2	4.7644e–06	1.1055	4.03
2^{-5}		4.5062e–04	0.7261	1.36		2.1831e–06	1.1259	7.95
2^{-6}		2.6918e–04	0.7433	2.66		9.8961e–07	1.1415	16.04
2^{-2}		3.2061e–04	–	0.56		1.4155e–07	–	1.74
2^{-3}		1.7125e–04	0.9047	1.10		5.0125e–08	1.4978	3.32
2^{-4}	2	9.0232e–05	0.9244	1.97	4	2.5470e–08	0.9767	6.49
2^{-5}		4.7019e–05	0.9404	3.91		2.7468e–08	–	13.07
2^{-6}		2.4284e–05	0.9532	7.85		3.6517e–08	–	26.02

- **Case I)** Let $\lambda = -0.2$, $f(t, u(t)) = 0$ and $u_0 = 1$. The corresponding analytical solution is given by $u(t) = E_\alpha(-0.2t^\alpha)$, where $E_\alpha(t)$ represents the Mittag-Leffler function (cf. Mainardi (2010)).
- **Case II)** Let $\lambda = -1$, $f(t, u(t)) = -0.1u^2 + g(t)$, $u_0 = 1$ and choose $g(t)$ such that the exact solution of (7.2.61) is given by $u(t) = 1 + t + t^2/2 + t^3/3 + t^4/4$.
- **Case III)** Let $\lambda = -1$, $f(t, u(t)) = 0.01u\left(1 - u^2\right) + 2\cos(2\pi t)$ and $u_0 = 1$.

We start with **Case I**, for which we consider $\Omega = (0, 40]$ and a tolerance $\epsilon^p = 10^{-7}$ for the Picard iteration, with varying number of correction terms M and $\alpha = 0.4$. The obtained results are presented in Table 7.7 for the methods IMEX(p), where the CPU time (CPU) measured in seconds

Table 7.8. *The global errors and convergence orders of the methods IMEX(p) for solving Case II with no correction term for $p = 0$ and $M = 2$ correction terms for $p = 1$, with $\sigma = \delta = \{1 - \alpha, 1\}$. Reprinted from the Journal of Applied Numerical Mathematics, 156, Zhou, Yongtao, Suzuki, Jorge L, Zhang, Chengjian, and Zayernouri, Mohsen, Implicit-explicit time integration of nonlinear fractional differential equations, 555–583, Copyright (2020), with permission from Elsevier and IMACS.*

IMEX(0)

	$\alpha = 0.2$		$\alpha = 0.5$		$\alpha = 0.8$	
h	err(h)	Order$_2$	err(h)	Order$_2$	err(h)	Order$_2$
2^{-6}	1.9855e–02	–	1.8557e–02	–	1.7421e–02	–
2^{-7}	9.9111e–03	1.0024	9.2800e–03	0.9998	8.7136e–03	0.9994
2^{-8}	4.9517e–03	1.0011	4.6425e–03	0.9992	4.3602e–03	0.9989
2^{-9}	2.4750e–03	1.0005	2.3226e–03	0.9991	2.1821e–03	0.9987
2^{-10}	1.2373e–03	1.0002	1.1619e–03	0.9992	1.0921e–03	0.9986

IMEX(1)

	$\alpha = 0.2$		$\alpha = 0.5$		$\alpha = 0.8$	
h	err(h)	Order$_2$	err(h)	Order$_2$	err(h)	Order$_2$
2^{-6}	5.3875e–04	–	4.4167e–04	–	3.7370e–04	–
2^{-7}	1.3610e–04	1.9849	1.1153e–04	1.9856	9.4359e–05	1.9856
2^{-8}	3.4206e–05	1.9924	2.8033e–05	1.9922	2.3740e–05	1.9909
2^{-9}	8.5751e–06	1.9960	7.0286e–06	1.9958	5.9679e–06	1.9920
2^{-10}	2.1498e–06	1.9960	1.7586e–06	1.9988	1.4862e–06	2.0056

represents the running time of the methods. We observe the linear convergence rate when using $M=2$ correction terms for IMEX(0). The convergence rates also improve for IMEX(1); however, we attain the accuracy limit of the scheme when using $M=4$. Such accuracy limit is determined by the value of $\epsilon = 5 \times 10^{-9}$ utilized for the fast inversion approach discussed in Section 7.2.6.

For **Case II**, we let $\Omega = (0, 1]$ and $\epsilon^p = 10^{-7}$. The obtained results are presented in Table 7.8, where we observe that both schemes achieve the theoretical convergence rates for the global error.

For **Case III**, we consider a long-time integration taking $\Omega = (0, 50]$, and we set a numerical tolerance $\epsilon^p = 10^{-6}$ for the Picard iteration. Figures 7.4 and 7.5 illustrate the obtained highly oscillatory solutions. We also perform

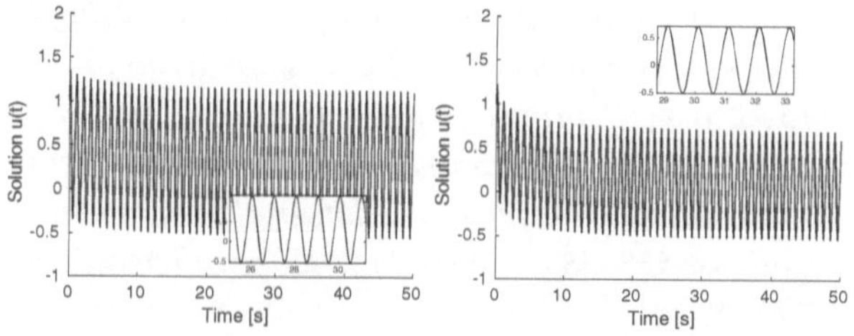

Figure 7.4 Solution versus time t for Case III using $h = 0.005$ and $M = 1$ correction term. Left: $\alpha = 0.2$; Right: $\alpha = 0.5$. Source: Reprinted from Zhou et al. (2020b) with permission from Elsevier.

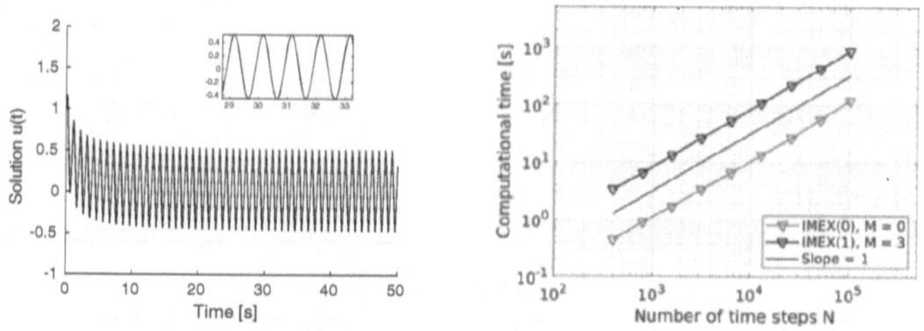

Figure 7.5 Left: Solution versus time t for Case III with $\alpha = 0.7$ using $h = 0.005$ with $M = 1$ correction term. Right: Computational time versus number of time-steps N for the developed IMEX schemes. Source: Reprinted from Zhou et al. (2020b) with permission from Elsevier.

a convergence analysis utilizing a benchmark solution with $h = 2^{-11}$ and $M = 3$ correction terms and evaluate the global error. The obtained results are presented in Table 7.9, where the expected first- and second-order convergence rates are obtained, respectively, with $p = 0$ without correction terms, and $p = 1$ using $M = 3$ correction terms. The computational times for the developed IMEX schemes are illustrated in Figure 7.5 including the initial phase for computation of correction weights. We observe the computational complexity of $\mathcal{O}(N \log N)$ for the developed schemes even for nonlinear problems, with a small difference between the first- and second-order schemes.

We also compare the developed IMEX(0) scheme with the direct, fully implicit L1 approach developed by Zeng et al. (2018a) with order $\mathcal{O}(h^{3-\alpha})$. Therefore, we use the same problem setup and parameters and employ a Newton sub-iteration for the nonlinear problem. We compute the errors and convergence order without correction terms and varying fractional orders. For $\alpha = \{0.2, 0.5, 0.7\}$, we obtained, respectively, err($h = 2^{-7}$) = {2.4581e–02, 1.5633e–02, 5.6208e–03} and $\text{Order}_2 = \{0.1233, 0.4930, 0.7524\}$. We

Table 7.9. *Convergence results for the IMEX(p) scheme solving Case III without corrections for $p = 0$ and $M = 3$ correction terms for $p = 1$, with $\sigma = \delta = \{\alpha, 2\alpha, 1 + \alpha\}$. Reprinted from the Journal of Applied Numerical Mathematics, 156, Zhou, Yongtao, Suzuki, Jorge L, Zhang, Chengjian, and Zayernouri, Mohsen, Implicit-explicit time integration of nonlinear fractional differential equations, 555–583, Copyright (2020), with permission from Elsevier and IMACS.*

IMEX(0)

	$\alpha = 0.2$		$\alpha = 0.5$		$\alpha = 0.7$	
h	err(h)	Order_2	err(h)	Order_2	err(h)	Order_2
2^{-3}	4.4883e–01	–	3.3552e–01	–	2.5798e–01	–
2^{-4}	2.2466e–01	0.9984	1.6681e–01	1.0082	1.3311e–01	0.9546
2^{-5}	1.1217e–01	1.0021	8.3553e–02	0.9975	6.7629e–02	0.9769
2^{-6}	5.6354e–02	0.9931	4.2141e–02	0.9875	3.4015e–02	0.9915
2^{-7}	2.8246e–02	0.9965	2.1163e–02	0.9937	1.7091e–02	0.9929

IMEX(1)

	$\alpha = 0.2$		$\alpha = 0.5$		$\alpha = 0.7$	
h	err(h)·	Order_2	err(h)	Order_2	err(h)	Order_2
2^{-3}	3.3534e-01	–	1.9664e-01	–	2.5260e-01	–
2^{-4}	9.5452e-02	1.8128	8.1541e-02	1.2699	7.1427e-02	1.8223
2^{-5}	2.2387e-02	2.0921	2.0062e-02	2.0230	2.0503e-02	1.8007
2^{-6}	5.4892e-03	2.0280	4.5814e-03	2.1306	4.8812e-03	2.0705
2^{-7}	1.8634e-03	1.5586	1.0402e-03	2.1389	1.1154e-03	2.1296

observe that although the L1 approach achieves similar error levels than the IMEX(0) scheme (Table 7.9, top), it is at most $\mathcal{O}(h^{\alpha})$ accurate for the tested values of h. This loss of accuracy could be due to the presence of nonlinearities in the solution, which is assumed to be $C^3(\Omega)$-continuous in the adopted L1-type discretization without corrections terms. On the other hand, the IMEX(0) scheme is first-order accurate without the need of any correction terms and α-independent.

The main advantages of the developed IMEX schemes in comparison to other works are: larger stability regions when compared to the IMEX schemes developed by Cao et al. (2016); and also a fast solution alternative when compared to the original fractional Adams–Bashforth/Moulton methods developed by Zayernouri and Matzavinos (2016), and the IMEX schemes by

Cao et al. (2016). When compared to the matrix-based fast solver for FODEs developed by Lu et al. (2015), the developed framework in this work handles the numerical solution of nonlinear and singular FODEs instead of only linear ones.

The developed schemes could be used for, e.g., fractional viscoelastic models under complex loading conditions and long-time integration (Jaishankar and McKinley, 2013). Regarding additional constitutive effects, such as fractional visco-elasto-plastic models (Suzuki et al., 2016; Suzuki, 2017), the developed methods could be potentially applied under simple monotone loads. Furthermore, the introduction of additional sets of correction terms motivates the use of data-infused self-singularity-capturing approaches (Suzuki and Zayernouri, 2021), which would decrease the number of correction terms per set.

7.3 An Auto-Singularity-Capturing Scheme

In this section, we introduce a two-stage computational framework for robust, accurate, and long-time integration of multi-term linear/nonlinear fractional differential equations. In the first stage, we formulate a self-singularity-capturing scheme, given available/observable data for diminutive time, experimentally obtained or sampled from an approximate numerical solution utilizing a fine grid near the initial time. The fractional differential equation provides the necessary knowledge/insight on how the hidden singularity can bridge between the initial and the subsequent short-time solution data. In the second stage, we utilize the multi-singular behavior of the solution in a variety of numerical methods, without resorting to making any ad hoc/uneducated guesses for the solution singularities. Particularly, we employ an implicit finite-difference method, where the captured singularities, in the first stage, are taken into account through some Lubich-like correction terms, leading to an accuracy of order $\mathcal{O}(\Delta t^{3-\alpha})$. We show that this novel framework can control the error even in the presence of strong multi-singularities.

The main contribution of the present section is to develop a two-stage data-infused computational framework for accurate time-integration of FODEs. In Stage-I, we formulate a self-singularity-capturing framework where:

- The scheme utilizes available data for initial diminutive time (application-oriented), and self-captures/determines multi-singular behaviors of the solution through the knowledge introduced by the FODE and its corresponding fractional operators. *No knowledge of the analytical solution is required a priori*, and some of the numerical tests throughout this work sample from analytical solution points only for verification purposes.

- We develop a new (finite-difference based) algorithm for automatic determination of the underlying power-law singularities near the initial data, employing gradient descent optimization. The singularities are introduced through Lubich's correction method (Lubich, 1986).
- We introduce the capturing scheme for M correction terms and construct a hierarchical, self-capturing framework. We test the framework for the particular case of up to $S = 3$ singularities and $M = 3$ correction terms.
- The self-capturing procedure makes use of two stopping criteria for the error minimization, namely ϵ (for solution error) and ϵ_1 (for gradient norm error). Here, numerical convergence with respect to ϵ_1 corresponds to the minimum error for a fixed number of correction terms. On the other hand, numerical convergence with tolerance ϵ defines our error control criterion, and determines if additional correction terms are needed to capture the true solution singularities.

In Stage-II we utilize the captured multi-singularities for full time-integration of the FODE in question, where a variety of the aforementioned numerical methods could be, in general, employed. At this stage, no knowledge of the solution data is required, but only the usual forcing terms. Our approach is explained as the following:

- To handle the numerical solution of the FODE using multiple correction terms, we develop an implicit finite-difference method, where we solve a linear system of size M for the first M time-steps. Therefore, we incorporate the captured singular behavior up to the desired precision ϵ, without the need of using a fine time-grid.
- We numerically demonstrate that the developed methodology is of order $\mathcal{O}(\Delta t^{3-\alpha})$;

We perform a set of numerical tests, where we demonstrate that the developed scheme is able to capture multiple singularities using a small number of time-steps, which can later be utilized for efficient time-integration of FODEs with relatively large time-step size Δt, being much more accurate than ad hoc choices of singularities when the regularity of the solution is unknown. The successful capturing of singularities motivates the development of kernel- and knowledge-based refinement of time-grids near $t = 0$ (Stynes et al., 2017), as well as self-construction of basis function spaces using Müntz polynomials (Esmaeili et al., 2011; Hou and Xu, 2017) for spectral element methods for FODEs/FPDEs (Kharazmi and Zayernouri, 2019).

Let G be an open set in $\mathbb{R}$, and let a function $f(t,u(t))\colon (a,b] \times G \to \mathbb{R}$ such that $f(t,u(t)) \in C_\gamma(\Omega)$ and is Lipschitz continuous with respect to any $u(t) \in G$. Let the fractional Cauchy problem of interest be given by

$$ {}^{\mathrm{C}}_{a}D^{\alpha}_{t} u(t) = f(t,u(t)), \quad u(a) = u_0. \tag{7.3.1}$$

Given the aforementioned conditions, there exists a unique solution for (7.3.1) in the following space of functions:

$$ \boldsymbol{C}^{\alpha}_{\gamma}(\Omega) = \left\{ u(t) \in C(\Omega) \, : \, {}^{\mathrm{C}}_{a}D^{\alpha}_{t} u(t) \in C_\gamma(\Omega) \right\}. \tag{7.3.2}$$

The corresponding solution with respect to $u(t)$ is equivalent to the following Volterra integral equation of the second kind:

$$ u(t) = u_0 + \frac{1}{\Gamma(\alpha)} \int_a^t \frac{f(s,u(s))}{(t-s)^{1-\alpha}} \, \mathrm{d}s, \quad a \le t \le b.$$

We remark that for other particular forms of (7.3.1), the corresponding Volterra equation of the second kind contains Mittag-Leffler functions $E_\alpha(z)$, which are defined through infinite sums and thus are not computationally friendly. Therefore, we choose to work with form (7.3.1), since the self-singularity-capturing framework can be extended for other FODEs in a straightforward way, without resorting to computationally expensive hypergeometric functions. Furthermore, although the existence and uniqueness of Cauchy problems has been investigated (Samko et al., 1993; Delbosco and Rodino, 1996; Kilbas et al., 2006; Diethelm and Ford, 2002), yet, there is no comprehensive framework to understand the singular behavior of the solution given any general $f(t,u(t)) \in C_\gamma(\Omega)$. The developed formulation in this work is suitable for problems where a few initial discrete data points are known from $u(t)$ and $f(t,u(t))$.

Due to the employed finite-difference discretization over the RL derivative in this work (see Section 7.3.2), we choose to rewrite (7.3.1) to obtain

$$ {}^{\mathrm{RL}}_{a}D^{\alpha}_{t} u(t) = f(t,u(t)) - \frac{u_0}{\Gamma(1-\alpha)(t-a)^{\alpha}}. \tag{7.3.3}$$

As will be shown in the next sections, the developed scheme is able to capture singularities with a minimal number of time-steps, i.e., with restricted data for a short period of time.

7.3.1 Two-Stage Time-Integration Framework

In the following sections, we develop the two-stage framework for efficient time-integration of FODEs. We start with Stage-I, where the self-singularity-capturing scheme is presented for S singularities, computed over a small

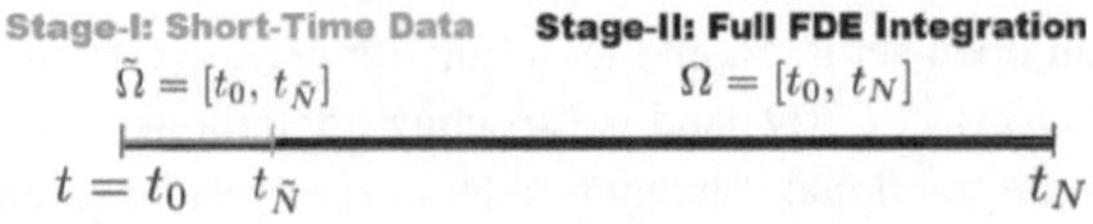

Figure 7.6 In Stage-I, the scheme captures the singularities for the short time-domain $\tilde{\Omega}$, given the short initial data. In Stage-II, the framework uses the singularities obtained in Stage-I for full time-integration of the FODE over Ω. Source: Suzuki and Zayernouri (2021), reprinted by permission of the publisher (Taylor & Francis Ltd, www.tandfonline.com).

number of initial data points denoted by $\tilde{N}$. Then, after the multi-singular solution behavior is learned, we present the developed solution for FODEs in Stage-II, where the subsequent full time-integration of the FODE is carried out over N time-steps (see Figure 7.6).

7.3.2 Stage-I: Self-Singularity-Capturing Stage

We develop the self-singularity-capturing framework, starting with M correction terms for the initial time-steps $\tilde{N}$, at which data is available. We then utilize the developed algorithm in a self-capturing approach, through a hierarchical and iterative fashion.

Minimization Method

Let $t \in \tilde{\Omega}$, with $\tilde{\Omega} = [t_0, t_{\tilde{N}}]$, where $\tilde{N}$ denotes the number of initial short data points, such that $\tilde{N} \geq M$, and let $\sigma \in \mathbb{R}^M$ be the following correction-power (singularity) vector:

$$\sigma = (\sigma_1, \sigma_2, \ldots, \sigma_M)^T. \tag{7.3.4}$$

We define an error function $E : \mathbb{R}^M \to \mathbb{R}^+$, given by

$$E(\sigma) = \sum_{n=1}^{\tilde{N}} \left(u_n^{data} - u_n^N(\sigma)\right)^2, \tag{7.3.5}$$

where $u_n^N(\sigma)$ denotes the σ-dependent numerical solution of $u(t)$ at $t = t_n$, and u_n^{data} represents the known initial short-time data. Figure 7.6 illustrates the integration domain $\tilde{\Omega}$ for Stage-I, where the error function (7.3.5) is evaluated for a short time.

We apply an iterative gradient descent scheme in order to find σ^* that minimizes (7.3.5). Let σ^k be known at the k-th iteration. We compute the updated value σ^{k+1} for iteration $k+1$ in the following fashion (Boyd and Vandenberghe, 2004):

$$\sigma^{k+1} = \sigma^k - \gamma^k \nabla E(\sigma^k), \tag{7.3.6}$$

where γ^k denotes the following two-point step size for the k-th iteration, given by Barzilai and Borwein (1988):

$$\gamma^k = \frac{\left(\sigma^k - \sigma^{k-1}\right)^T \left[\nabla E(\sigma^k) - \nabla E(\sigma^{k-1})\right]}{\|\nabla E(\sigma^k) - \nabla E(\sigma^{k-1})\|^2_{L^2(\mathbb{R}^M)}}. \tag{7.3.7}$$

The gradient $\nabla E\colon \mathbb{R}^+ \to \mathbb{R}^M$ of the error function is given by $\nabla E(\sigma) = (\partial E/\partial\sigma_1, \partial E/\partial\sigma_2, \ldots, \partial E/\partial\sigma_M)^T$, where, instead of obtaining a closed form for the derivatives (which would implicitly involve the differentiation of u_i^N), we utilize complex-step differentiation. Therefore, let $\boldsymbol{e}_j = (0, 0, \ldots, 1, \ldots, 0)^T$ be a vector in $\mathbb{R}^M$ of zeros, with a unit value in the j-th entry. Therefore, we have

$$\nabla E(\sigma)_j \approx \frac{\Im(E(\sigma + \mathrm{i}\Delta\sigma \boldsymbol{e}_j))}{\Delta\sigma}, \quad j = 1, 2, \ldots, M, \quad \mathrm{i} = \sqrt{(-1)}, \tag{7.3.8}$$

which only requires the additional evaluation of the error function perturbed by $\Delta\sigma$, which can be taken, e.g., as 10^{-14}. Given two numerical tolerances ϵ and ϵ_1, we iterate and find a new σ^{k+1} while both criteria $E(\sigma^{k+1}) > \epsilon$ and $\|\nabla E(\sigma^k)\| > \epsilon_1$ are true. The latter criterion is introduced to minimize our error function, while the former is used for error control in our self-capturing scheme. As will be shown in the next sections, the criterion $\|\nabla E(\sigma^k)\| < \epsilon_1$ is satisfied for small ϵ_1 even when $M < S$, but the criterion $E(\sigma^{k+1}) < \epsilon$ will only be satisfied when we use enough correction terms (see Section 7.3.3) to fully capture the S number of power-law singularities.

Numerical Scheme for Short-Time Integration

In the singularity-capturing scheme, given σ, in order to compute the error in (7.3.5) and the gradient (due to the error perturbation) (7.3.8), we need to compute the numerical solution u_n^N for $n = 1, \ldots, \tilde{N}$ initial data points over time-intervals Δt. For this purpose, we employ the finite-difference discretization with corrections presented in Lubich (1986) for the fractional RL derivative. Including correction terms, the discretization of the left-sided RL fractional derivative of order α, evaluated at time $t = t_n$, with initial time $a = 0$, has the following form:

$${}^{\mathrm{RL}}_{\;\;0}D_t^{\alpha} u(t)\big|_{t=t_n} \approx {}^{\mathrm{RL}}_{\;\;0}D_t^{(\alpha,M)} u^N(t)\big|_{t=t_n}. \tag{7.3.9}$$

The above approximation is augmented by the so-called Lubich's correction terms, which is given by

$$ {}^{RL}_{0}D_t^{(\alpha,M)} u^N(t)\big|_{t=t_n} = {}^{RL}_{0}D_t^{\alpha} u^N(t)\big|_{t=t_n} + \sum_{j=1}^{M} W_{j,n}(\sigma)\left(u_j^N - u_0\right), \quad (7.3.10) $$

with $n = 1, \ldots, \tilde{N}$. The term ${}^{RL}_{0}D_t^{\alpha} u^N(t)\big|_{t=t_n}$ is usually denoted in the convolution quadrature form $\Delta t^{-\alpha} \sum_{j=0}^{n} w_{n-j} u_j^N$ with weights w_{n-j}. However we keep the former notation in this work to be consistent with the adopted discretization for fractional RL derivatives (Zeng et al., 2018a) presented later in this section. The term $W_{j,n}(\sigma)$ denotes the correction weights, which depend on the correction powers $\sigma \in \mathbb{R}^M$. We assume a power-law solution singularity about $t = 0$, driven by the power-law kernel of the RL fractional derivative, i.e., $u\big|_{\substack{near \\ t=0}} \approx \sum_{k=1}^{M} c_k t^{\sigma_k}$. If $\sigma = \{\sigma_k\}_{k=1}^{M}$ are known, the RL fractional derivative of each singular term, t^{σ_k}, can be obtained by

$$ {}^{RL}_{0}D_t^{\alpha}(t^{\sigma_k})\big|_{t=t_n} + \sum_{j=1}^{M} W_{j,n} t_j^{\sigma_k} = \frac{\Gamma(1+\sigma_k)}{\Gamma(1+\sigma_k-\alpha)} t_n^{\sigma_k-\alpha}, \quad k = 1, \ldots, M, \quad (7.3.11) $$

where the first term on the left-hand side of (7.3.11) denotes the discretized RL fractional derivative of t^{σ_k}, while the right-hand side represents the exact fractional derivative of t^{σ_k}, evaluated at $t = t_n$. Equation (7.3.11) can be written as

$$ \sum_{j=1}^{M} V_{k,j} W_{j,n} = \frac{\Gamma(1+\sigma_k)}{\Gamma(1+\sigma_k-\alpha)\Delta t^{\alpha}} n^{\sigma_k-\alpha} - \frac{1}{\Delta t^{\sigma_k}} \, {}^{RL}_{0}D_t^{\alpha}(t^{\sigma_k})\big|_{t=t_n}, \quad (7.3.12) $$

with $k = 1, \ldots, M$, where $V_{k,j} = j^{\sigma_k}$ denotes a Vandermonde matrix with size $M \times M$. Therefore, to obtain the starting weights $W_{j,n}$, the above linear system has to be solved for all $n = 1, \ldots, \tilde{N}$.

Substituting (7.3.10) into (7.3.3), we obtain the following discrete form of the FODE for $n = 1, \ldots, \tilde{N}$:

$$ {}^{RL}_{0}D_t^{\alpha} u^N(t)\big|_{t=t_n} + \sum_{j=1}^{M} W_{j,n}(\sigma)\left(u_j^N - u_0\right) + \frac{u_0}{\Gamma(1-\alpha) t_n^{\alpha}} = f_n^{data}, \quad (7.3.13) $$

where f_n^{data} denotes the known short-time right-hand side at time t_n associated with u^{data}, which can be obtained experimentally or sampled through a fine solution grid near $t = 0$.

In order to discretize ${}^{RL}_{0}D_t^{\alpha} u^N(t)$, we follow the difference scheme developed in Zeng et al. (2018a), which is based on a second-order interpolation of

$u(t)$ for the fractional RL integral, similar to the L1-2 scheme for fractional Caputo derivatives developed in Gao et al. (2014). Therefore, when evaluating the left fractional integral at $t = t_n$, we split it into local $\mathcal{L}$ and history $\mathcal{H}$ parts as follows:

$$ {}^{\mathrm{RL}}_{\ 0}I^{\alpha}_{t} u^N(t)|_{t=t_n} = \underbrace{\int_{t_{n-1}}^{t_n} k(t_n - s) u^N(s)\, \mathrm{d}s}_{\mathcal{L}(u,t_n)} + \underbrace{\sum_{k=0}^{n-2} \int_{t_k}^{t_{k+1}} k(t_n - s) u^N(s)\, \mathrm{d}s}_{\mathcal{H}(u,t_n)}, \tag{7.3.14} $$

with the kernel $k(t) = t^{\alpha-1}/\Gamma(\alpha)$. The function $u^N(t)$ is approximated in an implicit fashion using Lagrange interpolants $l_j^{(p)}$ of order p:

$$ u^N(t_n) \approx I^{(p)} u_n^N = \sum_{j=0}^{p} l_j^{(p)} u_{n+j-p}^N. \tag{7.3.15} $$

Substituting (7.3.15) into (7.3.14), and evaluating the convolution integrals, we obtain the following approximations for the local and history parts, respectively, as

$$ \mathcal{L}^{(n,\alpha)} u^N = \sum_{j=0}^{p} d_j^{(p)} u_{n+j-p}^N, \tag{7.3.16} $$

$$ \mathcal{H}^{(n,\alpha)} u^N = \sum_{j=0}^{n-p} \left(b_{n-1-j}^{(1)} u_j^N + b_{n-1-j}^{(2)} u_{j+1}^N + b_{n-1-j}^{(3)} u_{j+2}^N \right), \tag{7.3.17} $$

where the corresponding α- and Δt- dependent coefficients $d_j^{(p)}$ and $b_j^{(1)}, b_j^{(2)}, b_j^{(3)}$ are presented in Section 7.3.3.

7.3.3 Discretization Coefficients for Fractional Derivatives

We present the α- and Δt-dependent coefficients for the local (7.3.16) and history (7.3.17) approximations for the RL fractional derivative, which are given by Zeng et al. (2018a):

$$ \begin{cases} d_0^{(1)} = -\frac{\alpha \Delta t^{\alpha}}{\Gamma(2+\alpha)}, \quad d_1^{(1)} = \frac{\Delta t^{\alpha}}{\Gamma(2+\alpha)}, & p = 1, \\ d_0^{(2)} = -\frac{\alpha \Delta t^{\alpha}}{2\Gamma(3+\alpha)}, \quad d_1^{(2)} = \frac{\alpha(3+\alpha)\Delta t^{\alpha}}{\Gamma(3+\alpha)}, \quad d_2^{(2)} = \frac{(4+\alpha)\Delta t^{\alpha}}{2\Gamma(3+\alpha)}, & p = 2, \end{cases} \tag{7.3.18} $$

and

$$b_j^{(1)} = \frac{\Delta t^{\alpha}}{2\Gamma(\alpha)}\left[a_j^{(\alpha+2)} - (2j-1)\,a_j^{(\alpha+1)} + j\,(j-1)\,a_j^{(\alpha)}\right], \tag{7.3.19}$$

$$b_j^{(2)} = -\frac{\Delta t^{\alpha}}{\Gamma(\alpha)}\left[a_j^{(\alpha+2)} - 2ja_j^{(\alpha+1)} + (j+1)\,(j-1)\,a_j^{(\alpha)}\right], \tag{7.3.20}$$

$$b_j^{(3)} = \frac{\Delta t^{\alpha}}{2\Gamma(\alpha)}\left[a_j^{(\alpha+2)} - (2j+1)\,a_j^{(\alpha+1)} + j\,(j+1)\,a_j^{(\alpha)}\right], \tag{7.3.21}$$

$$a_j^{(\alpha)} = \frac{1}{\alpha}\left[(j+1)^{\alpha} - j^{\alpha}\right]. \tag{7.3.22}$$

In our computations, we make use of $p = 1$ (piecewise linear approximation) for the local part $\mathcal{L}^{(1,\alpha)}u$, when $n = 1$ (first time-step), and $p = 2$ for $n > 1$ (subsequent time-steps). The fractional RL derivative can be obtained from the above discretization by setting $-1 < \alpha < 0$. Therefore,

$$^{\mathrm{RL}}_{\;\;0}D_t^{\alpha} u^N(t)\big|_{t=t_n} \approx \mathcal{L}^{(n,-\alpha)}u^N + \mathcal{H}^{(n,-\alpha)}u^N, \tag{7.3.23}$$

which has accuracy of $\mathcal{O}(\Delta t^{3-\alpha})$ when $p = 2$. Finally, substituting (7.3.23) into (7.3.13), and recalling (7.3.16), we obtain the discretized form for our FODE:

$$\begin{aligned}\sum_{j=0}^{p} d_j^{(p)} u_{n+j-p}^N + \mathcal{H}^{(n,-\alpha)}u^N + &\frac{u_0}{\Gamma(1-\alpha)t_n^{\alpha}} \\ &+ \sum_{j=1}^{M} W_{j,n}(\sigma)\left(u_j^N - u_0\right) = f_n^{data},\end{aligned} \tag{7.3.24}$$

with $n = 1, \ldots, \tilde{N}$.

Remark 7.17 Among a variety of available schemes, we have adopted the discretization of fractional operators introduced in Zeng et al. (2018a) using Lubich's correction terms. However, in Zeng et al. (2018a), the authors did not perform a fully implicit computation of the second term on the left-hand side of (7.3.13) when $M > 1$. Instead, they obtained the solutions for $n = 1, \ldots, M$ using a fine, uniform time-grid with time-step size $\tau = \Delta t^2$ and $M = 1$ correction term. We observe that using such a fine initial time-grid might be computationally expensive and ineffective when strong singularities are present (e.g., $u(t) = t^{\sigma^*}$, with $0 < \sigma^* < 1 - \alpha$). In our developed scheme, we treat such term in a fully implicit fashion, where we solve a small linear system with M unknowns to obtain u_n^N for $n = 1, \ldots, M$. This ensures a proper inclusion of all M singularities in all time-steps near $t = 0$.

We present here the developed finite-difference scheme to solve the discretized FODE (7.3.24) for the particular case of $M = 3$. Here, we solve a small linear system for the fully-implicit computation at the first 3 time-steps, that is, $t = \{\Delta t,\ 2\Delta t,\ 3\Delta t\}$, which is obtained by expanding (7.3.24) for each of the time-steps, where we use $p = 1$ for the first one, with $\mathcal{H}^{(1,-\alpha)}u^N = 0$, and $p = 2$ for the subsequent ones. The expansions are presented as follows:

First time-step $t = \Delta t$:

$$\underbrace{(d_1^{(1)} + W_{1,1})}_{A_{11}} u_1^N + \underbrace{(W_{2,1})}_{A_{12}} u_2^N + \underbrace{(W_{3,1})}_{A_{13}} u_3^N = f_1^{data} - r_1 u_0, \tag{7.3.25}$$

with

$$r_1 = d_0^{(1)} - \left(W_{1,1} + W_{2,1} + W_{3,1} + \frac{\Delta t^{-\alpha}}{\Gamma(1-\alpha)}\right). \tag{7.3.26}$$

Second time-step $t = 2\Delta t$:

$$\begin{aligned} &\underbrace{(d_1^{(2)} + b_0^{(2)} + W_{1,2})}_{A_{21}} u_1^N + \underbrace{(d_2^{(2)} + b_0^{(3)} + W_{2,2})}_{A_{22}} u_2^N \\ &\qquad + \underbrace{W_{3,2}}_{A_{23}} u_3^N = f_2^{data} - r_2 u_0, \end{aligned} \tag{7.3.27}$$

with

$$r_2 = d_0^{(2)} + b_0^{(1)} - \left(W_{1,2} + W_{2,2} + W_{3,2} + \frac{(2\Delta t)^{-\alpha}}{\Gamma(1-\alpha)}\right). \tag{7.3.28}$$

Third time-step $t = 3\Delta t$:

$$\begin{aligned} &\underbrace{(d_0^{(2)} + b_0^{(1)} + b_1^{(2)} + W_{1,3})}_{A_{31}} u_1^N + \underbrace{(d_1^{(2)} + b_0^{(2)} + b_1^{(3)} + W_{2,3})}_{A_{32}} u_2^N \\ &\qquad + \underbrace{(d_2^{(2)} + b_0^{(3)} + W_{3,3})}_{A_{33}} u_3^N = f_3^{data} - r_3 u_0, \end{aligned} \tag{7.3.29}$$

with

$$r_3 = b_1^{(1)} - \left(W_{1,3} + W_{2,3} + W_{3,3} + \frac{(3\Delta t)^{-\alpha}}{\Gamma(1-\alpha)}\right). \tag{7.3.30}$$

Therefore, from (7.3.25)–(7.3.30), we solve the following linear system for $n \le 3$:

$$\begin{bmatrix} A_{11} & A_{12} & A_{13} \\ A_{21} & A_{22} & A_{23} \\ A_{31} & A_{32} & A_{33} \end{bmatrix} \begin{bmatrix} u_1^N \\ u_2^N \\ u_3^N \end{bmatrix} = \begin{bmatrix} f_1^{data} \\ f_2^{data} \\ f_3^{data} \end{bmatrix} - \begin{bmatrix} r_1 \\ r_2 \\ r_3 \end{bmatrix} u_0, \quad \text{if } n \le 3. \tag{7.3.31}$$

Remark 7.18 To obtain the solutions for $M = 2$, we only need to remove the third row/column of the coefficient matrix $\boldsymbol{A}$, and set $W_{1,3} = W_{3,1} = W_{2,3} = W_{3,2} = 0$ in (7.3.25)–(7.3.30). Also, we observe that the correction terms lead to a full matrix of coefficients, which is reduced to a lower-triangular form for the uncorrected case.

For the remaining time-steps ($n > 3$), u_1^N, u_2^N, u_3^N are known and we solve for u_n^N in the usual (but still implicit) time-stepping fashion, as follows:

$$u_n^N = \left(f_n^{data} - \sum_{j=0}^{1} d_j^{(2)} u_{n-2+j}^N - \mathcal{H}^{(n,-\alpha)} u^N - \sum_{j=1}^{M} W_{j,n}(u_j^N - u_0) - \frac{u_0 t_n^{-\alpha}}{\Gamma(1-\alpha)}\right) / d_2^{(2)}, \tag{7.3.32}$$

with $3 < n \leq \tilde{N}$.

Remark 7.19 The current formulation can be extended for a larger number of correction terms; however, using $M > 9$ will incur an ill-conditioned system for (7.3.11). This fact was first analyzed by Diethelm et al. (2006) and later numerically shown by Zeng et al. (2017), which would incur significant errors when computing the weights $W_{n,j}$ and consequently propagate the errors to the operator discretization. In that sense, we choose $M = 3$ to capture the most critical singularities while keeping a small condition number and small residual for the linear system (7.3.11).

Stage-I Algorithm

The Stage-I framework is described in Algorithm I-1, which learns about the singularities of the solution using small $\tilde{N}$. The M captured singularities σ are utilized to initialize an FODE solver in Stage-II (see Section 7.3.4).

Algorithm I-2 summarizes the main steps for the numerical scheme to capture the singularities of u^{data}, f^{data} for diminutive times at fixed M.

Computational Complexity of Stage-I

We recall that in the presented scheme, the error function (7.3.5) is evaluated N_{it} times, which is the number of iterations until convergence of the minimization scheme. The complexity of this error function is dominated by the time-integration of the numerical solution u^N over $\tilde{N}$ time-steps. For the first M time-steps, we use (7.3.31) and solve the corresponding linear system, which costs $\mathcal{O}(M^3)$. For the remaining $\tilde{N} - M$ time-steps we use (7.3.32), where the dominant cost is due to the direct numerical evaluation of the fractional derivatives, that is, $\mathcal{O}((\tilde{N} - M)^2)$. Therefore, the computational complexity of the entire scheme is $\mathcal{O}((M^3 + (\tilde{N} - M)^2)N_{it})$. However, here we use $M \leq 3$, and therefore the asymptotic complexity becomes $\mathcal{O}(\tilde{N}^2 N_{it})$. Furthermore, $\tilde{N}$

Algorithm I-1 Stage-I Self-Singularity-Capturing Scheme for FODEs.

```
Known data about u^data, f^data for Ñ time-steps of size Δt.
Set M = 1, initial guess σ^(0) = 0 and numerical tolerances ε, ε_1.
while M ≤ 3 do
   Estimate σ_i^(k), i = 1, ..., M using Algorithm I and obtain the error E(σ^(k)).
   if E(σ^(k)) < ε then
      The singularities were captured within the specified tolerance ε.
      break
   else
      Additional correction terms needed: Set M = M + 1.
      if M == 2 then
         Set initial guess σ^(0) = [0, σ_1^(k)]^T.
      else if M == 3 then
         Set initial guess σ^(0) = [σ_1^(k), σ_2^(k), (σ_1^(k) + σ_2^(k))/2]^T.
      end if
   end if
end while
return M, σ, E(σ).
```

Algorithm I-2 Stage-I Singularity-Capturing Algorithm (with M correction terms).

```
Initial guess σ^0, γ^0, compute W_{n,j} using (7.3.11) and E(σ^0) using (7.3.5), (7.3.31), (7.3.32).
while E(σ^k) > ε and ||∇E(σ^k)|| > ε_1 do
   Compute the perturbation σ^k + iΔσ e_j for j = 1, 2, ..., M.
   Compute ∇E(σ^k) using (7.3.8), (7.3.11), (7.3.5), (7.3.31), (7.3.32).
   Compute γ^k using (7.3.7) and update σ^{k+1} = σ^k − γ^k ∇E(σ^k).
end while
return σ, E(σ).
```

is assumed to be small due to the short-time data, and we show in Section 7.3.4 that we are able to capture singularities with $\tilde{N} = \mathcal{O}(M)$, which makes the presented scheme numerically efficient, as long as a large number of iterations for convergence is not required.

Remark 7.20 We observe that since $\tilde{N}$ is small, there is no need to use fast time-stepping schemes in Stage-I, since the break-even point between fast and direct schemes usually lies in a range of moderate number of time-steps (e.g., about $\mathcal{O}(10^4)$ for the fast scheme in Zeng et al. (2018a)). Therefore, for small $\tilde{N}$, fast schemes would decrease the performance of Stage-I, and would only be beneficial for Stage-II.

7.3.4 Stage-II: Integration of FODEs with the Captured Singularities

Once the multi-singular behavior of the solution is captured through the framework presented in Section 7.3.3 and Algorithm I, a variety of numerical methods for FODEs can be implemented (e.g., other finite-difference and fast convolution schemes), incorporating the captured singularities. Here, to solve the FODE (7.3.3), we use the developed implicit finite-difference scheme presented in Section 7.3.2, and given by expressions (7.3.31) and (7.3.32) for $t \in \Omega = [t_0, t_N]$, using N time-steps of size Δt (not necessarily the same time-step size as Stage-I). The solution procedure is described in Algorithm II.

The computational complexity of Stage-II depends exclusively on the employed numerical discretization for the FODE. We utilize the same time-integration scheme as in Stage-I, for N time-steps. Therefore, recalling (7.3.31) and (7.3.32), we have a complexity of $\mathcal{O}(M^3 + (N-M)^2)$. However, since $M << N$, the asymptotic computational complexity of Stage-II becomes $\mathcal{O}(N^2)$.

Numerical Results

We start with Stage-I with a computational analysis of the correction weights (see Section 7.3.4), followed by the capturing scheme for the particular case of a single time-step and correction term (see Section 7.3.5). Then, we test the singularity-capturing Algorithm I-2 for $M = 1, 2$ and the self-capturing Algorithm I for $M = 3$ (see Section 7.3.7). We also utilize the two-stage framework with random singularities and compare the accuracy of the entire time-integration framework with the captured singularities against ad hoc choices through a comparison between the obtained error functions $E(\sigma)$ (see Section 7.3.8). In all aforementioned tests, if not stated otherwise, we use the method of fabricated solutions, that is, we take $u^{data} = u^{ext}$ and $f^{data} = f^{ext}$, where we assume the following solution:

$$u^{ext}(t) = \sum_{j=1}^{S} t^{\sigma_j^*}, \quad \sigma_j^* \in \mathbb{R}^+, \tag{7.3.33}$$

Algorithm II Complete solution framework for FODEs.

Stage-I: Self-Singularity-Capturing
Given data for u^{data}, f^{data} for $\tilde{N}$ time-steps, capture σ using Algorithm I.
Stage-II: Numerical solution of the FODE
Let N be the total number of time-steps. Compute $W_{n,j}$ using (7.3.12).
Solve u_n^N for $n = 1, \ldots, M$ using (7.3.31).
for $n = 4$ until N **do**
  Solve for u_n^N using (7.3.32).
end for

with homogeneous initial condition $u^{ext}(0) = 0$. The term S denotes the number of singularities/terms in $u^{ext}(t)$, and σ_j^* represents prescribed singularities to be captured. From the defined analytical solution, we obtain the forcing term using a direct calculation, given by Podlubny (1999):

$$f^{ext}(t) = \sum_{j=1}^{S} \frac{\Gamma(1+\sigma_j^*)}{\Gamma(1+\sigma_j^* - \alpha)} t^{\sigma_j^* - \alpha}. \tag{7.3.34}$$

The numerical examples in this work are presented for verification of the developed framework, where u^{ext} and f^{ext} represent synthetic data for the Cauchy problem. In practice, only Stage-I of the framework utilizes experimental data for diminutive time to capture the multi-singularities, and Stage-II utilizes the captured singularities for the full time-integration of the FODE.

Furthermore, we test Stage-I of the framework for a multi-term FODE (see Section 7.3.8) and the two-stage framework for a one-term FODE with a *singular-oscillatory* solution (see Section 7.3.8). Finally, we test the developed framework against a nonlinear FODE and demonstrate its effectiveness when the solution data is sampled from a fine time-grid near the initial time (see Section 7.3.8). The developed algorithms were implemented in MATLAB R2018b, and were run on a laptop with Intel Core i7-8650U CPU with 1.90 GHz, 16 GB RAM and Windows 10 operating system.

Numerical Behavior of Correction Weights

We investigate the behavior of the initial correction weights $W_{n,j}$ and the condition number for the Vandermonde matrix V presented in (7.3.12). Figure 7.7a illustrates the condition numbers for $V_{M\times M}$ using different choices for σ_k, namely $\sigma_k = \alpha k$ when the regularity of $u(t)$ is known, and $\sigma_k = 0.1k$ when it is unknown. We also illustrate the behavior of the correction weights using $M = 1$ with respect to the time-step size Δt and fractional order α

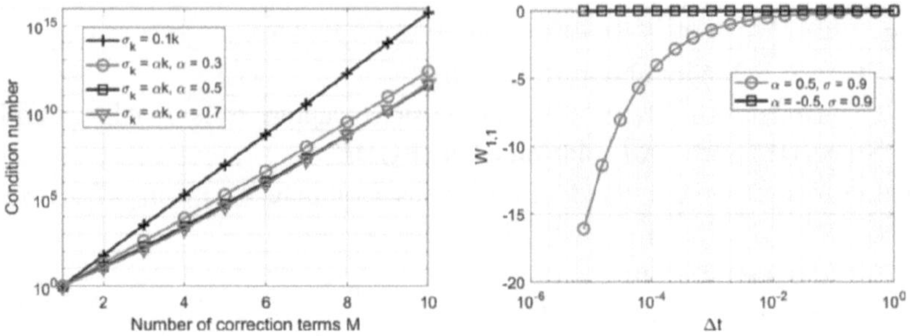

(a) Condition number of V vs M. (b) First correction weight $W_{1,1}$ vs Δt.

Figure 7.7 (a) Condition numbers for different choices of powers σ_k. We observe that the choice $\sigma_k = 0.1k$ for unknown regularity of $u(t)$ (Zeng et al., 2018a) leads to the highest condition number, where at most $M = 9$ correction terms can be used with double precision arithmetic. (b) First correction weight $W_{1,1}$ with respect to Δt. Source: Suzuki and Zayernouri (2021), reprinted by permission of the publisher (Taylor & Francis Ltd, www.tandfonline.com).

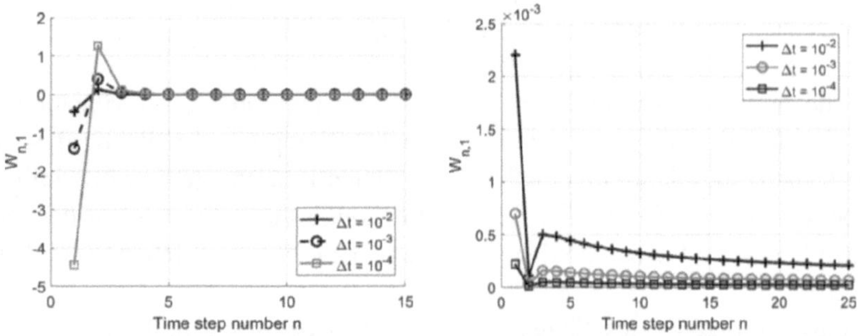

(a) $\alpha = 0.5$, $\sigma_k = 0.9$. (b) $\alpha = -0.5$, $\sigma_k = 0.9$.

Figure 7.8 Initial weights $W_{1,n}$ for different time-step sizes Δt, with corrections for fractional differentiation/integration (a) $\alpha = 0.5$ (fractional differentiation), where the initial weights have a larger magnitude that quickly decreases in the first time-steps, indicating that they are most relevant near $t = 0$. (b) $\alpha = -0.5$ (fractional integral). Source: Suzuki and Zayernouri (2021), reprinted by permission of the publisher (Taylor & Francis Ltd, www.tandfonline.com).

(corresponding to fractional differentiation/integration) in Figures 7.7b and 7.8. We observe that for $\alpha = 0.5$ (fractional differentiation), $W_{1,1}$ increases in magnitude as Δt decreases. On the other hand, for $\alpha = 0.5$ (fractional integration), it starts with $W_{1,1} \approx 0.2$ for $\Delta t = 1$, and decreases with Δt. For fractional integration, only positive correction weights are observed, that decrease slower with respect to n, but with all values decreasing to zero as Δt decreases.

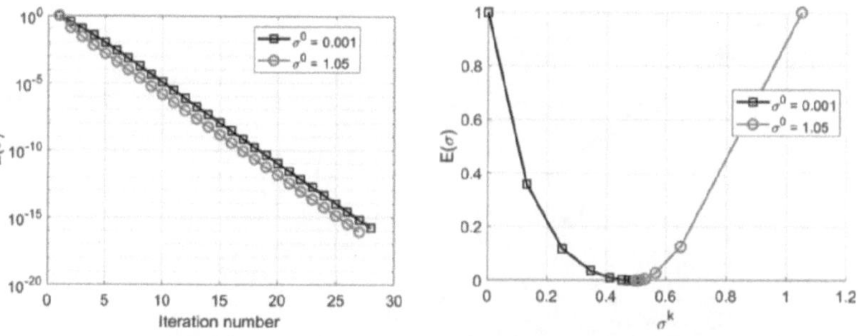

(a) Error vs iteration number. (b) Error vs iterated singularity

Figure 7.9 Convergence behavior with $\sigma^* = 0.5$ and initial guesses $\sigma^{(0)} = \{0.0001, 1.05\}$ and time-step size $\Delta t = 0.01$. Source: Suzuki and Zayernouri (2021), reprinted by permission of the publisher (Taylor & Francis Ltd, www.tandfonline.com).

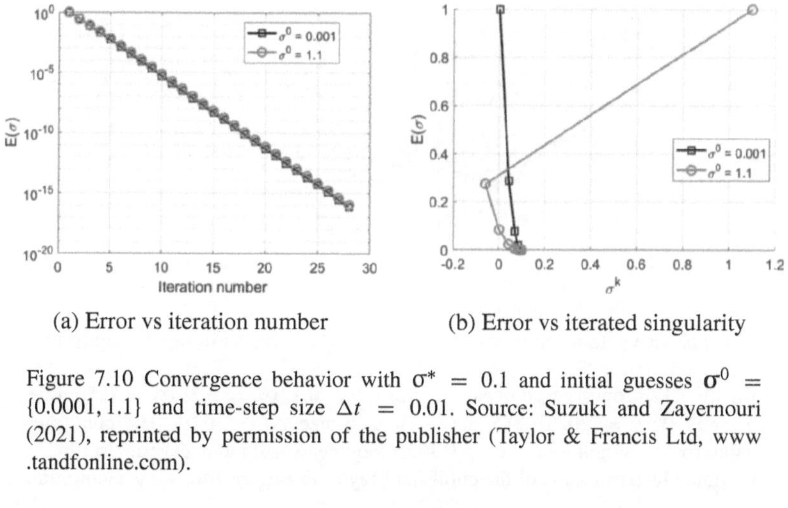

(a) Error vs iteration number (b) Error vs iterated singularity

Figure 7.10 Convergence behavior with $\sigma^* = 0.1$ and initial guesses $\sigma^0 = \{0.0001, 1.1\}$ and time-step size $\Delta t = 0.01$. Source: Suzuki and Zayernouri (2021), reprinted by permission of the publisher (Taylor & Francis Ltd, www.tandfonline.com).

7.3.5 Single Time-Step and Correction Term

We present the numerical results for a particular case of Stage-I in Section 7.3.6 for $S = 1$ singularity and $M = 1$ correction term, with a closed-form for the correction weights. Let the time domain $\tilde{\Omega} = [0, \Delta t]$, with fractional-order and time-step size kept constant, respectively, at $\alpha = 0.5$ and $\Delta t = 0.01$ in this section. Figure 7.9 presents the convergence behavior for $\sigma_1^* = 0.5$, and Figure 7.10 illustrates the obtained results using $\sigma_1^* = 0.1$. We observe that in both cases, the scheme captures the power σ_1^* from the analytical solution, with a relatively small number of iterations. Furthermore, we observe

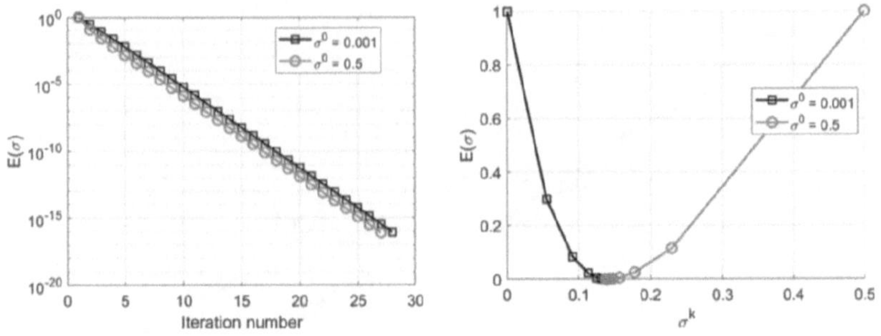

(a) Error vs iteration number (b) Error vs iterated singularity

Figure 7.11 Convergence behavior with $\sigma_1^* = 0.1$, $\sigma_2^* = 0.2$, initial guesses $\boldsymbol{\sigma}^{(0)} = \{0.001, 0.5\}$ and time-step size $\Delta t = 0.01$. The converged value for the singularity $\boldsymbol{\sigma} \approx 0.1377$ is between σ_1^*, σ_2^*. Source: Suzuki and Zayernouri (2021), reprinted by permission of the publisher (Taylor & Francis Ltd, www.tandfonline.com).

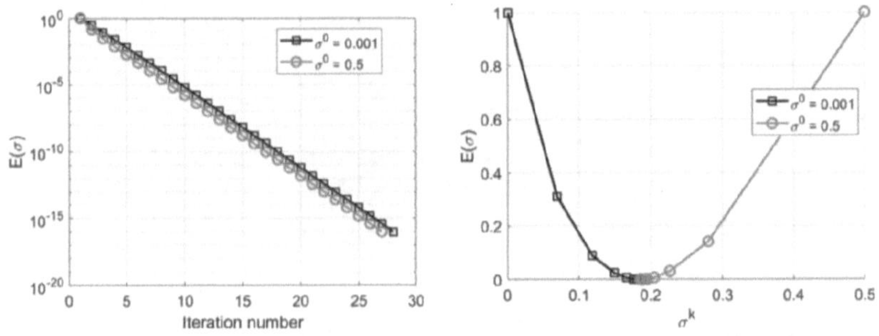

(a) Error vs iteration number (b) Error vs iterated singularity

Figure 7.12 Convergence behavior with $\sigma_1^* = 0.1$, $\sigma_2^* = 0.3$, $\sigma_3^* = 0.5$, initial guesses $\boldsymbol{\sigma}^{(0)} = \{0.001, 0.5\}$ and time-step size $\Delta t = 0.01$. The converged value for the singularity is $\sigma \approx 0.1856$. Source: Suzuki and Zayernouri (2021), reprinted by permission of the publisher (Taylor & Francis Ltd, www.tandfonline.com).

an overshooting in the iterative procedure for initial guess $\boldsymbol{\sigma}^{(0)} = 1.1$, but nevertheless, the scheme still converges within machine precision. Figures 7.11 and 7.12 show, respectively, the results for $S = 2$ with $\sigma_1^* = 0.1$, $\sigma_2^* = 0.2$ and $S = 3$ with $\sigma_1^* = 0.1$, $\sigma_2^* = 0.3$, $\sigma_3^* = 0.5$. We observe that the converged values for $\boldsymbol{\sigma}$ are closer to σ_1^*, which is the most critical singularity for the chosen $u^{ext}(t)$.

Since the converged values for $\boldsymbol{\sigma}$ captured intermediate values between the defined singularities $\sigma_1^*, \sigma_2^*, \sigma_3^*$ for the choice of $\Delta t = 0.01$, we analyze the

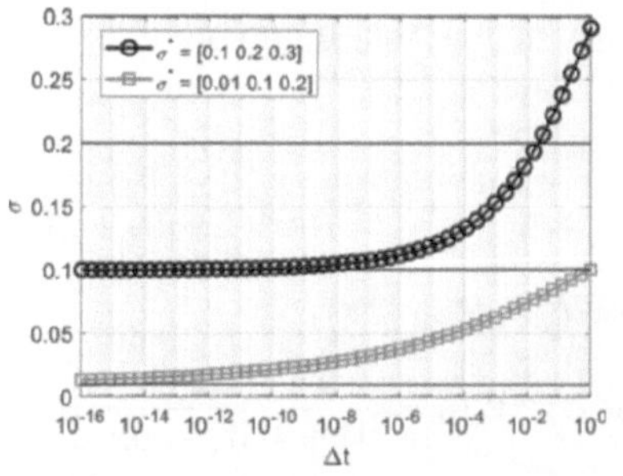

Figure 7.13 Converged values for the singularity σ vs time-step size Δt with two choices of $\sigma^* = \left[\sigma_1^*,\ \sigma_2^*,\ \sigma_3^*\right]^T$. Source: Suzuki and Zayernouri (2021), reprinted by permission of the publisher (Taylor & Francis Ltd, www.tandfonline .com).

convergence behavior of σ with respect to Δt. We present the obtained results in Figure 7.13 for two different sets of singular values, and we observe for the defined range of Δt, that σ lies between the singularities σ_1^*, σ_2^*, σ_3^*, and converge to the strongest singularity (in this case σ_1^*) as $\Delta t \to 0$.

7.3.6 Singularity-Capturing Formulation for a Single Correction Term and Time-Step

Let the time domain $\tilde{\Omega} = [0, \Delta t]$, and let $E\colon \mathbb{R} \to \mathbb{R}^+$ be the following quadratic error function over σ:

$$E(\sigma) = \|u^{data}(t) - u^N(t;\sigma)\|^2_{L^2(\tilde{\Omega})}, \tag{7.3.35}$$

where we assume $u^{data}(t)$ to be known, and $u^N(t;\sigma)$ represents the numerical approximation of $u^{data}(t)$, with $\sigma \in \mathbb{R}$ to be determined, such that $E(\sigma)$ is minimized. Therefore, let the case for $M = 1$ correction term and for the first time-step $t_1 = \Delta t$. We obtain the following cost function:

$$E(\sigma) = \left(u_1^{data} - u_1^N(\sigma)\right)^2 = \left(\Delta t^{\sigma^*} - u_1^N(\sigma)\right)^2, \tag{7.3.36}$$

where $u_1^N(\sigma)$ denotes the numerical solution for the FODE (7.3.3) at $t = \Delta t$, obtained through (7.3.13) and (7.3.34) as follows:

$${}^{\mathrm{RL}}_{\ 0}D_t^{\alpha} u^N(t)\big|_{t=\Delta t} + W_{1,1}\left(u_1^N - u_0\right) = \sum_{j=1}^{S} \frac{\Gamma(1+\sigma_j^*)}{\Gamma(1+\sigma_j^*-\alpha)} \Delta t^{\sigma_j^*-\alpha}. \tag{7.3.37}$$

Recalling (7.3.11), we obtain the initial correction weight $W_{1,1}$ by the equation

$$ {}^{\text{RL}}_{0}D^{\alpha}_{t}\left(t^{\sigma_1}\right)\big|_{t=\Delta t} + W_{1,1}\left(t_1^{\sigma}\right) = \frac{\Gamma(1+\sigma_1)}{\Gamma(1+\sigma_1-\alpha)} t_1^{\sigma_1-\alpha}. \tag{7.3.38} $$

Recalling (7.3.16) with $p = 1$,

$$ d_1^{(1)}\Delta t^{\sigma_1} + \Delta t^{\sigma_1} W_{1,1} = \frac{\Gamma(1+\sigma_1)}{\Gamma(1+\sigma_1-\alpha)}\Delta t^{\sigma_1-\alpha}, \tag{7.3.39} $$

therefore, we obtain the closed form for the correction weight:

$$ W_{1,1} = \frac{\Gamma(1+\sigma_1)}{\Gamma(1+\sigma_1-\alpha)}\Delta t^{-\alpha} - d_1^{(1)}. \tag{7.3.40} $$

Substituting (7.3.40) into (7.3.37), rewriting the RL derivative in its discretized form at $t = \Delta t$, and assuming homogeneous initial conditions $u_0 = 0$, we obtain

$$ d_1^{(1)}u_1^N + \left(\frac{\Gamma(1+\sigma_1)}{\Gamma(1+\sigma_1-\alpha)}\Delta t^{-\alpha} - d_1^{(1)}\right)u_1^N = f_1^{data}; \tag{7.3.41} $$

hence,

$$ u_1^N = \Delta t^{\alpha}\frac{\Gamma(1+\sigma_1-\alpha)}{\Gamma(1+\sigma_1)} f_1^{data}. \tag{7.3.42} $$

Substituting (7.3.42) into (7.3.36), we obtain, for $t = \Delta t$,

$$ E(\sigma_1) = \left(u_1^{data} - \Delta t^{\alpha}\frac{\Gamma(1+\sigma_1-\alpha)}{\Gamma(1+\sigma_1)} f_1^{data}\right)^2. \tag{7.3.43} $$

We assume that the above equation has a root at $\sigma_1 = \sigma^*$, that is, $E(\sigma^*) = 0$. Therefore, given an initial guess $\sigma_1^{(0)}$, we can apply a Newton scheme to iteratively obtain σ_1^{k+1} in the following fashion:

$$ \sigma_1^{k+1} = \sigma_1^k - \left\{E(\sigma_1)/\frac{\partial E(\sigma_1)}{\partial \sigma_1}\right\}\Big|_{\sigma_1=\sigma_1^k}, \tag{7.3.44} $$

where $\frac{\partial E(\sigma_1)}{\partial \sigma_1}$ can be obtained analytically, and is given by

$$ \frac{\partial E(\sigma_1)}{\partial \sigma_1} = -2\psi^*\Delta t^{\alpha}\frac{\Gamma(1+\sigma_1-\alpha)}{\Gamma(1+\sigma_1)}\left(u_1^{data} - \Delta t^{\alpha}\frac{\Gamma(1+\sigma_1-\alpha)}{\Gamma(1+\sigma_1)}\right), \tag{7.3.45} $$

with $\psi^* = \psi_0(1+\sigma_1-\alpha) - \psi_0(1+\sigma_1)$, where $\psi_0(z)$ denotes the polygamma function. We observe that the computational cost of the above procedure per iteration k is minimal, since there is no history computation when evaluating the corresponding time-fractional derivatives of u_1^N.

7.3.7 Singularity Capturing for M Correction Terms

We systematically test the capturing Algorithm I-2 for multiple correction terms $M = 1, 2, 3$ and multiple defined singularities $S = 1, 2, 3$ in u^{ext}. The tests are presented in an incremental fashion for M, where we show the capabilities of Algorithm I-2 to capturing/approximating the singularities. We then demonstrate how the self-capturing framework defined in Algorithm I successfully determines the singularities for $S = 3$ and $M = 3$, where we compare the obtained errors with ad hoc choices for σ using random strong singularities. Throughout this section, we keep the fractional order fixed at $\alpha = 0.5$, as well as short time-domain $\tilde{\Omega} = [0, 1]$, and perturbation $\Delta\sigma = 10^{-14}$ for the complex step differentiation. For all cases, we define a tolerance $\epsilon = 10^{-15}$ for $E(\sigma)$ and unless stated otherwise, $\epsilon_1 = 10^{-14}$ for $||\nabla E(\sigma)||$. We choose a smaller tolerance for the norm of the error gradient (since the $E(\sigma)$ is defined with the norm $||.||_2^2$) to make sure that $E(\sigma)$ is always minimized before we introduce additional correction terms in the self-capturing Stage-I. We also use $\gamma^0 = 10^{-3}$ for initialization in Algorithm I-2. For cases where $M = S$, we compute the component-wise relative error of the converged σ^k, which we define as

$$E_j^\sigma = \frac{|\sigma_j^* - \sigma_j^{(k)}|}{|\sigma_j^*|}, \quad j = 1, \ldots, M, \tag{7.3.46}$$

where k denotes the iteration number when convergence is achieved, i.e., $E(\sigma^{(k)}) < \epsilon$.

One Correction Term

We consider $M = 1$ and $\tilde{N} = 100$ initial data-points. Figures 7.14, 7.15 and 7.16 present the obtained results, respectively, for $S = 1, 2, 3$ singularities, using Algorithm I-2. We observe that, when $S = M$, we can capture singularities within machine precision. When $M < S$, we are still able to find a minimum for $E(\sigma)$ with an intermediate value for σ_1, similar to the results obtained for the simplified case in Section 7.3.5. Furthermore, when $S \geq 2$, or the initial guess $\sigma^{(0)}$ is far from the true singularity, the iterative procedure assumes the typical zig-zag behavior of the gradient descent scheme.

Two Correction Terms

We consider $M = 2$, where we first analyze the behavior of the error function $E(\sigma)$ using $S = 2$ singularities (see Figure 7.17). Then, we use Algorithm I-2 to capture the singularities $\sigma^* = [0.1, 0.3]^T$ (see Figure 7.18), where we observe a more pronounced zig-zag behavior for convergence, compared

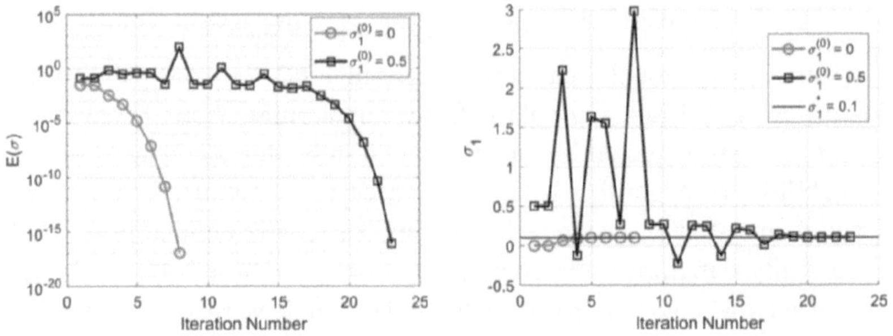

(a) Error vs iteration number (b) σ vs iteration number

Figure 7.14 Second-order asymptotic convergence behavior for $M = 1$ correction term, $S = 1$ singularity with value $\sigma_1^* = 0.1$. The obtained relative error for the converged σ_1 is $E_1^{\sigma} = 2.290 \times 10^{-8}$. Source: Suzuki and Zayernouri (2021), reprinted by permission of the publisher (Taylor & Francis Ltd, www.tandfonline .com).

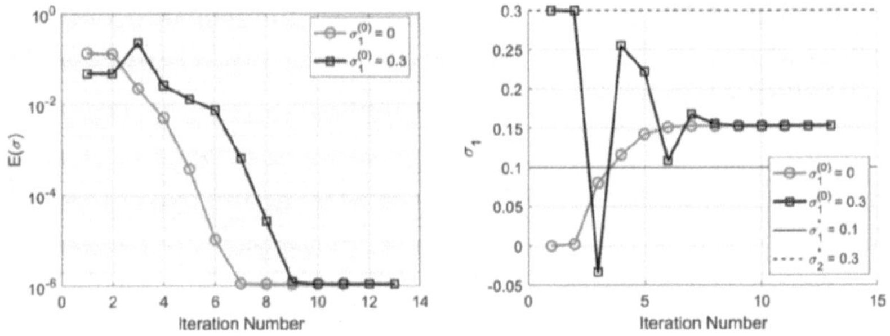

(a) Error vs iteration number. (b) σ vs iteration number.

Figure 7.15 Convergence behavior for $M = 1$, $S = 2$ with $\sigma^* = [0.1,\ 0.3]^T$. With the plateau in error and σ due to gradient convergence within the tolerance ϵ_1. The scheme converges to an "*intermediate*" value $\sigma \approx 0.153$, between the true singularities. Source: Suzuki and Zayernouri (2021), reprinted by permission of the publisher (Taylor & Francis Ltd, www.tandfonline.com).

to $M = 1$. We also obtain estimates of σ_1, σ_2 when dealing with $S = 3$ singularities, namely $\sigma^* = [0.1,\ 0.3,\ 0.5]^T$ (see Figure 7.19). Of particular interest, we observe that the obtained singularities lie in the range of the true singularities, which are useful observations to define initial guesses for cases with three correction terms and singularities.

Three Correction Terms

We test the case for $M = 3$, using $S = 1, 2, 3$. In this analysis, we use $\tilde{N} = 3$ initial data points. Figures 7.20 and 7.21 illustrate, respectively, the

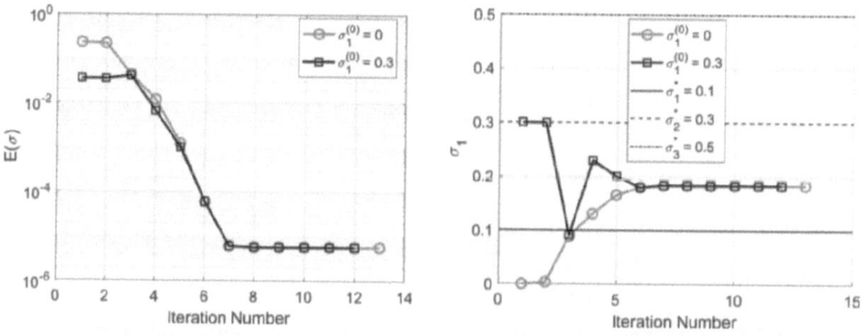

(a) $\boldsymbol{\sigma}$ vs iteration number. (b) $\boldsymbol{\sigma}$ vs iteration number.

Figure 7.16 Convergence behavior for $M = 1$, $S = 3$ with $\boldsymbol{\sigma}^* = [0.1,\ 0.3,\ 0.5]^T$. For both initial guesses, the scheme converges to an "*intermediate*" value $\boldsymbol{\sigma} \approx 0.184$, which is closer to the strongest true singularity $\sigma_1^* = 0.1$. Source: Suzuki and Zayernouri (2021), reprinted by permission of the publisher (Taylor & Francis Ltd, www.tandfonline.com).

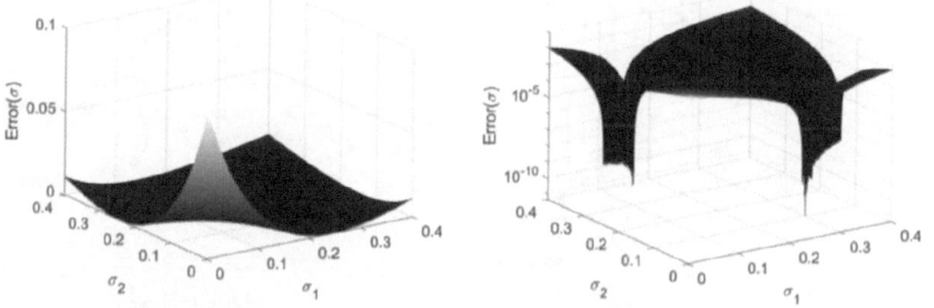

(a) Linear error-axis. (b) Logarithmic error-axis.

Figure 7.17 Error function $E(\boldsymbol{\sigma})$ for $M = 2$, $S = 2$, with $\sigma_1^* = 0.1$, $\sigma_2^* = 0.3$. Two minima occur at $\boldsymbol{\sigma}^* = [0.1,\ 0.3]^T = [0.3,\ 0.1]^T$, corresponding to the true singularities. We set $\sigma_1 \neq \sigma_2$ to avoid a singular matrix in linear system (7.3.11). Source: Suzuki and Zayernouri (2021), reprinted by permission of the publisher (Taylor & Francis Ltd, www.tandfonline.com).

obtained results for $S = 1, 2$ using Algorithm I-2. We observe a quadratic convergence rate for $S = 1$, similar to Figure 7.14, but with a few more iterations. Furthermore, for $M = 2$, we observed a small change in the final value of σ_3, but nevertheless, capturing the two singularities σ_1^*, σ_2^* is sufficient to minimize $E(\boldsymbol{\sigma})$ when $S = 2$.

The numerical convergence of Algorithm I-2 towards the correct singularities $\boldsymbol{\sigma}^*$ becomes much more difficult when using $S = M = 3$. Therefore, we make use of the self-capturing approach summarized in Algorithm I, which incrementally solves the minimization problem using $M = 1, 2, 3$ and use the

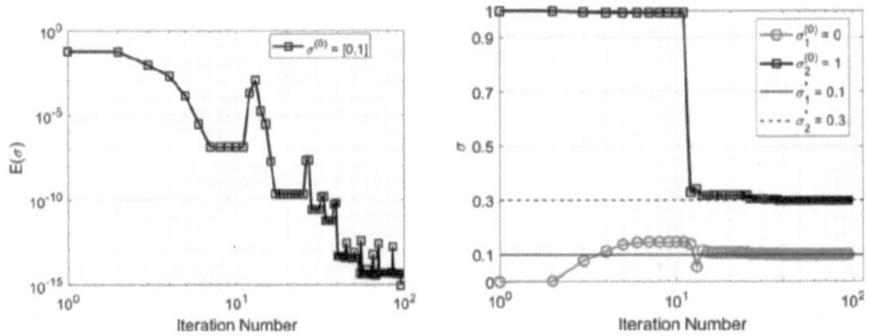

(a) Error vs iteration number. (b) $\boldsymbol{\sigma}$ vs iteration number.

Figure 7.18 Two correction terms $M = 2$ and two singularities $S = 2$, with $\boldsymbol{\sigma}^* = [0.1,\ 0.3]^T$. The obtained converged values are $\boldsymbol{\sigma} = [0.299974,\ 0.099990]^T$, which correspond to component-wise relative errors $E^{\sigma} = [8.61,\ 9.94]^T \times 10^{-5}$. Source: Suzuki and Zayernouri (2021), reprinted by permission of the publisher (Taylor & Francis Ltd, www.tandfonline.com).

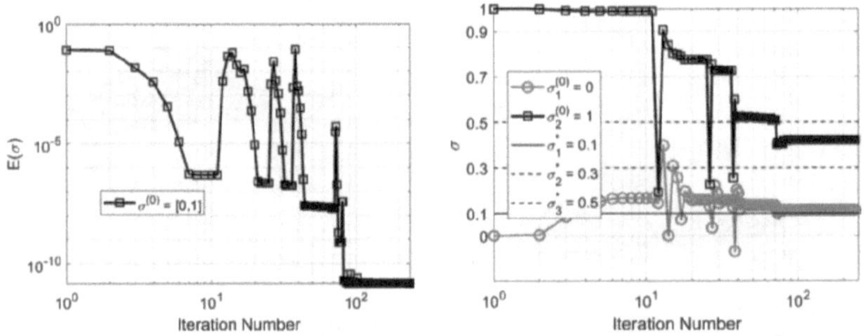

(a) Error vs iteration number. (b) $\boldsymbol{\sigma}$ vs iteration number.

Figure 7.19 Two correction terms $M = 2$ and three singularities $S = 3$, with $\boldsymbol{\sigma}^* = [0.1,\ 0.3]^T$ and $\epsilon_1 = 10^{-11}$. The converged values are $\boldsymbol{\sigma} = [0.112635,\ 0.420495]^T$, which are closer to the stronger singularities $\sigma_1 = 0.1$ and $\sigma_3 = 0.5$. The approximate singularities still lead to an approximation error of 1.47×10^{-11}. Source: Suzuki and Zayernouri (2021), reprinted by permission of the publisher (Taylor & Francis Ltd, www.tandfonline.com).

respective output singularities as initial guesses for the subsequent number of correction terms $M = M + 1$. We present the obtained results in Figure 7.22. We observe that the scheme converges to the true singularities with errors for σ up to 0.024, but nevertheless, they yield an approximation error for u^N of 10^{-15}.

We also test the influence of the number of initial data points $\tilde{N}$ in Stage-I over the number of iterations N_{it} and number of correction terms M. For this purpose, we set $S = 3$ singularities with values $\boldsymbol{\sigma}^* = [0.1,\ 0.3,\ 0.5]^T$

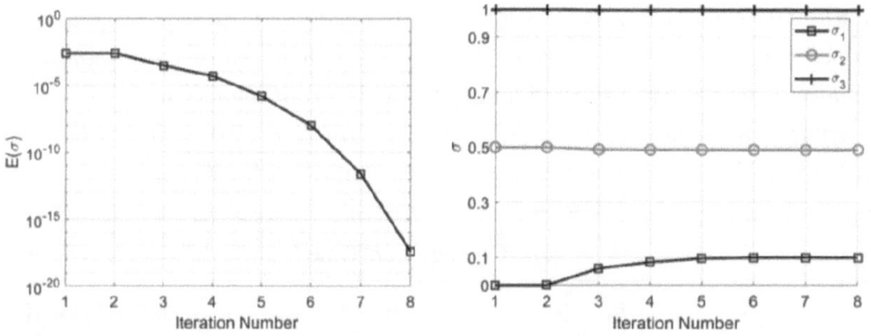

(a) Error vs iteration number. (b) σ vs iteration number.

Figure 7.20 Results for $M = 3$, $S = 1$. The obtained relative error for σ_1 is $E_1^\sigma = 4.869 \times 10^{-8}$. The components σ_1, σ_2 remained practically constant. Source: Suzuki and Zayernouri (2021), reprinted by permission of the publisher (Taylor & Francis Ltd, www.tandfonline.com).

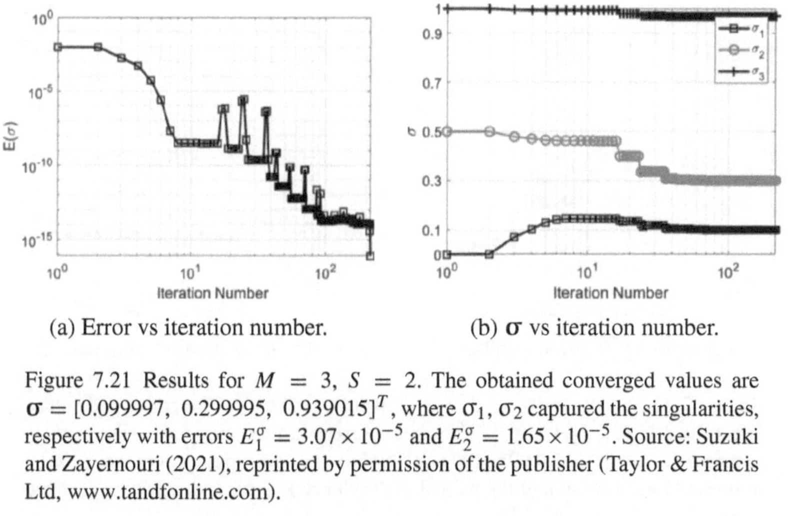

(a) Error vs iteration number. (b) σ vs iteration number.

Figure 7.21 Results for $M = 3$, $S = 2$. The obtained converged values are $\sigma = [0.099997,\ 0.299995,\ 0.939015]^T$, where σ_1, σ_2 captured the singularities, respectively with errors $E_1^\sigma = 3.07 \times 10^{-5}$ and $E_2^\sigma = 1.65 \times 10^{-5}$. Source: Suzuki and Zayernouri (2021), reprinted by permission of the publisher (Taylor & Francis Ltd, www.tandfonline.com).

and we fix $\Delta t = 0.1\,[s]$, $\alpha = 0.5$, $\epsilon = 10^{-13}$, $\epsilon_1 = 10^{-10}$. We then vary $\tilde{N}$ in Algorithm I and analyze the convergence behavior and computational time, which are presented in Table 7.10. We observe that the variation of $\tilde{N}$ did not affect the number of captured singularities M, but the choice $\tilde{N} = 3^7$ significantly increased the number of iterations. We also observe that despite the similar number of iterations N_{it} obtained for $3^1 \leq \tilde{N} \leq 3^6$, the computational time increases with $\tilde{N}$ due to the computational complexity of Stage-I. Therefore, we consider $\tilde{N} = \mathcal{O}(M)$ an appropriate choice, as the

Table 7.10. *Influence of $\tilde{N}$ in Stage-I of the developed framework. Source: Suzuki, Jorge L, and Zayernouri, Mohsen. 2021. A self-singularity-capturing scheme for fractional differential equations. International Journal of Computer Mathematics, 98(5), 933–960, reprinted by permission of the publisher (Taylor & Francis Ltd, http://www.tandfonline.com).*

$\tilde{N}$	M	N_{it}	Computation Time $[s]$	$E(\boldsymbol{\sigma})$
3^1	3	74	0.217	9.863015×10^{-14}
3^2	3	44	0.293	2.101287×10^{-14}
3^3	3	51	0.617	9.433562×10^{-14}
3^4	3	62	2.251	9.037342×10^{-14}
3^5	3	98	11.346	9.101754×10^{-14}
3^6	3	52	180.719	1.550662×10^{-14}
3^7	3	445	1275.168	8.823108×10^{-14}

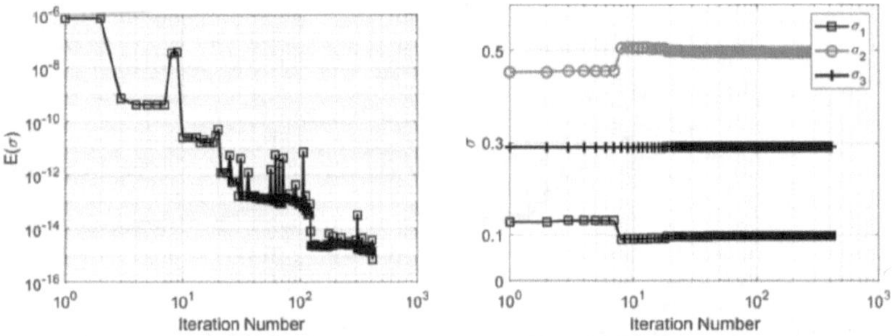

(a) Error vs iteration number. (b) $\boldsymbol{\sigma}$ vs iteration number.

Figure 7.22 Results for $M = 3$, $S = 3$ using $\epsilon = 10^{-15}$. The converged values are $\boldsymbol{\sigma} = [0.099,\ 0.498,\ 0.293]^T$, which correspond to component-wise relative errors $E^{\sigma} = [0.014,\ 0.005,\ 0.024]^T$. Source: Suzuki and Zayernouri (2021), reprinted by permission of the publisher (Taylor & Francis Ltd, www.tandfonline .com).

singularities $\boldsymbol{\sigma}^*$ are captured with less computational cost, which is suitable when diminutive initial data is available.

7.3.8 Several Random Singularities without Prior Knowledge

We perform two numerical tests involving random power-law singularities, where we define three strong singularities σ_1^*, σ_2^*, σ_3^* randomly sampled from a uniform distribution $\mathcal{U}(0, 1/2)$. We employ the self-singularity-capturing procedure in the context of single- and multi-term FODEs.

FODEs with Random Singularities

We test the two-stage framework for efficient time-integration of (7.3.3). We then compute $u^{ext}(t)$ and $f^{ext}(t)$, respectively using (7.3.33) and (7.3.34), and use the self-capturing framework presented in Algorithm I. We then compare the approximation error $E(\sigma)$ with the choice $\sigma_k = 0.1k$ defined in (Zeng et al., 2018a) when the singularities are unknown. Although our framework has no explicit information about the generated random singularities, we present them for verification purposes, which are given by

$$\sigma^* = [0.0172230402514543,\;\; 0.219372179828199,\;\; 0.190779228546504]^T.$$

We set $\tilde{N} = 3$ with $\Delta t = 1/3$, and therefore, $\tilde{\Omega} = [0, 1]$, $\epsilon = 10^{-15}$, and $\epsilon_1 = 10^{-13}$ to Algorithm II. Stage-I of the framework (Algorithm I) converges with $E(\sigma^{(k)}) = 2.96 \times 10^{-16}$ with $M = 2$ correction terms, with obtained values:

$$\sigma = [0.0187990387914248,\;\; 0.206944449676742]^T, \qquad (7.3.47)$$

with elapsed computational time of $0.169\,[s]$. After capturing the singularities within the desired precision ϵ, we enter Stage-II and compute the numerical solution $u^N(t)$ using the captured σ for multiple, longer time-integration domains Ω. We remark that when using an ad hoc choice of fixed $\sigma_k = 0.1k$ with 4 correction terms, that is,

$$\sigma = [0.1,\;\; 0.2,\;\; 0.3,\;\; 0.4]^T, \qquad (7.3.48)$$

we obtain an approximation error $E(\sigma) = 5.25 \times 10^{-5}$ over $N = 4$ time-steps. We use the captured powers (7.3.47) and the predefined ones (7.3.48) for longer time-integration. The obtained results are presented in Figure 7.23, where we observe how using 3 time-steps to capture the singularities leads to precise long time-integration with a large time-step size $\Delta t = 1/3$ (solid black curve). On the other hand, the errors for $\sigma_k = 0.1k$ are much larger, and also do not improve with smaller Δt due to the presence of very strong singularities.

Multi-Term FODEs with Random Singularities

In this section we show how the scheme in Stage-I can be applied for multi-term fractional differential equations. Let the multi-term FODE

$$\sum_{l=1}^{N_\alpha} {}^{\mathrm{RL}}_{0}D_t^{\alpha_l} u(t) = f(t, u(t)), \quad u(0) = u_0, \quad 0 < \alpha_l < 1, \qquad (7.3.49)$$

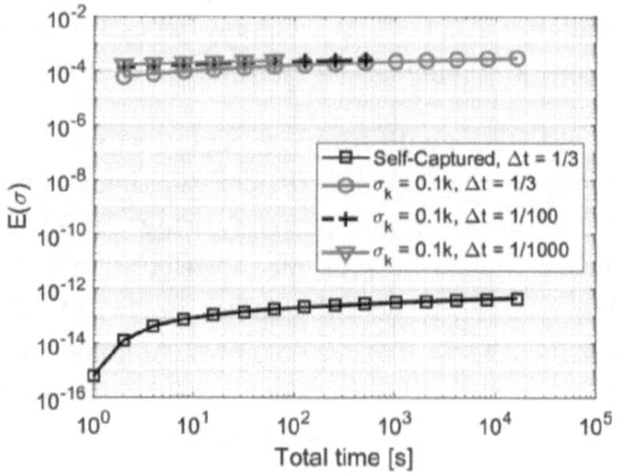

Figure 7.23 Comparison of error vs time for long time-integration between the captured singularities and $\sigma_k = 0.1k$ for different values of time-step size Δt. Source: Suzuki and Zayernouri (2021), reprinted by permission of the publisher (Taylor & Francis Ltd, www.tandfonline.com).

where α_l denotes the N_α multi-fractional orders with $l = 1, \ldots N_\alpha$. Using the approximation (7.3.10), and assuming the same powers σ for the correction weights for all fractional derivatives, we obtain

$$\sum_{l=1}^{N_\alpha} {}^{\text{RL}}_{0}D_t^{\alpha_l} u^N(t)\Big|_{t=t_n} + \sum_{j=1}^{M}\sum_{l=1}^{N_\alpha} W_{j,n}^{(\alpha_l)}(\sigma)\left(u_j^N - u_0\right) = f_n^\delta, \tag{7.3.50}$$

where we use the superscript (α_l) in $W_{j,n}^{(\alpha_l)}$ to denote the distinct sets of correction weights due to the fractional order α_l through (7.3.12). The entire minimization scheme is identical, since it depends on the solution data for $u(t)$. In order to solve (7.3.50) with the developed scheme in Stage-I, due to the linearity of (7.3.49), we only need to replace the initial correction weights $W_{n,j}$, and α-dependent coefficients d_j, b_j in (7.3.31) and (7.3.32) with the following summations:

$$\tilde{W}_{n,j} = \sum_{l=1}^{N_\alpha} W_{n,j}^{(\alpha_l)}, \quad \tilde{d}_j^{(p)} = \sum_{l=1}^{N_\alpha} d_j^{(p,\alpha_l)}, \quad \tilde{b}_j^{(k)} = \sum_{l=1}^{N_\alpha} b_j^{(k,\alpha_l)}, \quad k = 1, 2, 3. \tag{7.3.51}$$

We utilize the fabricated solution (7.3.33) in (7.3.49), which yields the following force term:

$$f^{ext}(t) = \sum_{\alpha_l}^{N_\alpha}\sum_{j=1}^{S} \frac{\Gamma(1+\sigma_j^*)}{\Gamma(1+\sigma_j^* - \alpha_l)} t^{\sigma_j^* - \alpha_l}. \tag{7.3.52}$$

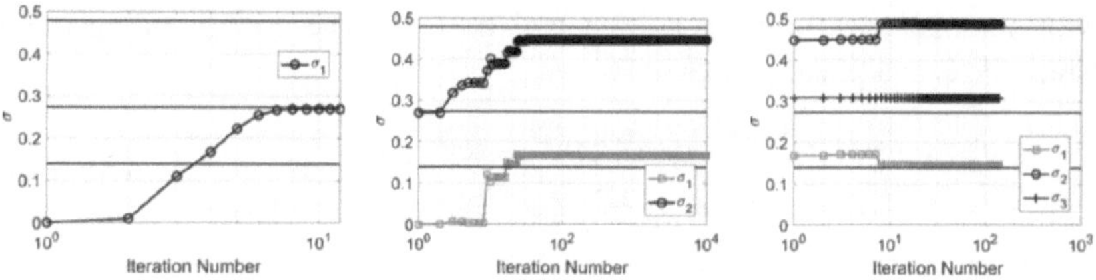

(a) $M = 1$ (b) $M = 2$ (c) $M = 3$

Figure 7.24 Singularities $\boldsymbol{\sigma}$ vs iteration number for a multi-term FODE with $N_\alpha = 3$ in Stage-I for each M in Algorithm I. (a) $M = 1$, with error $E(\boldsymbol{\sigma}) = 1.10 \times 10^{-5}$, (b) $M = 2$, with error $E(\boldsymbol{\sigma}) = 3.83 \times 10^{-13}$, and (c) $M = 3$ with convergence at $E(\boldsymbol{\sigma}) = 2.31 \times 10^{-15}$. The black horizontal lines correspond to the true random singularities in the solution. Source: Suzuki and Zayernouri (2021), reprinted by permission of the publisher (Taylor & Francis Ltd, www.tandfonline.com).

We set $N_\alpha = 3$ fractional orders $\{\alpha_1, \alpha_2, \alpha_3\} = \{0.3, 0.5, 0.7\}$, and identically to Section 7.3.8, we sample the following random singularities from $\mathcal{U}(0, 1/2)$:

$$\boldsymbol{\sigma}^* = [0.13924910943352420,\ 0.2734407596024919,\ 0.4787534177171488]^T.$$

Also, we set $\tilde{N} = 3$ with $\Delta t = 1/3$ and $\tilde{\Omega} = [0, 1]$, $\epsilon = 5 \times 10^{-15}$, and $\epsilon_1 = 10^{-14}$, with $\gamma^0 = 10^{-3}$ to Algorithm I, where the scheme converges with the following approximate singularities:

$$\boldsymbol{\sigma} = [0.1469249923105880,\ 0.4869203803691072,\ 0.3066386453671829]^T.$$

Figure 7.24 presents the obtained results for each M in the self-capturing procedure, where we observe that the scheme does not fully capture the true singularities, but provides sufficiently good approximations to obtain errors as low as $E(\boldsymbol{\sigma}) = 2.31 \times 10^{-15}$. The elapsed computational time is 10.65 $[s]$, due to the larger number of iterations N_{it} required for convergence of Stage-I.

Singular-Oscillatory Solutions

We also test the two-stage framework sampling data points from a fabricated solution for (7.3.3), defined as the following multiplicative coupling between a power-law and oscillatory parts:

$$u^{ext}(t) = t^{\sigma^*} \cos(\omega t), \qquad (7.3.53)$$

with homogeneous initial condition $u^{ext}(0) = 0$. The corresponding right-hand side is analytically obtained as

$$f^{ext}(t) = c_1 t^{\sigma^*-\alpha}\big[{}_p\tilde{F}_q(a^{(1)}; b^{(1)}; -\omega^2 t^2/4) \\ + c_3 \omega^2 t^2 {}_p\tilde{F}_q(a^{(2)}; b^{(2)}; -\omega^2 t^2/4)\big], \tag{7.3.54}$$

with

$$c_1 = -2^{\alpha-\sigma^*-4}\pi\Gamma(1+\sigma^*), \quad c_2 = 8\left(\alpha - \sigma^* - 1\right), \quad c_3 = (1+\sigma^*)(2+\sigma^*).$$

Also, ${}_p\tilde{F}_q$ represents the regularized, generalized hypergeometric function given by Weisstein (2003):

$${}_p\tilde{F}_q(a_1, \ldots, a_p; b_1, \ldots, b_q; z) = \frac{{}_pF_q(a_1, \ldots, a_p; b_1, \ldots, b_q; z)}{\Gamma(b_1)\ldots\Gamma(b_q)}, \tag{7.3.55}$$

where, for the particular case (7.3.54), we have

$$\begin{cases} a^{(1)} = \left\{\frac{(1+\sigma^*)}{2}, \frac{(2+\sigma^*)}{2}\right\}, & b^{(1)} = \left\{\frac{1}{2}, \frac{(2-\alpha+\sigma^*)}{2}, \frac{(3-\alpha+\sigma^*)}{2}\right\}, \\ a^{(2)} = \left\{\frac{(3+\sigma^*)}{2}, \frac{(4+\sigma^*)}{2}\right\}, & b^{(2)} = \left\{\frac{3}{2}, \frac{(4-\alpha+\sigma^*)}{2}, \frac{(3-\alpha+\sigma^*)}{2}\right\}. \end{cases}$$

We fix the fractional order $\alpha = 0.5$, frequency $\omega = 10\pi$ and randomly sample σ^* from a uniform distribution $\mathcal{U}(0, 1/3)$ and obtain a value $\sigma^* = 0.2426481954401539$. For Stage-I, we consider $\tilde{N} = 3$ data points with $\tilde{\Omega} = [0, 0.01]$, with $\Delta t = 3/100\,[s]$. Furthermore, we set $\epsilon = 10^{-11}$, $\epsilon_1 = 10^{-14}$ and $\gamma^0 = 10^{-3}$. Using Algorithm I with a slight initial guess modification for $\boldsymbol{\sigma}^{(0)}$ when $M = 2$ in line 11 to $\boldsymbol{\sigma}^{(0)} = [1, \sigma_1^{(k)}]^T$ (since the developed framework is mainly oriented to strong singularities), we capture the following two singularities:

$$\sigma = [0.2427452349772425, \;\; 2.220682758797950]^T,$$

where the first converged singularity is an approximation to the randomly sampled value for σ^*. To understand the second captured singularity σ_2, let the series expansion of (7.3.53) around ωt be

$$u^{ext}(t) = t^{\sigma^*}\cos(\omega t) = t^{\sigma^*}\left(1 - \frac{\omega^2}{2}t^2 + \mathcal{O}(t^4)\right) \\ = t^{\sigma^*} - \frac{\omega^2}{2}t^{2+\sigma^*} + \mathcal{O}(t^{4+\sigma^*}). \tag{7.3.56}$$

Therefore, the captured σ_2 is an approximation of the second term $2 + \sigma^*$ of the above series expansion. Although capturing the third singularity $4 + \sigma^*$ is

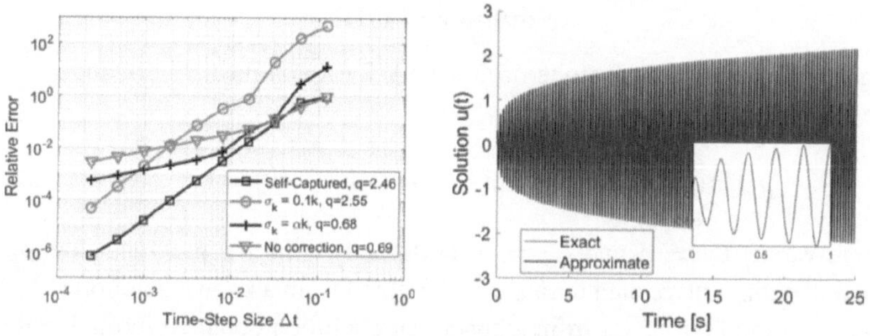

(a) Relative error vs time-step size. (b) Solutions for $u(t)$ vs time.

Figure 7.25 (a) Convergence analysis comparing the accuracy of the captured singularities in Stage-I with ad hoc choices for time-integration with $T = 1\,[s]$ in Stage-II. The term q denotes the convergence rate. The self-captured case uses $M = 2$ correction terms, while for $\sigma_k = 0.1k$ and $\sigma_k = \alpha k$ we use $M = 4$. (b) Exact and approximate solutions for $u(t)$ using the captured singularities for longer time-integration over $T = 25\,[s]$. Source: Suzuki and Zayernouri (2021), reprinted by permission of the publisher (Taylor & Francis Ltd, www.tandfonline .com).

possible, we would need to set higher values for the initial guesses in the self-capturing algorithm, and finding the corresponding minimum becomes more difficult, since such higher-order, weaker singularities have less influence in the accuracy of the solution.

After capturing the singularities in Stage-I, with elapsed computational time of $0.149\,[s]$, we apply them in Stage-II for a convergence test with $T = 1\,[s]$, with $\Omega = [0, T]$, and compare the approximate solution $u^N(t)$ with the exact one (7.3.56) through the relative error defined as

$$E_{L^2} = \frac{\|\boldsymbol{u}^N - \boldsymbol{u}^{ext}\|_{L^2(\Omega)}}{\|\boldsymbol{u}^{ext}\|_{L^2(\Omega)}}, \tag{7.3.57}$$

between the two-stage framework other ad hoc choices for σ, and also without corrections (see Figure 7.25a). We observe that capturing $M = 2$ singularities leads to the theoretical accuracy $\mathcal{O}(\Delta t^{3-\alpha})$ of the employed discretization, and also to errors about 2 orders of magnitude lower than the choice $\sigma_k = 0.1k$ with $M = 4$ for the singularities in the observed range for Δt. We also compare the exact and approximate solutions over $T = 25\,[s]$ with $N = 4096$ time-steps of size $\Delta t \approx 6.10 \times 10^{-3}$, and observe that we fully capture the singular and oscillatory behaviors of the solution, with elapsed computational time of $0.426\,[s]$.

Nonlinear FODE

Finally, we apply the developed framework for a nonlinear right-hand side of the FODE (7.3.3), now given by

$$ {}_{0}^{\mathrm{C}}D_{t}^{\alpha}u(t) = C_1 u(t)^2 \exp(-C_2\, t) + 1, \quad u(0) = 0, \tag{7.3.58}$$

where we set $C_1 = 1$ and $C_2 = 5$. Differently from the previous examples, where the diminutive data points were sampled from a known solution $u^{ext}(t)$, here, we sample $u^{data}(t)$ from a benchmark solution obtained through a fine-grid near $t = 0$. The right-hand side is evaluated explicitly. Our procedure follows:

- **I)** Set the short time-domain $\tilde{\Omega} = [0, 0.01]$ and numerically obtain the approximate solution of (7.3.58) using $\Delta t = 10^{-5}\,[s]$.
- **II)** Sample $\tilde{N} = 100$ benchmark solution data points and utilize the singularity-capturing algorithm (Stage-I) with $\Delta t = 10^{-4}$ and $\epsilon_1 = 10^{-14}$ over $\tilde{\Omega}$ to obtain σ. The value of ϵ is chosen according to the employed fractional order α.
- **III)** We utilize the captured singularities and the RHS of (7.3.58) evaluated explicitly for full time-integration (Stage-II) over $\Omega = [0, 1]$, using $\Delta t = 10^{-4}\,[s]$, and compare the results with the benchmark solution integrated over Ω.

We perform two tests: In the first, we set $\alpha = 0.2$, $\epsilon = 7.5 \times 10^{-5}$ in Stage-I, and capture $\sigma = 7.236455491751606 \times 10^{-2}$. In the second one, we set $\alpha = 0.25$, $\epsilon = 1.1 \times 10^{-5}$ and capture $\sigma = 7.602862102264105 \times 10^{-2}$. Figure 7.26 illustrates the obtained results for full time-integration. We observe that the self-captured solution matches the benchmark case for the entire time-integration domain, including near $t = 0$, demonstrating that the developed scheme is effective for nonlinear FODEs with unknown solutions. The obtained computational times for the self-capturing framework are: $0.07\,[s]$ for fine grid solution and sampling, $0.15\,[s]$ for Stage-I and $0.86\,[s]$ for Stage-II, yielding a total computational time of $1.08\,[s]$, compared to $48.58\,[s]$ for the benchmark solution. Therefore, the fine grid sampling at diminutive time had minimal influence on the computational cost.

In this section, we presented a two-stage framework for accurate and efficient solution of FODEs. It consisted of: **I)** A self-singularity-capturing scheme, where the inputs were limited data for diminutive time, and the output was the captured power-law singular values of the solution. The limited data can be either obtained experimentally or sampled from a fine time-grid near $t = 0$. We developed an implicit finite-difference algorithm

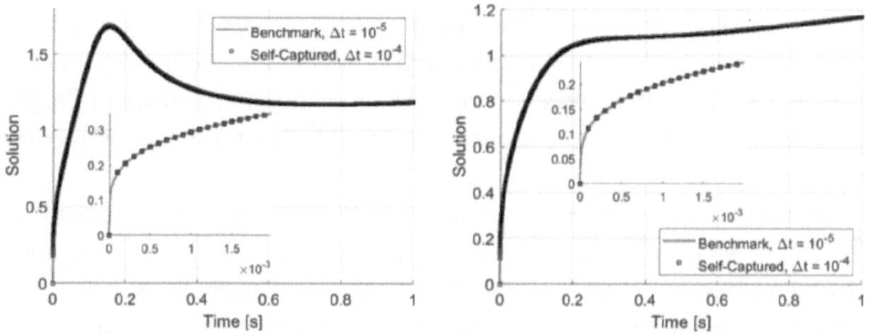

(a) $\alpha = 0.2$. (b) $\alpha = 0.25$.

Figure 7.26 Self-captured and benchmark solutions versus time for full time-integration of a nonlinear FODE. Source: Suzuki and Zayernouri (2021), reprinted by permission of the publisher (Taylor & Francis Ltd, www.tandfonline.com).

for the FODE solution and employed a gradient optimization scheme. **II)** The captured singularities were then inputs of an implicit finite-difference scheme for FODEs with accuracy $\mathcal{O}(\Delta t^{3-\alpha})$. Based on our numerical tests, we observed that:

- The scheme was able to fully capture S singularities using $M \geq S$ correction terms. When $S > M$, the scheme still obtained approximations of the M singularities, which was useful for error control.
- When $M = 1$, the most critical singularity was estimated by the scheme, regardless of S.
- The developed scheme was used in a self-singularity-capturing framework, which determined the singularity values up to the desired tolerance ϵ.
- Once the singularities are captured in the first stage, we can successfully employ any known numerical scheme for more accurate and efficient solution of the corresponding FODE, when compared to the ad hoc singularity choices done in (Zeng et al., 2018a).
- The scheme was able to capture two singularities from a solution using a strong power-law singularity coupled with an oscillatory smooth part, which allowed us to obtain the theoretical accuracy $\mathcal{O}(\Delta t^{3-\alpha})$ of the employed numerical discretization for the fractional derivative. Furthermore, by solving a nonlinear FODE, we demonstrate that the developed framework does not require the knowledge of any analytical solution.

The main advantage of the presented scheme over other works on correction methods (Lubich, 1986; Zeng et al., 2018a, 2017) is the self-capturing of strong multi-singularities σ of the solution data in Stage-I instead of ad hoc choices. This led to smaller solution errors while achieving the theoretical convergence rate of the employed numerical integration schemes in Stage-II, with algorithmically minimal correction terms M, and consequently requiring less computational cost.

The total computational complexity of the two-stage framework is $\mathcal{O}(\tilde{N}^2 N_{it} + N^2)$, since the number of correction terms M is small. Among a variety of numerical schemes, Stage-II of the framework could incorporate fast convolution frameworks (Lubich and Schädle, 2002; Zeng et al., 2018a) or a divide-and-conquer scheme (Fu et al., 2017). This would dramatically increase the efficiency for long time-fractional integration of anomalous materials under complex loading conditions (Suzuki et al., 2016, 2021, 2020; Barros de Moraes et al., 2021) and also further improve the numerical accuracy for nonlinear problems using fractional multi-step/IMEX schemes (Zayernouri and Matzavinos, 2016; Zhou et al., 2020b). The developed scheme can also be used to capture the power-law attenuations and creep-ringing (Jaishankar and McKinley, 2013) in highly-oscillatory vibration of anomalous systems. Also, discovering the multi-singular behavior of the solution can give insight on knowledge- and kernel-based grid refinement near $t = 0$ (Stynes et al., 2017). Furthermore, the presented scheme can leverage the development of self-constructing function spaces using Müntz polynomials (Esmaeili et al., 2011; Hou and Xu, 2017), increasing the efficiency of spectral element methods (Kharazmi and Zayernouri, 2019) applied to anomalous transport.

8
Applications of Anomalous Transport and Fractional Modeling

As highlighted in Chapter 1, anomalous transport phenomena can be observed in a wide variety of complex, multi-scale, and multi-physics systems such as: *sub-/superdiffusion in subsurface transport, kinetic plasma turbulence, aging polymers, glassy materials, in addition to amorphous semiconductors, biological cells, heterogeneous tissues, and fractal disordered media*, see e.g., Klages et al. (2008) and references therein. In this chapter, we focus on some selective applications of FPDEs and the methods presented in the previous chapters, reporting the scientific evidence of how and why fractional modeling naturally emerges in each case, along with a review of selected nonlocal mathematical models that have been proposed. The applications of interest are: (i) concentration transport in surface/subsurface dynamics, (ii) complex rheology and material damage, and (iii) fluid turbulence and geostrophic transport.

8.1 Concentration Transport in Surface/Subsurface Dynamics

The accurate prediction at large scales of contaminant transport in both surface and subsurface water is fundamental for the management of water resources and critical for environmental safety. However, the explicit description of the systems where transport takes place is extremely challenging, especially at large scales, due to the complexity of the medium. Such media feature heterogeneities that are either difficult or impossible to observe, and hence cannot be described with certainty at all relevant scales and locations. Even when the microstructure of the environment can be captured, simulations of appropriate differential models such as systems of advection–diffusion equations may be infeasibly expensive if conducted at fully-resolved small scales (Sun et al., 2020). In fact, the same types of equations that are accurate at small scales *do not* extrapolate and predict solutes' behavior at larger scales, due to the appearance of "anomalous", or "non-Fickian" behavior (Benson et al., 2000, 2001; Levy and Berkowitz, 2003; Neuman and Tartakovsky, 2009). At large scales, FPDEs are called for.

An upscaling model for groundwater flow and transport employs fractional derivatives. The fractional advection–dispersion equation (FADE) in one dimension is

$$ {}_{0}^{C}D_t^{\alpha} C(x,t) = -v\ \partial_x C(x,t) + \kappa\ {}_{-\infty}^{RL}D_x^{\beta} C(x,t). \tag{8.1.1} $$

Recall that the β-fractional derivative in space encodes anomalous superdiffusion, caused by long particle jumps. In particular, the proportion of jumps exceeding length L over a small time-step Δt falls off like $L^{-\beta}$ for some $1 < \beta < 2$. When $\alpha = 1$, this simple (constant coefficient) model captures the power-law downstream tail in the plume, see Figure 8.1. The parameter $\beta = 1.1$ in this model was fit using the available data, which also exhibits power-law statistics (Benson et al., 2001). The α-fractional derivative in time encodes anomalous subdiffusion, caused by long particle waiting times between movements. The proportion of resting times exceeding length T falls off like $L^{-\alpha}$ for some $0 < \alpha < 1$, see Schumer et al. (2003a) for an application to stream flow. In multiple dimensions, it has been shown by Schumer et al. (2003b) that the order of the fractional space derivative varies with the coordinate. This leads to an effective upscaling model in 2-D for the plume. Zhang et al. (2003) argued that a FADE model with variable parameters is needed to more effectively model the plume. One can systematically vary the fractional derivative orders to obtain a more faithful model. For example, there is enhanced anomalous dispersion further downstream, indicating a lower

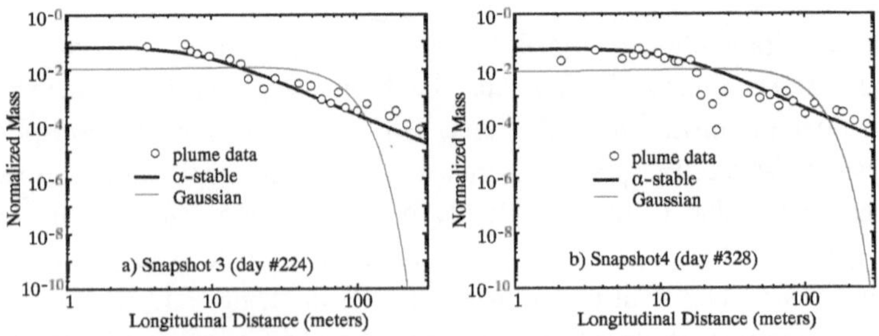

Figure 8.1 FADE application to ground water pollution. Source: Benson et al. (2001).

value of β (Clarke et al., 2005), so a variable order space-fractional derivative may be indicated.

In the previous sections, we provided evidences of the occurrence of anomalous behavior in subsurface transport even for a low degree of heterogeneity and we have shown that FADEs can be accurate models when properly tuned. However, the identification of an optimal fractional model for a specific setting (e.g., for specific hydraulic properties) is not trivial and has not been thoroughly explored in the literature. One of the main challenges in this context is the fact that the model parameters cannot be directly related to media properties, as carefully explained in Neuman and Tartakovsky (2009). Furthermore, oftentimes, it is hard or nearly impossible to collect solute measurements, so that only a very small set of data that are sparse in time and space and potentially affected by noise is available. Yet, in this context, FADEs have the advantage, compared to other models for subsurface transport, of having only a handful of parameters to tune, i.e., the identification problem consists in discovering a small set of parameters such as the diffusivity and the fractional order.

Only a few works in the literature have addressed this problem. In the context of highly heterogeneous settings, we mention the work by Pang et al. (2017) where the authors used multi-fidelity Bayesian optimization to discover *variable-order* fractional operators for the advection–diffusion equation from field data in the Macrodispersion Experiment (MADE) dataset in Columbus, Mississippi, see e.g., Barlebo et al. (2004) and references therein. Other recent works addressing a similar learning problem for fractional operators include optimization-based approaches such as the one used in Burkovska et al. (2021), fractional/nonlocal physics-informed neural network approaches such as Pang et al. (2019, 2020a) and operator-regression techniques such as the one developed in You et al. (2021). It is important to keep in mind that in all of these works, the computational cost may become prohibitive, due to the integral nature of the operators involved and to the strong singularities that require sophisticated (and expensive) quadrature rules. Thus, together with the development of new learning techniques or the extension of the current ones to more complex settings, it is mandatory to design more efficient discretization schemes and numerical solvers.

8.2 Complex Rheology and Material Damage

Accurate modeling of evolving material response and failure across multiple time- and length-scales is essential for life-cycle prediction and design of new materials. While the mechanical behavior of a number of standard engineering materials (e.g., metals, polymers, rubbers) is quite well understood, a significant modeling effort still needs to be conducted for complex materials, where microstructure heterogeneities, randomness and small scale physical mechanisms (such as collective behavior) lead to non-standard and, at times, counter-intuitive responses. Two examples are bio-tissues and natural materials (e.g., bio-polymers), which are multi-functional products of millions of years of evolution, locally optimized for their hosts and environment, and constrained by a limited set of building blocks and available resources (Imbeni et al., 2005; Wegst et al., 2015). These materials possess unprecedented properties at low densities, especially due to their hierarchical and multi-scale structure, leading to a wide spectrum of behaviors, such as power-law viscoelasticity, viscoplastic strains under hysteresis loading, materials failure/damage, material fatigue, fractal avalanche ruptures and self-healing mechanisms.

The main motivation for fractional materials modeling is the power-law fingerprint arising in microstructures undergoing anomalous diffusion, observed in a range of complex materials. Such microstructures often display a fractal nature with subdiffusive dynamics, e.g., of entangled polymer chains, and defect interactions such as dislocation avalanches, cracks and voids. Such non-exponential behavior cannot be accurately modeled by integer-order, linear viscoelastic models, which require arbitrary arrangements of Hookean/Newtonian elements and introduce a limited number of exponential (Debye) relaxation modes that, at most, represent a truncated power-law approximation (Bagley, 1989). While these approximations may be satisfactory for short times and engineering precision, they often result in high-dimensional parameter spaces and still lack predictability outside the experimental time/length-scales, often requiring re-calibration. In this context, fractional operators become appropriate and natural modeling choices, since their integro-differential operators naturally utilize power-law convolution kernels, coding self-similar microstructural features in a reduced-order mathematical language with smaller parameter spaces (similar to the case of anomalous transport, see Section 8.1). This fact allows accurate and predictive modeling, in an efficient manner, of bio-tissues (Suki et al., 1994; Craiem et al., 2008; Djordjević et al., 2003; Puig-de Morales-Marinkovic et al., 2007; Nicolle et al., 2010; Meral et al., 2010; Magin and Royston, 2010) and polymers (Ketz et al., 1988; Winter and Mours, 1997; Sollich, 1998; Baghdadi et al., 2005) for multiple time scales.

In this section, we review fractional models for materials undergoing power-law behaviors, termed *anomalous materials*, in a range of non-equilibrium and path-dependent responses. We start with linear viscoelasticity, introducing the basic modeling building block, known as the *Scott-Blair element*, which models a single power-law response and can be combined to incorporate more complex behaviors. In harmony with the previous sections, we will emphasize potential multi-scale connections, stochastic processes, and thermodynamic consistency. After providing evidence of cases where fractional behavior/power-laws appear as intrinsic qualities in a number of systems, we will report on the state-of-the-art models incorporating multiple physical mechanisms.

Fractional Viscoelasticity: Rheological Building Blocks. We start with the Boltzmann superposition integral for linear viscoelasticity, obtained from the linear superposition of infinitesimal step strains $\delta\varepsilon(t)$ applied to a viscoelastic material (Mainardi and Spada, 2011):

$$\sigma(t) = \int_{-\infty}^{t} G(t-\tau)\dot{\varepsilon}(\tau)\,d\tau, \tag{8.2.1}$$

where $\dot{\varepsilon}$ and $\sigma(t)$ denote, respectively, the strain rate and stress. The convolution kernel $G(t)$, is a relaxation function, directly related to stress relaxation experiments under step strains. It is traditionally modeled through combinations of Hookean springs and Newtonian dashpots, yielding a multi-exponential relaxation in the form

$$G(t) = \sum_{i=1}^{N} C_i \exp(-t/\tau_i).$$

For this particular choice of kernel, (8.2.1) is equivalent to a multi-term ordinary differential equation (ODE).

Relaxation experiments across multiple time- and frequency-scales indicate that anomalous materials exhibit memory effects in time for stress/strain responses, which translates into a single power-law scaling in the form $G(t) \propto t^{-\alpha}$, with $\alpha \in (0,1)$. This indicates that, contrary to exponential relaxation forms, there is a spectrum of relaxation time arising from the material microstructure (Jaishankar and McKinley, 2013), for which standard ODE models (e.g., generalized Maxwell model in creep/relaxation representations) would require a large number of parameters.

The fundamental fractional rheological building block element, termed the *Scott-Blair* (SB) element, is obtained by substituting the power-law kernel $G(t) = Et^{-\alpha}/\Gamma(1-\alpha)$ into (8.2.1), leading to the form:

$$\sigma(t) = {}_{-\infty}^{\;\;C}D_t^{\alpha}\varepsilon(t) = \frac{E}{\Gamma(1-\alpha)}\int_{-\infty}^{t}(t-\tau)^{-\alpha}\dot{\varepsilon}(\tau)\,d\tau, \tag{8.2.2}$$

which is equivalent to the Riemann–Liouville fractional derivative ${}^{RL}_{-\infty}D_t^{\alpha}\varepsilon(t)$ if the function $\varepsilon(t)$ is sufficiently well behaved at $t \to -\infty$ (Podlubny, 1999). While this equivalence is satisfied for semi-infinite domains, the choice of Riemann–Liouville and Caputo definitions matters when we introduce a causal strain history and switch the lower limit of (8.2.2) from $-\infty$ to 0, which leads to two different fractional Cauchy problems. For the Caputo definition, we have (Mainardi and Spada, 2011)

$$\sigma(t) = E\, {}^{C}_{0}D_t^{\alpha}\varepsilon(t), \quad t > 0, \quad 0 < \alpha < 1, \quad \varepsilon(0) = \varepsilon_0. \tag{8.2.3}$$

On the other hand, when employing Riemann–Liouville derivatives, we obtain

$$\begin{aligned} &\sigma(t) = E\, {}^{RL}_{0}D_t^{\alpha}\varepsilon(t), \quad t > 0, \quad 0 < \alpha < 1, \\ &{}^{RL}_{0}D_t^{\alpha-1}\varepsilon(t)\Big|_{t=0} = \varepsilon_0, \end{aligned} \tag{8.2.4}$$

where we remark that problem (8.2.3) is more commonly adopted due to the appearance of integer-order initial conditions (ICs), while both aforementioned problems are equivalent in the presence of homogeneous ICs.

The SB element provides a constitutive interpolation between a Hookean spring ($\alpha \to 0$) and a Newtonian dashpot ($\alpha \to 1$). The unique parameter pair ($E\,[Pa.s^{\alpha}]$, α) encodes snapshots of a dynamic process instead of an equilibrium state of the system (Jaishankar and McKinley, 2013). Consequently, these properties are only associated with equilibrium states in the limit cases for the fractional order α.

Remark 8.1 We remark that although (8.2.3) utilizing the Caputo definition is widely employed to represent the SB element in the literature, the pioneering works on anomalous rheology modeling are attributed to Gerasimov (1948) (and partially to Rabotnov (1948)) in 1948, introducing a similar power-law convolution operator as (8.2.2), which may be referred to as the Gerasimov–Caputo derivative in the literature (Valério et al., 2014). We refer the reader to Valério et al. (2014); Mainardi (2012) for more details on the historical context of fractional derivatives in viscoelasticity.

Mechanistic and Thermodynamic Interpretations. Apart from the Boltzmann integral representation (8.2.1), characterized by an integro-differential nature, the SB element can also be obtained through a continuous arrangement of canonical, Hookean and Newtonian elements, both from their constitutive and free-energy levels (Schiessel and Blumen, 1993; Lion, 1997), making the notion of SB elements intrinsically incorporating an infinite number of relaxation time more evident. In Schiessel and Blumen (1993), a hierarchical ladder-like structure of standard Maxwell viscoelastic elements was employed.

This structure led to a coupled system of ODEs, which had an infinite continued fraction (a recursion of fractions) representation in terms of the Maxwell model constants in the Laplace domain. Then, applying the inverse Laplace transform, a fractional stress–strain relationship was recovered for homonegeous initial conditions, therefore equivalent to both forms (8.2.3) and (8.2.4). In Lion (1997), an isothermal Helmholtz free-energy density was derived for the SB element from the elastic energies of a discrete-to-continuum arrangement of standard Maxwell elements/branches, obtaining the following form for the free-energy ψ as a function of the strain:

$$\psi(\varepsilon) = \frac{1}{2}\int_0^{\infty} h(z)\left[\int_0^t \exp\left(-\frac{t-s}{z}\right)\dot{\varepsilon}(s)ds\right]^2 dz, \tag{8.2.5}$$

$$h(z) = \frac{Ez^{-1-\alpha}}{\Gamma(\alpha)\Gamma(1-\alpha)}, \tag{8.2.6}$$

where $h(z)$ denotes the relaxation spectrum. Therefore, (8.2.5) represents the amount of available elastic energy to perform work from the SB element in the time domain, which cannot be directly inferred from (8.2.3) and (8.2.4). Naturally, the two limit cases for α are $\psi(\varepsilon) \to E\varepsilon^2/2$ when $\alpha \to 0$, and $\psi(\varepsilon) \to 0$ when $\alpha \to 1$. Furthermore, under suitable thermodynamic constraints, it is shown that the SB element is thermodynamically admissible and that the Caputo representation of (8.2.4) can be derived from (8.2.5) under continuum mechanics arguments.

Energy Decoupling in the Frequency Domain. Similar to the aforementioned representations, power-law structures also appear in viscoelastic dynamic properties and rheological experiments in the frequency domain (Jaishankar and McKinley, 2013), such as the complex shear modulus, defined as the ratio between the Fourier transform of stresses and strains:

$$G^*(\omega) := \frac{\mathcal{F}[\sigma](\omega)}{\mathcal{F}[\varepsilon](\omega)} = G'(\omega) + iG''(\omega), \tag{8.2.7}$$

where ω denotes the frequency. The term G' is the storage modulus, and G'' denotes the loss modulus, i.e., the stored and dissipated energy per cycle, respectively. Employing (8.2.7) into (8.2.4), the dynamic modulus of the Scott-Blair element is obtained (Schiessel, 1995):

$$G'(\omega) = \Re(G^*) = E\omega^{\alpha}\cos\left(\frac{\alpha\pi}{2}\right), \tag{8.2.8}$$

$$G''(\omega) = \Im(G^*) = E\omega^{\alpha}\sin\left(\frac{\alpha\pi}{2}\right), \tag{8.2.9}$$

which provides a clear storage/loss decomposition, with the value of α determining whether the material of interest is predominantly dissipative for certain frequency ranges. Here, $\Re(\cdot)$ and $\Im(\cdot)$ represent the real and imaginary parts, respectively.

Relationships to Material Microstructure and Stochastic Processes. The mechanistic origins of macroscopic power-law behaviors in complex materials are due to spatio-temporal anomalous subdiffusive processes (Metzler and Klafter, 2000) in fractal microstructures. We focus on the temporal case, in which the MSD of microstructural constituents follows a nonlinear scaling (in time) in the form $\langle \Delta x \rangle^2 \propto t^{\alpha}$. Bagley and Torvik (1983) provided a relationship between the complex shear modulus obtained from the Rouse theory of polymer dynamics. They started with the result of Rouse's theory for the shear modulus, i.e.,

$$G'(\omega) = nkT \sum_{p=1}^{N} \frac{\omega^2 \tau_p^2}{1 + \omega^2 \tau_p^2}, \tag{8.2.10}$$

$$G''(\omega) = \omega \mu_s + nkT \sum_{p=1}^{N} \frac{\omega \tau_p}{1 + \omega^2 \tau_p^2}, \tag{8.2.11}$$

where n denotes the number of molecules per unit volume, N is the number of monomers in the polymer chain, T represents the absolute temperature, and k is the Boltzmann's constant. The term τ_p denotes the relaxation time of the solution, which was approximated as

$$\tau_p \approx \tau_1 / p^2 = 6(\mu_0 - \mu_s)/(p^2 \pi^2 nkT),$$

which is valid when the number of submolecules N is large. The terms μ_0 and μ_s denote, respectively, the steady-flow viscosities of the solution and solvent. They further worked on Rouse's results, and by assuming the polymer chains and $\omega\tau_1$ to be sufficiently large, obtained the following power-law form for the dynamic shear modulus

$$G^*(\omega) = \mathrm{i}\omega\mu_s + \left[\frac{3}{2}(\mu_0 - \mu_s)nkT\right]^{1/2} (\mathrm{i}\omega)^{1/2}. \tag{8.2.12}$$

After applying the inverse Fourier transform, the above relationship leads to a Riemann–Liouville representation between stresses-strains with $\alpha = 1/2$. Similar observations were also reported for $\sigma(t)$ utilizing a Zimm model, where the inclusion of hydrodynamic interactions leads to the fractional order $\alpha = 2/3$. Glöckle and Nonnenmacher (1993) showed that fractional relaxation can be modeled by a special type of CTRW describing a trapping problem due to entanglements of polymer chains, thus slowing down the relaxation process. In their work, the random walkers, i.e., the particles, are considered as packages of free volume that allow conformational reorientations of chain segments, thus leading to relaxation. They obtained a waiting time distribution of such particles through a Fox–Wright representation in the form

$$\chi(t) \sim \frac{A}{\bar{\tau}} \sum_{k=0}^{\infty} \frac{(-1)^k}{\Gamma(-\beta k - \beta)} \left(\frac{\bar{\tau}}{t}\right)^{\beta k + \beta + 1}, \tag{8.2.13}$$

where $\bar{\tau}$ denotes the characteristic relaxation time, and the leading term indicates that the CTRW waiting time corresponding to fractional relaxation exhibits a Lévy-type decay in the form $\chi(t) \sim t^{-\beta-1}$.

Connecting Dynamic Viscoelasticity across Scales. A connection between power-laws propagating from micro- to macro-rheology was proposed in Mason (2000), with the use of a generalized Stokes–Einstein relation (GSER) for spheres undergoing generalized Langevin dynamics in a viscoelastic medium:

$$|G^*(\omega)| \approx \frac{kT}{\pi a \langle \Delta r^2(1/\omega)\rangle \Gamma(1+\alpha(\omega))}, \quad \text{where} \tag{8.2.14}$$

$$\alpha(\omega) \equiv \left.\frac{\mathrm{d}\ln\langle \Delta r^2(1/\omega)\rangle}{\mathrm{d}\ln t}\right|_{t=1/\omega},$$

which is valid for a sphere of radius a comparable to the length-scale of the embedding medium. Here, the dynamic shear modulus $G^*(\omega)$ is related to a velocity memory function from Langevin dynamics. Among a variety of representations for the GSER, (8.2.14) assumes a power-law structure of the MSD with exponent α, which approaches zero when the sphere is confined by elastic structures present in the complex fluid. Such power-law representation also reduces errors near the frequency extremes when employing Laplace and Fourier transforms.

Physical Interpretation of Fractional Orders. Despite existing connections between micro- and macro-rheological properties, the physical interpretation of the emerging fractional orders has been elusive. A connection between the fractional order and the fractal dimension of the material microstructure was made by Mashayekhi et al. (2019), where the authors extended the Zimm theory of polymer dynamics to fractal media as a bridge between the meso- and macro-scales. They showed that the fractional order is a rate-dependent material property that is strongly correlated with the fractal and spectral dimensions in fractal media.

Evidence of Fractional-Order Behavior

We provide a few examples of fractional/power-law behaviors in viscoelasticity and micro/macro-scale plasticity. We start with two examples in viscoelasticity of solid-like and fluid-like nature in which fractional modeling is more appropriate, both with better fits and a reduced number of model parameters.

Viscoelastic Rheology. Jaishankar and McKinley (2013) calibrated classical and fractional Maxwell models to the four orders-of-magnitude relaxation data for highly anomalous butyl rubber data from Blair et al. (1947) (Figure 8.2 (a)), and observed that the three-parameter fractional Maxwell model provided an excellent fit to the experimental data, while a multi-exponential, integer-order Maxwell model required six parameters to provide a satisfactory fit. Moreover, using the calibrated fractional relaxation parameters they obtained an accurate prediction of the creep compliance for the same material, especially for long-time behavior. The second experiment from Jaishankar and McKinley (2013) concerns the dynamic properties of acacia gum, a commonly used food preservative. In this case, they compared a four-parameter fractional Maxwell model with a single mode (three-parameter) standard Maxwell model (Figure 8.2(b)) and demonstrated that while the fractional Maxwell model captures a complex Cole–Cole behavior, its integer-order counterpart is unable to even estimate the qualitative response.

We note that other factors, such as material heterogeneity, can introduce multiple power-law relaxation regimes.

Stamenović et al. (2007) measured the complex shear modulus $G^*(\omega)$ of cultured human airway smooth muscle and observed two distinct power-law regimes separated by an intermediate plateau. Kapnistos et al. (2008) found an

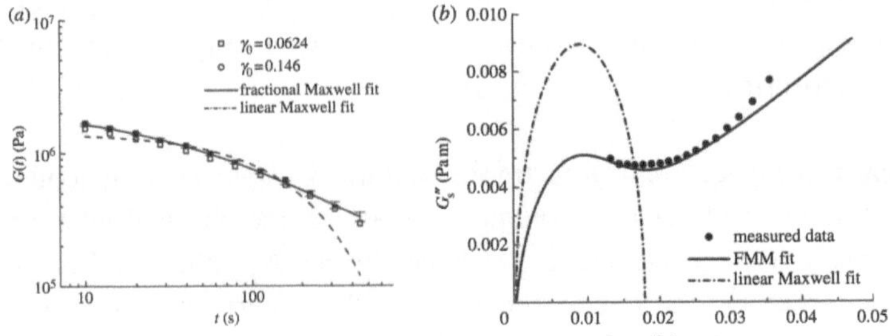

(a) (b)

Figure 8.2 Comparison between standard and fractional order models. (a) Relaxation behavior of butyl rubber using experimental data from Scott-Blair. (b) Cole–Cole plot (G' versus G'') for the dynamic properties of acacia gum. Source: Used with permission of The Royal Society (UK), from Jaishankar and McKinley (2013); permission conveyed through Copyright Clearance Center, Inc.

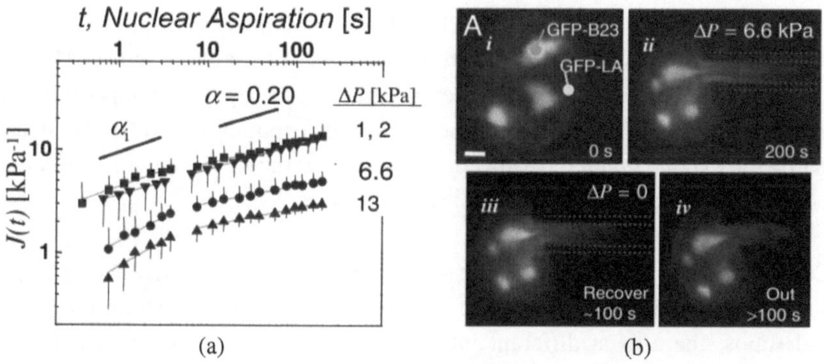

Figure 8.3 (a) Scale-free creep of ESCs nuclei under aspiration. Low applied stresses ΔP yield a single power-law creep scaling. For large stresses, a plastic transition is observed at $\tau_{plastic} \approx 8 - 10\,[s]$, with a creep exponent $\alpha \approx 0.2$, independent of stress values. (b) Different stages of nucleus aspiration, showing a viscoelastic recovery (ii)–(iii), followed by irreversible plastic deformation (iv). Source: Pajerowski et al. (2007), Copyright (2007) National Academy of Sciences, USA.

unexpected tempered power-law relaxation response of entangled polystyrene ring polymers, compared to the usual relaxation plateau of linear chain polymers. Such behavior was interpreted through self-similar conformations of double-folded loops of ring polymers, instead of the reptation observed in linear chains.

Power-Law Plasticity. The creep behavior of human embryonic stem cells (ESCs) under differentiation was studied by Pajerowski et al. (2007) through micro-aspiration experiments at different pressures. The cell nucleus demonstrated distinguished visco-elasto-plastic power-law scalings, with $\alpha = 0.2$ for the plastic regime, independent of the applied pressure. It is discussed that such a low power-law exponent arises due to the fractal arrangement of chromatin inside the cell nucleus.

Studies on force-induced mechanical plasticity of mouse embryonic fibroblasts were performed by Bonadkar et al. (2016). It was found that the viscoelastic relaxation and the permanent deformations followed a stochastic, normally-distributed, power-law scaling $\beta(\omega)$, with values ranging from $\beta \approx 0$ to $\beta \approx 0.6$. The microstructural mechanism of plastic deformation in the cytoskeleton is due to the combination of permanent stretching and buckling of actin fibers.

As for evidence of power-laws in failure of crystalline materials, Richeton et al. (2005) investigated the emergence of intermittency and dislocation avalanches in polycrystalline plasticity through acoustic emission experiments on ice under creep compression. Their findings demonstrate that unlike the scale-free, close-to-critical dislocation dynamics of single crystals (Miguel et al., 2001), the introduction of average grain sizes $\langle d \rangle$ from the polycrystal microstructure led to a tempered power-law distribution of avalanche sizes. While the exponential tempering cutoff changes with $\langle d \rangle$, the authors observed a constant power-law scaling for all samples.

Connections to Stochastic Processes. Although a subdiffusive MSD coefficient $0 < \alpha < 1$ is observed in a variety of studies for complex materials and fluids, there exist different interpretations on the underlying stochastic processes linked to the subdiffusive physics, e.g., crowding or caging effects in cells and polymers. Szymanski and Weiss (2009) utilized fluorescence correlation spectroscopy (FCS) of proteins immersed in crowded dextran solutions and reported a distribution of MSD coefficients with average $\langle \alpha \rangle \approx 0.82$, and compared this experimental finding with recovered distributions of MSD coefficients for simulated fractional Brownian motion (fBm), obstructed diffusion (OD), and CTRWs. Their findings indicated that the recovered distributions for fBm and OD matched the experiments, while the recovered distributions for CTRW-induced diffusion, with average $\langle \alpha \rangle \approx 0.59$ did not agree well with the data due to ergodicity breaking. Weber et al. (2010) studied the subdiffusion of bacterial chromosomal loci in viscoelastic cytoplasm and further concluded fractional Langevin motion to be more likely than CTRW and OD, due to the presence of ergodicity and a negative velocity autocorrelation function. Regarding polymers, Wong et al. (2004) studied the thermal motion of colloidal tracer particles in entangled actin filament (F-actin) networks, under different concentrations and network mesh sizes. They observed a subdiffusive behavior when the tracer particles' radii were comparable to the network mesh size, and suggested that such anomalous behavior happens due to intermittent caging behavior, followed by sudden infrequent jumps with a power-law distribution of caging times τ_c in the form $P(\tau_c) = \tau_c^{-1.33}$.

State of the Art: Anomalous Materials Modeling

As observed in Section 8.2, experimental evidence suggests that complex material behavior may possess more than a single power-law scaling in the viscoelastic regime, particularly in multi-fractal structures, which are characteristic of cells (Stamenović et al., 2007) and biological tissues (Vincent, 2012), due to their complex, hierarchical and heterogeneous microstructure. For such cases, a single SB element is not sufficient to capture the observed behavior, even if linear viscoelasticity holds. Furthermore, material nonlinearity due to large strains and additional physics such as plasticity, damage/failure requires more advanced rheological models, which could have full or partial fractional nature. In this section we review a class of fractional models in the literature, classified by rheology type and nature of the corresponding FDEs. We acknowledge that rheology is a vast field with a large number of different types of material behavior, and here we limit our review to visco-elasto-plasticity, damage mechanics and failure.

Viscoelasticity

Linear Viscoelasticity. We start by introducing two natural extensions of the SB viscoelastic model through serial and parallel combinations. The first one is the fractional Kelvin–Voigt (FKV) model, which is given by a parallel combination of SB elements, and relates the stresses $\sigma(t)$ and strains $\varepsilon(t)$ in the following additive form (Schiessel and Blumen, 1993):

$$\sigma(t) = E_1 \, {}_0^{\mathrm{C}}D_t^{\alpha_1} \varepsilon(t) + E_2 \, {}_0^{\mathrm{C}}D_t^{\alpha_2} \varepsilon(t), \quad t > 0, \quad \varepsilon(0) = 0, \tag{8.2.15}$$

where the fractional orders $0 < \alpha_1, \alpha_2 < 1$, and $E_1\,[Pa.s^{\alpha_1}]$, $E_2\,[Pa.s^{\alpha_2}]$ are the associated pseudo-constants. The corresponding relaxation function also assumes an additive form of two SB elements:

$$G^{\mathrm{FKV}}(t) := \frac{E_1}{\Gamma(1-\alpha_1)} t^{-\alpha_1} + \frac{E_2}{\Gamma(1-\alpha_2)} t^{-\alpha_2}, \tag{8.2.16}$$

where contrary to the scale-free relaxation behavior of a single SB element, since we assume $\alpha_2 > \alpha_1$, the FKV model possesses two time-scale dependent power-law regimes, given by $G^{\mathrm{FKV}} \sim t^{-\alpha_2}$ as $t \to 0$ and $G^{\mathrm{FKV}} \sim t^{-\alpha_1}$ as $t \to \infty$, which characterizes a transition from faster to slower relaxation regimes. We note that this quality allows the FKV model to describe materials that reach an equilibrium behavior for large times when $\alpha_1 \to 0$, which is intuitive from the mechanistic standpoint as one of the SB elements becomes a Hookean spring.

Through a serial combination of SB elements, we obtain the fractional Maxwell (FM) model (Jaishankar and McKinley, 2013), given by

$$\sigma(t) + \frac{E_2}{E_1} \, {}_0^{\mathrm{C}}D_t^{\alpha_2-\alpha_1} \sigma(t) = E_2 \, {}_0^{\mathrm{C}}D_t^{\alpha_2} \varepsilon(t), \quad t > 0, \tag{8.2.17}$$

with $0 < \alpha_1 < \alpha_2 < 1$, and two sets of initial conditions for strains $\varepsilon(0) = 0$, and stresses $\sigma(0) = 0$. We note that in the case of non-homogeneous initial conditions, compatibility conditions are required (Mainardi and Spada, 2011) between stresses and strains at $t = 0$. The corresponding relaxation function for this building block model assumes the more complex Miller–Ross form (Jaishankar and McKinley, 2013):

$$G^{\mathrm{FM}}(t) := E_1 t^{-\alpha_1} E_{\alpha_2-\alpha_1,1-\alpha_1}\left(-\frac{E_1}{E_2} t^{\alpha_2-\alpha_1}\right), \tag{8.2.18}$$

where $E_{a,b}(z)$ denotes the two-parameter Mittag-Leffler function, defined as (Mainardi and Spada, 2011)

$$E_{a,b}(z) = \sum_{k=0}^{\infty} \frac{z^k}{\Gamma(ak+b)}, \quad \Re(a) > 0, \quad b \in \mathbb{C}, \quad z \in \mathbb{C}. \tag{8.2.19}$$

Interestingly, the presence of a Mittag-Leffler function in (8.2.18) produces a stretched exponential relaxation for smaller time scales and a power-law behavior for larger time scales. The asymptotic behaviors are given by $G^{\mathrm{FM}} \sim t^{-\alpha_1}$ as $t \to 0$ and $G^{\mathrm{FM}} \sim t^{-\alpha_2}$ as $t \to \infty$, indicating that, contrary to the FKV model, the FM model has a constitutive transition from slower to faster relaxation. We refer the reader to McKinley and Jaishankar (2013) and Bonfanti et al. (2020) for a number of applications of the aforementioned models. Additionally, we note that both FKV and FM models are able to recover the SB element with a convenient set of pseudo-constants, or naturally reveal the necessity of standard rheological elements according to available data. Furthermore, we also outline more complex building block models that produce more flexible responses, including three to four fractional orders, such as the fractional Kelvin–Zener (FKZ), fractional Poynting–Thomson (FPT), and fractional Burgers (FB) models. We refer the reader to Bonfanti et al. (2020); Schiessel and Blumen (1993) for more details on such models.

Nonlinear Viscoelasticity. Fractional linear viscoelastic models are suitable candidates to describe the anomalous dynamics of a number of materials undergoing small strains. However, under large strains, material nonlinearities induce stress/strain dependencies on the relaxation behavior. One alternative to incorporate such a nonlinearity is through quasi-linear viscoelasticity (QLV) (Fung, 2013), which replaces $G(t)$ by a multiplicative decomposition between a reduced relaxation function $g(t)$ and an instantaneous, nonlinear elastic tangent response:

$$\sigma(t, \varepsilon) = \int_0^t g(t-s) \frac{\partial \sigma^e(\varepsilon)}{\partial \varepsilon} \dot{\varepsilon}(s)\, ds, \tag{8.2.20}$$

with $g(0^+) = 1$. Fractional approaches to QLV were developed by Doehring et al. (2005) for arterial valve cusp and by Craiem et al. (2008) for arterial wall viscoelasticity. In the latter, a reduced, fractional Kelvin–Voigt-type relaxation function $g(t) = C + Dt^{-\alpha}$ was employed, with pseudo-constant $D\,[s^\alpha]$, and nonlinear exponential form $\sigma^e(\varepsilon) = A\left(e^{B\varepsilon} - 1\right)$, with constant $A\,[Pa]$. Therefore, the fractional QLV formulation is able to capture not only the linear/nonlinear instantaneous stress response, due to the rearrangement and alignment of fibers with the load direction, but also the anomalous power-law relaxation of the fractal microstructure. We also mention nonlinear models that take into account the Mittag-Leffler-type relaxation dynamics, such as the fractional QLV model in Doehring et al. (2005) and the fractional K-BKZ model introduced by Jaishankar and McKinley (2014).

Visco-Elasto-Plasticity

Several works employed fractional calculus to account for the viscoplastic regimes of several classes of materials. We outline three of them: time-fractional, space-fractional and stress-fractional.

Time-fractional approaches focus on introducing memory effects into internal variables (Suzuki et al., 2016; Xiao et al., 2017; Suzuki et al., 2021), and consequently modeling power-laws in both viscoelastic and viscoplastic regimes. This is of interest for polymers, cells, and tissues. In this context, fractional visco-elasto-plastic models provide a constitutive interpolation between rate-independent plasticity and Perzyna's viscoplasticity by introducing a SB model acting the plastic regime (Suzuki et al., 2016) that utilizes a rate-dependent yield function in the form

$$f(\sigma, q) := |\sigma| - \left[\sigma^Y + K\, {}_0^{\mathrm{C}}D_t^{\alpha_K} q(t) + Hq(t)\right], \quad 0 < \alpha_K < 1, \quad (8.2.21)$$

where σ^Y and q denote, respectively, the yield stress and the accumulated plastic strain, with pseudo-constant $K\,[Pa.s^{\alpha_K}]$ and Hookean constant H. The above form for the yield function was later proved to be thermodynamically consistent in a further extension of the model to account for continuum damage mechanics (Suzuki et al., 2021).

A three-dimensional space-fractional approach to elastoplasticity was also developed by Sumelka (2014a) to account for spatial nonlocalities. The model is based on rate-independent elastoplasticity, and nonlocal effects are accounted for through a fractional continuum mechanics approach, where the strains are defined by a space-fractional Riesz–Caputo derivative of displacements $u(x)$ in the form

$$ {}_a^{\mathrm{RC}}D_b^{\alpha} u(x) = \frac{\Gamma(2-\alpha)}{2}\left({}_a^{\mathrm{C}}D_x^{\alpha} u(x) + (-1)^n\, {}_x^{\mathrm{C}}D_b^{\alpha} u(x)\right) \quad (8.2.22)$$

for left- and right-sided fractional Caputo derivatives (Sumelka, 2014a) with $n = \lceil \alpha \rceil$.

Finally, stress-fractional models for plasticity have found applicability in soil mechanics and geomaterials that follow non-associated plastic flow (Sumelka, 2014b; Sun and Sumelka, 2019), i.e., the yield surface expansion in the stress space does not follow the usual normality rule, and may be non-convex. The work by Sumelka (2014b) proposed a three-dimensional fractional viscoplastic model, where a fractional flow rule with order $0 < \alpha < 1$ in the stress domain naturally models non-associative plasticity. Interestingly, this model recovers the classical Perzyna viscoplasticity as $\alpha \to 1$, and the effect of the fractional flow rule can be a compact descriptor of microstructure anisotropy. Recently, a similar stress-fractional model was developed (Sun and Sumelka, 2019), and successfully applied to soils under compression. We refer the reader to the detailed review work by Sun et al. (2018) on fractional calculus in plasticity.

Damage Mechanics, Aging and Failure

There have also been recent efforts to represent damage, aging and failure effects via fractional calculus frameworks. Existing formulations are focused on either adding classical failure frameworks into existing fractional constitutive laws, or by developing fractional failure mechanisms. Here, we mostly focus on the latter and start with the work by Caputo and Fabrizio (2015), which developed a variable-order viscoelastic model in the form

$$\sigma(x,t) = g(\alpha(x,t))\; A(x)\; {}^{\mathrm{C}}_{t_0}D_t^{\alpha(x,t)}\, \varepsilon(x,t), \tag{8.2.23}$$

where

$$g(\alpha(x,t)) := (\alpha_C - \alpha(x,t))^2/4$$

denotes a material degradation function with critical damage α_C, $A(x)$ represents a space-dependent pseudo-property, and $0 < \alpha(x,t) < \alpha_C$ is the variable fractional order, also interpreted here as damage. The variable-order Caputo derivative is defined in (1.3.69). Interestingly, this mixed interpretation for $\alpha(x,t)$ makes it a multi-physics descriptor for anomalous damage, viscosity, and material aging. The evolution of $\alpha(x,t)$ is described by an integer-order phase-field equation, and the resulting model is proved to be thermodynamically admissible.

A key aspect to develop failure models relies on consistent forms of damage energy release rates, i.e., on obtaining the compatible operator for the loss of elastic energy, which is a nontrivial task even for the simplest fractional constitutive law (8.2.3). This has been achieved by employing the concept of fractional free-energy densities (Lion, 1997; Fabrizio, 2013; Alfano and Musto, 2017). Alfano and Musto (2017) developed a cohesive zone, damaged fractional viscoelastic Kelvin–Zener model, and studied the influence of integer and fractional damage energy release rates on damage evolution. In this case, integer-order energy loss considers Hookean-type rheology to compute the damage energy release rates, which may be justified when Hookean elements are present in the viscoelastic constitutive law, but incompatible for fully-fractional cases (an arrangement of Scott-Blair elements). The corresponding free-energy for the SB element is given by

$$\psi^{\mathrm{SB}}(t) = \frac{E}{2\Gamma(1-\alpha)} \int_0^t \int_0^t (2t - \tau_1 - \tau_2)^{-\alpha}\, \dot{\varepsilon}(\tau_1)\dot{\varepsilon}(\tau_2)\, d\tau_1 d\tau_2, \quad (8.2.24)$$

with $0 < \alpha < 1$, which clearly carries a power-law behavior over time. Among their findings, the authors obtained a rate-dependence of the fracture energy in terms of the fractional-order α, opening interesting directions towards failure of anomalous viscoelastic media such as polymers. In Suzuki et al. (2021) this idea was extended to plasticity, where a fractional visco-elasto-plastic model with memory-dependent damage was developed along with an isotropic damage evolution equation, given by Lemaitre's approach (Lemaitre, 1996):

$$\dot{D}(t) = \frac{\dot{\gamma}(t)}{1 - D(t)} \left(-\frac{Y^{ve}(t)}{S} \right)^s, \quad (8.2.25)$$

with material damage parameters $s,\ S \in \mathbb{R}^+$, plastic slip $\dot{\gamma}$ and damage energy release rate $Y^{ve}(t) = -\psi^{SB}(t)$, where $0 \leq D(t) < 1$. We note that although (8.2.25) is a nonlinear ODE, the memory is introduced through the power-law form of Y^{ve} (8.2.24). In this formulation, the viscoelastic and viscoplastic fractional orders introduce a competition between rate-dependent hardening and damage-induced softening, which could open interesting directions for modeling localized hardening in failing anomalous media. Sumelka et al. (2020) also developed the idea of memory-dependent damage for soft materials through a stress-driven time-fractional hyperelastic damage model, with evolution equation in the fractional nonlinear Cauchy form

$${}^{\mathrm{RL}}_{t-l_t}D^{\alpha}_{t}\, D(x,t) = \frac{1}{T^{\alpha}} \Phi \left\langle \frac{I_D}{\tau_D} - 1 \right\rangle_M, \quad (8.2.26)$$

where Φ represents an overstress function in terms of a stress intensity I_D, threshold stress τ_D for damage evolution, and a ramp function in Macaulay brackets notation $\langle \cdot \rangle_M$ (Sumelka et al., 2020). The memory length is driven by a time scale l_t, which was taken as a fraction of the total time T. This model was applied with an Ogden hyperelastic law to patient-specific three-dimensional abdominal aortic aneurysm (AA) for critical zone identification, with obtained fractional order $\alpha = 0.75$.

Additional work on variable-order models in the context of fractional damage, aging and failure include the following contributions. In Beltempo et al. (2018) a variable-order viscoelastic creep model was developed, where the evolution of the fractional order $\alpha(t)$ dictates the process of concrete aging. The variable-order viscoelastic model developed in Meng et al. (2019) employed a piecewise constant order followed by two linear decreasing functions for $\alpha(t)$ and successfully described the initial viscoelasticity, softening and hardening of amorphous glassy polymers under compression. Finally, variable-order operators were also proved to be useful mathematical tools to determine the onset of fracture. Patnaik and Semperlotti (2021) employed a variable fractional-order activation function for damage, where the sharp power-law activation threshold induced by the fractional operator was successfully employed to determine crack propagation and branching of brittle materials. We refer the reader to the recent review works on the use of variable-order (Patnaik et al., 2020) and distributed-order (Ding et al., 2021) fractional models in viscoelasticity and structural mechanics. In the distributed-order case, fractional derivatives are integrated with respect to a distribution of fractional orders within a certain range of values.

Anomalous Microstructural Defect Evolutions in Crystalline Materials The presence and evolution of defects that appear in the manufacturing processes also play a vital role in the failure mechanisms of engineering materials. In particular, the collective behavior of material dislocation dynamics (which can be thought of as the deviation of material microstructure from its ideal arrangement) at the mesoscale leads to avalanche, strain bursts, intermittent energy spikes, and nonlocal interactions producing anomalous features across different time- and length-scales, directly affecting plasticity, void and crack nucleation. Discrete dislocation dynamics (DDD) simulations are often used at the meso-level, but the cost and complexity increase dramatically with simulation time.

Dislocation dynamics is intrinsically connected to plasticity (Bulatov et al., 1998) and material failure, emitted from crack tips (Zhu et al., 2004), and piling-up leading to fatigue crack initiation (Tanaka and Mura, 1981). The long-range interaction of dislocation stress-fields leads to collective motion characterized by avalanches, intermittency, and power-law scaling in energy and velocity distributions (Miguel et al., 2001; Zaiser, 2006; Hähner et al., 1998). Numerical simulations have successfully reproduced those features from DDD models (Arsenlis et al., 2007). From a continuum perspective, early attempts of proposing evolution laws led to overly phenomenological models (Holt, 1970; Walgraef and Aifantis, 1985). Stochastic approaches have been proposed to account for uncertainties during dislocation motion (Hähner, 1996; Kapetanou et al., 2015). Data-driven surrogate modeling of dislocation glide for computation of mobility estimates with uncertainty was proposed in de Moraes et al. (2021). Lately, continuum dislocation dynamics (CDD) emerged as another alternative for the continuous modeling of dislocation lines (Hochrainer et al., 2014; Hochrainer, 2015), yet still focused on explicit modeling of dislocation–dislocation interactions. A meaningful representation of the collective dynamics of dislocations that highlights the nonlocal, stochastic, and anomalous behavior of dislocation ensembles in a fluid-limit continuous model is still missing. The use of nonlocal vector calculus for continuous modeling of dislocation dynamics is a natural, yet novel alternative.

To further understand how the anomalous features propagate to the continuum, de Moraes et al. (2023) developed a probabilistic model for dislocation motion constructed from the position statistics obtained from DDD simulations. They obtained the continuous limit of discrete dislocation dynamics through a PDF for the dislocation motion, and proposed a nonlocal transport model for the PDF. This approach connects the anomalous nature of DDD to the origin of its corresponding nonlocal operator at the continuum, facilitating the integration of dislocation dynamics into multi-scale, long-time material failure simulations. The striking observation in this study being related to the kernel discovery is the correspondence between the kernel fractional parameter/index and the scaling of velocity distribution tail from Figure 8.4.

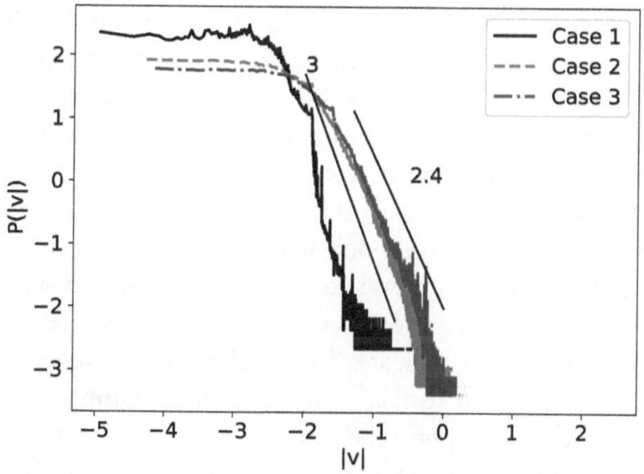

Figure 8.4 Probability density function of dislocation velocity for Case 1: the external load of $\sigma_{ext} = 0.0125\ \sigma_0$ without dislocation multiplication; Case 2: the load of $\sigma_{ext} = 0.0125\ \sigma_0$ with dislocation multiplication; Case 3: the load of $\sigma_{ext} = 0.0250\ \sigma_0$ with dislocation multiplication. A power-law scaling of order $\beta = 2.4$ when multiplication is present, and a sharper power-law for Case 1 with even no multiplication. Source: Reprinted from de Moraes et al. (2023) with permission from Elsevier.

The empirical scaling observed here and among other works in the literature matches the values of α found by the data-driven kernel learning algorithm. This is not surprising, as we can take the velocity distribution to be jump size distributions that define a particular Lévy measure, essential when transforming a stochastic process into differential equations through the semigroup theory (Chen et al., 2012).

Future Directions in Modeling Anomalous Materials

Although there exists a large spectrum of fractional models in the context of materials science, solid mechanics and rheology, these models are mostly characterized by constant-order fractional operators, for which a significant number of fast time-integration schemes is available. Yet, there is still a need for efficient numerical methods for variable- and distributed-order operators. In fact, although fractional models lead to a compact physical description with reduced number of material parameters, the computational cost is still high when calibrating the models with large experimental data sets. Furthermore, although there exists an increasing number of distributed-order operators in the context of viscoelasticity, structural mechanics, and anomalous diffusion (Ding et al., 2021), further validation against experimental data is needed.

We point out interesting research directions that could involve the use of variable- and distributed-order differential equations in the multi-scale modeling of materials. Nano-scale simulation studies on trapping of nanoparticles in hydrogel networks indicated a time-temperature dependency of the MSD in the evolution of anomalous diffusion regimes, where a subdiffusion regime has been found to be of transitional nature at intermediate time scales, with ballistic/normal diffusion dynamics for short/long time scales (Wang et al., 2020). This motivates the study of variable-order models in time to compactly describe the macroscopic rheological evolution of such polymer networks. Furthermore, the observation of distributions of power-law scaling parameters in micro-rheology creep experiments on cells (Bonadkar et al., 2016; Hecht et al., 2015) indicates the presence of microstructure-induced randomness in rheological response. In this sense, distributed-order models may arise as interesting approaches to naturally incorporate the stochastic parametric data into the differential operator (Kharazmi et al., 2017).

8.3 Fluid Turbulence and Geostrophic Transport

Richard P. Feynman described fluid turbulence as the most important unsolved problem in classical physics (Eames and Flor, 2011), a problem that stands today. By "turbulence", we refer to the three-dimensional and highly vortical fluid motions characterized by stochastic perturbations in pressure and flow velocity, and caused by excessive kinetic energy in areas of fluid flow that overcome the "damping effects" of the fluid's viscosity. The onset of turbulence can be predicted by the dimensionless Reynolds number *Re*, a ratio of kinetic energy to viscous damping in the fluid flow. Yet, the question remains of what mathematical equations govern the evolution of a turbulent flow and whether it is feasible to fully simulate turbulent flows by means of numerical methods.

Emmons (1970) reviewed the possibilities for computational fluid dynamics, concluding: "*... the problem of turbulent flows is still the big holdout. This straightforward calculation of turbulent flows – necessarily three-dimensional and unsteady – requires a number of numerical operations too great for the foreseeable future*". After almost a decade, however, the field of direct numerical simulation (DNS) of turbulence was established with successful numerical simulations of wind-tunnel flows at moderate *Re* by Orszag and Patterson Jr (1972); Kim et al. (1987); Karniadakis et al. (1993); Hussaini and Voigt (2012). These early computational developments were based on employing a Newtonian fluid assumption and applying the principles of conservation of mass, momentum, and energy to an infinitesimally small fluid element or parcel; see e.g., Chorin et al. (1990); Temam (2001); White and Majdalani (2006). In this context, assuming some proper (random) initial/boundary conditions, one can discretize the governing equations and solve for the "total number of degrees of freedom of turbulence" in the physical and parametric (stochastic) space.

The great challenge is that, in practice, DNS becomes prohibitively expensive, especially at high *Re*, more so in complex geometries. Hence, one of the main goals in *turbulence modeling* has been to systematically lower the total number of degrees of freedom to a manageable level, at the cost of reducing the accuracy of turbulence predictions. This approach has been mainly centered around the overarching theme of ensemble averaging the set of PDEs representing the various scalar and vector turbulence fields; see e.g., Lumley and McMichael (1995); Wilcox et al. (1998). This gives rise to new mathematical terms in the averaged or filtered governing equations, known as *turbulence closure* terms that can only be modeled as they are essentially unknown. When the entire time and length scales of turbulence are averaged, an operation denoted by $\omega \mapsto \overline{\omega}$, the Reynolds averaged Navier–Stokes (RANS) equations are obtained, solely describing the *mean-flow dynamics* of turbulence. Alternatively, if one applies a mathematically well-defined low-pass filter to the Navier–Stokes equations, an operation denoted by $\phi \mapsto \tilde{\phi}$, the resulting filtered governing equations describe the *large-eddy* dynamics of turbulence, where only small-scale subgrid dynamics need be modeled; this is referred to as large eddy simulation (LES). The common approach in the literature for modeling closure terms of any kind has been based on the use of classical local differential operators. More specifically, the majority of turbulence models have been constructed based on Boussinesq's turbulent viscosity concept (Pope, 2001), in which one assumes that the turbulent stress tensors are proportional to the *local gradient* of mean velocity at any point. The proportionality coefficient, referred to as *turbulent viscosity*, is to be inferred from data.

In the context of fractional turbulence modeling, Chen (2006b) made the first attempt to use the $2/3$-order fractional Laplacian modeling of Kolmogorov $-5/3$ scaling of fully developed turbulence and enhanced diffusing movements of random turbulent particles. Subsequently, Song and Karniadakis (2018) proposed a variable-order fractional model for the mean wall-bounded turbulent flow. They represented the Reynolds stresses with a nonlocal fractional derivative of variable-order that decays with the distance from the wall. Interestingly, they found that this variable fractional order has a *universal form* for all *Re* and for three different flow types, i.e., channel flow, Couette flow, and pipe flow. In addition to the aforementioned fully-developed flows, they modelled turbulent boundary layers and discussed how the streamwise variation affects the universal curve (see also Song and Karniadakis (2021) for a follow-up work). Later, Pang et al. (2020b) proposed a nonlocal truncated operator with spatially variable order, which is suitable for modeling wall-bounded turbulence, e.g., turbulent Couette flow. They showed that nonlocal physics-informed neural networks (nPINNs) can jointly infer the variable order, exhibiting a universal behavior with respect to *Re*, a finding that can contribute to better understanding of nonlocal interactions in wall-bounded turbulence. In terms of memory effects (i.e., nonlocality in time), Parish and Duraisamy (2017) developed a dynamic SGS model for LES, based on the Mori–Zwanzig (MZ) formalism. This closure model was constructed by exploiting similarities between two levels of coarse-graining via the Germano identity of fluid mechanics and by assuming that memory effects have a finite-temporal support. This work suggests future studies on using time-fractional derivatives in turbulence models.

The impetus for the fractional models we describe in this section is that the small-scale dynamics of turbulence are statistically anomalous, i.e., non-Markovian and non-Fickian, so that nonlocal closure models emerge as appropriate tools. Employed at the continuum level, fractional models therefore capture anomalous features in the small-scale stochastic subgrid dynamics of turbulence. The mathematical modeling of turbulence must address the fact that nonlinear interactions between the turbulence structures and motions create statistically complex phenomena that lead to a variety of anomalous features, including multi-power-law scalings in space-time, rare events, short-to-long range coherent motions, and enhanced turbulent mixing. These features urge better and novel understanding of the underlying nonlocal closure terms that appear as a result of the ensemble averaging or filtering of the governing equations. The nonlocal mode of thinking has the potential to shift the turbulence modeling paradigm and achieve a new level of physical and statistical consistency compared to classical approaches. This is especially true at high *Re*, for which a proper and efficient framework that unites computational, mathematical, and statistical aspects was not available until recently.

Evidence of Fractional Behavior in Turbulence

An intuitive concept of nonlocality and memory effects was established by Eringen and Wegner (2003), where a point within a fluid field (medium) is influenced by all points of the body at all past times. Coherent random motions and the spatially turbulent spot structures inherently give rise to *intermittent* signals with *self-similarities*, *sharp peaks*, *heavy-skirts*, and *skewed* distributions of velocity increments. Such statistical features have been well observed experimentally even in the context of most canonical problems, e.g., grid turbulence, in which the skewness factor negatively appears and the Kurtosis factor strongly exceeds three, emphasizing the non-Gaussian character of statistics (see e.g., Davidson (2015)). Moreover as demonstrated by Egolf and Hutter (2020, page 92), nonlocal effects appear even in the context of turbulent fields obtained numerically solving the Navier–Stokes equations. Such widespread statistical measures indicate the non-Markovian and non-Fickian nature of turbulence, and they are the consequence of nonlinear and coherent vortical effects that occur in a wide spectrum of length and time scales. Therefore, nonlocal interactions cannot be ruled out of modern turbulence physics. These considerations are particularly timely; in fact, we can now benefit from the spectrum of modern nonlocal and fractional modeling tools, equipped with well-established mathematical/statistical theories, that enable us to take such nonlocal/history effects into account with physical consistency and mathematical rigor.

Furthermore, averaging entire spatial scales as in RANS models or applying a spatial filter to the Navier–Stokes and energy equations as in LES models would make the underlying physical nonlocality in the corresponding closure terms in RANS models and the subgrid turbulent fluctuations in LES models even more pronounced. This sheds light on why turbulence modeling is a nonlocal task and further motivates the development of "nonlocal closure models" that can properly address and incorporate the underlying memory and long-range effects. Specifically, in what follows, we present a DNS study, presented by Akhavan-Safaei et al. (2021), which introduces new statistical measures and highlights the nonlocal character of subgrid scale dynamics in the context of scalar turbulence.

The Case of Scalar Turbulence Subgrid Dynamics

An ideal LES is such that the true, filtered turbulent intensity is captured accurately through a robust subgrid scale (SGS) modeling that is physically and mathematically expressive. In fact, the LES equations include closure terms that directly link the correct evolution in terms of turbulent intensity to the nature of the SGS closure and its modeling. Here, as a canonical problem, we consider the advection–diffusion (AD) equation

$$\frac{\partial \phi}{\partial t} + \frac{\partial}{\partial x_i} (\phi \, V_i) = -\theta \, V_2 + \mathcal{D} \frac{\partial^2 \phi}{\partial x_i \partial x_i}, \qquad i = 1, 2, 3, \tag{8.3.1}$$

in which $\mathcal{D}$ denotes the molecular diffusion coefficient of the passive scalar, and the imposed mean scalar gradient is taken to be uniform as $(0, \theta, 0)$, where θ is a real-valued constant. In the LES representation of the scalar turbulence, multiplying both sides of the filtered advection–diffusion equation by $\widetilde{\phi}$, the filtered scalar field ϕ, yields the time-evolution of the filtered turbulent *intensity*:

$$\frac{1}{2}\frac{\partial}{\partial t} (\widetilde{\phi}\,\widetilde{\phi}) + \widetilde{\phi}\, \frac{\partial}{\partial x_i} (\widetilde{\phi}\, \widetilde{V}_i) = -\theta\, \widetilde{\phi}\, \widetilde{V}_2 + \mathcal{D}\widetilde{\phi}\, \frac{\partial^2 \widetilde{\phi}}{\partial x_i \partial x_i} - \widetilde{\phi}\, \frac{\partial q_i^R}{\partial x_i}. \tag{8.3.2}$$

Here, q_i^R denotes the i-th component of the residual, SGS scalar flux defined as $q_i^R = \widetilde{\phi V_i} - \widetilde{\phi}\widetilde{V}_i$. Employing the filtered continuity equation $\nabla \cdot \widetilde{V} = 0$ and the chain rule for differentiation, we obtain

$$\frac{1}{2}\frac{\partial}{\partial t} (\widetilde{\phi}\,\widetilde{\phi}) + \widetilde{\phi}\, \widetilde{V}_i \frac{\partial \widetilde{\phi}}{\partial x_i} = -\theta\, \widetilde{\phi}\, \widetilde{V}_2 + \mathcal{D} \frac{\partial}{\partial x_i}\left(\widetilde{\phi}\, \frac{\partial \widetilde{\phi}}{\partial x_i}\right) - \mathcal{D} \frac{\partial \widetilde{\phi}}{\partial x_i} \frac{\partial \widetilde{\phi}}{\partial x_i}$$
$$- \frac{\partial}{\partial x_i}\left(\widetilde{\phi}\, q_i^R\right) + q_i^R \frac{\partial \widetilde{\phi}}{\partial x_i}. \tag{8.3.3}$$

Applying the ensemble-averaging operator, $\langle \cdot \rangle$, on (8.3.3) returns a transport equation for the *filtered scalar variance*, $\langle \widetilde{\phi}\, \widetilde{\phi} \rangle$. Akhavan-Safaei et al. (2021) considered the case of homogeneous turbulent velocity and scalar fields, in which

$$\left\langle \frac{\partial}{\partial x_i}(\cdot) \right\rangle = \frac{\partial}{\partial x_i}\langle(\cdot)\rangle = 0.$$

Defining the filtered scalar gradient as $\widetilde{\boldsymbol{G}}(\boldsymbol{x}) = \nabla\widetilde{\phi}(\boldsymbol{x})$, the time-evolution of the filtered scalar variance takes the following form

$$\frac{1}{2}\frac{d}{dt}\langle\widetilde{\phi}\,\widetilde{\phi}\rangle = -\widetilde{\mathcal{T}} + \widetilde{\mathcal{P}} - \widetilde{\chi} + \Pi, \tag{8.3.4}$$

where

$$\widetilde{\mathcal{T}} = \langle\widetilde{\phi}\,\widetilde{V}_i\,\widetilde{G}_i\rangle, \quad \widetilde{\mathcal{P}} = -\theta\,\langle\widetilde{\phi}\widetilde{V}_2\rangle,$$
$$\widetilde{\chi} = \mathcal{D}\,\langle\widetilde{G}_i\,\widetilde{G}_i\rangle, \quad \Pi = \left\langle q_i^R\,\widetilde{G}_i\right\rangle.$$

In (8.3.4), $\widetilde{\mathcal{T}}$ denotes the turbulent transport of filtered scalar variance while $\widetilde{\mathcal{P}}$ represents the production of resolved scalar variance by the uniform mean scalar gradient, and $\widetilde{\chi}$ is the resolved scalar variance dissipation due to the molecular diffusion. Unlike these three terms, Π (representing the SGS production of resolved scalar variance) is the only contributing term in (8.3.4) that contains the effects of the SGS scalar flux. Therefore, as pointed out earlier, understanding the true statistical nature of $\boldsymbol{q}^R \cdot \widetilde{\boldsymbol{G}}$ is essential for the SGS modeling and precise evaluation of the resolved scalar variance in the LES. This examination of $\boldsymbol{q}^R \cdot \widetilde{\boldsymbol{G}}$ might be viewed both from single-point and two-point statistics as discussed by Meneveau (1994) in the context of the LES for homogeneous isotropic turbulent flows. We focus on the two-point statistics of the SGS production of resolved scalar variance. This quantity is well represented in terms of the following normalized two-point correlation function

$$\mathcal{C}(q_i^R, \widetilde{G}_i) = \frac{\langle q_i^R(\boldsymbol{x})\,\widetilde{G}_i(\boldsymbol{x}+\boldsymbol{r})\rangle}{\langle q_i^R(\boldsymbol{x})\,\widetilde{G}_i(\boldsymbol{x})\rangle}, \tag{8.3.5}$$

where $\boldsymbol{r} = (r_1, r_2, r_3)$ denotes the spatial shift from the location $\boldsymbol{x}$. Moreover, the probability density function (PDF) of the SGS production of scalar variance normalized by its L_2-norm, i.e., $\boldsymbol{q}^R \cdot \widetilde{\boldsymbol{G}}/\|\boldsymbol{q}^R \cdot \widetilde{\boldsymbol{G}}\|$, is a novel statistical measure for studying the statistical behavior of Π and yielding a more comprehensive insight into the SGS modeling.

Let T_{LE} be the eddy turnover time. By taking a large sample space over $10\,T_{\mathrm{LE}}$ of this stationary process (after resolving the passive scalar field for $15\,T_{LE}$), the PDF of the normalized SGS production of filtered scalar variance is computed for four different filter widths, $\Delta/\eta = 8, 20, 41, 53$. These computations, shown in Figure 8.5(a), demonstrate that as Δ becomes larger, the PDF exhibits broader tails. Emergence of this tail behavior implies that as

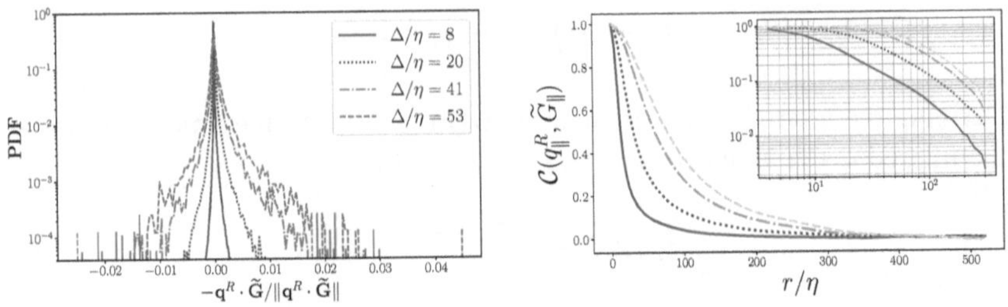

(a) (b)

Figure 8.5 Statistics of true subgrid-scale contribution to the filtered scalar variance rate. (a) PDF of normalized SGS dissipation of filtered scalar variance, $-\boldsymbol{q}^R \cdot \widetilde{\boldsymbol{G}}$, computed over a sample space of $10\,T_{LE}$ of statistically stationary turbulence. (b) Time-averaged two-point correlation function (8.3.5) between $q_{\|}^R$ and $\widetilde{G}_{\|}$ with $r = |\boldsymbol{r}_{\perp}|$. Source: Reprinted from Akhavan-Safaei et al. (2021) with permission from Elsevier.

the filter width increases, long-range spatial interactions become stronger and more pronounced (Akhavan-Safaei et al., 2020). Motivated by this observation, a two-point diagnosis of the SGS scalar production of the filtered variance as defined in (8.3.5) would be another statistical measure shedding light on the long-range interactions in addition to the filter width effects. Considering $\|$ as the direction along the imposed mean scalar gradient and $\perp$ representing the directions perpendicular to the imposed mean gradient, we focus on the evaluation of $\mathcal{C}(q_{\|}^R, \widetilde{G}_{\|})$.

Here, one case takes $\boldsymbol{r} = (r_1, 0, 0)$ and $\boldsymbol{r} = (0, 0, r_3)$ and takes the average of the resulting two-point correlation functions. Due to the statistically stationary turbulence, such procedure is performed for 20 data snapshots that are uniformly spaced over $10\,T_{LE}$ (on the same spatio-temporal data, used to compute the PDFs); hence, the time-averaged value of $\mathcal{C}(q_{\|}^R, \widetilde{G}_{\|})$ is obtained. Figure 8.5(b) illustrates this two-point correlation function extending over a wide range of spatial shift, $r = |\boldsymbol{r}|$, and evaluated at four filter widths similar to the ones utilized in Figure 8.5(a). This plot quantitatively and qualitatively reveals that as Δ increases, greater correlation values between the SGS scalar flux $q_{\|}^R(\boldsymbol{x})$, and filtered scalar gradient $\widetilde{G}_{\|}(\boldsymbol{x} + \boldsymbol{r})$ are observed at a fixed $\boldsymbol{r}$. These spatial correlations are significant both in the *dissipation* and also *inertial* subranges.

This confirms substantial nonlocal effects in the true SGS dynamics, which need to be carefully addressed in the SGS modeling for LES. A popular and fairly simple approach for modeling the SGS scalar flux is eddy-diffusivity modeling (EDM). In EDM, the main assumption is that the SGS scalar flux is

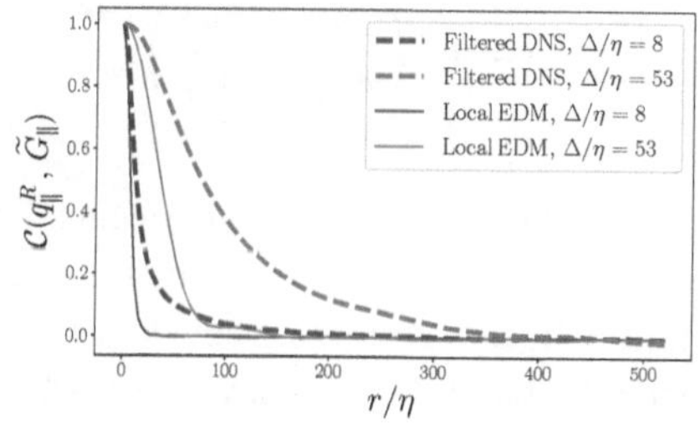

Figure 8.6 Comparison between the true values of the two-point correlation function given in (8.3.5) and the ones obtained from the local eddy-diffusivity modeling of the SGS scalar flux given in (8.3.6). The evaluations are performed at two filter widths of $\Delta/\eta = 8$, 53. Source: Reprinted from Akhavan-Safaei et al. (2021) with permission from Elsevier.

proportional to the resolved scale scalar gradient (i.e., the conventional locality assumption) as

$$\boldsymbol{q}^R(\boldsymbol{x}) \approx -\mathcal{D}_{\text{ED}}\, \widetilde{\boldsymbol{G}}(\boldsymbol{x}), \tag{8.3.6}$$

and $\mathcal{D}_{\text{ED}}$ is the proportionality coefficient. Obviously, EDM is a *local* modeling approach by construction. Computing $C(q_{\|}^R, \widetilde{G}_{\|})$ while $q_{\|}^R$ is approximated with EDM, one can compare it with its true value as shown in Figure 8.5(b). Figure 8.6 illustrates such a comparison for two filter widths, $\Delta/\eta = 8$, 53, and it reveals that in both cases local EDM substantially fails to predict the conspicuous long-range spatial correlations observed in the true two-point correlation values. *This evidence strongly suggests the adoption of a more sophisticated, nonlocal mathematical modeling tool that goes beyond conventional SGS modeling.*

State of the Art: Fractional Turbulence Modeling

In what follows, we present the history and state of the art in fractional turbulence modeling, including nonlocal RANS, fractional eddy-viscosity modeling, fractional scalar turbulence LES modeling, in addition to tempered fractional LES SGS modeling for turbulence.

Nonlocal RANS Models: A Narrative Survey

Recall that most turbulence models are built upon Boussinesq's turbulent viscosity concept. Thus, one conventionally assumes that the turbulent stress tensors τ_{ij}^R are proportional to the symmetric part of the local mean velocity gradient at any point (i.e., strain-rate tensor). Hence, the corresponding proportionality coefficient, known as the turbulent viscosity, emerges as the unknown turbulence model parameter μ_T in

$$\tau_{ij}^R = \mu_T \left(\frac{\partial \bar{u}_i}{\partial x_j} + \frac{\partial \bar{u}_j}{\partial x_i} \right), \tag{8.3.7}$$

where $\bar{u}_i$ represents the local mean velocity components. Prandtl (1942) aimed to move beyond this local constraint by introducing the *extended* mixing length concept for the first time. The corresponding new model was a great migration from locality to nonlocality, but did not achieve a remarkable success as it did not significantly improve accuracy. Afterward, he parametrized the primitive model in a way that the mixing length was taken to be greater than the (differential) length-scale of the problem, including a higher (second-order) Taylor expansion term. This strategy was analogous to adding a "weak sense of (short-range) nonlocality" to the model. This was regarded as a *weak nonlocal model* in the sense that the stress term was still in the form of Boussinesq's and the relation with the strain rate tensor in the same point was collinear. However, von Karman insisted on the consideration of the common local mixing length, which is generated by the local flow conditions and suggested considering the mixing length in terms of two succeeding derivatives (Egolf and Hutter, 2020). Bradshaw (1973) showed that Boussinesq's hypothesis fails over curved surfaces and noted that the form of the stress–strain relations was the main cause of this failure. It should be mentioned that there were some important developments mostly based on polynomial series, compared to the Boussinesq-type modeling including the works done by Spencer and Rivlin (1958, 1959), Lumley (1970) and Pope (1975); however, a noticeable lack of accuracy both in terms of physics and mathematics emerged as additional second- (and higher) order tensor series developments were demanded, where an interplay between predictability and practicality remained an open question.

As earlier indicated, Brownian motion can serve as a statistical model for the spread of a cloud of particles the continuum limit of which is a parabolic integer-order diffusion equation. Generalizing this approach to heavy-tailed processes such as Lévy processes can model the intermittency in turbulent flow signals and through a heavy-tailed central limit theorem converge to an anomalous diffusion equation with fractional derivatives in space and/or time. This suggests that employing fractional-based Reynolds stresses would be a proper model for the turbulent diffusion term. Interestingly and as a pioneering work, we see a nonlocal expression for the gradient of the transported field in a novel approach by Kraichnan (1964) for the *scalar quantity transport*. A decade later, Hinze et al. (1974) described the memory effect in a turbulent boundary layer flow. They utilized the experimental data produced downstream of a hemispherical cap, attached to the lower wall of channel geometry. They demonstrated that when one computes eddy-viscosity using Boussinesq's theory in the lateral gradient of the mean flow and turbulence shear-stresses, there is a huge non-uniform distribution that exists in the outer region of the boundary layer. Afterward, fractional-order models based on the RANS approach were offered in Hamba (1995, 2005); Chen (2006a); Egolf and Hutter (2017); Epps and Cushman-Roisin (2018). Most of these works use Green's functions based on the residual velocity to provide the expression for the Reynolds stress or scalar fluxes.

One of the main contributions to the development of nonlocal models was by Egolf and Hutter (2017; 2020). They started from Lévy flight statistics and generalized the zero-equation local Reynolds shear stress expression to a nonlocal and fractional type. The method is based on Kraichnan's convolution-integral approach and utilizing different weighting functions. Using the appropriate weighting functions, one can make a bridge between the first-order gradient of the common eddy-diffusivity models and the mean velocity difference term. The proposed model is based on the four distinct steps that can be followed conveniently to replace a local operator with a nonlocal one. In reality, the final proposed model is a more general and extended version of Prandtl's zero-equation mixing length and shear-layer turbulence models. The proposed model is called the *difference-quotient turbulence model* (DQTM), given by

$$\overline{u_2' u_1'} = -\sigma \chi_2 [\overline{u}_1(x_1, x_2) - \overline{u}_{1,\min}(x_1)] \frac{\overline{u}_{1,\max}(x_1) - \overline{u}_1(x_1, x_2)}{x_{2,\max} - x_2}, \quad (8.3.8)$$

where $\bar{\cdot}$ denotes the ensemble averaged quantity. Although well motivated and presented as somewhat of a generalization of classical models, such models have not been thoroughly tested against established integer-order models, and their practical efficiency in addressing the nonlocalities has not been adequately examined. A series of remarkable developments in fractional turbulence modeling gave a new and practical perspective on employing fractional calculus in turbulence, and introducing new statistical measures that directly reveal where classical approaches have room for modernization and enhancement. In what follows, we review such cutting-edge approaches.

Fractional Eddy-Viscosity Models

Di Leoni et al. (2021) developed a new nonlocal eddy-viscosity-based model (see (8.3.9) below) that can be applied in both isotropic and anisotropic turbulent flows. They obtained a proper two-point stress–strain-rate correlation structure for a priori testing the developed model and performed tests based on the high-resolution DNS data set for the homogeneous isotropic turbulence (HIT) and the channel flow canonical test cases.

The investigation of the model performance is set based on the necessary conditions for any LES approach in providing the accurate two-point statistics of the filtered quantities in the terms of correlations and spectra. The proposed model is given by

$$\tau_{ij}^{\alpha} = -\nu_T \left(D_i^{\alpha} \bar{u}_j + D_j^{\alpha} \bar{u}_i \right), \tag{8.3.9}$$

where the derivative operators D_i^{α} and D_j^{α} are both of order $0 < \alpha < 1$ respectively in i and j directions; however, they are employed as the *truncated* Caputo derivative variations, still being the convolution of the first derivative of velocity with respect to an inverse power-law kernel with index α, however, over a truncated (compact) integral support, forming a finite nonlocality horizon, for the purpose of lowering the computational cost.

Several numerical tests conducted in Di Leoni et al. (2021) indicated that the new model provides a better correlation between the filtered strain rate and subgrid scale stress tensor. Specifically, this model predicts the long tails in the ground-truth subfilter stress–strain-rate correlation functions. However, other conventional local eddy-viscosity-based models like classical Smagorinsky, which corresponds to $\alpha = 1$, miss this important feature as they decay faster (see Figure 8.7).

In addition to the significantly better capability in the prediction of the long-tail interactions in the new model, the probability density functions of the dissipation quantities for the HIT flow using the box filtering approach, are matching much better than the local model. The local model predictions are purely dissipative and with no tail behavior, which is in contrast with the ground-truth DNS data sets. Moreover, effects of the different parameters in

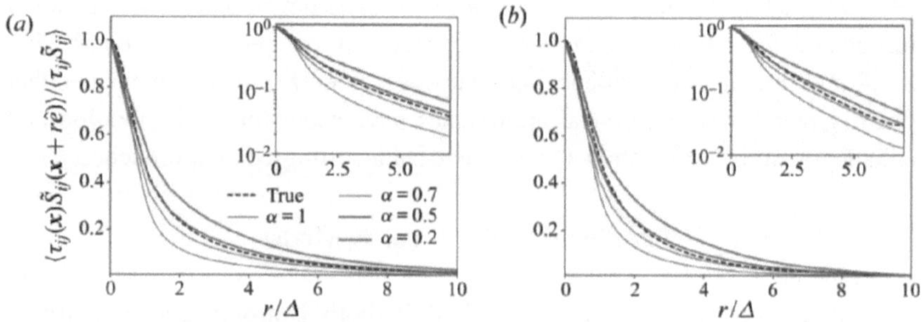

Figure 8.7 Two-point correlations between SGS stress and filtered rate of strain rate in different scenarios ($\alpha = 1$ corresponds to the local model) at the filter size $\Delta = 31\eta$ (a), and $\Delta = 53\eta$ (b). Source: Di Leoni et al. (2021), reproduced with permission.

the LES procedure have been analyzed including filter size, filter type, wall distance for the channel flow case, and integration radius.

Alternatively, in studying the turbulent transport and mixing, Boltzmann kinetic theory has shown a rich and promising ground based upon principles of statistical mechanics, which by construction is well suited to the stochastic description of turbulence at microscopic level (Harris, 2004). In the following, the fundamental sources of nonlocal closure and the SGS modeling for the residual passive scalar flux are studied at the Boltzmann kinetic transport framework. Our objective is to derive a nonlocal eddy-diffusivity SGS model at the continuum level. In what follows we present three recent developments of LES SGS where the SGS small motions are modeled by the Bhatnagar–Gross–Krook (BGK) kinetics transport.

Fractional LES SGS Modeling for Scalar Turbulence

Statistical description of LES is well represented through incorporating a filtering procedure into the Boltzmann kinetic transport. For the purpose of passive scalar transport, applying a spatially and temporally invariant filtering kernel, $\mathcal{G} = \mathcal{G}(\boldsymbol{r})$, onto the distribution function $g(t, \boldsymbol{x}, \boldsymbol{u})$ linearly decomposes that into the filtered, $\widetilde{g} = \mathcal{G} * g$, and the residual, $g' = g - \widetilde{g}$, components. Therefore, filtering the BGK equation results in the following filtered Boltzmann transport equation (FBTE) for the passive scalar:

$$\frac{\partial \widetilde{g}}{\partial t} + \boldsymbol{u} \cdot \nabla \widetilde{g} = -\frac{\widetilde{g} - \widetilde{g^{\mathrm{eq}}(\mathcal{B})}}{\tau_g}, \tag{8.3.10}$$

where $\mathcal{B}$ represents the generic *Boltzmann filter size*. As elaborated by Girimaji (2007), the nonlinear nature of the collision operator, $C_{\mathrm{BGK}}(g)$, prohibits the filtering kernel to commute with; thus, it initiates a source of closure at the kinetic level in FBTE (8.3.10). Defining $\widetilde{\mathcal{B}} := (\boldsymbol{u} - \widetilde{V})^2/c_T^2$, this closure problem is manifested in the following inequality:

$$\widetilde{g^{\mathrm{eq}}(\mathcal{B})} = \frac{\Phi \, \widetilde{\exp(-\mathcal{B}/2)}}{(2\pi)^{3/2}\, c_T^3} \neq \frac{\widetilde{\Phi} \, \exp(-\widetilde{\mathcal{B}}/2)}{(2\pi)^{3/2}\, c_T^3} = g^{\mathrm{eq}}(\widetilde{\mathcal{B}}). \tag{8.3.11}$$

The identified closure requires proper means of modeling so that one can numerically solve the FBTE (8.3.10). A common practice is to approximate this closure problem with a modified relaxation time approach that is described in detail in Sagaut (2010). Despite the success of this approach in some applications, it is not physically consistent with the filtered turbulent transport dynamics (Girimaji, 2007). Nevertheless, here we manage to adjust this inconsistency by looking at the nonlocal effects arising from filtering the Maxwell distribution function, $g^{\mathrm{eq}}(\mathcal{B})$, and model them with proper mathematical tools.

Considering the spatial filtering kernel $\mathcal{G}(\boldsymbol{r})$ with the filter-width Δ, and applying it on the Maxwell equilibrium distribution as

$$\widetilde{g^{\mathrm{eq}}(\mathcal{B})} = \mathcal{G} * g^{\mathrm{eq}}\big(\mathcal{B}(t,\boldsymbol{u},\boldsymbol{x})\big) = \int_{R_f} \mathcal{G}(\boldsymbol{r})\, g^{\mathrm{eq}}\big(\mathcal{B}(t,\boldsymbol{u},\boldsymbol{x}-\boldsymbol{r})\big)\,\mathrm{d}\boldsymbol{r}, \quad (8.3.12)$$

where $R_f = [-\Delta/2\,,\Delta/2]^3$. Subsequently, by rewriting the right-hand side of the passive scalar FBTE (8.3.10) into the following form

$$-\frac{1}{\tau_g}\big(\widetilde{g} - \widetilde{g^{\mathrm{eq}}(\mathcal{B})}\big) = \underbrace{-\frac{1}{\tau_g}\big(\widetilde{g} - g^{\mathrm{eq}}(\widetilde{\mathcal{B}})\big)}_{\text{closed}} + \underbrace{\frac{1}{\tau_g}\big(\widetilde{g^{\mathrm{eq}}(\mathcal{B})} - g^{\mathrm{eq}}(\widetilde{\mathcal{B}})\big)}_{\text{unclosed}}, \quad (8.3.13)$$

the unclosed part is structurally multi-exponentially distributed and may be approximated by a power-law distribution model as we propose

$$\widetilde{g^{\mathrm{eq}}(\mathcal{B})} - g^{\mathrm{eq}}(\widetilde{\mathcal{B}}) \approx g^{\alpha}(\widetilde{\mathcal{B}}) = \frac{\widetilde{\Phi}}{c_T^3}\, F^{\alpha}(\widetilde{\mathcal{B}}), \quad (8.3.14)$$

where $F^{\alpha}(\widetilde{\mathcal{B}})$ denotes an α-stable Lévy distribution that is mathematically designed based on heavy-tailed stochastic processes and replicates the power-law behavior (Applebaum, 2009; Meerschaert and Sikorskii, 2011). The corresponding macroscopic continuum variables associated with the filtered (8.3.1) are obtained in terms of the filtered distribution functions, $\widetilde{f}$ and $\widetilde{g}$, as

$$\widetilde{\Phi} = \int_{\mathbb{R}^d} \widetilde{g}(t,\boldsymbol{x},\boldsymbol{u})\,\mathrm{d}\boldsymbol{u}, \quad (8.3.15)$$

$$\widetilde{V}_i = \frac{1}{\rho}\int_{\mathbb{R}^d} u_i\, \widetilde{f}(t,\boldsymbol{x},\boldsymbol{u})\,\mathrm{d}\boldsymbol{u}, \quad i = 1,2,3. \quad (8.3.16)$$

According to the microscopic reversibility of the particles that assumes the collisions occur *elastically*, the right-hand side of (8.3.10) equals zero (Saint-Raymond, 2009). Therefore,

$$\frac{\partial\widetilde{\Phi}}{\partial t} + \nabla\cdot\int_{\mathbb{R}^d} \boldsymbol{u}\,\widetilde{g}\,\mathrm{d}\boldsymbol{u} = 0. \quad (8.3.17)$$

Since we are working with spatial filtering kernels, $\mathcal{G} = \mathcal{G}(\boldsymbol{r})$,

$$\int_{\mathbb{R}^d} \boldsymbol{u}\,\widetilde{g}\,\mathrm{d}\boldsymbol{u} = \int_{\mathbb{R}^d} (\boldsymbol{u} - \widetilde{\boldsymbol{V}})\,\widetilde{g}\,\mathrm{d}\boldsymbol{u} + \int_{\mathbb{R}^d} \widetilde{\boldsymbol{V}}\,\widetilde{g}\,\mathrm{d}\boldsymbol{u}. \quad (8.3.18)$$

By plugging (8.3.18) into (8.3.17), we obtain that

$$\frac{\partial\widetilde{\Phi}}{\partial t} + \nabla\cdot\big(\widetilde{\Phi}\,\widetilde{\boldsymbol{V}}\big) = -\nabla\cdot\boldsymbol{q}, \quad (8.3.19)$$

where

$$q_i = \int_{\mathbb{R}^d} \left(u_i - \widetilde{V}_i\right) \widetilde{g}\, \mathrm{d}\boldsymbol{u}. \tag{8.3.20}$$

The corresponding filtered passive scalar flux is obtained through a sequence of step-by-step derivations as

$$\widetilde{\boldsymbol{q}} = -\mathcal{D}\, \nabla \widetilde{\Phi}, \tag{8.3.21}$$

and the divergence of residual scalar flux is derived as the fractional Laplacian of the filtered total scalar concentration,

$$\nabla \cdot \boldsymbol{q}^R = -\mathcal{D}_\alpha\, (-\Delta)^\alpha\, \widetilde{\Phi}, \quad \alpha \in (0, 1], \tag{8.3.22}$$

where

$$\mathcal{D}_\alpha := \frac{C_\alpha (c_T\, \tau_g)^\alpha}{\tau_g}\, (\alpha + 2)\, \Gamma(\alpha)$$

is a model coefficient with the unit $[L^\alpha/T]$. The filtered AD equation for the total passive scalar concentration, developed from the filtered kinetic BTE with an α-stable Lévy distribution model, yields a fractional-order SGS scalar flux model at the continuum level. The aforementioned filtered AD equation reads

$$\frac{\partial \widetilde{\Phi}}{\partial t} + \frac{\partial}{\partial x_i}\left(\widetilde{\Phi}\, \widetilde{V}_i\right) = \mathcal{D}\, \Delta \widetilde{\Phi} + \mathcal{D}_\alpha (-\Delta)^\alpha\, \widetilde{\Phi}. \tag{8.3.23}$$

Through a proper choice for the fractional Laplacian order α, the developed model optimally works in an LES setting. Applying the Reynolds decomposition and considering the passive scalar with imposed uniform mean gradient, (8.3.23) fully recovers the filtered transport equation for the transport of the filtered scalar fluctuations, $\widetilde{\phi}$.

Nonlocal Spectral Transfer Model and Scaling Law for Scalar Turbulence

Akhavan-Safaei and Zayernouri (2023a) revisited the spectral transfer model for the turbulent intensity in passive scalar transport, and proposed a physically meaningful modification to the scaling of scalar variance cascade, given by

$$E_\phi(k) \sim \chi\, \varepsilon^{-1/3}\, k^{-2/3}\, (k^2 + \mathcal{C}_\alpha\, k^{2\alpha})^{-1/2}, \tag{8.3.24}$$

in which χ represents the rate of spectral flux function, ε denotes the dissipation rate of turbulent kinetic energy (TKE) and k is the Fourier wave number. This generalizes the $-5/3$ law, which corresponds to $\alpha = 0$. The comparison between the classic scaling law and this generalized model is depicted in Figure 8.8. We begin with redefining the corresponding length-scale of the scalar transport, being traditionally approximated only as $1/k$. While this way

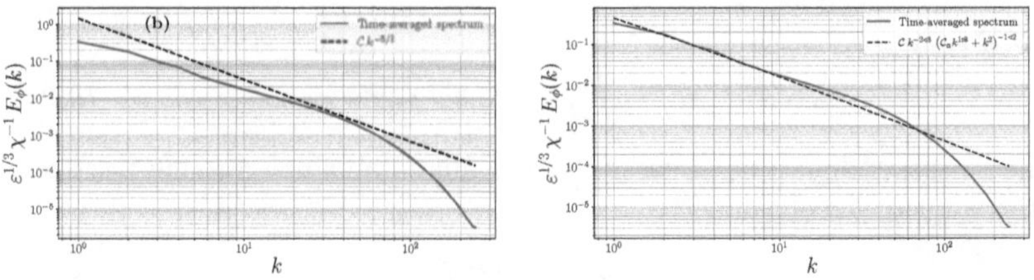

Figure 8.8 (Left) Turbulent scalar intensity $E_\Phi(k) \sim \chi\,\varepsilon^{-1/3}\,k^{-5/3}$, versus (Right) the modified (generalized) scaling law $E_\Phi(k) \sim \chi\,\varepsilon^{-1/3}\,k^{-2/3}\,(k^2 + \mathcal{C}_\alpha\,k^{2\alpha})^{-1/2}$, obtained from the DNS data. Source: Akhavan-Safaei and Zayernouri (2023a), reproduced with permission.

of thinking is quite consistent with the Brownian motion model at small scales (considering small jumps of finite variance) the authors argued using several experimental studies that the scalar turbulence (i.e., both the scalar increments and the scalar fluctuations) do not obey the K41 local-isotropy hypotheses, and they are anomalous and lead to nonlocal behavior. Hence, they modified the corresponding scalar variance length-scale in a way that it additionally included a new scale-free term to directly take the corresponding self-similar large jumps into account. This new inclusive length-scale, combined with A. N. Kolmagorov's velocity-scale and obtained from TKE, defined a new scalar time scale that leads to a new nonlocal power-law scaling for the cascade of scalar variance. From the generalized spectral transfer model, the authors obtained back a new fractional-order scalar transport model, which can be viewed as a re-derivation of their earlier fractional LES work, originally derived from the filtered Boltzmann transport equation in Akhavan-Safaei et al. (2021).

Fractional/Tempered-Fractional LES Models for Fluid Turbulence
For some pedagogical purposes, we first presented the case of fractional LES SGS modeling for scalar turbulence. However, this new paradigm in LES modeling actually began prior to Akhavan-Safaei et al. (2021). Samiee et al. (2020) developed the first ever fractional LES model for homogeneous isotropic turbulent flows as

$$(\nabla.\tau^R) = \mu_\alpha(-\Delta)^{(\alpha)}\bar{V}, \tag{8.3.25}$$

based on the derivation of the fractional Laplacian closure term in the spatially filtered Navier–Stokes equations when employing a Lévy stable distribution as the equilibrium model in the filtered BGK kinetic transport equation. To this end, a parallel computational–statistical framework for simulation of scalar and fluid turbulence with applications to data-driven fractional SGS modeling has been provided in Akhavan-Safaei and Zayernouri (2023b). In Samiee et al. (2022), they later developed a generalized version of this earlier model (suitable for incorporating data with tailored/truncated tails). Employing rather a *tempered Lévy stable* distribution in the kinetic level this time gave rise to the formulation of the tempered fractional LES closure term as

$$(\nabla.\tau^R) = \nu_\alpha \sum_{k=0}^{\kappa} \phi_k^\kappa (\Delta + \lambda_k)^{(\alpha)} \bar{V}, \tag{8.3.26}$$

forming a novel, data-friendly and expressive tempered fractional Laplacian SGS model for turbulence. They also showed that the newly developed nonlocal models can better recover the non-Gaussian statistics of subgrid-scale stress motions while they are being employed at the continuum level.

Dynamic Nonlocal LES Modeling

The recent developments in Samiee et al. (2020); Akhavan-Safaei et al. (2021); Samiee et al. (2022) offer a novel LES modeling paradigm for modeling the stochastic SGS motions. By employing fractional and tempered fractional Laplacian operators as additional linear terms to the filtered Navier–Stokes equations, this paradigm directly takes the superdiffusive nature of turbulence into account. However, all of the aforementioned models feature a static order of the fractional derivative that does not vary in time. To dynamically calculate the corresponding optimal fractional indices and tempering parameters throughout the flow simulation, dynamic nonlocal LES models have recently been formulated in Seyedi and Zayernouri (2022); Seyedi et al. (2022) along with the corresponding a priori and a posteriori tests. Such automated dynamic nonlocal LES modeling has been performed in the context of both flow and scalar turbulence, and has significantly enhanced the capability of the SGS LES models in prediction of TKE back-scattering and in properly addressing the notion of intermittency in the SGS dynamics. These features subsequently lead to sharp peaks and heavy tails of distributions in the small-scale motion statistics, e.g., in the dissipation of TKE and turbulent scalar variance.

Chaotic Transport in Quasi-geostrophic Flows

Long-term particle tracking has been used to study chaotic transport experimentally in laminar, chaotic, and turbulent flows in an annular tank that rotates sufficiently rapidly to insure two-dimensionality of the flow. For the laminar and chaotic velocity fields, the flow consists of flow regimes; tracer particles stick for long times to remnants of invariant surfaces around the vortices, then

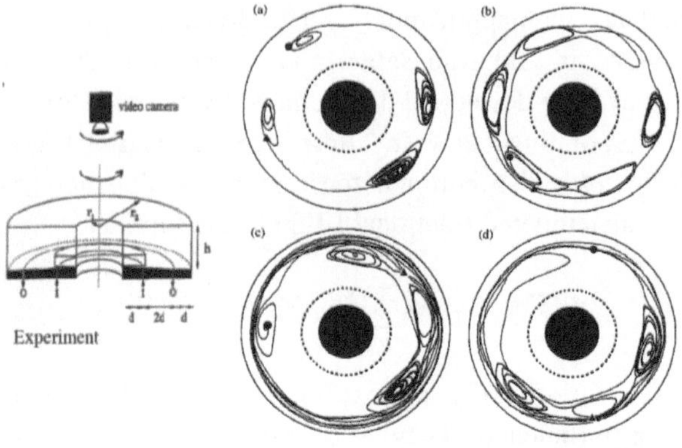

Figure 8.9 Chaotic transport in quasi-geostrophic flows: (Left) Cross section of the rotating annulus used by Solomon et al. (1994); (Right) the tracer particles stick for long times to remnants of invariant surfaces around the vortices, then make long excursions ("flights") in the jet regions. The mixing of an ensemble of tracer particles appear to be superdiffusive. Source: Reprinted from Solomon et al. (1994) with permission from Elsevier.

make long excursions ("flights") in the jet regions. The probability distributions for the flight time durations exhibit *power-law* rather than exponential decays, indicating that the particle trajectories are described mathematically as Lévy flights (i.e., the trajectories have infinite mean square displacement per flight). Sticking time probability distributions are also characterized by power-laws, as found in previous numerical studies. The mixing of an ensemble of tracer particles is superdiffusive: the variance of displacement grows with time as t^{λ} with $1 < \lambda < 2$. The dependence of diffusion exponent λ and the scaling of probability distributions are investigated for periodic and chaotic flow regimes, and the results are found to be consistent with theoretical predictions relating Lévy flights and anomalous diffusion, see Figure 8.9.

Future Directions in Fractional Turbulence Modeling

Laval et al. (2001) analyzed the effects of the local and nonlocal interactions on the intermittency corrections in the scaling properties in three-dimensional turbulence. They observed that nonlocal interactions are responsible for the creation of the intense vortices and on the other hand, local interactions are trying to dissipate them. Inspired by the mentioned observations, they came up with a new turbulence model that accounts for both the local and nonlocal interactions for the study of intermittency. In their proposed model, the large and small scales are being coupled by nonlocal interactions using a multiplicative process and additive noise along with a turbulent viscosity model for the local interactions. The results of the new model qualitatively cover the previously observed anomaly and intermittency aspects.

The aforementioned developments are practically interesting, mathematically exciting, and algorithmically robust. They encourage the field of research in turbulence to gradually open up towards a whole new wealth of recent mathematical developments in both theory and practice of fractional modeling. Inevitably, further systematic studies and developments of nonlocal turbulence models are needed (both numerically and experimentally) in order to achieve the charming blend of enhanced accuracy and lowered cost in realistic applications. On this note, we end this section by emphasizing that the non-Markovian/non-Fickian nature does not relax in compressible flows (i.e., variable density problems). Therefore the idea of developing generalized fractional turbulence models for transonic-to-hypersonic flows is a new and nourishing venue for research, in which the existing sense of thermodynamics can become fundamentally non-equilibrium and nonlocal.

References

Abel, Niels Henrik. 1823. Oplösning af et par opgaver ved hjelp af bestemte integraler. *Magazin for naturvidenskaberne*, **2**(55), 2.

Abramowitz, M., and Stegun, I.A. 1972. *Handbook of mathematical functions with formulas, graphs and mathematical tables*. New York: Dover.

Acosta, G., and Borthagaray, J. 2017. A fractional Laplace equation: Regularity of solutions and finite element approximations. *SIAM Journal on Numerical Analysis*, **55**, 472–495.

Acosta, Gabriel, Bersetche, Francisco M., and Borthagaray, Juan Pablo. 2017. A short FE implementation for a 2D homogeneous Dirichlet problem of a fractional Laplacian. *Computers & Mathematics with Applications*, **74**(4), 784–816.

Acosta, Gabriel, Borthagaray, Juan Pablo, and Heuer, Norbert. 2018. Finite element approximations of the nonhomogeneous fractional Dirichlet problem. *IMA Journal of Numerical Analysis*, **39**(3), 1471–1501.

Adams, R.A. 1975. *Sobolev Spaces*. New York: Academic Press.

Adda, F. B. 1998. Geometric interpretation of the differentiability and gradient of real order. *Comptes Rendus de l'Academie des Sciences Series I Mathematics*, **8**(326), 931–934.

Adler, Robert, Feldman, Raya, and Taqqu, Murad. 1998. *A practical guide to heavy tails: statistical techniques and applications*. Springer Science & Business Media.

Agarwal, Ravi P, Meehan, Maria, and O'Regan, Donal. 2001. *Fixed Point Theory and Applications*. Cambridge University Press.

Agranovich, M.S. 2015. *Sobolev Spaces, Their Generalizations and Elliptic Problems in Smooth and Lipschitz Domains*. Berlin: Springer.

Ainsworth, Mark, and Glusa, Christian. 2017. Aspects of an adaptive finite element method for the fractional Laplacian: A priori and a posteriori error estimates, efficient implementation and multigrid solver. *Computer Methods in Applied Mechanics and Engineering*, **327**, 4–35.

Ainsworth, Mark, and Glusa, Christian. 2018a. Hybrid finite element–spectral method for the fractional Laplacian: approximation theory and efficient solver. *SIAM Journal on Scientific Computing*, **40**(4), A2383–A2405.

Ainsworth, Mark, and Glusa, Christian. 2018b. Towards an efficient finite element method for the integral fractional Laplacian on polygonal domains. Pages 17–57 of: *Contemporary computational mathematics—a celebration of the 80th birthday of Ian Sloan. Vol. 1, 2.* Springer, Cham.

Ainsworth, Mark, and Mao, Zhiping. 2017a. Analysis and approximation of a fractional Cahn–Hilliard equation. *SIAM Journal on Numerical Analysis*, **55**(4), 1689–1718.

Ainsworth, Mark, and Mao, Zhiping. 2017b. Well-posedness of the Cahn-Hilliard equation with fractional free energy and its Fourier Galerkin approximation. *Chaos Solitons & Fractals*, **102**, 264–273.

Ainsworth, Mark, and Oden, J Tinsley. 1997. A posteriori error estimation in finite element analysis. *Computer Methods in Applied Mechanics and Engineering*, **142**(1-2), 1–88.

Akagi, Goro, Schimperna, Giulio, and Segatti, Antonio. 2016. Fractional Cahn-Hilliard, Allen-Cahn and porous medium equations. *Journal of Differential Equations*, **261**(6), 2935–2985.

Akhavan-Safaei, Ali, and Zayernouri, Mohsen. 2023a. A non-local spectral transfer model and new scaling law for scalar turbulence. *Journal of Fluid Mechanics*, **956**, A26.

Akhavan-Safaei, Ali, and Zayernouri, Mohsen. 2023b. A parallel computational-statistical framework for simulation of turbulence: Applications to data-driven fractional modeling. *Fractal and Fractional*, **7**(6), 488.

Akhavan-Safaei, Ali, Seyedi, S Hadi, and Zayernouri, Mohsen. 2020. Anomalous features in internal cylinder flow instabilities subject to uncertain rotational effects. *Physics of Fluids*, **32**(9), 094107.

Akhavan-Safaei, Ali, Samiee, Mehdi, and Zayernouri, Mohsen. 2021. Data-driven fractional subgrid-scale modeling for scalar turbulence: A nonlocal LES approach. *Journal of Computational Physics*, 110571.

Al-Maskari, M., and Karaa, S. 2019. Numerical approximation of semilinear subdiffusion equations with nonsmooth initial data. *SIAM Journal on Numerical.Analysis*, **57**(3), 1524–1544.

Alfano, Giulio, and Musto, Marco. 2017. Thermodynamic derivation and damage evolution for a fractional cohesive zone model. *Journal of Engineering Mechanics*, **143**(7), D4017001.

Andrews, G.E., Askey, R., and Roy, R. 1999. *Special functions, encyclopedia of mathematics and its applications, Vol. 71.* Cambridge: Cambridge University Press.

Antil, Harbir, and Rautenberg, Carlos N. 2019. Sobolev spaces with non-Muckenhoupt weights, fractional elliptic operators, and applications. *SIAM Journal on Mathematical Analysis*, **51**(3), 2479–2503.

Antil, Harbir, Pfefferer, Johannes, and Rogovs, Sergejs. 2017. Fractional operators with inhomogeneous boundary conditions: Analysis, control, and discretization. *arXiv preprint arXiv:1703.05256.*

Antil, Harbir, Verma, Deepanshu, and Warma, Mahamadi. 2019. External optimal control of fractional parabolic PDEs. *arXiv preprint arXiv:1904.07123.*

Aomoto, Kazuhiko, Kita, Michitake, Kohno, Toshitake, and Iohara, Kenji. 2011. *Theory of Hypergeometric Functions.* Springer.

Applebaum, David. 2009. *Lévy processes and stochastic calculus.* Cambridge University Press.

Arsenlis, A., Cai, W., Tang, M., Rhee, M., Oppelstrup, T., Hommes, G., Pierce, T.G., and Bulatov, V.V. 2007. Enabling Strain Hardening Simulations with Dislocation Dynamics. *Modeling and Simulation in Materials Science and Engineering*, **15**(6).

Askey, R. 1975. *Orthogonal Polynomials and Special Functions*. Society for Industrial and Applied Mathematics, Philadelphia, PA.

Askey, R., and Fitch, J. 1969. Integral Representations for Jacobi Polynomials and Some Applications. *Journal of Mathematical Analysis and Applications*, **26**, 411–437.

Atanackovic, T.M., Oparnica, L., and Pilipović, S. 2009. Distributional framework for solving fractional differential equations. *Integral Transforms and Special Functions*, **20**(3-4), 215–222.

Atkinson, K.E. 1997. *The numerical solution of integral equations of the second kind*. Cambridge Monographs on Applied and Computational Mathematics, vol. 4. Cambridge University Press, Cambridge.

Bachelier, Louis. 1900. Théorie de la spéculation. Pages 21–86 of: *Annales scientifiques de l'École normale supérieure*, vol. 17.

Baeumer, B., and Meerschaert, M. M. 2010. Tempered stable Lévy motion and transient super-diffusion. *Journal of Computational and Applied Mathematics*, **233**(10), 2438–2448.

Baeumer, Boris, Kovács, Mihály, Meerschaert, Mark M., and Sankaranarayanan, Harish. 2018a. Boundary conditions for fractional diffusion. *Journal of Computational and Applied Mathematics*, **336**, 408–424.

Baeumer, Boris, Luks, Tomasz, and Meerschaert, Mark M. 2018b. Space-time fractional Dirichlet problems. *Mathematische Nachrichten*, **291**(17-18), 2516–2535.

Baffet, Daniel, and Hesthaven, Jan S. 2017. High-Order Accurate Adaptive Kernel Compression Time-Stepping Schemes for Fractional Differential Equations. *Journal of Scientific Computing*, **72**(3), 1169–1195.

Baghdadi, H.A., Sardinha, H., and Bhatia, S.R. 2005. Rheology and gelation kinetics in laponite dispersions containing poly(ethylene oxide). *Journal of Polymer Science Part B: Polymer Physics*, **43**(2), 233–240.

Bagley, Ronald L. 1989. Power law and fractional calculus model of viscoelasticity. *AIAA journal*, **27**(10), 1412–1417.

Bagley, R.L., and Torvik, P.J. 1983. A theoretical basis for the application of fractional calculus to viscoelasticity. *Journal of Rheology*, **27**, 201–210.

Bagley, RL, and Torvik, PJ. 2000. On the existence of the order domain and the solution of distributed order equations-Part I. *International Journal of Applied Mathematics*, **2**(7), 865–882.

Balakrishnan, A. V. 1960. Fractional powers of closed operators and the semigroups generated by them. *Pacific Journal of Mathematics*, **10**, 419–437.

Banjai, L., and López-Fernández, M. 2019. Efficient high order algorithms for fractional integrals and fractional differential equations. *Numer. Math*, **141**(2), 289–317.

Barlebo, Heidi Christiansen, Hill, Mary C, and Rosbjerg, Dan. 2004. Investigating the Macrodispersion Experiment (MADE) site in Columbus, Mississippi, using a three-dimensional inverse flow and transport model. *Water Resources Research*, **40**(4).

Barles, Guy, Chasseigne, Emmanuel, Georgelin, Christine, and Jakobsen, Espen R. 2014. On Neumann type problems for nonlocal equations set in a half space. *Transactions of the American Mathematical Society*, **366**, 4873–4917.

Barros de Moraes, Eduardo A., Zayernouri, Mohsen, and Meerschaert, Mark M. 2021. An integrated sensitivity-uncertainty quantification framework for stochastic phase-field modeling of material damage. *International Journal for Numerical Methods in Engineering*, **122**(5), 1352–1377.

Barzilai, J, and Borwein, J.M. 1988. Two-Point Step Size Gradient Methods. *IMA Journal of Numerical Analysis*, **8**, 141–148.

Bas, E., and Metin, F. 2012. Spectral properties of fractional Sturm-Liouville problem for diffusion operator. *arXiv preprint arXiv:1212.4761*, 1–11.

Bateman, H. 1909. The solution of linear differential equations by means of definite integrals. *Trans. Cambridge Phil. Soc*, **21**, 171–196.

Bateman, Harry. 1953. *Higher transcendental functions [Volumes I-III]*.

Becken, W., and Schmelcher, P. 2000. The analytic continuation of the Gaussian hypergeometric function ${}_2F_1(a, b; c; z)$ for arbitrary parameters. *Journal of Computational and Applied Mathematics*, **126**(1-2), 449–478.

Beghin, L. 2015. On fractional tempered stable processes and their governing differential equations. *Journal of Computational Physics*, **293**, 29–39.

Beltempo, Angela, Zingales, Massimiliano, Bursi, Oreste S, and Deseri, Luca. 2018. A fractional-order model for aging materials: An application to concrete. *International Journal of Solids and Structures*, **138**, 13–23.

Bencherif-Madani, A., and Pardoux, E. 2009. A probabilistic formula for a Poisson equation with Neumann boundary condition. *Stochastic Analysis and Applications*, **27**(4), 739–746.

Benson, D.A., Wheatcraft, S.W., and Meerschaert, M.M. 2000. Application of a fractional advection-dispersion equation. *Water Resources Research*, **36**(6), 1403–1412.

Benson, D.A., Schumer, R., Meerschaert, M.M., and Wheatcraft, S.W. 2001. Fractional Dispersion, Lévy Motion, and the MADE Tracer Tests. *Transport in Porous Media*, **42**, 211–240.

Bernardi, C., and Maday, Y. 1997. Spectral methods. *Handbook of Numerical Analysis*, **5**, 209–485.

Biagini, Francesca, Hu, Yaozhong, Øksendal, Bernt, and Zhang, Tusheng. 2008. *Stochastic calculus for fractional Brownian motion and applications*. Springer Science & Business Media.

Blair, GW Scott, Veinoglou, BC, and Caffyn, JE. 1947. Limitations of the Newtonian time scale in relation to non-equilibrium rheological states and a theory of quasi-properties. *Proceedings of the Royal Society of London. Series A. Mathematical and Physical Sciences*, **189**(1016), 69–87.

Blumenthal, RM, Getoor, RK, and Ray, DB. 1961. On the distribution of first hits for the symmetric stable processes. *Transactions of the American Mathematical Society*, **99**(3), 540–554.

Bogdan, K., Burdzy, K., and Chen, Z.-Q. 2003. Censored Stable Process. *Probability Theory and Related Fields*, **127**, 89–152.

Bogdan, Krzysztof, Byczkowski, Tomasz, Kulczycki, Tadeusz, Ryznar, Michal, Song, Renming, and Vondracek, Zoran. 2009. *Potential analysis of stable processes and its extensions*. Springer Science & Business Media.

Boggio, Tommaso. 1905. Sulle funzioni di Green d'ordinem. *Rendiconti del Circolo Matematico di Palermo (1884-1940)*, **20**(1), 97–135.

Bonadkar, N., Gerum, R., Kuhn, M., Sporer, M., Lippert, A., Schneider, W., Aifantis, K.E., and Fabry, B. 2016. Mechanical plasticity of cells. *Nature Materials*, **15**, 1090–1094.

Bonfanti, Alessandra, Kaplan, Jonathan Louis, Charras, Guillaume, and Kabla, Alexandre J. 2020. Fractional viscoelastic models for power-law materials. *Soft Matter*.

Boniece, B. C., Sabzikar, F., and Didier, G. 2018. Tempered fractional Brownian motion: Wavelet estimation and modeling of turbulence in geophysical flows. Pages 174–178 of: *2018 IEEE Statistical Signal Processing Workshop (SSP)*. IEEE.

Bonito, Andrea, and Pasciak, Joseph E. 2015. Numerical approximation of fractional powers of elliptic operators. *Mathematics of Computation.*, **84**(295), 2083–2110.

Bonito, Andrea, Borthagaray, Juan Pablo, Nochetto, Ricardo H, Otarola, Enrique, and Salgado, Abner J. 2018. Numerical methods for fractional diffusion. *Computing and Visualization in Science*, 1–28.

Bonito, Andrea, Lei, Wenyu, and Pasciak, Joseph E. 2019. On Sinc quadrature approximations of fractional powers of regularly accretive operators. *Journal of Numerical Mathematics*, **27**(2), 57–68.

Borthagaray, Juan Pablo, Del Pezzo, Leandro M, and Martínez, Sandra. 2018. Finite element approximation for the fractional eigenvalue problem. *Journal of Scientific Computing*, **77**(1), 308–329.

Boyd, S., and Vandenberghe, L. 2004. *Convex optimization*. Cambridge University Press.

Bradshaw, P. 1973. AGARDograph, No. 169. *Nato Science and Technology Organisation, USA*.

Brosamler, G. 1967. A Probabilistic Solution of the Neumann Problem. *Mathematica Scandinavica*, **38**, 137–147.

Brunner, Hermann. 1986. Polynomial spline collocation methods for Volterra integrodifferential equations with weakly singular kernels. *IMA Journal of Numerical Analysis*, **6**(2), 221–239.

Brunner, Hermann. 1990. On the numerical solution of nonlinear Volterra–Fredholm integral equations by collocation methods. *SIAM Journal on Numerical Analysis*, **27**(4), 987–1000.

Brunner, H. 2017. *Volterra Integral Equations*. Cambridge: Cambridge University Press.

Bucur, Claudia. 2016. Some observations on the Green function for the ball in the fractional Laplace framework. *Communications on Pure & Applied Analysis*, **15**, 657.

Bucur, Claudia, and Valdinoci, Enrico. 2016. *Nonlocal Diffusion and Applications*. Vol. 20. Springer.

Bulatov, Vasily, Abraham, Farid F, Kubin, Ladislas, Devincre, Benoit, and Yip, Sidney. 1998. Connecting atomistic and mesoscale simulations of crystal plasticity. *Nature*, **391**(6668), 669–672.

Burkovska, Olena, and Gunzburger, Max. 2018. Regularity and approximation analyses of nonlocal variational equality and inequality problems. *arXiv preprint arXiv:1804.10282*.

Burkovska, Olena, Glusa, Christian, and D'Elia, Marta. 2021. An optimization-based approach to parameter learning for fractional type nonlocal models. *Computer and Mathematics with Applications*. to appear.

Caffarelli, Luis, and Silvestre, Luis. 2007. An extension problem related to the fractional Laplacian. *Communications in Partial Differential Equations*, **32**(8), 1245–1260.

Caffarelli, Luis A, and Stinga, Pablo Raúl. 2016. Fractional elliptic equations, Caccioppoli estimates and regularity. Pages 767–807 of: *Annales de l'Institut Henri Poincare (C) Non Linear Analysis*, vol. 33. Elsevier.

Cagliero, L., and Koornwinder, T.H. 2015. Explicit matrix inverses for lower triangular matrices with entries involving Jacobi polynomials. *Journal of Approximation Theory*, **193**, 20–38.

Canuto, C., Hussaini, M. Y., Quarteroni, A., and Zang, T. A. 2006. *Spectral methods*. Scientific Computation. Springer-Verlag, Berlin. Fundamentals in Single Domains.

Canuto, Claudio, Hussaini, M Yousuff, Quarteroni, Alfio, and Zang, Thomas A. 2007. *Spectral Methods: Fundamentals in Single Domains*. Springer Science & Business Media.

Cao, W., Zeng, F., Zhongqiang, Z., and Karniadakis, G.E. 2016. Implicit-Explicit Difference Schemes for Nonlinear Fractional Differential Equations with Nonsmooth Solutions. *SIAM Journal on Scientific Computing*, **38**, A3070–A3093.

Cao, Yanzhao, Herdman, Terry, and Xu, Yuesheng. 2003. A hybrid collocation method for Volterra integral equations with weakly singular kernels. *SIAM Journal on Numerical Analysis*, **41**(1), 364–381.

Caputo, M. 1995. Mean fractional-order-derivatives differential equations and filters. *Annali dell'Universita di Ferrara*, **41**(1), 73–84.

Caputo, M. 2001. Distributed order differential equations modelling dielectric induction and diffusion. *Fractional Calculus and Applied Analysis*, **4**(4), 421–442.

Caputo, Michele, and Fabrizio, Mauro. 2015. Damage and fatigue described by a fractional derivative model. *Journal of Computational Physics*, **293**, 400–408.

Carr, P., Geman, H., Madan, D.B., and Yor, M. 2002. The fine structure of asset returns: An empirical investigation. *The Journal of Business*, **75**(2), 305–333.

Cartea, Á., and del Castillo-Negrete, D. 2007. Fluid limit of the continuous-time random walk with general Lévy jump distribution functions. *Physical Review E*, **76**(4), 041105.

Chechkin, A. V., Gorenflo, R., and Sokolov, I. M. 2002. Retarding subdiffusion and accelerating superdiffusion governed by distributed-order fractional diffusion equations. *Physical Review E*, **66**(4), 046129.

Chechkin, A. V., Klafter, J., and Sokolov, I. M. 2003. Fractional Fokker-Planck equation for ultraslow kinetics. *Europhysics Letters*, **63**(3), 326.

Chechkin, A. V., Gonchar, V. Y., Gorenflo, R., Korabel, N., and Sokolov, I. M. 2008. Generalized fractional diffusion equations for accelerating subdiffusion and truncated Lévy flights. *Physical Review E*, **78**(2), 021111.

Chen, Minghua, and Deng, Weihua. 2015. Discretized fractional substantial calculus. *ESAIM: Mathematical Modelling and Numerical Analysis*, **49**(2), 373–394.

Chen, S., Shen, J., and Wang, L. L. 2016. Generalized Jacobi functions and their applications to fractional differential equations. *Mathematics of Compututation*, **85**(300), 1603–1638.

Chen, Sheng, and Shen, Jie. 2022. Log orthogonal functions: approximation properties and applications. *IMA Journal of Numerical Analysis*, **42**(1), 712–743.

Chen, Sheng, Shen, Jie, Zhang, Zhimin, and Zhou, Zhi. 2020. A spectrally accurate approximation to subdiffusion equations using the log orthogonal functions. *SIAM Journal on Scientific Computing*, **42**(2), A849–A877.

Chen, Sheng, Shen, Jie, and Wang, Li-Lian. 2018. Laguerre functions and their applications to tempered fractional differential equations on infinite intervals. *Journal of Scientific Computing*, **74**, 1286–1313.

Chen, Wen. 2006a. A speculative study of 2/ 3-order fractional Laplacian modeling of turbulence: Some thoughts and conjectures. *Chaos: An Interdisciplinary Journal of Nonlinear Science*, **16**(2), 023126.

Chen, Wen. 2006b. A speculative study of fractional Laplacian modeling of turbulence. *IFAC Proceedings Volumes*, **39**(11), 442–447.

Chen, W., and Holm, S. 2004. Fractional Laplacian time-space models for linear and nonlinear lossy media exhibiting arbitrary frequency power-law dependency. *The Journal of the Acoustical Society of America*, **115**(4), 1424–1430.

Chen, Wen, and Pang, Guofei. 2016. A new definition of fractional Laplacian with application to modeling three-dimensional nonlocal heat conduction. *Journal of Computational Physics*, **309**, 350–367.

Chen, Wen, Fu, Zhuo-Jia, and Chen, Ching-Shyang. 2014. *Recent Advances in Radial Basis Function Collocation Methods*. Springer.

Chen, Y.-M., Wei, Y.-Q., Liu, D.-Y., and Yu, H. 2015. Numerical solution for a class of nonlinear variable order fractional differential equations with Legendre wavelets. *Applied Mathematics Letters*, **46**, 83–88.

Chen, Zhen-Qing, and Song, Renming. 2005. Two-sided eigenvalue estimates for subordinate processes in domains. *Journal of Functional Analysis*, **226**(1), 90–113.

Chen, Z.Q., and Kumagai, T. 2008. Heat kernel estimates for jump processes of mixed types on metric measure spaces. *Probability Theory and Related Fields*, **140**(1-2), 277–317.

Chen, Zhen-Qing, Meerschaert, Mark M, and Nane, Erkan. 2012. Space–time fractional diffusion on bounded domains. *Journal of Mathematical Analysis and Applications*, **393**(2), 479–488.

Chihara, T. S. 1955. *Generalized Hermite polynomials*. Purdue University.

Chihara, T. S. 1978. *An introduction to orthogonal polynomials*. Gordon and Breach Science Publishers, New York-London-Paris. Mathematics and its Applications, Vol. 13.

Chorin, Alexandre Joel, Marsden, Jerrold E, and Marsden, Jerrold E. 1990. *A mathematical introduction to fluid mechanics*. Vol. 168. Springer.

Ciaurri, Óscar, Roncal, Luz, Stinga, Pablo Raúl, Torrea, José L, and Varona, Juan Luis. 2018. Nonlocal discrete diffusion equations and the fractional discrete Laplacian, regularity and applications. *Advances in Mathematics*, **330**, 688–738.

Clarke, D. D., Meerschaert, M. M., and Wheatcraft, S. W. 2005. Fractal travel time estimates for dispersive contaminants. *Groundwater*, **43**(3), 401–407.

Constantin, Peter, and Vicol, Vlad. 2012. Nonlinear maximum principles for dissipative linear nonlocal operators and applications. *Geometric And Functional Analysis*, **22**(5), 1289–1321.

Constantin, Peter, and Wu, Jiahong. 1999. Behavior of solutions of 2D quasi-geostrophic equations. *SIAM Journal on Mathematical Analysis*, **30**(5), 937–948.

Cont, Rama, and Tankov, Peter. 2003. *Financial modelling with jump processes*. CRC Press.

Cont, Rama, and Tankov, Peter. 2004. *Financial modelling with jump processes*. Chapman & Hall/CRC Financial Mathematics Series. Chapman & Hall/CRC, Boca Raton, FL.

Costabile, F.A., and Longo, E. 2010. A Birkhoff interpolation problem and application. *Calcolo*, **47**(1), 49–63.

Craiem, Damian, Rojo, Francisco J, Atienza, Jose Miguel, Armentano, Ricardo L, and Guinea, Gustavo V. 2008. Fractional-order viscoelasticity applied to describe uniaxial stress relaxation of human arteries. *Physics in Medicine & Biology*, **53**(17), 4543.

Cuesta, E., Lubich, C., and Palencia, C. 2006. Convolution quadrature time discretization of fractional diffusion-wave equations. *Mathematics of Computation*, **75**(254), 673–696.

Cusimano, N., del Teso, F., Gerardo-Giorda, L., and Pagnini, G. 2018. Discretizations of the spectral fractional Laplacian on general domains with Dirichlet, Neumann, and Robin boundary conditions. *SIAM Journal on Numerical Analysis*, **56**(3), 1243–1272.

Dai, F., and Xu, Y. 2013. *Approximation theory and harmonic analysis on spheres and balls*. Springer Monographs in Mathematics. Springer, New York.

Darve, Eric, D'Elia, Marta, Garrappa, Roberto, Giusti, Andrea, and Rubio, Natalia L. 2021. On the fractional Laplacian of variable order. *arXiv preprint arXiv:2109.01060*.

Davidson, Peter Alan. 2015. *Turbulence: an introduction for scientists and engineers*. Oxford University Press.

Davis, P.J. 1975. *Interpolation and Approximation*. New York: Dover Publications.

de Moraes, Eduardo A Barros, Suzuki, Jorge L, and Zayernouri, Mohsen. 2021. Atomistic-to-meso multi-scale data-driven graph surrogate modeling of dislocation glide. *Computational Materials Science*, **197**, 110569.

de Moraes, Eduardo A Barros, D'Elia, Marta, and Zayernouri, Mohsen. 2023. Machine learning of nonlocal micro-structural defect evolutions in crystalline materials. *Computer Methods in Applied Mechanics and Engineering*, **403**, 115743.

de Pablo, Arturo, Quirós, Fernando, Rodríguez, Ana, and Vázquez, Juan Luis. 2011. A fractional porous medium equation. *Advances in Mathematics*, **226**(2), 1378–1409.

DeBlassie, R. 1990. The first exit time of a two-dimensional symmetric stable process from a wedge. *Annals of Probability*, **18**, 1034–1070.

Defterli, O. and D'Elia, M. and Du, Q. and Gunzburger, M. and Lehoucq, R. and Meerschaert, M. M. 2015. Fractional diffusion on bounded domains. *Fractional Calculus and Applied Analysis*, **18**(2), 342–360.

Delbosco, D., and Rodino, L. 1996. Existence and Uniqueness for a Nonlinear Fractional Differential Equation. *Journal of Mathematical Analysis and Applications*, **204**(2), 609–625.

D'Elia, Marta, and Glusa, Christian. 2021. A fractional model for anomalous diffusion with increased variability. Analysis, algorithms and applications to interface problems. *arXiv preprint arXiv:2101.11765*. accepted in NMPDEs.

D'Elia, Marta, and Gunzburger, Max. 2013. The fractional Laplacian operator on bounded domains as a special case of the nonlocal diffusion operator. *Computers & Mathematics with Applications. An International Journal*, **66**(7), 1245–1260.

Deng, Weihua. 2007. Short memory principle and a predictor-corrector approach for fractional differential equations. *Journal of Computational and Applied Mathematics*, **206**(1), 174–188.

Di Leoni, Patricio Clark, Zaki, Tamer A, Karniadakis, George, and Meneveau, Charles. 2021. Two-point stress–strain-rate correlation structure and non-local eddy viscosity in turbulent flows. *Journal of Fluid Mechanics*, **914**.

Diethelm, Kai. 1997. Generalized compound quadrature formulae for finite-part integrals. *IMA Journal of Numerical Analysis*, **17**(3), 479–493.

Diethelm, K. 2010. *The analysis of fractional differential Equations, Lecture Notes in Math., Vol. 2004*. Berlin: Springer.

Diethelm, K., and Ford, N. J. 2002. Analysis of Fractional Differential Equations. *Journal of Mathematical Analysis and Applications*, **265**(2), 229–248.

Diethelm, K., Ford, J.M., Ford, N.J., and Weilbeer, M. 2006. Pitfalls in fast numerical solvers for fractional differential equations. *Journal of Computational Applied Mathematics*, **186**, 482–503.

Diethelm, Kai, Ford, Neville J, and Freed, Alan D. 2004. Detailed error analysis for a fractional Adams method. *Numerical Algorithms*, **36**(1), 31–52.

Ding, W., Patnaik, S., Sidhardh, S., and Semperlotti, F. 2021. Applications of distributed-order fractional operators: A review. *Entropy*, **23**, 110.

Ding, Zhonghai. 1996. A proof of the trace theorem of Sobolev spaces on Lipschitz domains. *Proceedings of the American Mathematical Society*, **124**(2), 591–600.

Dipierro, Serena, Ros-Oton, Xavier, and Valdinoci, Enrico. 2017. Nonlocal problems with Neumann boundary conditions. *Revista Matemática Iberoamericana*, **33**, 377–416.

Djordjević, V.D., Jarić, J., Fabry, B., Fredberg, J.J., and Stamenović, Dimitrije. 2003. Fractional Derivatives Embody Essential Features of Cell Rheological Behavior. *Annals of Biomedical Engineering*, **31**(6), 692–699.

Doehring, T.C., Freed, A.D., Carew, E.O., and Vesely, I. 2005. Fractional order viscoelasticity of the aortic valve cusp: an alternative to quasilinear viscoelasticity. *Journal of Biomechanical Engineering*, **127**(4), 700–708.

Doornik, Jurgen A. 2015. Numerical evaluation of the Gauss hypergeometric function by power summations. *Mathematics of Computation*, **84**(294), 1813–1833.

Du, Q., Gunzburger, M., Lehoucq, R., and Zhou, K. 2012. Analysis and approximation of nonlocal diffusion problems with volume constraints. *SIAM Review*, **54**, 667–696.

Du, Qiang, Yang, Jiang, and Zhou, Zhi. 2020. Time-fractional Allen–Cahn equations: analysis and numerical methods. *Journal of Scientific Computing*, **85**(2), 42.

Duan, Beiping, Jin, Bangti, Lazarov, Raytcho, Pasciak, Joseph, and Zhou, Zhi. 2017. Space-time Petrov–Galerkin FEM for fractional diffusion problems. *Computational Methods in Applied Mathematics.*

Duan, Beiping, Jin, Bangti, Lazarov, Raytcho, Pasciak, Joseph, and Zhou, Zhi. 2018. Space-time Petrov–Galerkin FEM for fractional diffusion problems. *Computational Methods in Applied Mathematics*, **18**(1), 1–20.

Duan, J. S., and Baleanu, D. 2018. Steady periodic response for a vibration system with distributed order derivatives to periodic excitation. *Journal of Vibration and Control*, **24**(14), 3124–3131.

Duo, Siwei, Wang, Hong, and Zhang, Yanzhi. 2019. A comparative study on nonlocal diffusion operators related to the fractional Laplacian. *Discrete & Continuous Dynamical Systems - B*, **24**, 231.

Duoandikoetxea, J. 2001. *Fourier Analysis.* Graduate Studies in Mathematics, vol. 29. American Mathematical Society, Providence, RI. Translated and revised from the 1995 Spanish original by David Cruz-Uribe.

Dyda, Bartłlomiej. 2012. Fractional calculus for power functions and eigenvalues of the fractional Laplacian. *Fractional Calculus and Applied Analysis*, **15**(4), 536–555.

Eames, Ian, and Flor, Jan-Bert. 2011. *New Developments in Understanding Interfacial Processes in Turbulent Flows.*

Egolf, Peter W, and Hutter, Kolumban. 2017. Fractional turbulence models. Pages 123–131 of: *Progress in Turbulence VII.* Springer.

Egolf, Peter William, and Hutter, K. 2020. Nonlinear, nonlocal and fractional furbulence. *Graduate Studies in Mathematics. Springer.*

Einstein, Albert. 1905. On the motion of small particles suspended in liquids at rest required by the molecular-kinetic theory of heat. *Annalen der Physik*, **17**(549-560), 208.

Emmons, Howard W. 1970. Critique of numerical modeling of fluid-mechanics phenomena. *Annual Review of Fluid Mechanics*, **2**(1), 15–36.

Engheta, N. 1998. Fractional curl operator in electromagnetics. *Microwave and Optical Technology Letters*, **17**(2), 86–91.

Epps, Brenden P, and Cushman-Roisin, Benoit. 2018. Turbulence modeling via the fractional Laplacian. *arXiv preprint arXiv:1803.05286.*

Eringen, A Cemal, and Wegner, JL. 2003. Nonlocal continuum field theories. *Appl. Mech. Rev.*, **56**(2), B20–B22.

Ern, Alexandre, and Guermond, Jean-Luc. 2013. *Theory and practice of finite elements.* Vol. 159. Springer Science & Business Media.

Ervin, V, Heuer, Norbert, and Roop, J. 2018. Regularity of the solution to 1-D fractional order diffusion equations. *Mathematics of Computation*, **87**(313), 2273–2294.

Ervin, V. J., and Roop, J. P. 2006. Variational formulation for the stationary fractional advection dispersion equation. *Numerical Methods for Partial Differential Equations*, **22**(3), 558–576.

Esmaeili, S., Shamsi, M., and Luchko, Y. 2011. Numerical solution of fractional differential equations with a collocation method based on Müntz polynomials. *Computers & Mathematics with Applications*, **62**(3), 918–929.

Ethier, Stewart N, and Kurtz, Thomas G. 2009. *Markov processes: Characterization and convergence.* Vol. 282. John Wiley & Sons.

Fabrizio, M. 2013. Fractional rheological models for thermomechanical systems. Dissipation and free energies. *Fractional Calculus and Applied Analysis*, **17**(1), 206–223.

Falconer, Kenneth. 2003. *Fractal geometry*. Second edn. Hoboken, NJ: John Wiley & Sons Inc. Mathematical foundations and applications.

Fischer, M. 2019. Fast and parallel Runge-Kutta approximation of fractional evolution equations. *SIAM Journal on Scientific Computing*, **41**(2), A927–A947.

Fix, GJ, and Roof, JP. 2004. Least squares finite-element solution of a fractional order two-point boundary value problem. *Computers & Mathematics with Applications*, **48**(7), 1017–1033.

Freidlin, Mark. 1985. *Functional Integration and Partial Differential Equations*. Annals of Mathematics Studies, vol. 109. Princeton University Press, Princeton, NJ.

Friedlander, Friedrich Gerard, Joshi, Mark Suresh, Joshi, M, and Joshi, Mohan C. 1998. *Introduction to the Theory of Distributions*. Cambridge University Press.

Fu, H., Ng, M.K., and Wang, H. 2017. A divide-and-conquer fast finite difference method for space-time fractional partial differential equation. *Computers & Mathematics with Applications*, **73**, 1233–1242.

Fung, Yuan-cheng. 2013. *Biomechanics: Mechanical Properties of Living Tissues*. Springer Science & Business Media.

Galé, José E, Miana, Pedro J, and Stinga, Pablo Raúl. 2013. Extension problem and fractional operators: semigroups and wave equations. *Journal of Evolution Equations*, **13**(2), 343–368.

Galeone, Luciano, and Garrappa, Roberto. 2006. On multistep methods for differential equations of fractional order. *Mediterranean Journal of Mathematics*, **3**(3), 565–580.

Galeone, Luciano, and Garrappa, Roberto. 2009. Explicit methods for fractional differential equations and their stability properties. *Journal of Computational and Applied Mathematics*, **228**(2), 548–560.

Gao, G-h., Sun, Z-z., and Zhang, H-w. 2014. A new fractional numerical differentiation formula to approximate the Caputo fractional derivative and its applications. *Journal of Computational Physics*, **259**, 33–50.

Garrappa, Roberto. 2010. On linear stability of predictor–corrector algorithms for fractional differential equations. *International Journal of Computer Mathematics*, **87**(10), 2281–2290.

Garrappa, R. 2015. Trapezoidal methods for fractional differential equations: Theoretical and computational aspects. *Mathematics and Computers in Simulation*, **110**, 96–112.

Gatto, Paolo, and Hesthaven, Jan S. 2015. Numerical approximation of the fractional Laplacian via Hp-finite elements, with an application to image denoising. *Journal of Scientific Computing*, **65**(1), 249–270.

Gautschi, Walter. 2002. Gauss quadrature approximations to hypergeometric and confluent hypergeometric functions. *Journal of Computational Applied Mathematics*, **139**(1), 173–187.

Gautschi, W. 2004. *Orthogonal polynomials: computation and approximation*. New York: Oxford University Press.

Gerasimov, A. 1948. A generalization of linear laws of deformation and its applications to problems of internal friction. *Prikladnaia Matematika i Mekhanica (Journal of Applied Mathematics and Mechanics)*, **12**(3), 251–260.

Getoor, R. 1961. First passage times for symmetric stable processes in space. *Transactions of the American Mathematical Society*, **101**, 75–90.

Girimaji, Sharath S. 2007. Boltzmann kinetic equation for filtered fluid turbulence. *Physical Review Letters*, **99**(3), 034501.

Glöckle, Walter G, and Nonnenmacher, Theo F. 1993. Fox function representation of non-Debye relaxation processes. *Journal of Statistical Physics*, **71**(3), 741–757.

Gottlieb, David, and Orszag, Steven A. 1977. *Numerical analysis of spectral methods: theory and applications*. SIAM.

Gradshteyn, I.S., and Ryzhik, I.M. 2015. *Table of integrals, series and products, 8th Ed.* New York: Academic Press.

Grubb, Gerd. 2015. Fractional Laplacians on domains, a development of Hörmander's theory of μ-transmission pseudodifferential operators. *Advances in Mathematics*, **268**, 478–528.

Grubb, Gerd. 2016. Regularity of spectral fractional Dirichlet and Neumann problems. *Mathematische Nachrichten*, **289**(7), 831–844.

Guan, Qing-Yang, and Ma, Zhi-Ming. 2005. Boundary problems for fractional Laplacians. *Stochastics and Dynamics*, **5**(03), 385–424.

Guan, Qing-Yang, and Ma, Zhi-Ming. 2006. Reflected symmetric α-stable processes and the regional fractional Laplacian. *Probability Theory and Related Fields*, **134**, 649–694.

Gulian, Mamikon. 2019. *Discovering and solving fractional-order partial differential equations: Machine learning and Monte Carlo methods*. Ph.D. thesis, Brown University.

Gulian, Mamikon, and Pang, Guofei. 2018. Stochastic solution of elliptic and parabolic boundary-value problems for the spectral fractional Laplacian. *arXiv preprint arXiv:1812.01206*.

Guo, B. Y. 1998. *Spectral methods and their applications*. World Scientific.

Guo, B. Y., Wang, L. L., and Wang, Z. Q. 2006a. Generalized Laguerre interpolation and pseudospectral method for unbounded domains. *SIAM Journal on Numerical Analysis*, **43**(6), 2567–2589.

Guo, B.Y., Shen, J., and Wang, L.-L. 2006b. Optimal spectral-Galerkin methods using generalized Jacobi polynomials. *Journal of Scientific Computing*, **27**(1-3), 305–322.

Guo, B.Y., Shen, J., and Wang, L.-L. 2009. Generalized Jacobi polynomials/functions and their applications. *Applied Numerical Mathematics*, **59**(5), 1011–1028.

Haas, Markus, and Pigorsch, Christian. 2009. Financial economics, fat-tailed distributions. *Encyclopedia of Complexity and Systems Science*, **4**(1), 3404–3435.

Hähner, P. 1996. A theory of dislocation cell formation based on stochastic dislocation dynamics. *Acta materialia*, **44**(6), 2345–2352.

Hähner, Peter, Bay, Karlheinz, and Zaiser, Michael. 1998. Fractal Dislocation Patterning During Plastic Deformation. *Physical Review Letters*, **81**(12), 2470–2473.

Hamba, Fujihiro. 1995. An analysis of nonlocal scalar transport in the convective boundary layer using the Green's function. *Journal of Atmospheric Sciences*, **52**(8), 1084–1095.

Hamba, Fujihiro. 2005. Nonlocal analysis of the Reynolds stress in turbulent shear flow. *Physics of Fluids*, **17**(11), 115102.

Harris, Stewart. 2004. *An introduction to the theory of the Boltzmann equation*. Courier Corporation.

Hecht, Fabian M, Rheinlaender, Johannes, Schierbaum, Nicolas, Goldmann, Wolfgang H, Fabry, Ben, and Schäffer, Tilman E. 2015. Imaging viscoelastic properties of live cells by AFM: power-law rheology on the nanoscale. *Soft matter*, **11**(23), 4584–4591.

Hermann, Richard. 2014. *Fractional calculus: An introduction for physicists*. 2 edn. World Scientific. 500pp.

Hesthaven, Jan S, Gottlieb, Sigal, and Gottlieb, David. 2007. *Spectral methods for time-dependent problems*. Vol. 21. Cambridge University Press.

Hinze, Julius Oscar, Sonnenberg, RE, and Builtjes, Peter Johann Henry. 1974. Memory effect in a turbulent boundary-layer flow due to a relatively strong axial variation of the mean-velocity gradient. *Applied Scientific Research*, **29**(1), 1–13.

Hochrainer, Thomas. 2015. Multipole expansion of continuum dislocations dynamics in terms of alignment tensors. *Philosophical Magazine*, **95**(12), 1321–1367.

Hochrainer, Thomas, Sandfeld, Stefan, Zaiser, Michael, and Gumbsch, Peter. 2014. Continuum dislocation dynamics: towards a physical theory of crystal plasticity. *Journal of the Mechanics and Physics of Solids*, **63**, 167–178.

Holt, David L. 1970. Dislocation Cell Formation in Metals. *Journal of Applied Physics*, **41**(8), 3197–3201.

Hormander, L. 2003. *The analysis of linear partial differential operators I: Distribution theory and Fourier Analysis*. 2 edn. Springer-Verlag.

Hou, D., and Xu, C. 2017. A fractional spectral method with applications to some singular problems. *Advances in Computational Mathematics*, **43**(5), 911–944.

Hsu, (Elton) P. 1984. Reflecting Brownian motion, boundary local time and the Neumann problem. *Dissertation Abstracts International Part B: Science and Engineering*, **45**(6).

Huang, C., and Stynes, M. 2016. A spectral collocation method for a weakly singular Volterra integral equation of the second kind. *Advances in Computational Mathematics*, **42**(5), 1015–1030.

Huang, Can, Jiao, Yujian, Wang, Li-Lian, and Zhang, Zhimin. 2016. Optimal fractional integration preconditioning and error analysis of fractional collocation method using nodal generalized Jacobi functions. *SIAM Journal on Numerical Analysis*, **54**(6), 3357–3387.

Huang, Yanghong, and Oberman, Adam. 2014. Numerical methods for the fractional Laplacian: A finite difference-quadrature approach. *SIAM Journal on Numerical Analysis*, **52**(6), 3056–3084.

Hussaini, M Yousuff, and Voigt, Robert G. 2012. *Instability and Transition: Materials of the workshop held May 15–June 9, 1989 in Hampton, Virginia Volume 2*. Springer Science & Business Media.

Imbeni, V, Kruzic, JJ, Marshall, GW, Marshall, SJ, and Ritchie, RO. 2005. The dentin–enamel junction and the fracture of human teeth. *Nature materials*, **4**(3), 229–232.

Jaishankar, Aditya, and McKinley, Gareth H. 2013. Power-law rheology in the bulk and at the interface: quasi-properties and fractional constitutive equations. *Proceedings of the Royal Society A: Mathematical, Physical and Engineering Sciences*, **469**(2149), 20120284.

Jaishankar, Aditya, and McKinley, Gareth H. 2014. A fractional K-BKZ constitutive formulation for describing the nonlinear rheology of multiscale complex fluids. *Journal of Rheology*, **58**(6), 1751–1788.

Jiao, Y.J., Wang, L.L., and Huang, C. 2016. Well-conditioned fractional collocation methods using fractional Birkhoff interpolation basis. *Journal of Computational Physics*, **305**, 1–28.

Jiao, Z., Chen, Y. Q., and Podlubny, I. 2012. Distributed-order dynamic systems: Stability, simulation, applications and perspectives. *London: Springer*.

Jin, B., and William, R. 2012. An inverse Sturm-Liouville problem with a fractional derivative. *Journal of Computational Physics*, **231**, 4954–4966.

Jin, B. T., Lazarov, R., Pasciak, J., and Zhou, Z. 2015a. Error analysis of semi-discrete finite element methods for inhomogeneous time-fractional diffusion. *IMA Journal of Numerical Analysis*, **35**(2), 561–582.

Jin, B. T., Lazarov, R., Pasciak, J., and Rundell, W. 2015b. Variational formulation of problems involving fractional order differential operators. *Mathematics of Computation*, **84**(296), 2665–2700.

Jin, B. T., Lazarov, R., and Zhou, Z. 2016a. Two fully discrete schemes for fractional diffusion and diffusion-wave equations with nonsmooth data. *SIAM Journal of Scientific Computing*, **38**(1), A146–A170.

Jin, B., Lazarov, R., Thomée, V. and Zhou, Z. 2017a. On nonnegativity preservation in finite element methods for subdiffusion equations. *Mathematics of Computation*, **86**(307), 2239–2260.

Jin, B. T., Li, B. Y., and Zhou, Z. 2017b. Correction of high-order BDF convolution quadrature for fractional evolution equations. *SIAM Journal of Scientific Computing*, **39**(6), A3129–A3152.

Jin, B. T., Li, B. Y., and Zhou, Z. 2018. Numerical analysis of nonlinear subdiffusion equations. *SIAM Journal on Numerical Analysis*, **56**(1), 1–23.

Jin, B. T., Li, B. Y., and Zhou, Z. 2019. Subdiffusion with a time-dependent coefficient: analysis and numerical solution. *Mathematics of Computation*, **88**(319), 2157–2186.

Jin, Bangti, Lazarov, Raytcho, and Zhou, Zhi. 2016b. A Petrov–Galerkin finite element method for fractional convection-diffusion equations. *SIAM Journal on Numerical Analysis*, **54**(1), 481–503.

Jin, Bangti, Li, Buyang, and Zhou, Zhi. 2020. Subdiffusion with time-dependent coefficients: improved regularity and second-order time stepping. *Numerische Mathematik*, **145**(4), 883–913.

Kac, M. 1949. On Distributions of Certain Wiener Functionals. *Transactions of the American Mathematical Society*, **65**, 1–13.

Kapetanou, Olga, Koutsos, Vasileios, Theotokoglou, Efstathios, Weygand, Daniel, and Zaiser, Michael. 2015. Statistical analysis and stochastic dislocation-based modeling of microplasticity. *Journal of the Mechanical Behavior of Materials*, **24**(3-4), 105–113.

Kapnistos, M., Lang, M., Vlassopoulos, D., Pyckhout-Hintzen, D., Richter, D., Cho, D., Chang, T., and Rubinstein, M. 2008. Unexpected power-law stress relaxation of entangled ring polymers. *Nature Materials*, **7**, 997–1002.

Karniadakis, George Em, Orszag, Steven A, et al. 1993. Nodes, modes and flow codes. *Physics Today*, **46**(3), 34–42.

Karniadakis, George Em, Karniadakis, George, and Sherwin, Spencer. 2005. *Spectral/hp element methods for computational fluid dynamics*. Oxford University Press on Demand.

Kelly, James F, Li, Cheng-Gang, and Meerschaert, Mark M. 2018. Anomalous diffusion with ballistic scaling: A new fractional derivative. *Journal of Computational and Applied Mathematics*, **339**, 161–178.

Kelly, James F., Sankaranarayanan, Harish, and Meerschaert, Mark M. 2019. Boundary conditions for two-sided fractional diffusion. *Journal of Computational Physics*, **376**, 1089–1107.

Ketz, R. J., Prud'homme, R. K., and Graessley, W. W. 1988. Rheology of concentrated microgel solutions. *Rheologica Acta*, **27**(5), 531–539.

Kharazmi, E., and Zayernouri, M. 2018. Fractional pseudo-spectral methods for distributed-order fractional PDEs. *International Journal of Computer Mathematics*, **95**(6-7), 1340–1361.

Kharazmi, E., and Zayernouri, M. 2019. Fractional Sensitivity Equation Method: Applications to Fractional Model Construction. *Journal of Scientific Computing*, **80**, 110–140.

Kharazmi, Ehsan, Zayernouri, Mohsen, and Karniadakis, George Em. 2017. Petrov–Galerkin and spectral collocation methods for distributed order differential equations. *SIAM Journal on Scientific Computing*, **39**(3), A1003–A1037.

Kilbas, A. A., Srivastava, H. M., and Trujillo, J. J. 2006. *Theory and Applications of Fractional Differential Equations, Volume 204 (North-Holland Mathematics Studies)*. New York, NY, USA: Elsevier Science Inc.

Kim, John, Moin, Parviz, and Moser, Robert. 1987. Turbulence statistics in fully developed channel flow at low Reynolds number. *Journal of Fluid Mechanics*, **177**, 133–166.

Kim, Panki, Song, Renming, and Vondracek, Zoran. 2020. On the boundary theory of subordinate killed Lévy processes. *Potential Analysis*, **53**(1), 131–181.

Klafter, J, and Silbey, R. 1980. Derivation of the continuous-time random-walk equation. *Physical Review Letters*, **44**(2), 55.

Klafter, Joseph, and Sokolov, Igor M. 2005. Anomalous diffusion spreads its wings. *Physics World*, **18**(8), 29.

Klages, R., Radons, G., and M Sokolov, I. 2008. *Anomalous transport: foundations and applications*. John Wiley & Sons.

Klein, C., Sparber, C., and Markowich, P. 2014. Numerical study of fractional nonlinear Schrödinger equations. *Proceedings of The Royal Society of London. Series A. Mathematical, Physical and Engineering Sciences*, **470**(2172), 20140364, 26.

Klimek, M., and Agrawal, O. P. 2012. On a regular fractional Sturm-Liouville problem with derivatives of order in $(0, 1)$. In: *Proceedings of 2012, 13th International Carpathian Control Conference (ICCC), 978-1-4577-1868.*

Kloeden, Peter E, and Platen, Eckhard. 1992. Stochastic differential equations. Pages 103–160 of: *Numerical Solution of Stochastic Differential Equations*. Springer.

Kochubei, A. N. 2008. Distributed order calculus and equations of ultraslow diffusion. *Mathematical Analysis and Applications*, **340**, 252–281.

Kochubei, A. N. 2011. General fractional calculus, evolution equations, and renewal processes. *Integral Equations and Operator Theory*, **71**(4), 583–600.

Kolokoltsov, Vassili N. 2011. *Markov processes, semigroups and generators*. De Gruyter Studies in Mathematics, vol. 38. Walter de Gruyter & Co., Berlin.

Koponen, Ismo. 1995. Analytic approach to the problem of convergence of truncated Lévy flights towards the Gaussian stochastic process. *Physics Review E*, **52**(Jul), 1197–1199.

Kopteva, Natalia, and Stynes, Martin. 2017. Analysis and numerical solution of a Riemann-Liouville fractional derivative two-point boundary value problem. *Advances in Computational Mathematics*, **43**(1), 77–99.

Kraichnan, Robert H. 1964. Direct-interaction approximation for shear and thermally driven turbulence. *The Physics of Fluids*, **7**(7), 1048–1062.

Kwaśnicki, M. 2017. Ten equivalent definitions of the fractional Laplace operator. *Fractional Calculus and Applied Analysis*, **20**(1), 7–51.

Kyprianou, Andreas E, Osojnik, Ana, and Shardlow, Tony. 2018. Unbiased "walk-on-spheres" Monte Carlo methods for the fractional Laplacian. *IMA Journal of Numerical Analysis*, **38**(3), 1550–1578.

Landkof, Naum. 1972. *Foundations of modern potential theory*. Vol. 180. Springer.

Laskin, Nikolai. 2000. Fractional quantum mechanics and Lévy path integrals. *Physics Letters A*, **268**(4), 298 – 305.

Laval, JP, Dubrulle, B, and Nazarenko, S. 2001. Nonlocality and intermittency in three-dimensional turbulence. *Physics of Fluids*, **13**(7), 1995–2012.

Lawler, Gregory F. 2010. *Random walk and the heat equation*. Vol. 55. American Mathematical Soc.

Lemaitre, J. 1996. *A course on damage mechanics*. Springer.

Letnikov, A.V., and Chernykh, V.A. 2011. *The Foundation of fractional calculus (with applications to the theory of oil and gas production, underground hydrodynamics and dynamics of biological systems)*. Neftegaz.

Levy, Melissa, and Berkowitz, Brian. 2003. Measurement and analysis of non-Fickian dispersion in heterogeneous porous media. *Journal of Contaminant Hydrology*, **64**, 203–226.

Li, C., and Deng, W.H. 2016. High order schemes for the tempered fractional diffusion equations. *Advances in Computational Mathematics*, **42**(3), 543–572.

Li, Changpin, and Cai, Min. 2017. High-order approximation to Caputo derivatives and Caputo-type advection–diffusion equations: Revisited. *Numerical functional analysis and optimization*, **38**(7), 861–890.

Li, Changpin, and Chen, An. 2018. Numerical methods for fractional partial differential equations. *International Journal of Computer Mathematics*, **95**(6-7), 1048–1099.

Li, Changpin, Zeng, Fanhai, and Liu, Fawang. 2012. Spectral approximations to the fractional integral and derivative. *Fractional Calculus and Applied Analysis*, **15**(3), 383–406.

Li, X., and Xu, C. 2009. A space-time spectral method for the time fractional diffusion equation. *SIAM Journal on Numerical Analysis*, **47**(3), 2108–2131.

Li, X., and Xu, C. 2010. Existence and uniqueness of the weak solution of the space-time fractional diffusion equation and a spectral method approximation. *Communications in Computational Physics*, **8**(5), 1016.

Li, Y., and Chen, Y. Q. 2014. Lyapunov stability of fractional-order nonlinear systems: A distributed-order approach. Pages 1–6 of: *ICFDA'14 International Conference on Fractional Differentiation and Its Applications 2014*. IEEE.

Li, Y., Sheng, H., and Chen, Y. Q. 2011. On distributed order integrator/differentiator. *Signal Processing*, **91**(5), 1079–1084.

Liemert, A., Sandev, T., and Kantz, H. 2017. Generalized Langevin equation with tempered memory kernel. *Physica A: Statistical Mechanics and its Applications*, **466**, 356–369.

Lin, Yumin, and Xu, Chuanju. 2007. Finite difference/spectral approximations for the time-fractional diffusion equation. *Journal Computational Physics*, **225**(2), 1533–1552.

Lion, Alexander. 1997. On the thermodynamics of fractional damping elements. *Continuum Mechanics and Thermodynamics*, **9**(2), 83–96.

Lions, P. L., and Sznitman, A. S. 1984. Stochastic differential equations with reflecting boundary conditions. *Communications on Pure and Applied Mathematics*, **37**(4), 511–537.

Liouville, J. 1832a. Mémoire sur le calcul des différentielles à indices quelconques. *Journal d l'Ecole Polytechnique*, **13**, 71–162.

Liouville, Joseph. 1832b. *Mémoire sur quelques questions de géométrie et de mécanique, et sur un nouveau genre de calcul pour résoudre ces questions*. tome XIII, XXIe cahier.

Lischke, Anna, Pang, Guofei, Gulian, Mamikon, Song, Fangying, Glusa, Christian, Zheng, Xiaoning, Mao, Zhiping, Cai, Wei, Meerschaert, Mark M, Ainsworth, Mark, et al. 2020. What is the fractional Laplacian? A comparative review with new results. *Journal of Computational Physics*, **404**, 109009.

López, José L., and Temme, Nico M. 2013. New series expansions of the Gauss hypergeometric function. *Advances in Computational Mathematics*, **39**(2), 349–365.

Lorenzo, C. F., and Hartley, T. T. 2002. Variable order and distributed order fractional operators. *Nonlinear Dynamics*, **29**(1-4), 57–98.

Lu, Xin, Pang, HongKui, and Sun, HaiWei. 2015. Fast approximate inversion of a block triangular Toeplitz matrix with applications to fractional sub-diffusion equations. *Numerical Linear Algebra with Applications*, **22**(5), 866–882.

Lubich, C. 1986. Discretized Fractional Calculus. *SIAM Journal on Mathematical Analysis*, **17**, 704–719.

Lubich, C. 1988. Convolution quadrature and discretized operational calculus. I. *Numerische Mathematik*, **52**(2), 129–145.

Lubich, C., and Schädle., A. 2002. Fast convoluion for nonreflecting boundary conditions. *SIAM Journal on Scientific Computing*, **24**, 161–182.

Lubich, Ch. 1983. On the stability of linear multistep methods for Volterra convolution equations. *IMA Journal of Numerical Analysis*, **3**(4), 439–465.

Lumley, John L. 1970. Toward a turbulent constitutive relation. *Journal of Fluid Mechanics*, **41**(2), 413–434.

Lumley, John L, and McMichael, James M. 1995. *Turbulence modeling*. Tech. rept. Sibley School of Mechanical and Aerospace Engineering, Ithaca NY.

Lynch, R.E., Rice, J.R., and Thomas, D.H. 1964. Direct solution of partial difference equations by tensor product methods. *Numerische Mathematik*, **6**, 185–199.

Ma, Heping, and Sun, Weiwei. 2001. Optimal error estimates of the Legendre–Petrov–Galerkin method for the Korteweg–de Vries equation. *SIAM Journal on Numerical Analysis*, **39**(4), 1380–1394.

Ma, S.N., Li, H.Y., and Zhang, Z. 2019. Novel spectral methods for Schrödinger equations with an inverse square potential on the whole space. *Discrete and Continuous Dynamical Systems. Series B. A Journal Bridging Mathematics and Sciences*, **24**(4), 1589–1615.

Maday, Y. 1990. Analysis of spectral projectors in one-dimensional domains. *Mathematics of Computation*, **55**, 537–562.

Magin, R., and Royston, T.J. 2010. Fractional-order elastic models of cartilage: A multi-scale approach. *Communications in Nonlinear Science Numerical Simulation*, **15**, 657–664.

Mainardi, F. 2010. *Fractional calculus and waves in linear viscoelasticity*. Imperial College Press.

Mainardi, Francesco. 2012. An historical perspective on fractional calculus in linear viscoelasticity. *Fractional Calculus and Applied Analysis*, **15**(4), 712–717.

Mainardi, Francesco. 2020. Why the Mittag-Leffler function can be considered the queen function of the fractional calculus? *Entropy*, **22**(12), 1359.

Mainardi, Francesco, and Spada, Giorgio. 2011. Creep, relaxation and viscosity properties for basic fractional models in rheology. *The European Physical Journal Special Topics*, **193**(1), 133–160.

Mainardi, Francesco, Mura, Antonio, Pagnini, Gianni, and Gorenflo, Rudolf. 2007. Sub-diffusion equations of fractional order and their fundamental solutions. Pages 23–55 of: *Mathematical methods in engineering*. Springer.

Mantegna, R. N., and Stanley, H. E. 1994. Stochastic process with ultraslow convergence to a Gaussian: the truncated Lévy flight. *Physical Review Letters*, **73**(22), 2946.

Mao, Z., and Karniadakis, G. 2018. A spectral method (of exponential convergence) for singular solutions of the diffusion equation with general two-sided fractional derivative. *SIAM Journal on Numerical Analysis*, **56**(1), 24–49.

Mao, Z.P., and Shen, J. 2017. Hermite spectral methods for fractional PDEs in unbounded domains. *SIAM Journal on Scientific Computing*, **39**(5), A1928–A1950.

Mao, Zhiping, and Shen, Jie. 2016. Efficient spectral–Galerkin methods for fractional partial differential equations with variable coefficients. *Journal of Computational Physics*, **307**, 243–261.

Mao, Zhiping, Chen, Sheng, and Shen, Jie. 2016. Efficient and accurate spectral method using generalized Jacobi functions for solving Riesz fractional differential equations. *Applied Numerical Mathematics*, in press.

Maroni, P., and da Rocha, Z. 2008. Connection coefficients between orthogonal polynomials and the canonical sequence: an approach based on symbolic computation. *Numererical Algorithms*, **47**(3), 291–314.

Mashayekhi, Somayeh, Hussaini, M Yousuff, and Oates, William. 2019. A physical interpretation of fractional viscoelasticity based on the fractal structure of media: Theory and experimental validation. *Journal of the Mechanics and Physics of Solids*, **128**, 137–150.

Masjed-Jamei, M., and Koepf, W. 2012. Two classes of special functions using Fourier transforms of generalized ultraspherical and generalized Hermite polynomials. *Proceedings of the American Mathematical Society*, **140**(6), 2053–2063.

Mason, Thomas G. 2000. Estimating the viscoelastic moduli of complex fluids using the generalized Stokes–Einstein equation. *Rheologica Acta*, **39**(4), 371–378.

McKinley, G.H., and Jaishankar, A. 2013. *Critical gels, Scott Blair and the fractional calculus of soft squishy materials*. Presentation.

Meerschaert, M, Sabzikar, F., and Chen, J. 2015. Tempered fractional calculus. *Journal of Computational Physics*, **293**, 14–28.

Meerschaert, M. M., Benson, D. A., and Baeumer, B. 1999. Multidimensional advection and fractional dispersion. *Physical Review E*, **59**(5), 5026.

Meerschaert, M. M., Mortensen, J., and Wheatcraft, S. W. 2006. Fractional vector calculus for fractional advection–dispersion. *Physica A: Statistical Mechanics and its Applications*, **367**, 181–190.

Meerschaert, M. M., Zhang, Y., and Baeumer, B. 2008. Tempered anomalous diffusion in heterogeneous systems. *Geophysical Research Letters*, **35**(17).

Meerschaert, Mark M, and Scheffler, Hans-Peter. 2001. *Limit distributions for sums of independent random vectors: Heavy tails in theory and practice*. Vol. 321. John Wiley & Sons.

Meerschaert, Mark M, and Sikorskii, Alla. 2011. *Stochastic models for fractional calculus*. Vol. 43. Walter de Gruyter.

Meerschaert, Mark M, and Sikorskii, Alla. 2019. *Stochastic models for fractional calculus*. de Gruyter.

Meerschaert, Mark M, Mortensen, Jeff, and Scheffler, Hans-Peter. 2004. Vector Grünwald formula for fractional derivatives. *Fractional Calculus and Applied Analysis*, **7**(1), 61–82.

Meneveau, Charles. 1994. Statistics of turbulence subgrid-scale stresses: Necessary conditions and experimental tests. *Physics of Fluids*, **6**(2), 815–833.

Meng, Ruifan, Yin, Deshun, and Drapaca, Corina S. 2019. Variable-order fractional description of compression deformation of amorphous glassy polymers. *Computational Mechanics*, **64**(1), 163–171.

Meral, FC, Royston, TJ, and Magin, R. 2010. Fractional calculus in viscoelasticity: an experimental study. *Communications in nonlinear science and numerical simulation*, **15**(4), 939–945.

Metzler, R., and Klafter, J. 2000. The random walk's guide to anomalous diffusion: a fractional dynamics approach. *Physics Reports*, **339**(1), 1–77.

Miguel, M.-Carmen, Vespignani, A., Zapperi, S., Weiss, J., and Grasso, J-R. 2001. Intermittent dislocation flow in viscoplastic deformation. *Nature*, **410**, 667–671.

Molchanov, S., and Ostrovskii, E. 1968. Symmetric stable processes as traces of degenerate diffusion processes. *Theory of Probability and its Applications*, **14**(1), 128–131.

Monegato, G. 2009. Definitions, properties and applications of finite-part integrals. *Journal of Computation and Applied Mathematics*, **229**(2), 425–439.

Montroll, E. W., and Weiss, G. H. 1965. Random walks on lattices, II. *Journal of Mathematical Physics*, **6**(2), 167–181.

Muller, Mervin E. 1956. Some continuous Monte Carlo methods for the Dirichlet problem. *Ann. Math. Statist.*, **27**(3), 569–589.

Musina, Roberta, and Nazarov, Alexander I. 2016. On fractional Laplacians – 2. *Annales de l'Institut Henri Poincare (C) Non Linear Analysis*, **33**(6), 1667–1673.

Mustapha, K. 2018. FEM for time-fractional diffusion equations, novel optimal error analyses. *Mathematics of Computation*, **87**(313), 2259–2272.

Naqvi, Q. A., and Abbas, M. 2004. Complex and higher order fractional curl operator in electromagnetics. *Optics Communications*, **241**(4-6), 349–355.

Neuman, Shlomo P., and Tartakovsky, Daniel M. 2009. Perspective on theories of non-Fickian transport in heterogeneous media. *Advances in Water Resources*, **32**, 670–680.

Nezza, E. Di, Palatucci, G., and Valdinoci, E. 2012. Hitchhiker's guide to the fractional Sobolev spaces. *Bulletin des Sciences Mathematiques*, **136**, 521–573.

Nicolle, S., Vezin, P., and Palierne, J.-F. 2010. A strain-hardening bi-power law for the nonlinear behaviour of biological soft tissues. *Journal of Biomechanics*, **43**(5), 927–932.

Nochetto, Ricardo H., Otàrola, Enrique, and Salgado, Abner J. 2015. A PDE approach to fractional diffusion in general domains: A priori error analysis. *Foundations of Computational Mathematics*, **15**(3), 733–791.

Nolan, John P. 1998. Parameterizations and modes of stable distributions. *Statistics & Probability Letters*, **38**(2), 187–195.

Nolan, John P. 2020. *Univariate stable distributions: Models for heavy tailed data*. Springer Nature.

Oldham, Keith, and Spanier, Jerome. 1974. *The fractional calculus theory and applications of differentiation and integration to arbitrary order*. Elsevier.

Orszag, Steven A, and Patterson Jr, GS. 1972. Numerical simulation of three-dimensional homogeneous isotropic turbulence. *Physical Review Letters*, **28**(2), 76.

Pajerowski, J David, Dahl, Kris Noel, Zhong, Franklin L, Sammak, Paul J, and Discher, Dennis E. 2007. Physical plasticity of the nucleus in stem cell differentiation. *Proceedings of the National Academy of Sciences*, **104**(40), 15619–15624.

Pang, G., Perdikaris, P., Cai, W., and Karniadakis, G.E. 2017. Discovering variable fractional orders of advection–dispersion equations from field data using multi-fidelity Bayesian optimization. *Journal of Computational Physics*, **348**, 694–714.

Pang, G., Lu, L., and Karniadakis, G. E. 2019. fPINNs: Fractional physics-informed neural networks. *SIAM Journal on Scientific Computing*, **41**, A2603–A2626.

Pang, G., D'Elia, M., Parks, M., and Karniadakis, G. E. 2020a. *nPINNs: nonlocal physics-informed neural networks for a parametrized nonlocal universal Laplacian operator. Algorithms and applications*. arXiv:2004.04276.

Pang, Guofei, Chen, Wen, and Sze, KY. 2013. Gauss–Jacobi-type quadrature rules for fractional directional integrals. *Computers and Mathematics with Applications*, **66**(5), 597–607.

Pang, Guofei, Chen, Wen, and Fu, Zhuojia. 2015. Space-fractional advection–dispersion equations by the Kansa method. *Journal of Computational Physics*, **293**, 280–296.

Pang, Guofei, D'Elia, Marta, Parks, Michael, and Karniadakis, George E. 2020b. nPINNs: nonlocal physics-informed neural networks for a parametrized nonlocal universal Laplacian operator. Algorithms and applications. *Journal of Computational Physics*, **422**, 109760.

Parish, Eric J, and Duraisamy, Karthik. 2017. A dynamic subgrid scale model for large eddy simulations based on the Mori–Zwanzig formalism. *Journal of Computational Physics*, **349**, 154–175.

Patnaik, Sansit, and Semperlotti, Fabio. 2021. Variable-order fracture mechanics and its application to dynamic fracture. *npj Computational Materials*, **7**(1), 1–8.

Patnaik, Sansit, Hollkamp, John P, and Semperlotti, Fabio. 2020. Applications of variable-order fractional operators: a review. *Proceedings of the Royal Society A*, **476**(2234), 20190498.

Pearson, John W, Olver, Sheehan, and Porter, Mason A. 2017. Numerical methods for the computation of the confluent and Gauss hypergeometric functions. *Numerical Algorithms*, **74**(3), 821–866.

Podlubny, I. 1999. *Fractional differential equations: An introduction to fractional derivatives, fractional differential equations, to methods of their solution and some of their applications*. Mathematics in Science and Engineering, vol. 198. San Diego, CA: Academic Press Inc.

Pope, SBj. 1975. A more general effective-viscosity hypothesis. *Journal of Fluid Mechanics*, **72**(2), 331–340.

Pope, Stephen B. 2001. *Turbulent flows*.

Popov, G. Ya. 1982. *Elastic stress concentration near punches, cuts, thin inclusions, and stiffeners*. Nauka.

Pozrikidis, C. 2016. *The fractional Laplacian*. CRC Press, Boca Raton, FL.

Prandtl, Ludwig. 1942. Bemerkungen zur Theorie der freien Turbulenz. *ZAMM-Journal of Applied Mathematics and Mechanics/Zeitschrift für Angewandte Mathematik und Mechanik*, **22**(5), 241–243.

Privault, Nicolas. 2008. Potential theory in classical probability. Pages 3–59 of: *Quantum potential theory*. Lecture Notes in Math., vol. 1954. Springer, Berlin.

Puig-de Morales-Marinkovic, M., Turner, K.T., Butler, J.P., Fredberg, J.J., and Suresh, S. 2007. Viscoelasticity of the human red blood cell. *American Journal of Physiology - Cell Physiology*, **293**(2), C597–C605.

Rabotnov, Yu N. 1948. The equilibrium of an elastic medium with after-effect. *Prikl. Mat. Mekh.*, **12**, 53–62.

Razminia, A., Dizaji, A. F., and Majd, V. J. 2012. Solution existence for non-autonomous variable-order fractional differential equations. *Mathematical and Computer Modelling*, **55**(3-4), 1106–1117.

Reed, Michael. 2012. *Methods of modern mathematical physics: Functional analysis*. Elsevier.

Revuz, Daniel, and Yor, Marc. 2013. *Continuous martingales and Brownian motion*. Vol. 293. Springer Science & Business Media.

Richeton, T., Weiss, J., and Louchet, F. 2005. Breakdown of avalanche critical behaviour in polycrystalline plasticity. *Nature Materials*, **4**.

Riemann, Bernhard. 1876. Versuch einer allgemeinen Auffassung der Integration und Differentiation. *Gesammelte Werke*, **62**(1876).

Roncal, Luz, and Stinga, Pablo Raúl. 2014. Transference of fractional Laplacian regularity. Pages 203–212 of: *Special functions, partial differential equations, and harmonic analysis*. Springer.

Roncal, Luz, and Stinga, Pablo Raúl. 2016. Fractional Laplacian on the torus. *Communications in Contemporary Mathematics*, **18**(03), 1550033.

Roop, J. P. 2004. *Variational solution of the fractional advection-dispersion equation, PhD Dissertation*. Ph.D. thesis, Clemson University, Department of Mathematical Sciences.

Ros-Oton, X., and Serra, J. 2014. The Dirichlet problem for the fractional Laplacian: Regularity up to the boundary. *Journal de Mathematiques Pures et Appliquees*, **101**, 275–302.

Rösler, M. 1998. Generalized Hermite polynomials and the heat equation for Dunkl operators. *Communications in Mathematical Physics*, **192**(3), 519–542.

Rudin, W. 1991. *Functional analysis*. 2 edn. McGraw-Hill.

Sabzikar, F., Meerschaert, M. M., and Chen, J. 2015. Tempered fractional calculus. *Journal of Computational Physics*, **293**, 14–28.

Sagaut, Pierre. 2010. Toward advanced subgrid models for lattice-Boltzmann-based large-eddy simulation: theoretical formulations. *Computers & Mathematics with Applications*, **59**(7), 2194–2199.

Saint-Raymond, Laure. 2009. *Hydrodynamic limits of the Boltzmann equation*. Springer Science & Business Media.

Sakamoto, K., and Yamamoto, M. 2011. Initial value/boundary value problems for fractional diffusion-wave equations and applications to some inverse problems. *Journal of Mathematical Analysis and Applications*, **382**(1), 426–447.

Samiee, M., Zayernouri, M., and Meerschaert, M. M. 2019. A Unified Spectral Method for FPDEs with Two-sided Derivatives; part II: Stability, and Error Analysis. *Journal of Computational Physics*, **385**, 244–261.

Samiee, Mehdi, Akhavan-Safaei, Ali, and Zayernouri, Mohsen. 2020. A fractional subgrid-scale model for turbulent flows: Theoretical formulation and a priori study. *Physics of Fluids*, **32**(5), 055102.

Samiee, Mehdi, Kharazmi, Ehsan, Meerschaert, Mark M., and Zayernouri, Mohsen. 2021. A unified Petrov-Galerkin spectral method and fast solver for distributed-order partial differential equations. *Communications on Applied Mathematics and Computation*, **3**, 61–90.

Samiee, Mehdi, Akhavan-Safaei, Ali, and Zayernouri, Mohsen. 2022. Tempered fractional LES modeling. *Journal of Fluid Mechanics*, **932**.

Samko, S. G., Kilbas, A. A., and Marichev, O. I. 1993. *Fractional integrals and derivatives: Theory and applications*. Switzerland: Gordon and Breach Science Publishers.

Samorodnitsky, Gennady, and Taqqu, Murad S. 1994. *Stable non-Gaussian random processes: Stochastic models with infinite variance*. Stochastic Modeling. London: Chapman Hall.

Sandev, T., Chechkin, A. V., Korabel, N., Kantz, H., Sokolov, I. M., and Metzler, R. 2015. Distributed-order diffusion equations and multifractality: Models and solutions. *Physical Review E*, **92**(4), 042117.

Sandev, T., Metzler, R., and Chechkin, A. 2018. From continuous time random walks to the generalized diffusion equation. *Fractional Calculus and Applied Analysis*, **21**(1), 10–28.

Scalas, Enrico, Gorenflo, Rudolf, and Mainardi, Francesco. 2004. Uncoupled continuous-time random walks: Solution and limiting behavior of the master equation. *Physical Review E*, **69**(1), 011107.

Schädle, A., López-Fernández, M., and Lubich, C. 2006. Fast and Oblivious Convolution Quadrature. *SIAM Journal on Scientific Computing*, **28**(2), 421–438.

Scher, Harvey, and Lax, Melvin. 1973a. Stochastic transport in a disordered solid. I. Theory. *Physical Review B*, **7**(10), 4491.

Scher, Harvey, and Lax, Melvin. 1973b. Stochastic transport in a disordered solid. II. Impurity conduction. *Physical Review B*, **7**(10), 4502.

Schiessel, H. 1995. Generalized viscoelastic models: their fractional equations with solutions. *Journal of Physics A: Mathematical and General*, **28**, 6567–6584.

Schiessel, H., and Blumen, A. 1993. Hierarchical analogues to fractional relaxation equations. *Journal of Physics A: Mathematical and General*, **26**, 5057–5069.

Schneider, R, Reichmann, O, and Schwab, Christoph. 2010. Wavelet solution of variable order pseudodifferential equations. *Calcolo*, **47**(2), 65–101.

Schumer, R., Benson, D.A., Meerschaert, M.M., and Wheatcraft, S.W. 2001. Eulerian derivation of the fractional advection-dispersion equation. *Journal of Contaminant Hydrology*, **48**, 69–88.

Schumer, R., Benson, D. A., Meerschaert, M. M., and Baeumer, B. 2003a. Fractal mobile/immobile solute transport. *Water Resources Research*, **39**(10).

Schumer, R., Benson, D. A., Meerschaert, M. M., and Baeumer, B. 2003b. Multiscaling fractional advection-dispersion equations and their solutions. *Water Resources Research*, **39**(1).

Seyedi, S Hadi, and Zayernouri, Mohsen. 2022. A data-driven dynamic nonlocal subgrid-scale model for turbulent flows. *Physics of Fluids*, **34**(3), 035104.

Seyedi, S Hadi, Akhavan-Safaei, Ali, and Zayernouri, Mohsen. 2022. Dynamic nonlocal passive scalar subgrid-scale turbulence modeling. *Physics of Fluids*, **34**(10).

Shen, J. 1994. Efficient spectral-Galerkin method. I. Direct solvers of second- and fourth-order equations using Legendre polynomials. *SIAM Journal on Scientific Computing*, **15**, 1489–1505.

Shen, J., and Sheng, C. T. 2019. An efficient space-time method for time fractional diffusion equation. *Journal of Scientific Computing*.

Shen, J., Tang, T., and Wang, L. L. 2011a. *Spectral methods: Algorithms, analysis and applications*. Series in Computational Mathematics, vol. 41. Springer-Verlag, Berlin, Heidelberg.

Shen, Jie, Tang, Tao, and Wang, Li-Lian. 2011b. *Spectral methods: algorithms, analysis and applications*. Vol. 41. Springer Science & Business Media.

Shen, Jie, Sheng, Changtao, and Wang, Zhongqing. 2015. Generalized Jacobi spectral-Galerkin method for nonlinear Volterra integral equations with weakly singular kernels. *Journal of Mathematical Study*, **48**(4), 315–329.

Sheng, Changtao, Ma, Suna, Li, Huiyuan, Wang, Li-Lian, and Jia, Lueling. 2021. Nontensorial generalised Hermite spectral methods for PDEs with fractional Laplacian and Schrödinger operators. *ESAIM: Mathematical Modelling and Numerical Analysis*, **55**(5), 2141–2168.

Sokolov, Igor M, and Metzler, Ralf. 2003. Towards deterministic equations for Lévy walks: The fractional material derivative. *Physical Review E*, **67**(1), 010101.

Sokolov, IM, Chechkin, AV, and Klafter, J. 2004. Fractional diffusion equation for a power-law-truncated Lévy process. *Physica A: Statistical Mechanics and its Applications*, **336**(3), 245–251.

Sollich, P. 1998. Rheological constitutive equation for a model of soft glassy materials. *Physics Review E*, **58**(1), 738–759.

Solomon, TH, Weeks, Eric R, and Swinney, Harry L. 1994. Chaotic advection in a two-dimensional flow: Lévy flights and anomalous diffusion. *Physica D: Nonlinear Phenomena*, **76**(1), 70–84.

Song, Fangying, and Karniadakis, George Em. 2018. A universal fractional model of wall-turbulence. *arXiv preprint arXiv:1808.10276*.

Song, Fangying, and Karniadakis, George Em. 2021. Variable-order fractional models for wall-bounded turbulent flows. *Entropy*, **23**(6), 782.

Song, Fangying, Xu, Chuanju, and Karniadakis, George Em. 2017. Computing fractional Laplacians on complex-geometry domains: Algorithms and simulations. *SIAM Journal of Scientific Computing*, **39**(4), A1320–A1344.

Song, R., and Vondracek, Z. 2003. Potential theory of subordinate killed Brownian motion in a domain. *Probab. Theory Relat. Fields*, **125**, 578–592.

Spencer, Anthony James Merrill, and Rivlin, Ronald S. 1958. The theory of matrix polynomials and its application to the mechanics of isotropic continua. *Archive for Rational Mechanics and Analysis*, **2**(1), 309–336.

Spencer, Anthony James Merrill, and Rivlin, Ronald S. 1959. Further results in the theory of matrix polynomials. *Archive for Rational Mechanics and Analysis*, **4**(1), 214–230.

Stamenović, Dimitrije, Rosenblatt, Noah, Montoya-Zavala, Martín, Matthews, Benjamin D, Hu, Shaohua, Suki, Béla, Wang, Ning, and Ingber, Donald E. 2007. Rheological behavior of living cells is timescale-dependent. *Biophysical Journal*, **93**(8), L39–L41.

Stein, Elias M. 1970. *Singular integrals and differentiability properties of functions*. Vol. 2. Princeton University Press.

Stinga, P. 2010. *Fractional powers of second order partial differential operators: extension problem and regularity theory*. Ph.D. thesis, Universidad Autonoma de Madrid.

Stinga, Pablo Raúl, and Torrea, José Luis. 2010. Extension problem and Harnack's inequality for some fractional operators. *Communications in Partial Differential Equations*, **35**(11), 2092–2122.

Stinga, Pablo Raúl, and Volzone, Bruno. 2015. Fractional semilinear Neumann problems arising from a fractional Keller–Segel model. *Calculus of Variations and Partial Differential Equations*, **54**(1), 1009–1042.

Stynes, M., O'Riordan, E., and Gracia, J.L. 2017. Error analysis of a finite difference method on graded meshes for a time-fractional diffusion equation. *SIAM Journal on Numerical Analysis*, **55**(2), 1057–1079.

Suki, B., Barabasi, A.L., and Lutchen, K.R. 1994. Lung tissue viscoelasticity: a mathematical framework and its molecular basis. *Journal of Applied Physiology*, **76**(6), 2749–2759.

Sumelka, W, Łuczak, B, Gajewski, T, and Voyiadjis, GZ. 2020. Modelling of AAA in the framework of time-fractional damage hyperelasticity. *International Journal of Solids and Structures*, **206**, 30–42.

Sumelka, Wojciech. 2014a. Application of fractional continuum mechanics to rate independent plasticity. *Acta Mechanica*, **225**(11), 3247–3264.

Sumelka, Wojciech. 2014b. Fractional viscoplasticity. *Mechanics Research Communications*, **56**, 31–36.

Sun, HongGuang, Zhang, Yong, Baleanu, Dumitru, Chen, Wen, and Chen, YangQuan. 2018. A new collection of real world applications of fractional calculus in science and engineering. *Communications in Nonlinear Science and Numerical Simulation*, **64**, 213–231.

Sun, Liwei, Qiu, Han, Wu, Chuanhao, Niu, Jie, and Hu, Bill X. 2020. A review of applications of fractional advection–dispersion equations for anomalous solute transport in surface and subsurface water. *WIREs Water*, **7**(4).

Sun, Yifei, and Sumelka, Wojciech. 2019. Fractional viscoplastic model for soils under compression. *Acta Mechanica*, **230**(9), 3365–3377.

Suzuki, J.L. 2017. *Aspects of fractional-order modeling and efficient bases to simulate complex materials using finite element methods*. Ph.D. thesis, University of Campinas, Brazil.

Suzuki, J.L., Zayernouri, M., Bittencourt, M.L., and Karniadakis, G.E. 2016. Fractional-order uniaxial visco-elasto-plastic models for structural analysis. *Computer Methods in Applied Mechanics and Engineering*, **308**, 443–467.

Suzuki, Jorge, Kharazmi, E., and Zayernouri, Mohsen. 2020. Nonlinear vibration of fractional viscoelastic cantilever beam: Towards structural health monitoring. *Journal of Sound and Vibration*.

Suzuki, Jorge, Zhou, Yongtao, D'Elia, Marta, and Zayernouri, Mohsen. 2021. A thermodynamically consistent fractional visco-elasto-plastic model with memory-dependent damage for anomalous materials. *Computer Methods in Applied Mechanics and Engineering*, **373**, 113494.

Suzuki, Jorge L., and Zayernouri, Mohsen. 2021. A self-singularity-capturing scheme for fractional differential equations. *International Journal of Computer Mathematics*, **98**(5), 933–960.

Susuki, Jorge L., Gulian, Mamikon, Zayernouri, Mohsen, and D'Elia, Marta. 2023. Fractional modeling in action: A survey of nonlocal models for subsurface transport, turbulent flows, and anomalous materials. *Journal of Peridynamics and Nonlocal Modeling*, **5**(3), 392–459.

Szegő, G. 1975. *Orthogonal Polynomials, 4th Ed.* Providence, RI: Amer. Math. Soc.

Szymanski, Jedrzej, and Weiss, Matthias. 2009. Elucidating the origin of anomalous diffusion in crowded fluids. *Physical review letters*, **103**(3), 038102.

Tanaka, K., and Mura, T. 1981. A Dislocation Model for Fatigue Crack Initiation. *Journal of Applied Mechanics*, **48**(1), 97.

Tang, T., Yuan, H.F., and Zhou, T. 2018. Hermite spectral collocation methods for fractional PDEs in unbounded domains. *Communications in Computational Physics*, **24**(4), 1143–1168.

Tang, Tao. 1993. A finite difference scheme for partial integro-differential equations with a weakly singular kernel. *Applied Numerical Mathematics*, **11**(4), 309–319.

Tarasov, V. E. 2008. Fractional vector calculus and fractional Maxwell's equations. *Annals of Physics*, **323**(11), 2756–2778.

Tarasova, V. V., and Tarasov, V. E. 2018. Concept of dynamic memory in economics. *Communications in Nonlinear Science and Numerical Simulation*, **55**, 127–145.

Temam, Roger. 2001. *Navier-Stokes equations: theory and numerical analysis*. Vol. 343. American Mathematical Soc.

Thomée, Vidar. 2006. *Galerkin finite element methods for parabolic problems*. Second edn. Springer Series in Computational Mathematics, vol. 25. Springer-Verlag, Berlin.

Tian, WenYi, Deng, Weihua, and Wu, Yujiang. 2014. Polynomial spectral collocation method for space fractional advection–diffusion equation. *Numerical Methods for Partial Differential Equations*, **30**(2), 514–535.

Torrejon, Diego, and Emelianenko, Maria. 2018. Generalized master equations for random walks with time-dependent jump sizes. *SIAM Journal on Applied Mathematics*, **78**(3), 1330–1349.

Treeby, Bradley E., and Cox, B. T. 2010. Modeling power law absorption and dispersion for acoustic propagation using the fractional Laplacian. *The Journal of the Acoustical Society of America*, **127**(5), 2741–2748.

Umarov, S., and Steinberg, S. 2006. Random walk models associated with distributed fractional order differential equations. Pages 117–127 of: *High dimensional probability*. Institute of Mathematical Statistics.

Vabishchevich, Petr N. 2015. Numerically solving an equation for fractional powers of elliptic operators. *Journal of Computational Physics*, **282**, 289–302.

Valério, Duarte, and Da Costa, Jose Sa. 2011. Variable-order fractional derivatives and their numerical approximations. *Signal Processing*, **91**(3), 470–483.

Valério, Duarte, Machado, José Tenreiro, and Kiryakova, Virginia. 2014. Some pioneers of the applications of fractional calculus. *Fractional Calculus and Applied Analysis*, **17**(2), 552–578.

Vázquez, Juan-Luis. 2014. Recent progress in the theory of nonlinear diffusion with fractional Laplacian operators. *Discrete and Continuous Dynamical Systems - S*, **7**(4), 857–885.

Vázquez, Juan Luis. 2017. The mathematical theories of diffusion: Nonlinear and fractional diffusion. Pages 205–278 of: *Nonlocal and Nonlinear Diffusions and Interactions: New Methods and Directions*. Springer.

Vincent, Julian. 2012. *Structural biomaterials*. Princeton University Press.

Walgraef, Daniel, and Aifantis, Elias C. 1985. Dislocation patterning in fatigued metals as a result of dynamical instabilities. *Journal of Applied Physics*, **58**(2), 688–691.

Wang, Hong, Yang, Danping, and Zhu, Shengfeng. 2016. Accuracy of finite element methods for boundary-value problems of steady-state fractional diffusion equations. *Journal on Scientific Computing*.

Wang, L.L., Samson, M., and Zhao, X.D. 2014. A well-conditioned collocation method using pseudospectral integration matrix. *SIAM Journal of Scientific Computing*, **36**, A907–A929.

Wang, Yi, Li, Zhen, Ouyang, Jie, and Karniadakis, George Em. 2020. Controlled release of entrapped nanoparticles from thermoresponsive hydrogels with tunable network characteristics. *Soft Matter*, **16**(20), 4756–4766.

Warma, Mahamadi. 2015. The fractional relative capacity and the fractional Laplacian with Neumann and Robin boundary conditions on open sets. *Potential Analysis*, **42**(2), 499–547.

Weber, Stephanie C, Spakowitz, Andrew J, and Theriot, Julie A. 2010. Bacterial chromosomal loci move subdiffusively through a viscoelastic cytoplasm. *Physical Review Letters*, **104**(23), 238102.

Wegst, Ulrike GK, Bai, Hao, Saiz, Eduardo, Tomsia, Antoni P, and Ritchie, Robert O. 2015. Bioinspired structural materials. *Nature Materials*, **14**(1), 23–36.

Weisstein, E.W. 2003. *Regularized Hypergeometric Function.*

White, Frank M, and Majdalani, Joseph. 2006. *Viscous fluid flow*. Vol. 3. McGraw-Hill New York.

Wilcox, David C, et al. 1998. *Turbulence modeling for CFD*. Vol. 2. DCW industries La Canada, CA.

Winter, H.H., and Mours, M. 1997. Rheology of polymers near liquid-solid transitions. *Advances in Polymer Science*, **134**, 165–234.

Wong, I.Y., Gardel, M.L., Reichman, D.R., Weeks, E.R., Valentine, M.T., Bausch, A.R., and Weitz, D.A. 2004. Anomalous diffusion probes microstructure dynamics of entangled F-actin networks. *Physical Review Letters*, **92**.

Wu, X., Deng, W., and Barkai, E. 2016. Tempered fractional Feynman-Kac equation: Theory and examples. *Physical Review E*, **93**(3), 032151.

Xiao, Rui, Sun, HongGuang, and Chen, Wen. 2017. A finite deformation fractional viscoplastic model for the glass transition behavior of amorphous polymers. *International Journal of Non-Linear Mechanics*, **93**, 7–14.

Xiu, Dongbin, and Karniadakis, George Em. 2002. The Wiener–Askey polynomial chaos for stochastic differential equations. *SIAM Journal on Scientific Computing*, **24**(2), 619–644

Xu, Qinwu, and Hesthaven, Jan S. 2014. Stable multi-domain spectral penalty methods for fractional partial differential equations. *Journal of Computational Physics*, **257**, 241–258.

Yamamoto, Masakazu. 2012. Asymptotic expansion of solutions to the dissipative equation with fractional Laplacian. *SIAM Journal on Mathematical Analysis*, **44**(6), 3786–3805.

Yan, Chanhao, Cai, Wei, and Zeng, Xuan. 2013. A parallel method for solving Laplace equations with Dirichlet data using local boundary integral equations and random walks. *SIAM Journal on Scientific Computing*, **35**(4), B868–B889.

Yan, Y. B., Khan, M., and Ford, N. J. 2018. An analysis of the modified L1 scheme for time-fractional partial differential equations with nonsmooth data. *SIAM Journal on Numerical Analysis*, **56**(1), 210–227.

Yang, Qianqian, Liu, Fawang, and Turner, I. 2010. Numerical methods for fractional partial differential equations with Riesz space fractional derivatives. *Applied Mathematical Modelling*, **34**(1), 200–218.

Yosida, Kôsaku. 1980. *Functional analysis*. Sixth edn. Grundlehren der Mathematischen Wissenschaften [Fundamental Principles of Mathematical Sciences], vol. 123. Springer-Verlag, Berlin-New York.

You, Huaiqian, Yu, Yue, Trask, Nathaniel, Gulian, Mamikon, and D'Elia, Marta. 2021. Data-driven learning of robust nonlocal physics from high-fidelity synthetic data. *Computer Methods in Applied Mechnics and Engineering*, **374**, 113553.

Yu, Y., Perdikaris, P., and Karniadakis, G.E. 2016. Fractional modeling of viscoelasticity in 3D cerebral arteries and aneurysms. *Journal of Computational Physics*, **323**, 219–242.

Yuste, S.B., and Joaquín, Q.-M. 2012. A finite difference method with non-uniform timesteps for fractional diffusion equations. *Computer Physics Communications*, **183**, 2594–2600.

Zaburdaev, V, Denisov, S, and Klafter, J. 2015. Lévy walks. *Reviews of Modern Physics*, **87**(2), 483.

Zaiser, Michael. 2006. Scale invariance in plastic flow of crystalline solids. *Advances in Physics*, **55**(1-2), 185–245.

Zaky, M. A., and Machado, J. A. T. 2017. On the formulation and numerical simulation of distributed-order fractional optimal control problems. *Communications in Nonlinear Science and Numerical Simulation*, **52**, 177–189.

Zayernouri, M., and Karniadakis, G.E. 2013. Fractional Sturm-Liouville eigen-problems: theory and numerical approximation. *Journal of Computational Physics*, **252**, 495–517.

Zayernouri, M., and Karniadakis, G.E. 2014a. Fractional spectral collocation method. *SIAM Journal of Scientific Computing*, **36**(1), A40–A62.

Zayernouri, Mohsen, and Karniadakis, George Em. 2014b. Exponentially accurate spectral and spectral element methods for fractional ODEs. *Journal of Computational Physics*, **257**(part A), 460–480.

Zayernouri, Mohsen, and Matzavinos, Anastasios. 2016. Fractional Adams–Bashforth/Moulton methods: an application to the fractional Keller–Segel chemotaxis system. *Journal of Computational Physics*, **317**, 1–14.

Zayernouri, M., Ainsworth, M., and Karniadakis, G.E. 2015. A unified Petrov–Galerkin spectral method for fractional PDEs, *Journal of Computer Methods in Applied Mechanics and Engineering*, **283**, 1545–1569.

Zayernouri, Mohsen, Ainsworth, Mark, and Karniadakis, George Em. 2015a. Tempered fractional Sturm–Liouville eigen-problems. *SIAM Journal on Scientific Computing*, **37**(4), A1777–A1800.

Zeng, F., Zhang, Z., and Karniadakis, G.E. 2015a. A Generalized Spectral Collocation Method with Tunable Accuracy for Variable-Order Fractional Differential Equations. *SIAM Journal on Scientific Computing*, **37**(6), A2710–A2732.

Zeng, F., Zhang, Z., and Karniadakis, G.E. 2017. Second-order numerical methods for multi-term fractional differential equations: smooth and non-smooth solutions. *Computer Methods in Applied Mechanics and Engineering*, **327**, 478–502.

Zeng, F., Turner, I., and Burrage, K. 2018a. A stable fast time-stepping method for fractional integral and derivative operators. *Journal of Scientific Computing*, **77**(1), 283–307.

Zeng, Fanhai, and Li, Changpin. 2014. Fractional differentiation matrices with applications. *arXiv preprint arXiv:1404.4429*.

Zeng, Fanhai, Li, Changpin, Liu, Fawang, and Turner, Ian. 2015. Numerical algorithms for time-fractional subdiffusion equation with second-order accuracy. *SIAM Journal on Scientific Computing*, **37**(1), A55–A78.

Zeng, Fanhai, Mao, Zhiping, and Karniadakis, George Em. 2017. A generalized spectral collocation method with tunable accuracy for fractional differential equations with end-point singularities. *SIAM Journal on Scientific Computing*, **39**(1), A360–A383.

Zeng, Fanhai, Turner, Ian, Burrage, Kevin, and Karniadakis, George Em. 2018b. A new class of semi-implicit methods with linear complexity for nonlinear fractional differential equations. *SIAM Journal on Scientific Computing*, **40**(5), A2986–A3011.

Zhang, C., and Guo, B.Y. 2012. Domain decomposition spectral method for mixed inhomogeneous boundary value problems of high order differential equations on unbounded domains. *Journal of Scientific Computing*, **53**(2), 451–480.

Zhang, Y., and Meerschaert, M.M. 2011. Gaussian setting time for solute transport in fluvial systems. *Water Resources Research*, **47**(8).

Zhang, Y., Benson, D.A., Meerschaert, M.M., LaBolle, E. M., and Scheffler, H.P. 2003. Lagrangian characterization of contaminant transport through multidimensional heterogeneous media with limited heterogeneity information. *Conference Proceedings for MODFLOW and MORE 2006: Managing Ground-Water Systems*, 639–643.

Zhang, Y.-N., Sun, Z.-Z., and Liao, H.-L. 2014. Finite difference methods for the time fractional diffusion equation on non-uniform meshes. *Journal of Computational Physics*, **265**, 195–210.

Zhang, Zhijiang, Deng, Weihua, and Karniadakis, George Em. 2018. A Riesz basis Galerkin method for the tempered fractional Laplacian. *SIAM Journal on Numerical Analysis*, **56**(5), 3010–3039.

Zhang, Zhongqiang, Zeng, Fanhai, and Karniadakis, George Em. 2015. Optimal error estimates of spectral Petrov–Galerkin and collocation methods for initial value problems of fractional differential equations. *SIAM Journal on Numerical Analysis*, **53**(4), 2074–2096.

Zhao, X. D., Wang, L. L., and Xie, Z. Q. 2013. Sharp error bounds for Jacobi expansions and Gegenbauer-Gauss quadrature of analytic functions. *SIAM Journal on Numerical Analysis*, **51**(3), 1443–1469.

Zheng, X., and Wang, H. 2020a. An Optimal-Order Numerical Approximation to Variable-order Space-fractional Diffusion Equations on Uniform or Graded Meshes. *SIAM Journal on Numerical Analysis*, **58**(1), 330–352.

Zheng, X., and Wang, H. 2020b. Wellposedness and regularity of a variable-order space-time fractional diffusion equation. *Analysis and Applications*, **18**(04), 615–638.

Zheng, Xiangcheng, Li, Yiqun, Cheng, Jin, and Wang, Hong. 2020. Inverting the variable fractional order in a variable-order space-fractional diffusion equation with variable diffusivity: analysis and simulation. *Journal of Inverse and Ill-posed Problems*, 000010151520190040.

Zhou, Yijing, and Cai, Wei. 2016. Numerical solution of the Robin problem of Laplace equations with a Feynman-Kac formula and reflecting Brownian motions. *Journal of Scientific Computing*, **69**(1), 107–121.

Zhou, Yijing, Cai, Wei, and Hsu, Elton. 2017. Computation of the local time of reflecting Brownian motion and the probabilistic representation of the Neumann problem. *Commun. Math. Sci.*, **15**(1), 237–259.

Zhou, Yongtao, Suzuki, Jorge L, Zhang, Chengjian, and Zayernouri, Mohsen. 2019. Fast IMEX time integration of nonlinear stiff fractional differential equations. *arXiv preprint arXiv:1909.04132*.

Zhou, Yongtao, Zhang, Chengjian, and Brugnano, Luigi. 2020a. An implicit difference scheme with the KPS preconditioner for two-dimensional time–space fractional convection–diffusion equations. *Computers & Mathematics with Applications*, **80**(1), 31–42.

Zhou, Yongtao, Suzuki, Jorge L, Zhang, Chengjian, and Zayernouri, Mohsen. 2020b. Implicit-explicit time integration of nonlinear fractional differential equations. *Applied Numerical Mathematics*, **156**, 555–583.

Zhou, Yongtao, Zhang, Chengjian, and Brugnano, Luigi. 2020c. Preconditioned quasi-compact boundary value methods for space-fractional diffusion equations. *Numerical Algorithms*, **84**(2), 633–649.

Zhu, Ting, Li, Ju, and Yip, Sidney. 2004. Atomistic study of dislocation loop emission from a crack tip. *Physical Review Letters*, **93**(2), 025503.

Zhuang, P., Liu, F., Anh, V., and Turner, I. 2009. Numerical methods for the variable-order fractional advection-diffusion equation with a nonlinear source term. *SIAM Journal on Numerical Analysis*, **47**(3), 1760–1781.

Index